ANNUAL REVIEW OF BIOCHEMISTRY

EDITORIAL COMMITTEE (1987)

ANNUAL REVIEW OF BIOCHEMISTRY

VOLUME 56, 1987

CHARLES C. RICHARDSON, *Editor*

Harvard Medical School

PAUL D. BOYER, *Associate Editor*

University of California, Los Angeles

IGOR B. DAWID, *Associate Editor*

National Institutes of Health

ALTON MEISTER, *Associate Editor*

Cornell University Medical College

ANNUAL REVIEWS INC.   4139 EL CAMINO WAY   P.O. BOX 10139   PALO ALTO, CALIFORNIA 94303-0897

ANNUAL REVIEWS INC.
Palo Alto, California, USA

International Standard Serial Number: 0066–4154
International Standard Book Number: 0–8243–0856-5
Library of Congress Catalog Card Number: 50–13143

Annual Review and publication titles are registered trademarks of Annual Reviews Inc.

Annual Reviews Inc. and the Editors of its publications assume no responsibility for the statements expressed by the contributors to this *Review*.

Typesetting by Kachina Typesetting Inc., Tempe, Arizona; John Olson, President Typesetting coordinator, Janis Hoffman

PRINTED AND BOUND IN THE UNITED STATES OF AMERICA

Annual Review of Biochemistry
Volume 56, 1987

CONTENTS

(continued)   v

SOME RELATED ARTICLES IN OTHER *ANNUAL REVIEWS*

From the *Annual Review of Biophysics and Biophysical Chemistry*, Volume 16 (1987)

Structure and Assembly of Coated Vesicles, B. M. F. Pearse and R. A. Crowther
Structural Studies of Halophilic Proteins, Ribosomes, and Organelles of Bacteria Adapted to Extreme Salt Concentrations, Henryk Eisenberg and Ellen J. Wachtel
The Thermodynamic Stability of Proteins, John A. Schellman
Biophysical Chemistry of Metabolic Reaction Sequences in Concentrated Enzyme Solution and in the Cell, D. K. Srivastava and Sidney A. Bernhard

From the *Annual Review of Cell Biology*, Volume 3 (1987)

Constitutive and Regulated Secretion, Teresa Lynn Burgess and Regis B. Kelly
Ubiquitin-Mediated Pathways for Intracellular Proteolysis, Martin Rechsteiner
Cell Surface Receptors for Extracellular Matrix Molecules, Clayton A. Buck and Alan F. Horwitz
Cell Transformation by the Viral src Oncogene, Hidesaburo Hanafusa and Richard Jove

From the *Annual Review of Genetics*, Volume 20 (1986)

Ribosomal Genes in Escherichia coli, Lasse Lindahl and Janice M. Zengel
Mechanism of Bacteriophage Mu Transposition, Kiyoshi Mizuuchi and Robert Craigie
Mammalian Urea Cycle Enzymes, Marian J. Jackson, Arthur L. Beaudet, and William E. O'Brien
Mismatch Repair in Escherichia coli, Miroslav Radman and Robert Wagner

From the *Annual Review of Immunology*, Volume 5 (1987)

Activation of the First Component of Complement, Verne N. Schumaker, Peter Zavodszky, and Pak H. Poon
Lipid Mediators Produced Through the Lipoxygenase Pathway, Charles W. Parker
Vaccinia Virus Expression Vectors, Bernard Moss and Charles Flexner
Biophysical Aspects of Antigen Recognition by T Cells, Tania H. Watts and Harden M. McConnell

From the *Annual Review of Medicine*, Volume 38 (1987)

Platelet-Derived Growth Factor, Russell Ross

From the *Annual Review of Microbiology,* Volume 41 (1987)

Physiology, Biochemistry, and Genetics of the Uptake Hydrogenase in Rhizobia, Harold J. Evans, Alan R. Harker, Hans Papen, Sterling A. Russell, F. J. Hanus, and Mohammed Zuber
Biosynthesis of Peptide Antibiotics, H. Kleinkauf and H. von Döhren
The Biosynthesis of Sulfur-Containing β-Lactam Antibiotics, J. Nüesch, J. Heim, and H.-J. Treichler
Export of Protein: A Biochemical View, L. L. Randall, S. J. S. Hardy, and Julia R. Thom

From the *Annual Review of Neuroscience,* Volume 10 (1987)

Molecular Biology of Visual Pigments, Jeremy Nathans

From the *Annual Review of Pharmacology and Toxicology,* Volume 27 (1987)

Biochemical and Molecular Genetic Analysis of Hormone-Sensitive Adenylyl Cyclase, Gerald F. Casperson and Henry R. Bourne

From the *Annual Review of Physical Chemistry,* Volume 38 (1987)

Three-Dimensional X-Ray Crystallography of Membrane Proteins: Insights into Electron Transfer, David E. Budil, Peter Gast, and James R. Norris
Molecular Modeling, Peter Kollman

From the *Annual Review of Physiology,* Volume 49 (1987)

Lateral Diffusion of Proteins in Membranes, Ken Jacobson, Akira Ishihara, and Richard Inman
Kinetics of the Actomyosin ATPase in Muscle Fibers, Yale E. Goldman

ANNUAL REVIEWS INC. is a nonprofit scientific publisher established to promote the advancement of the sciences. Beginning in 1932 with the *Annual Review of Biochemistry*, the Company has pursued as its principal function the publication of high quality, reasonably priced *Annual Review* volumes. The volumes are organized by Editors and Editorial Committees who invite qualified authors to contribute critical articles reviewing significant developments within each major discipline. The Editor-in-Chief invites those interested in serving as future Editorial Committee members to communicate directly with him. Annual Reviews Inc. is administered by a Board of Directors, whose members serve without compensation.

ANNUAL REVIEWS OF		SPECIAL PUBLICATIONS
Anthropology	Materials Science	
Astronomy and Astrophysics	Medicine	Annual Reviews Reprints:
Biochemistry	Microbiology	Cell Membranes, 1975–1977
Biophysics and Biophysical Chemistry	Neuroscience	Immunology, 1977–1979
Cell Biology	Nuclear and Particle Science	
Computer Science	Nutrition	Excitement and Fascination
Earth and Planetary Sciences	Pharmacology and Toxicology	of Science, Vols. 1 and 2
Ecology and Systematics	Physical Chemistry	
Energy	Physiology	Intelligence and Affectivity,
Entomology	Phytopathology	by Jean Piaget
Fluid Mechanics	Plant Physiology	
Genetics	Psychology	Telescopes for the 1980s
Immunology	Public Health	
	Sociology	

For the convenience of readers, a detachable order form/envelope is bound into the back of this volume.

Konrad Bloch

Ann. Rev. Biochem. 1987. 56:1–19

SUMMING UP

Konrad Bloch

James Bryant Conant Laboratories, Harvard University, Cambridge, Massachusetts 02138

To look ahead, to think of research problems still to be solved, comes more naturally to most scientists than to reflect on the past. Yet the opportunity to write a retrospective account like this has not been the chore I expected it to be. On reflection one sees the Why and the How of one's upbringing and career more clearly and from a perspective that is missing or has no place in the professional publication record. That circumstances and good fortune rather than deliberate decisions on my part have played a major role in my career is now much clearer to me. To begin at the beginning:

I was born and grew up in Neisse, Germany, in the province of Silesia, the second child of a middle class family in comfortable circumstances. My father had studied law but to his regret never practiced it. Instead, as a dutiful son he took over the family factory, but he never enjoyed being in business. Neisse was a midsized country town noted for several Gothic and Baroque churches, a beautiful Renaissance town hall, and a 300-foot-high watchtower. Hidden in the hills surrounding Neisse were fortifications dating to the era of the Prussian king Frederick the Second, who annexed Silesia from the Austrian Empress Maria Theresia. To her, we were told in school, Silesia was the most precious jewel in her crown. Perhaps for this reason we took little pride in being Prussians. My upbringing at home was fair and understanding yet strict, in line with customs and practices of the time. Only in one instance did I come near to rebelling against parental authority. On the occasion of my Bar mitzvah a generous great-uncle asked me to choose a present, the choices being either a cello or a canoe. For a 13-year-old the decision was obvious. Canoeing would be pure pleasure while practicing a musical instrument meant hours of drudgery. Thinking otherwise, my parents insisted that it was the cello I really wanted—and they won. It took many years before I came to

1

0066-4154/87/0701-0001$02.00

terms with the cello, but eventually I enjoyed playing it. That the canoe was really not for me I learned a few years later. After graduating from high school, I spent the summer of 1930 in a steel works near Lübeck at the Baltic Sea to obtain the practical experience required for an engineering degree, my goal at the time. It was a beautiful day when a friend asked me to join him on a canoe trip. I accepted eagerly but at the end of the day I was burned to a crisp and had to be hospitalized. Later in this account I will relate a second incident that bore out the wisdom of the parental decision.

I have only good memories of the schools I attended, in spite of the fact that most of the gymnasium teachers were martinets, some to the point of brutality. Learning by rote was considered an important, even essential educational principle. In retrospect one wonders whether there is a superior one, at least for adolescents. No doubt the system laid a sound basis for higher education even if in my case it failed to stimulate any special interests. Surely the chemistry teacher could not claim to have influenced my career. He succeeded in making the subject matter totally unattractive. As for my future, there was no family tradition to follow and therefore no parental pressure to enter one career or another. There were physicians among my forebears, but no scientists. I knew next to nothing about academic careers, and Einstein was probably the only living scientist whose name I knew. As for my interests as an adolescent, they tended towards the natural sciences or engineering. The few books I brought along from Europe as sentimental baggage included a set of slim volumes, published by Sammlung Göschen, on subjects such as mineralogy, crystallography, stereochemistry, and metallurgy. The last subject intrigued me most, and led me to read the authoritative textbook of the time, *Chemistry and Physics of Metals and Their Alloys* by Gustav Tamann. Stainless steel had just been invented, and there was much optimism that the time had come for a rational metallurgy, the design of alloys with a wide range of desired properties. I enrolled at the Technische Hochschule in Munich, but the uninspired course in metallurgy soon dampened my enthusiasm for the subject. The chemistry courses were a different matter. I no longer quite understand why I enjoyed the introductory courses, which emphasized quantitative and qualitative analysis. Obviously I must have derived much satisfaction from chemical manipulations and their visual aspects. Of lasting impact, however, was Hans Fischer's organic chemistry course. As he presented it, the subject matter was fascinating, the organization superb, and the delivery monotonous. To play to the galleries or to raise his voice for emphasis was not Fischer's style. At the end of my second year I knew I had found my field. Natural products chemistry then flourished in Germany, and Hans Fischer was one of the prominent figures in a stellar group that included Richard Willstaetter, Heinrich Wieland, and Alfred Windaus. Their lectures to the "Munchener Chemische Gesellschaft" have remained vivid memories. I remember especially Wieland's seminar in 1934 on butterfly pigments. He had discovered these novel structures and named them pteridines. His lecture ended on an emotional note. Wieland had isolated the pigments from some 200,000 butterflies schoolchildren all over Germany had collected for a penny a piece. In concluding, Wieland remarked, "I will be unable to continue this research since the Government regards the collecting of butterflies as cruelty to animals, incompatible with the ethics of the National Socialist party. I can only hope"—he added—"that this research will be continued by my colleagues abroad." To make this rather mildly critical statement in front of numerous brown-shirted students took considerable courage. To me Heinrich Wieland has remained a model, a great human being as well as a great scientist. The next two years I had to spend essentially full time in the organic lab. Some 100 "Gatterman" compounds had to be delivered to the Assistent, and after they passed the test, one advanced to so-called literature preps that required more skill and stamina. The latter were either pyrroles (in kg quantities) or porphyrins (in gram quantities) to be passed on as starting materials for the synthetic work of PhD students. H. Fischer had just completed the total synthesis of hemin, and chlorophyll was the next goal. This was perhaps the most physically demanding period in my life. To give an example, porphyrins were separated by distribution between HCl and ether in thick-walled and therefore heavy 10-liter separatory funnels.

In early 1934 the dean of the Technische Hochschule notified me that I was ineligible to continue in chemistry since Professor Fischer "had declined to accept me as a graduate student." This was a lie but in line with the impending racial laws. With a naiveté that seems incomprehensible in retrospect, I explored opportunities for doing graduate work elsewhere in Europe. Fortunately I received negative replies to inquiries I had sent to A. Butenandt, then at the Technische Hochschule of the "Free City of Danzig," and to F. Kögl at Utrecht. Events moved fast; Danzig was incorporated into the German Reich a few months later and Holland overrun by the Nazis early in World War II. To have been denied admission to the Utrecht department proved a lucky escape for yet another reason. In the mid-1930s the Utrecht laboratory claimed two spectacular discoveries: 1. "Heteroauxin," supposedly a new plant hormone, isolated from thousands of liters of horse urine and 2. the occurrence of D amino acids in tumor proteins. Both "discoveries" proved to be fabrications by one of Kögl's coworkers. Only after the war had ended did the truth come out.

Anxious to help me, H. Fischer, an ardent patriot but not a Nazi and impeccably fair to all of his students, arranged my appointment as research assistant at the Schweizerisches Höhenforschung's Institut in Davos. This delightful mountain resort had two claims to fame. Its ski runs were the longest in the Alps, and it had the largest number of tuberculosis sanatoria in the world. Thomas Mann had chosen it as the setting for *The Magic Mountain*[1], his most famous novel.

Frederic Roulet, a pathologist and head of the Institute, directed research appropriate to the locale. In his enlightened view, the cure for tuberculosis would ultimately come from biochemical studies, specifically the analysis of mycobacterial lipids. My predecessor at the Institute claimed to have found cholesterol among the lipids of human tubercle bacilli, contradicting an earlier report by Erwin Chargaff (1935). Worried about the reputation of the Institute, Roulet asked me to sort things out. This was my first research experience, a trying one for a chemistry undergraduate who had no one to turn to for advice. My inexperience showed during my first day in the laboratory. Whatever chemical operations one undertook 50 years ago, the first essential task was to purify all organic solvents. No matter how often I distilled acetone, I was unable to raise its boiling point above 52°C. Eventually it dawned on me that at an altitude of 5000 feet this is what it should be and not 56°C. I was now ready to repeat my predecessor's experiments. From lipid extracts of tubercle bacteria he had been able to obtain an insoluble digitonide, the classical test for cholesterol, and I confirmed him up to this point. However, pyridine cleavage of the precipitate failed to yield the expected, easily crystallizable cholesterol. It was apparent that a branched chain aliphatic hydrocarbon and not cholesterol was responsible for the

---

[1] While writing this chapter I happened to take another look at the novel in order to refresh the memory of my Davos years. This effort was rewarded when I came upon the following prophetic passage:

> "Nothing different.—Oh, well, the stuff to-day was pure chemistry," Joachim unwillingly condescended to enlighten his cousin. It seemed there was a sort of poisoning, an auto-infection of the organisms, so Dr. Krokowski said; it was caused by the disintegration of a substance, of the nature of which we were still ignorant, but which was present everywhere in the body; and the products of this disintegration operated like an intoxicant upon the nerve-centres of the spinal cord, with an effect similar to that of certain poisons, such as morphia, or cocaine, when introduced in the usual way from outside.
> "And so you get the hectic flush," said Hans Castorp. "But that's all worth hearing. What doesn't the man know! He must have simply lapped it up. You just wait, one of these days he will discover what that substance is that exists everywhere in the body and sets free the soluble toxins that act like a narcotic on the nervous system; then he will be able to fuddle us all more than ever. Perhaps in the past they were able to do that very thing. When I listen to him, I could almost think there is some truth in the old legends about love potions and the like."

To my surprise none of the Thomas Mann experts nor any of the neurobiologists I consulted were aware of this remarkable prediction of endorphins and enkephalins—made more than 70 years ago! It was of course pure accident that I reread *The Magic Mountain* now and not 10 years earlier, before the discovery of "the body's own opiates." My attempts to find the sources for Mann's inspiration have not yet yielded any clues.

insoluble digitonide. Roulet was not happy with the result but was persuaded to accept my negative findings when an American paper by R. J. Anderson and R. Schoenheimer confirmed the absence of cholesterol in mycobacteria as Chargaff had reported earlier. This was my earliest encounter with cholesterol. That I would find myself in the same laboratory with Erwin Chargaff at Columbia a few years later and that R. J. Anderson and Rudolf Schoenheimer would play crucial roles in my career I had no reason to suspect.

With the cholesterol problem out of the way I was next asked to investigate the phospholipid chemistry of the same microbes. It was known that the phospholipid fraction of the human tubercle bacillus, when injected subcutaneously, causes tissue changes (tubercular granulomas and giant cells) indistinguishable from those elicited by live bacteria. Roulet felt that the species specifity (human versus bovine) of the purified "phosphatide" fractions was worth investigating. I had not heard of phosphatides before, and learned everything then known about these substances from a monograph by Thierfelder and Klenk. I soon discovered that R. J. Anderson, professor and chairman of biochemistry at Yale University, was the foremost authority on the phospholipids of acid-fast bacteria. To my request for reprints he responded promptly. I followed his directions for preparing the so-called phosphatide A-3 fraction, but distressingly my elementary analyses of this substance showed a higher phosphorus content and the complete absence of nitrogen, at variance with Anderson's analytical data (1, 2).

With all the deference that seemed proper for a German undergraduate when writing to a senior professor, I informed Anderson of my results: "Can you advise me what I might have done wrong?" The reply from Yale was prompt and what it said astonishing: "Looking over our data, it seems to me quite possible that your preparation is purer than ours." Phosphatide A-3 actually proved to be phosphatidic acid[2]. Apart from boosting my morale, Anderson's letter gave me the courage to turn to him for help. My residential permit in Switzerland was about to expire without any chance of extension. To emigrate to the United States had long been my hope, but had remained a dream for want of American relatives who might provide the required affidavit of support. Again the response from overseas was prompt. Two letters arrived and the news was rarely, if ever, more welcome. "I have the pleasure to inform you that you have been appointed Assistant in Biological Chemistry, School of Medicine, Yale University." So wrote Michael Winternitz, the

[2]Only youthful enthusiasm explains that I volunteered to serve as the experimental animal for testing the biological activity of the two phosphatide fractions. The material from the human strain was injected into my left and the bovine phosphatide into my right forearm. Only the human fraction gave a positive response. Roulet was greatly pleased and I was left with two-inch scars still highly visible today. I should add that in my innocence I had paid little attention to the observation that tubercle bacilli remained viable after repeated washings with acetone.

dean. Anderson's separate letter was sober: "I hope the Dean's appointment letter will be of help to you. Unfortunately no funds for a stipend are available to go with the appointment." The American consul in Frankfurt—perhaps a Yale man—issued the immigration visa forthwith. I had not shown him Anderson's letter.

My morale needed a boost for another reason. In Roulet's opinion the publications resulting from my work in Davos (1, 2) might be sufficient for a PhD thesis. I enrolled at the University of Basel, and in due time submitted my thesis. Dekan Professor Bernoulli, a descendant of the famous physicist, was most encouraging. Yet a month later a curt letter arrived stating that the Naturwissenschaftliche Fakultaet had unanimously rejected my thesis as "ungenügend." While visiting the Basel Biozentrum a few years ago, I could not refrain from beginning my seminar by commenting on my dismal experience in 1936. My host's curiosity was aroused. From the university archives he extracted the information that one member of the faculty committee had rejected my thesis because I had failed to cite some important references (to his publications). According to the rules at that time one dissenting vote made the rejection unanimous. Very much later I wondered whether or not this was another occasion when fate was kind to me.

I arrived in the United States in December 1936 with great hopes but barely enough funds to support me for a month. While still in Munich I had not dared to aspire to an academic career. I expected to end up in the chemical industry like most of Fischer's students. At the time financial independence was a prerequisite for a "Habilitation," at least in practice. The Privatdozent, holder of the lowest academic rank, received no fixed salary and was expected to subsist on the meager tuition fees paid by the few students who enrolled in the special topics course a Privatdozent was allowed to teach. The large, remunerative courses were reserved for the professor. The larger the number of students, the greater the professorial income. Understandably the larger universities were also the best.

My initial euphoria of having made it to the United States began to subside, and reality demanded that I support myself. After scanning the ad section of the *New York Times* I applied without much enthusiasm for an opening in a small dye factory in New Jersey. Yet I felt that in spite of my distressing experience in Basel, I must at the very least explore the possibility of going on to a PhD degree. In the interim, however, I had not failed to pay my respects to Rudolph Anderson. Without his help I might not have made it to these shores. Besides, while I was still in Switzerland, he had offered to take me on for graduate work. Yet he now discouraged me from coming to Yale: "You won't learn very much here; why don't you work with Hans Clarke at P & S?" (College of Physicians and Surgeons, Columbia University). This modesty hardly conformed with my image of a Herr Professor but it was in keeping with Anderson's readiness to help some unknown youngster who had earlier dared to criticize his experiments, if ever so politely. This and other early contacts with senior scientists in the United States impressed me deeply and laid the basis for my undiminished regard and continuing admiration for the open and democratic spirit that prevails in American academia.

From abroad I had brought with me two letters of recommendation, one from Richard Willstaetter, whose counsel I had asked before leaving Munich, and a much shorter but very effective one from Hans Fischer. In essence his letter said no more than "Herr Bloch ist gut." Fischer had suggested that I seek advice from Max Bergmann at what was then the Rockefeller Institute. My visit to the institute proved to be a crucial turn of events. Bergmann agreed with Anderson that to work in Hans Clarke's department was my best bet. An interview was granted forthwith, and when the departmental secretary took me to his lab, Clarke asked me to wait a minute until he had finished taking a melting point. Since the compound melted sharply, Clarke was in good humor. He obviously had a high regard for the training students received in Hans Fischer's lab, and moreover his sympathies for refugees and his efforts to help them were widely known. The friendly interview and the probing of my background ended with the question, "By the way, do you play a musical instrument?" In good conscience I could answer "yes." I was not asked how well I played the cello. End of interview and admission as graduate student. Once again I had reason to be grateful for my parents' foresight. Elated by the outcome I reported the acceptance to Bergmann, who immediately offered to find financial support: "I will introduce you to a friend who might help." The friend was Leo Wallerstein, head of the Wallerstein Laboratories on Staten Island, New York, consultants to the brewing industry. An early immigrant from Germany, Wallerstein held a lucrative patent for controlling the turbidity of Lager beer by addition of proteolytic enzymes. His foundation assisted refugee scholars, beginning as well as established. With financial support assured for a year I could go to work.

Clarke, unlike the Basel faculty, did not find my published Davos papers wanting and generously accepted them as partial fulfillment for the PhD requirements. Still, to satisfy formalities I had to carry out some research at Columbia, but the problem Clarke assigned to me was straightforward and not demanding. For my thesis problem I was to synthesize a number of N-alkylcysteine derivatives and to examine their sulfur lability. The work went smoothly and according to plan except for the failure of N-methylcysteine hydrochloride to crystallize. Eventually, however, some crystals appeared and spread steadily. I took the round-bottom flask to Clarke, who smilingly told me about Adolph von Bayer's habit of expressing his appreciation for a student's success. On such occasions, when making the rounds, von Bayer would lift the Calabrese hat he wore indoors and outdoors. Clarke's symbolic gesture made this a sweet moment. It became even sweeter for the graduate student when Clarke suggested I call up "Dee" (Vincent du Vigneaud,

Professor and Chairman of Biochemistry at Cornell) to offer him a seed crystal—du Vigneaud's preparation of N-methylcysteine had not yet crystallized. du Vigneaud himself came to 168th Street and was most gracious. Nowadays a sharp chromatographic peak may be a superior criterion for purity, but as an aesthetic experience, it does not compare with the sight of crystals one is the first to see. Only X-ray crystallographers still experience this visual pleasure. After a year and a half, Clarke decided that along with my earlier papers, I had sufficient material for a thesis and for publication. The resulting *J. Biol. Chem.* paper (3) was a mere 12 pages long at a time when the journal format was one half of what it is today, the margins were wide, and the print at least twice as large.

In 1938 the Depression was still severe, and I found myself with a degree but without a job in sight. Max Bovarnick, a physician working for his Ph.D. in the same laboratory, needed an organic chemist in order to test his idea that a diodophenyl derivative of thyroxin should be of interest. A molecule containing six iodines instead of four, and three phenyl rings instead of two, was bound to be a superhormone. I managed to synthesize the monster, which proved to be totally inactive biologically. Its only distinction was complete insolubility in water. The outcome taught me an early lesson: Avoid chemical reasoning for predicting biological activity! This interlude came to an end one day when Rudolf Schoenheimer asked me to join his group. Later he confessed that he had hesitated hiring me because my thesis was so "thin," as indeed it was. Up to this point I had a modicum of experience in preparative organic chemistry but my knowledge of biology or biochemistry was nil. I accepted Schoenheimer's offer at once and soon realized that biochemistry was going to be my field from now on. How brilliantly Schoenheimer seized and exploited the opportunity of using stable isotopes for tracing metabolic pathways is a matter of record. It is tragic that Schoenheimer died too young to reap the rewards due to him.

Harold Urey's generous donations of stable isotopes and David Rittenberg's collaboration played essential roles in this pioneering research. To develop the tracer methodology required novel or specially adapted synthetic and analytical procedures and above all the insight to identify problems that could be solved uniquely with the aid of isotopes. A seminar by Harold Urey that Schoenheimer attended had convinced him that isotopes were the answer to a biochemist's prayer, especially for elucidating biosynthetic transformations. Reactions that are uphill energetically were essentially unknown in the 1930s. Besides, the tracer method was noninvasive and therefore gave information on metabolic events under normal physiological conditions. Almost overnight it superseded the classical balance method, which had inherent limitations even though it had furnished remarkably accurate information in some instances. Schoenheimer was fond of dramatizing the advance by comparing the balance method to a Coca-Cola dispenser. Feeding compound

A to an animal and isolating an increased amount of B from the tissues or excreta no more proved the conversion of A to B than the observation that a coin dropped into the vending machine is converted into a bottle of coke. One balance experiment that should have worked but did not was the conversion of muscle creatine into urinary creatinine. To settle this issue with the aid of $^{15}N_2$ was my first assignment. It was straightforward enough except that all earlier and inconclusive nonisotopic experiments had been done with humans or dogs. $^{15}N_2$ was precious and the labeled creatine I synthesized barely sufficient for feeding it to a few rats. But the isolation of creatinine by direct precipitation with picric acid, which worked so well with human urine, failed with rat urine. After I had laboriously devised a chromatographic method for isolating creatinine from this source, the creatine-creatinine conversion was proven but it was hardly a surprising discovery. More interesting was the next project, to elucidate the mechanism of creatine synthesis. The question was whether the methylation step occurred early or late in the pathway. A priori two mechanisms seemed reasonable, one involving the methylation of glycine to sarcosine and the other the methylation of guanidoacetic acid. In either case, the guanido group of creatine might originate from arginine. The pathway glycine + arginine → guanidoacetic acid → creatine proved to be the one nature has chosen. H. Borsook and J. W. Dubnoff (4) had already contributed to this problem by demonstrating the methylation of guanidoacetic acid to creatine. In the course of this work we became aware of some elegant amino acid chemistry M. Bergmann and L. Zervas had described (5). Twelve years earlier they had observed the transfer of an acetylated amidine group of arginine to glycine ethyl ester to form guanidoacetic acid ester, a gratifying model for the enzyme-catalyzed reaction. In this case nature proceeds "logically," following chemical precedent. Transamination is one example of fulfilled expectations and E. Snell's model studies on transamination another, and perhaps the best known one. Much more often, however, the choices of nature are not those the chemist would predict. Perhaps for this reason some chemists of earlier times found the unexpected designs of nature unsettling and therefore biochemistry without appeal, at least as subject for their own research. Fortunately this is no longer true, so much so that chemists have now taken over much of the territory that used to be the biochemists' preserve.

My first apprenticeship with Rudi and membership in his congenial and dedicated group were immensely enjoyable and gave me a superb education. I saw no reason to look for a job elsewhere. Yet when an offer came from Mt. Sinai Hospital in New York to join a Cancer Research team at twice the stipend I received at Columbia, I could not resist. I was anxious to marry and did not want to do so on a minimal budget. The interlude at Mt. Sinai was brief. After less than a year, Schoenheimer invited me to return to Columbia and offered to match the Mt. Sinai salary. Fortunately the hospital was understanding. On my return to Columbia Schoenheimer suggested that I investigate the origin of the hydroxyl oxygen in cholesterol. Was water or molecular oxygen the source? In retrospect, it is truly remarkable that these two alternatives should have occurred to Schoenheimer. In 1940 oxygen was widely believed to serve exclusively as a terminal electron acceptor in respiration but not as a source of carbon-bound oxygen in organic compounds. The much earlier proposal of Bach of peroxidase-catalyzed oxygenations had fallen into disrepute for lack of experimental evidence. At any rate, I failed to make progress with this problem because no method existed for the mass spectrometric analysis of stably bound oxygen in complex organic compounds. I was to develop such a method, but on this as well later occasions I demonstrated my ineptitude and lack of enthusiasm for designing analytical procedures. Had I been more skillful, the first oxygenase reaction might have been encountered some 15 years before Howard Mason and Osamu Hayaishi independently demonstrated the direct introduction of O_2 into aromatic systems. Their discovery of oxygenases opened one of the most remarkable new chapters in biochemical research.

Schoenheimer's untimely death in 1941 left his associates without the leader and the inspired leadership they so admired. We feared that we might have to look for jobs elsewhere, but Hans Clarke encouraged us to continue as heirs to the wealth of projects Schoenheimer had begun and developed. From now on we were to proceed on our own and to choose problems as we saw fit. This created somewhat of a problem since none of us had proprietary claims on a given subject. We shared the methodology but had no explicit commitment to any of the major areas then under investigation. How the division of "spoils" came about I do not recall—it may have been by drawing lots. At any rate, David Shemin "drew" amino acid metabolism, which led to his classical work on heme biosynthesis. David Rittenberg was to continue his interest in protein synthesis and turnover, and lipids were to be my territory.

At the time I was struggling with the cholesterol oxygen problem but quickly lost all interest to go on with it. A major turning point came with the arrival of a paper by Sonderhoff & Thomas (6), which had been delayed by the war. It reported that "The nonsaponifiable fraction of yeast grown in a medium supplemented with D-acetate had a Deuterium content so high that a direct conversion of acetic acid to sterols has to be postulated." This result was gratifying to Rittenberg, since his and Schoenheimer's experiments with D_2O had indicated that animal cholesterol is synthesized from small molecules, most likely intermediates of fat or carbohydrate metabolism. The obvious experiment to do, and it was done the next day, was to feed labeled acetate to rats and mice. Incorporation of deuterium into fatty acids and cholesterol was substantial, an outcome that set the stage for my long-lasting interest in the biosynthesis of cholesterol and fatty acids. Gratifying as it was, this result did not tell us how many of the 27 sterol carbon atoms were supplied by acetic acid. Feeding of a labeled metabolite leads necessarily to isotope dilution of unknown magnitude. Rough estimates of the endogenous acetate pool suggested, however, that acetate was a major and perhaps specific carbon and hydrogen source for all moieties of the sterol molecule although other precursors were not ruled out. The definitive answer came 10 years later with the arrival of microbial mutants. After I had moved to Chicago, I became aware of an acetateless mutant of *Neurospora crassa* that Ed Tatum, then at Yale, has isolated. The mutant seemed ideal for our purposes. Joining forces with the Yale investigators, I. Zabin could show that in this mutant, the sterol synthesized derived all its carbon atoms from exogenous acetate (7). Perhaps this result should have stimulated me to turn to microbes as experimental systems, but for such a drastic change I was not yet prepared. To establish the origin of individual sterol carbon atoms from either acetate carboxyl or methyl by traditional chemistry was to take several more years. This problem and the elucidation of biosynthetic pathways in general that occupied biochemists in the 1940s and 1950s could have been solved in a fraction of the time with the technologies available today. Yet the slow pace of getting there had rewards of its own.

My remaining years at Columbia were spent showing the suspected precursor role of cholesterol for bile acids and steroid hormones, if proof was needed. However, structural similarities, no matter how suggestive, do not prove biochemical relationships. Experience has taught biochemists to approach all such problems with an open mind. To demonstrate the origin of progesterone from cholesterol, which I thought needed to be shown, raised both technical and logistic problems. First of all, to prepare the requisite labeled cholesterol in sufficient quantity proved to be a major effort. Labeled compounds of any kind were not available commercially in the early 1940s. Much time was spent introducing deuterium into cholesterol by platinum-catalyzed exchange in heavy water–acetic acid mixtures. This method was destructive, but yielded gram quantities containing 4–5 atom percent D. Secondly, human pregnancy urine was the only practical source for isolating the progesterone metabolite pregnanediol in sufficient quantity. My request to the P & S department of obstetrics and gynecology for permission to administer labeled cholesterol to one of its patients was brusquely denied. Fortunately, however, I found sympathy and willingness to cooperate at home. The experiment gave the expected result, but ever since I have felt somewhat guilty that the essential collaborator was not a coauthor and mentioned only anonymously in the experimental section of the *J. Biol. Chem.* paper (8).

During the war years universities had suspended new appointments and promotions. Some of my contemporaries and I grew increasingly restive about the temporary nature of our appointments even though Hans Clarke allowed us total freedom in the choice of our research. The first break came early in 1946 when I was invited to Salt Lake City to be interviewed for an assistant professorship in the biochemistry department. During the discussion following my seminar, I rather curtly responded to a comment from the audience, not realizing that the very youthful questioner was someone very high up in the administration. I never learned whether my interview went well, but I suspect it did not. Breaking my return journey in Chicago, I visited Earl Evans, the recently appointed chairman of the biochemistry department of the University of Chicago, who was recruiting actively. As graduate students at Columbia Earl and I had worked at adjacent benches and had become good friends, sharing tastes in literature and music. Without any preliminaries Earl asked me whether I was interested in joining his department as an assistant professor. To accept was an easy decision, even if it meant sacrificing the prospect of skiing the wide open slopes of the Wasatch mountains in Utah. To join Evan's department was especially attractive because an isotope laboratory with a mass spectrometer that functioned more than half of the time had already been set up by Herbert Anker, my first graduate student at Columbia.

Once independent, an investigator may choose to play it safe by continuing earlier, ongoing research or he or she may venture in new directions. The decision is a matter of temperament and imagination, not necessarily of intelligence. I decided to temporize and to play it safe, although not entirely. Sterol biosynthesis, far from being finished, seemed too challenging a problem to abandon. At the same time, the temptation to branch out and to explore some other biosynthetic pathways was difficult to resist. In the early 1950s one of the major unresolved problems that clearly deserved a new look was protein biosynthesis. Several laboratories approached it, in retrospect naively, by examining the formation of peptide or amide bonds. Yet the underlying hypothesis that biosynthetic reactions require energy, specifically ATP, proved ultimately correct and an important first step. John Speck, my neighbor in Chicago, studied glutamine synthesis from glutamate and ammonia. Fritz Lipmann, who had earlier proposed that ATP serves as a general energy carrier for group activation, approached the problem by investigating the acetylation of sulfanilamide, and thus discovered coenzyme A. I chose glutathione synthesis as a model system and carried out some preliminary experiments myself. Robert Johnston took on the problem for his graduate work and could demonstrate that the tripeptide is synthesized in cell-free extracts and that ATP stimulates the process. Several years of hard and skillful work by John Snoke then led to the isolation and partial purification of the two component enzymes for the formation of the tripeptide via γ-glutamylcysteine (9). A few years later (1955), when Paul Zamecnik and Mahlon Hoagland launched the modern era of protein biosynthesis, it became obvious that apart from the ATP requirement, glutathione synthesis had no bearing at all on the mechanism of polypeptide formation. Yet, as my laboratory's first experience in enzyme purification it was not a wasted effort. This and other forays were undertaken by a group of unusually talented and mature graduate students (C. Gilvarg, R. Johnston, I. Zabin, R. Langdon, O. Reiss) eager to work on problems of their own rather than as members of a team. In one such project Charles Gilvarg explored the still unknown origin of the aromatic rings of phenylalanine and tyrosine. Might they originate from acetic acid like the cyclohexane rings of sterols? This proved not to be the case. Instead, the isotopic data seemed compatible with a "condensation of a triose unit with products of glucose metabolism leading possibly to a 7 carbon sugar" (10). When B. Davis discovered shikimic acid as an intermediate in aromatic biosynthesis (1951) with the aid of *Escherichia coli* mutants our traditional approach seemed obsolete. Gilvarg saw the light and joined B. Davis's laboratory contributing to the classical investigations that elucidated the biosynthesis of aromatic amino acids. In the meanwhile studies on the mechanism of cholesterol biosynthesis continued but progressed slowly. The goal was to establish the origin of each of the 27 individual carbon atoms of cholesterol, from acetate methyl or acetate carboxyl, respectively. This formidable task was eventually completed by parallel and complimentary efforts in the laboratories of Cornforth and Popjak and our own (11, 12). Happily, it was a friendly, agreed-upon division of labor, not a race. But the road from acetate to sterol was obviously long, and the chemistry unpredictable. Few if any of the hypotheses that steroid chemists had proposed over the years gave useful guidance or were experimentally testable. Yet one of the clues and speculative ideas, not widely quoted, seemed worth pursuing. My next door laboratory neighbor Thomas Gallagher reminded me of a report by the British nutritionist H. J. Channon showing that feeding of the shark oil hydrocarbon squalene increases the cholesterol content of animal tissues (13). To Channon the chemical structures of neither precursor nor presumed product were known and his evidence based on balance experiments suggestive at best. However, his paper had the virtue of stimulating Robert Robinson to draw a scheme for the mode of cyclization in the hydrocarbon-sterol conversion (14). Sir Robert's discussion of some of the mechanistic problems his scheme raised included the curious but for the times perhaps not atypical statement: "We must not allow the biogenetic tail to wag the chemical dog." In attempts to prove conversion with radioactive biosynthetic squalene, I spent a frustrating but otherwise most enjoyable summer at the Biological Station in Bermuda. All I was able to learn was that sharks of manageable length are very difficult to catch and their oily livers impossible to slice. Much labeled acetate was wasted in these experiments. Fortunately, Robert Langdon, working on his PhD thesis in Chicago, used the simpler, more sensible approach of feeding radioactive acetate to rats along with unlabeled squalene, experiments that proved the squalene-cholesterol conversion.

In 1952 I presented our latest results at an organic colloquium in the Harvard chemistry department, commenting on Robinson's hypothetical scheme for the cyclization of squalene. If my memory is correct, I made the point that the 30-carbon-containing lanosterol would be an attractive first cyclization product but for the fact that the site of the side chain attachment to steroid ring D, whether at C_{15} or C_{17}, was still unsettled. Later in the evening, Robert Woodward asked me whether I had seen the issue of *Helvetica Chimica Acta* that had just arrived. I had not and was delighted to learn that Ruzicka and his associates had now definitive evidence for the C_{17} alternative, proving lanosterol to be a 4,4',14 trimethyl derivative of cholesterol. Woodward and I proceeded to draw a cyclization scheme (B) with the object of accommodating the lanosterol structure as an intermediate on the way to cholesterol. That some methyl groups had to be shifted around in the process did not disturb us. The idea was testable since the predicted but only partly established distribution pattern of acetate-methyl and acetate-carboxyl in labeled cholesterol would distinguish between Robinson's scheme (A) and ours (B). Specifically, the arrangement of 4 of the 27 acetate carbons in cholesterol would be altered. On my return to Chicago I proceeded to degrade an appropriately labeled sample. The origin of C_{13}, one of the critical carbon atoms, conformed with scheme B but not A (15). One year later, during my sabbatical stay in Ruzicka's Institute for Organische Chemie at the ETH in Zurich in 1953, I was able to add to the evidence by establishing the origin of another of the four critical carbon atoms (C_7). This was the last time all of the experimental work was my own. I had decided to work with my own hands in spite of Ruzicka's skepticism. On arrival I had told him my research plans whereupon he asked me when I had received my PhD. To my reply that it was fifteen years ago, he responded, "Then you no longer know how to do experiments." I refused to be deterred, and when six months later I could tell Vlado Prelog that I had verified the prediction, his comment, with a twinkle in his eyes, was "Oh, how dull!" This damper reminded me of Enrico Fermi's statement: "Experimental confirmation of a prediction is merely a measurement. An experiment disproving a prediction is a discovery." I do not necessarily share this view.

The biggest remaining gap still to be filled in the biosynthetic pathway concerned the steps intervening between acetate and squalene. It was by chance that the key intermediate mevalonic acid was encountered in the course of studies on bacterial growth factors. This discovery (Folkers et al 1956) paved the way for the identification of the squalene precursor isopentenyl-pyrophosphate, the long-sought "biological isoprene unit" (16, 17). An account of these developments and the essential contributions of other laboratories (Cornforth & Popjak, Lynen) is given elsewhere (11). Another 20 years passed before I left the intriguing and still unfinished problem of sterol biosynthesis.

In 1954 our family moved east with anticipation that was tempered by regrets to part with friends and colleagues. Ever since I had arrived in Chicago, Earl Evans had given me all possible encouragement. He had become a close family friend. We had no reason to leave except for the yearning to live and bring up our children in a less urban environment. Going to Harvard meant one major change among others. As a member of the chemistry department I was expected to conform with the prevailing teaching requirements. My teaching load, now 3–4 times what it had been in Chicago, seemed inordinately heavy, but in retrospect I have no regrets. Teaching an entire undergraduate course forces one to read outside one's field of specialization.

By the 1950s, microbes and microbial mutants had become popular and proven in many respects superior to rat liver as systems for studying biosynthetic pathways that are often though not always shared by animals and microorganisms. The elegant elucidation of aromatic biosynthesis already mentioned was an outstanding example of the power of the mutant technique. In order to ease the effort of adjusting from familiar to unfamiliar biological systems, I enrolled in the famed microbiology summer course taught by C. B. van Niel at the Hopkins Marine station in Pacific Grove. To qualify for the course, one's record had to be satisfactory in the sense that one was not tainted by prior educational exposure to microbiology elsewhere. The exceedingly demanding course, a bravura performance and a model of pedagogy, taught me important lessons that were to influence much of my later research. First of all, the student was made aware of the enormous variety of microorganisms and their diverse life-styles. Second, I learned from van Niel that Nature allows the investigator to choose organisms uniquely suited for studying a specific biological phenomenon. The awareness that such choices exist is a valuable asset perhaps no longer emphasized sufficiently in the teaching of biology. I benefited especially from one piece of information van Niel mentioned casually during one of his lectures. The class was told that *Saccharomyces,* the organism that led to Pasteur's discovery of "la vie sans air," is in fact microaerophilic, not a strict anaerobe. This bit of information came from a paper by Andreasen & Stier (18), who had noted that in the strict absence of oxygen yeast fails to grow unless supplied with a sterol and an unsaturated fatty acid. I was pleased I could offer to the class a partial biochemical rationale since our laboratory had just proven the origin of the sterol hydroxyl group from molecular oxygen (19). In experiments stimulated by the proposals of Ruzicka (20) and Eschenmoser et al (21) that an electrophilic species of oxygen (OH^+) initiates the cyclization of squalene to sterol, T. T. Tchen was able to show that the OH group of cholesterol is indeed derived from O_2 not from water. It was the same experiment Schoenheimer had asked me to do 15 years earlier. T. T. Tchen had traveled to New York and returned with a glass balloon containing what was probably the world's total supply of $^{18}O_2$ gas. David Rittenberg had generously donated it. The positive result had important ramifications, stimulating my interest in oxygen as an essential biosynthetic reagent, and led me to investigate some consequences for aerobic versus anaerobic patterns of metabolism. To examine the oxygen requirement for the synthesis of oleic acid, implied by the nutritional requirement of anaerobic yeast, became D. Bloomfield's research project after I had returned east. Working with yeast microsomes he demonstrated that oxygen was indeed essential for the stearoyl-CoA→oleoyl-CoA conversion (22). It was an obvious question to ask next, obvious at least to someone indoctrinated by van Niel, how unsaturated fatty acids arise in obligate anaerobes as they do. Possible choices were (a) an electron acceptor other than O_2, or (b) an entirely different mechanism. Guided by studies on the structures of olefinic acids in lactobacilli (K. Hofman), we could eventually formulate and document a bacterial mechanism for olefin formation, novel in that it involved dehydration of medium-chain β-hydroxy acids rather than oxidative dehydrogenation of the corresponding saturated acid (23). The requisite enzyme, β-hydroxydecanoyl thioester dehydrase, that we isolated from *E. coli*, catalyzed the postulated olefin-generating dehydration step. Fortunately the enzyme also functioned as a reversible $\beta,\gamma \rightarrow \alpha,\beta$ C_{10}-enoyl thioester isomerase. Conveniently for enzyme assay but also of major consequence later on, a peak at 263 nm appears in the β,γ-α,β enoate isomerization. In crude *E. coli* extracts these transformations required the presence of a heat-stable protein (24), with properties reminiscent of a similar factor Vagelos and coworkers had just shown to be essential for fatty acid synthesis in *Cl. Kluyveri* (25). Before proceeding further with our studies Roy Vagelos and I had arranged to get together during the 1961 federation meeting in Atlantic City. Over cocktails, Roy and his coworkers Al Alberts and Peter Goldman, and Ann Norris, Bill Lennarz, and I from my laboratory, discussed the possible role of the heat-stable protein, known since as ACP, or acyl carrier protein (R. Vagelos, S. Wakil). Our two groups readily agreed not to compete but to pursue their separate objectives that had led to the chance discovery of ACP. Roy's laboratory was primarily interested in the mechanism of chain elongation, while we wanted to know how unsaturated fatty acids are formed anaerobically. I believe neither side had reasons later on to regret this amicable arrangement.

The substrate for dehydrase was unavailable commercially, a fortunate circumstance as it turned out. We prepared it by Raney-Ni reduction of 3-decynoic acid to the corresponding 3-olefin and subsequent thioesterification with N-acetylcysteamine (NAC). One of several identically prepared batches of 3-decenoyl NAC, luckily not the one made first, totally failed in the optical test with dehydrase. David Brock's detective work provided the first clue. In the run in question Raney-Ni reduction was incomplete, leaving

5% of the acetylenic decynoic acid unchanged. Subsequent thioesterification therefore afforded a mixture containing 95% olefinic thioester along with 5% 3-decynoyl-NAC. Was this contaminant to blame for the failed assay? The answer seemed to be yes since addition of this impure substrate to a dehydrase assay system completely inhibited the isomerization of authentic 3-decenoyl NAC. When pure 3-decynoyl thioester was prepared and tested, it inhibited dehydrase noncompetitively and irreversibly with a K_I of 1×10^{-7} M (26). Complete enzyme inactivation occurred with 1 mol of inhibitor per mole of enzyme and led to covalent modification of a single histidine residue at the active site (27). To be effective, the inhibitor had to be structurally identical with the substrate in all respects except for substitution of the olefinic bond by a triple bond. During one of the daily afternoon coffee hours, a cherished tradition I inherited from my days in Schoenheimer's lab and have continued ever since, the chemistry of the acetylene thioester—enzyme histidine interaction was the subject of lively discussion. It produced the important bit of information that acetylenes readily isomerize to the corresponding allenes. Intriguingly, George Helmkamp had already seen a gradual spectral change when decynoyl-NAC was kept in ethanolic solution. A new peak absorbing at 263 nm appeared and increased with time. Evidently the acetylenic thioester isomerized spontaneously in a reaction analogous mechanistically to the normal β,γ-α,β enoate transformation that dehydrase catalyzes. To our surprise the inhibitory potency of the solution increased rather than declined. Indeed, the pure allene, 2,3 decadienoyl-NAC, inhibited dehydrase much more rapidly than the acetylenic isomer, proving that the allene was the true inhibitory species (28). Still to be answered was the question of whether the spontaneous acetylene-allene transformation was sufficiently rapid to account for the result or whether the isomerization was enzyme-catalyzed. This issue was resolved by comparing kinetic isotope effects during dehydrase inhibition by 2-dideuterio 3-decynoyl-NAC and 2,3 deuterio decadienoyl-NAC. A k_H/k_D of 2.3 was found for the acetylenic thioester, while it was unity for the allenic analogue. Enzyme-catalyzed removal of the proton from the acetylenic thioester proved to be the rate-limiting step in the interaction between inhibitor and enzyme. Since the same enzyme that performs the physiological enoyl thioester isomerization during unsaturated fatty acid synthesis in *E. coli* also catalyzes the acetylene-allene isomerization, we were dealing with the rather unique case of an enzyme promoting its own destruction; i.e. the catalyzed transformation of a substrate analogue generated an active site probe of extreme chemical reactivity. It is fortunate that in this instance, as in many others, enzyme specificity is a matter of degree, not absolute. Today the phenomenon is more popularly, but perhaps not quite aptly, known as "enzyme suicide." I have argued, not very successfully, that this anthropomorphical term connotes a deliberate act on the part of the affected enzyme, but this is not strictly the case. Dehydrase falls prey to trickery because it fails to distinguish between the C_α protons of the acetylenic and the olefinic substrates, an error that seals its fate. "Mechanism-based enzyme inactivation" is gaining ground for describing the phenomenon, but whether it will catch on remains to be seen. I have always been quite indifferent on matters of terminology as long as the meaning is clear.

At least in principle, 3-decynoyl-NAC comes one step closer to Paul Ehrlich's concept of a "magic bullet." The substance inhibits the growth of *E. coli* and various other bacteria (29), an effect that is fully relieved by supplying the organism with oleic or vaccenic acids, the ultimate products of the inhibited reaction. The inhibitor seeks out a single susceptible enzyme. Animal cells that generate oleic acid by an entirely different mechanism lack a comparable target and therefore 3-decynoyl NAC is nontoxic to animals. Unfortunately the compound failed the most critical test a potential antibiotic must pass, i.e. to protect an organism against infection. The reasons are not known, but conceivably the lifetime of the thioester is too brief. Metabolic instability or side effects cannot be anticipated, and therefore the hopes for rationally designed therapeutic agents must remain guarded.

The subject of anaerobic versus aerobic life-styles in the context of evolution has continued to hold my interest ever since it came to my attention in van Niels' course. An example is the essential role of oxygen in the conversion of squalene to sterol. This requirement necessarily dates the appearance of cholesterol and cholesterol-derived metabolites to an era that followed or coincided with the arrival of aerobic cells during evolution. I pursued this line of reasoning in an essay contributed to a "Festschift" for Severo Ochoa on the occasion of his 70th birthday (30). Here I considered the structural and functional consequences arising from the stepwise enzymatic modification of the lanosterol molecule during conversion to cholesterol. Speculating on the motives of nature, I postulated that along with the sequential departure of the three nuclear methyl groups, in the order 14α-methyl, 4α-methyl, and 4β-methyl, the fitness of the molecule improves, reaching perfection with cholesterol. If this could be documented, functional improvement should parallel the sequence of steps the contemporary pathway employs. The results of subsequent research in recent years, mainly by Jean and Charles Dahl, both with model membranes and sterol auxotrophs, support this notion. Fitness for biological function, not chance, appears to be the driving force for structural modifications of a biomolecule (31). As for the function of sterols in membranes, the chemist could not have improved on nature—or perhaps has not so far. I am intrigued by the notion that hierarchies or Darwinian evolution are manifest at the level of small molecules as well as at the organismic and genomic level. In the course of these studies we have been led to a trail we are still following. There are indications for a novel function of the sterol molecule, involving control of phospholipid biosynthesis. Perhaps cholesterol started out as a hormonelike molecule serving as signal for the assembly of certain cell membranes. I hope to pursue this research a while longer.

One thought comes to mind once the hectic pace of teaching and research becomes a memory. Whatever the motives, whether curiosity or ambition—usually a combination of both—only near the end does one fully appreciate the rewards and privileges that go with a career in science. So much the better if the results should prove to have some degree of permanence. Science is indeed a glorious enterprise, and has been for me, I admit, glorious entertainment.

Literature Cited

1. Bloch, K. 1936. *Biochem. Z.* 285:372–85
2. Bloch, K. 1936. *Z. Physiol. Chem.* 244:1–13
3. Bloch, K. 1938. *J. Biol. Chem.* 125:275–82
4. Borsook, H., Dubnoff, J. W. 1940. *Science* 91:551–55
5. Bergmann, M., Zervas, L. 1927. *Z. Physiol. Chem.* 172:277–88
6. Sonderhoff, R., Thomas, H. 1937. *Ann. Chem.* 530:195–213
7. Ottke, R. C., Tatum, E. L., Zabin, I., Bloch, K. 1950. *J. Biol. Chem.* 189:419–33
8. Bloch, K. 1945. *J. Biol. Chem.* 157:661–66
9. Snoke, J. E., Bloch, K. 1954. In *Glutathione*, pp. 129–37. New York: Academic
10. Gilvarg, C., Bloch, K. 1952. *J. Biol. Chem.* 199:689–98
11. Bloch, K. 1965. *Science* 150:19–28
12. Cornforth, J. W., Hunter, G. D., Popjak, G. 1953. *Biochem. J.* 54:590–97
13. Channon, H. J. 1926. *Biochem. J.* 20:400–8
14. Robinson, R. 1934. *J. Chem. Soc. Ind.* 53:1062–63
15. Woodward, R. B., Bloch, K. 1953. *J. Am. Chem. Soc.* 75:2023
16. Bloch, K. 1958. *Proc. 4th Int. Congr. Biochem.* 4:50–55
17. Chaykin, S., Law, J., Phillips, A. H., Tchen, T. T., Bloch, K. 1958. *Proc. Natl. Acad. Sci. USA* 44:998–1004
18. Andreasen, A. A., Stier, T. J. B. 1954. *J. Cell. Comp. Physiol.* 15:119–25
19. Tchen, T. T., Bloch, K. 1956. *J. Am. Chem. Soc.* 78:1516–17
20. Ruzicka, L. 1953. *Experientia* 9:359–67
21. Eschenmoser, A., Ruzicka, L., Jeger, D., Arigoni, D. 1955. *Helv. Chim. Acta* 38:1890–904
22. Bloomfield, D., Bloch, K. 1960. *J. Biol. Chem.* 235:337–45
23. Bloch, K. 1969. *Acc. Chem. Res.* 2:193–202
24. Lennarz, W. J., Light, R. J., Bloch, K. 1962. *Proc. Natl. Acad. Sci. USA* 48:840–46
25. Goldman, P., Alberts, A. W., Vagelos, P. R. 1961. *Biochem. Biophys. Res. Commun.* 5:280–85
26. Brock, D. J. H., Kass, L. R., Bloch, K. 1967. *J. Biol. Chem.* 242:4432–40
27. Helmkamp, G., Bloch, K. 1969. *J. Biol. Chem.* 244:6014–22
28. Endo, K., Helmkamp, G. M., Bloch, K. 1970. *J. Biol. Chem.* 245:4293–96
29. Kass, L. R. 1968. *J. Biol. Chem.* 243:3223–28
30. Bloch, K. 1976. In *Reflections in Biochemistry*, ed. A. Kornberg, B. Horecker, L. Cornudella, J. Oro, pp. 143–50. New York: Pergamon
31. Bloch, K. 1983. *CRC Crit. Rev. Biochem.* 14:47–92

Ann. Rev. Biochem. 1987. 56:21–42
Copyright © 1987 by Annual Reviews Inc. All rights reserved

FRACTIONATION AND STRUCTURAL ASSESSMENT OF OLIGOSACCHARIDES AND GLYCOPEPTIDES BY USE OF IMMOBILIZED LECTINS

T. Osawa and T. Tsuji

Division of Chemical Toxicology and Immunochemistry, Faculty of Pharmaceutical Sciences, University of Tokyo, Bunkyo-ku, Tokyo 113, Japan

CONTENTS

INTRODUCTION

A number of recent studies have suggested that glycoconjugates on the cell surface constitute a great number of surface markers of the cells in various differentiation stages and with various functions. Furthermore, it has become apparent that these cell-surface markers control and determine cell-ligand and cell-cell interactions in many important cellular phenomena. In order to

21

clarify the molecular mechanisms of these cellular phenomena and to study the changes in cell-surface characteristics during differentiation, elucidation of the structure of the sugar chains of these membrane glycoconjugates is indispensable. However, in most cases, membrane glycoconjugates are very difficult to isolate in sufficient quantities for structural studies by conventional methods. Moreover, these glycoconjugates usually exhibit microheterogeneity in the structure of their carbohydrate moieties.

Lectins are sugar-binding proteins that agglutinate cells or precipitate glycoconjugates (1). Each lectin binds specifically to a certain sugar sequence in oligosaccharides and glycoconjugates. Therefore, these lectins have been widely used for histochemical detection of sugar chains on the cell surface (2, 3), staining and structural estimation of electrophoretically separated membrane glycoconjugates (4–7), separation of immunocyte subsets (2, 8) and the cells in different differentiation stages (9–11), and isolation of membrane glycoconjugates on a preparative scale by affinity chromatography (12).

Recently, many attempts have been made to fractionate and purify oligosaccharides and glycoconjugates by affinity chromatography by the use of immobilized lectin columns. The use of a series of different lectin columns whose binding specificities have been precisely elucidated enables us to fractionate a very small amount of radiolabeled oligosaccharides or glycopeptides (ca 10 ng depending on the specific activity) into structurally distinct groups, and to obtain general pictures of their structures, which make the subsequent structural study much easier.

In this review we summarize the recent advances in the application of immobilized lectin columns to the fractionation and the structural study of various oligosaccharides and glycopeptides.

PREPARATION OF IMMOBILIZED LECTIN COLUMNS

Various methods for coupling proteins to insoluble adsorbents can be used for the preparation of immobilized lectins. However, since some lectins are unstable under the conditions employed for the coupling, careful examination of the coupling conditions is always necessary for each lectin to avoid impairment of the binding activity of the lectin. In most cases, the coupling reaction is carried out in the presence of a haptenic sugar for the lectin to protect the binding site. The most popular method is to link lectins to agarose gel by the cyanogen bromide (CNBr) method (13–15). Agarose derivatives are easy to prepare and have ideal properties for affinity chromatography: namely, good flow rate and a very loose porous network that permits entry and exit of large molecules.

However, the following disadvantages can be pointed out for the immobilized lectins prepared by the CNBr method: 1. The pH of the eluent should not be alkaline, because, under alkaline conditions, some of the lectin molecules coupled to agarose gel exhibit slight but constant leakage from the solid matrix (16). 2. Since the N-substituted isourea formed on coupling of amino groups of a lectin to CNBr-activated agarose retains the weak basicity of the amino group, the column works as an ion-exchange column when the ionic strength of the eluent is low (17, 18). These disadvantages can be overcome by introducing linear polyacrylic hydrazide as a spacer between the solid matrix and the lectin molecule (19). The lectin column thus prepared is free of ion-exchange properties and, moreover, this column is stable because polyacrylic hydrazide is coupled to agarose at numerous positions. The agarose derivatives prepared by the epoxy method using bisoxirane (20) or epichlorohydrin (21) are also free of charge (22) and quite stable under alkaline conditions (21). However, this method is not adequate for the preparation of lectin columns, because the coupling to epoxy-activated agarose requires a high temperature (25–45°C), a high pH (9–14), and a large excess of the ligand. Matsumoto et al (23) converted epoxy-activated agarose into amino derivatives by using ammonia solution, and then the amino derivatives of agarose were succinylated with succinic anhydride and activated with N-hydroxysuccinimide according to the method of Cuatrecasas & Parikh (24). Now the reaction of coupling lectins to the activated agarose thus prepared can be performed under mild conditions. The adsorbent thus obtained has no charged groups in the linkage region between the lectin and agarose, reducing the nonspecific adsorption, and the bonds formed are stable even in an alkaline medium. Furthermore, a long spacer between the solid matrix and lectin molecules can be introduced by this method, minimizing steric interference between the lectin molecule and interacting substances.

Ito et al (25) converted epoxy-activated agarose into formyl-agarose by amination and subsequent reductive amination of the amino-agarose (23) with glutaraldehyde and NaCNBH$_3$. Various lectins were then successfully immobilized on formyl-agarose under mild conditions (pH 7.0, 4°C) during a short reaction period by reductive amination with NaCNBH$_3$. The amount of concanavalin A (Con A) immobilized on agarose by this method was found to be five times that of Con A immobilized on agarose by the CNBr method.

THE TYPE OF SUGAR CHAIN EACH LECTIN CAN RECOGNIZE ON THE CELL SURFACE

The interaction of lectins with cells can generally be inhibited by simple sugars. Mäkelä divided simple sugars into four groups based on the configuration of the hydroxyl groups at C-3 and C-4, and indicated the lectins that could be inhibited by each group of sugars (26). Since these lectins react with different cell surface structures, they have been used as molecular probes for investigations of the architecture of the cell surface. However, in these investigations, the results are often correlated simply with the specificities of lectins disclosed by hapten inhibition assays with simple sugars used as inhibitors. However, the results of hapten inhibition assays using simple sugars as hapten inhibitors cannot always be correlated with the structure of the binding sites on the cell surface.

We can divide the structure of the sugar chains of cell surface glycoproteins into two groups termed serine (or threonine)-linked sugar chains (mucin-type sugar chains) and asparagine-linked sugar chains. The latter group is further subdivided into three groups termed the high-mannose-type, complex-type, and hybrid-type. Typical examples of these sugar chains (27–30) are shown in Figure 1.

From the results of hemagglutination or precipitation inhibition assays using various simple sugars and oligosaccharides as hapten inhibitors (31–37), the lectins are divided into three groups based on the type of sugar chains to which they preferentially bind on the cell surface as shown in Table 1.

Even among the same group of lectins in Mäkelä's classification, differences can be seen in the type of sugar chains to which they preferentially bind. For example, in the case of galactose-binding lectins, *Agaricus bisporus* (mushroom) lectin, *Arachis hypogaea* (peanut) lectin, and *Bauhinia purpurea* lectin primarily bind to mucin-type sugar chains, whereas *Ricinus communis* lectin binds primarily to complex-type or hybrid-type sugar chains. It is also of interest that all mitogenic lectins belong to the group of lectins that bind preferentially to asparagine-linked sugar chains on the cell surface.

Recent investigations on binding specificities of various lectins indicate that even the lectins in the same group in Table 1 often bind to different sugar sequences in the same sugar chain. Detailed information on the structure of the sugar sequence that a lectin preferentially recognizes can be obtained by the use of affinity chromatography on the column on which the lectin is immobilized.

ELUCIDATION OF THE BINDING SPECIFICITIES OF INDIVIDUAL LECTINS BY THE USE OF THE IMMOBILIZED LECTIN COLUMN

Lectins that Bind to Asparagine-Linked Sugar Chains

CONCANAVALIN A (CON A)   Goldstein and his coworkers (38–40) were the first to clearly demonstrate through hapten inhibition of the precipitation of dextran with Con A that unmodified hydroxyl groups at the C-3, C-4, and C-6 positions of a glucopyranosyl residue are essential for the interaction of Con A, whereas certain modifications at the C-2 position are permitted for the binding of the lectin. Thereafter, the binding specificity of Con A was studied

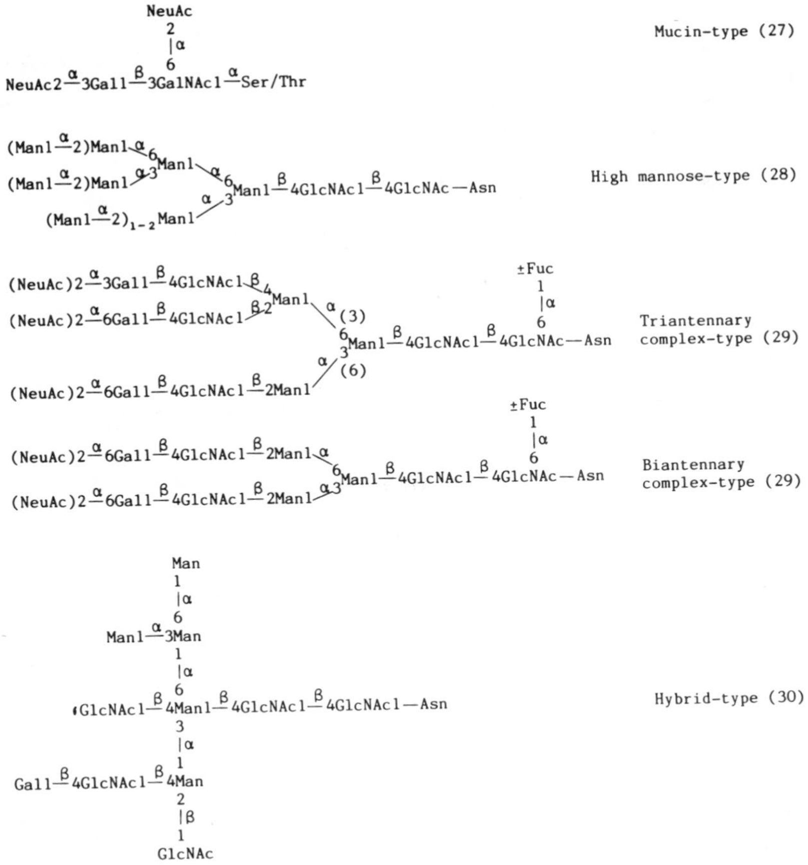

```
        NeuAc
          2                                          Mucin-type (27)
          |α
          6
NeuAc2─α─3Gal1─β─3GalNAc1─α─Ser/Thr


(Man1─α─2)Man1─α
                  ⁶Man1
(Man1─α─2)Man1─α─³      α       β        β
                  ⁶Man1─4GlcNAc1─4GlcNAc—Asn    High mannose-type (28)
                α─³
  (Man1─α─2)₁₋₂Man1


                                                    ±Fuc
(NeuAc)2─α─3Gal1─β─4GlcNAc1─β                          1
                          ⁴Man1                        |α
(NeuAc)2─α─6Gal1─β─4GlcNAc1─β─2   α  (3)              6      Triantennary
                              ⁶Man1─β─4GlcNAc1─β─4GlcNAc—Asn  complex-type (29)
                            α─³ (6)
(NeuAc)2─α─6Gal1─β─4GlcNAc1─β─2Man1


                                                    ±Fuc
                                                      1
(NeuAc)2─α─6Gal1─β─4GlcNAc1─β─2Man1─α                 |α
                                    ⁶Man1─β─4GlcNAc1─β─4GlcNAc—Asn   Biantennary
(NeuAc)2─α─6Gal1─β─4GlcNAc1─β─2Man1─α─³               6      complex-type (29)
```

```
              Man
               1
               |α
               6
      Man1─α─3Man
               1
               |α
               6
 ₁GlcNAc1─β─4Man1─β─4GlcNAc1─β─4GlcNAc—Asn    Hybrid-type (30)
               3
               |α
               1
Gal1─β─4GlcNAc1─β─4Man
               2
               |β
               1
             GlcNAc
```

Figure 1   Various types of sugar chains of glycoproteins.

by many investigators by a variety of approaches using simple sugars and oligosaccharides as inhibitors. The results of these studies were reviewed by Goldstein & Hayes (41).

Kornfeld & Ferris (42) measured the potency of the glycopeptides obtained from immunoglobulins as hapten inhibitors of ^{125}I-Con A binding to guinea pig erythrocytes. The most potent hapten had a branched chain oligosaccharide with two GlcNAc$\beta1\rightarrow$2Man$\alpha1\rightarrow$3(6)Man in the core. Removal of the GlcNAc residues from the chain somewhat reduced the activity of the glycopeptide. Debray et al (43) carried out hemagglutination inhibition assays using various asparagine-linked glycopeptides as hapten inhibitors, and demonstrated that Con A has a great affinity for the trimannosidic core Man$\alpha1\rightarrow$6(Man$\alpha1\rightarrow$3)Man substituted by two GlcNAc residues, but the

Table 1 The type of sugar chains to which an individual lectin preferentially binds

Type of sugar chains	Lectins	Simple sugar specificity
Ser/Thr-linked sugar chains (mucin-type chains)	*Agaricus bisporus* (mushroom)	Gal
	Arachis hypogaea (peanut)	Gal
	Bauhinia purpurea	Gal
	Iberis amara	Gal
	Maackia amurensis (hemagglutinin)	—
	Maclura pomifera	GalNAc
	Vicia graminae	Gal
	Vicia villosa	GalNAc
Asparagine-linked sugar chains	Concanavalin A[a]	Man
	Datura stramonium	GlcNAc
	Lens culinaris (lentil)[a]	Man
	Maackia amurensis (mitogen)[a]	—
	Phaseolus vulgaris[a]	GalNAc
	Phytolacca americana (pokeweed)[a]	GlcNAc
	Pisum sativum (pea)[a]	Man
	Ricinus communis	Gal
	Vicia faba (fava)[a]	Man
	Wistaria floribunda (mitogen)[a]	GalNAc
Either type of sugar chains	*Sophora japonica*	Gal
	Wheat germ	GlcNAc
	Wistaria floribunda (hemagglutinin)	GalNAc

[a]Potent mitogenic activity for human lymphocytes is observed.

affinity is reduced when these GlcNAc residues are substituted by galactosyl residues.

The structural basis of the interaction of immobilized Con A with oligosaccharides and glycopeptides was first investigated by Ogata et al (44). From the results of binding of ^{3}H-labeled oligosaccharides and glycopeptides to the Con A–agarose column, they concluded that at least two nonsubstituted or 2-*O*-substituted α-mannopyranosyl residues are required to be retained on the Con A–agarose column (Figure 2). Krusius et al (45) also chromatographed *N*-[^{3}H]-acetylated acidic *N*-glycosidically linked glycopeptides on a Con A–agarose column, and bound and unbound fractions were analyzed by methylation. They found that glycopeptides possessing two peripheral NeuAc-Gal-GlcNAc branches linked to the core (biantennary complex-type) were bound by the lectin, whereas glycopeptides with three branches (triantennary complex-type) were not (Figure 2). These findings are in good agreement with those of Ogata et al (44). More recently, Narasimhan et al (46, 47) further tested the binding of *N*-[^{3}H]-acetylated biantennary glycopeptides to a column of Con A–agarose. They found that, of two peripheral mannose residues, the Manα1–6 branch plays a more decisive role than the Manα1–3 branch in controlling interaction with Con A–agarose. Thus, a terminal GlcNAc group on the Manα1→6 branch favors very tight binding, but the substitution of this GlcNAc group with a Gal group greatly weakens the binding to Con A–agarose. Furthermore, they found that the presence of a "bisecting" GlcNAc group, linked 1–4 to the β-Man residue of the trimannosyl core (Figure 2), markedly weakens the interaction with Con A–agarose.

Baenziger & Fiete (48) carried out quantitative assays of the binding of various ^{125}I-labeled glycopeptides to Con A, and the ratios of the bound glycopeptide to the unbound glycopeptide were plotted according to Scatchard (49) to calculate the association constants of those glycopeptides. These association constants were compared with the ability of the glycopeptides to be retained on a Con A–agarose column. Some of their results are shown in Figure 3. Glycopeptides with association constants (K_a) in the range of $4.5 \times 10^6 \ M^{-1}$ to $25 \times 10^6 \ M^{-1}$ were retained by a Con A–agarose column, while glycopeptides with association constants in the range of $0.3 \times 10^6 \ M^{-1}$ to $4.0 \times 10^6 \ M^{-1}$ were not retained by but retarded on a Con A–agarose column. This study confirmed that the presence of the two α-linked mannose residues with unmodified hydroxyl groups at C-3, C-4, and C-6 is essential for interactions with the association constants of $4.5 \times 10^6 \ M^{-1}$ or greater. Their study also showed that 1. sequential removal of peripheral sugars on branches arising from C-2 of the outer mannose residues results in a progressive increase in the association constants, 2. the presence of a β1→4-linked GlcNAc residue on the β-linked mannose residue (bisecting GlcNAc residue) decreases the association constant to $2.0 \times 10^6 \ M^{-1}$ or less, and 3. the presence of a fucose residue on the innermost GlcNAc residue in the core does not significantly affect the association constant of a glycopeptide.

Ohyama et al (50) investigated the interaction of glycopeptides obtained from ovalbumin with a Con A–agarose column by frontal affinity chromatography. In this method, the retardation of the elution front of the labeled glycopeptide was measured and the dissociation constants for the glycopeptides were calculated (51–53). From their results they concluded that Con A specifically binds at high affinity to glycopeptides that contain the trimannose structure Manα1→6(Manα1→3)Man. The anomeric configura-

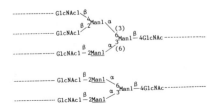

Figure 2 Structural requirements for the binding to Con A–agarose.

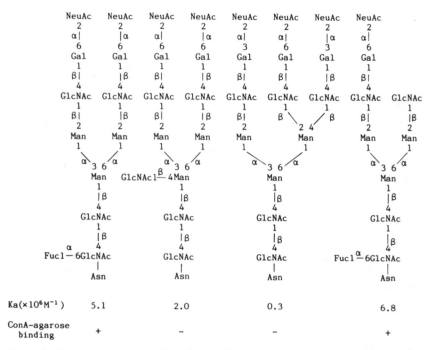

Figure 3 Association constants with and the ability of binding to Con A–agarose of various complex type glycopeptides.

tion of the central mannose of the trimannose structure is not critical. The C-3 hydroxyl group of the C-6-linked mannose and the C-4 hydroxyl group of the C-3-linked mannose should be unmodified. Substitution at these positions even by a mannose residue markedly weakens the binding. The C-2 positions of both the C-3- and C-6-linked mannose residues need not necessarily be free. The glycopeptides containing the structure that satisfies the above conditions can bind strongly to Con A–agarose with dissociation constants below 3.4×10^{-7} M. They also observed that the substitution of C-6-linked mannose at the C-2 position by an α-mannose residue or a β-GlcNAc residue enhances the binding to Con A–agarose.

LENTIL, PEA, AND FAVA LECTINS   These lectins are inhibited by Mäkelä's group III sugars (41, 54), and have sugar-binding specificities similiar to that of Con A. However, subtle differences in the sugar-binding specificities were observed among these lectins in inhibition assays with simple sugars used as hapten inhibitors. For example, the best inhibitor among the monosaccharide derivatives for lentil, pea, and fava lectins, but not for Con A, is methyl

2,3-di-*O*-methyl-α-D-glucopyranoside (54). Furthermore, the lentil lectin was shown to bind preferentially to the sequence GlcNAcβ1-2Man (32, 55, 56). Debray et al (43) demonstrated that the presence of an α1→6-linked fucose at the innermost GlcNAc residue in the core is essential for the maximum binding of a glycopeptide to the lentil lectin, but not for binding to Con A.

The sugar-binding specificities of immobilized lentil and pea lectins were first studied by Kornfeld et al (57). They showed that, in both cases, two α-mannosyl residues were required for glycopeptide binding. Substitution of the α-mannosyl residues at C-2 or substitution of one of the α-mannosyl residues at C-2 and C-6 did not prevent the binding to the two immobilized lectins. They also showed that the presence of a fucose residue attached to the innermost GlcNAc residue in the core of glycopeptides is essential for high-affinity binding to the immobilized lentil and pea lectins.

Yamamoto et al (58) compared the structural requirements of various glycopeptides for binding to the immobilized Con A, lentil, and pea lectins. Their results indicate that an intact GlcNAc residue at the reducing end of a complex-type oligosaccharide is essential for high-affinity binding to lentil lectin–agarose but not to Con A–agarose and that even an asparagine residue is required for high-affinity binding to pea lectin–agarose. In addition, interaction of a complex-type oligosaccharide with lentil lectin–agarose was enhanced by exposure of nonreducing terminal GlcNAc groups, whereas interaction with pea lectin–agarose was enhanced only after exposure of nonreducing, terminal α-mannopyranosyl groups.

The sugar-binding specificity of the immobilized fava lectin is similar to that of the immobilized pea lectin (59). The immobilized fava lectin can interact only with biantennary glycopeptides possessing an α-L-fucosyl group attached to the innermost GlcNAc residue in the core. An asparagine residue is essential for high-affinity binding to fava lectin–agarose. In addition, the enzymic exposure of a nonreducing terminal α-mannopyranosyl group enhances the interaction of the glycopeptide with the immobilized fava lectin.

These structural requirements for the binding of various oligosaccharides and glycopeptides to the immobilized lentil, pea, and fava lectins are shown in Figure 4 in comparison with those for binding to Con A–agarose.

WHEAT GERM AGGLUTININ (WGA)   The binding specificity of WGA has been studied in several laboratories by numerous methods (for review see Ref. 41). GlcNAc and its β1→4 oligomers have been found to be potent inhibitors of WGA in agglutination (60–63), in mitogenic stimulation of lymphocytes (64), and in precipitation of glycoconjugates (63, 65). Several workers have shown that the binding of WGA to cells or glycopeptides is decreased after treatment with neuraminidase (66–69). *N*-Acetylneuraminic acid residues were therefore implicated as important factors in WGA interactions (60, 66,

ELUTION RETARDED BY

	CON A	LENTIL	PEA	FAVA
Galβ→4GlcNAcβ→2Manα ⟍₆ Manβ→4GlcNAcβ→4GlcNAc — Asn (Fuc α1→6) ; Galβ→4GlcNAcβ→2Manα ⟋³	+	+	+	+
Galβ→4GlcNAcβ ⟍₄ Manα ; Galβ→4GlcNAcβ ⟋² ⟍₆ Manβ→4GlcNAcβ→4GlcNAc — Asn (Fuc α1→6) ; Galβ→4GlcNAcβ→2Manα ⟋³	—	—	—	—
Galβ→4GlcNAcβ→2Manα ⟍₆ Manβ→4GlcNAcβ→4GlcNAcol (Fuc α1→6) ; Galβ→4GlcNAcβ→2Manα ⟋³	+	—	—	—
Galβ→4GlcNAcβ ⟍₄ Manα ; Galβ→4GlcNAcβ ⟋² ⟍₆ Manβ→4GlcNAcβ→4GlcNAcol (Fuc α1→6) ; Galβ→4GlcNAcβ→2Manα ⟋³	—	—	—	—
Galβ→4GlcNAcβ→2Manα ⟍₆ Manβ→4GlcNAcβ→4GlcNAc (Fuc α1→6) ; Galβ→4GlcNAcβ→2Manα ⟋³	+	+	—	—
Galβ→4GlcNAcβ→2Manα ⟍₆ Manβ→4GlcNAcβ→4GlcNAc — Asn ; Galβ→4GlcNAcβ→2Manα ⟋³	+	—	—	—

Figure 4 Interaction of various oligosaccharides and glycopeptides with lentil-agarose, pea-agarose, and fava-agarose.

68, 70, 71). However, the inhibitory effect of N-acetylneuraminic acid is weaker than that of GlcNAc (66), and the presence of clustering sialyl residues may be necessary for the strong interaction of sialoglycoconjugates with WGA (72).

WGA-agarose has been widely used for the purification of many glycoproteins and glycopeptides (72–74), but structural requirements for the binding of oligosaccharides and glycopeptides to immobilized WGA were poorly understood. Yamamoto et al (75) tested the binding of chitin oligosaccharides, which were reduced with NaB^3H_4, and several N-[^3H]-acetylated asparagine-linked glycopeptides to WGA-agarose. Although N,N'-diacetylchitobiose alditol was not bound to WGA-agarose, N,N',N''-triacetylchitotriose alditol and N,N',N'',N'''-tetraacetylchitotetraose alditol were retained on the column, indicating that at least an intact N,N'-diacetylchitobiose moiety is required for strong binding to WGA-agarose. Furthermore, the affinity of chitin oligosaccharide alditol for WGA-agarose was found to increase with the increasing number of GlcNAc residues in agreement with the data of the precipitation inhibition experiments reported by Goldstein et al (65). When a high-mannose-type glycopeptide obtained from porcine thyroglobulin was applied to a column of WGA-agarose, it was recovered without any retardation in spite of the fact that it contained an intact N,N'-diacetylchitobiose structure in the core portion. A biantennary or triantennary complex-type

$(NeuAc)2\overset{\alpha}{-}6Gal1\overset{\beta}{-}4GlcNAc1\overset{\beta}{-}2Man1\searrow_{\alpha}$

$GlcNAc1\overset{\beta}{-}4$

$(NeuAc)2\overset{\alpha}{-}6Gal1\overset{\beta}{-}4GlcNAc1\overset{\beta}{-}2Man\nearrow^{\alpha}$

Fuc
1
$|\alpha$
6
$Man1\overset{\beta}{-}4GlcNAc1\overset{\beta}{-}4GlcNAc-Asn$

Figure 5 The complex-type glycopeptide obtained from human erythrocyte glycophorin A.

glycopeptide obtained from porcine thyroglobulin was also eluted in the void volume. On the other hand, hybrid-type glycopeptides prepared from ovalbumin were found to be retarded on a WGA-agarose column. These hybrid-type glycopeptides were modified by glycosidase treatment, Smith periodate degradation, acetolysis, and hydrazinolysis, and the products were tested for binding to WGA-agarose. The results showed that the GlcNAcβ1$\rightarrow$4Manβ1$\rightarrow$4GlcNAcβ1$\rightarrow$4GlcNAc-Asn structure is essential for tight binding of glycopeptides to a WGA-agarose column. An asialo-complex-type glycopeptide with a bisecting GlcNAc residue (Figure 5), which was prepared from human erythrocyte glycophorin A (76), was not retained by a WGA-agarose column, but after removal of an α-fucosyl residue attached to C-6 of the innermost GlcNAc residue, it was retarded by the column, indicating that the α-fucosyl residue in the core inhibits the interaction of the glycopeptide with WGA-agarose by steric hindrance.

Gallagher et al (77) investigated the interaction of sialylated glycopeptides isolated from the surface membrane of a murine hemopoietic cell line (416B) with WGA-agarose. After removal of the glycopeptides that were retained by columns of Con A–agarose and lentil lectin–agarose, the remaining glycopeptides were fractionated into three fractions (WGA-W, WGA-I, and WGA-S), which had weak, intermediate, and strong affinity for WGA-agarose, respectively. It was found that the binding of the WGA-W fraction was dependent on *N*-acetylneuraminic acid, whereas the WGA-S fraction interacted with the immobilized WGA by an *N*-acetylneuraminic acid–independent mechanism. Degradation of the WGA-S fraction with various exo- and endo-glycosidases showed that this fraction consisted of poly(*N*-acetyllactosamine)-type glycans (Figure 6). Since this type of sugar chain is included in cell-surface glycoproteins of various types of mammalian cells (78–82), such as a band-3 glycoprotein of human erythrocyte membranes (78, 79), these findings may have revealed what is the major binding site for WGA on the surface of various mammalian cells.

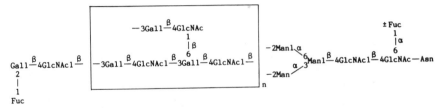

Figure 6 Poly(*N*-acetyllactosamine)-type glycopeptide.

POKEWEED MITOGENS   Börjeson and coworkers (83, 84) were the first to isolate a hemagglutinating and mitogenic lectin from pokeweed *(Phytolacca americana)* roots. Waxdal (85) and Yokoyama et al (86) separated five mitogenic lectins from pokeweed roots, of which one (Pa-1) was mitogenic for both B and T cells and the others (Pa-2, 3, 4, and 5) were mitogenic only for T cells (86, 87). The sugar-binding specificities of two major lectins (Pa-1 and 2) in pokeweed roots were studied by means of hemagglutination inhibition and the quantitative inhibition assay for the binding of ^{125}I-labeled lectins to human erythrocytes, with various oligosaccharides, glycopeptides, and glycoproteins used as inhibitors (36). Of the inhibitors used, chitin oligosaccharides and the glycopeptides and glycoproteins that bear asparagine-linked sugar chains, particularly PAS-1 and band-3 glycoproteins of human erythrocyte membranes, showed strong inhibitory activity. The inhibitory constants of the band-3 glycoprotein for the binding of Pa-1 and Pa-2 to human erythrocytes were found to be very close to the association constants for binding of the lectins to the cell, indicating that the band-3 glycoprotein is the major binding site for the lectins. Actually, when the solubilized membranes of human erythrocytes were subjected to affinity chromatography with immobilized Pa-1 and Pa-2 used as specific adsorbents, the band-3 glycoprotein was found to bind most strongly.

Irimura & Nicolson (88) immobilized Pa-1, Pa-2, and Pa-4 on agarose gels and determined their ability to bind to various glycopeptides. It was found that the glycopeptide bearing poly(*N*-acetyllactosamine)-type sugar chains prepared from the band-3 glycoprotein of human erythrocyte membranes bound to all three immobilized lectins, but the glycopeptides bearing biantennary or triantennary complex-type sugar chains did not.

Katagiri et al (89) studied the interaction of high-mannose-type and complex-type glycopeptides obtained from porcine thyroglobulin (28, 29) with an immobilized Pa-2 column. They showed that some of the high-mannose-type glycopeptides that have a common structure, Manα1→2Manα1→2Manα1→3Manβ1→4GlcNAcβ1→4GlcNAc-Asn, can be significantly retarded on the column, but none of the complex-type glycopeptides showed significant interaction with the column. However, when the high-mannose-type glycopeptides were treated with endo-β-*N*-acetylglucosaminidase, they lost the ability to interact with Pa-2–agarose, indicating that the *N*, *N'* -diacetylchitobiose moiety in the core is essential for the interaction.

RICINUS COMMUNIS LECTINS (RCA I AND RCA II)   The castor bean *(Ricinus communis)* contains two galactose-binding lectins, RCA I and RCA II (90). RCA I is a hemagglutinin, and RCA II is a strong toxin with hemagglutinating activity. Both lectins bind primarily to the terminal Galβ1→4GlcNAc sugar sequence (33) and much more weakly to the Galβ1→3GalNAc sugar sequence (91). However, sugar-binding specificity as tested with monosaccharides and simple oligosaccharides differs somewhat between the two lectins (92; for review see Ref. 41).

Baenziger & Fiete (93) and Debray et al (43) studied the interaction of various glycopeptides with the two lectins. These two lectins have almost identical association constants for triantennary complex-type glycopeptides, whereas association constants for mucin-type glycopeptides bearing the Galβ1→3GalNAc sugar sequence of RCA II were some two- to threefold higher than those of RCA I. Both lectins bind to terminal β-galactosyl residues of sugar chains, but RCA II can also bind to terminal GalNAc residues with a significant association constant ($>$ 10^6) when terminal galactose residues are removed from mucin-type sugar chains. Impairment of the association by sialylation at galactose residues is much greater when the galactose residues are substituted by sialic acid at C-3 rather than C-6. The ability of an immobilized RCA I or RCA II column to retain glycopeptides generally correlates with association constants greater than about 5 × 10^6 M^{-1}.

Narasimhan et al (47) tested the binding of various complex-type glycopeptides to a column of RCA I–agarose. They observed that the interaction with RCA I–agarose depends directly on the number of terminal β-galactosyl groups on the sugar chain and that the presence of a bisecting GlcNAc group enhanced binding of glycopeptides to RCA I–agarose. This indicates that the steric hindrance of the bisecting GlcNAc group affects the GlcNAcβ1→ 2Manα1→3Manβ1→4 portion as in the case of Con A–agarose, but does not extend to the terminal β-galactosyl group of this branch.

PHASEOLUS VULGARIS LECTINS (E-PHA AND L-PHA)   *Phaseolus vulgaris* seeds contain five isolectins. The individual isolectins are tetramers comprising various proportions of E and L subunits (94–98). The E subunit accounts for the erythroagglutinating activity of the lectin, while the L subunit is responsible for the mitogenic activity of the lectin (99, 100). The E$_4$ lectin (E-PHA) is strongly hemagglutinating and its mitogenic activity is relatively weak, whereas the L$_4$ lectin (L-PHA) is devoid of hemagglutinating activity and binds to lymphocytes, giving rise to strong mitogenesis.

None of the simple sugars inhibit E- and L-PHAs except GlcNAc, which inhibits hemagglutinating and mitogenic activities of PHAs at high concentrations (101–103). Kornfeld & Kornfeld (104) isolated a complex-type glycopeptide as a binding site for E-PHA from human erythrocyte membranes and showed that a galactose-containing sugar sequence is an important determinant for binding of the lectin. Kaifu & Osawa (33) synthesized Galβ1 →4GlcNAcβ1→2Man and demonstrated that this oligosaccharide is a strong inhibitor of E-PHA binding. As for the sugar-binding specificity of L-PHA, very little was known, but GalNAc has been used to dissociate L-PHA bound to cells (105). However, porcine thyroglobulin, which lacks GalNAc residues (29), inhibited the mitogenic activity of L-PHA (56), and L-PHA was retarded on a column prepared by coupling porcine thyroglobulin to agarose while E-PHA was retained by the column (106).

The first clue for the elucidation of the structure of the binding sites for E-PHA on human erythrocyte membranes was obtained by Irimura et al (76). They showed that a complex-type sugar chain of human erythrocyte glycophorin A is a biantennary sugar chain with a bisecting GlcNAc residue (Figure 5), and that this bisected biantennary complex-type sugar chain interacts with high affinity with E-PHA. They also demonstrated that, when terminal galactose residues were removed, the glycopeptide lost its ability to bind to E-PHA–agarose, indicating that the peripheral Galβ1→4GlcNAc β1→2Man sugar sequence is essential for the binding.

Later, Cummings & Kornfeld (107) carried out a systematic study of the binding of a panel of glycopeptides to immobilized E-PHA and L-PHA. They showed that a bisecting GlcNAc residue that links β1→4 to the β-linked mannose residue in the core is an important determinant for high-affinity binding to E-PHA–agarose. Without the bisecting GlcNAc residue, a complex-type glycopeptide could not be retained by an E-PHA–agarose column. On the other hand, the complex-type glycopeptide with a bisecting GlcNAc residue did not interact with L-PHA–agarose but, interestingly, triantennary and tetraantennary complex-type glycopeptides that have at least one of the α-linked mannose residues substituted at positions C-2 and C-6 with β-GlcNAc residues interacted with high affinity with L-PHA–agarose, but not with E-PHA–agarose. The interaction of this type of glycopeptides with L-PHA was completely abolished by the removal of the β-galactosyl residue in the peripheral portion, suggesting that the Galβ1→4GlcNAcβ1→2Man sugar sequence is essential for the binding, but the binding was not affected by the removal of a fucose residue at the innermost GlcNAc residue in the core. Hammarström et al (108) obtained the same results for the sugar-binding specificity of L-PHA by using quantitative precipitation and precipitation inhibition.

Yamashita et al (109) and Narasimhan et al (47) further investigated the sugar-binding specificity of immobilized E-PHA. Their results indicated that the Galβ1→4GlcNAcβ1→2Manα1→6 chain should not be sialylated, at least at C-6 of its terminal galactose residue, while the galactose residue in the outer chain, which is linked to the Manα1→3 residue, can be sialylated or even removed, still leaving the proximate GlcNAc residue as an important part of the determinant for E-PHA. Furthermore, Yamashita et al (109) showed that the tetraantennary complex-type sugar chain with a bisecting GlcNAc residue does not interact with E-PHA–agarose, because the Man

$\alpha1{\rightarrow}6$ residue is substituted by another outer chain. From these results, they proposed the most complementary structure for E-PHA (Figure 7), where R_1 and R_2 represent H or sugars and R_3 represents GlcNAc→Asn or (Fuc$\alpha1{\rightarrow}6$)GlcNAcOH.

ALEURIA AURANTIA LECTIN (AAL)   A fucose-binding lectin was isolated from fruiting bodies of *Aleuria aurantia* by Kochibe & Furukawa (110). Yamashita et al (111) investigated the sugar-binding specificity of immobilized AAL. The complex-type oligosaccharides with an α-fucosyl residue at the innermost GlcNAc residue interacted with the immobilized lectin most strongly, irrespective of the presence or the number of outer chain moieties. Even the trimannosyl portion was not required for binding. Furthermore, the interaction was not altered by the presence of a bisecting GlcNAc residue. The oligosaccharides with Fuc$\alpha1{\rightarrow}2$Gal$\beta1{\rightarrow}4$GlcNAc or Gal$\beta1$ $\rightarrow4$(Fuc$\alpha1{\rightarrow}3$)GlcNAc groups interacted with AAL-agarose, but less strongly than complex-type oligosaccharides with a fucosylated core. Lacto-*N*-fucopentaitol II, which has a Gal$\beta1{\rightarrow}3$(Fuc$\alpha1{\rightarrow}4$)GlcNAc group, interacted less strongly than the above two groups, and the oligosaccharides with Fuc$\alpha1{\rightarrow}2$Gal$\beta1{\rightarrow}3$GlcNAc or Gal$\beta1{\rightarrow}4$GlcNAc$\beta1{\rightarrow}3$Gal$\beta1{\rightarrow}4$(Fuc$\alpha1$ $\rightarrow3$)GlcNAc groups showed almost no interaction with the immobilized lectin.

Lectins that Bind to Mucin-Type Sugar Chains

Sueyoshi et al (112) quantitatively analyzed binding specificities of five immobilized lectins that preferably bind to serine/threonine-linked mucin-type sugar chains, by means of frontal affinity chromatography (52, 53). These five lectins were *Agaricus bisporus* (mushroom) lectin [ABA-I (113)], *Arachis hypogaea* (peanut) lectin [PNA (114)], *Bauhinia purpurea* lectin [BPA (115)], *Glycine max* (soybean) lectin [SBA (116)], and *Vicia villosa* lectin [VVA-B$_4$ (117, 118)]. The sugar-binding specificities of these lectins had been studied by hemagglutination-inhibition, precipitation-inhibition, or binding-inhibition assays (31, 91, 113, 118–122).

Sueyoshi et al (112) found that the lectins mentioned above could be divided into two groups with respect to the reactivities of the immobilized lectins with typical mucin-type glycopeptides, Gal$\beta1{\rightarrow}3$GalNAc$\alpha1$ →Ser/Thr and GalNAc$\alpha1{\rightarrow}$Ser/Thr. One group, which consists of ABA-I, PNA,

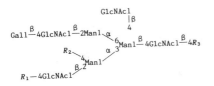

Figure 7 Structural requirements for the interaction of oligosaccharides and glycopeptides with E-PHA–agarose (109).

and BPA, preferentially binds to Galβ1$\rightarrow$3GalNAcα1$\rightarrow$Ser/Thr, and the other, which consists of SBA and VVA-B$_4$, shows higher affinity for GalNAcα1$\rightarrow$Ser/Thr than GalNAcGalβ1$\rightarrow$3GalNAcα1$\rightarrow$Ser/Thr. Among the immobilized lectin columns tested, only ABA-I–agarose retained a sialy-lated glycopeptide (123), which was prepared from human erythrocyte gly-cophorin A, and contains three tetrasaccharide chains with the structure shown in Figure 1. The association constant of ABA-I–agarose was 1.5×10^5 M^{-1}. The other four lectin columns did not significantly interact with this sialylated glycopeptide, and their association constants for the glycopeptide were below 3.5×10^3 M^{-1}. However, after desialylation, the resulting glycopeptide bearing Galβ1$\rightarrow$3GalNAcα1$\rightarrow$Ser/Thr chains was retarded on the immobilized PNA and BPA columns with association constants of 8.6×10^4 M^{-1} and 2.4×10^5 M^{-1}, respectively. On the other hand, the interactions of the desialylated glycopeptide with the immobilized SBA and VVA-B$_4$ were still very weak (K_a: 1.2×10^4 M^{-1} and 2.5×10^3 M^{-1}, respectively). Further removal of galactose residues from the desialylated glycopeptide resulted in significant decreases in the association constants of the immobilized ABA-I and PNA, but the absence of the β-galactosyl residue did not markedly affect the interaction with the immobilized BPA (K_a: 9.4×10^4 M^{-1}), and signifi-cantly enhanced the interaction with the immobilized SBA and VVA-B$_4$ (K_a: 9.4×10^4 M^{-1} and 2.5×10^5 M^{-1}, respectively). The difference between SBA-agarose and VVA-B$_4$–agarose was that the former preferentially in-teracted with the GalNAcα1$\rightarrow$3Galβ1$\rightarrow$3GalNAc-R sugar sequence, while the latter was highly specific for the GalNAcα1$\rightarrow$Ser/Thr structure (Tn antigenic structure). It was also shown on these immobilized lectin columns specific for mucin-type sugar chains that the glycopeptides with association constants of about 5.0×10^5 M^{-1} or greater were generally bound by the lectin columns depending on the amount of active lectin immobilized on column.

FRACTIONATION OF OLIGOSACCHARIDES AND GLYCOPEPTIDES BY THE USE OF IMMOBILIZED LECTINS

As mentioned in the introduction, cell membrane glycoproteins are in many cases very difficult to obtain in sufficient quantities for structural study. The sugar chains linked to asparagine residues of membrane glycoproteins seem to play a particularly important role in various biological interactions that lead to the change in cellular metabolism. This is suggested by the inhibition of various biological events with tunicamycin (124), which is a specific inhibitor of asparagine-linked sugar chains (125), and also by the fact that most mitogenic or cytotoxic lectins belong to the second group of lectins in Table

1, which preferentially bind to asparagine-linked sugar chains on the cell surface. However, there are several types of asparagine-linked sugar chains and in many cases a mixture of a variety of these sugar chains exists in the same glycoprotein molecule. Therefore, efficient methods for separating these asparagine-linked sugar chains are indispensable for elucidation of the structural and functional aspects of these sugar chains.

It is evident from the preceding discussion that various immobilized lectins can be successfully used for fractionation and for structural studies of asparagine-linked sugar chains of glycoproteins. This method needs less than 10 ng of a radiolabeled oligosaccharide or glycopeptide prepared from a glycoprotein by hydrazinolysis or by digestion with endo-β-N-acetylglucosaminidase or a protease. The fractionation and the structural assessment through the use of immobilized lectins make the subsequent structural studies much easier.

As originally proposed by Cummings & Kornfeld (126), a mixture of asparagine-linked sugar chains can be systematically fractionated by serial lectin–agarose affinity chromatography. An example is illustrated in Figure 8. To make the separation simpler, prior treatment of oligosaccharides or glycopeptides with sialidase can generally be recommended.

A mixture of asialo-asparagine-linked oligosaccharides is N-[^{3}H]-acetylated and subjected to affinity chromatography on a column of Con A–agarose. The glycopeptides with tri- and tetraantennary complex-type sugar chains pass through the column without any retardation, whereas the

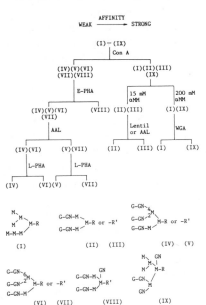

Figure 8 An example of the fractionation of various types of asparagine-linked oligosaccharides by the use of immobilized lectin columns. G, Gal; GN, GlcNAc; M, Man; R, -GlcNAc-GlcNAc-Asn; R', -GlcNAc-(Fuc-)GlcNAc-Asn.

glycopeptides with biantennary complex-type, hybrid-type, and high-mannose-type sugar chains are retained by the column. The glycopeptides with a bisected biantennary complex-type sugar chain are usually in the passed-through fraction (47).

Among the glycopeptides bound to the column, the biantennary gly-copeptides are usually removed by elution with 15-mM methyl α-D-mannopyranoside, while the high-mannose-type and hybrid-type gly-copeptides are obtained by elution with higher concentrations of methyl α-D-mannopyranoside (> 200-mM). The separation of complex-type oligo-saccharides from high-mannose-type oligosaccharides can also be carried out by affinity chromatography on a column of RCA I–agarose (127, 128), because this column can interact with complex-type oligosaccharides even when they are sialylated (93).

Hybrid-type glycopeptides and high-mannose-type glycopeptides can best be separated by the use of a WGA-agarose column. Since the sugar sequence GlcNAcβ1→4MANβ1→4GlcNAcβ1→4GlcNAcα1-Asn firmly binds to the WGA-agarose column, the hybrid-type glycopeptide with a bisecting GlcNAc residue can be retained by or at least retarded on the column, while the high-mannose-type glycopeptide is recovered without any retardation. Though E-PHA–agarose can also interact with some bisected asparagine-linked oligosac-charides, most hybrid-type glycopeptides cannot interact with the E-PHA–agarose column, because they have a substitution on the Manα1→6 residue that prevents interaction with the E-PHA–agarose (109).

The biantennary glycopeptides are further separated into two groups based on the presence or absence of a fucose residue on the innermost GlcNAc residue in the core. Since a fucose residue at the innermost GlcNAc residue is essential for binding to the immobilized lentil, pea, or fava lectin, any one of these lectin columns can be used for this separation. However, an intact innermost GlcNAc residue is required for tight binding to the lentil lectin–agarose and even an asparagine residue is necessary for tight binding to pea or fava lectin–agarose. Therefore, for the separation of radioactive oligosaccha-rides that are prepared by hydrazinolysis of glycoproteins followed by reduc-tion with NaB^{3}H$_4$, an immobilized AAL column should be used with attention to the fact that AAL-agarose can also bind complex-type oligosaccharides with an α-fucosyl residue on the outer chain.

Among the glycopeptides not bound to the Con A–agarose column, the glycopeptides with a poly(N-acetyllactosamine)-type sugar chain can be bound by Pa-2–agarose. Then, bisected biantennary glycopeptides are re-moved by a column of E-PHA–agarose. E-PHA–agarose binds bisected biantennary glycopeptides irrespective of the presence or absence of a fucose residue at the innermost GlcNAc residue in the core, whereas WGA-agarose cannot bind these glycopeptides with the fucose residues (75, 76). Since hybrid-type glycopeptides in most cases do not contain a fucose residue in the core, the WGA-agarose column is suitable for the separation of hybrid-type glycopeptides.

The fraction that does not bind to the columns of Pa-2–agarose and E-PHA–agarose still contains various types of triantennary and tetraantennary glycopeptides. When this fraction is applied to a column of AAL-agarose, tri- and tetraantennary glycopeptides with a core fucose residue are bound to the column. From the AAL-bound fraction, triantennary glycopeptides that contain an α-mannose residue substituted at C-2 and C-6 are separated by affinity chromatography on a column of the immobilized lentil, pea, or fava lectin. The unbound fraction is then applied to a column of L-PHA–agarose, and tetraantennary glycopeptides with the C-2- and C-6-substituted mannose residue can be obtained. However, in the case of oligosaccharides obtained by hydrazinolysis of a glycoprotein, the AAL-bound fraction should be directly applied to an L-PHA–agarose column, and the bound tri- and tetraantennary oligosaccharides with the C-2- and C-6-substituted α-mannose residue are then separated by gel filtration on a column of Biogel P-4. The remaining triantennary and tetraantennary glycopeptides or oligosaccharides with a core fucose residue but without the C-2- and C-6-substituted α-mannose residue are separated from each other by gel filtration on a column of Biogel P-4.

The glycopeptides that do not contain a core fucose residue but contain the outer α-mannose residue substituted at positions C-2 and C-6 can be separated by using L-PHA–agarose. The bound and unbound fractions are then each separated into triantennary and tetraantennary oligosaccharides by gel filtration on a column of Biogel P-4.

Serial lectin–agarose affinity chromatography is very sensitive and rapid. This method is being widely used for the analysis and separation of asparagine-linked oligosaccharides of various cellular glycoproteins (127–135) and their biosynthetic intermediates (136–139).

Literature Cited

1. Goldstein, I. J., Hughes, R. C., Monsigny, M., Osawa, T., Sharon, N. 1980. *Nature* 285:66
2. Lis, H., Sharon, N. 1977. *The Antigens,* ed. M. Sela, 4:429–529. New York: Academic. 582 pp.
3. Schrével, J., Gros, D., Monsigny, M. 1981. *Prog. Histochem. Cytochem.* 14(2):1–269
4. Irimura, T., Nicolson, G. L. 1983. *Carbohydr. Res.* 115:209–20
5. Irimura, T., Nicolson, G. L. 1983. *Carbohydr. Res.* 120:187–95
6. Irimura, T., Nicolson, G. L. 1984. *Cancer Res.* 44:791–98
7. Nicolson, G. L., Irimura, T. 1984. *Biol. Cell* 51:157–64
8. Sharon, N. 1983. *Adv. Immunol.* 34:213–98
9. Reisner, Y., Gachelin, G., Dubois, P., Nicolas, J. F., Sharon, N., Jacob, F. 1977. *Dev. Biol.* 61:20–27
10. Brabec, R. K., Peters, B. P., Bernstein, I. A., Gray, R. H., Goldstein, I. J. 1980. *Proc. Natl. Acad. Sci. USA* 77:477–79

40   OSAWA & TSUJI

11. Watanabe, M., Muramatsu, T., Shirane, H., Ugai, K. 1981. *J. Histochem. Cytochem.* 29:779–90
12. Lotan, R., Nicolson, G. L. 1979. *Biochim. Biophys. Acta* 559:329–76
13. Porath, J., Axen, R., Ernback, S. 1967. *Nature* 215:1491–92
14. Cuatrecasas, P., Anfinsen, C. B. 1971. *Methods Enzymol.* 22:345–78
15. Porath, J. 1974. *Methods Enzymol.* 34:13–30
16. Axén, R., Ernback, S. 1971. *Eur. J. Biochem.* 18:351–60
17. Rood, J. I., Wilkinson, R. G. 1974. *Biochim. Biophys. Acta* 334:168–78
18. Mega, T., Matsushima, Y. 1977. *J. Biochem.* 81:571–78
19. Wilchek, M., Miron, T. 1974. *Methods Enzymol.* 34:72–76
20. Sundberg, L., Porath, J. 1974. *J. Chromatogr.* 171:87–98
21. Matsumoto, I., Mizuno, Y., Seno, N. 1979. *J. Biochem.* 85:1091–98
22. Murphy, R. F., Conlon, J. M., Inman, A., Kelley, G. J. C. 1977. *J. Chromatogr.* 135:427–33
23. Matsumoto, I., Seno, N., Golovtchenko-Matsumoto, M., Osawa, T. 1980. *J. Biochem.* 87:535–40
24. Cuatrecasas, P., Parikh, I. 1972. *Biochemistry* 11:2291–99
25. Ito, Y., Seno, N., Matsumoto, I. 1985. *J. Biochem.* 97:1689–94
26. Mäkelä, O. 1957. *Ann. Med. Exp. Biol. Fenn.* 35:Suppl. 11. 133 pp.
27. Thomas, D. B., Winzler, R. J. 1969. *J. Biol. Chem.* 244:5943–46
28. Tsuji, T., Yamamoto, K., Irimura, T., Osawa, T. 1981. *Biochem. J.* 195:691–99
29. Yamamoto, K., Tsuji, T., Irimura, T., Osawa, T. 1981. *Biochem. J.* 195:701–13
30. Yamashita, K., Tachibana, Y., Kobata, A. 1978. *J. Biol. Chem.* 253:3862–69
31. Irimura, T., Kawaguchi, T., Terao, T., Osawa, T. 1975. *Carbohydr. Res.* 39:317–27
32. Kaifu, R., Osawa, T., Jeanloz, R. W. 1975. *Carbohydr. Res.* 40:111–17
33. Kaifu, R., Osawa, T. 1976. *Carbohydr. Res.* 52:179–85
34. Duk, M., Lisowska, E., Kordowicz, M., Wasniowska, K. 1982. *Eur. J. Biochem.* 123:105–12
35. Sarkar, M., Wu, A. M., Kabat, E. 1981. *Arch. Biochem. Biophys.* 209:204–18
36. Yokoyama, K., Terao, T., Osawa, T. 1978. *Biochem. Biophys. Acta* 538:384–96
37. Crowley, J. F., Goldstein, I. J., Arnarp,

J., Lönngren, J. 1984. *Arch. Biochem. Biophys.* 231:524–33
38. Goldstein, I. J., Hollerman, C. E., Smith, E. E. 1965. *Biochemistry* 4:876–83
39. So, L. L., Goldstein, I. J. 1967. *J. Immunol.* 99:158–63
40. Goldstein, I. J., Iyer, R. N., Smith, E. E., So, L. L. 1967. *Biochemistry* 6:2373–77
41. Goldstein, I. J., Hayes, C. E. 1978. *Adv. Carbohydr. Chem. Biochem.* 35:127–340
42. Kornfeld, R., Ferris, C. 1975. *J. Biol. Chem.* 250:2614–19
43. Debray, H., Decout, D., Strecker, G., Spik, G., Montreuil, J. 1981. *Eur. J. Biochem.* 117:41–55
44. Ogata, S., Muramatsu, T., Kobata, A. 1975. *J. Biochem.* 78:687–96
45. Krusius, T., Finne, J., Rauvala, H. 1976. *FEBS Lett.* 71:117–20
46. Narasimhan, S., Wilson, J. R., Martin, E., Schachter, H. 1979. *Can. J. Biochem.* 57:83–96
47. Narasimhan, S., Freed, J. C., Schachter, H. 1986. *Carbohydr. Res.* 149:65–83
48. Baenziger, J. U., Fiete, D. 1979. *J. Biol. Chem.* 254:2400–7
49. Scatchard, G. 1949. *Ann. NY Acad. Sci.* 51:660–72
50. Ohyama, Y., Kasai, K., Nomoto, H., Inoue, Y. 1985. *J. Biol. Chem.* 260:6882–87
51. Oda, Y., Kasai, K., Ishii, S. 1981. *J. Biochem.* 89:285–96
52. Kasai, K., Ishii, S. 1978. *J. Biochem.* 84:1051–60
53. Kasai, K., Ishii, S. 1978. *J. Biochem.* 84:1061–69
54. Allen, A. K., Desai, N. N., Neuberger, A. 1976. *Biochem. J.* 155:127–35
55. Kornfeld, S., Rogers, J., Gregory, W. 1971. *J. Biol. Chem.* 246:6581–86
56. Toyoshima, S., Fukuda, M., Osawa, T. 1972. *Biochemistry* 11:4000–5
57. Kornfeld, K., Reitman, M. L., Kornfeld, R. 1981. *J. Biol. Chem.* 256:6633–40
58. Yamamoto, K., Tsuji, T., Osawa, T. 1982. *Carbohydr. Res.* 110:283–89
59. Katagiri, Y., Yamamoto, K., Tsuji, T., Osawa, T. 1984. *Carbohydr. Res.* 129:257–65
60. Burger, M. M., Goldberg, A. R. 1967. *Proc. Natl. Acad. Sci. USA* 57:359–66
61. Allen, A. K., Neuberger, A., Sharon, N. 1973. *Biochem. J.* 131:155–62
62. Krug, U., Hollenberg, M. D., Cuatrecasas, P. 1973. *Biochem. Biophys. Res. Commun.* 52:305–12

63. Lotan, R., Sharon, N., Mirelman, D. 1975. *Eur. J. Biochem.* 55:257–62
64. Brown, J. M., Leon, M. A., Lightbody, J. J. 1976. *J. Immunol.* 117:1976–80
65. Goldstein, I. J., Hammarström, S., Sundblad, G. 1975. *Biochim. Biophys. Acta* 405:53–61
66. Bhavanandan, V. P., Katlic, A. W. 1979. *J. Biol. Chem.* 254:4000–8
67. Chandrasekaran, E. V., Davidson, E. A. 1979. *Biochemistry* 18:5615–20
68. Monsigny, M., Roche, A. C., Sene, C., Maget-Dana, R., Delmotte, F. 1980. *Eur. J. Biochem.* 104:147–53
69. Bhavanandan, V. P., Katlic, A. W., Banks, J., Kemper, J. C., Davidson, E. A. 1981. *Biochemistry* 20:5586–94
70. Cuatrecasas, P. 1973. *Biochemistry* 12:1312–23
71. Peters, B. P., Ebisu, S., Goldstein, I. J., Flashner, M. 1979. *Biochemistry* 18:5505–11
72. Bhavanandan, V. P., Umemoto, J., Banks, J. R., Davidson, E. A. 1977. *Biochemistry* 16:4426–37
73. Adair, W. L., Kornfeld, S. 1974. *J. Biol. Chem.* 249:4696–704
74. Lotan, R., Nicolson, G. L. 1979. *Biochim. Biophys. Acta* 559:329–79
75. Yamamoto, K., Tsuji, T., Matsumoto, I., Osawa, T. 1981. *Biochemistry* 20:5894–99
76. Irimura, T., Tsuji, T., Tagami, S., Yamamoto, K., Osawa, T. 1981. *Biochemistry* 20:560–66
77. Gallagher, J. T., Morris, A., Dexter, T. M. 1985. *Biochem. J.* 231:115–22
78. Fukuda, M., Fukuda, M. N., Hakomori, S. 1979. *J. Biol. Chem.* 253:3700–3
79. Tsuji, T., Irimura, T., Osawa, T. 1980. *Biochem. J.* 187:677–86
80. Fukuda, M., Fukuda, M. N. 1981. *J. Supramol. Struct. Cell. Biochem.* 17:313–24
81. Muramatsu, T., Gachelin, G., Nicolas, J. G., Condamine, H., Jakob, H., Jacob, F. 1978. *Proc. Natl. Acad. Sci. USA* 75:2315–19
82. Muramatsu, T., Gachelin, G., Dammonneville, M., Delabre, C., Jacob, F. 1979. *Cell* 18:188–91
83. Börjeson, J., Reisfeld, R., Chessin, L. N., Welsh, P. D., Douglas, S. D. 1966. *J. Exp. Med.* 124:859–72
84. Chessin, L. N., Börjeson, J., Welsh, P. D., Douglas, S. D., Cooper, H. L. 1966. *J. Exp. Med.* 124:873–84
85. Waxdal, M. J. 1974. *Biochemistry* 13:3671–77
86. Yokoyama, K., Yano, O., Terao, T., Osawa, T. 1976. *Biochim. Biophys. Acta* 427:443–52
87. Waxdal, M. J., Basham, T. Y. 1974. *Nature* 251:163–64
88. Irimura, T., Nicolson, G. L. 1983. *Carbohydr. Res.* 120:187–95
89. Katagiri, Y., Yamamoto, K., Tsuji, T., Osawa, T. 1983. *Carbohydr. Res.* 120:283–92
90. Nicolson, G. L., Blaustein, J., Etzler, M. E. 1974. *Biochemistry* 13:196–204
91. Kaifu, R., Osawa, T. 1979. *Carbohydr. Res.* 69:79–88
92. Olsnes, S., Saltvedt, E., Pihl, A. 1974. *J. Biol. Chem.* 249:803–10
93. Baenziger, J. U., Fiete, D. 1979. *J. Biol. Chem.* 254:9795–99
94. Allan, D., Crumpton, M. J. 1971. *Biochem. Biophys. Res. Commun.* 44:1143–48
95. Oh, Y. H., Conrad, R. A. 1972. *Arch. Biochem. Biophys.* 152:631–37
96. Weber, T. W., Aro, H., Nordman, C. T. 1972. *Biochim. Biophys. Acta* 263:94–105
97. Yachnin, S., Svenson, R. H. 1972. *Immunology* 22:871–83
98. Miller, J. B., Noyes, C., Heinrikson, R., Kingdon, H. S., Yachnin, S. 1973. *J. Exp. Med.* 138:939–51
99. Leavitt, R. D., Felsted, R. L., Bachur, N. R. 1977. *J. Biol. Chem.* 252:2961–66
100. Egorin, M. J., Bachur, S. M., Felsted, R. L., Leavitt, R. D., Bachur, N. R. 1979. *J. Biol. Chem.* 254:894–98
101. Borberg, H., Woodruff, J., Hirschhorn, R., Gesner, B., Miescher, P., Silber, R. 1966. *Science* 154:1019–20
102. Borberg, H., Yesner, I., Gesner, B., Silber, R. 1968. *Blood* 31:747–57
103. Dahlgren, K., Porath, J., Lindahl-Kiessling, K. 1970. *Arch. Biochem. Biophys.* 137:306–14
104. Kornfeld, R., Kornfeld, S. 1970. *J. Biol. Chem.* 254:2536–45
105. Perlés, B., Flanagan, M. T., Auger, J., Crumpton, M. J. 1977. *Eur. J. Immunol.* 7:613–19
106. Osawa, T., Terao, T., Matsumoto, I., Imbe, K. 1973. *Méthodologie de la Structure et du Métabolisme des Glycoconjugues*, ed. J. Montreuil, 2:725–31. Paris: CNRS
107. Cummings, R. D., Kornfeld, S. 1982. *J. Biol. Chem.* 257:11230–34
108. Hammarström, S., Hammarström, M.-L., Sundblad, G., Arnarp, J., Lönngren, J. 1982. *Proc. Natl. Acad. Sci. USA* 79:1611–15
109. Yamashita, K., Hitoi, A., Kobata, A. 1983. *J. Biol. Chem.* 258:14753–55
110. Kochibe, N., Furukawa, K. 1980. *Biochemistry* 19:2841–46

111. Yamashita, K., Kochibe, N., Ohkura, T., Ueda, I., Kobata, A. 1985. *J. Biol. Chem.* 260:4688–93
112. Sueyoshi, S., Tsuji, T., Osawa, T. 1987. *Carbohydr. Res.* In press
113. Sueyoshi, S., Tsuji, T., Osawa, T. 1985. *Biol. Chem. Hoppe-Seyler* 366: 213–21
114. Lotan, R., Sharon, N. 1978. *Methods Enzymol.* 50:361–67
115. Osawa, T., Irimura, T. 1978. *Methods Enzymol.* 50:367–72
116. Lis, H., Sharon, N. 1972. *Methods Enzymol.* 28:360–68
117. Tollefsen, S. E., Kornfeld, R. 1983. *J. Biol. Chem.* 258:5165–71
118. Tollefsen, S. E., Kornfeld, R. 1983. *J. Biol. Chem.* 258:5172–76
119. Presant, C. A., Kornfeld, S. 1972. *J. Biol. Chem.* 247:6937–45
120. Pereira, M. E. A., Kabat, E. A., Lotan, R., Sharon, N. 1976. *Carbohydr. Res.* 51:107–18
121. Prigent, M. J., Bencomo, V. V., Sinäy, P., Cartron, J. P. 1984. *Glycoconjugate J.* 1:73–80
122. Pereira, M. E. A., Kabat, E. A. 1974. *Carbohydr. Res.* 37:89–102
123. Prohaska, R., Koerner, T. A. W., Armitage, I. M., Furthmayr, H. 1981. *J. Biol. Chem.* 256:5781–91
124. Yamada, K. M., Olden, K. 1982. *Tunicamycin,* ed. G. Tamura, pp. 119–44. Tokyo: Japan Sci. Soc. Press. 220 pp.
125. Takatsuki, A., Tamura, G. 1982. See Ref. 124, pp. 35–70
126. Cummings, R. D., Kornfeld, S. 1982. *J. Biol. Chem.* 257:11235–40
127. Yamamoto, K., Tsuji, T., Tarutani, O., Osawa, T. 1984. *Eur. J. Biochem.* 143:133–44
128. Yamamoto, K., Tsuji, T., Tarutani, O., Osawa, T. 1985. *Biochim. Biophys. Acta* 838:84–91
129. Reitman, M. L., Trowbridge, I. S., Kornfeld, S. 1982. *J. Biol. Chem.* 257:10357–63
130. Cowan, E. P., Cummings, R. D., Schwartz, B. D., Cullen, S. E. 1982. *J. Biol. Chem.* 257:11241–48
131. Stanley, P., Vivona, G., Atkinson, P. H. 1984. *Arch. Biochem. Biophys.* 230:363–74
132. Chapman, A. J., Gallagher, J. T., Beardwell, C. G., Shalet, S. M. 1984. *J. Endocrinol.* 103:117–22
133. Cummings, R. D., Kornfeld, S. 1984. *J. Biol. Chem.* 259:6253–60
134. Stiles, G. L., Benovic, J. L., Caron, M. G., Lefkowitz, R. J. 1984. *J. Biol. Chem.* 259:8655–63
135. Krusius, T., Fukuda, M., Dell, A., Ruoslahti, E. 1985. *J. Biol. Chem.* 260:4110–16
136. Narasimhan, S. 1982. *J. Biol. Chem.* 257:10235–42
137. Cummings, R. D., Trowbridge, I. S., Kornfeld, S. 1982. *J. Biol. Chem.* 257:13421–27
138. Campbell, C., Stanley, P. 1984. *J. Biol. Chem.* 259:13370–78
139. Gleeson, P. A., Schachter, H. 1983. *J. Biol. Chem.* 258:6162–73

Ann. Rev. Biochem. 1987. 56:43–61

DYNAMICS OF MEMBRANE LIPID METABOLISM AND TURNOVER

E. A. Dawidowicz

Department of Physiology, Tufts Medical School, Boston, Massachusetts 02111

CONTENTS

INTRODUCTION AND PERSPECTIVES

Membrane biogenesis is an essential feature of cellular development and growth. Although some membrane proteins are synthesized on free polysomes (1) and de novo phospholipid biosynthesis can occur in the Golgi apparatus (2), the endoplasmic reticulum is the principal site for the initial assembly of membrane phospholipids and proteins. Cholesterol, a major membrane component in animal cells, appears to be synthesized on the endoplasmic reticulum (3) or is derived from low-density lipoprotein via receptor-mediated endocytosis (4).

The various organelle membranes in a eukaryotic cell differ widely in both protein and lipid composition, indicating that specific sorting and transport

43

0066-4154/87/0701-0043$02.00

processes must exist that ensure the correct ultimate location of these molecules within the cell. Although our understanding of these events for membrane proteins has increased over recent years (5), the intracellular movement of membrane lipids is less well understood and often overlooked in reviews and discussions of membrane biogenesis. However, it is quite clear that movement of a hydrophobic membrane protein within a cell must be accompanied by lipid. Furthermore, because examples of specific lipid requirements for protein function have been described (6–8), we know that failure to transport a given lipid to its correct destination could result in a malfunction of certain cellular processes.

At least three possible routes for the intracellular movement of lipids are considered in this review: spontaneous lipid movement, protein-facilitated lipid movement, and vesicular transport. Lipid assembly into membranes of prokaryotes is not discussed since it has recently been exhaustively covered (9). The reader is also directed to other reviews on intracellular movement of lipids (10–12), and to a recent review by the present author on the spontaneous and protein-facilitated movement of lipids between membranes (13).

Lipid turnover and degradation are vital for normal cellular function. This is manifested by the existence of lipid storage diseases where lipid degradation is impaired (14). However, investigations on the transport of lipids to their sites of degradation in intact cells are few. Although turnover of polyphosphoinositides is an important example of lipid turnover, the vast literature on this subject is outside the scope of the present review. The reader is referred to a recent review on this subject (15).

INTRACELLULAR LOCATION OF LIPID BIOSYNTHESIS

The various pathways and intracellular locations of the enzymes of phospholipid biosynthesis have been discussed in detail (16–19). However, it is important to consider these topics briefly before discussing the sorting and transport of lipids. In the subsequent discussion of the distribution of the lipids within the cell, it will become apparent that some of these molecules do not appear to leave the organelle in which they are assembled.

Phospholipids

The intracellular locations of the major pathways of phospholipid biosynthesis in animal cells have been determined by subcellular fractionation. A potential problem with this approach is that in several studies location has been defined by appearance in a microsomal fraction. Although derived mainly from the endoplasmic reticulum, microsomes can contain fragments of Golgi apparatus, plasma membrane, and small vesicles depending on the conditions that are used to homogenize the starting tissue (20), details that are not always adequately described.

The primary site for the syntheses of phosphatidylcholine and phosphatidylethanolamine by the choline- and ethanolamine-phosphotransferase pathways is located in the microsomal fraction of rat liver and other mammalian tissues (16–19). Further investigations demonstrated that these enzymes are located in both the endoplasmic reticulum and the Golgi apparatus of rat liver (2). However, over 90% of total cellular activity could be accounted for by the endoplasmic reticulum and only 1% by the Golgi. Purified plasma membranes lack these enzymes. Synthesis of phosphatidylinositol occurs in the microsomal fraction of liver and brain (19), whereas the syntheses of phosphatidylglycerol and cardiolipin take place in mitochondria of animal cells, where these lipids are almost exclusively located (19).

In addition to the pathways of de novo biosynthesis, transformation of existing lipids can occur. Fatty acid composition can be modified by deacylation-reacylation reactions (17), and polar head groups can be altered by base-exchange (19). The synthesis of phosphatidylserine in animal cells occurs almost exclusively by this latter pathway, involving the exchange of L-serine for ethanolamine in phosphatidylethanolamine, in the endoplasmic reticulum (21, 22). Conversion of phosphatidylserine to phosphatidylethanolamine in mitochondria (22, 23) and the synthesis of sphingomyelin by transfer of the phosphocholine group from phosphatidylcholine to ceramide in the plasma membrane (24–26) utilize phospholipid substrates that are synthesized in the endoplasmic reticulum. This indicates that intracellular transport must be directly involved in these latter processes (9).

Cholesterol

Cholesterol biosynthesis involves multiple steps (27). The rate-limiting enzyme in this pathway, HMG CoA reductase, is clearly located in the endoplasmic reticulum (28). Other enzymes involved in the biosynthesis of cholesterol from squalene are localized in microsomes (27, 29). This intracellular location for the synthesis of cholesterol and cholesterol ester has been more clearly defined in rat liver as the endoplasmic reticulum (3). These data have recently been challenged in a study (30) that will be described in detail later.

Glycolipids

The glycosylation of gangliosides by the sequential addition of monosaccharides from sugar nucleotides to an acceptor is catalyzed by glycosyltransferases located in the Golgi apparatus (31, 32). The sequential glycosylation of one ganglioside to the next higher homologue appears to involve a small pool of intermediates separate from the bulk of these molecules, which are located in the plasma membrane (33).

INTRACELLULAR LOCATION OF LIPIDS

Phospholipids, sterols, and glycolipids are not uniformly distributed amongst the various organelles of eukaryotic cells (16, 33–41). The reader is referred to the aforementioned references for details. Some trends are apparent if one examines the lipid composition of the organelles from rat liver (31, 37) as an example. Phosphatidylcholine, phosphatidylethanolamine, and phosphatidyl-inositol are present in all membranes. The highest concentration of phosphatidylethanolamine appears in the mitochondria, whereas phosphatidylcholine is slightly enriched in the endoplasmic reticulum. Cardiolipin is almost exclusively located on the inner mitochondrial membrane. Sphingomyelin appears highest in the plasma membrane, with a significant concentration in both Golgi and lysosomal membranes and almost complete absence in endoplasmic reticulum and mitochondria.

Added complexities to a complete description of the phospholipid distribution in animal cells are variations in fatty acid composition (34, 37), possible asymmetry across a given membrane (42, 43), and differences in the composition of plasma membrane domains in polarized cells (44). The degree of unsaturation of the fatty acyl chains increases from the plasma membrane to the Golgi to the endoplasmic reticulum (45). Evidence for lipid asymmetry is most compelling for the human erythrocyte (46). However, in the case of organelle membranes the data are less clear. In microsomal vesicles, derived from rat liver, the distribution of phospholipids that has been reported by three separate groups varies significantly and seems to depend on the method employed (47–49). This matter has not been satisfactorily resolved, and thus measurements of lipid sidedness in microsomal vesicles must still be viewed with caution. It had been suggested, from ^{31}P-NMR studies of intact microsomes (50), that a sizeable fraction of the phospholipids in this membrane is not organized in a classical bilayer structure, but experiences isotropic motion, which not only allows for a fast transbilayer movement of lipids, "but also makes studies on lipid localization, as a static phenomenon, irrelevant" (42). It is possible that the observed shift in the NMR spectrum is not due to isotropic motion or 'non-bilayer' structure, but results from the small size of the microsomal vesicles (51). Interpretation of the initial NMR has recently been questioned (52).

The distribution of cholesterol amongst the organelles of rat liver is quite striking (37). In the endoplasmic reticulum, the apparent site of cholesterol synthesis, the cholesterol:phospholipid molar ratios are 0.1 for the rough endoplasmic reticulum and 0.24 for the smooth endoplasmic reticulum. This ratio has a value of 0.1 in the mitochondrial membrane, 0.15 in the Golgi, and close to 0.8 in the plasma membrane. Similar values have been reported for the cholesterol distribution in cultured baby hamster kidney cells (38). Experiments with filipin have indicated an absence of filipin-sterol complexes from large coated pit regions on cell surfaces (53). Results from filipin staining have further suggested a heterogeneous distribution of cholesterol across the cisternae of the Golgi (54); the reticulum-related (or *cis*) cisternae are poor in cholesterol, whereas the secretory-granule (or *trans*) cisternae are rich in cholesterol. However, confirmation of these latter observations must await the detection of cholesterol in the Golgi by more direct methods, since it is well known that filipin can give a false negative result in cholesterol-rich membranes (55).

It has been suggested that the high cholesterol : phospholipid ratio correlates with the sphingomyelin content of the various organelles (56). This notion was supported by the finding that, in liposomes, phospholipids have different affinities for cholesterol in the order sphingomyelin > phosphatidylcholine > phosphatidylethanolamine (57). However, this does not appear to be the sole factor determining the intracellular cholesterol distribution in animal cells. The equilibrium partitioning of cholesterol between vesicles composed of egg lecithin-cholesterol and vesicles prepared from the lipid extract of either plasma membrane, mitochondria, or endoplasmic reticulum of LM cells showed only qualitative agreement with the sterol content of these membranes in vivo (45). From these data it was concluded that the distribution of sterol within cells is not an equilibrium phenomenon.

The eukaryotic cell *Acanthamoeba castellanii* does not contain any sphingomyelin; nevertheless the intracellular distribution of sterols is very similar to that described for cholesterol in mammalian cells (41). The highest sterol : phospholipid ratio is found in the plasma membrane of *Acanthamoeba,* whereas the sterol content of microsomes is significantly lower. These data further indicate our lack of understanding of the parameters that determine sterol distribution in eukaryotic cells.

A method for analyzing the cholesterol distribution in intact cells has recently been described (58). The data obtained using this approach have provocative implications. The method involves a determination of the fraction of cellular cholesterol that can be oxidized to cholest-4-en-3-one in the presence of cholesterol oxidase added to the outside of cells. Since cholesterol oxidase is membrane impermeable (59), it was inferred that this assay detects only plasma membrane cholesterol. One caveat associated with this procedure is that plasma membrane cholesterol is not susceptible to cholesterol oxidase under physiological conditions (60–63). Cholesterol is rendered susceptible to the enzyme after either warming the cells at low ionic strength or fixation with glutaraldehyde (64). The results obtained were as follows. Most of the cellular cholesterol was oxidized in three cell types: 94% in fibroblasts, 92% in Chinese hamster ovary cells, and 80% in hepatocytes with half-times of less than 30 seconds (58). It was concluded that the transmembrane movement of cholesterol across the plasma membranes of these cells is extremely rapid, in agreement with previous measurements of this process in human red cells and lipid vesicles using cholesterol oxidase (65, 66). The possibility exists that these very rapid movements could be enhanced by the oxidation process. The fraction of cellular cholesterol oxidized in the presence of the enzyme was taken to represent the percentage of cellular cholesterol present in the plasma membrane (58).

A reasonable value for the cholesterol:phospholipid molar ratio in whole cells is 0.3–0.4 (37, 38), whereas in plasma membranes this value does not exceed 0.8–0.9 (37, 38). In fact, a cholesterol:phospholipid molar ratio of greater than 1.0 is only found in human red blood cells from patients with spur cell anemia (67).

A prediction can be made from these data as follows. If 80% of the total cellular cholesterol is present in the plasma membrane of the hepatocyte (whole cell cholesterol:phospholipid molar ratio 0.3–0.4), then it follows that 27–40% of the total cellular phospholipid must also be present in the plasma membrane of this cell in order to achieve a cholesterol:phospholipid ratio of 0.8–0.9. If 90% of the cellular cholesterol is in the plasma membrane, this would require that 30–45% of the cellular phospholipid is also present in this membrane. These values are significantly higher than the value of 1–2% of total cellular lipid mass stated to be in the plasma membrane of eukaryotic cells (10, 120). However, in marked contrast, it has been indicated that 30% of the total cellular phospholipid is present in the plasma membrane of Chinese hamster ovary cells (68). A stereological study on the hepatocyte indicates that the plasma membrane represents 8–10% of the total membrane surface area (69).

The methods described above to measure the fraction of cellular phospholipid and cholesterol present in the plasma membrane have been used to study the transport of these molecules to the cell surface. It is thus imperative that all of the data are internally consistent. Furthermore, since reasonable estimates exist for the phospholipid:protein ratio in plasma membranes, one could predict the fraction of total cellular membrane protein present in the plasma membrane based on any phospholipid distribution. Direct chemical measurement of the total plasma membrane protein could then be compared with the predicted value.

ASSEMBLY OF THE LIPID BILAYER IN THE ENDOPLASMIC RETICULUM

The major intracellular site for the initial assembly of membrane lipids is the endoplasmic reticulum, and it is from this organelle that lipids must be transported throughout the cell. In a careful study, it has been demonstrated that the enzymes involved in the final step in the biosynthesis of phosphatidyl-choline and phosphatidylethanolamine are exclusively located on the cytoplasmic surface of microsomal vesicles derived from rat liver (70). Later studies further demonstrated that phosphatidylserine synthase and phospha-tidylinositol synthase are also located on this surface (71). Since the phospho-lipids in the endoplasmic reticulum are organized in a bilayer structure (52), the question arises as to how the newly synthesized lipids are translocated to the lumenal surface to assemble the growing membrane. Studies on the transbilayer movement of phospholipids across a variety of membranes pro-vide a clue as to how this translocation step could occur. These latter studies have recently been reviewed by the present author in (13), where the reader is referred for details.

Studies of transmembrane movement of phospholipids can be divided into three categories. In the first group are studies on pure phospholipid vesicles in which translocation across the bilayer is not detectable; half-times for this process have been estimated to be >11 days at 37°C (72–74). The second group includes studies on red cells and viral membranes, where the transmem-brane movement of phospholipids occurs with a half-time of 2–4 hours (75–77). In the third group are two membranes, microsomal vesicles derived from rat liver and *Bacillus megaterium,* where the measured half-times for the transmembrane movement of phospholipids are a few minutes or less (78–80). Interestingly, the extremely rapid translocation appears to occur in membranes capable of de novo biosynthesis. Attempts have been made to induce transmembrane movement of phospholipid in lipid vesicles. However, half-times of only 2–4 hours have been achieved (13). Transmembrane movement of phospholipids across vesicles prepared from a total lipid extract of rat liver microsomes was not detectable (79), which raises the question of whether the translocation process across the microsomal membrane is medi-ated by a protein "flippase" as postulated by Bretscher (81). A comparison of the rates of transmembrane movement for a diglyceride (extremely rapid) with a corresponding phosphatidylglycerol (extremely slow) (82) led to the sugges-tion that the hindrance to the movement of a phospholipid across a lipid bilayer was provided by the polar head group (83). It was further reasoned that a "flippase" would facilitate transfer of the polar head group across the microsomal membrane. With this in mind, Bishop & Bell made the exciting discovery of a transporter for water-soluble short-chain phospholipids in rat liver microsomal vesicles (83). Transport of these lipids was not detected into red blood cells or pure lipid vesicles. Transport was sensitive to proteolysis and treatment with N-ethylmaleimide, and appeared to be able to distinguish between the sn-1,2 phosphatidylcholine and the sn-2,3 isomer. In a pre-liminary communication, a successful reconstitution of a "flippase" activity from rat liver microsomes has been reported (84). This activity, which was not found in red cells, facilitates transmembrane movement of a long-chain bilayer-forming phosphatidylcholine in contrast to the transporter of short-chain water-soluble lipids described by Bishop & Bell (83). Discovery of such proteins is not without precedent. Evidence has been presented for the existence of a membrane protein involved in the permeation of long-chain fatty acids into adipocytes (85). It also now appears that transmembrane movement of phospholipids can occur in a specific manner across the plasma membrane of living cells. A fluorescent phosphatidylethanolamine analogue undergoes a rapid transmembrane movement across the plasma membrane of lung fibroblasts (86), whereas the translocation of the corresponding fluorescent phosphatidylcholine is not detectable (87). Recent studies have indicated that a rapid redistribution of the amino-containing phospholipids can occur across the erythrocyte membrane in an ATP-dependent manner (88–90). The physiological role of this latter process may be an involvement in the maintenance of lipid asymmetry.

INTRACELLULAR TRANSPORT OF LIPIDS

The approaches used to investigate intracellular lipid transport have involved following the fate of either radiolabeled or fluorescent lipids within the cell. For membrane growth to occur it is essential that these molecules are transferred in a net manner rather than in a one-for-one exchange process. This important distinction has not always been made in the studies that will be discussed.

Possible mechanisms that could be involved in these transport processes include protein-mediated transfer (91, 92) or a vesicular process (93). The criteria that have been used to distinguish between these possibilities have not been rigorous. For example, it has been assumed that if protein-mediated transfer occurs, the transported lipid will be bound to a carrier protein and thus detectable in the $100,000 \times g$ supernatant of a cell homogenate. This is not necessarily the case. It has been pointed out that proteins can facilitate intermembrane lipid transfer without binding lipid simply by increasing the off rate from the donor membrane (94). Attempts have been made to rule out a vesicular-mediated transfer by examining the effects of agents such as vinblastine (antimicrotubule), or cytochalasin B (antimicrofilament). It is tacitly assumed that if vesicles are involved, their movement would somehow be affected by the organization of the microfilaments or microtubules. More precise methods are required to elucidate the mechanisms of intracellular lipid transport.

Cholesterol

Maintenance of the cholesterol distribution in animal cells presents an enigma. Cholesterol is capable of spontaneous transfer between membranes by a mechanism that does not involve membrane-membrane contact (95, 96). This net transfer, which takes place down a membrane-cholesterol gradient, can occur with a half-time of 1–2 hours depending on the nature of the membranes involved (13). In addition the transmembrane movement of cholesterol across a variety of membranes is at least as rapid and may be extremely rapid (half-time of 1 minute or less) (58, 65, 66). With these facts in mind it is difficult to understand how the cholesterol gradient is maintained between the organelle membranes in the intact cell, unless the movement of cholesterol between membranes in vitro differs from the in vivo process.

Cholesterol transfer factors (97), cholesterol binding proteins (98), and sterol carrier proteins (99) have been detected in rat liver cytosol. It is possible that these proteins may be involved in the movement of cholesterol within the cell and are somehow regulated by surface characteristics of the various organelles. However, their role in vivo remains unclear. Transfer of cholesterol to the mitochondria during steroidogenesis in rat adrenal cells is inhibited by the antimicrofilament agent cytochalasin B, and by the anti-microtubule agent vinblastine (100). These results indicate an involvement of the cytoskeletal meshwork in cholesterol movement in adrenal cells, suggesting vesicular transport.

Transfer of radiolabeled cholesterol from the endoplasmic reticulum to the plasma membrane of Chinese hamster ovary cells has been followed using rapid membrane isolation (101). Newly synthesized radiolabeled cholesterol appeared in the plasma membrane with a 10-minute lag at 37°C when compared with the appearance of labeled sterol in the intact cell. This lag was interpreted as the transit time for the newly synthesized cholesterol from the endoplasmic reticulum to the plasma membrane. Transport could be arrested at 0°C and was inhibited by fairly high concentrations of energy poisons. Transfer of labeled glycoprotein to the cell surface occurred with the characteristic 20–30-minute lag, indicating a marked difference in the kinetics of intracellular transport of glycoproteins and cholesterol. Arrest of cholesterol transport at 0°C suggested a method for trapping newly synthesized cholesterol during transit to the cell surface and also indicated that possible scrambling of cholesterol by exchange during subcellular fractionation is not significant at 0°C. In a subsequent study the transit time of 10 minutes was confirmed (102). Cholesterol transport was blocked by fairly high concentrations of the energy poisons KCN and KF. However, these poisons did not appear to affect the intracellular distribution of cholesterol, suggesting that energy is only required in the assembly of the cholesterol gradient in the cell. Cytochalasin B, colchicine, monensin, NH_4Cl, and cycloheximide were without effect on cholesterol transport, suggesting that the transfer process does not appear to require an intact Golgi apparatus and is not affected by the organization of the cytoskeletal meshwork. As in the earlier study, cholesterol transfer was affected by temperature, apparently arrested or proceeding with a very slow rate at 15°C. An interesting finding was that following incubation of the cells with ^{3}H-acetate at 15°C and subsequent subcellular fractionation, a light fraction (appearing in the sucrose gradient of a density of less than 1.087) was enriched in newly synthesized cholesterol. After warming to 37°C, newly synthesized cholesterol was no longer enriched in this fraction. The authors suggested that this could indicate isolation of an intermediate in the transport process. Although no data were shown, it was indicated that electron microscopy of this fraction showed it to consist of smooth membrane vesicles. This result serves to emphasize the cautions expressed earlier in this review. Even though disruptors of the cytoskeletal network were without effect on the intracellular movement of cholesterol in these cells, the more direct approach has provided evidence in favor of vesicular transport. This would suggest that vesicular traffic may not always be affected by the organization of the cytoskeletal network.

Transfer of newly synthesized cholesterol to the cell surface in intact cells has been followed using the enzyme cholesterol oxidase to monitor the ^{3}H-cholesterol in the plasma membrane (103). It was reported that cholesterol synthesized from ^{3}H-acetate in both human fibroblasts and Chinese hamster ovary cells is transferred to the cell surface with a half-time of 1–2 hours at 37°C. The effects of temperature and possible inhibitors were not described.

To reconcile the difference in the reported transit times for cholesterol movement to the plasma membrane (101–103), Kaplan & Simoni (102) have proposed the following explanation. In order to be detected by cholesterol oxidase outside the cell, labeled cholesterol must be directly incorporated into the plasma membrane. In the cell fractionation approach the labeled cholesterol need only be associated with the plasma membrane to be detected. Thus if cholesterol transfer is mediated by a vesicular pathway, subcellular fractionation would detect vesicles arriving at the plasma membrane. Detection by cholesterol oxidase would require subsequent fusion. A weakness in the argument is the lack of indication of what factors would limit this fusion process; such indication is required to explain differences between the two sets of data.

Using cholesterol oxidase, Lange & Steck (30) have presented further data suggesting that newly synthesized cholesterol and lanosterol accumulate in an intracellular fraction of human fibroblasts, which appears to be separable from the endoplasmic reticulum by density gradient centrifugation. Since the nascent sterols did not appear in the same membrane fraction as HMG-CoA reductase, it was suggested that cholesterol biosynthesis may not be carried to completion in the endoplasmic reticulum, but that this could occur on a specialized, as-yet-undefined, intracellular membrane, which is also involved in the transport of cholesterol to the cell surface. Some of the enzymes involved in the conversion of lanosterol to cholesterol have been purified, and in a review of this field it is stated that this biosynthetic pathway is catalyzed by enzymes that are tightly bound to the endoplasmic reticulum (29). Further experimentation is required to critically evaluate the model proposed by Lange & Steck.

Transfer of newly synthesized sterols to the plasma membrane has been followed in the eukaryotic cell *A. castellanii* (104), using subcellular fractionation. The distribution of sterols in these cells is similar to that in other animal cells. These free-living amoeba do not require serum for growth; therefore the de novo synthetic pathway is the only means by which cellular sterol is acquired. Transport of newly synthesized sterols to the cell surface at 23°C occurred with a lag time of 30 minutes, which was interpreted as the transit time from the site of synthesis to the plasma membrane. Control experiments indicated that no scrambling of labeled sterols occurs during membrane isolation at 4°C.

Once the newly synthesized sterol has been transferred to the cell surface, the question of its subsequent fate must be considered. A suggestion has been made that the coated pit region of the plasma membrane could act as a molecular filter to prevent molecules such as cholesterol from leaving the plasma membrane by endocytosis (105). In support of this idea it was reported that cholesterol introduced into the plasma membrane of fibroblasts by exchange from lipid vesicles did not equilibrate with endogenously synthesized cholesterol, and that only this latter pool became esterified (106). Further studies indicated that cholesterol inserted into the plasma membrane does not appear to equilibrate with intracellular membranes (103). However, cholesterol introduced by exchange into the plasma membrane of either smooth muscle cells (107) or lung fibroblasts (108) did apparently enter the cells and become esterified after 1–4 hours at 37°C. In a recent study (109), a slow movement of cholesterol from the cell surface to the cell interior, as monitored by decreased cholesterol biosynthesis, decreased LDL binding, and increased cholesterol esterification, was detected 24 hours after exchange of cholesterol into the cell surface.

It is conceivable that in these above-mentioned studies, insensitive assays have been chosen to determine whether plasma membrane cholesterol enters the cell. Endocytic coated vesicles, purified from rat liver, have a cholesterol:phospholipid molar ratio of 0.84, characteristic of plasma membrane (110), clearly indicating that plasma membrane cholesterol is internalized by endocytosis. Pulse-chase experiments in *Acanthamoeba* have lead to a similar conclusion (104). A parallel turnover of sterol and phospholipid from the plasma membrane indicated that endocytic vesicles in *Acanthamoeba* have a sterol:phosopholipid ratio similar to the plasma membrane.

To which intracellular location are these sterol-containing endocytic vesicles transported? Although this is not known, it can be suggested that they might be directed to the *trans* Golgi cisternae, where a high cholesterol content has been detected (54). Plasma membrane components, including cholesterol, could then recycle back from the *trans* Golgi to the cell surface. Esterification of cholesterol would require transport to the endoplasmic reticulum (3). In the model proposed above, cholesterol in the plasma membrane would enter the cell, but significant amounts might not be transferred to the endoplasmic reticulum and thus not be detectable by cholesterol ester formation.

Phospholipids

The sorting and intracellular movement of phospholipids is a complex problem. In addition to possible interconversion by base-exchange enzymes and deacylation-reacylation reactions (16–19), a given phospholipid is usually distributed throughout the cell and, with the exceptions noted earlier, does not appear to be located in a particular organelle. In addition, transport to the various organelles could occur by different mechanisms.

Spontaneous transfer of phospholipids between membranes is a very slow process (111, 112), occurring with a half-time of at least 12 hours. By comparison, the transfer of newly synthesized phospholipids to mitochondria in rat liver occurs in minutes (113). These latter results lead to the proposal and subsequent isolation of phospholipid transfer proteins from the cytosol of many cells (13). Although several transfer proteins that catalyze the exchange and net transfer of phospholipids between isolated membranes in vitro have been purified (91, 92), their function in vivo remains unclear. As mentioned earlier, transfer proteins need not necessarily act as lipid carriers to promote phospholipid transfer (94).

A test for the possible role for transfer proteins in intracellular phospholipid movement has been devised (114). In this approach the relative rates of transfer for newly synthesized phospholipids to mitochondria in vivo were compared with their relative rates of transfer in vitro as stimulated by a partially purified transfer protein from the cell. In rat hepatocytes these rates corresponded, consistent with an essential function for the rat liver transfer protein. However, in baby hamster kidney cells the in vivo and in vitro rates did not correspond. This latter finding was not considered to be consistent with the postulate that soluble phospholipid transfer proteins are responsible for the rapid movement of phospholipids from microsomes to mitochondria in living cells.

An alternative method for the transfer of newly synthesized phospholipids involves vesicular transport. Chlapowski & Band (115) provided some of the first evidence in support of this mechanism for the transport of newly synthesized phospholipids to the cell surface in *Acanthamoeba palestinensis*.

These authors also demonstrated that during a pulse-chase experiment with ^{3}H-glycerol, the labeled phospholipids left the plasma membrane and appeared within the cell presumably via an endocytic pathway. In *Dictyostelium discoideum* phospholipid transfer to the surface membrane was investigated for cells in a developmental stage during which there is a significant increase in both the cellular phospholipid content (116) and outer surface area (117). This transfer was inhibited by colchicine, a known disrupter of the intracellular microtubule network, suggesting vesicular transport. A low-density phospholipid-rich vesicular fraction isolated from the cells was shown to be involved in the transfer of newly synthesized phospholipids to the plasma membrane in vivo. These vesicles had a phospholipid composition characteristic of the plasma membrane, yet were enriched in a marker enzyme characteristic of the smooth endoplasmic reticulum and Golgi. Vesicles seen in the electron microscope during this developmental stage in *Dictyostelium* were absent at other times. It is therefore not clear whether the vesicular-mediated transport of phospholipids occurs during all the developmental stages of this cell. In *Amoeba proteus,* electron microscopic autoradiography revealed intense patterns of radiolabeled lipids over vacuolar structures, which were suggested as possible intermediates in the transfer of choline phosphatides to the plasma membrane (118). Incorporation of radiolabeled precursors into the lipids of the endoplasmic reticulum in *Tetrahymena pyriformis* is rapid, whereas appearance in the surface membrane takes six hours (119). However, lipid transfer by a vesicular route was not suggested in this cell.

Transport of newly synthesized phosphatidylethanolamine to the cell surface of Chinese hamster V79 lung cell fibroblasts is not blocked by energy poisons, or by disrupting either the cytoskeletal meshwork or the Golgi apparatus (120). Appearance of phosphatidylethanolamine in the plasma membrane of intact cells was followed by derivatization with the membrane-impermeable amino reagent 2,4,6-trinitrobenzene sulfonic acid (TNBS) present outside the cells. Careful controls clearly indicated that only phosphatidylethanolamine in the outer leaflet of the plasma membrane was modified. Newly synthesized phosphatidylethanolamine appeared in the outer leaflet of the plasma membrane almost immediately after the addition of either ^{3}H-ethanolamine or various other labeled precursors, suggesting that both intracellular transport and subsequent transmembrane movement of phosphatidylethanolamine at the cell surface are very rapid processes. The possibility that these findings were a result of exchange rather than net transfer was not reported. Kinetics of protein secretion in these cells revealed that secreted proteins and newly synthesized phosphatidylethanolamine appear to be transported independently to the cell surface. Since the proteins are transferred by a vesicular route, it was suggested that these data were not consistent with a vesicular pathway for phosphatidylethanolamine transport. This conclusion was supported by the ineffectiveness of cytoskeletal disrupting agents and monensin in blocking transport of phosphatidylethanolamine. As part of this study, cells that had been prelabeled with $^{32}P_i$ for seven days were derivatized with TNBS. The extent of chemical modification demonstrated that only 3% of the total cellular phosphatidylethanolamine is located in the plasma membrane. This value should be compared with the predictions made earlier in this review for the fraction of total cellular phospholipid that is present in the plasma membrane.

Ganglioside transport from its site of synthesis in the Golgi to the outer leaflet of the plasma membrane requires 20 minutes (121), which is similar to the time required for glycoprotein transport (101). Newly synthesized phospholipids appear to be transported to the plasma membrane of A. castellanii with very similar kinetics to the transfer of newly synthesized sterols, suggesting cotransport, possibly by a vesicular pathway (104). Phospholipids and sterols turn over from the cell surface by endocytosis at similar rates, indicating that sterol in the plasma membrane is not excluded from an endocytic pathway. In contrast, newly synthesized phosphatidylcholine is transferred to the plasma membrane of Chinese hamster ovary cells with a rapid half-time of 2 minutes at 25°C (68), distinctly different from the transport of either cholesterol or glycoprotein to the surface of these cells described earlier (68, 101). It was not reported whether the detection of labeled phosphatidylcholine in the plasma membrane represented a net transfer. This transfer of phosphatidylcholine was not affected by disruption of the cytoskeleton, treatment with monensin, or depletion of metabolic energy. However, it was arrested at 0°C but continued at an appreciable rate at 15°C, when cholesterol transport in these cells is virtually halted. Any proposed mechanism for the transport of cholesterol, phospholipids, and glycoproteins to the cell surface will have to account for their dissimilar kinetics and energy requirements.

Voelker (122) has utilized intracellular biosynthetic reactions to follow transfer of newly synthesized phospholipid to the mitochondria in the intact cell in an innovative manner. Mammalian cells in culture derive a major fraction of phosphatidylethanolamine from phosphatidylserine, which requires transport of phosphatidylserine from its site of synthesis (endoplasmic reticulum) to the mitochondria, the intracellular location of phosphatidylserine decarboxylase (23, 24). Transfer of phosphatidylserine to the mitochondria by net transport can therefore be monitored by conversion to phosphatidylethanolamine in intact cells. This transport was blocked by the energy poisons NaN_3 and NaF, causing labeled phosphatidylserine to accumulate in the microsomal fraction. Since isolated phospholipid transfer proteins can stimulate transfer of phospholipids between membranes in vitro in the absence of ATP (91, 92), it was argued that this mode of action does not reflect the situation in the intact cell (122). Voelker concluded that the energy requirement of the transfer process in vivo does not necessarily rule out protein-mediated transport, but does place constraints on reconstitution of lipid transport systems.

Pagano and his colleagues have pioneered a most promising approach to investigate lipid transport in living cells using fluorescent lipid derivatives. A detailed summary of these studies has been reported (11). The method is based on the finding that synthetic phospholipids containing the fluorescent acyl residue N-4-nitrobenzo-2-oxa-1,3-diazole amino caproic acid at the sn-2 position (which will be abbreviated as NBD-phospholipid) are spontaneously transferred from phospholipid vesicles to the plasma membrane of intact cells at 2°C (123). From this initial location the fate of these molecules is followed by fluorescence microscopy. Although it is clear that the fluorescent lipids are not physiological, it has been argued that they are good analogues of their endogenous counterparts in mimicking metabolism and intracellular movement (10–12). One important advantage in their use is that the short-chain NBD-fatty acid cannot be reutilized by the cells (124); therefore any ambiguities that could be introduced by deacylation-reacylation reactions or de novo biosynthesis are eliminated. A minor disadvantage is that intracellular cholesterol movement cannot be investigated.

NBD-phosphatidylcholine and NBD-phosphatidylethanolamine, introduced into the plasma membrane of fibroblasts at 2°C, enter the cell by endocytosis after warming to 37°C, eventually accumulating in the Golgi apparatus, where they appear to be trapped on the lumenal surface (86, 87). This intracellular location for these two phospholipids appears to be a direct consequence of their initial site of insertion into the plasma membrane. A comparison of the rates of internalization of NBD-phosphatidylcholine and a fluorescent protein bound to the cell surface revealed that these components of the plasma membrane were transferred to different intracellular sites (87). It was not possible to determine where their segregation occurred. NBD-phosphatidylethanolamine is also transferred to the nuclear envelope and mitochondria in a manner independent of endocytosis, initiated by transmembrane movement at the plasma membrane (86). When NBD-phosphatidic acid is transferred to cells at 2°C, rapid fluorescent labeling of the mitochondria, endoplasmic reticulum, and nuclear envelope is observed while the temperature is maintained at 2°C (124). This rapid intracellular labeling results from the formation of NBD-diglyceride at the cell surface followed by transmembrane movement, and from facilitated translocation to intracellular membranes where rephosphorylation can occur (125). NBD-ceramide is detectable at similar intracellular locations, while the cells are maintained at 2°C, following insertion of this lipid into the plasma membrane (126). Since

NBD-ceramide is not metabolized at this stage of incubation, these data would imply a rapid transmembrane movement of ceramide across the plasma membrane. On warming the cells containing the fluorescent ceramide, conversion to NBD-cerebroside and NBD-sphingomyelin occurs accompanied by changes in the intracellular location of fluorescence (127). Initially intense fluorescence is seen in the Golgi apparatus followed by transfer to the plasma membrane. This transfer is apparently blocked by monensin, which does not appear to inhibit the synthesis of sphingomyelin. It was therefore concluded that the intracellular site of sphingomyelin biosynthesis is in the Golgi apparatus, in contrast to the studies presented earlier (24–26).

TURNOVER OF LIPIDS

Phospholipid turnover occurs rapidly in most animal cells, where almost half of the total phospholipid is degraded every one or two cell divisions (17, 128). It has been proposed that this turnover may be connected with the maintenance of cellular viability, and somehow involved in the 'repair' and maintenance of the membranes in the living cell (129). Impairment of the turnover and degradation of cellular lipids are manifested in disease states (14). The dynamics of lipid transfer to their site of degradation within the cell are not well understood.

Although phospholipases are distributed throughout the cell, many studies indicate that the role of the extralysosomal phospholipases is to function together with lysophospholipid acyltransferases in remodeling intracellular phospholipids (17). This is manifested in the detection of apparently different rates for turnover for the polar and nonpolar portions of phospholipids (130). The major intracellular site for the degradation of sphingolipids appears to be lysosomal (131). In Niemann-Pick's disease, for example, an accumulation of sphingomyelin can be accounted for by the lack of a lysosomal sphingomyelinase (131). Acquired phospholipid storage diseases can be experimentally induced by drugs that concentrate in lysosomes (132). How are lipids normally targeted to this organelle for degradation? Studies previously described demonstrate that lipids in the plasma membrane can enter cells by endocytosis and become distributed throughout the cell. However, transfer to lysosomes was not described. Few studies have addressed this latter question. An attempt has been made to follow this pathway by incorporating NBD-sphingomyelin into the plasma membrane of cultured human skin fibroblasts either from normal patients or those with Niemann-Pick's disease (133) using the methods described earlier (11). Surprisingly, it was reported that radiolabeled sphingomyelin was taken up by the cells at the same rate as the fluorescent analogue. After incubation for several hours at 37°C, an increase in labeled ceramide was detected in the normal cells. In contrast, no increase in the labeled ceramide was detected in the Niemann-Pick cells. These data are consistent with the notion that in normal cells plasma membrane sphingomyelin is transferred to and subsequently degraded in lysosomes. However, in another study (134), it was shown that Niemann-Pick Type A fibroblasts were able to degrade endogenously produced sphingomyelin at near normal rates despite a >96% reduction in the lysosomal sphingomyelinase activity in these cells. Two additional pathways for sphingomyelin modification were proposed. These were either hydrolysis by a cellular sphingomyelinase or the utilization of the intact phosphoryl choline moiety from sphingomyelin for the synthesis of phosphatidylcholine. The reasons for these reported differences are not clear. However, they demonstrate the need for more investigations in this important area of cellular lipid degradation.

ACKNOWLEDGMENT

Work described in this review from the author's laboratory was supported by grants from the NSF (DCB 8503856) and NIH (GM 37104).

Literature Cited

1. Lodish, H. F. 1973. *Proc. Natl. Acad. Sci. USA* 70:1526–30
2. Jelsema, C. L., Morré, D. J. 1978. *J. Biol. Chem.* 253:7960–71
3. Chesterton, C. J. 1968. *J. Biol. Chem.* 243:1147–51
4. Goldstein, J. L., Anderson, R. G. W., Brown, M. S. 1979. *Nature* 279:679–85
5. Pfeffer, S. R., Rothman, J. E. 1987. *Ann. Rev. Biochem.* 56:000–00
6. Sandermann, H. 1978. *Biochim. Biophys. Acta* 515:209–37
7. Hoover, R. L., Dawidowicz, E. A., Robinson, J. M., Karnovsky, M. J. 1983. *J. Cell Biol.* 97:73–80
8. Whetton, A. D., Houslay, M. D. 1983. *FEBS Lett.* 157:70–74
9. Voelker, D. R. 1985. In *Biochemistry of Lipids and Membranes,* ed. D. E. Vance, J. E. Vance, pp. 475–502. Menlo Park, Calif: Cummings
10. Pagano, R. E., Longmuir, K. J. 1983. *Trends Biochem. Sci.* 8:157–61
11. Pagano, R. E., Sleight, R. G. 1985. *Science* 229:1051–57
12. Pagano, R. E., Sleight, R. G. 1985. *Trends Biochem. Sci.* 10:421–25
13. Dawidowicz, E. A. 1987. *Curr. Top. Membr. Transp.* 29: In press
14. Glew, R. H., Basu, A., Prence, E. M., Remaley, A. T. 1985. *Lab. Invest.* 53:250–69
15. Downes, C. P., Michell, R. H. 1985. In *Molecular Mechanisms of Transmembrane Signalling,* ed. P. Cohen, M. D. Houslay, pp. 3–56. Amsterdam: Elsevier
16. Longmuir, K. J. 1987. *Curr. Top. Membr. Transp.* 29: In press
17. Esko, J. D., Raetz, C. R. H. 1983. In *The Enzymes,* ed. P. D. Boyer, 16:207–53. New York: Academic
18. Vance, D. E. 1985. See Ref. 9, pp. 242–70
19. Kennedy, E. P. 1986. In *Lipids and Membranes: Past, Present and Future,* ed. J. A. F. Op den Kamp, B. Roelofsen, K. W. A. Wirtz, pp. 171–206. Amsterdam: Elsevier
20. DePierre, J., Dallner, G. 1976. In *Biochemical Analysis of Membranes,* ed. A. H. Maddy, pp. 79–131. New York: Wiley
21. Kanfer, J. N. 1980. *Can. J. Biochem.* 58:1370–80
22. Voelker, D. R. 1984. *Proc. Natl. Acad. Sci. USA* 81:2669–73
23. Dennis, E. A., Kennedy, E. P. 1972. *J. Lipid Res.* 13:263–67
24. Voelker, D. R., Kennedy, E. P. 1982. *Biochemistry* 21:2753–59
25. Margraff, W. D., Anderer, F. A., Kanfer, J. N. 1981. *Biochim. Biophys. Acta* 664:61–73
26. van den Hill, A., van Heusden, P. H., Wirtz, K. W. A. 1985. *Biochim. Biophys. Acta* 833:354–57
27. Bloch, K. 1966. *Science* 150:19–28
28. Anderson, R. G. W., Orci, L., Brown, M. S., Garcia-Segura, L. M., Goldstein, J. L. 1983. *J. Cell Sci.* 63:1–20
29. Trzaskos, J. M., Bowen, W. D., Fisher, G. J., Billheimer, J. T., Gaylor, J. L. 1982. *Lipids* 17:250–56
30. Lange, Y., Steck, T. L. 1985. *J. Biol. Chem.* 260:15592–97
31. Keenan, T. W., Morré, D. J., Basu, S. 1974. *J. Biol. Chem.* 249:310–15

32. Fishman, P. H., Brady, R. O. 1976. *Science* 194:906–15
33. Miller-Podraza, H., Bradley, R. M., Fishman, P. H. 1982. *Biochemistry* 21: 3260–65
34. White, D. A. 1973. In *Form and Function of Phospholipids*, ed. G. B. Ansell, J. N. Hawthorne, R. M. C. Dawson, pp. 441–82. Amsterdam: Elsevier
35. McMurray, W. C. 1973. See Ref. 34, pp. 205–51
36. Keenan, T. W., Morré, D. J. 1970. *Biochemistry* 9:19–24
37. Colbeau, A., Nachbaur, J., Vignais, P. M. 1971. *Biochim. Biophys. Acta* 249:462–92
38. Renkonen, O., Gahmberg, C. G., Simons, K., Kaariainen, L. 1972. *Biochim. Biophys. Acta* 255:66–78
39. Green, C. 1977. In *International Review of Biochemistry, Biochemistry of Lipids II*, ed. T. W. Goodwin, 14:101–52. Baltimore: Univ. Park Press
40. Ashworth, L. A. E., Green, C. 1966. *Science* 151:210–11
41. Ulsamer, A. G., Wright, P. L., Wetzel, M. G., Korn, E. D. 1971. *J. Cell Biol.* 51:193–215
42. Op den Kamp, J. A. F. 1979. *Ann. Rev. Biochem.* 48:47–71
43. Etemadi, A-H. 1980. *Biochim. Biophys. Acta* 604:423–75
44. van Meer, G., Simons, K. 1982. *EMBO J.* 1:847–52
45. Wattenberg, B. W., Silbert, D. F. 1983. *J. Biol. Chem.* 258:2284–89
46. Verkleij, A. J., Zwaal, R. F. A., Roelofsen, B., Comfurius, P., Kastelijn, D., van Deenen, L. L. M. 1973. *Biochim. Biophys. Acta* 323:178–93
47. Higgins, J. A., Dawson, R. M. C. 1977. *Biochim. Biophys. Acta* 470:342–56
48. Nilsson, O. S., Dallner, G. 1977. *J. Cell. Biol.* 72:568–83
49. Sundler, R., Sarcione, S. L., Alberts, A. W., Vagelos, P. R. 1977. *Proc. Natl. Acad. Sci. USA* 74:3350–54
50. de Kruijff, B., Rietveld, A., Cullis, P. R. 1980. *Biochim. Biophys. Acta* 600: 343–57
51. Burnell, E. E., Cullis, P. R., de Kruijff, B. 1980. *Biochim. Biophys. Acta* 603: 63–69
52. van Meer, G. 1986. *Trends. Biochem. Sci.* 11:194–95
53. Montesano, R., Perrelet, A., Vassalli, P., Orci, L. 1979. *Proc. Natl. Acad. Sci. USA* 76:6391–95
54. Orci, L., Montesano, R., Meda, P., Malaisse-Lagae, F., Brown, D., et al. 1981. *Proc. Natl. Acad. Sci. USA* 78:293–97
55. Steer, C. J., Bisher, M., Blumenthal, R., Steven, A. C. 1984. *J. Cell Biol.* 99:315–19
56. Patton, S. 1970. *J. Theor. Biol.* 29:489–91
57. Demel, R. A., Jansen, J. W. C. M., van Dijck, P. W. M., van Deenen, L. L. M. 1977. *Biochim. Biophys. Acta* 465:1–10
58. Lange, Y., Ramos, B. V. 1983. *J. Biol. Chem.* 258:15130–34
59. Gottlieb, M. H. 1977. *Biochim. Biophys. Acta* 466:422–28
60. Patzer, E. J., Wagner, R. R., Barenholz, Y. 1978. *Nature* 274:394–95
61. Barenholz, Y., Patzer, E. J., Moore, N. F., Wagner, R. R. 1978. *Adv. Exp. Med. Biol.* 101:45–56
62. Pal, R., Barenholz, Y., Wagner, R. R. 1980. *J. Biol. Chem.* 255:5802–6
63. De Martinez, S. G., Green, C. 1979. *Biochem. Soc. Trans.* 7:978–79
64. Lange, Y., Cutler, H. B., Steck, T. L. 1980. *J. Biol. Chem.* 255:9331–37
65. Lange, Y., Dolde, J., Steck, T. L. 1981. *J. Biol. Chem.* 256:5321–23
66. Backer, J. M., Dawidowicz, E. A. 1981. *J. Biol. Chem.* 256:586–88
67. Cooper, R. A. 1970. *Semin. Hematol.* 7:296–322
68. Kaplan, M. R., Simoni, R. D. 1985. *J. Cell Biol.* 101:441–45
69. Blouin, A., Bolender, R. P., Weibel, E. R. 1977. *J. Cell Biol.* 72:441–55
70. Coleman, R., Bell, R. M. 1978. *J. Cell Biol.* 76:245–53
71. Bell, R. M., Ballas, L. M., Coleman, R. A. 1981. *J. Lipid Res.* 22:391–403
72. Rothman, J. E., Dawidowicz, E. A. 1975. *Biochemistry* 14:2809–16
73. Johnson, L. W., Hughes, M. E., Zilversmit, D. B. 1975. *Biochim. Biophys. Acta* 375:176–85
74. Roseman, M. A., Litman, B. J., Thompson, T. E. 1975. *Biochemistry* 14:4826–30
75. Shaw, J. M., Moore, N. F., Patzer, E. J., Correa-Freire, M. C., Wagner, R. R., Thompson, T. E. 1979. *Biochemistry* 18:538–43
76. van Meer, G., Poorthuis, B. J. H. M., Wirtz, K. W. A., Op den Kamp, J. A. F., van Deenen, L. L. M. 1980. *Eur. J. Biochem.* 103:283–88
77. Crain, R. C., Zilversmit, D. B. 1980. *Biochemistry* 19:1440–47
78. Zilversmit, D. B., Hughes, M. E. 1977. *Biochim. Biophys. Acta* 469:99–110
79. van den Besselaar, A. M. H. P., de Kruijff, B., van den Bosch, H., van Deenen, L. L. M. 1978. *Biochim. Biophys. Acta* 510:242–55
80. Rothman, J. E., Kennedy, E. P. 1977. *Proc. Natl. Acad. Sci. USA* 74:1821–25
81. Bretscher, M. S. 1973. In *The Cell Sur-*

face in Development, ed. A. A. Mosco- na, pp. 17–27. New York: Wiley
82. Ganong, B. R., Bell, R. M. 1984. *Biochemistry* 23:4977–83
83. Bishop, W. R., Bell, R. M. 1985. *Cell* 42:51–60
84. Backer, J. M., Dawidowicz, E. A. 1985. *J. Cell Biol.* 101:261a
85. Abumrad, N. A., Park, J. H., Park, C. R. 1984. *J. Biol. Chem.* 259:8948–53
86. Sleight, R. G., Pagano, R. E. 1985. *J. Biol. Chem.* 260:1146–54
87. Sleight, R. G., Pagano, R. E. 1984. *J. Cell Biol.* 99:742–51
88. Seigneuret, M., Devaux, P. F. 1984. *Proc. Natl. Acad. Sci. USA* 3751–55
89. Daleke, D. L., Huestis, W. H. 1985. *Biochemistry* 24:5406–16
90. Tilley, L., Cribler, S., Roelofsen, B., Op den Kamp, J. A. F., van Deenen, L. L. M. 1986. *FEBS Lett.* 194:21–27
91. Wirtz, K. W. A. 1974. *Biochim. Biophys. Acta* 344:95–117
92. Zilversmit, D. B. 1983. *Methods Enzymol.* 98:565–73
93. Morré, D. J., Kartenbeck, J., Franke, W. W. 1979. *Biochim. Biophys. Acta* 559:71–152
94. Brown, R. E., Stephenson, F. A., Markello, T., Barenholz, Y., Thompson, T. E. 1985. *Chem. Phys. Lipids* 38:79–93
95. Backer, J. M., Dawidowicz, E. A. 1981. *Biochemistry* 20:3805–10
96. McLean, L. R., Phillips, M. C. 1981. *Biochemistry* 20:2893–900
97. Bell, F. P. 1975. *Biochim. Biophys. Acta* 398:18–27
98. Srikantaiah, M. V., Hansbury, E., Loughran, E. D., Scallen, T. J. 1976. *J. Biol. Chem.* 251:5496–504
99. Erickson, S. K., Meyer, D. J., Gould, R. G. 1978. *J. Biol. Chem.* 253:1817–26
100. Crivello, J. F., Jefcoate, C. R. 1980. *J. Biol. Chem.* 255:8144–51
101. DeGrella, R. F., Simoni, R. D. 1982. *J. Biol. Chem.* 257:14256–62
102. Kaplan, M. R., Simoni, R. D. 1985. *J. Cell Biol.* 101:446–53
103. Lange, Y., Mathies, H. J. G. 1984. *J. Biol. Chem.* 259:14624–30
104. Mills, J. T., Furlong, S. T., Dawidowicz, E. A. 1984. *Proc. Natl. Acad. Sci. USA* 81:1385–88
105. Bretscher, M. S., Thompson, J. N., Pearse, B. M. F. 1980. *Proc. Natl. Acad. Sci. USA* 77:4156–59
106. Poznansky, M. J., Czekanski, S. 1982. *Biochim. Biophys. Acta* 685:182–90
107. Slotte, J. P., Lundberg, B. 1983. *Biochim. Biophys. Acta* 750:434–39
108. Slotte, J. P., Lundberg, B., Bjorkerud, S. 1984. *Biochim. Biophys. Acta* 793:423–26
109. Robertson, D. L., Poznansky, M. J. 1985. *Biochem. J.* 232:553–57
110. Helmy, S., Porter-Jordan, K., Dawidowicz, E. A., Pilch, P., Schwartz, A. L., Fine, R. E. 1986. *Cell* 44:497–506
111. Roseman, M. A., Thompson, T. E. 1980. *Biochemistry* 19:439–44
112. Nichols, J. W., Pagano, R. E. 1981. *Biochemistry* 20:2783–89
113. McMurray, W. C., Dawson, R. M. C. 1969. *Biochem. J.* 112:91–108
114. Yaffe, M. P., Kennedy, E. P. 1983. *Biochemistry* 22:1497–507
115. Chlapowski, F. J., Band, R. N. 1971. *J. Cell Biol.* 50:634–51
116. De Silva, N. S., Siu, C-H. 1980. *J. Biol. Chem.* 255:8489–96
117. De Silva, N. S., Siu, C-H. 1981. *J. Biol. Chem.* 256:5845–50
118. Flickinger, C. J., Read, G. A. 1982. *Cell Tissue Res.* 222:523–30
119. Thompson, G. A. Jr., Nozawa, Y. 1977. *Biochim. Biophys. Acta* 472:157–61
120. Sleight, R. G., Pagano, R. E. 1983. *J. Biol. Chem.* 258:9050–58
121. Miller-Podraza, H., Fishman, P. H. 1982. *Biochemistry* 21:3265–70
122. Voelker, D. R. 1985. *J. Biol. Chem.* 260:14671–76
123. Struck, D. K., Pagano, R. E. 1980. *J. Biol. Chem.* 255:5404–10
124. Pagano, R. E., Longmuir, K. J., Martin, O. C. 1983. *J. Biol. Chem.* 258:2034–40
125. Pagano, R. E., Longmuir, K. J. 1985. *J. Biol. Chem.* 260:1909–16
126. Lipsky, N. G., Pagano, R. E. 1983. *Proc. Natl. Acad. Sci. USA* 80:2608–12
127. Lipsky, N. G., Pagano, R. E. 1985. *J. Cell Biol.* 100:27–34
128. van den Bosch, H. 1980. *Biochim. Biophys. Acta* 604:191–246
129. Dawson, R. M. C. 1973. *Sub-Cell. Biochem.* 2:69–89
130. Omura, T., Siekevitz, P., Palade, G. E. 1967. *J. Biol. Chem.* 242:2389–96
131. Sweeley, C. C. 1985. See Ref. 9, pp. 361–403
132. Hostetler, K. Y. 1985. In *Phospholipids and Cellular Regulation*, ed. J. F. Kuo, 1:182–206. Boca Raton, Fla: CRC
133. Sutrina, S. L., Chen, W. W. 1984. *Biochim. Biophys. Acta* 793:169–79
134. Spence, M. W., Clarke, J. T. R., Cook, H. W. 1983. *J. Biol. Chem.* 258:8595–600

Ann. Rev. Biochem. 1987. 56:63–87
Copyright © 1987 by Annual Reviews Inc. All rights reserved

TOPOGRAPHY OF GLYCOSYLATION IN THE ROUGH ENDOPLASMIC RETICULUM AND GOLGI APPARATUS

Carlos B. Hirschberg

Department of Biochemistry, University of Massachusetts, Medical Center, Worcester, Massachusetts 01605

Martin D. Snider

Department of Biochemistry, Case Western Reserve University School of Medicine, Cleveland, Ohio 44106

CONTENTS

63

0066-4154/87/0701-0063$02.00

PERSPECTIVES AND SUMMARY

A great deal has been learned about the synthesis of N- and O-linked oligosaccharides, proteoglycans, and glycolipids (see reviews 1–5). Many of the reactions have been studied in detail, and a number of the glycosyltransferases involved have been purified. In addition, the subcellular locations of these reactions have been intensely studied. It is now known that nearly all of these steps occur in the Golgi apparatus (GA). Among the exceptions are the early stages of N-linked oligosaccharide synthesis, which occur in the rough endoplasmic reticulum (RER).

However, comparable understanding has not been available for the topographical aspects of glycosylation. The topographical problem is based on the fact that sugar residues must move across membranes as part of these reactions. While the nucleotide sugar substrates are made either in the cytoplasm or, in the case of CMP-NeuAc, in the nucleus (6–8), the macromolecular products are found in the lumen of the RER and GA. Moreover, because virtually none of these macromolecules are found in the cytoplasm, the glycosylation reactions must be organized to ensure the transmembrane movement of sugar residues.

Recent work has begun to uncover how this transmembrane movement occurs. In the GA, nucleotide sugars are transported across the Golgi membrane by specific transport systems (9, 10). They are then utilized by luminal glycosyltransferases. In contrast, during the assembly of asparagine-linked oligosaccharides in the RER, at least some sugar residues may enter the lumen via the transmembrane movement of lipid-linked intermediates (11–13).

The aim of this chapter is to review the recent evidence that has led to a better understanding of the topography of glycosylation. The reader will notice that our understanding of these events is not so complete for the ER as the GA. Nevertheless, we will present the results of the former organelle first in order to maintain a continuity with the sequence of reactions occurring in vivo, namely first RER and thereafter the GA.

TOPOGRAPHY OF ASPARAGINE GLYCOSYLATION IN THE ROUGH ENDOPLASMIC RETICULUM

The initial stages of asparagine-linked oligosaccharide synthesis in the RER involve the precursor oligosaccharide $Glc_3Man_9GlcNAc_2$. It is assembled in pyrophosphate linkage to the lipid carrier, dolichol (Dol), by the stepwise addition of single sugar residues, and is then transferred en bloc to a peptide asparagine residue (1) (Figure 1). The first seven residues are added directly from UDP-GlcNAc and GDP-Man, while the final seven are added from Man-P-Dol and Glc-P-Dol. The monosaccharide-lipids are made from Dol-P and the corresponding nucleotide sugars.

During the assembly of the lipid-linked precursor oligosaccharide and its transfer to peptide acceptors, sugar residues cross the RER membrane. While the nucleotide sugar substrates are cytoplasmic, newly glycosylated proteins are found in the RER lumen. A number of groups have studied this process, with a view to understanding how, and at what stage of assembly, this transmembrane movement occurs. This section reviews these efforts, which have been aimed at determining the orientation of enzymes of oligosaccharide-lipid synthesis and transfer, the topography of the oligosaccharide-lipid intermediates themselves, and the ability of nucleotide sugars to cross the RER membrane.

All of these studies are performed with vesicles isolated from the ER that are sealed and have the correct orientation (cytoplasmic side out). Either microsomes, which contain both ER and GA components, or highly purified ER fractions have been used. In all cases, it is necessary to show that the orientation of the vesicles is preserved during the experiment. Typical tests include showing that the activity of a luminal enzyme such as mannose-6-phosphatase remains latent or proving that a luminal marker cannot be digested or inactivated by added protease.

Topography of Enzymes of Glc3Man9GlcNAc2-PP-Dolichol Synthesis and Transfer

OLIGOSACCHARIDE TRANSFERASE   A great deal of evidence has been obtained in support of the luminal location of oligosaccharide transfer to protein acceptors: (*a*) Welply et al (14) used the tripeptide acetyl-Asn-Leu-Thr-NHCH$_3$, which was glycosylated when it was incubated with hen oviduct microsomal vesicles. The glycopeptide product was found to be trapped within the vesicles. This result suggests that the hydrophobic tripeptide had crossed the vesicular membrane and was glycosylated at the luminal surface. The resulting hydrophilic glycopeptide was then unable to escape from the vesicles. (*b*) Hanover & Lennarz showed that peptides glycosylated in vitro are within microsomal vesicles (15). They incubated vesicles with labeled GDP-Man under conditions where label was incorporated into lipid-linked

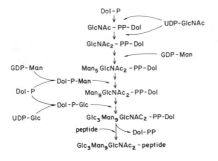

Figure 1   Pathway of asparagine-linked oligosaccharide synthesis in the rough endoplasmic reticulum. Reprinted by permission of John Wiley & Sons, Inc. from Snider, M. D. 1984. In *Biology of Carbohydrates*, ed. V. Ginsburg, P. W. Robbins, Vol. 2. pp. 169. ©

oligosaccharides and was then transferred to endogenous peptide acceptors. Using endoglycosidase H to probe the orientation of these products, they found that the glycopeptides were digested only if the vesicles were disrupted, suggesting that these acceptors resided within the RER lumen. (c) Glabe et al (16) studied the synthesis of ovalbumin, which has a single oligosaccharide. By determining the length of the shortest glycosylated nascent chain, they concluded that glycosylation cannot occur until there is a segment of 32 residues between the acceptor site and the ribosome. This segment would be long enough to extend across the ER membrane into the lumen. (d) Studies of in vitro translation have shown that peptides are glycosylated only if they are inserted into microsomal vesicles and segregated into the lumen (17, 18). (e) A number of secreted proteins are glycosylated posttranslationally in vivo, presumably when the polypeptides are sequestered within the RER lumen (19, 20). (f) The oligosaccharide donor for the reaction, $Glc_3Man_9GlcNAc_2$-PP-Dol, has been localized to the luminal face of the RER membrane (21) (see below).

ENZYMES OF $GLC_3MAN_9GLCNAC_2$-PP-DOLICHOL SYNTHESIS Several groups have attempted to obtain insight into the orientation of these enzymes by examining their protease sensitivity in microsomal vesicles. Typically, intact and disrupted microsomes are treated with protease and the effect of these treatments on enzyme activities assessed. In all cases, it must be demonstrated that protease does not penetrate the intact vesicles. This is usually accomplished by showing that a protein known to face the lumen is protease resistant in intact vesicles and protease sensitive in disrupted vesicles.

In a study with rat liver microsomes, Snider et al (22) found that a number of enzymatic activities were equally protease sensitive in intact and disrupted vesicles. This was true for the synthesis of Glc-P-Dol, Man-P-Dol, and $GlcNAc_2$-PP-Dol, as well as the incorporation of Glc residues from Glc-P-Dol into oligosaccharide-lipids. These conclusions have been confirmed by similar studies on Glc-P-Dol and $GlcNAc_2$-PP-Dol synthesis in hen oviduct microsomes (23) and on Glc-P-Dol synthesis in calf thyroid microsomes (24). These workers concluded that these enzymes have protease-sensitive sites facing the cytoplasmic side of the ER membrane.

Unfortunately, it was not possible to determine the orientation of the catalytic sites of these enzymes from these experiments. The results can be explained by (a) enzymes with cytoplasmic catalytic sites, (b) enzymes with transmembrane catalytic sites (e.g. ones that transfer a sugar from a cytoplasmic nucleotide sugar to a luminal acceptor), or (c) transmembrane enzymes with luminal active sites and cytoplasmic protease-sensitive sites.

While it is not possible to decide among these possibilities, the first two are the most likely, with the third considerably less so. There are a large number of membrane-bound enzymes in the RER and Golgi whose active sites are known to face the lumen. All of these enzymes are protease resistant in intact vesicles and protease sensitive in disrupted vesicles (25–27).

In these studies, some enzymatic activities were protease resistant in both intact and disrupted vesicles (22, 23). These include the synthesis of GlcNAc-P-Dol in both rat liver and hen oviduct microsomes and the synthesis of Man-P-Dol in hen oviduct microsomes. Presumably these enzymes are simply not sensitive to the proteolytic conditions used. As a result, no conclusion can be reached about their topography in the ER membrane.

Additional information about the orientation of Glc-P-Dol synthesis was obtained by Spiro & Spiro (24) from a study with the stilbene derivative DIDS. They showed that Glc-P-Dol synthesis was blocked when intact calf thyroid microsomes were treated with this inhibitor. Since DIDS probably cannot penetrate these vesicles, it is likely that a key part of the catalytic site of this enzyme faces the cytoplasmic side of the ER membrane, although it cannot be ruled out that DIDS actually inhibited the transport of UDP-Glc into the RER lumen for subsequent glucosylation of Dol-P.

The topography of enzymes of Dol-P metabolism has been examined by Adair and coworkers. They found that treatment of intact microsomes with proteases or mercury dextran derivatives inactivated two enzymes of Dol-P synthesis, long-chain prenyl transferase and dolichol kinase, suggesting that these enzymes have a cytoplasmic orientation (28). A second study showed that polyisoprenylphosphate phosphatase, which degrades Dol-P, is active with exogenous substrates in intact vesicles (29). This result is consistent with a cytoplasmic orientation of this enzyme, although translocation of the substrate to the site of a luminal enzyme could not be ruled out.

Topography of Oligosaccharide-Lipid Intermediates in the Rough Endoplasmic Reticulum Membrane

There are 16 lipid-linked intermediates involved in precursor oligosaccharide assembly: Man-P-Dol, Glc-P-Dol, and the 14 pyrophosphate-linked species, up to and including $Glc_3Man_9GlcNAc_2$-PP-Dol (1). Studies from three laboratories have provided information about the orientation of 12 of these species. Intermediates have been localized to both sides of the RER membrane, suggesting that oligosaccharide-lipid assembly occurs on both the cytoplasmic and luminal faces.

Hanover & Lennarz examined the orientation of $GlcNAc_2$-PP-Dol in hen oviduct microsomes (23, 30). Exogenous galactosyltransferase, which converts $GlcNAc_2$-PP-Dol into $GalGlcNAc_2$-PP-Dol in the presence of UDP-Gal, was used as a probe. Endogenous $GlcNAc_2$-PP-Dol was inaccessible to the probe in intact microsomes, but was accessible in disrupted vesicles, suggesting a luminal orientation. This was first demonstrated for GlcNAc$_2$-PP-Dol labeled by incubating microsomes with UDP-[^{14}C]GlcNAc in vitro. In a later experiment, oviduct slices were incubated with [^{3}H]glucosamine and microsomes were prepared. Then [^{14}C]GlcNAc$_2$-PP-Dol was added and the orientation of labeled species examined with galactosyltransferase. As expected, the added [^{14}C]GlcNAc$_2$-PP-Dol was accessible to the probe. However, [^{3}H] GlcNAc$_2$-PP-Dol synthesized in vivo was only accessible in disrupted vesicles. This experiment also supports a luminal orientation of this species and tends to rule out the possibility that endogenous GlcNAc$_2$-PP-Dol is present on the cytoplasmic face of intact vesicles, but is inaccessible to the probe in some way.

The topography of oligosaccharide-lipids in the RER membrane has been studied by Snider & Robbins (21) and Snider & Rogers (11). These workers used microsomal vesicles from Chinese hamster ovary cells labeled with [^{3}H]Man in culture. The cells were labeled under conditions where all the radioactivity was found in RER-derived vesicles. The lectin concanavalin A was used to probe oligosaccharide-lipid orientation. While oligosaccharide-lipid species are ordinarily extracted from membranes with organic solvents, when complexed with Con A, these molecules cannot be extracted using standard conditions. This property was used to measure the binding of Con A to different oligosaccharide-lipids in intact and disrupted vesicles.

This technique was first used by Snider & Robbins (21) to show that the largest lipid-linked oligosaccharide, Glc$_3$Man$_9$GlcNAc$_2$, was bound by Con A in disrupted, but not intact vesicles. This result suggests that this species resides on the luminal face of the ER membrane. This is in agreement with the luminal orientation of oligosaccharide transferase (see above) and is also supported by experiments of Hanover & Lennarz (23). These workers showed that storage of microsomal vesicles containing labeled Glc$_3$Man$_9$GlcNAc$_2$-PP-Dol resulted in the release of free oligosaccharide. This species was trapped within the vesicles, suggesting that it resulted from the degradation of luminal Glc$_3$Man$_9$GlcNAc$_2$-PP-Dol.

Snider & Rogers (11) extended the above studies with Con A to examine the orientation of nine other oligosaccharide-lipid species. They found that 50–70% of Man$_{3-5}$GlcNAc$_2$-PP-Dol was bound by Con A in intact vesicles. Similar binding was seen in disrupted vesicles, supporting the cytoplasmic orientation of these species. In contrast, larger species, including Man$_{6-9}$-GlcNAc$_2$-PP-Dol and Glc$_{1-3}$Man$_9$GlcNAc$_2$-PP-Dol, were bound by Con A in disrupted vesicles (50–80%), while <15% of each species was bound in intact vesicles, supporting a luminal orientation for these molecules. Two species, Man$_{1-2}$GlcNAc$_2$-PP-Dol, were not bound by Con A to a significant extent in either intact or disrupted vesicles. Apparently, the interaction of Con A with these molecules was not strong enough to make a determination about their orientation in the ER membrane.

Evidence suggesting that the synthesis of Man-P-Dol is accompanied by the transmembrane movement of Man residues has been obtained by Haselbeck & Tanner (12, 13). They used sealed phospholipid vesicles that contained Dol-P and purified Dol-P mannosyltransferase. When these vesicles were incubated with GDP-Man, Man-P-Dol was synthesized. When the reaction was done with GDP trapped within the vesicles, GDP-Man accumulated in the vesicles. This reaction occurred only with GDP as the trapped nucleotide and required Man-P-Dol synthesis, since GDP-Man accumulation occurred only when the vesicles contained Dol-P and active mannosyltransferase. In addition, experiments with GDP-Man radiolabeled with one isotope in the nucleotide and another in the mannose showed that the sugar but not the nucleotide was entering the vesicles. These results suggest that Man residues were transported across the vesicle membrane as part of Man-P-Dol synthesis; this was followed by resynthesis of GDP-Man within the vesicles in a reversal of the above reaction. The mechanism of this transport is not known. It could be part of the synthetic reaction or be due to Man-P-Dol flip-flop. In any case, these results suggest that cytoplasmic GDP-Man gives rise to luminal Man-P-Dol. Further support for a luminal orientation of Man-P-Dol is obtained from studies with yeast. Because Man-P-Dol serves as the donor for the direct mannosylation of secreted proteins in the RER of this organism (31, 32), this monosaccharide-lipid probably faces the RER lumen.

In conclusion, the above studies suggest that (a) $GlcNAc_2$-PP-Dol does not face the cytoplasmic side of the RER membrane, (b) most, if not all, $Man_{3-5}GlcNAc_2$-PP-Dol faces the cytoplasmic side of the RER membrane, (c) the intermediates from $Man_6GlcNAc_2$-PP-Dol to $Glc_3Man_9GlcNAc_2$-PP-Dol face the lumen, and (d) Man-P-Dol may face the RER lumen or flip-flop across the membrane.

Transport of Nucleotide Sugars into Rough Endoplasmic Reticulum Vesicles

The precursors of the sugars in $Glc_3Man_9GlcNAc_2$-PP-Dol are the corresponding nucleotide sugars UDP-Glc, GDP-Man, and UDP-GlcNAc. It is clear that the sugar residues must cross the RER membrane during oligosaccharide-lipid synthesis and transfer. One of the ways in which this might occur is via the specific transport of intact nucleotide sugars into the RER lumen. In this case, the organization of oligosaccharide-lipid synthesis would be similar to the topography of glycosylation in the GA, with a luminal pool of nucleotide sugars and both the glycosyltransferases and acceptors facing the lumen. Recent studies suggest that two of the nucleotide sugars involved in oligosaccharide-lipid assembly, UDP-GlcNAc and UDP-Glc, do in fact enter the RER via specific uptake systems, while GDP-Man does not. These experiments will be discussed briefly here, since nucleotide sugar uptake into the ER appears to be similar to this process in the GA. The reader is referred to the section on the GA for a more detailed review of nucleotide sugar transport and the methods used to study this process.

Perez & Hirschberg demonstrated the uptake of UDP-GlcNAc and UDP-Glc into vesicles from rat liver RER and SER in vitro (33, 34). By using a mixture of nucleotide sugars radiolabeled with one isotope in the sugar and a second one in the nucleotide, they found that the isotope ratio of the transported solute was similar to that of the substrates added to the medium, supporting the idea that intact nucleotide sugars are transported. Moreover, transport was found to be temperature dependent and saturable, with apparent K_ms of 3–4 μM for both nucleotide sugars. Preliminary evidence was also obtained suggesting that the transport of these nucleotide sugars involves a specific exchange with luminal UMP.

While UDP-GlcNAc and UDP-Glc are transported into RER and SER vesicles, GDP-Man is not. Two lines of evidence support this conclusion. First, incubation of RER vesicles with GDP-Man did not result in the accumulation of solutes within the vesicles (34). Moreover, when vesicles were incubated with GDP-Man labeled with one isotope in the sugar and a second isotope in the nucleotide, the isotope ratio in the pellet was quite different from the ratio in the incubation medium; the vesicles were enriched in radioactive sugar, suggesting that intact GDP-Man did not enter the vesicles. Finally, GMP was also not transported into the vesicles. Since the transport of most nucleotide sugars proceeds via exchange with the corresponding nucleoside monophosphate, the absence of GMP uptake also argues against a GDP-Man transport system.

Second, attempts to demonstrate a luminal pool of GDP-Man in RER-derived vesicles have been unsuccessful. Fleischer could not detect this nucleotide sugar in an analysis of compounds isolated with vesicles (35). In addition, when rat liver or hen oviduct slices were incubated with [3H]Man and microsomes were prepared, no labeled GDP-Man was associated with the vesicles (23, 36). A similar experiment with sialic acid has documented a pool of CMP-NeuAc within microsomes containing Golgi vesicles (23, 36). These experiments, therefore, argue against a pool of luminal GDP-Man. However, incubation of hen oviduct slices with [3H]glucosamine failed to demonstrate a microsomal pool of UDP-GlcNAc. Because this nucleotide sugar is known to be transported into the Golgi appartus, this result suggests that caution must be exercised in interpreting these negative results.

Taken together, these results suggest that GDP-Man is not translocated across the RER membrane. However, it is still possible that this transport occurs, followed by rapid utilization of the nucleotide sugar by glycosyltransferases and rapid efflux of the nucleotide product. If this is the case, then GDP-Man translocation is very different from all the other cases of nucleotide sugar and adenosine 3'-phosphate 5'-phosphosulfate (PAPS) transport into the RER and Golgi that have been described. In yeast, direct mannosylation of glycoproteins in the GA apparatus is known to occur (31, 32, 37). While the translocation of GDP-Man has not been tested in this organism, one might expect, by analogy with the topography of glycosylation in the vertebrate GA, that GDP-Man translocation does occur.

During their studies of RER vesicles, Perez & Hirschberg showed that incubation with radioactive UDP-GlcNAc and UDP-Glc resulted in the translocation of nucleotide sugars into RER vesicles (33, 34). In addition, during these incubations, nucleotide sugars were utilized to synthesize GlcNAc-PP-Dol, as well as Glc-P-Dol, oligosaccharide-lipid, and glycoproteins containing labeled GlcNAc and Glc residues. However, this experiment could not discern the relationship between these two events. Nucleotide sugar translocation could be a prerequisite for the synthetic reactions, or the two events could be unrelated.

To test the relationship between the uptake and utilization of UDP-GlcNAc and UDP-Glc in RER vesicles, Perez & Hirschberg (34) examined the effect of disrupting the vesicles on these two processes. This disruption resulted in a 70–90% decrease in solutes derived from UDP-GlcNAc and UDP-Glc in the vesicles, while no such effect was seen with GDP-Man, consistent with the lack of translocation of this nucleotide sugar. However, disruption had a different effect on incorporation. Disruption resulted in a 30% decrease in GlcNAc and Glc incorporation from the nucleotide sugars into endogenous acceptors, while a 13% increase of Man incorporation from GDP-Man into endogenous acceptors was observed. These results were interpreted as suggesting that the transfer of GlcNAc and Glc was probably linked to transport of the corresponding nucleotide sugars into the lumen of the vesicles. The stimulation of Man incorporation was interpreted as being the result of somewhat higher solubilization of the lipid substrates by the detergent used to break the vesicles. However, as discussed by the authors (34), this experiment is not conclusive because the effects on sugar incorporation into endogenous acceptors were relatively small, and detergents were the only agents used to disrupt the vesicles.

Model for the Topography of Glycosylation in the Rough Endoplasmic Reticulum

The data reviewed above can be combined to support the following model for the topography of glycosylation in the RER. The model is summarized in Figure 2 and the discussion will refer to the steps shown in the figure. All of the steps have previously been proposed by other workers and by ourselves (11, 13, 34, 38). It should be noted that many features of the model are supported only by a single type of experiment. Thus, it is likely that some aspects will be revised as more information becomes available.

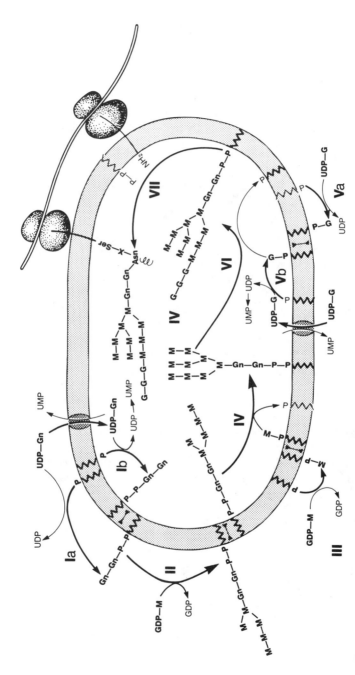

Figure 2 Model for the topography of asparagine glycosylation in the rough endoplasmic reticulum. P∿: Dol-P; P-P∿: Dol-PP; M: mannose; Gn: N-acetyl-glucosamine; G: glucose; GDP: guanosine 5'-diphosphate; GMP: guanosine 5'-monophosphate; UDP: uridine 5'-diphosphate; UMP: uridine 5'-monophosphate; ⟶ : translocase.

STEP I   Synthesis of GlcNAc$_2$-PP-Dol. Two arrangements for the topography of the addition of the two GlcNAc residues have been proposed. (I*a*) This species is synthesized on the cytoplasmic face of the ER by glycosyltransferases on this side of the membrane (22, 30). This is supported by the fact that the enzyme that adds the second GlcNAc residue has protease-sensitive sites on the cytoplasmic side of the membrane. (I*b*) UDP-GlcNAc is translocated into the RER lumen, where it is utilized to synthesize GlcNAc$_2$-PP-Dol on the luminal face. This is supported by the demonstration of UDP-GlcNAc translocation into RER vesicles (33), and by the finding that GlcNAc$_2$-PP-Dol probably resides on the luminal face of hen oviduct microsomes (23, 30). If the latter alternative is the correct one, then GlcNAc$_2$-PP-Dol must be translocated from the luminal to the cytoplasmic face after synthesis, since the synthetic steps that follow appear to occur on the cytoplasmic face.

STEP II   Synthesis of Man$_5$GlcNAc$_2$-PP-Dol. The addition of the first five Man residues from GDP-Man probably occurs on the cytoplasmic face. This is supported by the localization of Man$_{3-5}$GlcNAc$_2$-PP-Dol to this face of the membrane (11) and by the finding that GDP-Man is not transported into the ER (23, 34).

Flip-flop of Man$_5$GlcNAc$_2$-PP-Dol   This oligosaccharide-lipid has been localized to the cytoplasmic side of ER vesicles. However, all larger intermediates were found on the luminal face. This suggests that Man$_5$-GlcNAc$_2$-PP-Dol flips from the cytoplasmic to the luminal face, where it is elongated. Such flip-flop is probably protein mediated, since GlcNAc$_2$-PP-Dol (23) and spin-labeled Dol-P derivatives (39) have been shown not to flip in phospholipid vesicles. It should be noted that in yeast mutants defective in oligosaccharide-lipid synthesis, truncated oligosaccharides (e.g. Man$_{2-3}$-GlcNAc$_2$) are transferred to protein (40). Since these intermediates are thought to be made on the cytoplasmic face, this suggests that they can also flip to the luminal face where they act as donors in protein glycosylation.

STEP III   Synthesis of Man-P-Dol. This intermediate is probably synthesized on the cytoplasmic face and then flips to the luminal face where it serves as a Man donor in oligosaccharide elongation. This is supported by the demonstration that Dol-P mannosyltransferase may catalyze the flip-flop of this intermediate in reconstituted systems (12, 13). This view is also consistent with the findings that this enzyme has cytoplasmic protease-sensitive sites (22) and that GDP-Man is not transported into the RER (23, 34). Moreover, the addition of the last four Man residues to the oligosaccharide occurs on the luminal face (see Step IV).

STEP IV   Synthesis of $Man_9GlcNAc_2$-PP-Dol. The addition of the last four Man residues to the oligosaccharide from Man-P-Dol occurs on the luminal face. The demonstration that $Man_{6-9}GlcNAc_2$-PP-Dol face the lumen supports this conclusion (11).

STEP V   Synthesis of Glc-P-Dol. There are two possibilities for the topography of this reaction: (Va) Glc-P-Dol is synthesized on the cytoplasmic face and then flips to the luminal side, in a manner similar to that proposed for Man-P-Dol (Step III). The demonstration that the enzyme that synthesizes Glc-P-Dol has cytoplasmic protease-sensitive sites tends to support this possibility (22). (Vb) UDP-Glc penetrates the ER membrane and is used to synthesize Glc-P-Dol on the luminal face. This possibility is supported by the demonstration of UDP-Glc transport into the RER (33, 34). However, Parodi & coworkers (41) have shown that glycoproteins in the RER can be directly glycosylated from UDP-Glc. The UDP-Glc transport that has been described could serve to furnish substrate for this reaction.

STEP VI   Transfer of Glc residues to oligosaccharide-PP-Dol. These reactions probably occur on the luminal face. This conclusion is supported by the finding that $Man_9GlcNAc_2$-PP-Dol and $Glc_{1-3}Man_9GlcNAc_2$-PP-Dol are all found on this side of the membrane (11, 21). The finding by Snider et al (22) that the transfer of Glc residues to oligosaccharide-PP-Dol is inactivated by protease treatment of intact microsomes appears to conflict with this interpretation. However, it is possible that the glucosyltransferases have cytoplasmic protease-sensitive sites and luminal catalytic sites.

STEP VII   Transfer of $Glc_3Man_9GlcNAC_2$ to peptide acceptor. This step occurs on the luminal face of the RER. This conclusion is supported by localization to the luminal face of oligosaccharide transferase (14, 42), the oligosaccharide-lipid donor (21), and the glycopeptide products (14, 17, 18).

The Dol-PP product of oligosaccharide transferase is presumably released on the luminal face of the RER. While the carrier lipid is reutilized, the topography of this process is not understood. The orientation of the pyrophosphatase that converts Dol-PP to Dol-P is unknown, and the locations of the reactions that utilize Dol-P (Steps I, IV, and V) are not clear. Depending on the topography of these processes, either Dol-P or Dol-PP may flip across the RER membrane. In fact, flip-flop of one of these derivatives seems likely, since enzymes of Dol synthesis (28) and catabolism (29) have been localized to the cytoplasmic face of the RER, while Dol-PP is probably released on the luminal face.

This model has several significant features. First, it suggests that sugar residues cross the RER membrane by the flip-flop of lipid-linked intermediates. In fact, depending on which of the alternatives for Steps I and V are correct, all the sugar residues transferred to protein may cross the RER via such flip-flop. This is in sharp contrast to the topography of glycosylation in the Golgi apparatus, where sugar residues cross the membrane by the transport of intact nucleotide sugars. Second, this model offers an explanation for the use of two different Man donors in the synthesis of the oligosaccharide-lipid, namely that GDP-Man and Man-P-Dol act as donors on different sides of the RER membrane. The first five Man residues, which are derived from GDP-Man, would be added on the cytoplasmic face, while the final four, which are added from Man-P-Dol, would be added on the luminal face.

TOPOGRAPHY OF GLYCOSYLATION IN THE GOLGI APPARATUS

Glycosylation reactions in the GA are considerably more diverse than those in the RER. Golgi enzymes carry out the terminal stages of asparagine-linked oligosaccharide synthesis, as well as the assembly of proteoglycans, glycolipids, and O-linked oligosaccharides. The topography of these reactions is well understood. Nucleotide sugars are translocated across the Golgi membrane via specific carriers and utilized by glycosyltransferases in the Golgi lumen. The evidence supporting this model, as well as our knowledge of the nucleotide sugar carriers, will be reviewed in this section. This discussion will include studies with the sulfate donor PAPS, since the topography of sulfation reactions appears to be very similar to that of glycosylation.

Topography of Golgi Glycosyltransferases

There is ample evidence supporting a luminal orientation of Golgi glycosyltransferases: (a) The secretory and membrane glycoproteins that are substrates for glycosylation in the Golgi complex are known to face the lumen (15, 43). (b) Golgi glycosyltransferases in sealed vesicles in vitro are inactive with exogenous macromolecular acceptors that cannot penetrate the membrane. The vesicles must be disrupted to express the activity (25, 26). (c) These enzymes are protease resistant in sealed vesicles, but protease sensitive in disrupted vesicles (25, 26). (d) Many Golgi glycosyltransferases are glycosylated (44, 45). (e) Soluble glycosyltranferases are secreted from cells in active form. These enzymes are derived from the Golgi membrane-bound forms by proteolysis (45, 46). These data strongly support the conclusion that the active sites of Golgi glycosyltransferases face the Golgi lumen.

Detection of Luminal Pools of Nucleotide Sugars

The luminal orientation of Golgi glycosyltransferases suggests that there should be nucleotide sugars within the lumen to serve as substrates in glycosylation reactions. Three laboratories have presented evidence strongly suggesting the existence of such intraluminal pools. Carey et al (36) incubated mouse liver slices with [^{3}H]NeuAc and then isolated microsomes. CMP-[^{14}C]NeuAc was added to the slices prior to the isolation as a marker for adsorption of nucleotide sugars to microsomes during the subcellular fractionation. The isolated microsomes were enriched five- to eight-fold in CMP-[^{3}H]NeuAc synthesized in vivo over the added CMP-[^{14}C]NeuAc. Moreover, this enrichment was abolished when the vesicles were disrupted with detergent. These results strongly suggest that there was a pool of CMP-NeuAc synthesized in vivo that was isolated with the microsomes. Because detergent was required to liberate this pool and the sialylation of proteins and lipids occurs in the Golgi apparatus, this luminal pool was probably within Golgi vesicles in the microsomal fraction. Using a similar experimental design, Hanover & Lennarz (23) also detected an intraluminal pool of CMP-NeuAc in hen oviduct microsomes. However, these workers were unable to detect a similar pool of UDP-GlcNAc. Finally, Fleischer (35) found that the nucleotide and nucleotide sugar content of rat liver Golgi vesicles were very different from those of the cytoplasm. An HPLC analysis revealed AMP, UMP, CMP, UDP, and an unresolved mixture of UDP-Glc and UDP-Gal in the vesicles.

Nucleotide Sugar Translocation into Golgi Vesicles

The evidence reviewed above suggests that glycosylation occurs within the GA. Nucleotide sugar substrates, glycosyltransferases, and macromolecular acceptors all have a luminal orientation. However, nucleotide sugars are synthesized in the cytosol with the sole exception of CMP-NeuAc, which appears to be synthesized in the nucleus (6–8). This evidence led to the hypothesis that glycosylation in the GA involves the transport of nucleotide sugars from the cytosol into the Golgi lumen. The last 10 years have seen the direct demonstration of this translocation and the characterization of this process, which will be reviewed in this section.

The first evidence to support nucleotide sugar translocation was provided by Kuhn & White (47). Using a membrane fraction from rat mammary glands, they found that addition of glucose and UDP-galactose led to the synthesis of lactose within the vesicles. Lactose synthesis was inhibited by ovalbumin (a macromolecular substrate for the lactose synthase) with disrupted vesicles but not with intact ones. This suggested that lactose was a product of luminal galactosyltransferase and that lactose was not entering the vesicles subsequent to its synthesis in the medium. In addition, lactose synthesis was inhibited by UDP-Glc and UDP-GlcA in disrupted but not intact vesicles, suggesting that UDP-Gal was transported into vesicles by a system that did not recognize other nucleotide sugars (48). Based on these results, Kuhn & White proposed a uridine nucleotide cycle associated with lactose synthesis (49). In this model, UDP-Gal enters the Golgi apparatus by facilitated transport followed by galactose transfer to glucose catalyzed by luminal lactose synthase. The UDP formed is acted on by a UDPase to yield P_i and UMP, which then exits the vesicle. Many of the features of this cycle were later confirmed using other approaches (50, 51) and are the basis for the current view of nucleotide sugar transport described below.

An important advance towards the goal of direct measurement of nucleotide sugar transport into the GA was the demonstration by three groups that one could obtain Golgi vesicles that are highly purified, sealed, and of the same membrane topography as in vivo. Fleischer (25), using rat liver Golgi vesicles isolated from a sucrose gradient in the presence of D_2O, showed that galactosyl and sialyltransferases (using, respectively, ovalbumin and asialotransferrin as exogenous acceptors), were activated eightfold when the vesicles were disrupted. In addition, the enzyme activities could be destroyed by trypsin only if vesicles were first permeabilized, strongly suggesting that the active sites of these enzymes were facing the lumen of the vesicles. Carey & Hirschberg (26) labeled Golgi vesicles from mouse and rat liver in vitro with CMP-[^{14}C]NeuAc. They then showed that the incorporated NeuAc could only be removed with neuraminidase if the vesicles were first made leaky with detergents or by mechanical disruption. A similar finding was made by Creek & Morré (52). Carey & Hirschberg also showed that sialyltransferase activity could be destroyed by proteases only if vesicles were first made leaky (26).

Using these highly purified Golgi vesicles, two types of assays have been developed to directly measure the transport of nucleotide sugars (50a, 53). The first type is a centrifugation assay (53). Vesicles are incubated separately with radioactive standard penetrating and nonpenetrating solutes, centrifuged, and the amount of each solute in the pellets measured. The volumes accessible to these solutes in the vesicle pellets are then calculated from the specific activities. The volumes accessible to radioactive nucleotide sugars are also determined and compared to the volumes of the standards. Typically, volumes for nucleotide sugars are found to be even larger than those of the standard penetrators. The second assay is a filtration assay (50a). Vesicles are incubated with nucleotide sugars and filtered through a Millipore filter. The total solutes associated with vesicles are then determined. To correct for nonspecific absorption of solutes to vesicles, vesicles disrupted by filipin or low concentrations of detergents are used.

These assays, or slight variations, have been used with purified Golgi vesicles and microsomal preparations containing Golgi vesicles. Transport of the following compounds has been demonstrated (Table 1): CMP-NeuAc, GDP-fucose, UDP-Gal, UDP-GlcNAc, UDP-xylose, UDP-GalNAc, UDP-GlcA, and PAPS. Similar uptake of acetyl-CoA into rat liver Golgi vesicles

Table 1 Demonstration of nucleotide sugar and PAPS translocation into vesicles

Sugar nucleotide	Organelle	Tissue	Reference
UDP-Gal	Golgi	Rat mammary gland	48
	Golgi	Rat liver	33, 50, 50a, 55, 56
	Golgi	CHO cells	57, 58
	a	Mouse thymocytes	59
UDP-GlcNAc	Golgi	Rat liver	9, 33
	RER	Rat liver	33
	Golgi	CHO cells	57, 58
	a	Mouse thymocytes	60
	RER, SER	Calf thyroid	24
UDP-GalNAc	Golgi	Rat liver	61
	RER	Rat liver	61
	SER	Rat liver	61
	Golgi	CHO cells	58
UDP-xylose	Golgi	Rat liver	62
	RER	Rat liver	62
	SER	Rat liver	62
UDP-GlcA	Golgi	Rat liver	62
	RER	Rat liver	62
	SER	Rat liver	62
UDP-glucose	RER	Rat liver	34
	Golgi	CHO cells	58
GDP-fucose	Golgi	Rat liver	51, 63
	a	Mouse thymocytes	59
CMP-NeuAc	Microsomes (RER, SER, Golgi)	Rat liver	36
	Golgi	Rat liver	9, 52, 63, 64
	Golgi	CHO cells	57, 58
	Golgi	Hen oviduct	23
	a	Mouse thymocytes	59
PAPS	Golgi	Rat liver	64, 65, 70
	Microsomes (RER, SER, Golgi)	Chick chondrocytes	66

[a]Studies were done with permeabilized whole cells.

has also been reported (54). In the lumen, the acetyl CoA serves as a substrate in the acetylation of NeuAc residues.

The criteria used to demonstrate transport of nucleotide sugars and PAPS were the following:

1. Transport of nucleotide sugars is organelle specific. When vesicles derived from the rough and smooth endoplasmic reticulum and Golgi apparatus were assayed for their ability to mediate the transport of nucleotide sugars, it was found that CMP-NeuAc, GDP-fucose, PAPS, UDP-Gal, UDP-GalNAc, and UDP-GlcA were translocated into Golgi vesicles at rates at least 10-fold higher than into vesicles from other organelles.

2. The entire nucleotide sugar or PAPS is translocated into the lumen of the GA. This was demonstrated by using nucleotide sugars radiolabeled with one isotope in the sugar and a different one in the nucleotide. If the entire nucleotide sugar or nucleotide sulfate was being translocated into the lumen, then luminal molecules should have the same isotope ratio as the substrate added to the incubation medium. This was found for all the nucleotide sugars studied.

3. Transport is saturable. K_ms of between 1 and 10 μM have been measured for the nucleotide sugars and PAPS.

4. Transport is temperature dependent. In many instances, transport at 30°C was 8- to 10-fold more rapid than at 4°C.

5. Transport is competitively inhibited by the corresponding nucleoside mono-, di-, and triphosphates. Transport is not inhibited by free sugars.

6. Transport is protease sensitive. Treatment of intact Golgi vesicles with proteases abolished transport, but did not affect the activity of sialyltransferases, which face the lumen.

These data strongly suggest that nucleotide sugars are translocated into the Golgi lumen and argue that this translocation is catalyzed by transport proteins that are specific to the Golgi membrane. This conclusion is also consistent with studies on nucleotide sugar transport and intraluminal pools in permeabilized cultured thymocytes (59, 60).

Physiological Relevance of Nucleotide Sugar Translocation into Golgi Vesicles

Evidence strongly suggesting that nucleotide sugar translocation is of physiological importance has been obtained through analyses of mutant cultured cells. These Chinese hamster ovary cell lines, which belong to two complementation groups, were isolated through their resistance to the lectin wheat germ agglutinin by Stanley (67) and Briles et al (68). Cell lines of one complementation group (69) (Lec2 and 1021) had an 80–90% decrease in sialic acid in both glycoproteins and glycolipids. However, levels of CMP-NeuAc and sialyltransferase activity were not affected. Moreover, compositions of endogenous acceptors for sialylation appeared to be similar in mutant and wild-type cells (68). However, Golgi vesicles from Lec2 cells translocated CMP-NeuAc at only 2% of the rate of vesicles from wild-type cells (57). Translocation of PAPS, UDP-GlcNAc, and UDP-galactose, however, were comparable in vesicles derived from both cell lines. Cells of the second complementation group (69) (Lec8 and Clone 13) had an 80–90% reduction in both galactose and sialic acid in their glycoproteins and glycolipids. The levels of UDP-Gal and CMP-NeuAc, as well as activities of galactosyl and sialyltransferases, appeared to be normal, and endogenous acceptors for galatose in the mutant cell lines appeared to show no major differences with the corresponding parental cell lines (67, 68). However, the rate of UDP-Gal translocation into Golgi vesicles from the mutant cell lines was only 3% the rate of transport into vesicles isolated from the parental cell lines (58). Other nucleotide derivatives, such as UDP-GlcNAc, UDP-GalNAc, PAPS, and CMP-NeuAc were transported at comparable rates in vesicles derived from mutants and wild-type cells.

These results furnished strong evidence for the importance of nucleotide sugar translocation for glycosylation in the GA, since mutants that are unable to translocate individual nucleotide sugars are defective in the addition of these sugars to acceptors, even though the glycosyltransferases and acceptors are present. In addition, because the transport of a single nucleotide sugar was defective in each mutant, these results suggest that there are separate translocator proteins for each nucleotide sugar and PAPS. This is particularly significant for the uridine nucleotide sugars, which might have been thought to use a single translocator.

Properties of Nucleotide Sugar and PAPS Translocation

There appear to be separate transport proteins for each nucleotide sugar and for PAPS. This was established by the lack of competitive inhibition among different nucleotide sugars (64). This point is also supported by the genetic evidence discussed above.

INHIBITORS OF TRANSLOCATION   The effect of inhibitors on the translocation of nucleotide sugars and PAPS has been studied. Among the structural analogues of nucleotide sugars that were studied, the following conclusions were made (64).

1. The nucleotide base is recognized in the binding of nucleotide sugars and PAPS to the translocator. In general, the transport of pyrimidine nucleotide sugars is inhibited only by pyrimidine nucleotides, while purine nucleotides inhibit the translocation of purine nucleotide sugars. For example, UMP and CMP are competitive inhibitors of CMP-NeuAc translocation, while AMP and GMP do not inhibit. In these studies, the inhibition of CMP

was competitive and simple, while that of UMP was competitive, but complex, i.e. the plots of $1/V$ over $1/S$ did not give straight lines.

2. The position of the phosphate groups (of the nucleotides) appears to be critical for binding. The presence of a 5' phosphate group in uridine, guanosine, and cytidine nucleotides results in high inhibitory activity, while the 3' or 2' phosphate derivatives have virtually no inhibitory activity.

3. The number of 5' phosphate groups appears to be of less importance. Thus, the inhibitory activities of 5'-CTP, 5'-CDP, and 5'-CMP on CMP-NeuAc transport were similar.

4. The presence of a 2' hydroxyl group in the ribose moiety of the nucleotide sugar is not an important recognition feature since 2'-dCMP is as good an inhibitor of CMP-NeuAc translocation as CMP.

Since the nucleotide sugars and PAPS have net negative charges at physiological pH, the effect of inhibitors of anion transport on the translocation of nucleotide sugars and PAPS into Golgi vesicles was studied (70). The stilbene derivatives SITS and DIDS both inhibited translocation of CMP-NeuAc, GDP-fucose, and PAPS. Translocation was inhibited 50% by 10–20 μM DIDS or 100 μM SITS. DIDS was also used to demonstrate that the transfer of sulfate from PAPS occurred subsequent to translocation of the nucleotide sulfate into Golgi vesicles. When intact Golgi vesicles were incubated with DIDS, both the transport of PAPS and the incorporation of sulfate from PAPS into endogenous macromolecular acceptors within the vesicles were inhibited by 80%. However, when disrupted vesicles were treated with DIDS, no inhibition of sulfate transfer into endogenous macromolecules was detected. Unfortunately, when this experiment was performed with GDP-fucose and CMP-NeuAc, interpretable results were not obtained. In these cases, DIDS caused inhibition of incorporation in both intact and disrupted vesicles. These results, therefore, do not enable one to determine whether the inhibition of incorporation of sugars into endogenous acceptors with intact vesicles is caused by DIDS inhibition of the transporter, the transferase, or both.

Capasso & Hirschberg (70) also reported that the PAPS translocator in Golgi membranes appears to have structural features in common with the ADP/ATP translocator of mitochondria. They observed that palmitoyl coenzyme A, atractyloside, and carboxyatractyloside inhibited the translocation of PAPS into Golgi vesicles at concentrations that block ADP/ATP translocation in mitochondria. The inhibition of PAPS translocation was specific, since no inhibition of CMP-NeuAc and GDP-fucose translocation was observed. Neither coenzyme A nor free palmitate inhibited the translocation of PAPS; there was also no effect of bongkrekic acid. Since mitochondria were not found to be able to translocate PAPS, the conclusion was made that the PAPS and ADP/ATP translocators, while related, are not identical.

MECHANISM OF TRANSLOCATION: ANTIPORT WITH NUCLEOSIDE MONO-
PHOSPHATES   Nucleotide sugar translocation into Golgi vesicles in vitro
does not have an energy requirement (51). It was found that the addition of
ATP, phosphoenolpyruvate, or oligomycin had no effect, suggesting that
translocation occurs by facilitated diffusion. In addition, no effects were seen
with the ionophores nigericin, valinomycin, and monensin. Thus, the net
transport process appears to be electroneutral (51).

Initial studies by Kuhn & White (47), and Brandan & Fleischer (50),
examining the transport of UDP-Gal into Golgi vesicles, had studied the fate
of the nucleotide products. They obtained strong evidence that UMP exited
the lumen of the vesicles. This UMP was derived from the breakdown of UDP
by nucleoside diphosphatase facing the luminal side, following the transfer of
galactose to endogenous acceptors within the lumen.

One explanation of these results is that the entry of a nucleotide sugar is
coupled to the exit of the corresponding nucleoside monophosphate. Capasso
& Hirschberg tested this hypothesis for GDP-fucose (51). They incubated
Golgi vesicles with [^{3}H]GDP-fucose, which resulted in the accumulation of
^{3}H-solutes (primarily [^{3}H]GMP) within the vesicles. Addition of GDP-
[^{14}C]fucose resulted in exit of [^{3}H]GMP from the vesicles with a corresponding entry of GDP-[^{14}C]fucose. This stimulation of [^{3}H]GMP efflux was
specific, since it was not caused by PAPS or GDP-Man, a close structural
analogue of GDP-fucose. In a similar experiment, CMP-NeuAc also failed to
stimulate GMP efflux from Golgi vesicles. These results suggest that nucleotide sugar translocation occurs by an antiport mechanism involving the one for
one exchange of nucleotide sugars with the corresponding nucleoside
monophosphate. This idea is supported by direct measurements of the
stoichiometry of exchange. The ratio of entry of GDP-fucose and exit of GMP
was calculated to be 1.2. These calculations, however, are based on the
assumption that the specific activity of GMP within the vesicle lumen was
equal to that of the initial nucleotide sugar added to the vesicles at the
beginning of the incubation, which has not been proven.

Preliminary evidence was obtained that the translocation of other nucleotide sugars and PAPS also occurs by an antiport mechanism (51). Using an
approach similar to the one described above, evidence for the following
antiporters has been obtained: UDP-GlcNAc : UMP, CMP-NeuAc : CMP, and
PAPS : 3'-AMP. In these studies, the stoichiometry between entry of the
nucleotide sugar or nucleotide sulfate and exit of the corresponding nucleoside monophosphate was between 1.2 and 2.

For the reasons discussed above, these ratios are probably lower than the
calculated ones, supporting a one for one exchange. However, it cannot be
ruled out that the stoichiometry of exchange for some nucleotide sugars is not
one for one.

Model for the Topography of Glycosylation and Sulfation in the Golgi Apparatus

The experiments described above have led to the following model for the topography of glycosylation and sulfation in the GA. Figure 3 summarizes this model with particular emphasis on fucosylation from GDP-fucose. First, GDP-fucose binds through its guanosine moiety to the cytosolic domain of its antiporter. Then, the nucleotide sugar enters the Golgi lumen through its specific antiporter that recognizes the nucleotide and sugar (I). Fucose is transferred to luminal acceptors (II) in a reaction catalyzed by fucosyltransferases (III). GDP, a product of this reaction, is then cleaved to GMP by a luminal nucleoside diphosphatase (IV). This enzyme has been characterized biochemically and is probably the same as the thiaminepyrophosphatase that has been used extensively as a cytochemical marker for Golgi cisternae (71–75). The resulting GMP can then return to the cytoplasm by an equimolar exchange with a new molecule of GDP-fucose. An identical cycle is thought to occur for all other nucleotide sugars with the exception of CMP-NeuAc. In

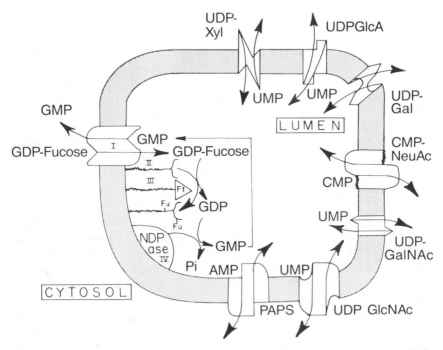

Figure 3 Model for the topography of glycosylation and sulfation in the Golgi apparatus. I: GDP-fucose/GMP antiporter; II: endogenous macromolecular acceptors for fucose; III: fucosyltransferases; IV: nucleosidediphosphatase.

this case, the action of the luminal nucleoside diphosphatase is not required, since CMP is the product of the glycosylation reaction. The fate of the luminal inorganic phosphate product is not known. Presumably, there is a phosphate transport system to allow this phosphate to return to the cytoplasm. Such transport has been described for the glucose-6-phosphatase system in the ER lumen (76).

FUTURE DIRECTIONS

Substantial progress has been made in understanding how sugar residues cross the ER and Golgi membranes during glycosylation in these organelles. However, there are a number of important problems that must remain the subject of future work.

While a plausible model for the topography of glycosylation in the ER has been proposed, it must be viewed as a working hypothesis. The uncertainties in the model must be resolved to prove how each of the sugars in the precursor oligosaccharide crosses the RER membrane. This will require more information about the topography of lipid-linked intermediates and the synthetic enzymes. Moreover, it has been proposed that the transmembrane movement of sugar residues occurs by the flip-flop of lipid-linked saccharides. Further experiments are needed to provide direct proof of this flip-flop, to study its mechanism, and to determine whether specific proteins catalyze this process.

A number of problems remain to be solved about the topography of glycosylation in the GA. The mechanism of nucleotide sugar:nucleoside monophosphate antiport has not been studied in detail. Moreover, while it has been shown that some nucleotide sugars are transported into both ER and Golgi, it is not known whether the translocator proteins for these molecules are the same in the two organelles. In addition, the exact location of the translocator proteins in the Golgi has not been studied. All of the glycosyltransferases studied to date are restricted to one or several cisternae in discrete parts of the GA. Nucleotide sugar translocators might be distributed throughout the GA or might be found only in those regions where the corresponding transferases are found. Purified translocator proteins and specific antibodies will allow all of these questions to be addressed. The topography of glycosphingolipid synthesis in the GA will need further examination particularly in view of a recent report suggesting that some of the reactions may occur on the cytoplasmic face of the GA membrane (77).

A number of new complex carbohydrates have been discovered in the last several years. Recently, several groups (78–83) have described O-linked GlcNAc residues on proteins. Most of these structures are on the cytosolic face of the nuclear envelope (78, 82), although they have also been reported on the cell surface (80) and other organelles (79). The topography of synthesis of these structures, and the way in which they are localized in a large number of different organelles, remain to be studied.

In addition, two new types of Man-containing structures have been identified recently. The first of these is a glycophospholipid that is attached to the carboxyl terminus of membrane proteins from vertebrates and trypanosomes (83–85). The second type of structure is O-linked oligosaccharides in brain chondroitin sulfate (86). Because GDP-Man is probably not translocated into the ER or GA of mammalian cells, questions remain about the transmembrane movement of Man residues during the synthesis of these structures. These structures may be synthesized on the cytoplasmic face of these organelles and then translocated to the lumen, or they may derive their Man residues from Man-P-Dol at the luminal face. In the case of the glycophospholipid, this latter alternative appears to be correct, since mutant cell lines defective in Man-P-Dol synthesis do not add this structure to glycoproteins (87).

Finally, the role of transmembrane movement of sugars in the regulation of glycoprotein synthesis remains to be addressed. It is possible that the ability of these residues to cross membranes is important in determining the structures of glycolipids and the oligosaccharides of glycoproteins and proteoglycans.

ACKNOWLEDGMENTS

We thank Claudia Abeijon and Enrique Brandan for helpful discussion and Peggy Kerner for skillful typing. Work in the authors' laboratories was supported by grants from the National Institutes of Health, GM 30365, 34396, and 38183.

Literature Cited

1. Kornfeld, R., Kornfeld, S. 1985. *Ann. Rev. Biochem.* 54:631–64
2. Höök, M., Kjellén, L., Johansson, S., Robinson, J. 1984. *Ann. Rev. Biochem.* 53:847–69
3. Lindahl, U., Feingold, D. S., Roden, L. 1986. *Trends Biochem. Sci.* 11:221–25
4. Basu, S., Basu, M., Kyle, J. W., Chon, H.-C. 1984. In *Ganglioside Structure, Function and Biomedical Potential,* ed. R. W. Ledeen, R. K. Yu, M. M. Rapport, K. Susuki, pp. 249–61. New York: Plenum
5. Kishimoto, Y. 1983. In *The Enzymes,* ed. P. D. Boyer, Vol. 16, pp. 358–408. New York: Academic 3rd ed.
6. Kean, E. L. 1970. *J. Biol. Chem.* 245:2301–8
7. Van Dijk, W., Ferwerda, W., van den Eijnden, D. H. 1973. *Biochim. Biophys. Acta* 315:162–75

8. Coates, S. W., Gurney, T., Sommers, L. W., Yeh, M., Hirschberg, C. B. 1980. *J. Biol. Chem.* 255:9225–29
9. Fleischer, B. 1983. *J. Histochem. Cytochem.* 31:1033–40
10. Perez, M., Hirschberg, C. B. 1986. *Biochim. Biophys. Acta.* 864:213–22
11. Snider, M. D., Rogers, O. C. 1984. *Cell* 36:753–61
12. Haselbeck, A., Tanner, W. 1982. *Proc. Natl. Acad. Sci. USA* 79:1520–24
13. Haselbeck, A., Tanner, W. 1984. *FEMS Micro. Lett.* 21:305–8
14. Welply, J. K., Shenbagamurthi, P., Lennarz, W. J., Naider, F. 1983. *J. Biol. Chem.* 258:11856–63
15. Hanover, J. A., Lennarz, W. J. 1980. *J. Biol. Chem.* 255:3600–4
16. Glabe, C. G., Hanover, J. A., Lennarz, W. J. 1980. *J. Biol. Chem.* 255:9236–42

17. Katz, F. N., Rothman, J. E., Lingappa, V. R., Blobel, G., Lodish, H. F. 1977. *Proc. Natl. Acad. Sci. USA* 74:3278–82
18. Lingappa, V. R., Lingappa, J. R., Prasad, R., Ebner, K. E., Blobel, G. 1978. *Proc. Natl. Acad. Sci. USA* 75:2338–42
19. Bergman, L. W., Kuehl, W. M. 1978. *Biochemistry* 17:5174–80
20. Roberts, J. L., Phillips, M., Rosa, P. A., Herbert, E. 1978. *Biochemistry* 17:3609–18
21. Snider, M. D., Robbins, P. W. 1982. *J. Biol. Chem.* 257:6796–801
22. Snider, M. D., Sultzman, L. A., Robbins, P. W. 1980. *Cell* 21:385–92
23. Hanover, J. A., Lennarz, W. J. 1982. *J. Biol. Chem.* 257:2787–94
24. Spiro, M. J., Spiro, R. G. 1985. *J. Biol. Chem.* 260:5808–15
25. Fleischer, B. 1981. *J. Cell Biol.* 89: 246–55
26. Carey, D. J., Hirschberg, C. B. 1981. *J. Biol. Chem.* 256:989–93
27. Grinna, L. S., Robbins, P. W. *J. Biol. Chem.* 1979. 254:8814–18
28. Adair, W. L. Jr., Cafmeyer, N. 1983. *Biochim. Biophys. Acta* 751:21–26
29. Keller, R. K., Adair, W. L. Jr., Cafmeyer, N., Simion, F. A., Fleischer, B., Fleischer, S. 1986. *Arch. Biochem. Biophys.* 249:207–14
30. Hanover, J. A., Lennarz, W. J. 1978. *J. Biol. Chem.* 254:9237–46
31. Sharma, C. B., Babczinski, P., Lehle, L., Tanner, W. 1974. *Eur. J. Biochem.* 46:35–41
32. Haselbeck, A., Tanner, W. 1983. *FEBS Lett.* 158:335–38
33. Perez, M., Hirschberg, C. B. 1985. *J. Biol. Chem.* 260:4671–78
34. Perez, M., Hirschberg, C. B. 1986. *J. Biol. Chem.* 261:6822–30
35. Fleischer, B. 1981. *Arch. Biochem. Biophys.* 212:602–10
36. Carey, D. J., Sommers, L. W., Hirschberg, C. B. 1980. *Cell* 19:597–605
37. Schekman, R. 1985. *Ann. Rev. Cell Biol.* 1:115–43
38. Hanover, J. A., Lennarz, W. J. 1981. *Arch. Biochem. Biophys.* 211:1–19
39. McCloskey, M. A., Troy, F. A. 1980. *Biochemistry* 19:2061–66
40. Huffaker, T. C., Robbins, P. W. 1983. *Proc. Natl. Acad. Sci. USA* 80:7466–70
41. Parodi, A. J., Mendelzon, D. H., Lederkremer, G. Z., Martin-Barrientos, J. 1984. *J. Biol. Chem.* 259:6351–57
42. Welply, J. K., Kaplan, H. A., Shenbagamurthi, P., Naider, F., Lennarz, W. J. 1986. *Arch. Biochem. Biophys.* 246:808–19
43. Rodriguez-Boulan, E., Kreibich, G., Sabatini, D. D. 1978. *J. Cell Biol.* 78:874–93
44. Schwyzer, M., Hill, R. L. 1977. *J. Biol. Chem.* 252:2338–45
45. Strous, G. J. A. M., Berger, E. G. 1982. *J. Biol. Chem.* 257:7623–28
46. Smith, C. A., Brew, K. 1977. *J. Biol. Chem.* 252:1294–98
47. Kuhn, N. J., White, A. 1975. *Biochem. J.* 148:77–84
48. Kuhn, N. J., White, A. 1976. *Biochem. J.* 154:243–44
49. Kuhn, N. J., White, A. 1977. *Biochem. J.* 168:423–33
50. Brandan, E., Fleischer, B. 1982. *Biochemistry* 21:4640–45
50a. Brandan, E., Fleischer, B. 1981. *Fed. Proc.* 40:681 (Abstr.)
51. Capasso, J. M., Hirschberg, C. B. 1984. *Proc. Natl. Acad. Sci. USA* 81:7051–55
52. Creek, K. E., Morré, D. J. 1981. *Biochim. Biophys. Acta* 643:292–305
53. Perez, M., Hirschberg, C. B. 1986. *Methods Enzymol.* 138:709–15
54. Varki, A., Diaz, S. 1985. *J. Biol. Chem.* 260:6600–8
55. Yusuf, H. K. M., Pohlentz, G., Sandhoff, K. 1983. *Proc. Natl. Acad. Sci. USA* 80:7075–79
56. Barthelson, R., Roth, S. 1985. *Biochem. J.* 225:67–75
57. Deutscher, S. L., Nuwayhid, N., Stanley, P., Briles, E. I. B., Hirschberg, C. B. 1984. *Cell* 39:295–99
58. Deutscher, S. L., Hirschberg, C. B. 1986. *J. Biol. Chem.* 261:96–100
59. Cacan, R., Cecchelli, R., Hoflack, B., Verbert, A. 1984. *Biochem. J.* 224:277–84
60. Cecchelli, R., Cacan, R., Verbert, A. 1985. *Eur. J. Biochem.* 153:111–16
61. Abeijon, C., Hirschberg, C. B. 1986. *Fed. Proc.* 45:1977 (Abstr.)
62. Nuwayhid, N., Glaser, J. H., Johnson, J. C., Conrad, H. E., Hauser, S. C., Hirschberg, C. B. 1986. *J. Biol. Chem.* 261:12936–41
63. Sommers, L. W., Hirschberg, C. B. 1982. *J. Biol. Chem.* 257:10811–17
64. Capasso, J. M., Hirschberg, C. B. 1984. *Biochim. Biophys. Acta* 777:133–39
65. Schwarz, J. K., Capasso, J. M., Hirschberg, C. B. 1984. *J. Biol. Chem.* 259:3554–59
66. Habuchi, O., Conrad, H. E. 1985. *J. Biol. Chem.* 260:13102–8
67. Stanley, P. 1980. *ACS Symp. Ser.* 128:214–21
68. Briles, E. B., Li, E., Kornfeld, S. 1977. *J. Biol. Chem.* 252:1107–16

69. Stanley, P. 1985. *Mol. Cell. Biol.* 5: 923–29
70. Capasso, J. M., Hirschberg, C. B. 1984. *J. Biol. Chem.* 259:4263–66
71. Ohkubo, I., Ishibashi, T., Tanigushi, N., Makita, A. 1980. *Eur. J. Biochem.* 112:111–18
72. Farquhar, M. G., Bergeron, J. J. M., Palade, G. E. 1974. *J. Cell Biol.* 60:8–25
73. Little, J. S., Widnell, C. C. 1975. *Proc. Natl. Acad. Sci. USA* 72:4013–17
74. Novikoff, A. B., Goldfischer, S. 1961. *Proc. Natl. Acad. Sci. USA* 47:802–10
75. Kuriyama, Y. 1972. *J. Biol. Chem.* 247:2979–88
76. Arion, W. J., Ballas, L. M., Lange, A. J., Wallin, B. K. 1976. *J. Biol. Chem.* 251:4901–7
77. Coste, H., Martel, M. B., Got, R. 1986. *Biochim. Biophys. Acta* 858:6–12
78. Davis, L. I., Blobel, G. 1986. *Cell* 45:699–709
79. Holt, G. D., Hart, G. W. 1986. *J. Biol. Chem.* 261:8049–57
80. Torres, C. R., Hart, G. W. 1984. *J. Biol. Chem.* 259:3308–17
81. Schindler, M., Hogan, M. 1984. *J. Cell Biol.* 99:99a (Abstr.)
82. Holt, G. D., Snow, C. M., Gerace, L., Hart, G. W. 1986. *J. Cell Biol.* (Abstr.) In press
83. Tse, A. G. D., Barclay, A. N., Watts, A., Williams, A. F. 1985. *Science* 230:1003–8
84. Ferguson, M. A. J., Haldar, K., Cross, G. A. M. 1985. *J. Biol. Chem.* 260: 4963–68
85. Haas, R., Brandt, P. T., Knight, J., Rosenberry, T. L. 1986. *Biochemistry* 25:3098–105
86. Krusius, T., Finne, J., Margolis, R. K., Margolis, R. U. 1986. *J. Biol. Chem.* 261:8237–42
87. Fatemi, S. H., Tartakoff, A. M. 1986. *Cell* 46:653–57

Ann. Rev. Biochem. 1987. 56:89–124

COMPLEXES OF SEQUENTIAL METABOLIC ENZYMES[1,2]

Paul A. Srere

Pre-Clinical Science Unit, Veterans Administration Medical Center and Department of Biochemistry, University of Texas Health Science Center, Dallas, Texas 75216

CONTENTS

[1]A note on terminology: A number of different expressions have been used to refer to multienzyme complexes. They include protein machines, clusters, supramolecular complexes, aggregates, and metabolons. I use these terms interchangeably. A multifunctional protein is one that contains two or more different catalytic centers on a single polypeptide chain. A multienzyme complex is an operational one in which the separate polypeptide chains remain tightly bound when usual isolation procedures are used.

These complexes carry out a series of sequential reactions on a substrate where the intermediates are considered to be out of diffusion equilibrium with identical molecules in the bulk phase of the same compartment of the cell. This process is referred to as channeling. Other terms that connote this phenomenon are coupling, processivity, direct transfer, and vectorial transfer. Two separate mechanisms may be included to result in channeling. In one mechanism the intermediates remain covalently bound to groups in the active sites, and in the other the intermediates are not covalently bound. In the latter mechanism one can observe either tight or leaky channeling.

Ambiguity is a term used by Wilson (120) to indicate the behavior of an enzyme whose location can be either bound to a structure or free in solution.

[2]The US Government has the right to retain a nonexclusive, royalty-free license in and to any copyright covering this paper.

SUMMARY AND PERSPECTIVE

Metabolic activities in cells are compartmented into various structures and organelles. In each compartment [outer cell membrane, Golgi, endoplasmic reticulum (rough and smooth), outer mitochondrial membrane, intermembrane space, inner mitochondrial membrane, matrix, lysosome, starch or glycogen particles, nuclear matrix, structural proteins, cytosol (aqueous cytoplasm, endoplasm)] one can locate metabolic enzymes concerned with individual metabolic pathways. There are some pathways distributed in two compartments. The data reviewed in this article deal with the hypothesis that the metabolic enzymes of a specific pathway within a single compartment are complexed in sequential or, at the very least, nonrandom arrangements.

Within some metabolic pathways there are stable multienzyme complexes and multifunctional enzymes that catalyze sequential reactions in that path. Recent evidence indicates that in addition to these more easily identifiable metabolic sequences there are specific interactions between many "soluble" sequential enzymes of metabolic pathways. Such complexes have been described in prokaryotes as well as eukaryotes. Additional evidence indicates that there are few, if any, free enzymes within cells. It appears also that the complexes of sequential metabolic enzymes are often bound to structural elements of the cell. I have listed those metabolic pathways for which there is evidence of the existence of complexes of sequential metabolic enzymes (metabolons) (Table 1). In this review I consider each of these metabolic pathways, citing only that evidence I think has been properly controlled. The table also lists nine different methodologies, each of which constitutes good evidence for organized enzyme systems. It is obvious from the table that the evidence is more compelling for some of the pathways than for others. I will indicate where possible where further studies are needed.

One corollary of this hypothesis is that many metabolites pass from one active site to another without complete equilibration with the bulk cellular fluid (channeling). It is obvious that there is leakiness in the channeling process, because there are metabolites present in the bulk aqueous phase of the cell. In this regard, there are two types of metabolic pathways, those that produce many multiusable intermediates and those in which only the end

Table 1 Complexes of sequential metabolic enzymes (metabolons)

Metabolic pathway	Evidence
DNA biosynthesis	A,B,C,E,F[a]
RNA biosynthesis	A,B,C,E,F
Protein biosynthesis	A,B,C,D,F
Glycogen biosynthesis	B,E
Purine biosynthesis	A,E
Pyrimidine biosynthesis	A,B,D,E,F
Amino acid metabolism	A,B,D,H
Lipid biosynthesis	B,C,F,H
Steroid biosynthesis	A,C,E
Glycolysis	A,B,C,D,E,I
Tricarboxylic acid cycle	B,C,D,G
Fatty acid oxidation	A,B,C,D
Electron transport	C,I
Antibiotic biosynthesis	A,E
Urea cycle	B,D
Cyclic AMP degradation	A,D,E

[a]A = channeling; B = specific protein-protein interactions; C = specific protein-membrane interactions; D = kinetic effects; E = isolation of complexes or multifunctional proteins; F = genetic evidence; G = model systems; H = existence of multifunctional or multienzymic proteins; I = physical chemical evidence. The evidence for most of these are in the text of this article.

products are usable. An example of the former would be glycolysis where glucose 6 phosphate and dihydroxyacetone phosphate are used in multiple ways. Protein synthesis is an example of the latter where the intermediates have no metabolic function except to become a complete protein.

A dramatic way of visualizing the multitude of reactions that constitute the intermediary metabolism of carbohydrate, amino acid, lipid, and nitrogen base metabolism has been presented by Alberts et al (1) (Figure 1). Instead of giving the individual reactions, as is usually done in metabolic maps, they introduced the simplification of representing each intermediate as a black dot and each enzyme as a line. Thus, the marvelous simplification is achieved by representing over 1000 different enzymes and substrates by just two elements. Their focus was on the fact that many intermediates had multiple fates and interactions. In order to emphasize the need of cells for the systems discussed in this review, I focus on a different aspect that becomes apparent upon the analysis of Alberts et al's schematic of metabolism.

The chart contains about 520 intermediates (dots). We can classify intermediates (dots) by the number of enzymes (lines) acting on them. Thus, a

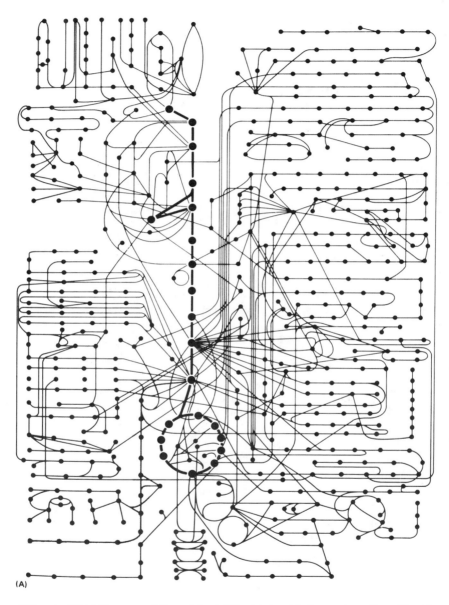

(A)

Figure 1   Schematic representation of about 500 related metabolic reactions. From (1) with permission.

dot connected to only one line is either a nutrient or an end product of metabolism. A dot connected to two lines is an intermediate with just one fate in metabolism. A dot with three lines has two metabolic fates and so on. An analysis of dots and the number of lines connected to them reveals the following:

lines	dots
1 or 2	410
3	71
4	20
5	11
6 or more	8

This indicates that about 80% of the shown metabolic intermediates have just one use in the cell. It is apparent that it would be a wasteful process if these intermediates each had to fill the water volume of the cell to attain its operating concentration. Atkinson (2) has pointed out that a limiting factor in cell structure is its solvation capacity, so that a viable strategy during evolution to overcome this difficulty would be the formation of sequential multienzyme complexes and their attendant channeling capabilities.

Another aspect one must consider is that some complexes (e.g. hexokinase binding to ADP + ATP translocator of the outer mitochondrial membrane) can be dissociated by metabolites. It is therefore possible that metabolic control can be exerted by control of the location of an enzyme in either a free or complex form (ambiquity)[1].

An additional corollary of this hypothesis is that enzymes are designed not only for their active sites, but also for their surface binding sites. These are not generalized surface sites for locating the enzymes in or out of membranes, which are also requirements for proteins, but more specific sites to interact with two or three other sequential enzymes or structural elements in the cellular compartment.

Two major criticisms have been made against this hypothesis. First, since diffusion is rapid and enzyme turnover slow, organization is not necessary. Second, the interactions of some isolated sequential metabolic enzymes are weak and usually require special (often considered nonphysiological) conditions to be demonstrated. These points have been discussed previously, but I summarize the evidence against them throughout the article.

I will not treat in depth all well-known multienzyme complexes or multifunctional enzymes, which have been reviewed elsewhere (3, 4). I concentrate on papers published within the last few years. Earlier work has been reviewed by others in a number of excellent books (5, 6), symposia (7–9), and reviews (10–15). It is the bias of this article that evidence for the existence of metabolic complexes is good. What remains to be unequivocally demonstrated is whether or not they are important in the regulation and operation of metabolic pathways in situ. This latter problem must be approached using newer techniques as I discuss later.

MACROMOLECULAR BIOSYNTHESES

DNA Replication

Alberts (16) has coined the term "protein machines"[1] to describe multienzyme complexes powered by nucleoside triphosphate hydrolysis. He has analyzed in detail the machine (metabolon) that is used for the synthesis of DNA of the T4 bacteriophage. A multienzyme complex can be reconstructed with the purified protein products of seven different genes. This complex moves along the DNA template using nucleoside triphosphate energy to synthesize a new DNA molecule. One of these proteins of the DNA synthesizing machine, a DNA polymerase, is believed to participate in several different types of discrete DNA polymerase complexes, each of which performs a different function such as repair or single-strand synthesis. The other polypeptides include a helix-destabilizing protein, three polymerase accessory proteins, DNA helicase, and two proteins for primer synthesis. At present we have no such detailed knowledge of the multienzyme DNA replicating complex of any eukaryotic system. It is interesting to note that the DNA-synthesizing multienzyme complex probably does not exist as such free from the DNA template. The DNA serves a dual purpose in this process, that of template and that of a structural framework. A structural component may be a feature of many metabolons, and I describe below multienzyme complexes that seem to aggregate on structural proteins of the cell, on membranes of the cell, and perhaps even on polysaccharide components of the cell.

One technique for showing the interaction between the individual polypeptides of the DNA replicating system is that of affinity chromatography (17). If the helix-destabilizing protein of T4 phage is bound to agarose and a T4-infected cell extract is placed on a column of the immobilized protein, one then finds 10 different T4-induced proteins bound to the column. Control columns with immobilized albumin in place of the helix-destabilizing protein do not show binding of these proteins so that specificity of the interaction is indicated.

Flanegan & Greenberg (18 and references therein) have shown that precursors of DNA are compartmentalized in replicating cells and proposed that the channeling observed in T4-infected *Escherichia coli* was due to a complex of enzymes that provided precursors directly into the DNA polymerizing complex. The rates of synthesis of DNA in vivo are high despite the fact that the concentrations of precursor pools in situ are lower than those required for comparable rates of synthesis in vitro.

Mathews and his colleagues (19) have isolated from T4-infected *E. coli* a complex of 10 enzymes responsible for the synthesis of the four deoxyribonucleotide triphosphates necessary for DNA synthesis. The complex isolated from lysozyme-treated cells was proposed to have the structure shown in Figure 2. Dihydrofolate reductase was shown to be part of the complex both physically and kinetically.

A transcription-replication enzyme complex has been isolated from T5-infected *E. coli* by Ficht & Moyer (20). The complex contains nuclease-protected double-stranded DNA. The enrichment of certain T5 DNA fragments in the complex and the enzyme activities present lead them to believe the complex is specific.

One component of the DNA replicating complex (termed "replisome" in bacteria and "replitase" in eukaryotes), namely the DNA polymerase, exists in several different forms. In mammals DNA polymerase alpha has several subunits, while DNA polymerases beta and gamma have only one subunit each. In *E. coli* DNA polymerase I has a single subunit and DNA polymerase III has seven subunits. It is not clear at present whether the multisubunit polymerases bind other polypeptides to become part of an even more complex multienzyme system.

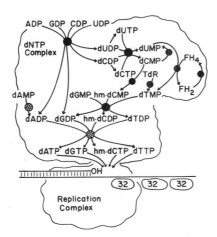

Figure 2 A speculative view of the T4 dNTP-synthesizing multienzyme complex. Closed circles denote phage-coded enzymes, and cross-hatched circles identify host cell enzymes [from (19) with permission]. The enzyme activities in the complex are: dCMP hydroxymethylase, dTMP synthase, FH$_2$ reductase, dNMP kinase, NDP kinase, dAMP kinase, dCTPase, dUTPase, dCMP deaminase, thymidine kinase, and rNDP reductase.

Reddy & Pardee (21) presented evidence that indicates channeling exists for DNA replication in a eukaryotic cell (Chinese hamster embryo fibroblasts). Earlier evidence had indicated that DNA replication in eukaryotic cells starts with DNA bound to the nuclear membrane. Similar membrane-DNA complexes had been reported for bacteria and for mitochondria.

Reddy & Pardee used the technique of cell permeabilization to study DNA replication in eukaryotic cells. This method enables one to study metabolic sequences without the massive disruption and dilution that occurs when metabolic studies are carried out on homogenized cells. These permeabilized cells showed rates of DNA replication equal to those observed in intact cells. These authors showed that in such cells under optimal conditions that NDPs could be used more efficiently for DNA synthesis than could NTPs. The most reasonable interpretation of their results would be a channeling phenomenon whereby exogenous NTPs could not enter the multienzyme complex easily. They further showed that during S phase (DNA synthesis), six enzymes associated with DNA replication, DNA polymerase, ribonucleotide reductase, thymidylate synthase, thymidine kinase, nucleoside diphosphate kinase, and dCMP kinase, cosedimented rapidly from nuclear extracts. Nuclear extracts from cells in the G phase did not behave in the same manner. Reddy & Pardee (22) have also shown that interactions of the enzymes of the DNA multienzyme complex make the thymidylate synthase of this complex susceptible to inhibition by hydroxyurea, novobiocin, and amphidicolin. This indicates a functional allosteric interaction when the complex is formed.

The DNA polymerase α_2 from HeLa cells has been shown to be a multiprotein consisting of eight polypeptides. Six separate binding and enzyme activities can be assigned (23) to this complex.

Although the evidence presented by Pardee and his coworkers for the mammalian system is certainly strong, it is not as clear-cut as that for the bacterial system. Recently Leeds et al (24) have shown that the dNTP pools in the nucleus are adequate to account for the observed DNA synthesis. The ribonucleotide reductase in these experiments could not be located in the isolated nucleus. Mathews & Slabaugh (25) point out that these results do not necessarily argue against a mammalian DNA synthesizing complex and that there still exists good indirect evidence for the existence of the complex.

In an excellent review, Nelson et al (25a) have stated that there is good evidence "that many important nuclear events occur not in solution but rather in association with relatively insoluble structural components that are bound to the nuclear matrix."

RNA Formation

The process of transcription of DNA has at least the same degree of complexity and processivity as does the replication of DNA. This topic has been reviewed a number of times (26 and references therein). In prokaryotes a single RNA polymerase with different σ subunits is responsible for RNA synthesis, while in eukaryotes there have been three different RNA polymerases described. The bacterial holoenzyme contains five subunits ($\alpha_2\beta\beta'\sigma$) with $M_r = 0.45 \times 10^6$. These subunits have different functions in the biosynthesis of RNA. Sigma factor binds to polymerase and causes tight binding to promoter sites on DNA. There is, in addition to this specific binding, loose binding, probably electrostatic, between the core enzyme ($\alpha_2\beta\beta'$) and DNA. After specific binding has occurred, synthesis proceeds a short way until σ factor is released, and only then can synthesis proceed to completion (27). According to McClure, most of the RNA polymerase within *E. coli* is bound in various states of activity both specifically and nonspecifically to DNA.

In eukaryotes three different RNA polymerases are known, each of which has a unique intranuclear location. RNA polymerase I is a nucleolar enzyme responsible for synthesizing RNA. This enzyme activity and the other two RNA polymerases are poorly characterized complexes of many proteins having molecular sizes in the region of 0.5×10^6 daltons. Each consists of two fairly large subunits (about 200 kd and 120 kd) and 10 or so smaller protein subunits. No reconstituted system has been produced from purified fractions from the cruder preparations.

Murooka & Lazzarini (28) reported the isolation and purification of a DNA-RNA polymerase complex from *E. coli* that exhibits as much synthetic activity as the original permeabilized cells. In another study on the RNA polymerases of *Neurospora,* Armaleo & Gross (29) isolated in a 700-kd complex of 12 polypeptides. This complex is associated with polymerase II and possibly polymerase I. Although no similar large complex has been isolated from other organisms, there are indications that preparations of polymerases from other cells are often "contaminated" with other proteins that have masses the same as the several polypeptides in the *Neurospora* complex.

It has been established that there is a sequential assembly of transcription factors on certain genes. Recently, however, RNA polymerase has been isolated from HeLa cells and found to contain transcription factors (30) in a functional complex in the absence of DNA.

Protein Biosynthesis

One of the most elaborate and complex "machines" for the biosynthesis of macromolecules is the one involved in protein biosynthesis (31). Its detailed structure and the number of different components required are well understood. The structure of the ribosome has been almost completely delineated (32) as to the three-dimensional arrangement of 4 RNAs and about 73 proteins, but we do not know the function of the individual RNAs and protein components of that structure. Protein synthesis occurs in three stages—initiation, elongation, and termination. At each stage the ribosome, mRNA, protein factors, and GTP are required. Protein complexes are formed between the components for each of the stages, and at the end of the stage the accessory proteins are dissociated from the rest of the ternary complex. Although in eukaryotes some protein synthesis occurs on free ribosomes, there is still a substantial portion of it that occurs while the ribosomes are attached to the endoplasmic reticulum. Some evidence exists that some mitochondrial protein synthesis occurs with free ribosomes and some with ribosomes attached to the outer membrane of the mitochondrion (33). In these latter cases the attachment to the membrane is important since the delivery of the newly synthesized protein is through the membrane.

An interesting observation has been reported concerning the interaction of amino acyl tRNA synthetases (34). A number of these enzymes can be isolated as a multienzyme complex. In higher eukaryotes as many as eight tRNA synthetases can comprise a complex (34). No special enzyme properties can be detected for the enzymes in complex compared to the free enzymes. Deutscher (35) has discussed the results of experiments dealing with the hydrophobic nature of the complex formation and has presented the following attractive hypothesis. The hydrophobic regions may allow the synthetase to be membrane bound in vivo. It has been shown that several amino acyl tRNA synthetases are bound to the endoplasmic reticulum. Since it is known that exogenous amino acids can be used for protein synthesis without mixing with endogenous pools, this membrane attachment may account for the channeling phenomenon. Deutscher also cites evidence indicating that there may be a pool of free synthetases using the endogenous pool of amino acids for the synthesis of specific proteins. Further, one might speculate that these so-called free synthetases are actually associated with cytoskeletal elements in the cell where Fulton et al (36) show one may find bound ribosomes.

Glycogen Metabolism

Fischer and his colleagues (37, 38) have shown that when glycogen particles are isolated from skeletal muscle, these particles are approximately one half protein. It was demonstrated that most of the proteins attached to glycogen were the enzymes of glycogen metabolism. These include phosphorylase b and its respective kinase and phosphatase, glycogen synthetase, and its respective kinase and phosphatase, and cAMP-dependent protein kinase. In addition, they have shown that all the glycolytic enzymes are present in the particle since these particles can form lactic acid from the glycogen. Elements of the sarcoplasmic reticulum are also present in the particles. In further studies on the protein composition of the complex, Cohen and his coworkers

(39, 40) have shown that debranching enzyme, glycogen synthase kinase 2, small amounts of protein phosphatase 2, inhibitor 1, and inhibitor 2 could be detected in the glycogen particle.

Electron microscopy has shown that the glycogen particles in skeletal muscle are closely associated with the sarcoplasmic reticulum near the thin filaments (41). The electron microscopic studies are thus in agreement with the experiments on the isolation of the complex.

Fischer and his coworkers also report (38) that the kinetic behavior of certain of the bound enzymes is different from that observed for the isolated enzyme. They have reported that phosphorylase phosphatase is active in particles but inactive in solution. This is due to presence of a bound in-activator of a phosphatase inactivator protein. Cycling of phosphorylase a to b, which would occur in solution, does not occur on the particle.

Radda and his associates (42) studied glycogen particles by NMR and ESR techniques. For phosphorylase a, they showed that ligand binding occurred on the particle but not in a simulated solution of the enzymes.

Glycogen particles are polydisperse, exhibiting a molecular weight range from 6×10^6 to 1600×10^6 (43). Turnover of glucose moieties occurs at the outer portions of the particle very rapidly, while the core glucose residues turn over slowly. Barber et al (44) showed that the different size glycogen particles separated by rate-zonal centrifugation had different distributions for phos-phorylase and for synthase. Phosphorylase activity was highest in the small particles, while the synthase activity was highest in the large particles.

In addition to these size and enzyme distribution heterogeneities, Geddes (43) has pointed out that glycogen is apparently synthesized along a protein backbone. All this information can be most easily interpreted in terms of two separate systems in the cytosol for glycolysis and for gluconeogenesis.

PRECURSOR MOLECULE METABOLISM

Purine and Pyrimidine Biosyntheses

The biosynthesis of uridine monophosphate occurs by means of multienzyme complexes and/or multifunctional enzymes [see Jones (45) for review]. In the conversion of HCO_3 to UMP, two cytosolic multifunction enzymes and a mitochondrial enzyme (outer surface of the inner membrane) are involved. The trifunctional enzyme (consisting of carbamyl phosphate synthetase, aspartate transcarbamylase, and dihydroorotase) converts bicarbonate, ammo-nia, and aspartate to dihydroorotate. This multifunctional enzyme is found in *Drosophila*, bullfrogs, mice, rats, and hamsters. Orotate is then produced by dihydroorotate dehydrogenase, which is on the outer surface of the inner mitochondrial membrane. Finally, a bifunctional enzyme containing orotate phosphoribosyl transferase and orotidylate decarboxylase produces uridine monophosphate. The concentrations of the five intermediates in the overall pathway in mammalian cells are held very low. This would indicate that both multifunction enzyme complexes are bound to the mitochondrion so that channeling can occur. Kaplan and his coworkers (46) have shown that channeling of intermediates occurs in the pyrimidine pathway in yeast. In yeast, the two catalytic proteins of uridine monophosphate synthase are on separate polypeptides. It has been postulated that mammalian cells must channel orotidylate monophosphate because of the very active pyrimidine nucleotidases in these cells. Since further utilization of the product of this pathway in eukaryotes occurs in the nucleus, it would appear that no direct coupling occurs from this system to the next metabolic sequence.

In the matter of purine biosynthesis there is good evidence for the existence of multienzyme complexes. There are 13 enzymes that can be considered part of this biosynthetic pathway (47). A number of years ago a paper reported that 10 enzymes of the pathway copurified from avian liver (48). In spite of this copurification no evidence for a large macromolecular complex was found when the purified complex was examined by column chromatography or sucrose gradient centrifugation in the absence of polyethylene glycol. This would indicate that the complex that was isolated with polyethylene glycol was held together by weak forces that are enhanced in the presence of a volume excluding molecule that simulates the high concentration of proteins in the cell cytosol. More recently the same group reported that they could also achieve partial copurification of some enzymes of the pathway from human lymphocytes (49). Henikoff et al (50) have shown that three activities of the de novo purine pathway in *Drosophila* are encoded at a single genetic locus. The activities of glycinamide ribonucleotide synthetase, glycinamide ribonucleotide transformylase, and aminoimidazole ribonucleotide synthetase are present together on a single protein, and in a *Drosophila* tissue culture cell line all are overproduced. The mRNA for the second enzyme (glycinamide ribonucleotide transformylase) seems to be capable of being produced separately. This multifunctional protein has been isolated by Daubner et al (51) from avian liver.

Another aspect of purine biosynthesis relates to the formation of the formyl groups used in the pathway. A multifunctional enzyme (C_1-tetrahydrofolate synthase) with three activities (10-formyl tetrahydrofolate synthetase, 5,10 methenyl tetrahydrofolate cyclohydrolase, and 5,10 methylene tetrahydrofolate dehydrogenase) is present in yeast (52), chicken, sheep, pig, and rabbit livers. Benkovic and his coworkers (53) have shown that multienzyme complexes can be isolated from chicken liver, which contains not only this trifunctional polypeptide but also glycineamide ribonucleotide (GAR) transformylase (a folate-requiring enzyme of purine biosynthesis), aminoimidazole carboxamide ribonucleotide transformylase, and serine transhydroxymethylase. The ratio of specific activities remains constant through many purification steps including a GAR affinity column and specific ATP elution. Resolution of the complex causes a loss of GAR transformylase activity, which can be recovered on reformation of the complex.

Channeling has been shown for the noncovalently bound intermediate in the bifunctional folate enzyme of formiminotransferase-cyclodeaminase (54). This channeling not only increases the overall efficiency of the enzyme, but also maintains a low concentration of a labile intermediate that has no other metabolic function.

In an interesting paper, Burns (55) has shown that gentle sonication of yeast cells stopped both purine and pyrimidine biosynthesis during the sonication period. When sonication was stopped, biosynthesis of purines and pyrimidines proceeded at normal rates and cells appeared to be normal with no damage from the sonication. Burns suggests that sonication disrupted multienzyme complexes, resulting in the interruption of these biosynthetic processes. It is clear that other studies on this system are necessary to show that the effect is due to disruption of complexes.

Amino Acid Metabolism

The biosynthesis of aromatic amino acids involves 13 enzymatic reactions and, depending on the cell type, shows a high degree of organization including both multifunctional proteins and multienzyme complexes (56–58). This pathway is absent in animals. In bacteria and higher plants it provides for the synthesis of phenylalanine, tyrosine, and tryptophan. It also provides for the biosynthesis of nicotinic acid, p-aminobenzoic acid, tetrahydrofolic acid, ubiquinone, and vitamin K in these cells. In plants, parallel pathways for aromatic amino acid synthesis exist, and a number of other substances such as lignin, alkaloids, and flavonoids are produced through this pathway. Seven enzyme activities catalyze the synthesis of chorismate from erythrose 4-P and phosphoenol pyruvate. In E. coli these activities exist on separate polypeptide chains. In Neurospora crassa and Saccharomyces cerevisiae, five of the activities are found on a single polypeptide. In addition to the existence of multifunctional proteins, interactions between sequential separate polypeptides have been found. A summary of these interactions is shown in Figure 3 (58). The multifunctional proteins exhibit the kinetic properties predicted for this system, i.e. a decrease in the transient time of the intermediates.

The most highly organized aromatic-tryptophan pathway has been found in Euglena where 10 of the 13 enzymes are found in two separate isolatable complexes. Although no organization for the first seven enzymes of the polyaromatic pathway has been detected generally in bacteria, there is evidence of some interactions of the tryptophan pathway enzymes in certain bacterial species.

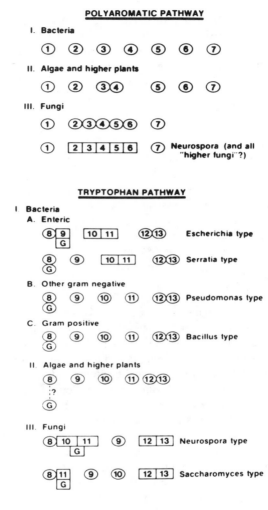

Figure 3 A phylogenetic comparison of the (in vitro) state of enzyme organization in the polyaromatic-tryptophan biosynthetic pathway. "Ovals" represent physically distinct enzyme activities. Conjoined "squares" represent multifunctional proteins. Conjoined "ovals" and conjoined "oval-square" assemblies represent multienzyme complexes. Numbers refer to respective pathway steps. The "G" subunit confers glutamine amidotransferase activity on step 8. From (58) with permission.

An interesting experiment has been reported by Manney (59) concerning the multifunctional protein tryptophan synthase of yeast. Manney utilized complementing diploids of indole-utilizing and indole-accumulating mutants. He found that these organisms grew at wild-type rates but only after a long lag period. The lag period could be shortened by adding indole. Thus, the multifunctional enzyme is able to work efficiently at low cellular indole concentrations, but high indole concentrations are needed when the enzymes are not able to channel substrate efficiently.

There are reports of multienzyme complexes involved in the metabolism of other amino acids. Thus, it has been reported that in *S. cerevisiae* there is channeling of histidinol in the histidine pathway (60). Bergquist et al (61) reported that five mitochondrial enzymes necessary for the biosynthesis of isoleucine and valine can be isolated as a stable multienzyme complex using a gel filtration technique. The complex could be isolated only from mitochondria that had been pretreated with ADP and phosphate.

In *Salmonella typhimurium*, Kredich et al (62) showed that cysteine synthetase is a multienzyme complex of the two enzymes serine transacetylase and *o*-acetylserine sulfhydrase in a 1:2 molar ratio. The isolated complex dissociates reversibly in the presence of *o*-acetylserine but associates in the presence of sulfide. Later studies on the kinetics of this multienzyme revealed, surprisingly, that the intermediate was not channeled, but was released to the surrounding medium, although a decrease in the transient time was detected (63).

Another apsect of amino acid metabolism is the formation or utilization of amino acids through the Krebs cycle. It is not surprising to find that specific interactions exist between enzymes of these two pathways. An important paper was that of Backman & Johansson (64), who showed that mitochondrial malate dehydrogenase and mitochondrial aspartate amino transferase could interact in the presence of derivatives of polyethylene glycol. Similarly, interaction between the pure cytosolic isozymes was demonstrated. They could detect no interaction when one of the pair of enzymes was cytosol in origin and the other was mitochondrial. Using enzymes immobilized on Sepharose, Beeckmans & Kanarek (65) confirmed the results on interactions obtained using countercurrent distribution techniques, but their results do not show as marked a specificity in the heterologous interactions. Two papers have appeared concerning the kinetics of this coupled enzyme system, one reporting anomalous kinetics and the other reporting no anomalous kinetics (66, 67). At present it is not possible to indicate whether the complex exhibits channeling behavior.

Mitochondrial aspartate aminotransferase and glutamate dehydrogenase also form a complex, and the mitochondrial aspartate amino transferase also complexes with carbamyl phosphate synthase I (68). The techniques used to measure these interactions are fluorescence, gel filtration equilibrium, crosslinking, and precipitation with polyethylene glycol. Fahien and his coworkers (69, 70) have reported more interactions between the transaminases, glutamate dehydrogenase, and malate dehydrogenase and how these interactions are affected by various metabolites. Salerno et al (71) did not observe the theoretically expected lag in the reaction rate at high concentrations of aspartate amino transferase and glutamate dehydrogenase if no interaction existed.

Hearl & Churchich have shown that the two mitochondrial enzymes that metabolize the neurotransmitter 4-aminobutyrate interact with each other (72). When the enzyme 4-aminobutyrate aminotransferase was immobilized through its antibodies to Sepharose, then the second enzyme, succinic semialdehyde dehydrogenase, would bind to the column. The addition of succinic semialdehyde dehydrogenase to fluorescently labeled aminobutyrate aminotransferase caused an increase in the polarization of fluorescence, which indicates protein-protein interaction. These enzymes could represent part of a bypass through alpha ketoglutarate in brain mitochondria.

Another mitochondrial multienzyme complex, glycine decarboxylase, consists of four proteins and has some similarity to the alpha keto acid dehydrogenase complexes. However, this complex is not as stable as the latter and can be dissociated easily. It has been studied in vertebrate liver, bacteria, and pea seedlings (73). One of the first descriptions of the channeling phenomenon was by Davis (74) in his studies on arginine metabolism. In *Neurospora* and *S. cerevisiae,* channeling of arginine and ornithine is due to the existence of vesicles and/or vacuoles that contain high concentrations of basic amino acids that are not in equilibrium with the cytosolic or mitochondrial pools of these amino acids. This form of channeling is therefore apparently not related to the multienzyme complex type discussed in this article.

Lipid Metabolism

Fatty acyl CoAs in animal tissues are synthesized by a pathway that includes ATP citrate lyase, acetyl CoA carboxylase, and fatty acid synthase. These three enzymes (along with several auxiliary enzymes) are coordinately expressed in animals and certain fibroblasts under nutritional and hormonal manipulations. When tissues containing these enzymes are disrupted by the usual biochemical procedures and then subjected to centrifugation, these three enzymes and other auxiliary enzymes appear in the $100,000 \times g$ supernatant fraction, which purportedly contains the "soluble" cytosolic enzymes. When the cytosol is subjected to sucrose gradient centrifugation, high-molecular-weight fractions are obtained containing all three enzyme activities (75). The total amount of enzyme in these high-molecular-weight fractions is variable and low, and these experiments have not been reproduced in other laboratories. We have shown that rat liver mitochondria binds small amounts of purified rat liver ATP citrate lyase (76). Janski & Cornell (77) have shown that rapid digitonin fractionation of rat hepatocytes did not release ATP citrate lyase in the same manner as the cytosolic marker enzyme lactate dehydrogenase. They interpreted their data as indicating that part of the ATP citrate lyase was bound to mitochondria. However, if ATP citrate lyase was bound to any intracellular membrane, the results they observed would have been the same. Linn & Srere (78) later showed the microsomal fraction of rat liver bound ATP citrate lyase. The binding capacity of this fraction was much greater than that of mitochondria and could be reversed with low concentrations of CoA.

Electron microscopy has indicated that acetyl CoA carboxylase is located on the endoplasmic reticulum of rat hepatocytes (79). Witters et al (80) have confirmed previous evidence that conventional fractionation of rat liver will yield microsomal acetyl CoA carboxylase. Its distribution is sensitive to the homogenization conditions. Acetyl CoA carboxylase can undergo a polymerization reaction so that cosedimentation of the polymerized enzyme with the microsomes is a possible explanation of the observations. Allred et al (81) have used radiolabeled avidin to locate biotinyl proteins in rat liver homogenate fractions, and report that one half of the total acetyl CoA carboxylase in the liver of fed rats is located in the mitochondrial fraction in an inactive form. They view this distribution as ambiquitous behavior of the enzyme. They detected no biotinyl proteins in the microsomal fraction.

Fatty acid synthase is an example of a complex of sequential enzymes (a metabolon). In most bacteria and plants the synthesis of fatty acids is catalyzed by six separate enzymes and an acyl carrier protein. There is no information available concerning possible complex formation between these enzymes in these systems. In fungal and animal cells the enzyme activities and acyl carrier proteins are part of multifunctional proteins. As pointed out earlier, in metabolic pathways in which the intermediates are not required for other metabolic purposes (either stoichiometric or regulatory), evolution seems to have proceeded toward the formation of covalent linkage of sequential metabolic reactions.

The other two enzymes of the fatty acid biosynthetic pathway, ATP citrate lyase and acetyl CoA carboxylase, could be considered multifunctional proteins. For ATP citrate lyase, two separate steps are involved: the formation of citryl CoA and the cleavage of citryl CoA. A similar reaction occurs in bacteria, the cleavage of malate via malyl CoA, but in this instance two separate polypeptides are involved. In certain bacteria the citrate lyase reaction, which is analogous to the eukaryotic ATP citrate lyase, is catalyzed by an enzyme containing three polypeptides; one is an acyl carrier protein (with bound dephospho CoA), one is an acyl transferase, and the third is a lyase.

Another example of evolutionary changes that yield covalently connected sequential enzyme activities is the catalysis by acetyl CoA carboxylase of two reactions, carboxylation of biotin and the transfer of CO_2 from biotin to acetyl CoA. In bacteria the enzyme is dissociable into three components, carboxyl carrier protein, biotin carboxylase, and carboxyl transferase. In animal tissues these three activities are on a single polypeptide chain.

Further metabolism of fatty acids is their conversion to complex lipids either as triglycerides or phospholipids. The synthesis of the latter occurs by way of the three enzymes located in the endoplasmic reticulum, a fatty acyl transferase, a phosphatase, and a choline phosphotransferase. A fourth enzyme, cytidylyl transferase, can be located in the cytosol or on the endoplasmic reticulum (82). This latter enzyme is controlled in several ways. It is activated by phospholipids and fatty acids, and it is inhibited by CTP and phosphocholine. Finally, it is an ambiquitous enzyme, that is, it can be either in the cytosol or endoplasmic reticulum depending on its state of phosphorylation. The phospho enzyme is cytosolic and relatively inactive, whereas the dephospho enzyme is attached to the membranes and active. Though this is a simplified version of the control of this system, the main point is that when this enzyme is in the membrane with the other enzymes of the pathway it is active. Part of this may be due to activation by the phospholipids of the membrane, but it is possible that a complex of enzymes forms within the membranes. It is known that all cellular membranes are synthesized in the endoplasmic reticulum and that membranes are asymmetric in their lipid distribution (83). This latter observation may be accounted for by a complex of enzymes on the cytosolic side of the lipid bilayer.

The available data supporting the existence of a complete complex of enzymes for phospholipid biosynthesis is thus suggestive but by no means conclusive. While the steps from fatty acyl CoA to phospholipid are located in the membrane, no experiments have been reported concerning the possible interactions of the enzymes in the membrane. For the first part of the pathway, the synthesis of fatty acids from acetate, there is evidence that two of the enzymes can bind to membranes. Further studies are needed to characterize the in vivo disposition of all the enzymes involved.

Another metabolic fate of fatty acids is their oxidation in the mitochondria. Early studies on this pathway in mitochondria had indicated that little if any of the intermediates could be detected, but soluble extracts of mitochondria could accumulate intermediates (84). A multienzyme complex present in the mitochondria and channeling of intermediates could be responsible for these observations. Sumegi & Srere have shown that thiolase, crotonase, and β-hydroxyacyl CoA dehydrogenase bind to the inner membrane of mitochondria (85). Kispal et al (86) have isolated the binding protein for β-hydroxyacyl CoA dehydrogenase from pig heart mitochondrial inner membranes. With a purified preparation they were able to show that liposomes containing the binding protein could bind β-hydroxyacyl CoA dehydrogenase but not malate dehydrogenase, citrate synthase, or fumarase. Earlier studies by Sumegi & Srere (87) have shown that several mitochondrial dehydrogenases, including β hydroxyacyl CoA dehydrogenase, bind to Complex I. It is not known whether the binding protein isolated for β hydroxyacyl CoA dehydrogenase is a subunit of Complex I.

In bacteria the multienzyme complex for the β oxidation of fatty acids contains all the enzymes except acyl CoA dehydrogenase (88). The complex has an $\alpha_2\beta_2$ structure, and the α subunit is a multifunctional protein and contains four enzyme activities of this pathway. Yang et al (89) have shown that the intermediate L-3-hydroxydecanoyl CoA is directly transferred from the active site of enoyl CoA hydratase of the multifunctional protein, to the 3-hydroxyacyl CoA dehydrogenase site on that protein. They could detect no equilibration with the bulk solvent since there was no lag time in the reaction, whereas a kinetic model of the unlinked enzymes indicated that a 30 sec lag time should occur.

The question of complex formation for the enzymes involved in steroido-genesis is the subject of an excellent review by Lieberman et al (90). These authors propose a scheme that includes the existence of multienzyme com-plexes for the synthesis of steroids such as estradiol, cortisol, and aldosterone. They refer to the separate multienzyme complexes as hormonads. The evi-dence for such organization in specific pathways includes the absence of certain stable expected intermediates, which indicates a channeling process. Thus, in the conversion of cholesterol to pregnenolone, the proposed in-termediates could not be trapped. A second line of evidence cited is the fact that there exist isozymes of certain processes that may exist for separate pathways. A third line of evidence presented indicates that different steroidal end products may have their own steroid precursors. Finally, evidence is presented that the late steps in the pathway of biosynthesis of different steroids are the ones that are regulated by trophic hormones. Burstein & Gut (91) have reported a complex isolated from adrenal acetone powders that converts cholesterol to pregnenolone. The interaction of the proteins involved in these conversions has not yet been demonstrated.

The biosynthesis of cholesterol itself may be a membrane-bound multi-enzymic system. First, it is known that many enzymes of cholesterol biosynthesis are located in the endoplasmic reticulum. Gaylor & Delwiche (92) have presented evidence for a multienzyme system in microsomes that converts lanosterol to cholesterol. It has been shown that most of the cholesterol of a cell occurs in the plasma membrane. In fibroblasts Lange & Steck (93) demonstrated that buoyant densities of membranes containing cholesterol and lanosterol were different from that of membranes with hy-droxymethylglutaryl reductase. They say that this result suggests that the biosynthesis of cholesterol is not a continuous process in the endoplasmic reticulum, and that a transfer step through the cytosol must occur.

Glycolysis

Experiments using isotopic glucose were interpreted as indicating the presence of two separate pools of certain glycolytic intermediates in a number of different systems. These included studies in *E. coli,* rat skeletal muscle, rat liver, rat diaphragm, Ehrlich ascite cells, and rat brain (12, 14, 94). Lynch & Paul (95) studied the metabolism of porcine vascular smooth muscle and concluded that glycolysis and glycogenolysis occurred in separate compartments. Postius (96) showed that in yeast cells endogenously produced pyruvate did not mix with exogenous pyruvate. On the other hand, Connett et al (97) studied the metabolism of *Tetrahymena* using isotopically labeled compounds and concluded that it was not necessary to propose two separated glycolytic pools to explain their data. Further, Clark et al (98), studying futile cycles in rat liver, found that interpretation of their isotope data could be made without the existence of several glycolytic pools.

Objections to the interpretation of isotopic data as indicating two separate substrate pools in the same compartment usually include 1. the heterogeneity of cells in the same tissue giving rise to separate pools (99) and 2. the possibility of isotope exchange reactions occurring. The first of these objections cannot be applied to studies on *E. coli,* Ehrlich ascite cells, or even vascular smooth muscle preparations. The second objection is also not applicable to the studies of Lynch & Paul (95) or the kinetic studies of Coe & Greenhouse (100), since both of these studies do not depend on isotopic evidence.

However, the results of isotope experiments are not the most convincing evidence concerning the existence of different complexes of glycolytic enzymes in the cytosol. Supporting evidence comes from studies that show specific interactions between sequential pairs of glycolytic enzymes, interactions between glycolytic enzymes and actin (101, 102), interactions between glycolytic enzymes and red blood cell membranes, isolation of complexes of glycolytic enzymes, electron microscopic evidence (103), and other evidence (see 12, 94, 101 for review). Further, it should be remembered that glycolysis is not as processive as nucleic acid or protein biosynthesis, since many of the intermediates serve several metabolic functions that may be reflected in formation of "looser" complexes that would be more difficult to isolate and detect.

Glyceraldehyde phosphate dehydrogenase binds to microsome fractions of skeletal muscle (104), and aldolase has been found in the microsomal fraction of rat liver (105). In *Tetrahymena pyriformis,* all of the lactate dehydrogenase (106), one half of the glyceraldehyde phosphate dehydrogenase (107), and about 75% of the phosphofructokinase (108) are bound to mitochondria. Paul and his coworkers (109, 110), who had shown by isotopic methods the existence of two compartments in smooth muscle cells, one for glycolysis and one for glycogenolysis, have recently demonstrated that plasma membranes and the contractile apparatus of this tissue bind glycolytic enzymes. Also, Bronstein & Knull (111) showed that aldolase, glyceraldehyde phosphate dehydrogenase, lactate dehydrogenase, and pyruvate kinase bound to columns containing covalently bound F-actin-tropomyosin complex. Huitorel & Pantoloni (112) have shown that glyceraldehyde phosphate dehydrogenases from brain, muscle, and erythrocyte bind to brain microtubules. Luther & Lee (113) have shown that phosphorylation of phosphofructokinase not only changes its kinetic behavior but also increases its binding to F-actin.

Glycolytic enzymes bind to red blood cell membranes (see 114 for review). The experiments of Steck and his coworkers (see 115 for review) and of Karadsheh & Uyeda (116) show that certain glycolytic enzymes can bind to the erythrocyte membrane. Thus glyceraldehyde phosphate dehydrogenase, aldolase, and phosphofructokinase can bind to protein band 3 of erythrocyte membrane. The binding of aldolase and glyceraldehyde phosphate dehydrogenase are inhibited by substrates. Studies by Friedrich's group (117) using labeled iodoacetate indicated that glyceraldehyde phosphate dehydrogenase was located near the membrane. Earlier crosslinking studies by Keokitichai & Wrigglesworth (118) also showed preferential location of this enzyme near the membrane. Several recent studies have reported that under slightly different experimental conditions there was no binding of glyceraldehyde phosphate dehydrogenase under putative in vivo conditions of ionic strength. The experiment by Rich et al (119) used a slow hemolysis rate and a long isolation procedure, both factors that may affect the binding.

Hexokinase in eukaryotes is often found bound to the outer mitochondrial membrane where it is bound to porin (120, 121). This interaction allows the ATP formed in mitochondria to be used directly by hexokinase without the ATP mixing with the cytosolic ATP pools. The binding of this enzyme to the membrane is modulated by its product glucose-6-PO_4. This ambiquitous behavior may represent a general method of regulation. Similar behavior was observed by Masters with certain glycolytic enzymes and actin (101).

Not only do glycolytic enzymes bind to structural elements and membranes, but there is good evidence that there are specific interactions between sequential enzymes (11). Early crystalline protein preparations from rabbit muscle (myogen A and myogen B) were found to contain mixtures of several glycolytic enzymes. Myogen A seems to be crystalline and homogeneous and contains both glyceraldehyde phosphate dehydrogenase and aldolase. Ovadi and her coworkers (122) and Batke & Tompa (123 and references therein) have shown this complex formation using a variety of techniques, including

kinetic studies, polarization of fluorescence, and active enzyme centrifugation. An interaction between aldolase and glyceraldehyde phosphate dehydrogenase has been demonstrated by fluorescence techniques (122) and by
the inactivation of glyceraldehyde phosphate dehydrogenase by aldolase
(124). Salerno & Ovadi (125) have shown an interaction between aldolase and
triose phosphate isomerase by their mutual protection against perchloric acid
denaturation. Ashmarina et al (126) showed by kinetic means an interaction
between glyceraldehyde phosphate dehydrogenase and phosphoglycerate
kinase. Cseke & Szabolcsi (127), using a molecular sieving property of
erythrocyte membrane, showed that the glycolytic enzymes catalyzing the
conversion of fructose 1,6 bisphosphate to phosphoenolpyruvate behave as if
they were in a large complex.

An interesting observation has been reported recently by Srivastava &
Bernhard (11). They showed first that direct transfer of 1,3 diphosphoglycerate occurred between 3-phosphoglycerate kinase and glyceraldehyde phosphate dehydrogenase. The tight binding of the metabolite to phosphoglycerate
kinase did not allow for a dissociation-diffusion mechanism of action. The
turnover rate of the coupled system was larger than the rate of dissociation of
1,3 diphosphoglycerate from the phosphoglycerate kinase. These workers
also showed that NADH was transferred directly from glyceraldehyde phosphate dehydrogenase to liver alcohol dehydrogenase. This experiment was
performed with concentrations of enzymes comparable to in situ enzyme
concentrations. Further work by this group has shown that a series of these
NADH transfers can occur between dehydrogenases, and in every case the
transfer occurs between dehydrogenases with opposite stereospecificity for
NADH. Direct transfer does not occur if both dehydrogenases have the same
chiral specificity for NADH. Further, a computer graphics analysis of the
electrostatic potential for the upper part of the active site region of glyceraldehyde phosphate dehydrogenase (B dehydrogenase) is positive and opposite
to that of corresponding active site areas on liver alcohol dehydrogenase and
lactate dehydrogenase (A dehydrogenases), to which NADH can be directly
transferred (128). This series of elegant experiments shows a possible
molecular basis for the interaction between sequential enzymes. I have
pointed out that the constancy of size of intracellular enzymes may well be
related to the need for specific surface sites to enable such interactions (15).

Another type of evidence for interactions of glycolytic enzymes comes
from experiments in which the direct isolation of such complexes can be
obtained by gel filtration of cell extracts. This has been reported for E. coli
(129) and for S. cerevisiae (130). In these experiments only a small portion of
the total content of glycolytic enzymes was found in a high-molecular-weight
fraction (1.6×10^6).

Kurganov (131), based on the binding data cited above, has constructed a model of a "glycolytic particle" consistent with the binding affinities and binding sites on the enzymes. A similar theoretical construct of glycolytic enzymes based on the existence of several isozymes for the glycolytic enzymes was proposed earlier by Ureta (132). In this latter model different isozymes served as parts of different metabolic pathways, e.g. glycolysis, gluconeogenesis.

In the trypanosome, an organelle, the glycosome, contains all the glycolytic enzymes (133). The metabolic rationale is that this organism must obtain all its energy glycolytically, and the concentration of these enzymes into an organelle may better achieve this. Additionally, sometimes a large complex of enzymes can be preserved when the membrane is removed. Crosslinking of these enzymes is easily achieved, but neither these experiments nor subsequent "channeling" experiments have given any indication of the structure, i.e. spatial arrangement of these enzymes within the organelle. At a total protein concentration of about 10 mM, the individual enzymes are in the mM range and the free substrate concentration is probably quite low. The main advantage of this system probably resides in the fact that the transient time for the system is greatly reduced (134, 135).

Particulate glycolysis in castor bean endosperm was located in the proplastid of that tissue (136). DeLuca & Kricka have shown that coimmobilizing the glycolytic enzymes results in an enormous increase in the rate of reaction (137) over that observed with the same quantity of enzymes free in solution.

Tricarboxylic Acid Cycle

The possibility that enzymes responsible for the oxidation of acetyl CoA to CO_2 and H_2O, the Krebs tricarboxylic acid cycle, form a multienzyme system, has been considered for many years. D. E. Green and his coworkers reported the isolation of a multienzyme system as an aggregate that could oxidize pyruvate and fatty acids to CO_2 and H_2O (138). It was shown, however, that they had isolated mitochondria, and for many years, the accepted concept has been that the electron transport system is localized in the inner mitochondrial membrane and most of the enzymes of the Krebs tricarboxylic acid cycle are free in the matrix of the organelle (see 139 for references).

Several kinetic observations, however, could not be explained by the existence of a completely unorganized tricarboxylic acid cycle. First, several isotope experiments indicated no mixing in the matrix of acetyl CoA generated by fatty acid oxidation and that generated by pyruvate oxidation (140). Von Glutz & Walter (141), using liver mitochondria oxidizing palmitate (^{14}C)

in the presence and absence of pyruvate, concluded that two acetyl CoA pools existed. There are data, however, that disagree with these results (142). Srere has shown that the high protein concentration of the matrix made local compartmentation of acetyl CoA possible in the matrix. Second, it was calculated that the apparent concentration of free oxaloacetate was not sufficient to account for the rate of oxidation observed in the cycle (142). Srere (143) postulated that an organized cycle could account for these apparent anomalies. Srere et al (144) showed that an immobilized system of malate dehydrogenase and citrate synthase showed a kinetic advantage over the free enzyme system. Further studies on oxidation rates in permeabilized mitochondria indicated that the enzymes of the cycle probably were not free in the matrix (145). A variety of other experiments using electron microscopy, crosslinking, histochemical techniques, and calculations from stereomorphological measurements on mitochondria all supported the concept of some level of organization of the Krebs tricarboxylic acid cycle enzymes within the mitochondrial matrix (see 139 for review).

One of the first direct experiments concerning interaction of sequential Krebs tricarboxylic acid cycle enzymes was the report that in the presence of polyethylene glycol, a volume excluding compound, a specific interaction between pig heart citrate synthase and pig heart mitochondrial malate dehydrogenase could be observed (146). These experiments were based on the earlier work of Backman & Johansson on the interaction of aspartate amino transferase and malate dehydrogenase (see *Amino Acid Metabolism*) (64). Specific interactions have now been shown between pyruvate dehydrogenase complex and citrate synthase (147), mitochondrial malate dehydrogenase and fumarase (148), α-ketoglutarate dehydrogenase complex and succinate thiokinase (149), citrate synthase and aconitase (150), and (NAD) isocitrate dehydrogenase and α-ketoglutarate dehydrogenase complex (151). In addition to these, interactions have been demonstrated between citrate synthase and thiolase (152), malate dehydrogenase and Complex I, pyruvate dehydrogenase complex and Complex I, and α-ketoglutarate dehydrogenase complex and Complex I (87). The techniques used to demonstrate these interactions included polarization of fluorescence, binding to an immobilized component, frontal elution analysis, and Hummel-Dreyer columns. In all cases the specificity was tested by using cytosolic isozymes or enzymes that were not sequential but had the same size or isoelectric point as the sequential partner.

For the pyruvate dehydrogenase complex-citrate synthase complex, it was observed that the K_m for CoA for pyruvate dehydrogenase complex decreases from 10 μM (free pyruvate dehydrogenase complex) to 1.5 μM (PDC-CS) (147). In the α-ketoglutarate dehydrogenase complex-succinate thiokinase complex, the K_m for succinyl CoA changed from 65 μM (free STK) to 1.5 μM (αKGDC-STK) (149).

Spivey and his colleagues have shown recently that a polyethylene glycol precipitate of mitochondrial malate dehydrogenase and citrate synthase could channel oxaloacetate (153). The coupled reaction was run in the presence of an aspartate amino transferase system in order to trap free oxaloacetate. When citrate synthase and mitochondrial malate dehdyrogenase were free in solution, the generated oxaloacetate was trapped so that citrate synthase could not form citrate. In the solid state (the polyethylene glycol precipitate of citrate synthase and mitochondrial malate dehydrogenase), no trapping of oxaloacetate occurred and citrate synthesis was only slightly inhibited. Thus an important functional test for this interaction is demonstrated.

Another level of organization of the Krebs cycle enzymes consists of their binding to the inner surface of the inner mitochondrial membrane (see 139 for review). Several tricarboxylic acid cycle enzymes are tightly bound to the inner membrane. Succinate dehydrogenase is extremely difficult to remove from membranes. Pyruvate dehydrogenase complex can be removed by treatment with Triton. α-Ketoglutarate dehydrogenase complex can be removed only with repeated washings. The other enzymes of the cycle are easily released as soluble enzymes when mitochondria were disrupted by sonication, freeze thawing, or by digitonin treatment of mitochondria. In heart mitochondria the inner membranes are so close to each other that almost all matrix proteins are near the inner membrane. In liver and other tissues this is not true, so that it is possible to picture that one level of organization of the Krebs cycle enzymes would be having them bound to inner membrane proteins. Crosslinking of liver mitochondria with glutaraldehyde indicated that fumarase, malate dehydrogenase, and citrate synthase were located next to the inner membrane (154). D'Souza & Srere (155) later showed that citrate synthase, malate dehydrogenase, and fumarase could bind to the inner surface of the inner mitochondrial membrane. The binding was specific since none was observed to occur to the outer surface of the inner membrane, the outer membrane, microsomes, red blood cell membranes, or to liposomes of various compositions, including those made of the lipids of the inner membrane. Protein components of the membrane were involved in the binding since they could be extracted and placed in liposomes and binding could then be observed. Recent experiments have shown that aconitase (the mitochondrial isozyme) and NAD isocitrate dehydrogenase also bind to the inner membrane (150). A number of cytosolic enzymes have been tried in the binding experiments, but no binding has been observed. Inner membranes from mitochondria of a wide variety of tissues exhibit this binding capacity, and the binding of heterologous Krebs cycle enzymes has been shown (156). The binding is prevented by moderate ionic strengths, and in the case of citrate synthase, it has been shown that very low concentrations of oxaloacetate prevents the binding.

Sumegi & Srere have shown the dehydrogenases of the Krebs tricarboxylic acid cycle can bind to Complex I either by itself or when it is incorporated into liposomes (152). Whether or not this is one protein or the only protein to which these enzymes bind in the inner membrane has not yet been demonstrated. It seems logical, however, that this complex, which is sequential to the dehydrogenases, contains binding sites for them. It should be remembered that purified Complex I contains about 20 subunits and only three or four have known catalytic functions.

The inner mitochondrial membrane is one of the most proteinaceous biological membranes. It has been estimated that 50% of its surface is occupied by protein, so that if there were a random distribution of proteins only one protein molecule diameter between them would exist. About 35% of the total protein of this membrane is the electron transport chain complexes and the ATPase (157). It has been proposed that multienzyme complexes of the individual electron transport components exist in the membrane. Recently, however, Hackenbrock and his coworkers (158) showed that the rate of diffusion of the complexes in inner membranes was greater than the rates of enzyme activity, and they concluded that there was no need for super complex formation. The rates of diffusion of the complexes (which are transmembrane proteins) were measured in isolated membranes, and it is likely that in situ the viscosity of the matrix will decrease these diffusion rates markedly. When Hochman et al (159) investigated the lateral diffusion of electron transport complexes in giant mitoplasts, their results yielded diffusion coefficients lower than those of Hackenbrock. They conclude that, although electron transfer can be achieved by diffusion, more rapid rates occur through transitory sequential aggregates of the electron transfer components. Berry & Trumpower (160) have isolated an enzyme complex with ubiquinol-cytochrome c oxidoreductase, cytochrome c reductase, and ubiquinol oxidase activities from *Paracoccus denitrificans*. Kell & Westerhoff (161) have argued strongly for the existence of electron transfer protein complexes in mitochondria based on calculations of the rates of diffusion and turnover number of the enzymes involved.

Robinson & Srere have recently isolated from sonicated mitochondria an easily sedimentable fraction that contains exposed activities of all the Krebs cycle enzymes (162). Citrate synthase, which is one of these exposed easily sedimented enzymes, is inhibited by its antibody so that it appears unlikely that the fraction consisted of free enzymes entrapped within a vesicular structure of altered permeability. The preparation is stable to ionic strengths as high as 400 mM KCl and to a broad range of pH values. The complex is unstable at intermediate concentrations of phosphate but stable at both high and low concentrations of this ion. Electron microscopy reveals what appear to be broken or open mitochondria. The isolation of this complex does not prove that the enzymes are arranged sequentially on the membrane. However, the fact that the complex is stable at high ionic strengths and that individual enzyme binding to the membrane is not stable under these conditions argues for cooperative interactions between the bound enzymes and the membrane. Recent experiments have shown a kinetic advantage for the multienzyme systems responsible for fumarate oxidation and for malate conversion to citrate of the slightly disrupted systems compared to the soluble systems.

OTHER METABOLIC PATHWAYS

A number of bioactive polypeptides, such as the antibiotic gramicidin, are synthesized on multifunctional proteins (see 163 for review). Thus, for gramicidin, two multifunctional synthetases catalyze the series of 17 reactions. Many of these enzyme systems contain 4-phosphopantetheine as the carrier for the activated carboxyl group. This reaction mechanism is similar to the fatty acid synthetase system. Recently Zocher et al (164) reported the isolation of a multifunctional protein that is involved in the synthesis of cyclosporin.

Electron microscopic studies of a unicellular green alga by Brown & Montezinos (165) indicated that cellulose microfibril biosynthesis occurred in association with a linear enzyme complex in the B face of the plasma membrane. This work is in agreement with other studies concerning the enzyme location of cellulose biosynthesis and with other electron microscopic studies in other cell types.

Harson et al (166) have shown that proteins released from the matrix of castor bean endosperm glyoxysomes can reassociate with the luminal side of glyoxysomal membranes. No identification of the binding proteins has been made. Bieglmayer et al (167) had shown earlier that osmotic shock of endosperm glyoxysomes gave rise to a membrane fraction that contained the major part of β-oxidation of fatty acids, malate synthase, citrate synthase, malate dehydrogenase, and antimycin A-insensitive NADH-cytochrome c reductase.

The secondary metabolites of plants are often biosynthesized by multienzyme systems (see 168 for review). Cutler et al (169) have shown that in the biosynthesis of the cyanogenic glucoside taxiphyllin, two intermediates are channeled in a cell-free preparation.

There are other metabolic pathways for which there is evidence of the existence of interacting sequential enzyme systems. Some of these systems are membrane associated. Most have in common the fact that the intermediates involved in the pathways do not have other functions in the cell, and thus represent tight processive sequences.

GENETIC CONSIDERATIONS

I have pointed out several times during the course of this review that several metabolic pathways occur as apparently free enzyme systems or multienzyme systems in those cellular forms that probably occurred early in evolution and appear as multifunctional enzymes in cells that arose later in evolution (see 170 for review of this aspect for fatty acid synthetase and 56 for review of aromatic amino acid biosynthesis).

One might expect that if the enzymes of a multienzyme system are unique and operative only in that one system, then control of their synthesis would be most easily accomplished if they were linearly and closely located on a chromosome. This is generally true except for the components of tight multienzyme complexes. Considering the penchant for DNA to move in a chromosome, this is not a surprising observation. However, it does entail that all the enzymes of a sequence that constitute a constant proportion group have the same promoter sequence so their synthesis can be coregulated.

McConkey (171) has shown for the proteins of HeLa and CHO cells, cells that had a common ancestor about 80×10^6 years ago, that at least half of the proteins are indistinguishable by two-dimensional gel electrophoresis and isoelectric focusing. This conservation of structure is interpreted by McConkey to indicate that interactions between proteins (quinary structure) impose severe limitations on the evolution of some classes of proteins. It is interesting to note that evolutionary changes in glycolytic enzymes are extremely slow (172).

Most metabolic enzymes are polymeric so that they have interacting surface sites. Sequential enzymes handle the same substrate so that they must have similar active sites. The interactions of sequential metabolic reactions could be postulated to have occurred by gene duplication and divergent evolution.

Fothergill-Gilmore (172) reviewed the status of the evolution of the glycolytic pathway. Although the three-dimensional structures of the glycolytic enzymes are similar, a careful analysis of these systems led her to tentatively propose that both convergent and divergent evolutionary factors were involved in development of the sequential enzymes. No similar analysis can be made as yet for other pathways.

BIOPHYSICAL AND CELLULAR ASPECTS

As mentioned in the first section of this review, one of the major criticisms of the hypothesis states that such complexes are not necessary since diffusion rates are in the range of μseconds and less, while enzyme turnover rates are in the range of mseconds. If this were so in a cell, to maintain a higher concentration of a product-substrate in the vicinity of the next enzyme in a sequence than exists in the bulk water phase, some energy input would be necessary. Since no mechanism for accomplishing this is immediately apparent, the assumption is that none exists and that substrates diffuse quickly throughout the cell to an equilibrium concentration. In a recent article, Srere (139) discussed all these factors in detail and showed how the simplistic diffusional calculations are far from describing the actual case that exists in cells. Experiments by Mastro & Keith (173) on diffusion coefficients in cells showed that these were from 1/5 to 1/2 those in H_2O for molecules smaller than $M_r = 700$. These effects may be due to viscosity or binding effects on the molecule studied.

For diffusion of large molecules ($M_r = 10,000$) in cells versus that in water, these factors range from 1/5 to 1/100 depending on the molecule studied. Recently Jacobson & Wojcieszyn (174) showed that diffusion of certain proteins in fibroblasts was slowed so markedly that one must assume binding of membrane protein to cytosolic elements has occurred. Their calculations showed that to account for their observations, 98% of the diffusing species must be bound at any one instance. Similar results were obtained by Lang et al (175) in hepatoma tissue culture cells. These results confirm for a few molecules the conclusions of earlier ultracentrifugal studies of Zalokar (176) in *Neurospora* and Kempner & Miller (177) in *Euglena*. These authors showed that no free proteins could be located in the cytosol of centrifuged intact cells.

Using the photobleaching recovery method, Brass et al (178) showed that the diffusion coefficient of the maltose binding protein in the periplasm of *E. coli* was 1/1000 that measured in water. They concluded that the periplasm contained little free water, confirming electron microscopic studies of Hobot et al (179). Since the mitochondrial matrix has a protein concentration similar to that of periplasm (139), it is possible that the proteins in that compartment have similar large restrictions in diffusional motion. It is interesting to note that reduction of dimensionality as a mechanism to reduce diffusion times in cells has recently been questioned as a viable explanation for a number of cellular phenomena (180).

A number of other studies are pertinent to the diffusion question. First, it is known that gradients for ATP and O_2 can exist within cells (181, 182). Second, in the frog oocyte, Horowitz & Miller (183) have demonstrated that H_2O, sodium, potassium, ATP, amino acids, and sugars are not uniformly distributed in the cytoplasm. Mollenhauer & Morre (184) have described electron microscopic studies that indicate that cytosol is not a uniform compartment but contains a large number of differentiated zones. It is also apparent that organelles, like mitochondria (139), are not randomly distributed in cells, but are located in functionally propitious sites. Thus, from physical chemical and cell biological studies, it is clear that many different cell types do not find intracellular diffusion rates sufficient for maintenance of life.

The second major criticism of studies on metabolic complexes is that they often cannot be demonstrated at presumed pH and ionic conditions of the cell. In some cases it has been necessary to add polyethylene glycol to show the interactions. When one isolates the components of a pathway to study individual interactions, the environment of the in vitro experiment is quite different from the in vivo circumstance. Thus, the components are usually at a lower concentration in vitro than in vivo and the total protein concentration is much lower. Therefore simulation of water exclusion effects of proteins in cells as described by Minton (185) is desirable, and polyethylene glycol often can serve this purpose. Since the interactions between proteins are often ionic, it is not surprising that they can be enhanced at low ionic strength. However, in vitro one is usually studying individual interactions, which is in contrast to the multiple interactions that occur in vivo along with the attendant cooperative binding effects. This is illustrated by the results of Robinson & Srere (162) with the Krebs cycle metabolon (see above).

One of the most important criteria in assessing the biological validity of in vitro studies on enzyme interactions is the specificity of the interactions. One must assess a whole spectrum of proteins to establish such specificity, and this was done for most of the cases cited in this article.

CONCLUSIONS AND FUTURE QUESTIONS

There have been a number of advantages listed for the existence of multi-functional proteins, multienzyme complexes, and metabolons. 1. If the substrates can be transferred directly from one active site to another (channeling) so that the substrate's concentration in the microenvironment of the second site is kept relatively high compared to the average bulk concentration of that substrate, then one spares the limited solvation capacity of cellular water. 2. In these situations one achieves high concentrations of substrates with fewer substrate molecules. 3. When the input to a sequence of metabolic reactions is changed, then the transient time for the attainment of a new steady state is reached faster for a multienzyme complex than for comparable enzyme activities free in solution. 4. If substrates are unstable in aqueous environments or if they can be acted on by other enzymes, then direct transfer to a successive active site is a mechanism for their preservation. 5. Interactions between the sequential proteins have been shown to have allosteric (thus regulatory) effects on the activities involved. 6. Such complexes, if they are attached to or are part of cellular structural elements or macromolecules, would have the advantage that the diffusion of their enzymatic components could possibly take place in two dimensions or one dimension rather than three dimensions. For RNA synthesis, diffusion in two dimensions must occur to account for the rapid rate of interaction of RNA polymerase with the appropriate promoter sequence on DNA.

Concurrent with the accumulation of the evidence presented in this article concerning the existence of interacting sequential enzymes, separate but related insights concerning cell structure have emerged. Among these are: 1. An elaborate structural network of proteins exists in the cytosol (186). 2. Protein concentrations in cells (139), certain organelles (139), and cell compartments are extremely high (178). 3. The location of organelles is specifically related to their function in the cell (139). 4. Few proteins seem to be free within cells (174–177).

One of the most important unsolved problems in metabolism is the question of whether or not sequential metabolic complexes exist within membranes, or membrane-bounded cellular compartments. If such complexes exist, a number of interesting corollaries must be considered. 1. Are structural cellular elements involved in these complexes? 2. Can the formation and dissolution of such complexes be involved in metabolic regulation? 3. If these complexes exist, then should we not use appropriate models for the kinetics and thermodynamic characteristics of solid state arrays rather than continuing to base our models on the solution properties of the enzymes that are obtained under conditions far from physiological conditions? 4. If these protein interactions are important for normal cell functions, should we not place even greater emphasis on the evolution and conservation of the surface binding sites on proteins?

It is now possible, using techniques of site-directed mutagenesis, to modify enzymes at sites involved in sequential interactions and to insert these modified enzymes into otherwise normal cells. Study of the change in phenotype that may occur could give more direct information concerning the role of metabolon occurrence in metabolism. Studies involving the use of NMR techniques on intact cells will prove useful in determining whether substrates are channeled and perhaps also whether the proteins are bound or free. The photobleaching techniques are useful for these problems. The use of electron microscopy for locating enzymes in situ may become more useful, and the use of multiple distinguishable labels may even give information about sequential arrangements of enzymes.

Extension of these concepts of interacting, metabolically related enzymes to interacting systems leads to the supramolecular structural features in a cell seen by electron microscopy. The structure of the protein synthetic system is thus seen as ribosomes on the endoplasmic reticulum, and the location of glycolytic enzyme systems may be responsible for the I-bands of muscle. It is plausible that the individual and different structures of the cristae seen in mitochondria from different cell types are due to the interactions of the mitochondrial metabolic complexes of the inner membrane described in this article. Further, it seems obvious that such gross features of cell structure as the cellular location of mitochondria next to energy-utilizing substructures simply reflect the same interactions described here at the next hierarchical level of cell structure. The description of the interactions responsible for the final structure of the cell will be difficult because they probably represent the sum of many weak interactions occurring in an environment that is not easily simulated in vitro. In addition, disruption of just one system may not cause a great perturbation in cell metabolism since the system was probably evolved partially to overcome the problem of solvation of thousands of intermediates. If just a few metabolites increase in concentration, the cell will certainly be able to handle this disruption so that phenotype changes will be difficult to find. Nonetheless, our current knowledge of protein structure and of the specific interactions seen at the level of metabolic complexes seems sufficient to provide an adequate basis of explanation of all that we see in the cell.

ACKNOWLEDGMENTS

I am indebted to many colleagues who discussed particular aspects of this review and sent me relevant material. These include Drs. M. E. Jones, F. Traut, C. Mathews, J. A. DeMoss, G. R. Welch, T. Steck, W. Dempsey, B. Sumegi, K. Uyeda, and A. Pardee. I thank Ms. Penny Perkins, who prepared the manuscript. My research was supported by the Veterans Administration Research Service, USPHS, and NSF.

Literature Cited

1. Alberts, B., Bray, D., Lewis, J., Raff, M., Roberts, K., Watson, J. D., eds. 1983. *Molecular Biology of the Cell.* New York: Garland
2. Atkinson, D. E. 1977. *Cellular Energy Metabolism and Its Regulation.* New York: Academic
3. Ginsburg, A., Stadtman, E. R. 1970. *Ann. Rev. Biochem.* 39:429–72
4. Kirschner, K., Bisswanger, H. 1976. *Ann. Rev. Biochem.* 45:143–66
5. Friedrich, P. 1984. *Supramolecular Enzyme Organization,* pp. 1–300. New York: Pergamon
6. Welch, G. R., ed. 1985. *Organized Multienzyme Systems: Catalytic Properties,* pp. 1–458. New York: Academic
7. Srere, P. A., Estabrook, R. W., eds. 1978. *Microenvironments and Metabolic Compartmentation,* pp. 1–455. New York: Academic
8. Welch, G. R., Clegg, J. S., eds. 1986. *Organization of Cell Metabolism.* New York: Plenum. In press
9. Nover, L., Lynen, F., Mothes, K., eds. 1980. *Cell Compartmentation and* *Metabolic Channelling,* pp. 1–523. Amsterdam: Elsevier North-Holland Biomed.
10. Welch, G. R. 1977. *Prog. Biophys. Mol. Biol.* 32:103–91
11. Srivastava, D. K., Bernhard, S. A. 1985. *Curr. Top. Cell. Regul.* 28:1–109
12. Clegg, J. S. 1984. *Am. J. Physiol.* 246:R133–R151
13. Wombacher, H. 1983. *Mol. Cell. Biochem.* 56:155–64
14. Srere, P. A., Mosbach, K. 1974. *Ann. Rev. Microbiol.* 28:61–83
15. Srere, P. A. 1984. *Trends Biochem. Sci.* 9:387–90
16. Alberts, B. M. 1985. *Trends Genet.* 1:26–30
17. Formosa, T., Burke, R. L., Alberts, B. M. 1983. *Proc. Natl. Acad. Sci. USA* 80:2442–46
18. Flanegan, J. B., Greenberg, G. R. 1977. *J. Biol. Chem.* 252:3019–27
19. Allen, J. R., Lasser, G. W., Goldman, D. A., Booth, J. W., Mathews, C. K. 1983. *J. Biol. Chem.* 258:5746–53

20. Ficht, T. A., Moyer, R. W. 1980. *J. Biol. Chem.* 255:7040–48
21. Reddy, G. P. V., Pardee, A. B. 1980. *Proc. Natl. Acad. Sci. USA* 77:3312–16
22. Reddy, G. P. V., Pardee, A. B. 1983. *Nature* 303:86–88
23. Vishwanatha, J. K., Coughlin, S. A., Wesolowski-Owen, M., Baril, E. F. 1986. *J. Biol. Chem.* 261:6619–28
24. Leeds, J. M., Slabaugh, M. B., Mathews, C. K. 1985. *Mol. Cell. Biol.* 5:3443–50
25. Mathews, C. K., Slabaugh, M. B. 1986. *Exp. Cell. Res.* 16:285–95
25a. Nelson, W. G., Pienta, K. J., Barrack, E. R., Coffey, D. S. 1986. *Ann. Rev. Biophys. Biophys. Chem.* 15:457–75
26. von Hippel, P. H., Bear, D. G., Morgan, W. D., McSwiggen, J. A. 1984. *Ann. Rev. Biochem.* 53:389–446
27. McClure, W. R. 1985. *Ann. Rev. Biochem.* 54:171–204
28. Murooka, Y., Lazzarini, R. A. 1973. *J. Biol. Chem.* 248:6248–50
29. Armaleo, D., Gross, S. R. 1985. *J. Biol. Chem.* 260:16174–80
30. Wingender, E., Jahn, D., Seifart, K. H. 1986. *J. Biol. Chem.* 261:1409–13
31. Moldave, K. 1985. *Ann. Rev. Biochem.* 54:1109–49
32. Wittmann, H. G. 1983. *Ann. Rev. Biochem.* 52:35–65
33. Kellems, R. E., Allison, V. F., Butow, R. A. 1975. *J. Cell Biol.* 65:1–14
34. Dang, C. V., Ferguson, B., Johnson-Burke, D., Garcia, V., Yang, D. C. H. 1985. *Biochim. Biophys. Acta* 829:319–26
35. Deutscher, M. P. 1984. *J. Cell Biol.* 99:373–77
36. Fulton, A. B., Wan, K., Penman, S. 1980. *Cell* 20:849–57
37. Meyer, F., Heilmeyer, L. M. G. Jr., Haschke, R. H., Fischer, E. H. 1970. *J. Biol. Chem.* 245:6642–48
38. Fischer, E. H., Heilmeyer, L. M. G. Jr., Haschke, R. H. 1971. *Curr. Top. Cell. Regul.* 4:236–47
39. Nimmo, H. G., Cohen, P. 1976. *Eur. J. Biochem.* 68:31–44
40. Cohen, P. 1978. *Curr. Top. Cell. Regul.* 14:183–93
41. Wanson, J. C., Drochmans, P. 1968. *J. Cell. Biol.* 38:130–50
42. Busby, S. J. W., Radda, G. K. 1976. *Curr. Top. Cell. Regul.* 10:89–160
43. Geddes, R. 1985. In *The Polysaccharides,* ed. G. O. Aspinall, pp. 283–336. New York: Academic
44. Barber, A. A., Orrell, S. A. Jr., Bueding, E. 1967. *J. Biol. Chem.* 242:4040–44
45. Jones, M. E. 1980. *Ann. Rev. Biochem.* 49:253–79
46. Aitken, D. M., Bhatti, A. R., Kaplan, J. G. 1973. *Biochim. Biophys. Acta* 309:50–57
47. Kornberg, A. 1982. *1982 Supplement to DNA Replication.* San Francisco: Freeman
48. Rowe, P. B., McCairns, E., Madsen, G., Sauer, D., Elliott, H. 1978. *J. Biol. Chem.* 253:7711–21
49. McCairns, E., Fahey, D., Sauer, D., Rowe, P. B. 1983. *J. Biol. Chem.* 258:1851–56
50. Henikoff, S., Keene, M. A., Sloan, J. S., Bleskan, J., Hards, R., Patterson, D. 1986. *Proc. Natl. Acad. Sci. USA* 83:720–24
51. Daubner, S. C., Schrimsher, J. L., Schendel, F. J., Young, M., Henikoff, S., et al. 1985. *Biochemistry* 24:7059–62
52. Appling, D. R., Rabinowitz, J. C. 1985. *J. Biol. Chem.* 260:1248–56
53. Smith, G. K., Mueller, W. T., Wasserman, G. F., Taylor, W. D., Benkovic, S. J. 1980. *Biochemistry* 19:4313–21
54. Paquin, J., Baugh, C. M., MacKenzie, R. E. 1985. *J. Biol. Chem.* 260:14925–31
55. Burns, V. W. 1964. *Science* 146:1056–58
56. Welch, G. R., Gaertner, F. H. 1980. *Curr. Top. Cell. Regul.* 16:113–62
57. DeMoss, J. A. 1986. *Metabolic Complexes,* ed. G. R. Welch, J. Clegg. New York: Plenum. In press
58. Welch, G. R., DeMoss, J. A. 1978. See Ref. 7, pp. 323–44
59. Manney, T. R. 1970. *J. Bacteriol.* 102:483–88
60. Bearden, L., Moses, V. 1972. *Biochim. Biophys. Acta* 279:513–26
61. Bergquist, A., Eakin, E. A., Murali, D. K., Wagner, R. P. 1974. *Proc. Natl. Acad. Sci. USA* 71:4352–55
62. Kredich, N. M., Becker, M. A., Tomkins, G. M. 1969. *J. Biol. Chem.* 244:2428–39
63. Cook, P. F., Wedding, R. T. 1977. *Arch. Biochem. Biophys.* 178:293–302
64. Backman, L., Johansson, G. 1976. *FEBS Lett.* 65:39–42
65. Beeckmans, S., Kanarek, L. 1981. *Eur. J. Biochem.* 117:527–35
66. Bryce, C. F. A., Williams, D. C., John, R. A., Fasella, P. 1976. *Biochem. J.* 153:571–77
67. Manley, E. R., Webster, T. A., Spivey, H. O. 1980. *Arch. Biochem. Biophys.* 205:380–87
68. Fahien, L. A., Kmiotek, E. H., Woldegiorgis, G., Evenson, M., Shrago, E.,

Marshall, M. 1985. *J. Biol. Chem.* 260:6069–79

69. Fahien, L. A., Kmiotek, E., Smith, L. 1979. *Arch. Biochem. Biophys.* 192:33–46

70. Fahien, L. A., Kmiotek, E. 1979. *J. Biol. Chem.* 254:5983–90

71. Salerno, C., Ovadi, J., Keleti, T., Fasella, P. 1982. *Eur. J. Biochem.* 121:511–17

72. Hearl, W. G., Churchich, J. E. 1984. *J. Biol. Chem.* 259:11459–63

73. Walker, J. L., Oliver, D. J. 1986. *J. Biol. Chem.* 261:2214–21

74. Davis, R. H. 1980. See Ref. 9, pp. 239–43

75. Gillevet, P. M., Dakshinamurti, K. 1982. *Biosci. Rep.* 2:841–48

76. Ranganathan, N. S., Srere, P. A., Linn, T. C. 1980. *J. Biol. Chem.* 203:52–58

77. Janski, A. M., Cornell, N. W. 1980. *Biochem. J.* 186:423–29

78. Linn, T. C., Srere, P. A. 1984. *J. Biol. Chem.* 259:13379–84

79. Yates, R. D., Higgins, J. A., Barrnett, R. J. 1969. *J. Histochem. Cytochem.* 17:379–85

80. Witters, L. A., Friedman, S. A., Bacon, G. W. 1981. *Proc. Natl. Acad. Sci. USA* 78:3639–43

81. Allred, J. B., Roman-Lopez, C. R., Pope, T. S., Goodson, J. 1985. *Biochem. Biophys. Res. Commun.* 129:453–60

82. Vance, D. E., Pelech, S. L. 1984. *Trends Biochem. Sci.* 9:17–20

83. Coleman, R., Bell, R. M. 1978. *J. Cell Biol.* 76:245–53

84. Stanley, K. K., Tubbs, P. K. 1975. *Biochem. J.* 150:77–88

85. Sumegi, B., Srere, P. A. 1984. *J. Biol. Chem.* 259:8748–52

86. Kispal, G., Sumegi, B., Alkonyi, I. 1986. *J. Biol. Chem.* In press

87. Sumegi, B., Srere, P. A. 1984. *J. Biol. Chem.* 259:15040–45

88. Pramanik, A., Pewar, S., Antonian, E., Schulz, H. 1979. *J. Bacteriol.* 137:469–73

89. Yang, S. Y., Bittman, R., Schulz, H. 1985. *J. Biol. Chem.* 260:2862–68

90. Lieberman, S., Greenfield, N. J., Wolfson, A. 1984. *Endocrine Rev.* 5:128–48

91. Burstein, S., Gut, M. 1973. *Ann. NY Acad. Sci.* 212:262–75

92. Gaylor, J. L., Delwiche, C. V. 1973. *Ann. NY Acad. Sci.* 212:122–38

93. Lange, Y., Steck, T. L. 1985. *J. Biol. Chem.* 260:15592–97

94. Ottaway, J. H., Mowbray, J. 1977. *Curr. Top. Cell. Regul.* 12:107–208

95. Lynch, R. M., Paul, R. J. 1983. *Science* 222:1344–46

96. Postius, S. 1981. *Z. Naturforsch. Teil C* 36:615–18

97. Connett, R. J., Wittels, B., Blum, J. J. 1972. *J. Biol. Chem.* 247:2657–61

98. Clark, D. G., Rognstad, R., Katz, J. 1973. *Biochem. Biophys. Res. Commun.* 54:1141–48

99. Jungermann, K., Katz, N. 1982. In *Metabolic Comparmentation,* ed. H. Sies, pp. 411–35. New York: Academic

100. Coe, E. L., Greenhouse, W. V. V. 1973. *Biochim. Biophys. Acta* 329:171–82

101. Masters, C. J. 1981. *CRC Crit. Rev. Biochem.* 11:105–43

102. Morton, D. J., Clarke, F. M., Masters, C. J. 1977. *J. Cell Biol.* 74:1016–23

103. Sigel, P., Pette, D. 1969. *J. Histochem. Cytochem.* 17:225–36

104. Caswell, A., Corbett, A. M. 1985. *J. Biol. Chem.* 260:6892–98

105. Foemmel, R. S., Gray, R. H., Bernstein, I. A. 1975. *J. Biol. Chem.* 250:1892–97

106. Eichel, H. J., Goldenberg, E. K., Rem, L. T. 1964. *Biochim. Biophys. Acta* 81:172

107. Conger, N. E., Fields, R. D., Feldman, C. J. 1971. *Fed. Proc.* 30:1158

108. Eldan, M., Blum, J. 1973. *J. Biol. Chem.* 248:7445–48

109. Paul, R. J., Wuytack, F., Raeymackers, L., Casteels, R. 1986. *Fed. Proc.* 45:766

110. Hardin, C., Paul, R. J. 1986. *Fed. Proc.* 45:767

111. Bronstein, W. W., Knull, H. R. 1981. *Can. J. Biochem.* 59:494–99

112. Huitorel, P., Pantaloni, D. 1985. *Eur. J. Biochem.* 150:265–69

113. Luther, M. A., Lee, J. C. 1986. *J. Biol. Chem.* 261:1753–59

114. Gillies, R. J. 1982. *Trends Biochem. Sci.* 7:41–42

115. Tsai, I. H., Murthy, S. N. P., Steck, T. L. 1982. *J. Biol. Chem.* 257:1438–42

116. Karadsheh, N. S., Uyeda, K. 1977. *J. Biol. Chem.* 252:7418–20

117. Solti, M., Bartha, F., Halasz, N., Toth, G., Sirokman, F., Friedrich, P. 1981. *J. Biol. Chem.* 256:9260–65

118. Keokitichai, S., Wrigglesworth, J. M. 1980. *Biochem. J.* 187:837–41

119. Rich, G. T., Pryor, J. S., Dawson, A. P. 1985. *Biochim. Biophys. Acta* 817:61–66

120. Wilson, J. E. 1980. *Curr. Top. Cell. Regul.* 16:1–44

121. Bessman, S. P., Geiger, P. J. 1980. *Curr. Top. Cell. Regul.* 16:55–86

122. Ovadi, J., Salerno, C., Keleti, T., Fasella, P. 1978. *Eur. J. Biochem.* 90:499–503

123. Batke, J., Tompa, P. 1986. In *Dynamics of Biochemical Systems*, ed. S. Danjanovich, T. Keleti, L. Tron. In press
124. Patthy, L., Vas, M. 1978. *Nature* 276:94–95
125. Salerno, C., Ovadi, J. 1982. *FEBS Lett.* 138:270–72
126. Ashmarina, L. I., Muronetz, V. I., Nagradova, N. K. 1985. *Eur. J. Biochem.* 149:67–72
127. Cseke, E., Szabolcsi, G. 1983. *Acta Biochim. Biophys. Acad. Sci. Hung.* 18:151–61
128. Srivastava, D. K., Bernhard, S. A., Langridge, R., McClarin, J. A. 1985. *Biochemistry* 24:629–35
129. Mowbray, J., Moses, V. 1976. *Eur. J. Biochem.* 60:25–36
130. Weitzman, P. J. D. 1986. In *Organization of Cell Metabolism*, ed. G. R. Welch, J. S. Clegg. Plenum. In press
131. Kurganov, B. I., Sugrobova, N. P., Milman, L. S. 1985. *J. Theor. Biol.* 116:509–26
132. Ureta, T. 1978. *Curr. Top. Cell. Regul.* 13:233–58
133. Opperdoes, F. R., Borst, P. 1977. *FEBS Lett.* 80:360–64
134. Aman, R. A., Kenyon, G. L., Wang, C. C. 1985. *J. Biol. Chem.* 260:6966–73
135. Hammond, D. J., Aman, R. A., Wang, C. C. 1985. *J. Biol. Chem.* 260:15646–54
136. Dennis, D. T., Green, T. R. 1975. *Biochem. Biophys. Res. Commun.* 64:970–75
137. DeLuca, M., Kricka, L. J. 1983. *Arch. Biochem. Biophys.* 226:285–91
138. Green, D. E., Loomis, W. F., Auerbach, V. H. 1948. *J. Biol. Chem.* 172:389–403
139. Srere, P. A. 1985. See Ref. 6, pp. 1–61
140. McKinley, M. P., Trelease, R. N. 1980. *Comp. Biochem. Physiol. B* 67:27–32
141. Von Glutz, G., Walter, P. 1975. *Eur. J. Biochem.* 60:147–52
142. Lopes-Cardoro, M., Klazingor, W., van den Bergh, S. G. 1978. *Eur. J. Biochem.* 83:635–40
143. Srere, P. A. 1972. In *Gluconeogenesis*, ed. R. W. Hanson, W. A. Mehlman, pp. 79–91. New York: Wiley
144. Srere, P. A., Mattiasson, B., Mosbach, K. 1973. *Proc. Natl. Acad. Sci. USA* 70:2534–38
145. Matlib, M. A., Shannon, W. A. Jr., Srere, P. A. 1977. *Arch. Biochem. Biophys.* 178:396–407
146. Halper, L. A., Srere, P. A. 1977. *Arch. Biochem. Biophys.* 184:529–34
147. Sumegi, B., Gyocsi, L., Alkonyi, I. 1980. *Biochim. Biophys. Acta* 616:158–66
148. Beeckmans, S., Kanarek, L. 1981. *Eur. J. Biochem.* 117:527–35
149. Porpaczy, Z., Sumegi, B., Alkonyi, I. 1983. *Biochim. Biophys. Acta* 749:172–79
150. Tyiska, R. L., Williams, J. S., Brent, L. G., Hudson, A. P., Clark, B. J., et al. 1986. *NATO Workshop: Organization of Cell Metabolism.* In press
151. Porpaczy, Z., Sumegi, B., Alkonyi, I. 1986. Submitted for publication
152. Sumegi, B., Gilbert, H. F., Srere, P. A. 1984. *J. Biol. Chem.* 259:15040–45
153. Datta, A., Merz, J. M., Spivey, H. O. 1985. *J. Biol. Chem.* 260:15008–12
154. D'Souza, S. F., Srere, P. A. 1983. *Biochim. Biophys. Acta* 724:40–51
155. D'Souza, S. F., Srere, P. A. 1983. *J. Biol. Chem.* 258:4706–9
156. Moore, G. E., Gadol, S. M., Robinson, J. B. Jr., Srere, P. A. 1984. *Biochem. Biophys. Res. Commun.* 121:612–18
157. Hatefi, Y. 1985. *Ann. Rev. Biochem.* 54:1015–69
158. Hackenbrock, C. R. 1981. *Trends Biochem. Sci.* 6:151–54
159. Hochman, J., Ferguson-Miller, S., Schindler, M. 1985. *Biochemistry* 25:2509–16
160. Berry, E. A., Trumpower, B. L. 1985. *J. Biol. Chem.* 260:2458–67
161. Kell, D. B., Westerhoff, H. V. 1985. See Ref. 6, pp. 63–139
162. Robinson, J. B. Jr., Srere, P. A. 1985. *J. Biol. Chem.* 260:10800–5
163. Kleinkauf, H., von Dohren, H. 1983. *Trends Biochem. Sci.* 8:281–83
164. Zocher, R., Nihira, T., Paul, E., Madry, N., Peeters, H., et al. 1986. *Biochemistry* 25:550–53
165. Brown, R. M. Jr., Montezinos, D. 1976. *Proc. Natl. Acad. Sci. USA* 73:143–47
166. Harson, M. M., Conder, M. J., Lord, J. M. 1983. *Planta* 157:143–49
167. Bieglmayer, C., Graf, J., Ruis, H. 1973. *Eur. J. Biochem.* 37:553–62
168. Luckner, M. 1980. See Ref. 9, pp. 255–62
169. Cutler, A. J., Hosel, W., Sternberg, M., Conn, E. E. 1981. *J. Biol. Chem.* 256:4253–58
170. McCarthy, A. D., Hardie, D. G. 1984. *Trends Biochem. Sci.* 9:60–63
171. McConkey, E. H. 1982. *Proc. Natl. Acad. Sci. USA* 79:3236–40
172. Fothergill-Gilmore, L. A. 1986. *Trends Biochem. Sci.* 11:47–51
173. Mastro, A., Keith, A. D. 1984. *J. Cell Biol.* 99:180S–187S
174. Jacobson, K., Wojcieszyn, J. 1984. *Proc. Natl. Acad. Sci. USA* 81:6747–51

175. Lang, I., Scholz, M., Peters, R. 1986. *J. Cell Biol.* 102:1183–90
176. Zalokar, M. 1960. *Exp. Cell Res.* 19:114–32
177. Kempner, E. S., Miller, J. H. 1968. *Exp. Cell Res.* 51:141–49
178. Brass, J. M., Higgins, C. F., Foley, M., Rugman, P. A., Birmingham, J., Garland, P. B. 1986. *J. Bacteriol.* 165:787–94
179. Hobot, J. A., Carleman, E., Villiger, W., Kellenberger, E. 1984. *J. Bacteriol.* 160:143–52
180. McCloskey, M. A., Poo, M. 1986. *J. Cell Biol.* 102:88–96
181. Kennedy, F. G., Jones, D. P. 1986. *Am. J. Physiol.* 250:C374–C383
182. Aw, T. Y., Jones, D. P. 1985. *Am. J. Physiol.* 249:C385–C392
183. Horowitz, S. B., Miller, D. S. 1984. *J. Cell Biol.* 99:172s–179s
184. Mollenhauer, H. H., Morre, D. J. 1978. *Sub–Cell. Biochem.* 5:327–59
185. Minton, A. P. 1981. *Biopolymers* 20:2093–2120
186. Wolosewick, J. J., Porter, K. R. 1979. *J. Cell Biol.* 82:114–39

Ann. Rev. Biochem. 1987. 56:125–58

AMINOACYL tRNA SYNTHETASES: GENERAL SCHEME OF STRUCTURE-FUNCTION RELATIONSHIPS IN THE POLYPEPTIDES AND RECOGNITION OF TRANSFER RNAs

Paul Schimmel

Department of Biology, Massachusetts Institute of Technology, Cambridge, Massachusetts 02139

CONTENTS

125

0066-4154/87/0701-0125$02.00

PERSPECTIVES AND SUMMARY

Background

Aminoacyl tRNA synthetases are canonical enzymes found in all life forms (1–5). They coevolved with the rules of the genetic code. Coding is based on the aminoacylation reaction in which specific amino acids are attached to tRNAs that have triplet anticodons cognate to the attached amino acids.

Early research concentrated on characterization of the 20 enzymes—one for each amino acid. Each enzyme catalyzes the esterification of its amino acid to a hydroxyl group at the 3'-end of its cognate transfer RNA. Most commonly this is accomplished through the synthesis of an aminoacyl adenylate (condensation of amino acid with ATP), followed by reaction of the enzyme-bound adenylate with tRNA.

The early work was highlighted by the discovery of an editing activity associated with isoleucine tRNA synthetase. This activity was shown to be manifested toward valyl adenylate bound to the isoleucine enzyme. Hydrolysis of valyl adenylate is induced by reaction of the complex with $tRNA^{Ile}$(6). A subsequent discovery showed that a misacylated tRNA(Val-$tRNA^{Ile}$) is a substrate for editing (7) and that, further, an editing deacylase site is general to this class of enzymes (8, 9). Editing reactions were then studied in depth (5, 10–12).

The early studies also established a paradox: in spite of the common catalytic features of these enzymes, the proteins are widely diverse in their subunit sizes and quaternary structures. While mechanisms for generating diversity (such as partial or full gene duplications followed by divergence of duplicated sequences) are easy to visualize, the selective pressure for advancing this process is difficult to rationalize. The paradox was reinforced because the particular quaternary structure and subunit size of at least some of the enzymes is roughly preserved throughout different organisms.

This is the background out of which research of the past seven years has emerged. Gene cloning, DNA sequencing, and more advanced methods of peptide sequencing have provided a technological base for the determination of a large number of primary structures of aminoacyl tRNA synthetases. At the same time, manipulation and dissection of synthetase genes and polypeptides have delineated structural and functional motifs within the proteins.

Concurrent with the research on the characteristics and mechanisms of the enzymes, the question of RNA recognition—the critical process on which hangs the basis of the genetic code—has been the central point of much of the interest in the enzymes. This problem has proved especially difficult, as the most simple and direct experiments failed to give a clear mechanism. Much less is known about protein-RNA interactions than about protein-DNA sequence recognition (13). The globular, folded structure of tRNA and of other RNA molecules adds a significant complication. As a consequence of folding, the potential is far greater for interactions that involve elements of the nucleic acid that are distant in the primary structure. In contrast, DNA-protein systems studied to date are relatively straightforward in that a limited number of consecutive base pairs generally define the recognition site in the nucleic acid.

The use of photo-crosslinking, nuclease digestion, and tritium labeling of synthetase-tRNA complexes established an operational model for the interaction (5, 14, 15). This model explained much of the aforementioned data by postulating that the enzymes bind along the inside of the three-dimensional L-shaped tRNA structure (16). It has been a useful, heuristic model, especially in view of the lack of rigorous structural information on a complex.

A General Scheme for the Structural Organization of Aminoacyl tRNA Synthetases

Notwithstanding the size and quaternary structure polymorphism of these enzymes, I review here the evidence for a common theme that runs through the organization of the structures of the entire class of enzymes. This is an arrangement of functional units along the amino acid sequence such that the synthesis of aminoacyl adenylate is determined by sequences located within the amino terminal halves of the proteins; RNA recognition is determined by these sequences and by those that are located on the carboxyl terminal side of sequences that are required for adenylate synthesis. This order along the sequence—adenylate synthesis followed by RNA recognition—is a recurrent theme in the enzymes, in spite of their many differences in primary and quaternary structures. I refer to the catalytic part of the structure as the core enzyme. To the core enzyme are joined sequences that are dispensable for catalysis. These sequences may be fused to the amino terminal or carboxyl terminal side of the core enzyme. The role of these sequences, which vary in size from enzyme to enzyme, is probably idiosyncratic. I suggest that, however, these sequences serve other biological functions such as regulation of expression or assembly of complexes (in eukaryotes). This means that the size and quaternary structure polymorphism of the enzymes reflects, at least in part, the different additional roles that are intrinsic to some of the enzymes. I note also that, in addition to the joining of sequences to the amino and carboxyl terminal side of the core enzymes, some of the extra sequences result from insertions into the catalytic domains. These sequences may be expected to protrude from the structure so as not to disturb the disposition of catalytic residues.

The sequences that are dispensable for catalysis have to be integrated with the core enzyme. This means that the core activity is not necessarily independent of the dispensable parts of the protein (17, 18). While the catalytic activity clearly resides with the core enzyme, that activity can be perturbed or modified by deletions or mutations in the dispensable sequences (17–19). This provides selective pressure to retain those sequences that are extraneous to the core enzyme.

Evidence is limited for a biological role for the sequences that are dispensable in the sense described above. There are examples of defined properties of certain synthetases (such as assembly of the quaternary structure) that are ascribable to these sequences. But proof that a biological function (for example, autoregulation of expression, of which there is evidence in some cases) requires dispensable polypeptide determinants has not yet been obtained.

Idiosyncratic Features of RNA Recognition

The study of RNA recognition has advanced through the manipulation of tRNA sequences and the testing of altered sequences as substrates for the relevant enzymes. These studies have reinforced and expanded on earlier ideas that recognition of tRNAs is as seemingly idiosyncratic as the sizes and quaternary structures of the various synthetases. The major advance has been in defining more fully the complexity of the problem and, at the same time, establishing more firmly that the identity of a tRNA extends beyond its anticodon.

GENERAL ORGANIZATION OF SEQUENCES AND STRUCTURES

Established Sequences

Table 1 lists 22 aminoacyl tRNA synthetases that have been sequenced. These sequences include examples from three bacterial species and from the yeast *Saccharomyces cerevisiae*. While bacterial organisms typically have just one aminoacyl tRNA synthetase for each amino acid, in yeast there is a cytoplasmic and a mitochondrial form of each aminoacyl tRNA synthetase. Both are nuclear encoded. (The tRNA substrates for the two cellular compartments, however, are encoded separately by the nuclear and mitochondrial genome.) The threonine enzyme is the first example in which both enzymes have been sequenced and shown to be encoded by separate nuclear genes (20, 21). The mitochondrial and cytoplasmic histidine tRNA synthetases, in contrast, are encoded by a single nuclear gene (22). The coding sequences for the cytoplasmic and mitochondrial histidine enzymes have 520 codons in common. The sequence for the mitochondrial enzyme has an additional 20 codons at its amino terminus (22).

In Table 1 there is at least one sequence for 12 of the enzymes. There are examples of all characterized quaternary structures of these enzymes: alpha,

Table 1   Completed sequences of aminoacyl tRNA synthetases

Enzyme	Organism	Quaternary structure	Size of polypeptide[a] (number of amino acids)	Reference
Alanine	E. coli	alpha-4	875	28, 29
Aspartate	yeast cytoplasmic	alpha-2	557	154, 155
Glycine	E. coli	alpha-2,beta-2	303 (alpha)	71
			689 (beta)	
Glutamine	E. coli	alpha	551	156
	yeast[b]	unknown	809	footnote c
Glutamic	E. coli	alpha	471	46
Histidine	E. coli	alpha-2	424	157
	yeast cytoplasmic	unknown	526	22
	yeast mitochondria	unknown	546	22
Isoleucine	E. coli	alpha	939	24
Methionine	E. coli	alpha-2	677	44
	yeast cytoplasmic	alpha-2	751	45
Threonine	E. coli	alpha-2	642	158
	yeast cytoplasmic	unknown	734	20
	yeast mitochondria	unknown	462	21
Phenylalanine	E. coli	alpha-2,beta-2	307 (alpha)	80, 159
			795 (beta)	
Tryptophan	E. coli	alpha-2	334	160
	B. stearothermophilus	alpha-2	327	23
	yeast mitochondria	alpha-2	374	161
Tyrosine	B. stearothermophilus	alpha-2	419	25
	B. caldotenax	alpha-2	419	26
	E. coli	alpha-2	424	162
	yeast mitochondria	unknown	492	footnote d

[a] The size is based on the nascent polypeptide that has the amino terminal methionine.
[b] The yeast S. cerevisiae.
[c] S. W. Ludmerer, P. Schimmel, in preparation.
[d] J. Hill, A. Tzagoloff, personal communication.

alpha-2, alpha-4, and alpha-2,beta-2. Note that the sizes of the polypeptides range from 327 [*Bacillus stearothermophilus* tryptophan enzyme (23)] to 939 [*Escherichia coli* isoleucine enzyme (24)] amino acids. Neither the long isoleucine tRNA synthetase polypeptide nor any of the others contains repeated sequences of any significance. It is unlikely, therefore, that sequence duplications are a part of the recent evolutionary history of these proteins.

Table 2 summarizes the extent to which sequences are conserved between the same enzymes from bacteria and yeast. The *B. stearothermophilus* and *Bacillus caldotenax* (both thermophilic strains) tyrosine enzymes differ by just four amino acids (25, 26). Between yeast and bacterial comparisons other than the histidine enzymes, the homologies range from 20% to 50%, in the portions that can be aligned. The homology between yeast and *E. coli* His-tRNA synthetase is minimal, and the alignment is obtained by placing many gaps in the *E. coli* His-tRNA synthetase sequence (22). Note that it is

Table 2   Amino acid sequence homologies between yeast and bacterial aminoacyl tRNA synthetases[a]

Enzyme	Source	Region (amino acids)	% Identity	Comments
His	S. cerevisiae	1–546	28.0	No region of strong
	E. coli	1–424		homology is shared
				the two enzymes
Gln	S. cerevisiae	254–732	47.9	Yeast enzyme contains
	E. coli	30–498		an N-terminal exten-
	S. cerevisiae	225–786	40.0	sion of 224 amino
	E. coli	1–551		acids
Met	S. cerevisiae	192–594	30.0	Yeast enzyme contains
	E. coli	1–401		an N-terminal exten-
	S. cerevisiae	192–751	20.0	sion of 191 amino
	E. coli	1–677		acids
Thr	S. cerevisiae (mitochondrial)	46–352	44.2	Yeast enzyme missing region corresponding
	E. coli	251–544		to the N-terminal 20
	S. cerevisiae (mitochondrial)	1–462	38.0	amino acids of the *E. coli* enzyme
	E. coli	251–642		
	S. cerevisiae	337–633	47.6	Yeast enzyme contains
	E. coli	251–544		an N-terminal exten-
	S. cerevisiae	50–725	38.0	sion of 49 amino
	E. coli	1–642		acids

Table 2 *(Continued)*

Enzyme	Source	Region (amino acids)	% Identity	Comments
	S. cerevisiae	347–633	51.5	Cytoplasmic enzyme
	S. cerevisiae (mitochondrial)	56–352		contains an N-terminal extension of 291 amino acids relative to the mitochondrial enzyme
	S. cerevisiae	292–725	42.0	
	S. cerevisiae (mitochondrial)	1–462		
Trp	*S. cerevisiae*	36–298	42.3	Yeast enzyme contains an N-terminal extension of 31 amino acids
	E. coli	5–251		
	S. cerevisiae (mitochondrial)	36–374	37.0	
	E. coli	5–334		
	S. cerevisiae (mitochondrial)	35–298	41.4	Yeast enzyme contains an N-terminal extension of 32 amino acids
	B. stearothermophilus	3–244		
	S. cerevisiae (mitochondrial)	33–374	39.0	
	B. stearothermophilus	1–327		
Tyr	*S. cerevisiae* (mitochondrial)	52–376	46.0	Yeast enzyme contains an N-terminal extension of 51 amino acids
	E. coli	1–307		
	S. cerevisiae (mitochondrial)	52–493	33.8	
	E. coli	1–423		
	S. cerevisiae	54–376	41.6	Yeast enzyme contains an N-terminal extension of 53 amino acids
	B. stearothermophilus	1–303		
	S. cerevisiae (mitochondrial)	54–493	34.6	
	B. stearothermophilus	1–419		
	S. cerevisiae (mitochondrial)	54–376	41.2	Yeast enzyme contains an N-terminal extension of 53 amino acids
	B. caldotenax	1–303		
Tyr	*S. cerevisiae* (mitochondrial)	54–493	33.9	
	B. caldotenax	1–419		

[a]References for specific sequences are given in Table 1. Adapted from Ludmerer, S. W. 1986. PhD thesis, MIT.

not known whether an enzyme from a bacterial organism will recognize in vivo specifically just its cognate tRNA in yeast, or vice versa. Some (probably a limited number) of the sequence differences between the enzymes for the same amino acid may be the constraints of specific recognition of tRNA sequences unique to a given organism.

The yeast cytoplasmic enzymes are larger than their bacterial counterparts. In the case of glutamine tRNA synthetase, the difference is 309 amino acids that are mostly due to an extension at the amino terminus (S. W. Ludmerer, P. Schimmel, in preparation). The large size difference of two synthetases for the same amino acid emphasizes the presence of sequences that are not associated with the aminoacylation function of these enzymes.

Bits and Pieces of an Aminoacyl tRNA Synthetase

The functional organization of a synthetase polypeptide has been studied by dissection and manipulation of the cognate gene. A large number of gene fragments have been created that encode pieces of the *E. coli* alanine tRNA synthetase polypeptide (17, 19). This enzyme is a tetramer of identical polypeptides of 875 amino acids (28, 29).

Figure 1 tabulates 18 partial deletions of *alaS* that have been investigated. These include a large number of C-terminal deletions as well as some internal deletions. In most cases, the deletions give rise to a product that is detectable in maxicell extracts. These products have been investigated for specific functional activities in vitro and in vivo (17, 19).

Removal of 490 amino acids from the C-terminus results in a stable protein fragment that is monomeric and has the full adenylate synthesis activity of the native protein. This establishes that adenylate synthesis activity does not depend on tetramer formation and that such activity is associated with each of the separate polypeptides. It also shows that the site for adenylate synthesis is positioned toward the amino terminal part of the protein.

Extension of the chain from 385 to 461 amino acids gives rise to aminoacylation activity, without assembly of the chain into a tetramer. Unlike the alanyl adenylate synthesis activity, the specific aminoacylation activity is reduced substantially from that of the native protein. When the gene fragment is present on a multicopy plasmid, however, the activity is sufficient to sustain growth of *E. coli* under circumstances where this is the only source of alanine tRNA synthetase activity (17, 30). This shows that the aminoacylation activity, while reduced, clearly resides with the amino terminal fragment. This activity must preserve the specificity of aminoacylation because significant misacylations should lead to cell death.

By using a similar approach, sequences important for tetramer formation have been located to amino acids 699 to 808 (19). Just as catalysis occurs with fragments that lack the tetramer formation capacity, the assembly of oligom-

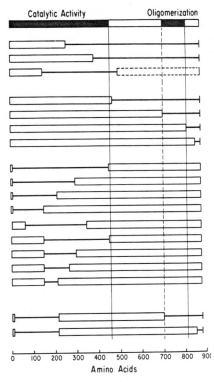

Figure 1 Gene deletions that map functional domains in an aminoacyl tRNA synthetase. The coding region of the 875 codons
of *E. coli alaS* is depicted. The gene for the
wild-type enzyme is depicted on top. Eigh-
teen different deletions are depicted below.
The deleted regions are designated by thin
lines and the regions left intact are shown as
wide bars. Regions of the polypeptide that
encode the catalytic and oligomerization
functions are shown by shading. Within the
catalytic domain, amino acids 1 to 368 are
sufficient to generate the full aminoacyl
adenylate synthesis activity and amino acids
1 to 461 encode a fragment that has
aminoacylation activity sufficient to sustain
cell growth when expressed from a multicopy plasmid. It has been shown that amino
acids 369 to 461 are especially critical for
specific binding of tRNA^{Ala} (48). Adapted
from Ref. 17.

ers does not require an intact or functional catalytic domain. These observations suggest that, to a zero-order approximation, functional domains are
arranged linearly along the sequence to give a modular structure to the
protein. This structure includes a core enzyme to which dispensable parts
have been fused. The core enzyme itself can be subdivided into a part that
executes adenylate synthesis and a part that has to be joined onto the domain
for adenylate synthesis in order to achieve aminoacylation.

Further studies with point mutants, and with enzyme fragments that interact
with point mutants, suggest that dispensable parts are coupled through the
three-dimensional structure to indispensable parts of the protein. Examples
are the *alaS4* and *alaS5* mutations in the dispensable C-terminal part of the
protein that affect the catalytic activity (18). These are discussed further in a
subsequent section.

These investigations have also established the importance of using null
alleles in the chromosome for testing for the function of an enzyme fragment.
The general system for creating and utilizing null alleles in essential proteins
is described by Jasin & Schimmel (30). Such alleles enable a rigorous test

of whether a fragment encoded by a multicopy plasmid can serve as the only source of enzyme activity. Strains with chromosomal point mutants, such as temperature-sensitive alleles, are dangerous to use as test strains. This is because the defective mutant protein may interact physically with a noncatalytic fragment that is introduced (17, 18). That interaction may give rise to a partial rejuvenation of the activity in the mutant protein that is encoded by the chromosome. The danger is that this activity is then interpreted as derived from the fragment.

Domain for Adenylate Synthesis is Located in the Amino Terminal Halves of the Enzymes

STRUCTURAL DATA   The pattern along the sequence of functional domains for alanine tRNA synthetase has been found, more or less, for several of the other enzymes. The ATP interaction site is well understood because of the high-resolution X-ray crystallographic information on the amino terminal parts of the *B. stearothermophilus* tyrosine and the *E. coli* methionine enzyme (31–35). Crystals of yeast aspartate (36–40) and *B. stearothermophilus* tryptophan tRNA synthetases (41, 42) have also been obtained, but these have not yet yielded high-resolution information.

The tyrosine and methionine enzymes form a characteristic mononucleotide binding fold that is comprised of beta strands and associated alpha helixes (31, 35). This alpha/beta mononucleotide binding fold starts near the amino terminus of each protein and extends over more than 200 amino acids. Although the two enzymes have little sequence homology over this region, there is considerable topological equivalence (35). The structural information has been the framework in which Fersht and coworkers have provided a detailed accounting of the energetics of binding and catalysis at the adenylate site of the tyrosine enzyme [(43) and see below].

A SIGNATURE SEQUENCE FOR SOME OF THE ENZYMES IS LOCATED IN THE AMINO TERMINAL PART   Webster et al discovered that several of the enzymes have a short, albeit compelling sequence homology in their amino terminal sections (24). This homology is now viewed as a signature sequence for at least some of the aminoacyl tRNA synthetases. This view resulted from the determination of the primary structure of the 939-amino-acid isoleucine tRNA synthetase. In a sequence of 11 consecutive amino acids, there are 10 identities and one conservative substitution between the *E. coli* isoleucine and methionine enzymes. This alignment of protein sequence is of amino acids 57 to 67 of the isoleucine enzyme with amino acids 14 to 24 of the methionine tRNA synthetase. Significantly, this part of the methionine enzyme's structure is especially close in conformation to that of the tyrosine enzyme. The alpha carbons of amino acids 14 to 24 of methionine tRNA synthetase have a root-mean-square deviation of 1.8 angstroms when placed on residues 38 to 52 of the *B. stearothermophilus* tyrosine enzyme (35). The latter two enzymes have only four amino acid identities, however, in this stretch of 11 amino acids.

The isoleucine/methionine enzyme homology is the strongest between two synthetases for different amino acids (24). Furthermore, it validates the weaker homologies among other enzymes in this region. A summary is given in Table 3. This shows an alignment of the isoleucine, methionine, glutamine, glutamic, and tyrosine enzymes. In some instances, sequences are available for the same enzyme from more than one source.

A characteristic feature is the HIGH sequence that is present in the *E. coli* methionine, isoleucine, glutamine, and tyrosine enzymes, as well as in *B. stearothermophilus* tyrosine and yeast glutamine enzyme sequences. (In some cases the I is replaced by an L.) In *B. stearothermophilus* tyrosine tRNA synthetase, His48 binds to ATP and His45 is believed to bind to the gamma phosphoryl group of ATP in the transition state. Substitution of Asn for His48 in the *B. stearothermophilus* tyrosine tRNA synthetase gives an enzyme that is indistinguishable in its kinetics from those of the wild-type protein (26). This argues that Asn's way of binding to ATP is equivalent to that of His. In

Table 3   A signature sequence for some aminoacyl tRNA synthetases[a]

Sequence	Enzyme
[254] T R F P P E P N G Y L H I G H	*S. cerevisiae* Gln-tRNA synthetase
[30] T R F P P E P N G Y L H I G H	*E. coli* Gln-tRNA synthetase
[5] T R F A P S P T G Y L H V G G	*E. coli* Glu-tRNA synthetase
[46] P P F A T G T P H Y G H	*S. cerevisiae* Ile-tRNA synthetase
[56] P P Y A N G S I H I G H	*E. coli* Ile-tRNA synthetase
[205] P Y V N N V P H L G N	*S. cerevisiae* Met-tRNA synthetase
[15] P Y A N G S I H L G H	*E. coli* Met-tRNA synthetase
[88] L Y C G V D P T A Q S L H L G N L V P L	*S. cerevisiae* mitochondrial Tyr-tRNA synthetase
[33] L Y C G F D P T A D S L H I G N L A A I	*B. caldotenax* Tyr-tRNA synthetase
[33] L Y C G F D P T A D S L H I G H L A T I	*B. stearothermophilus* Tyr-tRNA synthetase
[36] L Y C G F D P T A D S L H L G H L V P L	*E. coli* Tyr-tRNA synthetase

[a]The signature sequence was discovered by Webster et al (24). References to each of the listed sequences are given in Table 2. The location of the signature sequence of *S. cerevisiae* Ile-tRNA synthetase is tentative and is based on unpublished work of Z. Altboum, N. Clarke, S. W. Ludmerer, and P. Schimmel. Table is adapted from Ludmerer, S. W. 1986. PhD thesis, MIT.

this connection, note that the analogous site is a His in *E. coli* Met-tRNA synthetase (44) and an Asn in its yeast counterpart (45).

The analogous site in the *E. coli* glutamic tRNA synthetase, however, is a Gly, based on the alignment shown in Table 3 (46). This illustrates a deeper problem that arises in the searches to align sequences of synthetases. For the most part the sequences show few if any homologies amongst themselves. A number of the enzymes do not have the characteristic signature region of 10 to 15 amino acids that is shown in Table 3. Even when there is an obvious section of homology, however, significant departures from a consensus sequence nonetheless occur. Determination of how these departures are accommodated can only come from a detailed three-dimensional structural analysis of each of the proteins.

The 11 enzymes in Table 3 have polypeptides that range in size from 419 to 939 amino acids. In spite of this twofold size range, each has the signature sequence near the amino terminus. This is strong evidence for an amino terminal location for a nucleotide fold in the isoleucine, glutamine, and glutamic tRNA synthetases that is structurally like that found in the tyrosine and methionine tRNA synthetases. Note that, in the case of the alanine tRNA synthetase, there is no sequence that can be definitively assigned to the region of Table 3. In spite of this, the gene deletion analysis of this synthetase shows that the adenylate site is located well within the amino terminal half of the protein (17, 19).

Some Sites Critical for RNA Recognition are Located on the C-Terminal Side of the Domain for Adenylate Synthesis

The crystals of the tyrosine and methionine enzymes have not elucidated the tRNA binding site. There are crystals of yeast aspartate tRNA synthetase, free and complexed with its cognate tRNA, and these may in time yield detailed information (36–40). In the tyrosine enzyme, the C-terminal 99 amino acids are disordered and will not yield to structural analysis (31, 34, 35). Deletion of these 99 amino acids gives an enzyme fragment that executes tyrosyl adenylate synthesis but not aminoacylation or binding of tRNA (47). In this enzyme, therefore, some determinants critical for binding tRNA are located on the C-terminal side of the domain for adenylate synthesis.

Analysis of an extensive set of fragments of *E. coli* alanine tRNA synthetase shows that sequences essential for tRNA binding are located just beyond the adenylate synthesis domain. Especially significant are amino acids 368 to 385. A gene fragment that encodes amino acids 1 to 385 yields a protein that binds specifically to tRNA[Ala]. An amino terminal fragment that extends to Arg368 does not detectably bind, however (48). Both proteins are unimpaired in their adenylate synthesis activity. This suggests that they retain their native conformations. The results with the alanine enzyme raise the possibility that a region small relative to the size of the wild-type enzyme may be involved in RNA recognition.

Direct detection of regions in contact with bound tRNA has been attempted by crosslinking methods. The extensive early studies of photo-crosslinked enzyme-tRNA complexes delineated regions on the tRNAs that are close to or in contact with bound enzyme (5, 14). The recent investigations have coupled tRNAs to synthetases by one of two general chemical procedures. Most commonly, the 3'-end of a tRNA has been oxidized with periodate to give aldehydes that are reactive with lysine side chains (49, 50). Alternatively, Schulman has pioneered the use of a lysine-directed cleavable crosslinker that is joined to cytidines at various points in the tRNA structure (51–54).

A summary is given in Table 4.[1] By use of the lysine-directed cleavable

Table 5 Some of the characterized mutations in aminoacyl tRNA synthetases

Enzyme	Mutation	Phenotype/comments/reference
E. coli alanine alpha-4 tetramer	Gly674 → Asp (alaS4)	Temperature-sensitive lethal (75); mutation dissociates tetramer; little effect on adenylate synthesis activity; small changes in K_m's for substrates (alanine insignificant effect, ATP increased less than 5-fold, tRNA increased about 1-fold); k_{cat} for aminoacylation diminished by 20-fold (18). All measurements at permissive temperature of 30°C.
	Gly677 → Asp (alaS5)	Temperature-sensitive lethal; behaves similar to alaS4 mutant, main difference is that there is no effect on K_m for ATP (18).
	Ala409 → Val	Increased affinity for tRNA^{Ala}; elevation in k_{cat} for aminoacylation (77; L. Regan, L. Buxbaum, P. Schimmel, in preparation).
E. coli glycine alpha-2,beta-2 tetramer	Beta chain mutants Cys98 → Ala Cys395 → Ala Cys450 → Ala	There are 3 cysteines in beta subunit at positions 98, 395, and 450. Enzyme is inactivated by reaction of NEM with Cys395. Each Cys→Ala mutant is active in vitro and in vivo, as is the triple mutant (70).
	Cys395 → Gln	Aminoacylation activity is reduced to less than 10% of that of wild-type enzyme (70). Interpretation is that NEM labeling of Cys393 causes inactivation by a steric or conformational effect.
	Five subunit fusions that join the C-terminus of the beta chain to the amino terminus of the alpha chain, through a linker of six amino acids.	Fusions are stable in all extracts and are active in vivo and in vitro. Oligomerization of fusion chains occurs to give heterogenous aggregates. Because creation and manipulation of the linker region does not block activity, results imply that C-terminus of alpha chain and N-terminus of beta chain are not directly part of active site (83, 84).

[1]See page 140. Tables 4 and 5 are out of sequence.

Table 5   *(Continued)*

Enzyme	Mutation	Phenotype/comments/reference
B. stearothermophilus tyrosine alpha-2 dimer	Thr51 → Ala or Cys	Thr51 hydroxyl has a long H-bond contact with O-1 of ATP ribose. Ala substitution has small effect on K_m for ATP and on k_{cat} for adenylate synthesis and aminoacylation. Cys substitution significantly lowers K_m for ATP in aminoacylation (164, 165).
	Thr51 → Pro	16-fold decreases in K_m for ATP in adenylate synthesis (164); investigation of a series of double mutants suggests that the Pro substitution increases strength of interaction of His48 with alpha-phosphate of ATP (73). All three substitutions (Ala, Cys, Pro) are believed to improve transition state for formation of tyrosyl adenyate (72).
	one-by-one replacement of conserved (between *E. coli* and *B. stearothermophilus* enzyme) lysines and arginines with asparagines and glutamines, respectively.	K_m for tRNA is elevated 20-fold or more by substitution of any one of the following: Arg207, 368, 371, 407, 408; Lys208, 410, 411. k_{cat} for aminoacylation is diminished over 100-fold, with no effect on K_m for tRNA, by replacement of Lys151. All nine mutants have adenylate synthesis activity that is comparable to wild-type enzyme (55).
	Cys35 → Ser or Gly	Enzyme is active with either replacement. Modest elevation in K_m for ATP (57, 166).
	Phe164 → Asp	Phe164 is at subunit interface in wild-type enzyme. Dimer dissociates when pH is raised to ionize Asp side chain of mutant enzyme. Monomer has low activity. Dimer with Asp164 substitution is fully active (76).
	Thr40 → Ala and His45 → Gly	The double mutant lowers rate of adenylate synthesis by 3×10^5 and has small effect on K_m for tyrosine or ATP. Interpretation is that Thr40 and His45 contribute nothing to binding unreacted ATP, but make major contribution to binding of gamma-phosphate of transition state (74).

crosslinker, four lysines on *E. coli* methionine tRNA synthetase have been identified (54). All of the sites identified in this study are on the C-terminal side of the nucleotide-binding site of this enzyme (see above). One of these, Lys640, is known to result from crosslinking to the 5'-terminal C of tRNA$_i^{Met}$. This region of the enzyme can be removed by a gene deletion or by proteolysis, however, without substantial perturbation of the aminoacylation function of the enzyme.

The same system has been explored by the use of periodate-oxidized tRNA$_i^{Met}$. For this study an amino terminal tryptic fragment (approximately 550 amino acids) was used instead of the holoenzyme. Two sites are prominently labeled: Lys61 and Lys335 (49). The former is within the nucleotide binding fold and the latter is also. Because the 3'-end of bound tRNA must come close to the nucleotide site during catalysis, the labeling of a site in this region is plausible. Whether this particular site is especially significant is not clear, however.

The *E. coli* tyrosine enzyme is labeled by periodate-oxidized $tRNA^{Tyr}$ at three lysines (Lys229, 234, and 237) that are on the C-terminal side of the adenylate synthesis domain (50). This part of the enzyme is in the middle of a polypeptide segment that joins the end of the alpha/beta nucleotide fold to the first helix of a helical domain. It is perhaps noteworthy that Lys335 of the methionine enzyme is clustered with two other lysines to give a pattern in the sequence that has some resemblance to that around the crosslinked lysines on the *E. coli* tyrosine enzyme. There are sequences in *E. coli* isoleucine and tryptophan tRNA synthetases that can also be aligned with the crosslinking regions in the tyrosine and methionine enzymes (50).

The significance of these sequences is not yet certain. Replacement in *B. stearothermophilus* Tyr-tRNA synthetase of either Lys 229 or Lys 232 with asparagine results in reduction of adenylate synthesis activity (55). The effect of the mutations on binding to tRNA has not been assessed.

An Example where tRNA Binding Evidently Occurs Across Subunits and an Example where Binding is Entirely within a Single Polypeptide Chain

The dimeric *B. stearothermophilus* tyrosine tRNA synthetase shows half of the sites' reactivity. This raises the possibility that both subunits are required for formation of the active site (see also below). Carter et al have constructed heterodimers that are built from mutant subunits (56). One subunit is defective in tyrosyl adenylate synthesis and the other is defective in tRNA aminoacylation (for example, due to a weak interaction with tRNA). While the mutant homodimers are inactive, the heterodimers are active by virtue of binding of the tRNA across the subunits so that the subunit that donates a wild-type active adenylate site combines with the opposite subunit's contribution to the wild-type tRNA binding site to yield an active protein. More extensive analysis of mutants has confirmed and expanded this model (55, 56).

This is not a general model for how tRNAs interact with synthetases, however. Regan et al have investigated a set of amino terminal fragments of *E. coli* alanine tRNA synthetase (48). While the native enzyme and fragment 808N (encoded by an *alaS* gene fragment that extends to codon 808) are tetrameric, fragment 699N is a monomer. The three proteins bind to $tRNA^{Ala}$ with affinities that are the same within experimental accuracy. Regan et al have also established that binding of $tRNA^{Ala}$ does not induce oligomerization of a monomeric Ala-tRNA synthetase fragment. Further analyses have confirmed these observations (48).

To the variability of synthetase sequences, sizes, and quaternary structures, we can thus add the variability in their mode of interaction with tRNA. To determine sequences that are essential for recognition, there is an advantage, however, to a system that has all of the relevant sequences on a single polypeptide. This makes more straightforward the interpretation of mutations that affect binding, and the delineation of sequences that are relevant for binding.

CHARACTERIZED MUTATIONS IN RELATION TO STRUCTURE AND FUNCTION

The creation and analysis of point missense mutants in amino acyl tRNA synthetases have expanded considerably our ability to test specific hypotheses about conserved sequences, active site residues, and even the general structural organization of the proteins. Table 5 summarizes some of the mutants that have been investigated. Much effort has been directed at the *B. stearothermophilus* tyrosine enzyme, especially because of the structural model for the amino terminal nucleotide fold. This work, as mentioned, has expanded considerably our understanding of the energetics of binding and catalysis.

Table 4 Sites on aminoacyl tRNA synthetases that crosslink to tRNAs

Enzyme	tRNA site of coupling to enzyme	Sites labeled on protein	Reference
E. coli methionine (Tryptic amino terminal fragment of 550 amino acids)	3'-end of *E. coli* tRNA$_i^{Met}$ (periodate oxidized)	Lys61, Lys335	49
E. coli methionine (alpha-2 holoenzyme of 676 amino acids)	Lysine-directed cleavable cross-linker attached to 4 different C's located at 5'-end, 3'-terminal CCA, D-loop, and anticodon loop of tRNA$_i^{Met}$	Lys402, 439, 465, and 640 (coupled to 5'-terminal C)	53, 54
E. coli tyrosine (alpha-2 holoenzyme)	3'-end of *E. coli* tRNA^{Tyr} (periodate oxidized)	Lys234 is predominant site. Lys229, 237 are also labeled.	50
Yeast phenylalanine (alpha-2, beta-2 holoenzyme)	3'-end of yeast tRNA^{Phe} (periodate oxidized)	Specific Lys in the beta-subunit	167

I consider below a few examples that develop particular points about the design of aminoacyl tRNA synthetases.

Role of a Conserved Cys

In *B. stearothermophilus* tyrosine tRNA synthetase, Cys35 is hydrogen bonded to the 3'-hydroxyl of the ribose of bound tyrosyl adenylate. This cysteine is conserved in the *E. coli* and *B. caldotenax* tyrosine enzymes and the analogous position in the *E. coli* methionine enzyme is also a Cys. Replacement of Cys35 with Ser or Gly, in the *B. stearothermophilus* enzyme, does not block activity, although the mutant enzymes have reduced k_{cat}'s and increased K_m's for ATP (57, 58). The effects are not large (less than a factor of 10 for k_{cat}/K_m). This means that the Cys is not essential. Note that Cys35 flanks the signature sequence region shown in Table 3 and that, while it aligns with Cys12 of methionine tRNA synthetase, a cysteine is not found in several other enzymes that have an alignment in this region.

A Sulfhydryl Presumed Essential is not Required for Catalysis by an Aminoacyl tRNA Synthetase

Chemical modification of reactive sulfhydryl groups in aminoacyl tRNA synthetases commonly results in inactivation (59–66). These and other considerations have led to speculations about the role of a sulfhydryl in catalysis. For example, aminoacyl adenylates could react with thiols to give thioesters that, in turn, react with tRNA. Although aminoacyl adenylates are known to react with thiols in the course of gramacidin biosynthesis (67), there is no conclusive evidence for such a mechanism in the formation of aminoacyl tRNA. Evidence for a covalent enzyme-amino acid adduct has been presented in the case of the tryptophan enzyme, although a thioester linkage has not been established (68).

E. coli glycine tRNA synthetase is an alpha-2,beta-2 enzyme (69). This enzyme can be inactivated by treatment with the sulfhydryl reagent *N*-methyl maleimide (NEM). The NEM inactivation follows first order kinetics, and the target site is located on the beta subunit. The inactivation specifically blocks the tRNA-dependent step of aminoacylation and does not inhibit glycyl adenylate synthesis (65).

In the beta chain of the enzyme, there are cysteines located at positions 98, 395, and 450. Each of these has been separately changed to an alanine, and the triple mutant with three alanine substitutions has also been constructed. Each of these mutants is active in vitro and in vivo. This eliminates the possibility that any of these cysteines plays an essential role (70).

The target site for NEM inactivation is identified as Cys395. Because the Ala substitution has little effect on activity, the basis for inactivation by NEM is most likely a steric phenomenon. Support for this hypothesis comes from construction of a Cys395→Gln mutant. This mutant has severely reduced (more than 10-fold) catalytic activity (70).

The results with this system underscore the hazards of interpreting too freely the results of chemical modification or even site-directed mutagenesis experiments. The observed phenotype of a mutant may depend heavily on the amino acid substitution. Likewise, the sheer bulk of a bound chemical reagent can lead to stereochemical factors that severely affect activity.

In glycine tRNA synthetase there are five cysteines in the alpha subunit (71). None of these has been altered. The possibility that one of these is crucial to catalysis remains open.

Subtle Effects at the Active Site of Tyr-tRNA Synthetase

The crystal structure of the nucleotide binding part of the *B. stearothermophilus* tyrosine enzyme, and the three sequences of the enzyme that are available (from *E. coli, B. stearothermophilus, and B. caldotenax*), has invited extensive investigation of position 51. This position is occupied by Thr, Ala, and Pro in the *B. stearothermophilus, B. caldotenax, and E. coli* enzymes, respectively. This variation is in contrast with the relatively high overall degree of conservation between the three enzymes. More significantly, Thr51 in the crystal structure is proposed to form a hydrogen bond with the ribose ring oxygen. Many of the other polar side chains, which surround the bound tyrosyl adenylate, do not vary between the three enzymes.

The Thr51 hydrogen bond is viewed as nonoptimal, however, and capable, therefore, of improvement. Various substitutions have been made in the otherwise constant backbone of the *B. stearothermophilus* enzyme (72). The Thr51→Cys substitution is believed to improve the hydrogen bond to the ribose ring oxygen. This substitution uniformly lowers the free energy of all states in which the enzyme is bound to ATP. The dissociation constant for ATP in the adenylate synthesis reaction, for example, is lowered by 15-fold. The Ala51 and Pro51 substitutions lead to mixed effects, but the Pro51 replacement has the most dramatic overall consequences. These are manifested in a 25-fold increase in k_{cat}/K_m for ATP in adenylate synthesis. Introductions of other mutations have suggested that the effect of the Pro51 substitution is to alter the interaction of His48 with bound ATP (73).

Further studies have implicated Thr40 and His45 side chains as hydrogen bond donors to the gamma-phosphate of bound ATP in the transition state (74).

These investigations have probed deeply into the mechanisms and energies of enzyme-substrate interactions in the tyrosine tRNA synthetase system. The detailed interpretation of each mutant requires, in principle, the independent determination of the structure of each mutant protein. At the same time, the creation and investigation of mutants provide a heuristic approach to building molecular models of transition state complexes that are generally not deducible from an X-ray crystallographic analysis. The results demonstrate an extraordinary coupling of amino acid side chains within the three-dimensional structure. To a certain extent, the mutants call attention to interactions and effects that otherwise would escape notice in a more casual inspection of a structural model.

Mutations that Affect RNA Binding

Lysines and arginines have been substituted in *B. stearothermophilus* Tyr-tRNA synthetase, on the presumption that these groups are likely to interact with bound tRNA. Lysines were replaced with asparagines and arginines with glutamines. These investigations have identified nine positions in the sequence that span the region from Arg207 to Lys411. Each of the nine causes a sharp elevation in K_m for tRNA^{Tyr}. The locations of the nine sites suggest a model for how a tRNA associates with both subunits of the dimeric protein (55).

Coupling Between Dispensable and Indispensable Parts of an Aminoacyl tRNA Synthetase

The *alaS4* and *alaS5* mutations are temperature-sensitive lethal alleles of *E. coli* alanine tRNA synthetase (75). These alleles encode enzymes that have a low catalytic activity, even at the permissive temperature of 30°C (18). These mutations arise in the part of the protein that is dispensable for catalysis (Gly674→Asp and Gly677→Asp, respectively, for *alaS4* and *alaS5*). Each of these mutations dissociates the tetramer structure. [Note that the rate of adenylate synthesis and the affinity of the enzyme for tRNA are not dependent on the tetramer structure (see above).] These mutations are close to the region of the primary structure defined by deletion mutagenesis as essential for formation of the quaternary structure.

Inspection of kinetic parameters shows that the main effect of the mutations is on k_{cat} for aminoacylation, even though the mutations are outside the region required for the core catalytic activity. The K_m's for alanine, ATP, and tRNA are not greatly perturbed, nor is there much effect on the intrinsic adenylate synthesis activity (18). These observations suggest that the ligand interaction sites are not masked in the mutant enzyme. The sharp decrease in aminoacylation activity could reflect a dependence of the aminoacylation activity on the native quaternary structure. If this is true, then it is intriguing that there is not an effect of quaternary structure on the K_m's for the three substrates. A dependence of the intrinsic catalytic activity on quaternary structure would be an effective way to couple tightly the dispensable and indispensable parts of the structure.

These results can be compared with recent studies of the tyrosine enzyme, where the role of the dimer in relation to activity has also been explored (see below).

A Mutation that Causes pH-Dependent Dissociation of Tyrosine tRNA Synthetase

The phenyl side chain of Phe164 in *B. stearothermophilus* tyrosine tRNA synthetase interacts with its symmetrical partner in the dimeric native enzyme. The Phe164→Asp mutation leads to dissociation of the dimer at pH 7.8, where the carboxyl group is completely ionized. At pH 6, the carboxyl is protonated and the dimeric structure forms. This system facilitates a study of the dependence of activity on the quaternary structure (76).

The Asp164 dimer is essentially as active as the wild-type protein at pH 6. Dissociation of the dimer leads to a sharp diminution of both adenylate synthesis and aminoacylation activity. Dimerization is favored by addition of tyrosine and ATP, and independent binding measurements with tyrosine and tyrosyl adenylate show that these ligands have low affinity for the monomer (76). This is in contrast with the alanine tRNA synthetase system where there is no dependence of the adenylate synthesis activity on the quaternary structure (19, 28).

Selection of Amino Acid Replacements that Compensate for a Large Deletion in Ala-tRNA Synthetase

The amino terminal 461-amino-acid fragment of alanine tRNA synthetase has less than 0.5 percent of the aminoacylation activity of the wild-type protein. The affinity of this fragment for tRNA^{Ala}, under conditions of the tRNA binding assay, is reduced by 20-fold (48). The adenylate synthesis activity is indistinguishable from that of the wild-type enzyme. A selection has been designed to obtain point mutations that compensate for the large polypeptide deletion. This selection has yielded several replacements. One that has been characterized in some detail is an Ala409→Val substitution. This mutation results in a 5-fold enhancement of the k_{cat}/K_m for tRNA (77). Further studies suggest that this is, in part, due to an enhancement of tRNA binding to the fragment (L. Regan, L. Buxbaum, P. Schimmel, in preparation).

The magnitude of the enhancement of the activity that is obtained is limited by the selection. If a 5- to 10-fold increment in activity is sufficient to achieve cell viability in the selection, then that range of activity improvement will be included in and possibly predominate among the mutants that are obtained. The results show that the fragment has the capacity to be improved, and further mutagenesis might increase the activity even more than observed in the first experiments. While a capacity for improvement may be a general feature of many enzymes, in this instance it may also occur because the fragment is the prototypical alanine tRNA synthetase that is intrinsically capable of refinement.

Meaning of the Alpha-2,Beta-2 Structure

Glycine and phenylalanine tRNA synthetases in *E. coli* have alpha-2,beta-2 quaternary structures. In contrast, other characterized synthetases are monomeric or are oligomers of a single type of polypeptide chain (cf Table 1).

The genes for each of these proteins have been cloned from *E. coli*. In both instances the two coding regions are in tandem and are transcribed from a single promoter. The coding region for the smaller alpha subunit precedes that of the longer beta chain. In *glyS,* there are nine nucleotides between the stop of the alpha chain and the start of the beta subunit (79). There are 14 nucleotides in the spacer of *pheS* (80). Within the intersubunit coding region of each gene, there is a recognizable sequence for ribosome binding so that translation can reinitiate after the alpha chain's stop codon.

The rationale for the alpha-2,beta-2 structure is not clear. Studies of each isolated glycine tRNA synthetase subunit, together with reconstitution experiments, suggest that neither subunit alone has activity. Mixing of the separated chains gives both adenylate synthesis and aminoacylation activities (81). The isolated beta chain binds to tRNA, however (82). One explanation is that the active site is generated by the interface of subunits, perhaps at a junction formed between the alpha and beta chains. The two subunits could also represent functional domains that are tightly coupled in that activity is generated by conformational changes within the subunits that only occur on combining the chains.

Manipulation of the intersubunit coding region of *glyS* has resulted in synthesis of fused alpha and beta chains (83, 84). The fusions are stable and are active in vitro and in vivo. This demonstrates that the subunits do not have to be separated in order to generate activity. The apparent stability of the fusion protein is also noteworthy. The expected quaternary structure is a dimer of the two fused chains, based on the alpha-2,beta-2 structure of the native enzyme. A single, discrete dimeric species is not observed, however. Instead, more than one oligomeric form is detected in vitro. In spite of this disturbance of the quaternary structure, the fused protein is active (84).

The rationale for the alpha-2,beta-2 structure remains unknown. Whether there is an additional role for one or both of the free subunits is an open possibility.

STRUCTURAL ORGANIZATION IN RELATION TO COMPLEXES OF SYNTHETASES

It is now well established and generally accepted that, in higher eukaryotes, at least some aminoacyl tRNA synthetases assemble into high-molecular-weight complexes (85). Proteins other than synthetases may also be present. As discussed below, the assembly of these complexes appears to depend on the presence of sequences that are dispensable for catalysis.

The association of synthetases with each other, into a defined complex, is well proven by immunological methods. Antibodies raised against sheep liver methionine or lysine tRNA synthetase precipitate polypeptides for seven different synthetases (86). The polypeptide composition of these immune precipitates is the same as in the isolated complex. These and similar experiments, with immune precipitation protocols, give strong indication that these complexes are homogenous.

The seven enzyme polypeptides found originally in the sheep liver complex, in order of decreasing molecular weight, are isoleucine, leucine, methionine, glutamine, glutamic, lysine, and arginine tRNA synthetases (86–89). Subsequently, aspartyl tRNA synthetase has been assigned to one of the three previously unidentified polypeptides that is coprecipitated with the other enzymes (90). All or some of these eight synthetases are commonly found in the various studies of synthetase complexes. Proline tRNA synthetase is an additional enzyme that has been identified, in some studies, together with some or all of the aforementioned enzymes (90).

It is not definitively resolved whether purified complexes containing these eight or nine enzymes are themselves a breakdown product of a larger in vivo complex. It is reported that a complex of eight enzymes can break down progressively into a complex (or complexes) of fewer synthetases (91). Proteolysis is known to release active forms of enzymes from the complexes. The segment that is released in proteolysis is dispensable, evidently, for catalysis (85, 92). There is also evidence for a hydrophobic domain that is required for binding a synthetase into a complex (93, 94). This provides another rationale for the size polymorphism of synthetases. Those enzymes that assemble into a complex may require an extra adhesive domain that is not essential for catalysis, but that is the structural basis for building a complex.

There are also reports that at least some synthetases bind to components of the endoplasmic reticulum, if not to each other (95–97). The association is believed to be driven by ionic interactions. These associations provide another route for the appearance of synthetases as high-molecular-weight forms. These connections of synthetases with cellular components may not be independent of the multisynthetase complexes. One suggestion is that all synthetases are anchored to the endoplasmic reticulum and that, in addition, some are also associated with each other (96).

The biological significance of these complexes is not known. There is no evidence that the catalytic activity in vitro of a synthetase is radically altered when in the complex (98). Localization of the activity to defined places within the cell may be the main point about these complexes.

ENZYME STRUCTURE AND REGULATION OF AMINOACYL tRNA SYNTHETASE BIOSYNTHESIS

Regulation of the expression of these enzymes has long been investigated. The more recent studies have introduced two new concepts. One is that, in at least some systems, the enzymes themselves act directly as regulatory molecules that control their own biosynthesis. This constitutes a new biological activity that must be reflected in specialized additional structural motifs that are designed for a regulatory function. The other point is that there is no common mechanism that underlies the regulation of the levels of these enzymes. The mechanisms are as diverse as the enzymes themselves.

Different mechanisms of regulation pertain to alanine and threonine tRNA synthetases. In these cases, a role for additional sequences in regulation is plausible, although not proven. The levels of the threonine enzyme are controlled by a translation-based mechanism (99–101). The enzyme binds to a region in the 5'-part of the mRNA and thereby inhibits translation. Sequences in the mRNA that are important for interaction with the enzyme have been identified. These sequences have homologies with tRNA^{Thr}. Mutations in this region that relieve inhibition by the enzyme can be compensated by mutations in the enzyme that restore the regulatory phenotype. These mutations can be investigated to delineate parts of the enzyme that are important for interaction with mRNA.

The alanine enzyme was the first for which a molecular mechanism was suggested (102). The enzyme was shown to bind in vitro to its own promoter and block in vitro transcription of *alaS*. The amino acid alanine is a specific cofactor for the regulatory effect of the enzyme. Levels of alanine and of enzyme that are effective in vitro fall in the range of their in vivo concentrations. Controls have demonstrated that regulation of in vitro transcription does not occur with additions to the transcription mixture of other aminoacyl tRNA synthetases (102). In contrast to the threonine system, however, the regulation of the alanine enzyme has not been investigated in vivo. The reason for this is the lack of alanine auxotrophic strains that can be manipulated to control the levels of alanine. Because the interaction of the enzyme with its promoter depends strongly on the concentration of the amino acid, it is desirable to have a wide range of amino acid concentrations available so that the full extent of the regulated interaction can be investigated. Special gene constructions that may overcome these difficulties are now under consideration.

These results suggest that the alanine and threonine enzyme have specialized regions that enable them to function as regulatory molecules. Such additional sequences give a straightforward rationale for enzymes that are polymorphic in size. The delineation of such sequences within each enzyme has not yet been established, however.

Expression of phenylalanine tRNA synthetase is regulated by an attenuation mechanism (80). Transcription of the tandem alpha and beta subunit coding regions is driven by a single promoter. In front of these coding regions is a sequence for a 14-amino-acid peptide that encodes five phenylalanines. The characteristic stem-and-loop structures of an attenuator are also present. Extensive experimentation has documented the role of attenuation, in response to levels of aminoacylated tRNA^{Phe}, in controlling expression of phenylalanine tRNA synthetase (99, 103–107). The attenuation mechanism of control has no obvious connection with the particular quaternary structure of the phenylalanine enzyme.

The levels of the enzymes also are influenced by growth-rate-dependent or metabolic regulatory mechanisms (99, 108, 109). These mechanisms have not been elucidated. They presumably are superimposed on those that act specifically on individual synthetases, such as those described above.

RECOGNITION OF TRANSFER RNA

General Issues

Much of the research on aminoacyl tRNA synthetases has been motivated by a deep interest in the recognition problem, quite apart from the mechanism of aminoacylation or of other biological features of these proteins. The enzymes distinguish between RNA molecules that fold into the same basic three-dimensional structure that can accommodate the small variations in size that result from a variable number of nucleotides in the dihydrouridine and extra loops. The basis for specificity of tRNA recognition is subtle.

The association of the enzymes with their cognate tRNAs is not strong, by the standard of many specific protein-DNA interactions. At physiological pH, the dissociation constants are typically of the order of 0.1 to 1 micromolar. At pH 5 to 5.5, where the rate of aminoacylation is much slower, the interaction is generally stronger, and characterized by dissociation constants on the order of 1 to 10 nanomolar (5). DNA-protein interactions commonly have dissociation constants that are several orders of magnitude smaller, however. The relatively weak synthetase-tRNA interaction is a necessity, however. The enzymes must turn over rapidly in the aminoacylation reaction and for that reason cannot form tRNA or aminoacyl tRNA complexes with long lifetimes or, equivalently, with tiny equilibrium dissociation constants.

There is only one crystal of a synthetase tRNA complex: yeast Asp-tRNA synthetase and tRNA^{Asp} (36–38, 40). Detailed structural information on this crystal has not yet emerged. There is a long history of experiments that, however, attempt to define sites on the RNA that make contact with the protein. Based on much of this work, a model has been proposed for how most synthetases bind to tRNA. This involves contact of the enzyme along and around the inside of the L-shaped tRNA structure (16). Key regions in this area are the acceptor, dihydrouridine, and anticodon stems, and the anticodon itself. More recent experiments have reinforced this general picture for some of the complexes and suggested some variations (110–118).

Two Steps to Recognition and the Question of a Conformation Change

Recognition is believed to occur in two steps. One is the initial binding of tRNA to enzyme. The synthetases appear to bind selectively to their cognate tRNAs, although the degree of selectivity may not be sufficient to ensure accurate aminoacylation. The second step is after initial complex formation, perhaps in the transition state for aminoacylation. Binding of a noncognate tRNA does not lead to efficient aminoacylation, presumably because of the discrimination step after binding (5). There is evidence for a conformational change in the enzyme-tRNA complex that is part of the recognition process (5, 119–124).

An important example is an experiment by Ebel and coworkers (119). These authors attempted to aminoacylate the CCA trinucleotide sequence that is common to the 3'-end of tRNAs. The yeast phenylalanine enzyme was mixed with the trinucleotide, in the presence and absence of tRNA^{Phe}(-A). This tRNA species lacks the 3'-adenosine terminus and cannot be aminoacylated. The trinucleotide alone is not aminoacylated by the enzyme, but addition of tRNA^{Phe}(-A) triggers aminoacylation of CCA or even of adenosine. This is an important demonstration of the role of conformation of the complex for catalysis.

The Anticodon Problem

WHEN THE ANTICODON IS ESSENTIAL TO RECOGNITION   Efforts have continued to define determinants on the tRNAs that are crucial to recognition. These efforts have benefited from the recent technologies for tRNA nucleotide replacements in vitro and by the capability to make synthetic, mutant tRNA genes. These more powerful approaches have confirmed the complexity of the problem and have especially highlighted the idiosyncratic behavior of the various synthetase-tRNA systems.

Because all tRNAs are distinguished by their anticodons, this is the obvious site for discrimination by the aminoacyl tRNA synthetases (125). In spite of this, there is only one case where the anticodon appears essential to recognition. A series of experiments by Schulman and coworkers have established the anticodon as a critical parameter in the E. coli methionine tRNA synthetase system (126–129).

This has been demonstrated through systematic replacement of the anticodon nucleotides in tRNA$_i^{Met}$. The normal anticodon is CAU. Replacement of any of the three nucleotides results in serious diminution of the aminoacyla-tion rate by the cognate enzyme. The system is also sensitive to alterations in the anticodon loop size. Most significantly, the introduction of the amber suppressor CUA anticodon leads to misacylation by glutamine tRNA syn-thetase. The CUA-tRNA$_i^{Met}$ is a better substrate (by a factor of 1000) for the glutamine than for the methionine enzyme. Most of the deleterious effect of the CUA alteration for the methionine tRNA synthetase recognition is on the k_{cat} for tRNA, rather than the K_m. This emphasizes the importance of the catalytic step in recognition.

The misacylation with glutamine of the CUA-substituted tRNA$_i^{Met}$ is not the first example of a tRNA that can be altered so as to be misaminoacylated with glutamine. The amber-suppressing (CUA anticodon) tRNA^{Trp} is efficiently misacylated by this enzyme (130–132). An A→G mutation (at the fourth position from the 3'-end) of the amber suppressor su^{+3}tRNA^{Tyr} leads to its efficient misaminoacylation with glutamine (133).

Compared to other aminoacyl tRNA synthetases, the glutamine tRNA synthetase may have a more relaxed specificity, or a specificity that is more easily altered by mutation. Mutant Gln-tRNA synthetase with altered recogni-tion has been reported (134, 135).

The results establish the anticodon as a sensitive parameter for the *E. coli* methionine and glutamine tRNA synthetases. There may be additional nucle-otides that are critical to these systems, but it is unlikely that these are of such a nature as to make the anticodon irrelevant.

WHEN THE ANTICODON IS NOT THE ESSENTIAL PARAMETER   There are six codons for leucine and for serine. These codons are read by isoacceptor tRNAs with distinct anticodons. Each serine or leucine isoacceptor tRNA is specifically aminoacylated by only a single aminoacyl tRNA synthetase that is cognate to these tRNAs. This suggests that, at least for these systems, the anticodon is not the primary target for recognition.

This issue has been explored further by Normanly et al, who attempted to change the identity of an *E. coli* leucine isoacceptor (tRNA$_5^{Leu}$) to a serine-specific tRNA (136). For this purpose, sequences of six serine tRNAs were examined and the nucleotides common to these six were identified. (These six serine tRNAs are each specifically aminoacylated by *E. coli* serine tRNA synthetase.) Some of the shared nucleotides occur in the leucine isoacceptor and some are common to all tRNAs. Of the remainder, attention was given just to those that occur in the acceptor stem, and the dihydrouridine stem-and-loop. The stems have been prominent in studies that characterize contact points of synthetases on tRNAs.

Figure 2 summarizes the changes that were introduced into the amber suppressor (CUA anticodon) leucine isoacceptor. Twelve alterations, de-

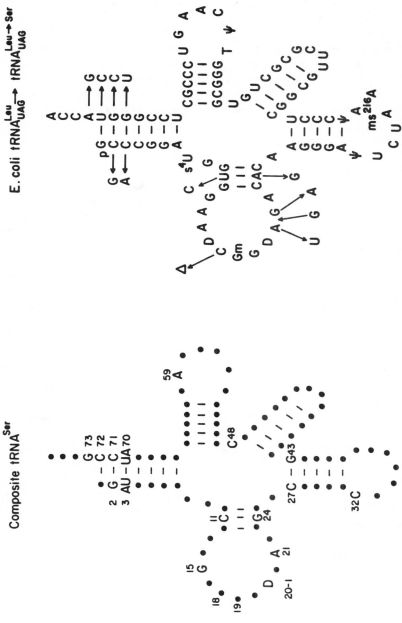

Figure 2 Shown on the left is a cloverleaf representation of conserved nucleotides amongst six *E. coli* serine tRNAs; these nucleotides are not found in tRNA$_5^{Leu}$. Shown on the right is a cloverleaf representation of tRNA$_5^{Leu}$ with changes that confer a serine acceptance specificity. Adapted from Ref. 136.

termined by the analysis of the six serine tRNAs, have been introduced. These alterations convert the leucine amber suppressor into a serine-inserting amber suppressor. Proof that serine is inserted at an amber codon included sequence analysis of a suppressed amber mutation of dihydrofolate reductase.

The results are especially noteworthy in the context of experiments that suggest that the amber suppressor CUA anticodon and a G at the fourth position from the 3'-end are critical factors in making a tRNA^{Trp}, tRNA^{Met}, and tRNA^{Tyr} susceptible to aminoacylation by glutamine tRNA synthetase (see above). Misaminoacylation with glutamine of the artificial tRNA$^{Leu \rightarrow Ser}$ has not been detected, even though the artificial species has a CUA anticodon and a G at the fourth position from the 3'-end (136).

This raises a crucial point. The assignment of sequences important for recognition cannot be done in isolation of the context in which those sequences have been defined as important. The tRNA backbone into which nucleotide substitutions are placed plays some role in determining whether those substitutions have an impact on recognition. This is a major complication in the interpretation of results of experiments where tRNA sequences are manipulated.

Other recent experiments have shown that tRNA^{Cys} and tRNA^{Ala} anticodons can be changed to the amber-suppressing CUA sequence without alteration of the fidelity of aminoacylation in vivo of those tRNAs (137). This result also differs from the sensitivity of tRNA^{Met} and tRNA^{Gln} to the anticodon sequence and emphasizes further the idiosyncratic, case-by-case theme of RNA recognition in the aminoacyl tRNA synthetase system.

The experiments described above have been done in vivo. The effectiveness in aminoacylation of the anticodon-substituted tRNAs has not been quantitatively investigated in vitro. Uhlenbeck and coworkers have constructed, by methods of RNA synthesis, anticodon variants of yeast tRNA^{Phe} and tRNA^{Tyr}. Significant effects of substitutions on the K_m for aminoacylation of tRNA^{Phe} have been found. Change of the anticodon of tRNA^{Tyr} from UGΨ to UGA results in detectable misacylation (with phenylalanine) of the tRNA^{Tyr} variant by Phe-tRNA synthetase. These results emphasize the importance of careful quantitative investigations to elucidate subtle, but significant, effects of anticodon substitutions on recognition (138). They also suggest that, while in some systems the anticodon is not the most essential part of the structure for recognition, it nonetheless may be part of the story.

CONCLUDING REMARKS

In spite of the limited three-dimensional structural information on aminoacyl tRNA synthetases, the extensive analysis of primary structures and genetic manipulations has led to a general scheme for the organization of the structures of this entire class of proteins. The arrangement of functional domains in a linear fashion along the sequence is a first order approximation of how the various proteins are assembled. There are no examples yet of a synthetase gene that is split by introns, however. The identification and characterization of a split gene is critical, to extend further our understanding of the assembly of domains.

While there are dispensable sequences identified in several of the synthetases, the biological significance of these sequences remains an open issue. The assembly of the quaternary structure is dependent on the presence of specific dispensable sequences, in certain instances. The assembly of large complexes of synthetases, in higher eukaryotes, evidently also requires the presence of sequences that can be removed without complete loss of enzymatic activity (see above). I pointed out that the examples of autoregulation of expression raise the possibility that some extra sequences may be used to accomplish the autoregulatory process, i.e. DNA or RNA binding. Some of the enzymes are phosphorylated, and there is some evidence that phosphorylation affects activity (139, 140). Sites for phosphorylation have not been identified, but it is reasonable to suggest that they might be associated with dispensable modules that are grafted to the core enzyme. Finally, several aminoacyl tRNA synthetases execute the synthesis of adenosine tetraphosphate adenosine (Ap_4A), which may be a pleiotrophic regulatory molecule (141–145). It is not known whether synthesis of Ap_4A can be executed by simply the adenylate synthesis domain, or whether additional specialized sequences are also required. An obvious experiment is to isolate a fragment, from the appropriate enzyme, that catalyzes adenylate synthesis and to test it for ability to make Ap_4A.

RNA recognition remains a challenge that is met with clever experimental approaches. Notwithstanding these approaches, there is no substitute for a detailed structural model based on an X-ray diffraction analysis. I am not convinced, however, that the iterative attempts to crystalize synthetase-tRNA complexes are the best investments of effort. There are several alternatives that would go far to provide a breakthrough in understanding. One is to solve the structure of the part of a free synthetase that interacts with tRNA. This structure could be combined with that of the known tRNA structure and the available data on the locations of critical sites (for the interaction) on both molecules. From these combined pieces of information, a heuristic model could be assembled and tested by directed mutagenesis of the tRNA and protein.

A second approach is to isolate the smallest part of a synthetase that is required for interaction with tRNA and to attempt crystalization of that piece in a complex with tRNA. If the polypeptide fragment can be reduced to under 100 amino acids, this might improve the chances for obtaining a crystal of a complex, because of the greater ease of manipulation of smaller components. Because of the reduced size, a small tRNA-binding fragment would clearly simplify the analysis of the crystal structure of a complex. At the least, the complex could be investigated more rigorously in solution by nuclear magnetic resonance, where the large molecular weights of the wild-type enzyme-tRNA complexes have severely hampered the ability to obtain high-resolution information. I note that the recent results obtained with a polypeptide derived from adenylate kinase show that the structure of a small segment (45 amino acids in this instance) bound to its ligand closely approximates the analogous complex with the wild-type protein (146).

Recent immunological studies show that certain patients with myositis (characterized by wasting of muscle tissue) have autoantibodies directed against either histidine, threonine, or alanine tRNA synthetase (147–149). In the latter case, the subjects also have autoantibodies directed against tRNA[Ala] (149). This raises the possibility that the anti-tRNA[Ala] antibodies are anti-idiotypic antibodies that are directed against the anti-Ala-tRNA synthetase antibodies. The anti-Ala-tRNA synthetase antibodies would, in this interpretation, be directed against the tRNA[Ala] binding site of the enzyme. This would suggest that the tRNA[Ala] binding site is antigenic.

In other studies, structural features of the *E. coli* and silk *B. mori* enzyme have been compared (150–152). In particular, an epitope in the *E. coli* enzyme cross-reacts with antibodies directed against silk *B. mori* alanine tRNA synthetase (153). The location of this epitope, in the bacterial enzyme, is at the C-terminal end of the domain for adenylate synthesis. This, in turn, marks the beginning of the sequences found to be critical for association of tRNA[Ala] with this enzyme (48). The results suggest that antibodies may be useful tools for identification and isolation of critical segments, such as the site for tRNA recognition, of aminoacyl tRNA synthetases.

ACKNOWLEDGMENT

I thank the National Institutes of Health (Grants GM 15539 and GM 23562) for support of research on aminoacyl tRNA synthetases.

Literature Cited

1. Söll, D., Schimmel, P. 1974. *The Enzymes* 10:489–538
2. Kisselev, L., Favorova, O. O. 1974. *Adv. Enzymol.* 40:141–238
3. Ofengand, J. 1977. *Molecular Mechanisms of Protein Biosynthesis*, ed. H. Weissbach, S. Pestka, pp. 7–79. New York: Academic
4. Ofengand, J. 1982. *Protein Biosynthesis in Eukaryotes*, ed. R. Perez-Bercoff, pp. 1–67. New York: Plenum
5. Schimmel, P. R., Söll, D. 1979. *Ann. Rev. Biochem.* 48:601–48
6. Baldwin, A. N., Berg, P. 1966. *J. Biol. Chem.* 241:839–45
7. Eldred, E. W., Schimmel, P. R. 1972. *Biochemistry* 11:17–23
8. Schreier, A. A., Schimmel, P. R. 1972. *Biochemistry* 11:1582–89
9. Yarus, M. 1972. *Proc. Natl. Acad. Sci. USA* 69:1915–19
10. Fersht, A. R. 1985. *Enzyme Structure*

and Mechanism. New York: Freeman. 2nd ed.
11. English, U., Gauss, D., Freist, W., English, S., Sternbach, H., von der Haar, F. 1985. Angew. Chem. Int. Ed. Engl. 24:1015–25
12. Freist, W., Pardowitz, I., Cramer, F. 1985. Biochemistry 24:7014–23
13. Pabo, C. O., Sauer, R. T. 1984. Ann. Rev. Biochem. 53:293–321
14. Schimmel, P. R. 1977. Acc. Chem. Res. 10:411–18
15. Schimmel, P. R. 1979. Adv. Enzymol. 49:187–222
16. Rich, A., Schimmel, P. R. 1977. Nucleic Acids Res. 4:1649–65
17. Jasin, M., Regan, L., Schimmel, P. 1984. Cell 36:1089–95
18. Jasin, M., Regan, L., Schimmel, P. 1985. J. Biol. Chem. 260:2226–30
19. Jasin, M., Regan, L., Schimmel, P. 1983. Nature 306:441–47
20. Pape, L. K., Tzagoloff, A. 1985. Nucleic Acids Res. 13:6171–80
21. Pape, L. K., Koerner, T. J., Tzagoloff, A. 1985. J. Biol. Chem. 260:15362–70
22. Natsoulis, G., Hilger, F., Fink, G. R. 1986. Cell 46:235–43
23. Winter, G. P., Hartley, B. S. 1977. FEBS Lett. 80:340–42
24. Webster, T. A., Tsai, H., Kula, M., Mackie, G., Schimmel, P. 1984. Science 226:1315–17
25. Winter, G., Koch, G. L. E., Hartley, B. S., Barker, D. G. 1983. Eur. J. Biochem. 132:383–87
26. Jones, M. D., Lowe, D. M., Borgford, T., Fersht, A. R. 1986. Biochemistry 25:1887–91
27. Deleted in proof
28. Putney, S. D., Sauer, R. T., Schimmel, P. R. 1981. J. Biol. Chem. 256:198–204
29. Putney, S. D., Royal, N. J., de Vegvar, H. N., Herlihy, W. C., Biemann, K., et al. 1981. Science 213:1497–501
30. Jasin, M., Schimmel, P. 1984. J. Bacteriol. 159:783–86
31. Blow, D. M., Brick, P. 1986. Biological Macromolecules and Assemblies, ed. F. Jurnak, A. MacPherson, 2:441–69. New York: Wiley
32. Risler, J. L., Zelwer, C., Brunie, S. 1981. Nature 292:384–86
33. Zelwer, C., Risler, J. L., Brunie, S. 1982. J. Mol. Biol. 155:63–81
34. Bhat, T. N., Blow, D. M., Brick, P., Nyborg, J. 1982. J. Mol. Biol. 158:699–709
35. Blow, D. M., Bhat, T. N., Metcalfe, A., Risler, J. L., Brunie, S., et al. 1983. J. Mol. Biol. 171:571–76
36. Giege, R., Lorber, B., Ebel, J.-P., Moras, D., Thierry, J.-C., et al. 1982. Biochimie 64:357–62
37. Moras, D., Lorber, B., Romby, P., Ebel, J.-P., Giege, R., et al. 1983. J. Biomol. Struct. Dyn. 1:209–23
38. Lorber, B., Giege, R., Ebel, J.-P., Berthet, C., Thierry, J.-C., et al. 1983. J. Biol. Chem. 258:8429–35
39. Lorber, B., Kern, D., Giege, R. 1984. Electrophoresis 84, ed. V. Neuhof, pp. 427–28. Germany: Verlag Chemie
40. Lorber, B., Giege, R. 1985. Anal. Biochem. 146:402–4
41. Coleman, D. E., Carter, C. W. Jr. 1984. Biochemistry 23:381–85
42. Carter, C. W. Jr., Coleman, D. E. 1984. Fed. Proc. 43:2981–83
43. Fersht, A. R., Shi, J.-P., Wilkinson, A. J., Blow, D. M., Carter, P., et al. 1984. Angew. Chem. Int. Ed. Engl. 23:467–73
44. Dardel, F., Fayat, G., Blanquet, S. 1984. J. Bacteriol. 160:1115–22
45. Walter, P., Gangloff, J., Bonnet, J., Boulanger, Y., Ebel, J.-P., et al. 1983. Proc. Natl. Acad. Sci. USA 80:2437–41
46. Breton, R., Sanfacon, H., Papayannopoulos, I., Biemann, K., Lapointe, J. 1986. J. Biol. Chem. 261:10610–17
47. Waye, M. Y., Winter, G., Wilkinson, A. J., Fersht, A. R. 1983. EMBO J. 2:1827–29
48. Regan, L., Bowie, J., Schimmel, P. 1986. Submitted for publication
49. Houtondji, C., Blanquet, S. 1984. Biochemistry 24:1175–80
50. Houtondji, C., Lederer, F., Dessen, P., Blanquet, S. 1986. Biochemistry 25:16–21
51. Schulman, L. H., Pelka, H., Reines, S. A. 1981. Nucleic Acids Res. 9:1203–17
52. Schulman, L. H., Valenzuela, D., Pelka, H. 1981. Biochemistry 20:6018–23
53. Valenzuela, D., Leon, O., Schulman, L. H. 1984. Biochem. Biophys. Res. Commun. 119:677–84
54. Valenzuela, D., Schulman, L. H. 1986. Biochemistry 25:4555–61
55. Bedouelle, H., Winter, G. 1986. Nature 320:371–73
56. Carter, P., Bedouelle, H., Winter, G. 1986. Proc. Natl. Acad. Sci. USA 83:1189–92
57. Wilkinson, A. J., Fersht, A. R., Blow, D. M., Winter, G. 1983. Biochemistry 22:3581–86
58. Wells, T. N. C., Fersht, A. R. 1986. Biochemistry 25:1881–86
59. George, H., Meister, A. 1967. Biochim. Biophys. Acta 131:165–74
60. Iaccarino, M., Berg, P. 1969. J. Mol. Biol. 42:151–69

61. Kuo, T., DeLuca, M. 1969. *Biochemistry* 8:4762–68
62. Bruton, C. J., Hartley, B. S. 1970. *J. Mol. Biol.* 52:165–78
63. Rouget, P., Chapeville, F. 1971. *Eur. J. Biochem.* 23:452–58
64. Waterson, R. M., Clarke, S. J., Kalousek, F., Koenigsberg, W. H. 1973. *J. Biol. Chem.* 248:4181–88
65. Ostrem, D. L., Berg, P. 1974. *Biochemistry* 13:1338–48
66. Chen, Z.-Q., Kim, J.-J.P., Lai, C.-S., Mehler, A. H. 1984. *Arch. Biochem. Biophys.* 233:611–16
67. Lipmann, F. 1973. *Acc. Chem. Res.* 6:361–67
68. Kisselev, L. L., Favorova, O. O., Kovaleva, G. K. 1979. In *Transfer RNA: Structure, Properties, and Recognition,* ed. P. R. Schimmel, D. Söll, J. N. Abelson, pp. 235–46. New York: Cold Spring Harbor Lab.
69. Ostrem, D. L., Berg, P. 1970. *Proc. Natl. Acad. Sci. USA* 67:1967–74
70. Profy, A. T., Schimmel, P. 1986. *J. Biol. Chem.* 261:15474–79
71. Webster, T. A., Gibson, B. W., Keng, T., Biemann, K., Schimmel, P. 1983. *J. Biol. Chem.* 258:10637–41
72. Ho, C. K., Fersht, A. R. 1986. *Biochemistry* 25:1891–97
73. Carter, P. J., Winter, G., Wilkinson, A. J., Fersht, A. R. 1984. *Cell* 38:835–40
74. Leatherbarrow, R. J., Fersht, A. R., Winter, G. 1985. *Proc. Natl. Acad. Sci. USA* 82:7840–44
75. Theall, G., Low, K. B., Söll, D. 1977. *Mol. Gen. Genet.* 156:221–27
76. Jones, D. H., McMillan, A. J., Fersht, A. R. 1985. *Biochemistry* 24:5852–57
77. Ho, C., Jasin, M., Schimmel, P. 1985. *Science* 229:389–93
78. Deleted in proof
79. Keng, T., Webster, T. A., Sauer, R. T., Schimmel, P. 1982. *J. Biol. Chem.* 257:12503–8
80. Fayat, G., Mayaux, J.-F., Sacerdot, C., Fromant, M., Springer, M., et al. 1983. *J. Mol. Biol.* 171:239–61
81. McDonald, T., Breite, L., Pangburn, K. L. W., Hom, S., Manser, J., et al. 1980. *Biochemistry* 19:1402–9
82. Nagel, G. M., Cumberledge, S., Johnson, M. S., Petrella, E., Weber, B. H. 1984. *Nucleic Acids Res.* 12:4377–84
83. Keng, T., Schimmel, P. 1983. *J. Biomol. Struct. Dyn.* 1:225–29
84. Toth, M. J., Schimmel, P. 1986. *J. Biol. Chem.* 261:6643–46
85. Deutscher, M. P. 1984. *J. Cell Biol.* 99:373–77
86. Mirande, M., Gache, Y., LeCorre, D., Waller, J.-P. 1982. *EMBO J.* 1:733–36
87. Kellermann, O., Tonetti, H., Brevet, A., Mirande, M., Pailliez, J.-P., et al. 1982. *J. Biol. Chem.* 257:11041–48
88. Mirande, M., Cirakoglu, B., Waller, J.-P. 1982. *J. Biol. Chem.* 257:11056–63
89. Mirande, M., Kellermann, O., Waller, J.-P. 1982. *J. Biol. Chem.* 257:11056–63
90. Mirande, M., Le Corre, D., Waller, J.-P. 1985. *Eur. J. Biochem.* 147:281–89
91. Dang, C. V., Yang, D. C. H. 1979. *J. Biol. Chem.* 254:5350–56
92. Vellekamp, F., Sihag, R. K., Deutscher, M. P. 1985. *J. Biol. Chem.* 260:9843–47
93. Sihag, R. K., Deutscher, M. P. 1983. *J. Biol. Chem.* 258:11846–50
94. Cirakoglu, B., Waller, J.-P. 1985. *Eur. J. Biochem.* 151:101–10
95. Dang, C. V., Yang, D. C. H., Pollard, T. D. 1983. *J. Cell. Biol.* 96:1138–47
96. Mirande, M., Le Corre, D., Louvard, D., Reggio, H., Pailliez, J.-P., et al. 1985. *Exp. Cell Res.* 156:91–102
97. Cirakoglu, B., Mirande, M., Waller, J.-P. 1985. *FEBS Lett.* 183:185–90
98. Mirande, M., Cirakoglu, B., Waller, J.-P. 1983. *Eur. J. Biochem.* 131:163–70
99. Grunberg-Manago, M. 1987. In *E. Coli and S. Typhimurium: Cellular and Molecular Biology,* ed. J. L. Ingraham, K. B. Low, B. Magasanik, F. C. Neidhardt, M. Schaechter, H. E. Umbarger. Washington, DC: Am. Soc. Microbiol. In press
100. Lestienne, P., Plumbridge, J. A., Grunberg-Manago, M., Blanquet, S. 1984. *J. Biol. Chem.* 259:5232–37
101. Springer, M., Plumbridge, J. A., Butler, J. S., Graffe, M., Dondon, J., et al. 1985. *J. Mol. Biol.* 185:93–104
102. Putney, S. D., Schimmel, P. 1981. *Nature* 291:632–35
103. Mayaux, J.-F., Fayat, G., Panvert, M., Springer, M., Grunberg-Manago, M., et al. 1985. *J. Mol. Biol.* 184:31–44
104. Plumbridge, J. A., Springer, M. 1982. *J. Bacteriol.* 152:661–68
105. Plumbridge, J. A., Springer, M. 1982. *J. Bacteriol.* 152:650–60
106. Springer, M., Mayaux, J.-F., Fayat, G., Plumbridge, J. A., Graffe, M., et al. 1985. *J. Mol. Biol.* 181:467–78
107. Springer, M., Plumbridge, J., Trudel, M., Grunberg-Manago, M., Fayat, G., et al. 1982. *Interaction of Translational and Transcriptional Controls in the Regulation of Gene Expression,* ed. M. Grunberg-Manago, B. Safer, pp. 25–47. New York: Elsevier Biomedical
108. Neidhardt, F. C., Parker, J., McKeever, W. B. 1975. *Ann. Rev. Microbiol.* 29:215–50

109. Neidhardt, F. C., Bloch, P. L., Pedersen, S., Reeh, S. 1977. *J. Bacteriol.* 129:378–87
110. Vlassov, V. V., Kern, D., Giege, R., Ebel, J.-P. 1981. *FEBS Lett.* 123:277–81
111. Vlassov, V. V., Kern, D., Romby, P., Giege, R., Ebel, J.-P. 1983. *Eur. J. Biochem.* 132:537–44
112. Wrede, P., Rich, A. 1985. *Endocyt. Cell Res.* 2:107–12
113. Favorova, O. O., Fasiolo, F., Keith, G., Vassilenko, S. K., Ebel, J.-P. 1981. *Biochemistry* 20:1006–11
114. Garret, M., Labouesse, B., Litvak, S., Romby, P., Ebel, J.-P., Giege, R. 1984. *Eur. J. Biochem.* 138:67–75
115. Gangloff, J., Jaozara, R., Dirheimer, G. 1983. *Eur. J. Biochem.* 132:629–37
116. Knorre, D. G., Vlassov, V. V. 1980. In *Molecular Biology Biochemistry and Biophysics,* ed. F. Chapeville, A.-L. Haenni, 32:278–300. Berlin/Heidelberg: Springer-Verlag
117. Renaud, M., Bacha, H., Dietrich, A., Remy, P., Ebel, J.-P. 1981. *Biochim. Biophys. Acta* 653:145–59
118. Romby, P., Moras, D., Bergdoll, M., Dumas, P., Vlassov, V. V., et al. 1985. *J. Mol. Biol.* 184:455–71
119. Renaud, M., Bacha, H., Remy, P., Ebel, J.-P. 1981. *Proc. Natl. Acad. Sci. USA* 78:1606–8
120. Fasiolo, F., Remy, P., Holler, E. 1981. *Biochemistry* 20:3851–56
121. Favre, A., Ballini, J. P., Holler, E. 1979. *Biochemistry* 13:2887–95
122. Butorin, A. S., Remy, P., Ebel, J.-P. 1982. *Eur. J. Biochem.* 121:587–95
123. Beresten, S., Scheinker, V., Favorova, O., Kisselev, L. 1983. *Eur. J. Biochem.* 136:559–70
124. Bacha, H., Renaud, M., Lefevre, J.-F., Remy, P. 1982. *Eur. J. Biochem.* 127:87–95
125. Kisselev, L. 1985. *Prog. Nucleic Acid Res. Mol. Biol.* 32:237–66
126. Schulman, L. H., Pelka, H., Susani, M. 1983. *Nucleic Acids Res.* 11:1439–55
127. Schulman, L. H., Pelka, H. 1983. *Proc. Natl. Acad. Sci. USA* 80:6755–59
128. Schulman, L. H., Pelka, H. 1984. *Fed. Proc.* 43:2977–79
129. Schulman, L. H., Pelka, H. 1985. *Biochemistry* 24:7309–14
130. Celis, J. E., Coulondre, C., Miller, J. H. 1976. *J. Mol. Biol.* 104:729–34
131. Yarus, M., Knowlton, R., Soll, L. 1977. *Nucleic Acid-Protein Recognition,* ed. H. J. Vogel, pp. 391–408. New York: Academic
132. Knowlton, R. G., Soll, L., Yarus, M. 1980. *J. Mol. Biol.* 139:705–20
133. Ghysen, A., Celis, J. E. 1974. *J. Mol. Biol.* 83:333–51
134. Hoben, P., Uemura, H., Yamao, F., Cheung, A., Swanson, R., et al. 1984. *Fed. Proc.* 43:2972–76
135. Inokuchi, H., Hoben, P., Yamao, F., Ozeki, H., Söll, D. 1984. *Proc. Natl. Acad. Sci. USA* 81:5076–80
136. Normanly, J., Ogden, R. C., Horvath, S. J., Abelson, J. 1986. *Nature* 321:213–19
137. Normanly, J., Masson, J.-M., Kleina, L. G., Abelson, J., Miller, J. H. 1986. *Proc. Natl. Acad. Sci. USA* 83:6548–52
138. Uhlenbeck, O. C., Bare, L., Bruce, A. G. 1984. *Gene Expression, Alfred Benzon Symp. 19,* ed. B. F. C. Clark, H. U. Petersen, pp. 163–77. Copenhagen: Munksgaard
139. Pendergast, A. M., Traugh, J. A. 1985. *J. Biol. Chem.* 260:11769–11774
140. Traugh, J. A., Pendergast, A. M. 1986. *Prog. Nucleic Acid Res. Mol. Biol.* 33:195–230
141. Brevet, A., Plateau, P., Cirakoglu, B., Pailliez, J.-P., Blanquet, S. 1982. *J. Biol. Chem.* 257:14613–15
142. Hilderman, R. H. 1983. *Biochemistry* 22:4353–57
143. Lee, P. C., Bochner, B. R., Ames, B. N. 1983. *J. Biol. Chem.* 258:6827–34
144. Varshavsky, A. 1983. *Cell* 34:711–12
145. Zamecnik, P. C., Rapaport, E., Baril, E. F. 1982. *Proc. Natl. Acad. Sci. USA* 79:1791–94
146. Fry, D. C., Kuby, S. A., Mildvan, A. S. 1985. *Biochemistry* 24:4680–94
147. Mathews, M. B., Bernstein, R. M. 1983. *Nature* 304:177–79
148. Mathews, M. B., Reichlin, M., Hughes, G. R. V., Bernstein, R. M. 1984. *J. Exp. Med.* 160:420–34
149. Bunn, C. C., Bernstein, R. M., Mathews, M. B. 1986. *J. Exp. Med.* 163:1281–91
150. Dignam, S. S., Dignam, J. D. 1984. *J. Biol. Chem.* 259:4043–48
151. Nishio, K., Kawakami, M. 1984. *J. Biochem.* 96:1867–74
152. Nishio, K., Kawakami, M. 1984. *J. Biochem.* 96:1875–81
153. Regan, L., Dignam, J. D., Schimmel, P. 1986. *J. Biol. Chem.* 261:5241–44
154. Amiri, I., Mejdoub, H., Hounwanou, N., Boulanger, Y., Reinbolt, J. 1985. *Biochimie* 67:607–13
155. Sellami, M., Chatton, B., Fasiolo, F., Dirheimer, G., Ebel, J.-P., Gangloff, J. 1986. *Nucleic Acids Res.* 14:1657–66
156. Hoben, P., Royal, N., Cheung, A., Fumiaki, Y., Biemann, K., Söll, D. 1982. *J. Biol. Chem.* 257:11644–50

157. Freedman, R., Gibson, B., Donovan, D., Biemann, K., Eisenbeis, S., et al. 1985. *J. Biol. Chem.* 260:10063–68
158. Mayaux, J.-F., Fayat, G., Fromant, M., Springer, M., Grunberg-Manago, M., Blanquet, S. 1983. *Proc. Natl. Acad. Sci. USA* 80:6152–56
159. Mechulam, Y., Fayat, G., Blanquet, S. 1985. *J. Bacteriol.* 163:787–91
160. Hall, C. V., van Cleemput, M., Muench, K. H., Yanofsky, C. 1982. *J. Biol. Chem.* 257:6132–36
161. Myers, A. M., Tzagoloff, A. 1985. *J. Biol. Chem.* 260:15371–77

162. Barker, D. G., Bruton, C. J., Winter, G. 1982. *FEBS Lett.* 150:419–23
163. Deleted in proof
164. Wilkinson, A. J., Fersht, A. R., Blow, D. M., Carter, P., Winter, G. 1984. *Nature* 307:187–88
165. Fersht, A. R., Wilkinson, A. J. 1985. *Biochemistry* 24:5858–61
166. Winter, G., Fersht, A. R., Wilkinson, A. J., Zoller, M., Smith, M. 1982. *Nature* 299:756–58
167. Renaud, M., Fasiolo, F., Baltzinger, M., Boulanger, Y., Remy, P. 1982. *Eur. J. Biochem.* 123:267–74

Ann. Rev. Biochem. 1987. 56:159–93

INOSITOL TRISPHOSPHATE AND DIACYLGLYCEROL: TWO INTERACTING SECOND MESSENGERS[1]

Michael J. Berridge

A.F.R.C. Unit of Invertebrate Physiology & Pharmacology, Department of Zoology, Downing Street, Cambridge CB2 3EJ, England

CONTENTS

[1]Abbreviations used: DG, diacylglycerol; ER, endoplasmic reticulum; G_p, G-protein responsible for stimulating phosphoinositidase; G_s, G-protein responsible for stimulating adenylate cyclase; G_i, G-protein responsible for inhibiting adenylate cyclase; 5-HT, 5-hydroxytryptamine; Ins, inositol; Ins1,4,5P$_3$, inositol 1,4,5-trisphosphate; cIns1:2,4,5P$_3$, inositol 1:2-cyclic 4,5-trisphosphate; PtdIns, phosphatidylinositol; OAG, 1-oleoyl-2-acetylglycerol; TXB$_2$, thromboxane B$_2$.

0066-4154/87/0701-0159$02.00

INTRODUCTION

External signals detected by surface receptors are translated into a limited repertoire of intracellular second messengers. Pre-eminent among these are inositol 1,4,5-trisphosphate (Ins1,4,5P$_3$) and diacylglycerol (DG), which constitute a bifurcating signal pathway that is attracting enormous interest because it is a central component in the control mechanisms of many different cells. Since the main features of this signaling system have been described in a number of recent reviews (1–9), I have chosen to concentrate on some of the latest developments concerning the way in which these two second messengers interact with each other to regulate cellular activity.

An inositol lipid located within the plasma membrane is the precursor used by the receptor mechanism to release Ins1,4,5P$_3$ to the cytosol, leaving DG within the plane of the membrane. The primary function of Ins1,4,5P$_3$ is to mobilize calcium from intracellular stores (1, 2) to constitute an Ins1,4,5P$_3$/Ca^{2+} pathway, whereas the other limb is controlled through DG, which stimulates protein kinase C (C-kinase) (4, 8) to form the DG/C-kinase pathway. Together the two limbs of this bifurcating signal pathway provide an exceptionally versatile signaling mechanism, which has been adapted to control short-term cellular responses such as contraction, secretion, and metabolism, and may also play a role in long-term events such as growth and perhaps even information storage in the brain. Most attention will be focused on the action of Ins1,4,5P$_3$, since this second messenger plays a central role through its ability to release calcium. It will be argued that the DG/C-kinase pathway functions primarily as an internal feedback system designed to modulate both calcium signaling and the activity of other receptor mechanisms. Such feedback interactions between second messenger pathways are assuming ever greater importance and may be the basis of the oscillatory phenomena that are being uncovered in many cell types. The formation, metabolism, and mode of action of Ins1,4,5P$_3$ and DG will be described first as a prelude to the discussion of these second messenger interactions.

FORMATION OF INOSITOL TRISPHOSPHATE

INOSITOL LIPID METABOLISM   Compared to all the other phospholipids in the membrane, PtdIns is unique in that it can be further phosphorylated. A PtdIns kinase transfers a phosphate from ATP onto the 4-position of the inositol headgroup to give phosphatidylinositol 4-phosphate (PtdIns4P). A separate PtdIns4P kinase adds another phosphate to the 5-position to give the phosphatidylinositol 4,5-bisphosphate (PtdIns4,5P$_2$), which is the substrate used by the receptor mechanism to generate Ins1,4,5P$_3$ (Figure 1). Only a small proportion (10–20%) of the inositol lipid pool is involved in signaling

(10–13). Further evidence for separate pools has come from experiments with Mn^{2+}, which stimulates an enzymic reaction leading to inositol headgroup exchange. This increases the incorporation of ^{3}H-inositol into a PtdIns pool that is insensitive to agonists (14, 15). Presumably Mn^{2+} stimulates incorporation of inositol into the hormone-insensitive pool.

The hormone-sensitive pool seems to be confined to the plasma membrane (13), which is where most of the PtdIns4,5P$_2$ is thought to reside (5). Even the PtdIns4,5P$_2$ within the plasma membrane, however, may not be a homogeneous pool (16, 17). In WRK-1 cells it has been estimated that only 60% of the PtdIns4,5P$_2$ is available for signaling, while the remainder is associated with the hormone-insensitive pool and remains stable for many hours (16). The PtdIns4,5P$_2$ in erythrocytes is similarly compartmentalized with a large metabolically inactive pool somehow kept apart from a smaller pool, which is turning over rapidly (17). It is difficult to imagine how PtdIns4,5P$_2$ might be compartmentalized within the plasma membrane, but a recent observation that polyphosphoinositides might be covalently linked to myelin basic protein (18) may explain how these lipids could be removed from the metabolic pool. There are clear indications, therefore, that the pool of PtdIns4,5P$_2$ available to the receptor may be a lot smaller than hitherto suspected.

The fact that Ins1,4,5P$_3$ is produced in many different cells provides unequivocal evidence that PtdIns4,5P$_2$ is one of the lipids used by the receptor

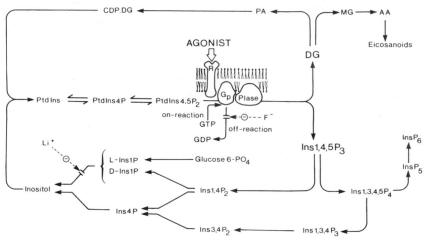

Figure 1   The bifurcating signal pathway begins with the hydrolysis of PtdIns4,5P$_2$ to give DG and Ins1,4,5P$_3$. When an agonist occupies its receptor (R), it activates a G-protein (G$_p$), which then binds GTP as part of an on-reaction leading to the stimulation of phosphoinositidase (PIase). Both second messengers (DG and Ins1,4,5P$_3$) can be metabolized via two separate pathways. See text for abbreviations and further details.

mechanism. There have been suggestions that the other two inositol lipids may also be used under certain circumstances, or by certain agonists. When platelets respond to thrombin, it has been argued that PtdIns4,5P$_2$ is used at the beginning and is followed later by a calcium-dependent hydrolysis of PtdIns (19). Vascular smooth muscle is also thought to switch over to hydrolyzing PtdIns during sustained stimulation (19a). Since PtdIns and PtdIns4P would release inositol phosphates that do not mobilize calcium, this would provide a mechanism for biasing the signal pathway towards protein kinase C and away from calcium. The possibility that the other inositol lipids might be hydrolyzed during signaling must not be ignored, but here we concentrate on PtdIns4,5P$_2$ as the precursor used to translate incoming signals into second messengers.

HYDROLYSIS OF PTDINS4,5P$_2$ An increased hydrolysis of PtdIns4,5P$_2$ to give DG and Ins1,4,5P$_3$ has been measured in many different cell types in response to a variety of stimuli including neurotransmitters, releasing factors, hormones, growth factors, fertilization, and even light (detailed lists of these stimuli are to be found in 1, 2, 5, 6, 9). Consistent with its role as a second messenger, the increase in the level of Ins1,4,5P$_3$ was found to precede the onset of calcium-dependent events in blowfly salivary gland (20) and in neutrophils (21). Measurements of intracellular calcium in different cell types have revealed that the increase in Ins1,4,5P$_3$ either preceded (22, 23) or coincided with the onset of the calcium signal (24–28). This evidence is consistent with the hypothesis that external signals act through a receptor mechanism to produce Ins1,4,5P$_3$, which then mobilizes calcium.

The transduction unit within the plasma membrane consists of three main components: 1. a receptor that detects the incoming signal, 2. a G-protein that serves to couple the receptor to the third component, and 3. the phosphodiesterase responsible for cleaving the lipid precursor (Figure 1). In order to avoid confusion with the nomenclature employed in the cyclic AMP field, where phosphodiesterase is used to describe the enzyme that degrades cyclic AMP to AMP, Downes & Michell (5) have introduced the term *phosphoinositidase* to describe this enzyme that hydrolyzes PtdIns4,5P$_2$.

Formation of cyclic Ins1,4,5P$_3$ Phosphoinositidase cleaves the phosphodiester bond between glycerol and phosphate, which requires the addition of a hydroxyl group. This group can enter as a free OH$^-$, causing the release of Ins1,4,5P$_3$, or the enzyme may use the resident hydroxyl located on the 2-position of the inositol ring, thus leading to the formation of inositol 1:2-cyclic 4,5-trisphosphate (cIns1:2,4,5P$_3$). The formation of cIns1:2, 4,5P$_3$ has been demonstrated in vitro following incubation of PtdIns4, 5P$_2$ with purified phosphoinositidase (29). There also is a report that cIns1:2,4,5P$_3$ exists in blood platelets, and that its concentration rises in response to thrombin (30), but others have failed to find this cyclic form (31). Considering the uncertainty surrounding cIns1:2,4,5P$_3$, I shall refer subsequently to the product of the receptor as Ins1,4,5P$_3$, bearing in mind that this could be either the cyclic form or a mixture of the two.

A role for G-proteins   A family of G-proteins are responsible for transducing signals across cell membranes (32). The inositol lipid signal transduction mechanism is another example where a G-protein (G$_p$) serves to couple receptors to the phosphoinositidase (33, 34). The subscript p (standing for phospholipid) has been used to distinguish this G-protein from that found in the other transduction mechanisms (35). The identity of G$_p$ and its relationship to other G-proteins is unknown. There is an intriguing possibility that G$_p$ may resemble or actually be the product of the *ras* oncogene (36–38). Expression of p21^{N-ras} in NIH 3T3 cells has no effect on the basal rate of inositol phosphate formation, but there is a large increase in their responsiveness to a number of growth factors, particularly bombesin (38). It would appear, therefore, that G$_p$ is limiting under normal conditions and that the expression of *ras* increases the ability of receptors to activate the phosphoinositidase.

The evidence that G$_p$ functions in signal transduction is based primarily on the observation that the breakdown of PtdIns4,5P$_2$ to Ins1,4,5P$_3$ and DG can be stimulated by the addition of nonhydrolyzable guanine-nucleotide analogues to either permeabilized cells or to isolated membranes. When studied in vitro, guanine nucleotide-binding proteins such as G$_s$, G$_i$, and transducin are activated by fluoride ions. The active form is an AlF$_4^-$ complex, which blocks the off-reaction by inhibiting the GTPase component, so stabilizing the G-proteins in their active configuration (Figure 1). It appears as if AlF$_4^-$ may also be capable of activating G$_p$ to stimulate the formation of Ins1,4,5P$_3$ both in vitro and in vivo (39, 40). This fluoride-induced formation of Ins1,4,5P$_3$ in intact liver cells was associated with an increase in intracellular calcium and the activation of phosphorylase. Fluoride also increased intracellular calcium levels in human platelets (41) and neutrophils (40). The ability of fluoride to stimulate the formation of Ins1,4,5P$_3$ and to elevate calcium levels in intact cells may explain how this ion can stimulate the release reaction of platelets (42), and can mimic the effect of light in *Limulus* photoreceptors (43).

Is there an inhibitory G-protein for the inositol lipid pathway?   Control of adenylate cyclase is exercised by both a stimulatory (G$_s$) and an inhibitory (G$_i$) protein. All the evidence described above pointed to G$_p$ being stimulatory, but there are suggestions that there may be an inhibitory pathway responsible for switching off the hydrolysis of inositol lipids. An inhibitory pathway in hippocampal neurons was proposed on the basis that acidic amino acids strongly suppress the formation of inositol phosphates induced by carbachol or histamine (44). Another example is found in the pituitary, where the release of hormones is inhibited by dopamine acting on D_2-receptors. While much of this inhibition probably occurs by lowering the level of cyclic AMP, dopamine may also act by inhibiting the breakdown of inositol lipids (45). Dopamine was able to suppress the stimulatory effect of angiotensin II on both prolactin secretion and the formation of inositol phosphates (45). Although these examples suggest the existence of an inhibitory pathway, more biochemical information is required before one can conclude that this pathway is mediated through a G-protein.

SECOND MESSENGER METABOLISM

Both of the second messengers formed by the hydrolysis of PtdIns4,5P$_2$ can be metabolized via two separate pathways. It has been known for some time that DG can either be phosphorylated to phosphatidic acid by a DG kinase or it can be hydrolyzed by a DG lipase to form monoacylglycerol, which is further hydrolyzed to release arachidonic acid (Figure 1). Since the latter is the precursor of the eicosanoids (prostaglandins, thromboxanes, and leukotrienes), the primary second messenger DG can give rise to additional messengers, which function as local hormones. The other primary second messenger, Ins1,4,5P$_3$, can also be metabolized via two separate pathways (Figure 1). Either it can be dephosphorylated to free inositol or it can enter an inositol tris/tetrakis pathway to form some newly identified inositol polyphosphates, which may have messenger functions.

DEPHOSPHORYLATION OF INS1,4,5P$_3$ One pathway for metabolizing Ins-1,4,5P$_3$ depends on a sequential series of dephosphorylation reactions that culminate in the formation of free *myo*-inositol (Figure 1). The first enzyme is Ins1,4,5P$_3$ 5-phosphatase, which has a somewhat precise positional selectivity in that it specifically removes the phosphate at the 5-position to give Ins1,4P$_2$ (46–48). Although the enzyme is predominantly particulate and associated with the plasma membrane (46–50), the enzyme found in blood platelets is soluble (48). This trisphosphatase is of key importance in that it terminates the second messenger action of Ins1,4,5P$_3$ because the product Ins1,4P$_2$ is incapable of releasing calcium. The enzyme is particularly sensitive to divalent cations in that it has an absolute requirement for magnesium and is markedly inhibited by silver, zinc, and cobalt (46, 47). Calcium inhibits the enzyme in blood platelets (48) but activates that from macrophages (51) and smooth muscle (52). The hydrolysis of Ins1,4,5P$_3$ is insensitive to lithium (47–50), but is blocked by spermine (50) and by 2,3-diphosphoglyceric acid (46, 50, 53).

Inhibition of inositol 1,4,5-trisphosphatase by 2,3-diphosphoglyceric acid could play a role in regulating the levels of Ins1,4,5P$_3$ in insulin-secreting β-cells (53). The latter are unusual cells in that they respond to glucose with an increase in Ins1,4,5P$_3$ formation even though they lack a glucose receptor (54). Since glucose must be metabolized via the glycolytic pathway in order to stimulate the release of insulin, Rana et al (53) reason that some of the glucose metabolites, particularly 2,3-diphosphoglyceric acid and fructose 1,6-bisphosphate, which increase markedly in the presence of glucose, may act to inhibit the enzyme that degrades Ins1,4,5P$_3$. It seems unlikely that the endogenous rate of Ins1,4,5P$_3$ production is fast enough to account for the rapid increased rate of formation that has been measured in response to glucose (54). Nevertheless, any decrease in the activity of the inositol tris-phosphatase would serve to potentiate their responsiveness to glucose. Since transformation of chick-embryo fibroblasts with v-*src* or v-*fps* results in an increase in the level of fructose 2,6-bisphosphate (55), it might be worth examining whether the increase in glycolysis, usually associated with cell transformation, may lead to inhibition of Ins1,4,5P$_3$ hydrolysis.

The next enzyme in the pathway is an inositol bisphosphatase with less precise positional selectivity in that it hydrolyzes Ins1,4P$_2$ to both Ins4P and Ins1P (47). The final step is to remove the phosphate from Ins1P and Ins4P through inositol monophosphatase(s). Most of the inositol phosphates within the cell are in the D-form, but there is a measurable quantity of L-Ins1P that originates from the isomerization of glucose 6-phosphate by an L-*myo*-inositol-1-phosphate synthase (56). A *myo*-inositol-1-phosphatase can hydro-lyze both D-Ins1P and L-Ins1P (57), but is incapable of dephosphorylating Ins2P (56). There is no information on whether or not this enzyme can hydrolyze Ins4P, so the metabolic route for this isomer is unknown. The Ins1P-phosphatase is inhibited by lithium with a K_i value of approximately 1 mM (57, 58), which is somewhat lower than that for the bisphosphatase (K_i 2.5 mM) (47).

The dephosphorylation of cIns1:2,4,5P$_3$ seems to occur along a similar pathway as for Ins1,4,5P$_3$. It is first dephosphorylated to cIns1:2,4P$_2$ and then to cIns1:2P (59). Cells have a D-*myo*-inositol 1:2-cyclic phosphate 2-inositol phosphohydrolase (60), which opens the cyclic ring to give Ins1P, which can then be dephosphorylated through the inositol 1-phosphatase. The appearance of cIns1:2P in pancreatic minilobules during stimulation with caerulein was used as evidence of direct phosphodiesteratic cleavage of PtdIns (61), but this cyclic phosphate may have originated from the step-wise dephosphorylation of cIns1:2,4,5P$_3$. The hydrolysis of cIns1:2,4,-5P$_3$ occurs very much more slowly than Ins1,4,5P$_3$, which may account for why the former was more effective when injected into *Limulus* photo-receptors (29).

PHOSPHORYLATION OF INS1,4,5P$_3$ The alternative pathway begins with a newly discovered Ins1,4,5P$_3$ 3-kinase that phosphorylates Ins1,4,5P$_3$ to inositol 1,3,4,5-tetrakisphosphate (Ins1,3,4,5P$_4$) (62). The kinase is a soluble enzyme first described in brain (62), but is apparently ubiquitous as it has been found in liver (63, 64), parotid (65), human T cells (66), and RINm5F cells (67). The kinase is an Mg^{2+}-requiring soluble enzyme, which transfers a phosphate from ATP specifically to the 3-position of Ins1,4,5P$_3$ to form Ins1,3,4,5P$_4$. The significance of the latter is that it is the precursor of inositol 1,3,4-trisphosphate (Ins1,3,4P$_3$) (Figure 1) (68). This new isomer was first described in parotid gland (69), and has since been found in many other cells (23, 54, 63–65, 70, 71).

A characteristic feature of this new Ins1,3,4P$_3$ isomer is that it has a somewhat slower turnover than Ins1,4,5P$_3$. In response to agonists, Ins-1,4,5P$_3$ builds up immediately, whereas there is a delay before Ins1,3,4P$_3$ begins to rise (23, 54, 63–65, 67, 70–72). When agonists are withdrawn, the Ins1,4,5P$_3$ isomer decays rapidly, and on the basis of in vitro experiments it has been estimated to have a half-life of approximately four seconds (47). On the other hand, the Ins1,3,4P$_3$ isomer has a much longer half-life (65). The latter is derived from Ins1,3,4,5P$_4$ by the specific removal of the 5-phosphate by an enzyme that is similar, if not identical, to the inositol trisphosphatase discussed earlier (68). The delay in the appearance of Ins1,3,4P$_3$ may depend on the time taken for Ins1,4,5P$_3$ to pass through this tris/tetrakis pathway, during which it undergoes a two-step isomerization using Ins1,3,4,5P$_4$ as an intermediary (Figure 1). The latter may also be the precursor for the formation of InsP$_5$ and InsP$_6$, which have been identified in various cells (71).

Like Ins1,4,5P$_3$, the new Ins1,3,4P$_3$ isomer is dephosphorylated through a sequence of dephosphorylation reactions that are beginning to be defined. Preliminary evidence suggests that one pathway depends on removal of the phosphate on the 1-position to give Ins3,4P$_2$ (R. F. Irvine, A. J. Letcher, D. J. Lander, J. P. Heslop, M. J. Berridge, unpublished observations), which is then dephosphorylated to either Ins3P or Ins4P. The latter is most likely since Ins4P has already been identified in brain (73). When cells are incubated in the presence of 10 mM Li$^+$ there is a selective enhancement of Ins1,3,4P$_3$ but no change in Ins1,4,5P$_3$ (64, 70).

In summary, the Ins1,4,5P$_3$ released from the membrane by the receptor can flow down two separate pathways. In one it is sequentially dephosphorylated to free inositol, whereas in the other pathway it is transformed into alternative polyphosphates before being dephosphorylated. Whenever one has such a bifurcation point in a metabolic pathway it is of interest to determine whether there is some control over the proportion of Ins1,4,5P$_3$ that is channeled down either pathway. There are indications that the DG/C-kinase pathway may act to stimulate the hydrolysis of Ins1,4,5P$_3$ (74, 75), whereas the increase in intracellular calcium may stimulate the kinase in RINm5F (67) and HL 60 cells (76), thus leading to an enhanced phosphorylation of Ins-1,4,5P$_3$ to Ins1,3,4,5P$_4$. This activation of both metabolic pathways may explain why the increase in Ins1,4,5P$_3$ formation that occurs following stimulation is so short-lived. Activation of the kinase by calcium will ensure the conversion of Ins1,4,5P$_3$ to Ins1,3,4,5P$_4$ and Ins1,3,4P$_3$, both of which may perform specific messenger functions within the cell (68, 69).

SECOND MESSENGER MODE OF ACTION

A key feature of this inositol lipid signaling system is that both products of PtdIns4,5P$_2$ hydrolysis function as second messengers, thus forming a bifurcating signal pathway for transferring information into the cell (Figure 2). While most attention here will be focused on how Ins1,4,5P$_3$ mobilizes calcium, it will be necessary to describe how DG stimulates protein kinase C, because the latter has a marked influence on all aspects of calcium signaling. While most attention has focused on the role of PtdIns4,5P$_2$ as a precursor for second messenger formation, it is worth considering the possibility that the hydrolysis and hence the removal of this lipid from the plasma membrane may in itself perform a 'messenger' role, particularly with regard to restructuring the cytoskeleton to alter cell shape.

Inositol Lipids and Cell Morphology

There are a number of observations indicating a functional relationship between PtdIns4,5P$_2$ and the cytoskeleton. In erythrocytes, PtdIns4,5P$_2$ appears to function in attaching the spectrin system to the plasma membrane by forming a link between glycophorin and protein 4.1 (77). Hydrolysis of this lipid component could possibly contribute to the remodeling of the cytoskeleton, which has been observed following stimulation of secretory cells (78). Hydrolysis of inositol lipids may also play a role in reorganizing actin; PtdIns4,5P$_2$ was found to dissociate profilactin, leading to the polymerization

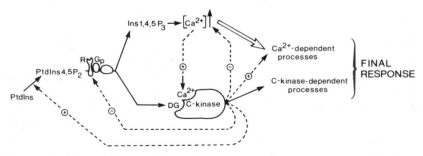

Figure 2 The dual signal hypothesis. The dashed lines represent the homologous interactions operating between the Ins1,4,5P$_3$/Ca^{2+} and DG/C-kinase pathways.

of actin (79). Following stimulation of blood platelets, there is a marked decrease in the G/F-actin ratio as actin polymerizes, and this might be driven by the hydrolysis of PtdIns4,5P$_2$. Another consequence of hydrolyzing PtdIns4,5P$_2$ is to produce DG, which has been shown to function in the attachment of α-actinin to the platelet membrane (80). The hydrolysis of PtdIns4,5P$_2$ may thus restructure microfilaments by enhancing both the polymerization of actin filaments and their attachment to the membrane.

The DG/C-Kinase Pathway

ACTIVATION OF PROTEIN KINASE C BY DG AND CALCIUM   The neutral DG that remains within the plane of the plasma membrane functions as a second messenger by activating protein kinase C (4, 8). In addition to DG, the enzyme requires calcium and phosphatidylserine (PS). When cells are activated there is usually an increase in calcium, which may contribute to the activation of protein kinase C perhaps by increasing the binding affinity of the enzyme for DG (81, 82). Experimentally the enzyme can be activated in intact cells either by adding a monoacyl-derivative of DG such as 1-oleoyl-2-acetylglycerol (OAG) or by means of phorbol esters (4, 8, 83). In some cells, such as the adrenal medulla, protein kinase C may be activated solely through calcium entering via voltage-dependent calcium channels (84). In most cells, however, where both limbs of the signal pathway are activated, DG and calcium probably act synergistically to stimulate the enzyme. An important aspect of the activation process appears to be a physical translocation of the enzyme from the cytosol into the membrane, which might be the role of calcium (85, 86). This calcium-mediated translocation into the membrane may then serve to 'prime' the enzyme to the stimulatory action of DG. A role for calcium in activating protein kinase C may be particularly relevant to the feedback interactions discussed later.

Calcium Signaling—the Ins1,4,5P$_3$/Ca^{2+} Pathway

SOURCE OF SIGNAL CALCIUM   For small cells with a large surface-to-volume ratio, entry of calcium across the plasma membrane, usually through voltage-dependent calcium channels, is sufficient for cell activation. There is no evidence for any role for inositol lipid hydrolysis in order to open such voltage-dependent calcium channels. This hydrolysis of inositol lipid is mainly confined to the action of calcium-mobilizing agonists, which bind to specific cell-surface receptors and gain access to both internal and external sources of calcium. The initial response to stimulation by calcium-mobilizing agonists is a release of internal calcium (Phase I), which is soon followed by entry of calcium across the plasma membrane (Phase II). Much of the calcium mobilized from the internal store during Phase I is pumped out of the cell (87), resulting in the calcium content of the cell declining by as much as 50%

(88–90). Removing external calcium or adding calcium entry antagonists usually has no effect on Phase I, but severely curtails Phase II (27, 81, 91). During Phase I, the level of intracellular calcium increases very fast, reaching a maximum value within seconds, but this high level is usually not maintained but decays back close to the resting level through the action of various negative feedback pathways that cooperate to curtail the calcium signal.

In considering how inositol lipid hydrolysis functions in calcium signaling, we must find answers to the following three questions. 1. How is calcium released from internal stores during Phase I? 2. How is calcium entry across the plasma membrane controlled during Phase II? 3. What feedback interactions exist to modulate this calcium signaling system?

MOBILIZATION OF INTERNAL CALCIUM BY INS1,4,5P₃   Of all the inositol phosphates that have been identified in cells, Ins1,4,5P₃ is the only one so far for which a clear second messenger role has been identified—it functions to release calcium from the endoplasmic reticulum (ER). Most attention has focused on its role in stimulating the release of calcium during cell activation, but of equal importance might be a role for Ins1,4,5P₃ in regulating the resting or basal level of calcium (92). Small variations in the endogenous turnover of inositol lipid may serve to adjust the resting level of Ins1,4,5P₃ and hence calcium. Even though activation of muscarinic receptors in adrenal chromaffin cells increases the levels of Ins1,4,5P₃ and calcium, there is no secretion, although responsiveness to nicotinic receptors was enhanced (93), apparently because of an increase in the resting level of calcium. Such alterations in responsiveness to incoming signals by resetting the resting level of calcium could be particularly important in modulating synaptic transmission in the nervous system.

The first direct evidence that Ins1,4,5P₃ functioned to mobilize intracellular calcium was obtained using permeabilized pancreatic cells (94). By incubating pancreatic cells in a low-calcium medium, the plasma membrane became leaky, thus allowing access to the internal membrane systems that sequester calcium. Using a calcium electrode it was possible to monitor the uptake of calcium into the ER and its subsequent release following the addition of Ins1,4,5P₃ (94). Similar observations have now been reported for many other permeabilized cells (Table 1). Based on all these studies, a detailed picture is beginning to emerge concerning this Ins1,4,5P₃-sensitive intracellular store of calcium. The identity of the store was initially based on the use of metabolic inhibitors, but subsequent cell fractionation experiments have confirmed that Ins1,4,5P₃ releases calcium from a membrane fraction derived solely from the ER (104, 123, 126, 131, 132). There is some degree of variability concerning the size of the Ins1,4,5P₃-sensitive pool (Table 1). In permeabilized liver cells, the ER contains 1 nM Ca^{2+}/mg protein, of which

0.5 nM is released by Ins1,4,5P$_3$, which is very similar to the amount released from intact cells in response to adrenaline (96). A kinetic analysis of $^{45}Ca^{2+}$ efflux from permeabilized hepatocytes estimated that approximately 35% of the nonmitochondrial calcium pool was sensitive to Ins1,4,5P$_3$ (133). In GH$_3$ cells, where Ins1,4,5P$_3$ releases approximately 40% of the stored calcium, addition of vanadate mobilizes the remainder, suggesting that this Ins1,4,5P$_3$-insensitive pool is stored within a separate ER membrane system (101). In the case of smooth muscle, electron microprobe X-ray analysis has revealed that agonists release a pool of calcium located close to the cell surface (134). It is possible, therefore, that the ER immediately below the plasma membrane is sensitive to Ins1,4,5P$_3$, whereas that lying deeper within the cell is insensitive. A similar conclusion was reached from experiments on *Xenopus* oocytes, where Ins1,4,5P$_3$ injected deep within the oocyte had less effect than when injected just below the cell surface (135).

Mechanism of release   Calcium is constantly cycling across the ER membrane, and all the available evidence points to Ins1,4,5P$_3$ acting to stimulate the passive efflux component while having no effect on the pump. Ins1,4,5P$_3$ had no effect on the calcium-sequestering system of adipocyte (129) or parotid gland ER (136). Once calcium has been accumulated within the ER, inhibiting the pump by adding vanadate or removing ATP results in a small release of calcium, which is enormously enhanced upon addition of Ins-1,4,5P$_3$ (97). The ability of Ins1,4,5P$_3$ to release calcium is independent of temperature (104, 114, 137, 137a). Since carrier-type antibiotics are markedly influenced by temperature, whereas ion channels are not, Smith et al (137) have argued by analogy that Ins1,4,5P$_3$ must act by opening a channel.

In order to release calcium, Ins1,4,5P$_3$ acts through a specific receptor, which may either be connected to or an integral part of the putative calcium channel. Of the inositol phosphates tested, only those having phosphates on the 4- and 5-position are capable of stimulating release (138, 139), and their order of potencies are Ins1,4,5P$_3$ > glycerophosphoinositol4,5P$_2$ = Ins-2,4,5P$_3$ > Ins4,5P$_2$. This sequence suggests that the two phosphates on the 4- and 5-position are essential to stimulate the release of calcium, whereas the phosphate on the opposite side of the molecule (1-position) functions to enhance the affinity of the molecule for its receptor. The product of Ins-1,4,5P$_3$ hydrolysis is Ins1,4P$_2$, which does not release calcium (94, 96, 137), nor do a number of other compounds including Ins1P, Ins2P, InsP$_6$, cIns1:2P, 2,3-diphosphoglyceric acid, fructose 1,6-bisphosphate, and *myo*-inositol.

First steps towards the isolation and characterization of the putative receptor are the identification of specific Ins1,4,5P$_3$ binding sites in various cells (140–142) and the development of a specific affinity label (143). A micro-

Table 1 Characteristics of (1,4,5)IP$_3$-induced calcium release from permeabilized cells and membrane fractions

Tissue	ER set-point (μM)	IP$_3$ concentration for half-maximal release (μM)	Amount of Ca^{2+} released (%)	Reference
Permeabilized cells				
Pancreas	0.4	0.4	—	94
Liver (rat)	0.45	0.1	—	95
Liver (guinea pig)	—	0.2	25	96
Neutrophil	0.2	0.6	—	97
Neutrophil	—	0.6	31	91
Insulinoma (Syrian hamster)	0.3–0.4	0.025	—	98
Insulinoma (RINm5F)	0.12	0.5	—	99
Insulinoma (RINm5F)	0.25	—	90	92
GH$_3$ cells	0.129	1.0	35	100
GH$_3$ cells	—	2.0	40	101
Swiss 3T3 cells	—	0.3	—	102
Porcine artery cells	—	0.7	—	103, 104
Rabbit mesenteric artery	—	1.0	40	105
Rat aortic smooth muscle	—	1.2	—	106
Dog tracheal smooth muscle	—	0.8	—	107
Macrophages	—	0.8	25	108, 109
Pancreatic islet	—	2.5	30	110
Parathyroid	0.60	0.3	—	111
Adrenal glomerulosa	—	0.7	~ 60	112
Mastocytoma	—	1.3	—	113
Blood platelet	—	1.1	65	114
Ehrlich ascites cells	—	0.4	—	115
Slime mold	—	—	70	116
N1E-115 neuronal cell line	—	0.5	30	117
Rat cortical kidney	0.55	—	—	118
Adrenal chromaffin cell	—	1.0	30	119
Lymphocytes	—	—	35	120
Membrane fractions				
Sea urchin homogenate	—	0.5–0.6	—	121
Liver microsomes	—	0.5	—	122
Insulinoma microsomes	0.1	3.0	—	123
Skeletal muscle	—	4–5	50	124
Hypocotyl microsomes	—	10–15	—	125
Bovine myometrium	—	—	40	126
Blood platelets	—	0.4	—	127
Blood platelets	—	0.25	40	128
Adipocytes	—	7.0	20	129
Sea urchin cortexes	—	0.4	—	130

somal fraction from bovine adrenal cortex had a single set of high-affinity Ins1,4,5P$_3$-binding sites with a K_d of 5.2 nM (140). Liver microsomes appeared to have two binding components with K_d values of 7.9 nM and 16 μM (141). Binding sites could also be demonstrated on permeabilized liver cells and neutrophils (142). The ability of different inositol phosphates to displace the ^{32}P-Ins1,4,5P$_3$ showed exactly the same specificity as for the release of calcium described earlier, i.e. Ins1,4,5P$_3$ > Ins2,4,5P$_3$ > Ins-4,5P$_2$. Spät et al (142) draw attention to the fact that the estimated receptor density is less than the K_d of the putative receptor, which means that there will be a close correspondence between Ins1,4,5P$_3$ formation and calcium mobilization. This point supports the view outlined earlier that Ins1,4,5P$_3$ may have a crucial role in regulating the level of intracellular calcium even when the cell is at rest.

An arylazide derivative of Ins1,4,5P$_3$ has been developed as a photoaffinity label of the receptor (143). When added to permeabilized cells and activated by light this label causes a specific inhibition of Ins1,4,5P$_3$-induced calcium release. This irreversible inhibition was prevented if the cells were incubated with a 10-fold excess of unmodified Ins1,4,5P$_3$. The development of this affinity label may provide a first step in the isolation of the Ins1,4,5P$_3$ receptor.

Ionic and nucleotide requirements of Ins1,4,5P$_3$-induced calcium release
Most of the studies using permeabilized cells or microsomes have been performed in media designed to resemble the cytoplasm, particularly in the high concentration of potassium. The latter turns out to be essential for both components of the ER calcium cycle (137a, 144). Of particular significance is that calcium release by Ins1,4,5P$_3$ requires potassium to function as a counterion to neutralize the buildup of charge resulting from the efflux of calcium. In addition to the calcium cycle, the ER has a potassium cycle based on a tetraethylammonium-sensitive potassium conductance and a furosemide-sensitive potassium/chloride cotransport pathway (144). The effect of Ins-1,4,5P$_3$ on calcium release seems to be insensitive to sodium (114), but is markedly enhanced when pH is increased from 6.7 to 7.5 (114, 121). When cells are stimulated, the pH often shifts to higher pH values, which would then enhance the calcium-mobilizing action of Ins1,4,5P$_3$ (114).

There is considerable uncertainty at present concerning the role of adenine and guanine nucleotides in Ins1,4,5P$_3$-induced calcium release. In order to maintain the calcium cycle, most experiments are conducted in the presence of ATP, but when the latter was removed from permeabilized neutrophils through the addition of glucose and hexokinase, Ins1,4,5P$_3$ was still able to release calcium (97). On the other hand, the release of calcium from per-meabilized smooth muscle cells required the presence of ATP (137) or was augmented by ATP (104). There was some degree of selectivity in that ADP and GTP were less effective, whereas AMP and CTP were inactive (137). ATP is apparently not required as a phosphate donor; the nonhydrolyzable ATP analogues AppNHp and AppCH$_2$p were equally active.

There is also some uncertainty concerning the role of GTP in the mode of action of Ins1,4,5P$_3$. Although Ins1,4,5P$_3$ can release calcium from microsomes, the effect is usually small and requires higher levels of Ins1,4,5P$_3$ than the corresponding permeabilized cells (Table 1). The Ins1,4,5P$_3$-induced release of calcium from liver microsomes was greatly enhanced by the addition of 10 μM GTP (145). This augmentation of release was not mimicked by the nonhydrolyzable GTP analogue GppNHp and was also dependent on having polyethylene glycol (PEG) in the incubation medium. When microsomes were incubated with ^{32}P-GTP, two proteins (M_r 38,000 and 17,000) were phosphorylated, and it is suggested that they play a role in the action of GTP (146). The requirement for PEG is explained by its ability to promote the binding of these proteins to the ER.

A role for GTP has also been identified in a neuronal cell line (N1E-115) where the release of calcium from a microsomal preparation by Ins1,4,5P$_3$ (1 μM) was greatly enhanced by the addition of GTP (1 μM) (117). Unlike the liver microsomes described above, GTP by itself released more calcium than Ins1,4,5P$_3$. GTP was also found to release calcium from parotid gland vesicles (136). Unlike Ins1,4,5P$_3$-induced release, this GTP-induced mobilization of calcium is specifically blocked by GDP or by lowering the temperature to 2°C (136, 147). Consistent with this evidence that Ins1,4,5P$_3$ and GTP operate through separate mechanisms, their effects were additive (136). Based on the fact that half-maximal release of calcium by GTP occurred at 0.3 μM, which is very much lower than the normal cellular level of this nucleotide, it is difficult to determine the physiological relevance of this release mechanism. Clearly, we need to learn more about the relationship between Ins1,4,5P$_3$ and GTP in the control of calcium release from the ER.

CALCIUM ENTRY ACROSS THE PLASMA MEMBRANE   During Phase II, when the cell switches over to using extracellular calcium, it is necessary to account for the enhanced entry of calcium induced by agonists. The term receptor-operated calcium channels was introduced in order to distinguish it from calcium entry mediated by voltage-dependent calcium channels. However, such receptor-operated calcium channels have never been detected in patch-clamp experiments, suggesting that a simple model of a channel being opened by the binding of an agonist may not be applicable. A number of hypotheses have been advanced to explain how the entry of external calcium may depend on the hydrolysis of inositol lipids.

A role for phosphatidic acid   One possibility is that the receptor mechanism may generate its own channel in the form of phosphatidic acid (PA) (148, 149). The idea is that the DG formed by the hydrolysis of PtdIns4,5P$_2$ is phosphorylated to PA, which then acts as a calcium ionophore before it is resynthesized back to PtdIns. Apart from the fact that it is formed during stimulation, the main evidence implicating an ionophoric role for PA is based on the fact that the addition of this lipid can stimulate various cells (148, 149). When tested on artificial membranes, however, PA failed to increase the flux of calcium (150), which seemed to argue that it may not act simply as a calcium ionophore. Recent studies on human A431 carcinoma cells reveal that PA increases intracellular calcium by stimulating the formation of Ins-1,4,5P$_3$ (151). Instead of acting as an ionophore, PA mimicked the action of other growth factors by stimulating the hydrolysis of PtdIns4,5P$_2$ (151).

A role for inositol phosphates   The other hypotheses for linking inositol lipids to calcium entry across the plasma membrane are both centered around a role for inositol polyphosphates. There apparently is no direct role for Ins1,4,5P$_3$ in regulating calcium entry across the plasma membrane; this second messenger had no effect on calcium release from plasma membrane vesicles (117, 129). However, there is preliminary evidence that Ins1,3,4,5P$_4$ may function to control calcium entry across the plasma membrane (63). The idea is that Ins1,4,5P$_3$ would release calcium from the ER during Phase I and then would be converted into Ins1,3,4,5P$_4$, which would function to control calcium entry across the plasma membrane during Phase II. Since the conversion of Ins1,4,5P$_3$ to Ins1,3,4,5P$_4$ is enhanced by calcium (67, 76), stimulation of influx by Ins1,3,4,5P$_4$ would be an example of calcium-induced calcium entry. The first direct evidence that Ins1,3,4,5P$_4$ acts to promote calcium entry has been obtained by injecting this molecule into eggs of the sea urchin *Lytechninus variegatus* (151a). Injection of submicromolar concentrations of Ins1,3,4,5P$_4$ caused an immediate elevation of the fertilization envelope, which required the presence of extracellular calcium. In addition, the action of Ins1,3,4,5P$_4$ was dependent on the coinjection of Ins2,4,5P$_3$, which is known to mobilize intracellular calcium (138, 139). In some way the action of Ins1,3,4,5P$_4$ seems to require the prior emptying of intracellular calcium stores. Such a mechanism contains the main element of another hypothesis (152) concerning the way in which inositol polyphosphates might act to regulate calcium entry.

Putney (152) has argued that Ins1,4,5P$_3$ may influence calcium entry indirectly through its effect on the mobilization of calcium from the ER. In many cells there is a component of the ER that is closely applied to the plasma membrane (153, 154). Once this pool has been depleted it can be reloaded with calcium entering from the outside without any change in the intracellular level of calcium (155). There appears to be a specialized pathway for calcium to move from outside the cell into the ER. It is this component of ER that is depleted of its calcium when smooth muscle is stimulated (134). The idea is that the permeability of the plasma membrane is somehow controlled by the calcium content of the underlying ER. When the latter is full of calcium, there is little calcium entry, but as $Ins1,4,5P_3$ begins to discharge this pool, calcium begins to flow in from the outside (152). Just how the ER dictates the permeability of the plasma membrane is still unclear. An attractive feature of this hypothesis is that, through a single site of action on the ER, $Ins1,4,5P_3$ can regulate both the release of internal calcium and the entry of external calcium.

Inhibition of calcium efflux   As for the ER, there is a constant cycling of calcium across the plasma membrane with passive influx being balanced by an active extrusion of calcium. There are indications from studies carried out on liver cells that calcium-mobilizing hormones may act by inhibiting the calcium pump (156). Since the calcium pump seems to require the presence of inositol lipids, the decrease in $PtdIns4,5P_2$ content that occurs during stimulation may result in a damping of the pump, thus enhancing the influx component of the plasma membrane calcium cycle. There is evidence to suggest that $Ins1,4,5P_3$ inhibits the calcium pump on the sarcolemma of coronary arteries (157).

Despite the elegance of these various proposals, we still do not know how the agonist-dependent hydrolysis of $PtdIns4,5P_2$ controls calcium entry across the plasma membrane.

SECOND MESSENGER ROLE FOR $INS1,3,4P_3$ Since the cell invests energy to convert $Ins1,4,5P_3$ to the new inositol polyphosphates, it is reasonable to suspect that they may have some second messenger function (68, 69). A possible role for $Ins1,3,4,5P_4$ in promoting calcium entry was mentioned earlier, but a function for $Ins1,3,4P_3$ has yet to be found. Recent experiments have shown that $Ins1,3,4P_3$ will release calcium from permeabilized Swiss 3T3 cells but with much lower efficacy (R. F. Irvine, A. J. Letcher, D. J. Lander, and M. J. Berridge, unpublished observation). However, the concentration of $Ins1,3,4P_3$ is known to rise far above that of $Ins1,4,5P_3$, so it is possible that the former may function to keep the internal pools of calcium discharged during Phase II, after the level of $Ins1,4,5P_3$ has returned close to its basal level.

DUAL SIGNAL HYPOTHESIS OF CELL ACTIVATION

The hallmark of this inositol lipid signaling system is that there is a bifurcation in the flow of information. The dual signal hypothesis concerns the way in which the $Ins1,4,5P_3/Ca^{2+}$ and DG/C-kinase signal pathways cooperate with each other to control a whole host of cellular processes. In many cells the two pathways appear to act synergistically (4, 8). The existence of a synergistic interaction was demonstrated by using pharmacological agents capable of stimulating each pathway independently of the other. Calcium ionophores (A23187 or ionomycin) were used to raise intracellular calcium, whereas phorbol esters or OAG could mimic the stimulatory effect of DG on protein kinase C. By combining a threshold concentration of an ionophore with a threshold concentration of an activator of protein kinase C, it is possible to activate many different cellular responses, including secretion of enzymes, hormones, and neurotransmitters; muscle contraction; and DNA synthesis. While such evidence suggests roles for both pathways, it provides no indication of their relative contributions to the final response. The dual signal hypothesis outlined in Figure 2 integrates information from many different cell types and summarizes how both limbs of the signal pathway function in the control of cellular activity:

It is proposed that the $Ins1,4,5P_3/Ca^{2+}$ pathway plays a major and direct role in initiating cellular responses. The DG/C-kinase pathway may also contribute directly to the final response, but its predominant role appears to be as a modulator of either the calcium-signaling pathway (homologous interactions) or of other signal pathways (heterologous interactions).

The Primary Role of Ins1,4,5P₃

The view that $Ins1,4,5P_3$ is the primary second messenger in initiating final responses is based mainly on the fact that it controls the level of intracellular calcium, which has been recognized as a major second messenger for stimulating a whole range of cellular processes such as contraction, secretion, and metabolism. Secondly, $Ins1,4,5P_3$ can trigger a number of complex physiological processes when added to permeabilized cells or when injected into intact cells (Table 2).

PERMEABILIZED CELLS   Addition of $Ins1,4,5P_3$ to permeabilized muscle cells can cause them to contract by stimulating calcium release from the sarcoplasmic reticulum. While there is general agreement that $Ins1,4,5P_3$ plays a central role in pharmacomechanical coupling in smooth muscle (105, 158, 159), there is less agreement concerning its role in excitation-contraction (E-C) coupling in skeletal muscle. Some groups have found that skinned skeletal muscle cells will contract in response to $Ins1,4,5P_3$ (124, 160), while others have found no effect (127, 182, 183). Another suggestion is that $Ins1,4,5P_3$ may act in skeletal muscle by increasing the sensitivity of the contractile mechanism to calcium (184). If $Ins1,4,5P_3$ is to function as a diffusible messenger between the cell surface and the sarcoplasmic reticulum (SR) membrane (185), it must be formed during the brief instant when the

Table 2 Summary of cellular responses elicited by adding Ins1,4,5P$_3$ to either permeabilized or intact cells

Cell type	Response	Reference
Permeabilized cells		
Vascular smooth muscle	Contraction	105, 158
Stomach smooth muscle	Contraction	159
Skeletal muscle	Contraction	124, 160, 161
Slime mold	Cyclic GMP formation	162
Slime mold	Actin polymerization	163
Sea urchin egg	Cortical reaction	121
Blood platelets	Protein phosphorylation	164–166
Blood platelets	Shape change	166, 167
Blood platelets	Aggregation	165–167
Blood platelets	5-HT secretion	114, 165, 166
Blood platelets	Formation of thromboxane B$_2$	165–167
Intact cells		
Limulus photoreceptors	Phototransduction and adaptation	168, 169
Limulus photoreceptors	Increased intracellular calcium	29, 170, 171
Salamander rods	Modulation of light response	172
Xenopus oocytes	Calcium mobilization	173
Xenopus oocytes	Membrane depolarization	174
Xenopus eggs	Membrane depolarization	135
Xenopus eggs	Cortical reaction	135, 175
Sea urchin eggs	Membrane depolarization	176
Sea urchin eggs	Cortical reaction	177, 178
NG 108-15 cells	Membrane hyperpolarization	179, 187
Smooth muscle	Increased potassium current	180
Lacrimal gland	Increased potassium current	181

t-tubule membrane is depolarized. It has been suggested that a G-protein may play a role in translating membrane depolarization into the activation of the phosphoinositidase that cleaves PtdIns4,5P$_2$ to release Ins1,4,5P$_3$ (186).

The chemotactic response of the slime mold *Dictyostelium discoideum* to pulses of cyclic AMP seems to be mediated through Ins1,4,5P$_3$, which can trigger both cyclic GMP formation (162) and actin polymerization (163) when added to permeabilized cells. The latter also responded to Ins1,4,5P$_3$ with a release of calcium (116). Finally, addition of Ins1,4,5P$_3$ to permeabilized blood platelets can trigger the complete heirarchy of normal cellular responses beginning with the change in shape and followed by aggregation, release of 5-HT, and the production of TXB$_2$ (114, 165–167). The induction of this panoply of effects was blocked by having EGTA in the bathing medium, indicating that $Ins1,4,5P_3$ triggered all these responses by mobilizing calcium (167). Other experiments have already shown that $Ins1,4,5P_3$ can release calcium from either platelet membrane vesicles (127, 128, 132) or from permeabilized platelets (29, 114). Taken at face value it would appear that $Ins1,4,5P_3$ acts to release calcium, which then triggers all these platelet responses. However, the sequence of reactions is a lot more complicated because the stimulatory effect of adding $Ins1,4,5P_3$ was markedly suppressed or, in some cases, abolished, by having indomethacin or aspirin in the bathing medium to block the formation of TXB_2 (165–167). The formation of TXB_2 may enable the $Ins1,4,5P_3/Ca^{2+}$ pathway to mediate its stimulatory effect by enhancing the hydrolysis of $PtdIns4,5P_2$ to generate additional mediators such as DG (165). Through this complex sequence of events, the addition of $Ins1,4,5P_3$ to permeabilized platelets may indirectly result in activation of the bifurcating signal pathway to produce the same spectrum of second messengers normally responsible for controlling shape change, aggregation, and the release reaction of intact cells.

INTACT CELLS   A primary messenger role for $Ins1,4,5P_3$ has been demonstrated by showing that it can trigger a larger number of physiological processes when injected into intact cells. *Limulus* photoreceptors respond to light with membrane depolarization resulting from the opening of sodium channels, an effect that can be exactly mimicked by injecting $Ins1,4,5P_3$ (168, 169). In addition, the adaptation that occurs in response to bright light is exactly mimicked by injecting $Ins1,4,5P_3$. The photosensitive R-lobe has a well-developed ER system located immediately below the invaginated cell surface containing the rhodopsin. Following injection with $Ins1,4,5P_3$ there is an increase in the intracellular level of calcium (170, 171) located specifically in the R-lobe (188).

Many of the physiological responses found in oocytes and eggs, including events associated with fertilization, can be exactly mimicked by injecting $Ins1,4,5P_3$. Immature oocytes of *Xenopus* have muscarinic receptors concentrated in the animal half of the cell, which respond to acetylcholine by stimulating the formation of $Ins1,4,5P_3$ (174). Injection of $Ins1,4,5P_3$ released calcium, which opens chloride channels to produce an electrophysiological response that closely resembles that induced by acetylcholine (173, 174). As these *Xenopus* oocytes mature in response to progesterone they lose their sensitivity to acetylcholine, but they retain all the elements of the inositol-lipid signal pathway, which is re-employed to carry out some of the early events of fertilization. The fusion of a sperm with an egg appears to be analogous to a hormone interacting with its receptor in that there is an increase in inositol lipid metabolism and a release of $Ins1,4,5P_3$ (189, 190). Injection of Ins1,4,5P$_3$ into eggs can mimic many of the early events of fertilization such as the release of cortical granules to form the fertilization membrane (135, 175, 177, 178) and the opening of ion channels resulting in the fertilization potential (135, 176). Injection of Ins1,4,5P$_3$ into *Xenopus* oocytes results in a prompt increase in intracellular calcium derived solely from internal stores, since removing external calcium had no effect (135).

Introducing Ins1,4,5P$_3$ into gland cells (181), smooth muscle (180), and neural cells (179, 187) can reproduce many of the electrophysiological responses elicited normally by the appropriate agonist. Many of these effects can be accounted for by Ins1,4,5P$_3$ mobilizing intracellular calcium, which then acts to open calcium-dependent potassium channels leading to membrane hyperpolarization. For example, the addition of bradykinin to a neuroblastoma/glioma hybrid cell line (NG 108-15) results in an initial hyperpolarization followed later by a longer-lasting depolarization (179, 191). Bradykinin is known to stimulate the hydrolysis of PtdIns4,5P$_2$ (191) with the release of Ins1,4,5P$_3$ (192) and a large increase in the intracellular level of calcium (193). Injection of Ins1,4,5P$_3$ into NG 108-15 cells reproduced exactly the early hyperpolarizing phase of the response, which is caused by the opening of Ca^{2+}-dependent potassium channels (179, 187). The subsequent depolarizing phase results from closure of a different voltage-dependent potassium current (M-current), which appears to be controlled by the DG/C-kinase limb of the signal pathway (187). The closing of potassium channels in response to protein kinase C activation has also been described in pyramidal hippocampal neurons (194) and *Hermissenda* photoreceptors (195) and may thus represent a general function of the DG/C-kinase pathway.

Modulatory Function of the DG/C-Kinase Pathway

Once protein kinase C has been activated through the concerted action of DG and calcium, it begins to phosphorylate specific proteins that are thought to contribute to the final response. Although protein kinase C can phosphorylate a large number of proteins when studied in vitro, it has proved extraordinarily difficult to identify the function of the smaller number of proteins that are phosphorylated in intact cells. Some of the identified substrates include vinculin, the epidermal growth factor (EGF) receptor, glycogen synthase, and lipocortin. Phosphorylation of the EGF receptor results in a decrease in both the affinity of the receptor for EGF and its tyrosine kinase activity, which is a good example of the modulatory action of protein kinase C.

Identification of the protein kinase C substrates mentioned above supports the hypothesis outlined earlier concerning the two main functions of protein kinase C. Liver provides a good example where the two signal pathways have two separate sites of action, both of which contribute to the final response of glycogen breakdown. The Ins1,4,5P$_3$/Ca^{2+} pathway controls the activity of phosphorylase kinase, whereas the DG/C-kinase pathway switches off gly-cogen synthase (7). In many other cases where the two pathways impinge on the same final effector system (e.g. control of exocytosis or transcription of the *myc* oncogene) their precise mode of action is unknown. In pancreas, the two pathways phosphorylate different sets of proteins, yet both can stimulate amylase release, to varying extents depending on cell type (196).

One way of trying to discern the contribution of protein kinase C is to block its activity either by adding the inhibitors H-7 and W-9 (197) or by down-regulating the enzyme by chronic treatment with a phorbol ester (198). When added to neutrophils, the inhibitor H-7 blocked the specific 50-kd protein phosphorylation induced by protein kinase C but failed to prevent the normal respiratory burst or secretory response (199, 200). Another example is found in fibroblasts where protein kinase C can be down-regulated by prolonged treatment with phorbol esters without seriously impairing their ability to respond to platelet-derived growth factor (PDGF) by increasing c-*myc* transcription and DNA synthesis (201). Although protein kinase C is capable of contributing to these various processes, it appears not to be essential, suggesting a degree of redundancy within the bifurcating signal pathway. The fact that c-*myc* can be activated by calcium ionophores (202, 203) provides further evidence that the $Ins1,4,5P_3/Ca^{2+}$ pathway represents one of the primary signal mechanisms for controlling early gene transcription during the action of growth factors. One way in which the DG/C-kinase pathway might contribute to the final response is to act indirectly to enhance the activity of this primary calcium-signaling pathway (Figure 2). Evidence for such a mechanism of 'sensitivity modulation' (204) has come from secretory cells, where the DG/C-kinase pathway can sensitize the exocytotic process to the stimulatory effect of calcium (84, 205). By sensitizing calcium-dependent processes, the DG/C-kinase limb of the signal pathway may be particularly important for maintaining cellular activity during Phase II (81, 206, 206a), and may explain how secretory responses in cells can be maintained at resting levels of intracellular calcium (207–209). As we learn more about protein kinase C, it becomes apparent that many of its functions are connected with modulating processes within the inositol lipid pathway (homologous in-teractions) or those connected with other signal pathways, particularly the cyclic AMP system (heterologous interactions).

HOMOLOGOUS INTERACTIONS RELATED TO CALCIUM SIGNALING   The mechanisms responsible for increasing the intracellular concentration of cal-cium are under stringent negative feedback control. In order to respond rapidly to external stimuli, there usually is an explosive increase in in-tracellular calcium during Phase I far exceeding that necessary to activate the cell. If unchecked, the calcium level may rise to levels that would damage the cell, which may explain the negative feedback interactions (Figure 2) that cooperate to curtail the increase in intracellular calcium levels, especially when cells are strongly stimulated.

Of particular importance with regard to curbing increases in intracellular calcium are the two negative feedback loops where protein kinase C acts both to inhibit the hydrolysis of PtdIns4,5P$_2$ and to reduce intracellular calcium by stimulating calcium pumps. Protein kinase C is also at the heart of the positive feedback interactions in that its activity is enhanced by calcium and it also seems to be responsible for stimulating the kinases that form PtdIns4,5P$_2$ (Figure 2). Protein kinase C occupies a pivotal position, because it can either enhance or inhibit calcium signaling; its net effect will depend on the degree to which these feedback mechanisms are expressed in each particular cell. For example, activation of protein kinase C with phorbol esters enhances the contractility of certain smooth muscle cells but relaxes others (194). Within the same muscle, agonist-dependent contractions were reduced, presumably due to the negative feedback effect on PtdIns4,5P$_2$ hydrolysis, whereas potassium-induced contractions were enhanced (210). From the large repertoire of second messenger interactions at its disposal, each cell selects those that enable it to respond appropriately to external signals.

Stimulation of PtdIns4,5P$_2$ formation   When cells are stimulated, especially with a high concentration of agonist, there initially is a fall in the level of PtdIns4,5P$_2$, but the level is soon restored and sometimes exceeds the resting level. This increase in PtdIns4,5P$_2$ formation might be a consequence of a positive feedback effect operated through protein kinase C because similar increases can be obtained when certain cells are stimulated with phorbol esters (211–214). In the case of thymocytes, the increase in PtdIns4,5P$_2$ labeling brought about by phorbol esters resulted in an increased formation of Ins-1,4,5P$_3$ when the cells were subsequently challenged with concanavalin A (213). The DG/C-kinase limb of the bifurcating signal pathway may thus have an important effect on inositol lipid metabolism by enhancing the formation of the PtdIns4,5P$_2$ used by the receptor mechanism.

Inhibition of calcium signaling   An important action of the DG/C-kinase pathway is to inhibit calcium signaling. The basic observation is that pretreatment with phorbol esters reduces agonist-induced increases in intracellular calcium in rat basophilic leukemia cells (215), blood platelets (216, 217), liver (218, 219), Swiss 3T3 cells (220, 221), neutrophils (222), vascular smooth muscle (223), adrenal glomerulosa cells (224), WRK-1 cells (225), and human A431 cells (226). The DG/C-kinase pathway can modulate calcium signaling either by inhibiting the hydrolysis of PtdIns4,5P$_2$, thus preventing the generation of Ins1,4,5P$_3$, or by activating the calcium pumps that remove calcium from the cytosol. When cells are pretreated with a phorbol ester or OAG to activate protein kinase C, there is a marked reduction in agonist-induced PtdIns4,5P$_2$ hydrolysis and inositol phosphate formation in cultured human 1321N1 astrocytoma cells (227), hippocampal slices (228), blood platelets (216, 229, 230), vascular smooth muscle (223), GH$_3$ cells (217), WRK-1 cells (225), and adrenal glomerulosa cells (224). There are a number of sites where the DG/C-kinase pathway could act to reduce the agonist-dependent hydrolysis of PtdIns4,5P$_2$. In liver, which has a number of receptors working through the inositol lipid pathway, phorbol esters seem to exert a selective inhibition of α_1-receptors, while the effects of vasopressin and angiotensin II were relatively unimpaired (218, 231). A similar situation exists in Swiss 3T3 cells where a phorbol ester inhibited inositol phosphate formation induced by bombesin but not that of PDGF (232). Such selective actions suggest that the DG/C-kinase pathway may exert its negative feedback effect by phosphorylating receptors rather than G$_p$ or the phosphoinositidase. In cultured smooth muscle cells, the phorbol ester–induced inhibitory effect on PtdIns metabolism mediated by α_1-agonists was associated with a rapid ($t_{1/2} = 2$ min) increase in the phosphorylation of the α_1-adrenergic receptor (233).

Another important negative feedback effect on calcium signaling exerted by the DG/C-kinase pathway is to stimulate the removal of calcium from the cytoplasmic compartment. Phorbol esters will reduce not only agonist-induced calcium rises but also those due to high potassium in GH$_3$ cells (217) or the calcium ionophore A23187 in neutrophils (234). Such observations could be explained by protein kinase C activating calcium pumps (235). Protein kinase C may also act to reduce calcium signaling by stimulating the enzyme that hydrolyzes Ins1,4,5P$_3$ (74, 75).

HETEROLOGOUS INTERACTIONS RELATED TO VOLTAGE-DEPENDENT CAL-CIUM CHANNELS AND MEMBRANE POTENTIAL   The inositol lipid transduction mechanism embedded within the plasma membrane is sensitive to membrane potential. Hydrolysis of inositol lipid in blowfly salivary gland by 5-HT was enhanced by membrane hyperpolarization but suppressed when the membrane was depolarized (236). Inhibition of receptor-activated inositol phosphate formation in hippocampus by excitatory amino acids may have resulted from their ability to depolarize the membrane. The situation becomes more complex when calcium-mobilizing receptors are found together with voltage-dependent calcium channels (e.g. in adrenal medulla, β-cells, smooth muscle, and synaptic endings). In such cases, membrane depolarization opens channels, allowing calcium to flood into the cell, and the concentration immediately below the plasma membrane may reach levels as high as 10^{-4} M, which is high enough to stimulate an agonist-independent breakdown of

PtdIns4,5P$_2$ (237). The large accumulation of calcium that occurs in pre-synaptic neurons during tetanic stimulation (238) may be sufficient to induce such lipid hydrolysis, and the resulting formation of DG could explain the activation of protein kinase C found in hippocampal neurons during the onset of long-term potentiation (239). The increased formation of inositol phos-phates following membrane depolarization reported in brain (240) and smooth muscle (241) were only partially blocked by adding calcium antagonists, suggesting that not all the increase is due to calcium. In skeletal muscle, where voltage-dependent calcium channels are absent, electrical stimulation resulted in an increased formation of inositol phosphates (124), suggesting that the hydrolysis of PtdIns4,5P$_2$ may be sensitive to voltage in some cells.

A reciprocal relationship has been uncovered in that the DG/C-kinase pathway modulates the activity of voltage-dependent calcium channels. In *Aplysia* neurons (242), *Hermissenda* photoreceptors (195), and rat adrenal medulla (243), phorbol esters facilitated the entry of calcium through voltage-dependent calcium channels, whereas in two mammalian cell lines it was reduced (244, 245).

HETEROLOGOUS INTERACTIONS RELATED TO CYCLIC NUCLEOTIDES   In addition to the interactions operating within the inositol lipid pathway, there are reciprocal interactions with other second messenger pathways, particular-ly the cyclic nucleotides (4, 8). There are numerous examples where the stimulation of receptors that hydrolyze inositol lipids strongly potentiate those receptors that form cyclic AMP. For example, the production of cyclic AMP in the pineal by β-receptors is enhanced by activation of α_1-receptors (246). Similarly, adenosine- or VIP-induced accumulation of cyclic AMP in brain is increased by the simultaneous activation of α_1-, 5-HT$_2$-, or H$_1$-receptors, all of which stimulate inositol lipid hydrolysis (247). This enhancement of cyclic AMP formation appears to be mediated through the DG/C-kinase pathway because phorbol esters augmented the accumulation of cyclic AMP by β-receptors in the pineal gland (246), frog erythrocytes (248), and S49 lympho-ma cells (249). The increase in cyclic AMP induced by adenosine in brain is also augmented by phorbol esters (250). Just where protein kinase C acts is unclear, but the evidence suggests that it may facilitate the interaction be-tween G$_s$ and adenylate cyclase (249). In blood platelets, protein kinase C phosphorylates G$_i$, so reducing the activity of the pathway, which inhibits adenylate cyclase while having no effect on the stimulatory pathway (251).

The reciprocal relationship between cyclic nucleotide and inositol lipid metabolism has not received much attention except in a few cell types such as blood platelets. Thrombin-induced platelet responses such as the increase in intracellular calcium, phosphorylation of the 40K protein, aggregation, and the release of 5-HT are all inhibited by agents that increase cyclic AMP

(252–254). This inhibitory effect of cyclic AMP is achieved by inhibiting the hydrolysis of PtdIns4,5P$_2$; there is a marked reduction in both DG (255) and Ins1,4,5P$_3$ formation (256, 257). Cyclic GMP plays a similar role in inhibiting agonist-induced hydrolysis of PtdIns4,5P$_2$ in platelets (257) and in smooth muscle (258). The site of action of the cyclic nucleotides is unclear, but there are suggestions that they may interefere with the enzymes that supply PtdIns4,5P$_2$ (259, 260).

Spatial and Temporal Aspects of Signaling

As more is learned about second messenger systems, it becomes apparent that we have to pay more attention to spatial and temporal aspects of signaling (89, 101). Spatial aspects concern the nonuniform spatial distribution of second messengers, whereas temporal aspects deal with growing evidence that second messenger levels may not be constant but may oscillate.

Spatial aspects   Various components of the second messenger pathway such as receptors, calcium stores, and the enzymes responsible for metabolizing Ins1,4,5P$_3$ are not equally distributed in the cell, which means that there may be localized concentrations of calcium. In *Xenopus* oocytes, for example, the muscarinic receptors that generate Ins1,4,5P$_3$ are localized at the animal pole (261). Compared to the vegetal pole, injections of calcium into the animal pole also gave much larger increases in chloride current (262), suggesting a localization of calcium-sensitive processes in one region of the cell. Another striking example is found in *Limulus* photoreceptors where the Ins1,4,5P$_3$-sensitive release of calcium is located specifically within the light-sensitive R-lobe (188).

Temporal aspects   There are many examples of oscillatory physiological processes such as the periodic release of cyclic AMP from the slime mold *Dictyostelium*, rhythmical cytoplasmic streaming in *Physarum*, pacemaker neurons in *Aplysia*, myogenic rhythms in smooth muscle, oscillations in membrane potential in insulin-secreting β-cells, anterior pituitary, and the insect salivary gland (263, 264). The oscillations recorded in the insect salivary gland are caused by periodic opening of calcium-dependent chloride channels, suggesting that the intracellular level of calcium was oscillating. The rhythmic membrane hyperpolarizations recorded from sympathetic ganglion cells (265) and hamster eggs (266) are thought to be driven by the periodic mobilization of calcium from an intracellular reservoir. Sustained oscillations in intracellular calcium have now been recorded in mouse oocytes (267), multinucleated L cells (268), cultured intestinal cells (269), and single hepatocytes (270). In the intestinal cells, the oscillations in membrane potential occurred in phase with the oscillations in intracellular calcium (269). In an earlier analysis, a model was put forward to explain such oscillations based on feedback interactions operating between cyclic AMP and calcium (263). The homologous feedback interactions operating within the inositol lipid pathway (Figure 2) suggest additional mechanisms for setting up such oscillations. A prediction from the earlier model (263) is that frequency of oscillations will increase if there is an increased input into any component of the oscillator. It is of some interest, therefore, to find that the frequency of oscillations in *Calliphora* salivary gland were enhanced by increasing the concentration of 5-HT (271). Similarly, the frequency of the periodic bursts of calcium in hepatocytes was accelerated by increasing vasopressin concentrations (270). Since inositol lipid hydrolysis is triggered by 5-HT in insect salivary gland (272) and by vasopressin in hepatocytes (7), it is reasonable to suppose that the second messengers associated with this receptor mechanism may be part of the biochemical oscillator. Such a possibility is supported by the observation that phorbol esters, which activate protein kinase C, induce oscillations in mouse oocytes (267). Large spontaneous transients in intracellular calcium have been recorded in rat mast cells, especially following perfusion with PS, which is a cofactor for protein kinase C (273). The frequency and amplitude of contractile oscillations in skinned cardiac fibers is increased following addition of Ins1,4,5P$_3$ (161), again suggesting a possible role for the inositol lipid pathway in either setting up or regulating intracellular oscillations in calcium. *Xenopus* oocytes show spontaneous fluctuations, which are sometimes cyclic in nature, and it was proposed that they represent the transient accumulation of a 'channel gating substance' (261). It is tempting to suggest that this substance might be Ins1,4,5P$_3$ functioning as part of a biochemical oscillator to give regular fluctuations in the intracellular level of calcium. In support of this suggestion, injection of Ins1,4,5P$_3$ into *Xenopus* oocytes resulted in oscillatory chloride currents, indicating that this second messenger has triggered oscillations in intracellular calcium (274).

In the insect gland there was also a clear relationship between oscillation frequency and the rate of fluid secretion, suggesting that control might be exercised through a frequency-dependent rather than an amplitude-dependent mechanism (271). Such a frequency-encoded control mechanism may also be responsible for regulating glycogenolysis in liver (270). What might be important, therefore, is not so much the absolute level of second messengers but rather the rate at which their concentrations fluctuate.

OTHER SECOND MESSENGERS

One of the outstanding problems in hormone action concerns the way in which insulin exerts its metabolic effects. A recent report indicates that it may employ an inositol phosphate as a second messenger (275, 276). In this case, the precursor lipid is a novel inositol-containing glycolipid that is cleaved in response to insulin to release a complex molecule consisting of an inositol phosphate glycan (which contains glucosamine) and diacylglycerol (276). The latter contains myristic acid and thus differs from that produced by hydrolyzing PtdIns4,5P$_2$. The fact that different forms of DG are produced by the two receptor mechanisms is particularly interesting in view of a recent report that there are at least three distinct forms of protein kinase C (277). The multiple forms of this enzyme may respond specifically to the DG formed from the different receptor mechanisms. The inositol phosphate glycan released by insulin functions as a second messenger to regulate cyclic AMP phosphodiesterase.

Inositol lipids may thus play a general role in signal transduction by functioning as precursors for second messengers. A role for PtdIns4,5P$_2$ as a precursor for Ins1,4,5P$_3$ is well established, and the possible function of a separate inositol-containing glycolipid as a precursor of a glycosylated inositol phosphate that may mediate the action of insulin is likely to increase interest in the role of inositol phosphates as intracellular messengers.

CONCLUSION

The receptor-mediated hydrolysis of PtdIns4,5P$_2$ yields two second messengers that control separate but interacting signal pathways. The dual signal hypothesis describes how the Ins1,4,5P$_3$/Ca^{2+} and DG/C-kinase pathways cooperate with each other to initiate a whole range of cellular processes. It is proposed that Ins1,4,5P$_3$ plays a primary role in this signaling system through its ability to mobilize intracellular calcium. Although the DG/C-kinase pathway may also contribute directly to these final responses, its primary function in many cells appears to be one of modulation both within the inositol lipid system as well as with other receptor mechanisms. The existence of feedback interactions operating between second messengers leads to a highly integrated and finely tuned signaling system capable of responding rapidly to external signals with appropriate changes in cellular activity.

The versatility of this signaling system is evident from the way it has been adapted to control widely divergent physiological processes, which can be separated into two major functions. First, it can bring about subtle changes in the sensitivity of the cell to other signal pathways, which may represent one of its major functions within the nervous system. Receptors operating through inositol lipid hydrolysis may potentiate synaptic transmission in several ways. By mobilizing intracellular calcium, Ins1,4,5P$_3$ could play a role in presynaptic facilitation by adjusting the resting level of calcium. Alternatively, the DG/C-kinase pathway may enhance transmitter release by either enhancing excitability through its ability to close potassium channels or by making the exocytotic process more sensitive to the stimulatory action of calcium.

The other major function of the inositol lipid signaling system is to provide the second messengers responsible for directly activating the cell. By releasing calcium, Ins1,4,5P$_3$ plays a direct role in stimulating a whole host of physiological processes, including the contraction of smooth muscle and perhaps also skeletal muscle; the breakdown of glycogen in liver; the release of amylase from pancreas; the opening of ion channels in parotid gland, smooth muscle, oocytes, and neural cells; some of the early events of fertilization; gene transcription in response to growth factors; and phototransduction in invertebrate photoreceptors. Many of these are short-term effects that are switched on and off quickly leaving the cell largely unchanged. However, there are two instances where the inositol lipid signal pathway may contribute to long-term changes such as occur when cells are stimulated to grow or during information storage in the nervous system. In both cases, signals detected at the cell surface must bring about changes in gene expression. The control mechanisms responsible for growth and the retention of long-term memory have much in common (278). In both cases, there are early second messenger events leading to rapid changes in ion fluxes and protein phosphorylation, which are somehow connected with the subsequent nuclear events responsible for growth or the acquisition of long-term memory. There already are indications that the bifurcating signal pathway based on Ins1,4,5P$_3$ and DG can activate gene transcription during the action of growth factors (202, 203). Second messengers of this kind could also play a role in activating the nuclear events responsible for memory (278). Unraveling the role of the inositol lipid signal pathway in such long-term changes in cellular activity represents a major challenge in the years ahead.

ACKNOWLEDGMENTS

I am indebted to the following for sending me copies of their unpublished papers: R. F. Irvine, P. W. Majerus, G. N. Europe-Finner, C. P. Downes, S. R. Nahorski, P. F. Blackmore, W. H. Moolenaar, R. Payne, M. Hirata, T. Biden, C. Wollheim, J. R. Williamson, M. V. Gershengorn, and H. Higashida.

Literature Cited

1. Berridge, M. J. 1984. *Biochem. J.* 220:345–60
2. Berridge, M. J., Irvine, R. F. 1984. *Nature* 312:315–21
3. Majerus, P. W., Neufeld, E. J., Wilson, D. B. 1984. *Cell* 37:701–3
4. Nishizuka, Y. 1984. *Nature* 308:693–98
5. Downes, C. P., Michell, R. H. 1985. In *Molecular Mechanisms of Transmembrane Signalling*, ed. P. Cohen, M. Houslay, pp. 3–56. New York: Elsevier
6. Hokin, L. E. 1985. *Ann. Rev. Biochem.* 54:205–35
7. Williamson, J. R., Cooper, R. H., Joseph, S. K., Thomas, A. P. 1985. *Am. J. Physiol.* 248:C203–C216
8. Nishizuka, Y. 1986. *Science* 233:305–12
9. Nahorski, S. R., Kendall, D. A., Batty, I. 1986. *Biochem. Pharm.* 35:2447–54
10. Fain, J. N., Berridge, M. J. 1979. *Biochem. J.* 180:655–61
11. Billah, M. M., Lapetina, E. G. 1982. *J. Biol. Chem.* 257:12705–8
12. Monaco, M. E., Woods, D. 1983. *J. Biol. Chem.* 258:15125–29
13. Rana, R. S., Kowluru, A., MacDonald,

M. J. 1986. *Arch. Biochem. Biophys.* 245:411–16
14. Gonzales, R. A., Crews, F. T. 1985. *J. Neurochem.* 45:1076–84
15. Horwitz, J., Tsymbalov, S., Perlman, R. L. 1985. *J. Pharm. Exp. Ther.* 233:235–41
16. Koreh, K., Monaco, M. E. 1986. *J. Biol. Chem.* 261:88–91
17. Müller, E., Hegewald, H., Jaroszewicz, K., Cumme, G. A., Hoppe, H., Frunder, H. 1986. *Biochem. J.* 235:775–83
18. Yang, J. C., Chang, P. C., Fujitaki, J. M., Chiu, K. C., Smith, R. C. 1986. *Biochemistry* 25:2677–81
19. Wilson, D. B., Neufeld, E. J., Majerus, P. W. 1985. *J. Biol. Chem.* 260:1046–51
19a. Griendling, K. K., Rittenhouse, S. E., Brock, T. A., Ekstein, L. S., Gimbrone, M. A., Alexander, R. W. 1986. *J. Biol. Chem.* 261:5901–6
20. Berridge, M. J., Buchan, P. B., Heslop, J. P. 1984. *Mol. Cell. Endocrinol.* 36:37–42
21. Dougherty, R. W., Godfrey, P. P., Hoyle, P. C., Putney, J. W. Jr., Freer, R. J. 1984. *Biochem. J.* 222:307–14
22. Drummond, A. H., Knox, R. J., Macphee, C. H. 1985. *Biochem. Soc. Trans.* 13:58–60
23. Wollheim, C. B., Biden, T. J. 1986. *J. Biol. Chem.* 261:8314–19
24. Beaven, M. A., Moore, J. P., Smith, G. A., Hesketh, T. R., Metcalfe, J. C. 1984. *J. Biol. Chem.* 259:7137–42
25. Thomas, A. P., Alexander, J., Williamson, J. R. 1984. *J. Biol. Chem.* 259:5574–84
26. Alexander, R. W., Brock, T. A., Gimbrone, M. A., Rittenhouse, S. E. 1985. *Hypertension* 7:447–51
27. Reynolds, E. E., Dubyak, G. R. 1985. *Biochem. Biophys. Res. Commun.* 130:627–32
28. Ramsdell, J. S., Tashjian, A. H. 1986. *J. Biol. Chem.* 261:5301–306
29. Wilson, D. B., Connolly, T. M., Bross, T. E., Majerus, P. W., Sherman, W. R., et al. 1985. *J. Biol. Chem.* 260:13496–501
30. Ishii, H., Connolly, T. M., Bross, T. E., Majerus, P. W. 1986. *Proc. Natl. Acad. Sci. USA* 83:6397–401
31. Hawkins, P. T., Berrie, C. P., Morris, A. J., Downes, C. P. 1986. *Biochem. J.* In press
32. Gilman, A. G. 1984. *Cell* 36:577–79
33. Litosch, I., Fain, J. N. 1986. *Life Sci.* 39:187–94
34. Taylor, C. W., Merritt, J. E. 1986. *Trends Pharmacol. Sci.* 7:238–42

35. Cockcroft, S., Gomperts, B. D. 1985. *Nature* 314:534–36
36. Chiarugi, V., Porciatti, F., Pasquali, F., Bruni, P. 1985. *Biochem. Biophys. Res. Commun.* 132:900–7
37. Fleischman, L. F., Chahwala, S. B., Cantley, L. 1986. *Science* 231:407–10
38. Wakelam, M. J. O., Davies, S. A., Houslay, M. D., McKay, I., Marshall, C. J., Hall, A. 1986. *Nature* 323:173–76
39. Blackmore, P. F., Bocckino, S. B., Waynick, L. E., Exton, J. H. 1985. *J. Biol. Chem.* 260:14477–83
40. Strand, C. F., Wong, K. 1985. *Biochem. Biophys. Res. Commun.* 133:161–67
41. Poll, C., Kyrle, P., Westwick, J. 1986. *Biochem. Biophys. Res. Commun.* 136:381–89
42. Mürer, E. H., Davenport, K., Siojo, E., Day, H. J. 1985. *Biochem. J.* 194:187–92
43. Fein, A., Corson, D. W. 1979. *Science* 204:77–79
44. Baudry, M., Evans, J., Lynch, G. 1986. *Nature* 319:329–31
45. Enjalbert, A., Sladeczek, F., Guillon, G., Bertrand, P., Shu, C., et al. 1986. *J. Biol. Chem.* 261:4071–75
46. Downes, C. P., Mussat, M. C., Michell, R. H. 1982. *Biochem. J.* 203:169–77
47. Storey, D. J., Shears, S. B., Kirk, C. J., Michell, R. H. 1984. *Nature* 312:374–76
48. Connolly, T. M., Bross, T. E., Majerus, P. W. 1985. *J. Biol. Chem.* 260:7868–74
49. Joseph, S. K., Williams, R. J. 1985. *FEBS Lett.* 180:150–54
50. Seyfred, M. A., Farrell, L. E., Wells, W. W. 1984. *J. Biol. Chem.* 259:13204–13208
51. Kukita, M., Hirata, M., Koga, T. 1986. *Biochim. Biophys. Acta* 885:121–28
52. Sasaguri, T., Hirata, M., Kuriyama, H. 1985. *Biochem. J.* 231:497–503
53. Rana, R. S., Sekar, M. C., Hokin, L. E., MacDonald, M. J. 1986. *J. Biol. Chem.* 261:5237–40
54. Turk, J., Wolf, B. A., McDaniel, M. L. 1986. *Biochem. J.* 237:259–63
55. Bosca, L., Mojena, M., Ghysdael, J., Rousseau, G. G., Hue, L. 1986. *Biochem. J.* 236:595–99
56. Eisenberg, F. 1967. *J. Biol. Chem.* 242:1375–82
57. Hallcher, L. M., Sherman, W. R. 1980. *J. Biol. Chem.* 255:1089–90
58. Takimoto, K., Okada, M., Matsuda, Y., Nakagawa, H. 1985. *J. Biochem.* 98:363–70

59. Connolly, T. M., Wilson, D. B., Bross, T. E., Majerus, P. W. 1986. *J. Biol. Chem.* 261:122–26
60. Ross, T. S., Majerus, P. W. 1986. *J. Biol. Chem.* 261:11119–23
61. Dixon, J. F., Hokin, L. E. 1985. *J. Biol. Chem.* 260:16068–71
62. Irvine, R. F., Letcher, A. J., Heslop, J. P., Berridge, M. J. 1986. *Nature* 320:631–34
63. Hansen, C. A., Mah, S., Williamson, J. R. 1986. *J. Biol. Chem.* 261:8100–103
64. Blackmore, P. F., Bocckino, S. B., Jiang, H., Pŕpic, V., Exton, J. H. 1986. *J. Biol. Chem.* 261:In press
65. Hawkins, P. T., Stephens, L., Downes, C. P. 1986. *Biochem. J.* 238:507–16
66. Stewart, S. J., Pŕpic, V., Powers, F. S., Bocckino, S. B., Isaaks, R. E., Exton, J. H. 1986. *Proc. Natl. Acad. Sci. USA* 83:6098–102
67. Biden, T. J., Wollheim, C. B. 1986. *J. Biol. Chem.* 261:11931–34
68. Batty, I. R., Nahorski, S. R., Irvine, R. F. 1985. *Biochem. J.* 232:211–15
69. Irvine, R. F., Letcher, A. J., Lander, D. J., Downes, C. P. 1984. *Biochem. J.* 223:237–43
70. Burgess, G. M., McKinney, J. S., Irvine, R. F., Putney, J. W. Jr. 1985. *Biochem. J.* 232:237–48
71. Heslop, J. P., Irvine, R. F., Tashjian, A. H., Berridge, M. J. 1985. *J. Exp. Biol.* 119:395–401
72. Irvine, R. F., Änggård, E. E., Letcher, A. J., Downes, C. P. 1985. *Biochem. J.* 229:505–11
73. Sherman, W. R., Munsell, L. Y., Gish, B. G., Honchar, M. P. 1985. *J. Neurochem.* 44:798–807
74. Molina y Vedia, L. M., Lapetina, E. G. 1986. *J. Biol. Chem.* 261:10493–95
75. Connolly, T. M., Lawing, W. J. Jr., Majerus, P. W. 1986. *Cell* 46:951–58
76. Lew, P. D., Monod, A., Krause, K-H., Waldvogel, F. A., Biden, T. J., Schlegel, W. 1986. *J. Biol. Chem.* 261:13121–27
77. Anderson, R. A., Marchesi, V. T. 1985. *Nature* 318:295–98
78. Perrin, D., Aunis, D. 1985. *Nature* 315:589–92
79. Lassing, I., Lindberg, U. 1985. *Nature* 314:472–74
80. Burn, P., Rotman, A., Meyer, R. K., Burger, M. M. 1985. *Nature* 314:469–72
81. Kojima, I., Kojima, K., Rasmussen, H. 1985. *J. Biol. Chem.* 260:9177–84
82. Dougherty, R. W., Niedel, J. E. 1986. *J. Biol. Chem.* 261:4097–100
83. Ashendel, C. L. 1985. *Biochim. Biophys. Acta* 822:219–42
84. Brocklehurst, K. W., Morita, K., Pollard, H. B. 1985. *Biochem. J.* 228:35–42
85. Wolf, M., LeVine, H., May, W. S., Cuatrecasas, P., Sahyoun, N. 1985. *Nature* 317:546–49
86. May, W. S., Sahyoun, N., Wolf, M., Cuatrecasas, P. 1985. *Nature* 317:549–51
87. Altin, J. G., Bygrave, F. L. 1985. *Biochem. J.* 232:911–17
88. Frantz, C. N. 1985. *Exp. Cell Res.* 158:287–300
89. Brown, R. D., Berger, K. D., Taylor, P. 1984. *J. Biol. Chem.* 259:7554–62
90. Mendoza, S. A., Schneider, J. A., Lopez-Rivas, A., Sinnett-Smith, J. W., Rozengurt, E. 1968. *J. Cell Biol.* 102:2223–33
91. Hamachi, T., Hirata, M., Koga, T. 1986. *Biochim. Biophys. Acta.* In press
92. Prentki, M., Corkey, B. E., Matschinsky, F. M. 1985. *J. Biol. Chem.* 260:9185–90
93. Forsberg, E. J., Rojas, E., Pollard, H. B. 1986. *J. Biol. Chem.* 261:4915–20
94. Streb, H., Irvine, R. F., Berridge, M. J., Schulz, I. 1983. *Nature* 306:67–69
95. Joseph, S. K., Thomas, A. P., Williams, R. J., Irvine, R. F., Williamson, J. R. 1984. *J. Biol. Chem.* 259:3077–81
96. Burgess, G. M., Godfrey, P. P., McKinney, J. S., Berridge, M. J., Irvine, R. F., Putney, J. W. Jr. 1984. *Nature* 309:63–66
97. Prentki, M., Wollheim, C. B., Lew, P. D. 1984. *J. Biol. Chem.* 259:13777–82
98. Joseph, S. K., Williams, R. J., Corkey, B. E., Matschinsky, F. M., Williamson, J. R. 1984. *J. Biol. Chem.* 259:12952–55
99. Biden, T. J., Prentki, M., Irvine, R. F., Berridge, M. J., Wollheim, C. B. 1984. *Biochem. J.* 223:467–73
100. Gershengorn, M. C., Geras, E., Purrello, V. S., Rebecchi, M. J. 1984. *J. Biol. Chem.* 259:10675–81
101. Biden, T. J., Wollheim, C. B., Schlegel, W. 1986. *J. Biol. Chem.* 261:7223–29
102. Berridge, M. J., Heslop, J. P., Irvine, R. F., Brown, K. D. 1984. *Biochem. J.* 222:195–201
103. Suematsu, E., Hirata, M., Hashimoto, T., Kuriyama, H. 1984. *Biochem. Biophys. Res. Commun.* 120:481–85
104. Suematsu, E., Hirata, M., Sasaguri, T., Hashimoto, T., Kuriyama, H. 1985. *Comp. Biochem. Physiol.* 82A:645–49
105. Hashimoto, T., Hirata, M., Itoh, T., Kanmura, Y., Kuriyama, H. 1986. *J. Physiol.* 370:605–18
106. Yamamoto, H., van Breemen, C. 1985.

Biochem. Biophys. Res. Commun.
130:270–74
107. Hashimoto, T., Hirata, M., Ito, Y. 1985. *Br. J. Pharmacol.* 86:191–99
108. Hirata, M., Suematsu, E., Hashimoto, T., Hamachi, T., Koga, T. 1984. *Biochem. J.* 223:229–36
109. Hirata, M., Kukita, M., Sasaguri, T., Suematsu, E., Hashimoto, T., Koga, T. 1985. *J. Biochem.* 97:1575–82
110. Wolf, B. A., Comens, P. G., Ackerman, K. E., Sherman, W. R., McDaniel, M. L. 1985. *Biochem. J.* 227:965–69
111. Epstein, P. A., Prentki, M., Attie, M. F. 1985. *FEBS Lett.* 188:141–44
112. Kojima, I., Kojima, K., Kreutter, D., Rasmussen, H. 1984. *J. Biol. Chem.* 259:14448–57
113. Muto, Y., Toymatsu, T., Yoshioka, S., Nozawa, Y. 1986. *Biochem. Biophys. Res. Commun.* 135:46–51
114. Brass, L. F., Joseph, S. K. 1985. *J. Biol. Chem.* 260:15172–79
115. Dubyak, G. R. 1986. *Arch. Biochem. Biophys.* 245:84–95
116. Europe-Finner, G. N., Newell, P. C. 1986. *Biochim. Biophys. Acta.* 887:335–40
117. Ueda, T., Chueh, S-H., Noel, M. W., Gill, D. L. 1986. *J. Biol. Chem.* 261:3184–92
118. Thévenod, F., Streb, H., Ullrich, K. J., Schulz, I. 1986. *Kidney Int.* 29:695–702
119. Stoehr, S. J., Smolen, J. E., Holz, R. W., Agranoff, B. W. 1986. *J. Neurochem.* 46:637–40
120. Suematsu, E., Nishimura, J., Hirata, M., Inamitsu, T., Ibayashi, H. 1985. *Biomed. Res.* 6:279–86
121. Clapper, D. L., Lee, H. C. 1985. *J. Biol. Chem.* 260:13947–54
122. Dawson, A. P., Irvine, R. F. 1984. *Biochem. Biophys. Res. Commun.* 120:858–64
123. Prentki, M., Biden, T. J., Janjic, D., Irvine, R. F., Berridge, M. J., Wollheim, C. B. 1984. *Nature* 309:562–64
124. Vergara, J., Tsien, R. Y., Delay, M. 1985. *Proc. Natl. Acad. Sci. USA* 82:6352–56
125. Drøbak, B. K., Ferguson, I. B. 1985. *Biochem. Biophys. Res. Commun.* 130:1241–46
126. Carsten, M. E., Miller, J. D. 1985. *Biochem. Biophys. Res. Commun.* 130:1027–31
127. Adunyah, S. E., Dean, W. L. 1985. *Biochem. Biophys. Res. Commun.* 128:1274–80
128. Authi, K. S., Crawford, N. 1985. *Biochem. J.* 230:247–53
129. Delfert, D. M., Hill, S., Pershadsingh,

H. A., Sherman, W. R., McDonald, J. M. 1986. *Biochem. J.* 236:37–44
130. Oberdorf, J. A., Head, J. F., Kaminer, B. 1986. *J. Cell Biol.* 102:2205–10
131. Streb, H., Bayerdorffer, E., Haase, W., Irvine, R. F., Schulz, I. 1984. *J. Membr. Biol.* 81:241–53
132. O'Rourke, F. A., Halenda, S. P., Zavoico, G. B., Feinstein, M. B. 1985. *J. Biol. Chem.* 260:956–62
133. Taylor, C. W., Putney, J. W. Jr. 1985. *Biochem. J.* 232:435–38
134. Bond, M., Kitazawa, T., Somlyo, A. P., Somlyo, A. V. 1984. *J. Physiol.* 355:677–95
135. Busa, W. B., Ferguson, J. E., Joseph, S. K., Williamson, J. R., Nuccitelli, R. 1985. *J. Cell Biol.* 101:677–82
136. Henne, V., Söling, H-D. 1986. *FEBS Lett.* 202:267–73
137. Smith, J. B., Smith, L., Higgins, B. L. 1985. *J. Biol. Chem.* 260:14413–16
137a. Joseph, S. K., Williamson, J. R. 1986. *J. Biol. Chem.* 261:14658–64
138. Burgess, G. M., Irvine, R. F., Berridge, M. J., McKinney, J. S., Putney, J. W. 1984. *Biochem. J.* 224:741–46
139. Irvine, R. F., Brown, K. D., Berridge, M. J. 1984. *Biochem. J.* 221:269–72
140. Baukal, A. J., Guillemette, G., Rubin, R., Spät, A., Catt, K. J. 1985. *Biochem. Biophys. Res. Commun.* 133:532–38
141. Spät, A., Bradford, P. G., McKinney, J. S., Rubin, R. P., Putney, J. W. 1986. *Nature* 319:514–16
142. Spät, A., Fabiato, A., Rubin, R. P. 1986. *Biochem. J.* 233:929–32
143. Hirata, M., Sasaguri, T., Hamachi, T., Hashimoto, T., Kukita, M., Koga, T. 1985. *Nature* 317:723–25
144. Muallem, S., Schoefield, M., Pandol, S., Sachs, G. 1985. *Proc. Natl. Acad. Sci. USA* 82:4433–37
145. Dawson, A. P. 1985. *FEBS Lett.* 185:147–50
146. Dawson, A. P., Comerford, J. G., Fulton, D. V. 1986. *Biochem. J.* 234:311–15
147. Gill, D. L., Ueda, T., Chueh, S-H., Noel, M. W. 1986. *Nature* 320:461–64
148. Salmon, D. M., Honeyman, T. W. 1980. *Nature* 284:344–45
149. Putney, J. W. Jr., Weiss, S. J., Van de Walle, C. M., Haddas, R. A. 1980. *Nature* 284:345–47
150. Holmes, R. P., Yoss, N. L. 1983. *Nature* 305:637–38
151. Moolenaar, W. H., Kruijer, W., Tilly, B. C., Verlaan, I., Bierman, A. J., de Laat, S. W. 1986. *Nature* 323:171–73
151a. Irvine, R. F., Moor, R. M. 1986. *Biochem. J.* 240:917–20

152. Putney, J. W. Jr. 1986. *Cell Calcium* 7:10–12
153. Henkart, M. P., Nelson, P. G. 1979. *J. Gen. Physiol.* 73:655–73
154. Charbonneau, M., Grey, R. D. 1984. *J. Cell Biol.* 102:90–97
155. Muallem, S., Fimmel, C. J., Pandol, S. J., Sachs, G. 1986. *J. Biol. Chem.* 261:2660–67
156. Charest, R., Prpic, V., Exton, J. H., Blackmore, P. F. 1985. *Biochem. J.* 227:79–90
157. Popescu, L. M., Hinescu, M. E., Musat, S., Ionescu, M., Pistritzu, F. 1986. *Eur. J. Pharmacol.* 123:167–69
158. Somlyo, A. V., Bond, M., Somlyo, A. P., Scarpa, A. 1985. *Proc. Natl. Acad. Sci. USA* 82:5231–35
159. Bitar, K. N., Bradford, P., Putney, J. W., Makhlouf, G. M. 1985. *Gastroenterology* 88:1326
160. Volpe, P., Salviati, G., Di Virgilio, F., Pozzan, T. 1985. *Nature* 316:347–49
161. Nosek, T. M., Williams, M. F., Zeigler, S. T., Godt, R. E. 1986. *Am. J. Physiol.* 250:C807–C811
162. Europe-Finner, G. N., Newell, P. C. 1985. *Biochem. Biophys. Res. Commun.* 130:1115–22
163. Europe-Finner, G. N., Newell, P. C. 1986. *J. Cell Sci.* 82:41–51
164. Lapetina, E. G., Watson, S. P., Cuatrecasas, P. 1984. *Proc. Natl. Acad. Sci. USA* 81:7431–35
165. Watson, S. P., Ruggiero, M., Abrahams, S. L., Lapetina, E. G. 1986. *J. Biol. Chem.* 261:5368–72
166. Israels, S. J., Robinson, P., Docherty, J. C., Gerrard, J. M. 1985. *Thromb. Res.* 40:499–509
167. Authi, K. S., Evenden, B. J., Crawford, N. 1986. *Biochem. J.* 233:707–18
168. Brown, J. E., Rubin, L. J., Ghalayini, A. J., Tarver, A. P., Irvine, R. F., et al. 1984. *Nature* 311:160–63
169. Fein, A., Payne, R., Corson, D. W., Berridge, M. J., Irvine, R. F. 1984. *Nature* 311:157–60
170. Brown, J. E., Rubin, L. J. 1984. *Biochem. Biophys. Res. Commun.* 125:1137–42
171. Payne, R., Corson, D. W., Fein, A., Berridge, M. J. 1986. *J. Gen. Physiol.* 88:127–42
172. Waloga, G., Anderson, R. E. 1985. *Biochem. Biophys. Res. Commun.* 126:59–62
173. Nadler, E., Grillo, B., Lass, Y., Oron, Y. 1986. *FEBS Lett.* 199:208–12
174. Oron, Y., Dascal, N., Nadler, E., Lupu, M. 1985. *Nature* 313:141–43
175. Picard, A., Giraud, F., Le Bouffant, F.,

Sladeczek, F., Le Peuch, C., Dofee, M. 1985. *FEBS Lett.* 182:446–80
176. Slack, B. E., Bell, J. E., Benos, D. J. 1986. *Am. J. Physiol.* 250:C340–C344
177. Whitaker, M., Irvine, R. F. 1984. *Nature* 312:636–39
178. Turner, P. R., Jaffe, L. A., Fein, A. 1985. *J. Cell Biol.* 102:70–76
179. Higashida, H., Streaty, R. A., Klee, W., Nirenberg, M. 1986. *Proc. Natl. Acad. Sci. USA* 83:942–46
180. Klöckner, U., Isenberg, G. 1985. *Pfluegers Arch.* 405:R61
181. Evans, M. G., Marty, A. 1986. *Proc. Natl. Acad. Sci. USA* 83:4099–103
182. Lea, T. J., Griffiths, P. J., Treager, R. T., Ashley, C. C. 1986. *FEBS Lett.* 207:153–61
183. Scherer, N. M., Ferguson, J. E. 1985. *Biochem. Biophys. Res. Commun.* 128:1064–70
184. Thieleczek, R., Heilmeyer, L. M. G. 1986. *Biochem. Biophys. Res. Commun.* 135:662–69
185. Volpe, P., Di Virgilio, F., Pozzan, T., Salviati, G. 1986. *FEBS Lett.* 197:1–4
186. Di Virgilio, F., Salviati, G., Pozzan, T., Volpe, P. 1986. *EMBO J.* 5:259–62
187. Higashida, H., Brown, D. A. 1986. *Nature.* 323:333–35
188. Payne, R., Fein, A. 1986. *J. Cell Biol.* In press
189. Kamel, L. C., Bailey, J., Schoenbaum, L., Kinsey, W. 1985. *Lipids* 20:350–56
190. Ciapa, B., Whitaker, M. 1986. *FEBS Lett.* 195:347–51
191. Yano, K., Higashida, H., Inoue, R., Nozawa, Y. 1984. *J. Biol. Chem.* 259:10201–207
192. Yano, K., Higashida, H., Hattori, H., Nozawa, Y. 1985. *FEBS Lett.* 181:403–6
193. Reiser, G., Hamprecht, B. 1985. *Pfleugers Arch.* 405:260–64
194. Baraban, J. M., Snyder, S. H., Alger, B. E. 1985. *Proc. Natl. Acad. Sci. USA* 82:2538–42
195. Farley, J., Auerbach, S. 1986. *Nature* 319:220–23
196. Burnham, D. B., Munowitz, P., Hootman, S. R., Williams, J. A. 1986. *Biochem. J.* 235:125–31
197. Hidaka, H., Inagaki, M., Kawamoto, S., Sasaki, Y. 1984. *Biochemistry* 23:5036–41
198. Rodriguez-Pena, A., Rozengurt, E. 1984. *Biochem. Biophys. Res. Commun.* 120:1053–59
199. Wright, C. D., Hoffman, M. D. 1986. *Biochem. Biophys. Res. Commun.* 135:749–55
200. Sha'afi, R. I., Molski, T. F. P., Huang,

C.-K., Naccache, P. H. 1986. *Biochem. Biophys. Res. Commun.* 137:50–60

201. Coughlin, S. R., Lee, W. M. F., Williams, P. W., Giels, G. M., Williams, L. T. 1985. *Cell* 43:243–51

202. Kaibuchi, K., Tsuda, T., Kikuchi, A., Tanimoto, T., Yamashita, T., Takai, Y. 1986. *J. Biol. Chem.* 261:1187–92

203. Moore, J. P., Todd, J. A., Hesketh, T. R., Metcalfe, J. C. 1986. *J. Biol. Chem.* 261:8158–62

204. Rasmussen, H., Waisman, D. M. 1982. *Rev. Physiol. Biochem. Pharmacol.* 95:111–48

205. Knight, D. E., Baker, P. F. 1983. *FEBS Lett.* 160:98–100

206. Martin, T. F. J., Kowalchyk, J. A. 1984. *Endocrinology* 115:1517–26

206a. Kolesnick, R. N., Gershengorn, M. C. 1986. *Endocrinology.* In press

207. Rink, T. J., Sanchez, A., Hallam, T. J. 1983. *Nature* 305:317–19

208. Pozzan, T., Gatti, G., Dozio, N., Vicentini, L. M., Meldolesi, J. 1984. *J. Cell Biol.* 99:628–38

209. Bruzzone, R., Pozzan, T., Wollheim, C. B. 1986. *Biochem. J.* 139–43

210. Menkes, H., Baraban, J. M., Snyder, S. H. 1986. *Eur. J. Pharmacol.* 122:19–27

211. De Chaffoy de Courcelles, D., Roevens, P., van Belle, H. 1984. *FEBS Lett.* 173:389–93

212. Halenda, S. P., Feinstein, M. B. 1984. *Biochem. Biophys. Res. Commun.* 124:507–13

213. Taylor, M. V., Metcalfe, J. C., Hesketh, T. R., Smith, G. A., Moore, J. P. 1984. *Nature* 312:462–65

214. Boon, A. M., Beresford, B. J., Mellors, A. 1985. *Biochem. Biophys. Res. Commun.* 129:431–38

215. Sagi-Eisenberg, R., Lieman, H., Pecht, I. 1985. *Nature* 313:59–60

216. Zavoico, G. B., Halenda, S. P., Sha'afi, R. I., Feinstein, M. B. 1985. *Proc. Natl. Acad. Sci. USA* 82:3859–62

217. Drummond, A. H. 1985. *Nature* 315:752–55

218. Cooper, R. H., Coll, K. E., Williamson, J. R. 1985. *J. Biol. Chem.* 260:3281–88

219. Lynch, C. J., Charest, R., Bocckino, S. B., Exton, J. H., Blackmore, P. F. 1985. *J. Biol. Chem.* 260:2844–51

220. Hesketh, T. R., Moore, J. P., Morris, J. D. H., Taylor, M. V., Rogers, J., et al. 1985. *Nature* 313:481–84

221. McNeil, P. L., McKenna, M. P., Taylor, D. L. 1985. *J. Cell Biol.* 101:372–79

222. Naccache, P. H., Molski, T. F. P., Borgeat, P., White, J. R., Sha'afi, R. I. 1985. *J. Biol. Chem.* 260:2125–31

223. Brock, T. A., Rittenhouse, S. E., Powers, C. W., Ekstein, L. S., Gimbrone, M. A., Alexander, R. W. 1985. *J. Biol. Chem.* 260:14158–62

224. Kojima, I., Shibata, H., Ogata, E. 1986. *Biochem. J.* 237:253–58

225. Monaco, M. E., Mufson, R. A. 1986. *Biochem. J.* 236:171–75

226. Moolenaar, W. H., Aerts, R. J., Tertoolen, L. G. J., de Laat, S. W. 1986. *J. Biol. Chem.* 261:279–84

227. Orellana, S. A., Solski, P. A., Brown, J. H. 1985. *J. Biol. Chem.* 260:5236–39

228. Labarca, R., Janowsky, A., Patel, J., Paul, S. M. 1984. *Biochem. Biophys. Res. Commun.* 123:703–9

229. Rittenhouse, S. E., Sasson, J. P. 1985. *J. Biol. Chem.* 260:8657–60

230. Watson, S. P., Lapetina, E. G. 1985. *Proc. Natl. Acad. Sci. USA* 82:2623–26

231. Corvera, S., Garcia-Sainz, J. A. 1984. *Biochem. Biophys. Res. Commun.* 119:1128–33

232. Sturani, E., Vicentini, L. M., Zippel, R., Toschi, L., Pandiella-Alonso, A., et al. 1986. *Biochem. Biophys. Res. Commun.* 137:343–50

233. Leeb-Lundberg, L. M. F., Cotecchia, S., Lomasney, J. M., DeBernardis, J. F., Lefkowitz, R. J., Caron, M. G. 1985. *Proc. Natl. Acad. Sci. USA* 82:5651–55

234. Rickard, J. E., Sheterline, P. 1985. *Biochem. J.* 231:623–28

235. Lagast, H., Pozzan, T., Waldvogel, F. A., Lew, P. D. 1984. *J. Clin. Invest.* 73:878–83

236. Berridge, M. J. 1981. In *Drug Receptors and Their Effectors*, ed. N. J. M. Birdsall, pp. 75–85. Bristol: Macmillan

237. Irvine, R. F., Letcher, A. J., Dawson, R. M. C. 1984. *Biochem. J.* 218:177–85

238. Connor, J. A., Kretz, R., Shapiro, E. 1986. *J. Physiol.* 375:625–42

239. Akers, R. F., Routtenberg, A. 1985. *Brain Res.* 334:147–51

240. Kendall, D. A., Nahorski, S. R. 1985. *Eur. J. Pharmacol.* 115:31–36

241. Best, L., Bolton, T. B. 1986. *Naunyn-Schmiedebergs Arch. Pharmacol.* 333:78–82

242. DeRiemer, S. A., Strong, J. A., Albert, K. A., Greengard, P., Kaczmarek, L. K. 1985. *Nature* 313:313–16

243. Wakade, A. R., Malhotra, R. K., Wakade, T. D. 1986. *Nature* 321:698–700

244. Di Virgilio, F., Pozzan, T., Wollheim, C. B., Vicentini, L. M., Meldolesi, J. 1986. *J. Biol. Chem.* 261:32–35

245. Harris, K. M., Kongsamut, S., Miller, R. J. 1986. *Biochem. Biophys. Res. Commun.* 134:1298–305

246. Sugden, D., Vanecek, J., Klein, D. C., Thomas, T. P., Anderson, W. B. 1985. *Nature* 314:359–361
247. Hollingsworth, E. B., Daly, J. W. 1985. *Biochim. Biophys. Acta* 847:207–16
248. Sibley, D. R., Jeffs, R. A., Daniel, K., Nambi, P., Lefkowitz, R. J. 1986. *Arch. Biochem. Biophys.* 244:373–81
249. Bell, J. D., Buxton, I. L. O., Brunton, L. L. 1985. *J. Biol. Chem.* 260:2625–28
250. Hollingsworth, E. B., Sears, E. B., Daly, J. W. 1985. *FEBS Lett.* 184:339–42
251. Katada, T., Gilman, A., Watanabe, Y., Bauer, S., Jacobs, K. H. 1985. *Eur. J. Biochem.* 151:431–37
252. Yamanishi, J., Takai, Y., Kaibuchi, K., Sano, K., Castagna, M., Nishizuka, Y. 1983. *Biochem. Biophys. Res. Commun.* 112:778–86
253. Rink, T. J., Sanchez, A. 1984. *Biochem. J.* 222:833–36
254. Zavoico, G. B., Feinstein, M. B. 1984. *Biochem. Biophys. Res. Commun.* 120:579–85
255. Imai, A., Hattori, H., Takahashi, M., Nozawa, Y. 1983. *Biochem. Biophys. Res. Commun.* 112:693–700
256. Watson, S. P., McConnell, R. T., Lapetina, E. G. 1984. *J. Biol. Chem.* 259:13199–203
257. Nakashima, S., Tohmatsu, T., Hattori, H., Okano, Y., Nozawa, Y. 1986. *Biochem. Biophys. Res. Commun.* 135:1099–104
258. Rapoport, R. M. 1986. *Circ. Res.* 58:407–10
259. Lapetina, E. G. 1986. *FEBS Lett.* 195:111–14
260. de Chaffoy de Courcelles, D., Roevens, P., van Belle, H. 1986. *FEBS Lett.* 195:115–18
261. Kusano, K., Miledi, R., Stinnakre, J. 1982. *J. Physiol.* 328:143–70
262. Miledi, R., Parker, I. 1984. *J. Physiol.* 357:173–83
263. Rapp, P. E., Berridge, M. J. 1977. *J. Theor. Biol.* 66:497–525
264. Berridge, M. J., Rapp, P. E. 1979. *J. Exp. Biol.* 81:217–80
265. Kuba, Y., Takeshita, S. 1981. *J. Theor. Biol.* 93:1009–31
266. Igusa, Y., Miyazaki, S-I. 1983. *J. Physiol.* 340:611–32
267. Cuthbertson, K. S. R., Cobbold, P. H. 1985. *Nature* 316:541–42
268. Ueda, S., Oiki, S., Okada, Y. 1986. *J. Membr. Biol.* 91:65–72
269. Yada, T., Oiki, S., Ueda, S., Okada, Y. 1986. *Biochim. Biophys. Acta* 887:105–12
270. Woods, N. M., Cuthbertson, K. S. R., Cobbold, P. H. 1986. *Nature* 319:600–2
271. Rapp, P. E., Berridge, M. J. 1981. *J. Exp. Biol.* 93:119–32
272. Berridge, M. J. 1983. *Biochem. J.* 212:849–58
273. Neher, E., Almers, W. 1986. *EMBO J.* 5:51–53
274. Parker, I., Miledi, R. 1986. *Proc. R. Soc. London Ser. B.* 228:307–15
275. Saltiel, A. R., Cuatrecasas, P. 1986. *Proc. Natl. Acad. Sci. USA* 83:5793–97
276. Saltiel, A. R., Fox, J. A., Sherline, P., Cuatrecasas, P. 1986. *Science* 233:967–72
277. Coussens, L., Parker, P. J., Rhee, L., Yang-Feng, T. L., Chen, E., et al. 1986. *Science* 233:859
278. Goelet, P., Castellucci, V. F., Schacher, S., Kandel, E. R. 1986. *Nature* 322:419–22

Ann. Rev. Biochem. 1987. 56:195–227

COMPLEX LOCI OF *DROSOPHILA*

Matthew P. Scott

Department of Molecular, Cellular and Developmental Biology, University of Colorado, Boulder, Colorado 80309-0347

CONTENTS

PERSPECTIVES AND SUMMARY

The purpose of this review is to present some of the information about gene structure that has been gathered through the combination of classical *Drosophila* genetics with molecular biology. What constitutes a complex locus? The definition has never been clearly established, but often the relevant characteristics are a plethora of apparently different phenotypes due to mutations in a single gene, responsiveness to a variety of different *trans*-acting

195

regulatory systems, and intragenic complementation. The mutations in a complex gene are sometimes called pseudoalleles.

A representative group of complex loci is discussed below; not all such loci can be included. Some complex genes, such as *Notch* and *white,* appear to be single transcription units controlled in an elaborate way. In other cases, such as the homeotic gene complexes, clusters of transcription units are affected in complex ways by mutations. The genetic complexity can be due not only to transcriptional regulation, but also to variable products. Alternative RNA splicing patterns are more the rule than the exception. In many cases, a family of related proteins is produced rather than a single protein. Sometimes only the untranslated parts of the mRNAs are altered by alternative splices, for reasons that are currently not understood for any gene. Yet another type of complexity is the result of multimeric protein structures, as is probably the case for the *rudimentary* and *Gart* loci. In a multiprotein complex, a function absent from one subunit may be provided by another, and although each subunit is defective in at least one function, the protein complex as a whole may still be able to provide all of the required activities. Different functional domains of a single protein may be able to act somewhat independently, and a mutation may affect only one domain of a protein. A cell containing two different mutant proteins (one from each allele) might have all of the wild-type functions, whether the proteins are associated in a multimeric complex or not. These sorts of complementation phenomena are primarily a matter of protein structure, not gene structure.

It is not possible to use *Drosophila* genetics to saturate a gene with mutations to the extent possible in *Escherichia coli* or lambda, and many genes that are not classified as complex would probably be genetically complex if analyzed in sufficient depth (e.g. 1–25). These include a large number of genes for which molecular analyses have revealed variable RNA and protein products and/or multiple *cis*-acting regulatory elements. Genes that are genetically complex but have not been analyzed molecularly, such as *dumpy* (26, 27) and *raspberry* (28) are beyond the scope of this review. Important molecular data about many loci have come from the relatively new technique of chromosome "walking," the isolation of overlapping cloned DNA segments (29). A second important advance has been the P-element transformation technique (30, 31), which allows genes to be modified in vitro and then reintroduced into the germline for testing. The P-element method allows a gene to be tested as a single copy insert into a chromosome, and to be tested for function during normal development.

Drosophila is estimated to have about 5000 genes based on saturation studies of several regions of the genome (32–34; reviewed in 35, 36). Measurements of mRNA complexity have led to somewhat higher estimates of about 10,000–15,000 different transcripts (37–39). Both of these estimates may seem disturbingly low, given the complexity of a fly. The problem, if there is one, can be at least partially solved by complex loci. The fly need not have a very large number of genes if each of them has multiple products and complex regulation. The molecular studies of complex loci have revealed three types of complexity: 1. combinatorial action of a large array of *cis*-acting regulatory elements, 2. multiple or variable products, 3. altered gene function due to the complex effects of transposon or other insertions. The first two of these may resolve the fly's apparent gene number paradox. Complex genetic effects due to the intricacies of transcriptional regulation, changes in protein structure, and transposon insertions are all observed at the first complex locus to be discussed, the *Notch* locus.

THE *NOTCH* LOCUS

Indications of genetic complexity at the *Notch* (*N*) locus come from the several phenotypes of *N* mutations and the recombinational size of the gene (40–42). The molecular analysis of the gene revealed a 40-kb transcription unit producing an approximately 10.5-kb RNA (43, 44). The *N* gene functions in early development to control the balance between the cells committed to forming neural ectoderm and those committed to forming epidermal ectoderm. In mutant embryos, there is a great excess of the former at the expense of the latter. The protein product is structurally related to epidermal growth factor and to blood-clotting factors such as factor IX (45, 46).

The wide variety of adult *N* phenotypes is an indication of the complexity of the gene (47). A null mutation heterozygous with a wild-type allele has notched wings and aberrant bristles. These are the dominant *N* alleles. They are dominant because two copies of the gene are required for wild-type development. In addition, such mutations are recessive lethals. Homozygous (dead) embryos have excess neural cells. Other types of *N* alleles have different phenotypes: The dominant *Abruptex* (*Ax*) alleles cause gaps in wing venation (some *Ax* alleles are recessive lethals, some are not); recessive *notchoid* (*nd*) alleles affect wing morphology very much like the dominant *N* alleles do; *glossy-like* (*facet^{glossy}*) alleles cause the development of smooth eye with mottled pigmentation; and *facet* (*fa*) alleles cause a rough eye phenotype with homogeneous pigmentation. The three latter classes of alleles are homozygous viable. Finally there are recessive lethal alleles that have no dominant effect; these are presumably alleles with too little embryonic function to survive as homozygotes, but with enough function during wing and bristle development to permit heterozygous (mutant/+) flies to develop with normal adult morphology.

What genetic data led to the grouping of mutations with such different phenotypes into one locus? The dominant *N* alleles are also recessive lethals, act like deletions of the gene, and fail to survive in heteroallelic combinations. [A fly carrying two different alleles of a gene in *trans,* a heteroallelic combination, tests whether the phenotype of either allele (or of both) is observed. This sort of complementation test can be used for dominant or recessive mutations.] The recessive visible alleles like *fa* and *nd* are grouped together with the dominant *N* alleles for two reasons. First, recombination mapping places the recessive alleles among the *N* alleles. Second, heteroallelic combinations of the *N* alleles with the recessive alleles give flies with the recessive phenotype as well as the *N* loss-of-function phenotype. For example, flies that are *N/fa* have both the dominant *N* phenotype and the *fa* phenotype. Thus, the dominant *N* alleles cannot provide *fa* function. If the *N* gene was unrelated to *fa,* no *fa* phenotype would be seen. The interpretation of these observations is that the *N* phenotype is due to a lack of *N* product, so *N* function is termed "haplo-insufficient." The *fa* phenotype is due to an alteration in the expression of *N* product. *fa* alleles are recessive because the *N* product can be provided by a wild-type allele in *trans* in the cells affected by *fa* mutations, but not by an *N* allele that fails to make the product. The other types of recessive alleles interact with *N* in the same way as *fa* alleles. However, different recessive alleles affect different aspects of *N* function, i.e. the different classes of recessive alleles can complement each other. A *fa/nd* fly looks wild-type, for example. The different effects of the recessive alleles are probably due in part to different effects of different transposon insertions, as will be discussed further below (see section on THE EFFECTS OF TRANSPOSON INSERTIONS).

One of the classes of mutations, the *Abruptex (Ax)* alleles, is not due to transposon insertions. The *Ax* alleles were originally defined as part of the *N* complex because they were mapped among the *N* alleles (48, 49). The *Ax* alleles are clustered in the major coding exon of *N* and appear to be point mutations that presumably alter the protein sequence (44). Some *Ax* mutations are recessive lethals; some are not. The alleles must be capable of providing enough *N* product so that there is not a haplo-insufficient (*N,* i.e. notched-wing) phenotype. Some *Ax* alleles enhance the phenotype of *N* mutations in *trans;* some suppress *N* mutations. The suppression effect suggests that the *Ax* alleles are overproducing *N* product or are producing hyperfunctional product that can make up for the lack of product produced by the *N* allele on the other chromosome. The same sort of logic applies to the suppression of the *Ax* phenotype by an *N* mutation in *trans,* which is usually observed. Therefore different *Ax* alleles have different amounts of wild-type function, and the range is from less than the wild-type amount of function to more than the wild-type amount of function. It should be noted, however, that a simple "hyperactive product" model is not adequate, since the *Ax* mutations simultaneously increase the level of some functions (e.g. the function required to prevent wing notching), and reduce other functions (e.g. those required for proper wing venation). Thus different functional domains of the protein may be differently affected.

Since the *Ax* alleles map within the coding region of the gene and appear to be point mutations, they are likely to cause altered function of the *N* protein. Genetic studies reveal some important features of the altered function. The *Ax* alleles exhibit negative complementation (48, 49). That is, heterozygous flies carrying two different *Ax* alleles may have more extreme phenotypes than flies homozygous for either mutation. The effect can be quite strong: some heterozygous combinations are lethal while flies homozygous for either mutation are completely viable.

Negative complementation can be accounted for by models in which a uniform, if aberrant, population of protein molecules is able to provide partial function, but a mixed population of two different types of aberrant proteins is less functional. Negative complementation has been observed among alleles of the *Drosophila* DOPA decarboxylase *(Ddc)* gene. In *Ddc* the negative complementation was interpreted as being indicative of a multisubunit protein, with nonfunctional subunits impairing the function of the functional subunits with which they assemble (50). Such a model might apply to *N* if, for example, the molecules on the surface of a cell need to associate in a certain way, and an *Ax* mutation alters the molecules in a way that still allows some protein associations. Two different sorts of alterations in the population of proteins might perturb the protein associations more severely, thus enhancing the phenotype.

THE *WHITE* LOCUS

One of the first *Drosophila* mutations described caused the normally red eyes to be white (51). The *white (w)* gene is on the X chromosome, is not essential for viability of the fly, and most of its alleles are recessive. Some *w* alleles eliminate all function ("bleached-white" phenotype), and some cause partial reductions in function (reviewed in 52; B. H. Judd, manuscript in preparation). Perhaps the most interesting *w* mutations are those that cause abnormal patterns of coloration rather than mere reductions in pigment (e.g. 53, 54). The gene controls pigmentation in the testes and the larval Malpighian tubules as well as in the eyes. The gene does not appear to encode any of the enzymes involved in synthesis of the eye pigments, and instead has been hypothesized to encode a function required for the integrity or for controlling the deposition of pigment granules. This hypothesis can explain how *w* can control the expression of both the red and brown pigments, despite the lack of common biochemical intermediates in the two biosynthetic pathways that form the pigments (52).

The w gene was cloned using "transposon tagging." The w DNA was isolated in association with a *copia* transposon that inserted into w in the w^c allele (55–58). Molecular probes permitted the confirmation of a number of genetic results, for example the assertion that the w^i allele was an intragenic duplication (59, 60). w^i had been proposed to be a duplication because cross-overs within w generated classes of progeny that were proposed to arise from unequal cross-overs within the duplicated part of w (61). The complete w gene has been shown to be contained within an 11.7-kb DNA fragment by germline transformation (54, 62). The 11.7 kb includes 1.95 kb of upstream sequence, and about 2.1 kb of downstream sequence. Insertions of the gene into novel chromosomal sites result in some striking abnormal patterns of eye coloration (differently pigmented vertical or horizontal zones) (63). All of the w gene has been sequenced (64). The w gene encodes a very rare 2.6-kb RNA in five exons distributed over about 6 kb (65–67). The protein-coding region begins in the first exon and continues into the fifth. A generally hydrophobic protein of unknown function is encoded in the w transcript.

Several features of w suggest complexity (52). First, the phenotypes are quite varied, giving many shades and patterns of eye coloration. Second, w interacts with the *zeste (z)* locus in a novel and interesting way. Third, different alleles differently affect dosage compensation, tissue-specific w expression, and z interactions. However, Judd commented in a previous review (52), "most of the evidence that points to complexity is quite indirect and inconclusive." The complexity of w may be due largely to the complex effects of transposon insertions, and in lesser degree to the separable *cis*-acting elements that control its expression.

Genetic studies showed that alleles causing bleached-white phenotypes tended to cluster at one end of the locus, while alleles that allowed partial function or altered function were at the other end. This latter "regulatory region" has turned out to be the 5' upstream region, and the coding exons are the sites of most of the bleached-white alleles. In general, insertions of transposons into w exons cause a bleached-white phenotype (68, 69). The insertions into introns, in contrast, often cause partial reductions of w function. Separable control elements involved in directing tissue-specific w expression are located between 0.22 and 0.4 kb and between 1.1 and 1.9 kb 5' of the cap site (70, 71). DNA involved in dosage compensation is located within a few hundred base pairs of the cap site, or within the transcription unit.

The $w^{spotted}$ (w^{sp}) alleles (53) are especially interesting for two reasons: the mutations affect the pattern of w expression as well as the amount of it, and four of the five alleles in this class are due to deletions of the 5' upstream region rather than to insertions (68, 69, 72). The effects of deletions, as opposed to the effects of transposon insertions, cannot be due to newly juxtaposed DNA; the effects of insertion mutations are often due to the properties of the transposons (see section on THE EFFECTS OF TRANSPOSON INSERTIONS below). The w^{sp} alleles all cause the eyes to be weakly pigmented in a mottled pattern, but the pigmentation of the Malpighian tubules is unaffected. The regulatory region defined by the molecularly mapped w^{sp} mutations is between 1270 and 590 bp upstream of the cap site. In this region there is a DNA sequence homology to the papovavirus transcriptional enhancer sequences (72). It has been suggested that the *cis*-acting region identified by the w^{sp} alleles acts as a transcriptional enhancer of *w* expression in the eyes. This enhancer model remains to be tested by germline transformation.

The *zeste (z)* locus is a regulatory locus that controls the expression of other genes and has effects that depend on the arrangement of the target genes on the chromosomes. Three loci, *w* (73), *Ultrabithorax (Ubx)* (74, 75), and *decapentaplegic (dpp)* (76), are known to interact with *z* (77). A wild-type allele of *z* improves the complementation of certain combinations of alleles in these target loci in *trans,* apparently by increasing the expression of the mutant genes. The enhancement is prevented if the alleles are not paired or in close association (a tandem duplication does permit enhancement, 78). In *Drosophila* the homologous chromosomes normally are paired, even in somatic cells.

The original *z* allele, z^1, is an example of a relatively rare form of *z* allele that is antimorphic in its effect; it interferes with the function of the wild-type allele. Instead of acting as an enhancer of gene function in *trans,* z^1 acts as a repressor (79). This was observed clearly in the first study of z^1 and its interactions with *w* (73). In males, which have only one X chromosome and therefore only one copy of the wild-type *w* gene, the presence of the z^1 mutation has no effect. In females homozygous for the z^1 mutation, the expression of their two (paired) wild-type *w* alleles is inhibited, causing a yellowish eye color. It appears that it is the pairing of the two *w* alleles that makes them susceptible to z^1 repression. The reduction in *w* function may be due to a head-specific reduction in *w* transcript levels (80). If a male is carrying two copies of *w* as a tandem duplication on the X chromosome, they also will be repressed by z^1 (78). However if the two *w* genes in a female are not paired, as can be the case if one of the chromosomes carries a rearrangement, then the z^1 alleles are not able to exert their repressive effect on *w*. Paired *w* genes introduced into other chromosomes can still be repressed by z^1 (54). The site of insertion of *w* into the chromosome has a strong effect on the *zeste-white* interaction (54): in one case, pairing of the *w* genes was not required for repression by z^1. The z^1 allele is recessive, but similar alleles that are strong enough to be dominant have been isolated (81). These stronger *z* alleles can repress *w* even when the *w* gene is present only in one copy and even when there is a wild-type *z* allele present. The *z* gene has been cloned (82), but nothing is yet known about the functions of its products.

The target of z^1 in the *w* locus has been mapped to the upstream regulatory region of *w* using germline transformation experiments. The part of the *w* gene that responds to *z* repression is located between -1850 and -1081 bp upstream of the cap site (70, 71). The papovavirus transcription enhancer homology is in the *z* target region, suggesting that *z* may act in *trans* on a transcription enhancer element (72). An insertion of a transposon far upstream (in the w^{DZL} allele) also can cause zeste-like effects, altering *w* function according to the pairing of the chromosomes (68, 83, 84).

The gene-pairing dependence of the *zeste-white* interaction is one example of a phenomenon named transvection by E. Lewis. Transvection is observed in the bithorax complex (see section below) as the chromosome-pairing–dependent expression of certain combinations of alleles (85). For example, the mutations bx^{34e} and *Ubx* cause only slight defects in the third thoracic segment when paired in a heterozygous fly, but the third thoracic segment is transformed into a second thoracic segment if the same two alleles are present in chromosomes that are unable to pair due to chromosome rearrangements. The rearrangements (for example, large inversions) need not directly alter the genes of the bithorax complex, they need only disrupt pairing. The effect has been used to screen for chromosome rearrangements that interfere with pairing (85). The *z* locus appears likely to encode part of the machinery that is involved in the transvection effect (75, 77). It has been speculated that transvection phenomena may be related to the effects of chromatin structure on gene expression. The transcription of a mutant *Drosophila* glue protein gene is increased by pairing of the mutant gene with a wild-type gene (85a), suggesting that at least in some cases transvection acts at the level of transcription or RNA stability.

TWO GENE CLUSTERS THAT ENCODE KNOWN ENZYMES

The rudimentary *Locus*

The enzymatic activities that carry out the first three steps in pyrimidine biosynthesis, carbamyl phosphate synthetase (CPSase), aspartate transcarbamylase (ATCase), and dihydroorotase (DHOase), are encoded by the sex-linked *rudimentary (r)* gene in *Drosophila* (52, 86). In prokaryotes the three enzymes are encoded by three genes (87), in yeast two of the enzymes are encoded by one gene and one by another (88), but in *Drosophila* and hamsters only a single gene is used (88, 89). A multifunctional protein of about 220,000 M_r carries all three activities. The *r* gene was discovered by Morgan in 1910 (90), and has been the focus of intensive genetic research (91). Recently, molecular analyses have revealed in some detail the structure of the gene (92, 93).

The primary *r* transcript is about 14 kb, and is processed into a 7.3-kb mRNA (93). There are seven exons and six introns. The relationship between the domains of the proteins and the arrangement of the coding sequences in the exons is not yet known. The order of the coding sequences for the three enzymes is the same in *Drosophila* as it is in hamsters, suggesting that the fusion of the three genes into one was an ancient event. The hamster gene, however, is 25 kb long rather than 14, and has 30–40 exons 0.1–0.4 kb long with 50–300-bp introns (94). The order of the coding regions in flies and hamsters is 5'-DHOase-CPSase-ATCase-3', while in yeast the order is 5'-CPSase-ATCase-3'. This suggests that the DHOase was added last to the complex, and to the 5' end of the gene. A fragment of the *Drosophila* sequence shows 35% homology with the protein sequence of the E. *coli* CPSase sequence; another section has 50% homology with the protein sequence of *E. coli* ATCase (93). Thus the protein sequences appear to be quite conserved.

One phenotype of *r* mutations is truncated wings in homozygous females and hemizygous males grown on normal rich media containing pyrimidine sources. This is a good example of how a mutation in a metabolic enzyme can have an apparently specific developmental effect even though the enzyme is required in many tissues. That the wing defect is not the whole story is revealed by the other two phenotypes of the *r* mutations: female sterility and pyrimidine auxotrophy (95, 96). Homozygous *r* females are sterile, or nearly so, if mated to mutant *r* males. If the male supplies a wild-type *r* allele the sterility is largely prevented. Zygotic gene activity is therefore adequate to rescue the progeny of homozygous mutant mothers. In the absence of an external pyrimidine source, homozygous flies die as larvae. Depending on the strength of the allele, the auxotrophy can be dominant.

The genetic complexity of *r* is revealed by complementation tests. Different studies have found different numbers of complementation groups (91, 96, 97). In one extensive study, Carlson (91) identified seven complementation "units" (inferred functional parts of the *r* gene) and 16 complementation groups (groups of alleles that behave identically in complementation tests; each such "group" is defective in a particular subset of the "unit" functions). Some mutations fail to complement any of the other mutations, and therefore act like deletions of all of the *r* functions. Other mutations complement some of the other alleles. This is not hard to understand in light of the tripartite structure of the gene product: mutations affecting all the enzyme functions are nulls (alleles that lack all functions, like deletions), and mutations affecting any one (or two) of the enzyme functions can complement mutations affecting only a different enzyme activity. This model can explain three complementation units, but not seven.

What is the basis for the complex complementation? It appears likely that different parts of the protein correspond to the different complementation groups (98, 99). Any mutation will fail to complement itself because no copy of the protein has that functional unit (or domain). Any mutation in the same functional unit will also fail to complement the first mutation. The many complementation groups found indicate that the functional unit does not always correspond to the whole protein (if it did, a single complementation group would be observed) or even a single enzymatic activity (in which case three complementation groups would be expected for r). The complementation data therefore suggest that a mutation in one part of the protein can in some cases complement a mutation in another part, even if both mutations are in the part of the protein responsible for one enzymatic function. One way in which this could work is if the protein functions as a multimer (which it almost certainly does; 88), and the seven complementation units described by Carlson correspond to protein domains. The r protein, based on this interpretation, is composed of at least seven genetically separable domains, and not just three domains that correspond to the three enzyme activities. The presence of a good domain in any one subunit protein molecule would suffice for the function of the whole multimer.

The Gart Locus

The *Gart* (glycinamide ribotide transformylase) locus also encodes three enzyme activities. The *Drosophila* version of the gene was first identified by complementation tests in yeast (100), in a search for *Drosophila* DNA segments that could complement the yeast *ade8* mutation. Only part of the *Drosophila* GART gene is required to complement *ade8* (101). It has been found that at least two other enzyme activities are associated with the GART products (102). All three enzymatic functions are in the purine nucleotide synthesis pathway. In yeast the three activities are encoded by two genes *ADE5,7* and *ADE8* (103). In both species, and probably others as well, the purine synthetic enzymes are probably associated in a multienzyme complex. *Drosophila Gart* and *rudimentary* appear to be cases of complexes encoding complexes.

The molecular analysis of the *Gart* locus has revealed the structure of the products (101). Two proteins are encoded by overlapping alternative mRNAs. One mRNA is a truncated version of the other, and is terminated by the use of an optional polyadenylation site. Only glycinamide ribotide synthetase (GARS) is encoded by the smaller transcript, while the larger transcript encodes GARS, aminoimidazole ribotide synthetase (AIRS), and glycinamide ribotide transformylase (GART). A 45-kd protein is encoded by the GARS (only) transcript; a 145-kd protein encoded by the larger transcript provides all three enzyme activities. In yeast, the GARS and AIRS activities are encoded by the *ADE5,7* locus, which encodes a single protein 802 amino acids long (103). Protein sequence homology between the two species extends through the entire *ADE5,7* yeast protein, and the entire *Drosophila* protein can be aligned with the tandemly joined yeast *ADE5,7* and *ADE8* (GART) proteins except that part of the *Drosophila* protein (the AIRS part) is internally duplicated. The organization of the *Drosophila* protein can be summarized as N terminus-GARS-AIRS-AIRS-GART-C terminus. The AIRS protein is suspected to function as a homomultimer in yeast, based on genetic analyses (99), and in *Drosophila* two AIRS "subunits" may be connected as parts of one protein.

Purine auxotroph mutants of *Drosophila* have been isolated, and one of them, *ade3¹*, has been shown to affect the *Gart* locus (104). A single base change was found in the coding region of the mutant allele. This case excepted, no genetic analysis has yet been done on the *Drosophila Gart* locus. However, the yeast genetics work on *ADE* genes provides a warning of what could be expected. The yeast gene ADE5,7 has a circular complementation map with up to 33 complementation groups (105, 106). All of the *ade5* (GARS) mutations map in one of the segments, so the genetic complexity is primarily due to the *ADE7* (probably AIRS) portion of the gene. Surprisingly, the *ADE7* part of the yeast protein is composed only of about 350 uninterrupted codons of sequence. The complexity is therefore probably due to complex interactions between protein subunits and not to complex gene structure. The interallelic complementation found may be due to mutations that block multimer assembly (and therefore function) when present on both AIRS subunits of, for example, a dimer. Dimers may, however, be able to form and function if the mutational change in one subunit is in a different location than the mutational change in the other (98, 99).

A circular complementation map with eight complementation classes has also been found for the xanthine dehydrogenase *(rosy)* gene of *Drosophila*, although the alleles exhibiting intragenic complementation only partially restore enzyme function (107). The *rosy* protein product functions as a dimer, and therefore the intragenic complementation probably reflects interactions between the two subunits, and between different parts within each subunit, and not complexity of regulation of gene expression. Another case of intragenic complementation involving an enzyme-coding locus is the DOPA decarboxylase gene (50).

Another twist has been added to the story of the *Gart* locus. Within the largest intron of *Gart* lies a gene transcribed in the opposite direction (108). The gene encodes a pupal cuticle protein that, so far as is known, is regulated quite independently of *Gart*. No other definite cases of genes within the introns of other genes have been reported. Although some insertions of transposons into introns appear to be tolerated (e.g. 109, 110), it is not certain that these particular copies of the transposons are active. The *fushi tarazu* segmentation gene appears to lie within the region involved in the function of the homeotic gene *Sex combs reduced* (109, 111), but the organization of the *Sex combs reduced* locus is not fully understood.

HOMEOTIC GENES

Mutations in homeotic genes cause the transformation of one part of the fly into another. Antennae may, for example, be transformed into legs. The homeotic genes have been intensively studied because they are among the best candidates for master regulatory genes that control development (112–118). They also serve as examples of complex gene structure and complex regulation of gene expression. Several general facts about homeotic genes can now be stated: 1. The position-specific effects of the genes on particular cells in the developing embryo are due to expression of the genes in only particular places in the embryo. It is therefore crucial to learn how *trans*-acting factors and *cis*-acting elements work to control position-specific expression of homeotic genes. The elaborate spatial and temporal patterns of expression indicate complex regulation. 2. The genetic complexity of the homeotic genes is due primarily to the intricacy of the gene structures and to the effects of DNA insertions, and not as much to a multiplicity of products. There is, however, growing evidence for variability in the encoded proteins although the variability is far more limited than that of, for example, immunoglobulins. 3. The homeotic genes are an interacting network that has attributes of both a hierarchical system and a combinatorial one.

The terminology of Muller (119) is frequently used in discussing *Drosophila* genes, especially homeotic genes, and although it can be confusing it is useful. A "hypomorph" is an allele of reduced function, and a "hypermorph" is an allele causing increased function. Hypermorph could refer to an increased amount of protein caused by a regulatory mutation or to a protein having increased activity. A "neomorph" is a dominant allele that causes the gene to function in abnormal ways, at abnormal times, or in abnormal places. Such a mutation may or may not allow normal function in addition to the novel function. Finally an "antimorphic" dominant allele is one that actively interferes with the function of a wild-type allele, such as a defective subunit that poisons a multimeric protein. Hypomorphic and null alleles are referred to as loss-of-function alleles, while hypermorphic and neomorphic alleles are referred to as gain-of-function alleles, a particularly confusing designation since the "gain" may not be quantitative but may instead allude to gene function in cells where the gene is normally quiescent.

The Antennapedia *Gene*

Antp is a member of the Antennapedia complex (ANT-C), a cluster of several homeotic genes and several other kinds of genes as well. The classic *Antp* phenotype is a dominant transformation of antennae into legs (120). Other dominant alleles transform dorsal head into thorax (121), or second and third legs into first legs (114, 122). Most of the dominant alleles are also recessive lethal alleles, and heteroallelic combinations of any of them fail to survive. In addition there are numerous recessive lethal alleles that have no dominant phenotype and fail to complement the recessive lethality of the dominant alleles (121, 123). The embryos that are homozygous for any combination of lethal alleles have their second and third thoracic segments transformed into hybrid segments having some characteristics of a first thoracic segment and some of a head segment (123).

A deletion of *Antp,* heterozygous with a wild-type allele, has no phenotype. The single wild-type copy of the gene is sufficient. Since the dominant *Antp* alleles, heterozygous with a wild-type allele, cause dramatic antenna-to-leg and other transformations, the dominant alleles must be gain-of-function mutations. The gain of function may be due to the expression of the mutant gene in tissues where the gene is normally silent or to expression at abnormal times during development. Screens for phenotypic reversion of the dominant *Antp* alleles (122, 124, 125) led to the isolation of new mutations that eliminate the abnormal function, for example by deleting the mutant allele or destroying it with an inversion breakpoint. Are the dominant *Antp* mutations hypermorphic, neomorphic, or antimorphic? Strong evidence in favor of viewing the antenna-to-leg mutants as neomorphs comes from experiments that demonstrate that *Antp* is normally not required in the head at all. The experiments used both transplantation of imaginal discs (126) and analysis of homozygous clones of cells (121, 125) to show that head structures develop normally without *Antp* function.

The structure of the *Antp* gene helps to reveal the basis of the genetic phenomena, though the gene is far from completely understood. *Antp* is the largest known *Drosophila* transcription unit, spanning about 103 kb (109, 110). Measurements of transcription rates in other insects suggest that it would take about 100 minutes to transcribe *Antp,* a substantial time in an organism that completes all of embryogenesis in only 22 hours. *Antp* has two promoters, one of which directs transcription of a (tiny!) 36-kb transcription unit nested within the large one (127–129). The protein-coding exons are all clustered within a 13-kb region near the 3' end of the gene, and are common to both transcription units. The major impact of having the two promoters is to attach partially different mRNA leader sequences to the coding parts of the mRNAs, and perhaps to allow different transcription completion times or independent control by two batteries of regulatory elements.

The structures of the *Antp* mRNAs are also unusual. The mRNAs have very long 5' leader sequences. Those from the promoter of the 103-kb transcription unit (P1) have a 1.5-kb leader sequence; those from the internal (P2) promoter have a 1.7-kb leader sequence. The *Antp* open reading frame is about 1.1 kb long, so only about 1% of the transcription unit is known to encode protein. The use of multiple alternative polyadenylation sites leads to heterogeneity at the 3' end of the mRNAs. Two major sites 1.5 kb apart, and a minor one between them, are used by transcripts from both promoters. The finished mRNAs are 3.4, 3.6, 4.9, and 5.1 kb long, the 3.4 and 4.9 from P1 and the others from P2. This sort of complexity in transcription and processing is not unique to complex homeotic loci. The actin 5C gene also has two promoters and three polyadenylation sites (19). The use of different polyadenylation sites in *Antp* is developmentally regulated, which is also true of the *serendipity* gene (10), a gene expressed during embryogenesis.

Most of the *Antp* alleles that cause the dominant antenna-to-leg phenotype are associated with chromosome rearrangements. The breakpoints of the rearrangements are located at different positions in the upstream 60 kb or so of the gene (109, 110). Two mechanisms can be envisioned to explain how such rearrangements could cause abnormal expression of *Antp* in the head. One possibility is that a negative control element that keeps *Antp* off in the head is removed by the rearrangements. The smaller (P2) transcription unit is still apparently intact and could be active in the absence of negative control. A second possibility is that control elements newly juxtaposed by the inversions or translocations turn on *Antp* in the head. These elements could be promoters or transcriptional enhancers, and could activate the P2 unit or generate new transcription units that would incorporate the *Antp* coding exons. Both a promoter fusion model and an enhancer fusion model rely on the observation that the *Antp* coding exons are unaffected by the rearrangements. The mutations could cause misregulation of the normal protein rather than a change in the structure of the protein.

The dominant mutations that transform second and third legs into first legs are also gain-of-function mutations that can be phenotypically reverted by deleting the mutant allele (122). However, one such allele, Scx^W, has a weaker phenotype in the presence of a duplication of the wild-type *Antp* gene (B. T. Wakimoto and T. C. Kaufman, personal communications). This suggests that Scx^W is an antimorph: the dominant allele interferes with wild-type function in some way, and the effect is ameliorated by adding more wild-type function. The Scx^W mutation is an inversion of about 50 kb, including the entire small (36-kb) transcription unit of *Antp* (109). The molecular basis of the effect of the Scx^W alleles is unknown, but one speculation is that the inversion in Scx^W leads to production of antisense RNA that could interfere with the processing or translation of the normal mRNA, an antimorphic effect. The only other *Scx* allele, Scx^I, contains an insertion near the 5' exon of *Antp*, but it is not known whether the insertion is the cause of the phenotype.

Some recessive *Antp* alleles complement each other completely (121, 130). The flies carrying the heteroallelic combination of *Antp* alleles look quite normal instead of dying as larvae with homeotically transformed thoracic segments. In other heteroallelic combinations, flies survive to late pupal stages ("pharate adults") or sometimes adulthood, but have defective legs or dorsal thoracic structures. It appears that different *Antp* mutations affect different subsets of the *Antp* functions. The structure of the gene immediately suggests how complementation may occur: mutations that affect only one of the transcription units could complement mutations that affect only the other transcription unit, and both groups of mutations would fail to complement mutations in coding exons or other parts of the gene that are common to both transcription units. This model appears to be consistent with two alleles, *Antp^{s1}* and *Antp^{s2}*, which complement completely (130). Both alleles are lethal when heterozygous with a deletion of *Antp*, although at different developmental stages. One allele, *Antp^{s2}*, is a chromosome inversion broken within *Antp* in the P1 transcription unit. The P2 unit is left apparently intact. The complementing allele, *Antp^{s1}*, contains a transposon insertion in the leader exon of the P2 transcription unit (109), an exon not used in the P1 unit. Not all cases of intragenic complementation are as complete as with the *Antp^{s1}* and *Antp^{s2}* alleles. In cases of partial complementation, one of the mutations may not leave either transcription unit fully functional.

The Ultrabithorax *and* bithoraxoid Units *of the Bithorax Complex*

The first part of the bithorax complex (BX-C) to be cloned was the *Ultrabithorax (Ubx)* gene and its surroundings (131). Several decades of elegant genetic analysis by E. Lewis (113, 132–139) had revealed the genetic complexity of the bithorax complex, and had also proven the importance to developmental biology of understanding how such gene complexes work. Like *Notch,* the *Ubx* part of the BX-C consists of a pseudoallelic series of mutations. *Ubx* alleles have a weak dominant effect (a slight swelling of the flight balancer organs, the halteres, indicative of a partial transformation of haltere to wing) and are recessive lethals. The homozygous *Ubx* larvae die with their third thoracic and first abdominal segments transformed into second thoracic segments. [The transformations actually appear to involve not segments but parasegments (140–142), but the details of the developmental aspects of BX-C gene function will not be described here.] The dominant effect of *Ubx* alleles is due to haplo-insufficiency of the gene, and the embryonic transformations involving the third thoracic and first abdominal segments, seen in *Ubx* homozygotes, are due to loss of *Ubx* function.

The genetic complexity of *Ubx* is observed with adult-viable mutations (113, 131). The mutations include *anterobithorax (abx)* alleles, which cause the transformation of the most anterior part of the third thoracic segment (and perhaps posterior second thoracic segment) into anterior second thoracic segment (and perhaps posterior first thoracic segment), *bithorax (bx)* alleles, which cause the transformation of anterior third thoracic segment into anterior second thoracic segment; *postbithorax (pbx)* alleles, which cause the transformation of posterior third thoracic segment into posterior second thoracic segment, and *bithoraxoid (bxd)* alleles, which cause the partial transformation of the third thoracic segment into a second thoracic segment and of the first abdominal segment into a third thoracic segment. Each of these types of allele is recessive. The recessive phenotype is also observed in flies heterozygous for the recessive allele and a *Ubx* mutation (recall *fa/N*). On the basis of these sorts of observations, the mutants were assigned to the *Ubx* unit of the BX-C (134, 135). All of these loss-of-function alleles cause transformations into more anterior structures, showing that the *Ubx* functions (*Ubx, abx, bx,* etc) are required to direct more posterior patterns of development. All *bx* alleles complement both of the *pbx* alleles and all of the *bxd* alleles. However *pbx* and *bxd* alleles generally fail to complement. Flies of the genotype *pbx/bxd* have a *pbx* phenotype. *bxd* mutations affect a function needed in the posterior part of the third thoracic segment, a function that is also inactivated by a *pbx* mutation.

The *Ubx* gene was the first homeotic gene cloned, and provided several surprises to its investigators. First, it was found to be (as of that time) the largest fly transcription unit, over 73 kb long (143–145). Large exons are located at each end of the 73-kb region and two 51-bp coding exons are located between the large exons. In contrast to *Antp,* the coding exons of *Ubx* are spread across the whole of the gene, and only one promoter has been identified. Second, and most usefully, nearly all of the mutations that were not associated with cytologically visible chromosome rearrangements were detectable at the DNA level because they were due to transposon insertions (131). The *Ubx* mutations were found to be associated with chromosome breaks that interrupt the 73-kb transcription unit at many points along its length. In addition, some *Ubx* mutations have been attributed to small deletions within the exons (146).

The *Ubx* transcription unit produces three major RNA species, 3.2, 4.3, and 4.7 kb in size (143). The 3.2- and 4.3-kb species both span the 73-kb region, and both have the same 5' ends, but little else is known about the 4.3-kb species. The 3.2-kb species contains protein-coding sequence from the exons at both ends of the 73-kb region and from the two 51-bp microexons in between. Most of the difference between the 3.2- and 4.3-kb RNA species may come from the use of alternative polyadenylation sites, since the 4.3-kb RNA terminates further downstream than the 3.2-kb RNA. The 4.7-kb transcript is quite different: 1. it is not polyadenylated, in contrast to the other two; 2. it is expressed only transiently at the very earliest stages of development, while the others are expressed a little later and then stay on; and 3. it starts near where the other species start but contains none of the downstream exons that make up the 3.2- and 4.3-kb RNAs. Instead the 4.7-kb RNA contains sequences located just downstream from the 5' exon. The time of expression of the 4.7-kb RNA suggests that it may be involved in an early function of *Ubx* in the posterior parts of the second- and third-thoracic segments (147).

Since the *Ubx* alleles inactivate a function that is common to the *bx* and *abx* functions, it is not surprising to find that the *Ubx* transcription unit spans the sites of the *bx* and *abx* mutations. The *bx* alleles are mostly insertion mutations, of at least four different transposons, *gypsy, I, Harvey,* and *412* (131, 148). The inserts are clustered in the largest *Ubx* intron about 30 kb downstream from the 5' *Ubx* exon. The *abx* alleles are all deletions, and all are located in the largest *Ubx* intron about 45 kb downstream from the 5' *Ubx* exon. The three *abx* deletions, of 1.5 kb, 6 kb, and 11 kb, overlap.

A fly that is *bxd*/*Ubx* has a *bxd* phenotype (transformation of third-thoracic segment to second-thoracic, and first abdominal segment to third-thoracic segment). A *pbx*/*Ubx* fly has the *pbx* phenotype (transformation of posterior third-thoracic segment to posterior second). Surprisingly, the *bxd* and *pbx* mutations do not map within the 73-kb *Ubx* transcription unit region. Instead, these classes of mutations map to an approximately 30-kb region upstream of the *Ubx* unit (131). The region has been named the *bxd* unit and has its own transcripts. Early in embryogenesis, the primary transcripts extend over 25 kb and are processed into finished molecules about 1.2 kb in size (143). In addition, there is a small late transcript from the center of the 25-kb *bxd* region. *bxd* mutations, like *bx* mutations, are often associated with *gypsy* inserts. Other *bxd* alleles are the result of chromosome breakpoints located in the *bxd* region. The closer the breakpoints are to the *Ubx* unit, the more severe the *bxd* phenotype (144). Both *pbx* alleles are deletions, and both delete the 5' region of the *bxd* unit. The two deletions overlap near the 5' exon of the *bxd* RNAs.

It has been shown that the *bxd* mutations act by affecting the expression of *Ubx* protein (149, 150). The *bxd* unit may therefore be a very large *cis*-acting regulatory region for *Ubx*. Analysis of *bxd* cDNA clones has shown that the RNA species represented in the clones do not encode proteins of significant length. Despite the elaborate and varied splicing patterns of the early *bxd* RNAs, they may not be related to the *bxd* functions (143). This startling possibility is based on the localization of the early transcripts, which is not what would be expected for *bxd* functions. *bxd* function is required in the posterior thorax and anterior abdomen, but the putative *bxd* transcripts are distributed over a broad region of the central and posterior abdomen. There is little of the RNA in the regions that are primarily affected by *bxd* mutations. Flies that are pbx^1/pbx^2 or pbx^1/pbx^1 make no early "*bxd*" RNA, but no *bxd*-like transformation of embryos or larvae is seen (W. Bender, personal communication). The function, if any, of the early transcripts remains mysterious. The late *bxd* RNA is made during metamorphosis, has a single exon, and is 0.8 kb long. It has a moderately sized open reading frame that has the codon usage expected for *Drosophila*. Its localization in the embryo has not been reported, and its function is unknown (143).

In addition to the loss-of-function mutations, the *Ubx* gene also has dominant gain-of-function alleles. One of these, *Contrabithorax (Cbx)*, is closely related to one of the *pbx* deletions because it arose from the same event, namely a transposition of 17 kb of *bxd* region DNA into *Ubx* (131). The *Cbx* phenotype is a transformation of posterior wing into posterior haltere, the transformation opposite to a *pbx* transformation. The deletion of the 17 kb causes the *pbx* phenotype, and the 17-kb insertion causes the *Cbx* phenotype, as was shown by molecular analysis (131) after E. Lewis (135) separated the two parts of the rearranged DNA by recombination. The 17 kb is inserted into one of the *Ubx* introns. The insert does not interfere markedly with normal *Ubx* function, since the *Cbx* mutation is homozygous viable and the homozygotes have about the same phenotype as the heterozygotes.

The *Cbx* effect appears to represent an activation of the normal *Ubx* function in posterior wing (second-thoracic segment), where *Ubx* should be off. *Ubx* normally functions in the third-thoracic segment to direct haltere development instead of wing development. Direct observation of *Ubx* products in *Cbx* larvae shows that *Ubx* products do indeed appear in the second-thoracic segment (151, 152). How does the 17-kb insertion activate *Ubx* expression in the wrong place? Either the insert disrupts a negative regulatory element that keeps *Ubx* off in the second-thoracic segment or the insert brings in a regulatory element that acts on the *Ubx* promoter to stimulate second-thoracic segment *Ubx* expression.

The Abdominal Region of the Bithorax Complex

The genes of the BX-C are arranged along the chromosome in approximately the order of the body segments they affect (113), which is also true for the genes of the Antennapedia complex (121). Whether the arrangement has functional importance or is an evolutionary accident remains to be determined. The genes in the centromere-distal portion of the BX-C are involved in controlling the differentiation of the abdominal segments of the fly, and have been less intensively investigated than the *Ubx* and *bxd* units. Genetic saturation analyses of the complex for lethal mutations revealed three lethal complementation groups: *Ubx, abd A,* and *Abd B* (153, 154). Earlier analyses by E. Lewis in which lethality was not the only criterion for gene identification had led to the detection of approximately one gene per abdominal segment (113, 155). He named the genes *infraabdominal 2–8 (iab2–iab8)*. The numbers of the *iab* loci refer to the most anterior abdominal segment that is affected, and all of the loss-of-function *(iab)* transformations cause segments to develop as though they are more anterior segments. An *iab2* mutation will transform the second through eighth abdominal segments into first abdominal segments; an *iab3* mutation will partially transform the third through sixth abdominal segments into second abdominal segments, and so on. The *iab* genes are arranged along the chromosome in the order of the segments they affect. The discrepancy in gene numbers does not now seem especially mysterious. There is one homeobox (indicating protein-coding sequences, see below) for each of the lethal complementation groups *Ubx, iab2* (which appears to be *abd A*), and *iab7* (which appears to be *Abd B*) (156). The other *iab* genes may act on the two protein-coding units to control their effects, much as *bxd* mutations act on *Ubx*. However it is also possible that the abdominal region BX-C genes are quite different from other homeotic genes, and there may be many other protein or RNA products in addition to the *iab2* and *iab7* products.

The abdominal genes are distributed across a region of about 200 kb (155). Together with the *Ubx-bxd* region the BX-C spans about 315 kb. The transcriptional map of the abdominal region has not yet been reported, but there has been extensive mapping of mutations (155). Many chromosome breakpoints affect more than one of the *iab* functions. Such *cis*-inactivation effects tend to be polar: a mutation in one *iab* gene also affects neighboring functions located to one side of the gene but not the other. The effects could be due to transcripts initiating in different *iab* regions, or to transcriptional enhancer-like effects of different *iab* genes on the expression of the *iab2* and *iab7* products, or to *cis*-active products of each of the *iab* units.

Dominant alleles affecting the abdominal region genes appear to exert their effects by causing inappropriate expression of one or more *iab* genes in an abnormal location. The *Tab* allele, for example, causes abdominal tissue to develop on the thorax, perhaps by causing a gene fusion of BX-C abdominal region sequences to a promoter that is active in the thorax (155).

The Homeobox and the Homeodomain: Protein Products of the Homeotic Loci

The 3' exons of *Antp, Ubx, Deformed (Dfd,* a homeotic gene active in the head), *Scr,* and probably other homeotic loci as well contain a conserved 180-bp sequence called the homeobox (157–159). The presence of a homeobox is an indicator of protein-coding sequence because the homeobox encodes a still more highly conserved 60-amino-acid domain called the homeodomain. There are at least three homeoboxes in the bithorax complex (one for each lethal complementation group) (156) and at least seven homeoboxes in the Antennapedia complex (one each for *Antp, Scr,* and *Dfd,* and others for segmentation genes in the complex) (156; A. Laughon, M. P. Scott, unpublished data). In addition, an ever growing number of homeoboxes is being discovered in genes located elsewhere in the genome.

An idea about the function of the homeodomain has come from the observation of a weak homology of part of the domain with certain bacterial DNA–binding proteins (16). The homology is with the region of the bacterial proteins that forms an α helix–β turn–α helix structure. In proteins such as lambda cro, one helix rests in the major groove of the DNA and makes sequence-specific contacts; the other helix lies over the first and stabilizes the interaction (160). The part of the homeodomain that is most highly conserved among the more than 25 sequences now known is the part that corresponds to the sequence-recognizing helix of the bacterial proteins (161). The homeodomain is also homologous to the protein products of the yeast mating type genes *MATa1* and *MATα2* (16, 16a). The *MATα2* product has been shown to be a sequence-specific DNA-binding protein (16b). These observations suggest that the homeodomain is involved in sequence-specific DNA binding, and that the sequences recognized by homeodomain-containing proteins are related just as the proteins are, thus explaining the stringent evolutionary conservation of the sequence-recognizing part of the protein. The first tests of the DNA-binding hypothesis were done using hybrid proteins composed of beta-galactosidase joined to part or nearly all of a *Drosophila* homeodomain-containing protein (162). The proteins were shown to be able to bind in a sequence-specific manner in vitro, though it is not known whether the detected binding sites play a role in vivo.

The homeodomain-containing proteins that have been characterized so far are about 40–60 kd in size. In five cases, *Ubx* (149, 150), *Antp* (163), *Scr* (P. D. Riley, S. B. Carroll, M. P. Scott, unpublished observations), *fushi tarazu* (164), and *engrailed* (165) (the latter two are segmentation genes), homeodomain-containing proteins have been localized with antibodies and found to be in the nucleus, in keeping with the proposed DNA-binding role. Apart from the homeodomain, the sequences of the proteins have little similarity, except that they all have poly–amino acid stretches. *Antp* protein, for example, has one section in which 31 out of 46 amino acids are glutamine (127–129). *Ubx* protein has 17 glycines in a row (150). The function of such sequences is unknown. *Ubx* and *Antp* produce multiple protein forms. In the case of *Ubx,* two alternative RNA splice donor sites at the 3' end of the 5' exon are located 27 bp (9 codons) apart (143, 166). *Antp* cDNA clones have been found that differ in the splicing pattern within the coding region so as to give rise to proteins that have or lack four amino acids just upstream of the homeodomain (J. R. Bermingham, M. P. Scott, unpublished data). The only obviously notable aspect of the optional *Ubx* and *Antp* sequences is the inclusion of a cysteine residue in each of them, raising the possibility that RNA splicing could affect disulfide bond formation.

It has been proposed that a cell's fate is determined by the array of homeotic gene products produced in that cell (113, 121, 167). Evidence from experiments employing immunofluorescence to detect homeotic protein products demonstrates that multiple homeotic genes can be expressed in the same cells (163; S. B. Carroll, S. DiNardo, R. A. H. White, P. H. O'Farrell, M. P. Scott, unpublished data). How could multiple homeotic proteins in the same cell direct the cell to follow one developmental pathway rather than another? A cell is presumably directed into a developmental pathway by the activation of a particular set of genes that encode the cell's structural components, enzyme activities, surface molecules, and so on. If the homeodomain is a DNA-binding domain, three possible models are: 1. The protein encoded by each gene independently binds to a certain set of target genes and activates or represses them (the same protein could conceivably activate at some sites and repress at others). The effects of the multiple homeotic proteins would be simply additive. 2. The proteins from some of the homeotic genes modulate the effects of other homeotic proteins, for example by competing for sites, by acting as cofactors, or by mediating interactions with molecules such as RNA polymerase. The effects of expressing multiple homeotic proteins would not be simply additive, since the presence of one protein could modulate the effect of the presence of another. 3. The homeotic proteins act as multimers, and function as, for example, homodimers or heterodimers. The heterodimers bind to a different array of target sites than any of the homodimers. The balance among the amounts of different heterodimers is determined by the relative amounts of monomers synthesized, by the affinity of different pairings of proteins, and by the stabilities of the different heterodimers. With this sort of model, the effects of expressing multiple homeotic proteins would not be simply additive, since new binding specificities would be generated by the formation of heterodimers.

OTHER COMPLEX GENES THAT CONTROL DEVELOPMENT

The Sex lethal *Locus*

The *Sex lethal (Sxl)* locus plays two roles, one in sex determination and one in dosage compensation (168, 169). *Sxl* activity is required in female cells 1. to direct and maintain differentiation along the female morphogenetic pathways and, 2. to prevent the heightened transcription of the X chromosome that is triggered to compensate for the presence of only one X chromosome in males. Thus chromosomally male cells can do without *Sxl* function, but the expression of *Sxl* is required to prevent dosage compensation in female cells. Since hypertranscription of two X chromosomes is lethal, homozygous null *Sxl* alleles are lethal in females. In females, most *Sxl* alleles are recessive; one wild-type copy of the gene is sufficient to provide both the sex determination and dosage compensation functions.

The multiple functions of *Sxl* are reflected in intragenic complementation. Some alleles affect primarily the initiation of the sexual differentiation decision, while other alleles affect primarily the maintenance and expression of the sexual differentiation decision. In some cases an allele from the first class can fully complement an allele from the second (168), so that a fly heterozygous for the two alleles does not have a detectable phenotype. Some *Sxl* alleles preserve the dosage compensation repression function, at least partially, but are unable to provide the female differentiation function. These alleles are female viable (in contrast to null alleles), allowing chromosomally female flies to survive but also allowing them to develop with male morphology.

The allele $Sxl^{M#1}$ is dominant and causes the death of males and male cells. This allele is likely to be a constitutively active form of the gene that is tolerated in females where *Sxl* would normally be active (170). In males, the inappropriate activity of the constitutive allele prevents dosage compensation, leading to insufficient X chromosome activity and killing the male flies. $Sxl^{M#1}$, being a dominant gain-of-function allele, can readily be reverted by selecting for male viability. The "revertant" alleles no longer have enough constitutive gene function to prevent dosage compensation in males, but the revertants also lack, partially or totally, the normal female function. Thus the common revertant alleles are not true revertants, since gene function is destroyed, not restored.

The *Sxl* gene has been cloned and initial investigations of the transcripts have been reported (168). The molecular structure is very complex, with at least 10 transcripts produced from a 32-kb transcribed region. There are three regions of homeobox homology (but they are not yet sequenced and proven to be homeoboxes), suggesting the possibility that the molecular functions of the *Sxl* protein products are related to those of the homeotic genes. As expected, at least one *Sxl* transcript is expressed only in females. What was not expected was an array of transcripts produced only in males or in both sexes. Some of the transcripts are also differentially expressed during development.

Many *Sxl* mutations are due to transposon insertions. *Sxl* was cloned using insertions of P-element transposons. Five other insertions are associated with female-specific lethal loss-of-function alleles. An insertion of the retrovirus-like transposon *B104* (also called *roo*) (171, 172), caused the dominant $Sxl^{M#1}$ mutation (168). The $Sxl^{M#1}$ insertion is within the transcribed region and has been proven to be the relevant change in the DNA by reversion studies: Rare incomplete excisions of the insert can restore wild-type function of the gene. Thus the insertion event itself does not seem to alter gene function; rather, the activity of the inserted element may be the causative factor. Three other alleles that are genetically similar to $Sxl^{M\#1}$ are caused by insertions of *B104* or another transposon in a region within 1 kb of the $Sxl^{M\#1}$ insert. In order to examine the effect of $Sxl^{M\#1}$ on transcription, a revertant derivative of the allele ($Sxl^{M\#1,fm3}$) was used, a double mutant in which the gain-of-function effect of the *B104* insert is reduced or eliminated by a second mutation that maps away from the *B104* insert. The reversion mutation might, for example, alter a protein-coding region, so that the *B104* insert misregulates a nonfunctional protein and is therefore harmless. The effect of the $Sxl^{M\#1,fm3}$ double mutation on transcription is surprising: the levels of at least seven *Sxl* transcripts is increased about 20-fold, while some others are decreased. The *B104* insert in $Sxl^{M\#1}$ may therefore act as a transcription enhancer for certain transcripts. Alternatively the heightened level of transcripts could be due to an alteration in their structure that stabilizes the RNA (the *fm3* change in the DNA has not been detected, and therefore appears to be a small change in the DNA).

Sxl is controlled by the *daughterless (da)* locus, which is expressed maternally (170, 173, 174). *da* product is provided during oogenesis; without it, *Sxl* is not activated and females die, presumably due to hypertranscription of their two X chromosomes. *da* probably participates in the mechanisms that read the X chromosome:autosome ratio to decide the pathway of sexual differentiation. The constitutive allele, $Sxl^{M\#1}$, does not require *da* product to be activated—it was isolated as a mutation that allowed the survival of female progeny of homozygous *da* females. The transcription of *Sxl* in mixed male and female embryos obtained from homozygous *da* females is detectably altered. There is an increase in the two male-specific transcripts and a decrease in two early embryo-specific transcripts (168). The single female-specific transcript does not appear to be affected. *Sxl* is also positively regulated by the *sisterless-a* gene, a gene probably also involved in measuring the X-to-autosome ratio (175).

An additional feature of *Sxl* function is its autoregulation: certain *Sxl* alleles can substitute for *da* product(s) in activating a wild-type *Sxl* allele (176). The experiments used *Sxl* alleles that provide dosage compensation repression functions sufficient for female survival but do not promote female differentiation. Chromosomally female flies bearing such alleles have the morphology of males. The alleles do not require activation by *da* to prevent the dosage compensation mechanism from being triggered; they are, in effect, constitutive repressors of dosage compensation. When two such alleles are in the same fly as a wild-type *Sxl* allele (which does require *da* product for activation), the wild-type allele functions independently of *da*. Female morphology instead of male morphology is observed, which is presumed to be due to the wild-type allele since the mutant alleles are not capable of providing the female differentiation function. The interpretation of these results is that the defective *Sxl* alleles activate, in *trans,* the wild-type allele, thus substituting for *da* and demonstrating positive autoregulation of *Sxl*. It remains to be demonstrated that autoregulation occurs in wild-type flies as well as in flies bearing the complex mutant alleles.

The doublesex *Locus*

In the regulatory hierarchy that controls sexual differentiation, *Sxl* has an early role, and the *doublesex (dsx)* locus has a later role (177, 178). *Sxl* acts on *dsx* through the intermediary genes *transformer* and *transformer 2*. *Sxl* controls both dosage compensation and sexual differentiation, while *dsx* is exclusively involved in sexual differentiation. *dsx* is believed to be a dual function negative regulator: in males it prevents female differentiation and in females it prevents male differentiation. Recessive null alleles of *dsx* transform both male and female flies into intersexes; the flies try to develop as both sexes at once. The complexity of *dsx* is revealed by alleles that affect only the male-specific or only the female-specific functions. One allele, for example, transforms males into intersexes but does not affect females (179). As in the case of *Sxl,* dominant alleles exist that appear to represent constitutively active forms of *dsx*. One such allele, dsx^D, does not affect males but transforms heterozygous females that are dsx^+/dsx^D into intersexes (177). The dsx^D allele may therefore constitutively express the male mode of *dsx* function.

dsx is as complex molecularly as the genetic data would suggest (179, 180). The gene has a minimum size of 27 kb based on breakpoints of chromosome rearrangements that affect *dsx* function. Transcripts spanning approximately the same 30-kb region were found, and, as with *Sxl,* more transcripts were found than had been anticipated. The transcripts overlap, apparently sharing some exons, and are expressed differentially in the two sexes and at different stages of development. Some pupal transcripts are male-specific, some pupal transcripts are female-specific, and the larval transcripts are common to both sexes. It is not yet known whether the male- or female-specific functions are due entirely to the expression of different RNA species, or if there are also differences in the position- or stage-specific expression of RNAs in males versus females.

The Achaete-Scute Complex

One cluster of genes that affects bristle and hair structure is the Achaete-scute (AS-C) complex (181). Deletions of the *achaete* and *scute* functions cause a complete absence of hairs and bristles, and deficiencies in the peripheral nervous system. The *scute (sc)* mutations prevent formation of subsets of the large bristles, each allele characteristically affecting certain bristles (120). The *achaete (ac)* mutations prevent formation of hairs and bristles to varying degrees. Both classes of mutations are recessive, and there is a continuum between the two classes: most alleles affect both the bristles and the hairs, but affect some more than others. Thus the *ac* and *sc* mutations may have fundamentally the same sort of effect on hairs and bristles, with the difference between the classes of mutations being the exact bristles or hairs that are most affected. Duplications of parts of the AS-C can lead to excessive formation of bristles or hairs, as can dominant alleles such as the *Hairy-wing* mutations.

The complex has been divided into four genetic regions: *achaete, scute α, lethal of scute (l'sc)* (182), and *scute β*. The *l'sc* function is required for development of the central nervous system; without *l'sc,* embryos die with degeneration of the neuroblasts of the ventral nervous system. All four genetic units are located within a 110-kb region that has been cloned (183, 184). *ac* mutations complement *sc* mutations, but widely separated (more than 40 kb) *sc* breakpoint mutations fail to complement each other completely [the phenotype of the heteroallelic flies is equal to that of the weaker allele (181, 183)]. Three *ac* mutations have been mapped to a 5-kb region at the left end of the AS-C. One is a transposon insert, one is a breakpoint, and one is a deletion. Next to the right are the *scute α* breakpoints, which are distributed over 30 kb. The *scute α* mutations are well separated from the *scute β* mutations; the latter are grouped in a region of about 22 kb at least 18 kb from the *scute α* breakpoints. Several *scute* mutations are due to insertions of transposable elements, especially of the element *gypsy*. Three *scute* alleles are due to deletions ranging in size from 1.2 kb to 18 kb. The position of a mutation seems to be more important than its nature: insertions or breakpoints in the same region have similar effects. Between the *scute α* and *scute β* regions should be the *l'sc* locus, but no breakpoints are available that clearly mark *l'sc*. The flanking mutations limit *l'sc* to a region of at most about 20 kb.

The large size of the complex, and the previous experiences with the BX-C and ANT-C genes, prompted a search for large transcription units, but the AS-C does not appear to have giant transcription units. Instead, six small transcription units have been found within the cloned region. One lies within the *l'sc* region and is found only in embryos (suggesting that it encodes the *l'sc* function), one is within the *scute α* region, one is within the *ac* region, one is within the *scute β* region, and two lie in flanking regions. The relationships between the transcripts and the genes are not fully understood. Little or no change in abundance or time of expression of any of the potentially relevant RNA species was observed in loss-of-function mutants (184), suggesting that the effects of the mutations are on spatial expression or a posttranscriptional step in gene expression. The closer *scute* breakpoints are to the putative *scute* α transcript, over a 50-kb region, the stronger their *scute* phenotype.

The decapentaplegic *Complex*

Flies lacking all of their extremities result from mutations in the *decapentaplegic (dpp)* complex (DPP-C) (185, 186). In addition to the the *dpp* phenotypes, two other effects have been attributed to mutations in DPP-C. One is the *shortvein (shv)* phenotype, which is an alteration in the wing venation pattern. The second is the *Haplo-insufficient near dpp (Hin-d)* phenotype, which is embryonic lethality accompanied by a transformation of structures that should be dorsal into ventral structures. The basis for assigning the three loci to a complex is that the *Hin-d* mutations appear to inactivate the *shv* and *dpp* functions, and *shv* alleles can inactivate *dpp* functions (187). Many *Hin-d* mutations are deletions of the entire complex, showing that the embryonic phenotype is a null phenotype.

A chromosome walk from a transposon allowed the isolation of DNA from the DPP-C. Many chromosome breakpoints (more than 40) were mapped onto the DNA (185). The mapping of the rearrangement mutations confirmed that the most severe *dpp* alleles map close to the *Hin-d* region, and less extreme to mild alleles map progressively further away. The *shv* mutations map on the opposite side of *Hin-d* from *dpp*. The breakpoints that delimit *Hin-d* show it to be within a 9-kb region. Within the 9 kb three size classes of RNA hybridize to each of two exon regions. The different transcripts vary in their 5' parts and in their pattern of expression during development. Two tRNA[tyr] genes are also located in or near *Hin-d*. The DPP-C may be a case of a single protein-coding region, or a few of them, governed by a complex set of regulatory elements.

THE EFFECTS OF TRANSPOSON INSERTIONS

If a transposon insertion into a gene causes a mutant phenotype, and they often do not (e.g. w^{i+a}, 59), does the phenotype reveal a normal aspect of gene regulation or an alteration of gene function due to the characteristics of the transposon? Both kinds of effects are observed (188). In some instances a functional part of the gene is interrupted; in other cases the transposon interrupts unimportant DNA, but the transposon's transcriptional activities or effects on chromatin structure cause changes in the "host" gene's function. The effect of a transposon on its host gene may be stage- or tissue-specific if the transposon's expression is developmentally regulated and if the activity of the transposon is what affects the host gene.

The *Notch* gene provides some well-studied examples of transposon effects. The recessive *facet (fa)*, *facet[glossy]*, and *glossy-like* alleles of *N* are often due to transposon insertions (189). All these alleles affect the eyes, but in distinguishable ways. All of the insertions were found to be in the second and third introns of the 38-kb gene. Each of the five *glossy-like* alleles is due to an insertion of the *flea* transposon, an element with at least some of the characteristics of a retrotransposon. Each *flea* insertion occurred at a different site, all in the same orientation so that *flea* transcription is opposite in direction to that of *N*. In contrast, the insertions that caused the *fa* phenotype (three cases) were due to three different transposons inserted at three different sites. Most strikingly, an insertion event that caused a *fa* phenotype occurred at exactly the same site as an insertion that caused a *glossy-like* phenotype. Therefore in this instance the different phenotypes must be due to the nature of the transposons. A different transposon insertion into the first intron (in a wild-type strain) does not interfere with gene function, so some insertions do not cause any phenotype. Therefore the interruption of the gene, or the change in the spacing of regulatory elements, is not in itself sufficient to cause a phenotype. A more attractive possibility is that transcription of the transposons leads to altered *N* function. Transcription of any of the *flea* inserts could lead to the synthesis of antisense *N* RNA, due to the opposing directions of transcription. The antisense RNA might inhibit *N* function by interfering with RNA processing or translation. The times and places that *flea* is transcribed during development would then control exactly how *N* function would be altered.

One way in which a transposon could inappropriately activate its host gene would be by acting as a portable promoter. The *white*^{DZL} allele (83) provides an example. The mutation is caused by the insertion of a transposon at a site about 6 kb upstream of the start of *w* transcription, outside of the *w* gene as it is defined by germline transformation (68, 80, 84). The insert is composed of two foldback elements flanking a single-copy 6.5-kb piece of DNA that is normally in another part of the genome. Since the mutation can be reverted by a deletion of part of the internal single-copy portion of the insert, some activity on the part of the single-copy DNA must be responsible for the phenotype. Molecular analysis revealed a novel transcript, found in the head, that initiates in the single-copy DNA of the insert and is spliced to the normal *w* exons (80). It is not clear how this transcript accounts for the reduction in *w* function. Surprisingly, the single-copy DNA of the insert is not normally transcribed at detectable levels in the head, suggesting that its transposition to the vicinity of *w* activated it in the head.

Transcription of the *Hairy-wing* alleles starts at the normal places in the AS-C DNA, but terminates within the transposons (190). The altered transcripts accumulate to 5–20 times the usual abundance, so the increased gene function is likely to be due to more stable transcripts resulting from either the AS-C-transposon fused transcription units, or from a transcription enhancement effect of the inserted transposons. The long terminal repeat is itself sufficient, in at least one case (*Hairy-wing*^{BS}), to cause abnormally high levels of AS-C transcripts, which might indicate the presence of a transcriptional enhancer-like element in the repeat (as expected from its homology with retroviruses; 191). Transposon insertions into *w* introns (59, 60, 68, 69) lead to transcripts that have the proper 5' ends, but most of them terminate within the transposons. Presumably termination is in the long-terminal repeats that are characteristic of the retrovirus-like transposons. Some of the transcripts may read through the inserts into the remainder of the *w* DNA, allowing RNA splicing to process out the transposon sequences. The low level of read-through and splicing may account for the low level of pigmentation seen in the eyes of these mutants. The inserts that have a polyadenylation site in the proper orientation may cause frequent termination of the readthrough transcripts, while those in the opposite orientation do not stop readthrough transcripts and can be spliced out efficiently.

Another effect of insertions into *w* can be to allow unequal crossing over, creating duplications and deficiencies. This effect has been demonstrated for two copies of the BEL transposon that are located 60 kb apart (192). Other cases of unequal crossing over at *w* are probably also due to internal repetitions or to transposons (193, 194).

Transposons inserted into the bithorax complex and into the achaete-scute complex cause disruptions that cannot be attributed entirely to interruptions of the DNA. The *bxd¹* allele is due to an insert of the *gypsy* transposon, resulting in a recessive phenotype (131). Two spontaneous reversion mutations have been observed, and in both cases a single 0.5-kb long-terminal repeat from *gypsy* remains at the original insertion site. A similar situation occurs in the *Hairy-wing* mutations in the AS-C (190). Insertions of *gypsy* are responsible for some of the dominant mutations, and three revertants of one of the mutations, *Hairy-wing⁺*, still contain insertions of a *gypsy* long-terminal repeat. In all of these cases, the activity of the transposon, and not the interruption of the DNA, must have been responsible for the original mutation.

Many transposon insertions have effects that can be modified by *trans*-acting regulatory loci (195–197). The locus *suppressor of Hairy-wing [su(Hw)]*, for example, when present in the homozygous mutant form, suppresses mutations at many loci that are caused by insertions of the *gypsy* transposable element (190, 198). The excessive accumulation of transcripts caused by the *Hairy-wing* mutation is partially reversed by the *su(Hw)* mutation, suggesting that the wild-type allele of the suppressor locus facilitates the transcriptional enhancer-like activity of the *gypsy* insert. The *su(wᵃ)* locus acts in a different way on a different insert. The *wᵃ* mutation is due to an insertion of the retrovirus-like transposon *copia* (55, 199). The insert prematurely terminates most of the *w* transcripts in the *copia* long-terminal repeat. In the presence of a *su(wᵃ)* mutation, the frequency of readthrough transcription increases, suppressing the w^a phenotype. Presumably the read-through transcripts can be spliced to give functional w mRNA.

CONCLUSIONS

Despite the progressive blurring of the distinction between simple and complex loci, it is still clear that some loci are far more complex than others. The basis for the complexity can include mechanisms for making multiple or variable products, mechanisms for expressing those products in intricate and precise spatial or temporal patterns, the often surprising effects of transposon insertions on gene function, and the complexities of protein structure and the interactions among multimer subunits. In the immediate future we may expect the rapid characterization of *cis*-acting regulatory elements to continue, leading to an understanding of how such elements must be arranged to function properly in directing tissue-, position-, or stage-specific expression of genes. The basis of chromosomal position effects and transvection may become clearer when *cis*-acting elements are better understood. In addition we may expect proteins that act upon the *cis*-acting DNA sequences to be purified and their functions determined. How multiple proteins interact while bound at neighboring *cis*-acting sites will be an especially interesting problem. The regulation of alternative RNA splicing patterns remains an almost un-approached problem of great importance. The roles of RNA 5' leader sequences and 3' trailer sequences in splicing, translational control, or other processes can be explored using germline transformation and should prove susceptible to the attack. Finally, the influences of transposons and other insertion mutations on gene function will continue to be an active area of research, with special emphasis on the ability of mutations in *trans*-acting genes to suppress or enhance the effects of the inserts. There is little evidence of serious limitations in the approaches now being used to explore gene structure, and the rate of progress will continue to depend on the care with which genes appropriate for certain problems are chosen for analysis.

ACKNOWLEDGMENTS

I thank my many colleagues for their donations of preprints, manuscripts, and generous advcie. Thanks to Margaret Fuller, Susan Dutcher, Sandra Sonoda, Welcome Bender, Claire Cronmiller, and Robert Boswell for criticizing the manuscript, and to Cathy Inouye for helping to prepare it. Thanks to Allen Laughon, Sean Carroll, and Pat O'Farrell for many helpful discussions about homeodomains and homeotic gene functions. Research in my laboratory is supported by the National Institutes of Health, the American Cancer Society, a Searle Scholar's Award, and a March of Dimes Basil O'Connor Grant.

224   SCOTT

Literature Cited

1. Savakis, C., Ashburner, M. 1985. *Cold Spring Harbor Symp. Quant. Biol.* 50:515–20
2. Spradling, A. C. 1981. *Cell* 27:193–202
3. Kafatos, F. C., Mitsialis, S. A., Spoerel, N., Mariani, B., Lingappa, J. R., et al. 1985. *Cold Spring Harbor Symp. Quant. Biol.* 50:537–48
4. Kuner, J. M., Nakanishi, M., Ali, Z., Drees, B., Gustaven, E., et al. 1985. *Cell* 42:309–16
5. Poole, S. J., Kauvar, L., Drees, B., Kornberg, T. 1985. *Cell* 40:37–43
6. Preiss, A., Rosenberg, U. B., Kienlin, A., Seifert, E., Jäckle, H. 1985. *Nature* 313:27–32
7. Searles, L. L., Jokerst, R. S., Bingham, P. M., Voelker, R. A., Greenleaf, A. L. 1982. *Cell* 31:585–92
8. Rozek, C. E., Davidson, N. 1983. *Cell* 32:23–34
9. Shermoen, A. W., Beckendorf, S. K. 1982. *Cell* 29:601–7
10. Vincent, A., Colot, V. H., Rosbash, M. 1985. *J. Mol. Biol.* 186:149–66
11. Davis, R. L., Davidson, N. 1986. *Mol. Cell. Biol.* 6:1464–70
12. Walldorf, U., Richter, S., Ryseck, R.-P., Steller, H., Edström, J. E., et al. 1984. *EMBO J.* 3:2499–504
13. Basi, G. S., Boardman, M., Storti, R. V. 1984. *Mol. Cell Biol.* 4:2828–36
14. Karlik, C. C., Fyrberg, E. A. 1985. *Cell* 47:57–66
15. Falkenthal, S., Parker, V. P., Davidson, N. 1985. *Proc. Natl. Acad. Sci. USA* 82:449–53
16. Laughon, A., Scott, M. P. 1984. *Nature* 310:25–31
16a. Shephard, J. C. W., McGinnis, W., Carrasco, A. E., DeRobertis, E. M., Gehring, W. J. 1984. *Nature* 310:70–71
16b. Johnson, A. D., Herskowitz, I. 1985. *Cell* 42:237–47
17. Hiromi, Y., Kuroiwa, A., Gehring, W. J. 1985. *Cell* 43:603–13
18. Bishop, J. M., Drees, B., Katzen, A. L., Kornberg, T. B., Simon, M. A. 1985. *Cold Spring Harbor Symp. Quant. Biol.* 50:727–32
19. Bond, B. J., Davidson, N. 1986. *Mol. Cell. Biol.* 6:2080–88
20. Walldorf, U., Richter, S., Ryseck, R.-P., Steller, H., Edström, J. E., et al. 1984. *EMBO J.* 3:2499–504
21. Garbe, J. C., Pardue, M. L. 1986. *Proc. Natl. Acad. Sci. USA* 83:1812–16
22. Cherbas, L., Schulz, R. A., Koehler, M. D., Savakis, C., Cherbas, P. 1986. *J. Mol. Biol.* 189:617–31

23. Schejter, E. D., Segal, D. Glazer, L., Shilo, B-Z. 1986. *Cell* 46:1091–101
24. Scholnick, S. B., Bray, S. J., Morgan, B. A., McCormick, C. A., Hirsh, J. 1986. *Science* In press
25. Morgan, B. A., Johnson, W. A., Hirsh, J. 1986. *EMBO J.* In press
26. Southin, J. L., Carlson, E. A. 1962. *Genetics* 47:1017–26
27. Grace, D. 1980. *Genetics* 94:647–62
28. Janca, F. C., Woloshyn, E. P., Nash, D. 1986. *Genetics* 112:43–64
29. Bender, W., Spierer, P., Hogness, D. S. 1983. *J. Mol. Biol.* 168:17–33
30. Spradling, A. C., Rubin, G. M. 1982. *Science* 218:341–47
31. Rubin, G. M., Spradling, A. C. 1982. *Science* 218:348–53
32. Judd, B. H., Shen, M. W., Kaufman, T. C. 1972. *Genetics* 71:139–56
33. Hochman, B. 1973. *Cold Spring Harbor Symp. Quant. Biol.* 38:581–89
34. Lefevre, G., Watkins, W. 1986. *Genetics* 113:869–95
35. Bishop, J. O. 1974. *Cell* 2:81–86
36. Spradling, A. C., Rubin, G. M. 1981. *Ann. Rev. Genet.* 15:219–64
37. Levy, B. W., Johnson, D. B., McCarthy, B. 1976. *Nucleic Acids Res.* 3:1777–89
38. Izquierdo, M., Bishop, J. O. 1979. *Biochem. Genet.* 17:473–97
39. Hough-Evans, B. R., Jacobs-Lorena, M., Cummings, M. R., Britten, R. J., Davidson, E. H. 1979. *Genetics* 95:81–94
40. Welshons, W. J., Von Halle, E. S. 1962. *Genetics* 47:743–59
41. Welshons, W. J. 1965. *Science* 150:1122–29
42. Welshons, W. J. 1971. *Genetics* 68:259–68
43. Artavanis-Tsakonas, S., Muskavitch, M. A. T., Yedvobnick, B. 1983. *Proc. Natl. Acad. Sci. USA* 80:1977–81
44. Kidd, S., Lockett, T. J., Young, M. W. 1983. *Cell* 34:421–33
45. Wharton, K. A., Johansen, K. M., Xu, T., Artavanis-Tsakonas, S. 1985. *Cell* 43:567–81
46. Kidd, S., Kelley, M. R., Young, M. W. 1986. *Mol. Cell. Biol.* 6:3094–108
47. Yedvobnick, B., Muskavitch, M. A. T., Wharton, K. A., Halpern, M. E., Paul, E., et al. 1985. *Cold Spring Harbor Symp. Quant. Biol.* 50:841–54
48. Foster, G. G. 1975. *Genetics* 81:99–120
49. Portin, P. 1975. *Genetics* 81:121–33
50. Wright, T. R. F., Black, B. C., Bishop,

C. P., Marsh, J. L., Pentz, E. S., et al. 1982. *Mol. Gen. Genet.* 188:18–26
51. Morgan, T. H. 1910. *Science* 32:120–22
52. Judd, B. H. 1976. In *The Genetics and Biology of Drosophila*, ed. M. Ashburner, E. Novitski, 1B:767–99. London: Academic
53. Lewis, E. B. 1956. *Genetics* 41:651 (Abstr.)
54. Hazelrigg, T., Levis, R., Rubin, G. M. 1984. *Cell* 36:469–81
55. Bingham, P. M., Levis, R., Rubin, G. M. 1981. *Cell* 25:693–704
56. Levis, R., Bingham, P. M., Rubin, G. M. 1982. *Proc. Natl. Acad. Sci. USA* 79:567–568
57. Goldberg, M. L., Paro, R., Gehring, W. J. 1982. *EMBO J.* 1:93–98
58. Pirrotta, V., Hadfield, C., Pretorius, G. H. J. 1983. *EMBO J.* 2:927–34
59. Karess, R., Rubin, G. M. 1982. *Cell* 30:63–69
60. Collins, M., Rubin, G. M. 1982. *Cell* 30:71–79
61. Rasmuson, B. 1962. *Hereditas* 48:587–611
62. Gehring, W. J., Klemenz, R., Weber, U., Kloter, U. 1984. *EMBO J.* 3:2077–85
63. Levis, R., Hazelrigg, T., Rubin, G. M. 1985. *Science* 229:558–61
64. O'Hare, K., Murphy, C., Levis, R., Rubin, G. M. 1984. *J. Mol. Biol.* 180:437–55
65. O'Hare, K., Levis, R., Rubin, G. M. 1983. *Proc. Natl. Acad. Sci. USA* 80:6917–21
66. Pirrotta, V., Brockl, Ch. 1984. *EMBO J.* 3:563–68
67. Fjose, A., Polito, L. C., Weber, U., Gehring, W. J. 1984. *EMBO J.* 3:2087–94
68. Zachar, Z., Bingham, P. M. 1982. *Cell* 30:529–41
69. Levis, R., O'Hare, K., Rubin, G. M. 1984. *Cell* 38:471–81
70. Pirrotta, V., Steller, H., Bozetti, M. P. 1985. *EMBO J.* 4:3501–8
71. Levis, R., Hazelrigg, T., Rubin, G. M. 1985. *EMBO J.* 4:3489–99
72. Davison, D., Chapman, C. H., Wedeen, C., Bingham, P. M. 1985. *Genetics* 110:479–94
73. Gans, M. 1953. *Bull. Biol. France Belg. (Suppl.)* 38:1–90
74. Kaufman, T. C., Tasaka, S. E., Suzuki, D. T. 1973. *Genetics* 75:299–321
75. Babu, P., Bhat, S. G. 1981. *Mol. Gen. Genet.* 183:400–2
76. Gelbart, W. M. 1982. *Proc. Natl. Acad. Sci. USA* 79:2636–40
77. Gelbart, W. M., Wu, C-T. 1982. *Genetics* 102:179–89
78. Green, M. M. 1961. *Genetics* 46:1555–60
79. Jack, J. W., Judd, B. H. 1979. *Proc. Natl. Acad. Sci. USA* 76:1368–72
80. Bingham, P. M., Zachar, Z. 1985. *Cell* 40:819–25
81. Lifschytz, E., Green, M. M. 1984. *EMBO J.* 3:999–1002
82. Mariani, C., Pirrotta, V., Manet, E. 1985. *EMBO J.* 4:2045–52
83. Bingham, P. M. 1980. *Genetics* 95:341–53
84. Levis, R., Rubin, G. M. 1982. *Cell* 30:543–50
85. Lewis, E. B. 1954. *Am. Nat.* 88:225–39
85a. Kornher, J. S., Brutlag, D. 1986. *Cell* 44:879–83
86. Rawls, J. M., Fristrom, J. W. 1975. *Nature* 255:738–40
87. O'Donovan, G. A., Neuhard, J. 1970. *Bacteriol. Rev.* 34:278–343
88. Jones, M. E. 1980. *Ann. Rev. Biochem.* 49:253–79
89. Wahl, G. M., Stern, M., Stark, G. R. 1979. *J. Biol. Chem.* 254:8679–89
90. Morgan, T. H. 1910. *Proc. Soc. Exp. Biol. Med.* 8:17–19
91. Carlson, P. 1971. *Genet. Res.* 17:53–81
92. Segraves, W. A., Louis, K., Tsubota, S., Schedl, P., Rawls, J. M., Jarry, B. P. 1984. *J. Mol. Biol.* 175:1–17
93. Freund, J. N., Vergis, W., Schedl, P., Jarry, B. P. 1986. *J. Mol. Biol.* 189:25–36
94. Padgett, R. A., Wahl, G. M., Stark, G. R. 1982. *Mol. Cell Biol.* 2:293–301
95. Norby, S. 1970. *Hereditas* 66:205–14
96. Jarry, B., Falk, D. 1974. *Mol. Gen. Genet.* 135:113–22
97. Rawls, J. M., Porter, L. A. 1979. *Genetics* 93:143–61
98. Kapuler, A. M., Bernstein, H. 1963. *J. Mol. Biol.* 6:443–51
99. Fincham, J. R. S. 1966. *Genetic Complementation.* New York: Benjamin
100. Henikoff, S., Tatchell, K., Hall, B. D., Nasmyth, K. A. 1981. *Nature* 289:33–37
101. Henikoff, S., Sloan, J. S., Kelly, J. D. 1983. *Cell* 34:405–14
102. Henikoff, S., Keene, M. A., Sloan, J. S., Bleskan, J., Hards, R., et al. 1986. *Proc. Natl. Acad. Sci. USA* 83:720–4
103. Henikoff, S. 1986. *J. Mol. Biol.* 190:519–28
104. Henikoff, S., Nash, D., Hards, R., Bleskan, J., Woolford, J. F., et al. 1986. *Proc. Natl. Acad. Sci. USA* 83:3919–23
105. Costello, W. P., Bevan, E. A. 1964. *Genetics* 50:1219–30
106. Dorfman, B. 1964. *Genetics* 50:1231–43

107. Chovnick, A., Gelbart, W., McCarron, M. 1977. *Cell* 11:1–10
108. Henikoff, S., Keene, M. A., Fechtel, K., Fristrom, J. W. 1986. *Cell* 44:33–42
109. Scott, M. P., Weiner, A. J., Polisky, B. A., Hazelrigg, T. I., Pirrotta, V. 1983. *Cell* 35:763–76
110. Garber, R. L., Kuroiwa, A., Gehring, W. J. 1983. *EMBO J.* 2:2027–34
111. Kuroiwa, A., Kloter, U., Baumgartner, P., Gehring, W. J. 1985. *EMBO J.* 4:3757–64
112. Garcia-Bellido, A. 1977. *Am. Zool.* 17:613–29
113. Lewis, E. B. 1978. *Nature* 276:565–70
114. Kaufman, T. C., Lewis, R., Wakimoto, B. 1980. *Genetics* 94:115–33
115. Lawrence, P. A., Morata, G. 1983. *Cell* 35:595–601
116. Scott, M. P. 1985. *Trends Genet.* 1:74–80
117. Scott, M. P., O'Farrell, P. H. 1986. *Ann. Rev. Cell Biol.* 2:49–80
118. Gehring, W. J., Hiromi, Y. 1986. *Ann. Rev. Genet.* 20:147–73
119. Muller, H. J. 1932. *Proc. 6th Int. Congr. Genet., Ithaca* 1:213–55
120. Lindsley, D. L., Grell, E. H. 1968. *Carnegie Inst. Washington Publ.* 627
121. Kaufman, T. C., Abbott, M. K. 1984. In *Molecular Aspects of Early Development,* ed. G. Malacinski, W. Klein, pp. 189–218. New York: Plenum
122. Hazelrigg, T., Kaufman, T. C. 1983. *Genetics* 105:581–600
123. Wakimoto, B. T., Kaufman, T. C. 1981. *Dev. Biol.* 81:51–64
124. Denell, R. E. 1973. *Genetics* 75:279–97
125. Struhl, G. 1981. *Nature* 292:635–38
126. Denell, R. E., Hummels, K. R., Wakimoto, B. T., Kaufman, T. C. 1981. *Dev. Biol.* 81:43–50
127. Schneuwly, S., Kuroiwa, A., Baumgartner, P., Gehring, W. J. 1986. *EMBO J.* 5:733–39
128. Laughon, A., Boulet, A. M., Bermingham, J. R., Laymon, R. A., Scott, M. P. 1986. *Mol. Cell Biol.* 6:4676–89
129. Stroeher, V. L., Jorgensen, E. M., Garber, R. L. 1986. *Mol. Cell Biol.* 6:4667–75
130. Abbott, M. K., Kaufman, T. C. 1986. *Genetics.* 114:919–42
131. Bender, W., Akam, M., Beachy, P. A., Karch, F., Peifer, M., et al. 1983. *Science* 221:23–29
132. Lewis, E. B. 1948. *Genetics* 33:113
133. Lewis, E. B. 1949. *Drosoph. Inform. Serv.* 23:59–60
134. Lewis, E. B. 1951. *Cold Spring Harbor Symp. Quant. Biol.* 16:159–74
135. Lewis, E. B. 1955. *Am. Nat.* 89:73–89
136. Lewis, E. B. 1963. *Am. Zool.* 3:33–56
137. Lewis, E. B. 1968. *Proc. 12th Int. Congr. Genet.* 2:96–97
138. Burger, M., ed. 1982. *Embryonic Development: Genes and Cells,* pp. 269–88. New York: Liss
139. Duncan, I., Lewis, E. B. 1982. *Developmental Order: Its Origin and Regulation,* pp. 533–54. New York: Liss
140. Hayes, P. H., Sato, T., Denell, R. E. 1984. *Proc. Natl. Acad. Sci. USA* 81:545–49
141. Struhl, G. 1984. *Nature* 308:454–57
142. Martinez-Arias, A., Lawrence, P. A. 1985. *Nature* 313:639–42
143. Hogness, D. S., Lipshitz, H. D., Beachy, P. A., Peattie, D. A., Saint, R. A., et al. 1985. *Cold Spring Harbor Symp. Quant. Biol.* 50:181–94
144. Bender, W., Weiffenbach, B., Karch, F., Peifer, M. 1985. *Cold Spring Harbor Symp. Quant. Biol.* 50:173–80
145. Akam, M. E., Martinez-Arias, A., Weinzier, R., Wilde, C. D. 1985. *Cold Spring Harbor Symp. Quant. Biol.* 50:195–200
146. Akam, M., Moore, H., Cox, A. 1984. *Nature* 309:635–37
147. Morata, G., Kerridge, S. 1981. *Nature* 290:778–81
148. Peifer, M., Bender, W. 1986. *EMBO J.* 6:2293–303
149. White, R. A. H., Wilcox, M. 1984. *Cell* 39:163–71
150. Beachy, P. A., Helfand, S. L., Hogness, D. S. 1985. *Nature* 313:545–51
151. White, R. A. H., Akam, M. E. 1985. *Nature* 318:567–69
152. Cabrera, C. V., Botas, J., Garcia-Bellido, A. 1985. *Nature* 318:569–71
153. Sanchez-Herrero, E., Vernos, I., Marco, R., Morata, G. 1985. *Nature* 313:108–13
154. Tiong, S., Bone, L. M., Whittle, R. S. 1985. *Mol. Gen. Genet.* 200:335–42
155. Karch, F., Weiffenbach, B., Pfeifer, M., Bender, W., Duncan, I., et al. 1985. *Cell* 43:81–96
156. Regulski, M., Harding, K., Kostriken, R., Karch, F., Levine, M., et al. 1985. *Cell* 43:71–80
157. Scott, M. P., Weiner, A. J. 1984. *Proc. Natl. Acad. Sci. USA* 81:4115–19
158. McGinnis, W., Garber, R. L., Wirz, J., Kuroiwa, A., Gehring, W. J. 1984. *Cell* 37:403–8
159. McGinnis, W., Hart, C. P., Gehring, W. J., Ruddle, F. J. 1984. *Cell* 38:674–80
160. Pabo, C. O., Sauer, R. T. 1984. *Ann. Rev. Biochem.* 53:293–321
161. Laughon, A., Carroll, S. B., Storfer, F. A., Riley, P. D., Scott, M. P. 1985.

Cold Spring Harbor Symp. Quant. Biol. 50:253–62

162. Desplan, C., Theis, J., O'Farrell, P. H. 1985. *Nature* 318:630–35

163. Carroll, S. B., Laymon, R. A., McCutcheon, M. A., Riley, P. D., Scott, M. P. 1986. *Cell.* 47:113–22

164. Carroll, S. B., Scott, M. P. 1985. *Cell* 43:47–57

165. DiNardo, S., Kuner, J. M., Theis, J., O'Farrell, P. H. 1985. *Cell* 43:59–69

166. Beachy, P. A. 1986. PhD Thesis. Stanford University School of Medicine. 113 pp.

167. Struhl, G. 1982. *Proc. Natl. Acad. Sci. USA* 79:7380–84

168. Maine, E. M., Salz, H. K., Schedl, P., Cline, T. W. 1985. *Cold Spring Harbor Symp. Quant. Biol.* 50:595–604

169. Cline, T. W. 1985. In *The Origin and Evolution of Sex,* pp. 301–27. New York: Liss

170. Cline, T. W. 1978. *Genetics* 90:683–98

171. Scherer, G., Tschudi, C., Perera, J., Delius, H., Pirrotta, V. 1982. *J. Mol. Biol.* 157:435–51

172. Meyerowitz, E. M., Hogness, D. S. 1982. *Cell* 28:165–76

173. Cline, T. W. 1979. *Dev. Biol.* 72:266–75

174. Cline, T. W. 1983. *Dev. Biol.* 95:260–74

175. Cline, T. W. 1986. *Genetics* 113:641–63

176. Cline, T. W. 1984. *Genetics* 107:231–77

177. Baker, B. S., Ridge, K. A. 1980. *Genetics* 94:383–423

178. Baker, B. S., Belote, J. M. 1983. *Ann. Rev. Genet.* 17:345–93

179. Belote, J. M., McKeown, M. B., Andrew, D. J., Scott, T. N., Baker, B. S. 1985. *Cold Spring Harbor Symp. Quant. Biol.* 50:605–14

180. McKeown, M., Belote, J. M., Andrew, D. J., Scott, T. N., Wolfner, M. F., et al. 1986. In *Gametogenesis and the Ear-*ly *Embryo,* ed. J. G. Gall, pp. 3–17. New York: Liss

181. Garcia-Bellido, A. 1979. *Genetics* 91:491–520

182. Muller, H. J. 1935. *Genetica* 17:237–52

183. Carramolino, L., Ruiz-Gomez, M., Guerrero, M. D., Campuzano, S., Modolell, J. 1982. *EMBO J.* 1:1185–91

184. Campuzano, S., Carramolino, L., Cabrera, C. V., Ruiz-Gomez, M., Villares, R., et al. 1985. *Cell* 40:327–38

185. Gelbart, W. M., Irish, V. F., St. Johnston, R. D., Hoffmann, F. M., Blackman, R., et al. 1985. *Cold Spring Harbor Symp. Quant. Biol.* 50:119–26

186. Spencer, F. A., Hoffmann, F. M., Gelbart, W. M. 1982. *Cell* 28:451–61

187. Segal, D., Gelbart, W. M. 1985. *Genetics* 109:119–43

188. Rubin, G. M. 1983. In *Mobile Genetic Elements,* ed. J. A. Shapiro, pp. 329–61. New York: Academic

189. Kidd, S., Young, M. W. 1986. *Nature* 323:89–91

190. Campuzano, S., Balcells, L., Villares, R., Carramolino, L., Garcia-Alonson, L., et al. 1986. *Cell* 44:303–12

191. Marlor, R. L., Parkhurst, S. M., Corces, V. G. 1986. *Mol. Cell. Biol.* 6:1129–34

192. Goldberg, M. L., Sheen, J.-Y., Gehring, W. J., Green, M. M. 1983. *Proc. Natl. Acad. Sci. USA* 80:5017–12

193. Green, M. M. 1959. *Heredity* 13:303–15

194. Judd, B. H. 1961. *Proc. Natl. Acad. Sci. USA* 47:545–50

195. Parkhurst, S. M., Corces, V. G. 1986. *BioEssays* 5:52–57

196. Chang, D-Y., Wisely, B., Huang, S-M., Voelker, R. A. 1986. *Mol. Cell. Biol.* 6:1520–28

197. Kubli, E. 1986. *Trends Genet.* 2:204–9

198. Modolell, J., Bender, W., Meselson, M. 1983. *Proc. Natl. Acad. Sci. USA* 80:1678–82

199. Mount, S. M., Rubin, G. M. 1985. *Mol. Cell. Biol.* 5:1630–38

Ann. Rev. Biochem. 1987. 56:229–62

ENVYMES OF GENERAL RECOMBINATION

Michael M. Cox

Department of Biochemistry, University of Wisconsin, Madison, Wisconsin 53706

I. R. Lehman

Department of Biochemistry, Stanford University School of Medicine, Stanford, California 94305

CONTENTS

HISTORICAL PERSPECTIVES AND SUMMARY

General recombination is the process by which DNA sequences are exchanged between homologous chromosomes at essentially any site. The first molecular model to account for such exchanges was suggested some 20 years ago by Robin Holliday to explain the patterns of gene conversion in the smut fungus *Ustilago maydis* (1). This model [and its subsequent variations (2)] has formed the conceptual framework for much of current research on the molecular mechanisms of general recombination (Figure 1). An important and

0066-4154/87/0701-0229$02.00

distinctive feature of the Holliday model is the covalent association of the two duplex DNA molecules that are to engage in recombination via a heteroduplex joint (the Holliday structure) that by virtue of its capacity to branch migrate can generate long regions of heteroduplex (3, 4). Such heteroduplex regions can account for the heterozygosity that is frequently observed in recombinant chromosomes (5, 6). Where mismatches result, their correction causes the unequal segregation of closely linked markers, i.e. gene conversion (7, 8). Because of its symmetry, the Holliday structure can be processed along either its horizontal or vertical axes to yield recombinant molecules in which the parental alleles bordering the potentially heterozygous regions are either conserved in their original linkage or reciprocally exchanged.

The isolation by Clark & Margulies (9) of strains of *Escherichia coli* defective in general recombination permitted for the first time the application of the powerful tools of genetics to a study of its molecular mechanism. Three loci were discovered initially: *recA, recB,* and *recC* (9). Additional genes, including *recE, recF, recG, recJ, recK, recL,* and *recN,* were subsequently identified (10–15). At the present time, only the products of the *recA, recB, recC,* and the recently discovered *recD* genes have been isolated in a functional form. *RecA* codes for the recombination enzyme of *E. coli,* the recA protein or recombinase (12–19); *recB, recC,* and *recD* (12, 20, 21) specify the three subunits of a multifunctional enzyme whose specific role in general recombination has yet to be established. By virtue of its nuclease activity, the recBCD enzyme could catalyze the processing of the DNA product of recA protein action to the final products of the recombinational exchange. [The *recF* and *recE* gene products probably catalyze processing by alternative pathways (22)]. The recBCD enzyme may also be involved at earlier stages in recombination. Another gene, *ssb,* codes for the single-stranded DNA-binding protein (23, 24). This protein, which plays a crucial role in DNA replication, also provides an important adjunctive function in recombination (25).

This review will primarily consider the products of the *recA, recB, recC, recD,* and *ssb* genes. The analysis of the structure and mechanism of these proteins has reached the stage where at least the outlines of an enzymatic pathway of homologous recombination can be formulated. It will not cover the enzymology of site-specific recombination exemplified by the bacte-

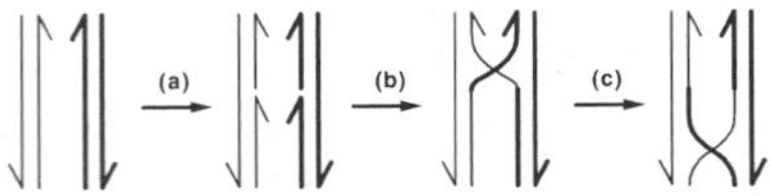

Figure 1.   The Holliday model for general recombination. (*a*) Two homologous DNAs are aligned and the apposing strands nicked. (*b*) The Holliday recombination intermediate is formed by reciprocal strand invasion followed by covalent connection of the two duplex DNA molecules. (*c*) Migration of the crossover point (branch migration) generates a long heteroduplex region.

riophage lambda–induced *int-xis* proteins, which has reached a high level of molecular resolution. For reviews of this subject, the reader is referred to Nash (26) and Sadowski (27).

REC BCD ENZYME

The recBCD enzyme (exonuclease V) of *E. coli* is a complex, multifunctional protein that catalyzes the hydrolysis of both linear duplex and single-stranded DNA, coupled to the hydrolysis of ATP (28–31). It is also an ATP-stimulated endonuclease that acts specifically on single-stranded DNA. In addition to its nuclease activity, exonuclease V is a DNA helicase, that is, it can use the energy of ATP hydrolysis to unwind linear duplex DNA (32–34). An enzyme with analogous activities has been purified from *Hemophilis influenzae* (35, 36). Exonuclease V had until recently been thought to consist of two subunits with molecular weights of approximately 140,000 and 130,000, the products of the *recB* and *recC* genes, respectively (31, 37, 38), although there were indications that it might possess a third subunit (39). It is now clear from the work of Smith and his colleagues (21) that the fully functional enzyme does indeed contain a third subunit with a molecular weight of 58,000, the product of the *recD* gene, hence its present designation as the recBCD enzyme.

A major impediment to an understanding of the mechanism of action of exonuclease V has been the availability of only very small amounts of pure enzyme, due in turn to its exceedingly low cellular concentration (31, 33, 37). However, cloning of the *recB* and *recC* genes either individually or together in high copy number vectors has permitted substantial overproduction of the enzyme (40–42), so that milligram quantities of quite pure enzyme are now readily available for analysis. Because *recD* is immediately adjacent to the *recB* gene, and, in fact, the *recB, C,* and *D* genes may constitute an operon, existing clones of the *recB* or *recB* and *recC* genes very likely contain the *recD* gene as well (21, 43).

The function of the individual subunits in the various activities associated with the recBCD enzyme has not yet been entirely clarified. However, a class of mutants, *recB*[‡], has been shown to lack nuclease activity, and correspondingly, to lack the recD subunit (44). Despite the absence of nuclease activity, *recB*[‡] mutants are recombination proficient, suggesting that it is the DNA helicase activity of the recBCD enzyme that functions in homologous recombination, although this has not yet been firmly established. The finding that the purified recB subunit retains DNA-dependent ATPase activity, but is lacking exonuclease, and that the recC subunit has neither of these activities (43), is consistent with the assignment of helicase activity to the recB and/or C subunits and nuclease to the recD subunit. Careful enzymatic analysis of each of the subunits together with the appropriate reconstitution experiments is obviously needed to establish with certainty the structure-function relationships of the three components of the recBCD complex.

In considering what role(s) the recBCD enzyme plays in homologous recombination, detailed information regarding its nuclease and helicase activities is obviously essential. In acting on linear duplex DNA, the recBCD enzyme initiates its attack at the ends of the DNA molecule and generates duplexes with long single-stranded tails and single-stranded DNA fragments ranging in size up to 1000 nucleotides (45). These are ultimately hydrolyzed to small oligonucleotides, with an average chain length of 4–5 residues (31, 46). However, in the presence of single-stranded DNA-binding protein (SSB), the amount of single-stranded fragments generated is greatly reduced (33). Similarly, exonucleolytic hydrolysis of linear single strands is inhibited by SSB. The generation of duplex DNA molecules with tails and long single-stranded fragments is also enhanced at high, but physiologically significant, ATP concentrations (3–5 mM) (31, 37, 45). Thus, under conditions similar to those to be expected in vivo, i.e. high concentrations of ATP and in the presence of SSB, the exonuclease activity of the recBCD enzyme is largely suppressed (although not eliminated) and the enzyme acts primarily as a helicase, whose activity is largely unaffected under these conditions. Exonuclease but not helicase activity is also inhibited completely by 1-mM Ca^{2+} in the presence of 1-mM Mg^{2+} (47). In this case, however, Ca^{2+} appears to act as a competitive inhibitor of Mg^{2+} and the inhibition of nuclease activity can be relieved by simply increasing the Mg^{2+} concentration (D. Julin, I. R. Lehman, unpublished).

The recBCD enzyme is most active on duplex DNA molecules with flush ends (49, 50). However, it will also attack circular duplexes containing gaps, provided that the gaps are ≥ 5 nucleotides in length (48, 49). This attack, which occurs at a 10-fold lower rate than at linear duplex ends, is very likely a consequence of endonucleolytic cleavage of the single strand at the gap to generate a linear duplex, which is then a substrate for exonucleolytic attack (48, 49). The requirement for gap sizes in excess of 5 residues presumably reflects the requirement by the enzyme for access to the cleavage site.

A detailed analysis of the helicase activity of the recBCD enzyme has shown that the optimal substrate is a linear duplex with nearly flush-ended 3' and 5' termini (50). Duplex molecules with single-stranded tails, particularly those in which the tails are >25 nucleotides in length, are unwound poorly, presumably because of the inability of the enzyme to bind such molecules. Circular duplex circles containing gaps are not unwound; however, the single-strand endonuclease activity of the recBCD enzyme can cleave the single-stranded region within the gap to produce linear molecules that can then be unwound at a rate dependent on the length of single-stranded tail generated by the cleavage (49).

Electron microscopic examination of the products formed from linear duplex DNA by the recBCD enzyme has revealed structures containing a single-stranded loop with two single-stranded tails, one emanating from each strand, and double loops of equal size, one on each strand of the duplex (34, 48). A plausible interpretation of the mechanism by which these structures are generated is that the recBCD enzyme binds to the ends of the duplex and unwinds the DNA to form a single-stranded loop and two single-stranded tails. These then give rise to a double loop structure by annealing of the two tails. The double loops are propagated along the duplex by continued unwinding and threading of one of the two strands past the enzyme. During this unwinding, which is coupled to ATP hydrolysis, the enzyme, by virtue of its nuclease activity, generates single-stranded DNA fragments.

As noted above, the recBCD enzyme possesses endonuclease activity that is specific for single-stranded DNA (31). Unlike the exonuclease activity, the endonuclease is not absolutely dependent on ATP; however, it is stimulated by ATP (some 7-fold). Since coating of single strands by SSB renders them largely insusceptible to the endonuclease (33), it is unlikely that the endonuclease represents an important activity of the recBCD enzyme in vivo. Although duplex DNA is not, in general, cleaved by the recBCD-associated endonuclease, it will catalyze the endonucleolytic cleavage of linear duplex DNA containing *chi* sites (51, 52). *Chi* sites are recombinational hot spots in *E. coli* and bacteriophage lambda DNA that enhance general recombination in their vicinity (53). *Chi*-dependent cleavage by the recBCD enzyme occurs on only one of the two strands of the duplex, that containing the *chi* sequence 5' G-C-T-G-G-T-G-G 3' (54). Cleavage does not occur within the *chi* sequence, but rather at sites located four to six nucleotides to the 3' side of *chi,* and may be the basis for the stimulatory effect of this sequence (52). Thus, mutations in the sequence that diminish *chi* activity in vivo also reduce cleavage by the recBCD enzyme in vitro. Similarly, mutants with enzymes that are defective in the recD subunit show no *chi* activity in vivo, and crude preparations of such enzymes do not show *chi*-dependent cleavage of duplex DNA (21, 44). Presumably, *chi*-specific endonuclease activity is associated with the recD subunit of the recBCD enzyme.

These findings have led to a model for the action of the recBCD enzyme in homologous recombination in which the helicase and *chi*-specific endonuclease activity of the recBCD enzyme act in the steps preceding strand exchange to generate a single strand, which can then serve as a substrate for recA protein–promoted D-loop formation (52, 55). According to this model the enzyme binds the flush ends of a duplex DNA molecule, and by virtue of its ATP-driven helicase activity unwinds it to produce first the single-stranded loop and then the double-looped structure described above. The double loop is then propagated along the duplex until the recBCD enzyme encounters a *chi* sequence in the correct orientation. Cleavage of the strand bearing the *chi* sequence followed by continued unwinding then generates a 3' terminated single strand, which can then be assimilated by an adjoining duplex, leading ultimately to the formation of a Holliday intermediate. Inasmuch as the recBCD enzyme requires the flush ends of a duplex DNA molecule to initiate its action, a double-strand cleavage of the circular *E. coli* chromosome is presumably required to provide an entry site for the recBCD enzyme. In the case of bacteriophage λ, the terminase enzyme that introduces a double-strand break at *cos,* the packaging origin, may provide the entry site (56).

Although the nuclease activities associated with the recBCD enzyme could in principle act in the resolution of Holliday intermediates, there is no evidence that this is the case (see below).

REC A PROTEIN

If one were to point to a single advance that triggered the very rapid development of our current understanding of the molecular mechanism of general recombination in *E. coli,* it would clearly be the construction of a lambda-transducing phage carrying the *recA* gene and the concurrent demonstration, nearly 20 years after the *recA* gene had first been identified, that its product was a protein with a molecular weight of approximately 40,000 (57, 58).

The discovery of the role of the recA protein in regulating the so-called SOS response of *E. coli* to DNA-damaging agents, and, in particular, autoregulation of the *recA* gene, led to a model that unified a mass of seemingly disparate data (59, 60). Knowledge of the way in which the *recA* gene is regulated also permitted the construction of strains of *E. coli* in which the *recA* gene was expressed constitutively at very high levels (61–63). Thus, gram quantities of homogeneous recA protein rapidly became available for the analysis of its structure and function. For a detailed treatment of the central role of recA protein in the regulation of SOS functions, the reader is referred to several excellent reviews (for example, 64). Activities of recA protein related to its role as a recombinase were reviewed in detail several years ago by Radding (65), McEntee & Weinstock (66), and by Dressler & Potter (67).

Structure of recA Protein

The recA protein is composed of 352 amino acids and has a molecular weight of 37,842 (68, 69). Other than a relatively low tyrosine and tryptophan content (seven and two residues, respectively), its amino acid content is unremarkable (68, 69). The amino terminal half of the recA protein contains sequences that are similar to peptide sequences at the active site of several of the serine proteases (69). However, the significance of this finding has diminished with the recent discovery that the recA protein enhances the intrinsic capacity of the lexA and phage λ repressors to undergo cleavage, but does not itself catalyze peptide bond hydrolysis (70).

Both tetragonal and hexagonal crystals of recA protein have been obtained at low pH (5 to 6). The hexagonal form is of appropriate size for X-ray crystallographic examination, and in a preliminary analysis the diffraction extended to 3.5 Å. In both forms the recA protein showed a space group of $P6_1$, indicating a helix of six recA protein monomers per unit cell (71). Predictions of secondary structure and the approximate tertiary folding of recA protein (72) suggest that it possesses a "nucleotide binding fold" and an alternating β strand–α helix pattern. Solution of the three-dimensional structure of recA protein, which hopefully is imminent, is an obvious prerequisite for a detailed understanding of how recA protein functions in general recombination.

Basic Activities of recA Protein

FORMATION OF REC A PROTEIN FILAMENTS   A striking feature of recA protein is its tendency to form aggregates or higher order polymers. This property of recA protein may be related to its action in stoichiometric rather than catalytic amounts in the DNA strand exchange reactions that will be considered later. In the presence of the nonhydrolyzable ATP analogue, ATPγS, and at pH 6.2, recA protein forms long filamentous structures easily visible in the electron microscope (73). More recent light-scattering studies have shown that recA protein can aggregate and self-assemble into filaments at neutral pH and in the absence of nucleotides (74–76). Formation of recA protein filaments is exceedingly sensitive to monovalent and divalent cation concentration. As an example, aggregation of recA protein as judged by light scattering is optimal at 10-mM Mg^{2+}, but is only marginally detectable at 5-mM Mg^{2+}, and virtually disappears at 40-mM Mg^{2+} (74). Similarly, the extent of aggregation declines precipitously between and 20- and 35-mM Tris (75). Nucleotides (ATP, ADP, GTP) have been observed to disrupt recA protein filaments in both electron microscopic and light-scattering studies (74).

A significant concern is the possibility that contaminating DNA in recA protein preparations may in some way potentiate filament formation. In fact, it has been observed that the very long filaments formed upon addition of Mg^{2+} and ATPγS to solutions of recA protein contain RNA, which is either present as a contaminant or generated by the action of a polynucleotide phosphorylase activity that appears to contaminate some preparations of recA protein (77). It is at present unclear to what extent, if any, RNA (or DNA) contamination can account for recA protein filamentation (75, 77).

The finding that various nucleoside triphosphates, i.e. ATP, at physiologically significant concentrations, can disrupt recA protein filaments would tend to suggest that these structures are not of significance in vivo. In fact, recent studies have indicated that recA protein filaments are structurally different from the filamentous complexes of recA protein formed on single-stranded DNA (see below). Indeed, formation of large, free recA protein filaments and binding of recA protein to single-stranded DNA to form nucleoprotein filaments that are active in strand exchange appear to be competing reactions (75). It is, however, possible that smaller aggregates (dimers, tetramers, etc) are intermediates in DNA binding. The uniform appearance of recA filaments in the electron microscope suggests that monomers within the filament are equivalent (76, 78–81). Whereas a monomer may be the active unit in the formation of filaments on DNA, the complexity of the aggregation reactions of recA protein has so far impeded efforts to determine if some other multimeric form of the protein plays a significant role.

BINDING OF REC A PROTEIN TO SSDNA   At neutral pH, recA protein polymerizes onto single-stranded DNA (ssDNA) in a highly cooperative manner (73, 75, 76, 78), coating the DNA completely and extending it significantly. The nucleoprotein filaments formed in this way are 12 nm in diameter and have contour lengths that are 60% that of protein-free DNA duplexes. The contour lengths of the recA protein-ssDNA complexes vary depending on factors such as the ionic strength and presence or absence of nucleotides. In the presence of ATP, the contour length can be as much as 150% of the length of the corresponding recA-free duplex DNA (81). These recA protein–ssDNA complexes are very likely the active species in DNA strand exchange promoted by the recA protein.

Measurements of the stoichiometry of binding of recA protein to ssDNA have yielded values that range from 3 to 6 nucleotides per recA monomer. A stoichiometry of 1 recA monomer per 4 nucleotides was reported by West et al based on the ratio of recA protein to single-stranded ϕX174 DNA in saturated recA protein–ssDNA complexes isolated by ultracentrifugation (82). A similar value was obtained by measurements of the amount of ssDNA protected from nuclease digestion by recA protein (83) and by light scattering (75). A value of 1 recA monomer per 4 nucleotides was also obtained in titrations using ATP hydrolysis as a measure of recA protein binding (84). In contrast, a value of 1 recA monomer per 6 nucleotides has been reported on the basis of titrations of the fluorescent etheno-derivative of ssDNA with recA protein (85, 86). The higher value for the size of the binding site may be related to the greatly enhanced affinity of recA protein for the modified residues in the etheno-DNA. The lower estimates for the binding site size (3 nucleotides per recA monomer) have been obtained with duplex DNA (see below).

Binding of recA protein to ssDNA is relatively stable as judged by its rate of equilibration with a challenging DNA ($t_{1/2}$ of approximately 30 min) (83). The binding is, however, strongly influenced by the type of anion present (87) and by the addition of nucleotide cofactors (83, 87). Addition of ATP to recA protein–ssDNA complexes stimulates the rate of equilibration so that the $t_{1/2}$ is approximately 3 min. In the presence of ADP, equilibration is even more rapid ($t_{1/2} \approx 0.2$ min). The slower equilibration with ATP suggests that a slow step leading to or including the ATP hydrolysis step precedes a rapid ADP-induced release from the DNA. Even in the presence of ADP, however, the overall equilibrium still favors almost complete association of recA protein with the ssDNA. In contrast to ATP, addition of ATPγS to recA protein–ssDNA complexes completely prevents movement of recA protein from the DNA (83). It would therefore appear that one function of the ATPase activity associated with the recA protein is to permit it to cycle on and off ssDNA, ATP being required for binding and ADP potentiating its dissociation (83, 87).

Transfer of recA protein from one DNA molecule to another appears to proceed via a ternary complex with no free recA protein intermediate (88). However, ATP hydrolysis is not tightly coupled to the transfer (89). Moreover, it is not tightly coupled to association or dissociation of recA monomers (at least when single-stranded DNA-binding protein is present), although exchange again occurs between adjacent recA-ssDNA complexes (90).

In contrast to the stoichiometry of 1 recA protein per 4 nucleotides of ssDNA observed in the absence of nucleotide, the stoichiometry in the presence of ATPγS is 1 recA protein per 8 nucleotides (83). The significance of the exact twofold increase in the stoichiometry of binding in the presence of ATPγS is not clear; however, one possibility is that a cryptic second DNA-binding site on the recA monomer is activated upon binding ATPγS. Some support for two binding sites for ssDNA per recA monomer has come from studies of recA protein–catalyzed annealing of complementary DNA strands described below.

The mutant recA protein (recA1) in which a glycine at position 160 is replaced by an aspartic acid residue (91) binds ssDNA cooperatively with a stoichiometry similar to the wild-type recA protein (1 monomer per 3.5 nucleotides). However, the mutant protein, which is completely lacking in ATPase activity (82, 91), is dissociated from ssDNA upon addition of ATP or ADP. A similar effect has been noted with wild-type recA protein in the presence of dTTP, a nucleotide not normally hydrolyzed by the recA protein (F. R. Bryant, I. R. Lehman, unpublished).

BINDING OF REC A PROTEIN TO DUPLEX DNA   Binding of recA protein to duplex DNA differs from the binding to ssDNA in two ways: (*a*) it is highly pH-dependent, with a pH optimum near 6.0 (73, 92), and (*b*) it occurs only in the presence of a ribonucleoside triphosphate cofactor (73, 93–96). The two binding reactions also differ in their sensitivity to ionic strength and in the extent of inhibition by ADP (73, 92).

Binding of recA protein generally results in unwinding of the duplex DNA (94–99). Unwinding in the presence of ATPγS produces a doubling in the number of base pairs per twist in the DNA helix (100). Although the unwinding observed with ATP appeared initially to occur to only a limited extent (94), it has recently been found that duplex DNA can be unwound by at least 28–30% in the presence of ATP (without a ssDNA cofactor), and that binding and unwinding are inseparably linked in this process (101).

All of the nucleoside triphosphates that are hydrolyzed by recA protein in the presence of ssDNA (see below) are also hydrolyzed in the presence of duplex DNA; a pH optimum of ~6.0 is observed in each case. Especially suitable cofactors include ATP, ATPγS, and UTPγS (73, 94), with both binding and unwinding observed in the presence of each. At pH 7.5, binding to duplex DNA in the presence of ATP is nearly undetectable; however, binding is facilitated by ATPγS (94–96, 101–103) if either homologous ssDNA or a low concentration of Mg^{2+} (1 mM) is added (94–99, 102).

In the presence of ATPγS, recA protein condenses on duplex DNA as thick rodlike filaments (73, 76, 78, 79, 104, 105). Close examination of these complexes by Koller and his colleagues (79, 100, 105) has shown the recA protein to be arranged along the DNA duplex as a helical structure. They observe that one helical turn with a pitch of approximately 100 Å and a diameter of 100 Å is formed upon interaction of 6.3 recA protein molecules with 18.6 base pairs of DNA. As a consequence, the DNA is unwound from the 10.5 base pairs per turn characteristic of the B-form to 18.6 base pairs per turn, i.e. 15° per base pair, and the DNA is stretched to 150% of its usual length (79, 100).

In contrast, Griffith and his collaborators (106) have found that binding of recA protein to supercoiled duplex DNA in the presence of Mg^{2+} and ATPγS generates filamentous structures with a repeat unit of 2 recA monomers per 17 base pairs. As a consequence, the DNA is unwound by 11.5° per base pair, a value significantly different from that reported by the Koller group. The discrepancy may be related to differences in methodology. In fact, considerable structural variation can occur depending upon ionic strength and the presence or absence of nucleotide cofactors (78, 80, 81, 107). Williams & Spengler have found that the helical pitch of the recA protein filaments is 55 Å in the absence of nucleotide cofactors, and 93–100 Å when ATPγS is added. In these experiments, the pitch in the presence of DNA was 72–75 Å (80).

The stoichiometries reported for recA protein binding to duplex DNA have again varied somewhat. The electron microscopic observations of Koller and colleagues (79) and the studies of Dombroski et al (108) yielded a value of 1 recA monomer per 3 base pairs. A stoichiometry of 1 monomer per 4 base pairs was obtained using light scattering (101) or ATP hydrolysis (J. Lindsley, M. M. Cox, unpublished) as probes of binding.

The weak binding of recA protein to duplex DNA at pH 7.5 observed in the presence of ATP is the result of a slow step in the association pathway rather than an unfavorable binding equilibrium (101). If this step is bypassed (e.g. by a pH shift from 6 to 7.5), binding at pH 7.5 is quite stable (101). The rate-limiting step in binding appears to be initiation of DNA unwinding. Once nucleation occurs, propagation of a recA protein filament on the duplex DNA is rapid (101). The ATP hydrolysis that accompanies this process occurs only after the DNA is bound and unwound, i.e. ATP hydrolysis is not coupled to unwinding. Slow binding and unwinding of DNA is mirrored by a long lag in ATP hydrolysis, which can be on the order of hours in binding experiments carried out at pH 7.5. Binding to duplex DNA is therefore a complex process leading to a recA-nucleoprotein filament on extensively unwound duplex DNA, which is capable of ATP hydrolysis at a rate approaching the rate observed with ssDNA as cofactor (k_{cat} 20–22 min^{-1}) (101). The unwinding of duplex DNA is especially relevant to (and indeed, it is required by) the DNA strand exchange reactions promoted by recA protein described below. RecA protein effectively holds the DNA at or near the transition state for the exchange of strands.

HYDROLYSIS OF NUCLEOSIDE TRIPHOSPHATES BY REC A PROTEIN   The first enzymatic activity to be associated with recA protein was its ssDNA-dependent ATPase (109, 110). In fact, both single- and double-stranded DNA stimulate ATP hydrolysis. The ssDNA-dependent reaction exhibits a broad pH optimum between 6.0 and 9.0 (92). The double-stranded DNA-dependent reaction, in contrast, exhibits a pH optimum near pH 6.0 (92, 111), paralleling double-stranded DNA binding (73). Values for the K_M for ATP range from 20 to 100 μM under various conditions (111, 112). The reaction proceeds with a turnover number of between 10 and 30 min^{-1}, again varying somewhat with conditions (84, 87, 101, 111, 112). The ssDNA-dependent ATPase activity has a Hill coefficient of 3.3 at pH 8 (as does the double-stranded DNA-dependent activity at pH 6.2), indicating that 3 or perhaps 4 recA monomers are required per hydrolytic cycle in a cooperative process in which binding of one ATP facilitates binding of further ATP molecules (111). It is therefore likely that the nucleoprotein filament described above is the active form of recA protein in the catalysis of ATP hydrolysis.

RecA protein–catalyzed hydrolysis of ATP is not restricted to the ends of these nucleoprotein complexes. Many or all recA monomers throughout the cooperatively assembled complex hydrolyze ATP. Thus, there is no correlation between the concentration of filament ends and either the initial rate or optimal turnover number for ATP hydrolysis (113). As described in a later section, this result also extends to the DNA strand exchange reaction.

In addition to ATP, hydrolysis of dATP, UTP, dUTP, and to a lesser extent CTP and dCTP is catalyzed by recA protein in the presence of single-stranded and duplex DNA (114, 115). In each instance hydrolysis stimulated by duplex DNA occurs only at acid pH (pH 5.5–6.5). In contrast to the *E. coli* recA protein, the recA protein from *Bacillus subtilis* has negligible ATPase activity. However, it can hydrolyze dATP at a rate comparable to that of the *E. coli* enzyme (116). Inasmuch as ATP can inhibit the dATPase activity of the *B. subtilis* recA protein analogue, it must be able to bind ATP. The significance of this stringent specificity for dATP is not known.

ADP and UTP are competitive inhibitors of the ATPase activity, indicating that there is only a single nucleoside triphosphate binding site per recA monomer (92, 115). Consistent with this idea is the finding that ATPγS, which is not hydrolyzed by the recA protein and is in fact a potent inhibitor of both the single- and double-stranded DNA-dependent ATPase (K_i = 0.6 μM), binds tightly to a single site per enzyme molecule (117). Modification of recA protein with the photoaffinity label 8-azido ATP both in the presence and absence of DNA results in covalent attachment of the azido ATP exclusively to Tyr-264 (118, 119). This residue is also the exclusive site of modification by another ATP analogue, 5'-P-fluorosulfonyl-benzoyladenosine (120). These and other findings suggest that Tyr-264 is located in the ATPase active site of recA protein and is positioned in close proximity to both the adenine ring and the triphosphate group of the ATP.

RecA protein does not catalyze any of the microscopic exchange reactions that are often associated with ATP hydrolysis. Experiments designed to detect [^{3}H]ADP⇌ATP, HPO_4^-⇌$H_2^{18}O_4$, and $HP^{18}O_4$⇌H_2O exchange were in each case negative, suggesting that ATP hydrolysis is irreversible (112).

RENATURATION OF COMPLEMENTARY SINGLE STRANDS   The alignment of complementary DNA sequences is the simplest DNA pairing activity associated with the recA protein (114), and may underlie the much more complex strand exchange reactions to be considered below. The renaturation reaction catalyzed by recA protein differs in several important respects from DNA renaturation promoted by the single-stranded DNA-binding protein, SSB, of *E. coli* (121) and the T4 gene 32 protein (122). (*a*) RecA protein, in contrast to SSB, is required in subsaturating amounts [maximal rates of renaturation are observed at a stoichiometry of 1 recA protein per 30 nucleotides of ssDNA, which is approximately 10–15% of saturation (see above)] (123). In fact, saturating levels of recA protein are inhibitory. (*b*) Renaturation promoted by SSB follows second order reaction kinetics, as does nonenzymatic renaturation. These reactions proceed by a rate-limiting collision between homologous sequences, followed by a rapid "zippering up" of the strands to form a DNA duplex (124). Catalysis by SSB is thought to involve unfolding of regions of secondary structure, resulting in increased rates of nucleation (122), although other factors may also be important (121). In contrast, recA protein–promoted renaturation proceeds by a first order rather than a second order process, suggesting that there is rapid non–rate determining formation of an intermediate prior to complete renaturation (123). (c) RecA protein–catalyzed renaturation is stimulated by ATP (114, 123, 125). There is no effect of ATP or other nucleoside triphosphate on SSB-promoted renaturation. The role of ATP hydrolysis in recA protein–catalyzed renaturation is complex. At low Mg^{2+} concentrations (10 mM) maximal rates of renaturation are obtained in the presence of ATP; at high recA protein levels, ATP is absolutely required for catalysis of renaturation. A lesser stimulation is observed with ADP as cofactor, whereas ATPγS, which induces the irreversible binding of recA protein to ssDNA, permits only low levels of recA protein–promoted renaturation (123, 125).

When the Mg^{2+} concentration is increased from 10 mM (the optimal concentration in the presence of ATP) to 30 mM, the rate of ATP-independent renaturation increases 2- to 3-fold and proceeds at approximately the same rate as the ATP-stimulated reaction. Mg^{2+} concentrations in excess of 40 mM are inhibitory (123).

RecA protein can catalyze the renaturation of a (+) circular ϕX174 DNA strand with its (−) linear complement to generate RFII molecules at 10-mM Mg^{2+} in the absence of ATP (F. R. Bryant, I. R. Lehman, unpublished). This reaction, which may be related to the three-strand exchange reaction involving a circular single strand and a homologous linear duplex, described below, may proceed by either of two mechanisms. In one, the DNA strands are brought together directly by means of two DNA-binding sites on the recA protein molecule (DNA·recA·DNA). In the second, the DNA strands are brought together by interaction between recA protein molecules bound to different strands (DNA·recA·recA·DNA). The increase in the stoichiometry of binding of recA protein to ssDNA from 1 recA monomer per 4 to 1 recA monomer per 8 nucleotides in the presence of ATPγS suggests that a recA monomer contains two DNA-binding sites (83). Further support for the two binding site model for DNA renaturation has come from experiments in which stable recA protein (+) circular DNA complexes ($t_{1/2}$ of dissociation ~30 min) are mixed with the complementary (−) linear strand. Under these conditions RFII formation occurs at nearly the same rate as that observed upon addition of recA protein to a mixture of the (+) and (−) DNA strands (F. R. Bryant, I. R. Lehman, unpublished).

The notion of two DNA-binding sites is also consistent with studies of the mechanism of transfer of recA protein from one ssDNA molecule to another. Transfer is slow in the absence of nucleotide cofactors, but is greatly stimulated by ATP (83). In both cases transfer appears to be a cooperative process in which many recA monomers, possibly in the form of clusters, are transferred from one ssDNA molecule to another in a single event. Cooperative transfer appears to proceed by the intermediate formation of a complex between a recA protein–ssDNA complex, and a second ssDNA molecule, followed by transfer of the recA protein from the first to the second strand (83, 88, 89). The formation of a two-stranded structure of this kind provides an intermediate that would appear to be well suited to bringing complementary DNA strands together so that pairing can occur. Stimulation of the reaction by ATP may result from the increase in the rate of transfer of recA protein, which may then serve to increase the cycling rate of recA protein between different DNA strands in the search for homology.

DNA Strand Exchange

The activities of recA protein described above—binding to DNA, ATP hydrolysis, and the pairing of complementary DNA strands—converge in the DNA strand exchange reaction, which represents a more complete experimental model for the action of recA protein in vivo. Given the appropriate substrates, recA protein can transfer a DNA strand from one homologous partner to another (126, 127). In the course of this reaction (when four strands are involved), an intermediate identical to that predicted by Holliday can readily be detected (128). A thorough understanding of this process should provide a chemical basis for the central stages of general recombination.

The substrates that can be utilized by recA protein for strand exchange are limited only by topological constraints (126) and by the requirement for a short region of ssDNA to initiate the reaction (126–129). The reactions can then take a variety of forms and lead to a variety of products (Figure 2). The need for well-defined substrates and products for kinetic analysis, and the recognition that recA protein can produce heteroduplex DNA thousands of base pairs in length (126, 127), have led to the development of a system that has become standard in the analysis of recA protein–promoted DNA strand exchange, the exchange between linear duplex and circular single-stranded phage DNAs (127) (Figure 2D). The substrates and products of this reaction, which occurs over a convenient time period (10–30 min), are well-defined and readily distinguished by a variety of methods. Nearly a dozen different assays have been employed to measure various aspects of the reaction. Since the single-stranded circles isolated from virions are (+) strands, they cannot pair with each other; the reaction therefore avoids interference by recA protein–promoted DNA renaturation. The availability of a great variety of homologous, heterologous, and chimeric substrates provides a flexibility that

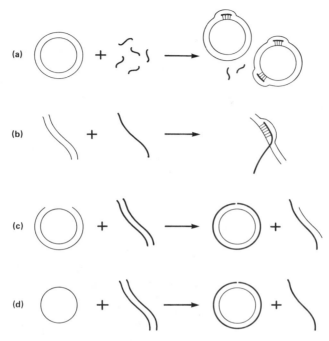

Figure 2. Strand exchange reactions promoted by recA protein.

has been exploited extensively. The reaction can be conveniently divided into three phases for study, each of which poses a different and interesting biochemical problem.

The first phase, complex formation or presynapsis, consists of the binding of stoichiometric amounts of recA protein to the available ssDNA (81, 130), resulting in the nucleoprotein filament described above that catalyzes ATP hydrolysis. The formation of this complex is related to the basic problem of how a protein binds nonspecifically and cooperatively to DNA. However, the ATP hydrolysis that occurs as a consequence of the binding complicates the analysis of this interaction. The product of the binding reaction is the active species in the subsequent phases of strand exchange. The structure and properties of the recA-ssDNA nucleoprotein filament are therefore central to any consideration of DNA strand exchange.

The second phase of the reaction involves alignment of the ssDNA within the nucleoprotein filament with complementary sequences in the duplex DNA (126, 131–133). The biochemical problem changes here. Binding to ssDNA effectively transforms recA protein into a duplex DNA-binding protein that now exhibits a sequence specificity dictated by the sequence of the ssDNA bound in the nucleoprotein filament. Several steps in this process have been identified.

Once homologous sequences are aligned, strand exchange begins. In the third phase of the reaction the (+) strand in the DNA duplex is displaced, and is replaced with the strand brought in by the nucleoprotein filament. This reaction is functionally equivalent to branch migration but differs from spontaneous branch migration (134) in that it requires ATP hydrolysis; in fact, it is the only phase of DNA strand exchange that requires ATP hydrolysis (127, 135), and it exhibits a unique polarity (136–138). Of special interest here is the coupling between a unidirectional reaction and chemical energy. The problem is a classical one in biochemistry, analogous in some respects to muscle contraction and ATP-driven ion pumps (139). Each of the three phases of the DNA strand exchange reaction will now be considered in detail.

FORMATION OF REC A—SSDNA NUCLEOPROTEIN COMPLEX   The central importance of ssDNA in the initiation of recA protein–promoted DNA strand exchange was recognized early (93, 97, 102, 114). As noted above, recA protein binds much more readily to ssDNA than to duplex DNA under conditions that are optimal for DNA strand exchange. Characterization of recA protein–ssDNA complex formation as a separate phase of DNA strand exchange has involved a variety of strategies. These include the use of DNA challenges (130), characterization of a lag in strand exchange (140), and isolation of the complex after its formation (81).

Formation of the complex requires the binding of stoichiometric amounts of recA protein to the ssDNA. As noted above, the values that have been reported for the binding site size for a recA monomer vary from 3 to 6 nucleotides. Secondary structure in the ssDNA impedes binding, imposing a barrier that can be circumvented by the addition of SSB (141, 84). Register & Griffith have recently found that the filaments assemble unidirectionally, $5' \rightarrow 3'$ along the ssDNA (142). This asymmetry may determine the polarity of the third phase of the reaction.

Formation of a complex active in strand exchange also requires ATP. Addition of ATP has several effects on the recA protein–ssDNA complex. These include a change in the range of nucleoside triphosphates that can be hydrolyzed by recA protein (F. R. Bryant, unpublished), as well as the change in the rate of transfer of recA protein between DNA molecules described above. These findings suggest that one or more ATP-induced changes in the state or conformation of recA monomers occurs in the complex. Structural evidence for such changes is presently limited to observations made by electron microscopy, and few details are available concerning changes at the molecular level. As already noted, characteristic differences can be observed by electron microscopy in recA protein–DNA complexes formed in the absence of nucleotide cofactors, in the presence of ATP, and in the presence of ATPγS. Leahy & Radding (143) have recently suggested that recA protein is in closest contact with the phosphate backbone of the DNA strand, leaving the bases free to participate in pairing reactions.

The rate of assembly of the active recA protein–ssDNA complex is affected by the nature of the ssDNA, the ionic conditions, and by the presence or absence of SSB. In the absence of SSB and with 10–15-mM $MgCl_2$, a discernible lag in strand exchange occurs, attributable to slow complex formation, which can be partially overcome by preincubation of recA protein with the ssDNA (140, 144). The lag is at least in part a function of secondary structure in the ssDNA. Thus, the rate of complex formation in the absence of SSB is enhanced by low Mg^{2+} concentrations, which eliminate most secondary structure in ssDNA (140, 141, 144), or by using polydT that lacks secondary structure (86, 87, 113, 145). When SSB is added, the rate of complex formation is greatly enhanced and the lag in strand exchange is abolished (81, 141). The role of SSB in the formation and maintenance of this complex is described in more detail below.

The recA-ssDNA nucleoprotein complex that results from this phase of the reaction is clearly the active species in the subsequent phases of strand exchange. Thus, there is a stoichiometric requirement for recA protein for the overall strand exchange reaction (146). Moreover, the rate and efficiency of strand exchange are functions of both the binding density of the recA protein in the complex and the stability of the complex (147). When SSB is added, the complexes have been shown to be kinetically competent as intermediates in strand exchange (130). Finally, when the complexes are separated from free recA protein and other components of the reaction, they are fully functional in strand exchange (81).

Once formed, the nucleoprotein complexes hydrolyze ATP with an apparent k_{cat} of about 30 min^{-1} at pH 7.5 and 37°C (84, 113). Hydrolysis occurs whether or not duplex DNA is added to initiate strand exchange, suggesting that much of the hydrolysis is irrelevant and not coupled to useful work. Alternatively, the ATP hydrolysis may reflect a system at idle, coupled to conformational changes that would result in strand exchange if homologous DNA were made available. The role of ATP hydrolysis will be considered in more detail below.

SEARCH FOR HOMOLOGY: SYNAPSIS   The mechanism by which proteins that bind specific sequences locate their binding sites is determined to a large extent by interactions with nonspecific DNA (148). As described above, binding of recA protein to duplex DNA under conditions optimal for strand exchange is highly dependent on the presence of homologous ssDNA. To put it another way, the recA-ssDNA nucleoprotein complex is a sequence-specific duplex DNA-binding entity, with the specificity determined by the sequence of the ssDNA. The complex is much larger and more complex than such well characterized specific DNA-binding proteins as RNA polymerase, the EcoRI restriction enzyme, and the lambda and lactose repressors. Mechanistically, however, the problem of how this complex searches for homologous sequences in duplex DNA can be analyzed to a first approximation in terms of interactions with nonspecific (heterologous) and specific (homologous) DNA sequences using approaches similar to those employed in the simpler systems. The problem is rendered more tractable by the demonstration that ATP hydrolysis is not required; the search for homology can be completed in the presence of ATPγS, which is not hydrolyzed by recA protein (127, 135).

A number of studies have provided evidence that recA protein–ssDNA complexes can bind nonspecifically to heterologous duplex DNA. Such interactions have been detected in the presence of both ATPγS (93, 94, 149) and ATP (150–152). The rate of formation of these complexes is consistent with the idea that they are intermediates in the search for homology (152). Radding and colleagues have shown that these nonspecific interactions are manifested in vitro by large networks that link together many recA protein–ssDNA complexes and duplex DNA molecules early in the strand exchange reaction (151, 152). These networks effectively reduce the volume in which the search for homology must occur and can be harvested by brief centrifugation. Conflicting results have been obtained in experiments designed to investigate the mechanism by which homologous sequences are aligned. Gonda & Radding found that the search for homology is facilitated by the presence of lengths of heterologous DNA attached to the homologous duplex, with rates of synapsis increasing with the length of the heterologous tail (131), suggesting a processive mechanism. Julin et al, in a similar set of experiments, found that heterologous tails had no effect on the rate of the reaction (132). Wherease the experiments differ in the use of SSB in the latter study (resulting in higher rates of synapsis), the discrepancy has not been explained satisfactorily.

In studies initiated by the Radding group, two types of synaptic structures have been described. These differ depending on whether the paired strands of DNA are interwound or not. Paranemic joints, in which the strands are paired but not interwound, are characterized by a lack of a requirement for a free homologous end (99, 126, 153) and their greater sensitivity to protein denaturants (154). They have been observed directly by electron microscopy (155). Paranemic joints are formed fast enough to be considered intermediates in the synapsis pathway, and are very likely precursors of the plectonemic joints in which the DNA strands are interwound (99, 132, 154).

The interaction with duplex DNA triggered by binding of recA protein to ssDNA generally results in the unwinding of the duplex. With ATPγS, unwinding has been detected in the presence of both heterologous and homologous ssDNA (149). The apparent unwinding in the presence of heterologous ssDNA may reflect in part a significant ssDNA-independent binding of duplex DNA, which is detected in the presence of ATPγS (79, 101, 103, 108). When ATP is used, the unwinding reaction exhibits a considerable degree of dependence on homology in the ssDNA (96–99), implying that the unwinding occurs after homologous alignment of the two DNAs has occurred. The extent of unwinding has not been precisely determined. Several studies indicate that 100–300 base pairs of duplex DNA are instantly unwound upon formation of a paranemic joint (140, 156, 157), a necessary prelude to the formation of a plectonemic joint molecule in which the (−) strand of the linear duplex is interwound with (+) ssDNA. More extensive unwinding has also been observed. Since it is dependent on ATP and homologous ssDNA, it appears to be triggered by the formation of paranemic joints (or D-loops) (96, 98). When the unwinding is topologically trapped in a closed-circular DNA molecule, it is manifested by a superhelical density significantly greater than that of RFI DNA. Formation of this extensively unwound species is slow, requiring 20–40 minutes to reach completion (96, 98). The unwound species can therefore not be an intermediate in the synapsis pathway, since the formation of a plectonemic joint generally requires less than a minute. The slow unwinding may reflect a direct binding of excess recA protein to the duplex DNA, which can occur following D-loop formation (96, 157) or strand exchange (B. C. Schutte, M. M. Cox, unpublished). As described above, extensive unwinding is a characteristic of the binding of recA protein to duplex DNA.

A reaction scheme consistent with these results is presented in Figure 3. The first two steps have been detected in the presence of ATPγS and do not require ATP hydrolysis. The second step, (b), includes both the homologous alignment of the two DNA molecules and the subsequent unwinding of a short region of the DNA duplex. The third step, (c), in which the incoming circular single strand is interwound with its complement to form a plectonemic joint, does require ATP hydrolysis (154). This final step, (d), initiates and is probably indistinguishable from the branch migration of the third phase. For purposes of discussion, we will define the end of the second phase, i.e. synapsis, as the product of the second step in the pathway shown in Figure 3. The product of synapsis by this definition is therefore a ternary complex containing the ssDNA-recA nucleoprotein filament and duplex DNA, which has been bound, homologously aligned with ssDNA, and unwound. Elucidation of this very complex reaction in mechanistic detail awaits more extensive structural and kinetic analysis.

REC A PROTEIN–PROMOTED BRANCH MIGRATION   Branch migration in solution under physiological conditions is a facile reaction. Rates measured in vitro were believed to account satisfactorily for the branch migration observed

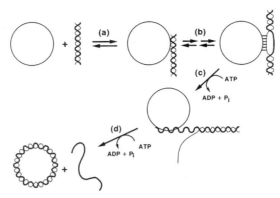

Figure 3.   Reaction pathway for the transfer of a strand from a linear duplex to a circular ssDNA showing (*a*) nonhomologous interactions leading to (*b*) formation of a paranemic joint. Formation of a plectonemic joint (*c*) is coupled to ATP hydrolysis. Branch migration (*d*) is shown as a step separate from plectonemic joint formation, but is most likely an extension of the same process.

during recombination in fungi and other organisms (134). There was no reason, therefore, to expect this reaction to be catalyzed. The observations that branch migration in the presence of recA protein required ATP hydrolysis (127) and proceeded with a unique polarity (138, 156), however, clearly implicated recA protein as a catalyst in this reaction. The polarity of the reaction between circular single-stranded and linear duplex DNA proceeds in the 5'→3' direction relative to the invading single strand (136–138). This is identical to the direction in which recA filaments assemble on ssDNA (142). With another pair of DNA substrates, linear single-stranded and circular duplex DNA, it has recently been found that strand exchange may advance in the 3'→5' direction relative to the linear single strand (158). In this case, however, it is not clear whether the polarity observed is at the level of synapsis or branch migration.

RecA protein contributes two important properties to branch migration. First, by providing a unique direction, it ensures that heteroduplexes thousands of base pairs in length are created efficiently. The recA protein–catalyzed reaction is actually much slower than the spontaneous one [5–20 bp s^{-1} (146) vs 6000 bp s^{-1} (134)]. However, by limiting the direction of the reaction, efficiency is improved relative to the spontaneous but random process. Whereas the spontaneous process provides a 1 in 3 chance of migrating 950 base pairs in 10 min (134), the recA protein–catalyzed reaction has a nearly 100% chance of proceeding linearly for 3000 base pairs in the same time period. The second property is probably closely related to the first. Unlike the spontaneous process, recA protein–promoted branch migration can proceed efficiently past short DNA mismatches and pyrimidine dimers (159,

160). It can even proceed past deletions or insertions hundreds of base pairs long (161). This property is probably of great utility to the cell (see ENERGET-ICS, below).

The active species in branch migration is the recA-ssDNA nucleoprotein filament formed in the first phase of the reaction. The mechanistic question then becomes: How is this filamentous structure employed to promote unidirectional branch migration coupled to ATP hydrolysis? Several basic characteristics of this reaction must be addressed in any consideration of mechanism. These include the location of recA protein at all stages of the reaction, the dynamic state of the filament, the location of the participating DNA strands relative to the filament, the specific role of ATP hydrolysis, and the finding by Bianchi et al that branch migration can proceed through extensive regions of nonhomology (161). Information about these basic features of DNA strand exchange is still very limited, as reflected in the variety of plausible models that can be proposed for the reaction (162–165).

The first question that can be asked about a filamentous system is whether the important events occur at ends or throughout the filament. An attractive model can be developed by drawing an analogy to two other filamentous systems, tubulin and actin. Hydrolysis of nucleoside triphosphates in these systems occurs at the ends of the filament and is coupled to a treadmilling reaction, which involves a net addition of subunits to one end of the filament at the expense of the other (166). Similarly, recA protein could promote unidirectional branch migration by coupling movement of the branch to association or dissociation of monomers at a filament end (162, 163). ATP hydrolysis by the recA nucleoprotein filament, however, is not restricted to the ends of the filament (113), and is not tightly coupled to complete association or dissociation of monomers (89, 90). Both observations provide evidence against treadmilling in recA filaments in the classical sense. An exchange of recA monomers between nucleoprotein complexes can be observed when the complexes are in transient contact (90). The role, if any, of this exchange of recA monomers in DNA strand exchange has yet to be determined.

A second and equally attractive model involves dissociation of recA protein at the branch point. RecA protein binds weakly to duplex DNA under conditions optimal for strand exchange. Inasmuch as the ssDNA in the recA nucleoprotein filament is converted to heteroduplex DNA as the branch point passes, dissociation of recA protein at the branch point might be expected. Direct observation of recA filaments during strand exchange by electron microscopy reveals that under some conditions dissociation might accompany movement of the branch (167). In fact, dissociation at the branch point has been incorporated into a model proposed by Howard-Flanders and colleagues (162). The electron microscopic observations, however, conflict with results obtained in solution and might reflect differences related to fixation of samples for microscopy. Results derived from patterns of ATP hydrolysis (113), DNase protection (168, 169), and the topology of the products of strand exchange (169) provide evidence that little net dissociation of recA protein occurs during the reaction. Instead they indicate that the recA protein is bound contiguously to the heteroduplex DNA well after strand exchange is complete. These findings do not conflict with the weak binding of recA protein to duplex DNA at neutral pH, since it has been demonstrated that the weak binding reflects a slow step in the association pathway rather than an unfavorable binding equilibrium (101). Strand exchange apparently bypasses this slow association step. These results taken together do not rigorously eliminate a role for association or dissociation in strand exchange, but they do indicate that the nucleoprotein filament remains relatively intact throughout the reaction.

A second question involves the location of the DNA strands undergoing exchange relative to the filament. One model has been suggested in which both DNA molecules are inside the spiral recA nucleoprotein filament (162). In another model, only one DNA molecule is inside the filament, with the other outside (164). The available information is again suggestive rather than definitive. The initiation of strand exchange results in an immediate decrease (up to 30%) in the rate of ATP hydrolysis that correlates precisely with the length of homology between the two DNA molecules involved (B. C. Schutte, M. M. Cox, unpublished). This finding suggests that some degree of homologous contact must occur almost immediately along the length of the homologous incoming DNA duplex. However, the duplex DNA acquires resistance to nuclease digestion with a time course that parallels the formation of heteroduplex DNA (169). Thus, the duplex DNA appears to be bound outside the filament until it is incorporated into heteroduplex. Much more detailed information is required about the structure of this complex and the space available within recA filaments for a more definitive answer to this question.

Fourier transform enhancements of electron micrographs of recA nucleoprotein filaments reveal a right-handed helical structure with a major groove (170). At least one DNA molecule lies within this complex, contacted primarily along the phosphate backbone (143). It is reasonable to suppose that the bases are exposed in the major groove of the filament, providing an active site for strand exchange. Inasmuch as DNA in its native form is a right-handed helix, rotation of the exchanging DNA molecules must occur to bring about unidirectional branch migration. This can occur, to a first approximation, in two ways, as illustrated in Figure 4. A model has in fact been proposed (164) in which the rotation shown in Figure 4b is coupled to strand exchange within the groove of the filament (Figure 5). Regardless of the

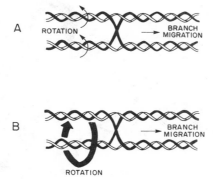

Figure 4.   Rotation of right-handed DNA to produce branch migration. Rotation of (*A*) both DNA molecules or (*B*) one about the other produces branch migration in the direction indicated.

actual mechanism of strand exchange, the rotation required to bring it about ensures that a topoisomerase is required to resolve topological strain induced by the reaction.

ENERGETICS OF DNA STRAND EXCHANGE: TO WHAT IS ATP HYDROLYSIS COUPLED?   RecA protein–promoted branch migration requires ATP hydrolysis. Indeed, a consideration of the principle of microscopic reversibility indicates that this reaction cannot be made unidirectional without an investment of chemical energy. ATP hydrolysis appears to be very inefficient, however, with over 100 ATPs hydrolyzed per base pair of heteroduplex formed under optimal conditions (112). All of the recA monomers in a nucleoprotein filament hydrolyze ATP in the absence of strand exchange. This is also true during strand exchange, with no correlation evident between rates of ATP hydrolysis and the number of migrating branch points (113). It is important to determine whether the apparent excess represents an idling mechanism, reflecting an activity that results in strand exchange when an appropriate substrate is available, or whether much of the ATP hydrolysis is not coupled to useful work.

The latter notion has in fact been suggested by Kowalczykowski and colleagues (112, 165). These workers noted that a favorable correlation can be drawn between ATP hydrolysis and strand exchange if it is assumed that most of the ATP hydrolysis is uncoupled from strand exchange and can therefore be subtracted from the total (165). Under most conditions, however, the ATP hydrolysis observed in the presence of ssDNA alone is significantly greater than the total ATP hydrolysis observed once strand exchange commences (B. C. Schutte, M. M. Cox, unpublished). In fact, the efficiency of the reaction can be improved to approximately 16 ATPs per base pair simply

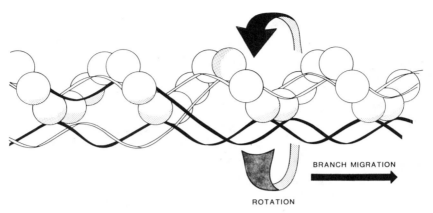

Figure 5.   A model for recA protein–promoted branch migration. A recA nucleoprotein filament formed on one DNA molecule promotes branch migration by rotating a second DNA molecule (linked to the first via a crossover junction) around the outside of the filament. The resulting rotation is that illustrated in Figure 4*b*. See Ref. 163 for details.

by adding low concentrations of ADP (112). High levels of ADP cause dissociation of recA protein from the DNA, and the improved efficiency may be related to a shortening of the filaments under these conditions (112). Radding and colleagues have observed that branch migration is much less sensitive to ADP than the earlier phases of the reaction (98, 140), possibly indicating that relatively short recA filaments are fully competent to promote extensive branch migration. Nevertheless, these results indicate that not all of the ATP hydrolysis that occurs during strand exchange is required for branch migration.

On the other hand, it is possible that most of the ATP hydrolysis is involved in strand exchange and that the molecular events in the process occur throughout the filament rather than uniquely at the branch point. That is, an important advantage may be gained by the apparent waste of chemical energy. This notion involves a consideration of the strategy of employing a filament to perform strand exchange. Unidirectional branch migration could in principle be carried out efficiently by an enzyme or small complex possessing appropriate helicase and reannealing activities. However, *E. coli* may have evolved a complex filamentous system to carry out this reaction. The inefficiency that might result if all of the recA monomers in the complex hydrolyze ATP is significant. However, even this expense is trivial when compared to the energy required by the protein synthetic machinery to assemble the 352 amino acids of every recA monomer in the filament. This investment by the cell can only be rationalized in terms of an essential reaction. Extensive branch migration promoted by the filamentous complex of recA protein may therefore have a significant impact on cell survival: (*a*) it could provide sufficient energy to bypass DNA lesions, (b) it could provide protection for the DNA branch point, preventing premature resolution, and (c) it might exclude other DNA-binding proteins that could block branch migration. The excess ATP hydrolysis may not be required thermodynamically to promote branch movement, however, it may be a price that is paid to ensure that the branch is protected and that extensive branch migration can occur. Possible roles for branch migration in vivo are discussed below.

FOUR-STRAND EXCHANGES   RecA protein promotes strand exchange between two duplex DNA molecules (82, 126, 128, 129, 171). To initiate the reaction, one of the molecules must have a single-stranded gap or tail and the second must have a free DNA end that can overlap the gap. Studies of these reactions have revealed the formation of classical Holliday intermediates by recA protein in vitro. These reactions are not as well characterized as the three-strand exchanges described above, reflecting in part the added difficulty in generating substrates and monitoring the reactions quantitatively. The recA nucleoprotein filament is again the active species in the reaction, and recA protein binding at the single-stranded region leads to a rapid and complete invasion of the adjacent DNA duplex (171; J. Lindsley, M. M. Cox, unpublished). The energetics of the four-stranded exchange reactions have not been investigated in detail, although efficiencies appear to be similar to those observed in the three-strand reaction (J. Lindsley, M. M. Cox, unpublished).

ROLE OF STRAND EXCHANGE IN VIVO   Approximately 1200 monomers of recA protein are present in an *E. coli* cell under normal growth conditions (172, 173). Upon induction of the SOS system, this number can increase by as much as two orders of magnitude (172, 173). The recA protein is involved in recombination, postreplication repair, and numerous other processes. In its role as a recombinase, recA protein is probably responsible for the steps in recombination that it catalyzes in vitro: pairing and branch migration. Some in vivo evidence mirrors the findings in vitro. Yancy & Porter (174) have found that low concentrations of mutant recA protein interfere with the activity of wild-type recA protein in the cell. A likely interpretation of this finding is that the mutant monomers interfere with recA filament formation, underlining the importance of the recA nucleoprotein complex in vivo. The polarity observed in recombination in vivo is also consistent with the polarity observed in vitro (175).

The direct role of recA protein in postreplication repair suggests that it is essentially a recombinational process (176). RecA protein is uniquely suited to play a role in the repair of pyrimidine dimers or other lesions occurring in ssDNA. Whereas these lesions may represent a major barrier to DNA polymerase action, recA protein–promoted branch migration can move past them easily. The role of recA protein in this instance is very likely the conversion of the lesion-containing ssDNA into a duplex via branch migration (176), facilitating repair. The incoming strand would be derived from DNA on the opposite side of the replication fork. Inasmuch as a lesion in a single-stranded region is unrepairable, and therefore lethal, the importance of branch migration becomes apparent. The use of a recA nucleoprotein filament and the enormous investment in energy that this represents, may serve to ensure that this and other related functions occur in vivo with high efficiency.

Role of SSB in Strand Exchange

SSB plays an important role in homologous recombination in *E. coli* (177–179). A role in moderating the nuclease activity of the recBCD enzyme has been noted above. Important effects of SSB were noted in some of the earliest work on recA protein–promoted pairing reactions in vitro. In these early studies, SSB appeared to act simply in a sparing role, reducing the amount of recA protein required by binding to excess ssDNA (180, 181). Subsequently it was demonstrated that DNA strand exchange is stimulated by SSB (127, 130). RecA protein is required in stoichiometric amounts for optimal activity, whether or not SSB is present (146).

SSB binds to ssDNA as a tetramer of 18,873-dalton subunits (182). Binding is rapid and stoichiometric, with binding densities ranging from 1 tetramer per 33–65 nucleotides depending on conditions (183, 184). The monomer contains four tryptophans, and binding is manifested by a large quenching of the tryptophan fluorescence of the protein (183, 184). This property has proved to be useful in binding studies. It has recently been demonstrated that SSB exists in at least two DNA binding modes (183–185). The "low salt" binding mode is prevalent at NaCl or Mg^{2+} concentrations of less than 10 or 1 mM, respectively. This mode is characterized by a high degree of cooperativity and a binding site size of 33 nucleotides. A relatively smooth filament of SSB is often observed in the electron microscope under these conditions (185). The "high salt" binding mode is prevalent above 200 mM NaCl or at 10 mM $MgCl_2$, and is characterized by a very low degree of cooperativity and a binding site size of approximately 65 nucleotides. This form is characterized by a "beads on a string" appearance when bound to ssDNA (185). At intermediate salt concentrations, the two forms can coexist (184). These findings explain a variety of observations regarding SSB obtained under different conditions (186–192).

When SSB exists in its "low salt" binding mode, binding of ssDNA by SSB and recA protein appears to be strictly competitive. SSB will displace recA protein almost entirely (147). Stimulation of recA protein–promoted DNA strand exchange is observed, however, only under conditions in which SSB exists predominantly in its "high salt," low-cooperativity binding mode. These conditions (10-mM Mg^{2+}) are optimal for the strand exchange reaction. Even here, SSB will rapidly displace recA protein from ssDNA in the absence of ATP (84). Upon addition of ATP, the steady-state binding equilibrium between the two proteins is displaced in favor of recA protein (84). If recA protein is added prior to SSB, the SSB does not displace the recA protein. Instead it plays a significant role in establishing a stable, stoichiometric, and highly active recA protein filament on the ssDNA (84, 146, 147). On circular ssDNA, the resulting complex is uniform and unbroken as viewed in the electron microscope (76, 80, 81). When SSB is added first, there is a long lag in the binding of recA protein (84). However, given sufficient time, recA protein can displace the SSB and the resulting complex is the same as that formed when SSB is added last (130, 146). This order of addition effect explains the finding that SSB inhibits recA protein–promoted ATP hydrolysis (180, 192). When SSB is added last, no inhibition occurs, and in fact an apparent enhancement of ATP hydrolysis is observed (84, 147, 193).

Stimulation of DNA strand exchange by SSB can be traced directly to an effect on the formation of recA-ssDNA complexes in the first phase of the reaction (146). Stimulatory effects of SSB on the formation of recA nucleoprotein filaments have been noted in a number of studies (76, 81, 185). Radding and colleagues have demonstrated that recA protein binding to ssDNA is impeded by DNA secondary structure (81, 141); this barrier to binding is removed upon addition of SSB (81, 84, 141).

Several explanations for the role of SSB in establishing a uniform recA-nucleoprotein filament on ssDNA have been suggested. The first and most straightforward is that SSB serves to denature secondary structure in the DNA (81, 141). The SSB is then displaced by recA protein to form the contiguous recA nucleoprotein filament. In support of the idea that SSB has only a transient action, Radding and coworkers have shown that when recA protein was bound to ssDNA under conditions that did not favor the formation of secondary structure (low Mg^{2+}), recA protein bound uniformly to the DNA. After further addition of Mg^{2+} up to concentrations optimal for strand exchange, the reactivity of the complexes was significantly improved relative to complexes that had not been subjected to Mg^{2+}-shift (81, 141).

Other studies, however, have demonstrated that SSB is not displaced from the recA protein–ssDNA complex (84). As measured by quenching of the intrinsic SSB fluorescence, an interaction of SSB with recA protein–ssDNA complexes was demonstrated. This interaction, which is continuous for periods greater than an hour, is ATP-dependent, and requires sufficient SSB to saturate the ssDNA. The association has the effect of increasing the binding density of recA protein and enhancing the apparent k_{cat} for recA protein–promoted ATP hydrolysis (84). The association is required to maintain the stability of recA protein–ssDNA complexes. In contrast, the stability and uniform binding of complexes formed via the Mg^{2+} shift do not persist.

Within a period of 20 min after the shift to high Mg^{2+} concentration, the complexes revert to a form equivalent in all respects to recA protein–ssDNA complexes formed at 10 mM Mg^{2+} in the absence of SSB (147). This deterioration does not occur in the presence of SSB (147).

Two models have been proposed to accommodate these results. In one, proposed by Kowalczykowski and coworkers (193, 193a), SSB prevents deterioration of the recA protein–ssDNA complexes by continually melting-out secondary structure and allowing rebinding of recA protein. Thus, SSB is used reiteratively to maintain the stability of the recA complex. In the second model, there is a direct interaction of SSB with the recA protein complexes (84). In this case it is the joint recA protein–SSB complex that denatures and binds to regions of secondary structure.

Data presently available suggest that SSB plays a role that may be more complicated than the transient or reiterative denaturation of secondary structure in DNA. Under conditions optimal for DNA strand exchange, the interaction between SSB and recA protein is not competitive. In fact, the amount of each protein in the complex is sufficient, separately, to saturate the ssDNA (84, 90). Maintaining high levels of SSB increases rather than decreases the stability of the recA complexes in dilution experiments (147). Moreover, the level of fluorescence quenching of SSB observed after complex formation is unaffected by the addition of excess recA protein or SSB (84). Exchange of recA protein between free and bound forms, which might be expected if it were cycling in and out of regions of secondary structure, is not observed in the presence of SSB (90). It has also been demonstrated that SSB that has participated in the formation of recA-ssDNA complexes is effectively sequestered and unable to participate immediately in further complex formation (D. A. Soltis, B. Stockman, M. M. Cox, I. R. Lehman, unpublished).

The "high salt" binding mode of SSB is characterized by a cooperativity parameter (ω) of approximately 50, which is low enough to suggest that little or no interaction occurs between SSB tetramers (183, 184). Lohman and coworkers have noted that the binding site size for SSB in this binding mode increases from 65 to 77 nucleotides when M13 ssDNA is substituted for polydT (184). They further suggested that this change might reflect blocking of SSB from regions of secondary structure. Under similar conditions, only 80% of a sample of ssDNA is protected from DNase digestion in the presence of excess SSB (169). These results suggest that neither recA protein nor SSB can bind to regions of significant secondary structure under these conditions. If this interpretation is correct, removal of secondary structure should involve a joint complex of the two proteins. However, attempts to isolate such complexes have been unsuccessful (194), nor have they been detected by electron microscopy. Some alterations in the structure of recA-ssDNA complexes formed in the presence of SSB have been noted, leading to the suggestion that mixed complexes may exist in which SSB and recA protein are not separated into domains (80). Finally, it should be noted that the stimulatory effect of single-stranded DNA-binding protein is not unique to *E. coli* SSB; the T4 gene 32 protein and the phage λ β protein can also stimulate recA protein–promoted DNA strand exchange (195). In sum, physical evidence for a direct interaction between recA protein and SSB is still lacking and is clearly required before the notion of such a complex can be accepted.

Whatever the mechanism of SSB action, these studies underline the importance of the recA nucleoprotein filament. The activity of recA protein in every instance correlates positively with an increase in the binding density of recA protein and the formation of unbroken filaments; interference with filament formation in every instance inhibits recA protein action. The active species in recA protein–promoted DNA strand exchange is clearly a uniform stoichiometric recA nucleoprotein filament.

RESOLUTION OF HOLLIDAY STRUCTURES

The final step in general recombination is the resolution of the Holliday intermediate. As described above, the recBCD enzyme could, by virtue of its nuclease activity, play a role in this reaction. Recently, however, an enzyme has been isolated from bacteriophage T4–infected cells, endonuclease VII, the product of T4 gene 49, that exhibits a demonstrable specificity for the cleavage of Holliday junctions (196–199). Cells infected with gene 49 mutants accumulate a highly branched multimeric form of T4 DNA called very-fast-sedimenting (VFS) DNA (196), and the infections are abortive. In vitro, the enzyme cleaves DNA specifically at the base of extruded cruciforms, which are used as analogues of the Holliday intermediate (198, 199). The cuts are symmetrically placed on both strands of the DNA, 2 or 3 nucleotides 5' to the end of the cruciform in each case (198). The resulting DNA molecule is linear with hairpin ends and single ligatable nicks at positions corresponding to the stem base of the cruciform (198, 199). A similar activity has recently been detected in yeast extracts (200, 201).

RECOMBINASES IN OTHER CELLS

Proteins closely related to recA protein of *E. Coli* appear to be widely distributed among bacteria. RecA-like proteins have been isolated from *B. subtilis* (116), *Salmonella typhimurium* (202), and *Proteus mirabilis* (203). Another recA-like protein, the product of the *uvsX* gene of bacteriophage T4, has been isolated from T4-infected cells (204–206). Although catalyzing many of the DNA pairing reactions catalyzed by recA protein, the uvs x protein hydrolyzes ATP at an approximately 20-fold greater rate than recA protein. Surprisingly, it produces AMP and PPi as well as ADP and Pi (206).

Only ssDNA functions as a cofactor for ATP hydrolysis. Notably, the T4 gene 32 protein strongly stimulates the activity of uvs x protein, possibly in a manner analogous to the effect of SSB on recA protein–promoted reactions (204–206). These enzymes are all very similar to recA protein, with the most interesting distinction being the hydrolysis of dATP but not ATP by the *B. subtilis* enzyme noted above.

Only one recA-like enzyme has been purified from eukaryotic cells, the rec1 protein of *Ustilago maydis* (207–209). This protein promotes many of the pairing and strand transfer activities associated with the *E. coli* recA protein. However, rec1 protein–promoted strand exchange exhibits the interesting difference that the polarity of branch migration is the opposite of recA protein (207). The rec1 protein also exhibits a pronounced Z-DNA binding activity, which may be related to the mechanism by which pairing and synapsis reactions are promoted by this enzyme (209).

A variety of recA-like activities have been detected in mammalian cell extracts (210–212), but to date none of these has been purified. In several instances the activity differs from recA protein in a basic property, in particular, the absence of a requirement for ATP (212). Nonetheless these studies suggest that studies of homologous recombination with purified enzymes from *E. coli* may be generally applicable to recombination in higher cells. On the other hand, the apparent absence of a gene with the pleiotropic effects of the *recA* gene in many organisms including yeast suggests that interesting alternative mechanisms remain to be discovered.

Many gaps remain in our understanding of the molecular mechanism of general recombination. The ready accessibility of the recA-recBCD-SSB system from *E. coli,* coupled to the importance of the biological and biochemical questions that remain to be addressed, ensures that this prototype will occupy a central place in studies of general recombination for years to come.

ACKNOWLEDGMENTS

Research cited from the authors's laboratories was supported by NIH Grants GM06196 to IRL and GM32335 to MMC and NSF Grant PCM7904638 to IRL.

Literature Cited

1. Holliday, R. 1964. *Genet. Res.* 5:282–304
2. Meselson, M. S., Radding, C. M. 1975. *Proc. Natl. Acad. Sci. USA* 72:358–61
3. Warner, R. C., Fishel, R. A., Wheeler, F. C. 1978. *Cold Spring Harbor Symp. Quant. Biol.* 43:957–68
4. Sigal, N., Alberts, B. 1972. *J. Mol. Biol.* 71:789–93
5. Fox, M. S., Dudney, C. S., Sodergren, E. J. 1978. *Cold Spring Harbor Symp. Quant. Biol.* 43:999–1007
6. Sodergren, E. J., Fox, M. S. 1979. *J. Mol. Biol.* 130:357–77
7. Fogel, S., Mortimer, R., Lusnak, K., Tavares, T. 1978. *Cold Spring Harbor Symp. Quant. Biol.* 43:1325–41
8. Radding, C. M. 1973. *Ann. Rev. Genet.* 7:87–111
9. Clark, A. J., Margulies, A. D. 1965. *Proc. Natl. Acad. Sci. USA* 53:451–59
10. Gottesman, M. M., Gottesman, M. E., Gottesman, S., Gellert, M. 1974. *J. Mol. Biol.* 88:471–87

11. Gillen, J. R., Karu, A. E., Nagaishi, H., Clark, A. J. 1977. *J. Mol. Biol.* 113:27–41
12. Bachmann, B. J., Low, K. B., Taylor, A. L. 1976. *Bacteriol. Rev.* 40:116–67
13. Clark, A. J., Volkert, M. R., Margossian, L. J. 1978. *Cold Spring Harbor Symp. Quant. Biol.* 43:887–92
14. Rothman, R. H., Clark, A. J. 1977. *Mol. Gen. Genet.* 155:279–86
15. Horii, Z.-I., Clark, A. J. 1973. *J. Mol. Biol.* 80:327–44
16. Castellazzi, M., Morand, P., George, J., Buttin, G. 1977. *Mol. Gen. Genet.* 153:297–310
17. Emmerson, P. T., West, S. C. 1977. *Mol. Gen. Genet.* 155:77–85
18. McEntee, K. 1977. *Proc. Natl. Acad. Sci. USA* 74:5275–79
19. Goudas, L. J., Mount, D. W. 1977. *Proc. Natl. Acad. Sci. USA* 74:5280–84
20. Tomizawa, J., Ogawa, H. 1972. *Nature New Biol.* 239:14–16
21. Amundsen, S. K., Taylor, A. F., Chaudhury, A. M., Smith, G. R. 1986. *Proc. Natl. Acad. Sci. USA* 83:5558–62
22. Clark, A. J. 1973. *Ann. Rev. Genet.* 7:67–86
23. Meyer, R. R., Glassberg, J., Kornberg, A. 1979. *Proc. Natl. Acad. Sci. USA* 76:1702–5
24. Johnson, B. F. 1977. *Mol. Gen. Genet.* 157:91–97
25. Glassberg, J., Meyer, R. R., Kornberg, A. 1979. *J. Bacteriol.* 140:14–19
26. Nash, H. 1981. *Ann. Rev. Genet.* 15:143–67
27. Sadowski, P. 1986. *J. Bacteriol.* 165:341–47
28. Buttin, G., Wright, M. 1968. *Cold Spring Harbor Symp. Quant. Biol.* 33:259–69
29. Oishi, M. 1969. *Proc. Natl. Acad. Sci. USA* 64:1292–99
30. Barbour, S. D., Clark, A. J. 1970. *Proc. Natl. Acad. Sci. USA* 65:955–61
31. Goldmark, P. J., Linn, S. 1972. *J. Biol. Chem.* 247:1849–60
32. Rosamond, J., Telander, K. M., Linn, S. 1979. *J. Biol. Chem.* 254:8646–52
33. MacKay, V., Linn, S. 1976. *J. Biol. Chem.* 251:3716–19
34. Taylor, A., Smith, G. R. 1980. *Cell* 22:447–57
35. Friedman, E. A., Smith, H. O. 1972. *J. Biol. Chem.* 247:2846–
36. Wilcox, K. W., Smith, H. O. 1976. *J. Biol. Chem.* 251:6127–34
37. Eichler, D. C., Lehman, I. R. 1977. *J. Biol. Chem.* 252:499–503
38. Hickson, I. D., Atkinson, K. E., Emmerson, P. T. 1982. *Mol. Gen. Genet.* 185:148–51
39. Lieberman, R. P., Oishi, M. 1974. *Proc. Natl. Acad. Sci. USA* 71:4816–20
40. Hickson, I. D., Emmerson, P. T. 1981. *Nature* 294:578–80
41. Umeno, M., Sasaki, M., Anai, M., Takagi, Y. 1983. *Biochem. Biophys. Res. Commun.* 116:1144–50
42. Dykstra, C. C., Prasher, D., Kushner, S. R. 1982. *J. Bacteriol.* 157:21–27
43. Hickson, I. D., Robson, C., Atkinson, K. E., Hutton, L., Emmerson, P. T. 1985. *J. Biol. Chem.* 260:1224–29
44. Chaudhury, A. M., Smith, G. R. 1984. *Proc. Natl. Acad. Sci. USA* 81:7850–54
45. MacKay, V., Linn, S. 1974. *J. Biol. Chem.* 249:4286–94
46. Karu, A. E., MacKay, V., Goldmark, P. J. 1973. *J. Biol. Chem.* 248:4874–84
47. Rosamond, J., Telander, K. M., Linn, S. 1979. *J. Biol. Chem.* 254:8646–52
48. Muskavitch, K. M. T., Linn, S. 1982. *J. Biol. Chem.* 257:2641–48
49. Prell, A., Wackernagel, W. 1980. *Eur. J. Biochem.* 105:109–16
50. Taylor, A. F., Smith, G. R. 1985. *J. Mol. Biol.* 185:431–43
51. Ponticelli, A. S., Schultz, D. W., Taylor, A. F., Smith, G. R. 1985. *Cell* 41:145–51
52. Taylor, A. F., Schultz, D. W., Ponticelli, A. S., Smith, G. R. 1985. *Cell* 41:153–63
53. Stahl, F. W. 1979. *Ann. Rev. Genet.* 13:7–24
54. Smith, G. R., Kunes, S. M., Schultz, D. W., Taylor, A., Triman, K. L. 1981. *Cell* 24:429–36
55. Smith, G. R., Amundsen, S. K., Chaudhury, A. M., Cheng, K. C., Ponticelli, A. S., et al. 1984. *Cold Spring Harbor Symp. Quant. Biol.* 49:485–95
56. Kobayashi, I., Stahl, M. M., Stahl, F. W. 1984. *Cold Spring Harbor Symp. Quant. Biol.* 49:497–506
57. McEntee, K., Hesse, J. E., Epstein, W. 1976. *Proc. Natl. Acad. Sci. USA* 73:3979–83
58. McEntee, K., Epstein, W. 1977. *Virology* 85:306–18
59. Witkin, E. M. 1976. *Bacteriol. Rev.* 40:869–907
60. Radman, M. 1975. In *Molecular Mechanisms for Repair of DNA*, ed. P. Hanawalt, R. B. Setlow, Vol. 5A, pp. 355–67. New York: Plenum
61. McEntee, K. 1977. *J. Bacteriol.* 132:904–11
62. Sankar, A., Rupp, W. D. 1979. *Proc. Natl. Acad. Sci. USA* 76:3144–48
63. Hickson, I. D., Gordon, R. C., Tomkins, A. E., Emmerson, P. T. 1981. *Mol. Gen. Genet.* 184:68–72
64. Little, J. W., Mount, D. W. 1982. *Cell* 29:11–22

65. Radding, C. M. 1982. *Ann. Rev. Genet.* 16:405–37
66. McEntee, K., Weinstock, G. M. 1981. *The Enzymes* 14:445–70
67. Dressler, D., Potter, H. 1982. *Ann. Rev. Biochem.* 51:727–61
68. Horii, T., Ogawa, T., Ogawa, O. 1980. *Proc. Natl. Acad. Sci. USA* 77:313–17
69. Sancar, A., Stachelek, C., Konigsberg, W., Rupp, W. D. 1980. *Proc. Natl. Acad. Sci. USA* 77:2611–15
70. Little, J. W. 1984. *Proc. Natl. Acad. Sci. USA* 81:1375–79
71. McKay, D. B., Steitz, T. A., Weber, I. T., West, S. C., Howard-Flanders, P. 1980. *J. Biol. Chem.* 255:6662–63
72. Blanar, M. A., Kneller, D., Clark, A. J., Karu, A. E., Cohen, F. E., et al. 1984. *Cold Spring Harbor Symp. Quant. Biol.* 49:507–11
73. McEntee, K., Weinstock, G. M., Lehman, I. R. 1981. *J. Biol. Chem.* 256:8835–44
74. Cotterill, S. M., Fersht, A. R. 1983. *Biochemistry* 22:3525–31
75. Morrical, S. W., Cox, M. M. 1985. *Biochemistry* 24:760–67
76. Flory, J., Radding, C. M. 1982. *Cell* 28:747–57
77. Register, J. C., Griffith, J. 1985. *Mol. Gen. Genet.* 199:415–20
78. Dunn, K., Chrysogelos, S., Griffith, J. 1982. *Cell* 28:757–65
79. DiCapua, E., Engel, A., Stasiak, A., Koller, Th. 1982. *J. Mol. Biol.* 157:87–103
80. Williams, R. C., Spengler, S. J. 1986. *J. Mol. Biol.* 187:109–18
81. Flory, J., Tsang, S. S., Muniyappa, K. 1984. *Proc. Natl. Acad. Sci. USA* 81:7026–30
82. West, S. C., Cassuto, E., Mursalim, J., Howard-Flanders, P. 1980. *Proc. Natl. Acad. Sci. USA* 77:2569–73
83. Bryant, F. R., Taylor, A., Lehman, I. R. 1985. *J. Biol. Chem.* 260:1196–202
84. Morrical, S. W., Lee, J., Cox, M. M. 1986. *Biochemistry* 24:1482–94
85. Silver, M. S., Fersht, A. R. 1982. *Biochemistry* 21:6066–72
86. Cazenave, C., Toulmé, J. J., Hélene, C. 1983. *EMBO J.* 2:2247–51
87. Menetski, J. P., Kowalczykowski, S. C. 1985. *J. Mol. Biol.* 185:281–95
88. Menetski, J. P., Kowalczykowski, S. C. 1987. *J. Biol. Chem.* In press
89. Menetski, J. P., Kowalczykowski, S. C. 1987. *J. Biol. Chem.* In press
90. Neuendorf, S. K., Cox, M. M. 1986. *J. Biol. Chem.* 261:8276–82
91. Kawashima, H., Horii, T., Ogawa, T., Ogawa, H. 1984. *Mol. Gen. Genet.* 193:288–92
92. Weinstock, G. M., McEntee, K., Lehman, I. R. 1981. *J. Biol. Chem.* 256:8829–34
93. Shibata, T., Cunningham, R. P., Das Gupta, C., Radding, C. M. 1979. *Proc. Natl. Acad. Sci. USA* 76:5100–4
94. McEntee, K., Weinstock, G. M., Lehman, I. R. 1981. *Progr. Nucleic Acid Res. Mol. Biol.* 26:265–81
95. Ohtani, T., Shibata, T., Iwabuchi, M., Watabe, H., Sino, T., Ando, T. 1982. *Nature* 299:86–89
96. Iwabuchi, M., Shibata, T., Ohtani, T., Natari, M., Ando, T. 1983. *J. Biol. Chem.* 258:12394–404
97. McEntee, K., Weinstock, G. M., Lehman, I. R. 1979. *Proc. Natl. Acad. Sci. USA* 76:2615–19
98. Wu, A. M., Bianchi, M., Das Gupta, C., Radding, C. M. 1983. *Proc. Natl. Acad. Sci. USA* 80:1256–60
99. Bianchi, M., Das Gupta, C., Radding, C. M. 1983. *Cell* 34:931–39
100. Stasiak, A., DiCapua, E. 1982. *Nature* (Lond.) 299:185–86
101. Pugh, B. F., Cox, M. M. 1987. *J. Biol. Chem.* In press
102. Shibata, T., Das Gupta, C., Cunningham, R. P., Radding, C. M. 1979. *Proc. Natl. Acad. Sci. USA* 76:1638–42
103. Volodin, A. A., Shepelev, V. A., Kisaganov, Y. N. 1982. *FEBS Lett.* 145:53–56
104. West, S. C., Cassuto, E., Mursalim, J., Howard-Flanders, P. 1980. *Proc. Natl. Acad. Sci. USA* 77:2569–73
105. Stasiak, A., DiCapua, E., Koller, Th. 1981. *J. Mol. Biol.* 151:557–64
106. Chrysogelos, S., Register, J. C., Griffith, J. 1983. *J. Biol. Chem.* 258:12624–31
107. Stasiak, A., Engelman, E. H. 1986. *Biophys. J.* 49:5–7
108. Dombroski, D. F., Scraba, D. G., Bradley, R. D., Morgan, A. R. 1983. *Nucleic Acids Res.* 11:7487–504
109. Ogawa, T., Wabiko, H., Tsurimoto, T., Horii, T., Masukata, H., Ogawa, H. 1978. *Cold Spring Harbor Symp. Quant. Biol.* 43:909–15
110. Roberts, J. W., Roberts, C. W., Craig, N. L., Phizicky, E. M. 1978. *Cold Spring Harbor Symp. Quant. Biol.* 43:917–20
111. Weinstock, G. M., McEntee, K., Lehman, I. R. 1981. *J. Biol. Chem.* 256:8845–49
112. Cox, M. M., Soltis, D. A., Lehman, I. R., DeBrosse, C., Benkovic, S. J. 1983. *J. Biol. Chem.* 258:2586–92
113. Brenner, S. L., Mitchell, R. S., Morrical, S. W., Neuendorf, S. K., Shutte, B.

C., Cox, M. M. 1986. *J. Biol. Chem.* In press
114. Weinstock, G. M., McEntee, K., Lehman, I. R. 1979. *Proc. Natl. Acad. Sci. USA* 76:126–30
115. Weinstock, G. M., McEntee, K., Lehman, I. R. 1981. *J. Biol. Chem.* 256:8856–58
116. Lovett, C. M., Roberts, J. W. 1985. *J. Biol. Chem.* 260:3305–13
117. Weinstock, G. M., McEntee, K., Lehman, I. R. 1981. *J. Biol. Chem.* 256:8850–55
118. Knight, K. L., McEntee, K. 1985. *J. Biol. Chem.* 260:867–72
119. Knight, K. L., McEntee, K. 1985. *J. Biol. Chem.* 260:10185–91
120. Knight, K. L., McEntee, K. 1985. *J. Biol. Chem.* 260:10177–84
121. Christiansen, C., Baldwin, R. L. 1977. *J. Mol. Biol.* 115:441–54
122. Alberts, B. M., Frey, L. 1970. *Nature* 227:1313–17
123. Bryant, F. R., Lehman, I. R. 1985. *Proc. Natl. Acad. Sci. USA* 82:297–301
124. Wetmur, J. G., Davidson, N. 1968. *J. Mol. Biol.* 31:349–70
125. McEntee, K. 1985. *Biochemistry* 24:4345–51
126. Das Gupta, C., Shibata, T., Cunningham, R. P., Radding, C. M. 1980. *Cell* 22:437–46
127. Cox, M. M., Lehman, I. R. 1981. *Proc. Natl. Acad. Sci. USA* 78:3433–37
128. Das Gupta, C., Wu, A. M., Kahn, R., Cunningham, R. P., Radding, C. M. 1981. *Cell* 25:507–16
129. West, S. C., Cassuto, E., Howard-Flanders, P. 1981. *Proc. Natl. Acad. Sci. USA* 78:2100–4
130. Cox, M. M., Lehman, I. R. 1982. *J. Biol. Chem.* 257:8523–32
131. Gonda, D. K., Radding, C. M. 1983. *Cell* 34:647–54
132. Julin, D. A., Riddles, P. W., Lehman, I. R. 1986. *J. Biol. Chem.* 261:1025–30
133. Tsang, S. S., Chow, S. A., Radding, C. M. 1985. *Biochemistry* 24:3226–32
134. Warner, R. C., Tessman, I. T. 1978. In *The Single-stranded DNA Phages*, ed. D. T. Denhardt, D. Dressler, D. S. Ray, p. 417. Cold Spring Harbor, N.Y.
135. Honigberg, S. M., Gonda, D. K., Flory, J., Radding, C. M. 1985. *J. Biol. Chem.* 260:11845–51
136. West, S. C., Cassuto, E., Howard-Flanders, P. 1981. *Proc. Natl. Acad. Sci. USA* 78:6149–53
137. Kahn, R., Cunningham, R. P., Das Gupta, C., Radding, C. M. 1981. *Proc. Natl. Acad. Sci. USA* 78:4786–90
138. Cox, M. M., Lehman, I. R. 1981. *Proc. Natl. Acad. Sci. USA* 78:6018–22
139. Jencks, W. P. 1980. *Adv. Enzymol.* 51:75–106
140. Kahn, R., Radding, C. M. 1984. *J. Biol. Chem.* 259:7495–503
141. Muniyappa, K., Shaner, S. L., Tsang, S. S., Radding, C. M. 1984. *Proc. Natl. Acad. Sci. USA* 81:2757–61
142. Register, J. C., Griffith, J. 1985. *J. Biol. Chem.* 260:12308–12
143. Leahy, M. C., Radding, C. M. 1986. *J. Biol. Chem.* 261:6954–60
144. Gonda, D. K., Shibata, T., Radding, C. M. 1985. *Biochemistry* 24:413–20
145. Cazenave, C., Chabbert, M., Toulme, J. J., Helene, C. 1984. *Biochem. Biophys. Acta* 781:7–13
146. Cox, M. M., Soltis, D. A., Livneh, Z., Lehman, I. R. 1983. *J. Biol. Chem.* 258:2577–85
147. Morrical, S. W., Cox, M. M. Submitted
148. Lohman, T. M. 1986. *CRC Crit. Rev. Biochem.* 19:191–245
149. Cunningham, R. P., Shibata, T., Das Gupta, C., Radding, C. M. 1979. *Nature* 281:191–95
150. Radding, C. M., Shibata, T., Cunningham, R. P., Das Gupta, C., Osber, L. 1980. In *Mechanistic Studies of DNA Replication and Genetic Recombination,* ed. B. Alberts, C. F. Fox, 19:863–70. New York: Academic
151. Tsang, S. S., Chow, S. A., Radding, C. M. 1985. *Biochemistry* 24:3226–32
152. Chow, S. A., Radding, C. M. 1985. *Proc. Natl. Acad. Sci. USA* 82:5646–50
153. Cunningham, R. P., Wu, A. M., Shibata, T., Das Gupta, C., Radding, C. M. 1981. *Cell* 24:213–23
154. Riddles, P. W., Lehman, I. R. 1985. *J. Biol. Chem.* 260:165–69
155. Christiansen, G., Griffith, J. 1986. *Proc. Natl. Acad. Sci. USA* 83:2066–70
156. Wu, A. M., Kahn, R., Das Gupta, C., Radding, C. M. 1982. *Cell* 30:37–44
157. Shibata, T., Makino, O., Ikawa, S., Ohtani, T., Iwabuchi, M., et al. 1984. *Cold Spring Harbor Symp. Quant. Biol.* 49:541–51
158. Konforti, B., Davis, R. 1986. *Proc. Natl. Acad. Sci. USA.* In press
159. Das Gupta, C., Radding, C. M. 1982. *Proc. Natl. Acad. Sci. USA* 79:762–66
160. Livneh, Z., Lehman, I. R. 1982. *Proc. Natl. Acad. Sci. USA* 79:3171–75
161. Bianchi, M. E., Radding, C. M. 1983. *Cell* 35:511–20
162. Howard-Flanders, P., West, S. C., Stasiak, A. 1984. *Nature* 309:215–20
163. Cox, M. M., Morrical, S. N., Neuendorf, S. K. 1984. *Cold Spring Harbor Symp. Quant. Biol.* 49:525–33
164. Cox, M. M., Pugh, B. F., Schutte, B. C., Lindsley, J. E., Lee, J., Morrical, S.

W. 1986. *Mechanisms of DNA Replication and Recombination*, Vol. 47, ed. T. Kelly, R. McMacken. New York: Liss

165. Roman, L. J., Kowalczykowski, S. C. 1986. *Biochemistry*. In press
166. Cleveland, D. N. 1982. *Cell* 28:689–91
167. Stasiak, A., Stasiak, A. Z., Koller, T. 1984. *Cold Spring Harbor Symp. Quant. Biol.* 49:561–70
168. Chow, S. A., Honigberg, S. M., Bainton, R. J., Radding, C. M. 1986. *J. Biol. Chem.* 261:6961–71
169. Pugh, B. F., Cox, M. M. 1987. Submitted
170. Howard-Flanders, P., West, S. C., Rusche, J. R., Egelman, E. H. 1984. *Cold Spring Harbor Symp. Quant. Biol.* 49:571–80
171. Cassuto, E., Howard-Flanders, P. 1986. *Nucleic Acids Res.* 14:1149–57
172. Salles, B., Paoletti, C. 1983. *Proc. Natl. Acad. Sci. USA* 80:65–69
173. Salles, B., Lang, M. C., Freund, A. M., Paoletti, C., Daune, M., Fuchs, R. P. 1983. *Nucleic Acids Res.* 11:5235–42
174. Yancey, S. D., Porter, R. D. 1984. *Mol. Gen. Genet.* 193:53–57
175. White, R. L., Fox, M. S. 1974. *Proc. Natl. Acad. Sci. USA* 71:1544–48
176. West, S. C., Cassuto, E., Howard-Flanders, P. 1981. *Nature* 294:659–62
177. Glassberg, J., Meyer, R. R., Kornberg, A. 1979. *J. Bacteriol* 140:14–19
178. Vales, L. D., Chase, J. N., Murphy, J. B. 1980. *J. Bacteriol* 143:887–96
179. Whittier, R. F., Chase, J. W. 1981. *Mol. Gen. Genet.* 183:341–47
180. McEntee, K., Weinstock, G. M., Lehman, I. R. 1980. *Proc. Natl. Acad. Sci. USA* 77:2606–10
181. Shibata, T., Das Gupta, C., Cunningham, R. P., Radding, C. M. 1980. *Proc. Natl. Acad. Sci. USA* 77:2606–10
182. Sancar, A., Williams, K. R., Chase, J. W., Rupp, W. D. 1981. *Proc. Natl. Acad. Sci. USA* 78:4274–78
183. Lohman, T. M., Overman, L. B. 1985. *J. Biol. Chem.* 260:3594–603
184. Lohman, T. M., Overman, L. B., Datta, S. 1986. *J. Mol. Biol.* 187:603–15
185. Griffith, J. D., Harris, L. D., Register, J. 1984. *Cold Spring Harbor Symp. Quant. Biol.* 49:553–59
186. Sigal, N., Delius, H., Kornberg, T., Gefter, M. L., Alberts, B. 1972. *Proc. Natl. Acad. Sci. USA* 69:3537–41
187. Weiner, J. H., Bertsch, L. L., Kornberg, A. 1975. *J. Biol. Chem.* 250:1972–80
188. Chrysogelos, S., Griffith, J. 1982.

Proc. Natl. Acad. Sci. USA 79:5803–7
189. Williams, K. R., Spicer, E. K., Lo Presti, M. B., Guggenheimer, R. A., Chase, J. N. 1983. *J. Biol. Chem.* 258:3346–55
190. Molineux, I. J., Pauli, A., Gefter, M. L. 1975. *Nucleic Acids Res.* 2:1821–37
191. Ruyechan, W. T., Wetmur, J. G. 1975. *Biochemistry* 14:5529–34
192. Cohen, S. P., Resnick, J., Sussman, R. 1983. *J. Mol. Biol.* 167:901–9
193. Kowalczykowski, S. C., Clow, J., Somani, R., Varghese, A. 1986. *J. Mol. Biol.* In press
193a. Kowalczykowski, S. C., Krupp, R. A. 1986. *J. Mol. Biol.* In press
194. Soltis, D. A., Lehman, I. R. 1983. *J. Biol. Chem.* 258:6073–77
195. Flory, S. S., Tsang, J., Muniyappa, K., Bianchi, M., Gonda, D., et al. 1984. *Cold Spring Harbor Symp. Quant. Biol.* 49:513–23
196. Kemper, B., Garabett, M. 1981. *Eur. J. Biochem.* 115:123–31
197. Jensch, F., Kemper, B. 1984. *EMBO J.* 5:181–89
198. Lilley, D. M. J., Kemper, B. 1984. *Cell* 36:413–22
199. Mizuuchi, K., Kemper, B., Hays, J., Weisberg, R. 1982. *Cell* 29:357–65
200. West, S. C., Körner, A. 1985. *Proc. Natl. Acad. Sci. USA* 82:6445–49
201. Symington, L. S., Kolodner, R. 1985. *Proc. Natl. Acad. Sci. USA* 82:7247–51
202. Pierre, A., Paoletti, C. 1983. *J. Biol. Chem.* 258:2870–74
203. West, S. C., Countryman, J. K., Howard-Flanders, P. 1983. *J. Biol. Chem.* 258:4648–54
204. Yonesaki, T., Ryo, Y., Minagawa, T., Takahashi, H. 1985. *Eur. J. Biochem.* 148:127–34
205. Griffith, J., Formosa, T. 1985. *J. Biol. Chem.* 260:4484–91
206. Formosa, T., Alberts, B. M. 1986. *J. Biol. Chem.* 261:6107–18
207. Kmiec, E. B., Holloman, W. K. 1983. *Cell* 33:857–64
208. Kmiec, E. B., Holloman, W. K. 1984. *Cell* 36:593–98
209. Kmiec, E. B., Holloman, W. K. 1986. *Cell* 44:545–54
210. Kenne, K., Ljungquist, S. 1984. *Nucleic Acids Res.* 12:3057–68
211. Kucherlapati, R. S., Spencer, J., Moore, P. D. 1985. *Mol. Cell. Biol.* 5:714–20
212. Hsieh, P., Meyn, M. S., Camerini-Otero, R. D. 1986. *Cell* 44:885–94

Ann. Rev. Biochem. 1987. 56:263–87

TRANSFER RNA MODIFICATION

Glenn R. Björk, Johanna U. Ericson, Claes E. D. Gustafsson,
Tord G. Hagervall, Yvonne H. Jönsson, and P. Mikael Wikström

Department of Microbiology, University of Umeå, S-901 87 Sweden

CONTENTS

1. PERSPECTIVES AND SUMMARY

Transfer RNA interacts with many different proteins, including elongation factors and aminoacyl-tRNA ligases, and with ribosomal RNA. This diversity of interactions for tRNA may be one reason for its complex content of

263

0066-4154/87/0701-0263$02.00

modified nucleosides. At present more than 50 different modified nucleosides have been characterized, all of which are derivatives of the normal nucleosides adenosine (A), guanosine (G), uridine (U), and cytosine (C) (1). Some are modified by the addition of a single methyl group to the base or to the 2'-hydroxyl group of the ribose, while others are formed by a complex sequence of reactions resulting in the hypermodified nucleosides. The synthesis of most modified nucleosides [except queuosine (Q) and inosine (I)] occurs at the polynucleotide level after the transcription of the tRNA genes (2, 3). tRNA from all organisms contains modified nucleosides (4). Most of them are not likely to be essential for viability, but play an important role in the fine tuning of tRNA activity. Not only does the presence of a modified nucleoside improve the efficiency of the tRNA in the decoding event, but it may also influence the fidelity of protein synthesis as well as sense the reading context, i.e. the nucleotides surrounding the codon. Furthermore, some modified nucleosides may also be involved in maintaining the reading frame. Therefore, the presence of different modified nucleosides is likely to have different impacts on the activity of tRNA. Due to its importance in the decoding steps, the level of tRNA modification also influences the expression of several operons through an inefficient reading of attenuator region of some operon transcripts. Thus, lack of a modified nucleoside may induce pleiotropic effects on cell physiology. Metabolic as well as genetic links/correlations exist between the synthesis of modified nucleosides in tRNA and intermediary metabolism, development, and cell cycle. Therefore, the level of tRNA modification has been suggested to be a regulatory device. This, as well as recent developments concerning the synthesis and function of modified nucleosides and the genetics of tRNA modification, is critically discussed in this review. Some other aspects of modified nucleosides in tRNA not covered in this review, like their possible role in tumor formation and their use as diagnostic tools, are discussed elsewhere (5). Furthermore, results not covered in detail here can be found in earlier reviews on this topic (1, 6–17).

II. PRESENCE OF MODIFIED NUCLEOSIDES IN TRANSFER RNA FROM DIFFERENT ORGANISMS

Figure 1 shows that cytosolic tRNA from all three kingdoms contains modified nucleosides. A few (Ψ13, Cm32, m^1G37, t^6A37, Ψ38, Ψ39, Ψ40, Ψ55, m^1A58) are present in tRNAs from all kingdoms, indicating a common evolutionary origin or a convergent evolution (4). Eukaryotic tRNA contains the largest variety and abundance of modified nucleosides, and consequently it contains several modified nucleosides specific for that kingdom. However, there are also specific modifications present in the other two kingdoms, e.g. mnm^5s^2U for eubacteria and $m^1\Psi$ for tRNA from archaebacteria (4). Most tRNAs from both eukaryotes and eubacteria contain m^5U54, while most tRNAs in archaebacteria contain $m^1\Psi$ in the same position. Since the position of the methyl group in $m^1\Psi54$ and m^5U54 is the same in relation to the ribose moiety and the phosphate-ribose backbone, an evolutionary convergence for the presence of a methyl group in this position of the tRNA has been suggested (18). Although not shown in Figure 1, there also exists species specificity of tRNA modification within a kingdom; e.g. mo^5U34 is present in tRNAs specific for val, ala, thr, and ser from gram-positive organisms, while the corresponding tRNAs from gram-negative organisms have cmo^5U34/ $mcmo^5U$.

Recently, selenium-containing modified nucleosides have been identified in tRNA from both bacteria and eukaryotes (19–21). One, mnm^5Se^2U, has been identified in position 34 of $tRNA^{Glu}$ from *Clostridium sticklandii* (22).

III. SYNTHESIS OF MODIFIED NUCLEOSIDES IN TRANSFER RNA

III.1 Transfer RNA–Modifying Enzymes: Their Recognition Signals and Locations

Genetic analyses have shown that different enzymes catalyze the synthesis of the same modified nucleosides in ribosomal RNA and in tRNA (23, 24). However, the $tRNA(m^5C)$methyltransferase from mammalian tissue shows activity towards *Escherichia coli* rRNA as well as towards synthetic polymers (25). Furthermore, different enzymes produce Ψ in the anticodon stem (positions 38, 39, and 40) and in positions 55 and 32 (26, 27). Two distinct $tRNA(m^2G)$methyltransferases from rat liver as well as two $tRNA(m^1$ G)methyltransferases from yeast with different site specificities have been characterized (28, 29). Thus, there are different enzymes catalyzing the synthesis of the same modified nucleoside not only in different nucleic acids but also in different positions of the tRNA.

Is the sequence surrounding the nucleotide to be modified part of the recognition signal for the tRNA-modifying enzymes? Transfer RNA reading codons starting with U usually have an i^6A derivative in position 37. The recognition signal for the corresponding enzyme is the sequence A36-A37-A38 as well as a five–base paired anticodon stem (30). For 11 other modified nucleotides, a correlation was pointed out between the modified nucleotide and the surrounding nucleotide sequence, suggesting the latter to be part of the recognition signal (30). Seryl-tRNA species I and V have no i^6A37 derivative, although they contain the sequence A36-A37-A38. However, these tRNAs do not have a five–base paired stem, supporting the notion that the tRNA-modifying enzymes catalyzing the formation of i^6A37 also require important recognition signals other than the primary sequence surrounding the

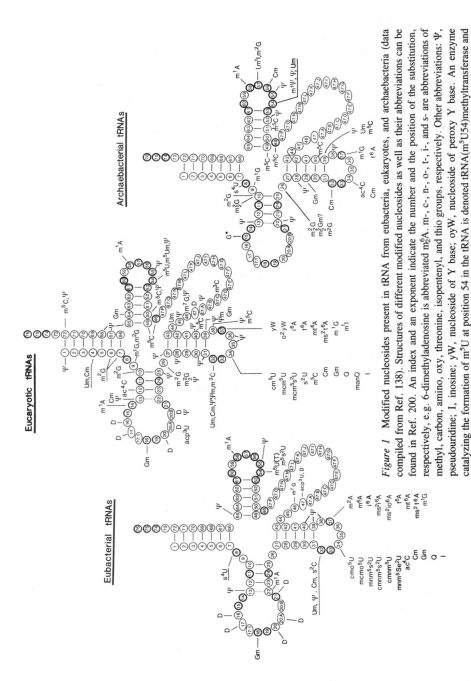

Figure 1 Modified nucleosides present in tRNA from eubacteria, eukaryotes, and archaebacteria (data compiled from Ref. 138). Structures of different modified nucleosides as well as their abbreviations can be found in Ref. 200. An index and an exponent indicate the number and the position of the substitution, respectively, e.g. 6-dimethyladenosine is abbreviated m$_2^6$A. m-, c-, n-, o-, t-, i-, and s- are abbreviations of methyl, carbon, amino, oxy, threonine, isopentenyl, and thio groups, respectively. Other abbreviations: Ψ, pseudouridine; I, inosine; yW, nucleoside of Y base; oyW, nucleoside of peroxy Y base. An enzyme catalyzing the formation of m^{5}U at position 54 in the tRNA is denoted tRNA(m^{5}U54)methyltransferase and likewise for other enzymes.

target nucleotide (31, 32). Other results suggest that the tertiary structure of the tRNA is also part of the recognition signal for tRNA-modifying enzymes (33–38). By injecting chimeric tRNAs into *Xenopus laevis* oocytes, the formation of Q34 in yeast tRNA^{Asp} was strongly influenced by the presence of U33-G34-U35 sequence (39). Furthermore, the reaction of A34 to I34 requires a Purine-35 and does not allow a U36 (40), whereas the tRNA(Gm34)methyltransferase is not dependent on the neighboring nucleotides but on the structure beyond the anticodon region (41). The tRNA (Ψ55)synthetase seems to have a strict requirement of a U54 (42). Thus, the highly specific tRNA-modifying enzymes have different requirements to recognize the target nucleotide. For some the sequence surrounding the nucleotide to be modified is crucial, but for others the three-dimensional structure may be a more determining factor.

In general, enzymes catalyzing the same reaction in both the cytoplasm and the mitochondria are coded for by different nuclear genes (43). However, based on biochemical experiments, Smolar & Svensson (44) suggested that the same nuclear gene might encode the enzyme catalyzing the formation of m$_2^2$G26 in cytoplasmic and mitochondrial tRNA, and this was later supported by genetic evidence (45). This was also shown to be true for the formation of m^{5}U54 as well as i^{6}A37 in yeast (45, 46).

By injecting the yeast tRNA^{Tyr} gene into *X. laevis* oocytes, it was established that all modifications, except at positions 34 and 37, take place in the nucleus before splicing occurs (47, 48). Modifications in the TΨC-loop preceded the processing of the 5'-leader sequence and the formation of Ψ in the anticodon stem, and Ψ35 preceded the splicing reaction. However, the formation of Q34 is one of the last steps in the maturation process and the responsible enzyme might be cytoplasmic (39, 49). So far, no modification has been shown to be a prerequisite for the splicing event (45, 50). Thus, during the maturation process, size reduction and modification of the tRNA are intimately related processes occurring in concert.

III.2 Regulation and Genetics of tRNA-Modifying Enzymes in Bacteria and Yeast

It can be estimated that at least 30 different modified nucleosides exist in bacterial tRNA. As stated above, there are different enzymes catalyzing the formation of the same modified nucleoside at different positions in the tRNA. Some modified nucleosides, like ms^{2}io^{6}A and mnm^{5}s^{2}U, have a complicated structure, and more than one enzyme is involved in their synthesis. From such considerations it can be inferred that at least 45 different tRNA-modifying enzymes exist in a eubacterial cell, which would require at least 45 kb of DNA. Thus, a substantial part of the genetic information (about 1% of the total DNA content in *E. coli*) is devoted to tRNA modification. However, each gene seems to be expressed at a low level, and therefore the energetic load for the cell is not too extensive.

Until recently, little was known about the regulation of the synthesis of the tRNA-modifying enzymes, and even less about the different processing enzymes. However, the level of tRNA(m^5U54)methyltransferase in *E. coli* increases with increasing growth rate, while the activity of the tRNA(m^1G37)-, tRNA(mnm^5s^2U34)methyltransferases, and the tRNA(Ψ38, 39, 40) synthetase is invariant with the growth rate (51, 52). Furthermore, the expression of tRNA(m^5U54)methyltransferase is stringently regulated, while the other two tRNA methyltransferases are not (53). In fact, the expression of tRNA(m^5U54)methyltransferase is regulated like stable RNA under all physiological conditions so far analyzed. Although the enzymes are all involved in the modification of tRNA, they are not coordinately regulated. Only the tRNA(m^5U54)methyltransferase is regulated as bulk tRNA (51). Temperature as well as the composition of the growth medium can specifically influence the expression of genes (54). The formation of Gm in *Bacillus stearothermophilus* as well as thiolation of m^5s^2U54 in *Thermus thermophilus* is stimulated by high temperature (55, 56). However, in the case of the formation of Gm18, m^1A, and m^5s^2U54, the activity but not the amount of the enzymes increased upon shifting to higher temperature (57).

Figure 2 shows the map locations of the 11 genes identified in *E. coli/ Salmonella typhimurium* likely to be structural genes for tRNA-modifying enzymes. It is clear that the genes are not clustered. In most cases one gene governs the synthesis of one modified nucleoside. However, the formation of s^4U and mnm^5s^2U requires more than one gene. The synthesis of mnm^5s^2U34 is governed by the *asuE* (25.3 min), the *trmE* (83 min), *trmF* (83 min), and the *trmC* (50 min) genes. The polypeptide synthesized from the *trmC* gene has two enzymatic activities, and two different mutations, *trmC1* and *trmC2*, accumulate two different derivatives, S2 and S1, respectively, in the tRNA (58, 59). The sequential formation of mnm^5s^2U34 is suggested to be:

$$U34 \xrightarrow{asuE} s^2U34 \xrightarrow{trmE} S2 \xrightarrow{trmC1} S1 \xrightarrow{trmC2} mnm^5s^2U34 \ .$$

The formation of s^4U8 requires at least two genes, *nuvA* and *nuvC*, which code for two polypeptides constituting the tRNA(s^4U8)synthetase (60). Since only 11 of the 45 potential structural genes for tRNA-modifying enzymes have been identified, it is obvious that more mutants defective in tRNA modification must be isolated before a complete picture of the complex biosynthesis as well as the function of the modified nucleosides in tRNA will emerge.

The genetic organization of a few eubacterial genes coding for tRNA-modifying enzymes has been established. The *trmA* gene, which codes for the tRNA(m^5U54)methyltransferase, is monocistronic and is transcribed coun

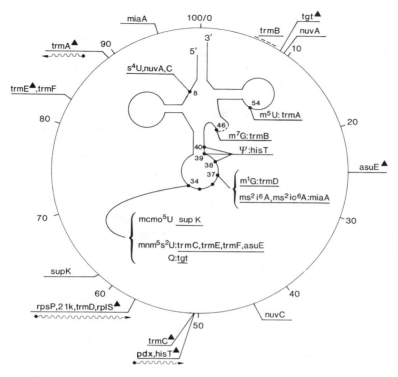

Figure 2 Location of potential structural genes for tRNA-modifying enzymes on the chromosomal map of *E. coli/S. typhimurium* (201). References to gene locations not found in Ref. 201: *miaA* (121, 202), *trmE* and *trmF* (144), *trmC* (203), *asuE* (145), *trmD* (204).

terclockwise on the *E. coli* chromosomal map (Figure 2; 61). The *trmA* promoter is homologous to the corresponding P1 promoter of rRNA genes, which may explain the similar regulatory behavior of the *trmA* gene and the rRNA genes (P. H. R. Lindström, unpublished results).

The *trmD* gene, which codes for the tRNA(m^{1}G37)methyltransferase, is part of a tetracistronic operon (Figure 2). Unexpectedly, the first gene and the last gene in the *trmD* operon code for ribosomal proteins S16(*rpsP*) and L19(*rplS*), respectively (62). These two ribosomal proteins are made in 100-fold larger amounts than the tRNA(m^{1}G)methyltransferase. Furthermore, the regulation of the *trmD* gene expression is not the same as that of the surrounding ribosomal protein genes. (P. M. Wikström, unpublished results). Under all physiological conditions tested, only one large transcript constituting the whole *trmD* operon was observed. (A. S. Byström, unpublished results). Thus, the mechanism(s) behind the differential expression as well as the noncoordinate regulation of expression of the genes within the operon operate at the posttranscriptional level.

The *hisT* gene, which codes for the tRNA(Ψ38,39,40)synthetase I, is downstream of a gene expressed 10–14-fold higher than the *hisT* gene. Thus it is, like the *trmD* gene, part of a differentially expressed multicistronic operon (63, 64). The upstream gene (*pdx*) is involved in the synthesis of vitamin B6 (65). Thus, a genetic link between pyridoxal phosphate biosynthesis and tRNA modification exists.

Eukaryotic tRNAs contain more modified nucleosides than tRNAs from eubacteria, and consequently more genes for tRNA-modifying enzymes must be present in e.g. yeast than in *E. coli*. In *Saccharomyces cerevisiae* the mutants *trm1* and *trm2* lack m_2^2G26 and m^5U54, respectively, in their tRNA (66, 45). The *sin1* gene in the fission yeast *Schizosaccharomyces pombe* and the *mod5-1* gene in *S. cerevisiae* govern the synthesis of i^6A37 (67, 68). The *mia* mutant of *S. cerevisiae* is deficient in dihydrouridine (69). Two anti-suppressors, *sin3* and *sin4*, of *S. pombe* were shown to be unlinked, but they influence the synthesis of the same nucleoside, mcm^5s^2U34 (70, 71). In a single mutant of *sin3* and *sin4* the level of mcm^5s^2U34 is reduced 6–13-fold, but in a double mutant the level of mcm^5s^2U34 was almost undetectable. However, since another unidentified modified nucleoside is also absent in the *sin3* and *sin4* mutants, it is possible that these genes are also involved in the synthesis of yet another modified nucleoside. Furthermore, another anti-suppressor mutant, *sin15*, reduces the efficiency of an opal suppressor, tRNA[Leu], and it is also deficient in ncm^5U34 (A. M. Grossenbacher, J. Kohli, University of Bern, and C. Gehrke, University of Missouri, personal communications). The yeast loci TRM1 and MOD5, which are involved in the synthesis of m_2^2G26 and i^6A37, respectively, have been cloned and sequenced (72, 73). Each is a structural gene for the corresponding enzyme, and as expected, each consists of a single cistron (74).

III.3 Metabolic and Developmental Aspects of the Synthesis of Modified Nucleosides

Changes in the metabolism of the bacterial cell that have no apparent relationship to any of the known direct precursors (methionine, cysteine, threonine, glycine, lysine, and carbonate) of modified nucleosides can still influence the synthesis of modified nucleosides, leading to undermodified tRNA in the cell. Starvation of leucine or arginine results in a severe unbalanced growth that induces specific undermodification of tRNA[Phe] and tRNA[Leu] (75). Unbalanced synthesis as such is probably not the reason for the appearance of undermodified tRNA; amino acid limitation under balanced growth also results in undermodified tRNA (76). A mutation in *ilvU* changes the concentrations of two tRNAs, isoaccepting tRNA[Val] and tRNA[Ile], and also influences the regulation of isoleucyl-tRNA ligase (77). The *ilvU*[+] product was suggested not only to allow derepression of isoleucyl-tRNA ligase but also to inhibit the modification of tRNAs specific for valine and isoleucine. Thus, changes in the intermediary metabolism of the cell may impose severe alterations in the modification of the tRNA, which thereby may change the level of expression of several operons (See below, Section IV).

Unexpectedly, three different kinds of auxotrophic mutants (Thi$^-$, Thy$^-$, Aro$^-$) are also defective in the modification of tRNA. Mutants of *E. coli* that require thiamine (Thi$^-$) for growth also lack s^{4}U8 in their tRNA (78). The mutation maps in the *nuvC* locus, and it was hypothesized that factor C of the tRNA(s^{4}U8)synthetase may also be a subunit of an unknown enzyme in the biosynthesis of thiamine. However, *nuvC* may be part of an operon that also encodes an enzyme involved in the biosynthesis of thiamine, and the *nuvC* mutation might affect another gene in the operon (cf *pdx*, *hisT* operon, and the *thyA* locus). A thymine-requiring mutant (Thy$^-$) suppresses nonsense as well as frameshift mutations, and it was suggested that some tetrahydrofolate-dependent methylation of tRNA may be influenced (79). However, recently the genetic organization of the *thyA* region has been established. An unidentified gene upstream of the *thyA* gene overlaps the latter gene, and therefore the suppressor phenotype of some Thy$^-$ mutations may be due to a mutation in the overlapping sequence for these two genes (80). If so, the coupling of tRNA modification and synthesis of thymidine may be a genetic rather than a metabolic link (cf *hisT*, *pdx* above). The three aromatic amino acids as well as four vitamins are synthesized by a common pathway in *E. coli* (Figure 3). Mutations in any of the *aroB*, *D*, *E*, *A*, and *C* genes involved in the synthesis of chorismic acid result in requirements for aromatic amino acids for growth (Aro$^-$), and give rise to a deficiency of cmo^{5}U34/mcmo^{5}U34 in the tRNA (81). The presence of shikimic acid in the growth medium for an *aroD* mutant restores the ability to synthesize cmo^{5}U34/mcmo^{5}U34 in the tRNA, while this capacity is lost in an *aroC* mutant. Mutations in any of the genes specific for the synthesis of tyrosine, phenylalanine, tryptophan, ubiquinone, folate, menaquinone, and enterochelin do not influence the synthesis of the modified nucleosides in tRNA (81; T. G. Hagervall, Y. H. Jönsson, G. R. Björk, unpublished results). Therefore chorismic acid itself or some unknown related metabolite thereof plays a key role in the formation of cmo^{5}U34/mcmo^{5}U34 in the tRNA.

One iron transport pathway in bacteria is mediated by enterochelin, an iron chelator that originates from chorismic acid (Figure 3). Bacteria growing under iron limitation produce an outer membrane receptor for the Fe^{3+}-enterochelin complex. Such cells contain i^{6}A37 instead of ms^{2}i^{6}A37 (82, 83). An identical effect on the tRNA modification occurs when cells of *E. coli* grow in body fluids where iron-binding host proteins are present (84). Such deficiency of the methyl thio group of ms^{2}i^{6}A stimulates the transport of the

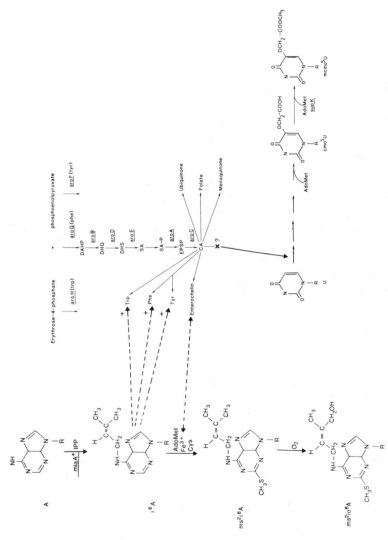

Figure 3 Links between the aromatic pathway and tRNA modifications in *E. coli/S. typhimurium*. Arrows to Phe, Tyr, and Trp denote the stimulation of transport of these amino acids in cells containing i⁶A37 instead of ms²i⁶A37 in the tRNA (85). The arrow between Fe^{3+} and enterochelin indicates the involvement of the latter in the transport of Fe^{3+}.

aromatic amino acids (85). The *miaA* mutant, which has an A37 instead of ms^2i^6A37, has an increased synthesis of enterochelin (86). This may indicate the presence of a regulatory circuit in which the transport of aromatic amino acids is stimulated to save more enterochelin to transport the limiting iron ions and at the same time stimulate the synthesis of the enterochelin. Thus, the synthesis and transport of the aromatic amino acids and enterochelin are intimately coupled to the formation of ms^2i^6A37 in *E. coli* and ms^2io^6A in *S. typhimurium*.

Facultative aerobic bacteria, like *S. typhimurium*, which are able to grow under both anaerobic and aerobic conditions, change their intermediary metabolism when shifted between these two conditions. The formation of one modified nucleoside, ms^2io^6A37, requires molecular oxygen for hydroxylation of the isopentenyl group to form ms^2io^6A37 from ms^2i^6A37. This reaction was suggested to be involved in the regulation of aerobiosis in *S. typhimurium* (86).

Bacillus subtilis is a differentiating eubacteria in which entry into the stationary growth phase is one way to start the sporulation process. In vegetative cells the formation of ms^2i^6A37 is not complete (87–90). During conditions that repress sporulation, the thiomethylation reaction is also inhibited, suggesting that this tRNA modification reaction may be coupled to the sporulation process (91). The pattern of $tRNA^{Lys}$ changes during the different growth phases, and a $tRNA^{Lys}$ species, deficient in ms^2t^6A37, is predominant in spores (92, 93). Thus, developmental changes as well as changes in the growth conditions strongly influence the modification of tRNA.

Queuosine-34 has been shown to influence the coding capacity of tRNA (see IV:2). Since it has been suggested that the level of tRNA modification may be a regulatory device (see V), changes in the level of Q34 are of particular interest. Such changes have been observed during the development of several organisms (94–99). Furthermore, tumor tissues completely lack Q34 in the tRNA (100). When *Dictyoistelium discodeum* is grown in a defined medium, the presence of Q34 in the tRNA is dependent on queuine in the growth medium (99). Although the presence or absence of queuine in the medium does not influence the growth rate in the vegetative growth phase, the presence of queuine—and thus the presence of Q34-containing tRNA—is a prerequisite for a normal development of the slime mold (101).

The cell cycle of mammalian cells comprises four successive phases. Following the M phase, the cells divide and enter the G_1 phase. In the G_1 phase the cells prepare themselves for the S phase, in which the cells synthesize DNA. One $tRNA^{Lys}$ isoaccepting species has been correlated to cell division (102–104). Purified growth factors stimulate specific tRNA-modifying enzymes involved in the synthesis of this $tRNA^{Lys}$ species (105,

106). A temperature-sensitive hamster cell line, which is blocked in the G_1 phase, shows an increase in the level of the undermodified tRNA^{Lys} species at nonpermissive temperature. The temperature-sensitive molecule in this mutant was suggested to be a tRNA-modifying enzyme involved in the modification of the anticodon (107). These results suggest that specific tRNA modification may be necessary for the commitment of cell division.

A relation between cell cycle and tRNA modification may also exist in yeast. The antisuppressor mutation *sin3*, which influences the synthesis of mcm^{5}s^{2}U34, also leads to increase of cell length (70). Furthermore, some allosuppressor strains, that also modulate the efficiency of nonsense suppressor tRNAs, have been shown to be allelic with the cell cycle control gene, *cdc25*, and also to possess an altered tRNA modification (108; C. Gehrke, University of Missouri, J. Kohli, University of Bern, personal communication).

IV. FUNCTION OF MODIFIED NUCLEOSIDES IN TRANSFER RNA

None of the modified nucleosides present in yeast tRNA^{Phe} appear to be essential to maintain the basic three-dimensional conformation (109). However, the presence of the modified nucleosides increases the surface area of the tRNA molecule by 20%, suggesting that the modified nucleosides are present to be recognized by various proteins/nucleic acids. Also, many participate in unusual hydrogen bonds, e.g. m^{1}A58-m^{5}U54 of tRNA^{Phe} (109), and the role of certain modified nucleosides is probably to bring about these interactions and thereby contribute to tRNA structure, especially where loops I, III, and IV interact. Since some modified nucleosides, like m^{7}G and m^{1}A, have a positive charge, this and not the methyl group per se is important for maintenance of the structure of tRNA as well as recognition sites for proteins (110). All tRNA modification mutants, except *supK* (111) and a mutant deficient in mnm^{5}s^{2}U34 (112), appear to be viable. Furthermore, some species of *Mycoplasma* have tRNA with a very low content of modified nucleosides (113). This suggests that most of the modified nucleosides in tRNA are nonessential for cell growth but still might have an important role in the fine tuning of the activity of the tRNA.

IV.1 Modified Nucleosides Next to the 3'-Side of the Anticodon (Position 37) Influence Translational Efficiency and Fidelity, Sensitivity of the tRNA to the Reading Context, and Reading Frame Maintenance

Position 37 is highly prone to be modified, and a correlation exists between the kind of modification present in this position and the coding capacity of the tRNA (Figure 4). Eukaryotic and eubacterial tRNAs reading UXX codons usually have a bulky hydrophobic modification at this position. However, such tRNAs from archaebacteria and *Mycoplasma* contain m^1G37, and therefore the suggestion that these bulky hydrophobic modifications are necessary to strengthen the weak A36-U interaction must be questioned. In fact, such tRNAs from eubacteria with an unmodified A37 are able to read codons starting with U (31, 32, 114, 115). However, when present in a tRNA, the ms^2 and ms^2i^6 modifications have been shown to stabilize anticodon-anticodon interaction (116, 117), and the ms^2 group to stabilize the secondary structure of a polynucleotide (118).

Transfer RNA^{Phe} from iron-starved *E. coli* contains i^6A37 instead of ms^2i^6A37 and is much less efficient in polyU-directed polyphenylalanine synthesis in vitro. However, no effect of the modification deficiency was observed using natural mRNA as template (119). Iron-limited media also derepress the tryptophan operon, presumably due to a lower ability to read the two contiguous tryptophan codons in the *trp* leader mRNA (120). The ms^2-deficient tRNA may therefore read contiguous codons, such as those present in polyU and in the attenuator region of the amino acid biosynthetic operon transcripts, less efficiently than codons not reiterated, as in genes coding for proteins. If so, the ms^2 modification is more important in some codon contexts than in others (see also below).

The *miaA* mutants of *E. coli/S. typhimurium* lack ms^2i^6A37/ms^2io^6A37 and

Codon [a] (nucleotide)			Modifications in position 37 in tRNA from:		
1st	2nd	3rd	Eucaryotes	Eubacteria	Archaebacteria
U	N	N	yW ; o^2yW ; i^6A ; m^1G	ms^2i^6A ; i^6A ; ms^2io^6A ; m^1G	m^1G
C	N	N	m^1G	m^1G ; m^2A	m^1G
A	N	N	t^6A ; mt^6A ; ms^2t^6A	t^6A ; mt^6A ; m^6A ; ms^2t^6A	t^6A
G	N	N	m^1G ; m^1I	m^2A ; m^6A	m^1G

[a] N= any of the four nucleotides

Figure 4 Presence of modified nucleosides at position 37 (3'-side of the anticodon) and the coding capacities of tRNAs from eubacteria, eukaryotes and archaebacteria.

have an unmodified A37 (120, 121). The *miaA1* mutant of *S. typhimurium* has a large (up to 50%) reduction in growth rate, and a reduced polypeptide chain elongation rate in vivo (121). Furthermore, regulation of several biosynthetic operons is affected (121). Lack of ms^2i^6A37/ms^2io^6A37 reduces the efficiency of nonsense suppression in vivo, of tRNA binding to the ribosome, and of UGA-suppression in vitro (116, 122–124). However, this modification is not essential for the activity of the tRNA, because the *miaA* mutant is viable (121, 125), and several functional tRNAs exist that have an unmodified A37 (31, 32). Thus, the extent of the effect of ms^2i^6A37 might vary with the tRNA species of which it is part; this has also been observed experimentally (124). Suppressor tRNAs lacking ms^2i^6A37/ms^2io^6A37 are more sensitive to the sequence surrounding the codon than the wild-type suppressor tRNA, suggesting that $ms^2i(o)^6$ modification is involved in determining the intrinsic codon context sensitivity of the tRNA (124). However, this modified nucleoside does not seem to be involved in the amino-acid-charging reaction (125, 126). The translational error level is reduced both in vivo and in vitro for ms^2i^6-deficient tRNA (116, 123, 124, 127). Since an improved accuracy requires energy (for review see Ref. 128), the decreased cellular yield and peptidyl release (129) may be reasonable. Combination of the *miaA* mutation and ribosomal mutations, which also increase the fidelity, reduces the growth rate, and in some cases the cell becomes streptomycin dependent, since this drug is known to suppress the proofreading flows (121, 123, 127). Thus, the $ms^2i(o)^6$ modification plays an important role in the efficiency and fidelity of translation as well as in the sensitivity of the tRNA to the sequence surrounding the codon.

The i^6A37 is present in some tRNAs from yeast that read UXX codons. Antisuppressor mutations of yeast, *sin1* and *mod5-1*, both contain an unmodified A37 instead of i^6A37 (67, 68). Lack of the i^6 modification reduces the activity of the serine-inserting UGA suppressor as well as the tyrosine-inserting suppressor. However, both mutants grow relatively well. Thus, like the $ms^2i(o)^6$-modification in eubacterial tRNA, the i^6 modification in tRNA from yeast is involved in the anticodon-codon interaction, suggesting that part of the effect observed in ms^2io^6A-deficient tRNA is due to the lack of the i^6 modification.

Removal of yW37 of yeast tRNA^{Phe} lowers its ability to bind to ribosomes and to form polyphenylalanine in a polyU-programmed translation system (130). Furthermore, tRNA^{Phe} deficient in yW37 shows variable efficiency at different sites in globin mRNA (131). This modified nucleoside has been suggested to intercalate between the two codon-anticodon triplet duplexes present in the A- and P-site on the ribosome (132). The model suggests that both in the A- and P-site the yW base is in contact with the mRNA and stabilizes the codon-anticodon interaction by stacking.

tRNA from all three kingdoms reading AXX codons usually contains a t^6A

derivative in position 37 (Figure 4). This conserved feature implies a common function in these groups of tRNA, such as the suggested strengthening of the weak binding of U36 to the A in the first position of the codon. Accordingly, *E. coli* tRNA^{Ile} and yeast tRNA$^{Arg}_{III}$, deficient in t^{6}A37, have reduced abilities to decode in vitro (133, 134). Eukaryotic tRNA^{Lys} deficient in t^{6}A was less efficient compared to fully modified tRNA^{Lys} in translating AAG codons in globin mRNA in vitro, and site-specific effects were observed for the hypo-modified tRNA^{Lys} (131). Measurements of anticodon-anticodon association showed that t^{6}A37 stabilizes U-A base pairs adjacent to the 5'-side of the modified nucleoside, most probably by an increased stacking interaction (134). tRNAs reading AXX codons contain a U36-t^{6}A37 sequence. Experiments using Upt^{6}A showed that t^{6}A stabilizes the stacking interaction compared with UpA, and also prevents wobble on the 3'-side of the anticodon (135).

Figure 4 shows that tRNAs that read codons starting with C or G usually have an unmodified purine nucleoside or a simple modification like m^{1}G, m^{2}A, etc, present in position 37. The conserved presence of m^{1}G37 in tRNAs that read codons starting with C is noteworthy. The energetically more stable GC-pairs compared to AU-pairs should have a lower requirement of stabilization of the anticodon-codon interaction. However, some of them, like m^{1}G and m^{6}A, destabilize the Watson-Crick double-helical structure by 1.0–1.8 Kcal/mol of methyl substituent (136). They may also prevent the formation of hydrogen bonds to nucleotides on the 5'-side of the codon as has been suggested from theoretical considerations (137). The presence of m^{1}G37 is not essential for the function of a tRNA, since a few tRNAs exist that have an unmodified G37, and chimeric tRNAs with an unmodified G37 are functional in vitro (115, 138). However, a mutant (*trmD3*) of *S. typhimurium,* in which m^{1}G37 is absent at 41°C but not at 30°C, has a dramatic reduction in growth rate at 41°C. Simultaneously, the mutant acquired a capacity to frameshift at runs of C at 41°C (P. M. Wikström, A. S. Byström, and G. R. Björk, unpublished observation). Probably the presence of a methyl group in position 1 of G influences the precision of the anticodon-codon interaction by preventing an interaction with the 5' neighboring nucleotide in the mRNA.

The modification at the 3'-side of the anticodon influences cell physiology in a profound way. More precisely, this modification influences not only the efficiency and the accuracy of the tRNA in its decoding function, but also the tRNA's sensitivity to the nucleotides surrounding the codon, and it may be involved in reading frame maintenance.

IV.2 Modified Nucleosides at the Wobble Position (Position 34) Influence Translational Fidelity and Codon Choice

The wobble position (position 34) is, like position 37, often modified. A correlation exists between the kind of modified nucleoside present and the coding capacity of the tRNA (1). The s^2U derivatives with different 5 substitutions (eubacteria and eukaryotes) are present at position 34 in tRNAs that read codons of the type NAA/G, where N can be C, A, or G. In triplet-dependent binding to the ribosome and in protein synthesis in vitro, s^2U derivatives recognize primarily A and much less efficiently G as the third letter of the codon (139, 140). Proton NMR analyses have shown that these derivatives almost exclusively have a conformation that allows the recognition of A but not of U and G (141). In such model experiments both the 5-substitution and the thio carbonyl group stabilize the conformation necessary for the proposed coding abilities. However, the presence of the thio carbonyl group was shown not to be involved in the preferential recognition of A but has been implicated to increase the stacking interaction within the anticodon (142–144).

Two mutants, *trmC1* and *trmC2*, of *E. coli* are both defective in the synthesis of mnm^5s^2U34, which is normally present in $tRNA^{Lys}_{UAA}$ (*supG*). The efficiency of $tRNA^{Lys}_{UAA}$ in reading both UAA and UAG codons is reduced in the *trmC* mutants, and the undermodified $tRNA^{Lys}_{UAA}$ is also more sensitive to the codon context than the normal $tRNA^{Lys}_{UAA}$ (58). The *trmC1* and *trmC2* mutants do contain the s^2 group, but also some unidentified substitution most likely at the 5 position. Mutants (*trmE, trmF*) having s^2U34 instead of mnm^5s^2U34 in their tRNA are more able to read UAA than UAG codons. Such undermodified tRNA also binds AAA better than AAG in vitro (145). A mutant, *asuE*, probably deficient in the thiolation of mnm^5s^2U34, has been isolated as an antisuppressor to *supL*, ($tRNA^{Lys}_{UAA}$) (146). Thus, alterations of the side chain at the 5 position as well as the thio group seem to be important for the decoding capacity of tRNAs possessing mnm^5s^2U34. These mutants are all viable, even though they all contain an undermodified derivative of mnm^5s^2U34. However, they all possess either the thio group or some kind of side chain. A temperature-sensitive mutant of *E. coli* has been isolated that is deficient in mnm^5s^2U34 (112). The chemical structure of the undermodified derivative present in this mutant has not been determined. If the temperature sensitivity of the mutant is due to the lack of mnm^5s^2U34, the reason might be that this mutant misread, because the mnm^5s^2 modification was suggested to restrict the ability to bind to U (141). The *trmE* and *trmC* mutants, which still have either a thio group or a side chain, are still restricted in this misreading capacity, thus they are still viable. However, they read codons ending either A or G with a reduced efficiency.

Yeast tRNAs that read split codon families have mcm^5s^2U34 or nm^5U34, and it has been suggested that the presence of these nucleosides explains why yeast ochre suppressors do not read UAG and why the nm^5U34-containing $tRNA^{Ser}_{UCA}$ cannot read UCG (147–149). The antisuppressors *sin3* and *sin4* of *Schizosaccharomyces pombe* inactivate serine-inserting ochre suppressor, and the double mutant has almost no mcm^{5}s^{2}U34 but an increased level of s^{2}U34 (70, 71). Independent of their effect on tRNA suppressors, the two mutations reduce the growth rate. In vivo decoding of the serine codon UCG by the UCA-reading serine tRNA is not promoted in the presence of mutations *sin3* and *sin4* (71). These results also support the restrictive function of the s^{2}U derivatives as suggested from early experiments in vitro as well as from model building, NMR analysis, and crystal structures of such modified nucleosides.

Another group of modified uridines (mo^{5}U, cmo^{5}U, mcmo^{5}U) has been found in position 34 of tRNAs specific for valine (GUN), serine (UCN), proline (CCN), threonine (ACN), and alanine (GCN). In triplet-dependent binding (150–154) and in vitro synthesis of MS2 coat protein (155, 156), these modified uridines can read not only codons ending with A or G, but also codons ending with U. From NMR analysis it was concluded that the -OCH$_2$- of the 5-substituent interacts with the 5'-phosphate to bring about the flexibility of the wobble nucleoside so as to recognize U, A, and G (141). These modified nucleosides are present in tRNAs that read families of four codons specifying the same amino acid. All these codon families are also read by at least one other tRNA species, which reads codons ending with U or C according to the wobble hypothesis. Furthermore, tRNA from mitochondria with an unmodified U in the wobble position and the only existing tRNA^{Gly}, tRNA^{Ala}, tRNA^{Pro}, and tRNA^{Val} from *Mycoplasma mycoides* are able to read all four codons in such a family (157, 158; T. Samuelsson, Y. S. Guindy, F. Lustig, T. Borén, U. Lagerkvist, Göteborg, Sweden, personal communication). Thus, it is not obvious why these modifications are present. However, an Aro$^-$ mutant of *E. coli* that lacks cmo^{5}U34/mcmo^{5}U34 in tRNA grows 20% slower than Aro$^+$ cells in rich medium, indicating that the presence of cmo^{5}U under some physiological conditions is important (G. R. Björk, unpublished results). This wobble base is within 4 Å of a pyrimidine in 16S rRNA when the tRNA is in the P-site on the ribosome (159). This may extend the anticodon stack into the 16S rRNA and stabilize the tRNA-ribosome interaction. The importance of the modification as such in this interaction has not been elucidated.

Recessive UGA suppressors (*supK*), which also suppress some frameshift mutations, have been isolated (111, 160). The *supK* mutants are deficient in a tRNA methyltransferase, which most likely catalyzes the formation of mcmo^{5}U in some tRNA species (161) (see Figure 3). These results suggest that a tRNA having cmo^{5}U34 instead of mcmo^{5}U34 is able to read UGA codons or to frameshift. However, the suppressing agent has so far not been identified as a tRNA species, and the mechanism behind this suppression is still unknown.

Elongator tRNA$_m^{Met}$ of *E. coli* contains ac^{4}C in position 34, while initiator tRNA$_f^{Met}$ contains a C. By chemically removing the ac^4 modification, it was shown that the tRNA$_m^{Met}$ lacking ac^{4}C34 binds to AUG-programmed ribosome almost twice as well as the tRNA$_m^{Met}$ containing ac^{4}C34 (162a). However, the presence of ac^{4}C decreases the misreading in vitro of AUA(Ile). Thus, the function of ac^4 modification appears to be primarily to reduce the misreading of the AUA(Ile) codon. This is achieved by a somewhat reduced efficiency to read the AUG(Met) codon.

In *S. cerevisiae* a leucine-inserting amber suppressor tRNA normally contains m^{5}C in the wobble position. By deleting the intron, the tRNA-(m^{5}C34)methyltransferase does not recognize the almost mature tRNA as substrate, which results in a tRNA$_{UAG}^{Leu}$ that has an unmodified C in the wobble position. Such undermodified tRNA is less efficient in suppression, which suggests that the presence of m^{5}C34 improves translation efficiency (162b).

The hypermodified nucleoside Q is present in position 34 of tRNAs specific for tyrosine, histidine, asparagine, and aspartic acid, which read codons ending with U or C. A mutant, (*tgt*), of *E. coli* that lacks Q34 dies in stationary phase and is unable to synthesize nitrate reductase under anaerobic conditions (163, 164). The Q modification was suggested to play a role in the expression of nitrate reductase (164). However, it was not ruled out that the mutant used harbors additional mutations, which may explain the observed pleiotropic effects. The presence of Q affects the codon choice, because tRNA^{His} containing G34 instead of Q34 prefers CAC to CAU, while low preference was observed with fully modified tRNA^{His} (165). tRNA^{Tyr} from *Drosophila melanogaster* or tobacco plants and containing G34 instead of Q34 is able to read UAG stop codons (166, 167), and it has been suggested that partial suppression of termination codons might be a regulatory device (168, 169). Thus, the modification of G to Q may be involved in regulation of gene expression (see Section III.3).

IV.3 Modified Nucleosides in the Anticodon Region other than Positions 34 and 37 Influence Translational Efficiency and Fidelity

A mutation in the *hisT* gene of *S. typhimurium* results in a Ψ deficiency in positions 38, 39, and 40 in many tRNA chains, among them tRNA^{His} (26, 170, 171). The growth rate, the polypeptide chain elongation rate, and the error level are all reduced in a *hisT* mutant (172, 173). Such mutants were isolated by their ability to derepress the histidine operon (174). The histidine leader mRNA contains seven histidine codons in a row, which are read inefficiently by the undermodified tRNA^{His}. This leads to derepression of the histidine operon (175). Since the anticodon stem of several tRNAs normally contain Ψ, a *hisT* mutation has a pleiotropic effect and accordingly influences the regulation of several amino acid biosynthetic operons, probably through an attenuation mechanism (171). Lack of Ψ in the anticodon stem also reduces the efficiency of suppression by *supE* (tRNA$_{UAG}^{Gln}$) and *supF* (tRNA$_{UAG}^{Tyr}$), but has no or only a minor effect in sensing the sequence surrounding the codon (176; G. R. Björk, unpublished results).

Deletion of the intervening sequence of a yeast tRNA$_{UAA}^{Tyr}$ ochre suppressor gene results in an inability of the modifying enzyme normally catalyzing the formation of Ψ in the middle of the anticodon (position 35) to perform its function. The resulting ochre suppressor tRNA$_{UAA}^{Tyr}$ has an unmodified U35 instead of Ψ35, and such tRNA has a much lower efficiency in suppression (177). These results suggest that Ψ35 is important in anticodon-codon interaction.

Upon chemical modification of s^{2}C32 of tRNA^{Arg}, it has been shown that the structure of the anticodon loop is affected. Such tRNA^{Arg} also suppresses the frameshifting that normally occurs in the translation of MS2 RNA in vitro (178).

IV.4 Modified Nucleosides Outside the Anticodon Region May Stabilize tRNA Conformation

Lack of m$_2^2$G26 in yeast tRNA^{Ser} reduces the in vitro charging ability by 20%, suggesting that some tRNA^{Ser} species might not be chargeable (179). Transfer RNA, deficient in dihydrouridine, supports polyphenylalanine synthesis in vitro with the same efficiency as normal tRNA (69). Chemical derivatization of acp^{3}U47 of tRNA^{Phe} does not influence the activity of the tRNA^{Phe} in the aminoacylation or polyphenylalanine synthesis in vitro (180). Chemical reduction of m^{7}G47 disrupts the C13-G22-m^{7}G47 base triple, which leads to a slightly less ordered tRNA structure (181). A mutant of *E. coli, trmB*, defective in the formation of m^{7}G47, grows slower than a *trmB$^+$* cell, supporting the idea that m^{7}G47 stabilizes the structure of the tRNA (182). Methylation in vitro producing m^{7}G47 or m^{2}G10 results in altered kinetics of the aminoacylation (183, 184). A mutant lacking s^{4}U8 in its tRNA shows identical growth characteristics as the wild-type cells (185, 186). Thus, modification in parts other than the anticodon region may be involved in the stabilization of the tRNA structure. Therefore more specific assays are necessary to reveal the precise function(s) of these modified nucleosides.

Ribothymidine (m^{5}U54, rT54) is one of the most abundant modified nucleosides, and because it is present in tRNA from both eukaryotes and eubacteria, its presence was thought to be essential for growth. However, a mutant (*trmA5*) of *E. coli*, which completely lacks m^{5}U54, is viable but is outgrown by a *trmA$^+$* cell in a mixed population experiment (187). The difference in the growth rates of the wild-type and the *trmA5* mutant is 4% (G. R. Björk, unpublished results). A yeast mutant (*trm2*) lacking m^{5}U54 grows normally (45). Furthermore, some bacterial species normally lack m^{5}U54 in their tRNA (188, 189). Deficiency of m^{5}U54 facilitates initiation of protein synthesis in *Streptococcus faecalis* with unformylated tRNA$_f^{Met}$, and a similar mechanism may operate in *E. coli* (17, 190–192). All these results show that m^5U54 is not essential for cell growth but may slightly enhance the activity of tRNA. Some eukaryotic tRNAs normally have U54 and are able to accept methyl group in vitro, which results in an m^5U54-containing tRNA. Using pairs of such tRNAs, only differing in the absence or presence of m^5U54, it was shown that depending on the species of tRNA used, the activity of the tRNA was increased or decreased when m^5U54 was present, suggesting that the degree of m^5U54 in some eukaryotic tRNA may regulate protein synthesis (193, 194). The content of m^5U54 seems to influence the elongation factor–directed A-site binding and the misincorporation of leucine in a polyU-directed polyphenylalanine-synthesizing system (195). Lack of m^5U54 augmented the intrinsic misreading capacity of $tRNA_4^{Leu}$, and does not induce other $tRNA^{Leu}$ species to misincorporate leucine (17). The stability of the tertiary structure of $tRNA^{Met}$ is increased by the presence of m^5U54, and it is further increased if the tRNA contains m^5s^2U54 (196). Therefore the presence of m^5U54 and m^5s^2U54 stabilizes the tRNA structure. The functional differences observed in vitro may explain the small but significant growth rate differences of the *E. coli trmA* mutant observed in vivo.

V. REGULATORY RÔLE OF THE SYNTHESIS OF MODIFIED NUCLEOSIDES IN tRNA

It has been suggested that the degree of modification may be a regulatory device (17, 58, 131, 171). The ratio of modified to unmodified tRNA may regulate the expression of specific operon(s) or gene(s) by the differential efficiency to decode either a leader mRNA in an attenuator-controlled operon or a structural gene or both. Since the tRNA is also decoding genes other than those to be regulated, an element of specificity must exist. Either the impact of the modified nucleoside in question is more important for the function of a certain specific tRNA molecule, or the modified nucleoside is involved in sensing the nucleosides close to the codon. Both these requirements have been shown to exist (31, 32, 124, 131). Thus, certain specific codon context sequences may have been evolved, e.g. in a leader mRNA sequence, making the function of the decoding tRNA extremely sensitive to the degree of modification. Since the formation of modified nucleosides so far analyzed is an irreversible reaction, such a regulatory device is likely to set a certain degree of expression rather than to operate quickly. Although lack of Ψ in the anticodon region of $tRNA^{His}$ leads to derepression of the histidine operon, it is difficult to see a metabolic connection between the synthesis of Ψ and histidine. The connection between ms^2i^6A37 and tryptophan synthesis is likewise difficult to reconcile (125). However, lack of iron results in i^6A37 instead of ms^2i^6A, which also increases the transport of aromatic amino acid (85). This may save more of chorismic acid to the synthesis of enterochelin (cf Figure 3), which explains why the degree of ms^2 modification may in fact be a regulatory device of iron metabolism.

Ames and collaborators (197) have pointed out that the histidine leader mRNA is in part homologous to $tRNA^{His}$. They suggested that some proteins, e.g. the tRNA-modifying enzymes, besides being involved in the biosynthesis of tRNA, may also influence the equilibrium between the different stem and loops that can form in the leader mRNA. Furthermore, tRNA-modifying enzymes have also been suggested to regulate the maturation of mRNA in some eukaryotic tissues (198). If so, some tRNA-modifying enzymes have two functions—one catalytic and one regulatory. However, no direct evidence for such a dual role has so far been presented. In this context one can also ask why some tRNA-modifying enzymes, e.g. the *trmD* enzyme and the *hisT* enzyme, are part of multicistronic operons (199).

The hypothesis that tRNA modification or tRNA-modifying enzymes are part of a regulatory device is attractive. Although parts of the requirements of the hypothesis have been fulfilled, so far no biosynthetic pathway, developmental changes, or cell cycle events have been directly shown to be regulated according to the hypothesis. Therefore, its generality awaits more knowledge about the function of the modified nucleosides, genetic organization of a regulated operon, and how intermediary metabolism in bacteria and development/cell cycle events in eucaryotes are linked to tRNA modification. Knowledge in these areas would allow us to evaluate the role of tRNA modification or tRNA-modifying enzymes in regulation of the metabolism of the cell.

ACKNOWLEDGMENTS

This work was supported by the Swedish Cancer Society (Proj. No. 680), Swedish National Science Foundation (BU-2930), and the Swedish Board for Technical Development (Grant No. 4206). Y. H. J. was supported by Grant DT 1459 from the Swedish National Science Foundation. The critical reading of the manuscript by Drs. K. Kjellin-Stråby and S. Normark, both of Umeå, Sweden, and P. Piper, of London, England, is gratefully acknowledged.

Literature Cited

1. Nishimura, S. 1979. In *Transfer RNA: Structure, Properties and Recognition,* ed. P. R. Schimmel, D. Söll, J. N. Abelson, pp. 59–79. New York: Cold Spring Harbor Lab.
2. Hankins, W. D., Farkas, W. R. 1970. *Biochem. Biophys. Acta* 213:77–89
3. Elliott, M. S., Trewyn, R. W. 1984. *J. Biol. Chem.* 259:2407–10
4. Björk, G. R. 1986. *Chem. Scr.* 26B:91–95
5. Nass, G., ed. 1983. *Modified Nucleosides and Cancer.* Berlin: Springer-Verlag
6. Borek, E., Srinivasan, P. R. 1966. *Ann. Rev. Biochem.* 35:275–98
7. Starr, J. L., Sells, B. H. 1969. *Physiol. Rev.* 49:623–69
8. Hall, R. H. 1971. *The Modified Nucleosides in Nucleic Acids.* New York: Columbia Univ. Press
9. Söll, D. 1973. *Science* 173:293–99
10. Nau, F. 1976. *Biochemie* 58:629–45
11. Agris, P. F., Söll, D. 1977. In *Nucleic Acid-Protein Recognition,* ed. H. Vogel, pp. 321–44. New York: Academic

12. Feldman, M. Ya. 1977. *Prog. Biophys. Mol. Biol.* 32:83–102
13a. Grosjean, H., Chantrenne, H. 1980. In *Chemical Recognition in Biology*, ed. F. Chapeville, A. C. Haenne, Berlin: Springer-Verlag
13b. Grosjean, H., Chantrenne, H. 1980. *Mol. Biol., Biochem. Biophys.* 32:347–67
14. Nishimura, S. 1983. *Prog. Nucleic Acid Res. Mol. Biol.* 28:49–73
15. Dirheimer, G. 1983. *Recent Results Cancer Res.* 84:15–46
16. Björk, G. R. 1984. In *Processing of RNA*, ed. D. Apirion, pp. 231–330. Boca Raton, Fla: CRC
17. Kersten, H. 1984. *Prog. Nucleic Acid Res. Mol. Biol.* 31:59–114
18. Pang, H., Ihara, M., Kuchino, Y., Nishimura, S., Gupta, R., et al. 1982. *J. Biol. Chem.* 257:3589–92
19. Wittwer, A. J. 1983. *J. Biol. Chem.* 258:8637–41
20. Ching, W.-M., Wittwer, A. J., Tsai, L., Stadtman, T. C. 1984. *Proc. Natl. Acad. Sci. USA* 81:57–60
21. Ching, W.-M. 1984. *Proc. Natl. Acad. Sci. USA* 81:3010–13
22. Ching, W.-M., Alzner-DeWeerd, B., Stadtman, T. C. 1985. *Proc. Natl. Acad. Sci. USA* 82:347–50
23. Björk, G. R., Isaksson, L. A. 1970. *J. Mol. Biol.* 51:83–100
24. Björk, G. R., Kjellin-Stråby, K. 1978. *J. Bacteriol.* 133:508–17
25. Keith, J. M., Winters, E. M., Moss, B. 1980. *J. Biol. Chem.* 255:4636–44
26. Singer, C. E., Smith, G. R., Cortese, R., Ames, B. N. 1972. *Nature New Biol.* 238:72–74
27. Green, C. J., Kammen, H. O., Penhoet, E. E. 1982. *J. Biol. Chem.* 257:3045–52
28. Kraus, J., Staehelin, M. 1974. *Nucleic Acids Res.* 1:1479–96
29. Smolar, N., Hellman, U., Svensson, I. 1975. *Nucleic Acids Res.* 2:993–1004
30. Tsang, T. H., Buck, M., Ames, B. N. 1983. *Biochem. Biophys. Acta* 741:180–96
31. Grosjean, H., Nicoghosian, K., Haumont, E., Söll, D., Cedergren, R. 1985. *Nucleic Acids Res.* 13:5697–706
32. Murgola, E. J., Pagel, F. T., Hijazi, K. A. 1984. *J. Mol. Biol.* 175:19–27
33. Kuchino, Y., Seno, T., Nishimura, S. 1971. *Biochem. Biophys. Res. Commun.* 43:476–83
34. Shershneva, L. P., Venkstern, T. V., Bayev, A. A. 1971. *FEBS Lett.* 14:297–300
35. Shershneva, L. P., Venkstern, T. V.,

Bayev, A. A. 1973. *Biochim. Biophys. Acta* 294:250–62
36. Gambaryan, A. S., Morozov, I. A., Venkstern, T. V., Bayev, A. A. 1979. *Nucleic Acids Res.* 6:1001–11
37. Prather, N. E., Murgola, E. J., Mims, B. H. 1981. *Nucleic Acids Res.* 9:6421–28
38. Matsumoto, T., Watanabe, K., Ohta, T. 1984. *Nucleic Acids Res. Symp. Ser.* 15:131–34
39. Carbon, P., Haumont, E., Fournier, M., de Henau, S., Grosjean, H. 1983. *EMBO J.* 7:1093–97
40. Haumont, E., Fournier, M., de Henau, S., Grosjean, H. 1984. *Nucleic Acids Res.* 12:2705–15
41. Droogmans, L., Haumont, E., de Henau, S., Grosjean, H. 1986. *EMBO J.* 5:1105–9
42. Drabkin, H. J., RajBhandary, U. L. 1985. *J. Biol. Chem.* 260:5580–87
43. Doonan, S., Barra, D., Bossa, F. 1984. *Inf. J. Biochem.* 16:1193–99
44. Smolar, N., Svensson, I., 1974. *Nucleic Acids Res.* 1:707–18
45. Hopper, A. K., Furukawa, A. H., Pham, H. D., Martin, N. C. 1982. *Cell* 28:543–50
46. Martin, N. C., Hopper, A. K. 1982. *J. Biol. Chem.* 257:10562–65
47. Melton, D. A., de Robertis, E. M., Cortese, R. 1980. *Nature* 284:143–48
48. Nishikura, K., de Robertis, E. M. 1981. *J. Mol. Biol.* 145:405–20
49. Nishikura, K., Kurjan, J., Hall, B. D., de Robertis, E. M. 1982. *EMBO J.* 1:263–68
50. Kohli, J. 1983. In *The Modified Nucleosides of Transfer RNA*, ed. P. F. Agris, R. A. Kopper, 2:1–10. New York: Liss
51. Ny, T., Björk, G. R. 1980. *J. Bacteriol.* 141:67–73
52. Arena, F., Ciliberto, G., Ciampi, S., Cortese, R. 1978. *Nucleic Acids Res.* 5:4523–36
53. Ny, T., Thomale, J., Hjalmarsson, K., Nass, G., Björk, G. R. 1980. *Biochim. Biophys. Acta* 607:277–84
54. Lemaux, P. G., Herendeen, S. L., Bloch, P. L., Neidhardt, F. C. 1978. *Cell* 13:427–34
55. Agris, P. F., Koh, H., Söll, D. 1973. *Arch. Biochem. Biophys.* 154:277–82
56. Watanabe, K., Shinma, M., Oshima, T., Nishimura, S. 1976. *Biochem. Biophys. Res. Commun.* 72:1137–44
57. Kumagai, I., Watanabe, K., Oshima, T. 1980. *Proc. Natl. Acad. Sci. USA* 77:1922–26
58. Hagervall, T. G., Björk, G. R. 1984. *Mol. Gen. Genet.* 196:194–200
59. Hagervall, T. G. 1984. *Biosynthesis and*

function of mnm⁵s²U in the wobble position of Escherichia coli tRNA. PhD thesis. Umeå Univ., Umeå, Sweden

60. Lipset, M. N. 1978. *J. Bacteriol.* 135:993–97
61. Lindström, P. H. R., Stüber, D., Björk, G. R. 1985. *J. Bacteriol.* 164:1117–23
62. Byström, A. S., Hjalmarsson, K. J., Wikström, P. M., Björk, G. R. 1983. *EMBO J.* 2:899–905
63. Marvel, C. C., Arps, P. J., Rubin, B. C., Kammen, H. O., Penhoet, E. E., Winkler, M. E. 1985. *J. Bacteriol.* 161:60–71
64. Arps, P. J., Marvel, C. C., Rubin, B. C., Tolan, D. A., Penhoet, E. E., Winkler, M. E. 1985. *Nucleic Acids Res.* 13:5297–315
65. Winkler, M. E., Arps, P. J. 1986. *ASM Ann. Meet. Abstr. H53*, p. 136
66. Phillips, J. H., Kjellin-Stråby, K. 1967. *J. Mol. Biol.* 26:509–18
67. Laten, H., Gorman, J., Bock, R. M. 1978. *Nucleic Acids Res.* 5:4329–42
68. Janner, F., Vögeli, G., Fluri, R. 1980. *J. Mol. Biol.* 139:207–19
69. Lo, R. Y. C., Bell, J. B., Roy, K. L. 1982. *Nucleic Acids Res.* 10:889–902
70. Heyer, W.-D., Thuriaux, P., Kohli, J., Ebert, P., Kersten, H., et al. 1984. *J. Biol. Chem.* 259:2856–62
71. Grossenbacher, A.-M., Stadelmann, B., Heyer, W.-D., Thuriaux, P., Kohli, J. 1986. Submitted
72. Ellis, S. R., Morales, M. J., Li, J.-M., Hopper, A. K., Martin, N. C. 1986. Submitted
73. Najarian, D., Ellis, S., Dihanich, M., Morales, M., Hopper, A., Martin, N. 1986. In *Yeast Cell Biology,* ed. A. Liss, J. Hicks. In press
74. Martin, N. C., Clark, R., Dihanich, M., Ellis, S. R., Li, J.-M., et al. 1985. *11th Int. tRNA Workshop.* Banz
75. Kitchingman, G. R., Fournier, M. J. 1977. *Biochemistry* 16:2213–20
76. Thomale, J., Nass, G. 1978. *Eur. J. Biochem.* 85:407–18
77. Fayerman, J. T., Vann, M. C., Williams, C. S., Umbarger, H. E. 1979. *J. Biol. Chem.* 254:9429–40
78. Ryals, J., Hsu, R.-Y., Lipset, M. N., Bremer, H. 1982. *J. Bacteriol.* 151:899–904
79. Herrington, M. B., Kohli, A., Lapchak, P. H. 1984. *J. Bacteriol.* 157:126–29
80. Belfort, M., Pedersen-Lane, J. 1984. *J. Bacteriol.* 160:371–78
81. Björk, G. R. 1980. *J. Mol. Biol.* 140:391–410
82. Wettstein, F. O., Stent, G. S. 1968. *J. Mol. Biol.* 38:25–40

83. Rosenberg, A. H., Gefter, M. L. 1969. *J. Mol. Biol.* 46:581–84
84. Griffiths, E., Humphreys, J. 1978. *Eur. J. Biochem.* 82:503–13
85. Buck, M., Griffiths, E. 1981. *Nucleic Acids Res.* 9:401–14
86. Buck, M., Ames, B. N. 1984. *Cell* 36:523–31
87. Keith, G., Rogg, H., Dirheimer, G. 1976. *FEBS Lett.* 61:120–23
88. Arnold, H. H., Raettig, R., Keith, G. 1977. *FEBS Lett.* 73:210–14
89. Arnold, H. H., Keith, G. 1977. *Nucleic Acids Res.* 4:2821–29
90. Vold, B. S. 1978. *J. Bacteriol.* 135:124–32
91. Buu, A., Menichi, B., Heyman, T. 1981. *J. Bacteriol.* 146:819–22
92. Smith, D. W. E., McNamara, A. L., Vold, B. S. 1982. *Nucleic Acids Res.* 10:3117–23
93. Vold, B. S., Keith, D. E. Jr., Buck, M., McCloskey, J. A., Pang, H. 1982. *Nucleic Acids Res.* 10:3125–32
94. White, B. N., Tener, G. M., Holden, J., Suzuki, D. T. 1973. *J. Mol. Biol.* 74:635–51
95. Hosbach, H. A., Kubli, E. 1979. *Mech. Ageing Dev.* 10:141–49
96. Lin, F.-K., Furr, T. D., Chang, S. H., Horwitz, J., Agris, P. F., Ortwerth, B. J. 1980. *J. Biol. Chem.* 255:6020–23
97. Dingermann, T., Ogilvie, A., Pistel, F., Muhlhofer, W., Kersten, H. 1981. *Hoppe-Seyler's Z. Physiol. Chem.* 362:763–73
98. Shindo-Okada, N., Terada, M., Nishimura, S. 1981. *Eur. J. Biochem.* 115:423–28
99. Ott, G., Kersten, H., Nishimura, S. 1982. *FEBS Lett.* 146:311–14
100. Okada, N., Shindo-Okada, N., Sato, S., Itoh, Y. H., Oda, K.-I., Nishimura, S. 1978. *Proc. Natl. Acad. Sci. USA* 75:4247–51
101. Schachner, E., Kersten, H. 1984. *J. Gen. Microbiol.* 130:135–44
102. Raba, M., Limburg, K., Burghagen, M., Katze, J. R., Simsek, M., et al. 1979. *Eur. J. Biochem.* 97:305–18
103. Conlon-Hollingshead, C., Ortwerth, B. J. 1980. *Exp. Cell Res.* 128:171–80
104. Smith, D. W. E., McNamara, A. L., Rice, M., Hatfield, D. L. 1981. *J. Biol. Chem.* 256:10033–36
105. Ortwerth, B. J., Wolters, J., Nahlik, J., Conlon-Hollingshead, C. 1982. *Exp. Cell Res.* 138:241–50
106. Lin, V. K., Ortwerth, B. J. 1983. *Biochem. Biophys. Res. Commun.* 115:598–605
107. Ortwerth, B. J., Lin, V. K., Lewis, J.,

Wang, R. J. 1984. *Nucleic Acids Res.* 12:9009–23
108. Nurse, P., Thuriaux, P. 1984. *Mol. Gen. Genet.* 196:332–38
109. Kim, S.-H. 1979. See Ref. 1, pp. 83–100
110. Agris, P. F., Sierzputowska-Gracz, H., Smith, C. 1986. *Biochemistry.* 25:5126–31
111. Reeves, R. H., Roth, J. R. 1971. *J. Mol. Biol.* 56:523–33
112. Colby, D. S., Schedl, P., Guthrie, C. 1976. *Cell* 9:449–63
113. Hayashi, H., Fisher, H., Söll, D. 1969. *Biochemistry* 8:3680–86
114. Litwack, M. D., Peterkofsky, A. 1971. *Biochemistry* 10:994–1001
115. Bruce, A. G., Atkins, J. F., Wills, N., Uhlenbeck, O., Gesteland, R. F. 1982. *Proc. Natl. Acad. Sci. USA* 79:7127–31
116. Vacher, J., Grosjean, H., Houssier, C., Buckingham, R. H. 1984. *J. Mol. Biol.* 177:329–42
117. Houssier, C., Grosjean, H. 1985. *J. Biomol. Struct. Dyn.* 3:387–408
118. Seela, F., Ott, J., Franzen, D. 1983. *Nucleic Acids Res.* 11:6107–20
119. Buck, M., Griffiths, E. 1982. *Nucleic Acids Res.* 10:2609–24
120. Eisenberg, S. P., Yarus, M., Soll, L. 1979. *J. Mol. Biol.* 135:111–26
121. Ericson, J. U., Björk, G. R. 1986. *J. Bacteriol.* 166:1013–21
122. Gefter, M. L., Russell, R. L. 1969. *J. Mol. Biol.* 39:145–57
123. Petrullo, L. A., Gallagher, P. J., Elseviers, D. 1983. *Mol. Gen. Genet.* 190:289–94
124. Bouadloun, F., Srichaiyo, T., Isaksson, L. A., Björk, G. R. 1986. *J. Bacteriol.* 166:1022–27
125. Yanofsky, C., Soll, L. 1977. *J. Mol. Biol.* 113:663–77
126. Goddard, J. P., Lowdon, M. 1981. *FEBS Lett.* 130:221–22
127. Diaz, I., Ehrenberg, M., Kurland, C. G. 1986. *Mol. Gen. Genet.* 202:207–11
128. Kurland, C. G., Ehrenberg, M. 1984. *Prog. Nucleic Acid Res. Mol. Biol.* 31:191–219
129. Petrullo, L. A., Elseviers, D. 1986. *J. Bacteriol.* 165:608–11
130. Thiebe, R., Zachau, H. G. 1968. *Eur. J. Biochem.* 5:546–55
131. Smith, D. W. E., Hatfield, D. L. 1986. *J. Mol. Biol.* 189:663–71
132. Fairclough, R. H., Cantor, C. R. 1979. *J. Mol. Biol.* 132:587–601
133. Miller, J. P., Hussain, Z., Schweizer, M. P. 1976. *Nucleic Acids Res.* 3:1185–201
134. Weissenbach, J., Grosjean, H. 1981. *Eur. J. Biochem.* 116:207–13

135. Watts, K. T., Tinoco, I. Jr. 1978. *Biochemistry* 17:2455–63
136. Engel, J. D., von Hippel, P. H. 1974. *Biochemistry* 13:4143–58
137. Pieczenik, G. 1980. *Proc. Natl. Acad. Sci. USA* 77:3539–43
138. Sprinzl, M., Moll, J., Meissner, F., Hartman, T. 1985. *Nucleic Acids Res.* 13:r1–r49
139. Sekiya, T., Takeishi, K., Ukita, T. 1969. *Biochim. Biophys. Acta* 182:411–26
140. Lustig, F., Elias, P., Axberg, T., Samuelsson, T., Tittawella, I., Lagerkvist, U. 1981. *J. Biol. Chem.* 256:2635–43
141. Yokoyama, S., Watanabe, T., Murao, K., Ishikura, H., Yamaizumi, Z., et al. 1985. *Proc. Natl. Acad. Sci. USA* 82:4905–9
142. Mazumdar, S. K., Saenger, W., Scheit, K. H. 1974. *J. Mol. Biol.* 85:213–29
143. Sen, G. C., Ghosh, H. P. 1976. *Nucleic Acids Res.* 3:523–35
144. Weissenbach, J., Dirheimer, G. 1978. *Biochim. Biophys. Acta* 518:530–34
145. Elseviers, D., Petrullo, L. A., Gallagher, P. J. 1984. *Nucleic Acids Res.* 12:3521–34
146. Sullivan, M. A., Cannon, J. F., Webb, F. H., Bock, R. M. 1985. *J. Bacteriol.* 161:368–76
147. Sherman, F., Ono, B., Stewart, J. R. 1979. In *Nonsense Mutations and tRNA Suppressors,* ed. J. E. Celis, J. D. Smith, pp. 133–53. New York: Academic
148. Piper, P. W. 1980. In *Transfer RNA: Biological Aspects,* ed. D. Söll, J. N. Abelson, P. R. Schimmel, pp. 379–94. New York: Cold Spring Harbor Lab.
149. Waldron, C., Cox, B. S., Wills, N., Gesteland, R. F., Piper, P. W., et al. 1981. *Nucleic Acids Res.* 9:3077–88
150. Oda, K., Kimura, F., Harada, F., Nishimura, S. 1969. *Biochim. Biophys. Acta* 179:97–105
151. Ishikura, H., Yamada, Y., Nishimura, S. 1971. *Biochim. Biophys. Acta* 228:471–81
152. Takeishi, K., Takemoto, T., Nishimura, S., Ukita, T. 1972. *Biochem. Biophys. Res. Commun.* 47:746–52
153. Takemoto, T., Takeishi, K., Nishimura, S., Ukita, T. 1973. *Eur. J. Biochem.* 38:489–96
154. Murao, K., Hasegawa, T., Ishikura, H. 1982. *Nucleic Acids Res.* 10:715–18
155. Mitra, S. K., Lustig, F., Åkesson, B., Axberg, T., Elias, P., Lagerkvist, U. 1979. *J. Biol. Chem.* 254:6397–401
156. Samuelsson, T., Axberg, T., Borén, T.,

Lagerkvist, U. 1983. *J. Biol. Chem.* 258:13178–84
157. Kilpatrick, M. W., Walker, R. T. 1980. *Nucleic Acids Res.* 8:2783–86
158. Heckman, J. E., Sarnoff, J., Alzner-Deweerd, B., Yin, S., RajBhandary, U. L. 1980. *Proc. Natl. Acad. Sci. USA* 77:3159–63
159. Ofengand, J., Liou, R., Kohut, J. III, Schwartz, I., Zimmermann, R. A. 1979. *Biochemistry* 18:4322–32
160. Atkins, J. R., Ryce, S. 1974. *Nature* 249:527–30
161. Reeves, R. H., Roth, J. R. 1975. *J. Bacteriol.* 124:332–40
162a. Stern, L., Schulman, L. H. 1978. *J. Biol. Chem.* 253:6132–39
162b. Strobel, M. C., Abelson, J. 1986. *Mol. Cell. Biol.* 6:2663–73
163. Noguchi, S., Nishimura, Y., Hirota, Y., Nishimura, S. 1982. *J. Biol. Chem.* 257:6544–50
164. Jänel, G., Michelsen, U., Nishimura, S., Kersten, H. 1984. *EMBO J.* 3:1603–8
165. Meier, F., Suter, B., Grosjean, H., Keith, G., Kubli, E. 1985. *EMBO J.* 4:823–27
166. Bienz, M., Kubli, E. 1981. *Nature* 294:188–90
167. Beier, H., Barciszewska, M., Sickinger, H.-D. 1984. *EMBO J.* 3:1091–96
168. Philipson, L., Andersson, P., Olshevsky, U., Weinberg, R., Baltimore, D., Gesteland, R. 1978. *Cell* 13:189–99
169. Geller, A. I., Rich, A. 1980. *Nature* 283:41–46
170. Cortese, R., Landsberg, R., Vonder Haar, R. A., Umbarger, H. E., Ames, B. N. 1974. *Proc. Natl. Acad. Sci. USA* 71:1857–61
171. Turnbough, C. L. Jr., Neill, R. J., Landsberg, R., Ames, B. N. 1979. *J. Biol. Chem.* 254:5111–19
172. Parker, J. 1982. *Mol. Gen. Genet.* 187:405–9
173. Palmer, D. T., Blum, P. H., Artz, S. W. 1983. *J. Bacteriol.* 153:357–63
174. Roth, J. R., Anton, D. N., Hartman, P. E. 1966. *J. Mol. Biol.* 22:305–23
175. Johnston, H. M., Barnes, W. M., Chumley, F. G., Bossi, L., Roth, J. R. 1980. *Proc. Natl. Acad. Sci. USA* 77:508–12
176. Bossi, L., Roth, J. R. 1980. *Nature* 286:123–27
177. Johnson, P. F., Abelson, J. 1983. *Nature* 302:681–87
178. Baumann, U., Fischer, W., Sprinzl, M. 1985. *Eur. J. Biochem.* 152:645–49
179. Björk, G. R., Kjellin-Stråby, K. 1977. In *The Biochemistry of Adenosylmethionine,* ed. F. Salvatore, E. Borek, V. Zappia, H. G. Williams-Asham, F. Schlenk, pp. 216–30. New York: Columbia Univ. Press
180. Friedman, S. 1979. *J. Biol. Chem.* 254:7111–15
181. Arcari, P., Hecht, S. M. 1978. *J. Biol. Chem.* 253:8278–84
182. Marinus, M. G., Morris, N. R., Söll, D., Kwong, T. C. 1975. *J. Bacteriol.* 122:257–65
183. Roe, B., Michael, M., Dudock, B. 1973. *Nature New Biol.* 246:135–38
184. Hoburg, A., Aschhoff, H. J., Kersten, H., Manderschied, U., Gassen, H. G. 1979. *J. Bacteriol.* 140:408–14
185. Ramabhadran, T. V., Jagger, J. 1976. *Proc. Natl. Acad. Sci. USA* 73:59–63
186. Thomas, G., Favre, A. 1980. *Eur. J. Biochem.* 113:67–74
187. Björk, G. R., Neidhardt, F. C. 1975. *J. Bacteriol.* 124:99–111
188. Johnson, L., Hayashi, H., Söll, D. 1970. *Biochemistry* 9:2823–31
189. Vani, B. R., RamaKrishnan, T., Taya, Y., Noguchi, S., Yamaizumi, Z., Nishimura, S. 1979. *J. Bacteriol.* 137:1084–87
190. Samuel, C. E., Rabinowitz, J. C. 1974. *J. Biol. Chem.* 249:1198–206
191. Delk, A. S., Rabinowitz, J. C. 1975. *Proc. Natl. Acad. Sci. USA* 72:528–30
192. Baumstark, B. R., Spremulli, L. L., RajBhandary, U. L., Brown, G. M. 1977. *J. Bacteriol.* 129:457–71
193. Marcu, K. B., Dudock, B. S. 1976. *Nature* 261:159–62
194. Roe, B. A., Tsen, H.-Y. 1977. *Proc. Natl. Acad. Sci. USA* 74:3696–700
195. Kersten, H., Albani, M., Männlein, E., Praisler, R., Wurmbach, P., Nierhaus, K. H. 1981. *Eur. J. Biochem.* 114:451–56
196. Davenloo, P., Sprinzl, M., Watanabe, K., Albani, M., Kersten, H. 1979. *Nucleic Acids Res.* 6:1571–81
197. Ames, B. N., Tsang, T. H., Buck, M., Christman, M. F. 1983. *Proc. Natl. Acad. Sci. USA* 80:5240–42
198. Sakamoto, K., Okada, N. 1985. *Nucleic Acids Res.* 13:7195–206
199. Björk, G. R. 1985. *Cell* 42:7–8
200. Nishimura, S. 1979. See Ref. 1, pp. 547–49
201. Bachmann, B. J. 1983. *Microbiol. Rev.* 47:180–230
202. Gallagher, P. J., Schwartz, I., Elseviers, D. 1984. *J. Bacteriol.* 158:762–63
203. Hagervall, T. G., Björk, G. R. 1984. *Mol. Gen. Genet.* 196:201–7
204. Byström, A. S., Björk, G. R. 1982. *Mol. Gen. Genet.* 188:440–46

Ann. Rev. Biochem. 1987. 56:289–315

FERRITIN: STRUCTURE, GENE REGULATION, AND CELLULAR FUNCTION IN ANIMALS, PLANTS, AND MICROORGANISMS

Elizabeth C. Theil

Department of Biochemistry, North Carolina State University, Raleigh, North Carolina 27695-7622

CONTENTS

It remaineth now in the next place to discourse of the mines of yron, a mettal which wee may well say is both the best and the worst implement used now in the world. . . . (1)

PERSPECTIVES

The history of civilization shows a long dependence on iron, particularly after the availability of iron for armaments increased (1). The evolution of organ-

289

0066-4154/87/0701-0289$02.00

isms also shows a long dependence on iron, even after the availability was diminished by dioxygen in the atmosphere (2). Ferritin maintains iron in an available, soluble form for use, e.g. in oxygen transfer, electron transfer, nitrogen fixation, and DNA synthesis (ribonucleotide reduction). The solubility of iron probably became a problem ca 2.5 billion years ago when H_2O began to be used as a source of hydrogen for photosynthesis (2). Dioxygen, the byproduct of such photosynthesis and probably the worst environmental pollutant of all time, created a dilemma for iron-dependent organisms: either move to environments devoid of dioxygen or accommodate to the low solubility of Fe(III) produced by dioxygen from Fe(II). [Fe(III) is ca $10^{-9} \times$ less soluble than Fe(II) (3). At concentrations greater than 10^{-18} M, hydrated Fe(III) forms insoluble, rustlike, hydrous ferric oxides.] The choice of maintaining and storing iron in a soluble form and accommodating to and using dioxygen has been the more successful one.

An illustration of the importance of ferritin in humans is the role of macrophage ferritin in recycling iron from old red blood cells: the amount of iron generated each day by such cells is ca 0.54 mmol. About 90% of the iron is converted by macrophages to ferritin and low-molecular-weight forms, from which the iron is slowly released to apotransferrin. Iron on transferrin is delivered to immature red cells, completing the cycle. An alternative to recycling, the excretion of the iron as a simple iron salt such as $FeCl_3$, would require ca 10^{13} liters of water each day to prevent precipitation of hydrous ferric oxide. For example, stabilizing the iron as a monoatomic chelate with citrate is also not feasible (4); five liters of orange juice would be needed for the daily burden of 0.54 mmol of iron, an amount likely to also alter the acid-base balance and remove other metals such as calcium. The advantages of recycling iron through temporary stores in macrophage ferritin are thus substantial. Other types of ferritin provide iron reservoirs for such uses as growth and cell replacement.

Ferritin is found in most cell types of humans and other vertebrates, and in invertebrates, higher plants, fungi, and bacteria. The role of ferritin in different cell types includes both specialized functions (e.g. recycling iron in macrophages, short- and long-term iron storage as in red cells of embryos or hepatocytes of adults) and intracellular housekeeping functions (providing a reserve of iron for cytochromes, nitrogenase, ribonucleotide reductase, hemoglobin myoglobin, etc, and possibly for detoxification, if excess iron enters the cell). Although all ferritins share structural properties, cell-specific variations in structure, function, and amount indicate the presence of cell-specific features of genetic regulation. (One of the more notable features of regulation is translational control of ferritin mRNA by iron in cells specialized for iron storage.)

The following discussion compares the constant and variable features of structure for the ferritin protein coat and iron core, of gene regulation (transcriptional vs translational control), and of ferritin function. Iron storage, the constant feature of ferritin function, is described in terms of three possible types: iron storage for other cells (specialized-cell ferritin), iron storage for intracellular needs (normal housekeeping ferritin), and iron storage for intracellular protection from iron overload (stress housekeeping ferritin). Cell-specific variations in ferritin function are considered as a basis for the observed variations in ferritin gene regulation and protein structure.[1]

FERRITIN STRUCTURE

Ferritin is a large protein, ca 12.0 nm in diameter, formed from a spherical protein coat (apoferritin), ca 1.0 nm thick with a mass of ca 450,000 (11), that surrounds a core of hydrous ferric oxide [Fe(III)O·OH]. Up to 4500 iron atoms can be included in the core along with variable amounts of phosphate (12). The protein coat, composed of 24 subunits (ca 20 kilodaltons), is in contact with the iron core at several points on the inner surface (13), forming an iron-protein interface. The iron-protein interface defines the site of core nucleation (14), and may be located where the protein subunit dimers interact. Iron passes into and out of the core in vivo and in vitro (reviewed in 5–9), most likely through the channels (15) in the protein. All ferritin molecules have the same general structure, which may reflect convergent evolution for *Escherichia coli* compared to mammals (16), but clearly results from conservative evolution among vertebrates (15, 17–20).

Apoferritin

THE GENERAL STRUCTURE OF APOFERRITIN AND VARIATIONS   The protein coat of ferritin, apoferritin, may be isolated from normal preparations of ferritin where the apoferritin fraction varies in amount, or may be prepared from ferritin by the reduction of the ferric iron in the core and chelation of the ferrous ion (21–23). Apoferritin thus prepared consists of assembled subunits, which can be used to reconstitute ferritin with Fe(II) and oxygen (reviewed in 6). Comparative studies of the morphology of crystals of apoferritin from mammals and *E. coli* indicate great similarity of packing, in spite of variations in primary structure (8).

Horse spleen apoferritin makes a useful standard against which to compare other types of ferritin, because it is composed (>90%) of identical subunits and because high-resolution X-ray crystallography by Harrison and col-

---

[1]Several earlier reviews of ferritin structure, function, and gene regulation are cited in references 5–10. Observations made since 1982 are emphasized here; annotation to older literature is made only in specific cases.

leagues has permitted the location of essentially all the amino acids (7, 15).
Each subunit of horse spleen apoferritin contains 174 amino acids (24) folded
into a bundle of two pairs of alpha-helixes joined by a long loop (Figure 1B).
The folded subunit has a tightly packed core of hydrophobic residues.

Interactions of the 24 subunits confer specific properties on the assembled
structure (Figure 1A) and depend on sequences of amino acids that are highly
conserved from amphibia to mammals (Figure 2; 18–20, 24–29). For example, a dimer pair is formed by interactions along the long axis of each
subunit with a groove 1.4 μm wide on the inner surface of the protein coat (7,
30); the grooves may be important in the nucleation of the iron core (31).
Interactions of three subunits form eight hydrophilic channels near the N end
of each subunit (Figure 1a); the detection of metal-binding sites in the
channels (7, 15) and the presence of conserved residues, Asp 127 and Glu
130, in the channels suggest that the channels could be important for transport
of metal in and out of the protein. Interactions of four subunits (helix E,
Figure 1A,B) form six channels (Figure 1A), 0.4–0.5 nm wide (7, 15, 30),
lined by hydrophobic residues such as Leu 165 (Figure 2). Although the
surrounding sequences are highly conserved, substitution of the hydrophobic
residues found in horse spleen apoferritin occurs in other types of ferritin

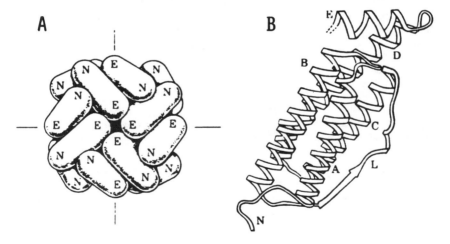

Figure 1 *A*. Schematic representation of the horse spleen ferritin molecule viewed down a
molecular four-fold axis, illustrating 4 3 2 symmetry. Each subunit is represented by a sausage-
shaped building brick. The N-terminal region of the polypeptide chain lies close to the end
labeled N; the E helix residues lie close to the end labeled E. *B*. Ribbon diagram of the alpha
carbon backbone of a horse spleen apoferritin subunit. The four long alpha helixes A, B, C, D are
comprised of residues 10–39, 45–72, 92–120, and 124–155, respectively; helix E contains
residues 160–169. L, a loop, connecting B and C, contains residues 73–91. Reprinted from Ref.
30 with permission.

(Figure 2); the function of the six channels on the four-fold axes is not known. In addition to the conserved amino acid sequences involved in defined subunit interactions, many other blocks of conserved sequence are found throughout the subunit; speculation about the structural or functional requirements that have maintained such sequences throughout evolution is difficult. However, the availability of cloned cDNA for a variety of ferritin subunits (18, 20, 27, 28) and their expression in *E. coli* (32, 33) provide the tools for illuminating investigations of modified ferritin.

Many types of apoferritin molecules are assembled from subunits of more than one structure (see, e.g., 34). Variations in ferritin subunits occur in the sequence of amino acids throughout the chain (Figure 2), in chain length [from 172 to 183 residues (Figure 2; 18, 20, 24, 27–29, 35, 36)], and in mass. The ramifications of such variety include heterogeneity of surface charge (pI ca 4–6 even within ferritin from a single cell type, e.g. 34, 35, 37; reviewed in 6 and 9), hydrophobicity (40, 41), and size (or shape) of the assembled molecule (42, 43). Nevertheless, when crystals were compared from apoferritin with different subunits, the X-ray diffraction patterns were very similar (8, 44), indicating that a large amount of structural homology is retained in spite of differences in primary structure.

Different combinations of constant and variable features of subunit folding may explain the similarity of assembled apoferritin molecules when subunits differ in primary structure. For example, the large L-type subunit of rat liver ferritin contains an internal octapeptide (36) that merely lengthens the chain connecting the alpha helixes at the DE turn, Figure 1*B*, with little effect on the regions involved in subunit-subunit interactions; the influence of the octapeptide was not detectable in difference electron density maps, which compared the X-ray diffraction patterns of rat liver and horse spleen apoferritin crystals (19).

Modifications of amino acid side chains in some ferritin subunits, including glycosylation (45) or in vitro phosphorylation of serine (39), also reflect differences in the primary structure of the subunits. Covalent crosslinks between subunit pairs (31) are a modification of ferritin that appears to be related to the physiological state of the cell and the amount of iron; whether formation of the crosslinks also reflects variations in the primary structure of subunits remains to be seen.

Many of the variations observed in ferritin structure are associated with particular cell types (e.g. 23, 39, 43, 46–49)[2] or cells in a particular physiological state (31) and coincide with functional differences as well (e.g. 31, 47, 48). While there is not yet enough information to correlate cell-specific differences in ferritin structure with differences in function, the existence of

---

[2]Cell-specific immunoreactivity (e.g. 46, 56) is superimposed on conservation of antigenicity within and between species and vertebrate classes (e.g. 23, 25).

```
1   5    10   15   20   25   30   35   40
SSQIRQNYSTEVEAAVNRLVNLYLYRASYTYLSLGFYFDRD  (1)Horse spleen L
M----------D------S------Q--------------   (2)Human liver L
MTTAST--V----HQDS---I--QI--E-Y---V---MSY-----  (3)Human liver H
MT--------------------H------------F----   (4)Rat liver L
MD--V---FHRDC---I--M--ME-Y-------MA------  (5)Tadpole red cell H
ME--V---FHQDC--GL--T---KFHS--V---MAS--N--  (6)Tadpole red cell L
MV--V----HSDC-------ML--E-Y------S-MYAF----  (7)Tadpole red cell X
MA--V----HSDC------ML--E-Y------S-MYAF----  (8)Tadpole pseudogene
```

```
41  45   50   55   60   65   70   75   80
DVALEGVCHFFRELAEEKREGAERLLKMQNQRGGRALFQD  (1)
-------S-------------Y-----------------   (2)
----KNFAKY-LHQSH-E--H--K-M-L-------IFL--  (3)
-------G-------------------L--E---------  (4)
-I---HN-AK--K-QSH-E--H--K-M-D--K----IVL--  (5)
----SNFAK----RS--EK-H--K-IEY-------VFL-S  (6)
----HN-AE--K-HSH-E--H--KFM-Y--K----VVL--  (7)
----HN-AE--K-HSH-E--H--KFM-Y--K----VVL--  (8)
```

```
81  85   90   95   100  105  110  115  120
LQKPSQDEWGTTLDAMKAAIVLEKSLNQALLDLHALGSAQ  (1)
IK--AE----K-P------MA---K-------------R   (2)
IK--DC-D-ESG-N--EC-LH----NV--S--E--K-ATDK  (3)
V---------K--E-----LA---N--------------   (4)
VK--ER----N--E--Q--LQ---TV--------KV--DK  (5)
VE--ER-D-ANG-E-LQT-LK-Q--V---------VAADK  (6)
IK--ER----N--E--Q--LQ---TV--------K-ATDK  (7)
IK--ER----N--E--Q--LQ---TV--------K-ATDK  (8)
```

```
                                     OCT
121 125  130  135  140  145  150  155/ 160
ADPHLCDFLESHFLDEEVKLIKKMGDHLTNIQRLVGSQAG  (1)
T---------T------------------LH--G-PE--   (2)
N--------I-T-Y-N-Q--A--EL---V--LRKMGAPES-  (3)
---------------K---------N----LR-WQ-P--S  (4)
V----------TEY-E-Q--S--QL--YI--LK--GLP-N-  (5)
S---MT------PY-S-S-ET---L---I-SLKK-WS-HP-  (6)
V----------EY-E-Q--D--RI--FI--LK--GLPEN-  (7)
V---------TEY-E-Q--D--RI--FI--LK--GLPEN-  (8)
```

```
161 165  170  175
LGEYLFERLTLKHD    (1)
--------------    (2)
-A----DKH--GDSDNES  (3)
--------------    (4)
M-----DKH-MGESS   (5)
MA----NKH--G      (6)
M-----DKHSV-ESS   (7)
M-----DKHSV-ESS   (8)
```

(OCT) - An octamer (QPAQTGVA) found
in rat liver ferritin, and
omitted in the figure for
purposes of comparison,
occurs between residues 157
and 158.

Figure 2 A comparison of the amino acid sequences of apoferritin subunits. The data are described in the following references: (*1*) horse spleen L (24); (*2, 3*) human liver L and H (28); (*4*) rat liver L (36); (*5*) tadpole red cell H (18); (*6*) tadpole red cell L (33); (*7, 8*) tadpole red cell X and tadpole pseudogene (62). Tadpole red cell X was sequenced from cloned red cell cDNA, and has 84% homology with tadpole red cell H and 99% homology with a processed pseudogene in the tadpole genome. The sequence for a human spleen subunit, not presented, is 82% homologous to horse spleen ferritin; differences occur at residues 11, 18, 25, 40, 48, 62, 81, 82, 85, 86, 145, 151, 154, 155, and 157 (29). All sequences except (*1*), horse spleen L, were determined by sequencing cloned DNA; (*2*), human liver L, corresponds to the sequence determined for the protein (35) except at residues 53 (E vs Q), 86 (E vs Q), and 175 (D vs N).

such structural variations within a species show that differential utilization of the genetic information for ferritin is clearly cell-specific.

THE H/L SUBUNIT STORY   The selection of horse spleen ferritin for study by early workers was serendipitous because abundance coincided with structural simplicity; but the studies also led to the assumption that all ferritins were composed of identical subunits. Subsequent examination of subunits from human heart and liver ferritin showed that the subunits from each tissue differed in mobility during electrophoresis in denaturing gels. The subunit types were designated H(eart) and L(iver); later, as more ferritins were examined and found to have similar variations in subunit mobility, and when both types of subunit were found in the same cell, the terminology shifted to H(eavy) and L(ight) (50). In addition to the H/L subunit variations, ferritins (and apoferritins) can differ in surface charge (isoelectric focusing), and immunoreactivity[2], while ferritin can also differ in iron content (density in sucrose gradients). The differences occur among normal cell types (e.g. 34, 39, 42, 48) and in pathological conditions such as hemochromatosis, transfusional or experimental iron overload, thalassemia, leukemia, and inflammation (copper poisoning) (e.g. 31, 50–55).

To accommodate the observed complexity of ferritin, Drysdale proposed that the structural features of ferritin arose from the combination in various ratios of two subunits, H and L, which differ in size, amino acid composition, surface charge, and immunoreactivity (50, 58). A corollary related differences in ferritin iron content to the functional efficiency of one of the two subunits for storing iron. Thus, in the human and the rat the H subunit was associated with a lower pI and lower iron content, and predominates in heart tissue, whereas the L subunit was associated with a higher pI and higher iron content, and predominates in the liver. Although there was little structural information about pure H and L subunits against which to test the predictions of the hypothesis, its appealing simplicity met widespread acceptance, the reservations of a large number of workers [and the poor correlations between horse and human ferritin (50)] notwithstanding.

Recent advances in the analysis of ferritin structure allow a more critical test of the two-subunit hypothesis. cDNAs encoding two distinct ferritin subunits have been isolated and sequenced for human liver (27, 28). One cDNA has been identified by hybrid-select translation and immunoprecipitation as an H chain (59), and the other has been identified as an L chain because of its differences from the H chain and its sequence similarity to the horse spleen ferritin L chain sequence (28); the masses predicted from the human liver ferritin cDNA sequences are 21,099 for the H chain (182 residues) and 19,766 for the L chain (175 residues), which correspond well to the masses of H and L subunits deduced from electrophoresis in SDS gels; eight sequences similar to the L subunit and 12 sequences similar to the H

Table 1   Percent homology of amino acid sequences among apoferritin subunits

	L vs horse spleen L	H vs L	H vs X
Horse spleen[a] (24)[b]	100	—	—
Human spleen[c] (24)	85	—	—
Rat liver[c] (36)	88	—	—
Human liver[c,d] (28)	87	55	—
Bullfrog red cell[c] (18, 33, 62)	52	61	84

[a]The sequence was determined from the protein.

[b]The numbers in parentheses are the references from which the data were obtained. The sequence of subunit X was determined from a tadpole red cell cDNA clone, and is 99% homologous to a processed pseudogene (62); the electrophoretic properties of the protein subunit (H or L) are not known.

[c]The sequence(s) were predicted from the cDNA sequence.

[d]The protein encoded by the human liver LcDNA was designated "L" because of its distinction from human liver H and its similarity to horse spleen L (28); electrophoretic properties were not determined.

subunit occur in the human genome (60). The two subunits have extensive sequence similarities (Figure 2) and are 55% homologous (Table 1).

To a first approximation, the two-subunit (H/L) hypothesis seems to have a structural basis if, as has been suggested (27), pseudogenes account for most of the extra hybridizable sequences observed in the genome. However, the sizes of L chains from human and rat liver ferritin differ from 174 residues (28) to 182 (36), even though the mobilities are similar for electrophoresis in denaturing gels. Moreover, for pig liver the L subunit has two different apparent masses, depending on the analytical method used (41), and can be resolved into three different fractions by reverse phase chromatography, each of which has a different primary structure (40). In addition, cDNAs encoding H and L subunits from frog red cells, identified by hybrid-select translation (18, 61), with apparently different masses of 20.0 and 22.8 kilodaltons determined by electrophoresis, have similar masses predicted from the cDNA sequence (20,536 and 19,941 daltons, respectively) (33) and contain similar numbers of residues (175 and 172 residues, respectively). Finally, mRNA encoding at least three ferritin subunits has been detected in HeLa cells, frog liver, and frog red cells (18, 33, 38, 62), and three distinct cDNAs have been characterized for frog red cells (18, 33). All apoferritin subunits share extensive homology (Figure 2, Table 1). In some cases the homology is extremely high, as in the case of mammalian L subunits or frog H and X subunits from liver and red cells. In other cases, such as mammalian and bullfrog L subunits, the homology is much lower (Table 1).

Clearly, the complexity of apoferritin subunit structure is greater than can be accounted for by the two-subunit hypothesis, because ferritin can be composed of at least three protein subunits, even in the same cell type (38, 61, 62). It is also clear that the H and L designations based on mobility in denaturing gels do not mean H(eavy) and L(ight) but reflect a property other than mass (33, 41, 61), such as amount of SDS bound. For example, alpha-tubulin (63a, 63b), Y_c-glutathione S-transferase (64), and cytoplasmic poly(A)-binding protein (65) all share with some H chains of apoferritin an electrophoretic mobility in SDS gels aberrantly slower than that predicted by the mass. The existence of more than two apoferritin subunits will explain with greater ease a variety of observations on housekeeping ferritins and ferritins in differentiating cells (e.g. 54, 66). Many other protein families, e.g. the cytochrome P-450s (67), appear to be even more complex than the ferritins. The challenge will be to analyze new information about ferritin for structural and functional relationships that accommodate naturally all the varieties being studied.

DISTINCTIVE FEATURES OF PLANTS AND MICROORGANISMS   Ferritin in plants has been examined mainly in spinach (68), bean leaves (69, 70), peas (71), soybeans (72), and soybean nodules (73). While the basic architecture of ferritin observed in animals appears to be preserved, the subunits can vary in size. In particular, the subunits appear to be synthesized as precursors of different sizes (electrophoresis in denaturing gels), which are subsequently processed during transport to the chloroplast of bean leaves (69) or during germination (soaking) in soybeans (72). Tryptic peptide maps (71) of pea ferritin suggest that some peptides may be common to those in animals, but much more information is needed before the structural (and evolutionary) relationship between ferritins in plants and animals can be established.

Iron storage molecules have only been examined in a few microorganisms, since iron storage in such cells is generally a transient phenomenon preceding rapid growth (74). Again the structure of a multisubunit protein coat around a core of iron is maintained and has been studied in *Azotobacter vinelandii* (75–77), *Pseudomonas aeruginosa* (12, 75), *Escherichia coli* (16, 79), and *Phycomyces blakesleeanus* (78). Apoferritin from bacteria has heme, as in cytochrome b557, often at 0.5/subunit (76, 79). The protein subunit sizes range from 16 to 19 kilodaltons (75–79) in analogy to apoferritin from animals. Immunological crossreactivity has been found between ferritin from *E. coli* and *A. vinelandii* but not with horse spleen ferritin (16). The available sequence information for the ferritin from *E. coli* (16) indicates no homology with horse spleen ferritin. Either the structural similarity of ferritin from animals and bacteria results from convergent evolution, or homology resides in those regions yet to be sequenced in the protein from *E. coli*.

The Iron-Apoferritin Interface

The iron core of ferritin has many properties in common with hydrous ferric oxides that form spontaneously. Some early views of the role of the apoprotein, therefore, envisioned little more than providing a coating for the iron core. Subsequent studies, using horse spleen apoferritin because of its relative simplicity and abundance, showed that the protein bound a variety of monoatomic metal ions including Fe(II) and Fe(III); a dimeric, Fe(II)-O-Fe(III) species; and a cluster of 3–4 Fe(III)-oxo atoms. Such observations indicate an important role for the protein in core formation that includes nucleation of the core.

Among the monoatomic metal ions that are bound by apoferritin are Cd(II), Zn(II), Mn(II), Tb(III), UO(III), and VO(IV) (21, 80; reviewed in 6–8). The facts that Fe(II) displaces Mn(II) (80) and VO(IV) (22) and that Zn(II) and Tb(III) alter the rate of iron core formation (81) show that the metal ions bind at protein sites that influence core formation. At least one of the sites appears to involve conserved carboxylate residues (Asp 127 and Glu 130) at the three-fold channels (Figures 1 and 2; 19, 80). Several observations suggest that binding sites in the three-fold channels may only represent the earliest stage of iron core formation. First, the stoichiometry of displacement of Mn(II) or VO(IV) by Fe(II) is complex (22, 80) and indicates that Fe(II) binding occurs at other types of sites in addition to those at the channels. Second, the oxidation of saturating amounts of Fe(II) in situ regenerates new Mn(II) and VO(IV) binding sites (22, 80). Third, if oxidation of Fe(II) to Fe(III) occurred in the three-fold channels, polynuclear Fe(III) complexes would plug the channels and prevent iron from reaching the inside of the protein shell. Polynuclear iron complexes would then form outside the protein when, in fact, the polynuclear iron core is inside the protein (12, 13). Taken together, the observations are compatible with the notion that Fe(II) binds first in the three-fold channel followed by oxidation and binding at sites on the interior of the protein. Esterification of carboxylate groups prevents oxidation of Fe(II) (57).

Ferritin can only be reconstituted in vitro from apoferritin with Fe(II); thus, oxidation must precede formation of the ferric iron core. An early stage, beyond monoatomic Fe(II), would be a dimer, Fe(II)-O-Fe(III); such dimers have been detected in several iron proteins, e.g. hemerythrin and uteroferrin (82, 83). The mixed valence dimeric iron species has also been detected in an iron-apoferritin complex using EPR spectroscopy (84). However, its location is not known and its low concentration, relative to the Fe(III) present, suggests that it is a transient species.

A still larger iron-apoferritin complex in which the iron environment is distinct from that in the core has been observed by EXAFS and Mössbauer spectroscopy; the Fe(III) atoms each appear to be attached to carboxylate-like ligands and linked to each other by oxygen (14). One type of metal-binding site commodious enough (7) and with sufficient conserved carboxylate ligands (Figure 2) to accommodate such an iron cluster is the subunit dimer interface. Such sites are also likely to be those affected by the dimer cross-links, which reduce iron uptake (31, 85). How many Fe(III) atoms the protein sites can accommodate before the clusters become indistinguishable from ferritin iron cores, remains to be determined. But it is certain that apoferritin is more than an inert cover for the iron core. The iron/protein interface contains sites important for formation of the iron core and may be important in relating variations in structure to cell-specific differences in function. Details of the site structures await the results of modification studies.

The Iron Core and Hemosiderin

The crystalline regions of the ferritin iron core have much in common with the natural mineral ferrihydrite (86), which also forms in the laboratory by heating solutions of $Fe(NO_3)_3$ (Ref. 6 describes ferrihydrite in more detail). Small amounts of phosphate have been observed associated with the core of mammalian ferritins, which have often been attributed to surface phosphate. However, because of the observed inaccessibility of some of the phosphate to ligand exchange and the effect phosphate would have on the regularity of hydrous ferric oxide structures, both the presence of phosphate throughout the iron core and its association with disordered regions were predicted (6). Recent observations indicate that ferritin cores can have phosphate through-out, that cores can have both ordered and disordered regions, and that the disorder increases when phosphate increases. For example, a model complex of Fe(III) and ATP (4:1) studied by EXAFS was found to be similar to polynuclear hydrous ferric oxides but with phosphate throughout the complex (87). In addition, natural cores were isolated with such a high phosphate content [Fe:P=1.7:1 instead of the 8:1 or 20:1 observed in mammalian ferritins (11, 12)] that the phosphate could not be accommodated on the core surface but had to be throughout the iron core; high-phosphate ferritin was characterized from two bacteria, *A. vinelandii* (12, 88) and *P. aeruginosa* (12, 75). Finally, the ferritin iron cores with high phosphate were highly disordered, with little or no long range order, by high resolution transmission electron microscopy, electron diffraction analysis (12), and Mössbauer spectroscopy (89). Ferritin cores from an invertebrate, the limpet *Patella vulgata*[3], were also studied by the same groups and found to be highly disordered (12, 89); however, it is not yet known whether the structure is due to phosphate or to some other variable.

Hemosiderin is another form of storage iron, often found in lysosomes. It is characterized functionally by insolubility in aqueous cellular extracts and by

[3]Limpets have a tongue, or radula, studded with teeth composed of iron oxide and silica that are regularly replaced. The amount of iron required is very high and appears to be stored in abundant ferritin in the hemolymph (90).

an iron/protein ratio higher than ferritin. The similarity of the iron environment in hemosiderin to the iron core of ferritin from the same tissue (e.g. 91–93) and the metabolism of hemosiderin and ferritin in lysomes (92) support the idea that hemosiderin is derived by the denaturation of ferritin.

Order in ferritin iron cores could be influenced both by the number of nuclei that form on the protein [availability of Fe(II)] and the structure of the nuclei. In a study with model iron cores formed in the presence of dextran or chondroitin sulfate (pharmaceutically important soluble complexes for the therapy of iron deficiency), both EXAFS and Mössbauer analyses indicated that sulfate appeared to nucleate highly ordered domains with the spectroscopic and X-ray diffraction properties of soluble hematite (Fe_2O_3); the hematite coexisted with the hydrous ferric oxide (94).

Considering both the effects of carbohydrate-bound sulfate and high levels of phosphate on the structure of soluble hydrous ferric oxides, variations in ferritin iron cores could be due to anions at nucleation sites as well in the cellular environment. Different iron core structures are associated with altered rates of iron release both in vitro and in vivo (94; discussion in 12). Thus, the specificity of protein/anion binding sites and cytoplasmic differences in anion concentrations provide additional possible explanations of cell-specific variations in ferritin function.

FERRITIN GENE REGULATION AND SYNTHESIS

Ferritin genes are expressed in most cells but the concentration of the protein among different cell types can vary 1000-fold (95). For many genes, changing mRNA concentration (transcription or stability) is the major mechanism of regulation, with changing the utilization of mRNA (storage or translational efficiency) providing modulation. In the case of ferritin, changing the utilization of mRNA is the major site of regulation, at least in cells specialized for iron storage where the potential accumulation of the protein is high, ferritin mRNA is abundant, and induction of ferritin synthesis can occur throughout the life of the cell. Iron can regulate the synthesis of ferritin at two posttranscriptional steps. For example, rates of synthesis 30–40× constitutive rates are achieved by recruitment of a stored ferritin mRNA, first observed by Munro (96), combined with competitive translation of the derepressed ferritin mRNA, first observed by us (97, 98). The synthesis of ferritin in cells specialized for iron storage is thus a paradigm for translational control mechanisms. However, because ferritin functions both as a housekeeping protein (iron stored for intracellular use) and a specialized-cell protein (iron stored for use by other cells), mechanisms of regulation could vary depending on the cell type; ferritin synthesis may depend more on mRNA concentration (transcriptional control or changed stability) in some cell types and more on mRNA utilization (translational control) in others. The mechanisms of ferritin synthesis induced by iron have been studied fairly extensively, but little is known about other agents that induce ferritin accumulations, e.g. infection, inflammation (99), and heat shock (100).

Translational Control

STORAGE OF FERRITIN mRNA   Studies of ferritin mRNA storage have taken advantage of the fact that iron induces ferritin synthesis. Many studies have used cells specialized for iron storage, such as the rat hepatocyte and the bullfrog red cell of the embryonic cell line.[4] In such cells, analysis of the amount of functional ferritin mRNA (translated in a cell-free system) and structural ferritin mRNA (hybridizable) showed that iron produces no detectable change in the amount of ferritin mRNA (18, 96, 98) for two types of subunits (61, 101); the constitutive amount of ferritin mRNA is also high (98, 102). Housekeeping ferritin mRNA is also stored, e.g. in reticulocytes of the adult cell line, but the amount is greatly reduced (61) compared to the embryonic cell type. Ferritin mRNA is stored in the cytoplasm (18), sedimenting at the top of sucrose gradients that fractionate polyribosomes (101). Although the iron-responsive repression is readily reversed during the isolation of poly(A[+]) RNA (98), the molecular basis for the repression is not yet known; mRNA storage occurs for a variety of other proteins, as well as ferritin, e.g. hemoglobin, general housekeeping proteins in heat-shocked cells (103), ornithine aminotransferase (104), and sterol-binding protein (105) in rat liver. No unifying reason for mRNA storage is apparent, although in the case of ferritin we have suggested (102) that the repeated entry of iron into the nucleus to activate transcription of ferritin genes could damage DNA by mechanisms similar to Fe-EDTA "nucleases" (106a,b); the storage and translational control of the abundant mRNA for the iron-containing subunit of ribonucleotide reductase (107, 108) may have a similar explanation.

COMPETITION AMONG mRNA   Formation of the initiation complex is the rate-limiting step in the translation of many mRNA molecules. Since inhibitors of elongation, such as cycloheximide, cause differential effects on the rate of synthesis of different proteins, message-specific variations occur in the efficiency of initiation (109). Low concentrations of cycloheximide, which had no detectable effect on total protein synthesis, reversed the induction of ferritin synthesis by iron in red cells specialized for iron storage (97), i.e. the

[4]Red cells of the embryonic cell line of many vertebrates, including humans, are specialized for iron storage and contain large amounts of ferritin; the specialized iron storage function is absent in red cells of the adult line (25, 95). Bullfrog tadpoles are a good source of red cells of the embryonic cell line because the animals are large and the cells are readily accessible.

embryonic red cell line (95). Such a result suggests that derepressed ferritin mRNA in such cells is a strong competitor for translation initiation. The results with cycloheximide in whole cells were confirmed by translating poly(A^+) RNA in a cell-free extract from wheat germ and demonstrating high translational efficiency for ferritin mRNA compared to the other cellular messages (98, 102). In iron-storing red cells, both derepression of ferritin mRNA and mRNA competition are necessary to obtain the observed rates of iron-induced ferritin synthesis in whole cells (102); in such cells the capacity to synthesize proteins declines during maturation (110). Hepatocytes, by contrast, exploit mRNA competition for ferritin synthesis only, apparently, when ribosomes and initiation factors have been reduced by starvation (compare Refs. 102 and 111). The two posttranscriptional sites for regulating ferritin synthesis available to cells programmed to store iron as a specialized function permit an exceedingly powerful and efficient response to iron.

Under normal conditions, any iron to be stored is delivered to the specialized iron storage cells by transferrin. During conditions of iron overload, excess iron is taken up by other cell types. However, the accumulations of ferritin may not reach the levels observed in cells specialized for iron storage, since less ferritin mRNA is stored; what ferritin mRNA is stored is nevertheless recruited for translation when iron is in excess (112). In the case of the iron-induced mouse fibroblast, ferritin mRNA appears to achieve a translational efficiency comparable to other mRNA (112). In contrast to the fibroblast, especially efficient translation of ferritin mRNA appears to occur in the iron-induced adult frog reticulocyte (61), since 20% the amount of the ferritin mRNA produces 80% as much ferritin protein, compared to the tadpole reticulocyte. The difference in competition between ferritin mRNA in fibroblasts and in adult erythroid cells probably reflects differences in the cellular concentration of initiation factors and ribosomes in the two cell types. Which structural features of ferritin mRNA are responsible for the high translational efficiency and which component(s) of the initiation complex recognize specific ferritin mRNA feature(s) are the subjects of current study. At this time the only notable features of ferritin mRNA structure are a relatively long (150–200 nucleotides) 5' untranslated region (18, 27, 119), heterogeneity at the 5' end (18), and regions of symmetry at the 3' end (27, 66); as yet none of the features have been definitively related to ferritin mRNA function.

Transcriptional Control or Changes in mRNA Stability

Changes in ferritin mRNA concentration have recently been observed in cells for which stored iron is used mainly for intracellular needs (housekeeping) during differentiation, development, and iron overload; cells that store iron for other cells (specialized-cell iron storage) are quite different. Differences occurred not only in the concentration of ferritin mRNA but also in the subunits encoded. The results for housekeeping ferritin contrast with those for cells specialized for iron storage (66, 113). Two cell types have been examined: first, promyelocytic leukemia cells (HL-60 from humans) during differentiation toward neutrophils or macrophages (66), and second, erythroid cells of the adult cell line during early maturation [Friend's erythroleukemia cells of the mouse (113)] or late maturation [circulating reticulocytes of the bullfrog (61)]. All the studies employed two types of ferritin cDNA as hybridization probes that had only 55–60% homology to each other and did not cross-hybridize under the conditions used. The cDNAs corresponded to mRNA encoding a slow-moving subunit in SDS gel electrophoresis (H) and a fast-moving subunit (L); homologous cDNAs were used for the human and bullfrog cells, while human liver cDNAs were used for the mouse erythroid cells.

Differentiation was accompanied by an increase in the relative concentration of ferritin mRNA subunit types (the H/L ratio), indicating either changes in transcription or mRNA stability to produce ferritin with a structure appropriate to the needs of each cell type. With regard to the total amount of ferritin mRNA, differentiation produced a transient increase in macrophages, a continuous (40×) increase in neutrophils (66), and a small (2–3×) increase in erythroblasts [Friend's cells, an early stage of maturation in the adult cell line (113)]. Reticulocytes of the adult red cell line, a later stage of erythroid maturation, have only ⅕ the amount of ferritin mRNA compared to the embryonic red cell line, which stores iron for other cells (61); the higher concentration of ferritin mRNA in the embryonic cell line suggests an even greater increase or stabilization of ferritin mRNA during maturation, compared to the adult red cell line. Clearly, changes in mRNA concentration (transcription or stability) are important in establishing the expression of ferritin genes appropriate to each cell type.

Iron can also cause changes in ferritin mRNA concentration for housekeeping ferritin, in contrast to specialized-cell ferritin (61, 70). In the case of bean leaves, excess iron increased the concentration of both ferritin mRNA and protein (69, 70); only one type of subunit was observed. For the iron-loaded reticulocyte of the adult red cell, the amount of ferritin mRNA for an L subunit increased 3×; the amount of the same L subunit in the protein also increased, but it is clear from the 33× increase in ferritin that excess iron caused all three subunit mRNAs to be recruited for translation (61). The change in both the amount and type of ferritin during experimental iron overload in cells that are not specialized for storage suggests the possibility that the ferritin synthesized might be part of a stress response to toxic levels of iron. Preliminary data show that ferritin also accumulates during the stress of heat shock in adult red cells (100).

Protein Stability

The opposing effects of synthesis and degradation determine the amount of ferritin in a cell. A number of investigators have measured the stability of synthesized ferritin, particularly during induction by iron, to determine if stabilization of the protein contributes to iron-induced accumulations (5, 52, 114–116). The studies have included specialized-cell iron storage, where iron is stored for other cells [rat hepatocytes (5, 52, 115, 116), and erythrocytes of the embryonic cell line from bullfrogs (97)] and housekeeping iron storage, where iron is stored for intracellular purposes [rat skeletal myoblasts (114)]. Observed half-lives of ferritin varied from 25 to 50 hours (114, 115). Other studies showed no loss of label during 17- to 24-hour chase periods (97, 116). Measurement of protein turnover is particularly difficult for ferritin because the protein is often partitioned between the cytosol and lysosomal vesicles, and because protein subunits in the ferritin of vesicles equilibrate very slowly with ferritin in the cytosol; for example, little if any label from [^{14}C]leucine appeared in lysosomal ferritin during a 24-hour period (117). Thus the calculation of turnover rates is influenced by the relative amount of ferritin in the cytosol and vesicles coupled with the differential recovery of ferritin from the two cellular compartments. Since the lifetime of stored iron can be weeks or months, the protein coat is probably more stable than indicated from the pulse-chase experiments in cells with both lysosomal and cytosolic ferritin. Although the subcellular distribution of ferritin may complicate measurements of protein stability, the subcellular partitioning has a virtue: the partitioning and the ability to observe individual ferritin molecules both by electron microscopy and immunocytology in subcellular fractions make ferritin a candidate for the investigation of certain forms of intracellular traffic, particularly in plants where it appears to be transported to leaf chloroplasts (69) and in insects where it is associated with the vacuolar system (123).

Transient changes in the measured rate of ferritin turnover (5, 114, 116) appear to be related to the lag between protein synthesis and the formation of the iron core. Once filled with iron, the measured turnover rate for the protein appears to be similar for constitutive and iron-induced synthesis (Figure 12 in Ref. 5; 116). Thus, accumulation of ferritin during induction by iron appears to be adequately explained by changes in the rate of ferritin synthesis, resulting from the recruitment and translational efficiency of ferritin mRNA and the stability of ferritin.

Organization of Ferritin Genes

The genomes of humans, rats, and bullfrogs contain multiple hybridizable sequences for the different ferritin subunits tested using cDNAs that do not cross-hybridize (18, 27, 118). The number of fragments encoding ferritin in genomic DNA ranges from 4 to 17. Processed ferritin pseudogenes have been isolated from humans, rats, and even bullfrogs (27, 62, 119), which could account for part of the apparent size of the ferritin gene family when cDNAs are used as hybridization probes. To date only one ferritin gene has been characterized (27), a human gene coding for an H type of subunit. There are three introns in the human H gene that interrupt the coding sequence of the gene at residues 34/35, 82/83, and 124/125. (The residue numbers are those used in Figure 2, where the residue numbers are aligned with horse spleen apoferritin.) In addition to sequences encoding the protein, exon 1 contains 208 bases of 5' untranslated sequences and exon 4 contains 160 bases of 3' untranslated sequences. Based on the single copy of the gene observed using a probe prepared from the 5' flanking region of the ferritin gene (TATA box through the transcription-initiation site), most of the genomic complexity was attributed to pseudogenes (27). However, multiple active genes could also exist that differ in the 5' flanking sequences.

The chromosomal location of ferritin genes has been examined for humans using both cell fusion and hybridization in situ. Such studies have been stimulated not only by a fundamental interest in ferritin but also by the genetic linkage between hemochromatosis, a defect in iron uptake, and the HLA-A antigens on chromosome 6. Each procedure used to determine chromosome location of ferritin genes has limitations. The limitation of the cell fusion studies resides in the requirement of immunogenicity of the expressed product. For example, a gene on a chromosome encoding a ferritin without immunoreactivity for the antiserum, or which is repressed or degraded in the host cell, would not be detected and an aberrantly low number of genes would result. On the other hand, chromosomal hybridization in situ, using cDNA as a probe, would not distinguish between pseudogenes and functional genes and would yield an erroneously high number of chromosomal sites for ferritin genes.

The results of cell fusion using antibodies with restricted specificities demonstrated ferritin genes at two chromosomal sites, one for an L type subunit on chromosome 19 and one for an H type subunit on chromosome 11; the antisera used were to spleen (mostly L type subunits) (120, 121) and heart (mostly H type subunits) (121). Hybridization in situ demonstrated the existence of H-type ferritin sequences on seven different chromosomes (1, 2, 3, 6, 14, 10, and X), in addition to chromosome 11. If any of the other human genes detected are functional genes, they are either repressed in rodent cells or are not recognized by antibodies to heart and spleen ferritins.

The most conservative interpretation of the available data on ferritin gene organization is the existence of two ferritin genes, which in humans are located on different chromosomes (120–122). Accounting for the existence of three translated ferritin mRNAs in HeLa cells (38) and three distinct cDNAs in bullfrog red cells (18, 33, 62; Figure 2) requires either three functional genes or alternate processing of the transcripts of one of two genes. In any case, the presence of two functional genes on separate chromosomes indicates that *trans*-acting elements may be required for coordinate expression of subunits. The presence of processed ferritin pseudogenes in organisms as diverse as mammals and amphibia (27, 62, 119), as well as multiple types of subunit mRNA and protein, complicates the genetic analysis, although no more so than for many other gene families. Because of the complexity of the ferritins, a great deal more information is needed before it is possible to know the complete organization of the genes, the mechanisms of producing mature mRNAs, and the stoichiometric relationship between the multiple ferritin subunit mRNAs and the genes that encode them.

CELLULAR FERRITIN FUNCTION

The feature of function common to all ferritins is the storage of iron in a soluble form. For convenience, three variations in ferritin function will be considered: (*a*) the storage of iron for other cells (specialized-cell ferritin); (*b*) the storage of iron for intracellular use (normal housekeeping ferritin); and (*c*) the storage of iron for detoxification (housekeeping stress ferritin).[5] Among the cells specialized for iron storage, rates of iron uptake and release depend on the purpose of the iron reservoir, e.g. long-term storage (hepatocyte), recycling (macrophages), or rapid consumption (red cells of embryos) (47–49, 124). Variations in iron turnover could also occur in normal housekeeping ferritin, depending on the rates of iron utilization. Ferritin iron uptake and release will be affected by both apoferritin structure and the environment (anions, metal cations and chelators, reductants, and subcellular distribution). It is premature to relate cell-specific variations in apoferritin structure only to differences in rates of iron turnover, although this hypothesis can be tested. Some structural differences may have a broader significance in the cell, such as compatibility with specific features of cellular architecture, but testing such an idea is not yet easy.

Mechanisms: Iron Uptake and Release

The structural aspects of iron uptake have been previously described (see STRUCTURE, *The Iron-Apoferritin Interface*). Apoferritin has the biggest effect on iron core formation at the beginning when nucleation occurs. Modification of the protein [carboxyl esterification (57), natural and synthetic subunit crosslinks (31)], the presence of other metals such as Zn (21, 22, 81), and the nature of the electron acceptor (125), all seem important when the iron

---

[5]In insects, the unique function of maintaining iron homeostasis for hemolymph iron has been attributed to pericardial cell ferritin (123).

content of the protein is low. Once the iron core has been nucleated, variations in the environment, e.g., appear to have less effect on the rate of iron uptake (125). The second phase of iron uptake can be attributed to a purely inorganic reaction (126) in which the final result depends more on the availability of iron (yielding one large or many small crystallites) or anions in the medium (producing high or low disorder) than on the structure of the protein. In fact, once core growth has begun, the nature of the iron donor is also less restricted and can be Fe(III) as well as Fe(II) (127). Iron entering the core is not uniform chemically and does not equilibrate for at least 24 hours (128). Although nothing is known of the structural basis for such chemical differences, the level of hydration is an obvious variable to consider and might be probed by comparing the effects of anions such as sulfate on the equilibration.

A number of experiments have sought to relate differences in the surface charge of ferritin or in the subunit composition to iron uptake. No general correlations have been found. Indeed, conflicting results exist from several laboratories (e.g. 52, 129). It is possible that apparent conflicts resulted from comparing different ferritins, e.g. macrophage vs hepatocyte or cells in different physiological states (constitutive, iron-induced, iron-intoxicated, etc). The feature observed that most consistently affects rates of reconstitution is the amount of iron originally present (129), which for some ferritins depends on the abundance of subunit dimer crosslinks (31).

Iron release from ferritin displays complex kinetics. In general, iron release is faster and more complete, in vitro, after conversion of Fe(III) to Fe(II) (6, 8, 130). However, either reduction or chelation could be used in vivo, depending on the quantity and rate of iron need. Reductants that are effective in vitro include reduced flavins, sulfhydryl compounds (thioglycolic acid, dithionite, dihydrolipoic acid) (6, 131, 132), Cu(II) plus ascorbate (133), and free radicals of oxygen (superoxide) and methylviologen (134–136).

Rates of reduction and iron release are altered by a variety of conditions, including pH, buffer ions (137), the age of the protein coat or iron core (39), and at least for synthetic iron cores, the disorder of the core (94). Posttranslational modifications of the protein coat also affect rates of iron release. For example, natural or synthetic (bis dinitrobenzene) crosslinks that form intramolecular dimers between subunits increase the rate of iron release in vitro, and are inversely correlated with reduced amounts of iron in vivo (31, 85).[6] Such modifications can explain, at least for some ferritins, the previously observed dependence of rates of iron release on the amount of iron/molecule

[6]Preliminary observations on crosslinked subunits on liver ferritin from iron-deficient rabbits (S. Bottomly, personal communication) appear to extend earlier observations on the inverse relationship between crosslinks and iron content.

(129). Phosphate in the core also affects reduction and iron release. The redox potential of *A. vinelandii* (AV) ferritin (Fe:P=1.7:1) is −420 mV compared to −310 mV for horse spleen ferritin (Fe:P=8:1) at pH 7 (88, 137). A curious feature of the electrochemical reduction of either AV or horse spleen ferritin is the apparent lag between the reduction of core iron, monitored by Mössbauer spectroscopy, and the release of Fe(II), i.e. the separation of protein and iron during anaerobic chromatography on Sephadex G-25. Possible explanations for the results include (*a*) the absence of an Fe(II) chelator, e.g. bipyridyl, which is present in most experiments on iron release from ferritin to enhance separation of Fe(II) from the protein; and (*b*) a kinetic lag between protonation or rehydration of the bridging Fe ligands and electron transfer to Fe(III), since protons are needed to reverse ferritin core formation. Preliminary data suggest that a separation between reduction and conversion of the ferritin iron core to monoatomic Fe(II) appeared to occur when X-ray absorption spectroscopy (138) was used to compare reduction by thioglycolate in the presence of Hepes or Tris.

Variations in Animal and Plant Cell Types and Microorganisms

It is convenient to consider at least three functional types of ferritin in multicellular organisms: specialized-cell ferritin, which stores iron for other cell types, normal housekeeping ferritin, which stores iron for usual intracellular purposes, and stress housekeeping ferritin, which sequesters iron during iron overload; within each type, variations in the kinetics of iron turnover may occur, depending on the dynamics of intracellular iron metabolism. Among the functional types of ferritin in hepatocytes, e.g., is a relatively small fraction of ferritin for normal housekeeping and a large amount of specialized-cell ferritin for long-term iron storage. Whether the two types of hepatocyte ferritin are structurally distinct is not known, but iron-induced changes in hepatocyte ferritin surface charge have been noted (52, 129); possibly the best time to examine normal housekeeping ferritin in hepatocytes would be in regenerating liver. Macrophages, in contrast to hepatocytes, could have three types of ferritin: normal housekeeping ferritin formed during differentiation such as that accumulated in 12-*O*-tetradecanoyl phorbol-3-acetate-treated HL-60 cells (66), specialized-cell ferritin for intermediate iron storage associated with ordinary erythrophagocytosis, and stress housekeeping ferritin associated with inflammatory conditions when iron and ferritin accumulate in macrophages (99). For example, the decrease in subunit cross-links that we observed in splenic macrophage ferritin during a shift from a normal to an inflammatory state in sheep was highly correlated with a lower rate of iron uptake and release in vitro and a higher steady-state iron content in vivo (31); the results suggest that ferritin formed during stress may play an active role in the retention of iron by macrophages. Ferritin produced by iron excess in cells like HeLa, fibroblast, and cultured muscle (38, 112, 114) may also function as stress ferritin, since such cells would normally be protected from excess iron in the whole organism where the transferrin/transferrin receptor system directs the flow of excess iron to the hepatocyte. Evaluating data on the properties of iron-induced ferritin in relation to the functional type of ferritin in the cell may, in the future, reduce the number of apparently conflicting results.

Ferritin in plants, at least in leaves (69) and leguminous nodules (73), is probably normal housekeeping ferritin, while that in seeds would appear to be specialized-cell ferritin, storing iron for other cell types. But so little is known about either the structure or the metabolism of ferritin in plants that it is too soon to know if different functional types actually exist. Similarly, ferritin in bacteria and fungal mycelia, which is important during growth (74, 139–141), seems to fulfill a normal "housekeeping" function, while ferritin in fungal spores would appear to be specialized-cell ferritin. Microorganisms rarely live in an environment with excess iron available. Moreover, the microbial transport of many forms of iron is tightly regulated. Thus, the problem of whether or not microorganisms have evolved a mechanism to detoxify iron, e.g. a stress ferritin, has attracted little, if any, attention. However, in *E. coli* the several genes identified that involve iron uptake are regulated by a single gene called Fur (142). Moreover, there appear to be appropriate mutants to use in the study of microbial iron overload and ferritin metabolism.

Other Possible Roles for Ferritin

Ferritin as isolated contains small amounts of other metals, e.g. Cu and Zn, usually about 4–10 atoms/molecule. Determination of whether the ferritin is a storage form for Cu and Zn in addition to metallothionein, as has been suggested (143), or whether the protein coat or the iron core of ferritin have bound the metal ions adventitiously during isolation, awaits the results of metabolic studies.

Fe(II) in the presence of dioxygen can catalyze the production of active oxygen species, e.g. superoxide, which accelerate peroxidation of lipids (134, 136). Ferritin has evolved to protect the cell from just such chemistry. However, activated oxygen species, such as superoxide, can reduce the iron in ferritin, producing Fe(II) (136). Indirectly, then, ferritin could appear to enhance lipid peroxidation (134–136). In vivo, only if superoxide levels were abnormally high, or if the ferritin coat were damaged, e.g. in a poisoned or diseased cell, or if the cell were loaded with iron in excess of that which could be stored, would ferritin no longer provide as complete protection from iron intoxication as in a normal cell.

Current concepts of iron uptake by ferritin involve the oxidation of Fe(II) and proton release from hydrated Fe(III). If each iron atom in ferritin followed the same pathway to the iron core, up to 4500 electrons and up to 11,250 H^+ would be produced. It is not known if the electrons generated by the oxidation are coupled to energy-producing reactions as they are during the oxidation of Fe(II) to Fe(III) by *Thiobacillus ferroxidans* (144). If the electrons produced from the oxidation of iron during ferritin formation were shown to be used to generate energy, a new role for ferritin would be defined.

CONCLUDING REMARKS

The storage of iron in a soluble form is achieved in aerobic cells and organisms by ferritin, which is composed of a large (ca 12 nm diameter), multisubunit protein coat surrounding a rustlike core of hydrous ferric oxide (and variable amounts of phosphate) containing up to 4500 iron atoms. Rates of iron deposition and release as well as nucleation of the iron core are all influenced by the protein coat. Ferritin is found in vertebrates, invertebrates, higher plants, fungi, and bacteria and has been needed since oxygen-evolving photosynthesis began, probably about 2.5 billion years ago. In vertebrates, the primary structure is highly conserved, but there is not enough information available currently to know if the similarity of tertiary structure observed among all life forms results from conservative or convergent evolution.

Superimposed on the common features of apoferritin structure are structural variations. As studied in vertebrates, the structural variations occur among cell types (e.g. hepatocytes, macrophages, myocytes, and erythroid cells) and within cell types in different physiological states (e.g. differentiating, iron-induced, or stressed). Such structural differences could reflect the functional type of ferritin characteristic of particular cell types or of a cell in a particular physiological state; the possible correlation is being examined. Functional differences in animals include specialized-cell ferritin, which stores iron for other cells over short, intermediate, or long time periods, normal housekeeping ferritin, which stores iron for normal intracellular use, and stress housekeeping ferritin, which sequesters iron during iron overload and heat shock. Functional differences could also be attributed to ferritin in plants, where it is found in leaf chloroplasts, seeds, and stressed (infected) nodules of legumes. But analysis of ferritin in plants is so limited that it is not even known yet if there are structural variations, let alone whether they are related to cell types. Variations in ferritin structure may relate not only to function, but also to specific features of cellular architecture. In vertebrates, variations in ferritin structure include the subunit sequence (as many as three different sequences have been observed in some cells, and the number may be higher), surface charge, intramolecular crosslinks between pairs of subunits, and, in the case of serum ferritin, glycosylation

Variations in the ferritin iron core have been studied by a variety of physical techniques including X-ray absorption (EXAFS), EPR and Mössbauer spectroscopy, high resolution electron microscopy, and X-ray diffraction. Differences in iron core structure include composition (amount of phosphate), size, and order that may reflect differences in the protein structure, anions in the environment, or both. Variation in ferritin core structure could contribute to functional differences.

Two posttranscriptional steps regulate ferritin synthesis in cells programmed for specialized iron storage, i.e. iron stored for other cells. In such cells, iron induces synthesis and accumulation of ferritin as much as 40× above constitutive levels by recruiting stored ferritin mRNA, which is a strong competitor for translation. Transcriptional control of ferritin gene expression (or alterations in mRNA stability) occurs during differentiation or in response to iron overload in cells that store iron for housekeeping (intracellular use); both total mRNA concentration and mRNA subunit type can change. Different combinations of both transcriptional (or mRNA stability) and translational control of ferritin synthesis may also occur. Such an example is the erythroid cell of the adult, which stores iron for housekeeping but needs large amounts of iron at one particular stage of maturation; changes in mRNA composition are associated with the early stages of maturation and mRNA storage at a later stage. mRNA storage allows the regulation of ferritin synthesis by iron in the cytoplasm. Keeping iron out of the nucleus may be an advantage to a specialized cell that is committed to handling large amounts of iron throughout its life, because of the potential for DNA degradation by iron. In contrast, transcriptional regulation of ferritin genes may suffice for housekeeping iron storage when the amount of storage iron needed is small and/or the duration of ferritin synthesis is short.

The presence of at least three protein and mRNA subunit types in HeLa cells and in amphibian red cells and the presence of 4–17 hybridizable sequences in the genomes indicate a large multigene ferritin family. However, only one ferritin gene has been characterized by cloning and only two chromosomal locations for ferritin have been identified so far. Moreover, the presence of processed pseudogenes, even in amphibia, places the problems of ferritin gene organization and transcript processing firmly in the category of unsolved. Other problems about ferritin for which solutions are currently being sought are the relationships between ferritin mRNA structure and function (storage and competition); between apoferritin structure (subunit and assembled), apoferritin function, and iron core structure; between cell specificity of structure and of function; between cell specificity of regulation and of function; and between eukaryotic and prokaryotic ferritin structure. Ferritin has specific subcellular locations, namely the cytosol and lysosomes in vertebrate cells, vacuoles in insects, or chloroplasts in green plants, but intracellular ferritin transport is largely unexplored. Ferritin not only has a magnetic appeal, but is also a model for translational control, intracellular trafficking, and cell specificity of gene regulation, of structure, and of function.

ACKNOWLEDGMENTS

The writing of the review and the work of the author cited was supported by the North Carolina Agricultural Research Service and NIH Grants AM20251 and GM34675. The author is grateful to E. Stuart Maxwell and Sunil Sreedharan for critically reading the manuscript and to Joann Fish for editorial assistance.

Literature Cited

1. Plinius, C. II. 1601. *The History of the World,* Tome I, Book 34, Chap. 14, p. 513. Transl. P. Holland. London
2. Walker, J. C. G. 1977. *The Evolution of the Atmosphere,* pp. 262–64. New York: Macmillan
3. Biedermann, G., Schindler, P. 1957. *Acta Chem. Scand.* 11:731–40
4. Spiro, T. G., Bates, G., Saltman, P. 1967. *J. Am. Chem. Soc.* 89:5559–62
5. Munro, H. N., Linder, M. C. 1978. *Physiol. Rev.* 58:317–96
6. Theil, E. C. 1983. *Adv. Inorg. Biochem.* 5:1–38
7. White, J. L., Smith, J. M. A., Harrison, P. M. 1983. *Adv. Inorg. Biochem.* 5:39–50
8. Clegg, G. A., Fitton, J. E., Harrison, P. M., Treffry, A. 1981. *Prog. Biophys. Mol. Biol.* 36:53–86
9. Aisen, P., Listowsky, I. 1980. *Ann. Rev. Biochem.* 49:357–93
10. Theil, E. C. 1986. In *Translational Control of Gene Expression in Eukaryotes,* ed. J. Ilan. New York: Plenum. In press
11. Harrison, P. M., Fischbach, F. A., Hoy, T. G., Haggis, G. H. 1967. *Nature* 216:1188
12. Mann, S., Bannister, J. V., Williams, R. J. P. 1986. *J. Mol. Biol.* 188:225–32
13. Massover, W. H., Cowley, J. M. 1973. *Proc. Natl. Acad. Sci. USA* 70:3847–51
14. Yang, C.-Y., Meagher, A., Huynh, B. H., Sayers, D. E., Theil, E. C. 1986. *Biochemistry.* In press
15. Banyard, S. H., Stammers, D. K., Harrison, P. M. 1980. *Nature* 288:298–300
16. Tsugita, A., Yariv, J. 1985. *Biochem. J.* 231:209–12
17. Macey, D. J., Smalley, S. R., Potter, I. C., Cake, M. H. 1985. *J. Comp. Physiol. B* 156:269–76
18. Didsbury, J. R., Theil, E. C., Kaufman, R. E., Dickey, L. F. 1986. *J. Biol. Chem.* 261:949–55
19. Harrison, P. M., White, J. L., Smith, J. M. A., Farrants, G. W., Ford, G. C., et al. 1985. In *Proteins of Iron Storage and Transport,* ed. G. Spik, J. Montreuil, R. R. Crichton, J. Mazurier, pp. 67–79. New York: Elsevier Science
20. Dorner, M. H., Salfeld, J., Will, H., Leibold, E. A., Vass, J. K., Munro, H. N. 1985. *Proc. Natl. Acad. Sci. USA* 82:3139–43
21. Macara, I. G., Hoy, T. G., Harrison, P. M. 1973. *Biochem. J.* 135:785–89
22. Chasteen, N. D., Theil, E. C. 1982. *J. Biol. Chem.* 257:7672–77
23. Crichton, R. R., Millar, J. A., Cumming, R. L. C., Bryce, C. F. A. 1973. *Biochem. J.* 131:51–59
24. Heustersreute, M., Crichton, R. R. 1981. *FEBS Lett.* 129:322–27
25. Theil, E. C., Brenner, W. E. 1981. *Dev. Biol.* 84:481–84
26. Addison, J. M., Treffry, A., Harrison, P. M. 1984. *FEBS Lett.* 175:333–36
27. Constanzo, F., Columbo, M., Staempfli, S., Santoro, C., Marone, M., et al. 1986. *Nucleic Acids Res.* 14:721–36
28. Boyd, D., Vecoli, C., Belcher, D. M., Byrd, S. K., Drysdale, J. W. 1985. *J. Biol. Chem.* 260:11755–61
29. Wustefeld, C., Crichton, R. R. 1982. *FEBS Lett.* 150:43–48
30. Ford, G. C., Harrison, P. M., Rice, D. W., Smith, J. M. A., Treffry, A., et al. 1984. *Philos. Trans. R. Soc. London Ser. B* 304:561–65
31. Mertz, J. R., Theil, E. C. 1983. *J. Biol. Chem.* 258:11719–26
32. Levi, S., Arosio, P., Cesareni, G., Cortese, O. R. 1985. *Proc. 7th Int. Conf.*

Proteins of Iron Metab., Lille, France (Abstr.)

33. Sreedharan, S. P., Didsbury, J. R., Theil, E. C., McKenzie, A. R., Kaufman, R. E. 1986. *Fed. Proc.* 45(6):1699 (Abstr.)
34. Otsuka, S., Maruyama, H., Listowsky, I. 1981. *Biochemistry* 20:5226–32
35. Addison, J. M., Fitton, J. E., Lewis, W. G., May, F., Harrison, P. M. 1983. *FEBS Lett.* 64:139–44
36. Leibold, E. A., Aziz, N., Brown, A. J. P., Munro, H. N. 1984. *J. Biol. Chem.* 259:4327–34
37. Lavoie, D. J., Ishikawa, K., Listowsky, I. 1978. *Biochemistry* 17:5448–54
38. Watanabe, N., Drysdale, J. W. 1983. *Biochim. Biophys. Acta* 743:98–105
39. Ihara, K., Maeguchi, K., Young, C. T., Theil, E. C. 1984. *J. Biol. Chem.* 259:278–83
40. Collawn, J. F. Jr., Fish, W. W. 1984. *Comp. Biochem. Physiol. B* 78:653–56
41. Collawn, J. F. Jr., Lau, P. L., Morgan, S. L., Fox, A., Fish, W. W. 1984. *Arch. Biochem. Biophys.* 233:260–66
42. Massover, W. H. 1985. *Biochim. Biophys. Acta* 829:377–86
43. Linder, M. C., Nagel, G. M., Roboz, M., Hungerford, D. M. Jr. 1981. *J. Biol. Chem.* 256:9104–10
44. Rice, D. W., Day, B., Smith, J. M. A., White, J. L., Ford, G. C., et al. 1985. *FEBS Lett.* 181:165–68
45. Worwood, M., Dawkins, S., Wagstaff, M., Jacobs, A. 1976. *Biochem. J.* 157:97–103
46. Stefanini, S., Chiacone, E., Arosio, P., Finazzi-Agro, A., Antonini, E. 1982. *Biochemistry* 21:2293–99
47. Yamada, H., Gabuzda, T. G. 1974. *J. Lab. Clin. Med.* 83:477–88
48. Brown, J. E., Theil, E. C. 1978. *J. Biol. Chem.* 253:2673–78
49. Van Wyck, C. P., Linder-Horowitz, M., Munro, H. N. 1971. *J. Biol. Chem.* 246:1025–31
50. Drysdale, J. W. 1977. *Ciba Found. Symp.* 51:41–57
51. Halliday, J. W., McKearing, L. V., Powell, L. W. 1976. *Cancer Res.* 36:4486–90
52. Bomford, A., Conlon-Hollingshead, C., Munro, H. N. 1981. *J. Biol. Chem.* 256:948–55
53. Cazzola, M., Dezza, L., Bergamaschi, G., Barosi, G., Bellotti, V., et al. 1983. *Blood* 62:1078–87
54. Wyllie, F., Jacobs, A., Waradanukul, K., Worwood, M., Wagstaff, M. 1984. *Leukemia Res.* 8:1095–101
55. Treffry, A., Lee, P. J., Harrison, P. M. 1984. *Biochem. J.* 220:717–22

56. Jones, B. M., Worwood, M. 1978. *Clin. Chim. Acta* 85:81–88
57. Wetz, K., Crichton, R. R. 1976. *Eur. J. Biochem.* 61:545–50
58. Arioso, P., Adelman, T. G., Drysdale, J. W. 1978. *J. Biol. Chem.* 253:4451–58
59. Boyd, D., Jain, S. K., Crampton, J., Barrett, K., Drysdale, J. W. 1984. *Proc. Natl. Acad. Sci. USA* 80:1265–69
60. Jain, S. K., Barrett, K. J., Boyd, D., Favreau, M., Crampton, J., et al. 1985. *J. Biol. Chem.* 260:11762–68
61. Dickey, L. S., Sreedharan, S., Theil, E. C., Didsbury, J. R., Wong, Y-H, Kaufman, R.E. 1987. *J. Biol. Chem.* In press
62. Dickey, L. F., Theil, E. C. 1986. *Fed. Proc.* 45(6):1883 (Abstr.)
63a. Sackett, L., Wolff, J. 1986. *J. Biol. Chem.* 261:9070–76
63b. Bryan, J. 1974. *Fed. Proc.* 33:152–57
64. Telakowski-Hopkins, C. A., Rodkey, J. A., Bennett, C. D., Lu, A. Y. H., Pickett, C. B. 1985. *J. Biol. Chem.* 260:5820–25
65. Sachs, A. B., Bond, M. W., Kornberg, A. 1986. *Cell* 45:827–35
66. Chou, C.-C., Gatti, R. A., Fuller, M. L., Concannon, P., Wong, A., et al. 1986. *Mol. Cell. Biol.* 6:566–73
67. Nebert, D. W. 1987. *Ann. Rev. Biochem.* 56:945–93
68. Seckbach, J. 1969. *Plant Physiol.* 44:816–20
69. Van der Mark, P., Van den Briel, W., Huisman, H. G. 1983. *Biochem. J.* 214:943–50
70. Van der Mark, P., Bienfait, F., Van den Ende, H. 1983. *Biochem. Biophys. Res. Commun.* 115:463–69
71. Crichton, R. R., Ponce-Ortiz, Y., Koch, M. H. J., Parfait, R., Stuhrmann, H. B. 1978. *Biochem. J.* 171:349–56
72. Sczekan, S., Joshi, J. 1986. *Fed. Proc.* 45(6):1602 (Abstr.)
73. Ko, M. P., Huang, P.-Y., Huang, J.-S., Barker, K. R. 1985. *Phytopathology* 75:159–64
74. McIntosh, M. A., Earhart, C. F. 1977. *J. Bacteriol.* 131:331–39
75. Moore, G. R., Mann, S., Bannister, J. V. 1986. *J. Inorg. Biochem.* In press
76. Stiefel, E. I., Watt, G. D. 1979. *Nature* 279:81–83
77. Li, J. D., Wang, J. W., Zhong, Z. P., Tu, Y., Dong, B. 1980. *Sci. Sinica* 23:897–904
78. La Bombardi, V. J., Pisano, M. A., Klavin, S. 1982. *J. Bacteriol.* 150:671–75
79. Yariv, J., Kalb, J., Sperling, R., Barringer, E. R., Cohen, S. G., et al. 1981. *Biochem. J.* 197:171–75
80. Wardeska, J. G., Viglione, B., Chas-

teen, N. D. 1986. *J. Biol. Chem.* 261:6677–83

81. Treffry, A., Harrison, P. M. 1984. *J. Inorg. Biochem.* 21:9–20

82. Wilkins, R. G., Harrington, P. C. 1983. *Adv. Inorg. Biochem.* 5:51–85

83. Antanaitis, B. C., Aisen, P. 1984. *J. Biol. Chem.* 259:2066–69

84. Chasteen, N. D., Antanaitis, B. C., Aisen, P. 1985. *J. Biol. Chem.* 260:2926–29

85. Theil, E. C. 1985. In *Frontiers in Bioinorganic Chemistry,* ed. A. V. Xavier, pp. 259–67. Weinheim/Deersfield Beach: VCH

86. Schwertman, V., Schulze, D. G., Murad, E. 1982. *J. Soil Sci. Soc. Am.* 46:869–75

87. Mansour, A. N., Thompson, C., Theil, E. C., Chasteen, N. D., Sayers, D. E. 1985. *J. Biol. Chem.* 260:7975–79

88. Watt, G. D., Frankel, R. B., Spartalian, K., Stiefel, E. I. 1986. *Biochemistry.* 25:4330–36

89. St. Pierre, T. G., Bell, S. H., Dickson, D. P. E., Mann, S., Webb, J., et al. 1986. *Biochim. Biophys. Acta* 870:127–34

90. Mann, S., Perry, C. C., Webb, J., Luke, B., Williams, R. J. P. 1986. *Proc. R. Soc. London Ser. B* 227:179–90

91. Bell, S. B., Weir, M. P., Dickson, D. P. E., Gibson, J. F., Sharp, G. A., et al. 1984. *Biochim. Biophys. Acta* 787:227–36

92. Richter, G. W. 1984. *Lab. Invest.* 50:26–35

93. Weir, M. P., Peters, T. J., Gibson, J. F. 1985. *Biochim. Biophys. Acta* 828:298–305

94. Yang, C.-Y., Bryan, A. M., Theil, E. C., Sayers, D. E., Bowen, L. H. 1986. *J. Inorg. Biochem.* 28:393–405

95. Theil, E. C. 1981. In *Hemoglobins in Development and Differentiation,* ed. G. Stamatoyannopoulos, A. Nienhuis, pp. 423–31. New York: Liss

96. Zahringer, J., Baliga, B. S., Munro, H. N. 1976. *Proc. Natl. Acad. Sci. USA* 73:857–61

97. Schaefer, F. V., Theil, E. C. 1981. *J. Biol. Chem.* 256:1711–15

98. Shull, G. E., Theil, E. C. 1982. *J. Biol. Chem.* 257:14187–91

99. Roeser, H. P. 1980. In *Iron in Biochemistry and Medicine,* ed. A. Jacobs, M. Worwood, 2:605–40. New York: Academic

100. Atkinson, B., Blaker, T. W., Dean, R. L. 1986. *J. Cell Biol.* (Abstr.) In press

101. Aziz, N., Munro, H. N. 1986. *Nucleic Acids Res.* 14:915–27

102. Shull, G. E., Theil, E. C. 1983. *J. Biol. Chem.* 258:7921–23

103. Lindquist, S. 1986. *Ann. Rev. Biochem.* 55:1151–91

104. Mueckler, M., Merrill, M., Pitot, H. C. 1985. *J. Biol. Chem.* 260:6109–14

105. McGuire, D. M., Chan, L., Smith, L. C., Towle, H. C., Dempsey, M. E. 1985. *J. Biol. Chem.* 260:5435–39

106a. Hertzberg, R. P., Dervan, P. B. 1982. *J. Am. Chem. Soc.* 104:313–15

106b. Dreyer, G. B., Dervan, P. B. 1985. *Proc. Natl. Acad. Sci. USA* 82:968–72

107. Standart, N. M., Bray, S. J., George, E. L., Hunt, T., Ruderman, J. N. 1985. *J. Cell Biol.* 100:1968–76

108. Sjoberg, B.-M., Eklund, H., Fuchs, J. A., Carlson, J., Standart, N. M., et al. 1985. *FEBS Lett.* 183:99–102

109. Lodish, H. F., Desalu, O. 1973. *J. Biol. Chem.* 248:3520–27

110. Theil, E. C. 1975. *Dev. Biol.* 46:343–48

111. Drysdale, J. W., Olafdottier, E., Munro, H. N. 1968. *J. Biol. Chem.* 243:552–55

112. Walden, W. E., Thach, R. E. 1986. *Biochemistry* 25:2033–41

113. Drysdale, J., Jain, S. K., Barrett, K. J., Vecoli, C., Belcher, D. M., et al. 1985. See Ref. 19, p. 346

114. Rittling, S., Woodworth, R. C. 1984. *J. Biol. Chem.* 259:5561–66

115. Linder, M. C., Moor, J. R., Scott, L. E., Munro, H. N. 1973. *Biochim. Biophys. Acta* 297:70–80

116. Treffry, A., Lee, P. J., Harrison, P. M. 1984. *FEBS Lett.* 165:243–46

117. Richter, G. 1984. *Lab. Invest.* 50:26–35

118. Brown, A. J. P., Leibold, E. A., Munro, H. N. 1983. *Proc. Natl. Acad. Sci. USA* 80:1265–69

119. Munro, H. N., Leibold, E. A., Vass, J. K., Aziz, N., Rogers, J., et al. 1985. See Ref. 19, pp. 331–41

120. Caskey, J. H., Jones, C., Miller, Y. E., Seligman, P. A. 1983. *Proc. Natl. Acad. Sci. USA* 80:482–86

121. Worwood, M., Brook, J. D., Cragg, S. J., Hellkuhl, B., Jones, B. M., et al. 1985. *Human Genet.* 69:371–74; 70:191 (correction)

122. Cragg, S. J., Drysdale, J., Worwood, M. 1985. *Human Genet.* 71:108–12

123. Locke, M., Leung, H. 1984. *Tissue Cell Res.* 16:739–66

124. Theil, E. C. 1980. *Br. J. Haematol.* 45:357–60

125. Treffry, A., Harrison, P. M. 1980. *FEBS Lett.* 100:33–36

126. Mayer, D. L., Rohrer, J. S., Schoeller, D. A., Harris, D. C. 1983. *Biochemistry* 22:876–80

127. Treffry, A., Harrison, P. M. 1979. *Biochem. J.* 181:709–16
128. Treffry, A., Harrison, P. M. 1984. *Biochem. J.* 220:857–59
129. Treffry, A., Lee, P. J., Harrison, P. M. 1984. *Biochim. Biophys. Acta* 785:22–29
130. Octave, J.-N., Schneider, Y.-J., Crichton, R. R., Trouet, A. 1983. *Biochem. Pharmacol.* 32:3413–18
131. Funk, F., Lenders, J.-P., Crichton, R. R., Schneider, W. 1985. *Eur. J. Biochem.* 152:167–72
132. Bonomi, F., Pagani, S. 1986. *Eur. J. Biochem.* 155:295–300
133. Bienfait, H. F., Van den Briel, M. L. 1980. *Biochim. Biophys. Acta* 631:507–10
134. Koster, J. F., Slee, R. G. 1986. *FEBS Lett.* 199:85–88
135. Thomas, C. E., Morehouse, L. A., Aust, S. D. 1985. *J. Biol. Chem.* 260:3275–80
136. Thomas, C. E., Aust, S. D. 1986. *J. Biol. Chem.* 261:13064–70
137. Watt, G. D., Frankel, R. B., Papaefthymiou, G. C. 1985. *Proc. Natl. Acad. Sci. USA* 82:3640–43
138. Theil, E. C., Sayers, D. E., Yang, C. Y., Fontaine, A., Dartyge, E. 1986. *Proc. 4th Int. EXAFS Conf., France.* In press
139. Klebba, P. E., McIntosh, M. A., Neilands, J. B. 1982. *J. Bacteriol.* 149:880–88
140. Neilands, J. B. 1983. *Adv. Inorg. Biochem.* 5:137–66
141. David, C. N. 1974. In *Microbial Iron Metabolism,* ed. J. B. Neilands, pp. 149–58. New York: Academic
142. Zimmerman, L., Hantke, K., Braun, V. 1984. *J. Bacteriol.* 159:271–77
143. Price, D., Joshi, J. G. 1982. *Proc. Natl. Acad. Sci. USA* 79:3116–19
144. Lundgren, D. G., Vestal, J. R., Tabita, F. R. 1974. See Ref. 141, pp. 457–73

Ann. Rev. Biochem. 1987. 56:317–32
Copyright © 1987 by Annual Reviews Inc. All rights reserved

IMPACT OF VIRUS INFECTION ON HOST CELL PROTEIN SYNTHESIS

Robert J. Schneider

Department of Biochemistry, New York University School of Medicine, 550 First Avenue, New York, NY 10016

Thomas Shenk

Department of Molecular Biology, Princeton University, Princeton, New Jersey 08544

CONTENTS

PERSPECTIVES AND SUMMARY

Viruses must compete with their host cell for macromolecular machinery at many levels subsequent to infection. Viruses utilize cellular gene products for DNA and RNA replication, transcription, processing and transport of mRNAs, protein synthesis, and posttranslational modification of polypeptides. Many viruses establish conditions within the infected cell that enable them to dominate various components of the cellular machinery. Translation is one area in which a wide variety of viruses usurp the cellular apparatus, and there appear to be many mechanisms through which this can be accomplished. The infected cell, for its part, can respond to virus infection by

317

implementing defensive strategies designed to inhibit viral translation, and, as one might expect, viruses have, in turn, evolved functions that frustrate the cellular defenses. Thus, virus-cell interactions at the level of protein synthesis appear especially complex.

Much effort has been devoted to the dissection of this interaction in a wide variety of viral systems. The problem has proven difficult but not intractable. There are a number of instances in which the principal players have been identified, and at least portions of the interaction can be described in molecular detail.

Here, we first overview the variety of mechanisms employed by viruses to dominate the cellular translational apparatus. Next, we discuss relatively recent work that has identified viral gene products able to antagonize antiviral effects normally induced by interferon treatment of cells. We have not attempted to reference every publication contributing to each topic under discussion. Additional references and more detailed accounts of many aspects of virus-host cell translational interactions can be found in recent reviews (1–3) and in a monograph edited by Fraenkel-Conrat & Wagner (4).

VIRUS-MEDIATED SHUT-OFF OF HOST CELL TRANSLATION

Although many viruses replicate without suppressing host cell protein synthesis, others partially or completely inhibit host cell translation subsequent to infection. Members of the adenoviruses, alphavirus subgroup of togaviruses, herpesviruses, picornaviruses, poxviruses, myxoviruses, reoviruses, and rhabdoviruses establish conditions within the infected cell that favor translation of viral mRNAs at the expense of cellular species. A surprising variety of mechanisms have been implicated in host cell shut-off. These include the degradation of host cell mRNAs, inactivation of translation factors, the production of factors that specifically inhibit cellular translation or facilitate viral translation, viral mRNAs that outcompete cellular mRNAs, and changes in the intracellular ionic environment that favor translation of viral mRNAs. There is no reason to expect that a virus would employ only one approach. In fact, this is almost certainly not the case. Vesicular stomatitis virus, a member of the rhabdovirus family, likely employs (a) competition by producing a large quantity of viral mRNAs (5, 6), which may initiate more efficiently than cellular mRNAs within the infected cell (7), and (b) partial inactivation of one or more initiation factors (8–10).

We first survey the variety of mechanisms employed to favor viral translation, and then examine the manner in which several viruses deal with host cell antiviral responses that impact at the level of translation.

Survey of Mechanisms

DEGRADATION OF HOST CELL MRNAS   Host cell mRNAs are rapidly degraded in both poxvirus (11) and herpes simplex virus–infected cells (12, 13). In the case of vaccinia virus, a poxvirus, a virus-coded product is required to prevent degradation of viral mRNAs late after infection (14). It is not clear whether this product protects against a virus-mediated degradation of mRNAs or a cellular antiviral response.

Shut-off of host cell translation subsequent to herpes simplex virus infection occurs in two stages. The first stage does not require viral gene expression and results in the dissociation of host cell mRNAs from polysomes. The second stage requires expression of the viral chromosome and results in degradation of host cell mRNAs (15). Mutants that are defective in stage-one shut-off are still competent for stage two, strongly suggesting the involvement of two different gene products (16).

The mechanism whereby the herpesvirus-specific nuclease discriminates host from viral mRNAs remains unclear. Conceivably, herpesvirus mRNAs are spared due to a signal encoded in their sequence or structure. Alternatively, differential compartmentalization may physically separate viral mRNAs from the nuclease.

INACTIVATION OF TRANSLATION FACTORS   The most thoroughly documented instance in which a normal translational initiation factor is inactivated is cleavage of the p220 component of the eIF-4F cap-binding complex subsequent to infection with poliovirus, a picornavirus. This makes good sense given the assumption that p220 is required for cap-binding activity, since poliovirus mRNA is uncapped and does not require the factor. The shut-off of cellular translation subsequent to poliovirus infection is discussed in detail below.

Another initiation factor, eIF-2, is partially inactivated by phosphorylation of its alpha subunit subsequent to infection with vesicular stomatitis virus (8, 10), reovirus (17), and possibly adenovirus (discussed below). It seems a safe bet that the viral mRNAs still require eIF-2 for initiation. Partial inactivation of the factor could, however, provide an opportunity for efficiently initiated viral mRNAs to outcompete less efficient cellular messages, and ultimately dominate the cellular translation apparatus.

FACTORS THAT INHIBIT CELLULAR TRANSLATION   Pensiero & Lucas-Lenard (18) found that cell-free translation extracts prepared from mengovirus-infected cells (mengovirus is a picornavirus) initiated translation more poorly than extracts prepared from uninfected cells. The defect was not relieved by addition of reticulocyte initiation factors, suggesting an initiation factor, such as eIF-2 or eIF-4F, was not simply inactivated. The inhibition could be relieved by washing infected-cell ribosomes with high salt, leading the investigators to suggest that mengovirus induces or activates a dominant inhibitor that binds to ribosomes. The origin and biochemical nature of the inhibitor are as yet unknown. It apparently does not discriminate between viral and cellular mRNAs, but since mengovirus mRNAs are inherently translated at very high efficiency (19, 20), mengovirus mRNAs are able to dominate residual translational capacity within the infected cell.

FACTORS THAT FACILITATE VIRAL TRANSLATION   An excellent example of this type of mechanism has been reported for frog virus 3 (21). This virus belongs to the iridovirus family, whose members infect either insects or amphibians. Late viral mRNAs are inefficiently traslated in rabbit reticulocyte lysates. A marked enhancement of protein synthesis was obtained by adding a ribosomal salt wash from infected but not uninfected cells. Apparently, frog virus 3 codes, induces, or activates a factor that facilitates translation of its late mRNAs.

Recently, there have been several suggestions of such positive-acting factors in the animal virus literature. Bernstein et al (22) have inferred the existence of a factor in poliovirus-infected cells that enables the virus to escape a nonspecific translational inhibition that occurs in certain cell types. Rosen et al (23) have reported that human T-lymphotropic virus type III (HTLV III or LAV) also actively enhances virus-specific protein synthesis. The HTLV III *trans*-activating gene product does not influence the level of viral cytoplasmic mRNAs, but enhances virus-specific protein levels more than 500-fold. Enhancement is dependent on the presence of a specific sequence near the 5' end of viral mRNAs. Whether the effect is due to direct action by the *trans*-activating gene product or to an indirect effect such as induction or activation of a factor that impacts on viral translation is, as yet, unclear.

COMPETITION   Viral mRNAs can gain control of cellular translation capacity by competing through simple abundance. Alternatively, they can compete through an intrinsic ability to initiate translation more efficiently than most cellular mRNAs. There are well-documented examples of both types of competition.

Competition results from simple abundance in both reovirus (24) and vesicular stomatitis virus–infected cells (5). Neither virus produces mRNAs that display unusually high efficiencies of initiation. Their efficiencies are similar to those of endogenous mRNAs, since, in both cases, the size of viral polysomes is the same or somewhat smaller than host cell polysomes that code for proteins of similar size. If initiation of viral mRNAs was more efficient, then viral polysomes would be larger assuming their rate of translational elongation remained constant. Both viruses produce great quantities of mRNAs, and the level of host cell shut-off tracks the concentration of intracellular vesicular stomatitis virus mRNA (6). Thus, the switch from host cell to viral translation must be governed at least in part by the abundance of viral mRNAs.

A number of viral mRNAs have been shown to initiate translation more efficiently than does the average cellular mRNA. These include mRNAs coded by several picornaviruses (encephalomyocarditis virus, 25,26; mengovirus, 19,20). In addition, mRNA coded by influenza virus shows above-average ability to compete for limited amounts of eIF-2 (27). On a molar basis, mengovirus mRNA competes 35 times more efficiently in cell-free translation extracts than does globin mRNA! It is not clear what makes one mRNA more efficient than another. Kozak (3) has suggested a highly efficient mRNA may simply lack such inhibitory features as 5' secondary structures or poor translational initiation consensus sequences.

ALTERED IONIC ENVIRONMENT   Carrasco (28) proposed that shut-off of host cell protein synthesis is often caused, at least in part, by an increase in intracellular sodium ion concentration. A number of viruses have been reported to alter cell membrane permeability, causing an imbalance in the *trans*-membrane sodium/potassium concentration gradient (reviewed in 29). In contrast to cellular mRNAs, whose translation is inhibited by high sodium, a number of viral mRNAs function at increased efficiency under these conditions (e.g. poliovirus, 30; encephalomyocarditis virus, 31; vesicular stomatitis virus, 7). An especially strong case can be made for a role of monovalent cations in encephalomyocarditis virus–induced host cell shut-off. The timing of monovalent cation influx correlates well with an overall decline in cellular protein synthesis (32), and host translation is restored when infected cells are placed in hypotonic medium (33).

It seems likely that salt affects translation at least in part by altering conformation of mRNAs, or by altering the ease with which an RNA's conformation can be changed. Along this line, Kozak (34) has produced an altered preproinsulin mRNA that is potentially able to form a hairpin (ΔG -30 kcal/mol) within its 5' noncoding region. While the mRNA was translated normally in standard medium, it was impaired in hypertonic medium.

Details of Picornavirus-Induced Shut-Off

The picornaviruses comprise a large but related group of RNA-containing viruses that include enteroviruses such as poliovirus and coxsackievirus and cardioviruses such as encephalomyocarditis virus, mengovirus, rhinovirus, and foot-and-mouth-disease virus (for a general review see 35). Picornavirus mRNAs lack the blocked, methylated cap structures found on most mRNAs that are required for efficient translation initiation (reviewed in 36, 37).

The picornaviruses epitomize the complexities and frustrations encountered in elucidating the manner in which viruses interact with and dominate protein synthetic machinery within infected cells. Although extensively related as a group, the viruses have seemingly evolved quite different mechanisms for usurping the host cell machinery. To some extent, this variety probably reflects evolutionary pressures deriving from differences in the types of cells infected by each virus and the attendant antiviral activities peculiar to each cell type. Poliovirus, encephalomyocarditis virus, and mengovirus are the best studied in terms of virus-mediated inhibition of host cell protein synthesis (reviewed in 38), and they effectively illustrate a variety of the takeover strategies discussed in the preceding section.

POLIOVIRUS   Shut-off of host cell translation is complete by two hours after infection with poliovirus. Cellular polysomes disaggregate and are replaced by polysomes containing poliovirus mRNA (39). The defect in translation of cellular mRNAs was first localized to the initiation step (40–42), and then, more specifically to the inability of host cell mRNAs to associate with 40S ribosomal subunits (43, 44). The nontranslated cellular mRNAs remain intact (40, 45, 46), and can be translated in cell-free extracts (47). Poliovirus can also interfere with the translation of a good variety of mRNAs coded by coinfecting viruses (e.g. herpesvirus, 48; vesicular stomatitis virus, 49, 50; adenovirus, 51) with similar kinetics to that observed for host cell mRNAs (49).

Several potential mechanisms can be readily ruled out as major players in poliovirus-induced shut-off. Intracellular cation levels do, indeed, increase in poliovirus-infected cells, and poliovirus mRNAs translate efficiently under these conditions (7, 28). However, the increase in intracellular cation concentration does not occur until several hours after inhibition of host cell protein synthesis (31, 52, 53). Competition for initiation on the basis of mRNA abundance can be excluded, since shut-off occurs before significant replication of poliovirus RNA that could produce a pool of competing mRNA molecules. Poliovirus mRNA does not compete on the basis of initiation efficiency. In fact, the viral mRNA is translated less efficiently than many of the cellular mRNAs that are inhibited (44, 54, 55). Finally, double-stranded RNA, which could be produced as a consequence of viral replication, strongly inhibits the initiation of translation (reviewed in 56, 57), and was considered a possible key to the shut-off phenomenon. However, double-stranded RNA shows no host versus virus specificity (58), and its accumulation does not fit with the kinetics of host cell shut-off (59).

Poliovirus-induced shut-off of host cell protein synthesis results at least in part from destruction of an initiation factor. Initial experiments produced conflicting results, suggesting either eIF-4B (54) or eIF-3 (60) was deficient in extracts of virus-infected cells. The confusion began to be resolved when it was shown that cap-binding protein (a 24-kd protein required for initiation by capped mRNAs; reviewed in 36) could be purified from preparations of both eIF-3 and eIF-4B (61). Although purified cap-binding protein alone failed to reactivate cap-dependent translation in poliovirus-infected cell extracts, a complex, termed cap-binding complex or eIF-4F (comprised of 24-kd, 46-kd, 73-kd, and 220-kd polypeptides), succeeded (62–64). In a key experiment, Etchison et al (65) employed immunoblotting procedures to demonstrate that the 220-kd polypeptide (p220) was not present in cap-binding complex prepared from infected cells. Rather, antigenically related polypeptides of 100–300 kd were found. The correlation between the kinetics of p220 cleavage and host shut-off was reasonably good. Thus, given the assumption that p220 is indeed a functional element of the cap-binding complex, poliovirus commandeers the host cell translational machinery by inactivating a factor required for initiation of capped host mRNAs but not uncapped viral mRNAs.

What is the origin of the proteolytic activity that cleaves p220? The viral mRNA comprises the entire genome and gives rise to a primary translation product termed the polyprotein. Poliovirus encodes at least one protease, called polypeptide 3C, that processes the viral polyprotein by cleavage at glutamine-glycine junctions (66). This protease was an attractive candidate for the activity mediating cleavage of p220. However, antiserum to protein 3C, which inhibits processing of the poliovirus polyprotein, does not inhibit cleavage of the 220-kd polypeptide in an in vitro assay (67). In a similar in vitro assay, it was determined that a second, putative viral protease, polypeptide 2A, also fails to cleave the 220-kd polypeptide (68). A poliovirus mutant incapable of inhibiting cellular translation has been described recently (22), and interestingly, it contains a single codon insertion in the 2A gene. The genetic data combined with the inability of 2A product to cleave p220 in vitro suggest the 2A product is required for the cleavage, but acts indirectly to mediate the process. Further, p220 cleavage products that comigrate with those formed subsequent to infection can be detected in small quantities within uninfected cells (22). It seems likely the virus activates a normal cellular activity responsible for p220 turnover.

One is left to question how uncapped mRNAs bypass the need for either a cap structure or a functional cap-binding complex. The complex is believed to melt secondary structure at the 5' end of mRNAs in an ATP-dependent reaction (69–72). Interestingly, inosine-substituted reovirus mRNA, which has reduced secondary structure due to the reduced stability of I·C as compared to G·C base-pairs, can form initiation complexes in poliovirus-infected cell extracts, while nonsubstituted mRNA cannot (73). Thus, the likely function of cap-binding complex in melting secondary structure, together with the inosine-substitution experiment, suggests that poliovirus might have evolved an mRNA with reduced secondary structure in its 5' domain, allowing it to bypass the need for a cap and the attendant functions of the cap-binding complex.

Inactivation of p220 may not be the whole story behind poliovirus-induced host cell shut-off. Significant alterations in the cytoskeletal framework occur as a result of virus infection (74), and the normal association of host cell mRNAs with this structure is disrupted (75). As yet it is difficult to say whether the cytoskeletal changes signal an additional viral strategy or simply constitute the inevitable result of host cell shut-off. In another area, an attenuated type 1 poliovirus has been shown to carry mutations upstream of the polyprotein's AUG and produce mRNAs that are translated less efficiently than wild-type mRNAs in cell-free extracts (76). Assuming the effect is due to the upstream mutations, it could prove informative to further pursue the role of the altered sequences on viral translation. Finally, Bernstein et al (22) have inferred the existence of a factor that enables poliovirus to escape a nonspecific translational inhibition that occurs subsequent to infection. There could well be much more to the poliovirus–host cell translational interaction than cleavage of p220.

It was surprising to find that encephalomyocarditis virus and mengovirus do not inactivate cap-binding complex (77). Extracts prepared from cells infected with either of these viruses will translate added cellular mRNA (19, 20, 78–80). Thus, different mechanisms must underlie shut-off in the case of these viruses.

ENCEPHALOMYOCARDITIS VIRUS   In contrast to the situation with poliovirus, host cell translation is not inhibited early after infection with encephalomyocarditis virus. In fact, there is little if any decline in total protein synthesis until relatively late in infection (80, 81).

The concurrent addition of viral and cellular mRNAs to uninfected cell translation extracts led to preferential usage of the viral mRNA (78). A similar study in which uninfected cell extracts were programmed with total mRNAs from virus-infected cells also demonstrated the predominant translation of viral messages despite the fact that they constituted only one-fifth the total mRNA population (26). These experiments suggest the viral mRNAs outcompete host cell species for host cell translation capacity. Consistent with this interpretation, supplementation of cell-free extracts with mixtures of initiation factors can relieve the competitive advantage of viral mRNAs (82–84).

As mentioned earlier in this review, the timing of encephalomyocarditis virus–induced host cell shut-off correlates well with an increase in intracellular cation concentrations (32). The case for a direct effect of cation concentration in shut-off is strong since host cell translation is restored when infected cells are placed in hypotonic medium (33). The ability of viral mRNAs to outcompete cellular species in extracts from uninfected cells may result from precisely the same characteristic that permits them to function more efficiently than cellular mRNAs in a hypertonic environment.

MENGOVIRUS   Like poliovirus, but unlike encephalomyocarditis virus, mengovirus induces rapid shut-off of host cell translation (85). So far, two elements of the mengovirus shut-off strategy have been identified. The first element of the strategy is a viral mRNA that is extremely efficient in initiating translation (19, 20, 86, 87). Lysates programmed with globin mRNA exhibit a half-maximal inhibition when 1/35 as much mengovirus mRNA is added. Mengovirus mRNA also binds eIF-2 about 30 times more efficiently than globin messages (87, 88). The second element of the viral takeover is a ribosome-associated translational inhibitor (18). The inhibitor can be removed from ribosomes by washing in 0.5-M KC1, restoring their activity. Its origin is not known.

Given the translational inhibitor plus highly efficient mRNA, it seems likely that mengovirus first inhibits overall translation and then outcompetes the host cell for residual translational capacity. This model fits well with the observation that the overall translational activity of a mengovirus-infected cell is reduced as compared to an uninfected cell (18).

SUMMARY   Each of the picornaviruses appears to employ a quite different strategy in order to dominate the infected cell's translational apparatus. Poliovirus and mengovirus both induce host cell shut-off rapidly after infection. However, poliovirus achieves this through inactivation of a factor required for initiation of capped mRNAs, while mengovirus encodes, induces, or activates a factor responsible for a general inhibition of protein synthesis and then outcompetes its host for residual translational capacity. Encephalomyocarditis virus does not induce host cell shut-off until quite late after infection. This virus likely relies on the ability of its mRNA to function efficiently in an ionic environment that is unsuitable for host mRNAs.

VIRAL RESPONSES TO CELLULAR ANTIVIRAL DEFENSES

The best understood antiviral responses of infected cells are those induced by interferons, and these impact at the level of translation. Two antiviral functions elicited by interferon are a protein kinase, termed the P1/eIF-2α kinase, and (2'-5') (A)$_n$ synthetase (reviewed in 89). Synthesis of the two enzymes is induced by interferon, and both are activated by double-stranded RNA. The activated protein kinase phosphorylates and inactivates a translational initiation factor, eIF-2, thereby inhibiting protein synthesis. Activated synthetase generates 2'-5'-linked oligoadenylates from ATP. These in turn activate a latent endonuclease, termed ribonuclease L, that cleaves single-stranded RNAs (e.g. mRNA and ribosomal RNA), again inhibiting translation. It has recently become apparent that several viruses have evolved gene products that antagonize these antiviral activities.

Poxvirus Factor

Poxviruses are the largest in size and structurally the most complex of animal viruses. Vaccinia virus has been studied as the family's prototype. It induces a rapid inhibition of host cell macromolecular synthesis (reviewed in 90). Cellular polysomes are disaggregated and cellular mRNAs degraded. Mechanisms underlying these events remain unclear. Extracts and crude preparations of initiation factors from vaccinia virus–infected cells are unable to support in vitro translation of encephalomyocarditis mRNA (91), and the block to initiation seems to occur at an early step in the initiation process (92). Products of early viral transcripts (93) and a fraction prepared from viral cores (94) have both been shown to inhibit translation in reticulocyte lysates.

The growth of vaccinia virus is insensitive to interferon in a variety of cell types (95). Not only is vaccinia itself insensitive, but it is able to protect coinfecting viruses from the effects of interferon (95–98). Both vesicular stomatitis virus and picornaviruses are normally inhibited by interferon, and both are protected when vaccinia virus is present. This protection appears to result from the ability of vaccinia virus to inhibit the interferon-induced, double-stranded RNA-dependent P1/eIF-2α kinase. Evidence is good that a factor coded or induced by the virus inhibits activity of the kinase (96, 98, 99). This activity has been termed specific kinase inhibitory factor, and appears to be protein in nature. It is not yet clear how it might function to prevent activation of the double-stranded RNA-dependent kinase. Perhaps the protein factor competes with kinase for an interaction with double-stranded RNA that is required for activation of the kinase. Further insight to its mechanism of action must await isolation and characterization of the specific kinase inhibitory factor.

Vaccinia virus also appears able to inhibit activation of ribonuclease L (100, 101). As described above, this nuclease is activated subsequent to production of 2'-5'-linked oligoadenylates by the interferon-induced (2'-5') $(A)_n$ synthetase. In vaccinia virus–infected cells, 2'-5'-linked oligoadenylates are produced in response to interferon, but fail to activate the nuclease. As yet, the factor responsible for this phenomenon has not been identified.

Herpes simplex viruses have also been reported to inhibit activation of ribonuclease L in interferon-treated cells (102). As was the case for vaccinia virus, $(2'-5')$ $(A)_n$ synthetase was activated, but the latent nuclease was not. Finally, encephalomyocarditis virus has been reported to inactivate the latent ribonuclease (103, 104).

Adenovirus VAI RNA

The human adenoviruses are a family of DNA tumor viruses. During the early phase of adenovirus infection host cell translation continues. Concomitant with the onset of the late phase of the infectious cycle, host cell protein synthesis is dramatically inhibited (e.g. 105). Host shut-off occurs at two levels (recently reviewed in 106). First, newly synthesized and processed cellular mRNAs do not accumulate in the cytoplasm (107). Two viral gene products, the E1B-55K (108–110) and the E4-34K polypeptide (111), which exist in a complex (112), are responsible for inhibition of mRNA accumulation. Available evidence suggests the complex likely blocks accumulation at the level of transport from nucleus to cytoplasm. There is also a block to translation of host cell mRNAs delivered to the cytoplasm prior to the halt of accumulation. These mRNAs remain intact (e.g. 113, 114) and apparently associated with a small number of ribosomes (115, 116). Nevertheless, they are not actively translated. The inhibition may result at least in part from the ability of late viral mRNAs to outcompete cellular mRNAs in the infected cell milieu. This seems likely in view of the fact that the majority of late mRNAs carry a 5' noncoding region that can enhance translation in late virus-infected cells of late mRNAs that normally do not carry the sequence (117, 118). This leader is not the whole story since many viral mRNAs that do not contain it are efficiently translated late after infection. Further, complex interactions with the interferon-induced P1/eIF-2α kinase are occurring late after infection, which may also influence virus-induced shut-off. This interaction and its implications will be discussed next.

Useful insights into adenovirus-host cell interactions at the level of translation have been generated by study of small, virus-coded RNAs termed VA RNAs (119). These are small transcripts of about 160 nucleotides synthesized in large amounts late after infection by RNA polymerase III (120). There are two VA RNA genes, designated VAI and VAII, located at about 30 map units on the adenovirus type 5 chromosome (121, 122). To explore the function of these RNAs, two variants were constructed, each of which fails to encode one of the VA species (123). Inability to produce VAII RNA generated no phenotype, while the mutant that failed to produce VAI RNA grew more poorly than its parent. Analysis of the defect indicated that VAI RNA is required for inefficient initiation of translation late after infection (123, 124).

The translational defect results from activation of the P1/eIF-2α kinase and subsequent loss of eIF-2 activity (125–127). The role of the kinase in loss of eIF-2 activity has been documented in several systems (reviewed in 56). During the initiation process, eIF-2, which is composed of three subunits (α,

β, γ), functions in a ternary complex with GTP and met-tRNA$_i$. GTP is hydrolyzed to GDP as the 80S initiation complex is formed and eIF-2 is released. The eIF-2-associated GDP must be exchanged for GTP before the factor can mediate a new round of initiation. This exchange is catalyzed by the GTP recycling factor, termed eIF-2B. It does not occur when the eIF-2α subunit has been phosphorylated. Rather, phosphorylated eIF-2 sequesters the limited quantities of eIF-2B in a tight complex, preventing the recycling reaction (reviewed in 57). Thus, GTP is no longer recycled, eIF-2 does not function catalytically, and the frequency of initiation is decreased.

As mentioned above, synthesis of the P1/eIF-2α kinase is induced by interferon, and it is activated by double-stranded RNA. VAI RNA prevents activation but not induction of the kinase in response to interferon (128). Further, VAI RNA can inhibit activation of the kinase by double-stranded RNA in extracts prepared from interferon-treated cells (128, 129).

How does VAI RNA inhibit activation of the kinase? To address this question we must consider the likely mechanism of kinase activation. First, the kinase binds to double-stranded RNA (130, 131). Second, activation of the kinase is associated with the phosphorylation of a polypeptide termed P1, which is very likely the kinase itself (89, 132). Third, although the kinase is activated by low levels of double-stranded RNA, high levels inhibit activation (133–136). Perhaps, then, activation occurs by binding of multiple kinase molecules to one double-stranded RNA molecule. The kinase molecules could then phosphorylate each other, leading to activation. The activated kinase would, in turn, phosphorylate the eIF-2α polypeptide, inhibiting translation. High levels of double-stranded RNA could prevent activation because each kinase moiety would bind to a different RNA molecule, and autophosphorylation, which is presumably required for activation, would not occur.

Given this model, it seems reasonable to suggest that VAI RNA functions by binding a single kinase molecule. This would, in theory, prevent auto-phosphorylation and subsequent kinase activation. Monovalent binding might occur simply because VAI RNA concentrations are high within infected cells (119), and multiple kinase molecules never find the same RNA molecule. Alternatively, VAI RNA may be designed to bind just one molecule of kinase irrespective of their relative concentrations. This idea fits well with the observation that VAI RNA cannot activate the kinase (128, 129), indicating that it does not function as a simple double-stranded RNA molecule. Mismatches in the VAI RNA molecule's duplex structure may be critical to its ability to interfere without activating. Double-stranded RNAs with one mismatch in eight base pairs fail to activate kinase (135).

One prediction of the model for VAI RNA function has already been met. VAI RNA binds to purified kinase (J. Kitajewski, T. Shenk, unpublished observation).

Not surprisingly, although growth of adenoviruses is relatively insensitive to interferon (137), propagation of a mutant adenovirus that fails to produce VAI RNA is inhibited by interferon (128). VAI RNA must play a key role in the ability of the virus to propagate in an infected host animal by antagonizing the antiviral activity of interferon.

Epstein-Barr virus encodes two small EBER RNAs that have secondary structures similar to VA RNAs (138, 139). EBER RNAs can partially substitute for VAI RNA in adenovirus-infected cells, suggesting they are functionally equivalent (140, 141). Similarly, hepatitis B virus encodes a small RNA polymerase III–transcribed RNA that facilitates translation subsequent to transfection of cultured cells with DNAs that encode the small viral transcript and a test mRNA (B. Aufiero, R. J. Schneider, unpublished observation). This assay scores for inhibition of the P1/eIF-2α kinase since the process of transfection with calcium phosphate-DNA precipitates generates kinase activity (142). Thus, three different viruses produce small RNAs with apparently similar functions. It is noteworthy that all three are able to maintain long-term infections of their animal hosts. Acquisition of this biological characteristic likely requires a solution to the interferon-induced P1/eIF-2α kinase.

Summary

There is good evidence that a variety of viruses encode functions that inhibit activation of either ribonuclease L or the P1/eIF-2α kinase. In general, this capability is evident among viruses that maintain long-term active or latent infections, and must therefore deal with interferon-induced antiviral responses of their hosts.

ACKNOWLEDGMENTS

We thank our colleagues who kindly provided us with reprints and preprints of their work, and gratefully acknowledge the competent secretarial assistance of Ms. Elena Chiarchiaro. Robert J. Schneider is the recipient of an Irma T. Hirschel Career Scientist Award, and Thomas Shenk is an American Cancer Society Research Professor.

Literature Cited

1. Shatkin, A. J. 1983. *Philos. Trans. R. Soc. London Ser. B* 303:167–76
2. Kääriäinen, L., Ranki, M. 1984. *Ann. Rev. Microbiol.* 38:91–109
3. Kozak, M. 1986. *Adv. Virus Res.* 31:229–92
4. Fraenkel-Conrat, H., Wagner, R. R., eds. 1984. *Comprehensive Virology* Vol. 19. New York: Plenum. 536 pp.
5. Lodish, H. F., Porter, M. 1980. *J. Virol.* 36:719–33
6. Lodish, H. F., Porter, M. 1981. *J. Virol.* 38:504–17
7. Nuss, D. L., Oppermann, H., Koch, G. 1975. *Proc. Natl. Acad. Sci. USA* 72:1258–62
8. Centrella, M., Lucas-Lenard, J. 1982. *J. Virol.* 41:781–91
9. Thomas, J. R., Wagner, R. R. 1983. *Biochemistry* 22:1540–46
10. Dratewka-Kos, E., Kiss, I., Lucas-Lenard, J., Mehta, H. B., Woodley, C.

L., Wahba, A. J. 1984. *Biochemistry* 23:6184–90
11. Rice, A. P., Roberts, B. E. 1983. *J. Virol.* 47:529–39
12. Nishioka, Y., Silverstein, S. 1977. *Proc. Natl. Acad. Sci. USA* 74:2370–74
13. Fenwick, M. L., McMenamin, M. M. 1984. *J. Gen. Virol.* 65:1225–28
14. Pacha, R. F., Condit, R. C. 1985. *J. Virol.* 56:395–403
15. Nishioka, Y., Silverstein, S. 1978. *J. Virol.* 27:619–27
16. Read, G. S., Frenkel, N. 1983. *J. Virol.* 46:498–512
17. Samuel, C. E., Duncan, R., Knutson, G. S., Hershey, J. W. B. 1984. *J. Biol. Chem.* 259:13451–57
18. Pensiero, M. N., Lucas-Lenard, J. M. 1985. *J. Virol.* 56:161–71
19. Abreu, S. L., Lucas-Lenard, J. 1976. *J. Virol.* 18:182–94
20. Hackett, P. B., Egberts, E., Traub, P. 1978. *Eur. J. Biochem.* 83:341–52
21. Raghow, R., Granoff, A. 1983. *J. Biol. Chem.* 258:571–78
22. Bernstein, H. D., Sonenberg, N., Baltimore, D. 1985. *Mol. Cell. Biol.* 5:2913–23
23. Rosen, C. A., Sodroski, J. G., Goh, W. C., Dayton, A. I., Lippke, J., Haseltine, W. A. 1986. *Nature* 319:555–59
24. Walden, W. E., Godelfroy-Colburn, T., Thach, R. E. 1981. *J. Biol. Chem.* 256:11739–46
25. Jen, G., Birge, C. H., Thach, R. E. 1978. *J. Virol.* 27:640–47
26. Svitkin, Y. V., Ginevskaya, V. A., Ugarova, T. Y., Agol, V. I. 1978. *Virology* 87:199–203
27. Katze, M. G., Detjen, B. M., Safer, B., Krug, R. M. 1986. *Mol. Cell. Biol.* 6:1741–50
28. Carrasco, L. 1977. *FEBS Lett.* 75:11–15
29. Carrasco, L., Lacal, J. C. 1983. *Pharmacol. Ther.* 23:109–45
30. Saborio, J. L., Pong, S. S., Koch, G. 1974. *J. Mol. Biol.* 85:191–211
31. Carrasco, L., Smith, A. E. 1976. *Nature* 264:807–9
32. Lacal, J. C., Carrasco, L. 1982. *Eur. J. Biochem.* 127:359–66
33. Alonso, M. A., Carrasco, L. 1981. *J. Virol.* 37:535–40
34. Kozak, M. 1986. *Proc. Natl. Acad. Sci. USA* 83:2850–54
35. Perez-Bercoff, R., ed. 1979. *The Molecular Biology of Picornaviruses.* New York: Plenum. 412 pp.
36. Shatkin, A. J. 1976. *Cell* 9:645–53
37. Banerjee, A. K. 1980. *Bacteriol. Rev.* 44:175–205
38. Ehrenfeld, E. 1984. *Compr. Virol.* 19:177–221
39. Penman, S., Scherrer, K., Becker, Y., Darnell, J. E. 1963. *Proc. Natl. Acad. Sci. USA* 49:654–61
40. Willens, M., Penman, S. 1966. *Virology* 30:355–67
41. Summers, D. F., Maizel, J. V. 1967. *Virology* 31:550–55
42. Leibowitz, R., Penman, S. 1971. *J. Virol.* 8:661–68
43. Ehrenfeld, E., Manis, S. 1979. *J. Gen. Virol.* 43:441–45
44. Brown, B. A., Ehrenfeld, E. 1980. *Virology* 103:327–39
45. Koschel, K. 1974. *J. Virol.* 13:1061–66
46. Fernandez-Munoz, R., Darnell, J. E. 1976. *J. Virol.* 18:719–26
47. Kaufmann, Y., Goldstein, E., Penman, S. 1976. *Proc. Natl. Acad. Sci. USA* 73:1834–38
48. Saxton, R. E., Stevens, J. C. 1972. *Virology* 48:207–20
49. Doyle, S., Holland, J. 1972. *J. Virol.* 9:22–28
50. Ehrenfeld, E., Lund, H. 1977. *Virology* 80:297–308
51. Bablanian, R., Russell, W. C. 1974. *J. Gen. Virol.* 24:261–79
52. Egberts, E., Hackett, P. B., Traub, P. 1977. *J. Virol.* 22:591–97
53. Nair, C. N., Stowers, J. W., Singfield, B. 1979. *J. Virol.* 31:184–89
54. Rose, J. K., Trachsel, H., Leong, K., Baltimore, D. 1978. *Proc. Natl. Acad. Sci. USA* 75:2732–36
55. Shih, D. S., Shih, C. T., Kew, O., Pallansch, M., Rueckert, R., Kaesberg, P. 1978. *Proc. Natl. Acad. Sci. USA* 75:5807–11
56. Jagus, R., Anderson, W. F., Safer, B. 1981. *Prog. Nucleic Acid Res. Mol. Biol.* 25:127–85
57. Safer, B. 1983. *Cell* 33:7–8
58. Celma, M. L., Ehrenfeld, E. 1974. *Proc. Natl. Acad. Sci. USA* 71:2440–44
59. Collins, F. D., Roberts, W. K. 1972. *J. Virol.* 10:969–78
60. Helentjaris, T., Ehrenfeld, E., Brown-Luedi, M. L., Hershey, J. W. B. 1979. *J. Biol. Chem.* 254:10973–78
61. Sonenberg, N., Morgan, M. A., Merrick, W. C., Shatkin, A. J. 1978. *Proc. Natl. Acad. Sci. USA* 75:4843–47
62. Tahara, S. M., Morgan, M. A., Shatkin, A. J. 1981. *J. Biol. Chem.* 256:7691–94
63. Grifo, J. A., Tahara, S. M., Morgan, M. A., Shatkin, A. J., Merrick, W. C. 1983. *J. Biol. Chem.* 258:5804–10
64. Edery, I., Lee, K. A. W., Sonenberg, N. 1984. *Biochemistry* 23:2456–62
65. Etchison, D., Milburn, S. C., Edery, I., Sonenberg, N., Hershey, J. W. B. 1982. *J. Biol. Chem.* 257:14806–10

66. Hanecak, R., Semler, B. C., Anderson, C. W., Wimmer, E. 1982. *Proc. Natl. Acad. Sci. USA* 79:3973–77
67. Lee, K. A. W., Edery, I., Hanecak, R., Wimmer, E., Sonenberg, N. 1985. *J. Virol.* 55:489–93
68. Lloyd, R. E., Toyoda, H., Etchison, D., Wimmer, E., Ehrenfeld, E. 1986. *Virology* 150:299–303
69. Kozak, M. 1980. *Cell* 19:79–90
70. Kozak, M. 1980. *Cell* 22:459–67
71. Morgan, M. A., Shatkin, A. J. 1980. *Biochemistry* 19:5960–66
72. Sonenberg, N., Guertin, D., Cleveland, D., Trachsel, H. 1981. *Cell* 27:563–72
73. Sonenberg, N., Guertin, D., Lee, K. A. W. 1982. *Mol. Cell. Biol.* 2:1633–38
74. Cervera, M., Dreyfuss, G., Penman, S. 1981. *Cell* 23:113–20
75. Lenk, R., Penman, S. 1979. *Cell* 16:289–301
76. Svitkin, Y. V., Maslova, S. V., Agol, V. I. 1985. *Virology* 147:243–52
77. Mosenkis, J., Daniels-McQueen, S., Janovec, S., Duncan, R., Hershey, J. W. B., et al. 1985. *J. Virol.* 54:643–45
78. Lawrence, C., Thach, R. E. 1974. *J. Virol.* 14:598–610
79. Svitkin, Y. V., Ugarova, T. Y., Ginevskaya, V. A., Kalinina, N. O., Scarlat, I. V., Agol, V. I. 1974. *Intervirology* 4:214–20
80. Jen, G., Detjen, B. M., Thach, R. E. 1980. *J. Virol.* 35:150–56
81. Jen, G., Thach, R. E. 1982. *J. Virol.* 43:250–61
82. Golini, F., Thach, S. S., Birge, C. H., Safer, B., Merrick, W. C., Thach, R. E. 1976. *Proc. Natl. Acad. Sci. USA* 73:3040–44
83. Tahara, S. M., Morgan, M. A., Grifo, J. A., Merrick, W. C., Shatkin, A. J. 1982. In *Interaction of Translational and Transcriptional Controls in the Regulation of Gene Expression*, pp. 359–72. New York: Elsevier
84. Ray, B. K., Brendler, T. G., Adya, S., Daniels-McQueen, S., Miller, J. K., et al. 1983. *Proc. Natl. Acad. Sci. USA* 80:663–67
85. Colby, D. S., Finnerty, V., Lucas-Lenard, J. 1974. *J. Virol.* 13:858–69
86. Kaempfer, R., Rosen, H., Israeli, R. 1978. *Proc. Natl. Acad. Sci. USA* 75:650–54
87. Rosen, H., DiSegni, G., Kaempfer, R. 1982. *J. Biol. Chem.* 257:946–52
88. Perez-Bercoff, R., Kaempfer, R. 1982. *J. Virol.* 41:30–41
89. Lengyel, P. 1982. *Ann. Rev. Biochem.* 51:251–82
90. Bablanian, R. 1984. *Compr. Virol.* 19:391–429
91. Schrom, M., Bablanian, R. 1979. *Virology* 99:319–28
92. Person, A., Beaud, G. 1980. *Eur. J. Biochem.* 103:85–93
93. Coppola, G., Bablanian, R. 1983. *Proc. Natl. Acad. Sci. USA* 80:75–79
94. Ben-Hamida, F., Person, A., Beaud, G. 1983. *J. Virol.* 45:452–55
95. Youngner, J. S., Thacore, H. R., Kelly, M. E. 1972. *J. Virol.* 10:171–81
96. Whitaker-Dowling, P. A., Youngner, J. S. 1983. *Virology* 131:128–36
97. Whitaker-Dowling, P. A., Youngner, J. S. 1984. *Virology* 137:171–81
98. Paez, E., Esteban, M. 1984. *Virology* 134:12–28
99. Rice, A. P., Kerr, I. M. 1984. *J. Virol.* 50:229–36
100. Paez, E., Esteban, M. 1984. *Virology* 134:29–35
101. Rice, A. P., Roberts, W. K., Kerr, I. M. 1984. *J. Virol.* 50:220–28
102. Cayley, P. J., Davies, J. A., McCullagh, K. G., Kerr, I. M. 1984. *Eur. J. Biochem.* 143:165–74
103. Cayley, P. J., Knight, M., Kerr, I. M. 1982. *Biochem. Biophys. Res. Commun.* 104:376–82
104. Silverman, R. H., Cayley, P. J., Knight, M., Gilbert, C. S., Kerr, I. M. 1982. *Eur. J. Biochem.* 124:131–38
105. Bello, L. J., Ginsberg, H. S. 1967. *J. Virol.* 1:843–50
106. Flint, S. J. 1984. *Compr. Virol.* 19:297–358
107. Beltz, G., Flint, S. J. 1979. *J. Mol. Biol.* 131:353–73
108. Babiss, L. E., Ginsberg, H. S., Darnell, J. E. 1985. *Mol. Cell. Biol.* 5:2552–58
109. Pilder, S., Moore, M., Logan, J., Shenk, T. 1986. *Mol. Cell. Biol.* 6:470–76
110. Karger, B. D., Ho, Y. S., Castiglia, C. L., Flint, S. J., Williams, J. 1986. *Cancer Cells* 4: In press
111. Halbert, D. N., Cutt, J. R., Shenk, T. 1985. *J. Virol.* 56:250–57
112. Sarnow, P., Hearing, P., Anderson, C. W., Halbert, D. N., Shenk, T., Levine, A. J. 1984. *J. Virol.* 49:692–700
113. Philipson, L., Pettersson, U., Lindberg, U., Tibbetts, C., Vennstrom, B., Persson, T. 1975. *Cold Spring Harbor Symp. Quant. Biol.* 39:447–56
114. Babich, A., Feldman, L. T., Nevins, J. R., Darnell, J. E., Weinberger, C. 1983. *Mol. Cell. Biol.* 3:1212–21
115. Khalili, K., Weinmann, R. 1984. *J. Mol. Biol.* 175:453–68
116. Katze, M. G., DeCorato, D., Krug, R. M. 1986. *J. Virol.* In press
117. Logan, J., Shenk, T. 1984. *Proc. Natl. Acad. Sci. USA* 81:3655–59

118. Bernkner, K. L., Sharp, P. A. 1985. *Nucleic Acids Res.* 13:841–57
119. Reich, P. R., Rose, J., Forget, B., Weissman, S. M. 1966. *J. Mol. Biol.* 17:428–39
120. Weinmann, R., Raskas, H. J., Roeder, R. G. 1974. *Proc. Natl. Acad. Sci. USA* 71:3426–30
121. Mathews, M. B. 1975. *Cell* 6:223–29
122. Pettersson, U., Philipson, L. 1975. *Cell* 6:1–4
123. Thimmappaya, B., Weinberger, C., Schneider, R. J., Shenk, T. 1982. *Cell* 31:543–51
124. Schneider, R. J., Weinberger, C., Shenk, T. 1984. *Cell* 37:291–98
125. Reichel, P. A., Merrick, W. C., Siekierka, J., Mathews, M. B. 1985. *Nature* 313:196–200
126. Schneider, R. J., Safer, B., Munemitsu, S. M., Samuel, C. E., Shenk, T. 1985. *Proc. Natl. Acad. Sci. USA* 82:4321–25
127. Siekierka, J., Mariano, T. M., Reichel, P. A., Mathews, M. B. 1985. *Proc. Natl. Acad. Sci. USA* 82:1959–63
128. Kitajewski, J., Schneider, R. J., Safer, B., Munemitsu, S. M., Samuel, C. E., et al. 1986. *Cell* 45:195–200
129. O'Malley, R. P., Mariano, T. M., Siekierka, J., Mathews, M. B. 1986. *Cell* 44:391–400
130. Hovanessian, A. G., Kerr, I. M. 1979. *Eur. J. Biochem.* 93:515–26
131. Lasky, S. R., Jacobs, B. L., Samuel, C. E. 1982. *J. Biol. Chem.* 257:11087–93
132. Berry, M. J., Knutson, G. S., Lasky, S. R., Munemitsu, S. M., Samuel, C. E. 1985. *J. Biol. Chem.* 260:11240–47
133. Hunter, T., Hunt, T., Jackson, R. J., Robertson, H. D. 1975. *J. Biol. Chem.* 250:409–17
134. Farrell, P. J., Balkow, K., Hunt, T., Jackson, R. J., Trachsel, H. 1977. *Cell* 11:187–200
135. Minks, M. A., West, D. K., Benvin, S., Baglioni, C. 1979. *J. Biol. Chem.* 254:10180–83
136. Miyamoto, N. G., Jacobs, B. C., Samuel, C. E. 1983. *J. Biol. Chem.* 258:15232–37
137. Stewart, W. E. 1979. *The Interferon System.* New York: Springer. 421 pp.
138. Lerner, M. R., Andrews, N. C., Miller, G., Steitz, J. A. 1981. *Proc. Natl. Acad. Sci. USA* 78:805–9
139. Rosa, M. D., Gottlieb, E., Lerner, M. R., Steitz, J. A. 1981. *Mol. Cell. Biol.* 1:785–96
140. Bhat, R. A., Thimmappaya, B. 1983. *Proc. Natl. Acad. Sci. USA* 80:4789–93
141. Bhat, R. A., Thimmappaya, B. 1985. *J. Virol.* 56:750–86
142. Akusjarvi, G., Svensson, C., Nygard, O. 1986. *Mol. Cell. Biol.* In press

Ann. Rev. Biochem. 1987. 56:333–64

INTRACELLULAR PROTEASES

Judith S. Bond and P. Elaine Butler

Department of Biochemistry and Molecular Biophysics, Virginia Commonwealth University, Richmond, Virginia 23298

CONTENTS

PERSPECTIVES AND SUMMARY

During the last decade there have been substantial advances in our understanding of cellular processes involving proteolysis, from the proteolytic processing of nascent polypeptide chains that accompanies protein synthesis to the extensive proteolysis that results in degradation of proteins to amino acids (1–4). Moreover, it has become evident that cellular proteolysis is a highly controlled, complex process that takes place in virtually all com-

333

0066-4154/87/0701-0333$02.00

partments of cells. There have been appreciable advances in the biochemical characterization of cellular proteinases, some of which are summarized in this review. It should be appreciated, however, that little is known about the physiological substrates of many cellular proteinases, what initiates and regulates proteolytic processes, or which proteinases are involved in specific functions.

Cellular proteinases range in size from approximately 20,000 to 800,000 daltons. Some of the well-characterized cellular proteinases, such as the lysosomal proteinases, fit the traditional concept of the small, monomeric proteases. However, we now know that many cellular proteinases have subunit molecular weights of 50,000 to 100,000 and many are oligomeric (see e.g. Tables 2, 3, and 5). The majority of polypeptides in cells (from *Escherichia coli* to HeLa cells) have subunit molecular weights of 30,000–50,000, and only 3–5% have subunit molecular weights greater than 80,000 (5, 6). In addition, most holoenzymes are smaller than 200,000 daltons (7). Hence, many proteinases are indeed larger than most cellular enzymes. There are probably good reasons for the synthesis of large proteinases that contain multiple subunits or domains. This may be critical for the selectivity and regulatability of cellular proteolysis. The large proteinases often have nonproteinaceous components; some are glycoproteins or lipoproteins, and some require divalent cations, ATP, or other small ligands. In addition, some have multicatalytic sites and cannot be easily categorized in one of the four major classes of endopeptidases (see section on endopeptidase classes). Many are unstable and difficult to purify. At least some of these qualities may be important for the synthesis, subcellular localization, regulation, and function of cellular proteinases.

As a general rule, it appears that many of the serine proteinases that are associated with mammalian cells are found in secretory granules or specialized granules (e.g. azurophil granules). Some are involved in specific protein processing reactions and others are general proteases that exert their action when they are secreted. Most of the cysteine proteinases have general proteinase activity, may be involved with the initial or terminal stages of extensive degradation of proteins, are found in the cytosol or in lysosomes, and exert their action within these compartments. The few aspartic proteinases found in cells are associated with secretory granules, membranes, endosomes, or lysosomes. The metallo-proteinases that have been described in mammalian cells are associated with endoplasmic reticulum, the plasma membrane, mitochondria, or the cytosol. It has been suggested that these enzymes are involved in removal of signal peptides and peptide processing, inactivation of biologically active peptides, and hydrolysis of proteins associated with membranes.

The subfield of cellular proteinases is in its infancy, especially compared with the area of extracellular proteases. Our knowledge of the structure, mechanisms, and function of the cellular proteinases is very limited compared to what we know about the kinetics, fine structure, and regulation of other enzymes. None of the mammalian cellular proteinases have been crystalized, amino acid sequence data are available for only a few, and the kinetic analyses of these enzymes are limited. Large quantities of purified cellular proteinases are not available and investigations of the enzymes have tended to be descriptive rather than mechanistic. We are still discovering which proteinases exist in cells and attempting to decipher how newly discovered enzymes relate to those already described. It is clear that many proteinases exist in high-molecular-weight complexes, either with specific substrates, or other polypeptide components. Compartments of mammalian cells, e.g. lysosomes, endosomes, secretory granules, membrane formations, transport vesicles, and mitochondria, interact and have multiple ways of controlling proteolytic processes. The substrates themselves, proteins and peptides, contain information that will determine their proteolytic susceptibility. With the new techniques that have emerged in cell and molecular biology, immunology, and protein chemistry, it is likely that we are on the brink of discoveries that will lead to a better understanding of cellular proteinases and cell proteolysis.

SCOPE OF THE REVIEW

All proteases are 'intracellular' at some stage in their existence. Some are clearly synthesized for export to extracellular spaces (secretory proteases), and they exert their biological action as discrete entities outside cells. There are many examples of extracellular proteases, from the small, well-characterized, pancreatic proteases trypsin and chymotrypsin, to the larger, more complex plasma proteases of the blood coagulation and complement systems. Other secretory proteases spend a longer span of time in cells after synthesis, usually stored in granules, before being secreted into the extracellular milieu; examples of these proteases are the mast cell proteases chymase and tryptase. The proteinases that we focus on in this review are mammalian cellular proteinases, those that exert their action as integral components of cells. These 'intracellular' proteinases may act on cellular or extracellular peptides or proteins (e.g. proteins brought into the cell by endocytosis), but they do so in association with a cell structure or compartment. Some of the cellular proteases may exclusively degrade resident cellular proteins and peptides (e.g. cytosolic and mitochondrial proteases); others may be active in the degradation of extracellular proteins (e.g. lysosomal proteases or plasma membrane proteases), in addition to cellular proteins; and yet others may act on proteins that will be secreted by the cell.

The term *protease* is synonymous with *peptide hydrolase*, and these terms include all enzymes that cleave peptide bonds (classified by the Enzyme Commission of the International Union of Biochemistry as EC 3.4). Proteases can be subdivided into *exopeptidases,* whose action is directed by the amino- or carboxy-terminus of the peptide (EC 3.4.11–19), or *endopeptidases,* enzymes that cleave peptide bonds internally in peptides and usually cannot accommodate the charged amino- or carboxyl-termini amino acids at the active site (EC 3.4.21–24, 99). It has been suggested that the term endopeptidase be used synonymously with *proteinase,* and we adopt that recommendation in this review (8).

The review discusses mammalian cellular proteinases first in the context of the well-defined mechanistic classes of proteinases. Where possible, evolutionary relationships and generalizations about these classes of proteinases are pointed out. Subsequently, the proteinases are discussed in terms of specific groups (grouped by functional or biochemical properties, or subcellular localization). Some proteinases are not discussed, and others are only mentioned briefly because of the constraints of space. The review emphasizes relatively well-characterized enzymes that illustrate the general characteristics and diversity of cellular proteinases.

ENDOPEPTIDASE CLASSES AND MECHANISMS

General Considerations

Endopeptidases are classified according to essential catalytic residues at their active sites (9). Four distinct classes of proteinases have been identified: serine (EC 3.4.21), cysteine (EC 3.4.22), aspartic (EC 3.4.23), and metalloproteinases (EC 3.4.24). Thus the serine proteinases are characterized by the presence of a uniquely reactive serine side chain at the active center, and the catalytic mechanism of these proteinases involves the covalent binding of substrates to this serine residue (e.g. 10). Cysteine proteinases (formerly called thiol proteinases) contain an essential cysteine residue that is involved in a covalent intermediate complex with substrates (11). Aspartic proteinases (formerly called acid proteinases) contain two aspartic residues at their active centers that are involved in catalysis. It is thought that general acid-base catalysis, rather than the formation of covalent enzyme-substrate intermediates, is operative in the mechanism of these enzymes (12). Metalloproteinases contain metal ions (usually zinc) at the active center. The metal ions are an integral part of their structures, and likely enhance the nucleophilicity of H_2O and polarize the peptide bond to be cleaved prior to nucleophilic attack (13, 14). Some proteinases of unknown catalytic mechanism have been assigned to a new, temporary subclass: EC 3.4.99. This subclass accommodates enzymes that either have not been sufficiently purified to allow assignation to one of the mechanistic classes or clearly do not fit one of the four classical groups. For example, some proteolytic enzymes do not appear to be inhibited by any of the classical protease inhibitors (e.g. 15), and these might represent a new mechanistic class of proteinases.

The class of a proteinase is usually determined according to the effects of protease inhibitors on enzyme activity (16). All serine proteinases are inhibited by diisopropyl fluorophosphate (DFP, DIFP, or diisopropyl phosphofluoridate, DipF), most by phenylmethanesulfonyl fluoride (PMSF), and some by chloromethyl ketones, e.g. TLCK (N-*p*-tosyl-L-lysine chloromethyl ketone) or TPCK (L-1-tosylamido-2-phenylethyl chloromethyl ketone). However, PMSF and the chloromethyl ketones will also inhibit some cysteine proteinases. Cysteine proteinases are inhibited by low concentrations of *p*-hydroxymercuribenzoate (pHMB, the hydrolysis product of *p*-chloromercuribenzoate, pCMB) and alkylating reagents (such as iodoacetate, iodoacetamide, and *N*-ethylmaleimide, NEM). Aspartic proteinases are inhibited specifically by pepstatins (acylated pentapeptides isolated from actinomycetes); diazoacetyl compounds, such as diazoacetyl-L-phe-methyl ester, also inhibit but will react with other proteinases as well. Metallo-proteinases are inhibited by chelating agents such as EDTA (ethylenediamine-tetraacetic acid) and 1,10-phenanthroline; some are inhibited by phosphoramidon (rhamnose-phosphate-leu-trp) (17). The serine and aspartic proteinases are probably the easiest enzymes to classify because DFP is a specific active site–directed inhibitor of the former class and pepstatin is a powerful and specific inhibitor of the latter class. Sulfhydryl reagents are generally not specific for cysteine residues at the active site, and because many intracellular proteinases contain reduced sulfhydryl groups somewhere in their structure, they are likely to be affected by sulfhydryl reagents. Some active site–specific cysteine reagents (e.g. E-64 and peptidyl diazomethyl ketones) are now available. E-64 [L-*trans*-epoxysuccinyl-leucylamido(4-guanidino)butane] reacts exclusively with active-site cysteine residues to form thioethers (18). Peptidyl diazomethyl ketones are also specific for cysteine proteinases under defined conditions, and modification of the peptide portion of these reagents may yield highly selective affinity reagents (19). Metal chelators may inhibit metal-activated proteinases in addition to metallo-proteinases. Metal-activated proteinases bind metals loosely and usually contain cysteine or serine as the catalytic residues. The metallo-proteinases can often be distinguished from the metal-activated proteinases by the tightness of the metal/enzyme association (dissociation constants for the enzyme-metal complex of 10^8 M^{-1} is a suggested dividing line), and by the profile of inhibition by a spectrum of inhibitors (14). Metals often have to be added during purification procedures for metal-activated enzymes.

Many protease inhibitors that are peptides and peptide analogues have been isolated from actinomycetes, and some are useful in classifying proteases. For example, bestatin inhibits only aminopeptidases. Some peptide aldehydes (leupeptin, antipain, chymostatin, elastinal) inhibit proteinases by forming hemiacetal or hemithioacetals with active-site serine or cysteine residues (20). The peptide aldehydes are potent inhibitors and have been used extensively as research tools. They may also have potential as therapeutic agents (21). However, they are not necessarily specific for particular classes of proteinases, e.g. leupeptin will inhibit both serine and cysteine proteinases.

Classification and assignation of EC numbers have been tremendously useful for organizing the large body of information generated as a considerable number of proteases have been discovered. The last published classification of proteases by the Nomenclature Committee of the International Union of Biochemistry (9) was in 1984; many cellular proteinases described herein do not appear in that listing as they have been characterized more recently.

Biochemical Characteristics/Evolutionary Relationships

SERINE PROTEINASES   The serine proteinases that have been characterized in prokaryotes and eukaryotes fall into two evolutionary superfamilies: the chymotrypsin family and the subtilisin family (22). Only the chymotrypsin family has been found in eukaryotes; this family includes many extracellular proteases such as trypsin, elastase, thrombin, plasma kallikrein, the plasma coagulation proteases, and cellular proteases (Table 1).

The most sophisticated work on the evolutionary relationships of the serine proteinases found in mammalian tissues is that on the proteases of the fibrinolytic and blood coagulation systems (23–25). These are clearly extracellular proteases, outside of the scope of this review; however, a brief note of the physicochemical characteristics of these proteases is relevant to the present review because some of the basic findings might be applicable to intracellular proteases. For instance, studies of the blood coagulation proteases have revealed that the large size of most of these proteases is probably a consequence of gene fusion. The catalytic domains of the proteases are clearly homologous to each other, as well as to other chymotrypsin-family enzymes, but the substrate specificities of the enzymes are very different and the multiple noncatalytic domains confer different regulatory properties upon these enzymes. Many of the plasma proteinases contain pre-pro-leader sequences, activation peptide domains, growth factor domains (homologous to EGF, epidermal growth factor), Ca^{2+}-binding domains, and 'kringle' domains. Kringles are triple-looped, disulfide-crosslinked domains that may appear once or in multiple copies in one protein; they are thought to play a role in binding mediators such as membranes, other proteins, or phospholipids and in the regulation of the proteolytic activity. There is little information presently available on the primary structure of the large cellular proteinases (calpain is an exception and will be discussed later). However, it is reasonable to expect that these too will contain simple proteolytic domains in addition to other domains that are responsible for the conversion of simple digestive enzymes into complex ones that are endowed with unique specificities and can be regulated.

Table 1 Proteinases found in mammalian cells[a]

That act as integral cellular components	That are present in 'secretory' or 'special' granules
Serine proteinases	
prolyl endopeptidase, cytosol	tissue kallikrein
ATP-dependent, mitochondria	submandibular proteinase A
processing, mitochondria & ER	nerve-growth-factor γ-subunit
trypsin-like, plasma membrane	β-nerve-growth-factor endopeptidase
enteropeptidiase, plasma membrane	acrosin, of sperm
chromatin proteinases, nuclei	cathepsin G, of neutrophils
cathepsin R, ER	elastase, of neutrophils
	tryase, of liver
	chymase, of mast cells
	tryptase, of mast cells
	chemotactic proteinase, of skin, lymphocytes
	plasminogen activator
Cysteine proteinases	
cathepsins B, L, H, N, S, M, T (lysosomal)	
calpains, cytosol	
ATP-dependent, cytosol	
multicatalytic, cytosol	
metal-dependent cysteine, cytosol	
Aspartic proteinases	
cathepsin D, E (lysosomal)	renin
red blood cell membrane	processing proteinase, pituitary
Metallo-proteinases	
endopeptidase 24.11, plasma membrane	
meprin, plasma membrane	
PABA-peptide hydrolase, plasma membrane	
procollagen C-peptidase, plasma membrane	
metallo-endopeptidase, cytosol	
signal peptidases, mitochondria & ER	

[a]Abbreviations: ER, endoplasmic reticulum; PABA-peptide, N-benzoyl-L-tyrosyl-p-aminobenzoic acid

There is a family of arginine-specific serine proteinases that are found in secretory or special granules in a number of cells and appear to have special processing functions (26–30). This group includes: tissue kallikrein (EC 3.4.21.35), the γ-subunit of nerve growth factor, β-nerve-growth-factor endopeptidase, tonin, rat urinary esterase A, EGF-binding protein, and sub-mandibular proteinase A (EC 3.4.21.40). These enzymes have many similar physicochemical and immunological characteristics and are all involved in the processing of precursors of polypeptide hormones, kinins, and growth factors. The enzymes derive from a multigene family: distinct mRNAs encode for the different enzymes and they are selectively expressed in different tissues (26). The exact relationship of this family of proteinases is in the process of being elucidated. It seems that submandibular proteinase A is identical with EGF-binding protein, and β-nerve-growth-factor endopeptidase is similar or identical to tissue kallikrein (27–29). Distinct differences in this family have also been found; e.g. while the γ-subunit of nerve growth factor is very similar to EGF-binding protein (they have approximately 70% sequence identity), the latter will not substitute for processing of nerve growth factor and the former will not substitute for liberation of EGF (30).

Some of the intracellular serine proteinases, listed in Table 1, will be discussed in sections on processing enzymes, ATP-dependent proteinases, and plasma membrane proteinases. Enteropeptidase (EC 3.4.21.9) is considered a plasma membrane proteinase, but it is loosely associated with duodenal brush border membranes and may actually act on trypsinogen in the lumen of the gut (31).

There is only one established example of a cytosolic serine proteinase, and that is prolyl endopeptidase (EC 3.4.21.26). It is a 70,000 dalton enzyme found in several tissues, including brain, liver, spleen, kidney, heart, lung, skeletal muscle, and pancreas. It cleaves -Pro-X- bonds in peptides but it will not degrade large proteins. It has been suggested that it is important in neuropeptide metabolism (32).

Three tissue-specific proteinases that are found in special granules are cathepsin G (EC 3.4.21.20), elastase (EC 3.4.21.37), and acrosin (EC 3.4.21.10). Cathepsin G and elastase are found in the azurophil granules of human neutrophil leukocytes, and they act extracellularly in inflammatory processes or in phagosomes in neutrophils (33). Acrosin, a 38,000-dalton proteinase that tends to self-associate to hexamers, is found in the acrosome, a modified lysosome, covering the anterior part of sperm heads. This enzyme is a peripheral membrane protein that is involved in fertilization (34).

Some of the serine proteinases found in cells stored in granules are extracellular proteinases, in that they are secreted and exert their action as discrete entities outside of cells. This group includes the mast cell proteinases (chymase and tryptase), chemotactic proteinase (35), and plasminogen activator (EC 3.4.21.31).

Chymase (EC 3.4.21.39), a chymotrypsin-like proteinase, and tryptase, a trypsin-like proteinase, are thought to be involved in inflammatory reactions and are present in fairly high concentrations in normal skeletal muscle, lung, and skin because of the mast cells in these tissues. They cannot be considered 'intracellular enzymes' as we have defined the term because the proteinases are active upon degranulation. However, they should be noted because they have been mistakenly thought to be associated with a number of different cell types and have been the source of considerable confusion. Thus the 'group-specific proteinases' (chymotrypsin-like enzymes, approximately 30,000 daltons) isolated from small intestine and from skeletal muscle, and originally thought to play a role in muscle protein catabolism, are now known to be of mast cell origin (36, 37). The major 'alkaline proteinase' activity in rat skeletal muscle tissue is mast cell chymase, and its activity increases in some muscles during starvation, diabetes mellitus, testosterone deficiency, and after glucocorticoid treatment (38). It is still unclear how mast cell proteinases might be involved in the degradation of muscle proteins, but there is some immunohistochemical evidence indicating that chymase may be present in certain muscle cells (39). Some confusion as to the origin of chymotrypsin-like enzymes has also occurred in nonmuscle tissues; a proteinase originally thought to be associated with rat liver mitochondria is now known to have a mast cell origin (40). In addition, there is some question as to whether the chromatin proteinases (that degrade histones), thought to be associated with nuclei isolated from various tissues (rat liver, thymus, uterus), are actually mast cell enzymes (41).

Tryptase has been purified from human pulmonary and lung mast cells. It has a molecular weight of approximately 140,000 and is composed of subunits of molecular weight 32,000 to 37,000 (42, 43). There have been several other 'trypsin-like proteinases' found associated with cells. For example, tryase, a 33,000-dalton, trypsin-like proteinase from rat liver tissue, is not inhibited by TLCK, as tryptase is (44). In addition, tryptase has not been identified in the granules of rat mast cells. The function of these proteinases is unknown, but proteinases that show a specificity for peptide bonds containing basic amino acid residues must be considered as possible protein-processing enzymes.

CYSTEINE PROTEINASES The cysteine proteinases from prokaryotes and eukaryotes characterized thus far fall into several evolutionarily related families; three of these are represented by a proteinase from *Streptococcus,* clostripain from *Clostridium histolyticum,* and papain from *Carica papaya* (22). The papain superfamily seems to be the predominant family in eukaryotes. Papain is a 23,400-dalton single polypeptide, and the structure and kinetics of the enzyme have been extensively investigated (45). The lysosomal cysteine proteinases, cathepsins B, H, L, and S, have a high degree of amino acid sequence homology with papain, making it likely that they all have evolved from a common ancestor (22). The amino acid residues around the active-site cysteine (Cys-25 in papain) and an essential histidine (His-159 in papain) have been highly conserved in the lysosomal cysteine proteinases. Calpain, a cytosolic proteinase present in a wide variety of eukaryotic cells, provides a fascinating example of a high-molecular-weight proteinase (110,000) that probably has evolved from the simple digestive proteinase, papain (see calpain section).

The cysteine proteinases are in general true 'intracellular proteinases' usually found in the cytosol or in lysosomes. The highly reducing environment in cells is probably important to their function. There are a few instances where lysosomal enzymes have been found in extracellular compartments, but this is usually in pathological situations. For example, some malignant tissues secrete a 'cathepsin B-like' cysteine proteinase that is thought to be involved in the destruction of the extracellular matrix as well as in facilitating cellular detachment from primary tumor masses and local invasion (46, 47). In addition, lysosomal enzymes from macrophages are released at sites of inflammation and may be damaging to the normal tissue and structural proteins in the area. Lysosomal proteinases are found in the synovium of arthritic joints, for example, and thought to be destructive to normal articular structures (48).

Most of the cysteine proteinases listed in Table 1 will be discussed later in this chapter in sections on lysosomal proteinases, calpains, ATP-dependent proteinases, and multicatalytic proteinases. There are a few proteinases that have been referred to as metal-dependent cysteine proteinases that deserve some mention. One of these is 'insulinase' (formerly EC 3.4.22.11, but removed from the 1984 listing). Insulin-degrading enzymes have been described in the cytosol of many tissues, including skeletal muscle, liver, kidney, brain, and erythrocytes (e.g. 49, 50, 50a). One form of 'insulinase' has a high affinity for insulin (K_m 2×10^{-8} M) and glucagon (K_m 3×10^{-6} M) and does not hydrolyze proinsulin or the separated A and B chains of insulin (51). Another form (termed 'neutral thiol peptidase') has the highest affinity for insulin, followed by insulin B chain and then glucagon (50a). All the insulin-degrading enzymes have essential cysteine residues, although the location of these residues at the active site has not been determined. The metal dependence of the enzymes is not at all clear. Most are inhibited by metal chelators and are referred to as 'metal-dependent cysteine proteinases'; zinc, manganese, and cobalt, but not calcium, prevent inactivation by chelators. Calcium activates the rat skeletal muscle enzyme (49). The rat liver insulin-degrading enzyme is assayed in the presence of EDTA at concentrations that would inhibit the metal-activated enzymes (52). Different forms of insulin-degrading enzymes have been reported to have molecular weights of 80,000, 110,000, and 180,000. Definitive characterization of the mechanistic class, physicochemical properties, and substrate specificities of the insulin-degrading proteinases will have to await further purification of these enzymes.

The insulin-degrading proteinases are distinct from other cytosol enzymes that hydrolyze insulin B chain, but not intact insulin (53, 54). The latter enzymes, isolated from human erythrocytes and mouse liver, have molecular weights of 300,000 and 190,000, respectively. They are also inhibited by metal chelators and low concentrations of pHMB. It is not yet clear whether these enzymes are cysteine or metallo-proteinases, or whether they represent a new mechanistic class.

ASPARTIC PROTEINASES   Aspartic proteinases have not been identified in prokaryotes but are present in eukaryotes. The structural data that are available indicate that the aspartic proteinases belong to one family (55). The enzymes generally have molecular weights in the range of 30,000 to 40,000, and are bilobal structures comprised of mainly β-sheet with a deep and extended cleft that contains the active site (56). It has been suggested that the two lobes and two domains per lobe of the enzymes have arisen as a result of gene duplication. These enzymes have a preference for peptide bonds flanked by hydrophobic amino acid residues.

In mammalian systems, several extracellular aspartic proteinases have been well studied: the pepsins (from the gastric juice of many species), chymosin (from calf stomach), and renin (from kidney and plasma of many species). Renin is stored in secretion granules in the kidney, and is liberated from cells in response to low blood pressure; it is highly specific for the cleavage of angiotensinogen to generate angiotensin I, which is subsequently hydrolyzed by angiotensin-converting enzyme (ACE) to angiotensin II, a powerful pressor peptide (57). Cathepsins D and E will be discussed in the section on lysosomal proteinases. Cathepsin D–like proteinases have also been found in association with endosomes (58), in neurosecretory granules (59), and red blood cell membranes (60).

METALLO-PROTEINASES   Metallo-proteinases are widely distributed in prokaryotes and eukaryotes. However, very little is known about evolutionary families of metallo-peptidases because there is little amino acid sequence data available. The primary amino acid sequence and three-dimensional structure of the bacterial enzyme thermolysin (EC 3.4.24.4) has been determined (61); the primary structure of only one mammalian metallo-endopeptidase, human fibroblast collagenase, has been determined (61a). The molecular weights of mammalian metallo-proteinases that have been characterized range from 17,000 to 800,000; most contain zinc as the essential cation (14). Some, such as meprin, also contain calcium. The role of the cations is not known for any of the mammalian enzymes, but by analogy with thermolysin it may be inferred that zinc is at the active site and plays a role in labilizing the peptide bond to be cleaved. Calcium ions, by contrast, are probably involved in enzyme stabilization (62).

Mammalian metallo-endopeptidases have been identified extracellularly, in cell membranes (plasma and endoplasmic reticulum), and in the cytosol (14). The extracellular enzymes include collagenases (EC 3.4.24.7), gelatinases, and procollagen N-proteinase (EC 3.4.24.14). Cell-associated enzymes include plasma membrane enzymes endopeptidase 24.11 (also called 'enkephalinase'), meprin, PABA-peptide hydrolase, and procollagen C–proteinase; endoplasmic reticulum and mitochondrial signal peptidases; and a soluble endopeptidase (EC 3.4.24.15). This class of proteinases, as the others, contains enzymes of high substrate-specificity (e.g. collagenases) as well as nonspecific enzymes (e.g. meprin). The plasma membrane proteinases and signal peptidases are discussed in later sections.

A few soluble metallo-endopeptidases have been described; one of the best characterized was identified by Orlowski and colleagues (63). This enzyme is present in many tissues but is at highest levels in brain, testis, and pituitary. It has a molecular weight of 67,000, and the metal-depleted enzyme can be reactivated by Zn^{2+}, Co^{2+}, and Mn^{2+}. The enzyme preferentially cleaves peptides with hydrophobic amino acid residues in the P_1, P_2, and P_3' position. The amino acid residues in the substrate are designated as P_1, P_2, P_3, etc in the N-terminal direction, and P_1', P_2', P_3' etc in the C-terminal direction from the bond undergoing cleavage (64). Studies with model substrates indicated the presence of an extended substrate binding site accommodating a minimum of five amino acid residues. It is capable of generating leu- and met-enkephalin from several larger peptide precursors as well as hydrolyzing other bioactive peptides. It is thought to function in the metabolism of neuropeptides.

It should be noted that most of the exopeptidases that have been characterized are metallo-proteins (65). They are generally high-molecular-weight (up to 300,000-dalton) glycoproteins.

SPECIFIC TYPES OF CELLULAR PROTEINASES

ATP-Dependent Proteinases

ATP-dependent proteinases that have been at least partially purified and characterized are listed in Table 2. The enzymes fall into two distinct proteinase classes: the first group is cytosolic and is composed of cysteine proteinases (inhibited by p-hydroxymercuribenzoate), and the second group is found in mitochondria or bacteria and is composed of serine proteinases (inhibited by DFP and PMSF). The two groups have some common features in that all the enzymes are high-molecular-weight, oligomeric proteins with alkaline pH optima. The major differences between the two groups lie in the active-site catalytic residue and the mechanism of action of ATP. The cysteine proteinases do not require magnesium ions for activity, are not inhibited by vanadate (an ATPase inhibitor), and are stimulated by both ATP and nonhydrolyzable analogues of the nucleotide. This indicates that ATP hydrolysis is not required for activity, but rather that ATP acts as an allosteric effector that regulates proteinase activity. The function of ATP is to stimulate a basal proteolytic activity (66, 67) or stabilize the proteinase against thermal denaturation (68, 69). Hence these enzymes are, more correctly, ATP-stimulated or ATP-stabilized proteinases.

By contrast, the serine proteinases require magnesium ions for activity, are inhibited by vanadate, and are not activated by nonhydrolyzable ATP analogues such as 5'-O-(3'-thiotriphosphate) (70). ATP hydrolysis is required for proteolytic activity; thus these enzymes are truly ATP-dependent proteinases. Protease La from $E.$ $coli$, the bacterial form of this proteinase, is the best characterized of this group (71, 72). Recent reports have suggested that while degradation of proteins requires ATP hydrolysis, degradation of small peptides does not (73). Furthermore, denatured protein substrates can stimulate the activity against small peptides, and small peptides can inhibit the ATPase activity of the proteinase. Complex interactions between active sites and regulatory sites are suggested to explain these observations.

Table 2 ATP-dependent proteinases[a]

Source	Class	M_r (kg/mol)	pH optimum	Inhibitors	Ref.
Rat skeletal muscle, cytosol	Cys	500	7.5	NEM, pHMBS	68
Rat skeletal muscle, cytosol	Cys	750	8–9	IAA, pHMBS	69
Rat liver, cytosol	Cys	550	7.5–9.5	IAA (partial)	67
Rat heart, cytosol	Cys	500	7–10	NEM, IAA	66
Mouse liver, cytosol	Cys	400	7.5–?	NEM, pHMB	79
Bovine adrenal cortex, mitochondria	Ser	650	8.2	PMSF, NEM, vanadate	80
Rat liver, mitochondria	Ser	550	7.5–10	DFP, PMSF, vanadate	70
$Escherichia$ $coli$ (Protease La)	Ser	370	9–9.5	DFP, pHMB, vanadate	71, 72
Mouse erythroleukemia cells, cytosol	?	600	?	NEM, hemin, DFP (partial), vanadate	74
Rabbit reticulocyte ubiquitin/ ATP, cytosol	?	600	7.8	NEM, EDTA, hemin	75

[a]Abbreviations: NEM, N-ethylmaleimide; IAA, iodoacetic acid; DFP, diisopropyl flourophosphate; PMSF, phenylmethylsulfonyl fluoride; pHMB, p-hydroxymercuribenzoate; pHMBS, p-hydroxymercuribenzene sulfonic acid.

One proteinase that appears to have properties common to both groups has been purified from mouse erythroleukemia cells (74). This enzyme is cytosolic, requires ATP hydrolysis for activity, is totally inhibited by the alkylating agent N-ethylmaleimide, and only partially inhibited by DFP. It is possible this represents a mixture of proteases.

A distinct ATP-dependent proteinase that degrades ubiquitin-protein conjugates has been partially purified from reticulocytes (75). This enzyme is involved in the ATP/ubiquitin-dependent proteolytic pathway that has been studied in the reticulocyte (see next section). The ubiquitin-conjugating enzymatic steps in this pathway have been extensively studied; the proteolytic steps are poorly characterized. The partially purified proteinase associated with this pathway has a high molecular weight (>600,000), a pH optimum of 7.8, and is inhibited by hemin, NEM, and partially by EDTA (75). It is possible that this is a mixture of proteases. In this pathway, ATP is required for both ubiquitin conjugation to proteins and for proteolysis.

ATP stimulation of other proteolytic activities, such as cathepsins L and D, has also been observed (76, 77). When the effects of ATP on cathepsin L were investigated, it was found that ATP interacted with the substrate (aldolase), rather than the proteinase, and made the substrate more vulnerable to proteolysis (78). These studies, and those above, illustrate that there are multiple ways in which ATP may stimulate proteolysis and that the mechanism of the stimulation must be evaluated before a protease can be designated 'ATP-dependent.'

Proteolytic Systems in Reticulocytes

Reticulocytes contain an ATP/ubiquitin-dependent proteolytic system composed of a high-molecular-weight conglomerate of 6–8 polypeptides (81). One component of the conglomerate is a heat-stable polypeptide ubiquitin, a 9,000-dalton peptide (82). It has been proposed that the conjugation of ubiquitin to protein substrates brands proteins for extensive degradation by proteases that specifically recognize the marked proteins. The conjugation of ubiquitin to polypeptides is a multistep, enzyme-catalyzed process that requires ATP and results in the formation of an isopeptide bond between the COOH-terminal glycine of ubiquitin and the epsilon amino groups, and possibly the alpha amino groups of the substrate (83).

An alternative hypothesis for the function of ubiquitin in ATP-dependent proteolysis has been put forward by Speiser & Etlinger, who propose that ubiquitin represses a protease inhibitor in reticulocytes (84). In addition, Rechsteiner and coworkers have demonstrated that lysozyme in which the amino groups have been blocked by guanidination, thereby preventing ubiquitin conjugation, is nevertheless degraded in an ATP-dependent, ubiquitin-activated manner (85). The fact that ubiquitin provides a fourfold stimulation of proteolysis in this system indicates that it has some other function in degradation besides direct marking of the substrate. In addition to multiple functions in degradation, ubiquitin probably plays other roles in cells, such as regulation of gene expression. This is indicated by the fact that ubiquitin is found conjugated to histones (86). In addition, ubiquitin is found on outer cell membranes where the proteins would be inaccessible to the cytosolic proteolytic system (87).

There is also considerable evidence for an ATP-independent pathway for protein breakdown in reticulocytes (88, 89). Depletion of the ATP in reticulocyte lysates, followed by dialysis and ammonium sulfate fractionation, results in the appearance of an ATP-independent protease activity. Rechsteiner and colleagues (75) have suggested that the treatment separates out a component (e.g. an inhibitor?) that is responsible for the ATP dependence of the reaction. It is proposed that the two activities (ATP-dependent and ATP-independent) are not present simultaneously, but are alternate forms of the same proteolytic complex. It is possible that the two forms are responsible for the degradation of different types of substrates. For example, Hipkiss and coworkers have demonstrated that shortened peptides produced by the action of puromycin or cyanogen bromide are degraded in an ATP-independent manner in reticulocyte lysates (89). In addition, Goldberg and colleagues have shown that oxidant-damaged hemoglobin is degraded in the absence of ATP in reticulocytes and erythrocytes (90). Conversely, 'abnormal' proteins of normal chain length, but substituted with the lysine analogue aminoethylcysteine, are degraded in an ATP-dependent process (91).

Obviously, there is a complex mechanism of protein breakdown in reticulocytes, with evidence for more than one ATP/ubiquitin-dependent pathway as well as an ATP/ubiquitin-independent pathway.

A number of possible roles of the proteolytic systems in reticulocytes have been suggested. As already mentioned, abnormal proteins and peptides are rapidly degraded in reticulocyte lysates (92), while a number of normal proteins are not readily degraded in the same system (93), indicating a specific role for the proteases in the removal of aberrant proteins. In addition, the proteases may be responsible for the maturational loss of reticulocyte organelles (94). This possibility is supported by the fact that most of the reticulocyte proteolytic activities described are lost in the mature erythrocyte (88, 95).

Although the ubiquitin proteolytic system has been well documented for the reticulocyte, it has been difficult to demonstrate its presence in other cell types. Indeed, if this system is concerned with reticulocyte-specific processes, it may not be active elsewhere. However, Haas and coworkers have shown that in rabbit liver a cysteine protease cleaves the C-terminal gly-gly from ubiquitin, thereby preventing conjugation to amino groups (96). Inhibition of this protease resulted in ATP/ubiquitin-dependent proteolysis. In addition, the system exists in mouse mammary carcinoma, and there is preliminary evidence that it exists in muscle and plant cells (97–99).

Multicatalytic Proteinases

In the past five years there have been a number of reports in the literature of high-molecular-weight complex proteinases that have been named multicatalytic proteinases (100–104). These enzymes have been purified from a number of sources, and some of their general properties are summarized in Table 3. They are cytosolic enzymes with alkaline pH optima, and they do not require ATP for activity. These properties distinguish them from both the lysosomal cathepsins and from the ATP-dependent proteinases described above. They are large proteins (600–700 kilodaltons) composed of numerous small subunits (21,000 to 34,000 daltons). They may tentatively be classified as cysteine proteinases, because total inhibition of activity is usually observed with cysteine reagents, such as p-hydroxymercuribenzoate, while dithiothreitol (DTT) and β-mercaptoethanol activate the enzymes. However, there is no definitive evidence that these reagents act exclusively on active-site cysteine residues. Interestingly, more than one type of peptide bond specificity has been observed for several of these proteinases. In general, these may be classified as trypsin-like (arg-X), chymotrypsin-like (phe-X), and peptidyl-glutamyl peptide bond–hydrolyzing activity (glu-X) (101). For the most part, each of these activities is differentially inhibited by peptide inhibitors such as leupeptin and chymostatin. For example, chymostatin inhibits the trypsin- and chymotrypsin-like activities in the rat skeletal muscle enzyme, but activates the glu-X activity (100). Similarly, leupeptin inhibits only the trypsin-like activity in the multicatalytic proteinase from bovine pituitary (101). Differential effects of serine proteinase inhibitors, such as DFP and PMSF, on the proteolytic activities have also been observed (100, 104). Such data (supported by the complex composition of the protein subunits) indicate that the proteinases contain several active sites that have different substrate specificities and inhibitor sensitivities and may actually belong to different mechanistic classes. Assignation of a particular proteolytic activity to a particular part of the proteinase complex has not yet been achieved.

Several important functions have been proposed for these proteinases, e.g. processing of prohormones in the pituitary (101), degradation of lens crystallins (102), and degradation of oxidized proteins in cells (105).

Calpains

The calpains (calcium-dependent papain-like proteinases) are a group of cysteine endopeptidases that require calcium ions for activity (for a recent

Table 3 Multicatalytic proteinases

Source	Molecular weight		Substrates	pH optimum	Ref.
	Enzyme (kg/mol)	Subunits (kg/mol)			
Rat skeletal mucle	650	25–32	arg-x	10.5	100
			phe-x	7.5	
			glu-x	9.0	
Bovine pituitary	700	24–32	arg-x	8.5	101
			leu-x	8.5	
			glu-x	8.5	
Bovine eye lens	700	24–32	leu-x	?	102
			glu-x		
			α,β-crystallins		
Rat and mouse liver	650	22–34	phe-x, insulin B chain, oxidized glutamine synthetase	9.0	103
Human erythrocytes	600	21–32	synthetic peptides, proteins	7.5	104

review see 106; 3). In the literature these enzymes are referred to by several names in addition to calpain (calcium-dependent proteinases, CDPs or CAPs; calcium-activated neutral proteinases, CANPs; calcium-activated factor, CAF), but it is clear that these are all the same enzyme. An observation that the amino acid sequence around the catalytic cysteine residue of calpain has about 33% homology with the papain sequence, whereas the sequences of cathepsins B, H, and L in this region are 90% homologous with papain, has led to the suggestion that calpains may be more closely related to one of the other families of cysteine proteinases; this lends support to those who would prefer names other than calpain (107). However, the Enzyme Commission of the International Union of Biochemistry has recommended the use of the term calpains (EC 3.4.22.17), and it seems reasonable that workers in the field adopt this recommendation for consistency. The calpains have been identified in most tissues and all appear to have similar properties. They are heterodimers of 110 kilodaltons composed of an 80-kilodalton catalytic subunit and a 30-kilodalton subunit of unknown function. Two different forms of the enzyme, calpain I and calpain II, have been isolated. These require μM and mM concentrations of calcium, respectively, for activity. The small subunit is the same in both enzymes, while the larger subunits are the products of different genes (108, 109). A tissue may contain either or both forms of the enzyme.

The primary structure of chicken calpain II has been deduced by Suzuki and coworkers, and the protein has been found to contain four domains (110). Domain II, the catalytic domain, has approximately 30% sequence homology with other cysteine proteinases such as cathepsin B and papain; the sequence of domain IV, a regulatory domain that binds Ca^{2+}, is homologous to other calcium-binding proteins such as calmodulin and troponin C. The functions of domains I and III are not known. It has been suggested that the domains resulted from gene fusion, and that this process has converted a simple papain-like proteinase into one that has a much more restricted substrate specificity and whose activity is regulated by calcium.

The calpains do not appear to have general proteolytic activity, but rather provide specific, limited cleavage of substrates giving rise to specific physiological responses. For example, the calpains have been implicated in myoblast fusion (111, 112). It is proposed that an influx of calcium into the cell activates calpain II thereby promoting the release of the surface glycoprotein, fibronectin. The membrane of the myoblast may then be rearranged to accommodate fusion (111). Similarly, in brain, membrane-associated calpain II will degrade fodrin when stimulated by an influx of calcium into a neuron (113, 114). This disrupts connections between the cytoskeleton and the membrane, causing a reorganization of cell structure. Such changes are thought to be ultimately involved in establishing long-term memory.

Lysosomal Proteinases

The lysosomal cathepsins (Table 4) are present in all mammalian cell types, with the exception of enucleated red blood cells. The name cathepsin is derived from a Greek term meaning 'to digest' (115). The concentration of lysosomes, and hence these proteinases, varies in different cells and tissues and is particularly high in liver, spleen, kidney, and macrophages. The properties of the lysosomal cathepsins from different species and cell types, however, are very similar. In general the lysosomal proteinases are small (20,000–40,000 daltons), optimally active at acidic pH values, and unstable at neutral and alkaline pH values. Most are glycoproteins and are active against a wide range of small peptide and large protein substrates. The best-characterized cathepsins are cathepsins B, L, H, and D, and there are several excellent detailed papers and reviews on these enzymes (e.g. 55, 116, 117).

Cathepsin B is generally active against synthetic substrates containing arginine in the P_1 position (e.g. Bz-arg-Nap, Bz-arg-arg-NMec), and has historically been thought of as a 'trypsin-like' enzyme with respect to substrate specificity. Interestingly, there is no evidence for this 'trypsin-like' activity with protein substrates. For example, when insulin B chain is digested with human liver cathepsin B there are 10 endoproteolytic cleavage points,

Table 4 Lysosomal proteinases[a]

Enzyme	M_r (g/mol)	Approx. pH optimum	Usual assay substrate	Inhibitors
cathepsin B (EC 3.4.22.1)	25,000	5	Z-arg-arg-NMec	thiol reagents
cathepsin L (EC 3.4.22.15)	24,000	5	azocasein	thiol reagents Z-phe-phe-CHN$_2$
cathepsin H (EC 3.4.22.16)	28,000	5	arg-NMec	thiol reagents
cathepsin M	30,000	5–7	aldolase	thiol reagents
cathepsin N	20,000	3.5	collagen	thiol reagents
cathepsin S	25,000	3.5	hemoglobin	thiol reagents
cathepsin T	35,000	6	TAT, azocasein	thiol reagents
cathepsin D (EC 3.4.23.5)	42,000	3.5	hemoglobin	pepstatin
cathepsin E	100,000	2.5	albumin	pepstatin

[a]Abbreviations: NMec, N-methyl coumarin; TAT, tyrosine aminotransferase; Z-phe-phe-CHN$_2$, benzoyloxy carbonyl-phe-phe-diazomethyl ketone.

but no evidence of specificity for basic amino acid residues (118). In addition, with glucagon or rabbit muscle fructose-1,6-bisphosphate aldolase as substrates, cathepsin B acts as a peptidyldipeptidase, cleaving dipeptides from the C-terminus, and again shows no endoproteolytic specificity for basic residues (119, 120). The fact that cathepsin B displays both endopeptidase and exopeptidase activity, depending on substrate, is unusual but not unique; cathepsin H, for instance, displays endopeptidase activity (e.g. against Bz-arg-NNap) and aminopeptidase activity (e.g. against arg-NMec) (116). In addition, the exopeptidase angiotensin-converting enzyme (peptidyldipeptidase A, EC 3.4.15.1) may also act as an endopeptidase (121). However, the action of proteinases on native protein structures cannot always be predicted on the basis of their action on synthetic substrates, and the structure of proteins and peptides determines to some extent the type of proteolytic action that will be expressed. Cathepsin B has been implicated as a processing enzyme for proinsulin and other proproteins on the basis of its specificity for dibasic pairs of residues in synthetic substrates. However, because of the discrepancy between the action of the enzyme on different substrates, it should not be assumed that cathepsin B is a processing enzyme.

An isozyme of cathepsin B has been isolated from porcine spleen that displays peptidyldipeptidase activity against several peptides but did not show endopeptidase activity (122). This observation may be a consequence of the substrates tested or species/tissue differences, because the endopeptidase activity of human, rat, and bovine liver cathepsin B is well established (16,

118). However, for the liver enzymes, higher concentrations of cathepsin B are required to show endopeptidase activity than are necessary for peptidyldipeptidase activity. Therefore, a range of enzyme concentrations must be tested against appropriate substrates before it can be concluded that porcine spleen cathepsin B has 'no endopeptidase activity.' Also of significance is the fact that cathepsin B exists in lysosomes at a very high concentration (approximately 1 mM; 123), and the enzyme would be expected to show potent endopeptidase activity in that location.

Cathepsin L is considered to be one of the most powerful lysosomal proteinases when assayed against protein substrates. It is routinely assayed with azocasein as substrate; Z-phe-arg-NMec (benzyloxycarbonyl-phe-arg-4-methyl 7-coumarylamide) has been recommended as an alternate substrate (116). A potent inhibitor of this enzyme is Z-phe-phe-CHN_2 (benzyloxycarbonyl-phe-phe-diazomethyl ketone), which has little effect on cathepsins B and H. The action of cathepsin L against insulin B chain indicates the enzyme has a preference for substrates with hydrophobic residues in the P_2 and P_3 positions.

The cysteine proteinases cathepsins M, N, S, and T are considered distinct from each other and from cathepsins B, H, and L. Cathepsin M from rabbit liver is distinguished from the others by its substrate specificity and sensitivity to inhibitors. In addition, at least a portion of this enzyme is associated with lysosomal membranes, unlike most, which are 'soluble' within the lysosome (124). The inactivation of fructose-1,6-bisphosphate aldolase is usually used to monitor cathepsin M activity, but it should be noted that this is not a specific substrate for the enzyme (125). Cathepsin N, sometimes called 'collagenolytic cathepsin,' cleaves N-terminal peptides of native collagen; it is similar to cathepsin L except that it has little activity against azocasein (126). Cathepsin S from beef spleen is very similar to cathepsin L, but the two may be distinguished on the basis of inhibition by Z-phe-phe-CHN_2 and ability to hydrolyze synthetic substrates (127). Cathepsin T from rat liver is distinguished by its ability to act on tyrosine aminotransferase (128).

Cathepsin D is a glycoprotein that resolves into several forms of similar molecular weight and different isoelectric points upon purification (55). A high-molecular-weight form (100,000 daltons), a 50,000-dalton single-chain form, and a two-chain form (34,000 and 12,000 daltons) of the porcine enzyme have been identified (129). There has been considerable work on the processing and transport of cathepsin D and the role of mannose phosphate in targeting the enzyme to the lysosome (130–133). Cathepsin D has limited action against native proteins but considerable activity against denatured proteins at pH 3.5–5. The enzyme preferentially attacks peptide bonds flanked by hydrophobic amino acids, phe-phe, phe-tyr, leu-phe (e.g. 134). It has been proposed that cathepsin D plays a role in pathological degradation of central nervous system proteins such as myelin basic protein (135). The inhibition of lysosomal proteolytic activity by pepstatin in vivo can largely be attributed to inhibition of cathepsin D activity (123).

Cathepsin E is found in bone marrow polymorphonuclear leukocytes and macrophages and appears to be distinct from Cathepsin D (136). It has an unusually high molecular weight for a lysosomal cathepsin and may be dimeric (41).

Signal and Processing Proteinases

Signal peptidases are responsible for the removal of signal sequences ('pre-sequences') from newly synthesized proteins (137). The signal peptide is necessary for the targeting of proteins to membranes, to extracellular spaces, or for their translocation into mitochondria (138, 139). Signal peptidases have not been extensively purified, and their activities have been mainly assayed in subcellular fractions enriched in endoplasmic reticulum and mitochondria (140, 141). The enzymes that have been investigated are metallo-proteinases, and they cleave peptide bonds at specific points in the polypeptide adjacent to small amino acid residues such as alanine, cysteine, glycine, and serine. A metallo-endopeptidase has been identified in pancreatic membranes that has the appropriate specificity for a signal peptidase (142). Whether this is its function in vivo is not known. Similarly, a metallo-proteinase has been partially purified from the matrix of both rat liver (140) and yeast mitochondria (143), which cleaves mitochondrial precursor proteins.

Processing enzymes convert inactive proproteins to mature, active polypeptides; both endo- and exopeptidases may be involved in processing (for reviews see 4, 144). These enzymes process many proteins, e.g. pro-albumin and prorenin (145, 146), and are involved in the maturation of peptide hormones and growth factors (144). In general, the endopeptidases have acidic or neutral pH optima and are located in secretory granules where they hydrolyze the proproteins before the active products are released at their site of action. The proteinases are generally specific for paired basic residues, the most common being lys-arg. However, some are also specific for monobasic residues, particularly single arginine residues (4). Depending on the specificity of the processing enzyme, both, one, or neither basic residue(s) may be removed from the peptide. Among the best-characterized processing enzymes are the arginine-specific family of serine proteinases, which include EGF-binding protein and nerve-growth-factor endopeptidases (26–30, 147; see also serine proteinase section). Nerve growth factor, for example, is stored and secreted as a multisubunit complex with a molecular weight of 140,000; containing two α-subunits, one β-dimer, two γ-subunits, and one or two zinc ions (148). The γ-subunits are processing endopeptidases; they have molecular weights of approximately 26,000, have no general proteinase activity, and are highly specific for the nerve-growth-factor proprotein. Tissue kallikreins (EC 3.4.21.35), serine endopeptidases with molecular weights of 25,000–48,000, are also found in secretory granules; they are thought to be involved in the processing of kininogens to kinins (e.g. lysyl-bradykinin) and possibly in the processing of proinsulin to insulin (149). A neutral serine endopeptidase that cleaves several prohormones at paired basic residues has been purified from porcine neurointermediate and anterior pituitary lobes (150). It is proposed that prohormones are processed within clathrin-coated vesicles. These vesicles mature into secretory granules with a concomitant decrease in the internal pH, which inhibits the processing enzyme (151). In addition, a serine endopeptidase has been implicated in the processing of somatostatin 28 to somatostatin 14 in rat brain cortex secretory granules (152) and in enkephalin maturation in bovine adrenal medullary secretory granules (153). One example of an aspartic endopeptidase involved in processing is proopiomelanocortin-converting enzyme, in the intermediary lobe of the pituitary (154). This is a 70-kilodalton protein that has an acidic pH optimum and is inhibited by pepstatin.

In general, if the processing enzyme removes the basic amino acid residues from the site of cleavage, the mature peptide is formed. If these residues are not removed, further processing by aminopeptidases and/or carboxypeptidases is required for the formation of the active peptide. One exopeptidase involved in processing is a carboxypeptidase B-like metallo-enzyme that has been identified in various endocrine organs (59, 155). It cleaves C-terminal arginine residues from enkephalins, a property that has earned it the name 'enkephalin convertase.' However, the enzyme will also process other peptide hormones such as ACTH and vasopressin (59).

Plasma Membrane Proteinases

The plasma membrane proteinases that have been identified are serine or metallo-proteinases (Table 5). Some of the enzymes are loosely associated with the plasma membrane (e.g. enteropeptidase, also found in the intestinal lumen), while others, such as meprin, are intrinsic membrane proteins that can only be removed from membranes by harsh treatments (166). The plasma membrane endopeptidases include enzymes that are very specific for substrates (e.g. enteropeptidase converts trypsinogen to trypsin; procollagen C-terminal proteinase specifically cleaves carboxy-terminal peptide bonds in procollagen) as well as those that will cleave peptide bonds in many substrates (e.g. meprin). Some of the plasma membrane enzymes will only hydrolyze peptide bonds in small peptides (up to 3,000 daltons) and will not act on large proteins; this includes the sheep red blood cell endopeptidase and endopeptidase 24.11. Others are active against large proteins and less so towards small peptides (e.g. meprin). They are generally large proteinases, both in terms of subunit and holoenzyme molecular weight. Interestingly, three of the plasma membrane proteinases, enteropeptidase, meprin, and PABA-peptide hydrolase, contain intersubunit disulfide bridges (156, 166). Intrasubunit disulfide bridges are common in proteinases, but intersubunit bridges are rare. Intersubunit bridges are a feature of many membrane proteins and receptors (e.g. 167).

Kidney brush border membranes have a very high content of proteases, which are likely involved in degradation of proteins and peptides from the glomerular filtrate. This is important for the function of the kidney in the retrieval of amino acids from proteins and peptides that are filtered. Many of the enzymes are exopeptidases, e.g. aminopeptidase N, ACE (168). Two metallo-endopeptidases have been highly purified and characterized: mouse

Table 5 Plasma membrane endopeptidases[a]

Enzyme	Source	Molecular weight		Substrates	Ref.
		Enzyme	Subunits		
Serine proteinases					
enteropetidase (EC 3.4.21.9; formerly enterokinase)	intestinal brush border, bovine	145,000	82,000 57,000	trypsinogen	156
trypsin-like	rat liver	120,000	30,000	peptides, protein	157
neutral endopeptidase (EC 3.4.21.24)	red blood cells, sheep	340,000	56,600	small peptides	158
proteinase 5 (neutral peptide-generating protease)	neutrophil	30,000	30,000	precursor	159
Metallo-proteinases					
endopeptidase 24.11	kidney, many tissues, species	93,000	93,000	neuropeptides, insulin B chain, small peptides	160
meprin	kidney, mouse, rat	320,000	85,000	insulin B chain, proteins, not small peptides	161, 166
neutral endopeptidase	kidney, rat	200,000		insulin, large polypeptides	162
	human	100,000		succ-AAA-NA	163
PABA-peptide hydrolase	intestine, human	200,000	100,000	PABA-peptide	164
procollagen C-terminal	connective tissue cells	100,000		procollagen	165

[a]Abbreviations: succ-AAA-NA, succinyl(alanyl)₃napthylamide; PABA-peptide, N-benzoyl-L-tyrosyl-p-aminobenzoic acid.

meprin and endopeptidase 24.11 from rabbit and pig. The relationship of neutral endopeptidases from rat and human kidney (162, 163) to meprin and endopeptidase 24.11 is not yet known. Endopeptidase 24.11 is a monomer of approximately 90 kilodaltons, while meprin is isolated and active as a tetramer with 85-kilodalton subunits. After sodium dodecyl sulfate polyacrylamide gel electrophoresis in the absence of β-mercaptoethanol, meprin migrates as a protein of 320 kilodaltons, indicating that the subunits are linked by reducible covalent bonds, probably disulfide bridges. Both meprin and endopeptidase 24.11 are extensively glycosylated with approximately 18% of the enzyme mass being attributable to carbohydrate residues, a common feature of membrane proteins (170). The active-site metal ion in both enzymes is zinc; meprin also contains 3 moles of calcium/subunit, a property that may contribute to the thermostability of this enzyme (171). The two metallo-endopeptidases are most clearly distinguished on the basis of their substrate specificity and inhibitor sensitivity. Meprin rapidly cleaves large proteins such as azocasein and azocoll and has no activity against peptides smaller than eight amino acids in length (172). By contrast, endopeptidase 24.11 will not hydrolyze proteins, but rapidly digests small peptides such as the enkephalins (169). In addition, nmolar concentrations of phosphoramidon will totally inhibit endopeptidase 24.11, while meprin is insensitive to this inhibitor. Meprin appears to be a kidney-specific enzyme and is a major component of the kidney brush border. Endopeptidase 24.11 is most abundant in kidney and is also present in lower concentrations in intestinal brush border (7.5% of kidney), lymph nodes (13%), and the central nervous system (1%). Endopeptidase 24.11 is present throughout the animal kingdom, while meprin has only been detected in mice and rats. An inherited deficiency of meprin has been discovered in inbred strains of mice (see section on synthesis and degradation of proteinases); activity of endopeptidase 24.11 is similar in low- and high-meprin activity strains, indicating that the enzymes are regulated independently.

REGULATION OF PROTEINASE ACTIVITY

Proteolytic activity in cells must be highly regulated to prevent inappropriate and uncontrolled degradation of proteins and to accomplish the aims of the many important processes in which proteases are involved. Regulation of cellular proteinase activity is affected in many ways.

Compartmentalization

The action of a proteinase may be limited by its subcellular localization. For example, the localization of the cathepsins in the lysosome provides them with an acidic environment, for activity and stability, while restricting their action to those proteins that enter this compartment. Leakage of the cathepsins from lysosomes would result in their inactivation by the higher pH of, and inhibitors in, the cytosol. Similarly, attachment of proteinases to membranes results in a loss of freedom, which (a) limits their accessibility to substrates and (b) may have the effect of increasing the concentration of specific substrates that interact with membrane components (e.g. receptors). The signal and processing proteinases are strategically localized in membranes or secretory granules, which enables access to specific proteins in specific conformations. The synthesis of processing proteases, along with proproteins and packaging of these complexes in secretory vesicles, ensures specific interactions between proteinase and substrate.

Synthesis and Degradation of Proteinases

There are many examples of tissue-specific proteinases and, thus, expression of these enzymes in different cell types is an important factor in regulation of proteinase activity. For those proteinases that are present in most mammalian cells, large variations in the concentrations of the proteinases are found in different tissues. Immunohistochemical techniques have confirmed that the concentrations of lysosomal proteinases, as well as activity of these enzymes, vary with cell type even within one tissue (e.g. liver, brain). The concentration of cell proteinases may be controlled by rates of synthesis and degradation, as for any enzyme. There have been few studies on the turnover of mammalian proteinases and how the rates of synthesis and degradation may change in different metabolic conditions. This may represent a rewarding area for future studies.

Interestingly, there are only a few examples of cellular proteinase deficiencies or synthesis of polymorphic forms of these enzymes in mammalian systems. This is particularly intriguing because there are many examples of lysosomal enzyme deficiencies (resulting in the 'lysosomal storage diseases'), but no lysosomal proteinase deficiency has been described (173). Only two mammalian cellular proteinase deficiencies have been described. One is a congenital deficiency of enteropeptidase in humans, associated with failure to thrive, diarrhea, anemia, and hypoproteinemic edema (174). The other is a heritable deficiency of meprin in inbred strains of mice (166, 175, 176). In the latter instance, the gene that determines whether mice will have high or low meprin activity has been located on chromosome 17 near the histocompatibility complex. This gene, termed Mep-1, may be a structural or regulatory gene for meprin (177). From the available data, it appears that the low-meprin phenotype is a consequence of improper posttranslational processing of the enzyme (178, 179).

Inhibitors and Activators

It is likely that endogenous intracellular inhibitors are important in the control of cellular proteinases. It is also possible that fluctuations in proteinase activities in cells are due to changes in inhibitor, rather than proteinase, concentrations. Two types of polypeptide inhibitors have recently been discovered in cells; they are the cystatins and stefins (180) that inhibit lysosomal cathepsins, and calpastatin (181), an inhibitor of calpains. The cystatin superfamily is comprised of three groups that are evolutionarily related: (a) the stefin family, small proteins (11 kilodaltons) that do not contain disulfide bonds, (b) the cystatin family, slightly larger proteins (13 kilodaltons) that have two disulfide loops, and (c) the kininogen family, more complex proteins (50–120 kilodaltons) that contain nine disulfide bonds (180). The kininogens are found extracellularly and are identical to alpha-1 cysteine proteinase inhibitors (182). The cystatins and stefins are found intracellularly and may function to prevent inappropriate proteolysis in the cytosol by lysosomal enzymes (183).

Calpastatin is a specific protein inhibitor of the calpains that has been isolated from a variety of mammalian and avian tissues (184–186). This inhibitor is equally effective in inhibiting calpains I and II and does not inhibit any other type of proteinase (186). The molecular weight of the inhibitor varies according to the source and the method of extraction (from 24,000 to 400,000). The higher-molecular-weight species are capable of binding several calpain molecules simultaneously (187). It is likely that the smaller calpastatin species are inhibitory domains derived from larger forms of the inhibitor (187). Calpastatin binds to the large subunit of calpain in the presence of high concentrations of calcium and is not cleaved by the proteinase. The mechanism of action of the inhibitor is not understood at present.

An activator of calpains (17–20 kilodaltons) has been isolated from brain tissue (188). The polypeptide stimulates calpain activity up to 25-fold, but does not act by altering the calcium sensitivity of either form of the proteinase.

Metabolites

The intracellular concentrations of small metabolites may affect proteinases in several ways. For example, calcium ions may stimulate calpain activity as well as increase autolysis of the enzyme and promote binding of the enzyme to calpastatin (106). Similarly, intracellular nucleotide concentration, especially of ATP, may affect the ATP-dependent proteinases as well as the energy-dependent pumps that allow acidification of endosomes, lysosomes, and other vesicles (189). Physiological concentrations of fatty acids (<100 μM) can activate the multicatalytic proteinase in skeletal muscle (190). This form of regulation may be important in diabetes where the intracellular concentration of fatty acids in skeletal muscle is increased (191).

Substrate Susceptibility

Degradation of intracellular proteins to amino acids and small peptides is at least partly determined by the structural characteristics of the protein substrates, and regulation may occur through modulation of protein structure by intrinsic and extrinsic factors (1, 2). The heterogeneity in half-lives observed for different proteins in a cell indicates that some proteins are more suscept-ible to degradation than are others. Recent evidence indicates that both the amino-terminal residue of proteins and specific amino acid sequences within proteins play some role in determining protein stability in vivo (191a,b). The observations that 'abnormal' protein structures are degraded more rapidly than normal proteins also indicate that protein structure is determinative in degradation rates. Substrates and allosteric activators can 'protect' a protein from proteolytic inactivation, while inhibitors and allosteric inactivators can render a protein more susceptible to the action of a protease (e.g. 195). In addition, covalent modifications such as ubiquitin-conjugation or oxidation may be important in branding proteins for degradation (81, 125, 192–194). At some point, proteinases or the proteolytic system must recognize and degrade the vulnerable and/or 'branded' proteins. Hence, proteolytic activity can be regulated by the availability of susceptible substrates.

FUNCTIONS OF CELLULAR PROTEINASES

There is such a diversity of processes in which cellular proteinases function that it is impossible to cover this aspect adequately. However, some in-dividual functions of proteinases have been suggested in previous sections, and some generalizations are offered here. Essentially, proteinases function to (a) create biologically active molecules, or (b) destroy biologically active proteins and peptides. The first group includes signal and processing enzymes as well as zymogen-activating enzymes, such as enteropeptidase. Cytosolic enzymes such as prolyl endopeptidase, or proteases involved in transmitting hormone signals, messages, or intracellular carriers, are also biogenic pro-teinases. The second group would include, for example, lysosomal cathepsins and ATP-dependent proteinases, that are involved in destroying or commit-ting active molecules to extensive hydrolysis. These catabolic proteinases have a role in the removal of defective, 'abnormal,' or normal polypeptides from cells. They control the concentrations of potent polypeptides and en-zymes in cells, have nutritive functions in the creation of amino acids from proteins, and permit the constant renewal of cellular contents. This extensive proteolysis allows flexibility in terms of the type and quantity of proteins that constitute cells, and thus enables cells to adapt to various needs imposed by diet, hormones, and other environmental factors.

The two general types of proteinase functions do not necessarily implicate, however, the multitude of processes in which cellular proteinases participate. These processes include reorganization of cytoskeleton, myoblast fusion and differentiation (94, 111), memory (113), protein synthesis (4), hormone action and inactivation (51), fertilization (34), growth and aging (196, 197), creation of immunologically recognizable molecules (135), and degradation of endocytosed material and necrosis (123, 198). In addition, cellular proteinases play an important role in diseases such as muscular dystrophy, diabetes, cachexia, cancer, and multiple sclerosis (2, 135, 199, 200). The irreversible modifications that proteinases effect, from the cleavage of a peptide bond, to the changes in protein conformation, to the determination of physiological processes, have profound consequences in biological systems.

ACKNOWLEDGMENTS

This work was supported by NIH Grant AM-19691. We thank Dr. Robert J. Beynon for many helpful discussions and critical reading of the manuscript.

Literature Cited

1. Beynon, R. J., Bond, J. S. 1986. *Am. J. Physiol: Cell Physiol.* 251:C141–52
2. Bond, J. S., Beynon, R. J. 1987. *Mol. Aspects Med.* In press
3. Khairallah, E. A., Bond, J. S., Bird, J. W. C., eds. 1985. *Intracellular Protein Catabolism.* New York: Liss
4. Schwartz, T. W. 1986. *FEBS Lett.* 200:1–10
5. Kiehn, E. D., Holland, J. J. 1970. *Nature* 226:544–45
6. Srere, P. A. 1984. *Trends Biochem. Sci.* 9:387–90
7. Righetti, P. G., Caravaggio, T. 1976. *J. Chromatogr.* 127:1–28
8. Barrett, A. J., McDonald, J. K. 1986. *Biochem. J.* 237:935
9. Webb, E. C. 1984. *Enzyme Nomenclature.* New York: Academic
10. Kraut, J. 1977. *Ann. Rev. Biochem.* 46:331–58
11. Polgar, L., Halasz, P. 1982. *Biochem. J.* 207:1–10
12. Tang, J. 1983. *Acid Proteinases: Structure, Function and Biology.* New York: Plenum
13. Holmes, M. A., Matthews, B. W. 1982. *Biochemistry* 20:6912–20
14. Bond, J. S., Beynon, R. J. 1985. *Int. J. Biochem.* 17:565–74
15. Wolfe, P. B., Silver, P., Wickner, W. 1982. *J. Biol. Chem.* 257:7898–902
16. Barrett, A. J., ed. 1977. *Proteinases in Mammalian Cells and Tissues.* Amsterdam: Elsevier
17. Komiyama, T., Suda, H., Aoyagi, T., Takeuchi, T., Umezawa, H., et al. 1975. *Arch. Biochem. Biophys.* 171:727–31
18. Hanada, K., Tamai, M., Adachi, T., Oguma, K., Kashiwagi, K., et al. 1983. See Ref. 21, pp. 25–36
19. Green, G. D. J., Shaw, E. 1981. *J. Biol. Chem.* 256:1923–28
20. Umezawa, H., Aoyagi, T. 1983. See Ref. 21, pp. 3–15
21. Katunuma, N., Umezawa, H., Holzer, H., eds. 1983. *Proteinase Inhibitors: Medical and Biological Aspects.* New York: Springer-Verlag
22. Barrett, A. J. 1986. In *Proteinase Inhibitors,* ed. A. J. Barrett, G. Salvesen, pp. 3–22. Amsterdam: Elsevier
23. Jackson, C. M., Nemerson, Y. 1980. *Ann. Rev. Biochem.* 49:765–811
24. Patthy, L. 1985. *Cell* 41:657–63
25. Neurath, H. 1984. *Science* 224:350–57
26. Ashley, P. L., MacDonald, R. J. 1985. *Biochemistry* 24:4520–27
27. Bothwell, M. A., Wilson, W. H., Shooter, E. M. 1979. *J. Biol. Chem.* 254:7287–94
28. Schenkein, I., Franklin, E. C., Frangione, B. 1981. *Arch. Biochem. Biophys.* 209:57–62
29. Taylor, J. M., Mitchell, W. M., Cohen, S. 1974. *J. Biol. Chem.* 249:2188–94
30. Thomas, K. A., Bradshaw, R. A. 1981. *Methods Enzymol.* 80:609–20
31. Rinderknecht, H., Nagaraja, M. R., Adham, N. F. 1978. *Digest. Dis.* 23:327–31
32. Wilk, S. 1983. *Life Sci.* 33:2149–57
33. Starkey, P. M. 1977. See Ref. 16, pp. 57–89
34. Muller-Esterl, W., Fritz, H. 1981. *Methods Enzymol.* 80:621–32
35. Hatcher, V. G., Oberman, M. S., Lazarus, G. S., Grayzel, A. I. 1978. *J. Immunol.* 120:665–70
36. Kominami, E., Kobayashi, K., Kominami, S., Katunuma, N. 1972. *J. Biol. Chem.* 247:6848–55
37. Woodbury, R. G., Everitt, M., Sanada,

Y., Katunuma, N., Lagunoff, D., Neurath, H. 1978. *Proc. Natl. Acad. Sci. USA* 75:5311–13
38. Dahlmann, B., Kuehn, L., Reinauer, H. 1983. *Biochim. Biophys. Acta* 761:23–33
39. Kay, J., Heath, R., Dahlmann, B., Kuehn, L., Stauber, W. T. 1985. See Ref. 3 pp. 195–205
40. Haas, R., Heinrich, P. C., Sasse, D. 1979. *FEBS Lett.* 103:168–71
41. Barrett, A. J., McDonald, J. K. 1980. *Mammalian Proteases: A Glossary and Bibliography:* Vol. 1, *Endopeptidases.* New York: Academic
42. Schwartz, L. B., Lewis, R. A., Austen, K. F. 1981. *J. Biol. Chem.* 256:11939–43
43. Smith, T. J., Hougland, M. W., Johnson, D. A. 1984. *J. Biol. Chem.* 259:11046–51
44. Saklatvala, J., Bond, J. S., Barrett, A. J. 1981. *Biochem. J.* 193:251–59
45. Dreuth, K. H., Kalk, H., Swen, H. M. 1976. *Biochemistry* 15:3731–38
46. Sloane, B. F., Dunn, J. R., Honn, K. V. 1981. *Science* 212:1151–53
47. Mort, J. S., Recklies, A. D. 1982. *Biochem. J.* 233:57–63
48. Dingle, J. T. 1984. *Clin. Orthop. Relat. Res.* 182:24–30
49. Ryan, M. P., Duckworth, W. C. 1983. *Biochem. Biophys. Res. Commun.* 116:195–203
50. Shii, K., Baba, S., Yokono, K., Roth, R. A. 1985. *J. Biol. Chem.* 260:6503–6
50a. Shroyer, L. A., Varandani, P. T. 1985. *Arch. Biochem. Biophys.* 236:205–19
51. Duckworth, W. C., Heineman, M., Kitabchi, A. E. 1975. *Biochim. Biophys Acta* 377:421–30
52. Brush, J. S., Nascimento, C. E. 1982. *Biochim. Biophys. Acta* 704:398–402
53. Kirschner, R. J., Goldberg, A. L. 1983. *J. Biol. Chem.* 258:967–76
54. Rosin, D. L., Bond, J. S., Bradley, S. G. 1984. *Proc. Soc. Exp. Biol. Med.* 177:112–19
55. Huang, J. S., Huang, S. S., Tang, J. 1980. In *Trends in Enzymology, Vol. 1: Enzyme Regulation and Mechanism of Action,* ed. P. Midner, B. Ries, 60:289–306. Oxford: Pergamon.
56. Blundell, T. L., Jones, H. B., Khan, G., Taylor, G., Sewell, B. T., et al. 1980. See Ref. 55, pp. 281–88
57. Ondetti, M. A., Cushman, D. W. 1985. *Crit. Rev. Biochem.* 16:381
58. Diment, S., Stahl, P. 1985. *J. Biol. Chem.* 260:15311–17
59. Hook, V. Y. H., Loh, Y. P. 1984. *Proc. Natl. Acad. Sci. USA* 81:2776–80
60. Tokes, Z. A., Chambers, S. M. 1975. *Biochim. Biophys. Acta* 389:325–38
61. Titani, K., Hermodson, M. A., Ericsson, L. H., Walsh, K. A., Neurath, H. 1972. *Nature* 238:35–37
61a. Goldberg, G. I., Wilhelm, S. M., Kronberger, A., Bauer, E. A., Grant, G. A., Eisen, A. Z. 1986. *J. Biol. Chem.* 261:6600–5
62. Feder, J., Garrett, L. R., Wildi, B. S. 1971. *Biochemistry* 10:4552–55
63. Chu, T. G., Orlowski, M. 1985. *Endocrinology* 116:1418–25
64. Schecter, I., Berger, A. 1967. *Biochem. Biophys. Res. Commun.* 27:157–62
65. McDonald, J. K., Barrett, A. J. 1986. *Mammalian Proteases: A Glossary and Bibliography:* Vol. 2, *Exopeptidases.* New York: Academic
66. DeMartino, G. N. 1983. *J. Mol. Cell. Cardiol.* 15:17–29
67. DeMartino, G. N., Goldberg, A. L. 1979. *J. Biol. Chem.* 254:3712–15
68. Ismail, F., Gevers, W. 1983. *Biochim. Biophys. Acta* 742:399–408
69. Dahlmann, B., Kuehn, L., Reinhauer, H. 1983. *FEBS Lett.* 160:243–47
70. Desautels, M., Goldberg, A. L. 1982. *J. Biol. Chem.* 257:11673–79
71. Charette, M. F., Henderson, G. W., Markovitz, A. 1981. *Proc. Natl. Acad. Sci. USA* 78:4728–32
72. Waxman, L., Goldberg, A. L. 1985. *J. Biol. Chem.* 260:12022–28
73. Goldberg, A. L., Voellmy, R., Chung, C. H., Menon, A. S., Desautels, M., et al. See Ref. 3, pp. 33–45
74. Waxman, L., Fagan, J. M., Tanaka, K., Goldberg, A. L. 1985. *J. Biol. Chem.* 260:11994–2000
75. Hough, R., Pratt, G., Rechsteiner, M. 1986. *J. Biol. Chem.* 261:2400–8
76. Pillai, S., Zull, J. E. 1985. *J. Biol. Chem.* 260:8384–89
77. Kirschke, H., Langer, J., Wiederanders, B., Ansorge, S., Bohley, P. 1977. *Eur. J. Biochem.* 74:293–301
78. McKay, M. J., Marsh, M. W., Kirschke, H., Bond, J. S. 1984. *Biochim. Biophys. Acta* 784:9–15
79. Rose, I. A., Warms, J. V. B., Hershko, A. 1979. *J. Biol. Chem.* 254:8135–38
80. Watabe, S., Kimura, T. 1985. *J. Biol. Chem.* 260:5511–17
81. Hershko, A., Ciechanover, A. 1982. *Ann. Rev. Biochem.* 51:335–64
82. Wilkinson, K. D., Urban, M. K., Haas, A. L. 1980. *J. Biol. Chem.* 255:7529–32
83. Hershko, A., Heller, H., Eytan, E., Kaklij, G., Rose, I. A. 1984. *Proc. Natl. Acad. Sci. USA* 81:7021–25

84. Speiser, S., Etlinger, J. D. 1983. *Proc. Natl. Acad. Sci. USA* 80:3577–80
85. Chin, D. T., Carlson, N., Kuehl, L., Rechsteiner, M. 1986. *J. Biol. Chem.* 261:3883–90
86. Goldknopf, I. L., Busch, H. 1977. *Proc. Natl. Acad. Sci. USA* 74:864–68
87. Siegelman, M., Bond, M. W., Gallatin, W. M., St. John, T., Smith, H. T., et al. 1986. *Science* 231:823–29
88. McKay, M. J., Daniels, R. S., Hipkiss, A. R. 1980. *Biochem. J.* 188:279–83
89. McKay, M. J., Hipkiss, A. R. 1982. *Eur. J. Biochem.* 125:567–73
90. Fagan, J. M., Waxman, L., Goldberg, A. L. 1986. *J. Biol. Chem.* 261:5705–13
91. Hershko, A., Eytan, E., Ciechanover, A. 1982. *J. Biol. Chem.* 257:13964–70
92. Etlinger, J. D., Goldberg, A. L. 1977. *Proc. Natl. Acad. Sci. USA* 74:54–58
93. Saus, J., Timoneda, J., Hernandez-Yago, J., Grisolia, S. 1982. *FEBS Lett.* 143:225–27
94. Rapaport, S., Daniel, W., Muller, M. 1985. *FEBS Lett.* 180:249–52
95. Speiser, S., Etlinger, J. D. 1982. *J. Biol. Chem.* 257:14122–27
96. Haas, A. L., Murphy, K. E., Bright, P. M. 1985. *J. Biol. Chem.* 260:4694–703
97. Ciechanover, A., Finley, D., Varshavsky, A. 1984. *Cell* 37:57–66
98. Etlinger, J. D., McMullen, H., Rieder, R. F., Ibrahim, A., Janeczko, R. A., Mamorstein, S. 1985. See Ref. 3, pp. 47–60
99. Vierstra, R. D., Langan, S. M., Haas, A. L. 1985. *J. Biol. Chem.* 260:12015–21
100. Dahlmann, B., Kuehn, L., Rutschmann, M., Reinauer, H. 1985. *Biochem. J.* 228:161–70
101. Wilk, S., Orlowski, M. 1983. *J. Neurochem.* 40:842–49
102. Ray, K., Harris, H. 1986. *FEBS Lett.* 194:91–95
103. Rivett, A. J. 1985. *J. Biol. Chem.* 260:12600–606
104. McGuire, M. J., DeMartino, G. N. 1986. *Biochim. Biophys. Acta.* 873:279–89
105. Rivett, A. J. 1985. *J. Biol. Chem.* 260:300–5
106. Pontremoli, S., Melloni, E. 1986. *Ann. Rev. Biochem.* 55:455–81
107. Metrione, R. M. 1986. *Trends Biochem. Sci.* 11:117–18
108. Sasaki, T., Yoshimura, N., Kikuchi, T., Hatanaka, M., Kitahara, A., et al. 1983. *J. Biochem.* 94:2055–61
109. Wheelock, M. J. 1982. *J. Biol. Chem.* 257:12471–74
110. Ohno, S., Emori, Y., Imajoh, S., Kawasaki, H., Kisagari, M., Suzuki, K. 1984. *Nature* 312:566–70
111. Schollmeyer, J. E. 1986. *Exp. Cell Res.* 162:411–12
112. Schollmeyer, J. E. 1986. *Exp. Cell Res.* 163:413–22
113. Lynch, G., Baudry, M. 1984. *Science* 224:1057–63
114. Siman, R., Baudry, M., Lynch, G. 1985. *Nature* 313:225–28
115. Willstatter, R., Bamann, E. 1929. *Z. Physiol. Chem.* 180:127–43
116. Barrett, A. J., Kirschke, H. 1981. *Methods Enzymol.* 80:535–61
117. Katunuma, N., Kominami, E. 1983. *Curr. Top. Cell. Regul.* 22:71–101
118. McKay, M. J., Offermann, M. K., Barrett, A. J., Bond, J. S. 1983. *Biochem. J.* 213:467–71
119. Aronson, N. N., Barrett, A. J. 1978. *Biochem. J.* 171:759–65
120. Bond, J. S., Barrett, A. J. 1980. *Biochem. J.* 189:17–25
121. Skidgel, R. A., Erdös, E. G. 1985. *Proc. Natl. Acad. Sci. USA* 82:1025–29
122. Takahashi, T., Dehdarani, A. H., Yonezawa, S., Tang, J. 1986. *J. Biol. Chem.* 261:9375–81
123. Dean, R. T., Barrett, A. J. 1976. *Essays Biochem.* 12:1–40
124. Pontremoli, S., Melloni, E., Salamino, F., Sparatore, B., Michetti, M., Horecker, B. L. 1982. *Arch. Biochem. Biophys.* 214:376–85
125. Offermann, M. K., McKay, M. J., Marsh, M. W., Bond, J. S. 1984. *J. Biol. Chem.* 259:8886–91
126. Evans, P., Etherington, D. J. 1979. *FEBS Lett.* 99:55–58
127. Kirschke, H., Locnikar, P., Turk, V. 1984. *FEBS Lett.* 174:123–27
128. Gohda, E., Pitot, H. C. 1981. *J. Biol. Chem.* 256:2567–72
129. Takahashi, T., Tang, J. 1981. *Methods Enzymol.* 80:565–81
130. Hasilik, A., Neufeld, E. F. 1980. *J. Biol. Chem.* 255:4946–50
131. Erickson, A. H., Conner, G. E., Blobel, G. 1981. *J. Biol. Chem.* 256:11224–31
132. Samarel, A. M., Worobec, S. W., Ferguson, A. G., Decker, R. S., Lesch, M. 1986. *Am. J. Physiol:Cell Physiol.* 250:C589–96
133. von Figura, K., Hasilik, A. 1986. *Ann. Rev. Biochem.* 55:167–93
134. Offermann, M. K., Chlebowski, J. F., Bond, J. S. 1983. *Biochem. J.* 211:529–34
135. Whitaker, J. N., Seyer, J. M. 1979. *J. Biol. Chem.* 254:6956–63
136. Yamamoto, K., Kamata, O., Katsuda, N., Kato, K. 1980. *J. Biochem.* 87:511–16

137. Walter, P., Gilmore, R., Blobel, G. 1984. *Cell* 38:5–8
138. Rodriguez-Boulan, E., Misek, D. E., Vega de Salas, D., Salas, P. J. I., Bard, E. 1985. *Curr. Top. Membr. Transp.* 24:251–94
139. Douglas, M. G., McCammon, M. T., Vassarotti, A. 1986. *Microbiol. Rev.* 50:166–78
140. Mori, M., Miura, S., Tatibana, M., Cohen, P. P. 1980. *Proc. Natl. Acad. Sci. USA* 77:7044–48
141. Jackson, R. C., Blobel, G. 1977. *Proc. Natl. Acad. Sci. USA* 74:5598–602
142. Mumford, R. A., Strauss, A., Powers, J. C., Pierzchala, P. A., Nishino, N., Zimmerman, M. 1980. *J. Biol. Chem.* 255:2227–30
143. Bohni, P. C., Daum, G., Schatz, G. 1983. *J. Biol. Chem.* 258:4937–43
144. Loh, Y. P., Brownstein, M. J., Gainer, H. 1984. *Ann. Rev. Neurosci.* 7:189–222
145. Russell, J. H., Geller, D. M. 1975. *J. Biol. Chem.* 250:3409–13
146. Panthier, J. J., Foote, S., Chambraud, B., Strosberg, A. D., Corvol, P., Rougeon, F. 1982. *Nature* 298:90–92
147. Frey, P., Forand, R., Maciag, T., Shooter, E. M. 1979. *Proc. Natl. Acad. Sci. USA* 76:6294–98
148. Thomas, K. A., Baglan, N. C., Bradshaw, R. A. 1981. *J. Biol. Chem.* 256:9156–66
149. Ole-Moi Yoi, O., Pinkus, G. S., Spragg, J., Austen, K. F. 1979. *N. Engl. J. Med.* 300:1289–94
150. Cromlish, J. A., Seidah, N. G., Chretien, M. 1986. *J. Biol. Chem.* 261:10850–58
151. Cromlish, J. A., Seidah, N. G., Chretien, M. 1986. *J. Biol. Chem.* 261:10859–70
152. Gluschankof, P., Morel, A., Gomez, S., Nicolas, P., Fahy, C., Cohen, P. 1984. *Proc. Natl. Acad. Sci. USA* 81:6662–66
153. Mizuno, K., Kojima, M., Matsuo, H. 1985. *Biochem. Biophys. Res. Commun.* 128:884–91
154. Loh, Y. P., Parish, D. C., Tuteja, R. 1985. *J. Biol. Chem.* 260:7194–205
155. Fricker, L. D., Snyder, S. H. 1982. *Proc. Natl. Acad. Sci. USA* 79:3886–90
156. Liepnieks, J. J., Light, A. 1979. *J. Biol. Chem.* 254:1677–83
157. Tanaka, K., Nakamura, T., Ichihara, A. 1986. *J. Biol. Chem.* 261:2610–15
158. Witheiler, J., Wilson, D. B. 1972. *J. Biol. Chem.* 257:2217–21
159. Coblyn, J. S., Austen, K. F., Wintroub, B. U. 1979. *J. Clin. Invest.* 63:998–1005
160. Kerr, M. A., Kenny, A. J. 1974. *Biochem. J.* 137:477–88
161. Beynon, R. J., Shannon, J. D., Bond, J. S. 1981. *Biochem. J.* 199:591–98
162. Varandani, P. T., Shroyer, L. A. 1981. *Biochim. Biophys. Acta* 661:182–90
163. Ishida, M., Ogawa, M., Mega, T., Ikenaka, T. 1983. *J. Biochem.* 94:17–24
164. Sterchi, E. E., Green, J. R., Lentze, M. J. 1983. *J. Pediatric Gastroenterol. Nutr.* 2:539–47
165. Hojima, Y., van der Rest, M., Prockop, D. J. 1985. *J. Biol. Chem.* 260:15996–6003
166. Bond, J. S., Beynon, R. J. 1986. *Curr. Top. Cell. Regul.* 28:263–90
167. Shin, J., Ji, T. H. 1985. *J. Biol. Chem.* 260:12828–31
168. Kenny, A. J. 1986. *Trends Biochem. Sci.* 11:40–43
169. Kenny, A. J., Gee, N., Matsas, R., Stewart, J., Bowes, M., Turner, A. 1985. See Ref. 3, pp. 175–84
170. Kenny, A. J., Maroux, S. 1982. *Physiol. Rev.* 62:91–128
171. Mulligan, M. T., Bond, J. S., Beynon, R. J. 1982. *Biochem. Int.* 5:337–43
172. Butler, P. E., McKay, M. J., Bond, J. S. 1987. *Biochem. J.* 241:229–35
173. Tager, J. M. 1985. *Trends Biochem. Sci.* 10:324–26
174. Hadorn, B., Tarlow, M. J., Lloyd, J. K., Wolff, O. H. 1969. *Lancet* 1:812–13
175. Bond, J. S., Shannon, J. D., Beynon, R. J. 1983. *Biochem. J.* 209:251–55
176. Beynon, R. J., Bond, J. S. 1983. *Science* 219:1351–53
177. Bond, J. S., Beynon, R. J., Reckelhoff, J. F., David, C. S. 1984. *Proc. Natl. Acad. Sci. USA* 81:5542–45
178. McKay, M. J., Garganta, C. L., Beynon, R. J., Bond, J. S. 1985. *Biochem. Biophys. Res. Commun.* 132:171–77
179. Beynon, R. J., Bond, J. S. 1985. See Ref. 3, pp. 185–94
180. Barrett, A. J., Fritz, H., Grubb, A., Isemura, S., Jarvinen, M., et al. 1986. *Biochem. J.* 236:312
181. Murachi, T. 1983. In *Calcium and Cell Function,* ed. W. Y. Cheung, pp. 377–410. New York: Academic
182. Muller-Esterl, W., Fritz, H., Kellerman, J., Lottspeich, F., Machleidt, W., Turk, V. 1985. *FEBS Lett.* 191:221–26
183. Turk, V., Brzin, J., Lenarcic, B., Locnikar, P., Popovic, T., et al. 1985. See Ref. 3, pp. 91–103
184. Melloni, E., Salamino, F., Sparatore, B., Michetti, M., Pontremoli, S., Horecker, B. L. 1984. *Arch. Biochem. Biophys.* 232:513–19

185. Takano, E., Kitahara, A., Sasaki, T., Kannagi, R., Murachi, T. 1986. *Biochem. J.* 235:97–102
186. DeMartino, G. N., Croall, D. E. 1984. *Arch. Biochem. Biophys.* 232:713–20
187. Nakamura, M., Inomata, M., Hayashi, M., Imahora, K., Kawashima, S. 1985. *J. Biochem.* 98:757–65
188. DeMartino, G. N., Blumenthal, D. K. 1982. *Biochemistry* 21:4297–303
189. Mellman, I., Fuchs, R., Helenius, A. 1986. *Ann. Rev. Biochem.* 55:663–700
190. Dahlmann, B., Rutschmann, M., Kuehn, L., Reinhauer, H. 1985. *Biochem. J.* 228:171–77
191. Garland, P. B., Randle, P. J. 1964. *Biochem. J.* 93:678–87
191a. Bachmair, A., Finley, D., Varshavsky, A. 1986. *Science* 234:174–86
191b. Rogers, S., Wells, R., Rechsteiner, M. 1986. *Science* 234:364–68
192. Rivett, A. J. 1986. *Curr. Top. Cell. Regul.* 28:291–337
193. McKay, M. J., Bond, J. S. 1985. See Ref. 3, pp. 351–61
194. Bond, J. S., Aronson, N. N. 1983. *Arch. Biochem. Biophys.* 227:367–72
195. Sinensky, M., Logel, J. 1983. *J. Biol. Chem.* 258:8547–49
196. McKay, M. J., Bond, J. S. 1986. In *Nutritional Aspects of Aging,* ed. L. H. Chen, pp. 173–94. Boca Raton, Fla:CRC
197. Stadtman, E. R. 1986. *Trends Biochem. Sci.* 11:11–12
198. deDuve, C. 1983. *Eur. J. Biochem.* 137:391–97
199. Duncan, W. E., Bond, J. S. 1981. *Am. J. Physiol.* 241:E151–59
200. Bird, J. W. C., Roisen, F. J. 1986. In *Myology,* ed. A. G. Engel, B. Q. Banker, pp. 745–68. New York: McGraw-Hill

Ann. Rev. Biochem. 1987. 56:365–94

THE STRUCTURE AND FUNCTION OF THE HEMAGGLUTININ MEMBRANE GLYCOPROTEIN OF INFLUENZA VIRUS

Don C. Wiley

Department of Biochemistry and Molecular Biology, Harvard University, Cambridge, Massachusetts 02138

John J. Skehel

National Institute of Medical Research, Mill Hill, London NW7 1AA, England

CONTENTS

365

0066-4154/87/0701-0365$02.00

PERSPECTIVES AND SUMMARY

The determination of the three-dimensional structure of the hemagglutinin membrane glycoprotein, HA, of influenza virus in 1981 (1, 2) provided a structural interpretation for a large amount of data accumulated before then and sparked an intensification of research on the structure-function relationship of this glycoprotein. Besides an important role as a model membrane glycoprotein, the HA has been studied primarily because of its three major activities in the virus's infectious cycle. 1. The HA binds to a sialic-acid-containing receptor on a target cell to initiate the virus-cell interaction. 2. The HA mediates the entry of the virus into the cytoplasm by a membrane-fusion event. 3. The HA is the major surface antigen of the virus against which neutralizing antibodies are produced and as a consequence undergoes antigenic variation leading to recurrent epidemics of respiratory disease.

This review describes progress in understanding the molecular mechanisms of these processes. The location and properties of the receptor-binding site are described as well as an atomic model for virus-cell binding based on the X-ray structure of an HA-receptor fragment complex. A conformational transition is described, probably activated in vivo by the low pH in endosomes, which results in the activation of membrane-fusion activity concomitant with a molecular rearrangement that exposes a hydrophobic 'fusion' peptide. The proposed location of antibody-binding sites on the HA and of regions recognized as processed peptides by thymic lymphocytes are reviewed as well as how variation in the sequence of these regions results in recurrent epidemics.

Reviews of the crystal structure determination of the HA (3), the structure of the neuramidase glycoprotein of influenza virus (4), virus-cell receptors (5), and viral membrane fusion (6) have appeared as well as several compendia of articles on influenza viruses (7–11).

INFLUENZA VIRUS MEMBRANE

Influenza virus, like a number of other enveloped viruses, contains a lipid membrane that it obtains during maturation by budding from the plasma membrane of an infected cell (for review, see 12). The membrane contains cellular lipids, but the membrane proteins are coded by the virus. These proteins are inserted into the membrane during biosynthesis by the same 'signal peptide'–mediated process employed by cellular membrane and secretory proteins. Influenza virus has two membrane glycoproteins, the hemagglutinin (HA) and the neuraminidase (NA). Both of their three-dimensional structures have been determined by X-ray diffraction (1, 4).

THE STRUCTURE OF THE HEMAGGLUTININ GLYCOPROTEIN

The HA of the 1968 influenza virus strain A/Aichi/2/68(H_3N_2) is synthesized as a single polypeptide chain of 550 amino acids, which is subsequently cleaved by removal of arginine 329 into two chains, HA_1 (36,334 daltons) and HA_2 (25,750 daltons). These chains are covalently attached by a disulfide bond from HA_1 position 14 to HA_2 position 137 (13), and the two-chain monomers are associated noncovalently to form trimers on the surface of membranes (14). Bromelain treatment of virus yields a soluble trimer, BHA, of the extracellular region, containing all of HA_1 and the first 175 of 221 amino acids of HA_2, but missing the hydrophobic membrane-anchoring peptide (15). The three-dimensional structure of this trimer has been determined from X-ray studies (2, 12).

Three-Dimensional Structure

Figure 1 shows a schematic drawing of a monomer of the HA. The membrane end of the molecule is at the bottom of the drawing, and a vertical line shows the location of the threefold symmetry axis that would relate two more monomers in the trimer. Both the amino terminus of HA_1 and the C-terminus of BHA_2 are found at the extreme membrane end of the molecule. The structure is long, projecting 135 Å off the membrane. The molecule appears to have a globular domain on top of an elongated stem region. The globular region at the top contains only part of HA_1, while the stem contains parts of HA_1 and all of HA_2. The conformation of the chain is unusually extended. From the amino terminus at the membrane, the first 63 amino acids of HA_1 reach in a nearly extended structure 96 Å up the molecule before the first compact folding occurs. The globular region at the top contains an eight-strand antiparallel beta sheet. This domain contains the receptor-binding site (see RECEPTOR BINDING). The remainder of HA_1 returns to the stem, running nearly antiparallel to the initial stretch of HA_1.

The amino terminus of HA_2 is 22 Å from the C-terminus of HA_1, indicating that a conformational change accompanied the cleavage of the two polypeptide chains. The hydrophobic amino terminal peptide that is involved in membrane fusion (see VIRAL ENTRY AND MEMBRANE FUSION) is tucked into the trimer interface about 35 Å from the bottom of the molecule. The major structural feature of the stem is a hairpin loop of two alpha helices (cylinders in Figure 1). The second helix is 80 Å long and forms the backbone of the stem region.

Independent local contacts stabilize the trimer in the globular and stem regions. The HA_1 globular domains make pairwise contacts near the upper

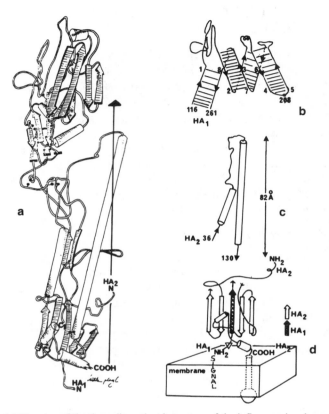

Figure 1 (*a*) Drawing of the three-dimensional structure of the influenza virus hemagglutinin. The terminal residues of HA_1 and HA_2 are labeled. The threefold symmetry axis relating molecules in the trimer is shown: the trimer is stabilized principally by packing of the long α-helixes (cylinders). Note that the amino terminus of HA_2, the 'fusion peptide,' tucks into the threefold contact. (*b*) The eight-stranded β-sheet structure and looped-out regions of the globular domain. (*c*) HA_2 residues 36–130 comprise an α-helical hairpin in the stem region of the monomer. Three of the long helixes, one from each monomer, pack together as a triple-stranded coiled-coil that stabilizes the trimer. (*d*) The membrane end of the molecule contains a five-stranded β-sheet. The central strand is the N terminus of HA_1, and the adjacent strands come from the C-terminal portions of the chain in HA_2. Broken lines suggest the path of the hydrophobic anchoring peptide removed by bromelain cleavage. The site of the oligosaccharide attached at HA_1 8 is shown as a triangle.

end of the molecule. A triple-stranded coiled-coil of alpha helixes formed from the top half of the long helix from each monomer provides the major contacts in the HA_2 region. These contacts are also stabilized by contacts between the second and third residues from each of the hydrophobic N-termini of HA_2 and a network of salt bridges in the lower part of the trimer.

Oligosaccharides

Six oligosaccharide chains are attached to asparagines of HA_1 (amino acid residues 8, 22, 38, 81, 165, 285), and one chain is attached to an asparagine of HA_2 (site number 154). All sites except 81 and 165 are processed, complex oligosaccharides (16). The most striking feature of the carbohydrate is its distribution (1). All sites except 165 are on the lateral surfaces of the molecule. One site, HA_1 8, is at the extreme membrane end of the molecule. The oligosaccharide at 165 appears to stabilize the oligomeric contacts between globular units at the top of the structure. The sequence of the oligosaccharides has been determined from one strain (17), however, no obligatory function has been assigned to them (1, 18, 19).

RECEPTOR BINDING

Sialic acid (N-acetyl neuraminic acid) is the only known essential component of cellular receptors for influenza type A virus. Oligosaccharides terminating in sialic acid are found on many cell-surface glycoproteins and glycolipids, and the removal of sialic acid from them by neuraminidase treatment destroys receptor binding, preventing infection (20). Furthermore, sialic-acid-containing glycolipids can serve as cellular receptors, restoring infectivity to neuraminidase-treated cells (21–23).

Receptor Site Structure

The influenza virus hemagglutinin's receptor-binding site is a pocket located on the distal end of the molecule (Figure 2a) and composed of amino acid side chains (tyr 98, his 183, glu 190, trp 153, leu 194) that are largely conserved in the numerous strains of the virus (1). Other conserved residues (cys 97, pro 99, cys 139, pro 147, tyr 195, arg 229) behind the pocket seem to stabilize the architecture of the site without being in a position to interact with the receptor. By contrast the perimeter of the surface of the pocket is composed of amino acid residues, which have varied during the antigenic drift that accompanies recurrent epidemics (2; Figure 6b).

The amino acid side chains forming the surface of the pocket, which are positioned such that they could make direct contact with a cellular receptor, are both polar and nonpolar (hydrophobic). Histidine 183, tyrosine 98, glutamic 190, and serine 136 have exposed side chain polar atoms that can act as hydrogen bond donors and acceptors. Leucine 194, tryptophan 153, and leucine 226 have exposed nonpolar side chains that have the potential to make van der Waal contacts with the hydrophobic portions of the receptor. In addition to these side chain atoms, sections of main chain from residues 225 to 228 and 131 to 137 have carbonyl oxygens and amide nitrogens of peptide bonds exposed with potential to interact with a receptor.

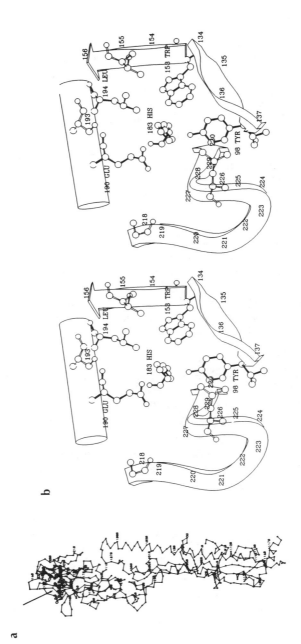

Figure 2   (*a*) Selected amino acids in and around the sialic acid receptor binding pocket are displayed on a schematic drawing of the alpha-carbon backbone of the hemagglutinin from the 1968 influenza virus X-31. (*b*) A detail of the same residues shown in a view of the top of the molecule in stereo. Leucine 226 projects into the pocket near the central residue Tyr 98. The site illustrated binds NeuAcα2→6Gal linkages, while glutamine at position 226 alters the site to bind preferentially NeuAcα2→3Gal linkages.

Topographically, the binding site is a depression, the bottom of which is formed by the phenolic hydroxyl of tyrosine 98 and the hydrophobic surface of tryptophan 153 (Figure 2b). The back wall of the site has histidine 183 projecting out to hydrogen bond with tyrosine 98 and both glutamic acid 190 and leucine 194 hanging down into the site from a short α-helix along the top of the back wall. Residues 224 to 228 form the 'left' edge and 133 to 137 the 'front' edge (Figure 2b). The right side of the site is an extended trough terminating at threonine 155.

In crystallographically refined electron density maps at 3 Å resolution (24), a number of water molecules are visible in the receptor-binding pocket that may mark positions where polar atoms of the receptor will bind [such patterns of bound water molecules being displaced by substrates have been observed in a number of cases in enzyme crystallography (e.g. see Ref. 25)].

Receptor Specificity Mutants

Confirmation that the pocket is the receptor-binding site was provided by the selection of single-amino-acid substitution mutants of the HA with altered receptor specificity. A comparison of the binding of different strains of influenza viruses of the H_3 subtype to erythrocytes derivatized with sialic acid in $\alpha2 \rightarrow 6$ and $\alpha2 \rightarrow 3$ linkages to galactose (NeuAc$\alpha2 \rightarrow$6Gal- or NeuAc$\alpha2 \rightarrow$3Gal-), had revealed distinct specificities (26). Preferential binding to sialic acid in $\alpha2 \rightarrow 6$ linkages, a trait of the 1968 HA, was found to correlate with a high sensitivity to neutralization of infection by glycoproteins (γ inhibitors) present in nonimmune horse serum (27). This allowed selection of receptor specificity variants by growth of virus in that serum (28). Five independently isolated mutants selected from two parental strains showed a reduction of affinity to $\alpha2 \rightarrow 6$ linkages and a marked increase in affinity for $\alpha2 \rightarrow 3$ linkages, so that four mutant viruses bind better to $\alpha2 \rightarrow 3$ than to $\alpha2 \rightarrow 6$ linkages, while one binds equally to both. In all five mutants, only a single-amino-acid substitution had occurred; in four cases glutamine was substituted for leucine at 226 and in one case methionine was substituted for leucine (28 and see Figure 2b), to produce the altered receptor specificity. Residue 226 is in the pocket described above (Figure 2b).

Structure of a Complex with Sialyl Lactose

At high concentration, the trisaccharide sialyl lactose inhibits the binding of influenza virus to erythrocytes (F. S. Escobar, D. C. Wiley, unpublished). To visualize directly the hemagglutinin's interaction with a receptor, .02 M $\alpha2 \rightarrow 6$ and $\alpha2 \rightarrow 3$ sialyl lactose were diffused into crystals of the wild-type 1968 HA (226 leu) and the specificity mutant (226 gln), respectively.

In both cases, difference Fourier electron density maps at 3 Å resolution show similar peaks in the receptor-binding site adjacent to position 226. Only the structure of the specificity mutant, 226 gln, complexed with $\alpha2\rightarrow3$ sialyl lactose, has been completed at this writing (W. Weis, J. C. Paulson, J. J. Skehel, D. C. Wiley, in preparation).

Model building and an exhaustive six-dimensional computer search of the fit of sialic acid to the difference electron density suggest two possible orientations, one of which is currently favored. This orientation places the glycerol side chain of sialic acid in the vicinity of tyrosine 98, glutamic 190, and serine 228. The methyl group of the N-acetyl moiety is located near tryptophan 153 and leucine 194, while the polar atoms of the amide link are near main chain polar atoms at residues 134 and 135. The carboxylate of sialic acid faces into the site near ser 136 and asn 137. The second possibility is in a similar position: C2 occupies about the same location, but the sugar ring is tilted and rotated 180° about an axis from C2 to C5 such that the glycerol and N-acetyl substituents switch locations and the carboxylate points up out of the site. Although the first alternative fits the density better, these models cannot be reliably distinguished at the current stage of the X-ray analysis.

No electron density is observed for the lactose moiety, indicating that it is spatially disordered, i.e. flexibly linked to the sialic acid.

Some alterations in the positions of side chains (< 1 Å) in the binding site are evident in the mutant's HA, which may be responsible for the altered specificity. Completing the X-ray structure analysis of the wild-type complex should help clarify this issue.

This description of binding to a receptor fragment now allows rationalization of some relative affinity measurements of various ligands and prepares the way for the design of inhibitors to block receptor binding and, therefore, infectivity of influenza virus.

Ligand-Binding Comparisons

The binding of influenza virus to cell membranes is effectively irreversible, presumably due to the statistical cooperativity of the large number of hemagglutinin molecules covering the surface of the virus. To date there are no reports of measurements of the intrinsic site affinity of the hemagglutinin for its receptor. Recently, however, the relative binding affinities of a series of ligands have been measured by their ability to inhibit the binding of virus to sialidase-treated erythrocytes, which were derivatized with so few sialic acid residues that the statistical cooperativity of virus binding was markedly reduced (29). Four points are suggested by the data in Figure 3: 1. The α anomer of sialic acid (with the carboxylate axial) has over 10-fold higher affinity than the natural mixture of α and β, which is dominated (20 to 1) by β anomer. 2. A methyl group esterified to the carboxylate of sialic acid reduces affinity about fourfold. 3. The α anomer of the monosaccharide (α-methyl glycoside of NANA) binds almost as tightly (factor of 2 or 3) as an $\alpha2\rightarrow6$-linked tetrasaccharide. 4. One hexasaccharide, doubly sialated, binds 10-fold tighter than the monosaccharide. Points 2 and 3 are consistent with the structural model of ligand binding suggested above. Point 2: Because the carboxylate appears to face toward the protein surface, esterification with a methyl group would be expected to effect the interaction with the protein. Point 3: The X-ray results that two saccharides of sialic lactose are spatially disordered and only the sialic acid is tightly bound are consistent with the affinity measurements indicating that tri and tetrasaccharides have only slightly higher affinity than an α monosaccharide.

Affinity Mutants

Variations in receptor-binding affinity have been observed in natural strains of influenza viruses, and the importance of this in the epidemiology of the disease has been discussed (5, 30–32). Relative affinities have been inferred

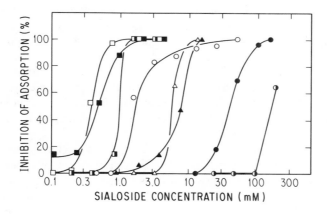

Figure 3 Relative ligand affinities deduced from the inhibition of viral absorbtion to derivatized erythrocytes (from Ref. 29). See text.

by measuring the ability of virus to agglutinate a series of erythrocyte preparations whose receptors have been modified by partial sialidase or periodate treatment.

In one study, mutants of the 1934 virus, A/Puerto Rico/8/38, were selected with enhanced 'affinity' for receptors as inferred by the receptor gradient described above (33). Selection was accomplished by growth of virus in the presence of a mixture of antihemagglutinin monoclonal antibodies at subneutralizing concentrations. In the three mutants selected, a single amino acid substituted in the HA at positions 185, 231, and 244 correlated with enhanced receptor 'affinity.' In the first two cases the amino acids are in the second shell of residues around the receptor-binding site, while the third would appear to require a conformational alteration across the trimer interface adjacent to the receptor-binding site.

Twenty-two monoclonal-antibody-selected antigenic mutants of the 1968 X31 virus with single-amino-acid substitutions in their HA, when tested on a series of erythrocytes treated with increasing amounts of periodate, exhibited altered receptor binding (34). Some single-amino-acid substitutions at 156, 189, 193, and 135 increased the apparent 'affinity,' while others at 144, 145, 158, 188, 189 decreased 'affinity' as measured by this procedure. This demonstrates that single amino acids responsible for antigenic changes can also cause altered receptor-binding characteristics. Whether variation in receptor affinities is important in generating or limiting the recurrence of epidemics is not yet clear.

In a recent study of receptor binding, a series of mutant viruses selected using a single monoclonal antibody were shown to exhibit alteration in both receptor affinity and specificity as well as changes in the activation pH for membrane fusion activity by the HA (35) (See Table 1). Amino acid substitutions at 218 (G to R or E) and 226 (L to P) and remarkably a deletion of 7 amino acids from 224 to 230 cause reduction in the affinity for 2→6-linked sialic acid as measured by binding to derivatized erythrocytes, an increase in affinity for 2,3-linked sialic acid, and at the same time, substantial (.2 to .53 pH units) changes in the pH dependence of membrane fusion activity. Residue 218 and the 224–230 loop form part of the trimeric interface between the globular domains at the top of the HA.

These observations, which indicate that alterations at the trimer interface can also influence membrane fusion, demonstrate the structural potential for information to be transmitted between the binding site and the trimer interfaces and, therefore, possible interrelationships between the process of binding to cells and the triggering of entry by membrane fusion. No evidence indicates that cell binding influences the pH optimum of membrane fusion, but whether some ligand, natural or designed, could cause such a change during normal entry or as an antiviral device to cause a premature conformational change is unexplored.

Table 1 Receptor binding–membrane fusion mutants

Variant	Amino acid at position						Receptor specificity					Change in fusion pH[d]
	145	158	193	218	219	226	Hemagglutination[a]		Absorption[b]		HI[c]	
							SAα2, 6Gal	SAα2, 3Gal	SAα2, 6Gal	SAα2, 3Gal	Eq α2M	
							HA titer		nmol/ml packed cells			
X-31	S	G	S	G	S	L	1024	0	13	>142	4096	—
63-E		N				P	512^t	2^t	30	93	0	+0.53
63-D						***	256^t	256^t	30	68	0	+0.20
63-3				E			1024	1024	10	54	1024	+0.43
v9A				R			256^t	0	32	>142	1024	+0.37
v68x			R				1024	512	24	42	0	0
X-31/HS						Q	0	1924	53	46	0	0

[a] Hemagglutination assays were performed using 12.5% (v/v) human erythrocytes enzymatically modified to contain 37-nmol NeuAc/ml packed cells for the SAα2, 6Galβ1, 4GlcNAc (SAα2, 6Gal) sequences and 107-nmol NeuAc/ml packed cells for SAα2, 3Galβ1, 3GalNAc (SAα2, 3Gal) sequences. All of the viruses agglutinated native (untreated) erythrocytes but did not agglutinate sialidase-treated cells.

[b] These values represent the amount of sialic acid (in nmol/ml packed erythrocytes) required to bind 50% of the applied virus. The actual values are an average of the 50% binding point determined by assaying the amount of virus bound to derivatized cells and the amount of virus remaining in the supernatant.

[c] Hemagglutination inhibition (HI) by equine α2 macroglobulin was determined at initial Eq α2M concentrations of 3.5 mg/ml.

[d] A reasonance energy transfer assay was used to measure hemagglutinin fusion. By this assay X31 exhibited 50% fusion at pH 5.5.

^t Indicates 'transient' agglutination where agglutination of derivatized erythrocytes was apparent after 30 min at room temperature, but no longer evident after 60 min.

*** These variants have deleted amino acids 224–230 inclusive (RGLSSRI).

VIRAL ENTRY AND MEMBRANE FUSION

The two glycopolypeptides of hemagglutinin subunits, HA_1 and HA_2, are derived from the primary translation product of HA mRNA by proteolytic processing, which involves removal of the amino-terminal signal sequence and cleavage by an arginine-specific protease to generate the C-terminus of HA_1 and the N-terminus of HA_2 (36–40). The identities of the enzymes involved in the cleavage reaction are not known, but their actions are required for the production of hemagglutinins active in membrane fusion; uncleaved HAs have receptor-binding activity but are unable to fuse membranes (41–43) and as a consequence, viruses containing such molecules are not infectious (44, 45). These observations and their similarity to the required proteolytic processing of the precursor of Sendai virus fusion glycoprotein (46, 47) were the first indications of the involvement of HA in membrane fusion, and were reinforced by the findings that the N-terminal sequence of the Sendai virus glycopolypeptide generated by the cleavage reaction is analogous to the conserved, hydrophobic amino-terminal sequence of HA_2 (40, 48–50). Subsequently, experiments with SV40-HA recombinant virus-infected cells in which HA was expressed at the cell surfaces directly demonstrated its role in membrane fusion (51), a role that was also supported by demonstrations of the ability of purified hemagglutinins to mediate membrane fusion in vitro (52, 53).

Unlike viruses such as Sendai virus, which fuse with the surface membranes of cells, many membrane viruses appear to enter cells by a process involving endocytosis and to fuse their membranes with the membranes of the endosomes (6, 54, 55). Consequently, the observations that endosomal pH was about pH 5.0 instigated a series of experiments in which a number of viruses including influenza virus were shown to lyse erythrocytes, fuse liposomes, or fuse cells in culture between pH 5.0 and 6.5, depending on the particular virus (56–58).

A Low-pH-Induced Conformational Change Activates Fusion Activity

For influenza viruses, these findings prompted analyses of hemagglutinin structure at low pH in attempts to understand the molecular basis of fusion activation. Soluble hemagglutinins released from viruses by digestion with bromelain (BHA) were observed to form aggregates of sedimentation coefficient approximately 30S containing about eight hemagglutinins specifically at the pH of fusion. Incubation of BHA at this pH in the presence of ^{3}H-TX-100 indicated its acquisition of the ability to bind detergent, and the low-pH-specific association of BHA with liposomes was shown in experiments involving liposome-BHA complex flotation (59, 60).

In addition to these observations, which indicate the exposure of hydrophobic regions of the hemagglutinin in its low-pH conformation, irreversible changes in structure were also shown by analyses of the proteolytic susceptibility of BHA, HA rosettes, and HA in virus particles after incubation at about pH 5.0 (59, 60). The resistance to proteolysis of native HA, clearly indicated by the quantitative isolation of BHA following extensive digestion of virus particles in high concentrations of bromelain (15), is in marked contrast to the susceptibility of HA in the low-pH conformation to digestion by trypsin, chymotrypsin, bromelain, proteinase K, and thermolysin (59–61). More detailed analysis of the tryptic digestion products shows that digestion in this case is restricted to the HA_1 glycopolypeptide yielding three characteristic fragments (Figure 4), one of which, HA_1 1–27, is linked to HA_2 by the single disulfide bond between HA_1 and HA_2 components of the subunit (HA_1 14S-S HA_2 137) and sediments in sucrose density gradients as a large aggregate. Thermolytic digestion of this aggregate converts it to a soluble product of sedimentation coefficient approximately 3S from which the N-terminal 23 residues of HA_2 have been removed (61). The conserved amino terminus of HA_2 appears, therefore, to be the hydrophobic region of the molecule involved in aggregate formation at low pH, its removal by thermolysin producing a soluble product. The other two tryptic peptides, HA_1 28–224 and HA_1 225–328, are recovered from the digestion products as a soluble complex of sedimentation coefficient about 2S, indicating that as a consequence of incubation at low pH, the membrane distal HA_1 domain is dissociated from the membrane proximal region of the molecule.

These observations were extended by the finding that HA_1 glycopolypeptides were quantitatively removed from viruses that had been incubated at pH 5 by cleavage of the disulfide bond between HA_1 and HA_2, which becomes susceptible to reduction specifically at low pH (62). Furthermore, the results of the sedimentation analysis and investigations of the subunit composition of the trypsin-solubilized HA_1 components by cross-linking with dimethyl suberimidate indicate that they are released as monomers and, therefore, that all of the interactions between individual HA_1 glycopolypeptides and other components of the hemagglutinin trimer are lost following incubation at low pH. Similar sedimentation analysis of hemagglutinin incubated at low pH indicates that at low concentrations in the presence of non-ionic detergents, the hemagglutinin completely dissociates (63, 64). At higher concentrations of hemagglutinin this appears not to be the case, and indeed cross-linking studies of the subunit composition of thermolysin-solubilized BHA_2 aggregates indicate that they remain trimeric (J. J. Skehel et al, unpublished).

The extensive nature of the changes in hemagglutinin conformation at low pH is also noted by electron microscopy, which appears to indicate that lengthening and dissociation of regions of the molecule occur (60, 65).

Changes in the antigenicity and immunogenicity of hemagglutinins resulting from incubation at low pH have also been reported and further indicate substantial changes in hemagglutinin conformation. The abilities of specific

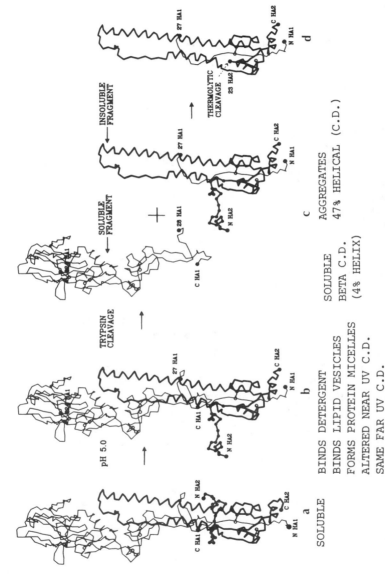

Figure 4 (a) A schematic representation of the effect of incubating BHA at low pH. (*a*) pH 7.0 monomer structure. (*b*) After low pH, the hydrophobic N-terminal peptide of HA₂ is exposed (see text). (*c*) The low pH conformation of BHA can be cleaved in vitro by trypsin at HA₁ 27, which leads to release of the 'top' globular domain from the stem, indicating a change in the tertiary interactions in the stem regions of the monomer. (*d*) The hydrophobic properties of the low pH form are lost after thermolytic removal of residues 1–23 of HA₂.

monoclonal antibodies to bind near intersubunit interfaces of hemagglutinins of the H_3 subtype were generally decreased, whereas the binding of antibodies recognizing sites A, E, and C (see *The Location of the Variable Antibody-Binding Sites*) was not affected, except for reactions with a number of antibodies specific for site C, which were enhanced (66–68). Studies with antibodies against A/PR/8/34 (H_1N_1) virus also indicated pH-dependent changes in hemagglutinin structure, although in these cases enhanced binding to sites A and E were the predominant effects (69).

In spectroscopic analyses, however, only small changes in fluorescence (52) and in near UV circular dichroism (CD) were noted (59). The results of far UV CD indicated that native HA and HA incubated at low pH were indistinguishable, implying that the changes in hemagglutinin conformation induced by low pH do not extend to large changes in secondary structure (59).

Overall, these analyses of the structure of hemagglutinin indicate that extensive rearrangement and loss of contact between components of the trimer that maintain their secondary structure occur as a consequence of incubation at low pH and are accompanied by or cause the extrusion of the hydrophobic amino terminus of HA_2 from its buried location in the trimer interface of the native molecule.

Membrane Fusion Mutants

Support for the proposed route of cell entry through endosomes has been obtained for a number of viruses by observing inhibitions of infections by reagents such as chloroquine, ammonium chloride, and amantadine, which elevate endosomal pH (eg. 70, 71). For influenza viruses in particular, high concentrations of amantadine have been shown to block cell entry (72, 73), and mutants that are resistant to this inhibition contain modifications specifically in their genes for hemagglutinin (74). The mutants retain membrane fusion activity, but their hemagglutinins assume the low pH conformation at higher pH than wild-type virus and mediate fusion at correspondingly higher pH. The HAs of the mutants have been sequenced to identify amino acid residues that influence the pH dependence of the change in conformation (74).

The structural locations of amino acid substitutions detected in the hemagglutinins of a collection of X-31 and Weybridge virus mutants are indicated in Figure 5a. The substitutions fall into two main groups on the basis of their proximity to the amino terminus of HA_2. Substitutions near the amino terminus involve amino acid residues that in the wild-type hemagglutinin stabilize the hydrophobic amino-terminal region in its buried position (Figure 5b). Residues that appear directly important in this context are HA_2 112-D and HA_1 17-H, which from hydrogen bonds with the terminal amino group and the amide nitrogens of residues 3, 4, and 5 of HA_2, and the carbonyl groups of residues 6 and 10 of HA_2, respectively. The amino acid substitutions at these two locations listed in Table 2 involving either changes in side chain charge or length decrease the extent of hydrogen bonding and consequently the stability of this region of the molecule. Less directly, substitutions in neighboring residues such as HA_2 114 also destabilize the location of the amino terminus of HA_2 by introducing a positively charged amino group in close proximity to it. Amino acid substitutions in the HA_2 amino terminus itself include HA_2 6 I→M, HA_2 2 L→F, and HA_2 9 L→F, all of which maintain the uncharged

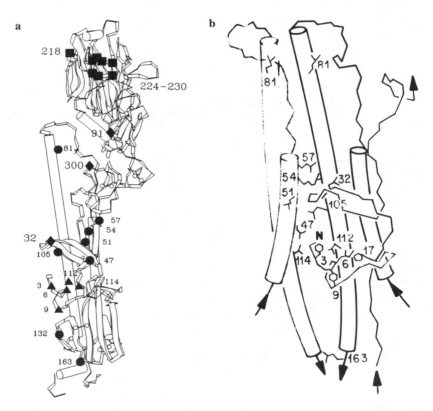

Figure 5 A schematic drawing of the HA monomer showing the location of single-amino-acid substitutions found in the membrane fusion mutants listed in Table 1. ■ indicates a substitution located in the HA_1-HA_1 interface, ● in the HA_2 trimer interface, ▲ in the vicinity of the N-terminal hydrophobic peptide of HA_2, and ♦ in the HA_1-HA_2 interface. Large number labels designate residues in HA_1. (*b*) An α carbon drawing of a detail from the HA trimer stem, showing the helical section of two HA_2 subunits (cylinders right and left) and a segment of HA_1 (thin line on right). The N-terminal peptide of HA_2, marked N, is tucked into the interface between two HA_2 subunits. The amino acid numbers show the positions of single-amino-acid substitutions in membrane fusion mutants.

Table 2 Amino acid substitutions in mutants that lyse erythrocytes at increased pH

	Residue			Amino acid substitution	ΔpH^b
X-31					
1			HA$_2$ 6	I→M	0.3
2	HA$_1$ 207	+	HA$_2$ 6	E→K; I→M	0.35
1a			HA$_2$ 112	D→G	0.4
2a			HA$_2$ 54	R→K	0.25
5a			HA$_2$ 114	E→K	0.6
6a			HA$_2$ 57+163	E→K; R→I	0.4
8a	HA$_1$ 102	+	HA$_2$ 6	V→M; I→M	0.5
4x			HA$_2$ 81	E→G	0.3
6x			HA$_2$ 9	F→L	0.4
12x			HA$_2$ 112	D→N	0.8
aa1			HA$_2$ 47	Q→R	0.35
aa2	HA$_1$ 17			H→Q	0.9
aa4			HA$_2$ 105	Q→K	0.3
aa9			HA$_2$ 112	D→E	0.25
Weybridge^a					
4			HA$_2$ 69+78	E→G; Q→D	0.15
12	HA$_1$ 91			R→Q	0.1
18			HA$_2$ 81	I→S	0.1
2Y	HA$_1$ 20+31			V→A; E→V	0.2
4Y			HA$_2$ 112	D→G	0.4
A1	HA$_1$ 113		HA$_2$ 82	R→I; N→D	0.3
A2	HA$_1$ 101		HA$_2$ 43	K→E; S→L	0.1
A4	HA$_1$ 32			R→G	0.2
A5	HA$_1$ 91			R→L	0.3
A6	HA$_1$ 300			R→S	0.3
A7	HA$_1$ 324	+	HA$_2$ 139	P→S; E→D	0.25
A8			HA$_2$ 54	R→K	0.3
A9	HA$_1$ 32	+	HA$_2$ 82	R→G; N→Y	0.25
B1			HA$_2$ 54+160	R→K; S→N	0.2
B5			HA$_2$ 51+91	K→R; R→Q	0.45
B7			HA$_2$ 114	E→K	0.5
B8			HA$_2$ 67+102	D→N; M→R	0.1
B9	HA$_1$ 221	+	HA$_2$ 114	P→S; E→K	0.45
C5			HA$_2$ 3	F→L	0.4
C6			HA$_2$ 54	R→G	0.1
C7			HA$_2$ 47	Q→L	0.45
C10			HA$_2$ 54	R→S	0.1
C13			HA$_2$ 51	K→N	0.2
C15	HA$_1$ 205	+	HA$_2$ 47	G→E; Q→K	0.3

^{a} The residue numbers for the Weybridge mutants are related to the X-31 hemagglutinin sequence.
^{b} ΔpH values are simple differences between the pHs at which hemolysis by the mutants and wild-type virus are 50% of maximum.

character of the region, but may influence its stability as a consequence of either changes in side chain size or orientation.

Mutant hemagglutinins in the second group contain substitutions in the interfaces between subunits in the trimer (Figure 5b). They all involve changes in charge and, therefore, influence local trimer stability by the loss of one or more salt bridges or hydrogen bonds. (See Table 2.) More importantly, they are located throughout the length of the molecule up to 100 Å distant one from another to indicate that different and widely separated residues influence the pH at which fusion activity is triggered. The regions included in this group have also been extended to HA_1 218 in the membrane distal HA_1-HA_1 interface by the observations that mutants selected with a specific anti-hemagglutinin monoclonal antibody that contain substitutions at this position also mediate fusion at higher pH than does wild-type virus (35). (See also Table 1.)

Mutants typical of both groups have also been isolated without the selective pressure of elevated endosomal pH. Viruses selected for the ability to repli-cate in MDCK cells without the addition of trypsin normally required to ensure precursor hemagglutinin cleavage were found to contain the substitu-tion HA_1 17 H→R (75), and a naturally occurring mutation contained HA_2 132 D→N (76).

Finally, mutant genes for hemagglutinin have been constructed and ex-pressed in SV40-HA recombinant virus-infected simian cells (77). All of the mutations reported were in the amino-terminal region of HA_2, involving residues HA_2 1, 4, and 11. Mutants containing the substitution HA_2 4 G→E mediated fusion at elevated pH, presumably as a consequence of destabilizing this region of the molecule in a similar fashion to the substitutions in the first group of mutants considered above. However, the ability of this mutant to mediate fusion abrogates an absolute requirement for conservation of the uncharged hydrophobic nature of this region, and in this regard is reminiscent of the HA_2 amino terminus in influenza C virus hemagglutinins, which contains aspartic acid residues at positions 5 and 6 (78, 79). Mutations at HA_2 1 G→E prevented membrane fusion and impaired lipid association, while the substitution HA_2 11 E→G had no apparent effect on membrane fusion per se but prevented polykaryon formation by cell-cell fusion. The molecular basis of these phenotypes is not clear at this stage, but such mutants may prove valuable in elucidating the mechanism of hemagglutinin-mediated membrane fusion.

The membrane fusion activities of a number of amantadine hydrochloride-selected mutants and of wild-type X-31 have also been investigated at various temperatures (53, 80). X-31, which at 37°C fuses membranes at pH 5.6, when incubated at 62°C fuses at pH 7.2. Mutants that at 37°C fuse at higher pH than the wild-type virus, fuse at lower temperatures, between 45°C and 60°C, at pH 7.2. Analysis of the heat-induced changes in structure that accompany fusion activation indicates similar but more extensive changes than those observed following low-pH activation of fusion at physiological temperature. Overall, however, the observations support the interpretation that mutations that elevate the pH of fusion decrease the stability of the hemagglutinin and the notion of a requirement for substantial changes in hemagglutinin structure for fusion activation.

Summary of Conformational Changes

The three-dimensional structure of the low-pH-induced fusion-active conformation of the HA is not known. However the experiments discussed above provide a partial picture of that active conformation. 1. The hydrophobic N-terminal peptide of HA_2 is exposed as indicated by the hydrophobic properties and thermolytic sensitivity, and the loss of the hydrophobic properties after thermolytic treatment removes HA_2 1–23. 2. Extended segments of HA_1 in the stem region of the molecule are rearranged. The region near HA_1 27 becomes exposed to tryptic digestion and a disulfide at HA_1 14 to reduction. HA_1 can dissociate from HA_2 after one proteolytic nick or after reduction, probably indicating that the conformation of the extended HA_1 regions that contact HA_2 have rearranged to make less extensive contacts than in Figure 1. 3. The HA_1-HA_1 trimer interfaces between the top, globular domains of the molecule appear to be dissociated based on the exposure of a tryptic site at HA_1 224 and the recovery of the tryptic glycopolypeptide HA_1 28–328 as a monomer rather than a trimer. It remains possible that the HA_1-HA_1 interface is not dissociated but has rearranged and that the observation of monomers of HA_1 after trypsin treatment results from subsequent dissociation at the protein concentration used in the experiment. This seems less likely based on the observed antigenic changes in the HA_1-HA_1 interface region and the destabilizing nature of the mutants (Table 1) in the HA_1-HA_1 interface that affect the pH of membrane fusion. 4. A rearrangement of the HA_2-HA_2 trimer interface along its whole length from the virus membrane to the top of the long helix is suggested by the location of amino acid substitutions in fusion mutants (Figure 5b). Chemical cross-linking of the thermolysin-solubilized BHA_2 aggregate indicates that the stem region of the molecule remains trimeric, although the rearranged interface appears weakened since low-pH BHA can dissociate at sufficiently low protein concentrations.

Proton-induced conformational changes have been observed in the switching of oxy to deoxy hemoglobin and the assembly of tobacco mosaic virus (TMV) (81, 82). The current description of the HA conformational change suggests that certain regions of the molecule, the N-terminal peptide of HA_2 and regions of HA_1 in the stem of the molecule, may undergo not just a change from one conformation to another, but may change from an ordered structure to a disordered ensemble of structures.

Membrane Fusion Mechanisms

The membrane fusion activity of the hemagglutinin has been measured both in vivo and in vitro and by a variety of techniques, including hemolysis (59, 83), polykaryon formation (84), resonance energy transfer (53, 85–87), liposome-cell fusion (88), spin-labeled phospholipid transfer (42), and electron microscopy (89, 90). In the main, determinations of the pH dependence of fusion by these procedures indicate correspondence between the pH of fusion and pH at which the characteristic changes in hemagglutinin structure occur. However, influenza-virus-mediated membrane fusion has also been reported to occur at neutral pH (86, 89, 90). In one case, liposomes containing negatively charged cardiolipin were used in the fusion assays, and the results obtained are probably not directly relevant to the hemagglutinin-mediated fusion process investigated using liposomes with more physiological lipid compositions (86, 87). The significance of other reports is, however, not clear. It may be that the efficiency of fusion in these systems is low, and simply results from a background level of fusion by viruses that fuse optimally at lower pH. Alternatively, the presence in virus preparations of small numbers of mutants able to fuse at elevated pH at 37°C may be involved, or it is possible that some ligands can induce the conformational change as discussed under *Affinity Mutants,* above. Nevertheless, the bulk of the evidence available at present indicates the requirement for a change in hemagglutinin conformation is triggered by low pH for influenza-virus-mediated membrane fusion.

What is the mechanism of this process? It is possible that the extruded HA_2 amino termini of neighboring molecules may associate to form hemagglutinin complexes active in membrane penetration or to expose areas of lipids in both virus and cellular membranes at which fusion may occur. However, the latter possibility seems inappropriate for fusion by viruses such as Sendai virus in which the protein involved in fusion is distinct from the receptor-binding protein and is therefore less likely as a general fusion mechanism. It is also possible in the absence of direct observations of membrane fusion in endosomes during influenza virus infections, that even though HA-mediated fusion can be demonstrated in vitro, the actual virus entry mechanism may result in membrane dissolution rather than in the formation of a continuous cellular-virus hybrid membrane. The favored scheme at present focuses on the conserved and hydrophobic nature of the amino terminus of HA_2 and suggests that as a result of the changes in hemagglutinin structure at low pH, direct interaction of this region of the molecule with an adjacent lipid bilayer is facilitated. Such an interaction may simply overcome the repulsive forces between the membranes, allowing their fusion, or may concomitantly destabilize the membrane and promote fusion as a consequence.

ANTIGENIC VARIATION

Antihemagglutinin antibodies neutralize virus infectivity, and as a consequence the antigenic variation for which influenza viruses are renowned involves extensive variation in the antigenic properties and structure of their hemagglutinin. For over 40 years the antigenicity of influenza isolates from outbreaks and epidemics has been monitored to ensure the inclusion in vaccines of variants closely related to the viruses in circulation (91), and more recently these comparative studies have been accompanied by nucleotide sequence analyses of the genes for the hemagglutinins of representative isolates (92–97).

The Location of the Variable Antibody-Binding Sites

The location of antibody-binding sites on the hemagglutinins of both subtypes of influenza A currently circulating (H_3 subtype viruses since 1968 and H_1 subtype since 1977) have been proposed based on the location of sequence variation in the HAs of natural isolates and on sites of single-amino-acid substitutions observed in antigenic mutants selected by growth in monoclonal antibodies (18, 30, 66, 98–103). Figure 6a shows the location on a schematic diagram of the HA of the naturally occurring sequence variation of the H_3 subtype HA during the last 18 years (five shaped symbols) and the single-site antigenic mutants selected from 1968 and 1972 parental strains with monoclonal antibodies (stars). (For a comparison with the H_1 subtype, see Ref. 103 and Figures 22 and 23 in Ref. 12). The variation is predominantly in the HA_1 glycopolypeptide, involving residues covering much of the surface of the distal domain of the molecule. The substitutions have been segregated into five regions designated A to E (2, 18), although some subdivision and overlap of these areas has been noted (66, 103).

It is currently thought that each of these regions is an antibody-binding area and that the HAs of new epidemic strains must include substitutions in each of these regions in order to escape the antisera produced during an earlier cycle of infection. The antigenic significance of the observed natural variation in regions A to E is strongly supported by the coincident mapping of monoclonal-antibody-selected variants into each of the regions (Figure 6a), because these variants are known to no longer bind the selecting antibody and as a consequence replicate efficiently in its presence. An additional indication that the natural variation is antigenically significant is the observation that the frequency of substitutions in HA_1, about .8% per year since 1968, is greater than that of silent mutation, .3%, indicating that it probably resulted from immune selective pressure.

The assumption that the location of a single-site substitution defines the site of antibody binding has been explored in two cases by crystallizing and determining the X-ray structure of the HAs from the monoclonal-antibody-

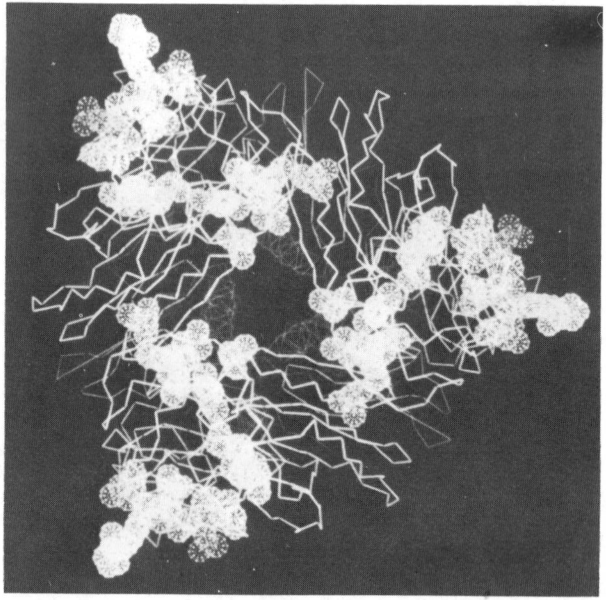

Figure 6 Natural variation since 1968 and monoclonal variants suggest the antibody-binding sites on the 1968 HA. (*a*) ●, Site A; ■, Site B; ▲, Site C; ◆, Site D; ▼, Site E. The symbols represent locations of natural sequence variation between 1968 and 1979. ★, single-site monoclonally selected variant, each star represents a separate variant; ★ plus site symbol, represents a site of natural variation that has also been observed in a monoclonally selected variant. Underlined amino acids in the list of amino acid substitutions were observed in monoclonally selected variants only, no underline indicates in natural variants only, first letter underlined indicates substitutions found in both natural and monoclonally selected variants. An asterisk indicates addition of an N glycosylation site, a minus indicates the loss of such a site. (*b*) Variable amino acid positions defining antibody-binding sites surround the conserved residues forming the receptor-binding pocket. The HA trimer is viewed from above, dotted spheres mark variable amino acid positions shown in Figure 6*a*. The center of the figure shows the α-helices of HA_2 extending 'into' the page. The receptor sites are inside the crescent of variable residues. The remaining position of HA_1, which is not in the site or variable, is covered by the oligosaccharide position 165 (not shown).

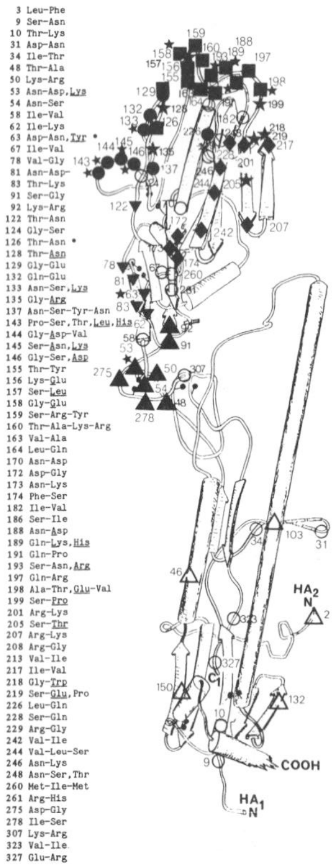

selected mutants 146 G→D and 188 N→D (sites A and B) (104; W. Weis et al, unpublished). In both cases, the structural changes between the HA of the antigenic mutant and wild-type were confined to the immediate vicinity of the amino-acid substitution, demonstrating that for the antibody to recognize this mutation, it must bind directly to that region and illustrating that rather small, local variations in structure suffice to allow escape from neutralization by antibodies (104). Evidence consistent with the conclusion that the locations of amino-acid substitutions define the sites of antibody binding on the HA and that most substitutions influence antibody binding directly rather than from a distance, has also been obtained from electron microscopy of monoclonal antibody–hemagglutinin complexes using antibodies recognizing regions A, B, and E (105). No such direct information is presently available for binding sites C and D.

Because of the proximity of antibody-binding sites and the receptor-binding pocket, certain amino-acid substitutions influence both the binding of specific monoclonal antibodies and the specificity of receptor recognition (105a). As discussed above (RECEPTOR BINDING), a number of antihemagglutinin antibodies have also been observed to select mutants with different receptor-binding specificities which are, however, not antigenic mutants, since they continue to bind the selecting antibodies (33, 35). The sites of binding such antibodies are not known, but they do not in these cases appear to correspond with the locations of the amino acid sequence changes.

Some of the observed variation results in the addition or loss of an N-linked oligosaccharide (asterisk and dash in the list of Figure 6a). In the case of the monoclonal-antibody-selected mutant HA_1 63 D→N, the addition of the oligosaccharide has been proven to be the cause of the antigenic change, as mutant virus grown in the presence of tunicamycin and lacking the new oligosaccharide at 63 was not antigenically distinguishable by the antibody (18). A similar experiment on the natural epidemic strain A/VIC/3/75, which is glycosylated at 63, indicated that it would bind antibody produced against the 1968 strain (two epidemics earlier) when grown in tunicamycin and lacking the oligosaccharide at 63. Thus, carbohydrate, which is host-specific, can mask surfaces of the hemagglutinin from the immune system. Similar masking is seen in the area of HA_1 165 when the proposed antigenic sites of H_3 (glycosylated at 165) and H_1 (unglycosylated at 165) are compared (61, 103).

Figure 6b shows that the proposed antibody-binding sites surround the conserved receptor binding site on the HA. A similar observation has been made on the influenza NA (4) and has been proposed for Rhinovirus and Poliovirus based on their recent X-ray structure determinations (106, 107). Thus, even if antibodies could be directed at the conserved residues in the concave binding site, because of the large footprint of an antibody (108), they would probably be dislodged by the variation that occurs readily on the rim of the binding site.

Recognition by Thymic Lymphocytes

Several studies of hemagglutinin immunogenicity and antigenicity have involved antibody production using synthetic peptides (109–113), and antibodies produced in this way have been of value in monitoring structural changes in the hemagglutinin and in studies of the molecular basis of antigenicity (114, 115). In the main, however, antipeptide antibodies have been inefficient in neutralizing virus infectivity and in binding to the native protein. This contrasts with the recognition of peptides by immune cells and the probable role of hemagglutinin fragments in stimulating cellular immunity.

Hemagglutinin-specific helper T-cell lines derived from peripheral blood lymphocytes from humans were found to be stimulated to proliferate by peptides HA_1 1–36, 105–140, 200–228, 306–328 of H_3 subtype hemagglutinin, and to a lesser extent, by peptides of similar sizes equivalent to all regions of HA_1. The most immunodominant peptide, HA_1 306–328, was recognized by both subtype-specific and cross-reactive clones isolated from these T-cell lines (116, 117).

Helper T-cells induced following infection of mice with A/Puerto Rico/8/ 34 (H_1N_1) were shown to recognize three regions of the HA, two of which were defined using peptides HA_1 109–120 and HA_1 290–310. The fine specificity of clones that recognized peptide HA_1 109–120 was analyzed using fragments of HA_1, various synthetic peptides from the region HA_1 109–120, a series of natural antigenic variants, and a monoclonal-antibody-selected variant to stimulate IL2 production (118–121). As a result, the importance for cellular recognition of residue HA_1 115, which is conserved in natural isolates of the H_1 subtype, was demonstrated. The third epitope defined in these studies involved recognition of HA_1 136.

Experiments with H_3 hemagglutinin-specific mouse cells, also involving the use of peptides, show the recognition of HA_1 48–68 and HA_1 128–148. In these studies, as with the studies of the H_1 subtype, clones of different specificity were differentiated by their abilities to recognize natural and monoclonal-antibody-selected antigenic variants, and also a fusion mutant (122, 123). The results suggest the importance of additional as yet undefined regions of HA_1 in helper T-cell recognition.

Overall hemagglutinin recognition by helper T-cells appears from these studies to be restricted to the HA_1 glycopolypeptide, including a number of regions that are recognized by antibodies. Peptides equivalent to these sites stimulate cells efficiently, which is consistent with the antigen processing roles proposed for antigen-presenting cells (124). In other studies, however, in which synthetic peptides (125) and isolated HA_1 and HA_2 glycopolypeptide chains (126) were used to stimulate proliferation, helper T-cells that recognize components of HA_2 were detected and evidence was presented that they are more subtype cross-reactive than cells that recognize HA_1.

Cytotoxic T-cells appear to recognize both HA_1 and HA_2 components of the hemagglutinin. For H_2 subtype–specific cells, fragments of hemagglutinin have been reported to be ineffective in directing cell recognition (127). In other experiments, however, induction of cytotoxic cells was observed with a CNBr fragment, HA_2 103–123 (128), and this was confirmed using an equivalent synthetic peptide (129). In addition, a peptide analogue of HA_1 181–204 was shown to induce cytotoxic cells at about 10 times greater molar efficiency than the HA_2 peptide. Both of these peptides were recognized by subtype-specific cytotoxic cells (129). Experiments with a fusion protein containing the HA_2 glycopolypeptide produced in *E. coli* have also indicated the ability of this region of the hemagglutinin to generate target cells recognized by H_1 subtype–specific cytotoxic T-cells (130).

During many influenza infections, cytotoxic T-cells specific for HA represent only a minority of the immune cells induced, and as a consequence HA-specific cell clones are infrequently isolated. Following infection with A/Japan/305/57, however, about 45% of the clones obtained were found to be HA specific. Among these, about half were completely strain specific, reacting only with target cells expressing A/Japan/305/57 HAs, about one third were subtype specific, and the remainder either recognized hemagglutinins of both H_1 and H_2 subtypes or of all three subtypes tested, H_1, H_2, and H_3 (131).

Mechanisms of Antigenic Variation

Antigenic variation in influenza viruses involves two processes frequently referred to as antigenic shift and antigenic drift. The former is responsible for the introduction into the human population of viruses containing hemagglutinins antigenically similar to those of viruses circulating in animals of other species and birds. The most recent antigenic shift, which resulted in the Hong Kong influenza epidemic in 1968, was caused by a recombinant virus containing the gene for the H_3 hemagglutinin and the seven other genes in the virus-genome from an H_2N_2 virus circulating at that time in humans (132). Sequence analysis of the hemagglutinins of viruses of the H_3 subtype isolated from ducks and horses before 1968 strongly suggests that the recombination involved an avian influenza virus, and occurred presumably in a bird or human simultaneously infected with a human H_2N_2 virus and an avian virus of the H_3 subtype (133–136). The amino acid sequence homology between the Hong Kong (H_3) hemagglutinin and Asian influenza (H_2) hemagglutinins is approximately 36% HA_1 and 50% HA_2 (137, 138).

Antigenic variants of the H_3 subtype isolated between 1968 and 1986, however, vary in sequence homology by about 0.8% per year, and the process involved in this variation is termed antigenic drift. Antigenic drift appears to involve the selection under antibody pressure of antigenic mutants with the ability to reinfect at least a proportion of the population. The precise mechanism of this selective process is unknown. The frequency at which antigenic mutants occur during selection with monoclonal antibodies is between 10^{-4} and 10^{-5} (139), and given that antibodies against any of the five antigenic regions neutralize virus infectivity, antigenic drift to an extent that would allow reinfection of the majority of the population would occur spontaneously at a very low frequency. Changes at each of the antigenic sites may, on the other hand, occur during reinfections by antigenic mutants changed in only one or two sites in individuals who develop only partial immunity during initial infections. Analysis of the variety of antibody specificities in post-infection human sera and of the restricted ability of human sera to neutralize monoclonal-antibody-selected mutants suggest that this is the case (140, 141). Information on the frequency of reinfections from more extensive serological surveys is required for assessment of this possibility.

CONCLUSION

The molecular models emerging for the major membrane activities of the HA glycoprotein provide a challenge for the future. Can measurements of intrinsic site affinities coupled with structural analysis of HA-receptor analogue complexes provide a description of the events and energetics of virus-to-cell binding? Can the binding potential of the HA receptor-binding site be sufficiently understood to allow binding inhibitors to be designed that could prevent infection? What is the mechanism of membrane fusion? Can a description of the conformational change required for membrane fusion activity be used as the basis for designing strategies to prevent the change, or to trigger it to occur prematurely, thus preventing membrane fusion required for infection?

Using current information and reagents—site-defined monoclonal antibodies, single-site antigenic mutants, extensive sequence information for the natural antigenic variants, and collections of post-infection antisera from a decade of recurrent influenza—can the spectrum of antihemagglutinin antibodies produced in individuals and their consequences be determined to a degree that allows a detailed path for the selection of new epidemic strains to be seen? What is a plausible explanation for the generation of recurrent epidemics?

ACKNOWLEDGMENTS

DCW acknowledges support from NIH AI-13654 and NSF DMB 85-02920.

Literature Cited

1. Wilson, I. A., Skehel, J. J., Wiley, D. C. 1981. *Nature* 289:368–73
2. Wiley, D. C., Wilson, I. A., Skehel, J. J. 1981. *Nature* 289:373–78
3. Wiley, D. C., Wilson, I. A., Skehel, J. J. 1984. In *Biological Macromolecules and Assemblies*, Vol. I: *Viral Structures*, ed. F. Jurnak, A. McPherson, pp. 299–336. New York: Wiley
4. Colman, P. M., Varghese, J. N., Laver, W. G. 1983. *Nature* 303:41–47
5. Paulson, J. C. 1985. In *The Receptors*, Vol. II, pp. 131–219. New York: Academic
6. White, J., Kielian, M., Helenius, A. 1983. *Q. Rev. Biophys.* 16:151–95
7. Laver, W. G., Chu, C. M., eds. 1985. *The Origin of Pandemic Influenza Viruses*. New York: Elsevier-North Holland
8. Nayak, D. P., ed. 1981. *ICN-UCLA Symp. Mol. Cell. Biol.* Vol. 21. New York: Academic
9. Palese, P., Kingsbury, D. W., eds. 1983. *Genetics of Influenza Viruses*. New York: Springer-Verlag
10. Laver, W. G., Air, G., eds. 1985. *Immune Recognition of Protein Antigens. In Current Communication in Molecular Biology*. New York: Cold Spring Harbor Lab.
11. Kendal, A. P., Patriarca, P. A., eds. 1985. *Options for the Control of Influenza Virus. UCLA Symp. Mol. Cell. Biol., Vol. 36.* New York: Liss
12. Wiley, D. C. 1985. In *Virology*, ed. B. N. Fields, pp. 45–68. New York: Raven
13. Ward, C. W. 1981. *Curr. Top. Microbiol. Immunol.* 94/95:1–17
14. Wiley, D. C., Skehel, J. J., Waterfield, M. 1977. *Virology* 79:446–48
15. Brand, C. M., Skehel, J. J. 1972. *Nature* 238:145–47
16. Ward, C. W., Dopheide, T. A. 1980. *Virology* 103:37–53
17. Keil, W., Geyer, R., Dabrowski, J., Dabrowski, U., Niemann, H., et al. 1985. *EMBO J.* 4:2711–20
18. Skehel, J. J., Stevens, D. J., Daniels, R. S., Douglas, A. R., Knossow, M., et al. 1984. *Proc. Natl. Acad. Sci. USA* 81:1779–83
19. Raymond, F. L., Caton, A. J., Cox, N. J., Kendal, A. P., Brownlee, G. G. 1983. *Nucleic Acids Res.* 11:7191–203
20. Gottschalk, A. 1959. In *The Viruses*, ed. F. V. M. Bumet, W. V. M. Stanley, 3:51–61. New York: Academic
21. Bergelson, L. D., Bukrinskaya, A. G., Prokazova, N. V., Shaposhnikova, G. I., Kocharov, S. L., et al. 1982. *Eur. J. Biochem.* 128:467–74
22. Suzuki, Y., Matsunaga, M., Matsumoto, M. 1985. *J. Biol. Chem.* 260:1362–65
23. Bukrinskaya, A. G., Kornilaeva, G. V., Vorkunova, N. K., Timofeeva, N. G., Shaposhnikova, G. I., Bergelson, L. G. 1982. *Vopr. Virusol.* 1982:661–66
24. Knossow, M., Lewis, M., Rees, D., Wilson, I. A., Wiley, D. C. 1986. *Acta Crystallogr.* In press
25. James, M. N. G., Sielecki, A. R. 1984. *Biochemistry* 24:3701–13
26. Rogers, G. N., Paulson, J. C. 1982. *Fed. Proc.* 41:5880
27. Rogers, G. N., Pritchett, T., Paulson, J. C. 1983. *Fed. Proc.* 42:2181
28. Rogers, G. N., Paulson, J. C., Daniels, R. S., Skehel, J. J., Wilson, I. A., Wiley, D. C. 1983. *Nature* 304:76–78
29. Pritchett, T., Brosmer, R., Paulson, J. C. 1986. Submitted for publication
30. Fazekas de St. Groth, S. 1978. In *Topics in Infectious Diseases,* ed. W. G. Laver, H. Bachmayer, R. Weil, 3:25–48. Vienna: Springer-Verlag
31. Underwood, P. A. 1982. *J. Gen. Virol.* 62:153–69
32. Underwood, P. A. 1985. *Arch. Virol.* 84:53–62
33. Yewdell, J. W., Caton, A. J., Gerhard, W. 1986. *J. Virol.* 57:623–28
34. Underwood, A., Skehel, J. J., Wiley, D. C. 1987. *J. Virol.* 61(1):206–8
35. Daniels, R. S., Jeffries, S. A., Yates, P. A., Schild, G. C., Rogers, G. N., et al. 1986. Submitted for publication
36. Elder, K. T., Bye, J. M., Skehel, J. J., Waterfield, M. D., Smith, A. E. 1979. *Virology* 95:343–50
37. Air, G. M. 1979. *Virology* 97:468–72
38. McCauley, J., Bye, J. M., Elder, K. T., Gething, M. J., Skehel, J. J., et al. 1979. *FEBS Lett.* 108:422–26
39. Garten, W., Bosch, F. X., Linder, D., Rott, R., Klenk, H.-D. 1981. *Virology* 115:361–74
40. Skehel, J. J., Waterfield, M. D. 1975. *Proc. Natl. Acad. Sci. USA* 72:93–97
41. Huang, R. T. C., Rott, R., Klenk, H.-D. 1981. *Virology* 110:243–47
42. Maeda, T., Kawasaki, K., Ohnishi, S. 1981. *Proc. Natl. Acad. Sci. USA* 78:4133–37
43. White, J., Matlin, K., Helenius, A. 1981. *J. Cell Biol.* 89:674–79
44. Klenk, H.-D., Rott, R., Orlich, M., Blodorn, J. 1975. *Virology* 68:426–39

45. Lazarowitz, S. G., Choppin, P. W. 1975. *Virology* 68:440–54
46. Homma, M., Ouchi, M. 1973. *J. Virol.* 12:1457–65
47. Scheid, A., Choppin, P. W. 1974. *Virology* 57:475–90
48. Gething, M. J., White, J., Waterfield, M. 1978. *Proc. Natl. Acad. Sci. USA* 75:2737–40
49. Scheid, A., Graves, M., Silver, S., Choppin, P. W. 1978. In *Negative Strand Viruses and the Host Cell,* ed. B. W. J. Mahy, R. D. Barry, pp. 181–93. London: Academic
50. Richardson, C. D., Scheid, A., Choppin, P. W. 1980. *Virology* 105:205–22
51. White, J., Helenius, A., Gething, M. J. 1982. *Nature* 300:658–59
52. Sato, S. B., Kawasaki, K., Ohnishi, S. I. 1983. *Proc. Natl. Acad. Sci. USA* 80:3153–57
53. Wharton, S. A., Skehel, J. J., Wiley, D. C. 1986. *Virology* 149:27–35
54. Simons, K., Garoff, H., Helenius, A. 1982. *Sci. Am.* 246:58–66
55. Helenius, A., Marsh, M., White, J. 1980. *Trends Biochem. Sci.* 5:104–6
56. Baananen, T., Kaariainen, L. 1979. *J. Gen. Virol.* 43:593–601
57. Matlin, K., Reggio, H., Simons, K., Helenius, A. 1982. *J. Mol. Biol.* 156:609–31
58. Lenard, J., Miller, D. K. 1981. *Virology* 110:479–82
59. Skehel, J. J., Bayley, P. M., Brown, E. B., Martin, S. R., Waterfield, M. D., et al. 1982. *Proc. Natl. Acad. Sci. USA* 79:968–72
60. Doms, R. W., Helenius, A., White, J. 1985. *J. Biol. Chem.* 260:2973–81
61. Daniels, R. S., Douglas, A. R., Skehel, J. J., Waterfield, M. D., Wilson, I. A., Wiley, D. C. 1985. See Ref. 7, pp. 1–7
62. Graves, P. N., Schulman, J. L., Young, J. F., Palese, P. 1983. *Virology* 126:106–16
63. Nestorowicz, A., Laver, W. G., Jackson, D. C. 1985. *J. Gen. Virol.* 66:1687–95
64. Doms, R. W., Agnew, W., Helenius, A. 1986. Submitted for publication
65. Ruigrok, R. W. H., Wrigley, N. G., Calder, L. J., Cusack, S., Wharton, S. A., et al. 1986. *EMBO J.* 5:41–49
66. Daniels, R. S., Douglas, A. R., Skehel, J. J., Wiley, D. C. 1983. *J. Gen. Virol.* 64:1657–62
67. Webster, R. G., Brown, L. E., Jackson, D. C. 1983. *Virology* 126:587–99
68. Jackson, D. C., Nestorowicz, A. 1985. *Virology* 145:72–83
69. Yewdell, J. W., Gerhard, W., Bachi, T. 1983. *J. Virol.* 48:239–48
70. Helenius, A., Marsh, M., White, J. 1982. *J. Gen. Virol.* 58:47–61
71. Miller, D. K., Lennard, J. 1981. *Proc. Natl. Acad. Sci. USA* 78:3605–9
72. Skehel, J. J., Hay, A. J., Armstrong, J. A. 1977. *J. Gen. Virol.* 38:97–110
73. Kato, N., Eggers, H. J. 1969. *Virology* 37:632–41
74. Daniels, R. S., Downie, J. C., Hay, A. J., Knossow, M., Skehel, J. J., et al. 1985. *Cell* 40:431–39
75. Rott, R., Orlich, M., Klenk, H.-D., Wang, M. L., Skehel, J. J., Wiley, D. C. 1985. *EMBO J.* 3:3329–32
76. Doms, R. W., Gething, M. J., Henneberry, J., White, J., Helenius, A. J. 1986. *Virology* 57:603–13
77. Gething, M. J., Doms, R. W., York, D., White, J. 1986. *J. Cell Biol.* 102:11–23
78. Nakada, S., Treager, R. S., Krystal, M., Aaronson, R. P., Palese, P. 1984. *J. Virol.* 50:118–24
79. Pfeifer, J. B., Compans, R. W. 1984. *Virus Res.* 1:281–96
80. Ruigrok, R. W. H., Martin, S. R., Wharton, S. A., Skehel, J. J., Bayley, P. M., Wiley, D. C. 1986. *Virology.* 155:484–97
81. Perutz, M. F., Fermi, G., Shih, T.-B. 1984. *Proc. Natl. Acad. Sci. USA* 81:4781–84
82. Butler, P. J. G., Durham, A. C. H., Klug, A. 1972. *J. Mol. Biol.* 72:1–18
83. Maeda, T., Ohnishi, S. I. 1980. *FEBS Lett.* 122:283–87
84. White, J., Helenius, A., Kartenbeck, J. 1983. *EMBO J.* 1:217–22
85. Wilschut, J., Hoekstra, D. 1984. *Trends Biochem. Sci.* 9:479–83
86. Stegmann, T., Hoekstra, D., Scherphof, G., Wilschut, J. 1985. *Biochemistry* 24:3107–33
87. Stegmann, T., Hoekstra, D., Scherphof, G., Wilschut, J. 1986. *J. Biol. Chem.* In press
88. Van Meer, G., Bavoust, J., Simons, K. 1985. *Biochemistry* 24:3593–602
89. Huang, R. T. C., Wahn, K., Klenk, H.-D., Rott, R. 1980. *Virology* 104:294–302
90. Haywood, A. M., Boyer, P. P. 1985. *Proc. Natl. Acad. Sci. USA* 82:4611–15
91. Pereira, M. S. 1982. In *Virus Persistence, Symp. 33, Soc. Gen. Microbiol.,* eds. B. W. J. Mahy, A. C. Minson, G. K. Darby, pp. 15–37. Cambridge Univ. Press
92. Sleigh, M. J., Both, G. W., Un-

derwood, P. A., Bender, V. J. 1981. *J. Virol.* 37:845–53

93. Both, G. W., Sleigh, M. J. 1981. *J. Virol.* 39:663–72

94. Both, G. W., Sleigh, M. J., Cox, N. J., Kendal, A. P. 1983. *J. Virol.* 48:52–60

95. Skehel, J. J., Daniels, R. S., Douglas, A. R., Wiley, D. C. 1983. *Bull. WHO* 61:671–76

96. Daniels, R. S., Douglas, A. R., Skehel, J. J., Wiley, D. C. 1985. *Bull. WHO* 63:273–77

97. Laver, W. G., Air, G. M., Dopheide, T. A., Ward, C. W. 1980. *Nature* 283:454–57

98. Gerhard, W., Webster, R. G. 1978. *J. Exp. Med.* 148:383–92

99. Moss, B. A., Underwood, P. A., Bender, V. J., Whittaker, R. G. 1980. In *Structure and Variations in Influenza Virus*, ed. W. G. Laver, G. M. Air, pp. 329–38. Amsterdam: Elsevier

100. Laver, W. G., Air, G. M., Webster, R. G., Gerhard, W., Ward, C. W., Dopheide, T. A. 1979. *Virology* 98:226–37

101. Gerhard, W., Yewdell, J., Frankel, M. E., Webster, R. 1981. *Nature* 290:713–17

102. Newton, S. E., Air, G. M., Webster, R. G., Laver, W. G. 1983. *Virology* 128:495–501

103. Caton, A. J., Brownlee, G. G., Yewdell, J. W., Gerhard, W. 1982. *Cell* 31:417–27

104. Knossow, M., Daniels, R. S., Douglas, A. R., Skehel, J. J., Wiley, D. C. 1984. *Nature* 311:678–80

105. Wrigley, N. G., Brown, E. B., Daniels, R. S., Douglas, A. R., Skehel, J. J., Wiley, D. C. 1983. *Virology* 131:308–14

105a. Daniels, R. S., Douglas, A. R., Skehel, J. J., Wiley, D. C., Naeve, C. W., et al. 1984. *Virology* 138:174–77

106. Rossmann, M. G., Arnold, E., Erickson, J. W., Frankenberger, E. A., Griffith, J. P., et al. 1986. *Nature* 317:145–53

107. Hogle, J. M., Chow, M., Filman, D. J. 1986. *Science* 229:1358–65

108. Amit, A. G., Mariuzza, R. A., Phillips, S. E. V., Poljak, R. J. 1986. *Science* 233:747–53

109. Green, N., Alexander, H., Olson, A., Alexander, S., Shinnick, T. M., et al. 1982. *Cell* 28:477–87

110. Muller, G. M., Shapira, M., Arnon, R. 1982. *Proc. Natl. Acad. Sci. USA* 79:569–73

111. Jackson, D. C., Murray, J. M., White, D. O., Fagan, C. N., Tregear, G. W. 1982. *Virology* 120:273–76

112. Shapira, M., Jibson, M., Muller, G., Arnon, R. 1984. *Proc. Natl. Acad. Sci. USA* 81:2461–65

113. Atassi, M. Z., Webster, R. G. 1983. *Proc. Natl. Acad. Sci. USA* 80:840–44

114. White, J., Wilson, I. 1986. Submitted for publication

115. Wilson, I. A., Niman, H. L., Houghten, R. A., Cherenson, A. R., Connolly, M. L., Lerner, R. A. 1984. *Cell* 37:767–78

116. Lamb, J. R., Eckels, D. D., Lake, P., Woody, J. N., Green, N. 1982. *Nature* 300:66–69

117. Lamb, J. R., Green, N. 1983. *Immunology* 50:659–66

118. Hackett, C. J., Dietzschold, B., Gerhard, W., Ghrist, B., Knorr, R., et al. 1983. *J. Exp. Med.* 158:294–302

119. Hurwitz, J. L., Haber-Katz, E., Hackett, C. J., Gerhard, W. 1984. *J. Immunol.* 133:3371–77

120. Gerhard, W., Hackett, C., Melchers, F. J. 1983. *J. Immunol.* 130:2379–85

121. Hackett, C. J., Hurwitz, J. L., Dietzschold, B., Gerhard, W. U. 1985. *Immunology* 135:1391–94

122. Mills, K. H. G., Skehel, J. J., Thomas, T. B. 1986. *J. Exp. Med.* 163:1477–90

123. Mills, K. H. G., Skehel, J. J., Thomas, T. B. 1986. *Eur. J. Immunol.* In press

124. Unanue, E. R. 1984. *Ann. Rev. Immunol.* 2:395–428

125. Attasi, M. Z., Kurisaki, J. I. 1984. *Immunol. Commun.* 13:539–51

126. Katz, J. M., Laver, W. G., White, D. O., Anders, D. M. 1985. *J. Immunol.* 134:616–22

127. Morrison, L. A., Lukacher, A. E., Braciale, V. L., Fan, D. P., Braciale, T. J. 1986. *J. Exp. Med.* 163:903–21

128. Wabuke-Bunoti, M. A. N., Fan, D. P. 1983. *J. Immunol.* 130:2386–91

129. Wabuke-Bunoti, M. A. N., Taku, A., Fan, D. P., Kent, S., Webster, R. G. 1984. *J. Immunol.* 133:2194–201

130. Yamada, A., Young, J. F., Ennis, S. A. 1985. *J. Exp. Med.* 162:1720–25

131. Braciale, T. J., Henkel, T. J., Lukacher, A., Braciale, V. L. 1986. *J. Immunol.* 137:995–1002

132. Scholtissek, C., Rohde, W., von Hoyningen, V., Rott, R. 1978. *Virology* 87:13–20

133. Laver, W. G., Webster, R. G. 1973. *Virology* 51:383–91

134. Ward, C. W., Dopheide, T. A. 1981. *Biochem. J.* 195:337–40

135. Fang, R., Min Jou, W., Huylebrook, D., Devos, R., Fiers, W. 1981. *Cell* 25:315–23

136. Daniels, R. S., Skehel, J. J., Wiley, D. C. 1985. *J. Gen. Virol.* 66:457–64
137. Gething, M. J., Bye, J., Skehel, J. J., Waterfield, M. D. 1980. *Nature* 287:301–6
138. Verhoeyen, M., Fang, R., Min Jou, W., Devos, R., Huylebrook, D., et al. 1980. *Nature* 286:771–76
139. Yewdell, J. W., Webster, R. G., Gerhard, W. U. 1979. *Nature* 279:246–48
140. Natali, A., Oxford, J. S., Schild, G. C. 1981. *J. Hyg.* 87:185
141. Wang, M. L., Skehel, J. J., Wiley, D. C. 1986. *J. Virol.* 57:124–28

Ann. Rev. Biochem. 1987. 56:395–433

INTRACELLULAR CALCIUM HOMEOSTASIS

Ernesto Carafoli

Laboratory of Biochemistry, Swiss Federal Institute of Technology (ETH), 8092 Zurich, Switzerland

CONTENTS

INTRODUCTION

Most of the Ca^{2+} of higher organisms is immobilized in the bones and teeth as hydroxyapatite [$Ca_{10}(PO_4)_6(OH)_2$]. A negligible amount of Ca^{2+} (in humans only a few grams out of a total of about 1250) is contained in the extracellular and intracellular fluids. The concentration in the former compartments, including blood plasma, is controlled by the movement of Ca^{2+} in and out of the bone deposits, and is fixed at about 3 mM, of which approximately half is ionized Ca^{2+}. Extracellular (plasma) Ca^{2+} derives its importance from its relationship to the intracellular Ca^{2+}, since the latter performs the fund-

395

0066-4154/87/0701-0395$02.00

amental task of carrying signals to a large number of biochemical activities in the various subcellular compartments. The relationship between extracellular Ca^{2+} and the signaling Ca^{2+} inside cells is not well understood, but it is evident that the extracellular pool provides a relatively large reservoir from which Ca^{2+} is drawn and made to flow into the cell. The maintenance of the extracellular Ca^{2+} within a narrow concentration range ensures that a constant source of Ca^{2+} will always be available to cells. The very high concentration (mM) of free Ca^{2+} in the extracellular pool as compared to that in the intracellular milieu (sub μM) (see below), and the resulting very large electrochemical force on Ca^{2+}, are particularly convenient to its role as an intracellular regulator, since even minor changes in the permeability of the plasma membrane to Ca^{2+} induced by physiological stimuli will produce very significant fluctuations in its cytosolic concentration. Such large fluctuations would naturally be more difficult to achieve if the electrochemical force on Ca^{2+}, i.e. its tendency to penetrate into the cell, were minor. It is important to mention that the difference in concentration between the extra and in-tracellular Ca^{2+} pools concerns the ionized portion of Ca. The total concen-tration of Ca^{2+} inside cells can indeed approach or surpass the mM level; if erythrocytes contain only 20 μM total Ca^{2+} (1, 2), axons contain 200–400 μM (1, 3), heart cells 4 mM (4), liver cells 1.6 mM (5, 6), and brain cells 1.5 mM (4). The essential difference is that in the extracellular pool about half of the total Ca^{2+} is ionized, whereas within cells only 0.1% or less of the total is. This conclusion stems from experiments carried out by Hodgkin & Keynes almost 30 years ago (7); they injected radioactive Ca^{2+} into the giant axon of a squid, and found that the diffusion of Ca^{2+} when a voltage gradient was applied was very limited over a period of hours. Based on the difference between the rate of diffusion of Ca^{2+} in solution and in the axoplasm, and assuming that the difference reflected the different proportions of free Ca^{2+} in the two situations, Hodgkin & Keynes (7) concluded that the ionized Ca^{2+} in the axoplasm was less than 0.022% of the total (since axons contain about 0.4 mM total Ca^{2+}, the ionized Ca^{2+} concentration was calculated to be less than 0.1 μM).

Any chemical designed to function as an intracellular messenger must undergo large fluctuations in concentration around the targets of the messen-ger function, and it is self-evident that significant concentration swings can only be achieved rapidly if the chemical in question is poised to exert the signaling function in a very low concentration range. Thus, once the decision to use Ca^{2+} as a messenger had been made, it became imperative for cells to develop ways and means to keep its background ionized internal concentra-tion very low. The concentration of other biological messengers like cyclic nucleotides, inositol trisphosphate, or diacylglycerol can be easily modified by metabolic synthesis and degradation. Since this is impossible for the case of Ca^{2+}, different mechanisms had to be developed to control it within cells.

Today, cells accomplish this by reversibly complexing Ca^{2+} to specific ligands. Since the binding must be very specific, i.e. privilege Ca^{2+} in the presence of much higher concentrations of other cations, structurally complex ligands are required. As is the rule for biological situations that demand a high degree of structural complexity, protein molecules are also used in this case.

HIGH-AFFINITY CYTOSOLIC CALCIUM-BINDING PROTEINS

Why was Ca^{2+} chosen over the other ions present in the biological environment as the intracellular messenger? There is no conclusive answer to this question, but a plausible key to the choice appears to be the need to bind the messenger tightly and specifically. Na^+ and K^+ (and the monovalent anions) appear unsuitable since they normally form only loose complexes with proteins, due to their large ionic radii and low charge. Equally unsuitable are the large polyatomic anions like phosphate and bicarbonate. Williams (8–10) has repeatedly discussed the binding chemistry of Ca^{2+} and Mg^{2+}, and has concluded that the latter ion, being much smaller than Ca^{2+}, imposes a greater physical constraint on the surrounding protein, assembling the coordinating oxygen atoms (usually six) into a regular octahedron of small dimensions. Since protein backbones are normally insufficiently flexible to provide regular, and small cavities, Mg^{2+} frequently satisfies part of its coordination requirements with water oxygens, thus greatly weakening the strength of its binding to the protein. Ca^{2+}, on the other hand, with its larger radius and more flexible (6 to 8) coordination number, while still binding tightly, imposes a smaller physical constraint on the surrounding protein, accepting amply variable distances to the coordinating oxygen atoms. The greater versatility of Ca^{2+} makes it ideally suited to be complexed by irregularly shaped protein cavities that would refuse Mg^{2+}, and it is likely that this has been the decisive advantage that has tilted the evolutionary choice in its direction. The choice, and the resulting possibility to maintain a very low intracellular free Ca^{2+} concentration, has had at least one other very important advantage. As first pointed out by Weber (11), low intracellular Ca^{2+} makes possible the widespread use of phosphate-containing compound as metabolic fuels. Since phosphorylated compounds are continuously degraded to liberate energy, and resynthesized to store it, a significant concentration of inorganic phosphate always exists in cells. The high concentration of inorganic phosphate would evidently be incompatible with a high concentration of Ca^{2+}; the maintenance of the intracellular free Ca^{2+} at very low concentration levels allows the phosphate-oriented metabolism characteristic of life as we know it.

To complex Ca^{2+} tightly and specifically while rejecting Mg^{2+} and other cations, evolution has developed, and by now probably perfected, especial proteins, the first of which, troponin C, was described by Ebashi about 20 years ago (12). The Ca^{2+}-binding proteins that are involved in the regulation and processing of the Ca^{2+} signal belong to two groups, the soluble (or, more precisely, nonmembranous) proteins, and the membrane-intrinsic proteins. Both types of proteins contribute to the buffering of cell Ca^{2+}, but it is evident that the amount of Ca^{2+} that can be buffered by the soluble cytosolic proteins is limited by their total amount. This quantitative limitation does not apply to the case of the membrane-intrinsic proteins, which bind Ca^{2+} on one side of a membrane (the plasma membrane, or the membrane of one of the organelles), transport it across, and "come back" uncomplexed to repeat the binding and transport cycle. In this way even minute amounts of binding proteins can handle large amounts of Ca^{2+}, and thus modify significantly its free concentration in the cytosol or within the intracellular organelles. Thus, it is now accepted that the movement of Ca^{2+} across membrane boundaries on specific proteins is the most efficient way to regulate its ionic concentration in the cell. The main role of the soluble binding proteins that complex Ca^{2+} reversibly in the cytoplasm, and of the nonmembraneous Ca^{2+}-binding proteins like troponin C, is thus not the buffering of Ca^{2+} but the processing of its signal. The essential ingredient in the processing of the signal is the conformational change of the proteins, which express hydrophobic sites and increase the α-helical content upon binding of Ca^{2+} (13–17). The structurally modified proteins now acquire the ability to interact with enzyme targets, and have thus been aptly called "Ca^{2+}-modulated proteins" (18). These proteins as a group exhibit a number of common properties. They have relatively small M_r (10,000–20,000), are negatively charged at neutral pH (pI 4–5), contain unusually high amounts of acidic amino acids, and bind two to four Ca^{2+} with high affinity.

The first Ca^{2+}-modulated protein to be studied in detail, thanks to the availability of its crystal structure, has been parvalbumin, a muscle protein that is particularly abundant in fishes, amphibia, and reptiles. In a series of now classical studies on parvalbumin, Kretsinger and his associates (19, and see Kretsinger & Nelson, 18, for a review) unraveled the mechanism by which parvalbumin binds Ca^{2+}, and established a series of general structural principles for the high-affinity binding by the Ca^{2+}-modulated proteins. In the years that followed Kretsinger's studies on parvalbumin, the primary structures of a number of other Ca^{2+}-modulated proteins became available, and it was immediately clear that they were compatible with the set of rules established by Kretsinger and his associates. More recently, the crystal structures of three other Ca^{2+}-modulated proteins (20–22) have directly verified the structural predictions from the work on parvalbumin.

The Ca^{2+}-modulated proteins contain repeat domains that bind Ca^{2+} with high affinity and selectively. Each domain consists of a loop of 10–12 amino acid residues, flanked by two α-helixes pointing in opposite directions, perpendicular to each other. The loop constitutes the Ca^{2+}-binding site proper, and contributes 6 to 8 oxygen atoms to the coordination of Ca^{2+}. Both carboxyl oxygens from the side chains of glutamic and aspartic acid residues, and carbonyl oxygens from the protein backbone take part in the binding. In the case of parvalbumin the helix-loop-helix arrangement is repeated three times, but only the loops between helixes C-D and helixes E-F bind Ca^{2+}. The loop between helixes A and B is not operational, the reason for this being the deletion of two amino acids from it. Evidently, the Ca^{2+}-binding cavity in proteins of this type has been perfected during evolution to a degree where even a relatively minor variation becomes incompatible with the binding of Ca^{2+}. The three other crystal structures of Ca^{2+}-modulated proteins now available are those of the steroid-dependent intestinal Ca^{2+}-binding protein (20), of skeletal muscle troponin C (21), and of calmodulin (22). Although they have confirmed the general principles projected from the structure of parvalbumin, one important difference has become apparent, at least in the case of calmodulin and troponin C. Unlike parvalbumin, which is a globular protein, in crystalline troponin C and calmodulin the exit helix of the second Ca^{2+}-binding domain runs without interruption into the starting helix of the third α-helical stretch. Calmodulin and troponin C thus contain a central 25-amino-acid α-helical stretch, which connects Ca^{2+}-binding domains A-B/C-D and E-F/G-H. This confers to the proteins an extended dumbshell-like appearance, very different from that of parvalbumin. Conceivably, this extended conformation enhances the structural flexibility of the protein [Herzberg & James have suggested (21) that a glycine in the middle of the long α-helix serves as a hinge around which the two halves can rotate, or even come into contact], and thus plays a role in the processing of the Ca^{2+} message (i.e. in the Ca^{2+}-induced conformational change of the protein that is essential for the interaction with the targets). A caveat, however, is in order on this point, since there is at the moment no conclusive evidence that the long connecting α-helix, which is certainly present in the crystal state, also exists in the protein in solution. It may be pertinent to note in this context that the crystals of calmodulin and troponin C were obtained at a pH considerably below physiological neutrality (between 5.0 and 5.6). Under these conditions Ca^{2+}-binding domains A-B and C-D of troponin C lack Ca^{2+} and have unorthodox inter-helix angles, the starting and exit helices being almost parallel.

CALCIUM BUFFERING BY THE CYTOSOLIC HIGH-AFFINITY CALCIUM-BINDING PROTEINS

The main message that should be extracted from the discussion above is that the role of Ca^{2+}-modulated proteins in the buffering of cell Ca^{2+} is of minor importance as compared to their function in the processing of the signal. This is a reasonable conclusion. In addition to the quantitative limitation allowed to above, it is difficult to see how Ca^{2+}-modulated proteins could respond rapidly, i.e. complex more or less Ca^{2+}, in response to physiological stimuli (hormonal or otherwise) that demand rapid variations of the level of ionized Ca^{2+} in the cytosol or in any of the other cell compartments. However, Ca^{2+}-binding proteins are present in the cytosol of cells, and it is therefore appropriate to discuss their contribution to the total Ca^{2+}-binding capacity of cells. Since no more than 0.1% of the total cell Ca^{2+} is ionized and since cells contain up to mM concentrations of total Ca^{2+} (see above), it follows that between 20 μM and several μM Ca^{2+} (see above) is permanently complexed to ligands in resting cells (or in the form of inert precipitates). As will be discussed in detail below, the ionized Ca^{2+} concentration in resting cells is thought to oscillate between 0.1 and 0.2 μM; this implies that ligands with K_d values in the mM range will have a very low percentage saturation with Ca^{2+}. The inorganic and small organic ligands, as well as nucleic acids and zwitterionic phospholipids, will have little or no Ca^{2+} bound at the sub-μM concentrations present in resting cells. Negatively charged phospholipids have higher affinity constants for Ca^{2+} (23), and could thus bind considerable amounts of it. This may be particularly significant for the case of cardiolipin in the inner leaflet of the inner mitochondrial membrane. Thus, biological membranes may be responsible for a sizable fraction of the total Ca^{2+}-buffering capacity of cells: Manery has calculated a value of 0.3 mM Ca^{2+} for the overall half-saturation of cell membranes with Ca^{2+} (24). It is clear, however, that a significant fraction of Ca^{2+} is complexed to high-Ca^{2+}-affinity proteins, which contribute significantly to the rapidly exchangeable pool of bound Ca^{2+}; Baker (25) has calculated that the pool in the cytosol of the giant axon of the squid amounts to about 10 μM Ca^{2+}. Calmodulin is probably the most important of the high-affinity Ca^{2+}-binding proteins, since it is present in considerable amounts in all eukaryotic cells. It is now known to interact with a large number of cellular enzymes, and to mediate their function: among them, one can quote adenylate cyclase and cyclic nucleotide phosphodiesterase, phosphorylase *b* kinase, phospholamban, the Ca^{2+}-pumping ATPase of plasma membrane (see 26 for a recent review). According to present knowledge, calmodulin is found in the cytosol, and bound to the plasma membrane. Among intracellular organelles, mitochondria are calmodulin free (27), and no information is available as yet on endo(sarco)plasmic reticulum. Other high-affinity Ca^{2+}-binding proteins are present only in selected tissues, for instance troponin C and parvalbumin in muscles. Troponin C, the first recognized Ca^{2+}-binding protein and mediator of the Ca^{2+} message (12), is one of the three subunits of the troponin complex. It shares considerable sequence homology with calmodulin and has four Ca^{2+} binding sites like the latter (in heart, however, there are only three sites).

In recent years, troponin C and calmodulin have been joined by a relatively large number of (putative) Ca^{2+}-binding proteins. As a group, they share some properties, e.g. the expression of hydrophobic sites in the presence of Ca^{2+}, and the Ca^{2+}-dependent association with biomembranes. Their metabolic function is, at the moment, a matter for speculation, but they could play a role in the processing of the Ca^{2+} message and/or in the buffering of cell Ca^{2+}. One interesting set of proteins present in several tissues have been called calcimedins (28), and differ from calmodulin in isoelectric point, DEAE-cellulose binding characteristics, and heat stability. They have apparent M_r of 67,000, 35,000, 33,000, and 30,000, and the first has been shown to interact in a Ca^{2+}-dependent manner with another intracellular protein (29). They do not activate the classical target of calmodulin action cyclic nucleotide phosphodiesterase, although pretreatment of the enzyme with calcimedins attenuates the activation by calmodulin (30). They have been shown to activate the Ca^{2+} ATPase activity of isolated liver microsomes (30). The properties of calcimedins are interesting, but it is well to remember that their ability to bind Ca^{2+} is based on still indirect experiments, e.g. UV circular dicroism changes following Ca^{2+} titration, or fluorescence measurements with Tb^{3+} titration (31). It is of interest that the Ca^{2+} concentration required for these effects is in the μM range.

Synexin is a protein first described in adrenal medulla chromaffin granules (32), but later found in other tissues as well. It is a 47,000-dalton protein that causes isolated chromaffin granules to aggregate and apparently promotes their fusion with the plasma membrane in the process of exocytosis. The granule aggregation is Ca^{2+} dependent, and coincides with the Ca^{2+}-promoted polymerization of synexin. It has been proposed that the granule membrane contains a synexin receptor protein (33), although phosphatidyl-inositol has also been considered as the receptor for synexin (34).

A different group of proteins with synexin-like activity have been found first in the electric organ of *Torpedo marmorata* (35) and then in several mammalian tissues (36). They have been termed calelectrins and exist in different states of aggregation (from 32,500 to 67,000 daltons). Like synexin, they become associated with membranes in a Ca^{2+}-dependent manner. Their ability to bind Ca^{2+} is suggested by the Ca^{2+} dependence of the membrane association, and is thus, as for the case of calcimedin, only presumptive. Chromobindins (37); synhibin (38); proteins I, II, and III from liver and intestine (39); lipocortin (40–42); endonexin (p 32.5), an adrenal Ca^{2+}-binding protein (46); and p 36, a Ca^{2+}-binding protein of intestine brush border (43), are some of the newly discovered proteins that also belong to this group. All these proteins are immunologically related to one another, and to calcimedin (44). [Calregulin, a major Ca^{2+}-binding protein of bovine tissues, is apparently immunologically unrelated to them, and it is of interest that it expresses hydrophobic sites upon binding of Zn^{2+} instead of Ca^{2+} (45)]. A recent study (46) has noted a 17-amino-acid consensus sequence, present one or more times in each protein, in Torpedo calelectrin, protein II, p. 36, endonexin (p 32.5), and lipocortin. This sequence may represent a portion of a new type of Ca^{2+}-binding site, or of a lipid-binding domain.

One can attempt to evaluate quantitatively the Ca^{2+} buffering capacity of the two most widely studied, and possibly most important Ca^{2+}-binding proteins, calmodulin and troponin C, since data are available on their content in a variety of tissues (calmodulin) and in heart and skeletal muscle (troponin C). In attempting the evaluation, however, one ought to bear in mind that the Ca^{2+}-binding properties, particularly of calmodulin, are influenced by a number of factors, among them the cooperativity between the first and the second binding site, the presence of Mg^{2+}, and the salt composition of the environment (47, 48). Thus, the extrapolation of the in vitro data to the in vivo situation can only be approximate. In mammalian tissues, the concentration of calmodulin is highest in brain (about 30 μM, 49–56), testis (about 28 μM, 13, 56, 57), and endocrine pancreas (about 24 μM, 58), and lowest in thyroid and skeletal muscles (about 2 μM, 59, 60). Interestingly, axons apparently contain much more calmodulin than the remainder of the nervous tissue (C. Klee, personal communication). On the basis of four Ca^{2+}-binding sites per molecule, and bearing in mind the limitations alluded to above, it can then be calculated that between 8 μM (thyroid, skeletal muscles) and 120 μM (brain) cell Ca^{2+} may be buffered by calmodulin in resting cells. Since the calmodulin content of heart and skeletal muscles is at the lowest end among mammalian cells, it is interesting that these two cell types contain large amounts of another high-affinity Ca^{2+}-binding protein, troponin C (about 2 and about 20 μM, respectively, 53). By contrast, smooth muscle cells, which appear not to contain troponin C, contain instead amounts of calmodulin that are almost as high as in brain or testis (about 25 μM in chicken gizzard, 17 μM in myometrium, 53).

THE MEASUREMENT OF INTRACELLULAR FREE CALCIUM

The measurement of intracellular free Ca^{2+}, obviously a problem of great importance, has seen a large number of efforts, and the use of a large number of procedures, over a period of decades. The ideal method should in principle be able to "see" Ca^{2+} specifically down to a concentration of about 10 nM, in the presence of a large excess of Mg^{2+} and of other physiological cations. It should be able to reveal transient changes in ionized Ca^{2+} that occur on a very fast time scale, e.g. milliseconds, in contracting muscles. It should reveal free Ca^{2+} intracellularly without damaging the functional properties of the cells, including their total Ca^{2+} buffering capacity and their morphological integrity. Lastly, if an indicator is used it should ideally not diffuse back into the medium across the plasma membrane, nor redistribute across intracellular membrane boundaries. No one of the methods used so far satisfies all of these demands, although in cells large enough to permit microinjection of suitable indicators (phosphoproteins, 61–64, metallochromic dyes, 65, 66) or impalement with Ca^{2+}-sensitive microelectrodes (67, 68), adequate measurements have been made, yielding values around 0.1 μM. Small cells, not amenable to microinjection or impalement, have become accessible to analysis in the last few years thanks to the use of esters of fluorescent Ca^{2+} chelators, introduced into cells by passive diffusion and blocked inside by removal of the ester moiety (69, 70). These fluorescent esters have now rapidly become the method of choice in most situations, even when large cells are the object. The principle of the method, which was applied for the first time to an acetoxymethyl tetraester of a Ca^{2+} chelator termed Quin 2 (69, 70), is both simple and ingenious. Being relatively apolar, the ester diffuses freely into cells across the plasma membrane, to be trapped inside by cytosolic esterases that unmask the four carboxyl groups of the molecule, and render it too hydrophilic to cross membrane barriers. Quin 2, despite its immediate and wide acceptance, is not ideal, chiefly because its low fluorescence yield requires very large concentrations inside cells, affecting in a significant way their total buffering capacity, and underestimating rapid Ca^{2+} transients. The new generation of fluorescent Ca^{2+} chelators, e.g. Fura 2 (71), is far more satisfactory, essentially because the higher fluorescent yield requires much lower intracellular concentrations. An additional advantage is the fact that the peak of fluorescence emission is now removed from the regions of the spectrum where intracellular fluorescent compounds might interfere. Fura 2 (71) is now well on its way to becoming the most popular, and most useful, indicator of intracellular free Ca^{2+}.

A large number of studies using fluorescent Ca^{2+} chelators in a number of cell types, and photoproteins like aequorin (61–63), and obelin (64, 72), metallochromic indicators like arsenazo III (65, 66, 73), Ca^{2+}-specific microelectrodes, especially microelectrodes employing neutral charge carriers (67, 74) in large cells that permit their use, point to a resting ionized Ca^{2+} concentration in most cells that oscillates between 0.1 and 0.2 μM.

MEMBRANE TRANSPORT OF CALCIUM

Eukaryotic cells contain Ca^{2+}-transporting systems in the plasma membrane, in mitochondria, and in endo(sarco)plasmic reticulum. Ca^{2+}-transporting systems (ATPases) have been described also in the Golgi vesicles (75) and in the lysosomes (76), but their characterization is not nearly as advanced as that of the systems in the other membranes mentioned. Typically, plasma membranes contain three systems: a specific ATPase (77), a Ca^{2+} channel (78), and a Na^+/Ca^{2+} exchanger (79, 80) (the Na^+/Ca^{2+} exchanger, however, is absent from mature erythrocytes). Mitochondria contain an electrophoretic uniporter that is used exclusively for the uptake of Ca^{2+} (81) and a Na^+/Ca^{2+} exchanger, different from that of the plasma membrane, used for the release of Ca^{2+} from the matrix to the cytosol (82). Some mitochondrial types may also contain a Ca^{2+}/H^+ exchange system. Sarco(endo)plasmic reticulum contains a specific ATPase, different from that of the plasma membrane (83, 84) for the uptake of Ca^{2+}, and a hitherto unknown system for the release of Ca^{2+} back to the cytoplasm.

These transport systems have different kinetic properties, poised to satisfy the different requirements of cells during the functional cycle. Indeed, there will be situations where Ca^{2+} must be regulated in the cytosol, or in other cell compartments, very rapidly and with utmost precision, e.g. the contraction-relaxation cycle of muscle, especially fast muscles. Conversely, other situations may require slower movements of bulk amounts of Ca^{2+}. The systems outlined above are diversified to do just that, since they have different affinities of interaction with Ca^{2+}, and different total Ca^{2+} handling capacity. In general, whenever the need arises to transport Ca^{2+} with high interaction affinity, ATPases are chosen, since this appears to be the only transport mode that confers to the system high Ca^{2+} affinity. As a result, cells rely solely on ATPases for the fine tuning of their Ca^{2+}. On the other hand, more options are open to situations that demand the movement of bulk amounts of Ca^{2+} with intermediate affinity. Exchangers, channels, and electrophoretic uniporters are all low-Ca^{2+}-affinity systems.

THE CALCIUM TRANSPORTING SYSTEMS OF PLASMA MEMBRANES

The Ca^{2+} Channel

Ca^{2+} action potentials, implying specific Ca^{2+} channels, were first recorded in 1958 in the crayfish muscle fiber membrane (78), and then in a number of other excitable tissues (see 85–89 for reviews). The essential characteristics of the channel, as derived originally from recordings of electrical currents in intact tissues and for cells, have been greatly refined and extended in more recent times using patch-clamp recordings of currents through single Ca^{2+} channels (90). The conductance of the channel is of the order of 15–25 pS, each channel allowing the passage of about 3×10^6 Ca^{2+} ions per second. The density of the Ca^{2+} channel varies in different excitable tissues: it is of the order of 0.1 to 1.0 per μm^2 in heart (91, 92), but much higher, for instance, in the T-tubular membranes of skeletal muscles (93–95). One important point in this context is the existence of Ca^{2+} channels in nonexcitable cells. Naturally, excitable cells offer great advantages in the study of channel-mediated Ca^{2+} currents, and this helps explain why almost all of the work on Ca^{2+} channels has been performed on excitable systems. In fact, the impression one often has from the literature is that Ca^{2+} channels are thought to be present only in these cell types. This is of course incorrect; the essential role of Ca^{2+} and the absolute necessity of all cells to carefully control its penetration demand that a controlled (proteinaceous) structure, i.e. a "channel," always be present. The gating system(s), the Ca^{2+} selectivity, and the pharmacology of the channel as described in excitable cells, may well be peculiar to these cell types. But no cell can afford to let Ca^{2+} diffuse inside without the precise control only gated channel structures can provide. Future experimental developments will undoubtedly see the extension of the concept of Ca^{2+} channels to all cells in general. Interesting reports on this have already begun to appear (96, 97).

The Ca^{2+} channels are gated by the electrical potential across the plasma membrane. Upon depolarization of the plasma membrane, inward Ca^{2+} currents become apparent at potentials around -40 mV and reach a maximum around 0 mV. The relationship linking voltage to the probability of channel opening saturates at less than one. The selectivity of the Ca^{2+} channel is not absolute, since Ba^{2+} and Sr^{2+} are in fact transported in preference over Ca^{2+}. It has been proposed (86) that the Ca^{2+} channel possesses a selectivity filter responsible for the exclusion of monovalent cations. When the filter becomes damaged, Na^+ is also allowed to cross the channel.

Ca^{2+} channels can be blocked by several organic compounds. Tertiary amines like verapamil and its derivatives and the benzothiazepine diltiazem block the channel from inside, after having entered it in the open state (98, 99). Therefore, their effect is voltage- and usage-dependent. Dihydropyridines, the most frequently used Ca^{2+} channel blockers, do not show voltage and usage dependance (99), although their potency is dependent on the membrane potential (100–103). They are thought to block the channel from the inner surface, or from the lipid phase of the plasma membrane, since their effect increases markedly with their lipophilicity (103, 104). Some dihydropyridines do not act as channel blockers (Ca^{2+} antagonists), but as Ca^{2+} agonists, since rather than favoring the inactivated state of the channel they block it in the open position (103–105).

One important aspect of the Ca^{2+} channel is its sensitivity to adrenergic neurotransmitters, first observed in isolated atria (106) and, most recently, in patch-clamp experiments on cultured heart myocytes (90) or single cardiac cells. cAMP or the catalytic subunit of the cAMP-dependent protein kinase (90, 107–109) enhance the channel opening probability, essentially by increasing the forward rate constants k_1 and k_2, and by decreasing slightly k_{-2} in a kinetic model that assumes two closed states C_1 and C_2, and one open state O (89).

$$C_1 \underset{k_{-1}}{\overset{k_1}{\rightleftharpoons}} C_2 \underset{k_{-2}}{\overset{k_2}{\rightleftharpoons}} O \hspace{4cm} 1.$$

It has recently become apparent that there are several types of Ca^{2+} channels, each one with different properties. Three types have been identified in cultured sensory neurons (110), two in mammalian hearts (111). In neurons three kinetically different unitary conductances have been observed using patch-clamping. The most common corresponds to the sq-called L-type channels, whose repeated openings produce currents of relatively long duration. The T-type channel (112) opens at more negative transmembrane potentials than the L-channels, and is responsible for short-duration currents. The third type, the N-channel, is not present in heart sarcolemma and is only opened by large depolarizations from very negative transmembrane potentials. Only the L-type channels are sensitive to dihydropyridines (111), and it has been postulated that they may be more abundant in heart (and smooth muscle) than in the central nervous system (113), since dihydropyridines have very evident effect on heart and smooth muscle, and minor effects on the central nervous system. It has also been observed (111) that only the T-type channels survive in isolated membrane patches, whereas L-channels do not.

Labeled dihydropyridines have been used to tag Ca^{2+} channel components during isolation attempts. Experiments on brain cells (114), rabbit skeletal muscle T-tubules (93, 94, 115), and guinea pig skeletal muscle (116) have produced a glycoprotein of approximately 210,000 daltons. Whether the total dihydropyridine receptor consists of three polypeptides of M_r about 142,000, about 56,000, and about 31,000 or of two polypeptides of M_r about 142,000 and about 31,000 is still controversial. It has been claimed (117) that the 56,000-dalton subunit, termed β-subunit, is selectively phosphorylated by the cAMP-dependent protein kinase in T-tubule membranes, but more recent work (118, 119) has shown that the phosphorylation of the 142,000-dalton subunit may be the one that is important functionally. One important recent development is the reconstitution of the purified dihydropyridine receptor complex. The complex has been incorporated into planar phospholipid bilayer membranes at the tip of glass patch pipettes (120). The reconstituted receptor forms a functional 20-pS Ca^{2+}-channel that retains the regulatory, biochemical, and pharmacological properties of L-type channels in the native membranes (119).

One can attempt to calculate whether the opening of Ca^{2+} channels leads to a significant increase in cytosolic free Ca^{2+} (121). If heart cells are taken as an example, one may assume that their transmembrane potential is depolarized by 50 mV and that all of the current necessary to produce this depolarization is carried by Ca^{2+}. The charge (Q) will be given by

$$Q = \frac{CV}{ZF},\qquad\qquad\qquad 2.$$

which corresponds to 2.8×10^{-13} moles cm^{-2} (assuming $C = 1\ \mu F$ cm^{-2}). A spherical cell of diameter 20 μm will have a surface area of 1.3×10^{-5}cm^2, corresponding to the influx of 3.64×10^{-18} moles of Ca^{2+}. In a cell volume of 4.2 pl this would correspond to a change in cytosolic Ca^{2+} concentration of about 1 μM, assuming uniform distribution and no removal of free Ca^{2+} from the cytosol by internal buffers, membranous or otherwise. In fact, the last assumption is likely to be incorrect, since it seems probable that the increase in Ca^{2+} is limited to a narrow zone beneath the plasma membrane where the concentration increase would then be much larger. The iternal Ca^{2+} buffers will certainly act on the penetrated Ca^{2+}. That the internal concentration of Ca^{2+} indeed increases upon activation has been experimentally demonstrated (albeit under somewhat extreme conditions) in squid giant axons using the photoprotein aequorin (85, 122).

The Na^{+}/Ca^{2+} Exchanger

The Na^{+}/Ca^{2+} exchange is particularly active in excitable plasma membranes, like those of nervous and heart cells. Long ago, its existence was indicated by the observation that heart muscle contracts when placed in a Na^{+}-free saline (while skeletal muscle, for example, does not). The most plausible explanation for the observation was that if external Na^{+} is reduced, Ca^{2+} from the saline medium can exchange for internal Na^{+}, thus providing the inward-directed flux of Ca^{2+} necessary for the contraction. Experiments in the late 1960s on cardiac muscle preparations and on the giant axon of the squid (79, 80, 123) documented the existence of the exchanger. It was shown that the exit of Ca^{2+} from the cell is dependent on external Na^{+}, and it was recognized at the outset that the system is reversible, i.e. that either raising the intracellular Na^{+} concentration, or lowering the Na^{+} concentration in the extracellular medium, induces Ca^{2+} influx (124, 125). It soon became clear that the exchanger, although undoubtedly most active in heart and brain cells, is also present in nonexcitable tissues like endocrine tissues (126), various other epithelial cells (74, 127–130), and bone cells (131). In fact, it appears that the erythrocyte may be the only cell where a Na^{+}/Ca^{2+} exchanger is completely absent. Dog erythrocytes contain a Na^{+}-coupled Ca^{2+} transport system, but it is not mediated by a direct Na^{+}/Ca^{2+} exchanger.

Logically, the role of the Na^{+}/Ca^{2+} exchanger should be the control of the intracellular free Ca^{2+} concentration. In the original postulate of a Na^{+}/Ca^{2+} exchange system in heart plasma membranes (79), an electrically silent exchange of 2 Na^{+} per 1 Ca^{2+} was suggested, in which the inwardly directed downhill movement of Na^+ provides the energy for the outwardly directed uphill movement of Ca^{2+} (in the case of the exchanger operating in the direction of Ca^{2+} influx into the cell it would of course be the inwardly directed downhill movement of Ca^{2+} that provides the energy for the uphill extrusion of Na^+). It was recognized very early, however, that the energy content of the transmembrane Na^+ gradient is not adequate to maintain the very large Ca^{2+} gradient across the plasma membrane if the system were to operate electroneutrally (123). In fact, a 3 Na^+ per 1 Ca^{2+} exchange stoichiometry had already been suggested in the original squid giant axons experiment (123, 124), and has recently been directly demonstrated for the case of heart in a series of experiments using isolated sarcolemmal vesicles (132). The latter system represents a decisive experimental advance, since it affords a quantitative definition of the kinetic parameters of the fluxes of Na^+ and Ca^{2+}, uncomplicated by other intracellular events involving the two cations. Concentrated suspensions of plasma membrane vesicles are passively loaded with large concentrations of Na^+, and are then diluted 200–400-fold at time zero in a medium containing Ca^{2+} but no Na^+. The Na^+ loading-plus-dilution procedure establishes an outwardly directed Na^+ gradient that is specifically discharged to drive the uptake of Ca^{2+}. Vesicle experiments have shown that the total content of Na^+/Ca^{2+} exchange activity varies considerably in different plasma membranes, being maximal (see above) in heart and nervous tissue. In heart, the exchanger is clearly a large system, moving under optimal conditions more that 30 nmoles of Ca^{2+} per mg of membrane protein per sec (133). Vesicle experiments have also been used to establish the affinity of the exchanger for Ca^{2+}, but have produced results that have opened some problems. This point will be considered later. Using vesicles, the demonstration of electrogenicity has come from direct isotopic measurements of Na^+ and Ca^{2+} (134), from measurements of the movements of lipophilic cations during the operation of the exchanger (135, 133), and from experiments in which the Na^+-driven uptake of Ca^{2+} into sarcolemmal vesicles has been shown to be modulated by artificially establishing negative or positive intravesicular potentials (136, 137). Thus, the energy contained in the chemical transmembrane gradient of Na^+ (or of Ca^{2+}) must be added algebraically to the energy contained in the transmembrane electrical potential.

If the electrochemical Na^+ gradient is the only source of energy for the operation of the Na^+-coupled Ca^{2+} movement, and assuming for the sake of simplicity that the Na^+/Ca^{2+} exchanger is the sole regulator of the intracellular free Ca^{2+} (however, see the next chapter, where the Ca^{2+} pump is discussed), the system ought to be able to reduce the cytosolic free Ca^{2+} to the values that have been found in several laboratories with a number of techniques, chiefly the fluorescent Ca^{2+} chelators. For the case of heart,

Reuter (138) has calculated that the system would lead to an equilibrium in which the cytosolic free Ca^{2+} concentration would be given by

$$[Ca^{2+}]_i = [Ca^{2+}]_o \frac{[Na]_i^3}{[Na]_o^3} \exp\left[\frac{(n-2)FV_m}{RT}\right], \qquad 3.$$

where V_m is the transmembrane potential, and F, R, and T have the usual values. If the outside Na^+ concentration is 140 mM, the corresponding cytosolic value is 6 mM, and the transmembrane potential is -80 mV, the system would establish a free Ca^{2+} equilibrium (at 37°C and with 1 mM external Ca^{2+}) of about 70 nM. In fact, the experimentally measured values in heart (139–141) are 2–5 times higher. There is little doubt, thus, that an electrogenic 3 Na^+ per 1 Ca^{2+} exchange system contains adequate energy to maintain the cytosolic free Ca^{2+} concentrations at the physiological levels. The electrogenicity of the system, however, means that it will operate as a current generator, the direction of the current being defined by the reversal potential, i.e. the potential where the net exchange current is zero. V_R is given by

$$V_R = \frac{n\,V_{Na} - 2\,V_{Ca}}{n-2} \text{ (138)}, \qquad 4.$$

where V_{Na} and V_{Ca} are the equilibrium potentials calculated from the Nernst equation for Na^+ and Ca^{2+}. To take again the case of heart as an example (138), it follows that at transmembrane potentials positive to V_R the current of the exchanger will be directed outwardly, whereas at potentials negative to V_R it will be directed inwardly. In other words, the Na^+/Ca^{2+} exchanger will lead to Ca^{2+} influx if the cell membrane is depolarized, and to Ca^{2+} efflux as the membrane repolarizes.

One important problem that has become apparent from studies of the exchanger in isolated plasma membrane vesicles (132) is its Ca^{2+} affinity. The apparent K_m values measured in different laboratories vary rather extensively, but in no case have they been shown to be lower than 2–5 μM. Several factors may influence the affinity of the exchanger: in dialyzed squid axons ATP increases the Ca affinity of the exchanger during Na_o-driven Ca efflux about tenfold, and the Na_o affinity about twofold (142). In the same preparation ATP and Ca_i are required for the Na_i-driven Ca influx (143), possibly due to the presence of a calmodulin-directed phosphorylation step

(144). A number of other treatments (reviewed in 145), among them the exposure to acidic phospholipids and a controlled proteolytic step, also activate the exchanger. Even with these treatments, however, the apparent K_m of the exchanger has never been reduced to the sub-μM level, which would seem necessary for the efficient interaction of the system with the very low (100–200 nM) free Ca^{2+} concentrations presumed to exist in the cytosol. At the moment there are no obvious explanations for this problem: one possibility that could be entertained is that the properties of the exchanger in isolated plasma membrane vesicles differ from those prevailing in vivo.

No specific inhibitor of the exchanger has so far been found, although compounds known to affect other reactions, e.g. doxorubicin and its de-rivatives (146) and amiloride and its derivatives (147), have been shown to possess inhibitory activity. The lack of an inhibitor to tag the exchanger has greatly hampered the attempts to purify it from plasma membranes. Some progress has nevertheless been recently made: it has been possible to solubil-ize plasma membranes with detergents, and to reassemble the exchanger, together with most other plasma membrane proteins, into liposomes. When the latter are passively loaded with Na^+ and then diluted in a Ca^{2+}-containing Na^+-free medium, Ca^{2+} is taken up, i.e. the exchanger has been reconsti-tuted. Several fractionation and purification schemes have been applied to detergent-solubilized plasma membranes, and have yielded reconstituted preparations in which the putative Na^+/Ca^{2+} exchanger has been tentatively identified with a protein of M_r 82,000 (148), 70,000 (149), and 33,000 (150).

The Ca^{2+} ATPase

The third Ca^{2+} transport system present in plasma membranes is a specific ATPase that exports Ca^{2+} from the cell. It was first described in erythrocytes (77), and later demonstrated in a number of other cells (see 151 for a recent comprehensive review). The plasma membrane used in most studies of the Ca^{2+} ATPase has traditionally been the erythrocyte, on which most of the properties of the enzyme have been established, and from which the enzyme has recently been purified to homogeneity (152). The pump is a high-affinity enzyme, which interacts with Ca^{2+} with a K_m well below 1 μM, but which transports Ca^{2+} with a low total capacity; figures of about 0.5 nmoles of Ca^{2+} transported per mg of membrane proteins per sec are typically measured (153). These values are certainly in excess of the Ca^{2+} extrusion requirements of cells like the erythrocyte, but most likely insufficient to satisfy those of other cells, particularly heart cells, where the Na^+Ca^{2+} exchanger may then operate in parallel as a bulk Ca^{2+}-extruding system.

The Ca^{2+} ATPase of plasma membranes is one of the so-called $E_1 E_2$ transport ATPases (see 151), which are postulated to exist in two different conformational states E_1 and E_2 in different moments of the reaction mechanism, and which conserve the ATP energy intramolecularly in the form of an acyl phosphate, most likely an aspartyl phosphate.

One important property of the ATPase is the stimulation by calmodulin (154, 155), which increases the Ca^{2+} affinity of the enzyme and its maximal transport rate. At variance with other targets where calmodulin acts by a phosphorylation mechanism, in this case no phosphorylation reaction is involved, and the stimulation is due to the direct interaction of calmodulin with the pump. The interaction with calmodulin has permitted the isolation of the enzyme on calmodulin affinity chromatography columns, first from erythrocytes (152), and then from at least four other plasma membranes (153, 156–158). The isolation procedure involves elution of the calmodulin column with EGTA in the presence of detergents, the most convenient of them being Triton X-100, and of phospholipids. The purified enzyme is a single polypeptide of M_r 138,000 (159), which reproduces the essential properties of the enzyme in situ, among them the transition to high Ca^{2+} affinity upon calmodulin addition, and the transport of Ca^{2+} across membranes in reconstituted systems. The reconstituted liposomal system has permitted the establishment that the stoichiometry between transported Ca^{2+} and hydrolyzed ATP approaches 1 (159, 160), and that the enzyme functions as an obligatory Ca^{2+}/H^+ exchanger, whose probable stoichiometry is 1 to 2 (161).

One interesting finding made on the purified enzyme, which extends observations made on the enzyme in situ (162–164), is that a number of treatments alternative to calmodulin activate the pump much in the same way as calmodulin does. Among them is the exposure of the enzyme to acidic phospholipids and long-chain polyunsaturated fatty acids (159, 165) and a controlled treatment with trypsin (165, 166). The trypsin treatment has permitted the mapping of some of the functionally important domains of the enzyme molecule (167), among them the calmodulin-binding region, which appears to be contained in a peripheral sequence of about 9000 daltons. Interestingly, up to about 60,000 daltons can be removed from the purified enzyme by controlled proteolysis with trypsin without essential impairment of the Ca^{2+}-transporting function in reconstituted systems (167, 168). This opens the problem of the role of the about 60,000-dalton component, which is apparently not involved in the reaction mechanism proper. One possibility is that this portion of the molecule is involved in some type of regulatory function. Recent experiments (170) have confirmed previous work in which the ATPase in the heart sarcolemma membrane was found to be activated by the cAMP-dependent protein kinase (169): the isolated ATPase can be phosphorylated, and apparently activated, by a kinase-directed phosphorylation step dependent on cAMP.

A number of hormonal effects on the plasma membrane Ca^{2+} ATPase have been reported. Both activation (171) and inhibition (172) of the erythrocyte enzyme have been claimed to be induced by the thyroid hormone, apparently dependent on the nutritional status of the animal. Oxytocin has been claimed to inhibit the enzyme in rat myometrial plasma membranes (173), and insulin in those of adipocytes (174). Also of interest is the description of protein activators (175, 176) and inhibitors (177) of the ATPase.

One last point on the plasma membrane Ca^{2+} ATPase concerns the case of liver, where the enzyme appears to have properties that set it apart from the enzyme in other plasma membranes (178). Distinctive characteristics are the lack of calmodulin effects, the poor nucleotide specificity, and the much higher affinity for Ca^{2+} (K_m in the nM range). Also of interest is the presence of an endogenous, membrane-bound protein inhibitor (179). Based on these special properties, it has even been suggested that the liver enzyme may in fact not be a Ca^{2+} pump (180).

CALCIUM TRANSPORT ACROSS INTRACELLULAR MEMBRANES

Endoplasmic and Sarcoplasmic Reticulum

Most of the work on this membrane system has been traditionally carried out on sarcoplasmic reticulum, but interest in endoplasmic reticulum is now rapidly mounting. The reason for this upsurge of interest is twofold: on the one hand mitochondria are now generally considered to play no major role in the cytosolic homeostasis of Ca^{2+}, leaving endoplasmic reticulum as the only alternative intracellular structure that could be active in the process. On the other hand, a recent finding, namely that the very popular second messenger inositol trisphosphate releases Ca^{2+} from an intracellular store that appears to be the endoplasmic reticulum (see below), has focused a great deal of attention on the Ca^{2+} handling ability of this membrane system.

Both isolated sarcoplasmic and endoplasmic reticulum accumulate Ca^{2+} from the ambient through a specific ATPase, which is a member of the $E_1 E_2$ class like the equivalent enzyme of the plasma membrane. Very little information is available on the ATPase of endoplasmic reticulum, but the assumption that its general properties repeat those of the better known enzyme of sarcoplasmic reticulum seems very reasonable. At variance with endoplasmic reticulum, where the content of Ca^{2+}-ATPase is minor, the sarcoplasmic reticulum ATPase is a very abundant protein: it may represent up to 90–95% of the total protein of sarcoplasmic reticulum, and even in the least favorable case, which is that of heart, it does not fall below 40–50%. Naturally, the availability of such large amounts of protein has oriented research towards it, and has facilitated its study.

A role of sarcoplasmic reticulum in the cellular homeostasis of Ca^{2+} in muscle was first indicated about 25 years ago by experiments of Hasselbach &

Makinose (184), in which a preparation of muscle microsomes, now known to derive from sarcoplasmic reticulum, was shown to remove significant amounts of Ca^{2+} from the medium in a process that was dependent on ATP. That sarcoplasmic reticulum accumulates Ca^{2+} in vivo has been demonstrated by treating skinned fibers with oxalate to precipitate Ca^{2+}, a procedure which leads to the formation of Ca^{2+} deposits within the cisternae (181). More recently, work using electron probe X-ray microanalysis (182, 183) has shown that sarcoplasmic reticulum in situ contains large amounts of Ca^{2+} (30–50 mmoles per kg of dry weight in smooth muscle) and has also shown that the Ca^{2+} content decreases considerably as the muscle contracts.

The Ca^{2+} uptake system in the sarcoplasmic reticulum of heart, skeletal, and smooth muscle has now been characterized in considerable detail, and its function in the removal of Ca^{2+} from muscle cytosol down to the levels that would induce dissociation of Ca^{2+} from troponin C, has become firmly established. The opposite reaction, i.e. the release of Ca^{2+} from sarcoplasmic retictulum to saturate the Ca^{2+}-binding sites of troponin C and to induce the association of actin and myosin, has on the other hand proven to be a more formidable experimental problem, whose solution has yet to come.

The ATPase that mediates the process of Ca^{2+} uptake has been purified in 1970 (184) as a protein of M_r about 100,000. It has been incorporated into artificial phospholipid vesicles and shown to pump Ca^{2+} in an ATP-dependent reaction in a number of laboratories (see 181 for a recent review). Studies on the interaction of Ca^{2+} with the purified enzyme (185–187) have confirmed indications from work on sarcoplasmic reticulum vesicles that an average of two Ca^{2+} ions are accumulated per ATP hydrolyzed (188). The enzyme interacts with Ca^{2+} with high affinity, as expected of the ATPase transport mode (see above); kinetic studies have shown half-maximal rates of Ca^{2+} influx into the vesicles with values of ionized Ca^{2+} varying between 0.1 and 1.0 μM (189–191). Upon addition of Ca^{2+} to the ATP (and Mg^{2+})-containing reaction medium, Ca^{2+} is taken up to maximal levels of 150–200 nmol per mg of reticulum protein (in fast skeletal muscles). Much higher levels of loading can be achived in the presence of the permeant anion oxalate, which precipitates Ca^{2+} inside the vesicles (184). Values for the maximum velocity of Ca^{2+} influx into fast skeletal muscle sarcoplasmic reticulum vary between about 0.5 and 4 μmol per mg of protein per min (189, 190, 192–194), but the maximal rate in heart is considerably lower (195, 196).

The cisternae of sarcoplasmic reticulum in the proximity of the transverse tubules, which inflect periodically from the sarcolemma at the level of the Z-lines, are anchored to the tubular system by feetlike projections (197), forming the so-called triadic junctions. Somehow, the depolarization of the cell membrane is transmitted through the transverse tubules to the interior of the muscle cell, thus causing efflux of Ca^{2+} from the neighboring cisternae of the reticulum. Evidently, a portion of the reticulum membrane acts as a sensor for the excitation wave spreading from the plasma membrane enveloping the cell to the transverse tubules: it is believed that the feetlike projections that connect the transverse tubules with the reticulum cisternae are physically involved in the transmission of the excitation to the interior of the cell, but the nature of the signal that will eventually direct the release of the Ca^{2+} from sarcoplasmic reticulum, whether mechanical, electrical, or chemical, is unknown. It has been suggested (198) that the feetlike projections would gate (i.e. open) putative Ca^{2+} channels in the terminal cisternae of the sarcoplasmic reticulum in response to the flow of charges in the tubular system. That the depolarization of the plasma membrane activates a mechanism that dramatically alters the Ca^{2+} permeability of the sarcoplasmic reticulum membrane is obvious, since the rate of Ca^{2+} release from Ca^{2+}-loaded, isolated vesicles of sarcoplasmic reticulum, even under the most favorable experimental circumstances, is orders of magnitude lower than that occurring during muscle excitation. A number of experimental procedures have been shown to induce release of Ca^{2+} from isolated sarcoplasmic reticulum vesicles, beginning with the reversal of the ATP-dependent uptake process in the presence of ADP under conditions of low extravesicular Ca^{2+} and ATP (199). The phenomenon, which is coupled to the resynthesis of ATP, is most certainly not the physiological mechanism for Ca^{2+} release in vivo. Two other interesting mechanisms for the release of Ca^{2+} have been proposed: in the first, manipulations of the extravesicular electrolyte composition, in an attempt to mimic the conditions used to study the depolarization-induced release of Ca^{2+} from skinned skeletal muscle fibers, have been used to produce transient potentials across the vesicular membrane. The procedure causes Ca^{2+} efflux from skeletal muscle reticulum vesicles (200, 201), but it has been argued that osmotic phenomena caused by the procedure may alter the permeability of the vesicular membrane (202). In addition, it is known that in some muscles (e.g. heart), the penetration of Ca^{2+} into the fiber across the Ca^{2+} channel is necessary to elicit contraction. This militates in favor of the second proposal for the release of Ca^{2+} from sarcoplasmic reticulum, the so-called Ca^{2+}-induced Ca^{2+} release (203). The phenomenon has been discovered and studied in skinned fibers, and it consists of inducing the release of massive amounts of Ca^{2+} from sarcoplasmic reticulum by introducing Ca^{2+} into the medium at concentrations that are below the threshold concentration (0.1 μM) that activates myofibrils. It now appears likely that the Ca^{2+}-induced Ca^{2+} release phenomenon, which nicely rationalizes the requirement for Ca^{2+} penetration into heart cells to elicit contraction, but whose molecular mechanism is still unknown, may be the physiological mechanism for sarcoplasmic reticulum Ca^{2+} release in heart. In smooth muscle the case may be different, since in this slowly contracting tissue the possibility of Ca^{2+} release in response to inositol trisphosphate (204) appears more realistic than in heart and other muscle types (205).

One aspect of the function of sarcoplasmic reticulum that may be of importance to the homeostasis of Ca^{2+} is the regulation of the activity of the Ca^{2+}-pumping ATPase by hormonally controlled processes. Sarcoplasmic reticulum from heart, slow skeletal muscle, and smooth muscle contains an acidic proteolipid that has been named phospholamban (206, 207). This component, whose M_r was originally estimated to be about 22,000 (207), is apparently contained in the membranes of sarcoplasmic reticulum in an approximate 1:1 stoichiometry to the ATPase and is the substrate of two protein kinases, one cAMP dependent, one Ca^{2+} plus calmodulin dependent (208). Phosphorylation of phospholamban by either of the two kinases, or by both, increases the rate of ATP hydrolysis and of the Ca^{2+} translocation that is coupled to it (208). Although the mechanism of the stimulation is not known, it is generally assumed that it involves the interaction of phospholamban with the ATPase. An interesting recent proposal suggests that nonphosphorylated phospholamban is permanently attached to the ATPase and inhibits it: phosphorylation would induce a conformational change in the molecule that would lead to its detachment from the ATPase and to its activation (209). Important new findings related to phospholamban are (a) the demonstration that the molecule consists of five probably identical subunits of M_r about 6000, producing a total M_r of about 25,000–28,000 (210–212), (b) the determination of its primary structure (213), and, (c) the identification of separate phosphorylation sites for the two kinases (210–214) [serine 16 and threonine 17 in the 6080-dalton monomer; also of potential interest is the recent finding that protein kinase C phosphorylates phospholamban (215)].

An important new development has been the recent determination of the primary structure of the Ca^{2+} ATPase from both fast skeletal muscles and heart sarcoplasmic reticulum (216). It appears that the enzyme contains 10 transmembrane stretches, and a very large portion protruding from the cytosolic side of the membrane. The latter portion is postulated to consist of several domains, containing the ATP-binding sequence, the aspartic acid residue which becomes phosphorylated, the portion of the molecule that transduces the ATP energy to the region that translocates Ca^{2+}, and the Ca^{2+}-binding domain proper. The latter has been ascribed to a region of the molecule next to the hydrophobic intramembrane α-helixes that contains an unusually high concentration of glutamic acid residues (216). The domain contains no indications for parvalbumin-type Ca^{2+} binding loops, showing that a different structural organization of the Ca^{2+} interacting region is evidently at work in membrane-intrinsic proteins. At this early stage generalizations are clearly dangerous, but the idea seems reasonable that the parvalbumin-type structure has evolved, and has now been probably optimized, only for proteins that must express Ca^{2+}-dependent hydrophobic sites to process the Ca^{2+} signal. Other structural organizations are evidently required for other types of Ca^{2+} interactions, in membranes and elsewhere, not directly linked to the processing of the Ca^{2+} signal.

One final point on sarcoplasmic reticulum concerns another Ca^{2+}-binding protein, named calsequestrin (217). This is an extrinsic, hydrophilic protein (M_r 45,000), which binds large amounts of Ca^{2+} (maximum binding capacity about 40 mol per mol) with low affinity (apparent K_d 0.8 mM). Interest in this protein has mounted after the discovery that most of it is found in the terminal cisternae of sarcoplasmic reticulum (218), which are supposed to contain the Ca^{2+} that is mobilized in response to the depolarization event (see above). It is presumed that calsequestrin plays a role in the complexing of Ca^{2+} within the cisternal space: in intact vesicles, it is protected against tryptic digestion, indicating its internal location (219). A high-affinity Ca-binding protein has been identified in sarcoplasmic reticulum preparations from fast skeletal muscle (220, 221). It binds one mol Ca/mol (M_r 65,000) with a K_d comparable to that of troponin C. The function of this protein is still unclear.

Endoplasmic reticulum in nonmuscle cells takes up Ca^{2+} through an $E_1 E_2$ ATPase: in liver, it has been solubilized from the membrane environment but not yet completely purified (222). Phosphorylation experiments with γ-labeled ATP (see for example 223) have indicated that the ATPase has the same M_r as the sarcoplasmic reticulum counterpart. The enzyme has high Ca^{2+} affinity (K_m, 1 μM), and loads relatively high amounts of Ca^{2+} into isolated reticulum vesicles in the presence of the trapping anion oxalate. Recent in situ work using electron probe X-ray microanalysis (224) has shown that endoplasmic reticulum indeed is a major intracellular store of Ca^{2+}. The regulation of the Ca^{2+} uptake by endoplasmic reticulum appears to be a complex phenomenon that may involve calmodulin (222, 225) and the operation of the cAMP-dependent protein kinase (226). A very interesting suggestion that may be relevant to the regulation of Ca^{2+} by (liver) endoplasmic reticulum has been made recently by Benedetti et al (226b). Glycogen mobilization, e.g. by α_1 adrenergic agonists, results in the production of glucose-6-phosphate. Simultaneously, the concentration of Ca^{2+} in the cytosol increases due to its mobilization from intracellular stores (see below). Benedetti et al (226b) have observed that glucose-6-phosphate, hydrolyzed by the glucose-6-phosphatase that is located on the internal face of the endoplasmic reticulum membrane, greatly increases the accumulation of Ca^{2+}. Evidently, the accumulation of phosphate from glucose-6-phosphate within the endoplasmic reticulum vesicles acts as a sink for the accumulation of Ca^{2+}. It has been suggested that the mechanism may be physiologically relevant, since it couples the removal of phosphate from glucose-6-phosphate to supply glucose to the blood with the reduction of the Ca^{2+} signal that had originally promoted the conversion of phosphorylase *b* to *a*; this is necessary to prevent the permanent activation of other Ca^{2+}-linked functions unrelated to glycogen mobilization. The suggestion is undoubtedly interesting. Even if it may have oversimplified the complex matter of Ca^{2+} regulation by (liver) endoplasmic reticulum (see for example 226c), it deserves further exploration.

Phosphoinositides appear to be involved in the release of Ca^{2+} from endoplasmic reticulum. The current phenomenal interest in the role of these compounds in the cellular homeostasis of Ca^{2+} has its origin in the observation made by Hokin & Hokin more than 30 years ago (227, 228) that the stimulation of enzyme secretion in pancreas by a number of agonists is linked to the increased incorporation of phosphate into phospholipids, specifically into phosphatidyl-inositol (PI) and phosphatidic acid (PA). Later on, a phosphatidyl-inositol/phosphatidate cycle was proposed (229) in which PI, upon stimulation by agonists, would break down to diacylglycerol (DG), which would then be rephosphorylated to PA. In the absence of the agonist, PA would be converted back to PI. In 1975, Michell (230) observed that the mobilization of cellular Ca^{2+} induced by a number of agonists is preceded by the breakdown of PI, and it was soon recognized that it is the breakdown of polyphosphoinositides (PIP_2), rather than that of PI, which plays a key role in the mobilization of cytosolic Ca^{2+} (231–234). It is undoubtedly the finding that Ca^{2+}-mobilizing agonists (e.g. vasopressin in liver) induce the rapid breakdown of polyphosphoinositides that is responsible for the recent upsurge of interest in the metabolism of the latter compounds. The correlation with Ca^{2+} mobilization initially indicated the possibility of entry of Ca^{2+} into the cells (Ca gating) (230). The hypothesis received some experimental support from experiments on blowfly salivary glands (235), where 5-hydroxytryptamine caused PI breakdown and entry of Ca^{2+}. However, the key experiment was performed by Streb et al (236) on permeabilized exocrine pancreatic cells, and it showed that one of the breakdown products of PIP_2, inositoltrisphosphate (IP_3), releases Ca^{2+} from a nonmitochondrial intracellular store when added to the medium in μM concentrations. The experiment has now been repeated on a number of other permeabilized cell types (see 237 for a review), and it has become clear that the nonmitochondrial Ca^{2+} store is the endoplasmic reticulum. More recently, IP_3-induced Ca^{2+} release has also been observed in several isolated endoplasmic reticulum preparations (238–242). The phenomenon is now clearly established, but a number of its aspects still await clarification, e.g. the potentiating effects of guanidino nucleotides and of polyethylene glycol (241, 242). A recent study (243) indicates that the effects of guanidino nucleotides and of polyethylene glycol on the IP_3-induced Ca^{2+} release may be the expression of two different phenomena. Also of interest is the observation that intrareticulum Ca^{2+} may modulate the releasing effect of IP_3 (243). The release of Ca^{2+} induced by IP_3 is rapid; although not on a time scale that would be compatible with the requirements for heart and skeletal muscle contraction, the rate of release could be adequate in the case of smooth muscle (204). Such release occurs at less than μM concentrations of IP_3; IC_{50} values as low as 0.025 μM have been measured in insulinoma cells (244). Ca^{2+} is released specifically by the 1,4,5-isomer of IP_3 (236, 245); the other two products of phosphoinositide breakdown, IP_2 and IP, are inactive as Ca^{2+} releasers. The release phenomenon is transient due to the rapid hydrolysis of IP_3, and is followed by Ca^{2+} reuptake into endoplasmic reticulum (236, 246). Based on calculations of the amount of PIP_2 hydrolyzed under the influence of agonists, it has been estimated that the cellular concentration of IP_3 could theoretically rise to 15 μM in 2 min. Very likely, the values reached in vivo are much lower, since the inositol trisphosphatase has been estimated to have a $t_{1/2}$ of 4 sec (in liver, 247).

The effect of IP_3 on intracellular (endoplasmic reticulum) Ca^{2+} stores appears well documented. The possibility that the activation of the phosphoinositide cycle will have an additional effect on the cellular homeostasis of Ca^{2+} by opening plasma membrane Ca^{2+} channels to the penetration of extracellular Ca^{2+} (see above) appears to have lost favor, and is indeed not supported by conclusive experimental evidence. It has been suggested that PA, one of the products of the cycle, could function as a plasma membrane Ca^{2+} ionophore, but its ionophoric action is relatively weak (reviewed in 237).

The Ca^{2+}-Transporting Systems of Mitochondria

The observation that isolated energized mitochondria could take up large amounts of Ca^{2+} is now more than 25 years old (81, 248, 249), and has been followed by an impressive number of studies that have established the essential properties of the process. For a long time mitochondria have been regarded as very important in the regulation of cytosolic Ca^{2+}: in fact, one could say that until recently the interest in mitochondria as cellular Ca^{2+} buffers had grown in parallel with the importance of Ca^{2+} as a universal cellular messenger. More recent experimental developments, two of which are of particular importance, have shown that the importance of mitochondria as cytosolic Ca^{2+} buffers had been markedly overestimated. The first development has been the precise definition of the kinetic parameters of the energy-linked Ca^{2+} uptake reactions (see for example 250), which showed that the affinity of mitochondria for Ca^{2+} (K_m about 10 μM) was insufficient for a role in the regulation of reactions that are modulated by sub-μM concentrations of Ca^{2+}. In addition, it became clear that the rate of Ca^{2+} uptake by mitochondria is at least one order of magnitude lower than that of sarcoplasmic reticulum [in heart, comparative measurements on the two organelles can be made. Under the most favorable, and most likely artificial, experimental conditions, mitochondria remove Ca^{2+} from the medium at a V_{max} of 0.6 μmol per mg of prot per min (251)]. The second development has been the demonstration by electron probe X-ray microanalysis experiments (252–254) that much less Ca^{2+} is associated with mitochondria in situ than had been previously assumed based on measurements on isolated mitochondria. Mitochondria in situ are now agreed to contain only between 1 and 2 nmoles of Ca^{2+} per mg of protein. It may be relevant to mention in this context the dense granules normally seen with the electron microscope within the mitochondrial profiles of osmium-treated specimens of a number of tissues. These granules have routinely been assumed to consist of precipitates of Ca^{2+} and phosphate, but the assumption, at least in normal cells, appears unwarranted. Direct correlation experiments between granules counts and mitochondrial Ca^{2+} (255), as well as more recent X-ray microanalysis experiments on a number of tissues (183, 224, 252, 256–258), have indicated that the granules contain phospholipid-bound phosphorous but not Ca^{2+}. The latter becomes associated with them only if the cell loses the Ca^{2+} impermeability barrier of the plasma membrane due to noxious effects. The point on the Ca^{2+} content of mitochondria in situ is important, since it says that the size of the pool of Ca^{2+} that mitochondria can contribute to the cytosol, even if all of the pool were rapidly mobile, is very limited. One can offer the example of heart, which contains about 100 mg of mitochondrial protein per gram of wet weight (259), corresponding to a total mitochondrial Ca^{2+} pool of between 100 and 200 nmol per gram of wet weight.

In parallel with the decrease of interest in mitochondria as cytosolic Ca^{2+} buffers, workers in the field have become increasingly aware of their importance as regulators of their own internal Ca^{2+} concentration. The trigger here has been the demonstration that three matrix dehydrogenases, pyruvate dehydrogenase, the NAD-dependent isocitric dehydrogenase, and α-oxoglutarate dehydrogenase (260–262), are modulated by oscillations of matrix Ca^{2+} in the μM range. Clearly, the arguments mentioned above against a (major) role of mitochondria in the regulation of cytosolic Ca^{2+} do not apply to the case of the intramitochondrial space: by necessity, only mitochondria can regulate their own matrix Ca^{2+} content. As a result, most of the emphasis of mitochondrial Ca^{2+} transport research has now deviated from cytosolic effects, and is being increasingly focused on intramitochondrial phenomena.

Several reviews have recently appeared on the topic of mitochondrial Ca^{2+} transport (see, e.g. 263, 264), and the reader is referred to them for comprehensive information. Here, only the aspects of the process that are relevant to the topic of Ca^{2+} homeostasis in the cell and in the intramitochondrial compartment will be discussed. The phenomenon of mitochondrial Ca^{2+} uptake is traditionally interpreted in chemiosmotic terms, i.e. Ca^{2+} penetrates across the inner membrane without charge compensation in response to the negative membrane potential established inside the inner membrane barrier by the activity of the respiratory chain (265). The polycation ruthenium red completely inhibits the uptake reactions (266). In the absence of inorganic phosphate in the medium, both the rate of uptake and the maximal extent of accumulation are low, even in the presence of saturating Ca^{2+} concentrations. This is the so-called "limited loading" phenomenon, in which Ca^{2+} is assumed to saturate anionic binding sites (e.g. cardiolipin, see above) inside the mitochondria. The extent of the increase in the matrix-free Ca^{2+} concentration under these conditions is difficult to evaluate, although it certainly occurs, as revealed by the increased activity of the Ca^{2+}-modulated matrix dehydrogenases (267). In the presence of the permeant anion inorganic phosphate, both the extent of accumulation and its rate are strongly stimulated, as was recognized almost from the beginning of the work in the area (268). This is due to the precipitation of insoluble Ca-phosphate deposits, probably amorphous hydroxyapatite, in the matrix ("matrix loading"). The deposits are visible as electron dense masses in the electron microscope (269), and there is no doubt that they contain the accumulated Ca^{2+}, at variance with the case of the normal dense granules routinely seen within mitochondria. The Ca-phosphate deposits are probably primed, or stabilized, by organic compounds, e.g. adenine nucleotides, which are also known to penetrate into the matrix under these conditions (270). It is remarkable that mitochondria appear to be relatively little disturbed, structurally and functionally, by the accumulation of Ca^{2+} and phosphate until very large levels of loading are reached. This points to the importance of the phenomenon of "matrix loading" with Ca^{2+} and phosphate as a safety device that enables mitochondria to buffer large amounts of extra Ca^{2+} when its concentration in the cytosol increases pathologically to levels sufficient to activate their low-affinity uptake system. In this way mitochondria buy precious time for the cell, enabling it to survive until the noxious agent that has caused the extra accumulation of Ca^{2+} in the cytosol subsides.

At the free Ca^{2+} concentrations presumed to exist in the cytosol, the low Ca^{2+} affinity uptake system is expected to function at a fraction of its maximal capability. Increased activity (e.g. increased affinity for Ca^{2+}) could conceivably be induced by endogenous factors, for example the polyamine spermine, which has recently been shown to increase the affinity of liver mitochondria for Ca^{2+} at physiological concentrations (271). However, physiological concentrations of Mg^{2+} severely inhibit the uptake reaction, shifting its affinity for Ca^{2+} to much lower values (K_m in the 30 μM range, 250, 259). One can thus assume that under in vivo conditions mitochondria take up Ca^{2+} at a marginal rate only. However, since the transmembrane potential that drives the uptake of Ca^{2+} is unlikely to ever collapse to values that would allow the reversal of the electrogenic uptake route, the latter functions essentially as a one-way system. As a result, no matter how slowly, mitochondria would still calcify in situ if no way existed either to completely block the slow uptake route, or to export the penetrated Ca^{2+} back to the cytosol despite the presence of the large negative membrane potential present inside. It was this consideration that led to the investigation of conditions that would permit the electroneutral release of Ca^{2+} from mitochondria; in 1975 Carafoli et al (82) discovered that the exposure of Ca^{2+}-loaded heart mitochondria to Na^+ specifically discharged the accumulated Ca^{2+}. The fact that the Na^+-promoted Ca^{2+} efflux operated in the presence of the blocker of the uptake pathway, ruthenium red, clearly established that the phenomenon occurs on a pathway that is independent of that used for the electrophoretic uptake of Ca^{2+}. The Na^+-promoted release route has now been characterized in detail in a number of studies (272–277) that have traced it back to the existence of a Na^+/Ca^{2+} exchanger that operates electroneutrally (it may even operate electrogenically in the direction of Ca^{2+} efflux, i.e. exchanging more than 2 Na^+ for 1 Ca^{2+} 272, 264), and has different inhibitor sensitivity with respect to the electrophoretic uptake pathway (278). The exchanger is particularly active in excitable tissues, chiefly heart and nervous tissue, and minimally in nonexcitable tissues like liver. Values of about 0.01–0.02 μmoles of Ca^{2+} released per mg of protein per min are typically measured in heart mitochondria at optimal Na^+ concentrations (10 mM), and it is interesting that the relationship linking the external Na^+ concentration to the rate of Ca^{2+} efflux is strongly sigmoidal, half-maximal activity being seen at 4–5 mM Na^+ (272). In principle, then, even a comparatively minor concentration change of extramitochondrial Na^+ could have a major effect on the rate of Ca^{2+} release. However, it appears unlikely that the in vivo changes of the Na^+ concentration in the cytosol are large enough under normal conditions to exploit the sigmoidal character of the relationship to confer significant regulatory properties to cytosolic Na^+. It is well possible, however, that pharmacological influences (e.g. digitalis in heart), increase the Na^+ concentration in the cytosol to a level sufficient to activate the exchanger (272).

The affinity of the exchanger for Ca^{2+} has been estimated at the external face of the inner membrane by operating the system as a Ca^{2+}-Ca^{2+} exchanger, and it has been found to be about 10 μM (273). The affinity of the exchanger for Ca^{2+} at the matrix side of the inner membrane is still only presumptive, since the measurement of intramitochondrial Ca^{2+} relies on indirect methods. The so-called null-point Ca^{2+} titration procedure in the presence of Ca^{2+} and the ionophore A23187 in the presence of ruthenium red (279) permits the approximate estimation of the relation between total and free-matrix Ca^{2+}. Using this procedure, K_m values of 6 μM (279) and 3–4 μM (280) have been obtained for the affinity of the exchanger for internal Ca^{2+}. It must also be considered that a competitive interrelation between Na^+ and Ca^{2+} has been noted at the outer face of the inner membrane, with Na^+ decreasing the binding affinity of the exchanger for Ca^{2+} quite considerably (281). If the exchanger were operating symmetrically, one would expect the same competitive inhibition also at the matrix side of the inner membrane, where Na^+ is to be found as a consequence of the equilibrium of the Na^+/H^+ antiporter. This would increase the internal K_m of the exchanger for Ca^{2+} to values that would greatly exceed the presumed free Ca^{2+} levels of the matrix (279), and probably greatly limit the exit of Ca^{2+} from mitochondria through the exchanger . For this reason, Crompton has speculated (264) that the exchanger may operate asymmetrically in energized mitochondria, with higher Ca^{2+} affinity at the matrix than at the external side and/or higher Na^+ affinity at the external than at the matrix side. Crompton has further speculated (264) that matrix ATP could impose asymmetry on the carrier, in analogy with the effect of ATP on the Na^+/Ca^{2+} exchanger of the plasma membrane (see above). One additional regulatory effect on the exchanger has been observed in liver, a mitochondrial type where the exchanger is very poorly active; indications have been obtained for an increase in the internal affinity for Ca^{2+} after glucagon or β-adrenergic treatment (282, 283). One recent interesting finding on the exchanger is its inhibition by a number of Ca^{2+} antagonists, already known to act on the plasma membrane Ca^{2+} channel (284, 285). The most effective among them has been found to be the benzothiazepine diltiazem. Benzodiazepines (diazepam) have also been reported to inhibit the exchanger (286).

Even in mitochondrial types where the Na^+-promoted route is very active, Ca^{2+} is released in the absence of Na^+ at a slow rate in the presence of the inhibitor of the electrophoretic uptake route, ruthenium red. This implies the existence of a Ca^{2+} release route different from that promoted by Na^+, and also different from the (reversed) electrophoretic uptake route, which is blocked by ruthenium red. In general, the Na^+-independent route is least active in mitochondria where the Na^+-promoted release is highest, and vice versa. Thus, in liver and kidney mitochondria the Na^+-independent Ca^{2+} release is much faster than in heart mitochondria. A Ca^{2+}/H^+ antiporter has been proposed by several authors (287–290), and now appears to have gained wide acceptance. It must be noted, however, that the existence of such a direct exchanger is questioned by recent evidence, for example by experiments in which the rate of Ca^{2+} efflux has been found to correlate inversely with intramitochondrial pH (291). Also the fact that no specific inhibitors of the Na^+-independent release route are known, while not a conclusive argument, raises doubts on the existence of a (proteinaceous) H^+/Ca^{2+} exchanger. In general, a pertinent question is whether the efflux of Ca^{2+} from mitochondria under Na^+-independent conditions may be due to alterations of the inner membrane induced by the accumulation of excess Ca^{2+} rather than due to activation of a carrier-mediated process. Although the mechanism of the Ca^{2+}-induced damage is not known, it has been known for a long time that Ca^{2+} induces swelling of mitochondria and loss of membrane potential. Also unclear is the mechanism by which some agents potentiate and others inhibit the damaging effects of Ca^{2+}. Based on some correlation between the Ca^{2+}-induced permeability increase and the activation of membrane-bound phospholipase A_2 (292–295), it has been proposed that stimulation of membrane-bound phospholipase A_2 by Ca^{2+} is responsible for the activation of a nonspecific (i.e. damage-linked) Ca^{2+} release pathway (292–295). Chemically different agents known to either potentiate (N-ethylmaleimide, diamide, inorganic phosphate, oxaloacteate) or inhibit (adenine nucleotides, Mg^{2+}, β-hydroxybutyrate) the deleterious effects of Ca^{2+} would do so by acting on the acylation/deacylation reactions of the phospholipids of the inner membrane. Conditions that induce reduction of mitochondrial pyridine nucleotides have been found to favor the retention of Ca^{2+} in liver mitochondria, whereas conditions that promote their oxidation favor its loss (296, 297). It is still unclear whether this phenomenon is also the expression of nonspecific permeability changes of the inner membrane. Although this would be indicated by the finding that protection against the loss of Ca^{2+} is seen with ATP (298), evidence has been presented in favor of the intactness of liver mitochondria during the phenomenon (299). Irrespective of this, it is of interest that the phenomenon is linked to the hydrolysis of pyridine nucleotides, to the reversible loss of nicotinamide from mitochondria, and to the ADP-ribosylation of an inner membrane protein (300).

The finding that the routes for Ca^{2+} uptake and Ca^{2+} release are independent, and that the Na^+-induced release route is either undisturbed, or even promoted by the negative internal membrane potential, form the basis for the proposal of the mitochondrial "Ca^{2+} cycle" (301), in which Ca^{2+} penetrates into the matrix driven by the transmembrane potential, and exits from them electroneutrally (or, possibly, further dissipating the transmembrane potential if more than 2 Na^+ exchange for 1 Ca^{2+}). This would imply that, in addition to the well known burst of respiratory energy dissipation observed upon pulsing mitochondria with Ca^{2+}, one should observe a permanent activation of resting respiration after the accumulation of the Ca^{2+} pulse. In fact the Ca^{2+} cycle should operate even without added Ca^{2+}, exploiting the endogenous matrix pool. That endogenous Ca induces permanent dissipation of respiratory energy has been experimentally verified by the inhibition of a portion of the resting (state 4) respiration of mitochondria by ruthenium red (302), which evidently prevents the use of energy (i.e. of the transmembrane potential) by the electrophoretic uptake leg of the Ca^{2+} cycle. It has also been directly demonstrated in a number of mitochondria from excitable tissues, by showing that the rate of state 4 respiration is increased by Na^+ in the presence of Ca^{2+}, the increase being abolished by ruthenium red (274). Considering the arguments presented above on the heavy damping of the process of Ca^{2+} uptake under in situ conditions, it is evident that the amount of energy dissipation by the cycle will be very limited in vivo. It has in fact been estimated to account for less than 0.2% of the total oxygen consumption of heart tissue (303).

The overall rate of Ca^{2+} cycling will naturally be determined by the rate-limiting step of the cycle, which contains, in addition to the Ca^{2+} uptake and release legs, a Na^+/H^+ antiporter (304) that mediates the efflux of the excess Na^+ that has penetrated into the mitochondria. The alternative of a H^+/Ca^{2+} exchanger in mitochondria that are poorly Na^+-sensitive would only require a two-leg cycle since the chemiosmotic H^+ would in this case reenter mitochondria in direct exchange for Ca^{2+}. Nicholls (276, 277) has proposed a kinetic regulation of the cycle based on the algebraic sum of an essentially constant Na^+-induced release rate (until about 60 μM free-matrix Ca^{2+}) and a rate of the uptake uniporter that will be dependent on the free Ca^{2+} concentration in the extramitochondrial medium. The rates of influx and efflux will be roughly equivalent at 1 μM external Ca^{2+}, which will then be the "set point" of the cycle. At higher external Ca^{2+}, concentrations the uptake pathway would predominate, leading to temporary Ca^{2+} loading of mitochondria, whereas at lower external Ca^{2+} concentrations the efflux pathway would take primacy, returning the external Ca^{2+} to the set point (and depleting mitochondria of Ca^{2+} in the process). A more recent analysis by Crompton (264) considers the cycle essentially from the angle of the internal "set point," and takes into account the value of about 1 μM internal Ca^{2+} estimated to exist in isolated respiring heart mitochondria in the presence of physiological Mg^{2+}, 0.5 μM external Ca^{2+}, and amounts of Na^+ that saturate the efflux pathway (279, 280). The value is in the optimal range for the modulation of the Ca^{2+}-controlled matrix dehydrogenases both in the isolated state and in vivo. Denton et al (305) have observed that coupled heart mitochondria suspended in media containing physiological concentrations of Na^+ and Mg^{2+} maintain the Ca^{2+}-dependent matrix dehydrogenases half-maximally activated at 0.1–0.5 μM external Ca^{2+}. The Ca^{2+} cycle, then, will modulate the matrix-free Ca^{2+} based on the external (cytosolic) Ca^{2+}, affording precise control of the activity of the Ca-dependent matrix dehydrogenases. Whether the increase in the activity of the latter will be automatically translated into the overall increase of the mitochondrial respiratory activity and ATP production will ultimately depend on the definition of the rate-limiting step(s) of the general process of oxidative phosphorylation in vivo.

CONCLUSIONS: AN INTEGRATED PICTURE

The following general points can be extracted from the material presented in this review:

1. Cells limit greatly the exchange of Ca^{2+} with the extracellular universe, and control it with great care. Perhaps it is worth returning to the concept mentioned above of the convenience of the exposure of cells to a very large gradient of Ca^{2+}. Under these conditions even minor changes of the permeability of the plasma membrane to Ca^{2+} will cause significant swings of its concentration in the cytosol. While useful, this highly dynamic situation places a great burden on the precise functioning of the Ca^{2+} influx-mediating Ca^{2+} channel, and helps explain why this structure is controlled by the sophisticated array of mechanisms described in the preceding sections.

2. The division of labor among the different structures, membranous and otherwise, that preside over the task of controlling intracellular Ca^{2+} ensures that the oscillations in its free concentration in the cytosol, while actually occurring, are transient in nature. The time scale of the Ca^{2+} oscillations varies with the cell, depending essentially on the relative abundance of slow (low-affinity) and rapid (high-affinity) buffering structures. Thus, in muscles containing large amounts of high-Ca^{2+}-affinity sarcoplasmic reticulum, the life-span of the Ca^{2+}-linked oscillations is of the order of 200–400 msec. In other cells, and/or in other situations, the time scale is longer, particularly so when mitochondria come into play to dispose of abnormal Ca^{2+} pulses that have penetrated. The experiments by Rose & Löwenstein on the epithelial cells of the Chironomous salivary glands (306) are particularly instructive: the time course for the (presumed) energy-linked absorption of Ca^{2+} pulses by mitochondria was of the order of minutes.

3. Nonmembranous Ca^{2+}-buffering structures (e.g. soluble proteins) probably complex a major portion of the total cell Ca^{2+}. Even if this de facto results in a Ca^{2+}-buffering role, the high-affinity Ca^{2+}-binding proteins interact with Ca^{2+} for another reason, i.e. to process the information it contains. Although additional, up-to-now unexplored mechanisms may exist, it appears that a common denominator for the processing of the Ca^{2+} signal (i.e. for the interaction with targets), by high-affinity proteins is the expression of hydrophobic sites upon Ca^{2+} complexation. Of particular interest in view of their subcellular location is the case of the proteins whose association with the membrane is regulated by Ca^{2+} (307). They could conceivably play an important role in mediating the transduction of the Ca^{2+} signal from the

plasma membrane to the cell interior, and in mediating the interaction between the cytoskeleton and membrane components.

4. The structural model for Ca^{2+} binding evolved from the parvalbumin studies (18, 19) has been verified in a number of soluble (or nonmembranous) Ca^{2+}-binding proteins, including the central mediator of the Ca^{2+} message calmodulin. It appears, however, that a novel site mediates the interaction of high-affinity membrane proteins with Ca^{2+}. Still another site may mediate the binding of Ca^{2+} with the cytoskeletal proteins reversibly associated with membranes (307).

5. The point made above on the limited exchanges of Ca^{2+} between the intracellular milieu and the extracellular spaces, together with existing knowledge of the amounts of Ca^{2+} known to be mobilized by cells during their functional cycle, leads to the conclusion that cells are essentially self-sufficient in terms of Ca^{2+}. Although the long-term control of the total cell Ca^{2+} obviously depends on the activity of the plasma membrane systems, cells satisfy the greatest quantitative portion of their Ca^{2+} demands using their own Ca^{2+} pool, cytosolic or in the organelles. However, the interplay of cell Ca^{2+} with Ca^{2+} in the extracellular spaces, however limited quantitatively, is of vital importance, particularly to some cells. The penetrating Ca^{2+} sets in motion cascades of events, sometimes even triggering the massive liberation of Ca^{2+} from intracellular stores (e.g. the sarcoplasmic reticulum). Without this interplay the functional activity of (some) cells would be impossible.

6. The picture of the role of the cellular organelles accepted today wants endo(sarco)plasmic reticulum to fine-tune cytosolic Ca^{2+} down to the resting levels of between 100 and 200 nM, and to do it rapidly. It also wants mitochondria to play a minor role, or no role at all, in the control of cytosolic Ca^{2+}, and to be mainly concerned with the modulation of their own matrix Ca^{2+}. This seems reasonable under normal physiological conditions. However, the mitochondrial low-affinity Ca^{2+} uptake machinery has an impressive total capacity for Ca^{2+} accumulation, due to its ability to coaccumulate phosphate and to precipitate hydroxyapatite deposits in the matrix space. Thus, if noxious agents interfere with the Ca^{2+} permeability barrier of the plasma membrane, and permit cytosolic Ca^{2+} to rise to levels where the balance between mitochondrial Ca^{2+} uptake and Ca^{2+} release routes is shifted in favor of the former, mitochondria will store away the excess cytosolic Ca^{2+}, and will do so for a relatively long time without intolerable disturbances to their function. This is a safety device of great importance to cells (308–310), in view of the fact that the penetration of excess Ca^{2+} into the cytosol is a frequent and early phenomenon in cell injury (311), and also considering that no other organelle has the Ca^{2+}-storing capacity of mitochondria. That mitochondria in injured cells accumulate massive amounts of Ca^{2+} is extensively documented by electron microscope studies in

which the mitochondrial profiles appear crowded with electron-opaque masses resembling those seen in isolated mitochondria after massive loading with Ca^{2+} and phosphate (see 308–310 for reviews of the topic). In some cases, e.g. liver after intoxication with CCl_4 (312) and kidney after uranium intoxication (313), the association of large amounts of Ca^{2+} with mitochondria has been demonstrated by direct measurements. If the injuring condition is eliminated before the point of no return is reached, i.e. before the level of mitochondrial Ca^{2+} loading has become intolerable to the organelle, the excess Ca^{2+} will be gradually dissolved from the matrix deposits and released from mitochondria at a rate compatible with the Ca^{2+}-exporting ability of the plasma membrane systems. Essentially, then, it will be the ability of mitochondria to cope with large amounts of Ca^{2+} that will decide whether or not injured cells will survive.

Literature Cited

1. Williams, R. J. P., Wacker, W. E. C. 1967. J. Am. Med. Assoc. 210:18–22
2. Long, C., Mouat, B. 1973. Biochem. J. 123:829–36
3. Keynes, R. D., Lewis, P. R. 1956. J. Physiol. 134:399–407
4. Widdowson, E. M., Dickerson, J. W. T. 1964. In Mineral Metabolism, ed. C. L. Comar, F. Bronner, 2:1–247. New York: Academic
5. Thiers, R. E., Vallee, B. L. 1957. J. Biol. Chem. 226:911–20
6. Bresciani, F., Auricchio, F. 1962. Cancer Res. 22:1284–89
7. Hodgkin, A. L., Keynes, R. D. 1957. J. Physiol. 138:253–81
8. Williams, R. J. P. 1970. Q. Rev. Chem. Soc. 24:331–65
9. Williams, R. J. P. 1974. Biochem. Soc. Symp. 39:133–38
10. Williams, R. J. P. 1976. Symp. Soc. Exp. Biol. 30:1–17
11. Weber, A. 1976. Symp. Soc. Exp. Biol. 30:445–000
12. Ebashi, S. 1963. Nature 200:1010
13. Dedman, J. R., Potter, J. D., Jackson, R. L., Johnson, J. D., Means, A. R. 1977. J. Biol. Chem. 252:8415–22
14. Richman, P. G., Klee, C. B. 1978. Biochemistry 17:928–35
15. Anderson, J. M., Charbonneau, H. J., Jones, H. P., McCann, R. O., Cormier, M. J. 1980. Biochemistry 19:3113–20
16. La Porte, D. C., Wierman, B. M., Storm, D. R. 1980. Biochemistry 19:3814–19
17. Levine, B. A., Mercola, D., Coffman, D., Thornton, J. M. 1977. J. Mol. Biol. 115:743–60
18. Kretsinger, R. H., Nelson, D. J. 1977. Coord. Chem. Rev. 18:29–124
19. Kretsinger, R. H., Nockolds, C. E. 1973. J. Biol. Chem. 248:3313–26
20. Szebenyi, D. M. E., Obendorf, S. K., Moffat, K. 1981. Nature 294:327–32
21. Herzberg, O., James, M. N. G. 1985. Nature 313:653–59
22. Babu, Y. S., Sack, J. S., Greenhough, T. J., Bugg, C. E., Means, A. R., Cook, W. J. 1985. Nature 315:37–40
23. Dawson, R. M. C., Hauser, H. 1970. In Calcium and Cellular Function, ed. A. W. Cuthbert, pp. 17–41. Cambridge: Cambridge Univ. Press
24. Manery, J. F. 1969. In Mineral Metabolism, ed. C. L. Comar, F. Bronner, 3:405–52. New York: Academic
25. Baker, P. F. 1972. Progr. Biophys. Mol. Biol. 24:177–223
26. Klee, C. B., Vanaman, T. C. 1982. Adv. Protein Chem. 35:213–321
27. Carafoli, E., Niggli, V., Malmström, K., Caroni, P. 1980. Ann. NY Acad. Sci. 356:258–66
28. Moore, P. B., Dedman, J. R. 1982. J. Biol. Chem. 257:9663–67
29. Moore, P. B., Dedman, J. R. 1984. Ann. NY Acad. Sci. 435:173–75
30. Moore, P. B., Kraus-Friedmann, N., Dedman, J. R. 1984. J. Cell Sci. 72:121–33
31. Moore, P. B., Dedman, J. R. 1985. In Calmodulin Antagonists and Cell Physiology, ed. H. Hidaka, D. J. Hartshorne, pp. 483–94. New York: Academic
32. Creutz, C. E., Paxoles, C. J., Pollard,

H. B. 1978. *J. Biol. Chem.* 253:2858–66

33. Dabrow, M., Zaremba, S., Hogue-Angeletti, R. A. 1980. *Biochem. Biophys. Res. Commun.* 96:1164–71
34. Hong, K., Duzgunes, N., Papahadjopoulous, D. 1982. *Biophys. J.* 37:297–305
35. Sudhof, T. C., Walker, J. H., Fritsche, V. 1985. *J. Neurochem.* 44:1302–7
36. Sudhof, T. C., Ebbecke, M., Walker, J. H., Fritsche, V., Bonstead, C. 1984. *Biochemistry* 23:1103–9
37. Cruetz, C. E., Dowling, L. G., Sando, J. J., Villar-Palasi, C., Whipple, J. H., Zaks, W. J. 1983. *J. Biol. Chem.* 258:14664–74
38. Pollard, H. B., Scott, J. 1982. *FEBS Lett.* 150:201–6
39. Shadle, P. J., Gerke, V., Weber, K. 1985. *J. Biol. Chem.* 260:16354–60
40. Flower, R. J., Blackwell, G. J. 1979. *Nature* 278:456–59
41. Hirata, F. 1980. *Proc. Natl. Acad. Sci. USA* 77:2533–36
42. Di Rosa, M., Persico, P. 1979. *Br. J. Pharmacol.* 66:161–63
43. Gerke, V., Weber, K. 1984. *EMBO J.* 3:227–33
44. Smith, V. L., Dedman, J. R. 1987. In press
45. Khanna, N. C., Tokuda, M., Waisman, D. M. 1986. *J. Biol. Chem.* 261:8883–87
46. Geisow, M. J., Fritsche, V., Hexham, J. M., Dash, B., Johnson, T. 1986. *Nature* 320:636–38
47. Crouch, T. H., Klee, C. B. 1980. *Biochemistry* 19:3692–98
48. Haiech, J., Klee, C. B., Demaille, J. G. 1981. *Biochemistry* 20:3890–94
49. Watterson, D. M., Harrelson, W. G. Jr., Keller, P. M., Sharief, F., Vanaman, T. C. 1976. *J. Biol. Chem.* 251:4501–13
50. Lin, Y. M., Lin, Y. P., Cheung, W. Y. 1974. *J. Biol. Chem.* 249:4943–54
51. Wolff, J., Poirier, P. G., Brostrom, C. O., Brostrom, M. A. 1977. *J. Biol. Chem.* 252:4108–17
52. Vandermeers, A., Vandermeers-Piret, M. C., Rothé, G., Kutzner, R., Delforge, A., Christophe, J. 1977. *Eur. J. Biochem.* 81:379–86
53. Grand, R. J. A., Perry, S. V., Weeks, R. A. 1979. *Biochem. J.* 177:521–29
54. Grand, R. J. A., Perry, S. V. 1980. *Biochem. J.* 189:227–40
55. Watterson, D. M., Mendel, P. A., Vanaman, T. C. 1980. *Biochemistry* 19:2672–76
56. Chafouleas, J. G., Dedman, J. R., Mun-jaal, R. P., Means, A. R. 1979. *J. Biol. Chem.* 254:10262–67
57. Autric, F., Ferraz, C., Kilhoffer, M. C., Cavadore, J. C., Demaille, J. G. 1980. *Biochim. Biophys. Acta* 631:139–47
58. Sugden, M. C., Christie, M. R., Ashcroft, S. J. 1979. *FEBS Lett.* 105:95–100
59. Kobayashi, R., Kuo, I. C. Y., Coffee, C. J., Field, J. B. 1979. *Metabolism* 28:169–82
60. Nairn, A. C., Perry, S. V. 1979. *Biochem. J.* 179:89–97
61. Ridgway, E. B., Ashley, C. C. 1967. *Biochem. Biophys. Res. Commun.* 29:229–34
62. Campbell, A. K., Daw, R. A., Luzio, J. P. 1979. *FEBS Lett.* 107:55–60
63. Campbell, A. K., Lea, T. J., Ashley, C. C. 1979. In *Detection and Measurement of Free Calcium in Cells*, ed. C. C. Ashley, A. K. Campbell, pp. 13–72. Amsterdam: Elsevier/North Holland
64. Hallett, M. B., Campbell, A. K. 1982. *Nature* 295:155–58
65. Scarpa, A., Brinley, F. J., Tiffert, T., Dubyak, G. 1978. *Ann. NY Acad. Sci.* 307:86–112
66. Scarpa, A. 1979. See Ref. 63, pp. 85–115
67. Amman, D., Meier, P. C., Simon, W. 1979. See Ref. 63, pp. 117–29
68. Brown, H. M., Pemberton, J. P., Owen, J. D. 1976. *Anal. Chim. Acta* 85:261–76
69. Tsien, R. Y. 1980. *Biochemistry* 19:2396–404
70. Tsien, R. Y. 1981. *Nature* 290:527–28
71. Grynkiewicz, G., Poenie, M., Tsien, R. Y. 1985. *J. Biol. Chem.* 260:3440–50
72. Ashley, C. C., Ridgway, E. B. 1970. *J. Physiol.* 209:105–30
73. Brown, J. E., Coles, J. A., Pinto, L. H. 1977. *J. Physiol.* 269:299–320
74. Lee, C. O., Taylor, A., Windhager, E. C. 1980. *Nature* 287:859–61
75. Virk, S. S., Kirk, C. J., Shears, S. B. 1985. *Biochem. J.* 226:741–48
76. Klempner, M. 1985. *J. Clin. Invest.* 76:303–10
77. Schatzmann, H. J. 1966. *Experientia* 22:364–68
78. Fatt, P., Ginsborg, B. L. 1958. *J. Physiol.* 142:516–43
79. Reuter, H., Seitz, N. 1968. *J. Physiol.* 195:451–70
80. Blaustein, M. P., Hodgkin, A. L. 1968. *J. Physiol.* 198:46–48P
81. Vasington, F. D., Murphy, J. 1961. *Fed. Proc.* 20:146
82. Carafoli, E., Tiozzo, R., Lugli, G., Crovetti, F., Kratzing, C. 1974. *J. Mol. Cell. Cardiol.* 6:361–71

83. Ebashi, S. 1958. *Arch. Biochem. Biophys.* 76:410–23
84. Hasselbach, W., Makinose, M. 1961. *Biochem. Z.* 333:518–28
85. Reuter, H. 1975. In *Calcium Movements in Excitable Cells,* ed. P. F. Baker, H. Reuter, pp. 55–97. New York: Pergamon
86. Kostyuk, P. G. 1981. *Biochim. Biophys. Acta* 650:128–50
87. Kostyuk, P. G. 1982. In *Membrane Transport of Calcium,* ed. E. Carafoli, pp. 1–40. London: Academic
88. Tsien, R. W. 1983. *Ann. Rev. Physiol.* 45:341–58
89. Reuter, H. 1984. *Ann. Rev. Physiol.* 46:473–84
90. Reuter, H., Stevens, C. F., Tsien, R. W., Yellen, G. 1982. *Nature* 297:501–4
91. Bean, B. P., Nowycky, M. C., Tsien, R. W. 1983. *Biophys. J.* 41:295A
92. Reuter, H. 1983. *Nature* 301:569–74
93. Borsotto, M., Barhanin, J., Norman, R. I., Lazdunski, M. 1984. *Biochem. Biophys. Res. Commun.* 122:1357–66
94. Curtis, B. M., Catterall, W. A. 1984. *Biochemistry* 23:2113–18
95. Ferry, D. R., Goll, A., Glossmann, H. 1983. *EMBO J.* 2:1729–32
96. Varecka, L., Carafoli, E. 1982. *J. Biol. Chem.* 257:7414–21
97. Neyses, L., Locher, R., Stimpel, M., Streuli, R., Vetter, W. 1985. *Biochem. J.* 227:105–12
98. Hescheler, J., Pelzer, D., Trube, G., Trautwein, W. 1982. *Pflügers Arch.* 393:287–91
99. Lee, K. S., Tsien, R. W. 1983. *Nature* 302:790–94
100. Sanguinetti, M. C., Kass, R. S. 1984. *Circ. Res.* 55:336–48
101. Bean, B. P. 1984. *Proc. Natl. Acad. Sci. USA* 81:6388–92
102. Sanguinetti, M. C., Kass, R. S. 1984. *J. Mol. Cell. Cardiol.* 16:667–70
103. Kokubun, S., Reuter, H. 1984. *Proc. Natl. Acad. Sci. USA* 81:4824–27
104. Affolter, H., Coronado, R. 1986. *Proc. 6th Conf. Cyclic Nucleotides, Calcium and Protein Phosphorylation, Bethesda,* A124
105. Hess, P., Lansman, J. B., Tsien, R. W. 1984. *Nature* 311:538–44
106. Grossmann, A., Furchgott, R. F. 1964. *J. Pharmacol. Exp. Ther.* 145:162–72
107. Irisawa, H., Kokubun, S. 1983. *J. Physiol.* 338:321–27
108. Osterrieder, W., Brum, G., Hescheler, J., Trautwein, W., Flockerzi, V., Hofmann, F. 1982. *Nature* 298:576–78
109. Trautwein, W., Taniguchi, J., Noma, A. 1982. *Pflügers Arch.* 392:307–14
110. Nowycky, M. C., Fox, A. P., Tsien, R. W. 1985. *Nature* 316:440–43
111. Nilius, B., Hess, P., Lansman, J. B., Tsien, R. W. 1985. *Nature* 316:443–46
112. Carbone, E., Lux, H. D. 1984. *Nature* 310:501–2
113. Reuter, H. 1985. *Nature* 316:391
114. Curtis, B. M., Catterall, W. A. 1983. *J. Biol. Chem.* 258:7280–83
115. Borsotto, M., Norman, R. I., Fosset, M., Lazdunski, M. 1984. *Eur. J. Biochem.* 14:449–55
116. Glossmann, H., Ferry, D. R. 1983. *Naunyn-Schmiedebergs. Arch. Pharmacol.* 323:279–91
117. Curtis, B. M., Catterall, W. A. 1985. *Proc. Natl. Acad. Sci. USA* 82:2528–32
118. Hosey, M. M., Borsotto, M., Lazdunski, M. 1986. *Proc. Natl. Acad. Sci. USA* 83:3733–37
119. Flockerzi, V., Oeken, H. J., Hofmann, F., Pelzer, D., Cavalié, A., Trautwein, W. 1986. *Nature* 323:66–68
120. Hanke, W., Methfessel, C., Wilmsen, U., Boheim, G. C. 1984. *Bioelectrochem. Bioenergetics* 12:329–39
121. Campbell, A. K. 1983. *Intracellular Calcium, its Universal Role as Regulator,* pp. 155–60. New York: Wiley
122. Baker, P. F., Meves, H., Ridgway, E. B. 1973. *J. Physiol.* 231:527–48
123. Blaustein, M. P., Hodgkin, A. L. 1969. *J. Physiol.* 200:496–527
124. Baker, P. F., Blaustein, M. P., Hodgkin, A. L., Steinhardt, R. A. 1969. *J. Physiol.* 200:431–58
125. Baker, P. F., Blaustein, M. P., Manil, J., Steinhardt, R. A. 1967. *J. Physiol.* 191:100–2P
126. Herchuelz, A., Sener, A., Malaisse, W. J. 1980. *J. Membr. Biol.* 57:1–12
127. Blaustein, M. P. 1974. *Rev. Physiol. Biochem. Pharmacol.* 70:33–82
128. Famulski, K., Carafoli, E. 1982. *Cell Calcium* 3:263–81
129. Gmaj, P., Murer, H., Kinne, R. 1979. *Biochem. J.* 178:549–57
130. Grinstein, S., Erlij, D. 1978. *Proc. R. Soc. London (Biol.)* 202:353–60
131. Krieger, S. Y., Tashijian, A. H. Jr. 1980. *Nature* 283:843–45
132. Reeves, J. P., Sutko, J. L. 1979. *Proc. Natl. Acad. Sci. USA* 76:590–94
133. Caroni, P., Reinlib, L., Carafoli, E. 1980. *Proc. Natl. Acad. Sci. USA* 77:6354–58
134. Pitts, B. J. R. 1979. *J. Biol. Chem.* 254:6232–35
135. Reeves, J. P., Sutko, J. L. 1980. *Science* 208:1461–64
136. Philipson, K. D., Nishimoto, A. Y. 1980. *J. Biol. Chem.* 255:6880–82

137. Bers, D. M., Philipson, K. D., Nishimoto, A. Y. 1980. *Biochim. Biophys. Acta* 601:358–71
138. Reuter, H. 1985. *Med. Res. Rev.* 5:427–40
139. Lee, C. O., Uhm, D. Y., Dresdner, K. 1980. *Science* 209:699–701
140. Sheu, S. S., Fozzard, H. A. 1982. *J. Gen. Physiol.* 80:325–51
141. Lado, M. G., Sheu, S. S., Fozzard, H. A. 1984. *Circ. Res.* 54:576–85
142. Blaustein, M. P. 1977. *Biophys. J.* 20:79–111
143. Di Polo, R., Beaugé, L. 1983. *Ann. Rev. Physiol.* 45:313–24
144. Caroni, P., Carafoli, E. 1983. *Eur. J. Biochem.* 132:451–60
145. Philipson, K. D. 1985. *Ann. Rev. Physiol.* 47:561–71
146. Caroni, P., Villani, F., Carafoli, E. 1981. *FEBS Lett.* 130:184–86
147. Siegl, P. K. S., Cragoe, E. J., Trumble, M. J., Kaczorowski, G. J. 1984. *Proc. Natl. Acad. Sci. USA* 81:3238–42
148. Hale, C. C., Slaughter, R. S., Ahrens, D. C., Reeves, J. P. 1984. *Proc. Natl. Acad. Sci. USA* 81:6569–73
149. Barzilai, L., Spanier, R., Rahamimoff, H. 1984. *Proc. Natl. Acad. Sci. USA* 81:6521–25
150. Soldati, L., Longoni, S., Carafoli, E. 1985. *J. Biol. Chem.* 260:13321–27
151. Schatzmann, H. J. 1982. See Ref. 87, pp. 41–108
152. Niggli, V., Penniston, J. T., Carafoli, E. 1979. *J. Biol. Chem.* 254:9955–58
153. Caroni, P., Carafoli, E. 1981. *J. Biol. Chem.* 256:3263–70
154. Gopinath, R. M., Vincenzi, F. F. 1977. *Biochem. Biophys. Res. Commun.* 77:1203–9
155. Jarrett, H. W., Penniston, J. T. 1977. *Biochem. Biophys. Res. Commun.* 77:1210–16
156. Hakim, G., Itano, T., Verma, A. K., Penniston, J. T. 1982. *Biochem. J.* 207:225–31
157. Wuytack, F., De Schutter, G., Casteels, R. 1981. *Biochem. J.* 198:265–71
158. Michalak, M., Famulski, K., Carafoli, E. 1984. *J. Biol. Chem.* 259:15540–47
159. Niggli, V., Adunyah, E. S., Penniston, J. T., Carafoli, E. 1981. *J. Biol. Chem.* 253:395–401
160. Clark, A., Carafoli, E. 1983. *Cell Calcium* 4:83–88
161. Niggli, V., Sigel, E., Carafoli, E. 1982. *J. Biol. Chem.* 257:2350–56
162. Ronner, P., Gazzotti, P., Carafoli, E. 1977. *Arch. Biochem. Biophys.* 179:578–83
163. Taverna, R. D., Hanahan, D. J. 1980. *Biochem. Biophys. Res. Commun.* 94:652–59
164. Sarkadi, B., Enyedi, A., Gardos, G. 1980. *Cell Calcium* 1:287–97
165. Niggli, V., Adunyah, E. S., Carafoli, E. 1981. *J. Biol. Chem.* 256:8588–92
166. Stieger, J., Schatzmann, H. J. 1981. *Cell Calcium* 2:601–16
167. Zurini, M., Krebs, J., Penniston, J. T., Carafoli, E. 1984. *J. Biol. Chem.* 259:618–27
168. Benaim, G., Zurini, M., Carafoli, E. 1984. *J. Biol. Chem.* 259:8471–77
169. Caroni, P., Carafoli, E. 1981. *J. Biol. Chem.* 256:9371–73
170. Neyses, L., Reinlib, L., Carafoli, E. 1985. *J. Biol. Chem.* 260:10283–87
171. Davis, P. J., Blas, S. D. 1981. *Biochem. Biophys. Res. Commun.* 90:1073–80
172. Galo, M. G., Unates, L. E., Farias, R. N. 1981. *J. Biol. Chem.* 256:7113–14
173. Soloff, M. S., Sweet, P. 1982. *J. Biol. Chem.* 257:10687–93
174. Pershadsingh, H. A., McDonald, J. M. 1979. *Nature* 281:495–97
175. Mauldin, D., Roufogalis, B. D. 1980. *Biochem. J.* 187:507–13
176. Maretzki, D., Klatt, D., Reimann, B., Rapoport, S. 1982. *Biochem. Int.* 4:323–29
177. Lee, K. S., Au, K. S. 1983. *Biochim. Biophys. Acta* 742:54–62
178. Lotersztajn, S., Hanoune, J., Pecker, F. 1981. *J. Biol. Chem.* 256:11209–15
179. Lotersztajn, S., Pecker, F. 1982. *J. Biol. Chem.* 257:6638–41
180. Lin, S. H. 1986. *J. Biol. Chem.* 260:10976–80
181. Constantin, L. L., Franzini-Armstrong, C., Podolsky, R. J. 1965. *Science* 147:158–60
182. Bond, M., Kitazawa, T., Somlyo, A. P., Somlyo, A. V. 1984. *J. Physiol.* 355:677–95
183. Kowarski, D., Shuman, H., Somlyo, A. P., Somlyo, A. V. 1985. *J. Physiol.* 366:153–75
184. MacLennan, D. H. 1970. *J. Biol. Chem.* 245:4508–18
185. Meissner, G. 1973. *Biochim. Biophys. Acta* 298:906–26
186. Ikemoto, N. 1975. *J. Biol. Chem.* 250:7219–24
187. Kanazawa, T., Yamada, S., Yamamoto, T., Tonomura, Y. 1971. *J. Biochem.* 70:95–123
188. Hasselbach, W. 1964. *Progr. Biophys. Chem.* 14:167–222
189. Weber, A., Herz, R., Reiss, I. 1966. *Biochem. Z.* 345:329–69

190. Worsfold, M., Peter, J. B. 1970. *J. Biol. Chem.* 245:5545–52
191. Makinose, M. 1969. *Eur. J. Biochem.* 10:74–82
192. Sreter, F. A. 1969. *Arch. Biochem. Biophys.* 134:25–33
193. Martonosi, A., Feretos, R. 1964. *J. Biol. Chem.* 239:648–58
194. Inesi, G., Scarpa, A. 1972. *Biochemistry* 11:356–59
195. Fanburg, B. L., Finkel, R. M., Martonosi, A. 1964. *J. Biol. Chem.* 239:2298–306
196. Besch, H. R., Schwartz, A. 1971. *Biochem. Biophys. Res. Commun.* 45:286–92
197. Franzini-Armstrong, C. 1980. *Fed. Proc.* 39:2403–9
198. Schneider, M. F., Chandler, W. K. 1973. *Nature* 242:44
199. Barlogie, B., Hasselbach, W., Makinose, M. 1971. *FEBS Lett.* 12:267–68
200. Inesi, G., Malan, N. 1976. *Life Sci.* 18:773–80
201. Kasai, M., Miyamoto, H. 1976. *J. Biochem.* 79:1053–66
202. Meissner, G., McKinley, D. 1976. *J. Membr. Biol.* 30:79–80
203. Fabiato, A., Fabiato, F. 1975. *J. Physiol.* 249:469–95
204. Somlyo, A. V., Bond, M., Somlyo, A. P., Scarpa, A. 1985. *Proc. Natl. Acad. Sci. USA* 82:5331–35
205. Fabiato, A., Putney, J. 1986. *Proc. 30th Int. Congr. Physiol., Vancouver,* S. 172.01
206. Tada, M., Kirchberger, M. A., Li, H. C. 1975. *J. Cyclic Nucleotides Res.* 1:329–38
207. Tada, M., Kirchberger, M. A., Li, H. C., Katz, A. M. 1975. *J. Biol. Chem.* 250:2640–46
208. Le Peuch, C. J., Haiech, J., Demaille, J. G. 1979. *Biochemistry* 18:5150–57
209. Inoui, M., Chamberlain, B. K., Saito, A., Fleischer, S. 1986. *J. Biol. Chem.* 261:1794–800
210. Imagawa, T., Watanabe, T., Nakamura, T. 1986. *J. Biochem.* 99:41–53
211. Wegener, H. L., Jones, L. R. 1984. *J. Biol. Chem.* 259:1834–41
212. Gasser, J. T., Chiesi, M. P., Carafoli, E. 1986. *Biochemistry.* In press
213. Fujii, J., Kadoma, M., Tada, M., Toda, H., Sakiyawa, F. 1986. *Biochem. Biophys. Res. Commun.* 138:1044–50
214. Fujii, J., Ueno, A., Kitano, K., Tanaka, S., Kadoma, M., Tada, M. 1986. Submitted for publication
215. Movsesian, M. A., Nishikawa, M., Adelstein, R. S. 1984. *J. Biol. Chem.* 259:8029–32
216. MacLennan, D. H., Brandl, C. J., Korczak, B., Green, N. M. 1985. *Nature* 316:696–700
217. MacLennan, D. H., Wong, P. T. S. 1971. *Proc. Natl. Acad. Sci. USA* 68:1231–35
218. Meissner, G. 1975. *Biochim. Biophys. Acta* 385:51–68
219. Stewart, P. S., MacLennan, D. H. 1974. *J. Biol. Chem.* 249:985–93
220. Ostwald, T. J., MacLennan, D. H., 1974. *J. Biol. Chem.* 249:974–79
221. Ostwald, T. J., MacLennan, D. H., Dorrington, K. H. 1974. *J. Biol. Chem.* 5867–71
222. Moore, P. B., Kraus-Friedmann, N. 1983. *Biochem. J.* 214:69–75
223. Parys, J. B., De Smedt, H., Vandenberghe, P., Borghgraef, R. 1985. *Cell Calcium* 6:413–29
224. Somlyo, A. P., Bond, M., Somlyo, A. V. 1985. *Nature* 314:622–25
225. Famulski, K. S., Carafoli, E. 1984. *Eur. J. Biochem.* 140:447–52
226. Bygrave, F. L., Tranter, C. J. 1978. *Biochem. J.* 174:1021–30
226b. Benedetti, A., Fulceri, R., Ferro, M., Comporti, M. 1986. *Trends Biochem. Sci.* 10:284–85
226c. Hers, H. G. 1986. *Trends Biochem. Sci.* 10:285
227. Hokin, L. E., Hokin, M. R. 1955. *Biochim. Biophys. Acta* 18:102–10
228. Hokin, L. E., Hokin, M. R. 1958. *J. Biol. Chem.* 233:805–10
229. Hokin, M. R., Hokin, L. E. 1967. *J. Gen. Physiol.* 50:793–811
230. Michell, R. H. 1975. *Biochim. Biophys. Acta* 415:81–147
231. Berridge, M. J. 1984. *Biochem. J.* 220:345–60
232. Downes, C., Michell, R. H. 1982. *Cell Calcium* 3:467–502
233. Michell, R. H. 1982. *Cell Calcium* 3:285–94
234. Michell, R. H., Kirk, C. J., Jones, L. M., Dawnes, C., Creba, J. A. 1981. *Philos. Trans. R. Soc. London Ser. B* 296:123–37
235. Fein, J. N., Berridge, M. J. 1979. *Biochem. J.* 178:45–58
236. Streb, H., Irvine, R. F., Berridge, M. J., Schulz, I. 1983. *Nature* 306:67–69
237. Sekar, C. M., Hokin, L. E. 1986. *J. Membr. Biol.* 89:193–210
238. Streb, H., Bayerdörffer, E., Haase, W., Irvine, R. F., Schulz, I. 1984. *J. Membr. Biol.* 81:241–53
239. Prentki, M., Biden, T. J., Janjic, D., Irvine, R. F., Berridge, M. J., et al. 1984. *Nature* 309:562–64
240. Dawson, A. P., Irvine, R. F. 1984.

Biochem. Biophys. Res. Commun. 120: 858–64

241. Muallem, S., Schoefield, M., Pandol, S., Sachs, G. 1985. *Proc. Natl. Acad. Sci. USA* 82:4433–37

242. Ueda, T., Cueh, S. H., Noel, M. W., Gill, D. L. 1986. *J. Biol. Chem.* 261:3184–92

243. Thiery, J., Klee, C. B. 1986. *J. Biol. Chem.* In press

244. Joseph, S. K., Williams, R. J., Corkey, B. E., Matschinsky, F. M., Williamson, J. R. 1984. *J. Biol. Chem.* 259:12952–55

245. Irvine, R. F., Brown, K. D., Berridge, M. J. 1984. *Biochem. J.* 221:269–72

246. Hirata, M., Suematsu, E., Hashimoto, T., Hamachi, T., Koga, T., 1984. *Biochem. J.* 223:229–36

247. Storey, D. J., Shears, S. B., Kirk, C. J., Michell, R. H. 1984. *Nature* 312:374–76

248. Vasington, F. D., Murphy, J. V. 1962. *J. Biol. Chem.* 237:2670–72

249. De Luca, H. F., Engström, G. W. 1961. *Proc. Natl. Acad. Sci. USA* 47:1744–50

250. Crompton, M., Sigel, E., Salzmann, M., Carafoli, E. 1976. *Eur. J. Biochem.* 69:429–34

251. Vercesi, A., Reynafarje, B., Lehninger, A. L. 1978. *J. Biol. Chem.* 253:6379–85

252. Somlyo, A. P., Somlyo, A. V., Shuman, H. 1979. *J. Cell Biol.* 81:316–35

253. Somlyo, A. P., Urbanics, R., Vadasz, G., Kovach, A. G. B., Somlyo, A. V. 1985. *Biochem. Biophys. Res. Commun.* 132:1071–78

254. Somlyo, A. P. 1985. *Circ. Res.* 57:497–507

255. Barnard, T. 1981. *Scanning Electron Microsc.* 1981(P2):419–33

256. Bond, M., Shuman, H., Somlyo, A. P., Somlyo, A. V. 1984. *J. Physiol.* 357:185–201

257. Somlyo, A. V., Gonzales Serratos, H., Shuman, H., McClellan, G., Somlyo, A. P. 1981. *J. Cell Biol.* 90:577–94

258. Somlyo, A. P., Walz, B. 1985. *J. Physiol.* 357:185–95

259. Scarpa, A., Graziotti, P. 1973. *J. Gen. Physiol.* 62:756–72

260. Denton, R. M., Randle, P. J., Martin, B. R. 1972. *Biochem. J.* 128:161–63

261. Denton, R. M., Richards, D. A., Chin, J. G. 1978. *Biochem. J.* 176:899–906

262. McCormack, J. G., Denton, R. M. 1979. *Biochem. J.* 180:533–44

263. Carafoli, E. 1982. See Ref. 87, pp. 109–39

264. Crompton, M. 1985. *Curr. Top. Membr. Transp.* 25:231–76

265. Rottenberg, H., Scarpa, A. 1974. *Biochemistry* 13:4811–19

266. Moore, C. L. 1971. *Biochem. Biophys. Res. Commun.* 42:405–18

267. McCormack, J. G., Denton, R. M. 1984. *Biochem. J.* 218:235–47

268. Rossi, C. S., Lehninger, A. L. 1963. *Biochem. Z.* 338:698–713

269. Greenawalt, J. W., Rossi, C. S., Lehninger, A. L. 1964. *J. Cell Biol.* 23:21–38

270. Carafoli, E., Rossi, C. S., Lehninger, A. L. 1965. *J. Biol. Chem.* 240:2254–61

271. Nicchitta, C. V., Williamson, J. R. 1984. *J. Biol. Chem.* 259:12978–83

272. Crompton, M., Capano, M., Carafoli, E. 1976. *Eur. J. Biochem.* 69:453–62

273. Crompton, M., Kunzi, M., Carafoli, E. 1977. *Eur. J. Biochem.* 79:549–58

274. Crompton, M., Moser, R., Lüdi, H., Carafoli, E. 1978. *Eur. J. Biochem.* 82:25–31

275. Crompton, M., Heid, I. 1978. *Eur. J. Biochem.* 91:599–608

276. Nicholls, D. G. 1978. *Biochem. J.* 170:511–22

277. Nicholls, D. G. 1978. *Biochem. J.* 176:463–74

278. Crompton, M., Heid, I., Baschera, C., Carafoli, E. 1978. *FEBS Lett.* 104:352–54

279. Coll, K. E., Joseph, S. K., Corkey, B. E., Williamson, J. R. 1982. *J. Biol. Chem.* 257:8696–704

280. Hansford, R. G., Castro, F. 1981. *Biochem. J.* 198:525–33

281. Hayat, L. H., Crompton, M. 1982. *Biochem. J.* 202:509–18

282. Goldstone, T. P., Crompton, M. 1982. *Biochem. J.* 204:369–71

283. Goldstone, T. P., Duddridge, R. J., Crompton, M. 1983. *Biochem. J.* 210:463–72

284. Vaghy, P. L., Johnson, J. D., Matlib, M. A., Wang, T., Schwartz, A. 1982. *J. Biol. Chem.* 257:6000–2

285. Wolkowicz, P. E., Michael, L. H., Lewis, R. M., McMillin-Wood, J. 1983. *Am. J. Physiol.* 244:H644–H651

286. Matlib, M. A., Lee, S. W., Depover, A., Schwartz, A. 1983. *Eur. J. Pharmacol.* 89:327–36

287. Akerman, K. 1978. *Arch. Biochem. Biophys.* 189:256–62

288. Fiskum, G., Cockrell, R. S. 1978. *FEBS Lett.* 92:125–28

289. Fiskum, G., Lehninger, A. L. 1979. *J. Biol. Chem.* 254:6236–39

290. Tsokos, J., Cornwell, T. F., Vlasuk, G. 1980. *FEBS Lett.* 119:297–300

291. Gunter, T. E., Chace, J. H., Puskin, J. S., Gunter, K. K. 1983. *Biochemistry* 22:6341–51
292. Beatrice, M. C., Palmer, J. W., Pfeiffer, D. R. 1980. *J. Biol. Chem.* 255:8663–71
293. Palmer, J. W., Pfeiffer, D. R. 1981. *J. Biol. Chem.* 256:6742–50
294. Epps, D. E., Palmer, J. W., Schmid, H. H. O., Pfeiffer, D. R. 1982. *J. Biol. Chem.* 257:1383–91
295. Beatrice, M. C., Stiers, D. L., Pfeiffer, D. R. 1982. *J. Biol. Chem.* 257:7161–71
296. Lehninger, A. L., Vercesi, A., Bababunmi, E. A. 1978. *Proc. Natl. Acad. Sci. USA* 75:1690–94
297. Lötscher, H. R., Winterhalter, K. H., Carafoli, E., Richter, C. 1979. *Proc. Natl. Acad. Sci. USA* 76:4340–44
298. Hofstetter, W., Mühlebach, T., Lötscher, H. R., Winterhalter, K. H., Richter, C. 1981. *Eur. J. Biochem.* 117:361–67
299. Baumhütter, S., Richter, C. 1982. *FEBS Lett.* 148:271–75
300. Richter, C., Winterhalter, K. H., Baumhütter, S., Lötscher, H. R., Moser, B. 1983. *Proc. Natl. Acad. Sci. USA* 80:3188–92
301. Carafoli, E. 1979. *FEBS Lett.* 104:1–5
302. Stücki, J. W., Ineichen, E. A. 1974. *Eur. J. Biochem.* 48:365–75
303. Crompton, M. 1984. *Biochem. Soc. Trans.* 8:261–62
304. Mitchell, P., Moyle, J. 1967. *Biochem. J.* 105:1147–62
305. Denton, R. M., McCormack, J. G., Edgell, N. J. 1980. *Biochem. J.* 190:107–17
306. Rose, B., Löwenstein, W. R. 1975. *Science* 190:1204–6
307. Owens, R. J., Crumpton, M. J. 1984. *Bio Essays* 1:61–63
308. Carafoli, E. 1974. *Biochem. Soc. Symp.* 39:89–113
309. Carafoli, E., Malmström, K., Jerusalem, F. 1983. In *Cellular Pathobiology of Human Disease*, ed. B. F. Trump, A. Laufer, R. T. Jones, pp. 99–122. New York-Stuttgart: Fischer
310. Carafoli, E. 1982. In *Pathophysiology of Shock, Anoxia, and Ischemia*, ed. R. A. Cowley, B. F. Trump, pp. 95–112. Baltimore-London: Williams & Wilkins
311. Schanne, F. A. X., Kane, A. B., Young, E. E., Farber, J. L. 1979. *Science* 206:700–2
312. Thiers, R. E., Reynolds, E. S., Vallee, B. L. 1960. *J. Biol. Chem.* 235:2130–33
313. Carafoli, E., Tiozzo, R., Ronchetti, I., Laschi, R. 1971. *Lab. Invest.* 25:516–27

Ann. Rev. Biochem. 1987. 56:435–66

DNA MISMATCH CORRECTION

Paul Modrich

Department of Biochemistry, Duke University Medical Center, Durham, North Carolina 27710

CONTENTS

PERSPECTIVES AND SUMMARY

Both prokaryotes and eukaryotes process mismatched base pairs, pairing errors in which the Watson-Crick bases occur in noncomplementary opposition within the DNA helix.[1] Such mispairs arise as a consequence of genetic recombination and as a result of DNA biosynthetic errors or the deamination

---

[1] For the purpose of this review the term mismatch will be used to refer to the transition mispairs G-T and A-C and the transversion mismatches G-G, A-A, G-A, C-C, T-T, and C-T. In addition, the term will be extended to include helix anomalies resulting from insertion or deletion of one or more nucleotides in one strand of the helix.

0066-4154/87/0701-0435$02.00

of 5-methylcytosine in a G-m^{5}C pair. The recognition and correction of mispairs generated in this manner has been of interest for several reasons. The processing of mismatches within recombination intermediates probably contributes to a number of marker effects associated with recombination, including gene conversion, coconversion, map expansion effects, localized negative interference, and postmeiotic segregation (1–6). On the other hand, mismatches resulting from DNA biosynthetic errors and G-T mispairs generated by the deamination of 5-methylcytosine are lesions that will be fixed as mutations unless corrected on the proper DNA strand. Wagner & Meselson (7) suggested that mutation avoidance via mismatch correction could be achieved by use of secondary signals within the helix to direct repair to the appropriate strand. This hypothesis, as initially proposed for the postreplication repair of biosynthetic errors, is illustrated in Figure 1. Mismatch correction systems capable of strand discrimination via secondary signals do indeed exist within bacteria (6), and perhaps in mammalian cells as well (8). As anticipated by Wagner & Meselson, the properties of such systems are in accord with the idea that they function in the maintenance of genetic stability.

Mismatched base pairs and the systems that process them are also of interest from the viewpoint of chemistry and mechanism. Although the concept of mismatch formation has played a major role in thoughts concerning the basis of spontaneous mutation (9, 10), information concerning the structural nature of mispaired DNA bases has become available only during the past four years. The past few years have also seen the development of

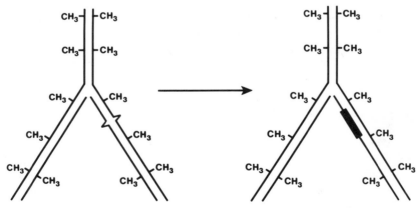

Figure 1 The Wagner-Meselson model for postreplication mismatch correction—Wagner & Meselson (7) suggested that mismatch correction might function in the elimination of DNA biosynthetic errors. They postulated that strand discrimination necessary for function of such a system could be based on a special relationship between the repair system and the replication complex or, since DNA methylation occurs postsynthetically, on the transient undermethylation of newly synthesized strands.

cell-free systems that support mismatch correction in vitro and isolation of the set of *Escherichia coli* proteins known to be required for mismatch correction in this organism. Thus, the nature of mismatch recognition and the mechanisms of repair can now be addressed at the level of protein-DNA interaction.

This review emphasizes the biochemical features of mismatch correction in *E. coli*, the organism that has been most extensively studied in this respect, and also considers available information concerning the structural nature of mismatched base pairs. The discussion that follows also alludes to correction in other organisms for comparative purposes, but it is by no means comprehensive in this respect. For alternate perspectives, the reader is referred to recent reviews by Marinus (11), Radman & Wagner (11a), Claverys & Lacks (12), and a particularly excellent account by Meselson (6).

BIOLOGY OF MISMATCH CORRECTION

Evidence for Mismatch Processing in Vivo

The term "mismatch correction" is used in this review to refer to processes in which recognition of a mismatched base pair elicits a specific response resulting in its repair. It is important that this definition be kept in mind, since in principle, rectification of a mispair can occur by reactions that do not depend on the presence of a mismatch within the DNA helix. Although evidence establishing the existence of mismatch-specific repair systems is summarized in the sections that follow, the major points supporting this view can be stated at the outset. These are: (*a*) Different mispairs are corrected with different efficiencies, implying that mismatch specificity is associated with correction. (*b*) Bacterial and yeast mutants have been identified that are selectively defective in mismatch correction. (*c*) In vitro experiments have demonstrated repair DNA synthesis that is provoked by the presence of a mismatch and have led to identification of a protein that binds to at least some mispairs.

Direct evidence for the intracellular processing of mismatched base pairs has been provided by transfection experiments that used artificially constructed DNA heteroduplexes, in which the two strands could be distinguished genetically. In such experiments the fate of heterozygotic markers, and hence the corresponding mismatches, was determined by analysis of the genotypes of virus particles emerging from single infective centers. The first definitive studies of this type in the *E. coli* system utilized multiply marked bacteriophage lambda heteroduplexes under conditions of replication or recombination block (7, 13, 14). These experiments demonstrated heteroduplex correction prior to the onset of replication and showed that the efficiency of correction of different mismatches can vary by as much as an order of magnitude. Furthermore, for those heteroduplexes where two or more closely spaced mismatches were repaired, correction was usually restricted to one DNA strand with repair events at the several mispairs being nonindependent (7, 14). Cocorrection events of this sort were interpreted in terms of an excision repair mechanism with an average excision tract length of about 3 kilobases (7).

The evidence for mismatch correction in *Streptococcus pneumoniae* is somewhat more indirect but equally compelling. Transformation of this organism involves assimilation of single-strand fragments of donor DNA into the chromosome of recipient bacteria to generate a heteroduplex region. A particularly striking feature of this process is an associated marker-specific variation in transformation efficiency (reviewed in Ref. 12). High-efficiency markers yield transformants with an efficiency approaching one per genome equivalent of donor DNA entering the cell. In contrast, the transformation efficiency of other markers is typically in the range of 0.05 to 0.5. It is now evident that marker discrimination in this system reflects mismatch repair on the donor strand within the heteroduplex recombination intermediate. Thus when mismatches within the heteroduplex are subject to repair, the genotype of the donor strand is corrected to that of the recipient and consequently lost. As in the case of the lambda system discussed above, cocorrection data and physical analysis of the fate of a transforming DNA fragment have indicated that correction in *S. pneumoniae* involves repair tracts in the range of 5 to 10 kilobases (15, 16).

Transformation and DNA injection methods have also been used to study heteroduplex processing in yeast and mammalian cells. Under conditions where reassortment of markers by recombination was excluded, heteroduplexes containing two insertion mismatches (8 and 12 bp in size) or two point mismatches were efficiently repaired in yeast (17). Furthermore, the two insertions were corrected with different efficiencies, suggesting that this class of mismatch can promote its own repair in this organism. In contrast to *E. coli* and *S. pneumoniae,* cocorrection efficiencies indicated that about half of the repair tracts provoked by an insertion mismatch in yeast were less than a kilobase in length. Similar findings in mammalian cells have also been interpreted in terms of mismatch repair (8, 18–20). However, inasmuch as recombination could have contributed to results obtained in these studies and since marker-specific effects have not yet been described, conclusions concerning mismatch repair in mammalian cells should be viewed with caution.

Postreplication Repair of Biosynthetic Errors

The initial impetus for study of mismatch correction was based on the idea that rectification of mispairs within recombination intermediates could explain certain marker effects associated with crossing over (21). Consideration of such effects is beyond the scope of this review, and the reader is referred elsewhere for discussion of this area (reviewed in Refs. 1–3, 6). However, alternate roles for mismatch repair were suggested in the mid-1970s. The mutator phenotype associated with *hex⁻* mutants of *S. pneumoniae* [*hex⁻* mutants are defective in mismatch correction and behave as high-efficiency recipients for transformation with all markers (22)], led Tiraby & Fox (23) to suggest that mismatch correction might function in mutation avoidance. A more explicit hypothesis was described in 1976 by Wagner & Meselson (7), who suggested that mismatch correction could serve to eliminate DNA biosynthetic errors from newly synthesized DNA. This hypothesis included the proposal that the strand discrimination necessary for function of such a system could be based on the undermethylation of newly synthesized DNA, or alternatively, could reflect a special relationship between the repair system and the replication apparatus.

The general features of the Wagner-Meselson model, which is outlined in Figure 1, have been confirmed, at least in bacterial systems. Indeed and as mentioned above, mismatch correction as monitored by transformation in *S. pneumoniae* is strand specific in the sense that processing is largely limited to the incoming donor strand within the heteroduplex region (24). Guild & Shoemaker (15) proposed 10 years ago that strand direction in this system is based on the presence of free ends on the donor DNA strand within the heteroduplex recombination intermediate, and this is still the favored explanation for strand discrimination in *S. pneumoniae*. A comprehensive discussion of mismatch correction in this organism can be found in a recent review (12).

Strand-directed mismatch correction has also been demonstrated in *E. coli*, and the system responsible displays the features anticipated by Wagner & Meselson (7) for methyl-directed, postreplication repair of biosynthetic errors (Figure 1). In this organism adenine methylation of d(G-A-T-C) sequences determines the strand on which repair occurs. The most direct evidence in support of this view has been provided by transfection with heteroduplexes in which the two DNA strands were in defined states of methylation at such sites (6, 25–29). With hemi-methylated heteroduplexes, which are methylated at d(G-A-T-C) sequences on only one DNA strand, repair is highly biased to the unmethylated strand, with the methylated strand serving as template for correction. Mismatch repair also occurs on heteroduplexes in which neither strand is methylated, but in this case correction shows little strand preference. Heteroduplexes that are fully modified at d(G-A-T-C) sites on both DNA strands are subject to repair at substantially reduced efficiency (26, 29), although exceptions to this rule exist (6). These exceptions, which apparently involve the action of alternate correction pathways, will be considered below.

Several of these transfection studies (27, 28) have utilized the small, single-stranded phages f1 and M13, in which the density of d(G-A-T-C) sequences is lower than anticipated on a statistical basis. For example, the f1 heteroduplexes studied by Lu et al (27) contain only four such sites, with the shortest distance between the mismatch and the nearest d(G-A-T-C) sequence being 1000 base pairs. Nevertheless, strand direction by methylation operates effectively on such molecules.

Application of the transfection assay for heteroduplex repair has led to identification of several E. coli mutants that are defective in mismatch repair. Mutant strains defective in uvrD (also called uvrE, mutU, or recL), mutH, mutL, or mutS function exhibit reduced levels of heteroduplex correction as judged by this biological assay (26, 30–32). Since mutations in these loci also confer high spontaneous mutability (33), the associated defects in mismatch correction are consistent with a role for mismatch repair in mutation avoidance.

A key feature of the methyl-directed repair hypothesis is the idea that newly synthesized DNA will exist in an undermethylated form for a period of time sufficient to allow mismatch correction to occur. The transfection experiments discussed above are in accord with this view since they imply that mismatch correction usually initiates prior to replication or methylation of unmethylated or hemi-methylated heteroduplexes. With one exception, attempts to directly assess the rate of in vivo methylation of newly synthesized DNA also indicate a significant delay between synthesis and methylation. Marinus, and Lyons & Schendel have found that d(G-A-T-C) sequences in newly synthesized DNA are undermethylated relative to those within the bulk chromosome (34, 35). In contrast, Szyf et al (36) reported that DNA synthesized during pulse labeling periods of 30 s to 5 min at 30°C was fully methylated at such sites. This led the latter authors to conclude that strand discrimination based on methylation of d(G-A-T-C) sites was very unlikely. However, this conclusion may be invalid for several reasons. The effectiveness of the quenching protocol (36) for termination of in vivo methylation was not evaluated, and since the method used to monitor the extent of methylation was not quantitative, significant levels of undermethylation could have gone undetected. It also can be estimated that even during the shortest pulse time studied, an E. coli replication fork would progress about 10,000 base pairs at 30°C. Consequently, these experiments were not particularly sensitive for newly synthesized DNA.

Genetic evidence has also indicated a role in mismatch correction for the E. coli dam methylase, the S-adenosylmethione-dependent activity responsible for modification of d(G-A-T-C) sequences in this organism (37, 38). Mutants deficient in this activity (dam⁻) are hypermutable (39), as are strains that overproduce the methylase more than 10-fold (40, 41). These findings are consistent with the transfection results cited above. The mutator phenotype of dam⁻ mutants can be understood in terms of a loss in strand bias for repair, while that associated with overproduction can be explained by more rapid methylation of newly synthesized DNA coupled with the reduced efficiency of correction on symmetrically modified regions. For the latter explanation to be valid, it is necessary that the biological rate of d(G-A-T-C) methylation be limited by the intracellular level of the *dam* enzyme. This has been shown to be the case (42).

The *E. coli dam* methylase has also been implicated in a pathway involving function of *mutH*, *mutL*, and *mutS* genes. In combination with *recA*, *recB*, *recC*, *recJ*, *lexA*, or *polA* mutations, lesions in the *dam* gene result in an inviable phenotype (11). Futhermore, *dam* mutants grow poorly in the presence of certain base analogues like 2-aminopurine (43). McGraw & Marinus (44) identified suppressor mutations that restored viability to *dam⁻ recA⁻*, *dam⁻ recB⁻* and *dam⁻ recC⁻* double mutants, and Glickman & Radman (43) isolated suppressor mutations that allowed *dam⁻* mutants to grow in the presence of 2-aminopurine. The majority of these suppressors were second site mutations that inactivated *mutH*, *mutL*, or *mutS* function, indicating that *dam* and these *mut* genes function in a common pathway. Glickman & Radman have explained these observations by suggesting that in the absence of methylation, mismatch correction may initiate on both DNA strands, leading to the generation of double-strand breaks (43). Under conditions of elevated mismatch correction provoked by base analogue mutagenesis or when double-strand break repair is blocked by *recA* or *recB* mutations (45, 46), lethal events ensue unless mismatch correction is blocked by *mutH*, *mutL*, or *mutS* mutations. This idea received experimental support with the recent demonstration that *dam⁻ recA^{ts}* and *dam⁻ recB^{ts}* strains accumulate double-strand breaks at 42°C, a temperature at which they are inviable, and that double-strand-break formation is suppressed and viability restored at this temperature by introduction of *mutL* or *mutS* mutations (47).

While it is clear that d(G-A-T-C) methylation can direct mismatch correction, Lacks and colleagues have proposed that this represents only a minor pathway for strand discrimination in *E. coli* (12, 48). They have suggested that the majority of correction events in *E. coli* are directed by DNA termini, as is the case in *S. pneumoniae*. This proposal is based on the relative mutability of *dam⁻* strains as compared to that of strains defective in *mutH*, *mutL*, or *mutS* function. Mutations in these *mut* loci result in a 100- to 1000-fold increase in the spontaneous mutation rate (33, 43), and all else being equal, *dam⁻* mutations would be expected to result in about half this increase. In fact methylase mutations increase spontaneous mutability only about 20- to 80-fold. This discrepancy in mutation rates suggests that the methyl-directed pathway may comprise only one component of a multifaceted *mutHLS*-dependent repair system.

As will be discussed below, alternate mismatch correction systems do exist in *E. coli,* but these represent low-efficiency pathways relative to *dam-*directed repair. In addition, there is no evidence to indicate the existence of a major correction pathway directed by termini as postulated by Lacks. On the contrary, attempts to detect such a system both in vivo and in vitro have yielded negative results. The presence of strand-specific scissions did not support strand-directed, *mutHLS*-dependent mismatch repair in either case, but the removal of d(G-A-T-C) sites by mutagenesis was found to result in dramatic reduction of correction, both in vivo and in vitro (49, 50, 50a). d(G-A-T-C) sites therefore play a major role in the *mutHLS* pathway of mismatch correction. Indeed, in vitro experiments indicate that one function of the *mutH* product is recognition of such sequences (K. Welsh, A.-L. Lu, P. Modrich, in preparation).

On the other hand, the discrepancy in spontaneous mutabilities of *dam*⁻ and *mut*⁻ strains can be explained without invocation of alternate pathways. Mismatch correction on unmethylated heteroduplexes can lead to double-strand breaks, and transfection experiments have demonstrated that unmethylated heteroduplexes suffer loss of biological activity in a reaction that is dependent on the presence of a mismatch and on *mut* gene function (29). The differences in mutabilities of *dam*⁻ and *mut*⁻ strains may therefore reflect a selection against mismatches, and hence mutation, in *dam*⁻ strains. It would thus appear that strand discrimination in *E. coli* is largely dictated by the state of methylation of d(G-A-T-C) sequences, while in *S. pneumoniae,* DNA termini provide the basis for strand direction.

As discussed above, the study of mismatch correction in eukaryotes is not as advanced as in bacterial systems. Nevertheless, Hare & Taylor have obtained evidence suggesting that DNA methylation and DNA termini may also contribute to strand direction of mismatch repair in mammalian cells (8). These experiments monitored the fate of G-T and A-C mismatches in SV40 heteroduplexes subsequent to their introduction into monkey CV-1 cells. The presence of strand breaks was found to affect the transmission of heteroduplex genotypes to progeny virus, an effect attributed to mismatch repair. With covalently closed heteroduplexes, there was little strand bias to repair. However, when covalently closed molecules were hemi-methylated at two *Hha*I sites (Gm⁵CGC) spanning the mismatches, repair was observed only on the unmethylated strand. Since the most commonly methylated dinucleotide in vertebrates is CpG, these findings suggest that methylation at such sites, as well as DNA termini, may function to direct mismatch correction in this class of organism.

dam-*Independent Mismatch Correction in* E. coli

Although activity of the *E. coli mutHLS*-dependent mismatch correction system is highly dependent on the presence of d(G-A-T-C) sequences in the unmethylated or hemi-methylated configuration, alternate repair systems exist that are independent of the state of methylation of such sites and probably independent of such sequences altogether. One such system is illustrated by the behavior of the two possible heteroduplexes constructed using wild-type and *Pam*80 lambda DNA (Peterson and Meselson cited in Ref. 6). The two possible heteroduplexes in this case are $P/+$ (l/h) and $+/P$ (l/h), where l and h designate the two phage strands. With hemi-methylated or unmethylated heteroduplexes, results obtained were those expected for the *dam*-directed pathway, namely correction on the unmethylated strand or repair with little strand bias, respectively. Furthermore, as observed in other studies (26, 29), symmetric methylation at d(G-A-T-C) sites resulted in a large reduction in correction of the $+/P$ (l/h) heteroduplex. However, in the symmetrically methylated configuration, the alternate $P/+$ (l/h) heteroduplex was efficiently repaired to wild type. Thus, the *Pam*80/+ and +/*Pam*80 mismatches are subject to methyl-directed repair, but when this pathway is blocked by symmetric methylation, the *Pam*80/+ (l,h) mismatch is subject to specific correction by a pathway that functions less efficiently than the *dam*-directed system.

The nature of this alternate pathway has been deduced by Lieb (51–53). Her analysis of the fine structure of the lambda *cI* gene has led to the identification of six exceptional mutations that yield excess recombinants in four factor crosses over short intervals. Two of these involve C→T transitions at the second position within the sequence d(C-C-A/T-G-G) (52), with the rest being C→T transitions in the related sequences d(C-A/T-G-G) and d(C-C-A/T-G) (53). Excess recombination of these markers has been attributed to sequence-specific, very short patch (VSP) mismatch correction, which involves excision tracts of 10 base pairs or less and acts unidirectionally in the sense that the d(C-C-A/T-G-G) or related sequence serves as template for repair (52, 53). In the *Pam*80 example cited above, VSP repair corrects the G-T mismatch in 5'-C-T-A-G-A/G-G-T-C-T-5' to C-C-A-G-A/G-G-T-C-T. The A-C mispair in the alternate heteroduplex is not subject to repair by this system (6).

The mismatch and sequence specificity of VSP repair is of particular interest since it suggests that this system corrects G-T mispairs that arise by deamination of 5-methylcytosine in G-m^{5}C base pairs. In *E. coli* K strains d(C-C-A/T-G-G) is methylated at the internal C, and such sequences are hot spots for mutation due to spontaneous deamination of the methylated base (54, 55). Although the uracil DNA glycosylase can recognize and eliminate the spontaneous deamination product of cytosine, this activity cannot excise thymine resulting from deamination of 5-methylcytosine (56).

VSP repair requires *mutL* and *mutS* gene products, proteins that are also required for methyl-directed mismatch correction, but in contrast to the latter system, VSP repair is independent of *mutH* and *uvrD* proteins (M. Lieb, personal communication). In addition, VSP repair requires the *dcm* gene product, the enzyme that methylates d(C-C-A/T-G-G) sequences in *E. coli* (11; M. Lieb, personal communication). The requirement for the *dcm* methylase lends additional credence to the idea that VSP repair functions to eliminate thymine resulting from 5-methylcytosine deamination. However, function of the *dcm* enzyme in correction does not appear to involve DNA methylation since G-T mismatches in d(C-A/T-G-G) and d(C-C-A/T-G) sequence contexts are also subject to VSP repair. Although subsets of the *dcm* recognition site, these sequences are not known to be modified by the *dcm* methylase.

Two additional low-efficiency methyl-independent pathways have been identified in *E. coli* by Kolodner and colleagues, who utilized transformation methods to study the fate of pBR322 plasmid heteroduplexes symmetrically modified at d(G-A-T-C) sites (57, 58). One pathway involves excision tracts in excess of a kilobase and does not require *mutH* or *mutL* function but was reduced by 50 to 60% in *mutS* or *uvrD* hosts. The second system, which is extremely weak, is characterized by short repair tracts (<300 base pairs) and stringent dependence on *recF* or *recJ* function. In contrast to *dam*-directed and VSP correction systems discussed above, repair by these two methyl-independent pathways can occur on either DNA strand. Fishel & Kolodner have suggested that these "error prone" systems may act on heteroduplex regions formed during recombination or generated as a consequence of physical damage (57).

Mismatch Specificity

As mentioned above, mispairs corresponding to distinct genetic markers can be corrected with different efficiencies, with the implied specificity representing a major argument that mismatches provoke their own repair. Correction efficiencies of defined mispairs have been evaluated in both *S. pneumoniae* and *E. coli,* and in both cases the results are remarkably similar. The G-T and A-C transition mismatches are good substrates for repair in both organisms (27, 28, 59–61). A-A, G-G, and T-T transversion mispairs are also subject to correction in both organisms (60, 61), as are the four possible mismatches corresponding to insertion/deletion of A, G, C, or T (59, 62), although this group of mismatches is recognized somewhat less well than the transition mispairs (28, 59, 60). Insertions/deletions of 10 nucleotides are also corrected in *E. coli* (58), but in those cases tested, mismatches involving larger regions of noncomplementarity (>30 base pairs in *S. pneumoniae,* 800 base pairs in *E. coli*) were found to be refractory to repair (29, 59, 60).

In contrast to the G-T, A-C, A-A, G-G, and T-T point mispairs, or mismatches corresponding to small insertions or deletions, the G-A, T-C, and C-C transversion mispairs are generally poor substrates for correction in the bacterial systems (28, 59–61). However, exceptional G-A and T-C mismatches, which are subject to repair, have been identified in both *S. pneumo-niae* and *E. coli* (49, 59, 60; M. Jones, R. Wagner, M. Radman, personal communication). This differential sensitivity to repair indicates that correction efficiency is sensitive to sequence environment, at least in the case of G-A and T-C mismatches.

A similar hierarchy of correction efficiencies has been inferred in yeast by Fogel and colleagues based on the postmeiotic segregation frequencies (4). Postmeiotic segregation (PMS), in which two alleles segregate during the first mitotic division of a haploid spore, has been attributed to the presence of uncorrected mismatches within regions of heteroduplex resulting from meiotic recombination. Hence, alleles that result in high PMS would correspond to a poorly repaired mismatch that persists within the heteroduplex, while alleles resulting in low PMS would correspond to a well-corrected mispair. In fact, PMS frequencies in yeast correlate extremely well, and in the expected manner, with efficiencies of correction of the different mismatches as determined in bacterial systems (4). Moreover, mutations have been isolated that confer elevated PMS, and like bacterial mutations that block mismatch correction, yeast *pms1* mutations confer a mitotic mutator phenotype (5).

STRUCTURES OF BASE PAIR MISMATCHES

A major feature of the Watson-Crick proposal was the idea that only purine-pyrimidine pairs would be readily accommodated within the structure of the helix, with the specificity of purine-pyrimidine pairing dictated by satisfaction of the hydrogen-bonding potential of the bases (63). This led to their suggestion that spontaneous mutation might reflect the occasional occurrence of a base in a rare tautomeric form that could form a well-fitting but nevertheless incorrect base pair. This sort of idea received additional impetus with the demonstration that bacterial DNA polymerases, which are thought to select incoming nucleotide precursors by virtue of their ability to form sterically acceptable base pairs (64), misincorporate incorrect nucleotides at detectable frequencies (reviewed in Ref. 65). Consequently, a number of mechanisms have been postulated to account for mispair formation, including elaboration of the tautomer hypothesis (9), formation of wobble pairs (66–69), involvement of ionized bases (68), and pairing schemes involving *anti-syn* isomerization about the *N*-glycosidic bond (9, 68, 70). Although such pairing schemes can account for the rare entry of an incorrect nucleotide into a DNA chain, the question at hand is concerned with the conformation of the noncomplementary base once it is stably incorporated within a stretch of helix. Is the resulting mismatch intrahelical under physiological conditions, stabilized by the stacking and hydrogen-bonding potential of bases involved, or alternatively, does it adopt an extrahelical conformation (67, 71)? The following discussion will consider this problem with respect to mismatches within DNA helixes, and as will be seen, the consensus of studies addressing this point is that at least some mispairs can adopt intrahelical conformations. It is tempting to view such structures in terms of the problem of mismatch recognition, and the discussion that follows does so. However, it should be kept in mind that such judgments may be premature at this point for several reasons. The currently available structures do not comprise the complete set of possible point mismatches and single base insertion/deletion mispairs. Secondly, correction of some mispairs depends on sequence environment (59; M. Jones, R. Wagner, M. Radman, personal communication), and it is known that neighboring sequences can affect the stability of a mismatch and in at least one case, possibly conformation as well (see below). Since the set of available structural information is small, the significance of sequence environment in determination of mismatch conformation cannot yet be evaluated. Lastly, the NMR experiments reviewed below demonstrate that mismatches are dynamic structures. Consequently, their recognition might involve interaction with minor species.

Extent of DNA Helix Disruption by Noncomplementary Bases: Melting Behavior and Single-Strand Nuclease Sensitivity

Early attempts to address the extent of helix disruption caused by the presence of mismatched bases relied on thermodynamic analysis of homopolymer duplexes in which one strand contained interspersed, noncomplementary nucleotides (67). In one of the first systematic studies of this type with DNA helixes, Dodgson & Wells (72, 73) examined two sets of heteroduplexes of the forms $d(G)_n \cdot d(C_{12}A_mC_x)$, where $m = 1$–6, and $d(G)_n \cdot d(C_{10}G_mC_x)$ where $m = 1$ and 3–5. Thermal melting analysis demonstrated that the presence of noncomplementary blocks of G-G or G-A mismatches did not disrupt cooperative interactions between the flanking regions of G-C base pairs. This finding, and the relative resistance of such structures to high levels of S1 and mung bean single-strand nucleases, led to the suggestion that G-G and A-G mispairs might be accommodated within the model heteroduplexes as stacked intrahelical forms.

Although S1 nuclease readily hydrolyzes heteroduplex loops as small as 30 nucleotides in size (74), the finding that single-base-pair mismatches are generally resistant to this activity has recently been extended to a large set of mispairs by Maniatis, Lerman and colleagues (75). These results do not, however, necessarily imply that the stable conformation of all mispairs is intrahelical in nature. Rather, they show that the significantly populated conformations available to a mispair are not sufficiently unstacked to be recognized by the DNA-binding site of the nuclease. This is pertinent since model-building studies (67, 71) demonstrate that rotation of noncomplementary bases out of the helix can be accomplished with limited backbone distortion to yield a structure in which the two Watson-Crick pairs on either side of the noncomplementary region may stack on each other. A similar idea was suggested in the study of block homopolymer heteroduplexes mentioned above. Dodgson & Wells (72) pointed out that their conclusions could be complicated by slippage of dC blocks along the $d(G)_n$ chain. In one scenario slippage would result in extrusion of a noncomplementary loop to bring $d(C)_n$ blocks spanning this region into register, thus yielding stacked $d(G)_n \cdot d(C)_n$ blocks that would melt in the cooperative fashion observed.

The advent of high-yield methods for defined oligodeoxyribonucleotide synthesis has permitted construction of model heteroduplexes, which are not subject to this sort of slippage problem, in quantities sufficient for high-resolution physical and structural analysis. Several comprehensive studies of the effects of single mismatched base pairs on the thermal stabilities of small DNA heteroduplexes are now available. Tinoco and colleagues (76) determined the thermodynamic parameters for formation of the 16 possible helixes of the form $d(CA_3XA_3G) \cdot d(CT_3YT_3G)$, where X and Y were A, G, T, and C. In similar experiments Werntges et al (77) examined helix-coil transitions of 16 octadecamer heteroduplexes, which were also substituted at a single position.

Both of these studies demonstrated that substitution of a G-C pair or an A-T pair by a mismatch reduces the stability of the helix, with the degree of destabilization depending on the mispair. For example, Aboul-Ela et al observed a hierarchy of stabilities that can be expressed approximately as G-T > G-G > G-A > C-T > A-A = T-T > A-C = C-C (76). A similar variation in stability was observed by Werntges et al (77), and by Patel et al (G-A = G-T > A-C = C-T, Ref. 78), with differences attributable to effects of sequence environment. The importance of the latter factor in determination of mismatch stability is exemplified by the finding that inversion of a mismatch can result in significant changes in helix stability (76, 77). For example, the study by Tinoco and colleagues demonstrated that replacement of T-G by G-T or G-A by A-G yielded heteroduplexes differing in free energy by 0.7 to 0.9 kcal/mol at 25°C, with differences in enthalpy and entropy of helix formation being much more dramatic (76).

The hierarchy of heteroduplex stability determined in such experiments does not correlate with the efficiencies of correction of the different mispairs. G-T and G-A mismatches are among the more stable mispairs, while A-C and C-C fall into the least stable class. By contrast, G-T and A-C transition mispairs are usually well repaired, while G-A and C-C transversion mismatches are generally poor substrates (28, 29, 59–61). In fact, the sequences of the 16 octadecamer duplexes studied by Werntges et al (77) corresponded to a set of M13 heteroduplexes used in a previous study to assess efficiencies of correction in vivo (28). No correlation was observed between in vivo correction efficiencies and melting temperatures of the corresponding octadecamer heteroduplexes (77).

In an attempt to relate thermodynamic properties to structural features of a mismatch that might affect correction efficiency, Werntges et al (77) evaluated the melting curves for their 16 octadecameric heteroduplexes in terms of a stack model for the helix-to-coil transition, which allowed for internal loop formation. Estimates for the enthalpy and entropy of melting of stacks involving mismatches were obtained by empirical fit of melting profiles to the theoretical expression for the partition function of the stack model. In contrast to heteroduplex stability, the enthalpy of melting of a mismatch stack estimated in this manner correlates to some extent with correction efficiency. This parameter was used to classify the 12 mismatches studied (4 of the 16 heteroduplexes contained Watson-Crick pairs). Mismatches with enthalpies of stack melting comparable to A-T or G-C pairs were defined as wobble pairs (T-G, G-G, C-A, A-A, and A-G), those with enthalpies about half that of A-T or G-C pairs were classified as weak (G-T, A-C, and G-A), while those with enthalpies of stack melting near zero were inferred to be unstacked or extrahelical (T-T, C-T, T-C, and C-C). Comparison of mismatch classification with correction efficiency revealed that wobble or weak mismatches were corrected with good to poor efficiency while open mispairs were corrected very poorly. This study therefore suggests that pyrimidine-pyrimidine mismatches adopt an extrahelical conformation and that such conformations are not recognized by the correction system.

G-T and A-C Transition Mismatches

The best-studied DNA base pair mismatch has been the G-T transition mispair. Extensive nuclear magnetic resonance (NMR) data was available documenting the conformation and dynamics of this mismatch in B-DNA in solution (69, 78–82), and single crystal X-ray structures have been solved for G-T mispairs in A (83, 84), B (85), and Z-DNA (85, 86). In every case, the G-T mismatch has been found to adopt the wobble conformation anticipated by Crick (66), which is stabilized by two imino proton-carbonyl hydrogen bonds. Intrahelical mismatch structures determined by solution and solid state methods are shown in Figure 2, with the model heteroduplexes used for structural determinations listed in Table 1.

The G-T wobble pair is accommodated within A, B, or Z-DNA helices with little effect on global helix structure or backbone conformation (82–86). However, formation of the wobble pair results in displacement of guanine and thymine bases relative to their positions in G-C or A-T pairs. In both A and B-helixes, guanine is shifted into the minor groove while thymine is displaced into the major groove (83–85). In the Z-structure, the guanine base is shifted into the groove and thymine away from the groove (86). Kennard and

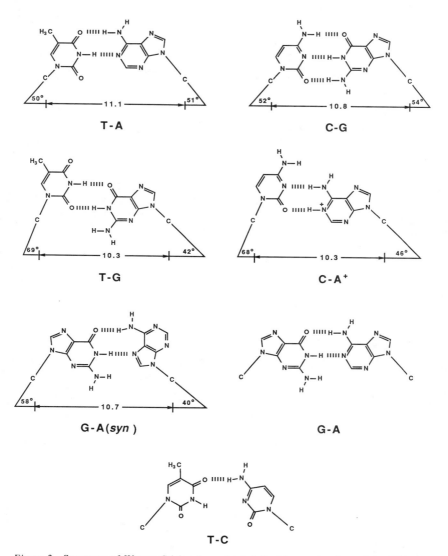

Figure 2 Structures of Watson-Crick pairs and intrahelical base pair mismatches—With the exception of adenine in the G-A*(syn)* pair, all bases are in the *anti* conformation. As discussed in the text, structures shown for T-G and C-A$^+$ have been determined by X-ray and solution NMR methods, the G-A*(syn)* structure by X-ray analysis, and the G-A structure by NMR. The tentative structure for T-C is also based on NMR measurements. Bridging water molecules, which are thought to contribute additional stability to T-G, C-A$^+$, and T-C pairs, are not shown. Note the symmetry of T-A and C-G pairs with respect to a pseudodyad located between the bases and the symmetrical disposition of their glycosidic bond angles relative to the Cl'-Cl' vector. Significant deviations from this symmetry are evident in T-G, C-A$^+$, and G-A*(syn)* mispairs (85, 87). Adapted from Hunter et al (87) and Patel et al (78).

Table 1 Oligonucleotide models for study of mismatch structure

Duplex	Conformation	Mismatch	Ref.
5'- C - G - t - G - A - A - T - T - C - g - C - G 3'- G - C - g - C - T - T - A - A - G - t - G - C	B	G-T wobble	79, 82
5'- C - G - C - g - A - A - T - T - t - G - C - G 3'- G - C - G - t - T - T - A - A - g - C - G - C	B	G-T wobble	85
5'- G - G - G - g - t - C - C - C 3'- C - C - C - t - g - G - G - G	A	G-T wobble	84
5'- G - G - g - G - C - t - C - C 3'- C - C - t - C - G - g - G - G	A	G-T wobble	83
5'- C - g - C - G - t - G 3'- G - t - G - C - g - C	Z	G-T wobble	86
5'- t - G - C - G - C - g 3'- g - C - G - C - G - t	Z	G-T wobble	85
5'- C - G - C - a - A - T - T - c - G - C - G 3'- G - C - G - c - T - T - A - A - a - C - G - C	B	A^+-C wobble	87
5'- C - G - c - G - A - A - T - T - C - a - C - G 3'- G - C - a - C - T - T - A - A - G - c - G - C	B	Probably A^+-C	88, 89

Sequence		Description		Ref.
5'-C-C-A-A-g-a-T-T-G-G 3'-G-G-T-T-a-g-a-A-C-C	B	G (*anti*)-A (*anti*)		90
5'-C-G-a-G-A-A-T-T-C-g-C-G 3'-G-C-g-C-T-T-A-A-G-a-G-C	B	G (*anti*)-A (*anti*)		91
5'-C-G-C-g-A-A-T-T-a-G-C-G 3'-G-C-G-a-T-T-A-A-g-C-G-C	B	G (*anti*)-A (*syn*)		92
5'-C-G-c-G-A-A-T-T-C-t-C-G 3'-G-C-t-C-T-T-A-A-G-c-G-C	B	T-C wobble?		78
5'-A-T-C-C-T-A-t-T-A-G-G-A-T 3'-T-A-G-G-A-T-t-A-T-C-C-T-A	B	T-T wobble?		93
5'-C-G-C-a-G-A-A-T-T-C- -G-C-G 3'-G-C-G- -C-T-T-A-A-G-a-C-G-C	B	Extra A intrahelical		94
5'-C-G-C-a-G-A-G-C-T-C- -G-C-G 3'-G-C-G- -C-T-C-G-A-G-a-C-G-C	B	Extra A intrahelical		95
5'-C-A-A-A-c-A-A-A-G 3'-G-T-T-T- -T-T-T-C	B	Extra C extrahelical		96

colleagues (84, 85, 87) have pointed out that this displacement renders the G-T pair devoid of elements of pseudosymmetry inherent to A-T or G-C base pairs. In the normal Watson-Crick pairs, the glycosidic bonds are symmetrically disposed relative to a vector joining the Cl' carbons of the two sugar residues. The pseudodyad relating the glycosidic bonds in A-T and G-C pairs also results in approximate equivalence of purine N3 and pyrimidine O2 minor groove hydrogen bond acceptors. Deviation of the G-T wobble pair from these elements of pseudosymmetry can be seen in Figure 2.

A particularly striking feature of the G-T wobble pair is the role of solvent in stabilizing base-base interactions in the mismatch. The keto O4 of thymine and the amino N2 of guanine participate in interbase hydrogen bonds in A-T and G-C pairs, but in the G-T wobble pair these functional groups are displaced into solvent. Visualization of first shell water molecules in crystals of the Z-DNA helix formed by d(C-G-C-G-T-G) (86) and crystals of the A-helix formed by d(G-G-G-G-T-C-C-C) (84) revealed that the wobble bases are bridged by a network of solvent molecules linking exposed functional groups. In both structures, the amino N2 of G was linked via a water molecule to the keto O2 of T, with the keto O4 of T and the keto O6 of G also bridged by solvent. The bridging waters thus satisfy the bonding potential of the wobble bases, and as pointed out by Kneale et al (84), in a sense result in stabilization of the G-T wobble pair by four hydrogen bonds rather than two. Rich and colleagues (86) have suggested that bridging water molecules may prove of general importance in stabilization of base-base mispairs in both DNA and RNA helixes.

Analysis of the set of model G-T heteroduplexes (Table 1) has demonstrated that the wobble pair is also stabilized by stacking interactions. Base pairing is maintained on either side of the wobble pair in B-DNA in solution at low temperature (79), and in the solid state in A (83, 84), B (85), and Z-helixes (85, 86). Base pair stacks visualized in crystals of G-T model heteroduplexes have been compared with those observed in isomorphous crystals of the corresponding parental helixes, which in each case contain a G-C pair instead of a mismatch. Although stacks involving the wobble pair deviate from those observed with the G-C pair due to the displacement of guanine and thymine mentioned above, perturbation of stacking interactions was found to be localized to stacks involving the mispair (84, 85). Furthermore, wobble pair stacks displayed significant base overlap, ranging from somewhat less to somewhat more than that observed for the G-C pair in parental helixes, depending on the nature of neighboring base pairs and helix conformation (84–86). Given the quality of stacking by the G-T pair, Ho et al (86) have suggested that helix destabilization by the wobble pair may at least partially reflect the increased hydrophobic surface of the helix resulting from displacement of the 5-methyl group of thymine into solvent. Destabilization by this mechanism would be entropic in nature, an idea consistent with the finding by Patel et al (79) that the enthalpy of melting of the G-T heteroduplex formed by d(C-G-T-G-A-A-T-T-C-G-C-G) is identical to that of the d(C-G-C-G-A-A-T-T-C-G-C-G) helix, despite the 20°C lower T_m of the former.

The dynamics of the G-T wobble pair have been addressed in the elegant studies of Patel and coworkers (78–81), who used NMR methods to examine the solution behavior of the B-form heteroduplex formed by the self-complementary dodecamer d(C-G-T-G-A-A-T-T-C-G-C-G). Thermal dependence of chemical shifts for nonexchangeable base protons revealed a common melting transition for the 10 nonterminal base pairs within the heteroduplex (52°C in 0.1 M phosphate pH 7.7; compare with 72°C for the parental helix containing a G-C pair), demonstrating that the wobble pair is included in the cooperative unit for the helix-to-coil transition (79). At temperatures below the T_m, the destabilizing effect of the G-T pair was evident in enhanced rates of base pair and helix opening as deduced by rates of imino proton exchange (80, 81), an effect localized to the mismatch and one base pair on either side. Such findings show that the G-T mismatch can be a much more dynamic entity than the conventional Watson-Crick pairs.

Unlike guanine and thymine, the major tautomers of A and C do not contain imino protons that can participate in interbase hydrogen bonds. Nevertheless, X-ray (87) and NMR (88, 89) studies of B-DNA heteroduplexes (Table 1) indicate that adenine and cytosine can form an intrahelical base pair without tautomerization of either base. As shown in Figure 2, the resulting wobble pair is thought to involve the N1-protonated form of adenine and to be stabilized by two hydrogen bonds.

The crystal structure of d(C-G-C-A-A-A-T-T-C-G-C-G) (Table 1 and Ref. 87) indicates that the A-C wobble pair shares several features in common with the G-T mismatch considered above. Like the G-T pair, the A-C mismatch is further stabilized by a water molecule bridging the N4 and N6 amino groups of cytosine and adenine, respectively. Moreover, formation of the A-C pair involves displacement of the cytosine into the major groove and adenine into the minor groove, resulting in loss of the pseudosymmetry characteristic of A-T and G-C pairs. Lastly, as observed in the case of the G-T mismatch, the A-C wobble pair is accommodated within B-DNA with only limited effects on local helix conformation, and the A-C pair was found to stack well with neighboring G-C and A-T base pairs.

However, effects of the A-C mismatch on helix stability and dynamics differ from those of the G-T pair. As mentioned above, the G-T mismatch is one of the more stable mispairs, while A-C is among the least stable. Measurement of rates of imino proton exchange has demonstrated that the A-C mismatch also results in greater kinetic destabilization of the helix than the G-T wobble pair (78, 88). In contrast to the highly localized effect of the

G-T mismatch on kinetics of helix opening, the A-C mispair results in an enhancement in the rate of helix opening, which is evident two to three base pairs removed from the mismatch.

Purine-Purine and Pyrimidine-Pyrimidine Transversion Mismatches

The only purine-purine mismatch that has been examined structurally is the G-A mispair. Unlike the other mismatches studied in this manner, G-A has been found to adopt two intrahelical conformations. Kan et al (90) and Patel et al (91) have utilized NMR methods to determine solution conformations of d(C-C-A-A-G-A-T-T-G-G) and d(C-G-A-G-A-A-T-T-C-G-C-G) heteroduplexes (Table 1). Both groups observed hydrogen-bonded NH—N resonances for the G-A mismatch at low temperature, and based on proton nuclear Overhauser enhancement experiments, both concluded that this mispair was of the form G(anti)-(anti), which is stabilized by two hydrogen bonds (Figure 2). In this mispair, the guanine and adenine bases are in their major tautomer forms and in the usual anti configuration with respect to the glycosidic bond. Consistent with this assignment was the observation of nuclear Overhauser effects between the H-2 proton of the mismatch adenine and the imino protons of the G-A and adjacent base pairs (90, 91). Failure to observe such effects between the G-A imino proton and the H-8 protons of adenine or guanine (90, 91) excluded the alternate G(anti)-A(syn) (67, 70) and G(syn)-A(anti, imino) (9) conformations.

In contrast, solution of the crystal structure of d(C-G-C-G-A-A-T-T-A-G-C-G) has shown that the G-A mismatches within this heteroduplex (Table 1) assume the alternate G(anti)-A(syn) conformation (92). Like the G(anti)-A(anti) pair, this conformation involves the major guanine and adenine tautomers and is stabilized by two hydrogen bonds, but in the G(anti)-A(syn) pair adenine adopts the unusual syn orientation with respect to the glycosidic bond (Figure 2). As observed for G-T and A-C mismatches in B-DNA, G(anti)-A(syn) is asymmetric with respect to glycosidic bond angles and results in only small perturbations of local and global helix parameters. However, the former effect is less pronounced for G(anti)-A(syn) than for the transition mispairs (Figure 2). The syn configuration of adenine in the G(anti)-A(syn) pair allows this mismatch to fit well in the B-helix (9, 92), but accommodation of a G(anti)-A(anti) mispair is expected to require significant distortion of the sugar-phosphate backbone. Unfortunately, information on backbone conformation in the latter case is not available.

Although details of the stacking properties of G(anti)-A(anti) and G(anti)-A(syn) have not been described, it is clear that base pairing is maintained on either side of these mismatches, at least at low temperature (91, 92). Furthermore, NMR analysis of the d(C-G-A-G-A-A-T-T-C-G-C-G) heteroduplex

(91) demonstrated that the adenine H-2 proton of the G-A mismatch resonates upfield from its unstacked value, with the chemical shift being independent of temperature between 0 and 40°C. This finding suggests that this adenine remains stacked in the *anti* configuration within the model helix over this temperature range.

It is not clear what factors determine whether G-A assumes the *anti-anti* or *anti-syn* conformation. Since the neighboring base pairs differed in each of the three studies considered above (Table 1), it is possible that sequence environment determines the conformation of the G-A pair. Alternatively, external factors, such as crystal forces, may also play a role in this respect. The demonstration of two distinct conformations for the G-A mismatch is of interest in view of the biological finding that some G-A mismatches are subject to repair in *S. pneumoniae* and *E. coli,* while others are not (49, 59–61; M. Jones, R. Wagner, M. Radman, personal communication).

Effects of the G-A mismatch on helix stability are similar to those found for the G-T wobble pair (above). In only one case, however, is it possible to relate such effects to the conformation assumed by G-A. Patel et al (91) demonstrated that the T_m of d(C-G-A-G-A-A-T-T-C-G-C-G), which contains two G(*anti*)-A(*anti*) mispairs, is almost identical to that of the related dodecamer containing two G-T pairs at the corresponding positions (Table 1). G(*anti*)-A(*anti*) is also similar to the G-T pair with respect to effects on helix dynamics. Presence of the G-A pair results in enhanced rates of helix opening, but as observed for the G-T wobble pair, this kinetic effect is localized to the mismatch and immediately adjacent base pairs (91).

Two of the three pyrimidine-pyrimidine mismatches have also been studied by proton NMR methods, but only in very preliminary fashion. Cornelis et al (93) have examined the tridecamer d(A-T-C-C-T-A-T-T-A-G-G-A-T) heteroduplex, which contains a central T-T mismatch (Table 1). Although nuclear Overhauser enhancement methods were not used to assign protons in this system, an imino proton resonance was attributed to the T-T pair. Since this resonance was shifted upfield relative to that of an unstacked thymine residue, it was inferred that the T-T mismatch was intrahelical and stacked. This resonance, and that attributed to imino protons of the adjacent A-T pairs, broadened rapidly with increasing temperature from 0 to 40°C, suggesting exchange with solvent and hence enhanced rates of helix opening in this region of the heteroduplex.

In their analysis of analogues of d(C-G-C-G-A-A-T-T-C-G-C-G), Patel and coworkers (78) have also examined proton NMR spectra of the d(C-G-C-G-A-A-T-T-C-T-C-G) heteroduplex containing two T-C mismatches (Table 1). As in the case of the T-T mispair, these experiments resulted in identification of an imino proton resonance, which was attributed to pairing of T and C bases at low temperature. A tentative pairing scheme, involving a single amino-keto hydrogen bond and a water bridge, was proposed to explain this observation (78; Figure 2). This structure was proposed as tentative since nuclear Overhauser effects could not be used to assign the imino proton due to rapid exchange with solvent (78). Indeed, effects of the T-C mismatch on rates of helix opening are dramatic. Lifetimes of imino protons one to three base pairs removed from the mismatch were found to be reduced by almost an order of magnitude relative to those observed for the parental helix (78). Although the thermal stability of the T-C heteroduplex is similar to that of the related A-C heteroduplex (Table 1), the degree of kinetic destabilization by the T-C pair is much more pronounced (78).

Insertion/Deletion Mispairs

The structural fate of an extra, noncomplementary base within right-handed DNA has also been addressed by NMR methods. Analysis of one-dimensional chemical shifts and nuclear Overhauser effects led Patel et al (94) to conclude that the two extra, noncomplementary adenines within the tridecamer d(C-G-C-A-G-A-A-T-T-C-G-C-G) heteroduplex stack within the helix (Table 1). Intercalation of an extra adenine into the helix has been confirmed by Hare et al (95), who utilized two-dimensional NMR and distance geometry methods to deduce the structure of the related tridecamer d(C-G-C-A-G-A-G-C-T-C-G-C-G) (Table 1). The presence of the stacked but unpaired adenine in such molecules reduces the thermal stability of the helix, but destabilization is somewhat less than that imparted by a G-T pair as judged by T_m measurement (78, 94). Furthermore, a single helix-to-coil transition was observed for the tridecamer d(C-G-C-A-G-A-A-T-T-C-G-C-G) heteroduplex as monitored by chemical shift of nonexchangeable base protons, with the stacked adenine melting as a component of the cooperative unit. Thus, the unpaired adenine base is reasonably stable in the intercalated state. However, the stacked A residue results in longer-range kinetic effects on helix opening than the G-T pair, with several-fold enhanced rates of imino proton exchange being evident three base pairs removed from the mismatch (80).

The conformation assumed by extra, unpaired pyrimidines in model heteroduplexes differs from that of an unpaired adenine. Using NMR methods, Tinoco and colleagues (96) concluded that the unpaired cytosine in the heteroduplex formed by d(C-A-A-A-C-A-A-A-G)•d(C-T-T-T-T-T-T-G) (Table 1) is outside the helix. The observation of a nuclear Overhauser effect between imino protons of the two A-T pairs spanning the extra C indicated that these protons are less than four angstroms apart, much less than the seven angstroms expected if the extra C were to stack between the two A-T pairs. The temperature dependence of the chemical shift of the unpaired C also indicated that the noncomplementary C was unstacked.

A similar conclusion has been drawn by Evans & Morgan (97) concerning the nature of extra, noncomplementary thymine residues. Mixing curves of $d(T-C)_n$ and $d(G-G-A)_n$ demonstrated maximal duplex formation between the copolymers at 1.33 pyrimidine nucleotide equivalents per purine equivalent. Since it was also shown that ultraviolet irradiation of such duplexes led to formation of both T-C and C-C dimers, Evans & Morgan concluded that these copolymers form a duplex in which every other thymine in the pyrimidine strand is extrahelical. It is not clear, however, whether this conclusion can be extrapolated to natural heteroduplexes, which are incapable of slippage.

Implications for Mismatch Recognition

The number of proteins involved in mismatch recognition is not known, but the size of the available set of *E. coli* mismatch repair mutants suggests that this number may be small. This idea is consistent with the finding that the *E. coli mutS* protein can recognize several different mispairs (Ref. 98 and below) and constrains possible mechanisms of recognition.

As discussed above, G-T and A-C transition mismatches are generally the best substrates for mismatch repair. A-A, G-G, and T-T point mispairs are also corrected, as are nonhomologies corresponding to small insertions or deletions. G-A, T-C, and C-C transversion mispairs are usually poor substrates for repair, but exceptional G-A and T-C mismatches are corrected. Comparison of correction efficiencies with heteroduplex stabilities or helix dynamics in the vicinity of the mismatch reveals little correlation between such parameters. G-T and G-A are among the more stable mismatches, while A-C and T-C are among the least stable. Although all mismatches examined result in enhanced rates of helix opening, it is also known that the Watson-Crick base pairs are subject to opening several times per second in vitro (80, 99). It therefore seems unlikely that mere adoption of an extrahelical conformation would be sufficient for recognition of a base associated with the rarely occurring mismatch. However, it has been shown that breathing of the normal DNA helix usually involves opening and closing of individual base pairs, while breathing at a mismatch reflects opening of several base pairs (80). A recognition mechanism based on this distinction can be imagined, but it is difficult to account for mismatch specificity by such a scheme.

With the possible exception of T or C insertions, all mispairs studied can adopt an intrahelical form. Kennard and colleagues (85, 87) and Werntges et al (77) have suggested that this is the conformation recognized during mismatch repair. This is an attractive hypothesis since occurrence of a mispair will result in variation in the number and/or placement of base functional groups available in major and minor grooves (Figure 2). Unfortunately, the structural data available are not yet sufficient to permit elaboration of a recognition scheme that is based on this idea and that can account for the observed specificity of mismatch correction.

MISMATCH CORRECTION IN VITRO

In vitro assays for mismatch repair have utilized heteroduplex constructs containing a mismatched base pair within the recognition sequence for a Type II restriction enzyme (27, 57). Such sites, which are resistant to cleavage by the endonuclease, are rendered sensitive to the enzyme by mismatch correction on the appropriate DNA strand. This method has been used to demonstrate the methyl-directed and methyl-independent pathways in cell-free extracts of *E. coli* (27, 57, 100), and more recently to detect mismatch repair in extracts of yeast (101). Since the *E. coli* methyl-directed pathway has been the most extensively studied in this respect, the remainder of this section is devoted to this system.

Requirements for Methyl-Directed Mismatch Correction in Vitro

The biochemistry of *dam*-directed correction has been addressed using substrates derived from bacteriophage f1 (6.4 kilobases), for which derivatives containing 0, 1, 2, or 4 d(G-A-T-C) sites are available (27, 50, 50a). Heteroduplex repair in this system is dependent on use of concentrated cell extracts, requires ATP, is reduced in the absence of exogenous dNTPs, and as observed in vivo, is directed by the state of methylation of d(G-A-T-C) sequences. Unmethylated heteroduplexes are corrected with little strand preference, repair of hemi-methylated molecules is highly biased to the unmethylated DNA strand, and symmetrical methylation results in substantial loss in substrate activity. Since heteroduplexes lacking a d(G-A-T-C) site are extremely poor substrates for mismatch correction, this sequence must have a direct role in the *dam*-directed reaction (50, 50a). The role of d(G-A-T-C) sequences in correction becomes even more striking when one considers that the state of d(G-A-T-C) methylation can control correction of a mismatch located more than 1000 base pairs distant and that the presence of a single d(G-A-T-C) site within a heteroduplex is sufficient to elicit this effect (27, 50, 50a).

Methyl-directed mismatch correction in *E. coli* extracts is also similar to intracellular repair in its protein requirements. Extracts derived from strains bearing functional defects in *mutH*, *mutL*, *mutS*, or *uvrD* gene products support heteroduplex repair at less than 10% the wild-type rate, but normal levels of mismatch correction can be restored by mixing mutant extracts (27, 100). This complementation method has permitted isolation of the *mut* gene products, the nature of which will be considered below. In addition, the in vitro assay has not only revealed a requirement for the *E. coli* single-strand binding protein (SSB), but has also shown that methyl-directed correction is independent of DNA polymerase I, *recBC* nuclease, and the *recF* gene product (50, 100). In vitro correction also occurs at normal rates in extracts of *dam-4* strains, suggesting that the only involvement of the *dam* methylase in repair is its role in modification of d(G-A-T-C) sites (K. Welsh, P. Modrich, unpublished experiments). This conclusion is in accord with transfection experiments that indicate that heteroduplex correction occurs normally in *dam-3* strains (26), but is in contrast to indirect arguments suggesting that the *dam* methylase may have a direct role in mismatch repair (102).

The complete set of mismatches has not been tested in vitro, but those examined suggest that the specificity of the cell-free system is similar to that deduced on the basis of transfection experiments (49, 61; M. Jones, R. Wagner, M. Radman, personal communication). Mismatches subject to *dam*-directed repair in *E. coli* extracts include two different G-T mispairs, two A-C's, one G-A, one A-A, and one T-T. One T-C mismatch has been found to be a weak substrate, and two G-A's mispairs were not subject to *mutHLS*-dependent correction (100; K. Au, R. Lahue, S.-S. Su, P. Modrich, unpublished).

Proteins Required for dam-*Directed Mismatch Repair*

Of the five proteins implicated in in vitro repair, two have been the subject of extensive study in several laboratories. The *uvrD* gene product has been shown to be DNA helicase II (103–105), an activity that catalyzes the ATP-dependent unwinding of duplex DNA (106, 107). *E. coli* SSB binds tightly and cooperatively to single-stranded DNA and stimulates the action of *E. coli* DNA polymerases II and III (reviewed in Ref. 108).[2]

The *mutH, mutL,* and *mutS* genes of *E. coli* and *Salmonella typhimurium* have been isolated, and in the case of the *E. coli* genes, overproducers have been constructed (98, 100, 109, 110). Availability of these overproducing strains and the in vitro complementation assay for mismatch repair has permitted isolation of near homogeneous, biologically active forms of the *E. coli mutH, mutL,* and *mutS* proteins (50, 98; K. Welsh, P. Modrich, in preparation). Although these three proteins together with DNA helicase II and SSB are not sufficient to mediate mismatch correction in a defined system, examination of the individual *mutH* and *mutS* gene products has suggested functional roles for these proteins in methyl-directed mismatch repair.

Footprinting methods have demonstrated that the purified *mutS* protein [subunit $M_r = 97,000$ (98, 110)] binds to at least some mismatched base pairs (98). Highest affinity was observed for a G-T mismatch, while the protein

[2]As stated in the text, DNA polymerase I is not required for repair of a mismatch by the methyl-directed system. Preliminary experiments suggest that DNA polymerase III may function in this respect (K. Au, R. Lahue, P. Modrich, unpublished). Involvement of the latter activity would be consistent with the requirement for SSB.

displayed lowest affinity for a T-C mispair. Affinities for A-C and G-A mismatches were intermediate. Moreover, affinity of *mutS* protein for the different mispairs was found to parallel the efficiency of their correction as determined by in vitro assay (98; S.-S. Su, P. Modrich, unpublished). This finding suggests that mismatch binding by the protein is significant in the context of the overall repair reaction. This study also revealed an unusual feature of *mutS* footprints: in every case the mispair was found to be acentric within the protected region. Since sequences bounding a mismatch were excluded as source of the effect, it was concluded that *mutS*-DNA interaction involves an asymmetry inherent to a mispair or alternatively, an asymmetry imposed upon the helix by the presence of a mismatch (98).

As in the case of the *mutS* gene product, a simple activity is associated with the *mutH* protein [subunit $M_r = 25,000$ (109)] that can account for its involvement in methyl-directed mismatch repair. This activity is a Mg^{2+}-dependent endonuclease that cleaves 5' to the dG of d(G-A-T-C) sites, generating 5'-phosphoryl and 3'-hydroxyl termini (K. Welsh, A.-L. Lu, P. Modrich, in preparation). Symmetrically methylated d(G-A-T-C) sequences are resistant to attack by this endonuclease, hemimethylated sites are cleaved on the unmethylated strand, and unmethylated sites are usually subject to scission on only one of the two DNA strands. However, this *mutH*-associated activity has several puzzling features. Site-specific hydrolysis at d(G-A-T-C) sequences does not require the presence of a mismatch within the DNA substrate, and the endonuclease activity is extremely weak, with an estimated turnover number of about one scission per hr per mol *mutH* protein. Nevertheless, several arguments indicate that the activity is not a simple contaminant of *mutH* preparations. d(G-A-T-C) cleavage activity copurifies with *mutH* complementing activity through multiple column steps without change in relative specific activities, and the endonuclease activity cannot be resolved from *mutH* protein by electrophoresis (K. Welsh, A.-L. Lu, P. Modrich, in preparation).

If it is assumed that d(G-A-T-C) cleavage activity is mediated by the *mutH* product, then this suggests a role for the protein in strand discrimination during methyl-directed correction. In fact, in vitro analysis of repair DNA synthesis associated with mismatch repair has indicated occurrence of strand scissions on the unmethylated strand in the vicinity of d(G-A-T-C) sites (100), and biological experiments have suggested that the *mutH* protein functions in the strand discrimination stage of the repair reaction (28). Several possible explanations for the peculiar properties of the *mutH*-associated endonuclease can be envisioned. For example, the free form of the protein may exist largely in an inactive conformation, or naked DNA may not represent the true substrate for hydrolysis. Activation of the *mutH*-associated activity in such models would occur subsequent to proper assembly of a set of repair proteins on the heteroduplex. However, direct evidence supporting this sort of idea is not available.

Repair DNA Synthesis Associated with Methyl-Directed Mismatch Correction

In vitro experiments have shown that methyl-directed mismatch correction is accompanied by repair DNA synthesis (27). As judged by several criteria, the majority of this synthesis is associated with heteroduplex repair. Thus, repair DNA synthesis is dependent on the presence of a mismatch and on the state of d(G-A-T-C) methylation. Synthesis on hemi-methylated molecules lacking a mismatch or on fully methylated molecules containing a mispair was only 30% of that on heteroduplexes that were both hemi-methylated and contained a mismatch. Furthermore, synthesis on hemi-methylated DNAs was largely confined to the unmethylated strand and was dependent on functional *mutH, mutL,* and *mutS* gene products (27, 100).

In the concentrated cell extracts required for mismatch correction (27), localization studies demonstrated that repair synthesis was distributed over much of the f1 heteroduplex (100), presumably reflecting large repair tracts of the type observed in vivo (7). However, repair DNA synthesis, which was similarly dependent on the presence of a mismatch, state of DNA methylation, and *mut* gene function, was also observed in dilute extracts that did not support efficient mismatch correction (100). In this case repair tracts were localized to the vicinity of d(G-A-T-C) sites, and it was suggested that these short repair tracts reflected initiation and premature termination of mismatch repair under dilute conditions (100). This localization effect led Lu et al (100) to suggest that *dam*-directed mismatch repair involves scission of the un-methylated strand in the vicinity of a d(G-A-T-C) site.

Models for Methyl-Directed Mismatch Correction

Although information concerning the mechanism of methyl-directed mis-match repair is still quite limited, available information suggests that an understanding of the process will require answers to two questions. Perhaps the simplest deals with site(s) of incision and directionalities of excision and resynthesis. As discussed above, it would appear that heteroduplex repair involves an incision event on the unmethylated strand of a d(G-A-T-C) site. A mechanism in which excision initiates at such a site is particularly attractive since it provides a simple means by which *dam* methylation can govern the strandedness of repair, although it is noteworthy that additional sites for incision have not been excluded. Speculation concerning various possible excision and resynthesis schemes have been described (100) and will not be further belabored here.

The second and more interesting question deals with the nature of signal transduction between a d(G-A-T-C) site and the mismatch provoking repair. That such signal transduction occurs is indicated by several lines of evidence. It is clear that mismatches promote their own repair (7), but it has also been shown that d(G-A-T-C) sequences are directly involved in the correction process (49, 50). Therefore, both elements must be recognized during the course of the reaction. The finding that *mutHLS*-dependent repair DNA synthesis requires both a mismatch and a d(G-A-T-C) sequence that is unmethylated on at least one strand is also in accord with this view (100).

Two types of signal transduction schemes that can account for the known features of mismatch correction are illustrated in Figure 3. For simplicity the mechanisms shown assume incision on the unmethylated strand at a d(G-A-T-C) site followed by an excision or strand displacement reaction that is presumed to occur with fixed directionality. The first class of mechanism, illustrated in examples A and B, is based on DNA transport in the manner of Type I restriction enzymes (111). In mechanism A, the repair system binds to a hemi-methylated d(G-A-T-C) sequence, with the state of methylation imposing an asymmetry on the complex. DNA to a particular side of the complex is then subject to transport through a second binding site. Entry of a mismatch into this site triggers DNA cleavage within the d(G-A-T-C)-binding site, and repair ensues. The difficulty with this mechanism is that directional DNA transport, and hence energy consumption, are unrelated to the occurrence of a mispair.

This problem is resolved in mechanism B, which invokes an analogous transport scheme, but which in this case is provoked by the mismatch and occurs in a bidirectional manner. Provided that the repair complex bound at the mismatch possesses elements of dyad symmetry, it can be seen that d(G-A-T-C) sequences on either side of the mismatch will enter transport sites in different orientations. Thus, cleavage of the unmethylated strand would occur only on that side of the mismatch consistent with the directionality of the excision reaction.

Mechanism C is similar to mechanism B, but differs in the manner of linear signal transduction along the helix. In this scheme the mismatch recognition protein is presumed to possess elements of dyad symmetry that permit it to act as a nucleation site for bidirectional polymerization of a second protein along the helix. If the second protein possesses elements of asymmetry, then it would impose different environments on hemi-methylated d(G-A-T-C) sites on either side of the mispair. As in the case of mechanism B, this would provide the basis for an incision reaction on only one side of the mismatch as necessary for unidirectional excision.

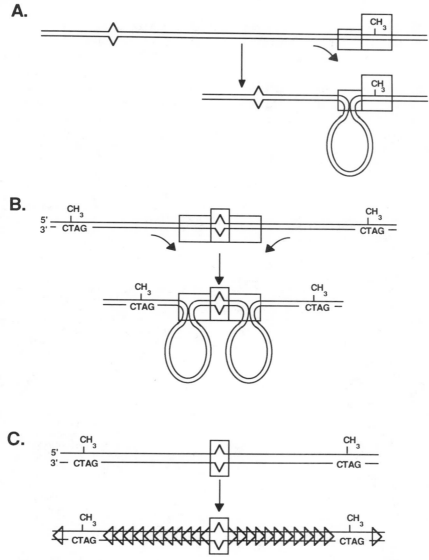

Figure 3 Possible mechanisms for signal transduction along the DNA helix during methyl-directed mismatch correction—A discussion of these mechanisms can be found in the text.

Of course such schemes are merely illustrative and at this point without basis in fact. Establishment of the mechanism of signal transduction in this system must await reconstitution of methyl-directed repair in a defined system so that true intermediates in the process may be identified.

ACKNOWLEDGMENTS

Work in our laboratory was supported by Grant GM23719 from the National Institute of General Medical Sciences. I am indebted to Karin Au, Susanna Clark, Bob Lahue, Mike Su, Brian Terry, Kate Welsh, and Beverly Yashar for many useful comments on the manuscript and for their contributions to the mismatch repair project. I am also grateful to Olga Kennard, Richard Kolodner, Peggy Lieb, Matthew Meselson, Dinshaw Patel, and Miro Radman for their willingness to share unpublished findings and for provision of manuscripts prior to publication.

Literature Cited

1. Holliday, R. 1974. *Genetics* 78:273–87
2. Fox, M. S. 1978. *Ann. Rev. Genet.* 12:47–68
3. Radding, C. M. 1978. *Ann. Rev. Biochem.* 47:847–80
4. White, J. H., Lusnak, K., Fogel, S. 1985. *Nature* 315:350–52
5. Williamson, M. S., Game, J. C., Fogel, S. 1985. *Genetics* 110:609–46
6. Meselson, M. 1987. In *The Recombination of Genetic Material,* ed. K. B. Low. Academic. In press
7. Wagner, R., Meselson, M. 1976. *Proc. Natl. Acad. Sci. USA* 73:4135–39
8. Hare, J. T., Taylor, J. H. 1985. *Proc. Natl. Acad. Sci. USA* 82:7350–54
9. Topal, M. D., Fresco, J. R. 1976. *Nature* 263:285–89
10. Drake, J. W., Baltz, R. H. 1976. *Ann. Rev. Biochem.* 45:11–37
11. Marinus, M. G. 1984. In *DNA Methylation, Biochemistry and Biological Significance,* ed. A. Razin, H. Cedar, A. D. Riggs, pp. 81–109. New York: Springer-Verlag
11a. Radman, M., Wagner, R. 1986. *Ann. Rev. Genet.* 20:523–38
12. Claverys, J.-P., Lacks, S. A. 1986. *Microbiol. Rev.* 50:133–65
13. White, R. L., Fox, M. S. 1975. *Mol. Gen. Genet.* 141:163–71
14. Wildenberg, J., Meselson, M. 1975. *Proc. Natl. Acad. Sci. USA* 72:2202–6
15. Guild, W. R., Shoemaker, N. B. 1976. *J. Bacteriol.* 125:125–35
16. Mejean, V., Claverys, J.-P. 1984. *Mol. Gen. Genet.* 197:467–71
17. Bishop, D. K., Kolodner, R. D. 1986. *Mol. Cell. Biol.* 6:3401–9
18. Miller, L. K., Cooke, B. E., Fried, M. 1976. *Proc. Natl. Acad. Sci. USA* 73:3073–77
19. Abastado, J.-P., Cami, B., Dinh, T. H., Igolen, J., Kourilsky, P. 1984. *Proc. Natl. Acad. Sci. USA* 81:5792–96
20. Folger, K. R., Thomas, K., Capecchi, M. R. 1985. *Mol. Cell. Biol.* 5:70–74
21. Holliday, R. A. 1964. *Genet. Res.* 5:282–304
22. Lacks, S. 1970. *J. Bacteriol.* 101:373–83
23. Tiraby, J.-G., Fox, M. S. 1973. *Proc. Natl. Acad. Sci. USA* 70:3541–45
24. Ephrussi-Taylor, H., Gray, T. C. 1966. *J. Gen. Physiol.* 49 (Pt. 2):211–31
25. Radman, M., Wagner, R. E., Glickman, B. W., Meselson, M. 1980. In *Progress in Environmental Mutagenesis,* ed. M. Alacevic, pp. 121–30. Amsterdam: Elsevier/North Holland
26. Pukkila, P. J., Peterson, J., Herman, G., Modrich, P., Meselson, M. 1983. *Genetics* 104:571–82
27. Lu, A.-L., Clark, S., Modrich, P. 1983. *Proc. Natl. Acad. Sci. USA* 80:4639–43
28. Kramer, B., Kramer, W., Fritz, H.-J. 1984. *Cell* 38:879–87
29. Wagner, R., Dohet, C., Jones, M., Doutriaux, M.-P., Hutchinson, F., Radman, M. 1984. *Cold Spring Harbor Symp. Quant. Biol.* 49:611–15
30. Nevers, P., Spatz, H. 1975. *Mol. Gen. Genet.* 139:233–43
31. Rydberg, B. 1978. *Mutation Res.* 52:11–24
32. Bauer, J., Krammer, G., Knippers, R. 1981. *Mol. Gen. Genet.* 181:541–47
33. Cox, E. C. 1976. *Ann. Rev. Genet.* 10:135–56
34. Marinus, M. G. 1976. *J. Bacteriol.* 128:853–54
35. Lyons, S. M., Schendel, P. F. 1984. *J. Bacteriol.* 159:421–23
36. Szyf, M., Gruenbaum, Y., Urieli-Shoval, S., Razin, A. 1982. *Nucleic Acids Res.* 10:7247–59
37. Marinus, M. G., Morris, N. R. 1973. *J. Bacteriol.* 114:1143–50
38. Herman, G. E., Modrich, P. 1982. *J. Biol. Chem.* 257:2605–12

39. Marinus, M. G., Morris, N. R. 1974. *J. Mol. Biol.* 85:309–22
40. Herman, G. E., Modrich, P. 1981. *J. Bacteriol.* 145:644–46
41. Marinus, M. G., Poteete, A., Arraj, J. A. 1984. *Gene* 28:123–25
42. Szyf, M., Avraham-Haetzni, K., Reifman, A., Shlomai, J., Kaplan, F., et al. 1984. *Proc. Natl. Acad. Sci. USA* 81:3278–82
43. Glickman, B. W., Radman, M. 1980. *Proc. Natl. Acad. Sci. USA* 77:1063–67
44. McGraw, B. R., Marinus, M. G. 1980. *Mol. Gen. Genet.* 178:309–15
45. Krasin, F., Hutchinson, F. 1977. *J. Mol. Biol.* 116:81–98
46. Wang, T. V., Smith, K. C. 1983. *J. Bacteriol.* 156:1093–98
47. Wang, T. V., Smith, K. C. 1986. *J. Bacteriol.* 165:1023–25
48. Mannarelli, B. M., Balganesh, T. S., Greenberg, B., Springhorn, S. S., Lacks, S. A. 1985. *Proc. Natl. Acad. Sci. USA* 82:4468–72
49. Langle-Rouault, F., Maenhaut-Michel, G., Radman, M. 1986. *EMBO J.* 5:2009–13
50. Lahue, R. S., Su, S.-S., Welsh, K., Modrich, P. 1986. *UCLA Symp. Mol. Cell. Biol.* 47: In press
50a. Lahue, R., Su, S.-S., Modrich, P. 1987. *Proc. Natl. Acad. Sci. USA* 84:In press
51. Lieb, M. 1983. *Mol. Gen. Genet.* 191:118–25
52. Lieb, M. 1985. *Mol. Gen. Genet.* 199:465–70
53. Lieb, M., Allen, E., Read, D. 1986. *Genetics* 114:1041–60
54. Coulondre, C., Miller, J. H., Farabaugh, P. J., Gilbert, W. 1978. *Nature* 274:775–80
55. Duncan, B. K., Miller, J. H. 1980. *Nature* 287:560–61
56. Lindahl, T. 1982. *Ann. Rev. Biochem.* 51:61–87
57. Fishel, R. A., Kolodner, R. 1984. *Cold Spring Harbor Symp. Quant. Biol.* 49:603–9
58. Fishel, R. A., Siegel, E. C., Kolodner, R. 1986. *J. Mol. Biol.* 188:147–57
59. Lacks, S. A., Dunn, J., Greenberg, B. 1982. *Cell* 31:327–36
60. Claverys, J.-P., Mejean, V., Gasc, A.-M., Sicard, A. M. 1983. *Proc. Natl. Acad. Sci. USA* 80:5956–60
61. Dohet, C., Wagner, R., Radman, M. 1985. *Proc. Natl. Acad. Sci. USA* 82:503–5
62. Dohet, C., Wagner, R., Radman, M. 1986. *Proc. Natl. Acad. Sci. USA* 83:3395–97
63. Watson, J. D., Crick, F. H. C. 1953. *Nature* 171:964–67
64. Kornberg, A. 1980. *DNA Replication.* San Francisco: Freeman
65. Loeb, L. A., Kunkel, T. A. 1982. *Ann. Rev. Biochem.* 51:429–57
66. Crick, F. H. C. 1966. *J. Mol. Biol.* 19:548–55
67. Lomant, A. J., Fresco, J. R. 1975. *Prog. Nucleic Acid Res. Mol. Biol.* 15:185–218
68. Topal, M. D., Fresco, J. R. 1976. *Nature* 263:289–93
69. Early, T. A., Olmsted, J., Kearns, D. R., Lezius, A. G. 1978. *Nucleic Acids Res.* 5:1955–70
70. Traub, W., Sussman, J. L. 1982. *Nucleic Acids Res.* 10:2701–8
71. Fresco, J. R., Alberts, B. M. 1960. *Proc. Natl. Acad. Sci. USA* 46:311–21
72. Dodgson, J. B., Wells, R. D. 1977. *Biochemistry* 16:2374–79
73. Dodgson, J. B., Wells, R. D. 1977. *Biochemistry* 16:2367–74
74. Shenk, T. E., Rhodes, C., Rigby, P. W. J., Berg, P. 1975. *Proc. Natl. Acad. Sci. USA* 72:989–93
75. Myers, R. M., Lumelsky, N., Lerman, L. S., Maniatis, T. 1985. *Nature* 313:495–98
76. Aboul-Ela, F., Koh, D., Tinoco, I. Jr., Martin, F. H. 1985. *Nucleic Acids Res.* 13:4811–24
77. Werntges, H., Steger, G., Riesner, D., Fritz, H.-J. 1986. *Nucleic Acids Res.* 14:3773–90
78. Patel, D. J., Kozlowski, S. A., Ikuta, S., Itakura, K. 1984. *Fed. Proc.* 43:2663–70
79. Patel, D. J., Kozlowski, S. A., Marky, L. A., Rice, J. A., Broka, C., et al. 1982. *Biochemistry* 21:437–44
80. Pardi, A., Morden, K. M., Patel, D. J., Tinoco, I. 1982. *Biochemistry* 21:6567–74
81. Patel, D. J., Pardi, A., Itakura, K. 1982. *Science* 216:581–90
82. Hare, D., Shapiro, L., Patel, D. J. 1986. *Biochemistry* 25:7445–56
83. Brown, T., Kennard, O., Kneale, G., Rabinovich, D. 1985. *Nature* 315:604–6
84. Kneale, G., Brown, T., Kennard, O., Rabinovich, D. 1985. *J. Mol. Biol.* 186:805–14
85. Kennard, O. 1985. *J. Biomol. Struct. Dyn.* 3:205–26
86. Ho, P. S., Frederick, C. A., Quigley, G. J., van der Marel, G. A., van Boom, J. H., et al. 1985. *EMBO J.* 4:3617–23
87. Hunter, W. N., Brown, T., Anand, N.

N., Kennard, O. 1986. *Nature* 320:552–55

88. Patel, D. J., Kozlowski, S. A., Ikuta, S., Itakura, K. 1984. *Biochemistry* 23:3218–26
89. Sowers, L. C., Fazakerley, G. V., Kim, H., Dalton, L., Goodman, M. F. 1986. *Biochemistry* 25:3983–88
90. Kan, L.-S., Chandrasegaran, S., Pulford, S. M., Miller, P. S. 1983. *Proc. Natl. Acad. Sci. USA* 80:4263–65
91. Patel, D. J., Kozlowski, S. A., Ikuta, S., Itakura, K. 1984. *Biochemistry* 23:3207–17
92. Brown, T., Hunter, W. N., Kneale, G., Kennard, O. 1986. *Proc. Natl. Acad. Sci. USA* 83:2402–6
93. Cornelis, A. G., Haasnoot, J. H. J., den Hartog, J. F., de Rooij, M., van Boom, J. H., Cornelis, A. 1979. *Nature* 281:235–36
94. Patel, D. J., Kozlowski, S. A., Marky, L. A., Rice, J. A., Broka, C., et al. 1982. *Biochemistry* 21:445–51
95. Hare, D., Shapiro, L., Patel, D. J. 1986. *Biochemistry* 25:7456–64
96. Morden, K. M., Chu, Y. G., Martin, F. H., Tinoco, I. 1983. *Biochemistry* 22:5557–63
97. Evans, D. H., Morgan, A. R. 1982. *J. Mol. Biol.* 160:117–22
98. Su, S.-S., Modrich, P. 1986. *Proc. Natl. Acad. Sci. USA* 83:5057–61
99. Mandal, C., Kallenbach, N. R., Eng-lander, S. W. 1979. *J. Mol. Biol.* 135:391–411
100. Lu, A.-L., Welsh, K., Su, S.-S., Modrich, P. 1984. *Cold Spring Harbor Symp. Quant. Biol.* 49:589–96
101. Muster-Nassal, C., Kolodner, R. 1986. *Proc. Natl. Acad. Sci. USA* 83:7618–22
102. Schlagman, S. L., Hattman, S., Marinus, M. G. 1986. *J. Bacteriol.* 165:896–900
103. Kumura, K., Sekiguchi, M. 1984. *J. Biol. Chem.* 259:1560–65
104. Kushner, S. R., Maples, V. F., Easton, A., Farrance, I., Peramachi, P. 1983. *UCLA Symp. Mol. Cell. Biol.* 11:153–59
105. Hickson, I. D., Arthur, H. M., Bramhill, D., Emmerson, P. T. 1983. *Mol. Gen. Genet.* 190:265–70
106. Abdel-Monem, M., Durwald, H., Hoffmann-Berling, H. 1977. *Eur. J. Biochem.* 79:39–46
107. Matson, S. W. 1986. *J. Biol. Chem.* 261:10169–75
108. Chase, J. W., Williams, K. R. 1986. *Ann. Rev. Biochem.* 55:103–36
109. Grafstrom, R. H., Hoess, R. H. 1983. *Gene* 22:245–53
110. Pang, P. P., Tsen, S.-D., Lundberg, A. S., Walker, G. C. 1984. *Cold Spring Harbor Symp. Quant. Biol.* 49:597–602
111. Yuan, R. 1981. *Ann. Rev. Biochem.* 50:285–315

Ann. Rev. Biochem. 1987. 56:467–95
Copyright © 1987 by Annual Reviews Inc. All rights reserved

ALTERNATIVE SPLICING: A UBIQUITOUS MECHANISM FOR THE GENERATION OF MULTIPLE PROTEIN ISOFORMS FROM SINGLE GENES

Roger E. Breitbart, Athena Andreadis, and Bernardo Nadal-Ginard

Laboratory of Molecular and Cellular Cardiology, Howard Hughes Medical Institute, Department of Cardiology, The Children's Hospital, and Department of Pediatrics, Harvard Medical School, Boston, Massachusetts 02115

CONTENTS

PERSPECTIVES AND SUMMARY

The regulated expression of structurally distinct, developmentally regulated, and cell type–specific protein isoforms is a fundamental characteristic of eukaryotic cell differentiation. The molecular mechanisms responsible for generating this protein diversity, although presently poorly understood, might be broadly categorized into two systems: those that select one gene among the members of a multigene family for expression in a particular cell, developmental stage, or physiological condition, as in myosin heavy chain and globin; and those that generate several different proteins from a single gene. The latter comprise DNA rearrangement and alternative RNA splicing, each leading to the differential expression of genomic sequences and so producing multiple protein isoforms from a single gene. DNA rearrangement appears to be restricted to a very limited set of genes for the immunoglobulins and T cell receptors. In contrast, increasing numbers of genes in organisms ranging from *Drosophila* to human, including their RNA and DNA viruses, are now known to use alternative RNA splicing. These genes encode proteins with myriad functions, and are particularly prevalent in muscle. The ubiquity of alternative splicing indicates that it is an important genetic regulatory mechanism in metazoan organisms and their viruses.

The mechanisms involved in the regulation of alternative splicing pathways are presently not understood. In recent years significant advances have been made in the elucidation of the biochemistry of splicing. Neither the experimental data available nor the hypotheses put forward for the removal of introns in constitutively spliced transcripts can account for the cell- and developmental-specific patterns of alternative splicing exhibited by some genes. Primary nucleotide sequences at or around the splice junctions are unlikely to be the only determinants of whether a given exon is constitutively or alternatively spliced since no consistent differences in the sequences of these two types of junctions have been found. Nevertheless, primary transcript sequences acting in *cis* are likely to be involved, particularly in transcripts with different primary structures. Specific *trans*-acting factors are essential for regulated alternative splicing of a single transcript.

In most cases, the pairs of exons that are presently alternatively spliced originated through duplications in ancestral genes. Therefore, alternative splicing could have appeared relatively early in metazoan evolution. The physiological basis for the selective advantage of encoding different protein isoforms in a single gene, rather than in separate genes, is not clearly understood. Beyond its role in increasing the coding power of the genome, the existence of alternative RNA splicing raises important questions concerning the function of different protein isoforms, the evolution of split genes and their proteins, and the mechanism of splicing per se.

MANIFESTATIONS OF ALTERNATIVE SPLICING

In contrast to their prokaryotic counterparts, most eukaryotic protein coding genes contain the sequences present in the corresponding mature mRNA in discontinuous DNA segments (exons) interspersed among sequences (introns) that do not form a part of the mature mRNA. Therefore, the primary transcripts of these genes also contain the sequences corresponding to the introns. These intron sequences are precisely excised by a nuclear multistep process known as pre-mRNA splicing, which represents an additional posttranscriptional level of regulation in eukaryotic gene expression (1). In the majority of instances studied so far, each and every one of the exons present in a gene are incorporated into one mature mRNA through the invariant ligation of consecutive pairs of donor and acceptor splice sites (see below) removing every intron. This type of splicing, termed *constitutive* splicing, yields a single gene product from each transcriptional unit even when its coding sequence is split into as many as 41 exons as is the case for a mammalian myosin heavy chain gene (2). In the context of this high precision, efficiency, and apparent universality of the splicing apparatus in higher eukaryotes, there are instances in which nonconsecutive exons (or splice sites) are joined in the processing of some, but not all, transcripts from a gene. This *alternative* pattern of pre-mRNA splicing can exclude individual exon sequences from the mature mRNA in some transcripts but include them in others (Figure 1). The use of such differential splicing patterns in transcripts from a single gene yields mRNAs with different primary structures. When the exons involved contain translated sequence, these alternatively spliced mRNAs will encode related but distinct protein isoforms.

More than 50 genes are currently known to generate protein diversity through the use of alternative splicing in organisms ranging from *Drosophila* to human (Table 1), including their RNA and DNA viruses. The genes that produce alternatively spliced mRNAs encode proteins with a wide variety of functions and distributions. These products include proteins of the contractile apparatus, extracellular matrix, and cytoskeleton, as well as membrane receptors, peptide hormones, and enzymes of intermediary metabolism and DNA transposition (see references in Table 1). There is every reason to believe that these represent only a minor sample of the extensive use of this mechanism.

The variation and complexity of the alternative splicing patterns exhibited by this heterogeneous group of genes are considerable. The number of different mRNAs, and hence protein isoforms, encoded by a given gene increases exponentially as a function of the number of exons that participate in alternative splicing pathways (see below). The capacity to generate different but closely related protein isoforms by alternative splicing increases significantly the phenotypic variability that can be obtained from single genes such

PATTERNS OF ALTERNATIVE RNA SPLICING

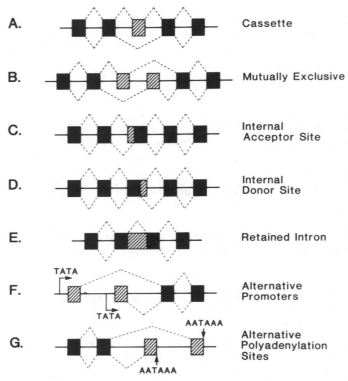

Figure 1 Patterns of alternative splicing. Constitutive exons (black), alternative sequences (striped), and introns (solid lines) are spliced according to different pathways (dotted lines), as described in the text. Alternative promoters (TATA) and polyadenylation signals (AATAAA) are indicated.

as fibronectin (3) or from gene families such as those of the contractile proteins (see 4 for review). The power of this mechanism to generate diversity is further increased when the products of alternatively spliced genes and gene families assemble to form protein complexes, as in the sarcomere (4).

SPLICE SITE SELECTION

The steps involved in the production of a primary gene transcript, including promoter selection, transcription termination, and 3' end formation, have been extensively reviewed (1, 5, 6, and references therein). This transcript, capped and polyadenylated, constitutes the substrate for the splicing system.

It becomes complexed with specific ribonucleoproteins to form a "spliceo-some." Introns are demarcated by invariant consensus sequences at their 5' (donor) and 3' (acceptor) boundaries. The pathway of splicing comprises cleavage at the donor site, formation of the lariat branchpoint, and cleavage at the acceptor site with concomitant ligation of the 5' and 3' exons. Small nuclear RNAs are implicated in the process (see Ref. 1 for review).

The central problem in pre-mRNA splicing, both constitutive and alterna-tive, is the selection of the correct pairs of donor and acceptor sites to be joined. The sequences at the splice sites, though conserved, are multiply repeated elsewhere in the transcript, and under the proper conditions they can become functional as evidenced by the recognition of cryptic equivalents by the splicing complex (7, 8). However, because of the size and lability of mRNA precursors, it has been impossible to investigate rigorously the role of RNA secondary structure in splicing. Tentative results, linking hairpins to splice junction choices, have emerged from studies with viruses (8a–10). In general, the silent phenotype of large intron deletions (11, 12) and the wide variation in intron length argue against strict higher order structure require-ments. On the other hand, recent in vitro and in vivo experiments indicate that sequestering of an exon on a loop of a very stable hairpin leads to its occasional exclusion from the mRNA (13).

Given that most genes contain many introns, and that all splice sites are formally equivalent, it is imperative that the splicing process be limited to the correct donor/acceptor pairs to ensure proper combinations of exons. One solution to this dilemma is that the splicing process be orderly, and that it proceed via a scanning mechanism that starts at one end of the transcript. In the β-globin gene, the first intervening sequence is almost always excised before the second (14). In contrast, the opposite directionality is exhibited in the E2A pre-mRNA of adenovirus (15). Furthermore, partially spliced mole-cules with different combinations of persisting introns accumulate in both the nucleus and in vitro systems (16–19). From these experiments the consensus is that there is no strict 5' to 3' or 3' to 5' order for the removal of multiple introns from pre-mRNA molecule.

The selection among several putative donors or acceptors flanking a single intron is not well understood. Experiments with chimeric constructs that contain tandem duplications of donors and acceptors have given conflicting results (20, 21) and, together with the results obtained with several exon truncation and substitution experiments, have led to the conclusion that exon sequences do not play a major role in splicing (1). This point of view is in contradiction with recent findings. The pattern of splice site selection in alternatively spliced viral pre-mRNAs (22), and possibly in a tropomyosin gene (23), is altered by mutation within an exon. In *cis*-competition assays it has been shown that utilization of 5' and 3' splice sites can be significantly

Table 1 Cellular genes exhibiting alternative RNA splicing

Gene[a]	Origin	Alternative splicing patterns[b]	Regulation[c]	References
Actin 5C	Drosophila	alternative 5'-terminal exons	stochastic	Bond & Davidson (89)
Alcohol dehydrogenase	Drosophila	alternative 5'-terminal exons	development	Benyajati et al (90)
Aldolase A	rat	alternative 5'-terminal exons	tissue	Joh et al (64a)
Amy 1[a] (α-amylase)	mouse	alternative 5'-terminal exons	tissue	Young et al (31)
Antennapedia	Drosophila	alternative 5'-terminal exons		Schneuwly et al (91)
Argininosuccinate synthetase	human	cassette		Freytag et al (92)
Brain gene (product unknown)	rat	not elucidated	stochastic	Tsou et al (93)
c-abl proto-oncogene	mouse	not elucidated		Ben-Neriah et al (94)
c-Ki-ras	human, mouse	cassette		Shimizu et al (95)
				McGrath et al (96)
				Capon et al (97)
				George et al (98)
				McCoy et al (99)
Calcitonin	rat, human	alternative 3'-terminal exons	tissue	Amara et al (68)
				Rosenfeld et al (68a, 68b)
				Jonas et al (69)
αA-Crystallin	mouse, hamster	cassette	stochastic	King & Piatigorsky (81, 100)
				van den Heuvel et al (100a)
Dash	Drosophila	retained intron	stochastic	Telford et al (101)
Dunce[+]	Drosophila	not elucidated	development	Davis & Davidson (102)
EH8	Drosophila	not elucidated	development	Vincent et al (103)
ERCC-1	human	cassette		van Duin et al (104)
γ-Fibrinogen	rat, human	retained intron (alternative polyadenylation site in human)	tissue (preliminary)	Crabtree & Kant (59)
				Fornace et al (60)
				Chung & Davie (61)

Protein	Type of alternative splicing	Regulation	Species	Reference
Fibronectin	internal acceptor sites	tissue	rat	Homandberg et al (62)
	cassette; internal donor and acceptor sites	tissue	human	Schwarzbauer et al (51)
				Tamkun et al (52)
				Kornblihtt et al (53, 54)
				Hynes (3)
FMRFamide neuropeptide	not elucidated		Aplysia	Schaefer et al (105)
Gastrin-releasing peptide	internal donor and acceptor sites		human	Sausville et al (106)
Glucocorticoid receptor	not elucidated		human	Hollenberg et al (107)
Granulocyte colony-stimulating factor	internal donor site		human	Nagata et al (108)
Growth hormone	internal acceptor site		human	DenNoto et al (109)
Histocompatability antigen H-2K	internal acceptor site (exon 8); internal donor and acceptor sites (exon 2)	stochastic	mouse	Kress et al (110)
				Transy et al (111)
Histocompatability antigen Qa/Tla	cassette (exon 7)		mouse	Brickell et al (112)
Histocompatability antigen Q10	cassette (exon 3)		mouse	Lalanne et al (113)
HMG-CoA reductase	internal donor sites		hamster	Reynolds et al (114)
Immunogobulin heavy chains μ, γ, δ, ε	alternative 3'-terminal exons, internal donor site	maturation	mouse	Early et al (70)
				Alt et al (71)
				Maki et al (72)
				Rogers et al (73)
				Moore et al (74)
				Tyler et al (75)
				Yaoita et al (76)
				Perlmutter & Gilbert (77)
Insulinlike growth factor precursor	alternative 3'-terminal exons		human	Rotwein et al (115)
Interleukin-2 receptor	cassette		human	Leonard et al (66, 67)
Myelin basic protein	cassettes (2)	development	mouse	de Ferra et al (55)
				Takahashi et al (56)

Table 1 (*Continued*)

Gene[a]	Origin	Alternative splicing patterns[b]	Regulation[c]	References
Myosin heavy chain	*Drosophila*	cassette	development and tissue	Rozek & Davidson (34, 35) Bernstein et al (36)
Myosin light chain 1/3	rat, mouse, chicken	alternative 5'-terminal exons, mutually exclusive cassettes (1 pair)	development	Periasamy et al (37) Robert et al (38) Nabeshima et al (39) Strehler et al (40)
Myosin alkali light chain	*Drosophila*	cassette	development	Falkenthal et al (79)
Nerve growth factor	mouse	cassette	tissue	Edwards et al (116)
Neural cell adhesion molecule	chicken	not elucidated	tissue	Murray et al (117) Hemperly et al (118)
(2'-5') Oligo A synthetase	human	alternative 3'-terminal exons	tissue	Benech et al (119) Saunders et al (120)
Ovomucoid	chicken	internal donor site	stochastic	Stein et al (121)
P transposable element	*Drosophila*	retained intron	tissue	Laski et al (64)
Plasminogen activator, urokinase-like	porcine	internal donor site		Nagamine et al (122)
Polymeric immunoglobulin receptor	rabbit	cassette		Deitcher & Mostov (123)
Prekininogen (bradykinin)	bovine	internal donor site, alternative 3'-terminal exons		Kitamura et al (124, 125)
Preprotachykinin	bovine	cassette	tissue	Nawa et al (82)
Prolactin	rat	internal acceptor site		Maurer et al (126)
Pro-opiomelanocortin	rat	internal acceptor site		Oates & Herbert (127)
Salivary proline-rich protein	human	retained intron, internal acceptor site		Maeda et al (128)
Serine protease homologue	mouse	internal acceptor site	maturation	Cook et al (129)
T cell marker *Lyt-2*	mouse	not elucidated		Zamoyska et al (130)
T cell receptor T3	human	cassette		Tunnacliffe et al (131)
T cell receptor chain	mouse	cassette		Behlke & Loh (132)

Gene	Species	Splicing pattern	Regulation	References
Thy-1.2 glycoprotein	mouse	alternative 5'-terminal exons		Ingraham & Evans (65)
α-Tropomyosin	rat	cassettes, mutually exclusive cassettes, alternative 3'-terminal exons	tissue	Ruiz-Opazo et al (41); Ruiz-Opazo & Nadal-Ginard (41a); D. Wieczorek et al, submitted
β-Tropomyosin	rat	cassettes, mutually exclusive cassettes, alternative 3'-terminal exons	tissue	Helfman et al (45a)
β-Tropomyosin	human	not elucidated	tissue	MacLeod et al (42)
Tropomyosin I (Tropomyosin 2)	Drosophila	cassette	development and tissue	Basi et al (43); Basi & Storti (44); Karlik & Fyrberg (23)
Tropomyosin II (Tropomyosin 1)	Drosophila	mutually exclusive cassettes (1 pair), alternative 3'-terminal exons	development and tissue	Karlik & Fyrberg (45)
Troponin T (skeletal fast)	rat, quail	combinatorial cassettes (5), mutually exclusive cassettes (1 pair)	development and tissue	Medford et al (46); Breitbart et al (47); Breitbart & Nadal-Ginard (48); Hastings et al (49)
Troponin T (cardiac)	chicken	cassette	development and tissue	Cooper & Ordahl (50)

[a] This list was compiled following an extensive computer-assisted search of the literature and, to the best of the authors' knowledge, was complete at the time of submission. Only those cellular genes in which alternative splice site usage (see text) has been clearly documented by cDNA and/or genomic sequencing, and/or nuclease protection mapping, have been included. Because many genes exhibit more than one pattern of alternative splicing, they are not readily grouped according to mechanism; hence, an alphabetical order is used for simplicity.

[b] Alternative splicing patterns are as described in the text.

[c] Where known, developmental and tissue-specific (including lymphocyte maturation) regulation of splicing is indicated. Documented stochastic alternative splice site usage is also specified.

affected by the sequences within the flanking exons as well as the relative proximity of competing 5' and 3' splice sites (24). Taken together, these results suggest that intron removal is likely to be kinetically or thermodynamically determined as if, contrary to what has been believed, not all donor and acceptor sites are equivalent. Evidence is accumulating (24, 25, and authors' unpublished observations) in support of the notion that combinations between certain donors and acceptors are preferred to others. This phenomenon of exon "compatibility" might play a fundamental role in exon selection in alternatively spliced transcripts and may function in combination with a scanning mechanism within the intron.

Despite the lack of experimental proof, the conclusion that splicing operates by some sort of processive mechanism is almost unavoidable. However, the apparatus may not recognize the intronic primary structure but, rather, a characteristic secondary feature (26). One of the most compelling reasons for a tracking mechanism is the absence of bona fide *trans* splicing (with the possible exception of the spliced common leader of Trypanosome RNAs; 27). Within the spliceosome, linearly contiguous donor and acceptor sites might not be in closer proximity than distant sites in the same or other molecules. The lack of a tracking system would lead to inter- or intramolecular *trans* splicing. The former would yield mRNAs with insertions and deletions very likely to result in nonfunctional proteins; the latter would alter the sequential order of the exons in the mRNA. No such molecules have been detected in vivo or in vitro. Nevertheless, intermolecular splicing can be produced in vitro but is favored by the formation of hybrid duplexes between the two reacting substrates (26, 28).

Whether splicing proceeds via scanning or not, it is clear that correct exon selection does not require the presence of the entire gene: chimeric gene constructs are correctly spliced (29–32). Only in the case of a globin intron engineered in the herpes simplex thymidine kinase gene has intron position in the transcript been correlated with its lack of excisability (33). Yet, if the recent experiments suggesting nonequivalence of different donor and acceptor sites are confirmed (24, 25), results obtained with chimeric genes will have to be interpreted with caution.

PATTERNS OF ALTERNATIVE PRE-mRNA SPLICING

As indicated throughout this review, alternative splicing is a prevalent mechanism for generating isoform diversity that is especially common and elaborate among the contractile protein genes. Differential splicing has been demonstrated in four out of the eight major sarcomeric proteins: myosin heavy chain (34–36), alkali myosin light chain (37–40), tropomyosin (23, 41–45a; D. Wieczorek et al, submitted) and skeletal (46–49) and cardiac (50) troponin

T. Alternative splicing is also particularly well developed and well studied in the extracellular matrix protein fibronectin (3, 51–54). Together, these genes display the full range of alternative splicing patterns described for all the genes in Table 1. These genes encode proteins that are relatively abundant, and that have well-defined biological functions. They can serve as paradigms for the study of the mechanisms involved in alternative splicing in diverse systems. For these reasons, and due to the familiarity of the authors with these genes, they will be used preferentially as examples to illustrate site usage.

Alternatively spliced genes may employ one or more patterns of differential splice site usage in combination with conventional constitutive exons (Table 1 and Figure 1). In principle, a primary gene transcript undergoes alternative RNA splicing if at least one pair of donor and acceptor sites that are joined in the formation of one mRNA, fail to be joined to each other in the formation of another mRNA. Each may remain unspliced, or be spliced instead to an alternative partner. The result, therefore, is at least two mRNAs with different primary sequences. Where that difference involves translated sequences, they will encode structurally distinct protein isoforms.

The alternative usage of splice sites in a primary gene transcript inevitably results either in the removal of potential coding (exon) sequence or in the incorporation of otherwise noncoding (intron) sequence in the mature mRNA. Therefore, the conventional distinction between exons and introns becomes obscured in alternatively spliced genes. Figure 1 diagrams the various patterns of alternative splicing exhibited by the genes in Table 1, and detailed below. The classification system indicated in the figure is convenient for descriptive purposes. In addition, it provides a basis for consideration of the mechanisms responsible for alternative splicing, which may well differ among the particular patterns.

Combinatorial Exons

Many alternatively spliced genes contain entire exons that are individually included or excluded from the mature mRNA (Table 1). When such an exon is retained, the splicing pattern resembles that for a constitutive gene in which all potential coding sequences are incorporated into the mature RNA (Figure 1A). When it is removed, it is presumably carried on a long intron that also contains its flanking noncoding sequences. Such alternatively spliced exons represent discrete *cassettes* of genetic information encoding peptide subsegments that are differentially incorporated into the mature gene product. Several genes contain more than one such cassette (Figure 2, Table 1). If n is the number of exons in a gene that may each be individually included or excluded in a *combinatorial* fashion, then there is a potential for up to 2^n different mRNAs to be encoded by the single gene.

The most remarkable example of this pattern occurs in the fast skeletal

FAST SKELETAL TROPONIN T GENE AND SPLICED mRNAs

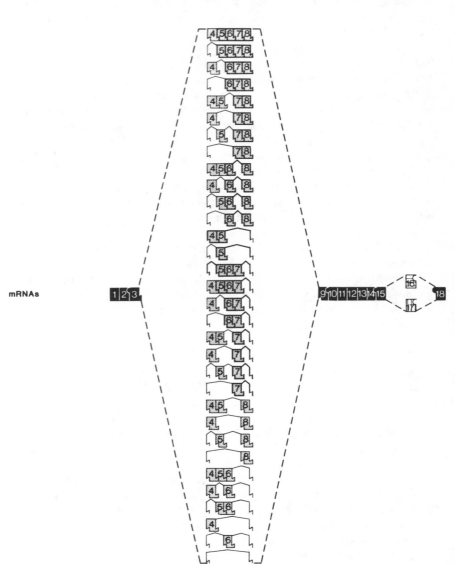

troponin T (TnT) gene, where five consecutive cassettes (exons 4–8) are spliced in a combinatorial fashion to yield as many as 32 (2^5) different sequences in the corresponding domain of the protein, subject to tissue-specific and developmental stage–specific regulation (Figure 2; 47). The combinatorial exons, each 12–18 nucleotides long, are among the smallest reported (the cardiac TnT gene has a six-nucleotide exon; 50). Another example of combinatorial splicing is found in the gene for the myelin basic protein, which contains two nonconsecutive cassettes, each of which is differentially incorporated, generating four (2^2) isoforms (55, 56).

Mutually Exclusive Exons

Several contractile protein genes contain pairs of consecutive cassette exons that are differentially spliced in a *mutually exclusive* fashion (Figure 1B). Here, one member or the other of the pair is invariably spliced into a given mRNA, but the exclusion or inclusion of both simultaneously does not occur. This stands in contrast to the combinatorial patterns described above in which the splicing of each cassette exon would appear to be independent of that of others in the gene. Each mutually exclusive cassette encodes an alternative version of the same protein domain in two distinct mRNAs. In theory, series of three or more mutually exclusive exons are possible, where the number of mRNAs generated would be equal to the number of such exons in the series; however, only mutually exclusive pairs have thus far been identified.

In fast skeletal TnT, the α and β isotype switch exons 16 and 17 are spliced in a mutually exclusive manner to yield two different internal peptides near the carboxyl terminus in different isforms, subject to developmental and tissue-specific control; (Figure 2; 46, 49). Thus, the potential for 32 TnT mRNAs with distinct patterns of the combinatorial exons 4–8 (see above) is doubled to 64 by these exchangeable but mutually exclusive exons (47). No other gene has been reported to exhibit such potential to generate protein diversity by alternative splicing.

Analogous pairs of mutually exclusive cassette exons in the rat α- and β-tropomyosin (TM) genes encode different internal domains that specify, in part, the skeletal and smooth or nonmuscle TM isoforms (Figure 3; 41, 41a,

---

Figure 2 Fast skeletal troponin T gene organization (rat) and 64 possible mRNAs. Exons (black, constitutive; gray, combinatorial; striped, mutually exclusive) are diagrammed to represent their terminal split codons (sawtooth boundaries lie between the first and second nucleotides of the codon, concave/convex between the second and third, and flush boundaries lie between intact codons). The promoter (TATA), polyadenylation signal (AATAAA), untranslated sequences (UT), and encoded amino acids (numbered below exons) are indicated. Each mRNA comprises the constitutive sequences, one of the 32 combinations of exons 4–8, and either exon 16 or 17.

45a). The *Drosophila* TM II gene also has a pair of mutually exclusive exons that are spliced in a regulated manner during development (45). The myosin light chain (MLC) 1/3 gene has mutually exclusive cassettes generating different amino-terminal domains in the two MLC isoforms (Figure 4; 37–40).

To date, mutually exclusive splicing has been detected only in genes for these three contractile proteins, but their analysis provides significant insight into the process of splicing. Where the pair of mutually exclusive exons is located between a common donor and a common acceptor, the sequence between them is a pseudointron; although this pseudointron contains appropriate and functional 5' and 3' splice sites, as evidenced by their capacity to pair with more distant junctions, it is never excised as a precise unit. This feature clearly demonstrates the nonequivalence among splice sites, with dramatically different affinities between otherwise perfectly functional donors and acceptors. In addition, mutually exclusive exons require either a strict directionality of splicing, or another mechanism that otherwise ensures that the joining of one exon of the pair to the common donor is not followed by the joining of the other to the common acceptor. Incorporation of both exons in the mRNA would generate a premature stop codon in the case of TnT (46) and a duplication of 42 amino acids in TM (41a). These mRNAs are either not produced or they do not accumulate to detectable levels.

Internal Donor and Acceptor Sites

Cassette exons are delimited by splice sites that lie at the boundaries between mRNA-coding and noncoding sequence. While the exon itself may or may not be incorporated, its immediate flanking introns are invariably excluded in the splicing process. There are numbers of genes with *internal* alternative splice sites which, in contrast, actually lie entirely within potential coding sequence. Splicing at such a site results in the exclusion of some fraction of an otherwise intact exon (Figure 1C and D). The same pattern may also be viewed conversely as incorporating immediately adjacent intron sequence into the mRNA. It is in such instances that the terms "exon" and "intron" become most ambiguous. This type of splicing is always characterized by the existence of a minimum of two functional donor sites that can be spliced to a single acceptor or vice versa. In this case the potential to generate diversity is equal to the number of duplicated splice sites. In rat fibronectin, for example, a single donor site may be spliced to any of three consecutive acceptor sites (i.e. not separated by additional donor sites) downstream (3, 51, 52). Two of the three isoforms, therefore, contain additional amino acids, representing the differences among the various subunits of the plasma protein. Analogous patterns of internal splicing operate in adenovirus (57) and SV40 (58).

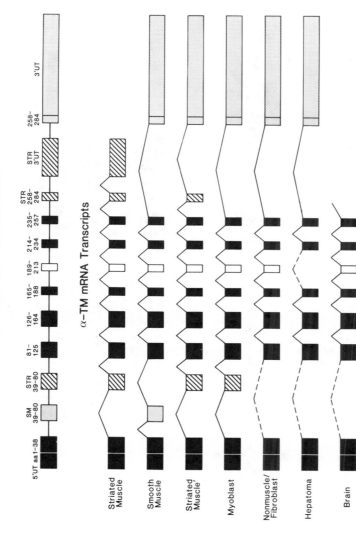

Figure 3 α-Tropomyosin gene organization (rat) and seven alternative splicing pathways. Exons (black, constitutive; gray, smooth muscle–specific; striped, striated muscle–specific; white, variable) are indicated with their encoded amino acids (numbered). Split codons are not diagrammed here. Experimentally documented splicing pathways (solid lines) and others (dotted lines) inferred from nuclease protection mapping are shown. The smooth (SM) and striated (STR) exons encoding amino acid residues 39–80 are mutually exclusive, and there are alternative 3'-terminal exons as well. (From D. Wieczorek et al, submitted.)

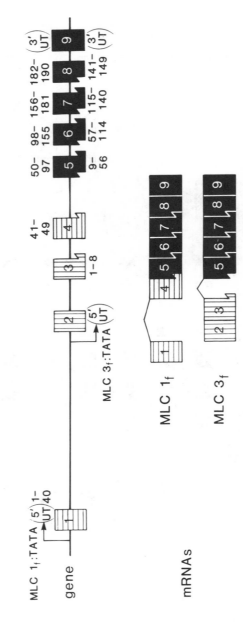

Figure 4 Myosin light chain (MLC) gene organization and two spliced mRNAs (MLC1$_f$ and MLC3$_f$). Constitutive (black), MLC1$_f$-specific (horizontal stripes), and MLC3$_f$ (vertical stripes) are diagrammed to show their split terminal codon structure (see legend to Figure 2) and encoded amino acids (numbered). The two isoform-specific promoters (TATA) are indicated. Exons 3 and 4 are mutually exclusive cassettes.

Retained Introns

Several other genes incorporate intron sequence into mRNA by failing to splice both members of a donor-acceptor pair altogether (Figure 1*E*). The *retained intron* necessarily maintains an intact translational reading frame (see below) and, in effect, creates a longer fusion exon. This is found in the alternative splicing of the rat γ-fibrinogen gene transcripts, which retain the complete seventh intron in 10% of the mRNAs, thereby adding an extra domain to the corresponding protein isoforms (59–62). In the retroviruses, the retention of such putative intron sequence represents the difference between the full-length viral RNA genome and the spliced mRNAs that encode the viral proteins (reviewed in 63). The unspliced transcripts are packaged into mature viral particles. Removal of the first intron from other primary transcripts, however, approximates the downstream *gag, pol,* and *env* translational reading frames to the common leader sequence. Retention of the third intron of the *Drosophila* P transposable element in somatic but not germ line tissues results in a truncated gene product lacking the transposase domain encoded by open reading frame 3 (64). Here, the retained intron presumably carries a new stop codon that precludes translation of that final open reading frame.

These patterns of alternative RNA splicing, which involve splice sites residing within potential coding sequences (Figure 1*C, D,* and *E*), may or may not proceed by a different mechanism than the cassette model (see below). It is clear, however, that the outcome for the mature mRNA and encoded protein sequence is the same whether the alternative sequence arises from some fraction of a longer exon, or as a separate, self-contained cassette. Both mechanisms provide a means for the differential incorporation of particular protein domains. The corollary is that, while the manifestations of alternative splicing may be inferred from comparisons of related mRNA or protein sequences, the particular patterns involved cannot be determined without knowledge of the sequence organization of the particular gene.

Alternative 5'- and 3'-Terminal Exons

In many systems, alternatively spliced mRNAs are associated with different primary transcripts of the same gene (Table 1 and references therein). Heterogeneous sites of transcription initiation and of 3' end formation necessarily result in transcripts having decidedly distinct primary structures. Different promoters and different polyadenylation sites may specify alternative 5' and 3' *terminal exons,* respectively. These exons are not cassettes in the sense defined above, in that each is flanked by a single splice site at its internal boundary alone. In some instances, alternative splice site usage is also manifest in exons internal to these heterogeneous termini.

MULTIPLE PROMOTERS   Two promoters in the mouse α-amylase gene, one active in salivary gland and the other in liver, initiate alternative first exons (Figure 1F; 31). The splicing of the shorter (liver) transcript, is essentially constitutive because the upstream exon 1 donor site is absent from the RNA and, hence, not available to the exon 3 acceptor. In the longer (salivary) transcript, the exon 2 donor site is present but remains unspliced. An entirely analogous pattern of alternative transcription initiation produces different leader exons in the gene for aldolase A (64a) and the *Thy1.2* cell surface glycoprotein (65).

The MLC1/3 gene is one of the best documented examples of mutually exclusive splicing regulated through the alternative use of two different promoters (Figure 4; 40). Vertebrate fast skeletal muscle contains two alkali MLC light chain isoforms (MLC1 and MLC3) that differ from each other at the amino-terminus (exons 1 and 2), and where they have additional isoform-specific internal sequences (exons 4 and 3, respectively). The structural organization of this gene in rat (37), mouse (38), and chicken (39) is identical, with two promoters separated by more than 10 kilobases. The isoform-specific exons are arranged in a sequence that necessitates alternative splicing (Figure 4). In the longer transcript, the exclusion of the MLC3-specific exon 2 can easily be explained by its lack of a consensus splice acceptor sequence. On the other hand, as in the other mutually exclusive exon pairs described, no known mechanism can account for the mutually exclusive use of the MLC3- and MLC1-specific exons 3 and 4 (see above). It is significant, however, that the wide separation of the transcription initiation sites for MLC1 and MLC3 gives rise to two mRNA precusors of significantly different size (20 and 10 kilobases, respectively) and sequence. These differences could play a role in determining the pattern of alternative splicing of isoform-specific exons (see below).

The gene for the human interleukin-2 receptor has two promoters that are quite closely spaced and initiate overlapping first exons (66, 67). Moreover, the fourth exon is a true cassette, encoding an internal protein domain, that is differentially incorporated. In contrast to the patterns for the MLC, however, it is not clear whether the alternative splicing of this fourth exon is linked to one or the other of its two transcripts, or whether it occurs independently. Promoter selection in this gene, therefore, may not determine the internal exon usage.

MULTIPLE POLYADENYLATION SITES   Primary transcripts with 3′ end heterogeneity arise from cleavage at different polyadenylation sites. When each polyadenylation site is located in a separate exon, their differential use will generate alternatively spliced mRNAs (Table 1 and references therein). Here as with alternative promoters, the structural differences in the poly-adenylated transcripts may dictate splicing patterns of internal exons as well.

Among the most remarkable examples of cellular genes combining alternative polyadenylation sites with alternatively spliced exons is the TM gene. In vertebrates, the genes for α- (41, 41a) and β-TM (45a) each encode striated and smooth/nonmuscle protein isoforms. The *Drosophila* TM I (23, 43, 44) and TM II (45) genes also encode multiple isoforms. The rat α-TM gene produces a minimum of seven distinct isoforms. (Figure 3; D. Wieczorek et al, submitted). Interestingly, these isoforms are all transcribed from a promoter that is active in all cell types. Therefore, the production of these different tissue-specific isoforms is regulated not at the transcriptional level but by a combination of alternative splicing and polyadenylation site selection. Both the final and penultimate exons contain polyadenylation sites. In addition, there are two mutually exclusive cassettes lying further upstream, as well as two combinatorial cassettes. When the longer transcript, bearing the more distant carboxyl terminus, is spliced to incorporate only the upstream alternative exon of this pair, it forms the mRNA for the smooth muscle isoform. When the shorter transcript is spliced instead to include only the downstream member of the mutually exclusive pair, it specifies the striated muscle isoform. Thus, for the mRNAs of these two major TM isoforms, there appears to be a clear correlation between a particular splicing pattern a long distance from the 3' end and the primary structure of the transcript produced by the alternative use of two polyadenylation sites separated by approximately six kilobases. Yet, polyadenylation cannot be the sole determinant of splicing pattern since there is a minimum of five different mRNAs that contain the same 5' and 3' ends (Figure 3).

Alternative splicing and polyadenylation are linked in a number of other cellular genes as well (see Table 1). In the calcitonin gene, tissue-specific polyadenylation at sites either in the fourth or the sixth (final) exon results in transcripts spliced to encode the calcitonin hormone in thyroid C cells or the calcitonin gene–related peptide in brain, respectively (68–69). Alternative polyadenylation sites are also associated with the generation of either the secreted or membrane-bound forms of the immunoglobulins by differential splicing (70–77). In several other genes, including that for vimentin (78), different polyadenylation sites are employed in the absence of alternative splice site usage: all splice junctions present in a given transcript are cleaved and religated in a wholly constitutive fashion.

READING FRAME CONSTRAINTS ON ALTERNATIVE SPLICING

A final consideration for the patterns of alternative splicing concerns the translational reading frame. Just as constitutive splicing must produce an intact reading frame, continuous across exon boundaries, so must alternative splicing. Most cassette exons, as well as retained introns, encode translated sequences only. The inclusion or exclusion of the alternative sequences in the mRNA, with the exception of mutually exclusive exons, results in an overall longer or shorter peptide, respectively. In each of three *Drosophila* contractile protein genes, however, namely myosin heavy chain (34–36), alkali MLC (79), and TM I (23, 43, 44), the penultimate exon carries in-frame a translational stop codon. Similarly, the retained third intron of the *Drosophila* P element (64) introduces a new stop codon (see above). In each of these cases, incorporation of the alternative sequence results in an alternative carboxyl terminus, preempting that encoded in the final exon lying further downstream. The final exon then constitutes 3'-untranslated sequence only.

With the exception of these alternative sequences in *Drosophila,* maintenance of the original reading frame requires that the boundaries of a combinatorial cassette exon comprise either intact triplet codons or complementary split codons (47, 80), such that the exon contains an integral multiple of three nucleotides. This must hold, as well, for retained introns, and for the alternative portions of exons with internal splice sites. The five combinatorial exons in the amino terminal region of TnT, for example, share identical split codon symmetry, and each comprises either 12, 15, or 18 nucleotides (Figure 2). Each may, therefore, be individually included or excluded from the mature mRNA with absolute preservation of the translational reading frame.

This symmetry of terminal codons is not necessary for constitutive exons, which are invariably incorporated, nor for cassette exons that are spliced in a mutually exclusive fashion. Indeed, the mutually exclusive exons in the MLC (37–39), TnT (46), and α-TM (41a) genes among others are bounded by asymmetric codons. Both members of each pair of exons, however, have the same asymmetry. Thus, they do not contain integral multiples of three nucleotides, requiring that one or the other, but never both, of each pair be included to preserve the reading frame.

MECHANISMS OF ALTERNATIVE SPLICE SITE SELECTION

Alternative splicing is the most clear-cut demonstration that consensus splice sites are not equivalent. With rare exception (81), canonical donor and acceptor site sequences are uniformly repeated in the introns of alternatively spliced genes. However, these sequence elements that confer the fidelity of splice site selection common to all exons must be reconciled with a means for differentiating those few subject to alternative splicing, but no such mechanism has yet been found. It is likely that the basis for the different affinities among donor and acceptor sites resides, at least in part, in the primary sequence in and around the splice sites. The role of these putative *cis*-acting elements is likely to vary in the different modes of splicing, and preliminary experiments with alternatively spliced exons suggest some testable hypotheses.

Primary and Secondary Structure of RNA Transcripts

The key to regulated alternative splicing must reside in information encoded in the gene transcript *(cis)*, but may require additional control by diffusible *(trans)* factors. In a number of genes, including skeletal (47) and cardiac (50) TnT and preprotachykinin (82), multiple regulated mRNAs are produced from structurally identical primary transcripts, initiated at one promoter and terminated at a single polyadenylation site. The execution of different developmental or tissue-specific alternative splicing patterns in the homogeneous transcripts of these genes, therefore, necessitates control by *trans*-acting factors (47). Possible candidates include small nuclear RNAs, which are developmentally regulated in *Xenopus* (83).

In contrast, several other alternatively spliced genes produce structurally heterogeneous primary transcripts, with either different promoters or polyadenylation sites (see above). The sequences of these different transcripts might each contain enough information to govern its own splicing pathway by *cis* mechanisms alone. The involvement of *trans* factors would be optional. Selection of an alternative promoter or terminator would make the subsequent splicing pattern obligatory. Thus transcriptional regulation at the level of initiation might predetermine the splicing of the two MLC transcripts, which differ in length by ten kilobases, as well as those of other genes with multiple promoters (Table 1 and references therein). Similarly, it is clear that selection between the two polyadenylation sites in the immunoglobulin μ gene, separated by about two kilobases, governs the ensuing splicing pathway leading to either the membrane-bound or secreted protein (84). The same mechanism may reasonably be expected to operate in other alternatively spliced genes with multiple 3'-terminal exons (Table 1 and references therein). In rat α-TM, there is a clear correlation in the major striated and smooth muscle isoforms between polyadenylation site selection and the splicing of the upstream mutually exclusive cassettes (41a). Processing at the 3' end, however, cannot be the sole determinant of the splicing pattern in TM since there are at least five different mRNAs with the same 3' (and 5') end (Figure 3; D. Wieczorek et al, submitted). Therefore, it is likely that this gene is regulated by a combination of the mechanisms outlined above for the TnT and MLC genes. Primary structure differences are probably important but, particularly in nonmuscle cells, not sufficient.

Beyond primary sequence, it is possible to envisage higher order structures of gene transcripts that may juxtapose particular exons and exclude others. In this manner certain splices would be favored by local conformation of the

RNA having the least free energy, either alone or stabilized by additional *cis* intramolecular secondary structures or by interaction with *trans* factors. Regions of the RNA molecule that may be spliced in more than one pattern, then, would assume different conformations subject to such stabilizing influences. These might be quite subtle; for example, fluctuating nuclear concentrations of a factor could modulate the relative affinities of different splicing sites for each other. Moreover, splices executed at one point of the RNA molecule, even in the constitutive regions, might generate conformational changes that affect splicing elsewhere. The findings that a mutant *Neurospora* mitrochondrial pre-mRNA with an anomalous 3' extension fails to be spliced (85), and that mutations within an exon affect alternative splicing patterns both in vitro (24) and in vivo (22, 23), lend validity to the concept that distant intramolecular changes might affect local splicing. The mechanism is not understood, but it could involve the production of different splicing intermediates that, by changing the affinity of the remaining splice sites, has a cascade effect on the remaining exons.

Combinatorial Splicing

Conformational mechanisms are particularly appropriate to the TnT combinatorial exons. Their small sizes may increase the likelihood that the splicing together of two of them, forming a "fusion exon" intermediate, might significantly alter the affinities of the remaining acceptor and donor for the sites of the as yet unspliced exons. Thus, the regulated execution of a single splice could trigger a particular cascade of subsequent splicing events to produce a given combination of these exons in the TnT mRNA.

Transfection assays with in vitro–modified constructs show that the combinatorial exons of TnT can be spliced in myogenic and nonmyogenic cells, but that the patterns are modulated depending on the cell type and the sequence environment in which the exons are placed (authors' unpublished observations). These findings confirm and extend the report that a gene hybrid, containing a portion of the human fibronectin gene that is spliced in two alternative modes in the native transcripts, is similarly spliced when expressed in HeLa cells (32). Taken together, these results indicate that alternative splicing is not wholly dependent on the structural integrity of the pre-mRNA transcript, but that the pattern of utilization might vary significantly depending on the sequence and cell contexts in which the exon is located.

Mutually Exclusive Splicing

All pairs of mutually exclusive exons so far detected appear to represent duplications of ancestral exons (see below). The most striking feature of these exons is the functional incompatibility of the 5' and 3' splice junctions between them. This incompatibility is not dependent on the integrity of the whole transcript, since in vitro–made constructs containing the mutually exclusive exons of TnT, MLC1/3, and α-TM, flanked either by their natural or foreign donor and acceptor sites, always incorporate one of the exons but not both when introduced to a variety of cell types (M. Gallego, C. Smith, and authors' unpublished observations). These results indicate the existence of cis-acting elements capable of conferring the mutually exclusive character to these exons in different cell and sequence environments. Interestingly, analyses of the TnT (48) and α-TM (41a) genomic sequences show that the primary transcripts of these two genes have the capacity to form alternative local stem-and-loop structures that sequester either one or the other alternatively spliced exons. In addition, homologous sequences in rat U3B and U1 small nuclear RNAs can, in theory, compete for base-pairing with the TnT intron sequences in these stems (48). The existence of such alternative structures in the nucleus, stabilized by *trans* factors, could satisfactorily explain why one or the other of the exons is always excluded. Recently, it has been shown that exons and splice sites incorporated in engineered RNA hairpin loops can be bypassed in the splicing process (13). This mode of regulation, combining *cis-* and *trans-*acting elements, remains to be experimentally proved for native gene transcripts.

Where mutually exclusive splicing is executed in transcripts with different 5' or 3' ends, a strict directionality of the process might provide a mechanism for its regulation. The most telling clues come from the behavior of exons 3 and 4 in the MLC gene (Figure 4; 37–39). In addition to their failure to join each other, their donor sites fail to compete for the common exon 5 acceptor in the two transcripts of this gene. These observations suggest that there is strong specificity in the relative affinities of individual donor and acceptor sites. The donor of exon 1 would be highly specific for the acceptor of exon 4, while exon 2 would be highly specific for exon 3. Thus, assuming a 5' to 3' direction of processing, it is argued that the donor site of the 5'-most exon of MLC1 or MLC3 is first activated and spliced to the acceptor site of its corresponding miniexon in a kinetically (or thermodynamically) favored reaction for the particular transcript. This splice results in a longer fusion exon, the donor site of which would be more "compatible" with the common acceptor of exon 5 than would be that of the alternative miniexon of the pair. This differential compatibility might be conferred by local RNA conformation (see above), or by the immediate primary sequence context. Recent results with duplicated donor and acceptor sites (24) and from mutants of the dihydrofolate reductase gene (25), as well as with in vitro–made mutant constructs of the MLC1/3 gene (M. Gallego, B. Nadal-Ginard, unpublished observations) lend support to this notion.

The mechanism postulated is dependent only on promoter selection and *cis*-acting elements to produce the two mRNAs. Similar but slightly modified mechanisms could explain alternative splicing of transcripts with different polyadenylation sites, and might also contribute to the differential splicing of identical transcripts in different cells.

Internal Acceptor and Donor Sites and Retained Introns

There is a significant body of data on natural (7, 25) and in vitro (8) mutations demonstrating that a single base substitution around a splice site can switch the splicing from a normal donor or acceptor to a cryptic one that competes successfully for the splicing factors. It is likely that most cases of alternative splicing involving internal donor or acceptor sites, as well as retained introns, are variations on this theme. In these cases, in-frame alternative splice sites might have evolved randomly, and if the altered proteins were advantageous, they would have been retained. If, through further mutation, a new splice site becomes significantly stronger than the original one, it could be used constitutively, leading to exon/intron junctional "sliding" (86). Here, the coding capacity of the exon is either increased by incorporating intronic sequences or lost when a portion of the exon is converted to intron sequence. The alternative use of these conditional sites might be regulated by specific factors, or by subtle changes in the concentrations of the constitutive splicing factors. Recent results in vitro clearly demonstrate that the concentration of the splicing extract affects the relative splicing efficiency of two splice sites that compete for a single partner (24).

EVOLUTIONARY AND BIOLOGICAL IMPLICATIONS OF ALTERNATIVE SPLICING

The role of introns in gene evolution, as well as the selective advantage that has maintained them in eukaryotes, has remained a puzzle. Most hypotheses envision introns as dispersing agents for exons in a process of exon shuffling (see Ref. 87). According to these theories, exons would be discrete functional units that code for structural or functional protein domains. The role of introns would then be facilitation of this exon exchange and creation of complex genes out of relatively simple precursors. Eukaryotic cells, which must adapt as members of multicellular organisms, would retain introns for future specialization, whereas prokaryotes would streamline their genomes by systematic intron deletion.

There is no evidence that constitutive splicing contributes to gene regulation. Alternative splicing, however, endows a gene with the capacity to generate more than one mRNA and the corresponding proteins. Because this mechanism can operate only in the presence of introns, it may constitute a major selective advantage for split genes in higher organisms.

Splicing appears to be an ancient process, and may have evolved from autocatalyzed RNA ligation (88). The remarkable intra- and interspecies conservation of intron positions among alternatively spliced genes indicates that at least some of their alternative exons were present in ancestral genes, in some cases even at the time of the radiation of the arthropods 600 million years ago (40, 41a, 48, 49). The outstanding question is whether alternative splicing represents a predecessor of constitutive splicing or, rather, a refinement of it, developed through exon and splice site duplication. It is abundantly clear that exon duplication has produced many if not all mutually exclusive pairs, as well as some combinatorial cassettes, which share size, sequence homology, and terminal split codon structure (37, 46–48). Although the evolutionary record does not indicate whether these exons were alternatively spliced in the ancestral genes, it is highly suggestive. In particular, alternative splicing would have to have coevolved with the duplication of exons with asymmetric terminal split codons (see above); otherwise, strictly constitutive inclusion of both would result in frameshifts.

The regulated production of different proteins in eukaryotes has been attributed primarily to promoter selection. The diversity achieved through transcriptional control is greatly augmented, however, by alternative splicing, which generates molecules with additional or interchangeable domains. Moreover, this mechanism permits stringent conservation of constant protein domains simultaneous with relatively rapid divergence of variable protein domains during evolution. In proteins that are multifunctional or operate as parts of a complex, this allows for increased specificity and efficiency of interaction.

Both alternative splicing and, for a limited set of genes, DNA rearrangement, are capable of generating related protein isoforms. Alternative splicing is particularly adept for terminally differentiated, long-lived cells that have lost DNA replicating capacity, and for cells that must respond rapidly to environmental stimuli (for example, nerve and muscle cells). Unlike DNA rearrangement, it has the added advantage of being reversible. At present the functional differences among alternatively spliced isoforms are generally unknown. Two notable exceptions are the immunoglobulins, where alternative heavy chain 3'-terminal domains target antibodies to a particular subcellular location (70–77), and the *Drosophila* P element, in which excision of an intron in the germ line, which is otherwise retained in somatic cells, allows transposase expression (64).

As growing numbers of genes are found to utilize alternative splicing, the importance of gaining detailed knowledge of this mechanism is clear. Structural analyses of these genes, in combination with in vitro experiments, have led to certain testable hypotheses regarding the process of alternative splicing. Further investigation will contribute to a better understanding of splicing, in particular, and gene regulation overall.

ACKNOWLEDGMENTS

The assistance of Rachel Beth Cohen in the preparation of this manuscript is greatly appreciated. This work was supported by grants from the National Institutes of Health, the American Heart Association, and the Muscular Dystrophy Association. R. E. B. is a recipient of a Physician Scientist Award from the National Heart, Lung and Blood Institute. A. A. is a postdoctoral research fellow of the Muscular Dystrophy Association. B. N.-G. is an established investigator of the Howard Hughes Medical Institute.

Literature Cited

1. Padgett, R. A., Grabowski, P. J., Konarska, M. M., Seiler, S., Sharp, P. A. 1986. *Ann. Rev. Biochem.* 55:1119–50
2. Strehler, E. E., Strehler-Page, M.-A., Perriard, J.-C., Periasamy, M., Nadal-Ginard, B. 1986. *J. Mol. Biol.* 190:291–317
3. Hynes, R. O. 1985. *Ann. Rev. Cell Biol.* 1:67–90
4. Nadal-Ginard, B., Breitbart, R. E., Strehler, E. E., Ruiz-Opazo, N., Periasamy, M., et al. 1986. In *Molecular Biology of Muscle Development*, ed. C. Emerson, D. Fischman, B. Nadal-Ginard, M. A. Q. Siddiqui, pp. 387–410. New York: Liss. 957 pp.
5. Leff, S. E., Rosenfeld, M. G., Evans, R. M. 1986. *Ann. Rev. Biochem.* 55:1091–117
6. Platt, T. 1986. *Ann. Rev. Biochem.* 55:339–72
7. Treisman, R., Orkin, S. H., Maniatis, T. 1983. *Nature* 302:591–96
8. Krainer, A. R., Maniatis, T., Ruskin, B., Green, M. R. 1984. *Cell* 36:993–1005
8a. Nussinov, R. 1980. *J. Theor. Biol.* 83:647–62
9. Munroe, S. H. 1984. *Nucleic Acids Res.* 12:8437–56
10. Seiki, M., Hikikoshi, A., Taniguchi, T., Yoshida, M. 1985. *Science* 228:1532–34
11. Wieringa, B., Hofer, E., Weissmann, C. 1984. *Cell* 37:915–25
12. van Santen, V. L., Spritz, R. A. 1985. *Proc. Natl. Acad. Sci. USA* 82:2885–89
13. Solnick, D. 1985. *Cell* 43:667–76
14. Lang, K. M., van Santen, V. L., Spritz, R. A. 1986. *EMBO J.* 4:1991–96
15. Gattoni, R., Stevenin, J., Devilliers, G., Jacob, M. 1978. *FEBS Lett.* 90:318–23
16. Ryffel, G. U., Wyler, T., Muellener, D. B., Weber, R. 1980. *Cell* 19:53–61
17. Roop, D. R., Nordstrom, J. D., Tsai, S.-Y., Tsai, M. J., O'Malley, B. W. 1978. *Cell* 15:651–85
18. Berget, S. M., Sharp, P. A. 1979. *J. Mol. Biol.* 129:547–65
19. Tsai, M.-J., Ting, A. C., Nordstrom, J. L., Zimmer, W., O'Malley, B. W. 1980. *Cell* 22:219–30
20. Kuhne, T., Wieringa, B., Reiser, J., Weissmann, C. 1983. *EMBO J.* 2:727–33
21. Lang, K. M., Spritz, R. A. 1983. *Science* 220:1351–55
22. Somasekhar, M. B., Mertz, J. E., 1985. *Nucleic Acids Res.* 13:5591–609
23. Karlik, C. C., Fyrberg, E. A. 1985. *Cell* 41:57–66
24. Reed, R., Maniatis, T. 1986. *Cell.* 46:681–90
25. Mitchell, P. J., Urlaub, G., Chasin, L. 1986. *Mol. Cell. Biol.* 6:1926–35
26. Solnick, D. 1985. *Cell* 42:157–64
27. Milhausen, M., Nelson, R. G., Sather, S., Selkirk, M., Agabian, N. 1984. *Cell* 38:721–29
28. Konarska, M. M., Padgett, R. A., Sharp, P. A. 1985. *Cell* 42:165–71
29. Chu, G., Sharp, P. A. 1981. *Nature* 289:378–82
30. Kaufman, R. J., Sharp, P. A. 1982. *Mol. Cell. Biol.* 2:1304–19
31. Young, R. A., Hagenbuchle, O., Schibler, U. 1981. *Cell* 23:451–58
32. Vibe-Pedersen, K., Kornblihtt, A. R., Baralle, F. E. 1984. *EMBO J.* 3:2511–16
33. Greenspan, D. S., Weissman, S. M. 1985. *Mol. Cell. Biol.* 5:1894–900
34. Rozek, C. E., Davidson, N. 1983. *Cell* 32:23–34
35. Rozek, C. E., Davidson, N. 1986. *Proc. Natl. Acad. Sci. USA* 83:2128–32
36. Bernstein, A. I., Hansen, C. J., Becker, K. D., Wassenberg, D. R. II, Roche, E. S., et al. 1986. *Mol. Cell. Biol.* 6:2511–19
37. Periasamy, M., Strehler, E. E., Garfin-

kel, L. I., Gubits, R. M., Ruiz-Opazo, N., et al. 1984. *J. Biol. Chem.* 259:13595–604

38. Robert, B., Daubas, P., Akimenko, M.-A., Cohen, A., Garner, I., et al. 1984. *Cell* 39:129–40

39. Nabeshima, Y., Fujii-Kuriyama, Y., Muramatsu, M., Ogata, K. 1984. *Nature* 308:333–38

40. Strehler, E. E., Periasamy, M., Strehler-Page, M.-A., Nadal-Ginard, B. 1985. *Mol. Cell. Biol.* 5:3168–82

41. Ruiz-Opazo, N., Weinberger, J., Nadal-Ginard, B. 1985. *Nature* 315:67–70

41a. Ruiz-Opazo, N., Nadal-Ginard, B. 1987. *J. Biol. Chem.* In press

42. MacLeod, A. R., Houlker, C., Smillie, L. B., Talbot, K., Modi, G., et al. 1985. *Proc. Natl. Acad. Sci. USA* 82:7835–39

43. Basi, G. S., Boardman, M., Storti, R. V. 1984. *Mol. Cell. Biol.* 12:2828–36

44. Basi, G. S., Storti, R. V. 1986. *J. Biol. Chem.* 261:817–27

45. Karlik, C. C., Fyrberg, E. A. 1986. *Mol. Cell. Biol.* 6:1965–73

45a. Helfman, D. M., Cheley, S., Kuismanen, E., Finn, L. A., Yamawaki-Kataoka, Y. 1986. *Mol. Cell Biol.* 6:3582–95

46. Medford, R. M., Nguyen, H. T., Destree, A. T., Summers, E., Nadal-Ginard, B. 1984. *Cell* 38:409–21

47. Breitbart, R. E., Nguyen, H. T., Medford, R. M., Destree, A. T., Mahdavi, V., et al. 1985. *Cell* 41:67–82

48. Breitbart, R. E., Nadal-Ginard, B. 1986. *J. Mol. Biol.* 188:313–23

49. Hastings, K. E. M., Bucher, E. A., Emerson, C. P. Jr. 1985. *J. Biol. Chem.* 260:13699–703

50. Cooper, T. A., Ordahl, C. P. 1985. *J. Biol. Chem.* 260:11140–48

51. Schwarzbauer, J. E., Tamkun, J. W., Lemischka, I. R., Hynes, R. O. 1983. *Cell* 35:421–31

52. Tamkun, J. W., Schwarzbauer, J. E., Hynes, R. O. 1984. *Proc. Natl. Acad. Sci. USA* 81:5140–44

53. Kornblihtt, A. R., Vibe-Pedersen, K., Baralle, F. E. 1984. *Nucleic Acids Res.* 12:5853–68

54. Kornblihtt, A. R., Umezawa, K., Vibe-Pedersen, K., Baralle, F. E. 1985. *EMBO J.* 4:1755–59

55. de Ferra, F., Engh, H., Hudson, L., Kamholz, J., Puckett, C., et al. 1985. *Cell* 43:721–27

56. Takahashi, N., Roach, A., Teplow, D. B., Prusiner, S. B., Hood, L. 1985. *Cell* 42:139–48

57. Nevins, J. R. 1983. *Ann. Rev. Biochem.* 52:441–66

58. Ziff, E. B. 1980. *Nature* 287:491–99

59. Crabtree, G. R., Kant, J. A. 1982. *Cell* 31:159–66

60. Fornace, A. J. Jr., Cummings, D. E., Comeau, C. M., Kant, J. A., Crabtree, G. R. 1984. *J. Biol. Chem.* 259:12826–30

61. Chung, D. W., Davie, E. W. 1984. *Biochemistry* 23:4232–36

62. Homandberg, G. A., Williams, J. E., Evans, D. B., Mosesson, M. W. 1985. *Thromb. Res.* 38:203–9

63. Temin, H. M. 1985. *Mol. Biol. Evol.* 2:455–68

64. Laski, F. A., Rio, D. C., Rubin, G. M. 1986. *Cell* 44:7–19

64a. Joh, K., Arai, Y., Mukai, T., Hori, K. 1986. *J. Mol. Biol.* 190:401–10

65. Ingraham, H. A., Evans, G. A. 1986. *Mol. Cell. Biol.* 6:2923–31

66. Leonard, W. J., Depper, J. M., Crabtree, G. R., Rudikoff, S., Pumphrey, J., et al. 1984. *Nature* 311:626–31

67. Leonard, W. J., Depper, J. M., Kanehisa, M., Kronke, M., Peffer, N. J., et al. 1985. *Science* 230:633–39

68. Amara, S. G., Jonas, V., Evans, R. M. 1982. *Nature* 298:240–44

68a. Rosenfeld, M. G., Mermod, J.-J., Amara, S. G., Swanson, L. W., Sawchenko, P. E., et al. 1983. *Nature* 304:129–35

68b. Rosenfeld, M. G., Amara, S. G., Evans, R. M. 1984. *Science* 225:1315–20

69. Jonas, V., Lin, C. R., Kawashima, E., Semon, D., Swanson, L. W., et al. 1985. *Proc. Natl. Acad. Sci. USA* 82:1994–98

70. Early, P., Rogers, J., Davis, M., Calame, K., Bond, M., et al. 1980. *Cell* 20:313–19

71. Alt, F. W., Bothwell, A. L. M., Knapp, M., Siden, E., Mather, E., et al. 1980. *Cell* 20:293–301

72. Maki, R., Roeder, W., Traunecker, A., Sidman, C., Wabi, M., et al. 1981. *Cell* 24:353–65

73. Rogers, J., Choi, E., Souza, L., Carter, C., Word, C., et al. 1981. *Cell* 26:19–28

74. Moore, K. W., Rogers, J., Hunkapiller, T., Early, P., Nottenburg, C., et al. 1981. *Proc. Natl. Acad. Sci. USA* 78:1800–4

75. Tyler, B. M., Cowman, A. F., Adams, J. M., Harris, A. W. 1981. *Nature* 293:406–8

76. Yaoita, Y., Kumagai, Y., Okumura, K., Honjo, T. 1982. *Nature* 297:697–99

77. Perlmutter, A. P., Gilbert, W. 1984. *Proc. Natl. Acad. Sci. USA* 81:7189–93

78. Zehner, Z. E., Paterson, B. M. 1983. *Proc. Natl. Acad. Sci. USA* 80:911–15

79. Falkenthal, S., Parker, V. P., Davidson,

N. 1985. *Proc. Natl. Acad. Sci. USA* 82:449–53

80. Sharp, P. A. 1980. *Cell* 23:643–46
81. King, C. R., Piatigorsky, J. 1983. *Cell* 32:707–12
82. Nawa, H., Kotani, H., Nakanishi, S. 1984. *Nature* 312:729–34
83. Forbes, D. J., Kirschner, M. W., Caput, D., Dahlberg, J. E., Lund, E. 1984. *Cell* 38:681–89
84. Danner, D., Leder, P. 1985. *Proc. Natl. Acad. Sci. USA* 82:8658–62
85. Garriga, G., Bertrand, H., Lambowitz, A. M. 1984. *Cell* 36:623–34
86. Craik, C. S., Rutter, W. J., Fletterick, R. 1983. *Science* 220:1125–29
87. Gilbert, W. 1985. *Science* 228:823–24
88. Cech, T. R., Bass, B. L. 1986. *Ann. Rev. Biochem.* 55:599–629
89. Bond, B. J., Davidson, N. 1986. *Mol. Cell. Biol.* 6:2080–88
90. Benyajati, C., Spoerel, N., Haymerle, H., Ashburner, M. 1983. *Cell* 33:125–33
91. Schneuwly, S., Kuroiwa, A., Baumgartner, P., Gehring, W. J. 1986. *EMBO J.* 5:733–39
92. Freytag, S. O., Beaudet, A. L., Bock, H. G. O., O'Brien, W. E. 1984. *Mol. Cell. Biol.* 4:1978–84
93. Tsou, A. P., Lai, C., Danielson, P., Noonan, D. J., Sutcliffe, J. G. 1986. *Mol. Cell. Biol.* 6:768–78
94. Ben-Neriah, Y., Bernards, M. P., Paskind, M., Daley, G. Q., Baltimore, D. 1986. *Cell* 44:577–86
95. Shimizu, K., Birnbaum, D., Ruley, M. A., Fasano, O., Suard, Y., et al. 1983. *Nature* 304:497–500
96. McGrath, J. P., Capon, D. J., Smith, D. H., Chen, E. Y., Seeburg, P. H., et al. 1983. *Nature* 304:501–6
97. Capon, D. J., Seeburg, P. H., McGrath, J. P., Hayflick, J. S., Edman, U., et al. 1983. *Nature* 304:507–13
98. George, D. L., Scott, A. F., Trusko, S., Glick, B., Ford, E., et al. 1985. *EMBO J.* 4:1199–204
99. McCoy, M. S., Bargmann, C. I., Weinberg, R. A. 1984. *Mol. Cell. Biol.* 4:1577–82
100. King, C., Piatigorsky, J. 1984. *J. Biol. Chem.* 259:1822–26
100a. van den Heuvel, R., Hendriks, W., Quax, W., Bloemendal, H. 1985. *J. Mol. Biol.* 185:273–84
101. Telford, J., Burckhardt, J., Butler, B., Pirrotta, V. 1985. *EMBO J.* 4:2609–16
102. Davis, R. L., Davidson, N. 1986. *Mol. Cell. Biol.* 6:1464–70
103. Vincent, A., O'Connell, P., Gray, M. R., Rosbash, M. 1984. *EMBO J.* 3:1003–13

104. van Duin, M., de Wit, J., Odijk, H., Westerveld, A., Yasui, A., et al. 1986. *Cell* 44:913–24
105. Schaefer, M., Picciotto, M. R., Kreiner, T., Kaldany, R.-R., Taussig, R., et al. 1985. *Cell* 41:457–67
106. Sausville, E. A., Lebacq-Verheyden, A. M., Spindel, E. R., Cuttitta, F., Gadzar, A. F., et al. 1986. *J. Biol. Chem.* 261:2451–57
107. Hollenberg, S. M., Weinberger, C., Ong, E. S., Cerelli, G., Oro, A., et al. 1985. *Nature* 318:635–41
108. Nagata, A., Tsuchiya, M., Asano, S., Yamamoto, O., Hirata, Y., et al. 1986. *EMBO J.* 5:571–81
109. DeNoto, F. M., Moore, D. D., Goodman, H. M. 1981. *Nucleic Acids Res.* 9:3719–30
110. Kress, M., Glaros, D., Khoury, G., Jay, G. 1983. *Nature* 306:602–4
111. Transy, C., Lalanne, J.-L., Kourilsky, P. 1984. *EMBO J.* 3:2383–86
112. Brickell, P. M., Latchman, D. S., Murphy, D., Willison, K., Rigby, W. J. 1983. *Nature* 306:756–60
113. Lalanne, J.-L., Transy, C., Guerin, S., Darche, S., Meulien, P., et al. 1985. *Cell* 41:469–78
114. Reynolds, G. A., Goldstein, J. L., Brown, M. S. 1985. *J. Biol. Chem.* 260:10369–377
115. Rotwein, P., Pollock, K. M., Didier, D. K., Krivi, G. G. 1986. *J. Biol. Chem.* 261:4828–32
116. Edwards, R. H., Selby, M. J., Rutter, W. J. 1986. *Nature* 319:784–87
117. Murray, B. A., Hemperly, J. J., Prediger, E. A., Edelman, G. M., Cunningham, B. A. 1986. *J. Cell Biol.* 102:189–93
118. Hemperly, J. J., Murray, B. A., Edelman, G. M., Cunningham, B. A. 1986. *Proc. Natl. Acad. Sci. USA* 83:3037–41
119. Benech, P., Mory, Y., Revel, M., Chebath, J. 1985. *EMBO J.* 4:2249–56
120. Saunders, M. E., Gewert, D. R., Tugwell, M. E., McMahon, M., Williams, B. R. G. 1985. *EMBO J.* 4:1761–68
121. Stein, J. P., Catterall, J. F., Kristo, P., Means, A. R., O'Malley, B. W. 1980. *Cell* 21:681–87
122. Nagamine, Y., Pearson, D., Grattan, M. 1985. *Biochem. Biophys. Res. Commun.* 132:563–69
123. Deitcher, D. L., Mostov, K. E. 1986. *Mol. Cell. Biol.* 6:2712–15
124. Kitamura, N., Takagaki, Y., Furuto, S., Tanaka, T., Nawa, H., et al. 1983. *Nature* 305:545–49
125. Kitamura, N., Kitagawa, H., Fukushima, D., Takagaki, Y., Miyata, T., et al. 1985. *J. Biol. Chem.* 260:8610–17

126. Maurer, R. A., Erwin, C. R., Donelson, J. E. 1981. *J. Biol. Chem.* 256:10524–28

127. Oates, E., Herbert, E. 1984. *J. Biol. Chem.* 259:7421–25

128. Maeda, N., Kim, H. S., Azen, E. A., Smithies, O. 1985. *J. Biol. Chem.* 260:11123–30

129. Cook, K. S., Groves, D. L., Min, H. Y., Spiegelman, B. M. 1985. *Proc. Natl. Acad. Sci. USA* 82:6480–84

130. Zamoyska, R., Vollmer, A. C., Sizer, K. C., Liaw, C. W., Parnes, J. R. 1985. *Cell* 43:153–63

131. Tunnacliffe, A., Sims, J. E., Rabbitts, T. H. 1986. *EMBO J.* 5:1245–52

132. Behlke, M. A., Loh, D. Y. 1986. *Nature* 322:379–82

Ann. Rev. Biochem. 1987. 56:497–534

INHIBITORS OF THE BIOSYNTHESIS AND PROCESSING OF N-LINKED OLIGOSACCHARIDE CHAINS

Alan D. Elbein

Department of Biochemistry, The University of Texas Health Science Center, San Antonio, Texas 78284

CONTENTS

PERSPECTIVES AND SUMMARY

In 1913, Paul Ehrlich wrote, "Parasites possess a whole series of chemorecep-tors which differ specifically from each other. Now if we were to succeed in discovering amongst these a receptor which was not represented in the organs

497

of the host, we would have the possibility of constructing an ideal medicament by selecting a haptophore group which fits exclusively this particular receptor of the parasite. A medicament provided with such a haptophore group would be entirely innocuous because it is not anchored by the organs, it would, however strike the parasites with full force and, in this sense, correspond to the immune substances which in the manner of magic bullets seek out the enemy." Although most of the inhibitors described in this chapter are not useful as chemotherapeutic agents, nevertheless the goal of having specific agents or "magic bullets" applies here also. That is, the more specific the site of action of a compound is, the more useful it is likely to be as a biochemical tool for studying functions of molecules or pathways.

The oligosaccharide chains of the N-linked glycoproteins are believed to confer biological specificity at the cell surface where they may be involved in cell-cell adhesion, differentiation, recognition, regulation, modulation of protein receptors, and so on. Thus, it is not surprising that there is such great interest in compounds that can prevent the glycosylation of N-linked glycoproteins or cause alterations in the structure of the carbohydrate chains. The biosynthesis of the N-linked glycoproteins involves a number of metabolic pathways, and inhibition of any of these pathways may affect the final product. For example, the initial biosynthesis of the "core" oligosaccharide involves the transfer of sugars to the lipid carrier, dolichyl-P, to form a large oligosaccharide-lipid, which ultimately donates oligosaccharide to protein. Inhibition of the transfer of GlcNAc-1-P to the lipid carrier by antibiotics such as tunicamycin can prevent glycosylation of the protein, whereas inhibition of other reactions in the formation of lipid-linked saccharides can modify the oligosaccharide structure. Alternatively, inhibition of the synthesis of the carrier, dolichyl-P, by agents such as compactin will also prevent glycosylation of the protein. Another way to prevent glycosylation is to modify the amino acids in Asn-X-Ser (Thr), the tripeptide that carries the carbohydrate chain. Amino acid analogues such as β-hydroxynorvaline and fluoroasparagine are effective in this regard. Once the "core" oligosaccharide is transferred to protein, the carbohydrate chain is subjected to a variety of processing reactions to produce the typical high-mannose, hybrid, and complex types of oligosaccharides. Several plant alkaloids, and sugar analogues with nitrogens in the ring instead of oxygen, block the processing pathway at various stages, giving rise to modified oligosaccharide structures. Finally, the transport of glycoproteins from the endoplasmic reticulum (ER) through the various Golgi stacks is inhibited by certain ionophores such as monensin, and this effectively stops terminal glycosylation by preventing exposure of glycoprotein to those glycosyl transferases. This review briefly covers these various means of interfering with N-linked glycosylation.

INTRODUCTION

Several recent reviews have nicely covered various aspects of the structure and biosynthesis of the oligosaccharide chains of the N-linked or asparagine-linked glycoproteins (1–5). Therefore this introduction only breifly discusses the major pathways that participate in the assembly of the complex, hybrid, and high-mannose types of oligosaccharides. Inhibitors of N-linked oligosaccharide formation (i.e. glycosylation inhibitors and inhibitors of glycoprotein processing) have not been covered previously in this series. However, several reviews (4, 6–9) have dealt with various aspects of these inhibitors.

The biosynthesis of the N-linked oligosaccharide involves the participation of lipid-linked saccharide intermediates as shown in Figure 1. In this pathway, dolichyl-P serves as the carrier, and the sugars GlcNAc, mannose, and glucose are transferred from their sugar nucleotide derivatives, or from lipid-linked monosaccharides, to produce the lipid-linked oligosaccharide $Glc_3Man_9(GlcNAc)_2$-PP-dolichol. In the final step of this pathway, the "core" oligosaccharide is transferred to protein while the protein is being synthesized on membrane-bound polysomes. The first step in the lipid-linked saccharide pathway, i.e. the transfer of GlcNAc-1-P to dolichyl-P, is inhibited by

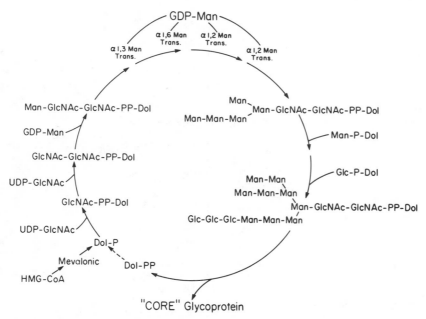

Figure 1 Reactions of the lipid-linked saccharide pathway leading to the formation of $Glc_3Man_9(GlcNAc)_2$-PP-dolichol.

tunicamycin, thus preventing the formation of any lipid intermediates. Figure 1 also shows that dolichyl-P is formed via the conversion of acetate to mevalonic acid, and thus its synthesis is apparently regulated by the enzyme hydroxymethylglutaryl CoA reductase. This enzyme is inhibited by compactin.

Once glycosylation of the protein has occurred, the oligosaccharide portion is modified by a series of glycosidases and glycosyl transferases, as shown in Figure 2, to give rise to various high-mannose, hybrid, and complex types of oligosaccharides. Two different glucosidases (I and II) that are located in the endoplasmic reticulum remove all three glucose residues to produce a glycoprotein with a $Man_9(GlcNAc)_2$ structure. The route from high-mannose structures to hybrid and complex chains involves the removal of 4–6 mannoses by specific α-mannosidases that reside in the ER and Golgi apparatus, and addition of GlcNAc and other sugars by Golgi-bound glycosyl transferases to produce an array of oligosaccharide structures. A number of glycosidase inhibitors are described that modify the processing pathway.

Processing of Glycoproteins

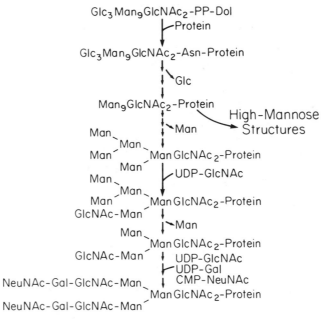

Figure 2 Reactions involved in the processing of the oligosaccharide chains of the N-linked glycoproteins.

INHIBITORS OF LIPID-LINKED SACCHARIDE FORMATION

Tunicamycin and Related Antibiotics

Tunicamycin was first identified and isolated from *Streptomyces lysosuperificus* by Takatsuki et al (10), by virtue of the fact that it inhibited replication of a number of enveloped viruses. Since initial work on the mechanism of action suggested that the antibiotic interfered with the formation of viral and cellular surface coats, it was given the name tunicamycin after the Latin word "tunica," for coats. A comprehensive survey of the literature on tunicamycin has been published (11), as well as a monograph on this antibiotic (12).

Tunicamycin is a nucleoside antibiotic whose structure was elucidated by the elegant studies of Tamura and associates (12). This compound is shown in Figure 3. The antibiotic is composed of uracil, a fatty acid, and two glycosidically linked sugars. The sugars are N-acetylglucosamine and an unusual 11-carbon aminodeoxydialdose, called tunicamine. The tunicamine is attached to uracil in an N-glycosidic bond and is itself substituted at two positions. At the anomeric carbon, the GlcNAc is linked in an O-glycosidic bond, while a long-chain fatty acid is bound in amide linkage to the amino group (13). Using high performance liquid chromatography (HPLC) with a reversed-phase column, the antibiotic was separated into as many as 10 different components that had molecular weights ranging from 802 to 858 (14–16). These differences in molecular mass were shown to be attributable to differences in the fatty acid components, which varied in chain length from C-13 to C-17, and in structure from normal to branched and from saturated to unsaturated. Recent studies dealing with the chemical synthesis of tunicamycin have made excellent progress. Initially, tunicamine was synthesized by the addition of a nitro sugar to a sugar aldehyde using KF as the catalyst, and was then condensed with uracil to form tunicaminyl uracil (17). The tunicaminyl uracil was then condensed with a GlcNAc derivative and fatty acid to form one of the tunicamycin homologues, tunicamycin V, formerly called tunicamycin A (18).

Five other antibiotics are reported to be structurally related to tunicamycin. These antibiotics are known as streptovirudin (16, 19, 20), mycospocidin (21), antibiotic 24010 (22), antibiotic MM 19290 (23), and corynetoxin (24, 25). All of these compounds have the same general structure as tunicamycin, but they differ in the nature of the fatty acid component. Thus, the streptovirudins have shorter chain fatty acids as evidenced by the fact that their molecular weights range from 790 to 816 (20). On the other hand, mycospocidin appears to contain components identical to those of tunicamycin, since an HPLC comparison of the two gave the same series of peaks (21). The

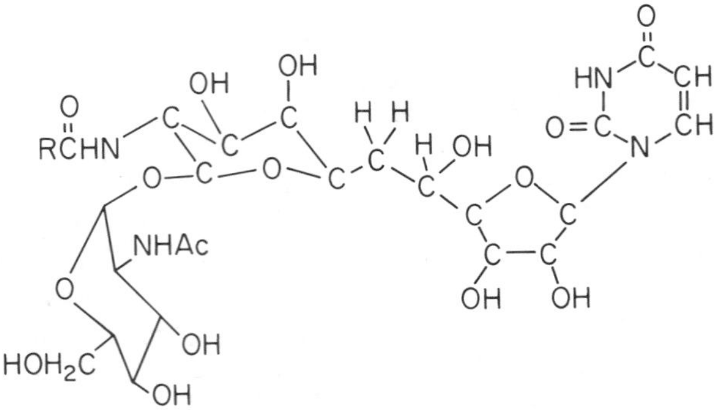

$$
\begin{array}{ll}
\text{I} : \text{R} = \text{CH}_3)_2\text{CH}(\text{CH}_2)_7\text{CH}=\text{CH}- & \text{VI} : \text{R} = (\text{CH}_3)_2\text{CH}(\text{CH}_2)_{11}- \\
\text{II} : \quad (\text{CH}_3)_2\text{CH}(\text{CH}_2)_8\text{CH}=\text{CH}- & \text{VII} : \quad (\text{CH}_3)_2\text{CH}(\text{CH}_2)_{10}\text{CH}=\text{CH}- \\
\text{III} : \quad \text{CH}_3(\text{CH}_2)_{10}\text{CH}=\text{CH}- & \text{VIII} : \quad \text{CH}_3(\text{CH}_2)_{12}\text{CH}=\text{CH}- \\
\text{IV} : \quad \text{CH}_3(\text{CH}_2)_{11}\text{CH}=\text{CH}- & \text{IX} : \quad \text{CH}_3(\text{CH}_2)_{13}\text{CH}=\text{CH}- \\
\text{V} : \quad (\text{CH}_3)_2\text{CH}(\text{CH}_2)_9\text{CH}=\text{CH}- & \text{X} : \quad (\text{CH}_3)_2\text{CH}(\text{CH}_2)_{11}\text{CH}=\text{CH}-
\end{array}
$$

Figure 3   Structure of tunicamycin and its various analogues.

corynetoxins represent an interesting series of tunicamycin-related compounds, since they have been implicated as the causative agent of ryegrass toxicity. In this condition, the developing seedheads of annual ryegrass may be invaded by a nematode, causing a gall to be formed. Some of these nematodes carry the bacterium, *Corynebacterium rathayi*, which produces the glycolipid(s) referred to as corynetoxin. These compounds are apparently highly toxic to sheep that ingest this infected ryegrass (24, 25). They are all potent inhibitors of N-linked glycosylation in vivo and in vitro.

The site of action of tunicamycin was initially demonstrated by Tkacz & Lampen using a microsomal enzyme preparation from calf liver. This antibiotic inhibited the first step in the lipid-linked saccharide pathway, i.e. the transfer of GlcNAc-1-P from UDP-GlcNAc to dolichyl-P to form dolichyl-PP-GlcNAc (26). Tunicamycin had the same site of action in membrane preparations of plants (27) and in chick oviduct microsomes (28), but the enzyme that adds the second GlcNAc to form dolichyl-PP-GlcNAc-GlcNAc was not affected by this antibiotic (29), nor were the GlcNAc transferases that add terminal GlcNAc residues to the mannose chains (28). In addition, tunicamycin did not inhibit the phospho-N-acetylglucosamine transferase that adds GlcNAc-1-P, from UDP-GlcNAc, to the terminal mannose residues on the high-mannose oligosaccharides of lysosomal hydrolases (30). Several microbial enzymes that catalyze GlcNAc-1-P (or closely related sugar phosphates) transfer to polyprenyl-P are also inhibited by tunicamycin. These include enzymes involved in the formation of undecaprenyl-PP-N-acetylmuramoyl-pentapeptide (31) and undecaprenyl-PP-GlcNAc (32), but not undecaprenyl-PP-galactose (11) or undecaprenyl-PP-N-acetylgalactosamine (33). Thus, tunicamycin appears to affect phosphotransferases that catalyze the translocation of an N-acetylhexosamine-1-P from a UDP-precursor to a polyprenyl-P. However, the sugar must be an N-acetylglucosamine, or chirally related derivative.

The mechanism of action of tunicamycin was examined with the solubilized GlcNAc-1-P transferase from pig aorta (34) or hen oviduct (35), and with the membrane-bound enzyme from chick embryo (36). With the aorta enzyme, competitive inhibition could not be demonstrated, presumably because of a very strong affinity of tunicamycin for the transferase (34). However, with streptovirudin, which has a lower affinity for the GlcNAc-1-P transferase, it was possible to reverse the inhibition with UDP-GlcNAc (37). Addition of UDP-GlcNAc, but not dolichyl-P, to the solubilized oviduct enzyme protected it from inactivation by tunicamycin (35). The suggestion was presented that tunicamycin acted as a tight-binding, reversible inhibitor and might be a substrate-product transition state analogue. Kinetic analysis with the transferase gave a K_m for UDP-GlcNAc of 2.9×10^{-6}M and an apparent K_i for tunicamycin of 5×10^{-8}, (36).

As indicated earlier, tunicamycin is produced as a complex of analogous compounds that differ only in the structure of the fatty acid moiety. The various homologues were separated by HPLC in several laboratories (14–16), and the individual components were compared in terms of their ability to inhibit the formation of dolichyl-PP-GlcNAc, in vitro and in vivo, and also in their ability to inhibit protein synthesis (see below). In one study, 17 tunicamycin peaks were isolated, and all of these were potent inhibitors of the solubilized GlcNAc-1-P transferase (16). Thus, 50% inhibition was achieved at 2–8 ng/ml of antibiotic. However, the homologues with longer fatty acid chains were generally better inhibitors, although other aspects of the fatty acid structure also appear to be involved. Another study also found differences between the various components, but these were not directly correlated with the length of the fatty acid component (38). One difference between these studies was the use in the latter experiments of a microsomal enzyme preparation that required significantly larger amounts of antibiotic for inhibition (0.2–1 μg/ml). Part of this increased requirement is probably due to the hydrophobic nature of tunicamycin and the fact that it interacts with membrane phospholipids. In fact, phosphatidylcholine or phosphatidylserine has been shown to prevent binding of tunicamycin to yeast protoplasts and to block antibiotic inhibition (39).

There are several cautions that should be considered in using tunicamycin (or any other inhibitor) for biochemical studies. In the first place, tunicamycin may inhibit protein synthesis (40–43), and therefore any observed effects on glycoprotein function or cellular physiology must be considered in that light. In various studies cited below, protein synthesis (i.e. ^{3}H-leucine, ^{3}H-proline incorporation into TCA precipitable material) was inhibited from 10% to 60%, at levels of tunicamycin (0.1–10 μg/ml) that inhibited glycosylation by 70–90%. However, not all cell systems are sensitive, and in some, no inhibition of protein synthesis was caused by antibiotic (44–47). It is not clear why some systems are much more susceptible than others. Furthermore, it is not known whether tunicamycin inhibits the synthesis of all proteins, or whether it only inhibits synthesis of certain kinds of proteins. That is, does tunicamycin inhibit the synthesis of glycoproteins to a greater extent than synthesis of other proteins? Several studies have suggested that glycosylation and protein synthesis are linked in such a way that when glycosylation is blocked, the synthesis of certain proteins is prevented. For example, when glycosylation of yeast carboxypeptidase Y (48), α-amylase of barley aluerone layer (49), thyroid thyroglobulin (50), and hydroxymethylglutaryl CoA reductase in CHO cells (51) were blocked with tunicamycin, the nonglycosylated proteins were not found in the cells, even though the carbohydrate-free protein appeared to be stable to proteolysis. At this time, there is not sufficient data to suggest how a block in glycosylation might regulate the synthesis of the polypeptide chain. However, in this regard, it is interesting to note that tunicamycin has also been reported to suppress the synthesis of mRNA for alkaline phosphatase (52). Again, no mechanism was presented to explain these results. Some of the tunicamycin homologues, separated by HPLC, have been reported to inhibit protein synthesis, whereas others do not (15, 16, 37). It is not clear whether inhibition is due to specific homologues (or specific fatty acid structures), or whether some of the HPLC fractions are contaminated with an inhibitor of protein synthesis.

Another consideration in the use of tunicamycin is the amount of antibiotic required to prevent glycosylation. The many studies cited in this review using cultured eukaryotic cells or tissue slices have generally used ranges of antibiotic from 0.1 to 10 μg/ml. Thus, it is clear that the concentration of tunicamycin necessary for 50% inhibition depends on the cell type, and probably on other factors such as the type of media used for growth, the age or growth phase of the cells, length of exposure to drug, etc. Certain cell types, notably chemically or virally transformed cell lines, and perhaps other tumor cells (53–55), appear to be especially sensitive to this antibiotic. Whether these differences reflect an increased sensitivity of, or decreased amounts of, the GlcNAc-1-P transferase, increased permeability of the cells to inhibitor, alterations in nutrient uptake, etc, remains to be established. At any rate, in view of these differences, it is advisable to test each cell type at various concentrations of tunicamycin before choosing an experimental dose.

Tunicamycin has been used in a great number of studies (over 650 citations since 1980) to examine the role of oligosaccharide in the properties and function of various secreted proteins, membrane receptors, glycoprotein enzymes, viral envelope proteins, developmentally related glycoproteins, tumor-associated proteins, transport proteins, and so on. Selected examples are presented to demonstrate the specific effects of this inhibitor. In almost every case examined, at the appropriate concentration, tunicamycin effectively prevented the attachment of the N-linked oligosaccharide as measured by the inhibition of $[2\text{-}^3H]$mannose incorporation into protein.

The consequences of the protein not being glycosylated may vary widely depending on the protein in question. For example, in terms of secreted proteins, tunicamycin had little or no effect on the following: secretion of procollagen from chick embryos (56) or chick cranial bones in organ culture (41); transferrin, the apo B component of very low-density lipoprotein or α-fetoprotein secretion by rat hepatocytes (42, 57, 58); interferon secretion by leukocytes (59) or interferon in L-cells induced by Newcastle virus (60); β-N-acetylhexosaminidase secretion by fibroblasts from normal individuals (61); secretion of glycoprotein hormone α-subunit by carcinoma cells (62); secretion of pro-opiomelanocortin by mammalian pituitary or cultured pituitary cells, or the secretion of the cleavage products, ACTH, and endorphin (63); prothrombin secreted from the livers of treated rats (64); lipoprotein lipase secretion from adipose cells (65). In this last case, the protein was secreted at the same rate as that of control cells, but it was enzymatically inactive. The above studies demonstrate that in many cases, the N-linked oligosaccharides do not play an essential role in the rate of synthesis, transport, or secretion of these extracellular proteins, although in some cases, carbohydrate may have a role in function (65).

However, the secretion of some proteins is profoundly altered when glycosylation is prevented. This may result in an absence of secretion, or in a greatly reduced rate of secretion. Thus the secretion of the immunoglobulins IgA and IgE by mouse and rat plasma cells was inhibited 85–100% at 0.5 μg/ml of tunicamycin (40). In various hybridoma cells that produce altered IgM chains (altered in the protein or in the number of carbohydrate chains), tunicamycin blocked IgM secretion in the wild type and in three different mutant cell lines, but caused less inhibition in two other cell lines (66). On the other hand, IgM secretion is not inhibited by this antibiotic in a mouse lymphoma cell, even at 5 μg/ml (47). Tunicamycin prevented the majority of IgM secretion but did not affect IgG secretion in hybridoma cells producing both isotypes (67), suggesting that the differential effects of this drug on IgM and IgG secretion are due to factors intrinsic to the respective heavy chain polypeptides themselves, rather than to other properties of the producing cells.

Interestingly enough, not all proteins are secreted from the same cell at the same rate even in the absence of inhibitors (68), and the inhibitor may affect certain proteins but not others. For example, in hepatomas, glycoproteins fall into three classes in terms of secretion rate, with retention times of 35' 77' and 115'. Of three rapidly secreted glycoproteins, tunicamycin inhibited only one (α_1 protease inhibitor), while of the intermediately secreted proteins, the antibiotic affected two of the three examined (α_2-macroglobulin and ceruloplasmin (69). Since it is generally believed that all secretory and plasma membrane glycoproteins migrate through the same Golgi vesicles during their path from the site of synthesis in the endoplasmic reticulum to their ultimate goal (70), it is not clear what signals affect the rate of transport or secretion. Recently, evidence has been obtained to indicate that newly synthesized membrane and secretory proteins exit the ER at different rates (68, 71, 72), with half times of transport to the Golgi varying as much as from 14 minutes to 137 minutes. To explain these differences in rate, it was postulated that the ER-to-Golgi step is a selective and rate-limiting step, and that newly synthesized proteins have (an) amino acid sequence(s) that bind(s) to (a) carrier protein(s) involved in transport (71, 72). One study suggested that removal of glucose might provide the signal for transport (73).

Tunicamycin also has dramatic effects on various other biological systems, such as membrane receptor activity and virus replication. Thus, the function of most receptors that have been examined is blocked or decreased by this inhibitor. Examples are β-adrenergic receptor of astrocytoma cells (74), acetylcholine receptor of muscle cells (75, 76), insulin receptor of 3T3-L1 adipocytes (77, 78), gonadotropin receptor (79), epidermal growth factor receptor of smooth muscle cells (80), or epidermoid carcinoma cells (81), nerve growth factor receptor of rat pheochromocytoma cells (82), and opiate receptor in neurotumor cells (45, 83). In the latter case, the use of tunicamycin allowed the authors to distinguish down-regulation of receptors from densensitization. Low-density lipoprotein receptors of human fibroblasts were also decreased by growth in tunicamycin (84, 85). However, the asialoglycoprotein receptor of hepatocytes was apparently not affected by this antibiotic (86). In most of these cases, tunicamycin caused a diminished number of receptor molecules at the cell surface, but it did not affect the binding affinity of those receptors that were present. However, in the case of the receptor for diphtheria toxin in CHO cells, tunicamycin did not alter the number of receptors, but it decreased their affinity for ligand (87). The decreased number of receptors at the cell surface is probably a combination of decreased rate of transport and increased rate of degradation (see below). In a number of these cases (43, 75, 82), as well as in some of the studies with secreted proteins, the effects of tunicamycin were reversible after removal of drug. In fact, with the insulin receptor, pulse chase studies with heavy (N^{15}) and light

(N^{14}) amino acids indicated that the nonglycosylated insulin receptor could be glycosylated after the removal of antibiotic (88). These studies indicate that under certain circumstances, N-linked glycosylation can occur after protein synthesis is complete.

Tunicamycin was originally identified by virtue of its antiviral activity against enveloped viruses. Thus, it is not surprising that it greatly decreases the infectivity of a number of viruses, including Semliki forest virus (89), Sindbis virus (90), Newcastle disease virus (91), Rauscher murine leukemia virus (92), Rous sarcoma virus (44), mouse leukemia virus (93), various strains of influenza virus (89), Hantaan virus (94), and rotavirus (95). The effect of tunicamycin on cells infected with various RNA-containing enveloped viruses is essentially similar for all viruses tested thus far, i.e. a reduction in infectivity and interference in the synthesis of glycopeptides. Whereas in a few systems noninfectious virus particles are made (Rous sarcoma and influenza), in other cases (Sindbis, VSV, fowl plaque) viral particles are almost undectable. The apparent reason for the absence of viral particles is that the nonglycosylated envelope proteins are subjected to rapid proteolysis and are therefore not available for particle formation. This was shown in one case by the use of the protease inhibitor N-α-p-tosyl-L-lysin-chloromethylketone (TLCK). Cells infected with fowl plaque virus and incubated in tunicamycin did not contain any detectable viral hemagglutinin. However, in the presence of TLCK, the cells accumulated the nonglycosylated hemagglutinin (89).

Various other glycoproteins are also more rapidly degraded when their nonglycosylated forms are produced in the presence of tunicamycin. Such is the case for fibronectin (6) for thyroid-stimulating hormone (96), for acetylcholine receptor (76), for the ACTH-endorphin common precursor (97), and for certain immunoglobulins (98), to cite a few examples. In some of these cases, the tunicamycin-induced disappearance of the glycoprotein in question could be prevented by the addition of protease inhibitors such as leupeptin or TLCK. These studies indicate that one role of the oligosaccharide portion of these proteins is to protect the glycoprotein from degradation, or to aid in the stability of the protein. It appears likely that as the protein is being synthesized, N-linked oligosaccharides are added to certain exposed asparagine residues, and these carbohydrate chains may have a pronounced effect on the subsequent folding of the protein, and thus on its final conformation. The end result is that certain proteins require carbohydrate to assume stable conformations. However, not all glycoproteins become unstable when they are in the nonglycosylated form, as shown for the hemagglutinin of influenza virus grown in MDCK cells in the presence of tunicamycin (99), and for various receptors and secretory proteins that are not degraded when produced in tunicamycin. Thus, the influence of carbohydrate on the conformation of the protein must depend ultimately on the amino acid sequence.

One extensive study that provides strong evidence for the effect of oligo-saccharide on protein conformation was done with various temperature-sensitive mutants of VSV (100). With one of these viral strains (San Juan), tunicamycin prevented viral replication more than 90% at either 38°C or 30°C. However, with the Orsay strain, replication was again inhibited 85–90% when the virus was grown at 38°C, but at 30°C tunicamycin inhibited only 30–50%. When the Orsay strain was grown at 30°C in tunicamycin, the nonglycosylated G protein could be detected at the host cell surface, indicating that it had been synthesized and transported in the normal manner. However, at 38°C, no G protein appeared at the cell surface. In addition, differences in the physical properties of the various G proteins produced in tunicamycin at 38°C or 30°C were apparent. Thus, the Orsay G protein produced at $30^{\circ C}$ in the presence of antibiotic could be readily solubilized in Triton X-100, but the protein produced at 38°C remained insoluble. The most likely explanation for these results involves the role of carbohydrate in providing a conformation that has increased solubility and/or stability (100).

Tunicamycin has been used to provide evidence for the presence of, or role of, N-linked oligosaccharides in various biological systems. Thus, the synthesis of keratan sulfate in chick embryo corneas indicated the presence of N-linked oligosaccharides (101), while its effects on various cartilage pro-teoglycans have suggested the presence of N-linked oligosaccharides in the link proteins (102). Tunicamycin also affects the synthesis of the gangliosides GM_1 and GM_2 in isolated Golgi vesicles (103). This inhibition was found to be due to a block in carrier-mediated transport of nucleotide sugars across the Golgi vesicles, since the transport of UDP-galactose into the Golgi vesicles was prevented. On the other hand, this antibiotic had no effect on the Golgi UDP-GlcNAc carrier system (104). These conflicting results are difficult to interpret, especially since tunicamycin bears a closer resemblance to UDP-GlcNAc than to UDP-galactose. Nutrient transport in cultured cells is affected by tunicamycin, probably because the plasma membrane transport carriers are N-linked glycoproteins. Thus, glucose transport in a variety of cells, such as chick embryo fibroblasts (105), Swiss 3T3 cells (106), erythrocytes (107), and malignant and nonmalignant cells (108), is inhibited, as were some of the amino acid transport systems of hepatocytes (109). However, other amino acid carrier systems were not affected by tunicamycin. Whether some of these transporters are glycoproteins and others are not, or whether the effect depends on conformation, remains to be determined. It should, of course, be remembered that the effect of tunicamycin on a system such as transport or development might be due to inhibition of a key enzyme necessary for synthesis of the protein, or to some such secondary effect.

To summarize the studies done with tunicamycin on the role of carbo-hydrate, one must conclude that there is no unifying concept that explains all situations, unless it is that carbohydrate affects the ultimate conformation of the protein. In that way, oligosaccharide may (a) affect the stability of the protein towards denaturation and proteolysis, (b) increase the solubility of the protein, (c) cause the exposure of sites on the protein that affect its transport, membrane interaction, and/or function, (d) itself play a role in biological recognition.

In terms of recognition reactions for various biological functions, it is clear that carbohydrate is involved in some cases [asiologlycoprotein receptor (110); mannose 6-phosphate receptor (111)], but in others, protein structure appears to be the determining factor. For example, certain animal cells, such as epithelial cells, segregate different sets of membrane proteins into two plasma membrane domains, apical and basolateral, and these are morphologically, physiologically, and biochemically distinct (112). Thus, in MDCK cells, the G protein of VSV is treated as a basolateral membrane protein, while the hemagglutinin of influenza virus is segregated to the apical membrane (113). Treatment of the infected cells with tunicamycin had no effect on this polarity, indicating that the transport mechanism that directs proteins to various surfaces must recognize certain aspects of the protein structure (114). In one interesting study, it was found that IgG that is secreted by certain hybridomas contained sulfated N-linked oligosaccharides. However, in the presence of tunicamycin, the protein contained four times as much sulfate in the heavy chain, but the sulfate was attached to tyrosine residues (115). Is this a case where the lack of carbohydrate results in an altered conformation that exposes tyrosine residues to an appropriate sulfotransferase?

Other Antibiotics That Inhibit Lipid-Linked Saccharide Formation

Several other antibiotics have been shown to inhibit the formation of lipid-linked saccharides in cell-free extracts of various tissues, and some of these compounds have proven useful for biochemical studies. In most cases, these inhibitors were initially tested on the "dolichol pathway" because they were known to inhibit bacterial cell wall synthesis, a series of reactions that also involves lipid-linked saccharide intermediates. For example, amphomycin and tsushimycin inhibit the phospho-N-acetylmuramoyl-pentapeptide transferase involved in peptidoglycan biosynthesis (116), while bacitracin inhibits the enzyme that catalyzes the dephosphorylation of undecaprenyl-PP (117). Diumycin, a member of the moenomycin group of antibiotics, also inhibits cell wall synthesis, but is exact site of action is not known (118). Showdomycin its a nucleoside antibiotic whose mechanism of action is probably related to the reaction of its maleimide group with sulfhydryl groups of proteins (119).

Amphomycin is an undecapeptide containing either 3-isododecanoic or

3-anteisododecanoic acid attached to an N-terminal aspartic acid in amide linkage (120). In cell-free extracts of pig aorta, amphomycin inhibited the formation of dolichyl-P-mannose but not the transfer of mannose from GDP-mannose or dolichyl-P-mannose to lipid-linked oligosaccharides. Thus, when microsomes were incubated with GDP-[^{14}C]mannose under conditions where dolichyl-P-mannose formation was inhibited, mannose was still transferred to lipid-linked oligosaccharide to form a Man$_5$(GlcNAc)$_2$-PP-dolichol (121). Such studies suggested that some of the mannose residues were donated directly from GDP-mannose. Similar results were obtained using extracts of oviduct tissue (122) and embryonic liver (123). Isolation of a Chinese hamster ovary cell line that is unable to synthesize dolichyl-P-mannose but still synthesizes the Man$_5$(GlcNAc)$_2$-PP-dolichol, confirmed the idea that GDP-mannose was the direct donor of the first five mannose residues (124). Purification of several of the mannosyl transferases involved in synthesis of Man$_5$(GlcNAc)$_2$-PP-dolichol also substantiated these results (125).

In extracts of brain, amphomycin inhibited the formation of dolichyl-P-glucose, dolichyl-P-mannose, and dolichyl-PP-GlcNAc, leading these investigators to suggest that amphomycin might form a complex with dolichyl-P, thereby inhibiting these reactions (126). In fact, amphomycin interfered with the extraction of endogenous dolichyl-P, further suggesting the formation of a complex. In one study, amphomycin was used to prove that elongation of the Man$_7$(GlcNAc)$_2$-PP-dolichol involved the transfer of mannose from dolichyl-P-mannose, rather than GDP-mannose (127), while in another study, the antibiotic was used to provide evidence that dolichyl-P is involved in sugar translocation through membranes (128). That study suggested a solution to the problem of how cytoplasmic sugar nucleotides can give rise to dolichyl-P-mannose that faces the lumen of the ER. Tsushimycin is another lipopeptide antibiotic that appears to have the same site of action as amphomycin (129). The unfortunate thing about these antibiotics is that they apparently cannot cross the membrane, and therefore are not useful for cell culture experiments. Perhaps incorporation of amphomycin into liposomes would overcome this problem.

Bacitracin is also a peptide antibiotic that inhibits lipid-linked saccharide formation. However, the exact site of action of this compound remains obscure. For example, hen oviduct membranes incubated with UDP-[^{14}C]GlcNAc in the presence of 1 mM bacitracin accumulated a trisaccharide-lipid, characterized as Manβ-GlcNAc-GlcNAc-PP-dolichol (130). This suggested that bacitracin interfered with the addition of the first α-mannosyl residue. On the other hand, 0.7–1 mM bacitracin was reported to inhibit the transfer of GlcNAc-1-P to dolichyl-P by calf pancreas microsomes, without affecting the synthesis of dolichyl-PP-(GlcNAc)$_2$, dolichyl-P-mannose, or dolichyl-P-glucose (131). On the other hand, in yeast membrane preparations, bacitracin inhibited the formation of dolichyl-PP-(GlcNAc)$_2$, but no inhibition in the synthesis of dolichyl-PP-GlcNAc was observed (132). These studies indicate at least three different sites of action for this inhibitor.

It seems likely that the action of bacitracin is due to its ability to form a complex with polyisoprenyl-phosphates, as was observed in the bacterial systems. In fact, when the formation of lipid-linked saccharides was examined in a microsomal preparation of aorta, bacitracin blocked the transfer of both mannose and GlcNAc from their sugar nucleotides into lipid-linked monosaccharides and lipid-linked oligosaccharides. Inhibition of dolichyl-P-mannose formation could be overcome by the addition of high concentrations of dolichyl-P, but this did not reverse the inhibition of dolichyl-PP-GlcNAc formation. Thus, bacitracin appears to affect all of the reactions in which polyprenyl-phosphates participate (133). It would be of interest to determine whether this antibiotic can bind all of the available dolichyl-P in these systems.

Diumycin has only been used in a few studies to inhibit lipid-linked saccharides, in vitro. Using a membrane preparation, or a solubilized enzyme fraction, from yeast, diumycin was found to inhibit dolichyl-P-mannose synthesis. Inhibition of this reaction was somewhat better with the solubilized enzyme. In the presence of antibiotic, transfer of manose from GDP-mannose to preformed dolichyl-PP-(GlcNAc)$_2$ still occurred, but mannose transfer from dolichyl-P-mannose to serine or threonine residues on the protein was also inhibited (134). With a solubilized enzyme preparation from *Acanthamoeba,* diumycin inhibited mannose and GlcNAc transfer from their sugar nucleotides into lipid-linked monosaccharides. The synthesis of dolichyl-P-glucose was only slightly inhibited. The antibiotic also blocked the transfer of the second GlcNAc to dolichyl-PP-GlcNAc, and this reaction was even more sensitive than the GlcNAc-1-P transfer (135). Since diumycin does not inhibit dolichyl-P-glucose formation, it may prove useful in defining or distinguishing reactions of the lipid-linked saccharide pathway. In fact, this antibiotic, as well as showdomycin, was used to define the role of dolichyl-P-mannose in the stimulation of the enzyme UDP-GlcNAc-dolichyl-P: GlcNAc-1-P transferase (136). Dolichyl-P-mannose, either synthesized from GDP-mannose by microsomes or added directly to microsomes, stimulated the formation of GlcNAc containing mono-, di-, and trisaccharide-lipids. The inhibition of dolichyl-P-mannose formation by these antibiotics prevented this stimulation, suggesting that dolichyl-P-mannose may play a role in regulation of the lipid-linked saccharide pathway. Another member of the moenomycin group of antibiotics is flavomycin, and this compound also interferes with the formation of lipid-linked monosaccharides in pig brain microsomes (137).

Showdomycin probably inhibits the lipid-linked saccharide pathway because of the reactivity of its maleimide group towards sulfhydryl groups on the sensitive enzymes. Consistent with this is the observation that the synthesis of dolichyl-P-glucose and dolichyl-PP-GlcNAc by membrane preparations of Volvox were equally sensitive to this antibiotic in the presence of 0.2% Triton X-100. This inhibition was lost when excess dithiothreitol was added, and further, the inactivation of the enzymes by N-ethylmaleimide was comparable to showdomycin inhibition (138). On the other hand, in aorta extracts, the formation of dolichyl-P-glucose and dolichyl-P-mannose as sensitive to showdomycin, whereas the formation of dolichyl-PP-GlcNAc was not very sensitive (139). This observation is difficult to reconcile with the finding that the partially purified GlcNAc-1-P transferase was sensitive to sulfhydryl reagents such as N-ethylmaleimide and p-chloromercuribenzene-sulfonate (140). Perhaps the lack of reactivity with showdomycin is due to accessibility. On the other hand, some of the mannosyl transferases are also sensitive to sulfhydryl reagents (141), and these enzymes are inhibited by this antibiotic.

Inhibition by Sugar Analogues

Just as in the case of tunicamycin, early interest in 2-deoxy-D-glucose and D-glucosamine was based on the fact that they inhibited multiplication of various enveloped viruses (142, 143). A number of sugar analogues, including 2-deoxy-2-fluroro-D-glucose, 2-deoxy-2-fluoro-D-mannose, 4-deoxy-4-fluoro-D-mannose, 6-deoxy-6-fluoro-D-glucose, mannosamine, etc, have been synthesized and studied, and results with some of these compounds have been extensively reviewed (7). Therefore, we cover sugar analogues only briefly here. It is now clear that the sugar analogues, 2-deoxyglucose, 2-deoxy-2-fluoroglucose, and 2-deoxy-2-fluoromannose (and probably other fluorosugars) have to be metabolized in order to exert their inhibitory effect (7). Thus, various studies have suggested that nucleotide esters of deoxyglucose or fluoroglucose (or mannose) are the inhibitory agents (144).

When cultured animal cells are incubated with 2-deoxyglucose, the sugar is converted to both UDP-2-deoxyglucose and GDP-2-deoxyglucose, as well as to dolichyl-P-2-deoxyglucose. Apparently the inhibition of protein glycosylation is not from depletion of sugar nucleotide pools, since the levels of GDP-mannose and UDP-GlcNAc are actually increased. The major compound involved in inhibition is GDP-2-deoxyglucose, since addition of mannose reversed the inhibition and decreased the levels of GDP-2-deoxyglucose (145). GDP-2-deoxyglucose also serves as a sugar donor for the formation of dolichyl-PP-(GlcNAc)$_2$-2-deoxyglucose, which cannot be further elongated (144). In the presence of both GDP-mannose and GDP-2-deoxyglucose, membrane preparations produced dolichyl-PP-(GlcNAc)$_2$-Man-2-deoxyglucose, which also could not be further elongated, nor was it transferred to protein. Thus, the accumulation of such abnormal lipid-linked saccharides within the cell may prevent glycosylation from occurring, and may also tie up all the available dolichyl-P. Interestingly enough, although the 2-deoxyglucose-containing oligosaccharides are not transferred to protein, 2-deoxyglucose can be incorporated into glycoproteins under noninhibitory conditions (146). In terms of the in vivo effects of 2-deoxy-D-glucose, they appear to be analogous to those seen with tunicamycin. However, because of the close resemblance of this sugar and its metabolic products to glucose, it probably has other side effects on various metabolic pathways. The fluorosugars also inhibit lipid-linked saccharides and protein glycosylation. The UDP- and GDP-derivatives of 2-deoxy-2-fluoromannose were found in treated cells, and GDP-2-deoxy-2-fluoromannose appeared to inhibit synthesis of Man-β-GlcNAc-GlcNAc-PP-dolichol (147). 4-Deoxy-4-fluoroglucose also inhibits lipid-linked saccharide synthesis and protein glycosylation (148).

Animal cells grown in glucosamine accumulate intracellular glucosamine and are inhibited in terms of protein glycosylation (149). The removal of glucosamine leads to a decrease in intracellular glucosamine and a reversal of inhibition, indicating that inhibition was due to glucosamine. Apparently glucosamine inhibits at an early stage in the assembly of the lipid-linked oligosaccharides, but the inhibition requires intact cells (150). Since glucosamine has been reported to affect the properties of the endomembrane system (151), its in vivo inhibition of lipid-linked saccharides and protein glycosylation could be a secondary effect. Nevertheless, when MDCK cells were incubated with 1 mM concentrations of glucosamine, there was a complete shift in the size of the lipid-linked oligosaccharide from the normal, $Glc_3Man_9(GlcNAc)_2$ to a $Man_7(GlcNAc)_2$. When the concentration of glucosamine was raised to 10 mM, the major oligosaccharide associated with the lipid was a $Man_3(GlcNAc)_2$. This effect of glucosamine was reversible, and after removal of this amino sugar, the cell could renew its synthesis of $Glc_3Man_9(GlcNAc)_2$-PP-dolichol (152). The site of inhibition causing this alteration is not clear.

The inhibition of MDCK cells by glucosamine was specific and could not be mimicked by other amino sugars such as galactosamine, mannosamine, or N-acetylglucosamine. However, mannosamine was also an inhibitor, although it inhibited at a different step than glucosamine, since the major lipid-linked oligosaccharides at 1–10 mM mannosamine were $Man_5(GlcNAc)_2$ and $Man_6(GlcNAc)_2$ (153). Strangely enough, the $Man_5(GlcNAc)_2$ structure was sensitive to endo-β-N-acetylglucosaminidase H, indicating that it was not the usual $Man_5(GlcNAc)_2$ biosynthetic precursor. Pulse chase studies indicated that the block was in the biosynthetic pathway and suggested that mannosamine was altering the sequence of addition of mannose residues, perhaps by its specific action on mannosyl transferases. The lipid-linked oligosaccharides produced in the presence of mannosamine were still transferred to protein, and these altered oligosaccharides were still processed to hybrid and complex chains (153).

One problem with the use of sugar analogues is that they may be metabolized by the cell, leading to alterations that indirectly affect protein glycosylation and lipid-linked saccharides. For example, glucosamine may cause depletion of UTP and/or ATP levels under certain conditions (154). The lack of UTP may result in inhibition of RNA synthesis, while energy depletion may in turn affect lipid-linked saccharide formation (see below). Thus, some observed effects could be of a secondary nature.

Effect of Energy Charge on Glucose Starvation

When cells are treated with 10 mM CCCP (carbonyl-cyanide m-chlorophenylhydrazone), an uncoupler of oxidative phosphorylation, the cells are unable to produce dolichyl-P-mannose, but still can synthesize dolichyl-P-glucose, dolichyl-PP-GlcNAc, and dolichyl-PP-(GlcNAc)$_2$. Cells incubated with CCCP produce a Glc$_3$Man$_5$(GlcNAc)$_2$-PP-dolichol, and transfer the oligosaccharide to protein (155). On the other hand, incubation of thyroid slices with CCCP leads to the disappearance of Glc$_3$Man$_9$(GlcNAc)$_2$-PP-dolichol and the appearance of Man$_9$(GlcNAc)$_2$-PP-dolichol and Man$_8$(GlcNAc)$_2$-PP-dolichol. There was also a decrease in the N-glycosylation of proteins. Several inhibitors of respiration such as N$_2$ or antimycin A gave similar effects (156).

Another condition that affects the lipid-linked saccharides of cultured cells is glucose starvation. Thus, when certain mammalian cells (such as CHO cells) are incubated in the absence of glucose, they no longer synthesize the Glc$_3$Man$_9$-(GlcNAc)$_2$-PP-dolichol, but instead accumulate the Man$_5$(GlcNAc)$_2$-PP-dolichol. This lipid can be glucosylated to the Glc$_3$Man$_5$(GlcNAc)$_2$-PP-dolichol and then transferred to proteins by what has been called the "alternate pathway" (157–159). In addition, some cells apparently accumulate smaller amounts of Man$_2$(GlcNAc)$_2$-PP-dolichol (160). In several cell lines, the effect of glucose starvation was observed within 20 minutes of glucose removal as long as cells were at low-to-moderate densities, but no effect was observed at high densities (161). This glucose starvation could be overcome by addition of glucose, but not of pyruvate, glutamine, galactose, inositol, or glycerol. In BHK cells infected with VSV virus and starved for glucose, the Glc$_3$Man$_5$(GlcNAc)$_2$ was formed and transferred to the G protein (162). Rat hepatoma cells starved for glucose produced a lower-molecular-weight form of α_1-acid glycoprotein (163). This reduction in molecular weight was due not only to a truncated oligosaccharide but also to a reduced number of oligosaccharide chains, perhaps as a result of inhibition of oligosaccharide synthesis. Since some uncouplers of oxidative phosphorylation have been reported to disrupt the recycling of the glucose transport carrier, these compounds (such as CCCP) may limit the entry of glucose into the cells, thus mimicking the glucose effect.

GLYCOPROTEIN PROCESSING INHIBITORS

During the past five or six years, a number of compounds have been identified that are specific and potent inhibitors of glycosidases. Since at least four of the enzymes in the processing pathway are glycosidases (but with a much stricter substrate specificity than lysosomal or other glycosidases), these processing inhibitors have specific sites of action in this pathway. The difficulty with inhibitors such as tunicamycin or compactin is that they completely prevent glycosylation, and therefore the nonglycosylated protein may have an altered conformation, causing it to be insoluble or to be rapidly degraded, and so on. The processing inhibitors, on the other hand, allow glycosylation to occur, but the N-linked oligosaccharides cannot be modified beyond a specific stage. This gives rise to glycoproteins having various types of altered oligosaccharide structures. A review on processing inhibitors has appeared (163), and some of these compounds have been discussed in another review (8).

Swainsonine

Certain species of wild plants that are found in the western United States and Australia are well known and notorious to ranchers, since animals that eat these plants develop severe abnormalities (called locoism) that eventually lead to death (164). In the United States, the toxicity is due to *Astragalus* sp., commonly called locoweed, while in Australia it is due to *Swainsona* sp. (165). These plants represent a major hazard to the livestock industry and probably cause millions of dollars in losses annually. The indolizidine alkaloid, swainsonine [(1S,2R,8R,8αR)-1, 2, 8 trihydroxyoctahydroindolizidine; Figure 4], was isolated from these plants (166–168) and shown to be a potent inhibitor of lysosomal α-mannosidase (169), and also a potent competitive inhibitor of jack bean α-mannosidase (170, 171). Fifty percent inhibiton of the jack bean enzyme occurred at about 1×10^{-7} M swainsonine. It was suggested that the inhibitory action of swainsonine results from the structural similarity of its protonated form to the mannosyl cation, since the glycosyl ion intermediate is presumably formed during hydrolysis of natural substrates by glycosidases (172). Swainsonine is also produced by the fungus *Rhizoctonia leguminicola,* and in fact this fungus, infecting red clover, may give symptoms like those of locoweed (173).

Lymph nodes of sheep that had ingested *Swainsona* (or *Astragalus*) sp. on the range contained high levels of mannose-rich oligosaccharides (174), which led to the postulation that consumption of *Swainsona* sp. produces a lysosomal storage disease that is biochemically and morphologically similar to genetically determined α-mannosidosis of man and sheep. Swainsonine was fed to pigs and its effects compared to that of feeding pigs the whole *Swainsona* plant. Both experimental groups showed signs of poisoning typical

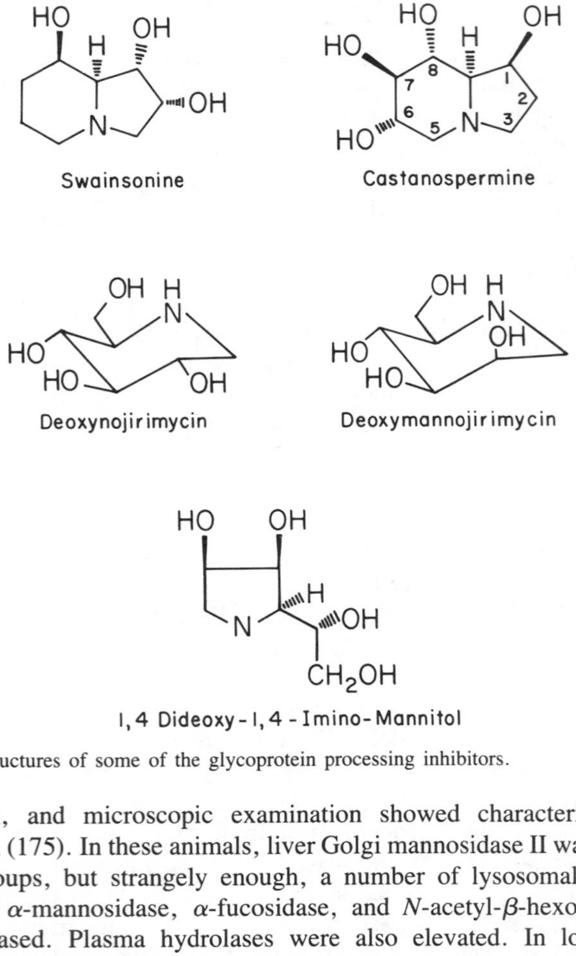

Swainsonine

Castanospermine

Deoxynojirimycin

Deoxymannojirimycin

1,4 Dideoxy-1,4-Imino-Mannitol

Figure 4   Structures of some of the glycoprotein processing inhibitors.

of locoism, and microscopic examination showed characteristic foamy vacuolation (175). In these animals, liver Golgi mannosidase II was decreased in both groups, but strangely enough, a number of lysosomal hydrolases (especially α-mannosidase, α-fucosidase, and N-acetyl-β-hexosaminidase) were increased. Plasma hydrolases were also elevated. In locoweed-fed sheep, high-mannose oligosaccharides were found in the urine. Two major oligosaccharides were identified by HPLC as $Man_4(GlcNAc)_2$ and $Man_5(GlcNAc)_2$ (176). The relative abundance of the various individual oligosaccharides changed over the course of the feeding studies (177). When locoweed feeding was discontinued, the amount of urinary oligosaccharides declined rapidly and reached a baseline within 12 days. Accumulation of mannose-rich oligosaccharides was also induced in cultured human fibroblasts by growth in swainsonine (178), and also in rats and guinea pigs by feeding them the alkaloid (179). Although all of the oligosaccharides were of the high-mannose type, there were differences in structure in the various systems, such as the presence of one or two GlcNAc residues. This probably relates to the presence or absence of an endo-β-N-acetylglucosaminidase. In general, the structures were analogous to the urinary oligosaccharides found in bovine, feline, and human α-mannosidosis (180).

Swainsonine was the first compound that was found to inhibit glycoprotein processing. In cultured cells grown in the presence of this alkaloid, there was a great decrease in the amount of [2-^{3}H]mannose incorporated into complex types of oligosaccharides and a considerable increase in mannose incorporation into oligosaccharides susceptible to digestion by endo-β-N-acetylglucosaminidase H (181). Since the major oligosaccharide released by endoglucosaminidase H sized like a Hexose$_9$GlcNAc, it was originally thought that swainsonine was inhibiting mannosidase I. However, when swainsonine was tested directly on mannosidase I and mannosidase II, it was found to be a potent inhibitor of mannosidase II, but was without effect on mannosidase I (182). Swainsonine also did not inhibit the endoplasmic reticulum α-mannosidase, nor the corresponding soluble α-mannosidase (183). In keeping with this site of action, swainsonine caused the formation of hybrid types of oligosaccharides having an oligomannosyl core [Man$_5$(GlcNAc)$_2$] characteristic of neutral oligosaccharides, and a (or several) NeuNAc-Gal-GlcNAc sequence(s) characteristic of complex chains. Thus, the oligosaccharide chains of the G protein of VSV were altered from complex to hybrid structures (184) as was the N-linked carbohydrate of fibronectin (185). BHK cells grown continuously in the presence of swainsonine had similar types of hybrid structures as did a ricin-resistant mutant, and in fact, the swainsonine-grown cells showed increased resistance to ricin (186).

Swainsonine has been used in a number of studies in order to determine whether changes in the structure of the N-linked oligosaccharides affect glycoprotein function. In most cases, swainsonine has little effect on the glycoprotein in question, which may indicate that a partial complex chain is sufficient for activity, and that protein conformation is not altered. Swainsonine did not impair the synthesis or export of thyroglobulin in porcine thyroid cells (187), surfactant glycoprotein A from Type II epithelial cells (188), H2-DK histocompatability antigens from macrophages (189), or von Willebrand protein in epithelial cells (190). In one study, the secretion of α_1-antitrypsin from primary rat hepatocytes was not altered by growth in swainsonine (191), but in another study using human hepatoma cells, the alkaloid increased the rate of secretion of transferrin, ceruloplasmin, α_2-macroglobulin, and α_1-antitrypsin (192). The authors suggest that these proteins traverse the Golgi more rapidly than their normal counterparts.

Swainsonine also did not affect the insertion or function of the insulin receptor (193), the epidermal growth factor receptor (194), or the receptor for asialoglycoproteins (195). The alkaloid did, however, block the receptor-mediated uptake of mannose-terminated glycoproteins by macrophages. This inhibition appeared to be due to formation of hybrid chains on glycoproteins present on the macrophage surface, which could then bind to and tie up the mannose receptors (196). Viral proteins are still assembled and processed in the presence of this drug, and infectious particles are produced (184, 197–199). Swainsonine also did not prevent the fucosylation (185, 191, 200) or the sulfation (201) of the N-linked oligosaccharide in several systems, including the influenza viral hemagglutinin. The evidence indicated that the fucosyl transferase that adds fucose in α1,6 linkage to the innermost GlcNAc residue acts after the addition of GlcNAc to the α1,3-linked mannose, as previously suggested by in vitro studies (5). In addition, these studies showed that addition of sulfate or fucose to the internal GlcNAc residues did not inhibit the action of endoglucosaminidase H.

On the other hand, some functions are affected by swainsonine. For example, the glucocorticoid stimulation of resorptive cells, which probably involves the attachment of cells such as osteoclasts to bone, is blocked by swainsonine (202). This alkaloid also reduced the interaction of *Trypanosoma cruzi* with peritoneal macrophages when either host or parasite was treated with the drug (203). There was a dramatic decline in the ability of B16–F10 melanoma cells to colonize the lungs of experimental animals in the presence of swainsonine (204). Lymphocytes can be stimulated to proliferate by treatment with Concanavalin A, and this stimulation is suppressed by an immunosuppressive factor isolated from the serum of mice bearing a tumor (sarcoma 180). The suppression caused by this factor is overcome by swainsonine (205), suggesting that the alkaloid could be valuable in immunosuppressive diseases. The turnover or degradation of endocytosed glycoproteins is inhibited by swainsonine, leading to an accumulation of the glycoproteins in the lysosomes. The block appears to be an inhibition of lysosomal α-mannosidase (and perhaps other mannosidases), and suggests that the oligosaccharide portion of the glycoprotein must be degraded before the protein (206). The above studies suggest some important roles for swainsonine, and the need for more experimentation with this alkaloid.

A number of laboratories have now accomplished the chemical synthesis of swainsonine. In most cases, the synthesis has started with a derivative of glucose or mannose, such as methyl 3-amino-3-deoxy-α-D-mannopyranoside (207, 208) or methyl 4,8-imino-2,3-O-isopropylidene-4,6,7,8-tetradeoxy-α-D-manno-octapyranoside (209). The synthesis of swainsonine was also accomplished from the four-carbon precursor, trans-1,4-dichloro-2-butene (210). Utilizing this synthesis, two other isomers of swainsonine, referred to as "Glc-Swainsonine" (1S, 2S, 8R, 8αR-trihydroxyindolizidine) and "Ido-Swainsonine (1S, 2S, 8S, 8αR-trihydroxyindolizidine) were produced. These isomers afforded an opportunity to determine how the chirality of these molecules affected their ability to inhibit various glycosidases. Interestingly enough, the "Glc-Swainsonine" did not inhibit jack bean α-mannosidase, but instead inhibited fungal α-glucosidase. However, this isomer still inhibited the glycoprotein processing mannosidase II, rather than glucosidase I or II (211). These results indicate that the indolizidine ring structure is not directly analogous to the pyranose ring of the sugar and therefore the chirality of the hydroxyl groups may not be the only factor in specificity of inhibition. One of the major advantages of being able to chemically synthesize the various inhibitors is that modifications in synthesis may produce a number of isomers. Comparisons of the biological activity of these isomers can provide a great deal of information about the requirements for inhibition of specific glycosidases. Several other isomers of swainsonine have also been synthesized, but nothing is yet known about their biological activity (212).

Castanospermine and Deoxynojirimycin

Castanospermine [(1S, 6S, 7R, 8R, 8αR)-1, 6, 7, 8-tetrahydroxyindolizidine] is another plant alkaloid (Figure 4) that was isolated from the nuts of the Australian tree, *Castanospermum australe,* also called Moreton Bay Chestnut (213). This alkaloid is also toxic to animals that eat these nuts and causes severe gastrointestinal upset, possibly leading to death. These symptoms may be related to its potent inhibitory effect on α-glucosidases, including intestinal sucrase and maltase (see below). Castanospermine was found to be a strong competitive inhibitor of a number of α-glucosidases, and it also inhibited β-glucosidase, although less effectively, in a competitive manner. On the other hand, it was inactive towards mannosidases, galactosidases, or hexosaminidases (214). Castanospermine also blocked glycoprotein processing by virtue of its inhibition of glucosidase I and glucosidase II (215). In cell culture, castanospermine inhibited complex oligosaccharide formation and caused the accumulation of oligosaccharides having $Glc_3Man_{7-9}(GlcNAc)_2$ structures (216–218).

Deoxynojirimycin (as well as nojirimycin) is a glucose analogue with an NH-group substituting for the oxygen atom in the pyranose ring (Figure 4). These compounds are synthesized by certain *Bacillus* species and were found to be potent inhibitors of a variety of α-glucosidases (219). Deoxynojirimycin inhibited the partially purified glucosidase I (I_{50}, 20 μM) from *Saccharomyces cerevisiae* as well as glucosidase II (I_{50}, 2 μ) (220). However, with the enzymes obtained from calf liver, deoxynojirimycin showed a preference for glucosidase I (50% inhibition at 3 μM) (221). It is not clear whether these differences reflect differences in the enzymes from yeast and liver, or whether they are due to changes in the assay conditions. In IEC-6 intestinal cells, this inhibitor (at 5 mM) caused a decrease in complex types of oligosaccharides and an increase in high-mannose types. Since these high-mannose structures were less susceptible to α-mannosidase, they were thought to contain glucose residues (220).

The inability of the cell to remove glucose from its glycoproteins may have profound effects on the transport, synthesis, and/or secretion of the protein in question. Thus, in Hep-G-2 cells grown in deoxynojirimycin, the rate of secretion of α_1-antitrypsin fell, but only marginal effects were seen on the glycoproteins, ceruloplasmin, and the C3 component of complement, and on the nonglycoprotein, albumin (73). Deoxynojirimycin also inhibited α_1 proteinase inhibitor (222), cathepsin D (223), and IgD (224). It was suggested that the presence of glucose in the oligosaccharide might retard transport of the protein, and in fact α_1-antitrypsin accumulated in the RER (73). That explanation, however, is not entirely satisfactory, since other glycoproteins were not retarded by deoxynojirimycin. Deoxynojirimycin has been reported to inhibit the formation of $Glc_3Man_9(GlcNAc)_2$-PP-dolichol in intestinal epithelial cells and to cause the formation of $Man_9(GlcNAc)_2$-PP-dolichol (225). Furthermore, at high concentrations of deoxynojirimycin, inhibition of glycosylation was observed, and α_1-antitrypsin forms with lower M_r were detected (222). Thus, the decline in the rate of secretion could be a result of decreased glycosylation. Or it is possible that for some proteins, the removal of glucose residues affects the protein structure and causes such proteins to be transported to the Golgi at a faster rate.

Some indication that glucose removal may cause changes in conformation was obtained in several viral systems. Deoxynojirimycin and castanospermine inhibited the formation of Sindbis virus in BHK cells and in several other cell lines. Growth of Sindbis was inhibited to a much greater extent at 37°C than at 30°C in BHK cells (226). Studies with the San Juan strain of VSV indicated that the formation of the virus becomes temperature sensitive when glucose residues are retained on the oligosaccharide chains of the G protein (227). However, in another study, deoxynojirimycin was reported to have no effect on the surface expression of VSV G protein, influenza hemagglutinin, and class I histocompatability antigens (228).

Castanospermine was injected into rats to determine its effects on glycogen metabolism. The alkaloid caused a marked decrease in liver α-glucosidase activity, and after injections for three or four days, glycogen disappeared from the cytoplasm of liver cells and accumulated in the lysosomes (229). The symptoms were somewhat like those found in Pompe's disease. Castanospermine is also a potent inhibitor of intestinal sucrase and maltase, and these enzyme activities are markedly reduced in animals treated with castanospermine.

The seeds of the Moreton Bay chestnut also contain another indolizidine alkaloid that was isolated and characterized as 6-epicastanospermine. This isomer was expected to be an α-mannosidase inhibitor since the 6-hydroxyl of the alkaloid should be analogous to the 2-hydroxyl of the sugar. Instead the 6-epimer differed from castanospermine only in that it no longer inhibited β-glucosidase. However, it still inhibited the fungal α-glucosidase as well as the processing glucosidases. It did not inhibit other glycosidases (230). Thus, as in the case of swainsonine analogues, the chirality of hydroxyl groups on the indolizidine ring is not sufficient to predict the inhibitory capacity of a given alkaloid. Perhaps with the availability of additional isomers, enough data will be obtained to provide a set of rules for predicting such phenomena.

Deoxymannojirimycin

Based on the inhibition of glucosidase I by deoxynojirimycin, the mannose analogue of this compound was synthesized chemically and shown to be a mannosidase inhibitor (231). This compound inhibited mannosidase IA/B and caused the accumulation of high-mannose oligosaccharides of the $Man_{8-9}(GlcNAc)_2$ structure in cultured cells (232, 233). Deoxymannojirimycin did not inhibit the secretion of IgD or IgM (while deoxynojirimycin did) in cultured cells (234). This compound also was without effect on the appearance of the VSV G protein, the influenza hemagglutinin, and the HLA-A,-B, and -C antigens (232), as well as α_1 proteinase inhibitor and α_1-acid glycoprotein (234).

In one interesting study, deoxymannojirimycin was used as a tool to study the recycling of membrane glycoproteins through the Golgi regions that contained mannosidase I. Thus, membrane glycoproteins were synthesized and labeled in the presence of deoxymannojirimycin, causing the proteins to have $Man_{9-8}(GlcNAc)_2$ structures. The label and inhibitor were then removed and the change in structure of the oligosaccharide chains was followed with time. The data indicated that the transferrin receptor and other glycoproteins were transported to the mannosidase I compartment and could be reprocessed during reculture (235).

Deoxymannojirimycin was also used to determine the role of the ER α-mannosidase in the processing of the HMG CoA reductase in UT-1 cells. In the absence of inhibitor, the predominant oligosaccharides on the reductase are single isomers of $Man_6(GlcNAc)_2$ and $Man_8(GlcNAc)_2$. However, in the presence of deoxymannojirimycin, the $Man_8(GlcNAc)_2$ accumulates, indicating that the ER α-mannosidase is responsible for the initial mannose processing (236). However, not all hepatocyte glycoproteins were found to be substrates for this ER α-mannosidase.

Pyrrolidine Alkaloids

The pyrrolidine alkaloid 2,5-dihydroxymethyl-3,4-dihydroxypyrrolidine (DMDP, Figure 4) was isolated from the plants *Lonchocarpus sericeus* and

Derris elliptica (237). DMDP was tested as an inhibitor of various glycosidases and found to be a potent inhibitor of almond emulsin β-glucosidase (K_i = 7 × 10^{-6} M), yeast α-glucosidase (K_i 6 × 10^{-6} M), and insect trehalase (K_i = 5.5 × 10^{-5} M) (238).

DMDP was found to inhibit the processing of the N-linked oligosaccharide of the influenza viral hemagglutinin (239). Thus, in the presence of 250 μg/ml of DMDP, more than 80% of the glycopeptides became susceptible to endoglucosaminidase H and were characterized as having mostly a Glc$_3$Man$_8$GlcNAc structure. These studies indicated that DMDP inhibited glucosidase I. However, in studies with IEC-6 intestinal cells, 5 mM DMDP also inhibited complex chain formation by 80%, but there was an increase in high-mannose (i.e. Man$_{7-9}$GlcNAc) as well as glucosylated oligosaccharides (240). These workers suggested that DMDP did not primarily inhibit glucosidase I in these cells. In another study, involving the synthesis of the oncogenic membrane protein V-erb B, DMDP caused the same results as deoxynojirimycin, indicating that it inhibited glucosidase I in this system also (241). When tested on the purified glucosidase I from plants, DMDP was found to be a reasonable inhibitor of the enzyme, although not as good as castanospermine (215). DMDP did not inhibit the plant processing mannosidase I.

Although DMDP is not nearly as effective as the other glucosidase I inhibitors, it is interesting in several respects. First of all, its activity indicates that six-membered ring structures are not absolutely necessary for glycosidase inhibition, although they may be more effective than five-membered rings. Secondly, these studies indicate that an NH-group in the ring in place of the oxygen, and probably three or more hydroxyl groups in the proper orientation, are necessary for activity. Finally, DMDP has been synthesized chemically (242), and this synthetic route should allow various isomers to be produced (see below). These compounds should be valuable additions to the already existing glycosidase inhibitors.

Chemically Synthesized Processing Inhibitors

1,4-dideoxy-1,4-imino-D-mannitol (DIM) was synthesized chemically from benzyl-α-D-mannopyranoside and shown to be a competitive inhibitor of jack bean α-mannosidase (243). DIM was also shown to inhibit the formation of complex chains in influenza virus–infected MDCK cells (244). The oligosaccharide chains of the viral hemagglutinin produced in the presence of DIM were mostly of the Man$_9$GlcNAc structure, suggesting that DIM inhibited mannosidase I. In keeping with this site of action, DIM also inhibited the action of rat liver mannosidase I on the Man$_9$GlcNAc substrate (244). Although DIM is not as effective an inhibitor as swainsonine or deoxymannojirimycin, it can be synthesized in good yield in a relatively straightforward manner. Thus, DIM, or other related isomers, should be valuable compounds for studies in animals to produce model systems for lysosomal storage diseases or other glycosidase deficiencies.

Conduritol epoxide and conduritol derivatives are inhibitors of glycosidases and have been used in various studies (245). This inhibitor administered to rats produced a model of Gaucher's disease by inhibiting β-glucosidase of brain, liver, and spleen and causing elevated levels of glucosylceramide in tissues (246). Bromoconduritol (6-bromo-3,4,5-trihydroxycyclohex-1-ene) is an active site–directed covalent inhibitor of glycosidases that was also chemically synthesized (247). Bromoconduritol caused the accumulation in virus-infected cells of endoglucosaminidase H–sensitive oligosaccharides that had the structures GlcMan$_9$GlcNAc, GlcMan$_8$GlcNAc, and GlcMan$_7$GlcNAc. This led the authors to suggest that this compound inhibited glucosidase II (248). However, inhibition of glucosidase II should be expected to give oligosaccharides with two glucoses, since this enzyme presumably removes the two α1,3-linked glucoses. One problem with the use of bromoconduritol is that it has a half-life in water of only 15 minutes and is therefore difficult to use in cultured cells. It is not clear whether its degradation products are toxic.

Glycosylmethyl-p-nitrophenyltriazenes irreversibly inactivate lysosomal glycosidases such as β-galactosidase and β-glucosidase in vitro (249). The mannosylpyranosyl-methyl-p-nitrophenyltriazene was tested on the biosynthesis and degradation of α_1-acid glycoprotein in liver cells. This compound prevented the conversion of high-mannose chains to complex oligosaccharides and gave mostly oligosaccharides of the Man$_{7-9}$GlcNAc structure. This suggested that this inhibitor acted at the mannosidase I stage. The compound also blocked the hydrolysis of endocytosed α_1-acid glycoprotein chains, probably by inhibiting lysosomal α-mannosidase (250).

COMPOUNDS THAT MODIFY PROTEIN STRUCTURE OR SYNTHESIS

β-Hydroxynorvaline and Fluoroasparagine

The amino acid sequence at the asparagine residue that becomes glycosylated is Asn-X-Ser (Thr), where X can be any amino acid except aspartic acid and probably proline (251). That sequence alone, however, is not sufficient for glycosylation (252), and it seems likely that the tripeptide sequence must be in a favorable conformation for glycosylation to occur. In fact, statistical studies on a number of naturally occurring glycoproteins suggest that most of the asparagines that are glycosylated are located in segments of the peptide that favor the formation of β-turns (253).

The importance of threonine in the tripeptide sequence was tested by studying the effect of the threonine analogue β-hydroxynorvaline on cotranslational glycosylation in ascites tumor lysates. This amino acid analogue inhibited the glycosylation of the α subunit of human chorionic gonadatropin and the β subunit of bovine luteinizing hormone. The inhibition could be overcome by adding threonine, indicating that β-hydroxynorvaline was acting through its incorporation into protein (254). In cultured fibroblasts, β-hydroxynorvaline caused the formation of cathepsin D–molecules having 2,1, or 0 oligosaccharide chain(s). The nonglycosylated form was a minor species and was degraded within 45 minutes of its synthesis, in keeping with studies on the results of tunicamycin inhibition of various proteins. Cathepsins with 2 or 1 oligosaccharide(s) were normally segregated into lysosomes, and their proteolytic maturation was not affected (255). Similar results were seen with α_1-acid glycoprotein in rat hepatocytes, where β-hydroxynorvaline caused the production of molecules having 0–6 oligosaccharide chains. The nonglycosylated species showed the slowest rate of secretion, while increasing the number of oligosaccharides increased the rate of secretion. However, the fully glycosylated molecule produced in the presence of the analogue exited the cell more slowly than the native protein, indicating that the effect of the amino acid analogue on peptide sequence as well as on glycosylation may affect the secretion rate (256). Apparently, glycosylation was not completely inhibited by this analogue, which may indicate that glycosylation can still occur, although less readily when this analogue replaces threonine. Or the incorporation of β-hydroxynorvaline into protein may be much less efficient than that of threonine, and therefore even small pools of threonine in the cells may be much more readily utilized. It would be interesting to know whether those glycoproteins with only a single oligosaccharide instead of the normal six, all are glycosylated at the same site, or whether this glycosylation is random.

Threo-β-fluoroasparagine is selectively toxic to asparagine-dependent cells when aspartic acid is included in the culture media. Although the mechanism of cytoxicity is not known, it is believed to be due to incorporation of the analogue into protein (257). In fact, threo-β-fluoroasparagine markedly inhibited glycosylation at a concentration of 1 mM, and this effect was blocked by asparagine, suggesting that incorporation of the analogue into protein was necessary for inhibition of glycosylation. On the other hand, the erythro-isomer was inactive (258). Short peptides containing N-acetyl-threo-β-fluoroasparagine were synthesized, and these were first tested as oligosaccharide acceptors with hen oviduct microsomes (259). The data suggested that incorporation of β-fluoroasparagine into cellular protein inhibits N-linked glycosylation by rendering the protein substrates ineffective for glycosylation. Thus, the toxicity of fluoroasparagine, like the toxicity of tunicamycin, might be due to the inability of the cell to glycosylate the asparagine-linked glycoproteins. Of course, incorporation of fluoroasparagine into protein may alter its conformation and properties. Since the fluoroasparagine prevents glycosylation at the transferase step, it will be of interest to determine whether it has any effect on the synthesis of lipid-linked saccharides in cell culture.

Inhibitors of Protein Synthesis

Several inhibitors of protein synthesis have been found to also affect the synthesis of lipid-linked oligosaccharides. When MDCK cells were grown in the presence of cycloheximide or puromycin to inhibit protein synthesis, there was also a substantial inhibition in the incorporation of $[2\text{-}^3\text{H}]$mannose into lipid-linked oligosaccharide. However, the formation of dolichyl-P-mannose was affected only slightly. Since cycloheximide did not affect mannose incorporation into lipid-linked saccharides in cell-free extracts and since the glycosyltransferases of the dolichol pathway are not greatly affected during the course of these studies, the results were thought to be due either to limitations in the amount of dolichyl-P available to serve as a carrier, or to a feedback control mechanism that affects synthesis of lipid-linked oligosaccharides. Since the formation of dolichyl-P-mannose was apparently not affected by cycloheximide, the feedback mechanism appeared more likely (260). Another study, using actinomycin D to depress levels of mRNA, or cycloheximide to inhibit protein synthesis, found that synthesis of lipid-linked oligosaccharides was proportional to the rate of protein synthesis. The regulated step appeared to be prior to the formation of $\text{Man}_5(\text{GlcNAc})_2\text{-PP-}$dolichol, leading these investigators to suggest that a likely control point was the availability of dolichyl-P (261). However, in LM cells inhibited with cycloheximide, the incorporation of mannose into lipid-linked oligosaccharides stopped, but there was no effect on the formation of dolichyl-PP-GlcNAc, dolichyl-PP-$(\text{GlcNAc})_2$, or dolichyl-P-mannose. Dolichyl-P addition to uninhibited cells caused a 300% stimulation of lipid-linked oligosaccharide formation but had no effect on inhibited cells, indicating that inhibition is probably not due to limitations in dolichyl-P (262). The authors suggest feedback regulation possibly caused by elevated levels of a metabolite such as GTP. Thus, these protein synthesis inhibitors may be novel tools for determining whether the formation of lipid-linked saccharides is linked to protein synthesis and if so, by what mechanism.

INHIBITORS OF DOLICHYL-P SYNTHESIS

Since dolichyl-P serves as an obligatory carrier in the biosynthesis of N-linked oligosaccharides, the prevention of dolichyl-P formation should result in inhibition of the synthesis of lipid-linked saccharides and thus in protein glycosylation. Dolichyl-P is synthesized by the same series of reactions from acetate to farnesyl-PP (i.e. acetate to hydroxymethylglutaryl CoA and mevalonic acid to farnesyl-PP) as are involved in cholesterol formation. Thus, at least one point in the regulation of dolichyl-P should be the hydroxy-methylglutaryl CoA reductase, which is the rate-limiting step in cholesterol biosynthesis. Several inhibitors of this enzyme, 25-hydroxycholesterol and compactin, have been shown to affect N-linked glycosylation.

When aortic smooth muscle cells are grown in 25-hydroxycholesterol, the incorporation of acetate into cholesterol and dolichyl-P is inhibited 85–90%, as is N-linked glycosylation. However, mevalonate incorporation into these lipids is not altered by 25-hydroxycholesterol, and in fact, mevalonate addition could overcome the inhibition of N-linked glycosylation (263). In L-cell cultures, hydroxycholesterol also affected dolichol and cholesterol synthesis, but in this case, conditions that caused large fluctuations in cholesterol synthesis had only small effects on dolichol (264). The authors postulate that other rate-limiting steps in addition to the reductase must control the levels of these two lipids. Feeding animals a diet high in cholestyramine, a compound that increases HMG CoA reductase activity, resulted in a great stimulation in acetate incorporation into cholesterol but not into dolichyl-P (265). This suggested that dolichyl-P synthesis might not be regulated at the reductase stage. On the other hand, animals fed a high-cholesterol diet, which should suppress cholesterol synthesis, show increased incorporation of mevalonate into dolichol and dolichol-linked oligosaccharides, and also increased activity of some of the glycosyltransferases involved in glycoprotein synthesis (266). Perhaps inhibiting the cholesterol branch may result in increased concentrations of various intermediates that are also involved in dolichol synthesis.

Compactin is a fungal metabolite that is a competitive inhibitor of HMG CoA reductase (267). This compound (and other closely related compounds such as mevinolin) has been intensively studied in regard to cholesterol metabolism because of its potential therapeutic value in lowering serum cholesterol levels. When rat hepatocytes were cultured in compactin for 24 hours, dolichyl-P synthesis was inhibited 77–91%, leading the authors to suggest that dolichyl-P was mainly synthesized via the de novo pathway and that CTP-mediated phosphorylation was of limited functional importance. However, another study with compactin indicated that the dolichol kinase and the dolichyl-P phosphatase probably play key roles in regulating the cellular level of dolichyl-P (268). When compactin was given to sea urchin embryos at 1–5 μM concentrations, it induced abnormal gastrulation. This effect appeared to be due to the inhibition of dolichyl-P synthesis, since these embryos had a decreased capacity to synthesize N-linked glycoproteins and dolichol-linked saccharides. Furthermore, the compactin effect could be overcome by supplementing the embryos with dolichyl-P, but not with cholesterol and/or coenzyme Q (269). Similar results were found with mouse embryos where development was arrested by compactin or an oxygenated sterol at the

32-cell stage, leaving the blastomeres decompacted (270). In this case also, N-linked glycosylation was inhibited and mevalonate could reverse this inhibition.

It is interesting to note that in UT-1 cells starved for cholesterol by growth in compactin, crystalloid ER is formed, which contains elevated levels of HMG CoA reductase (271). Since the reductase is a glycoprotein containing N-linked high-mannose chains, it is curious that compactin does not prevent its glycosylation. In fact, the oligosaccharide chain from the reductase of UT-1 cells has been characterized as Man_8 to $Man_6(GlcNAc)_2$ structures (236). At any rate, if compactin inhibits dolichyl-P synthesis completely in cells, it should cause the same alteration in N-linked glycoproteins as does tunicamycin. However, effects with compactin are likely to be somewhat more complex, since cholesterol and ubiquinones are also inhibited. Even if these two compounds are added back to cells, one must be concerned about their reaching the proper location and in the appropriate concentration. Finally, compactin probably inhibits isopentenyladenine synthesis as well, and this compound has been reported to affect DNA synthesis. In BHK cells, compactin does inhibit DNA replication, and this inhibition is overcome by adding isopentenyladenine (272).

INHIBITORS OF GLYCOPROTEIN MOVEMENT

During the synthesis and processing of the N-linked oligosaccharides, the glycoproteins are transported from their site of synthesis in the ER through the various Golgi stacks (from cis to trans). Although the signals involved in targeting these proteins to specific locations are not known, a number of compounds called ionophores have been found to perturb the movement of glycoproteins. Thus, the secretion of immunoglobulins by plasma cells was markedly inhibited by monensin, and striking alterations in the ultrastructural appearance of the Golgi were observed (273). The suggestion was made that depletion of Ca^{2+} levels by the ionophore rendered the Golgi vesicles incapable of fusing. Thus vesicular traffic between the ER and the Golgi is perturbed (274).

The movement of various other proteins, such as procollagen and fibronectin, is also inhibited by ionophores (275), as are various enveloped viruses (276), and membrane vacuoles could be visualized by electron microscopy. Immunofluorescence of these vacuoles showed that they contained large accumulations of procollagen and fibronectin. In one study, monensin was used to isolate those cisternae containing accumulated viral capsids, since they sedimented at a higher density (277). In the presence of monensin, terminal glycosylation reactions were inhibited, but incorporation of amino acids into protein and hydroxylation of proline were not affected. Thus, in monensin-treated cells, newly synthesized proteins accumulate in intracellular vacuoles that appear to be derived from the Golgi complex, and some of the posttranslational modifications are blocked. Apparently there exists an acidic subcompartment in the Golgi that may be important in the secretory process (278).

An antiviral antibiotic, brefeldin A, also strongly blocked secretion in rat hepatocytes without affecting protein synthesis. This compound inhibited the terminal glycosylation of α_1 protease inhibitor and haptoglobulin, and also blocked the proteolytic conversion of proalbumin to albumin. These two modifications occur in the trans Golgi. However, proteolytic processing of haptoglobulin, which occurs in the ER, was not affected. Brefeldin A had no effect on the endocytic pathway (279). Thus, this agent appears to impede protein transport from the endoplasmic reticulum to the Golgi by a mechanism different from that of monensin.

ACKNOWLEDGMENTS

The author expresses his sincere appreciation to Ms. Grace Everett for an excellent job of editing and typing. Research from the author's laboratory was supported by grants from the National Institutes of Health (HL 17783 and AM 21800) and the Robert A. Welch Foundation.

Literature Cited

1. Kobata, A. 1984. In *Biology of Carbohydrates*, ed. V. Ginsburg, P. W. Robbins, 2:87–161. New York: Wiley-Interscience
2. Vliegenthart, J. F. G., Dorland, L., van Halbeek, H. 1983. *Adv. Carbohydr. Chem. Biochem.* 41:209–373
3. Hubbard, S. C., Ivatt, R. J. 1981. *Ann. Rev. Biochem.* 50:555–83
4. Kornfeld, R., Kornfeld, S. 1985. *Ann. Rev. Biochem.* 54:631–64
5. Schachter, H., Narasimhan, S., Gleeson, P., Vella, G. 1983. *Can. J. Biochem. Cell Biol.* 61:1049–66
6. Olden, K., Parent, J. B., White, S. L. 1982. *biochim. Biophys. Acta* 650:209–32
7. Schwarz, R. T., Datema, R. 1982. *Adv. Carbohydr. Chem. Biochem.* 40:287–379
8. Elbein, A. D. 1984. *Crit. Rev. Biochem.* 16:21–47
9. Wagh, P. V., Bahl, O. P. 1981. *Crit. Rev. Biochem.* 14:307–77
10. Takatsuki, A., Arima, K., Tamura, G. 1971. *J. Antibiot.* 24:215–21
11. Tkacz, J. S. 1981. In *Antibiotics, Modes and Mechanisms of Microbial Growth Inhibitors*, 6:1–52. Berlin/New York: Springer-Verlag
12. Tamura, G., ed. 1982. *Tunicamycins.* Tokyo: Japan Sci. Soc. Press
13. Ito, T., Kodama, Y., Kawamura, K., Suzuki, K., Takatsuki, A., Tamura, G. 1977. *Agric. Biol. Chem.* 41:2303–5
14. Ito, T., Kodama, Y., Kawamura, K., Suzuki, K., Takatsuki, A., Tamura, G. 1979. *Agric. Biol. Chem.* 43:1187–95
15. Mahoney, W. C., Duksin, D. 1980. *J. Chromatogr.* 198:506–10
16. Keenan, R. W., Hamill, R. L., Occolowitz, J. L., Elbein, A. D. 1980. *Biochemistry* 70:2968–73
17. Suami, T., Fukuda, Y., Yamamoto, J., Saito, Y., Ito, M., Ohba, S. 1982. *J. Carbohydr. Chem.* 1:9–13
18. Suami, T., Sasai, H., Matsuno, K., Suzuki, N. 1985. *Carbohydr. Res.* 143:85–96
19. Eckardt, K., Wetzstein, H., Thrum, H., Ihn, W. 1980. *J. Antibiot.* 33:908–10
20. Elbein, A. D., Occolowitz, J. L., Hamill, R. L., Eckardt, K. 1981. *Biochemistry* 20:4210–16
21. Tkacz, J. S., Wong, A. 1978. *Fed. Proc.* 37:1766
22. Murazumi, N., Yamamori, S., Araki, Y., Ito, E. 1979. *J. Biol. Chem.* 254:11791–93

23. Kenig, M., Reading, C. 1979. *J. Antibiot.* 32:549–54
24. Vogel, P., Stynes, B. A., Coackley, W., Yeoh, G. T., Petterson, D. S. 1982. *Biochem. Biophys. Res. Commun.* 105:835–40
25. Frahn, J. L., Edgar, J. A., Jones, A. J., Cockrum, P. A., Anderton, N., Culvenor, C. C. J. 1984. *Aust. J. Chem.* 37:165–82
26. Tkacz, J. S., Lampen, J. O. 1975. *Biochem. Biophys. Res. Commun.* 65: 248–53
27. Ericson, M., Gafford, J., Elbein, A. D. 1977. *J. Biol. Chem.* 252:7431–33
28. Struck, D. K., Lennarz, W. J. 1977. *J. Biol. Chem.* 252:1007–13
29. Lehle, L., Tanner, W. 1976. *FEBS Lett.* 71:167–80
30. Reitman, M. L., Kornfeld, S. 1981. *J. Biol. Chem.* 256:4275–81
31. Bettinger, G. E., Young, F. E. 1975. *Biochem. Biophys. Res. Commun.* 67: 16–21
32. Bracha, R., Glaser, L. 1976. *Biochem. Biophys. Res. Commun.* 72:1098–98
33. Ward, J. B., Wyke, A. W., Curtis, C. A. 1980. *Biochem. Soc. Trans.* 8:164–66
34. Heifetz, A., Keenan, R. W., Elbein, A. D. 1979. *Biochemistry* 18:2186–92
35. Keller, R. K., Boon, D. Y., Crum, F. C. 1979. *Biochemistry* 18:36–64
36. Takatsuki, A., Tamura, G. 1982. See Ref. 12, pp.
37. Elbein, A. D., Gafford, J., Kang, M. S. 1979. *Arch. Biochem. Biophys.* 196: 311–18
38. Duksin, D., Mahoney, W. C. 1982. *J. Biol. Chem.* 257:3105–9
39. Kuo, S-C., Lampen, J. O. 1976. *Arch. Biochem. Biophys.* 172:574–81
40. Hickman, S., Kulczycki, A. Jr., Lynch, R. G., Kornfeld, S. 1977. *J. Biol. Chem.* 252:4402–8
41. Duksin, D., Bornstein, P. 1977. *J. Biol. Chem.* 252:955–62
42. Struck, D. K., Suita, P. B., Lane, M. D., Lennarz, W. J. 1978. *J. Biol. Chem.* 253:5332–37
43. Olden, K., Pratt, R. M., Yamada, K. M. 1978. *Cell* 13:461–73
44. Digglemann, H. 1979. *J. Virol.* 30:799–804
45. McLawhon, R. W., Cermak, D., Ellory, J. C., Dawson, G. 1983. *J. Neurochem.* 41:1286–96
46. Mizunaga, T., Noguchi, T. 1982. *J. Biochem.* 91:191–200
47. Sibley, C. H., Wagner, R. A. 1981. *J. Immunol.* 126:1868–73
48. Hasilik, A., Tanner, W. 1978. *Eur. J. Biochem.* 91:567–75
49. Schwaiger, H., Tanner, W. 1979. *Eur. J. Biochem.* 102:375–79
50. Seagar, M., Miquelis, R. D., Simon, C. 1980. *Eur. J. Biochem.* 113:91–96
51. Borgford, T. J., Hurta, R. A., Tough, D. F., Burton, D. N. 1986. *Arch. Biochem. Biophys.* 244:502–16
52. Ito, F., Chou, J. Y. 1984. *J. Biol. Chem.* 259:14997–99
53. Duksin, D., Bornstein, P. 1977. *Proc. Natl. Acad. Sci. USA* 74:3433–37
54. Olden, K., Pratt, R. M., Yamada, K. M. 1979. *Int. J. Cancer* 24:60–66
55. Irimura, T., Nicolson, G. L. 1981. *J. Supramol. Struct. Cell. Biochem.* 17: 325–36
56. Tanzer, M. L., Rowland, I. N., Murray, L. W., Kaplan, J. 1977. *Biochem. Biophys. Acta* 506:187–96
57. Ledford, B. E., Davis, D. F. 1983. *J. Biol. Chem.* 258:3304–8
58. Katz, N. R., Goldfarb, V., Liem, H., Müller-Eberhard, U. 1985. *Eur. J. Biochem.* 146:155–59
59. Mizrahi, A., O'Malley, J. A., Carter, W. A., Takatsuki, A., Tamura, G., Sulkowski, E. 1978. *J. Biol. Chem.* 253: 7612–15
60. Fujisawa, J., Iwakura, Y., Kawade, Y. 1978. *J. Biol. Chem.* 253:8677–79
61. Miller, A. L., Kress, B. C., Lewis, L., Stein, R., Kinnon, C. 1980. *Biochem. J.* 186:971–75
62. Cox, G. S. 1981. *Biochemistry* 20: 4893–900
63. Budarf, M. L., Herbert, E. 1982. *J. Biol. Chem.* 257:10128–35
64. Swanson, J. C., Suttie, J. W. 1985. *Biochemistry* 24:3890–97
65. Amri, E. Z., Vannier, C., Etienne, J., Ailhaud, G. 1986. *Biochem. Biophys. Acta* 875:334–43
66. Sidman, C., Potash, M. J., Kohler, G. 1981. *J. Biol. Chem.* 256:13180–87
67. Sidman, C. 1981. *J. Biol. Chem.* 256:9374–76
68. Strous, G., Lodish, H. F. 1980. *Cell* 27:709–17
69. Bauer, H. C., Parent, J. B., Olden, K. 1985. *Biochem. Biophys. Res. Commun.* 138:368–75
70. Farquhar, M. G. 1985. *Ann. Rev. Cell Biol.* 1:447–88
71. Fitting, T., Kabat, D. 1982. *J. Biol. Chem.* 257:14011–17
72. Scheele, G., Tartakoff, A. 1985. *J. Biol. Chem.* 260:926–31
73. Lodish, H. F., Kong, N. 1984. *J. Cell Biol.* 98:1720–29
74. Perkins, J. P., Toews, M. L., Harden, T. K. 1984. *Adv. Cyclic Nucleotide Protein Phosphoryl. Res.* 17:37–46
75. Merlie, J. P., Sebbane, R., Tzartos, S.,

Lindstrom, J. 1982. *J. Biol. Chem.* 257:2694–701

76. Prives, J. M., Olden, K. 1980. *Proc. Natl. Acad. Sci. USA* 77:5263–67

77. Deutsch, P. J., Rosen, O. M., Ruben, C. S. 1982. *J. Biol. Chem.* 257:5350–58

78. Reid, B. C., Ronnett, G. V., Lane, M. D. 1981. *Proc. Natl. Acad. Sci. USA* 78:2908–12

79. Schwartz, I., Hazum, E. 1985. *Endocrinology* 116:2341–46

80. Bhargawa, G., Makman, M. H. 1980. *Biochim. Biophys. Acta* 629:107–12

81. Slieker, L. J., Lane, M. D. 1985. *J. Biol. Chem.* 260:687–90

82. Baribault, T. J., Neet, K. E. 1985. *J. Neurosci. Res.* 14:49–60

83. Law, P. Y., Ungar, H. G., Horn, D. S., Loh, H. H. 1985. *Biochem. Pharmacol.* 34:9–17

84. Chattergee, S., Kwiterovich, P. O. Jr., Sekerke, C. S. 1979. *J. Biol. Chem.* 254:3704–7

85. Filipovic, I., von Figura, K. 1980. *Biochem. J.* 181:373–75

86. Breitfeld, P. P., Rup, D., Schwartz, A. L. 1984. *J. Biol. Chem.* 259:5310–15

87. Kranitzky, W., Durham, D. L., Hart, D. A., Eidels, L. 1985. *Infect. Immun.* 49:336–43

88. Ronnett, G. V., Lane, M. D. 1981. *J. Biol. Chem.* 256:4704–7

89. Schwarz, R. T., Rohrschneider, J. M., Schmidt, M. F. G. 1976. *J. Virol.* 19:782–91

90. Leavitt, R., Schlesinger, S., Kornfeld, S. 1977. *J. Virol.* 21:375–85

91. Takatsuki, A., Tamura, G. 1978. *Agric. Biol. Chem.* 42:275–78

92. Schultz, A. M., Oroszlan, S. 1979. *Biochem. Biophys. Res. Commun.* 86:1206–13

93. Polonoff, E., Machida, C. A., Kabat, D. A. 1982. *J. Biol. Chem.* 257:14023–28

94. Schmaljohn, C. S., Hasty, S. E., Rasmussen, L., Dalrymple, J. M. 1986. *J. Gen Virol.* 67:707–17

95. Petrie, B. L., Estes, M. K., Graham, D. Y. 1983. *J. Virology* 46:270–74

96. Weintraub, B. D., Stannard, B. S., Meyers, L. 1983. *Endocrinology* 112:1331–35

97. Loh, Y. P., Gainer, H. 1979. *Endocrinology* 105:474–87

98. Dulis, B. H., Kloppel, T. M., Grey, H. M., Kubo, R. T. 1982. *J. Biol. Chem.* 257:4369–74

99. Nakamura, K., Compans, R. W. 1978. *Virology* 84:303–19

100. Gibson, R., Schlesinger, S., Kornfeld, S. 1979. *J. Biol. Chem.* 254:3600–7

101. Hart, G., Lennarz, W. J. 1978. *J. Biol. Chem.* 253:5795–801

102. Lohmander, L. S., Fellini, S. A., Kimura, J. H., Stevens, P. L., Hascall, V. C. 1983. *J. Biol. Chem.* 258:12280–86

103. Yusuf, H. K. M., Pohlentz, G., Schwarzmann, G., Sandhoff, K. 1983. *Eur. J. Biochem.* 134:47–54

104. Cecchelli, R., Cacan, R., Verbert, A. 1985. *Eur. J. Biochem.* 153:111–16

105. Olden, K., Pratt, R. M., Jaworski, C., Yamada, K. M. 1979. *Proc. Natl. Acad. Sci. USA* 76:791–95

106. Kitagawa, K., Nishino, M., Iwashima, A. 1985. *Biochim. Biophys. Acta* 821:67–71

107. Haspel, H. C., Birnbaum, M. J., Wilk, E. W., Rosen, O. M. 1985. *J. Biol. Chem.* 260:7219–25

108. White, M. K., Bramwell, M. E., Harris, H. 1984. *J. Cell. Sci.* 68:257–70

109. Barber, E. F., Handlogten, M. E., Kilberg, M. S. 1983. *J. Biol. Chem.* 258:11851–55

110. Ashwell, G., Harford, J. 1982. *Ann. Rev. Biochem.* 51:531–54

111. von Figura, K., Hasilik, A. 1986. *Ann. Rev. Biochem.* 56:167–94

112. Cereijido, M., Robbins, E. S., Dolan, W. J., Rotunno, C. A., Sabatini, D. D. 1978. *J. Cell. Biol.* 77:853–80

113. Roth, M. G., Fitzpatrick, J. P., Compans, R. W. 1975. *Proc. Natl. Acad. Sci. USA* 76:6430–34

114. Green, R. F., Meiss, H. K., Rodriguez-Boulan, E. 1981. *J. Cell Biol.* 89:230–39

115. Bauerle, P. A., Huttner, W. B. 1984. *EMBO J.* 3:2209–15

116. Tanaka, H., Oiwa, R., Matsukura, S., Inokoshi, J., Omura, S. 1987. *J. Antibiot.* 35:1216–21

117. Wedgwood, J. F., Strominger, J. L. 1980. *J. Biol. Chem.* 255:1120–23

118. Rogers, H. J., Perkins, H. R., Ward, J. B. 1980. *Microbial Cell Walls and Membranes*, p. 298. London: Chapman & Hill

119. Visser, S. W., Roy-Burman, S. 1979. *Antibiotics* 5:363

120. Bodanszky, M., Sigler, G. F., Bodanszky, A. 1973. *J. Am. Chem. Soc.* 95:2352–57

121. Kang, M. S., Spencer, J. P., Elbein, A. D. 1978. *J. Biol. Chem.* 253:8860–66

122. Hayes, G. R., Lucas, J. J. 1983. *J. Biol. Chem.* 258:15095–100

123. Kyosseva, S. V., Zhivkov, V. I. 1985. *Int. J. Biochem.* 17:813–17

124. Chapman, A., Fujimoto, K., Kornfeld, S. 1980. *J. Biol. Chem.* 255:4441–46

125. Jensen, J. W., Schutzbach, J. S. 1981. *J. Biol. Chem.* 256:12899–904
126. Banerjee, D. K., Scher, M. G., Waechter, C. J. 1981. *Biochemistry* 20:1561–68
127. Katal, A., Prakash, C., Vijay, I. K. 1984. *Eur. J. Biochem.* 141:521–26
128. Haselbeck, A., Tanner, W. 1982. *Proc. Natl. Acad. Sci. USA* 79:1520–24
129. Elbein, A. D. 1981. *Biochem. J.* 193:477–84
130. Chen, W. W., Lennarz, W. J. 1976. *J. Biol. Chem.* 251:7802–9
131. Herscovics, A., Bugge, B., Jeanloz, R. W. 1977. *FEBS. Lett.* 82:800–04
132. Reuvers, F., Boer, P., Steyn-Parve, E. P. 1978. *Biochem. Biophys. Res. Commun.* 82:800–5
133. Spencer, J. P., Kang, M. S., Elbein, A. D. 1978. *Arch. Biochem. Biophys.* 190:829–32
134. Babczinski, P. 1972. *Eur. J. Biochem.* 112:53–58
135. Villemez, C. L., Carlo, P. L. 1980. *J. Biol. Chem.* 255:8174–178
136. Kean, E. L. 1983. *Biochem. Biophys. Acta* 752:488–90
137. Bause, E., Legler, G. 1982. *Biochem. J.* 201:481–87
138. Muller, T., Bause, E., Jaenicke, L. 1981. *FEBS Lett.* 128:208–12
139. Kang, M. S., Spencer, J. P., Elbein, A. D. 1979. *J. Biol. Chem.* 254:10037–43
140. Kaushal, G. P., Elbein, A. D. 1985. *J. Biol. Chem.* 260:16303–9
141. Kaushal, G. P., Elbein, A. D. 1986. *Arch. Biochem. Biophys.* 250:38–47
142. Kaluza, G., Schmidt, M. F. G., Scholtissek, C. 1973. *Virology* 54:179–89
143. Schwarz, R. T., Schmidt, M. F. G., Anwer, U., Klenk, H. D. 1977. *J. Virol.* 23:217–26
144. Datema, R., Lezica, R. P., Robbins, P. W., Schwarz, R. T. 1981. *Arch. Biochem. Biophys.* 206:65–71
145. Datema, R., Schwarz, R. 1978. *Eur. J. Biochem.* 90:505–16
146. Steiner, S., Courtney, R. J., Melnick, J. L. 1973. *Cancer Res.* 33:2402–7
147. McDowell, W., Datema, R., Romero, P. A., Schwarz, R. T. 1985. *Biochemistry* 24:8145–52
148. Grier, J. J., Rasmussen, J. R. 1984. *J. Biol. Chem.* 259:1027–30
149. Koch, H. U., Schwarz, R. T., Scholtissek, C. 1979. *Eur. J. Biochem.* 94:515–22
150. Datema, R., Schwarz, R. T. 1979. *Biochem. J.* 184:113–23
151. Friedman, S. J., Skeham, P. 1980. *Proc. Natl. Acad. Sci. USA* 77:1172–76
152. Pan, Y. T., Elbein, A. D. 1982. *J. Biol. Chem.* 257:2795–801
153. Pan, Y. T., Elbein, A. D. 1985. *Arch. Biochem. Biophys.* 242:447–56
154. Schlostissek, C. 1975. *Curr. Top. Microbiol. Immunol.* 70:101–19
155. Datema, R., Schwarz, R. T. 1981. *J. Biol. Chem.* 256:11191–98
156. Spiro, R. G., Spiro, M. J., Bhoyroo, V. D. 1983. *J. Biol. Chem.* 258:9469–76
157. Turco, S. 1980. *Arch. Biochem. Biophys.* 205:330–39
158. Sefton, B. M. 1977. *Cell* 10:659–69
159. Rearick, J. J., Chapman, A., Kornfeld, S. 1981. *J. Biol. Chem.* 256:6255–61
160. Gershman, H., Robbins, P. W. 1981. *J. Biol. Chem.* 256:7774–80
161. Turco, S. J., Pickard, J. L. 1982. *J. Biol. Chem.* 257:8674–79
162. Baumann, H., Jahreis, G. P. 1983. *J. Biol. Chem.* 258:3942–49
163. Fuhrmann, U., Bause, E., Ploegh, H. 1985. *Biochim. Biophys. Acta.* 825:95–110
164. Nielsen, D. B., James, L. F. 1985. In *Plant Toxicology*, ed. A. A. Seawright, M. P. Hegartz, L. F. James, R. F. Keeler, pp. 24–31. Yeerongpilly, Australia: Queensland Poisonous Plant Committee
165. Hartley, W. J. 1978. In *Effects of Poisonous Plants on Livestock*, ed. R. F. Keeler, K. R. Van Kampen, L. F. James, pp. 363–95. New York: Academic
166. Colegate, S. M., Dorling, P. R., Huxtable, C. R. 1979. *Aust. J. Chem.* 32:2257–64
167. Molyneux, R. J., James, L. F. 1982. *Science* 216:190–92
168. Davis, D., Schwarz, P., Hernandez, T., Mitchell, M., Warnock, B., Elbein, A. D. 1984. *Plant Physiol.* 76:972–75
169. Dorling, P. R., Huxtable, C. R., Colegate, S. M. 1980. *Biochem. J.* 191:649–51
170. Kang, M. S., Elbein, A. D. 1983. *Plant Physiol.* 71:551–54
171. Tulsiani, D. P. R., Broquist, H. P., Touster, O. 1985. *Arch. Biochem. Biophys.* 236:427–34
172. Lalegerie, P., Legler, G., Yon, J. M. 1982. *Biochemie* 64:977–1000
173. Broquist, H. P. 1985. *Ann. Rev. Nutr.* 5:391–409
174. Dorling, P. R., Huxtable, C. R., Vogel, P. 1978. *Neuropathol. Appl. Neurobiol.* 4:285–95
175. Tulsiani, D. P. R., Broquist, H. P., James, L. F., Touster, O. 1984. *Arch. Biochem. Biophys.* 232:76–85
176. Sadeh, S., Warren, C. D., Daniel, P. F., Bugge, B., James, L. F., Jeanloz, R. W. 1983. *FEBS Lett.* 163:104–19
177. Daniel, P. F., Warren, C. D., James, L. F. 1984. *Biochem. J.* 221:601–7

178. Di Bello, I. C., Dorling, P., Winchester, B. 1983. *Biochem. J.* 215:693–96
179. Abraham, D. J., Sidebotham, R., Winchester, B. G., Dorling, P. R., Dell, A. 1983. *FEBS Lett.* 163:110–13
180. Abraham, D., Blakemore, W. F., Jolly, R. D., Sidebotham, R., Winchester, B. 1983. *Biochem. J.* 215:573–79
181. Elbein, A. D., Solf, R., Vosbeck, K., Dorling, P. R. 1981. *Proc. Natl. Acad. Sci. USA* 78:7393–97
182. Tulsiani, D. P. R., Harris, T. M., Touster, O. 1982. *J. Biol. Chem.* 257:7936–39
183. Bischoff, J., Kornfeld, R. 1986. *J. Biol. Chem.* 261:4758–65
184. Kang, M. S., Elbein, A. D. 1983. *J. Virology* 46:60–69
185. Arumughan, R. G., Tanzer, M. L. 1983. *J. Biol. Chem.* 258:11883–89
186. Foddy, L., Feeney, J., Hughes, R. C. 1986. *Biochem. J.* 233:697–706
187. Franc, J. L., Houseplan, S., Fayet, G., Bauchilloux, S. 1986. *Eur. J. Biochem.* 157:225–32
188. Whitsett, J. A., Ross, G., Weaver, T., Rice, W., Dion, C., Hull, W. 1985. *J. Biol. Chem.* 260:15273–79
189. Le, A. Y., Doyle, D. 1985. *Biochemistry* 24:6238–45
190. Wagner, D. D., Mayadas, T., Urban-Pickering, M., Lewis, B. H., Marder, V. J. 1985. *J. Cell Biol.* 101:112–20
191. Gross, V., Tran-Thi, T-A., Vosbeck, K., Heinrich, P. C. 1983. *J. Biol. Chem.* 258:4032–36
192. Yeo, T. K., Yeo, K. T., Parent, J. B., Olden, K. 1985. *J. Biol. Chem.* 260:2565–69
193. Duronio, V., Jacobs, S., Cuatrecasas, P. 1986. *J. Biol. Chem.* 261:970–75
194. Soderquist, A. M., Carpenter, G. 1984. *J. Biol. Chem.* 259:12586–94
195. Breitfeld, P. P., Rup, D., Schwartz, A. L. 1984. *J. Biol. Chem.* 259:10414–21
196. Chung, K. M., Shepard, V. L., Stahl, P. 1984. *J. Biol. Chem.* 259:14637–41
197. Elbein, A. D., Dorling, P. R., Vosbeck, K., Horisberger, M. 1982. *J. Biol. Chem.* 257:1573–76
198. Repp, R., Tamura, T., Boschek, C. B., Wege, H., Schwarz, R. T., Niemann, H. 1985. *J. Biol. Chem.* 260:15873–79
199. Hadwiger, A., Niemann, H., Kablisch, A., Bauer, H., Tamura, T. 1986. *EMBO J.* 5:689–94
200. Schwartz, P., Elbein, A. D. 1985. *J. Biol. Chem.* 260:14452–58
201. Merkle, R., Elbein, A. D., Heifetz, A. 1985. *J. Biol. Chem.* 260:1083–89
202. Bar-Scavit, Z., Kahn, A. J., Pegg, L. E., Stone, R. R., Teitelbaum, S. L. 1984. *J. Clin. Invest.* 73:1277–83
203. Villalta, F., Kierszenbaum, F. 1985. *Mol. Biochem. Parasitol.* 16:1–10
204. Humphries, M. J., Matsumoto, K., White, S. L., Olden, K. 1986. *Proc. Natl. Acad. Sci. USA* 83:1752–56
205. Hino, M., Nakayama, O., Tsurami, Y., Adachi, K., Shibata, T., et al., 1985. *J. Antibiot.* 38:926–35
206. Winkler, J. R., Segal, H. L. 1984. *J. Biol. Chem.* 259:1958–62
207. Suami, T., Tadano, K. I., Iimura, Y. 1985. *Carbohydr. Res.* 136:67–75
208. Ali, M. H., Hough, L., Richardson, A. C. 1984. *J. Chem. Soc. Chem. Commun.* 1984(7):447–48
209. Yasuda, N., Tsutsumi, H., Takaya, T. 1984. *Chem. Lett.* 1984:1201–4
210. Adams, C. E., Walker, F. T., Sharpless, K. B. 1985. *J. Org. Chem.* 50:420–24
211. Elbein, A. D., Szumilo, T., Sanford, B. A., Sharpless, K. D., Adams, C. E. 1987. *Biochemistry.* In press
212. Yasuda, N., Tsutsumi, H., Takaya, T. 1985. *Chem. Lett.* 1985(1):31–34
213. Hohenschutz, L. D., Bell, E. A., Jewess, P. J., Leworthy, D. P., Pryce, R. J., et al. 1981. *Phytochemistry* 20:811–14
214. Saul, R., Chambers, J. P., Molyneux, R. J., Elbein, A. D. 1983. *Arch. Biochem. Biophys.* 221:265–75
215. Szumilo, T., Kaushal, G. P., Elbein, A. D. 1986. *Arch. Biochem. Biophys.* 247:261–71
216. Pan, Y. T., Hori, H., Saul, R., Sanford, B. A., Molyneux, R. J., Elbein, A. D. 1983. *Biochemistry* 22:3975–84
217. Palamarczyk, G., Elbein, A. D. 1985. *Biochem. J.* 227:795–804
218. Sasak, V. W., Ordovas, J. M., Elbein, A. D., Berninger, R. W. 1985. *Biochem. J.* 232:759–66
219. Frommer, W., Junge, B., Mueller, L., Schmidt, D., Truscheit, E. 1979. *Planta Med.* 35:195–207
220. Saunier, B., Kilker, R. D., Tkacz, J. S. Jr., Quaroni, A., Herscovics, A. 1987. *J. Biol. Chem.* 257:14155–61
221. Hettkamp, H., Bause, E., Legler, G. 1982. *Biosci. Rep.* 2:899–906
222. Gross, V., Andus, T., Tran-Thi, T-A., Schwarz, R. T., Decker, K., Heinrich, P. C. 1983. *J. Biol. Chem.* 258:12203–9
223. Lemansky, P., Grieselmann, V., Hasilik, A., von Figura, K. 1984. *J. Biol. Chem.* 259:10129–35
224. Peyrieras, N., Bause, E., Legler, G., Vasilov, R., Claesson, L., et al. 1983. *EMBO J.* 2:823–32

225. Romero, P. A., Friedlander, P., Herscovics, A. 1985. *FEBS Lett.* 183:29–32
226. Schlesinger, S., Koyama, A. H., Malfer, C., Gee, S. L., Schlesinger, M. J. 1985. *Virus Res.* 2:139–49
227. Schlesinger, S., Malfer, C., Schlesinger, M. J. 1984. *J. Biol. Chem.* 259:7597–601
228. Burke, B., Matlin, K., Bause, E., Legler, G., Peyrieras, N., Ploegh, H. 1984. *EMBO J.* 3:551–56
229. Saul, R., Ghidoni J. J., Molyneux, R. J., Elbein, A. D. 1985. *Proc. Natl. Acad. Sci. USA* 82:93–97
230. Molyneux, R. J., Roitman, J. N., Dunnheim, G., Szumilo, T., Elbein, A. D. 1986. *Arch. Biochem. Biophys.* 251:450–57
231. Legler, G., Julich, E. 1984. *Carbohydr. Res.* 128:61–72
232. Fuhrmann, U., Bause, E., Legler, G., Ploegh, H. 1984. *Nature* 307:755–58
233. Elbein, A. D., Legler, G., Tlusty, A., McDowell, W., Schwarz, R. T. 1984. *Arch. Biochem. Biophys.* 235:579–88
234. Gross, V., Steube, K., Tran-Thi, T-A., McDowell, W., Schwarz, R. T., et al. 1985. *Eur. J. Biochem.* 150:41–46
235. Snider, M. D., Rogers, O. C. 1986. *J. Cell Biol.* 103:265–75
236. Bischoff, J., Liscum, L., Kornfeld, R. 1986. *J. Biol. Chem.* 261:4766–74
237. Welter, A., Jadot, J., Dardenne, G., Marlier, M., Casimir, J. 1976. *Phytochemistry* 15:747–49
238. Evans, S. V., Fellows, L. E., Bell, E. A. 1985. *Phytochemistry* 22:768–70
239. Elbein, A. D., Mitchell, M., Sanford, B. A., Fellows, L. E., Evans, S. V. 1984. *J. Biol. Chem.* 259:12409–13
240. Romero, P. A., Friedlander, P., Fellows, L. E., Evans, S. V., Herscovics, A. 1985. *FEBS Lett.* 184:197–201
241. Schmidt, J. A., Beug, H., Hayman, M. J. 1985. *EMBO J.* 4:105–12
242. Fleet, G. W. J., Smith, P. W. 1985. *Tetrahedron Lett.* 26:1469–72
243. Fleet, G. W. J., Smith, P. W., Evans, S. V., Fellows, L. E. 1984. *J. Chem. Soc. Chem. Commun.* 1985:1240–41
244. Palamarczyk, G., Mitchell, M., Smith, P. W., Fleet, G. W. J., Elbein, A. D. 1985. *Arch. Biochem. Biophys.* 243:35–45
245. Legler, G. 1977. *Methods Enzymol.* 46:368–81
246. Kamfer, J. N., Legler, G., Sullivan, J., Rughavan, S. S., Mumford, R. A. 1975. *Biochem. Biophys. Res. Commun.* 67:85–90
247. Legler, G., Lotz, W. 1973. *Hoppe Seyler's Z. Physiol. Chem.* 354:243–54
248. Datema, R., Romero, P. A., Legler, G., Schwarz, R. T. 1982. *Proc. Natl. Acad. Sci. USA* 79:6787–91
249. Van Diggelen, O. P., Galjaard, H., Sinnott, M. L., Smith, P. J. 1980. *Biochem. J.* 188:337–43
250. Docherty, P. A., Kuranda, M. J., Aronson, N. N. Jr., Be Miller, J. N., Myers, R. W., Bohn, J. A. 1986. *J. Biol. Chem.* 261:3457–63
251. Marshall, R. D. 1972. *Ann. Rev. Biochem.* 41:673–702
252. Bause, E., Legler, G. 1981. *Biochem. J.* 195:639–44
253. Aubert, J. P., Biserte, G., Loucheux-Lefebore, M. H. 1976. *Arch. Biochem. Biophys.* 175:410–18
254. Hortin, G., Boime, I. 1980. *J. Biol. Chem.* 255:8007–10
255. Hentze, M., Hasilik, A., von Figura, K., 1984. *Arch. Biochem. Biophys.* 230:375–82
256. Docherty, P. A., Aronson, N. N. Jr. 1985. *J. Biol. Chem.* 260:10847–55
257. Stern, A. M., Foxman, B. M., Tashjian, A. H. Jr., Abeles, R. H. 1982. *J. Med. Chem.* 25:544–50
258. Hortin, G., Stern, A. M., Miller, B., Abeles, R. H., Boime, I. 1983. *J. Biol. chem.* 261:6461–50
259. Rathod, P. K., Tashjian, A. H. Jr., Abeles, R. H. 1986. *J. Biol. Chem.* 261:6461–69
260. Schmidt, J., Elbein, A. D. 1979. *J. Biol. Chem.* 254:12291–94
261. Hubbard, S. C., Robbins, P. W. 1980. *J. Biol. Chem.* 255:11782–93
262. Grant, W., Lennarz, W. J. 1983. *Eur. J. Biochem.* 134:575–83
263. Mills, J. T., Adamany, A. M. 1978. *J. Biol. Chem.* 253:5270–73
264. James, M. J., Kandutsch, A. A. 1979. *J. Biol. Chem.* 254:8442–46
265. Keller, R. K., Adair, W. L. Jr., Ness, G. C. 1979. *J. Biol. Chem.* 254:9966–69
266. White, D. A., Middleton, B., Pawson, S., Bradshaw, J. P., Clegg, R. J., et al. 1981. *Arch. Biochem. Biophys.* 208:30–36
267. Goldstein, J. L., Brown, M. S., Anderson, R. G. W., Russell, D. W., Schneider, W. J. 1985. *Ann. Rev. Cell Biol.* 1:1–39
268. Astrand, I. M., Fries, E., Chojnacki, T., Dallner, G. 1986. *Eur. J. Biochem.* 155:447–52
269. Carson, D. P., Lennarz, W. J. 1979. *Proc. Natl. Acad. Sci. USA* 76:5709–13
270. Surani, M. A. H., Kimber, S. J., Osborn, J. C. 1983. *J. Embryol. Exp. Morphol.* 75:205–23

271. Orci, L., Brown, M. S., Goldstein, J. L., Garcia-Legura, L. M., Anderson, R. G. 1984. *Cell* 36:835–45

272. Huneeus, V. Q., Wiley, M. H., Siperstein, M. D. 1980. *Proc. Natl. Acad. Sci. USA* 77:5842–46

273. Tartakoff, A. M., Vassalli, P., Detraz, M. 1977. *J. Exp. Med.* 146:1332–45

274. Uchida, N., Smilowitz, H., Tanzer, M. L. 1979. *Proc. Natl. Acad. Sci. USA* 76:1868–72

275. Tartakoff, A. M. 1983. *Cell* 32:1026–28

276. Johnson, D. C., Schlesinger, M. J. 1980. *Virology* 103:407–9

277. Quinn, P., Griffiths, G., Warren, G. 1983. *J. Cell Biol.* 96:851–56

278. Anderson, R. G. W., Pathak, R. K. 1985. *Cell* 40:635–43

279. Misumi, Y., Miki, K., Takatsuki, A., Tamura, G., Ikehara, Y. 1986. *J. Biol. Chem.* 261:11398–403

Ann. Rev. Biochem. 1987. 56:535–65

THE NUCLEUS: STRUCTURE, FUNCTION, AND DYNAMICS

John W. Newport and Douglass J. Forbes

Department of Biology, University of California at San Diego, La Jolla, California 92093

CONTENTS

I. PERSPECTIVES AND SUMMARY

The enclosure of DNA within its own compartment or organelle, the nucleus, was a major evolutionary event that separated eukaryotic from prokaryotic organisms. The development of the nucleus was likely the result of increased organizational and regulatory problems associated with the large genomes of more complex organisms. The unique organization of the nucleus solves these problems in several ways. The nuclear envelope separates DNA from cytoplasmic processes, while selective transport through the nuclear pores

535

0066-4154/87/0701-0535$02.00

creates a unique biochemical environment within the nucleus. The nuclear lamina located between the inner nuclear membrane and chromatin acts to support nuclear shape and serves perhaps as a rigid framework for chromatin attachment and organization. Within the nucleus the DNA is further organized into loop domains. All of these elements undoubtedly contribute to the remarkable efficiency and fidelity of nuclear functions, such as DNA replication and recombination, and RNA transcription and processing.

In the last five years significant progress has been made in understanding both the function of specific nuclear components and the way in which these components interact in the nucleus. Macromolecular traffic into and out of the nucleus is now known to be regulated by an active transport system involving the nuclear pores and specific signal sequences. Transport of certain proteins into the nucleus appears to be developmentally regulated. This raises the interesting possibility that minor changes in transport selectivity may have major effects on nuclear structure and function. Progress towards understanding the mechanics of nuclear transport at the molecular level has been significantly enhanced by the identification of the first two pore-specific proteins and by development of in vitro transport systems.

The composition and regulation of the nuclear lamina is, in many cases, now well understood. Human lamins A and C have been cloned, and sequence analysis reveals that they are related to intermediate filaments. This relationship supports a possible nucleoskeletal role for the lamins. The provocative, albeit unproven, theory that lamins also mediate nuclear membrane attachment to chromosomes has stimulated intense and ongoing research interest in the lamin proteins.

Since the discovery of the packaging of DNA into nucleosomes in the early 1970s, rapid progress has been made towards understanding the higher organization of chromatin fibers in the nucleus. Chromatin is organized into 30–100-kilobase-pair loop domains. Whole chromosomes are attached to the nuclear envelope at defined points. In addition, advances have been made in defining both specific DNA sequences and proteins involved in maintaining these aspects of chromosome organization.

Until recently, dynamic aspects of nuclear structure such as nuclear growth, assembly, and disassembly have been difficult to study. The development of in vitro systems in which nuclear assembly and disassembly occur now allows detailed study of nuclear dynamics at the molecular level. The ability to manipulate these events experimentally should greatly facilitate eventual understanding of nuclear dynamics.

A comprehensive review of nuclear structure in the space allotted would be impossible. We have therefore limited ourselves to reviewing in detail those aspects of the nucleus in which recent experiments have significantly increased our understanding. These include nuclear transport, composition and regulation of the nuclear lamina, chromatin organization within the nucleus, and cell-free systems for studying nuclear dynamics at the biochemical level. Other aspects of the nucleus such as the transcription and processing of messenger and ribosomal RNA, in which much progress has been made, have been left for a more specialized review. Similarly, we do not attempt to present data related to the nuclear matrix or DNA replication, since these have been recently reviewed (1). Previous reviews on all aspects of nuclear structure are cited throughout the text.

II. NUCLEAR ENVELOPE STRUCTURE AND FUNCTION

The Nuclear Membranes

The most prominent feature of the nucleus as visualized in the light microscope is the nuclear envelope. The envelope is composed of two bilayer membranes, the outer and inner nuclear membranes, separated by a perinuclear cisternal space. Nuclear pores perforate the double nuclear membrane and provide a link between cytoplasm and nucleoplasm. Recent reviews have presented a compendium of work to date on these membranes (2, 3, 8), and only a brief synopsis is offered here.

Together the nuclear membranes act as a passive barrier separating nucleus from cytoplasm. Little progress has been made in defining additional functions for the membrane components of the nuclear envelope. Although numerous studies have identified enzymatic activities associated with the nuclear membranes [see Franke (2) for a comprehensive review], these studies have been complicated by the fact that the outer nuclear membrane is continuous at many points with the endoplasmic reticulum (ER) (2, 3). Indeed, most enzymes associated with the nuclear membranes closely resemble those of the ER (see 4–6 for exceptions).

The nuclear membranes and the space that they enclose can perform at least some of the major functions of the endoplasmic reticulum. The outer nuclear membrane clearly contains ribosomes on its cytoplasmic surface (2, 3), as does the rough ER. Puddington et al (7) determined that the outer nuclear membrane is a major site for both synthesis and posttranslational processing of the vesicular stomatitis virus G protein, a secreted protein. In Friend erythroleukemia cells, the nuclear envelope is the site of 40% of G protein synthesis and initial glycosylation (7). Furthermore, both signal peptidase and the appropriate glycosyltransferase activities are present in the nuclear envelopes.

It is not known whether the inner and outer nuclear membranes are identical in protein and lipid composition. Unlike the outer membrane, the inner membrane does not contain attached ribosomes and instead lies closely apposed to the proteinaceous nuclear lamina (see below), which lines the inner nuclear envelope (see 2, 3, 8 for reviews). To biochemically distinguish between the inner and outer nuclear membranes, a separation protocol is necessary. Although it was originally thought that Triton X-100 extraction removed the outer membrane but left the inner membrane intact, Aaronson & Blobel (9) have since shown that this treatment removes all nuclear lipids. Recently, Schindler et al (10) reported that citraconic anhydride removes the outer membrane preferentially. Such differential extration should be useful in identifying biochemical and functional differences between the two nuclear membranes.

The Nuclear Pores

In contrast to the passive barrier of the nuclear membranes, the nuclear pores actively control the nuclear entry and exit of macromolecules. A number of reviews have been written previously on the nuclear envelope, nuclear pores, and nuclear transport (2, 3, 8, 11–25). The present review briefly covers these subjects, and concentrates on recent studies of pore structure and function.

NUCLEAR PORE STRUCTURE   The existence of nuclear pores was suggested by Hertwig in 1876 (26). Having observed a punctate pattern on nuclear envelopes with the light microscope, he proposed that the structures represented channels linking cytoplasm to nucleoplasm. With the advent of the electron microscope, Callan in 1949 (27, 28) first observed the ultrastructure of nuclear pores. Using this technique, a number of labs (see 2, 3 for reviews) resolved specific structures within the pore complex with little variation observed between different organisms. In a recent study, Unwin & Milligan (29) examined nuclear pores at 90 Å resolution using Fourier averaging methods to reveal fine details of pore structure. Their results, like those of previous studies (see 2, 3, 8), describe the pore complex as consisting of two rings of eight globular subunits, one ring affixed to each side of the nuclear envelope. Each ring has an inner diameter of 800 Å and an outer diameter of 1200 Å, the latter being identical to the diameter of the complete pore complex. The subunits have been termed annular granules in previous studies (see 2, 3). Eight spokes extend from the periphery of the pore lumen toward the center of the pore. In cross section these spokes comprise the diaphragm of the pore. These pore structures are built around a precise circular hole in the nuclear membranes of 900 Å diameter. At the edge of this hole the two nuclear membranes are clearly seen to fuse. A large central granule ~350 Å in diameter is observed to occupy the lumen of the pore, but this central granule is not always present. Although thought to be an integral part of pore structure, it has also been hypothesized that the central granule is a ribonucleoprotein particle in transit through the pore (see 2, 3, 30). Lastly, Unwin & Milligan (29) observed particles of 220 Å in diameter, varying in number from 0 to 8 per pore, affixed to the annular granules on the cytoplasmic rim of the pore. They suggest that these resemble ribosomes and may represent a functional part of pore structure.

To summarize, the waist of the pore, defined by the circular membrane opening, is occupied by eight spokes and the central granule. Above and below this waist are eight annular granules connected to the spokes by an equal number of links. On the cytoplasmic surface of the pore are often additional granules that may be part of the pore structure. Different fixation techniques have led to different interpretations of pore structure, with some reports describing filaments extending into the nucleoplasm and/or cytoplasm (29; see 2, 3 for reviews) instead of or appended to the annular granules. Such filaments have been hypothesized to act as tracks that lead molecules to or through the pore or to be ribonucleoprotein fibrils.

Essentially identical pore complexes, as judged by electron microscopy, have been described by Kessel in "annulate lamellae" (30; see 2, 3, 30 for reviews). These cytoplasmic structures contain closely packed pores arranged in 1–100-membrane stacks (30). [Occasionally single pores have been observed in ER cisternae and intranuclear annulate lamellae (2, 3, 30)]. Annulate lamellae are found most frequently in oocytes and transformed cells and less often in other cell types, although their origin and function are unknown. Recent work in *Drosophila* embryos suggests that annulate lamellae result from a transient overproduction of nuclear pore components (31), a hypothesis consistent with the observed storage of many nuclear components in oocytes for use later in development.

The number of pores per nucleus varies over a wide range (100–5×10^7 per nucleus) (3). The density of pores per μm^2 of nuclear envelope also varies greatly, from a low of $3/\mu m^2$ in highly differentiated but metabolically inactive cells such as nucleated red blood cells and lymphocytes, to $50/\mu m^2$ in amphibian oocytes and Acetabularia. Somatic cells that are differentiated but highly active (brain, liver, kidney) contain 15–20 pores/μm^2 (3). Work by Maul and colleagues (32) indicates that the pore number/nucleus in HeLa cells doubles in S phase in preparation for subsequent mitosis.

During mitosis, the defined structures of the nuclear pore disappear. In cases where mitosis has been followed closely with electron microscopy, the pore structures are seen to become indistinct in prometaphase, then to disappear entirely in metaphase (33–36). In *Drosophila,* where the nuclear membrane does not break down in mitosis, the pores disappear (33) in a manner identical to that of higher eukaryotes. In late anaphase and telophase in all organisms examined the process is essentially reversed, with the pore appearing first as an indistinct structure and then gradually achieving its former and precise eightfold symmetry. Protein synthesis is not necessary for pore reformation (32). Feldherr (38, 39) has observed a greater pore permeability following mitosis in amoebae, possibly indicative of the presence of an immature pore structure with an altered lumen. Since pores have never been observed in the cytoplasm in mitosis, and since the pore appears to undergo a gradual structural dimunition followed by a subsequent accretion at the end of mitosis, it is probable that the pore is dissociated into its component parts at mitosis.

NUCLEAR PORE PROTEINS   The nuclear pore is estimated to have a molecular mass of 10^8 daltons (40). If one assumes that each pore protein has a mass of 200,000 daltons, then 500 such proteins would be required to construct a pore (or 500/16 = 31 proteins per annular granule if all the mass were in these prominent pore structures). Some authors have suggested that pores have RNA components (see 2, 3, 30 for reviews), which would reduce the number of proteins required. In either case, the pore is a very large structure and may be composed of multiple copies of a few proteins, as is characteristic of cytoskeletal structures, or fewer copies of many different proteins, as is the case for the ribosome.

Initial attempts to identify pore proteins by SDS-PAGE focused on the residual protein matrix or "pore complex-lamina" that remains after high salt–detergent extraction of nuclear envelopes [41–48; see Franke (2) and Fisher (24) for additional references]. These pore complex-lamina preparations have been derived most often from rat liver or *Xenopus* oocyte nuclear envelopes, the latter of which is highly enriched in nuclear pores. The major proteins observed are surprisingly few, and include the nuclear lamins (60–70 kilodalton) (see below) and gp 190 (see below; 44–46). Of the proteins that make up the pore, only 2 (and possibly a third; 24, 24a) have been characterized.

The 190-kilodalton pore glycoprotein (gp 190)   To search for pore proteins, Gerace and colleagues (44) fractionated rat liver nuclear envelopes into peripheral and integral membrane proteins. A single major 190 kilodalton (kd) integral membrane protein was found. This Con A–binding glycoprotein remains associated with pore complex-lamina preparations even when the nuclear membranes are removed with Triton X-100, and is not present in isolated ER. The protein, designated gp 190, is the predominant Con A–binding protein of the nuclear envelope. Fluorescent Con A staining of the nuclear envelope had been seen previously by a number of groups (50–54). Virtanen (54) had also observed two Con A–binding proteins of 180 kd and 34 kd in rat liver nuclear membranes.

Antibodies prepared against gp 190 (44) were seen to specifically label the cytoplasmic surface of nuclear pores as determined by immunoferritin electron microscopy. No staining was seen on the lamina, nor was staining seen if the nuclear membranes were present, again indicating an integral membrane location. Anti-190-kd antiserum cross-reacts with a similarly sized protein in cow, dog, and chicken. Rough calculations suggest ~25 copies per pore, with complete dispersion throughout the cell in mitosis. Gerace et al (44) proposed that the carbohydrate portion of gp 190 lies in the perinuclear space and that the protein may anchor the pore within the nuclear membranes. As such, gp 190 would play a primarily structural role in the pore.

Fisher and colleagues (45, 48) have purified a related 188-kd protein from *Drosophila*. Digestion with endoglycosidase H reduces the molecular weight of this protein by 10 kd. Although anti-rat antisera do not recognize the *Drosophila* gp 188, monoclonal antisera against the *Drosophila* protein (48) cross-react with rat gp 190, and similarly sized proteins from *Xenopus* chicken, opossum, and guinea pig. Fisher (24) finds that the gp 190 of *Xenopus* oocytes is highly protease sensitive, perhaps explaining why this pore protein has not appeared as a prominent band in isolated oocyte nuclear envelopes.

The 62-kilodalton pore protein   Unlike gp 190, which can be visualized on gels as a Coomassie staining band, the second known pore protein identified was found by several less direct means. In screening monoclonal antibodies prepared against rat liver pore complex-lamina, Davis & Blobel (56), using immunofluorescence, found one antibody that stained the periphery of nuclei in a distinctly punctate manner. The antigen dispersed throughout the cytoplasm in mitosis. When immunogold electron microscopy was performed on extracted pore complex-lamina preparations, staining of the nuclear pore was observed. A major protein band of ~62 kd was identified on protein blots using this antiserum, as was a smaller cytoplasmic precursor that could be chased into the larger nuclear form. When p62 was isolated by antibody affinity chromatography, it was found to bind the lectin, wheat germ agglutinin (WGA), indicating that p62, the second pore protein identified, is also a glycoprotein.

In a separate approach, Finlay et al (57) found that a WGA-binding protein that is localized in the nuclear pore is involved in nuclear transport. Baglia & Maul (58) had previously observed that WGA inhibits the efflux of RNA from rat liver nuclei in vitro. Using an in vitro nuclear transport system composed of a fluorescently labeled nuclear protein (nucleoplasmin) and isolated nuclei added to an extract of *Xenopus* eggs, Finlay et al (57, 59) observe nuclear transport that closely parallels in vivo transport. A number of lectins were tested for nuclear binding and/or inhibition of nuclear transport. Only WGA (0.1 mg/ml) had an effect, being found to inhibit transport completely. The authors found that fluorescently labeled WGA bound to rat liver nuclei in a punctate manner (57). By electron microscopy, ferritin-labeled WGA was found to bind exclusively to the cytoplasmic face of nuclear pores in isolated rat liver nuclei (up to 15 copies of WGA per pore). Competing sugar (*N*-acetylglucosamine) relieved all inhibition of transport and abolished ferritin-WGA binding to the pores, indicating a specific lectin-ligand interaction at the pore. A single major band of 63–65 kd was observed on protein blots when probed with iodinated WGA (57). This protein appears to be identical to that seen by Blobel & Davis (56). Thus, the WGA-binding protein or proteins are the first pore proteins identified that when modified affect nuclear protein transport.

In retrospect, peripheral binding to nuclei by WGA had been seen previously (10, 51–53), although a punctate pattern was not observed. Furthermore, several groups when probing protein blots of isolated nuclei observed a single prominent WGA-binding band of ~65 kd (60–62). The 62-kd pore protein belongs to a new class of glycoproteins that contain single additions of *O*-linked *N*-acetylglucosamine residues as shown in an elegant study by Holt & Hart (63). The majority of these residues in the nucleus are present on protein the size of the 62-kd pore protein (63). Recent work by Holt et al (64) indicates that there is a family of proteins in the nuclear pore, all of which contain *O*-linked *N*-acetylglucosamine residues.

Nucleocytoplasmic Exchange

Traffic between the nucleus and cytoplasm has been assumed to occur through the nuclear pores. This has been definitively shown by electron microscopy in only a few cases (38, 39, 66, 67). In these studies, electron-opaque particles are seen to pass through the central channel of the pore (38, 39, 67). The pore appears not to be simply an open channel, but has the properties of a molecular sieve [68–83; see Dingwall & Laskey (25) for a recent review]. By determining the size of proteins, dextrans, or gold particles that can enter the nucleus when injected into the cytoplasm, the apparent pore diameter has been estimated to be 90 Å for *Xenopus* oocytes (70) and *Drosophila* salivary gland nuclei (72), 100–110 Å for hepatocytes (77) and hepatoma cells (80, 81), and 120 Å for amoebae (84).

SIGNAL SEQUENCES FOR NUCLEAR LOCALIZATION OF PROTEINS   Many nuclear proteins are too large to be accommodated by the apparent diameter of the pore. For these, as well as some small nuclear proteins (85–88), it is becoming clear that there exist signal sequences on the proteins that allow rapid entry into the nucleus (25, 89–105; T. Burglin, personal communication). [Certain large proteins are able to enter the nucleus in the absence of signal sequences by an unknown mechanism (106–108)]. The sequence characteristics and homologies between different signal sequences have recently been reviewed by Dingwall & Laskey (25). Briefly, nuclear localization signals, i.e. nuclear signal sequences, differ from signal sequences that target proteins to the ER or mitochondria in that the nuclear signals are a permanent part of the protein and are not removed upon transport. A nuclear protein can thus enter the nucleus not only after synthesis, but also after each mitosis. Signal sequences can map to the amino or carboxy termini or to the central region of the protein. Such sequences not only allow nuclear entry, but also promote entry that is much faster than could be accounted for by diffusion (86, 109).

Identification of signal sequences has been accomplished by both genetic and biochemical approaches. The *Xenopus* nuclear protein, nucleoplasmin, has an extremely high nuclear affinity when injected into oocytes. Dingwall and colleagues (91) find that if the carboxy terminal third of nucleoplasmin is removed by protease digestion, the N-terminal core is unable to enter the nucleus. The separated carboxy tail of nucleoplasmin, in contrast, quickly enters the nucleus. It has since been shown that complexing nucleoplasmin to large nonnuclear molecules such as the immunoglobulin, IgG (110), an algal protein, phycoerythrin (81), or even to large gold particles (67, 111) confers rapid entry into the nucleus. Feldherr (67) injected nucleoplasmin-coated gold particles into *Xenopus* oocytes, and observed that particles up to 200 Å in size enter the nucleus through the central channel of the pore, whereas particles coated with proteins lacking signal sequences remain cytoplasmic.

Hall et al (92) first used genetic techniques to study nuclear signal sequences. They fused various fragments of the yeast $\alpha 2$ gene to a gene encoding a large nonnuclear protein, and found that the amino terminal 13 amino acids of $\alpha 2$ are sufficient to allow nuclear localization of the fusion protein. Genetic manipulation has similarly allowed definition of signal sequences in other proteins (93–102). The most definitive studies of this sort have identified a signal sequence in the SV40 large T-antigen protein (93–95). A seven-amino-acid stretch of T antigen is sufficient to act as a signal sequence when genetically fused to cytoplasmic pyruvate kinase (95). Mutation of a central lysine residue in this sequence renders both the T-antigen protein (93, 94) and the seven-amino-acid signal sequence present in fusion proteins (95) incapable of nuclear localization. Extending this work, the seven-amino-acid signal sequence was synthesized chemically by Goldfarb et al (104) and Lanford et al (105). The synthetic peptide was covalently crosslinked to large nonnuclear proteins (~10 peptides/protein) ranging in size from ovalbumin (43 kd; diameter 54 Å) to ferritin (465 kd; diameter 94 Å) (104, 105). The peptide was found to confer nuclear localization on all these proteins, while covalent addition of the mutant peptide sequence did not allow accumulation within the nucleus. The rate of transport is influenced by the number of peptides per cytoplasmic protein and to some extent by the size of the protein (105). Presence of excess free signal peptide reduces nuclear transport of the peptide-protein (104), suggesting possible saturation of a receptor protein within the pore.

Antisera raised against the SV40 T-antigen signal sequence peptide (Pro-Lys-Lys-Lys-Arg-Lys-Val) has been found to cross-react with 12 nuclear proteins from a human lymphocyte cell line, indicating the existence of a family of proteins bearing homologous nuclear localization signals (104). Since the signal sequences identified in other nuclear proteins often share little sequence homology with the T-antigen sequence (25), there are presumably a number of families of signal sequences in each cell. Recognition of the various signal sequences may involve several different "receptor" molecules or a single "receptor" able to recognize various signal sequences. At present it is unknown where signal sequence recognition occurs, although it precedes or accompanies passage through the pore (67; see 25 for a review). The most likely location for signal sequence is the nuclear pore itself, but there are no data that rule out the existence of a cytoplasmic "carrier" protein that would recognize proteins bearing signal sequences and ferry them to the pore.

DEVELOPMENTAL CONTROL OF NUCLEAR TRANSPORT   There are several instances where proteins are nuclear or cytoplasmic depending on the developmental stage of the organism or alternately on the growth conditions of the cell. In *Xenopus* oocytes, the majority of small nuclear RNP (snRNP) proteins are cytoplasmic (112) and remain so during fertilization and early embryogenesis. At 12th cleavage synthesis of their companion small nuclear RNAs (snRNAs) occurs and the snRNP proteins migrate into the nucleus— presumably a signal sequence is exposed or activated by the binding of snRNA molecules to these proteins. A short RNA sequence of snRNA U2 is required for snRNP protein binding and nuclear localization of U2 snRNA (113).

Using monoclonal antibodies, Dreyer and colleagues (114–116) have identified a separate class of *Xenopus* proteins that are interesting in that they are nuclear in the *Xenopus* oocyte, cytoplasmic in the early embryo, and nuclear at later stages of development. These proteins have been termed "late-migrating nuclear antigens." It is possible that a reversible modification of signal sequences is occurring, the proteins being capable of nuclear transport in one state and not in the other, although no such change has yet been observed (114–116). Richter et al (98) have found that a modification of the adenoviral protein E1a appears to be required for nuclear accumulation of E1a.

In another example of differential control of nuclear transport, several groups have observed (117–121) that both the catalytic and regulatory subunits of cAMP-dependent protein kinase type II are cytoplasmic in the absence of cAMP, but that the catalytic subunit rapidly becomes nuclear in the presence of cAMP. Woffendin and colleagues (121), studying the distribution of these subunits in *Dictyostelium discoideum,* find that the catalytic subunit is cytoplasmic in vegetative amoebae (low cAMP) but nuclear when cells became aggregated into the migrating slug (high cAMP). Other examples of possible control of nuclear transport include (*a*) herpes virus, where mutation in one viral protein eliminates its ability to localize to the nucleus and similarly effects two other viral nuclear proteins (122) and (*b*) *Xenopus* oocytes, where Dabauvalle & Franke (123) have identified proteins that are diffusible in the oocyte and theoretically small enough to pass through the pores but that remain cytoplasmic. Such "karyophobic" proteins (123) and the proteins identified by Dreyer (114–116) suggest nuclear transport may involve more than pore size and the presence or absence of signal sequences.

RNA IMPORT AND EXPORT   The pores are also sites for RNA exchange between the nucleus and cytoplasm; newly synthesized RNAs must exit the nucleus, while other RNAs, such as the small nuclear RNAs, must reenter the nucleus after complexing with proteins in the cytoplasm [see De Robertis (18) for a review]. Experimental approaches to in vitro RNA export from nuclei have been recently reviewed by Agutter (23) and Clawson and colleagues (22). Only a few in vivo studies of RNA transport have been done to date. Zasloff et al (124–126) have identified regions of an initiator tRNA which, if mutated, block transport from the nucleus to the cytoplasm in *Xenopus* oocytes. Dworetzky & Feldherr (127) have injected large gold particles coated with methionine tRNA or 5S RNA into the nucleus of oocytes and observed rapid transport of the particles through the pores and into the cytoplasm. If injected into the cytoplasm, the particles do not enter the nucleus, mimicking the normal vectorial nature of transport of these RNAs. BSA- or ovalbumin-coated gold particles injected into the nucleus fail to enter the pores or cytoplasm, suggesting that the nucleoplasmic surface of the pores specifically recognizes and exports tRNA and 5S RNA molecules. In similar experiments using tRNA– and nucleoplasmin-coated gold particles of different size, it has been possible to show that a single nuclear pore is capable of simultaneous bidirectional transport, with RNA-coated gold exiting the nucleus and nucleoplasmin-coated gold entering the nucleus through the same pore (127; C. Feldherr, personal communication).

IN VITRO NUCLEAR TRANSPORT   Although rapid progress is being made in determining the nature of signal sequences using in vivo systems, dissecting the molecular mechanism of nuclear transport will require an in vitro transport system that can be experimentally manipulated. Such systems are now available (57, 59, 116, 128, 129). Extracts of *Xenopus* eggs have been found to excel at maintaining nuclei in and/or restoring nuclei to an intact state (59, 81,

116). This property would be predicted to be an absolute requirement for any in vitro transport system (80, 81). Presumably the ability of *Xenopus* egg extracts to repair and/or maintain integrity of added nuclei stems from the stabilizing presence of large amounts of nuclear components stored in the eggs for use later in development. A rapid in vitro nuclear transport system has been developed (57, 59) using the fluorescently labeled nuclear protein, nucleoplasmin, *Xenopus* egg extracts, and added nuclei (see also 129). With this system, fluorescent nucleoplasmin was observed to accumulate up to 17-fold in rat liver nuclei within 30 minutes. As first shown for radioiodinated nucleoplasmin, transport and accumulation are completely dependent on the presence of ATP in the extract, ceasing when ATP is depleted (59, 129) and resuming when ATP is added back (59). Nucleoplasmin accumulation is also temperature dependent in vitro, showing a maximum at ~30° C and being abolished at 0° C (59), as is characteristic of an energy-requiring process and as is found in vivo (91). Accumulation of nucleoplasmin is dependent on an intact nuclear envelope (59), as suggested by Peters (81) and in contradiction to earlier studies on other proteins (18, 73, 74). In similar egg extracts, Dreyer and colleagues (116) find that added nuclei (sperm and erythrocyte) rapidly accumulate nuclear antigens endogenous to the extract. It has also been found that nuclei reconstituted from bacteriophage DNA (116, 128–131) and several types of normal nuclei are equally capable of fluorescent nucleoplasmin accumulation in vitro (59).

With such systems, it has been possible to approach questions of nuclear transport. For example, the late-migrating nuclear antigens (114–116) fail to accumulate in nuclei added to *Xenopus* egg extracts (116)—such extracts may thus prove powerful in determining the point at which entry of these antigens into the nucleus is controlled. As described in a previous section, through use of an in vitro transport system the lectin wheat germ agglutinin was shown to bind to the pore and inhibit completely the transport of nucleoplasmin (56). These systems should prove fruitful for the study of other aspects of protein and RNA transport, as well as for understanding the molecular mechanism of transport and the nature of its developmental control.

The Nuclear Lamina

Underlying the inner nuclear membrane is a polymeric network, the lamina, composed of 1–4 different proteins called lamins (60–70 kd) (for review see 8, 132–134). It has been proposed that the lamins play important roles in stabilizing the nuclear envelope and serving as an attachment site for chromatin to the envelope (132, 135–138). However, definitive experiments demonstrating a role of the lamina in either of these two capacities are still lacking. Direct visualization of the lamina by electron microscopy indicates that it is a discrete layer of 30–100 nM in thickness, located between the inner nuclear membrane and chromatin (41, 42, 135). The lamina can be isolated intact from nuclei following nuclease digestion and extraction with nonionic detergents and high salt (41, 42, 135, 139). The residual matrix after extraction often retains the same shape and size as the original nucleus and is composed almost exclusively of the lamin proteins, indicating that the lamins form a highly stable polymeric network. The physical properties of the lamina are consistent with its proposed structural role within the nucleus.

The lamin proteins have been purified from a wide variety of animals (for review see 134). Identification, characterization, and comparison between lamins isolated from different organisms and cell types have usually involved either analysis of peptide maps after proteolytic digestion or comparison of conserved antigenic epitopes using either poly- or monoclonal antibodies (41, 42, 135, 140–154). In mammals the lamina is composed of three lamin proteins (lamins A, B, C) (135, 141, 146, 148) and a fourth minor species (153). In avians there are two major (A and B_2) (146, 149, 150, 153) and one minor lamin (B_1) proteins (153). In amphibians there are four lamins L_I–L_{IV} (134, 145, 146, 155), in *Drosophila* two lamins (144, 151, 152), and in the germinal vesicle of the mollusc *Spisula* a single species (147).

Based on similarities between peptide maps, antibody cross-reactivity, and isoelectric point, many of the different lamins can be segregated into two different classes: mammalian lamins A and C, avian A and to some extent B_2, and amphibian L_{II}–L_{IV} in one class and mammalian lamin B, avian B_1, and amphibian L_I in a second class (134). It is not clear whether these two lamin classes are functionally distinct (see below)

Sequence analysis of cDNA clones encoding human lamins A and C indicates that these two proteins are identical in amino acid sequence except for the presence of an additional 133 amino acids at the carboxy end of the lamin A proteins (156, 157). Southern blots using the lamin cDNA clone as probe suggest that a single gene encodes both lamin A and C and that the different proteins are the result of differential processing of the mRNA. Similarly, in *Drosophila* the two lamins are encoded by a single gene (Y. Gruenbaum et al, personal communication). The cDNA-derived amino acid sequence of human lamins A and C indicates that the lamins share extensive regions of homology with the intermediate filament family of proteins (156, 157). This homology is primarily restricted to those unique regions of the intermediate filament proteins thought to be alpha-helical and involved in forming coiled-coil polymers. The strong similarity between lamins and intermediate filaments suggests that the lamins, like intermediate filaments, might form filaments arranged in long fibers or in a dense network bound to the nuclear membrane.

The composition of the lamin proteins within interphase nuclei varies between cell types in the same organism. This has been best characterized in the amphibian *Xenopus laevis* (134, 143, 150). In *Xenopus* the large ooctye germinal vesicle nucleus and the nuclei in the first 4000 cells made following fertilization of the egg contain a single lamin protein (L_{III}). At the mid-blastula transition, a new lamin is made and incorporated into the nucleus (L_I), while a third lamin (L_{II}) is synthesized beginning at gastrulation. In the mature adult the lamina of most nuclei is composed of lamins L_I and L_{II}, but the nuclei of certain terminally differentiated cell nuclei (myocardial cells and neurons) contain L_I, L_{II}, and L_{III}. *Xenopus* spermatid nuclei contain a lamina composed exclusively of another lamin protein, L_{IV}. How or whether these changes in lamin composition regulate nuclear structure or function is current-ly unknown. Additional regulation of lamin structure occurs during the cell cycle and is discussed below.

III. THE ORGANIZATION OF DNA WITHIN THE NUCLEUS

The Organization of Chromosomes within the Nucleus

The work of Sedat and colleagues has provided a high-resolution model of the three-dimensional organization of chromosomes within interphase nuclei (158–162). They have mapped the conformation, orientation, and arrange-ment of the large polytene chromosomes within nuclei of unfixed *Drosophila* salivary gland cells. To do this, fluorescently labeled chromosomes were optically sectioned in situ to minimize distortion and the optical images recorded and refined electronically. The folding pathway of polytene chromo-somes within the nucleus was reconstructed by electronically superimposing the refined optical images. A statistical comparison of chromosome organiza-tion within 24 different nuclei reconstructed by these techniques has revealed several generalizations about chromosome folding and packaging. First, the chromosomes are oriented with centromeres attached to the nuclear envelope at one pole of the nucleus and telomeres attached to the envelope at the opposite pole. This orientation most likely occurs during anaphase of mitosis and is preserved during interphase (158). Second, the prominent chromosome coiling observed is strongly chiral with right-handed turns predominating. Third, the chromosome arms are maintained in separate spatial domains. Thus, although the chromosomes are highly contorted and very closely packed within the nucleus they do not loop around each other. Sedat's group suggests that the nonintertwined pattern of chromosome packing within nuclei is established at the end of mitosis. Thus, when the mitotic chromosomes bind to the nuclear envelope they do so in a nonintertwined configuration. Mainte-nance of these initial contacts during subsequent envelope growth ensures that the chromosomes remain within separate spatial domains. These observations suggest that the initial chromatin envelope attachments are stable over very long periods of time. Consistent with this is the important finding that certain specific chromosomal loci are frequently found attached to the nuclear envelope (158, 159). Such loci occur almost exclusively at positions of intercalary heterochromatin, suggesting that these sequences likely play an important role in defining chromatin-envelope attachment sites. These sites are relatively evenly distributed along each chromosome arm and occur on the average every 10^6 bp. As indicated above, the conserved envelope attachment sites could play a pivotal role in maintaining chromosome orientation within the nucleus. Furthermore, as discussed below, such sites could act as sites for regulating nuclear envelope growth at the end of mitosis. Identification of the DNA bands located at attachment regions should prove extremely useful for identifying sequences important for envelope attachment.

A statistical comparison of both the positions of individual chromosomes and chromosomes relative to each other did not reveal any strikingly conserved folding pattern. Individual chromosomes folded into a wide variety of different configurations, and chromosome-chromosome interactions were largely random. It appears from these results that although individual chromosomes in different nuclei share some structural similarities (for example, right-hand gyration and envelope attachment sites), nonetheless chromosome geometry within the nucleus is not precisely determined.

Whether the conclusions about chromosome organization of polytene chromosomes hold true for normal diploid nuclei has yet to be determined. Some evidence indicates that the chromosomes contained in diploid nuclei are partially ordered (163–165, 212). However, high-resolution studies of chromosome organization in these nuclei are currently restricted by technological limitations.

The Organization of DNA into Chromosomes

LAMPBRUSH CHROMOSOMES   The earliest studies on the organization of DNA within the chromosome were done using lampbrush chromosomes (166, 167). Results from these studies have provided a remarkably accurate model for the organization of DNA within the somatic cell nucleus, as determined by later biochemical studies. Lampbrush chromosomes are present at the diplotene stage of meiosis in the oocytes of almost all animals. They are 10–100 times longer in axial length than mitotic chromosomes and easily observed in the light microscope under phase optics. Generally, the lampbrush chromosomes are enclosed within a nucleus 2–3 orders of magnitude larger than a typical somatic nucleus. It is not clear whether the chromosomes are visible in the light microscope because they are in a unique stage of condensation or in part because such a large nuclear volume allows them to separate into individual units. If the latter is true, then the inability to observe individual chromosomes within somatic nuclei is not due to their decondensation below the limits of resolution of light microscopy, but rather to the fact that in these much smaller nuclei the chromosomes are more densely packed and therefore not resolvable.

When the nuclei containing lampbrush chromosomes are manually dissected from oocytes and the envelope removed, a large number of DNA loops can be observed to extend perpendicularly from a central chromosomal axis. This was observed in shark oocytes by Ruchert in 1892 (168) and was the first indication that DNA is organized into loop domains within the nucleus. Callan found that when a lampbrush chromosome was stretched, single loops separated at the base to form a single fiber (167). This simple experiment showed that a loop is part of the long molecular axis that runs the length of the lampbrush chromosome. Gall subsequently demonstrated by kinetic experiments using DNAase digestion that this central axis is composed of a single strand of duplex DNA (169).

Direct measurements of DNA loop size in lampbrush chromosomes found that loops range from 1 to over 100 μm (166). Assuming 1 μm of chromatin is equal to 3000 base pairs (bp) of DNA, loops contain between 3000 and 300,000 bp of DNA. Both the number of visible loops and the average size of the loops vary with the state of oocyte development (170). These changes in loop number and size correlate with the stage of transcriptional activity of the oocyte, suggesting that loops represent actively transcribed regions of DNA. This was confirmed by electron microscopy using the Miller technique (for a review see 166). Furthermore, these experiments demonstrated that there may be several transcription units of similar or different lengths within a single loop and that neighboring transcription units may be of the same or opposite polarity (171, 172). Thus, a loop does not represent a single gene and transcription within a loop is not restricted to one direction. During those stages of oogenesis when transcription is at its highest, RNA polymerase is maximally packed along the DNA every 20–25 nm and loops are fully extended. When transcription levels decrease, either due to oocyte maturation or addition of actinomycin-D, RNA polymerase packing is reduced and the loops retract into a more condensed state (166). It is possible that a modulation of supercoiling density regulates loop extension and retraction. Observations on lampbrush chromosomes are fully consistent with recent studies both on the organization of chromatin structure within somatic cell interphase nuclei and models for gene activation within these nuclei. Because chromatin loops in lampbrush chromosomes are visible at the light microscope level they provide a unique opportunity for directly studying factors responsible for maintaining loop structure and correlating changes in loop topology with function.

ORGANIZATION OF LOOP DOMAINS IN THE INTERPHASE NUCLEUS   The first clear demonstration that the lateral loop organization of chromatin found in lampbrush chromosomes was a universal feature of chromosome organization in somatic cells came from experiments of Cook and coworkers in 1975 (173–175). They found that after removal of the majority of proteins (including histones) from isolated somatic cell nuclei by extraction with Triton X-100 and high salt (2-M NaCl), the DNA remained bound to a residual structure that resembled the original nucleus. When such protein-depleted nuclei (or "nucleoids") were sedimented through sucrose gradients containing different concentrations of the DNA-intercalating agent ethidium bromide, their sedimentation properties changed in a manner similar to that observed with supercoiled circular plasmid DNA (173–176). These studies demonstrated that the long linear DNA within somatic cell nuclei is physically restricted and, therefore, organized into loop domains (173–175, 177, 178). Based on sedimentation and direct DNA measurements, the average size of a loop ranges from 80 to 150 kbp of DNA, well within the range of loop sizes found in lampbrush chromosomes. Because these techniques are not as sensitive to small loops as they are to larger loops, the average value is most likely an overestimate. In direct observation of lampbrush chromosomes, a large percentage of the loops in chromosomes are substantially smaller than 80,000 bp. From these data, the current working model for organization of DNA in the interphase nucleus is that the DNA is organized into loop domains of various sizes via attachment to the nuclear periphery and/or an internal matrix of the nucleus.

Laemmli and coworkers have demonstrated that the organization of chromatin into loop domains in interphase nuclei is preserved when the chromatin condenses into mitotic chromosomes (179–182). When isolated mitotic chromosomes are extracted with high salt to remove the majority of proteins, the DNA remains anchored to a residual proteinaceous scaffold that retains the basic morphological organization of an unextracted metaphase chromosome. The scaffold is composed primarily of two proteins: ScI, a 170-kd protein, and ScII, a 135-kd protein. Together these proteins represent 1–2% of the total chromosomal protein. The DNA is attached to this residual scaffold at numerous sites, forming loops. Direct electron microscope measurements show that these loops are the same size as those found in interphase nuclei. Thus the organization of DNA into loop domains is a structural feature of both interphase and metaphase chromosomes.

DNA SEQUENCE-SPECIFIC ATTACHMENT SITES   The organization of chromatin into loop domains is a stable and conserved feature of lampbrush chromosomes, as demonstrated by the finding that characteristic loops can be identified from generation to generation (166). Similarly, the banding pattern of *Drosophila* salivary gland chromosomes is stably inherited. These findings suggest that the organization of chromatin into loop domains is determined by the interactions of specific DNA sequences with nuclear proteins. Although many attempts to identify such sequences were made, the results were often not clear cut (for a review see 1). Recently Laemmli and coworkers have provided convincing evidence that such sequences do exist (183, 184). The key element in their experiments was the development of a method for extracting proteins from interphase nuclei that maintains specific protein-DNA interactions. The method uses low concentrations of the salt lithium diiodosalicylate (LIS) for protein extraction. Apparently, extraction of nuclei with high salt weakens ionic interactions between the proteins that maintain loops and the specific DNA sequences that interact with these proteins. The high salt–induced reduction in specific binding affinity allows the DNA to slide such that specificity is no longer observed. The general procedure for identifying specific DNA sequences as nuclear attachment sites involves extraction of nuclei with low concentrations of LIS, followed by restriction digestion of the DNA, and separation of solubilized (unattached) DNA from bound DNA by centrifugation. Sequences specific to the base of the loop remain bound to the nucleus during this fractionation, whereas other loop sequences unattached to nuclear structures are released. DNA from each fraction is then separated on a gel, blotted, and probed with a labeled cloned fragment.

Using this procedure, specific nuclear-binding sequences have been identified in the *Drosophila* histone gene repeat (183), *hsp70* gene (183), alcohol dehydrogenase gene (184), mouse Kappa immunoglobulin gene (185), and ribosomal RNA genes (186). These specific sequences are spaced along the genome from 5000 to 112,000 bp apart (187). Different sites are not homologous in sequence, but are generally similar in that they are approximately 200 bp long and A-T rich (70%). Because these sites remain bound to the nucleus after protein extraction and restriction digestion they have been named either scaffold attachment regions (SAR) or matrix attachment regions (MAR).

The best-characterized SAR is located in the *Drosophila* histone repeat in the nontranscribed spacer region between histone genes H1 and H3 (183). DNAase protection experiments show that the region is composed of two 200-bp nuclease-resistant sequences separated by a 100-bp DNAase-accessible region. Because nucleosomes are in register on either side of the SAR site, it is likely that this 200-bp region stably interacts with one or more proteins that then block nucleosome formation in the region.

In the case of both the *hsp70* gene and the mouse immunoglobulin gene, attachment of SAR sites to the nucleus is independent of transcriptional activity (183, 185). This suggests that previous studies indicating that activation of transcription is correlated with attachment of the regions to nuclear sites may need to be reexamined (188–192). This is particularly important in light of the demonstration that the *hsp70* gene shows transcriptionally dependent attachment to the nucleus when nuclei are extracted with 2-M NaCl but not when extracted with LIS (183).

Cockerill & Garrard found that the in vitro–defined SAR sequences bind specifically to nuclear scaffolds (185), when salt-extracted, DNAase-digested nuclei were incubated with cloned SAR sites, as defined by the Laemmli procedure. Using this technique with nuclei from a variety of tissues they found that SAR sequences were not tissue specific. They further found that *Drosophila* SAR sequences competitively inhibit the binding of the mouse kappa SAR sequence to nuclei. This competition demonstrates that although SAR sequences are not homologous in sequence they interact with some of the same proteins within the nucleus. Current estimates based on saturation hybridization of labeled SAR sequences to salt-extracted nuclear matrix indicate that there are at least 10,000 SAR-binding sites per nucleus. In summary, it appears that the SAR sequences represent an extended family of similar sequences that play a pivotal role in the formation and maintenance of chromatin loop domains within the nucleus and metaphase chromosome.

TOPOISOMERASE II AND LOOP ORGANIZATION   Analysis of SAR sequences reveals that they contain numerous sites similar or identical to sites cleaved by topoisomerase II (185, 193). This, in conjunction with the finding that topoisomerase II is a major component of the nucleus after high salt extraction and DNAase digestion, strongly suggests that topoisomerase II plays a major structural and enzymatic role in loop formation (194). An enzyme that can deconcatenate loops and possibly regulate the topology of a loop by regulating the supercoiling density would likely play an important role in maintaining loop organization and regulating function (185, 195–199; for reviews see 200, 201). In support of this, Earnshaw and colleagues have demonstrated that the major metaphase chromosome scaffold protein, ScI, is topoisomerase II (202, 203). They have further shown by immunostaining of metaphase chromosomes that ScI is located at the base of loops. ScI is present in metaphase chromosomes at one copy per 23,000 bp (204). In hybridization experiments, the histone SAR sequence binds to metaphase scaffolds. Thus, based on chromosomal location, correct stoichiometry, and possible interaction with identified loop attachment sequences, it is likely that topoisomerase II plays a major role in the organization of chromatin into loops both in metaphase chromosomes and interphase nuclei.

ARE LOOPS DYNAMIC?   Almost all studies on the organization of chromatin into loop domains have involved nuclei isolated from proliferating cells (173–176, 183, 185, 187). These nuclei are active in both DNA replication and RNA transcription. If there is a correlation between chromatin structure and function, then it would be expected that nuclei with different activities would organize their chromatin differently, i.e. the organization of loop domains would be dynamic. Several lines of evidence suggest that this is the case. In experiments using chicken red blood cell nucleoids, Cook & Brazell found that as the nuclei became inactive during the course of development, their sedimentation rate approached the value expected for nuclei in which loops were no longer present (173, 174). Further evidence for the possible dynamic nature of loop domains is suggested by DNA replication studies. In somatic cells the average length of a replicon measured by fiber autoradiography is the same as the estimated loop domain size, indicating that a replicon unit is defined by the DNA contained in a loop (205–207). However, during early embryogenesis in *Drosophila* and *Xenopus,* where DNA replication is extremely rapid, the average replicon length decreases at least fourfold relative to the somatic replicon size (205, 208). In amphibian spermatocytes, replicon length increases at least fourfold relative to the somatic unit (206). If replicon length and loop size are tightly coupled, these results suggest that the number and size of loops must change dramatically between nuclei of different activity. Consistent with this proposal, Heck & Earnshaw, in a detailed study, have demonstrated that the amount of topoisomerase II present in cells at different stages of the cell cycle can vary over 3–4 orders of magnitude (209; see also 210), being present at 10^6 copies per cell during S, G2, and M phases and undetectable in nonproliferating cells. They also found that topoisomerase II disappeared from red blood cells at a rate that was consistent with the loss of loop structure observed by Cook & Brazell (174). These results suggest that loop domains are dynamic at different stages of the cell cycle as well as during development. It is possible that the family of SAR sequences is composed of sequences with different binding affinities for topoisomerase II and/or other unidentified loop-forming proteins. Competition between these DNA sites under conditions when loop proteins are present in limited amounts could efficiently regulate both loop size and number.

INTERACTIONS BETWEEN CHROMATIN AND THE NUCLEAR ENVELOPE
Relatively little is known about the associations between chromatin and elements of the nuclear envelope. As discussed above, Sedat and colleagues have mapped several unique interactions between polytene chromosomes and the envelope in salivary gland nuclei (158–161). These include the centromeric region, telomeric region, and several unique heterochromatic regions within each chromosome. As many as 1500 such attachments have been estimated to exist in interphase nuclei (211). By using antibodies against centromere antigens it has also been demonstrated that centromeres bind to the nuclear envelope of interphase somatic cell nuclei (212). These results indicate that envelope attachment, like loop formation, may depend on specific DNA sequences.

Lebkowski & Laemmli have shown that when salt-extracted nuclei are treated with metal-chelating agents there is a further unfolding of the nucleoid as measured by a decreased sedimentation rate (136, 137, 213). They have proposed that this treatment disrupts the associations that maintain loops and that the remaining associations holding the nucleoid together are due to interactions between chromatin and the nuclear lamina. Such periodic interactions between chromatin and the nuclear envelope would thus constitute a second level of nuclear folding, involving loops significantly larger than the 100-kbp domains discussed above. These interactions would provide a solid foundation for the bulk organization of the chromosome within the nucleus and could act as initiation sites for the reassembly of the nuclear envelope at the end of mitosis.

IV. DYNAMIC REARRANGEMENT OF NUCLEAR STRUCTURE DURING MITOSIS

Changes in nuclear size occur continually in response to increases in DNA content, variations in transcriptional activity, and different stages of the cell cycle and development. For example, during S phase of the cell cycle nuclear volume roughly doubles as the amount of DNA doubles. Such changes in nuclear size demonstrate that nuclear components such as the lamina and nuclear membranes are dynamic. The most dramatic rearrangement of nuclear components occurs during mitosis. At the onset of mitosis RNA transcription is inactivated, chromatin condenses, the nuclear lamina depolymerizes, and in most higher eukaryotes the nuclear membrane vesicularizes. The end of mitosis involves a reversal of these processes. The molecular processes regulating nuclear structure during mitosis may also, in an attenuated form, regulate processes such as nuclear membrane and lamina growth, as well as chromatin rearrangement, in nuclei during interphase. Hence, a molecular understanding of nuclear rearrangements during mitosis could provide insight into the dynamics of interphase nuclei.

Assembly and Disassembly of the Nuclear Membrane

The extent to which the nuclear membrane breaks down during mitosis varies in different organisms (for reviews see 214–218). Many lower eukaryotes undergo "closed mitosis" in which the nuclear membrane remains intact throughout mitosis, while the majority of higher eukaryotes undergo "open mitosis" in which the extent of membrane breakdown ranges from partial to complete (33, 219). In other organisms, membrane breakdown also varies with the developmental stage. For example, in *Physarum* and related slime molds membrane breakdown occurs during the cellular stage of the life cycle but does not occur during the multinucleate plasmodial phase (220–222). The observed variation in extent of nuclear membrane breakdown during mitosis, in conjunction with the observation that some cells actively control this process, suggests that membrane breakdown is a regulated process. Little is currently known of the biochemical nature of this regulatory process. The continuity between the outer nuclear membrane and ER has, however, led Warren (223) to propose that nuclear membrane breakdown at mitosis may involve a hyperactivity of the mechanisms that normally regulate budding of the endoplasmic reticulum membrane during intracellular transport of vesicles between the ER and Golgi (223, 224).

Assembly and Disassembly of the Nuclear Lamina

The polymeric lamina of interphase nuclei is depolymerized in mitosis (133, 225–227). Sucrose sedimentation studies of lysates of *Xenopus* and *Drosophila* eggs demonstrate that the lamin proteins are both soluble and present in the cytoplasm in a monomeric or dimeric form during the mitotic stage of the cell cycle (152). Using Chinese hamster ovary (CHO) mitotic cell lysates, Gerace & Blobel have found that lamins A and C are converted to a soluble form, whereas lamin B remains largely attached to membrane vesicles (141; see also 228). They have hypothesized that the tight association of lamin B with nuclear membrane vesicles during mitosis facilitates reassociation of the membrane vesicles with chromosomes during envelope assembly at the end of mitosis. However, in *Xenopus* and *Drosophila* embryos (143, 152), where lamins do not associate with membrane vesicles during mitosis, alternative mechanisms must be active for directing the reassociation of membranes around chromosomes.

Gerace and colleagues have demonstrated that depolymerization of the lamina at mitosis is likely regulated by phosphorylation of the lamin proteins (225, 226). They have found that at mitosis the level of lamin phosphorylation increases 4–7-fold above the low level present in interphase and that lamin dephosphorylation occurs at telophase. Phosphorylation is observed at multiple serine residues, a subset of these phosphorylation sites being mitosis-specific. Phosphorylation is the only detectable charge-altering modification associated with lamin disassembly at mitosis (226). This phosphorylation-dephosphorylation cycle is thought to control the monomer-polymer equilibrium regulating lamin breakdown and formation during mitosis.

Because the lamina is located between chromatin and the nuclear membrane, it has been suggested that lamina depolymerization may act as a trigger for disassembly of the nuclear membrane in mitosis (133). However, using both indirect immunofluorescence and electron microscopy, Stick & Schwarz have shown that the nuclear lamina of chick and *Xenopus* oocytes disappears in meiosis without altering the structure of the nuclear membranes or the nuclear pores (229). This finding strongly suggests that although lamina disassembly may be required for nuclear membrane breakdown, it is not sufficient.

Formation of Synthetic Nuclei

In cells that undergo "open mitosis," the chromosome acts as a template for the assembly of new nuclear envelope at the end of mitosis. This is most clearly observed in rapidly dividing cells in which a nuclear envelope forms around each chromosome, resulting in numerous small nuclei or "karyomers." These karyomers subsequently fuse to form a single large nucleus (230, 231). It has been found that when bacteriophage lambda DNA is injected into *Xenopus* eggs that this initially protein-free DNA acts as a template for the assembly of nuclei that are normal in morphology (128, 130, 232, 233; see 234 for similar experiments in *Drosophila*). The reconstituted nuclei contain a lamina, double nuclear membrane, and nuclear pores. This finding indicates that the assembly of nuclear envelope components does not require specific eukaryotic DNA sequences. Furthermore, it demonstrates that in the absence of a DNA template many nuclear components can remain unassembled in the cytoplasm and require only the addition of DNA to trigger their assembly.

In Vitro Systems for Studying Nuclear Assembly

Until recently most studies on nuclear assembly and disassembly were carried out using one of several in vivo systems. Those systems included cells synchronized at mitosis using inhibitors of microtubule assembly, fusion of cells at different stages of the cell cycle (235–237), and injection of nuclei into large eggs naturally arrested at one stage of the cell cycle (238–243). Although these techniques have proved useful for answering many questions of nuclear dynamics, they have been of limited use for answering others. Recently, several cell-free systems have been developed for investigating nuclear assembly and disassembly at the biochemical level.

Nuclear assembly in vitro has been achieved using several templates of differing complexity. Lohka & Masui find that demembranated sperm nuclei can act as a template for nuclear envelope reassembly when added to an extract of amphibian eggs (244–246). Burke & Gerace, using a lysate of mitotic CHO cells, find that metaphase chromosomes also serve as efficient templates for assembly of nuclear envelopes (247). Newport and colleagues find that even protein-free bacteriophage DNA can act as a template for nuclear assembly in vitro when added to *Xenopus* egg extracts (128, 131, 232). In all cases the nuclei assembled in vitro are morphologically similar to normal interphase nuclei in terms of their membrane, pore, and lamina composition. The assembly of nuclear structure occurs spontaneously when DNA is added to a mixture of unassembled nuclear components (128, 131,

140, 232). Newmeyer et al showed that nuclear assembly requires ATP (129). In the amphibian systems, the components used for nuclear assembly in vitro are derived from the excess pool of components made and stored in the egg during oogenesis, while in the CHO system, soluble membrane and lamin components are derived from the disassembled nuclei of mitotic cells.

Assembly of the nuclear envelope occurs by a process involving (*a*) the binding of 100–200-nM diameter membrane vesicle to chromatin, followed by (*b*) the fusion of these vesicles to form a continuous double membrane, as shown by electron microscopy of nuclear assembly using sperm nuclei, metaphase chromosomes, or protein-free DNA (131, 245, 247). The assembly of nuclear lamina and membranes around metaphase chromosomes does not require ATP hydrolysis and is, in fact, inhibited by high concentrations of ATP (247). The energy stored in lamins by phosphorylation at the onset of mitosis may be used to drive envelope formation at the end of mitosis (247).

The amphibian assembly extracts can be fractionated into soluble and particulate fractions using differential centrifugation (246). Both are required for nuclear reconstitution around demembranated sperm nuclei (246). The necessary particulate components obviously include membrane vesicles, while the necessary soluble components presumably include the lamins and factors required for chromatin decondensation. Nuclear assembly extracts can also be depleted of particular nuclear proteins to test theories of nuclear assembly. By incubating CHO cell extracts with lamin antibodies coupled to Staph A, Burke & Gerace depleted extracts of either lamins A and C or lamin B (247). In extracts depleted of lamin B, lamins A and C bind to chromosomes; however, envelope formation is blocked. In extracts depleted of lamins A and C, lamin B binding to chromosomes is significantly reduced and envelope assembly is blocked. These depletion results support the proposal that lamins A and C bind to chromatin independently of membrane formation and that the interaction of membrane-bound lamin B with chromatin-bound lamins A and C facilitates formation of a new nuclear membrane at the end of mitosis (133).

The formation of nuclei is an ordered process with discrete intermediates, as shown by Newport using protein-free DNA templates in an amphibian cell-free system (128, 131, 232). The DNA is first assembled into nucleosomes, then rearranged to form a distinctive highly condensed sphere. The condensation process resulting in this sphere apparently involves formation of a scaffold, perhaps similar to that present in metaphase chromosomes, since the chromatin within the condensed intermediate retains its spherical organization following extraction of histones with 2-M NaCl. Topoisomerase II inhibitors block formation of the condensed intermediate suggesting that topoisomerase II is actively involved in the condensation process. Only after the chromatin is fully condensed do lamin and membrane components begin to assemble around the DNA, as judged by immunological techniques and electron microscopy. In contrast, lamins and membrane vesicles begin binding to metaphase chromosomes almost immediately after addition to such an extract (131). This suggests that envelope formation may be mediated by specific chromatin-bound receptors. These putative receptors may bind to protein-free DNA only after chromatin formation and full condensation, but be already present on the added metaphase chromosomes. Using monoclonal antibodies, several groups have identified proteins that are bound to the periphery of metaphase chromosomes and could function as chromatin-bound receptors (248, 249). Such hypothetical receptors may function during envelope formation as membrane-binding sites or as sites for initiating lamin polymerization (131, 248, 249). Clearly, systems for investigating nuclear assembly in vitro have made biochemical analysis of nuclear structures and dynamics much more accessible.

In Vitro Systems for Studying Nuclear Disassembly

The development of in vitro systems for nuclear disassembly has largely stemmed from in vivo studies of amphibian eggs. Unfertilized Xenopus eggs are arrested in metaphase. Interphase nuclei injected into unfertilized eggs undergo nuclear envelope breakdown and chromosome condensation, demonstrating that the mitotic activities needed for nuclear disassembly are active in these eggs (238–240, 250). A specific protein factor called maturation promoting factor (MPF), which is present not only in unfertilized eggs but also in somatic cells in mitosis, is thought to play a major role in maintaining these mitotic components in an active form (239, 240, 250–253). Both Miake-Lye & Kirschner (254) and Lohka & Maller (255) have shown that addition of partially purified MPF to a Xenopus egg nuclear assembly extract causes activation of nuclear disassembly factors. Thus, addition of MPF to a nuclear assembly extract induces nuclear envelope breakdown, lamin phosphorylation and depolymerization, and chromosome condensation (224, 254, 255).

The transition from the assembly to disassembly state induced by MPF is transient, lasting 20–30 minutes (254). To develop a nuclear disassembly system that would remain stable long enough for biochemical analysis, Newport & Spann (224) have made cell-free extracts directly from unfertilized eggs, using a combination of high concentrations of EGTA and beta-glycerophosphate to stabilize the mitotic state. The mitotic activities in such extracts are stable for five hours at room temperature and indefinitely at $-70°C$ (128, 224, 232). Since these stable extracts are composed exclusively of soluble factors, particulate components are not required for nuclear disassembly. That cell-free nuclear disassembly extracts can be made directly from amphibian eggs arrested in mitosis indicates it should be possible to make similar extracts from somatic cells arrested in mitosis.

A lamin kinase has been proposed to be important in regulating lamin depolymerization during mitosis (133, 225, 226). Consistent with this, labeling experiments have demonstrated that a lamin kinase is active in the amphibian disassembly extract (224, 254). In disassembly extracts the rate of lamin depolymerization as a function of the concentration of added nuclei displays simple first order saturation kinetics (224), suggesting that a lamin kinase is solely responsible for lamin depolymerization.

Using the amphibian nuclear disassembly extracts, it has been shown that lamin phosphorylation and depolymerization precede nuclear membrane breakdown (224, 255). Furthermore, if large numbers of nuclei are added to a disassembly extract, at a concentration of 40,000 per μl of extract, membrane breakdown stops but lamin release continues (224). This indicates that membrane vesicularization during mitosis is, in part, independent of lamin solubilization. It also suggests that membrane vesicularization is an active process that requires proteins that are titrated out at high nuclear concentrations. One possibility is that membrane breakdown is dependent on a clathrin-like protein that actively vesicularizes the membrane during mitosis (223, 224). If so, regulation of the levels of such a protein might account for the variation in the extent of nuclear membrane breakdown observed in different organisms.

The use of in vitro systems, in conjunction with in vivo techniques, has made careful biochemical analysis of nuclear organization and dynamics possible. Using these techniques it should be possible to investigate both the properties of individual nuclear components such as the pores, membranes, and lamina, as well as how these elements interact with chromatin and one another to form a functionally intact nucleus.

ACKNOWLEDGMENTS

The authors thank K. Wilson, P. Hartl, B. Dunphy, D. Finlay, D. Newmeyer, T. Spann, and S. Lonergan for reading drafts of the manuscript. We thank S. Allen and D. Newmeyer for aid in preparing the manuscript. Work in the authors' laboratories was supported by grants to J. N. from the NIH (GM33523), March of Dimes (GRMOD5-471), and Searle Foundation, and to D. F. from the NIH (GM33279) and the Pew Memorial Trust.

Literature Cited

1. Nelson, W. G., Pienta, K. J., Barrack, E. R., Coffey, D. S. 1986. *Ann. Rev. Biophys. Biophys. Chem.* 15:457–75
2. Franke, W. W. 1974. *Int. Rev. Cytol. Suppl.* 4:71–236
3. Maul, G. G. 1977. *Int. Rev. Cytol. Suppl.* 6:75–186
4. Fahl, W. E., Jefcoate, C. R., Kasper, C. B. 1978. *J. Biol. Chem.* 253:3106–13
5. Matsuura, S., Masuda, R., Omori, K., Negishi, M., Tashiro, Y. 1981. *J. Cell Biol.* 91:212–20

6. Richardson, J. C. W., Maddy, H. 1980. *J. Cell Sci.* 43:269–77
7. Puddington, L., Lively, M. O., Lyles, D. S. 1985. *J. Biol. Chem.* 260:5641–47
8. Franke, W. W., Scheer, U., Krohne, G. Jarasch, E.-D. 1981. *J. Cell Biol.* 91:39s–50s
9. Aaronson, R. P., Blobel, G. 1974. *J. Cell Biol.* 62:746–54
10. Schindler, M., Holland, J. F., Hogan, M. 1985. *J. Cell Biol.* 100:1408–14

11. Feldherr, C. 1972. *Adv. Cell Mol. Biol.* 2:273–307
12. Kay, R. R., Johnston, I. R. 1973. *Sub-Cell. Biochem.* 2:127
13. Fry, D. J., Boston, M. A. 1976. In *Mammalian Cell Membranes,* ed. G. A. Jamieson, D. M. Robinson, 2:197–265. London: Butterworths
14. Harris, J. R. 1978. *Biochem. Biophys. Acta* 515:55–104
15. Bonner, W. M. 1978. In *The Cell Nucleus, Chromatin, Part C,* ed. H. Busch, 6:97–148. New York: Academic
16. Deleted in proof
17. Paine, P. L., Horowitz, S. B. 1980. In *Cell Biology: A Comprehensive Treatise,* ed. D. M. Prescott, L. Goldstein, 4:299–338. New York: Academic
18. De Robertis, E. M. 1983. *Cell* 32:1021–25
19. Dingwall, C. 1985. *Trends Biochem. Sci.* 10:64–66
20. Rine, J., Barnes, G. 1985. *BioEssays* 2:158–61
21. Feldherr, C. M. 1985. *BioEssays* 3:52–55
22. Clawson, G. A., Feldherr, C. A., Smuckler, E. A. 1985. *Mol. Cell. Biochem.* 67:87–100
23. Agutter, P. S. 1983. *Biochem. J.* 214:915–21
24. Fisher, P. A. 1987. In *Chromosomes and Chromatin Structure,* ed. K. W. Adolph. Boca Raton: CRC In press
24a. Berrios, M., Fisher, P. A. 1986. *J. Cell Biol.* 103:711–24
25. Dingwall, C., Laskey, R. A. 1986. *Ann. Rev. Cell Biol.* 2:365–88
26. Hertwig, R. 1876. *Morphol. Jahrb.* 2:63–82
27. Callan, H. G., Randall, J. R., Tomlin, S. G. 1949. *Nature* 163:280
28. Callan, H. G., Tomlin, S. G. 1950. *Proc. R. Soc. London Ser. B* 137:367–78
29. Unwin, P. N. T., Milligan, R. A. 1982. *J. Cell Biol.* 93:63–75
30. Kessel, R. G. 1983. *Int. Rev. Cytol.* 82:181–303
31. Stafstrom, J. P., Staehelin, L. A. 1984. *J. Cell Biol.* 98:699–708
32. Maul, G. G., Maul, H. M., Scogna, J. E., Lieberman, M. W., Stein, G. S., et al. 1972. *J. Cell Biol.* 55:433–47
33. Stafstrom, J. P., Staehelin, L. A. 1984. *Eur. J. Cell Biol.* 34:179–89
34. Zatsepina, O. V., Polyakov, V. Y., Chentsov, Y. S. 1977. *Cytobiologie* 16:130–44
35. Roos, U.-P. 1973. *Chromosoma* 40:43–82
36. Maul, G. G. 1977. *J. Cell Biol.* 74:492–500
37. Deleted in proof
38. Feldherr, C. M. 1968. *Nature* 218:184–85
39. Feldherr, C. M. 1968. *J. Cell Biol.* 39:49–54
40. Blobel, G. 1985. *Proc. Natl. Acad. Sci. USA* 82:8527–29
41. Aaronson, R. P., Blobel, G. 1975. *Proc. Natl. Acad. Sci. USA* 72:1007–11
42. Dwyer, N., Blobel, G. 1976. *J. Cell Biol.* 70:581–91
43. Krohne, G., Franke, W. W., Scheer, U. 1978. *Exp. Cell Res.* 116:85–102
44. Gerace, L., Ottaviano, Y., Kondor-Koch, C. 1982. *J. Cell Biol.* 95:826–37
45. Filson, A. J., Lewis, A., Blobel, G., Fisher, P. A. 1985. *J. Biol. Chem.* 260:3164–72
46. Berrios, M., Filson, A. J., Blobel, G., Fisher, P. A. 1983. *J. Biol. Chem.* 258:13384–90
47. Harris, J. R. 1985. *Micron Microsc. Acta* 16:89–108
48. Fisher, P. A., Blobel, G. 1983. *Methods Enzymol.* 96:589–96
49. Deleted in proof
50. Feldherr, C. M., Richmond, P. A., Noonan, K. D. 1977. *Exp. Cell Res.* 107:439–44
51. Seve, A. P., Hubert, J., Douvier, D., Masson, C., Geraud, G., Bouteille, M. 1984. *J. Submicrosc. Cytol.* 16:631–41
52. Virtanen, I., Wartiovaara, J. 1976. *J. Cell Sci.* 22:335–44
53. Virtanen, I., Wartiovaara, J. 1978. *Cell Mol. Biol.* 23:73–79
54. Virtanen, I. 1977. *Biochem Biophys. Res. Commun.* 78:1411–17
55. Deleted in proof
56. Davis, L. I., Blobel, G. 1986. *Cell* 45:699–709
57. Finlay, D. R., Newmeyer, D. D., Price, T. M., Forbes, D. J. 1987. *J. Cell Biol.* 104:189–200
58. Baglia, F. A., Maul, G. G. 1983. *Proc. Natl. Acad. Sci. USA* 80:2285–89
59. Newmeyer, D. D., Finlay, D. R., Forbes, D. J. 1986. *J. Cell Biol.* 103:2091–102
60. Glass, W. F., Briggs, R. C., Hnilica, L. S. 1981. *Anal. Biochem.* 115:219–24
61. Schindler, M., Hogan, M. 1984. *J. Cell Biol.* 99:99a
62. Guinivan, P., Noonan, N. E., Noonan, K. D. 1980. *J. Cell Biol.* 87:205a
63. Holt, G. D., Hart, G. W. 1986. *J. Biol. Chem.* 261:8049–57
64. Holt, G. D., Snow, C. M., Gerace, L., Hart, G. W. 1986. *J. Cell Biol.* 103:320a
65. Deleted in proof
66. Stevens, B. J., Swift, H. 1966. *J. Cell Biol.* 31:55–77
67. Feldherr, C. M., Kallenbach, E.,

Schultz, N. 1984. *J. Cell Biol.* 99:2216–22

68. Bonner, W. M. 1975. *J. Cell Biol.* 64:421–30
69. Bonner, W. M. 1975. *J. Cell Biol.* 64:431–37
70. Paine, P. L., Moore, L. C., Horowitz, S. B. 1975. *Nature* 254:109–114
71. De Robertis, E., Longthorne, R. F., Gurdon, J. B. 1978. *Nature* 272:254–56
72. Paine, P. L. 1975. *J. Cell Biol.* 66:652–57
73. Feldherr, C. M., Pomerantz, J. 1978. *J. Cell Biol.* 78:168–75
74. Feldherr, C. M., Ogburn, J. A. 1980. *J. Cell Biol.* 87:589–93
75. Reynolds, C. R., Tedeschi, H. 1984. *J. Cell Biol.* 70:197–207
76. Einck, L., Bustin, M. 1984. *J. Cell Biol.* 98:205–13
77. Peters, R. 1984. *EMBO J.* 3:1831–36
78. Peters, R. 1983. *J. Biol. Chem.* 258:11427–29
79. Lang, I., Peters, R. 1984. In *Information and Energy Transduction in Biological Membranes,* ed. C. L. Bolis, E. J. M. Helmrich, H. Passow, pp. 377–86. New York: Liss
80. Lang, I., Scholz, M., Peters, R. 1986. *J. Cell Biol.* 102:1183–90
81. Peters, R., Lang, I., Scholz, M. Schulz, B., Kayne, F., 1986. *Biochem. Soc. Trans. London.* 14:821–22
82. Jiang, L. W., Schindler, M. 1986. *J. Cell Biol.* 102:852–58
83. Schindler, M., Jiang, L. W. 1986. *J. Cell Biol.* 102:859–62
84. Feldherr, C. 1965. *J. Cell Biol.* 25:43–53
85. Rechsteiner, M., Kuehl, L. 1979. *Cell* 16:901–8
86. Dingwall, C., Allan, J. 1984. *EMBO J.* 3:1933–37
87. Wu, L. H., Kuehl, L., Rechsteiner, M. 1986. *J. Cell Biol.* 103:465–74
88. Kuehl, L., Rechsteiner, M., Wu, L. 1985. *J. Biol. Chem.* 260:10361–68
89. Mills, A. D., Laskey, R. A., Black, P., De Robertis, E. M. 1980. *J. Mol. Biol.* 139:561–68
90. Dabauvalle, M.-C., Franke, W. W. 1982. *Proc. Natl. Acad. Sci. USA* 79:5302–6
91. Dingwall, C., Sharnick, S. V., Laskey, R. A. 1982. *Cell* 30:449–58
92. Hall, M. N., Hereford, L., Herskowitz, I. 1984. *Cell* 36:1057–65
93. Lanford, R. E., Butel, J. S. 1984. *Cell* 37:801–13
94. Kalderon, D., Richardson, W. D., Markham, A. F., Smith, A. E. 1984. *Nature* 311:33–38
95. Kalderon, D., Roberts, B. L., Richard-son, W. D., Smith, A. E. 1984. *Cell* 39:499–509
96. Moreland, R. B., Nam, H. G., Hereford, L. M., Fried, H. M. 1985. *Proc. Natl. Acad. Sci. USA* 82:6561–65
97. Silver, P. A., Keegan, L. P., Ptashne, M. 1984. *Proc. Natl. Acad. Sci. USA* 81:5951–55
98. Richter, J. D., Young, P., Jones, N. C., Krippl, B., Rosenberg, M., et al. 1985. *Proc. Natl. Acad. Sci. USA* 82:8434-38
99. Richardson, W. D., Roberts, B. L., Smith, A. E. 1986. *Cell* 44:77–85
100. Davey, J., Colman, A., Dimmock, N. J. 1985. *J. Gen. Virol.* 66:2319–34
101. Davey, J., Dimmock, N. J., Colman, A. 1985. *Cell* 40:667–75
102. Welsh, J. D., Swimmer, C., Cocke, T., Shenk, T. 1986. *Mol. Cell. Biol.* 6:2207–12
103. Schickedanz, J., Scheidtmann, K. H., Walter, G. 1986. *Virology* 148:47–57
104. Goldfarb, D. S., Gariepy, J., Schoolnik, G., Kornberg, R. D. 1986. *Nature* 322:641–44
105. Lanford, R. E., Kanda, P., Kennedy, R. C. 1986. *Cell* 46:575–82
106. Stacey, D. W., Allfrey, V. G. 1984. *Exp. Cell Res.* 154:283–92
107. Barnes, G., Rine, J. 1985. *Proc. Natl. Acad. Sci. USA* 82:1354–58
108. Rine, J., Barnes, G. 1986. *UCLA Symp. Mol. Cell. Biol.* 33:395–413
109. Feldherr, C., Cohen, R. J., Ogburn, J. A. 1983. *J. Cell Biol.* 96:1486–90
110. Sugawa, H., Imamoto, N., Wataya-Kaneda, M., Uchida, T. 1985. *Exp. Cell Res.* 159:419–29
111. Dingwall, C., Burglin, T. R., Kearsey, S. E., Dilworth, S., Laskey, R. A. 1986. In *Proc. Workshop Nucleocytoplasmic Transport, Heidelberg)* ed. R. Peters, M. F. Trendelenburg. In press
112. Fritz, A., Parisot, R., Newmeyer, D., De Robertis, E. M. 1984. *J. Mol. Biol.* 178:273–85
113. Mattaj, I. W., De Robertis, E. M. 1985. *Cell* 40:111–18
114. Dreyer, C., Singer, H., Hausen, P. 1981. *Wilhelm Roux's Arch.* 190:197–207
115. Dreyer, C., Scholz, E., Hausen, P., Glaser, B., Muller, U., Siegel, E. 1982. *Wilhelm Roux's Arch.* 191:228–33
116. Dreyer, C., Stick, R., Hausen, P. 1986. See Ref. 111
117. Kuettel, M. R., Schwoch, G., Jungmann, R. A. 1984. *Cell Biol. Int. Rep.* 11:949–57
118. Squinto, S. P., Kelley-Geraghty, D. C., Kuettel, M. R., Jungmann, R. A. 1985.

J. Cyclic Nucl. & Protein Phosphoryl. Res. 10:65–73

119. Kuettel, M. R., Squinto, S. P., Kwast-Welfeld, J., Schwoch, G., Schweppe, J. S., et al. 1985. *J. Cell Biol.* 101:965–75

120. Nigg, E. A., Hilz, H., Eppenberger, H. M., Dutly, F. 1985. *EMBO J.* 4:2801–6

121. Woffendin, C., Chambers, T. C., Schaller, K. L., Leichtling, B. H., Rickenberg, H. V. 1986. *Dev. Biol.* 115:1–8

122. Knipe, D. M., Smith, J. L. 1986. *Mol. Cell. Biol.* 6:2371–81

123. Dabauvalle, M.-C., Franke, W. W. 1986. *J. Cell Biol.* 102:2006–14

124. Zasloff, M., Rosenberg, M., Santos, T. 1982. *Nature* 30:81–84

125. Tobian, J. A., Drinkard, L., Zasloff, M. 1985. *Cell* 43:415–22

126. Zasloff, M. 1983. *Proc. Natl. Acad. Sci. USA* 80:6436–40

127. Dworetzky, S. I., Feldherr, C. M. 1987. *J. Cell Biol.* 103:320a

128. Newport, J., Forbes, D. 1985. *Banbury Report 20: Genetic Manipulation of the Early Mammalian Embryo*, ed. F. Constantini, R. Jaenisch, pp. 243–50. Cold Spring Harbor: Cold Spring Harbor Lab. Press

129. Newmeyer, D. D., Lucocq, J. M., Burglin, T. R., De Robertis, E. M. 1986. *EMBO J.* 5:501–10

130. Forbes, D. J., Kirschner, M. W., Newport, J. N. 1983. *Cell* 34:13–23

131. Newport, J. W. 1987. *Cell.* 48:205–17

132. Gerace, L., Blobel, G. 1982. *Cold Spring Harbor Symp. Quant. Biol.* 46:967–78

133. Gerace, L., Comeau, C., Benson, M. 1984. *J. Cell Sci. Suppl.* 1:137–60

134. Krohne, G., Benavente, R. 1986. *Exp. Cell Res.* 162:1–10

135. Gerace, L., Blum, A., Blobel, G. 1978. *J. Cell Biol.* 79:546–66

136. Lebkowski, J. S., Laemmli, U. K. 1982. *J. Mol. Biol.* 156:325–44

137. Lebkowski, J. S., Laemmli, U. K. 1982. *J. Mol. Biol.* 156:309–24

138. Burke, B., Gerace, L. 1986. *Cell* 44:639–52

139. Krohne, G., Franke, W. W., Ely, S., D'Arcy, A., Jost, E. 1978. *Cytobiologie* 18:22–38

140. Scheer, U., Kartenbeck, J., Trendelenburg, M. F., Stadler, J., Franke, W. W. 1976. *J. Cell Biol.* 69:1–18

141. Gerace, L., Blobel, G. 1980. *Cell* 19:277–87

142. Deleted in proof

143. Benavente, R., Krohne, G., Franke, W. W. 1985. *Cell* 41:177–90

144. Fuchs, J. P., Giloh, H., Kuo, C., Saumweber, H., Sedat, J. 1983. *J. Cell Sci.* 64:331–49

145. Krohne, G., Dabauvalle, M. C., Franke, W. W. 1981. *J. Mol. Biol.* 151:121–41

146. Krohne, G., Debus, E., Osborn, M., Weber, K., Franke, W. W. 1984. *Exp. Cell Res.* 150:47–59

147. Maul, G. G., Baglia, F. A., Newmeyer, D. D., Ohlsson-Wilhelm, B. M. 1984. *J. Cell Sci.* 67:69–85

148. McKeon, F. D., Tuffanelli, D., Fukuyama, K., Kirschner, M. 1983. *Proc. Natl. Acad. Sci. USA* 80:4374–78

149. Shelton, K. R., Higgins, L. L., Cochran, D. L., Ruffolo, J. J., Egle, P. M. 1980. *J. Biol. Chem.* 255:10978–84

150. Stick, R., Hausen, P. 1985. *Cell* 41:191–202

151. Fisher, P. A., Berrios, M., Blobel, G. 1982. *J. Cell Biol.* 92:674–86

152. Smith, D. E., Fisher, P. A. 1984. *J. Cell Biol.* 99:20–28

153. Lehner, C., Kurer, V., Eppenberger, H., Nigg, E. 1986. *J. Biol. Chem.* 261:13293–301

154. Maul, G. G., French, B. T., Bechtol, K. B. 1986. *Dev. Biol.* 115:68–77

155. Benavente, R., Krohne, G. 1985. *Proc. Natl. Acad. Sci. USA* 82:6176–81

156. McKeon, F., Kirschner, M., Caput, D. 1986. *Nature* 319:463–68

157. Fisher, D., Chaudhary, N., Blobel, G. 1986. *Proc. Natl. Acad. Sci. USA* 83:6450–54

158. Sedat, J. W. 1986. *J. Cell Biol.* 102:112–23

159. Mathog, D., Hochstrasser, M., Gruenbaum, Y., Saumweber, H., Sedat, J. 1984. *Nature* 308:414–20

160. Agard, D. A., Sedat, J. W. 1983. *Nature* 302:676–81

161. Zakian, V. A. 1984. *Nature* 308:406

162. Hochstrasser, M., Mathog, D., Gruenbaum, Y., Saumweber, H., Sedat, J. W. 1986. *J. Cell Biol.* 102:112–23

163. Murray, A., Davies, H. 1979. *J. Cell Sci.* 35:59–66

164. Avivi, L., Feldman, M., Brown, M. 1982. *Chromosoma* 86:1–16

165. Avivi, L., Feldman, M., Brown, M. 1982. *Chromosoma* 86:17–26

166. Callan, H. G. 1982. *Proc. R. Soc. London Ser. B* 214:417–48

167. Callan, H. G. 1963. *Int. Rev. Cytol.* 15:1–34

168. Ruchert, J. 1892. *Anat. Anz.* 7:107–58

169. Gall, J. 1963. *Nature* 198:36–38

170. Scheer, U., Sommerville, J., Bustin, M. 1979. *J. Cell Sci.* 40:1–20

171. Angelier, N., Lacroix, J.-C. 1975. *Chromosoma* 51:323–35

172. Scheer, U., Franke, W. W., Trendelenburg, M. F., Spring, H. 1976. *J. Cell Sci.* 22:503–19

173. Cook, P. R., Brazell, I. A. 1976. *J. Cell Sci.* 22:287–302
174. Cook, P. R., Brazell, I. A. 1976. *J. Cell Sci.* 22:303–24
175. Warren, A. C., Cook, P. R. 1978. *J. Cell Sci.* 30:211–26
176. Wang, J. 1969. *J. Mol. biol.* 43:263–72
177. Benyajati, C., Worcel, A. 1976. *Cell* 9:393–407
178. Vogelstein, B., Pardoll, D. M., Coffey, D. S. 1980. *Cell* 22:79–85
179. Adolph, K., Cheng, S., Laemmli, U. 1977. *Cell* 12:805–16
180. Paulson, J. R., Laemmli, U. K. 1977. *Cell* 12:817–28
181. Earnshaw, W. C., Laemmli, U. K. 1983. *J. Cell Biol.* 96:84–93
182. Lewis, C. D., Lebkowski, J. S., Daly, A. K., Laemmli, U. K. 1984. *J. Cell Sci. Suppl.* 1:103–22
183. Mirkovitch, J., Mirault, M.-E., Laemmli, U. K. 1984. *Cell* 39:223–32
184. Gasser, S. M., Laemmli, U. K. 1986. *EMBO J.* 5:511–18
185. Cockerill, P. N., Garrard, W. T. 1986. *Cell* 44:273–82
186. Keppel, F. 1986. *J. Mol. Biol.* 187:15–21
187. Mirkovitch, J., Spierer, P., Laemmli, U. K. 1986. *J. Mol. Biol.* 190:255–58
188. McCready, S. J., Cook, P. R. 1984. *J. Cell Sci.* 70:189–96
189. Jackson, D. A., Caton, A. J., McCready, S. J., Cook, P. R. 1982. *Nature* 296:366–68
190. Smith, H. C., Puvion, E., Buchholtz, L. A., Berezney, R. 1984. *J. Cell Biol.* 99:1794–802
191. Cook, P. R., Brazell, I. A. 1980. *Nucleic Acids Res.* 8:2895–905
192. Small, D., Nelkin, B., Vogelstein, B. 1982. *Proc. Natl. Acad. Sci. USA* 79:5911–15
193. Udvardy, A., Schedl, P., Sander, M., Hsieh, T. 1985. *Cell* 40:933–41
194. Berrios, M., Osheroff, N., Fisher, P. A. 1985. *Proc. Natl. Acad. Sci. USA* 82:4142–46
195. Lynn, R., Giaever, G., Swanberg, S. L., Wang, J. C. 1986. *Science* 233:647–49
196. Goto, T., Wang, J. C. 1984. *Cell* 36:1073–80
197. Hsieh, T. S., Brutlag, D. 1980. *Cell* 21:115–25
198. Halligan, B. D., Edwards, K. A., Liu, L. F. 1985. *J. Biol. Chem.* 260:2475–82
199. Glikin, G. C., Blangy, D. 1986. *EMBO J.* 5:151–55
200. Gellert, M. 1981. *Ann. Rev. Biochem.* 50:879–910
201. Wang, J. 1985. *Ann. Rev. Biochem.* 54:665–97
202. Earnshaw, W. C., Halligan, B., Cooke, C. A., Heck, M. M. S., Liu, L. F. 1985. *J. Cell Biol.* 100:1706–15
203. Earnshaw, W. C., Heck, M. M. S. 1985. *J. Cell Biol.* 100:1716–25
204. Gasser, S. M., Laroche, T., Falquet, J., Boy de la Tour, E., Laemmli, U. K. 1986. *J. Mol. Biol.* 188:613–29
205. Blumenthal, A. B., Kriegstein, H. J., Hogness, D. S. 1973. *Cold Spring Harbor Symp. Quant. Biol.* 38:205–23
206. Callan, H. G. 1973. *Cold Spring Harbor Symp. Quant. Biol.* 38:195–203
207. Huberman, J., Riggs, A. 1968. *J. Mol. Biol.* 32:327–40
208. Buongiorno-Nardelli, M., Micheli, G., Carri, M. T., Marilley, M. 1982. *Nature* 298:100–2
209. Heck, M. M. S., Earnshaw, W. C. 1987. *J. Cell Biol.* 103:2569–81
210. Duguet, M., Lavenot, C., Harper, F., Mirambeau, G., DeRecondo, A. M. 1983. *Nucleic Acids Res.* 11:1059–75
211. Skaer, R. J., Whytock, S., Emmines, J. P. 1976. *J. Cell Sci.* 21:479–96
212. Moroi, Y., Hartman, A. L., Nakane, P. K., Tan, E. M. 1981. *J. Cell Biol.* 90:254–59
213. Lewis, C. D., Laemmli, U. K. 1982. *Cell* 29:171–81
214. Heath, B. 1980. *Int. Rev. Cytol.* 64:1–80
215. Kubai, D. 1975. *Int. Rev. Cytol.* 43:167–227
216. Pickett-Heaps, J. 1974. *BioSystems* 6:37–48
217. Fuller, M. S. 1976. *Int. Rev. Cytol.* 45:113–53
218. Goode, D. 1975. *BioSystems* 7:318–25
219. Zeligs, J. D., Wollman, S. H. 1979. *J. Ultrastruct. Res.* 66:53–77
220. Aldrich, H. 1969. *Am. J. Bot.* 56:290–99
221. Hinchee, A., Haskins, E. 1980. *Protoplasma* 102:235–52
222. Hinchee, A., Haskins, E. 1980. *Protoplasma* 102:117–30
223. Warren, G. 1985. *Trends Biochem. Sci.* 10:439–43
224. Newport, J., Spann, T. 1987. *Cell.* 48:219–30
225. Gerace, L., Blobel, G. 1980. *Cell* 19:277–87
226. Ottaviano, Y., Gerace, L. 1985. *J. Biol. Chem.* 260:624–32
227. Jost, E., Johnson, R. T. 1981. *J. Cell Sci.* 47:25–53
228. Lebel, S., Raymond, Y. 1984. *J. Biol. Chem.* 259:2693–96
229. Stick, R., Schwarz, H. 1983. *Cell* 33:949–58

230. Ito, S., Dan, K., Goodenough, D. 1981. *Chromosoma* 83:441–53
231. Hernandez-Verdum, D., Gregaire, M., Labidi, B., Bouteille, M. 1986. *Exp. Cell Res.* 164:243–50
232. Newport, J., Spann, T., Kanki, J., Forbes, D. 1985. *Cold Spring Harbor Symp. Quant. Biol.* 50:651–55
233. Shiokawa, K., Sameshima, M., Tashiro, K., Muira, T., Nakakura, N., Yamana, K. 1986. *Dev. Biol.* 116:539–42
234. Steller, H., Pirrotta, V. 1985. *Dev. Biol.* 109:54–62
235. Johnson, R. T., Rao, P. N. 1970. *Nature* 226:717–22
236. Hanks, S. K., Rao, P. N. 1980. *J. Cell Biol.* 87:285–91
237. Gollin, S. M., Wray, W., Hanks, S. K., Hittelman, W. N., Rao, P. N. 1984. *J. Cell Sci. Suppl.* 1:203–21
238. Gurdon, J. 1976. *J. Embryol. Exp. Morphol.* 36:523–40
239. Masui, Y., Markert, C. 1971. *J. Exp. Zool.* 177:129–46
240. Newport, J., Kirschner, M. 1984. *Cell* 37:731–45
241. Balakier, H., Masui, Y. 1986. *Dev. Biol.* 113:155–59
242. Czolowska, R., Modlinski, J. A., Tarkowski, A. K. 1984. *J. Cell Sci.* 69:19–34
243. Clarke, H., Masui, Y. 1986. *J. Cell Biol.* 102:1039–46
244. Lohka, M., Masui, Y. 1983. *Science* 220:719–21
245. Lohka, M., Masui, Y. 1984. *J. Cell Biol.* 98:1222–30
246. Lohka, M., Masui, Y. 1984. *Dev. Biol.* 103:434–42
247. Burke, B., Gerace, L. 1986. *Cell* 44:639–52
248. Chaly, N., Bladon, T., Setterfield, G., Little, J. E., Kaplan, J. G., Brown, D. 1984. *J. Cell Biol.* 99:661–71
249. McKeon, F. D., Tuffanelli, D. L., Kobayashi, S., Kirschner, M. W. 1984. *Cell* 36:83–92
250. Miake-Lye, R., Newport, J., Kirschner, M. 1983. *J. Cell Biol.* 97:81–91
251. Wasserman, W. J., Smith, L. D. 1978. *J. Cell Biol.* 78:R15–R22
252. Wu, M., Gerhart, J. 1980. *Dev. Biol.* 79:465–77
253. Gerhart, J., Wu, M., Kirschner, M. 1984. *J. Cell Biol.* 98:1247–55
254. Miake-Lye, R., Kirschner, M. 1985. *Cell* 41:165–75
255. Lohka, M., Maller, J. 1985. *J. Cell Biol.* 101:518–23

Ann. Rev. Biochem. 1987. 56:567–613

PROTEIN SERINE/THREONINE KINASES

Arthur M. Edelman

Department of Pharmacology and Therapeutics, State University of New York at Buffalo, Buffalo, New York 14214

Donald K. Blumenthal

Department of Biochemistry, University of Texas Health Center at Tyler, Tyler, Texas 75710

Edwin G. Krebs

Howard Hughes Medical Institute, University of Washington, Seattle, Washington 98195

CONTENTS

0066-4154/87/0701-0567$02.00

PERSPECTIVES AND SUMMARY

Phosphorylation and dephosphorylation of proteins catalyzed by protein kinases and protein phosphatases are recognized as major processes for regulating cellular functions. First implicated more than 30 years ago as a means of regulating glycogen phosphorylase activity, protein phosphorylation is now known to affect a vast array of other proteins. Protein phosphorylation is particularly prominent for the role that it serves in signal transduction. Signals impinging on cells have their effects amplified and disseminated by a network of protein phosphorylation-dephosphorylation reactions, the complexity of which we are only beginning to appreciate (reviewed in 1). The protein kinases and the protein phosphatases are both subject to control. The present review, however, is limited to the protein kinases that catalyze the phosphorylation of serine and threonine residues in proteins (for a review of protein tyrosine kinases, see 2). Primary emphasis is given to the physicochemical properties of the kinases, to the specific reactions that they catalyze, and to the molecular mechanisms by which they are regulated. Of necessity, the physiological aspects of the processes in which the individual kinases are involved are discussed only briefly. In addition, because of space limitations, the important topic of sequence homology between protein kinases is not treated (reviewed in 2).

CYCLIC AMP–DEPENDENT PROTEIN KINASES

In most eukaryotes cyclic AMP (cAMP) is thought to act by binding to the regulatory (R) subunit of the cAMP-dependent protein kinase (cAK). This results in the phosphorylation of specific target proteins by the catalytic (C) subunit of the enzyme. The structure and function of cAK have recently been reviewed (3–6). cAK has a phylogenetic distribution that extends from mammals to yeast (reviewed in 3, 7), but there is no compelling evidence for the existence of a cAK in prokaryotes (7), and in higher plants, cAK may be quite different from the mammalian enzyme(s) (8).

There are two major forms (Types I and II) of mammalian cAK that are distinguished on the basis of DEAE-chromatography, cAMP analogue specificity, and their ability to undergo autophosphorylation. The C subunits of these forms appear to be essentially identical in terms of their gross physical and enzymatic properties, although there are indications of slight structural differences (9). Uhler et al (10) have recently isolated cDNA clones coding for a protein that appears to represent a second isoform of C. The role of this second form is not clear at present although it does not appear to be a Type I- or Type II-specific form. The two major isozymic forms of regulatory subunit, R_1 and R_{11}, appear to confer most of the observed structural and functional differences to the two types of cAK. R_1 and R_{11} differ with respect to tissue distribution, amino acid sequence, interactions with C, cAMP analogue-binding properties, and the ability to be autophosphorylated. Species and tissue heterogeneity of R_{11} has been observed, and this may also be true of R_1 (11–13).

The holoenzyme form of cAK exists as an inactive tetrameric complex, R_2C_2. When cAMP binds to R, the complex dissociates to form R_2cAMP_4 and two active C's. Purified mammalian cAK holoenzyme has an apparent native M_r ranging from 150,000 to 170,000, R-subunit dimers (R_2's) have apparent native M_r's ranging from about 92,000 to 108,000, and monomeric C has an apparent native M_r of 38,000. Hydrodynamic data indicate that the holoenzyme and the R-subunit dimer are highly asymmetric, whereas C is relatively symmetrical and globular (3, 14, 15).

The amino acid sequences of R_1 from bovine skeletal muscle (16) and R_{11} (17) and C (18) from bovine cardiac muscle have been determined. In addition, the sequences of mouse C (19) and bovine testes R_1 (20) have been determined from cDNA clones. The M_r's of R_1, R_{11}, and C from amino acid sequence analysis are 42,804; 45,004; and 40,580; respectively. The M_r of C is in good agreement with the values obtained by SDS-PAGE and by other techniques, but the molecular weight of the two R's were substantially less than the values indicated by SDS-PAGE; it has been suggested that the unusual electrophoretic behavior may be due to the asymmetries of the dimerization domains (17). An anomalous shift in electrophoretic mobility on SDS-PAGE resulting from phosphorylation of some forms of R_{11} at Ser-95 may be a related phenomenon (3).

The C subunit from bovine cardiac muscle was the first protein kinase catalytic domain to be sequenced (18); the predicted amino acid sequence of mouse C obtained from cDNA clones shows a 98% sequence homology to the bovine protein (19). The bovine enzyme has 349 amino acids; the mouse protein has 351. Affinity labeling with the ATP analogue p-fluorosulfonyl-benzoyladenosine (FSBA) modifies Lys-72 in bovine cardiac C (18, 21, 22). Affinity labeling with a 3-nitro-2-pyridinesulfenyl(Npys) analogue of a peptide substrate (Leu-Arg-Arg-Ala-Cys(Npys)-Leu-Gly) modifies Cys-199 (23), but this may be fortuitous since the corresponding 2-pyridinesulfenyl peptide analogue labels both Cys-199 and Cys-343. The C subunit is $N(\alpha)$-myristylated (22); the possible role of this unusual modification is not presently understood. Crystals of C that refract to 3.5 Å have recently been grown (24).

The two R isozymes from bovine muscle show significant sequence homology (17). They are each organized into at least four functional domains: two cAMP-binding domains, a dimerization domain, and a C interaction (inhibitory) domain. The two cAMP-binding domains in each subunit show high internal sequence homology as well as significant homology with specific regions of the cGMP-dependent protein kinase (cGK) and the cAMP-binding catabolite gene activator protein (CAP) from *Escherichia coli* (17). The amino terminal regions of R_1 and R_{11} (residues 1–134) contain the C subunit interaction and R-R dimerization domains, but show only marginal sequence homology with each other and no homology with CAP or cGK (17). The R-R dimerization domain resides within the first 45 residues of the amino terminus of R_{11} (25). The C subunit interaction domain is not well defined; in R_{11} this domain includes the autophosphorylation site (Ser-95) and must extend past residue 99, since a proteolytic fragment consisting of residues 1–99 is not inhibitory to C (D. K. Blumenthal, unpublished observation). Residues 1–45 (R_{11}) do not appear to be critical for interaction with C (25). The tertiary structures of the cyclic nucleotide–binding domains of R_1 and R_{11} have recently been predicted on the basis of the X-ray crystal structure of CAP using interactive computer modeling (5, 26). Data from photolabeling and from binding of cAMP analogues were used in developing the models. Crystals of R_1 that diffract to at least 3.5 Å have recently been obtained (27).

cAK has a broad protein substrate specificity in vitro and in vivo (reviewed in 28–30). Synthetic peptide substrates have been used extensively to study structural determinants of specificity. With many substrates, primary structural determinants appear to be dominant, since short peptides (such as those containing the sequence Arg-Arg-Xxx-Ser-Xxx) have kinetic properties ($K_m \simeq 10$–20 μM; $V_{max} \simeq 8$–20 μmol/min/mg) comparable to the corresponding intact proteins (31, 32). However, some synthetic peptides are very poor substrates compared to the intact protein, e.g. troponin I (33) and protein phosphatase inhibitor-1 (34), suggesting that higher order determinants can be important. Peptide substrate binding induces conformational changes in C; substrate-induced conformational changes require that the peptide possess the aforementioned basic residues, a seryl hydroxyl, and the ability to assume an appropriate conformation (35). The structure of the peptides, Leu-Arg-Arg-Ala-Ser-Leu-Gly and Arg-Arg-Ala-Ser-Leu, have been studied by NMR spectroscopy in the presence of C, Co-$(NH_3)_4$AMP-PCP, and Mn^{2+}; the data are consistent with these peptides assuming an extended coil conformation when bound to the enzyme (36).

MgATP is the preferred metal-nucleotide substrate of cAK (37). Studies using ATP analogues (38) and NMR spectroscopy (6, 39) indicate that the nucleotide substrate binds to the enzyme in the *anti* conformation. A number of intersubstrate and intermetallic distances have been determined by NMR (40). A random, ping-pong Bi-Bi mechanism based on studies using histone H1 as substrate has been proposed (41, 42), whereas a sequential reaction mechanism has been indicated by kinetic studies using peptide substrate (6, 43–45). Cook et al (44) proposed a mechanism involving the random addition of substrates and ordered release of products. Whitehouse et al (45) proposed a steady-state ordered Bi-Bi mechanism in which an $(E \cdot ATP)^*$ complex forms before the peptide/protein substrate binds. This complex might also exist as $(E \cdot ADP \cdot P^*)$, where P^* denotes a possible metaphosphate intermediate (40). Arguments against a ping-pong mechanism have been made (43, 45).

The activity of cAK is primarily regulated by the interaction of C with R. In the absence of cAMP, R_2 binds to and inhibits C with a $K_i = 0.1$–1.0 nM (46–48). Addition of cAMP to the inactive R_2C_2 complex results in binding of cAMP to R ($K_d \sim 10$ nM) (6, 46) with the formation of metastable ternary complexes. A ternary complex consisting of $R_2 \cdot C \cdot cAMP_2$ has recently been demonstrated (49); the composition of this complex suggests that the two members of the R-subunit dimer may act independently of one another. Upon binding cAMP, a conformational change in R occurs that lowers R-C affinity by about 10^4 such that at physiological concentrations of the enzyme essentially complete dissociation of R and C would be expected (40, 50, 51). Work with cAMP analogues specific for one or the other of the intrachain cAMP-binding sites in R_1 and R_{11} indicates that both sites are involved in kinase activation (52, 53). The physiological role of a small heat-stable protein kinase inhibitor (PKI) is not well understood, although it may function to suppress cAK activity stimulated by basal levels of cAMP (54). The R subunits (16, 17) and PKI (55, 56) all contain segments of sequence that closely resemble the sequences of preferred substrates and thus appear to act as "pseudo-substrate" inhibitors (40, 57); additional determinants that contribute to their high affinity ($K_d \simeq 0.1$ nM) for C must also be present (58, 59; reviewed in 3, 5).

Each subunit of cAK is known to be phosphorylated at multiple sites. Purified C contains endogenous phosphate at Thr-197 and Ser-338 (18, 22), but the significance of these phosphorylations is not known. R_{11} can be phosphorylated at Ser-44, -47, -74, -76, and -95 (17, 60, 61, 63, 64), but the only site with demonstrable importance is Ser-95, the "autophosphorylation" site. R_{11} as purified contains significant amounts of phosphate at Ser-74 and/or Ser-76 (60), and these sites can also be phosphorylated in vitro by casein kinase II (60, 61). Ser-44 and Ser-47 are phosphorylated in vitro by glycogen synthase kinase-3 (61). The phosphorylation state of Ser-95 is altered in intact tissues by interventions that change cAMP levels (62). Autophosphorylation is an intramolecular reaction (64, 65) that results in reduced rates of R_{11} reassociation with C (66). R_1 contains one site that is phosphorylated in situ ("in vivo") and one that is phosphorylated by the cGMP-dependent protein kinase in vitro (67, 68). The latter site is Ser-99 (16, 69); phosphorylation at this site results in loss of inhibitory activity towards C and loss of one cAMP-binding site, but it is not clear whether phosphorylation of this site occurs physiologically (70). Phosphorylation of the "in vivo" site has no apparent effect on R_1 function, and its position within the sequence is not known (70).

CYCLIC GMP–DEPENDENT PROTEIN KINASE

A protein kinase that was about 50-fold more sensitive to activation by cyclic GMP (cGMP) than by cAMP was reported in 1970 (71). This enzyme, termed cGMP-dependent protein kinase (cGK), has been purified to homogeneity from several sources (reviewed in 3, 28, 72–74), but its physiological functions are still not well understood (72). The enzyme is found in appreciable quantities in smooth muscle, lung, heart, and Purkinje cells of the cerebellum (76–78). It appears to be primarily cytosolic in many tissues (71, 72, 77, 78), although a significant amount of the enzyme in platelets (85–90%; 79) and aortic smooth muscle (25%; 80) is associated with the particulate fraction. A distinct membrane-specific form of cGMP-dependent protein kinase from intestinal brush border has been described (81).

The native M_r of the soluble form of cGK from bovine lung and heart is about 155,000 (82–85); it is composed of two identical subunits (86). The M_r of the N(α)-acetylated monomer is 76,331, as determined by sequence analysis (86), which agrees well with estimates of 74,000–82,000 from SDS-PAGE under reducing conditions (82–85). The native enzyme is rather asymmetric (83, 85, 87). The two subunits are thought to be linked by disulfide bonds (85), although this may be an artifact of purification (88); arginyl residues also appear to be important in dimer formation (89). The two subunits of cGK are thought to be arranged in an antiparallel manner with the regulatory domain of one subunit inhibiting the catalytic domain of the other (89–91).

The 670 residues of the cGK sequence can be divided into six segments corresponding to four functional domains (86). The N-terminal segment contains the dimerization domain (86, 88, 91), the "hinge" region (88), and the major site of autophosphorylation (86, 88). Although this segment is functionally similar to the N-terminal regions of the regulatory subunits of cAK, there is little or no significant sequence homology between cGK and the two types of cAK in this region (86). The second and third segments show high sequence homology with the cAMP-binding domains of the R subunits of cAK and CAP from *E. coli*. Together the fourth, fifth, and sixth segments represent the catalytic domain of the enzyme. The fourth and fifth segments display high homology with all members of the protein kinase family, whereas the sixth segment is homologous only with the catalytic subunit of cAK. The ATP affinity analogue FSBA labels Lys-389 in the fourth segment (86, 92). Thus, cGK is "chimeric," being composed of regulatory (cyclic nucleotide–binding) and catalytic domains derived from two separate protein families (86). A distinct cyclic nucleotide–dependent protein kinase, which also contains both the catalytic and regulatory domains in the same chain, has been purified from a developing insect (93). This enzyme can be distinguished from cGK by its similar affinity for cAMP and cGMP and may represent an ancestral form of cGK.

cGK exhibits a substrate specificity towards proteins and peptides in vitro that is similar but not identical to the specificity shown by cAK (reviewed in 28, 72). Both enzymes require multiple basic residues amino terminal to the phosphorylation site and both prefer seryl residues as phosphate acceptors. A basic residue immediately C-terminal and a prolyl residue immediately N-terminal to the phosphorylation site appear to be positive determinants for cGK, whereas these amino acids are neutral or negative determinants for cAK (94, 95). PKI (see preceding section) has no effect on cGK activity (82, 96–98), and synthetic peptide inhibitors show markedly different inhibition patterns with the two enzymes (96, 99), thus providing additional evidence that the two kinases recognize different substrate-specificity determinants. It is thought that each enzyme phosphorylates different substrates in vivo and that cGK has the narrower substrate specificity (28, 72). A number of proteins that are selectively phosphorylated in tissue homogenates by cGK have been described. One such protein from rabbit cerebellum termed G-substrate (100) has been well characterized (101, 102a).

Ser-32 in histone H2B is preferentially phosphorylated by cGK, as is its corresponding synthetic peptide, Arg-Lys-Arg-Ser-Arg-Lys-Glu (28, 102b). This peptide and an analogue containing Ala instead of Ser have been used in studies that show that the mechanism of the reaction catalyzed by cGK is sequential (103). Product and dead-end inhibition patterns indicate an ordered Bi-Bi mechanism in which MgATP binds first and MgADP leaves last. Initial-velocity patterns obtained using histone H2B as substrate suggested a ping-pong mechanism, but product and dead-end inhibition by MgADP and MgAMP-PNP, respectively, ruled out this possibility (103). These differences in kinetic patterns, which depend on whether histone or peptide substrate is used, are reminiscent of the patterns obtained with cAK (see above). The preferred metal-nucleotide substrate for cGK is MgATP (72).

The activity of cGK is regulated by nanomolar levels of cGMP, but cAMP and cIMP at micromolar concentrations can also activate the enzyme (72). Binding of cGMP to cGK results in enzyme activation, but, unlike cAK, the subunits do not dissociate. The two intrasubunit cGMP-binding sites differ with respect to their K_d values, dissociation rates, and analogue specificity (104–108). Both sites appear to be involved in enzyme activation (104, 107). Comparison of analogue specificity between the cGK and cAK indicates that both cyclic nucleotides are bound by their respective kinase by similar forces in the 2', 3', and 5' regions, that the *syn* conformation of cyclic nucleotide is universally preferred, and that selectivity (between intrachain sites and between the two kinases) is determined by features in the base moiety (107). The specificities of the slowly exchanging binding sites in cGK and cAK are strikingly similar; comparable similarities in specificity were proposed for the rapidly exchanging sites (107).

Purified cGK contains slightly more than 1 mol covalently bound phosphate per mol subunit (110), suggesting that the enzyme is phosphorylated in vivo, but the sites and the kinase(s) responsible are not known. In vitro studies indicate that 2.1–2.5 mol phosphate can be incorporated autocatalytically per mol subunit (68, 91, 110). Thr-58 is the primary site phosphorylated autocatalytically (86, 88, 111), but phosphorylation of five additional sites is also seen, particularly when autophosphorylation occurs in the presence of cAMP (111). Thr-58 is probably phosphorylated in an "intramolecular" reaction (89). The rate of autophosphorylation is increased several fold by micromolar concentrations of cAMP or cIMP, whereas cGMP has little effect or is slightly inhibitory (73, 89, 109, 110). Autophosphorylation (in the presence of cAMP) of cGK is much slower ($k = 0.07$/min) than autophosphorylation of cAK (109). The K_a for cAMP is decreased as much as 10-fold by autophosphorylation (to approximately 0.8–7 μM; 109, 110, 112), apparently the result of an increase in the affinity of cAMP for the high-affinity (slowly exchanging) binding site (112, 113). There is little or no effect of autophosphorylation on cGMP interactions with the enzyme (109, 110). If autophosphorylation takes place in vivo, this might allow the enzyme to be activated by cAMP as well as by cGMP.

PHOSPHORYLASE KINASE

Phosphorylase kinase (PhK) is a key regulatory enzyme in the metabolism of glycogen (reviewed in 114, 115). Its primary substrate is phosphorylase b, although the enzyme may catalyze the phosphorylation of other proteins. The kinase purified from rabbit muscle (where it represents approximately 0.5% of the soluble protein) has been studied most extensively, but it has also been purified from skeletal muscle of other species, and from bovine cardiac muscle (116) and rat liver (117). The various forms from different tissues and species share many physical and biochemical properties (reviewed in 114). Knowledge of the phylogenetic distribution and variation of PhK is incomplete; the properties of the dogfish enzyme are generally similar to those of the rabbit enzyme, whereas yeast appear to have a distinct form of PhK. PhK from rabbit skeletal muscle has a native molecular weight of approximately 1.3×10^6 and is composed of four different subunits, termed α and α', β, γ, δ. The α and α' subunits represent isozymic variants, with the α subunit predominating in fast-twitch skeletal muscle and the α' subunit predominating in cardiac and slow-twitch skeletal muscle. The reported M_r values (as determined by SDS-PAGE) range from 118,000 to 145,000 for the α subunit, from 133,000 to 140,000 for the α' subunit, and from 108,000 to 128,000 for the β subunit. The M_r's of the γ subunit and δ subunit (as determined by sequence analysis) are 44,673 (118) and 16,680 (119), respectively. The holoenzyme is generally thought to have the subunit structure, $[\alpha \cdot \beta \cdot \gamma \cdot \delta]_4$, although other stoichiometries have been reported. Electron microscopy shows two major forms, a "butterfly" form (120, 121) and a "chalice" form (122), both forms being about 20×20 nm in overall dimension. A possible spatial arrangement of subunits in the holoenzyme complex has been proposed (123).

The function of each subunit of PhK is not completely understood. The δ subunit (124) is identical to calmodulin (CaM). Its role as a regulatory subunit through which Ca^{2+} acts to increase catalytic activity of the holoenzyme is therefore reasonably certain. Moreover, the γ subunit has been isolated in a catalytically active form, indicating that it is responsible for at least part of the catalytic activity of the holoenzyme (125). Support for this is provided by the high sequence homology between the γ subunit and the catalytic domains of other protein kinases (118). Isolation of active partial complexes consisting of $\alpha \cdot \gamma \cdot \delta$ and $\gamma \cdot \delta$ indicate that Ca^{2+} regulation occurs by direct interaction of the δ and γ subunits (126). There is indirect evidence that the β subunit and perhaps the α subunit also contain catalytic sites, but partial complexes that lack both α and β subunits have been shown to have the same molecular catalytic activity as the holoenzyme (126). Evidence that activation of the enzyme can be effected by phosphorylation or by limited proteolysis of the α and β subunits, or by their dissociation from the holoenzyme complex, suggests that these subunits regulate activity by an inhibitory mechanism.

Several substrates of possible physiological significance in addition to phosphorylase b have been described, including glycogen synthase, phospholamban, sarcolemmal Na^+/K^+-ATPase, and sarcoplasmic reticulum Ca^{2+}-dependent ATPase (reviewed in 114). There is also a report that PhK exhibits phosphatidylinositol kinase activity (127). Synthetic peptides corresponding to the phosphorylation sequences in phosphorylase b and glycogen synthase have been used to study the substrate specificity of PhK (114, 128). None of the synthetic peptides used have had V_{max}/K_m ratios approaching that of the intact protein substrates (K_m = 270 μM; V_{max} = 15 μmol/min/mg, using phosphorylase b and nonactivated PhK; 128). The shortest peptides with reasonable kinetic properties ($K_m \simeq$ 1 mM; $V_{max} \simeq$ 0.5–3.0 μmol/min/mg) had the sequence (Arg)-Lys-Gln-Ile-Ser-Val-Arg-Gly-(Leu). Basic residues are required near the site of phosphorylation, but the optimum location is not apparent. Threonine cannot substitute for serine in the phosphorylase b peptides, although threonine is known to be phosphorylated in vitro in troponin I (129). MgATP appears to be the only metal-nucleotide substrate used by the the enzyme (38). Free ATP is an inhibitor of the enzyme, whereas ADP is an allosteric activator that binds the holoenzyme with K_d values ranging from 0.26 to 17 μM and with a stoichiometry of 8 mol ADP per mol holoenzyme (130). ATP affinity-labeling analogues react with sites on both the β and γ subunits (131, 132). The site of γ subunit labeling by FSBA is predicted from sequence homology with other protein kinases (118), but it is not clear whether the nucleotide-binding sites in the β subunit are allosteric or catalytic. The K_m values for phosphorylase b vary markedly depending on pH and the phosphorylation state of PhK (from 18 to 370 μM), whereas the K_m value for ATP (0.24–0.38 mM) and the V_{max} value (10–20 μmol/min/mg) remain relatively constant under these conditions (114, 126). The kinetic mechanism has been studied under initial rate conditions using a synthetic tetradecapeptide corresponding to the phosphorylation site in phosphorylase b. The results of these studies suggest a sequential random Bi-Bi mechanism (133).

The activity of PhK is subject to regulation by a number of allosteric effectors, as well as by mechanisms such as protein phosphorylation. Allosteric effectors that have been identified include Ca^{2+}, CaM, ADP, glycogen, and Mg^{2+}. Regulation of enzyme activity by Ca^{2+} is probably the most important of these mechanisms, since catalytic activity is almost completely dependent on micromolar concentrations of this metal ion, at least with skeletal muscle PhK (134, 135). Extrinsic CaM, referred to as the δ' subunit, can stimulate the V_{max} of the enzyme 2- to 10-fold (135, 136). Cross-linking experiments indicate the δ' subunit binds to the α and β subunits (137). Extrinsic CaM does not activate α' subunit isozymes (138, 139). Troponin C has also been shown to activate phosphorylase kinase in vitro, although the K_d ($\simeq$ 1 μM) is much higher than that for CaM ($K_d \simeq$ 2–15 nM) (136). Effectors such as ADP, glycogen, and Mg^{2+} also stimulate enzyme activity in vitro at physiological concentrations, but their roles are not well understood.

PhK is phosphorylated in vitro by a number of different protein kinases including PhK itself, the cyclic nucleotide–dependent protein kinases, the Ca^{2+} (CaM)-dependent protein kinase(s), and perhaps protein kinase C. In addition, PhK purified from unstimulated muscle contains 7 to 19 moles of "intrinsic" alkali-labile phosphate per mole $[\alpha \cdot \beta \cdot \gamma \cdot \delta]_4$. The cAK-catalyzed phosphorylation of the α (α') and β subunits has been well studied in vitro and in situ. This enzyme phosphorylates a specific site on each of these subunits; the sequences around these sites have been determined (140). The β subunit is phosphorylated approximately 5- to 10-fold faster than the α (α') subunit (141, 142). Recent work indicates that α (α') subunit phosphorylation cannot occur until the β subunit has been phosphorylated to the extent of at least 1 mol phosphate/mol holoenzyme and that once this stoichiometry has been reached additional β subunit phosphorylation is inhibited (141). The presence of extrinsic CaM (δ') also inhibits β subunit phosphorylation (143). Phosphorylation of both α and β subunits results in enzyme activation through a marked decrease in K_m for phosphorylase b, but there is not a simple relationship between phosphorylation and activation, particularly in the case of α (α') subunit phosphorylation (114, 141). It has been suggested that α (α') subunit phosphorylation only regulates activity when the β subunit is in a phosphorylated state (114). PhK is also subject to autocatalytic phosphorylation of the α (α') and β subunits, although it is not clear whether the reactions are inter- or intra-molecular, or both (144). Autophosphorylation is associated with enzyme activation, perhaps to a greater extent than that obtained by cAK-catalyzed phosphorylation. It is not known whether the autophosphorylation sites are identical to those phosphorylated by cAK, but tryptic maps of phosphopeptides suggest that this is the case for the β subunit (145a). cGK appears to phosphorylate the same sites phosphorylated by cAK (as indicated by tryptic maps of phosphopeptides), but phosphorylates the α subunit more rapidly (87, 145b).

The physiological and pharmacological regulation of PhK activity in situ has been studied extensively in heart, skeletal muscle, and liver (114); in smooth muscle (146); and in neutrophils (147). The results of these studies are in basic accord with in vitro studies that indicate PhK activity is primarily controlled by changes in intracellular Ca^{2+} and cAMP concentrations.

MYOSIN LIGHT CHAIN KINASES

The myosin light chain kinases (MLCKs) are distinguished by their high degree of substrate specificity and calmodulin (CaM) dependence. MLCK was first identified and purified from rabbit skeletal muscle (148). The enzyme(s) has now been purified from cardiac and smooth muscle, as well as several nonmuscle tissues (for reviews see 149 and 150). The phylogenetic distribution of MLCKs is still sketchy, but the enzyme has been purified from muscle of the invertebrate, *Limulus* (151). Striated muscle MLCKs appear to be primarily cytosolic, whereas smooth muscle MLCKs require extensive extraction to release them from myofibrils (reviewed in 152). The role of myosin light chain (MLC) phosphorylation in cell function differs depending on the specific cell type. In smooth muscle, MLC phosphorylation is required for the initiation of contraction (reviewed in 150 and 153), whereas in striated muscle it appears to modulate the degree of tension produced during contraction (149). The role of MLC phosphorylation in nonmuscle cells is poorly understood, but is thought to be involved in motile processes such as cell shape changes, secretion, movement, phagocytosis, receptor capping, etc (150).

MLCKs from different species and different tissues display large differences in M_r (as determined by SDS-PAGE), ranging from 37,000 for the enzyme from *Limulus* muscle (151) to 155,000 for several of the smooth muscle and nonmuscle enzymes (149, 154, 155). Skeletal (and cardiac) muscle MLCKs are generally smaller, ranging in size from M_r 68,000 (human) to 150,000 in the chicken (149, 152, 156a, 156b). The best characterized enzymes are those from rabbit skeletal muscle and avian gizzard.

Rabbit skeletal muscle MLCK was reported to have M_r values from 77,000 to 94,000 based on SDS-PAGE (149), but the M_r determined by sequence analysis is 65,040 (157, 158). The anomalous electrophoretic behavior probably results from the highly asymmetric shape and unusual amino acid composition of the amino-terminal "tail" region (158–160), since the isolated C-terminal "head" region appears to exhibit more typical behavior (157, 161). The native M_r of MLCK is between 70,000 and 80,000, with one report of 103,000 (149), indicating that the enzyme is monomeric. The enzyme has 603 residues with an acetylated N-terminus (158), and is organized into three distinct domains: a catalytic domain, a CaM-binding domain, and an N-terminal domain of unknown function. The catalytic and CaM-binding domains are located in the C-terminal 360-residues (157, 161); the CaM-binding domain lies at the extreme C-terminus (157, 161, 162). It has been suggested that the N-terminal domain, which comprises nearly half of the entire molecule, might function to maintain a tertiary structure necessary for optimal catalytic activity (161) or to localize the enzyme in the myofilament (159).

MLCK purified from avian gizzard has an M_r of about 130,000 as determined by SDS-PAGE (149) and a native M_r of 124,000; the enzyme is asymmetric (163). MLCK partially purified from bovine adrenal medulla has an M_r of 150,000 and is also asymmetric (155). Limited proteolysis has been used to determine the organization of functional domains in the gizzard enzyme. In one study, the catalytic site was placed near the center of the molecule and the two sites phosphorylated by cAK were positioned within 24.5 kd of the N-terminus (164). The CaM-binding domain was placed between these two regions. In a different arrangement, the catalytic domain was placed at the center of the molecule, the phosphorylation sites toward the

C-terminus, and the CaM-binding domain in between (165). The latter arrangement is similar to that of the skeletal muscle enzyme, except that the skeletal muscle enzyme may not have homologous phosphorylation sites and its CaM-binding domain falls at the extreme C-terminus (see above). Recently the sequence of a 20- or 21-residue peptide containing one of the phosphorylation sites was determined (166). This peptide may also represent the CaM-binding domain.

As mentioned earlier the MLCKs have a narrow substrate-specificity. A specific seryl residue near the amino terminus of the phosphorylatable- or P-light chain of myosin is preferentially phosphorylated. The sequence of this site in the P-light chain of different tissues is not identical. In general, MLCKs from smooth muscle and nonmuscle tissues "prefer" smooth and nonmuscle light chains, whereas striated muscle MLCKs phosphorylate P-light chains from different tissues with similar kinetics. Studies using synthetic peptides have shown that basic residues at three specific positions towards the N-terminal side of the phosphorylated residue are important specificity determinants, with the two positions proximal to the phosphorylation site being most important for the smooth muscle kinases and the distal position being more critical for the skeletal muscle kinases (reviewed in 149). The smooth muscle MLCKs may also be capable of slowly phosphorylating an additional site in the P-light chain sequence (167–170). This site appears to be at a threonine residue near the first site of phosphorylation. The physiological importance of this second phosphorylation site is not clear.

Gizzard myosin is phosphorylated by gizzard MLCK with K_m and V_{max} values of 15 μM and 15 μmol/min/mg, respectively; similar values are observed with isolated gizzard myosin P-light chains (149). Early studies suggested that the two P-light chains in each molecule of filamentous gizzard myosin are phosphorylated in an ordered manner in vitro (171–173). More recent studies, however, indicate random phosphorylation of smooth muscle myosin (174–176). There is evidence for strong product inhibition by phosphorylated myosin (174). The kinetic constants for the phosphorylation of rabbit skeletal muscle myosin by skeletal muscle MLCK are 19 μM and 47 μmol/min/mg (K_m and V_{max} values, respectively), and similar constants obtain for the isolated skeletal muscle P-light chains (149). The two P-light chains of skeletal muscle myosin appear to be phosphorylated in a random manner (177).

Early kinetic studies with the skeletal muscle enzyme indicated a random Bi-Bi catalytic mechanism (178). A recent investigation of the kinetic mechanism of the skeletal muscle enzyme under steady-state conditions and using forward reaction, product inhibition, and reverse reaction data, also indicates a rapid-equilibrium random Bi-Bi catalytic mechanism (179). The data indicate that the enzyme forms a dead-end complex with ADP and light chain, and that the forward reaction is strongly inhibited by both products. The equilibrium constant for the reaction has a value of about 60. With intact smooth and skeletal muscle, the rate of myosin light chain phosphorylation observed following electrical stimulation is in good agreement with the rate of phosphorylation predicted based on the V_{max} of purified MLCK and the MLCK content of the muscle (152, 180).

MgATP is the metal-nucleotide substrate used by the MLCKs; other metal ions or nucleoside triphosphates are poor substrates. K_m values for MgATP range between 50 and 400 μM depending on the source of enzyme, with smooth muscle and nonmuscle MLCKs exhibiting the lower values. The location of the nucleotide-binding site in the sequence of rabbit skeletal muscle MLCK has been predicted from sequence homology data (157).

Most MLCKs require the presence of Ca^{2+} and CaM for activity, but there are reports of CaM-independent forms of MLCK, although the latter might be proteolytic degradation products. In CaM-regulated forms of MLCK, catalytic activity requires the formation of a 1 : 1 complex between MLCK and CaM in which all four Ca^{2+}-binding sites on CaM are occupied (149). The complex formed has a nanomolar or subnanomolar dissociation constant. The rate of enzyme inactivation upon lowering free Ca^{2+} concentration is approximately 1/sec in vitro and in situ (152). The binding of calmodulin to MLCK results in conformational changes in MLCK that are indicated by changes in its intrinsic fluorescence (181, 182) and circular dichroism spectra (159).

MLCKs from skeletal (183), cardiac (156b), and smooth muscle (184) have been observed to undergo autophosphorylation reactions, but these reactions are slow and their importance is not clear. Autophosphorylation appears to have little effect on enzymatic properties, but recently Geuss et al (179) showed that autophosphorylated skeletal muscle MLCK was more susceptible to product inhibition by phosphorylated light chain. MLCKs from several different tissues are also phosphorylated by other protein kinases, most notably cAK (150). With the skeletal (183) and cardiac (156b) muscle enzymes, phosphorylation by cAK has little effect on catalytic activity. With smooth muscle and platelet MLCKs, cAK-catalyzed phosphorylation can cause a marked change in the affinity of CaM for the enzyme (150). Although this phosphorylation appears to occur in situ (185), there is debate as to its physiological importance (4, 149, 153).

OTHER CALCIUM (CALMODULIN)-DEPENDENT PROTEIN KINASES

The isolation and characterization of Ca^{2+}/calmodulin(CaM)-dependent protein kinases from rat brain (186–194), rabbit liver (195, 196), and skeletal muscle (197), revealed the presence of an enzyme or class of enzymes distinct from PhK or MLCK. This enzyme has been variably referred to as the multifunctional Ca^{2+}/CaM-dependent protein kinase or as Ca^{2+}/CaM-dependent protein kinase II. Here it will be referred to simply as kinase II. Kinase II has been purified from a wide range of other organisms and tissues (reviewed in 149, 198). In brain, which contains a very high concentration of the enzyme (199–201), it is partly bound to membranes or cytoskeletal structures and partly soluble (200–210). By immunocytochemistry (204), it was found in nerve endings, axons, dendrites, neuronal somata, and postsynaptic densities.

Kinase II utilizes numerous substrates (149, 198), among which the best include synapsin I, tryptophan hydroxylase, skeletal muscle glycogen synthase, and the microtubule-associated proteins Tau and MAP-2 with K_m's of 0.3–3000 nM (186, 187, 189–191, 193, 211, 212). Sites in protein substrates phosphorylated by kinase II may either be identical to those phosphorylated by cAK, PhK, or MLCK, as for example, Ser-7 of glycogen synthase or Ser-19 of smooth muscle myosin light chain (189, 195–197; A. M. Edelman, M. K. Bennett, M. B. Kennedy, E. G. Krebs, unpublished results), or distinct, as in some of the sites phosphorylated in MAP-2, pyruvate kinase, synapsin I, and tyrosine hydroxylase (199, 211–217). Skeletal muscle kinase II was found to phosphorylate synthetic peptides based on skeletal muscle glycogen synthase and smooth muscle myosin light chain (218) with kinetic constants comparable to those obtained using proteins as substrates. An arginine three residues N-terminal to the phosphorylated serine or threonine was critical. The enzyme did not readily phosphorylate synthetic peptide substrates of cAK, PhK, and histone H4 kinase. Kinase II has a low K_m for ATP (7–109 μM) and high K_a (10–123 nM) for CaM (186, 187, 191–197) when compared with MLCK and PhK. Neuronal cytoskeletal kinase II (but not the soluble enzyme) displays positive cooperativity in its activation by CaM although CaM binding is noncooperative (219a).

Kinase II consists of a 49,000–55,000-dalton subunit (α) and often a 60,000/58,000-dalton (β/β') doublet (186–197). The α subunit in some preparations also appears as a doublet. This heterogeneity may be due to proteolysis or phosphate incorporation into the subunits. Both α and β subunits undergo Ca^{2+}/CaM-dependent, intramolecular autophosphorylation. Both subunits are competent to autophosphorylate when separated by SDS-PAGE and renatured (219b), are labeled by a photoreactive ATP derivative (221), and are able to bind calmodulin (188, 190, 191, 193, 194, 221). Thus catalytic and regulatory sites exist on both subunits. Autophosphorylation results in Ca/CaM-independent activity reversible to Ca/CaM-dependency on dephosphorylation (222–226a, b) and either no change (195, 197, 224, 225), inhibition (219b, 223, 226a, b), or in one case enhancement (221) of activity assayed in the presence of Ca/CaM. The inhibition was reportedly due to a decreased V_{max} (223) or decreased CaM binding (227). The extent of inhibition is temperature-dependent (224), but since it was reported to be reversible (223) or nonreversible (228) on dephosphorylation, it is not clear if it is due to thermal denaturation of the phosphoenzyme. Furthermore, the sites and functional consequences of autophosphorylation are reported to vary with ATP concentration (226b). Immunoblotting and peptide mapping experiments suggest that α and β/β' subunits are structurally similar yet distinct (188–191, 193, 194, 206). The size of the holoenzyme is 500,000–700,000 daltons in most tissues except liver where it is about 300,000 daltons (195, 196). By electron microscopy the skeletal muscle enzyme appears to be a dodecamer composed of two stacked hexameric rings (197). The relative ratio of subunits in kinase II depends on the source of kinase (190, 191). Kinase II from rat forebrain and cerebellum is composed of subunits in ratios of $3:1$ and $1:4$ ($\alpha:\beta/\beta'$), respectively. Skeletal muscle kinase II is similar to the cerebellar form in this respect (197). More of the forebrain enzyme is associated with postsynaptic densities than the cerebellar form (191), consistent with the finding that the α subunit is the major postsynaptic density protein (203, 205, 206). The hypothesis that kinase II plays a role in neurotransmission is supported by the effects on neurotransmission of microinjection of this kinase (229).

Two other Ca(CaM)-dependent protein kinases termed I and III have been described. CaM kinase-I (230) is found in high concentrations in brain, is monomeric (M_r 37,000–42,000), and has a restricted substrate specificity, phosphorylating efficiently a neuron-specific substrate (termed protein III) and a site in synapsin I distinct from that phosphorylated by kinase II. CaM kinase-III (231a) is found in high concentrations in pancreas and skeletal muscle and phosphorylates readily only a 100,000-dalton endogenous substrate.

PROTEIN KINASE C

Protein kinase C (PK-C) was originally described as a "proenzyme" requiring proteolysis for expression of its activity (231b). Later it was shown that in the absence of proteolysis, it has activity that is dependent on Ca^{2+} and membranes, or phospholipid, and enhanced by diacylglycerol (DAG) (231c–234). PK-C is widely distributed in nature (234). In mammals, platelets, spleen, vas deferens, and brain have the highest, and heart, fat, and skeletal muscle the lowest levels (234–236). The subcellular localization of PK-C is hormonally regulated (see below) and varies with the tissue (234–236).

PK-C purified from diverse sources has similar properties [reviewed in (237–241)]. Its carboxyl terminal region shows significant sequence homology to other protein kinases (242–244). A family of PK-C-related genes encode three highly homologous sequences (244, 245), which may be related to multiple PK-C isozymes identified in rat brain (246). PK-C is monomeric with an $M_r = 77,000$. Ca^{2+}-dependent proteolysis yields a catalytically active M_r ~50,000 form (PK-M) unable to interact with Ca^{2+} and membranes or phospholipid (231b, 247, 248), suggesting a hydrophobic regulatory domain, separable (249, 250) from a hydrophilic catalytic domain. Another active fragment of M_r 65,000 is largely unresponsive to Ca^{2+} and phospholipid (235).

PK-C phosphorylates a broad range of cellular proteins (240). Although PK-C phosphorylates many of the same substrates as cAK and sometimes the same sites (251), it does have a specificity clearly distinct from that of cAK (253). In a number of substrates for PK-C, such as the EGF (254, 255) and IL-2 (256) receptors, HMG17 (257), and myelin basic protein (MBP) (251), the phosphorylated residue is flanked on its amino and carboxyl sides by basic residues. However, Ser-115 in MBP, which has no basic residues closely C-terminal, is also a major PK-C phosphorylation site (258). Synthetic peptide studies also suggest a requirement for basic residues on either the N- (258–260) or C- (261) sides of the phosphorylatable serine (or threonine). PK-C utilizes ATP ($K_m \simeq 5\ \mu M$) but not GTP (236, 262).

The activation of PK-C by Ca^{2+} is specific except for Ba^{2+} and Sr^{2+} at high concentrations (231c, 263). In the presence of Ca^{2+} purified phospholipids activate PK-C with order of potency a function of the physical state of the phospholipid and the Ca^{2+} concentration (231c, 263–266). In the presence of DAG and 2 μM Ca^{2+}, phosphatidylserine (PS) is the only effective phospholipid (264). Various sonicated phospholipids were reported to potentiate or inhibit PS-dependent activation (264), but this was not found with mixed micelles (266) or defined phospholipid vesicles (265). Other lipophilic compounds, e.g. unsaturated fatty acids, have been reported to substitute for phospholipids (267–273); however their role in the physiological activation of PK-C is unclear. In the presence of Ca^{2+} and sonicated phospholipids, the activity of PK-C is enhanced by DAG (232–234); however with Ca^{2+} and mixed micelles there is close to absolute dependence on DAG (266). Sn-1,2-DAGs stereospecifically activate PK-C (232, 233, 274, 275). The 3-OH group and 1, 2-carbonyls are required for activation (276, 277). A long-chain fatty acid is required for at least one (either) of the 1,2 positions (278) with activity increasing from 3 to 11 carbons (266). DAG increases the affinity of the enzyme for phospholipid 2–3-fold and decreases the K_a for Ca^{2+} ~100-fold (233, 263, 266). Conversely both phospholipid and Ca^{2+} increase the affinity of the enzyme for DAG (266). A model has been proposed (239) in which a "primed" $(PK-C)_1$ $(phosphopholipid)_4$ $(Ca^{2+})_1$ complex is formed to which one molecule of DAG binds to activate the enzyme.

Phorbol esters (279) and other tumor promoters (275, 280) with structural similarities (237, 281, 282) to DAG activate PK-C. Interaction with phorbol esters is regulated by Ca^{2+} and phospholipid and occurs at the same site as DAG (284, 287–289). Copurification of kinase activity and phorbol ester binding (283, 285, 290–292) and other evidence (241) confirm that PK-C is a major phorbol ester receptor in cells. Binding data (284) and the photoaffinity labeling of phospholipids with a phorbol derivative (293) support the above model of activation (239). PK-C is subject to autophosphorylation (236, 294, 295), which may increase its affinity for Ca^{2+} and phorbol esters (295).

Many extracellular signals transiently stimulate the hydrolysis of phosphatidylinositol bisphosphate (PIP_2) to DAG and inositol trisphosphate (296). The latter has been proposed to function by mobilizing intracellular Ca^{2+}. Several lines of evidence suggest Ca^{2+} and DAG promote the activation of PK-C by strengthening its association with membranes. Phorbol esters and cell-permeable DAGs applied extracellularly result in PK-C activation (279, 297, 298) and produce decreases in cytosolic and increases in particulate PK-C (299–302). Ca^{2+} in homogenization buffers enhances recovery of PK-C in particulate fractions (231c, 248, 285). Ca^{2+} ionophores or high external Ca^{2+} concentrations cause PK-C translocation to membranes (303, 304). Ca^{2+} and phorbol esters potentiate each other's ability to stabilize PK-C association with membranes (305, 306). A number of extracellular signals whose mechanisms of action appear to involve PIP_2 hydrolysis (307–314) cause translocation of PK-C to membranes, although other agents may cause membrane-to-cytosol translocation (315, 316). The interaction of presumably inactive cytosolic PK-C with membrane phospholipid in the presence of Ca^{2+} and DAG thus allows synergistic activation of the enzyme (317). Ca^{2+} elevation is also reported to stimulate the binding to membranes of Ca^{2+}-dependent protease (248), which converts PK-C to 65,000- and 50,000-dalton fragments in a reaction enhanced by phospholipid and DAG (247) and stimulated in intact cells by phorbol esters (318–321). The active fragments are released back into the cytoplasm (248, 318, 320, 321), where they may then phosphorylate a new subset of protein substrates. Proteolysis may account for the observed down-regulation of PK-C in cells treated with phorbol ester (322, 323).

HEME-REGULATED PROTEIN KINASE

The heme-regulated protein kinase, also referred to as the heme-controlled repressor (HCR), functions in controlling the initiation of protein synthesis (reviewed in 324–327). Phosphorylation of eIF-2α, the 38,000-dalton subunit of eIF-2, impairs formation of the ternary complex of eIF-2, GTP, and Met-tRNA$_i$. The guanine nucleotide exchange factor (GEF) is also important with respect to the function of the kinase (324–329). Briefly, for continuous formation of the ternary complex during protein synthesis, GTP must replace

GDP in the eIF-2 · GDP complex that is formed in the eIF-2 recycling process. GEF serves as a catalyst for this exchange. When eIF-2 is phosphorylated, however, GEF gets tied up in a stable complex and cannot circulate to nonphosphorylated eIF-2 molecules. Since more eIF-2 than GEF is present in the cell, GEF is rendered ineffective with less than stoichiometric phosphorylation of eIF-2α.

Most of what is known about the heme-regulated protein kinase is based on work in reticulocytes. The enzyme has been enriched about 2000-fold from the postribosomal supernatant fraction of rabbit reticulocyte lysates and is essentially homogeneous at this stage (330, 331). The purified kinase is 95,000 daltons by SDS-PAGE and 140,000 to 160,000 daltons under nondenaturing conditions (330, 332). Estimates of the size of the enzyme by gel filtration give values in the range of 300,000–350,000 daltons (331, 332), which are high because the protein is asymmetric, having frictional ratio of 1.73 (332). The kinase "prefers" ATP over GTP as its nucleoside triphosphate substrate (333). The only known protein substrate for the heme-regulated protein kinase is eIF-2α, which is phosphorylated at a single site near its amino terminus (327, 334–336), having the sequence Leu-Leu-Ser[48]-Glu-Leu-Ser (Ser-48 is the phosphorylatable serine). The apparent K_m for eIF-2α is 0.5 μM (337). Now that the phosphorylation site in eIF-2α has been elucidated (336), it may be possible to develop synthetic peptide substrates for the kinase; this should greatly facilitate its study.

The precise mechanisms involved in control of the activity of the heme-regulated protein kinase are unknown (324–328). The enzyme undergoes multisite autophosphorylation by an intramolecular reaction in which 3–5 mol P per mol of kinase are introduced (327, 332, 333, 338). One study supports the concept that autophosphorylation causes enzyme activation (339), but this relationship is not always apparent (331, 338, 340). Activation of the kinase by autophosphorylation constitutes an attractive hypothesis to account for what has been referred to as the conversion of a proenzyme (pro HCR) to an active form (HCR), but the process may be more complicated (340, 341). Regardless of the relationship between autophosphorylation and activation of the kinase, it is clear that heme inhibits this process and the eIF-2 phosphorylation reaction itself. The concentration of heme required for half-maximal inhibition is 5 μM (331). Direct binding of heme to the kinase has been demonstrated (339).

For the heme-regulated protein kinase to remain in an inactive state in the presence of heme, it would appear that its sulfhydryl (-SH) groups must be in a reduced or unmodified state (324, 327, 342). The enzyme is irreversibly activated by N-ethylmaleimide in the presence of heme and can also be activated reversibly by GSSG under these conditions; reversal requires NADPH (343). It is hypothesized that the kinase is in an inactive configuration when certain of its sulfhydryl groups are reduced and in an active configuration when they are oxidized or modified. In an extensive study of this problem using reticulocyte lysates (342), one group has recently obtained support for a system in which NADPH can furnish the reducing power to keep protein -SH groups in the reduced state provided that a functional thioredoxin/thioredoxin reductase system is operating. It is likely that additional factors may also be important for control of the heme-regulated protein kinase. For example, interaction of the enzyme in situ with other proteins, such as spectrin, has been considered (344). In addition, the substrate, eIF-2, exists in a complex with GEF and this may affect its phosphorylation by the kinase (345).

DOUBLE-STRANDED RNA-DEPENDENT PROTEIN KINASE

For recent reviews of the double-stranded RNA-dependent protein kinase (ds RNA–dep. PK), also called ds RNA–activated translational inhibitor (DAI), see (324, 327, 346). This enzyme is found in reticulocyte lysates and is also present in nonerythroid cells. In the latter it is inducible by treatment with α, β, or γ interferons. The only known physiological substrate for the kinase is eIF-2, which is phosphorylated on its α subunit at the same site as in the reaction catalyzed by the heme-regulated protein kinase and with the same functional effect (327, 347, 348). Highly purified ds RNA–dep. PK has also been reported to phosphorylate histones (324, 349, 350), but see (351). Although the inhibition of protein synthesis by ds RNA is readily demonstrable in reticulocyte lysates, there is no indication as to the physiological role in this type of cell. In virus-infected cells, however, the induction of the kinase, its activation by ds RNA, and the resulting inhibition of protein synthesis is believed to be one of the mechanisms involved in the antiviral action of interferons. Interferon increases the level of eIF-2α phosphorylation in intact cells (352). A recent report suggests that the host protein required for in vitro replication of poliovirus might be the ds RNA–dependent protein kinase (353).

The ds RNA–dep. PK has been purified from reticulocytes (354–356) and other cell types (349–351). These studies, as well as recent work utilizing monoclonal antibodies (357, 358), have supported the concept that a phosphorylatable M_r 68,000 protein, which is enriched in parallel with enzyme activity, is the kinase itself. The M_r 68,000 protein possesses a ds RNA–dependent ATP binding site (359). Highly phosphorylated "M_r 68,000 protein," which may contain as many as 10 mol of phosphate per mol, displays an apparent gel molecular weight increase of 1500 over the nonphosphorylated form (360) on SDS- PAGE. The apparent M_r of the protein under nondenaturing condition is essentially the same as that obtained by SDS-

PAGE (354, 356, 361). Values for a $S_{20,w}$ of 3.2 (354) or 3.7 (327) have been reported. One recent study, however, indicates that the M_r 68,000 protein exists in a complex with an M_r 48,000 protein that is the actual catalytic moiety (362).

The ds RNA–dep. PK is activated by preincubation with ds RNA in the presence of ATP, a process that is accompanied by autophosphorylation (reviewed in 324, 327). No activation of the kinase occurs when it is preincubated with high levels of ds RNA, but once activated it is not inhibited under these conditions (333). This provides a convenient method for studying the activation process, and most of the evidence supports the idea that activation is caused by autophosphorylation. One model for the auto-phosphorylation-activation process (363) involves the mandatory binding of multiple molecules of the kinase to one ds RNA molecule, which is then followed by an "intermolecular" phosphorylation in which one bound molecule phosphorylates its neighbor. This could explain why activation of the enzyme does not occur at high levels of ds RNA. The mechanism of activation of the ds RNA–dep. PK assumes added importance, because certain viruses produce substances that "rescue" them from the inhibitory effects of interferon, and these substances appear to act at the level of the activation step (363, 364). One substance in this category is VAI RNA produced by adenovirus (364–366); it is suggested that this RNA interferes with activation by binding only a single molecule of the kinase.

PYRUVATE DEHYDROGENASE KINASE

Pyruvate dehydrogenase (PDH) kinase is a specific mitochondrial protein kinase found in eukaryotic cells that is involved in the regulation of PDH activity by a phosphorylation-dephosphorylation cycle [reviewed in (367, 368)]. The specific target for the kinase is the α subunit (E1α) of the PDH component (E1) of the PDH complex, which has an $\alpha_2\beta_2$ structure. E1α is phosphorylated on three different serine residues, as identified in two tryptic peptides, having the structures:

Site 1   Site 2
Tyr-His-Gly-His-Ser(P)-Met-Ser-Asp-Pro-Gly-Val-Ser(P)-Tyr-Arg

Site 3
Tyr-Gly-Met-Gly-Thr-Ser(P)-Val-Glu-Arg

It is now agreed that the residue at position 8 in the first peptide is aspartic acid, rather than asparagine as had been thought at one time (368, 369); work involving synthetic peptide substrates has shown that this residue is important as a specificity determinant (370). The rate of phosphorylation of site 1 by the kinase is much greater than that of site 2, which is equal to or slightly greater than that of site 3 (369, 371, 372). Moreover, the phosphorylation of site 1 correlates with the decrease in enzyme activity that accompanies phosphorylation of the dehydrogenase (369, 373). Total loss of activity of PDH occurs when only one phosphate has been introduced into the $\alpha_2\beta_2$ structure (369). Although the phosphorylation of sites 2 and 3 does not affect the activity of PDH, under certain conditions occupancy of these sites may affect the rate of dephosphorylation and reactivation of the dehydrogenase catalyzed by PDH phosphatase (374). Both partial reactions catalyzed by the E_1 component of PDH, i.e. the decarboxylation of pyruvate and the reductive acetylation of lipoic acid, appear to be regulated by phosphorylation-dephosphorylation (375).

PDH kinase is tightly bound to the dehydrolipoamide acetyltransferase or "core component" (E2) of the PDH complex and is present to the extent of 2–3 copies (368, 376) per core, which contains 60 copies of the acetyltransferase. It is believed that these kinase molecules remain fixed to the E2 core and catalyze the phosphorylation of 20 (or 30) dehydrogenase molecules by a mechanism that does not require their dissociation from the complex (377). The kinase has been separated from the acetyltransferase and purified 2700-fold from extracts of bovine kidney mitochondria (378). It is made up of one α and one β subunit with M_r's of 48,000 and 45,000, respectively, as measured by SDS-PAGE. The $S_{20,w}$ of the native enzyme is 5.5S (368), consistent with an $\alpha\beta$ structure. Partial proteolysis experiments support the concept that the α subunit is the catalytic subunit. The β subunit may have a regulatory function (378). The kinase is strongly inhibited by N-ethylmaleimide and certain disulfides; inhibition by the latter can be reversed by thiols (379). The turnover number (k_{cat}) of pyruvate dehydrogenase kinase is only 32 min^{-1} (368). The enzyme appears to be specific for ATP as the phosphate donor (380); an apparent K_m for MgATP was estimated to be 25 μM (371).

The activity of the PDH complex is subject to end-product inhibition by NADH and acetyl CoA and is regulated by phosphorylation-dephosphorylation as indicated above. With respect to the latter process attention has been equally divided between the kinase and the Ca^{2+}-sensitive phosphatase, (367, 368). PDH kinase activity is enhanced by acetyl CoA and NADH and inhibited by ADP and high concentrations of pyruvate. Although the metabolite effects on kinase activity have been appreciated for many years, there is still no universal agreement as to the molecular mechanisms involved. In an isolated model system utilizing highly purified PDH kinase and the dephospho form of the phosphorylation site tetradecapeptide (see above) as substrate, the metabolite effects were clearly demonstrable in support of the concept that they interact directly with the kinase (381).

However, other evidence indicates that stimulation of the kinase by metabolites may be mediated through the reduction and acetylation of lipoyl moieties covalently bound either to E2 or to a newly characterized protein, protein X, in the PDH complex (377, 382, 383). It has been suggested that protein X serves to anchor the kinase to the core of the complex (377, 383). In addition to being regulated by metabolites, PDH kinase may also be under the control of what is referred to as the kinase activator protein, the amount of which is increased by starvation or diabetes (384).

BRANCHED-CHAIN α KETOACID DEHYDROGENASE KINASE

The branched-chain α ketoacid dehydrogenase (BCKAD) is regulated by phosphorylation-dephosphorylation in a manner analogous to the regulation of PDH (385–388). BCKAD, which catalyzes the oxidation of 4-methyl-2-oxopentanoate (ketoleucine), L-3-methyl-2-oxopentanoate (ketoisoleucine), and 3-methyl-2-oxobutyrate (ketovaline), is rate limiting in the oxidation of the three essential branched-chain amino acids. Like PDH, BCKAD consists of a complex of three components, E1 (the dehydrogenase or decarboxylase), E2 (the acyl transferase), and E3 (dihydrolipoamide dehydrogenase). As isolated, however, the complex has lost its E3. The exact stoichiometry for the various components in BCKAD is unknown. Phosphorylation occurs on the α subunit of E1, which has an $\alpha_2\beta_2$ structure. (For a recent comprehensive review of the BCKAD complex see Ref. 389.)

The BCKAD complex has been isolated from several different tissues in a form that contains "intrinsic" BCKAD kinase (390–394), which is bound to the E2 component (395). No information is available concerning the molecular properties of the enzyme. The reaction catalyzed by BCKAD kinase consists of the phosphorylation of two serine residues in the 46,000–49,000–dalton α subunit of E1. These serines are close to each other in a tryptic peptide obtained from E1α (396–398):

Site 1		Site 2	

Ile-Gly-His-His-Ser(P)-Thr-Ser-Asp-Asp-Ser-Ser-Ala-Tyr-Arg-Ser(P)-Val-Asp-Glu-Val-Asn-

Tyr-Trp-Asp-Lys

BCKAD kinase prefers ATP over GTP as the nucleoside triphosphate substrate (389). The K_m for ATP has been estimated to be 12.6 μM or 25 μM for the beef kidney (389) or rabbit liver enzyme (391), respectively.

Phosphorylation of BCKAD causes inactivation of the enzyme. Of the two sites in the α subunit of E1 that are phosphorylated (see above), occupancy of

Site 1 correlates reasonably well with inactivation (397, 399, 400). The role (if any) of Site 2 phosphorylation is unknown. BCKAD that has been inactivated by phosphorylation of E1 can be reactivated by the action of a specific phosphatase (394, 401) or by the addition of an activator protein that can be isolated from the high-speed supernatant fractions obtained from liver and kidney mitochondria but not from heart or skeletal muscle mitochondria (402–404). The activator has been shown to be dephospho E1 (403, 404), but it is not known whether this represents a loosely bound physiologically significant regulatory fraction of E1 or whether it is simply an isolation artifact (389).

Information on the regulation of BCKAD kinase is based on work with isolated BCKAD complex (391, 397, 405–407), intact mitochondria (408), and with whole cells, organs, or intact animals (409–411), since it has not been possible thus far to study the phosphorylation reaction in an isolated system with pure kinase and E1. The kinase reaction is inhibited by ADP and by each of the branched-chain ketoacids, the relative potency being ketoleucine > ketoisoleucine > ketovaline. In addition, BCKAD kinase is inhibited by thiamine pyrophosphate (bovine kidney kinase) and acetoacetyl CoA (rabbit liver kinase). Inhibition by the three branched-chain ketoacids is believed to be especially important with respect to physiological adaptation to dietary variations in the levels of leucine, isoleucine, and valine (390).

CASEIN KINASES I AND II

Casein kinases are messenger-independent protein kinases operationally defined by their preferential utilization of acidic proteins such as casein or phosvitin as substrates (reviewed in 412, 413).

Casein kinase I (CK I), so named for its elution ahead of casein kinase II (CK II) on DEAE cellulose (414), is widely distributed within the plant and animal kingdoms (412). In the rat, the kidney, spleen, and liver contain the highest levels of activity and skeletal muscle and fat, the lowest (415). In rat liver the distribution of CK I activity is: cytosol (72%), microsomes (18%), mitochondria (10%), and nuclei (1%). CK I purified from calf thymus (416), rabbit reticulocytes (414), skeletal muscle (417), and liver (418), and diverse other sources (412, 419–424) is similar in molecular properties. The enzyme is monomeric of M_r ~37,000. It undergoes autophosphorylation without any apparent effect on activity (418). It utilizes ATP well ($K_m \simeq 7$–22 μM) but GTP poorly ($K_m \gtrsim 1$ mM). The major contributions to the nucleotide-binding affinity are from the adenine base and triphosphate (particularly the β-phosphoryl group) (423). Inactivation of the enzyme by NEM is prevented by phosvitin, but not by ATP, suggesting the participation of thiols at the protein substrate binding site (422). Mg^{2+} in excess of the amount required for formation of the MgATP complex increases activity (425); however, the substrate used influences the extent of stimulation. Similarly, the stimulation of activity by monovalent cations (100–300 mM) is dependent on the substrate (417, 421).

Although a variety of protein substrates have been reported for CK I (412, 420, 422, 426–434), in relatively few cases have functional changes as a result of phosphorylation been noted. Exceptions include phosphorylase kinase (431) and hepatoma poly (A) polymerases (428), which are activated, and muscle glycogen synthase (417, 418) and aminoacyl tRNA synthetases (434), which are inhibited by phosphorylation by CK I. Sites in genetic variants of casein that are phosphorylated by CK I are characterized by a cluster of acidic residues N-terminal to a phosphorylated serine (435, 436). SerP appears to function as an acidic residue, since dephosphorylation of casein leads to significantly lower rates of phosphorylation by CK I (419, 435, 436); also, multiple serine residues in close proximity are phosphorylated in muscle glycogen synthase by CK I (437). In general, CK I "prefers" serine to threonine residues (412, 438). CK I is relatively insensitive to heparin, which is a potent inhibitor of CK II (reviewed in 412). No physiological regulator for CK I is known, although evidence that insulin or glucagon treatment leads to CK I activation in vivo was presented (439–441).

CK II, like CK I, is widely distributed among eukaryotic organisms (412). In the rat, the highest levels of activity are found in spleen and testis, and the lowest in fat and skeletal muscle (415). In rat liver CK II activity is predominantly cytosolic (90%), with the remaining activity divided among the nuclear, mitochondrial, and microsomal fractions. CK II has been purified and extensively characterized from most of the sources used for purification of CK I (414, 416, 420, 421, 442–445, 448), and other tissues from both higher and lower eukaryotes (451–458). The enzyme is composed of two types of subunits: α of M_r 37,000–44,000 and β of M_r 24,000–28,000, although a subunit of M_r 100,000 was reported for the porcine liver nuclear enzyme (443, 459). The α subunit often runs as a doublet (α/α') on SDS-PAGE. This may be due to proteolysis during purification (412, 444, 448), however α' appears to be present in calf thymus in vivo (460). Recently a heterogeneity of α due to charge differences was detected (461a). CK II has an $\alpha_2\beta_2$ structure ($\sim$ 130,000 daltons) resistant to dissociation by high ionic strength, detergents, and limited proteolysis (448). The kinase from lower eukaryotes has an α (or α/α') subunit(s) but no apparent β subunit (420, 445, 458, 449), although in yeast a 41,000-dalton species cross-reactive with β of Drosophila CK II was detected (461b). At low ionic strength CK II forms large aggregates seen by electron microscopy to involve the formation of linear filaments that then aggregate side-to-side (461c). The α subunit contains the catalytic site as shown by affinity labeling with FBSA, and by the isolation of an enzymatically active α subunit (462–464). Moreover, α exhibits sequence homology with the catalytic domains of other protein kinases (K. Takio, personal communication). Reconstituting separated α and β subunits leads to greater activity than that exhibited by α alone, reaching a maximum at an $\alpha:\beta$ ratio of 1:1, suggesting that β has a role in the production of optimal activity (464). CK II incorporates, intramolecularly, 2 mol of phosphate almost exclusively into the β subunit without an effect on activity (414, 416, 444, 456), although effectors (e.g. spermine) that block autophosphorylation also lower the K_m for protein substrates, suggesting a negative influence of autophosphorylation on activity (448). A distinguishing feature of CK II is its ability to utilize GTP ($K_m \simeq 30 \ \mu$M) almost as well as ATP ($K_m \simeq 10 \ \mu$M). The adenine base and the triphosphate (particularly the β phosphoryl group) make the major contributions to nucleotide-binding affinity (465). CK II is stimulated by free Mg^{2+} and monovalent cations, the extent of stimulation being dependent on the substrate.

CK II phosphorylates many of the same substrates as CK I (412, 426, 427, 429, 430, 444, 450, 466–470) and a variety of others (413). In several cases, notably DNA topoisomerase II (471) and hepatoma RNA polymerases I and II (469), phosphorylation by CK II was reported to regulate activity. The phosphorylation of several proteins by CK II is known to potentiate their phosphorylation by glycogen synthase kinase-3/Fa (see below). The sites phosphorylated by CK II in protein substrates, for example, casein variants (436, 472–474), R_{11} (60, 61), protein phosphatase inhibitor-2 (475), and high-mobility group protein 14 (476), are characterized by clusters of acidic residues C-terminal to the phosphorylated serine or threonine. One of the requirements for optimal phosphorylation may be that these sites be located within β turns (472). CNBr fragments of caseins containing the presumed recognition sequences are phosphorylated at lower rates (477), suggesting an influence of higher-order determinants of structure. This is supported by the finding that synthetic peptides have K_m's 1–2 orders of magnitude higher than the parent protein substrates (478–480). Systematic studies employing a synthetic peptide substrate, ser-(glu)$_5$, indicated that a glutamate at every position enhanced the rate of phosphorylation, but those 5 and 3 residues removed from the serine were the most critical (480). Addition of an N-terminal arginine significantly inhibited phosphorylation of this peptide (478). Serine was preferred to threonine as the phosphorylatable residue (480).

A number of compounds have been proposed to regulate CK II through inhibition (heparin, other glycosaminoglycans, 2,3-diphosphoglycerate) or activation (polyamines) (412, 446, 447). Although the enzyme is often quite sensitive to some of these compounds in vitro (K_i for heparin $\simeq 1$ nM), it remains to be seen whether fluctuation in their intracellular concentrations controls CK II activity in vivo.

RHODOPSIN KINASE AND β-ADRENERGIC RECEPTOR KINASES

Rhodopsin kinase (RK) catalyzes the phosphorylation of serine and threonine residues located primarily in the hydrophilic C-terminal cytosolic domain of rhodopsin (481–484). Five of the serines and threonines that have been specifically identified as being phosphorylated (485) are underlined in the following sequence, which represents the 19 C-terminal amino acids of ovine rhodopsin: Asp-Asp-Glu-Ala-Ser-Thr-Thr-Val-Ser-Lys-Thr-Glu-Thr-Ser-Gln-Val-Ala-Pro-Ala. A synthetic peptide corresponding to the last nine amino acids in this sequence can serve as a substrate for the enzyme (486). The role of rhodopsin phosphorylation in the visual excitation cycle has been reviewed recently (482). Briefly, photolyzed rhodopsin, acting in a catalytic manner, facilitates formation of a Gα·GTP complex, which activates cyclic GMP phosphodiesterase. Phosphorylation of rhodopsin (488–490) leads to its tight binding to a 48,000-dalton protein (retinal S-antigen), which competes with the binding of the Gα (491, 492) and suppresses diesterase activation.

Although RK is tightly bound to bleached rhodopsin, it can be readily extracted from dark-adapted photoreceptor membranes and has been purified to homogeneity (493, 494). The enzyme exhibits an $M_r = 67,000$–$69,000$ under denaturing or nondenaturing conditions (493, 495). It has no activity toward phosvitin, casein, mixed histones, or protamine but does undergo autophosphorylation (496). ATP ($K_m \simeq 8~\mu M$, $V_{max} = 40$ nmol/min/mg) is a better phosphate donor than GTP ($K_m \simeq 400~\mu M$, $V_{max} = 2$ nmol/min/mg) (494). The kinase is 90% inhibited in the presence of 0.1 M Na$^+$ (494).

No specific regulators of RK have been described. Instead, rhodopsin phosphorylation is regulated at the "substrate level" by factors that influence the exposure of the phosphorylation sites on rhodopsin, the major one being photoactivation (497, 498). This comes about through a light-induced conformational change in rhodopsin (499). The critical conformational change appears to occur at the meta-rhodopsin II stage (499). Another factor that affects the exposure of the phosphorylation sites on rhodopsin is its interaction with transducin (G-protein) as evidenced by the fact that GTP stimulates the phosphorylation of rhodopsin (500, 501).

It would not be surprising if RK has functions other than the phosphorylation of rhodopsin. The enzyme is present in the pineal gland at a concentration close to that of the retina and has also been detected in other tissues (502). It was postulated (502) that the kinase may function in the phosphorylation of receptors other than rhodopsin. In this connection it is noteworthy that a β-adrenergic receptor kinase, which phosphorylates the receptor when it is agonist-occupied, has recently been identified (503). Phosphorylation results in a functional down-regulation of the receptor. The kinase translocates from the cytosol to the plasma membrane under these conditions, i.e. in a manner analogous to the binding of rhodopsin kinase to rhodopsin when it is activated (504). The cloning of the gene and cDNA for the β-adrenergic receptor has revealed evidence for homology with rhodopsin (505).

HYDROXYMETHYLGLUTARYL-CoA REDUCTASE KINASE

Strong evidence now exists that the activity of hydroxymethylglutaryl-CoA (HMG-CoA) reductase is regulated by phosphorylation-dephosphorylation (reviewed in 506, 507). Phosphorylation of the reductase may also affect the rate of its degradation and thus constitute a factor in control of the level of this rate-limiting enzyme of cholesterol biosynthesis (508, 509). Phosphorylation of HMG-CoA reductase, which inactivates the enzyme, is catalyzed by several protein kinases including protein kinase C (510), a calcium (calmodulin)-dependent kinase (511), and a messenger-independent kinase that is generally referred to as HMG-CoA reductase kinase.

HMG-CoA reductase kinase (HMG-CoA RK) has not been well characterized. The kinase, which is soluble or at least readily extracted by 250 mM NaCl from microsomes, has been purified to apparent homogeneity from rat liver (512, 513). M_r's of 58,000 (512) or 105,000 (513), as determined by SDS-PAGE, and 380,000 (512) or 205,000 (513), as determined by gel filtration, have been reported. The kinase catalyzes the phosphorylation of serine residues (514) in the reductase; two different phosphorylation sites may be involved (514, 515). The K_m for ATP (presumably ATP is used in preference to GTP) has been estimated to be 0.2 mM in the presence of 10 mM MgCl$_2$ (516). An unusual property of the kinase is its activation by ADP (517, 518), or AMP (519).

Regulation of the reversible phosphorylation of HMG-CoA reductase and the role of HMG-CoA RK are poorly understood. The kinase has been reported to undergo modest activation as a result of autophosphorylation (513) and more striking activation as a result of phosphorylation by an HMG-CoA RK kinase (507, 516). The latter kinase, like HMG-CoA RK, is messenger-independent. The lack of distinct regulatory properties for the kinase and the kinase kinase have prompted investigators to look carefully at possible control mechanisms that may be operating in the dephosphorylation of HMG-CoA reductase. Nonetheless, it is probable that regulation also occurs at the kinase level as evidenced by the report of Beg et al (520) that the administration of mevalonate to animals, which is known to cause inhibition of HMG-CoA reductase activity, appears to cause activation of HMG-CoA RK through activation of the kinase kinase.

GROWTH-ASSOCIATED HISTONE H1 KINASE

During cell growth and division histone H1 undergoes extensive phosphorylation, which may be involved in the initiation of mitosis and chromosome condensation (reviewed in 521). This phosphorylation is catalyzed by a chromatin-bound protein kinase (522) commonly referred to as the growth-associated histone H1 kinase or the mitosis-associated kinase (523). The kinase appears to be highly specific for H1 and closely related proteins. Four to five phosphorylation site sequences for the enzyme, which are of the type Lys-Ser/Thr-Pro-Lys or Lys-Ser/Thr-Pro-X-Lys (524), have been identified in calf thymus H1. In trout histone H1, all of the identified phosphorylation sites have the sequence Lys-Ser-Pro-Lys (525). Phosphorylation sites similar to those in H1 have also been identified in chicken erythrocyte histone H5 (526). Although several laboratories have carried out partial purification of the growth-associated histone H1 kinase from a variety of sources (527–532) only limited data are available with respect to its enzymic and physical properties. Estimates of the M_r of the native kinase range from 80,000 to 96,000 (527, 528, 531). Recently the M_r of the enzyme, or a catalytic subunit, labeled using an ATP affinity analogue, was found to be 67,000 by SDS-PAGE (532). Properties of the growth-associated kinase that have been described include its relatively high tolerance for salt (526, 528) and its possible preference for GTP over ATP (529, but see 532). K_m values of 58 μM for ATP and 1.4 μM for GTP were reported for the kinase from plasmocytoma (529). The kinase displays micromolar affinity for its substrate, H1. The growth-associated kinase activity in dividing cells increases from the G_1/S boundary to late G_2/early M phase and then falls (521, 523, 532); these fluctuations are probably due to activation and inactivation rather than variation in the amount of enzyme (533, 534).

GLYCOGEN SYNTHASE KINASE 3 (F_A)

Glycogen synthase kinase 3 (GSK 3), also called F_A, was identified independently by two groups, one of whom recognized it as a glycogen synthase kinase (535) and the other as an activating factor (F_A) for an ATP-Mg^{2+}-dependent protein phosphatase (536–538). Protein phosphatase inhibitor-2 was found to be a component of the phosphatase (539), and it was found (540) that this protein, in addition to glycogen synthase, is a substrate for GSK 3. A third substrate for the kinase (541) is Type II regulatory subunit of cAK (R_{11}). The phosphorylation site sequences for GSK 3 in each of its three protein substrates are known, and they bear some resemblance to each other in that they all contain a number of proline residues and several arginines (475). A peptide fragment from glycogen synthase, which contains serines that can be phosphorylated by GSK 3, is a very poor substrate for the enzyme (542). Evidence that the substrate specificity of GSK 3 may be more complicated than that of many protein kinases comes from observations that prior phosphorylation of glycogen synthase (474, 543a, 543b), inhibitor-2 (544), and R_{11} (541) by casein kinase II makes these proteins better substrates for GSK 3. In each instance the casein kinase II phosphorylation sites(s) are C-terminal to the GSK 3 sites.

GSK 3 has been purified to near homogeneity (538, 542, 545). Approximately 0.2 mg of enzyme can be isolated from 5 kg rabbit skeletal muscle after a 67,000-fold enrichment; a relative molecular mass of 47,000 was found by sedimentation-equilibrium and a value of 51,000 by SDS-PAGE (542). The kinase utilizes GTP ($K_m \simeq 400 \ \mu M$) or ATP ($K_m \simeq 20 \ \mu M$) as the nucleoside triphosphate substrate; $V_{GTP}/V_{ATP} = 0.65$ (545). The K_m for glycogen synthase as the protein substrate is 0.3 mg/ml (546). GSK 3 is not inhibited by heparin. The enzyme has no known physiological regulators but in all probability is nonetheless involved in important aspects of hormonal control. For example, GSK 3 is known to phosphorylate the same sites in glycogen synthase that are altered by insulin (546). In addition, the ATP-Mg^{2+}-dependent protein phosphatase, which is regulated as a result of the phosphorylation of inhibitor-2 by GSK 3, is involved in reactions that are subject to regulation.

HORMONE-STIMULATED RIBOSOMAL PROTEIN S6 KINASE

The phosphorylation of ribosomal protein S6 by cyclic AMP–dependent and cyclic AMP–independent processes has been known for many years (reviewed in 547). Among the enzymes that catalyze the phosphorylation of S6 is one protein kinase whose activity rises significantly when quiescent cells are stimulated with serum or growth factors (439, 549–555) and also during the oncogenic transformation of cells (552); this enzyme will be referred to here simply as "S6 kinase." Protein kinase C catalyzes the phosphorylation of S6 at sites similar to those phosphorylated by S6 kinase (294, 556, 557), but the two enzymes are distinct entities (550–552, 554, 558, 559). Protein kinase C is nonetheless involved in one pathway for activation of the latter enzyme (551, 558); but S6 kinase can also be activated by a protein kinase C–independent pathway (561). Traugh and coworkers have suggested that S6 kinase is identical to protease-activated protein kinase II, which was originally detected in reticulocytes (562, 563). The S6 kinase has also been identified as the H4-specific protease-activated protein kinase (564). Finally, an S6 kinase has also been purified to homogeneity from *Xenopus* eggs (565), which may be the enzyme responsible for catalyzing S6 phosphorylation in oocytes undergoing maturation (566).

S6 kinase catalyzes the phosphorylation of up to 5 mol of phosphate per mol of S6 (reviewed in Ref. 563). Some of the phosphorylation sites are located near the carboxyl terminus of S6 in an 18-residue segment having the structure (567): Arg-Arg-Leu-Ser-Ser-Leu-Arg-Ala-Ser-Thr-Ser-Lys-Ser-Glu-Glu-Ser-Gln-Lys. This sequence contains the principle (Ser-4) and secondary (Ser-5 and Ser-9) cAMP-dependent protein kinase phosphorylation sites as well as S6 kinase sites (567). A synthetic peptide having a structure corresponding to the first eight residues of the above sequence can be phosphorylated on Ser-5 by a protease-activated protein kinase from liver (568). The peptide has also been used in studying S6 kinase in fibroblast growth factor or phorbol ester–stimulated 3T3 cells (555).

S6 kinase has been difficult to purify (439, 551) and relatively little is known about its physical and enzymic properties. Partially purified enzyme characterized by Rosen and her group (551) has a M_r of 50,000–60,000, a pH optimum between 8 and 9, utilizes ATP but not GTP as a phosphate donor, and is inhibited by Ca^{2+} (0.5 mM) and NaF (30 mM). The *Xenopus* egg S6 kinase (565) has a molecular weight of 92,000. The mechanism of activation of S6 kinase has not been determined, but it has been shown (569) that the microinjection of purified insulin receptor into *Xenopus* oocytes causes activation of an S6 kinase; this could occur as a result of protein-tyrosine phosphorylation.

ISOCITRATE DEHYDROGENASE KINASE

Although several prokaryotic protein kinases have been identified (reviewed in 570), the only one that will be discussed here is the isocitrate dehydrogenase (IDH) kinase of *E. coli* and *Salmonella typhimurium*, which is involved in the adaptation to growth using acetate as the carbon source. This adaptation requires the induction of the enzymes of the glyoxylate bypass (isocitrate lyase and malate synthase) and a mechanism to regulate the flow of isocitrate through this pathway (IDH kinase/phosphatase). The genes for these several enzymes reside in the same operon and are thus under the same genetic control (571). In *E. coli* and *S. typhimurium* IDH becomes phosphorylated upon shifting the organism from glucose to acetate (572, 573). In *E. coli* phosphate is incorporated into a single seryl residue in the sequence: Ser-Leu-Asn-Val-Ala-Leu-Arg (574). Data from phosphorylation of IDH in vitro are also consistent with a stoichiometry of one mole phosphate per mole IDH monomer (575, 576). Stoichiometric phosphorylation of IDH results in its inhibition, causing an increased flow of isocitrate through the glyoxylate bypass. One of the unusual features of IDH kinase is its association with IDH phosphatase (576, 577). IDH kinase and phosphatase activities copurify and are coded for by a single gene (577, 578), suggesting that the two activities reside on the same M_r 66,000 polypeptide chain. The kinase exhibits K_m values for MgATP and IDH of 88 μM and 0.35 μM, respectively (579). The turnover number of the kinase is 0.3/min (V_{max} = 4.5 nmol/min/mg; Ref. 575). It prefers MgATP as nucleotide substrate (579). Interestingly, IDH phosphatase activity requires ATP or ADP (570, 577). A number of metabolites have been identified as inhibitors of the kinase; some of these, such as isocitrate and 3-phosphoglycerate, simultaneously activate IDH phosphatase activity (575, 579), resulting in enhanced sensitivity of IDH phosphorylation via the "multistep" effect (575). The phosphorylation of IDH is also subject to "zero-order ultrasensitivity" (575).

THYLAKOID MEMBRANE KINASE(S)

Our understanding of the various roles that protein phosphorylation might play in plant physiology is limited (reviewed in 580). A number of different plant protein kinases have been partially-purified, but only a few have been purified to homogeneity (581–586). The best characterized enzyme is a thylakoid membrane kinase that phosphorylates the protein component of the light-harvesting chorophyll a/b complex (LHC). Phosphorylation of LHC is thought to be involved in maximizing the efficient capture of light energy (reviewed in 587–591). The LHC kinase is activated under conditions in which the chloroplast is exposed to low levels of red light (~640 nm); these conditions cause a buildup of reduced plastoquinone, which stimulates the kinase by an unknown mechanism (587). LHC is phosphorylated on one or perhaps both of the threonyl residues in the sequence: Ser-Ala-Thr-Thr-Lys-Lys, which occurs at the amino terminus of LHC protein (592, 593). In the dark, or when the chloroplast is illuminated with low-intensity far-red light (>680nm), the plastoquinone pool is largely oxidized; the kinase is inactive under these conditions and LHC is dephosphorylated by a thylakoid-associated protein phosphatase (592).

A protein kinase activity that may represent the LHC kinase has recently been purified to homogeneity from spinach thylakoid membranes (585). It represents >80% of the total thylakoid membrane histone kinase activity and is apparently the only purified thylakoid kinase that is active towards LHC. The kinase exhibits an M_r of 64,000 by SDS-PAGE and undergoes an autophosphorylation reaction of unknown significance. The solubilized enzyme has a specific activity of 30–50 nmol/min/mg using histone (III-S) as substrate. The K_m for MgATP is 35 μM, in good agreement with values obtained in situ (594). Although the kinase appears to phosphorylate the same LHC threonyls that are phosphorylated in situ (595), the activity of the purified kinase towards purified LHC is much slower than that expected based on in situ rates of phosphorylation. The reason for the low rate may be an artifact of the detergent-solubilization procedures used to prepare the enzyme and its substrate (585).

OTHER PROTEIN SERINE/THREONINE KINASES

A number of protein serine/threonine kinases have been described in addition to those that have been discussed above. Included in this group are two additional enzymes that phosphorylate glycogen synthase. They include *glycogen synthase kinase 4* (GSK 4) purified from skeletal muscle (545, 467) and a new multifunctional kinase originally detected as a liver *ATP-citrate lyase kinase* (596–598). GSK 4 catalyzes the phosphorylation of "site 2" in glycogen synthase (545), and the ATP-citrate lyase kinase catalyzes the phosphorylation of "sites 2 and 3" (598). The latter enzyme, which was found to have a M_r of 36,000 under nondenaturing conditions, also phosphorylates acetyl-CoA carboxylase.

An increasing number of reports of membrane-bound kinases are appearing. Examples are the *histone-activated casein kinase* obtained from placental and Ehrlich ascites tumor cell membranes (599), the cAMP and Ca^{2+}-independent *protein kinase from neurofilaments* (600), and the *protein kinase(s) from coated vesicles* (for recent papers see Refs. 601–605).

Histone kinase II, which has been purified from calf thymus, is a cyclic nucleotide–independent protein kinase that phosphorylates histone H1 on Ser 103 (524, 606). The sequence of amino acids at the phosphorylation site is Ala-Ala-Ala-Ser-Phe-Lys-Ala. A synthetic peptide containing this sequence can serve as a substrate for the enzyme (606). Another "histone kinase" that has been identified is a 65,000-dalton protein (gel filtration) referred to as *histone kinase III,* which preferentially phosphorylates histone H2B (607).

Protease-activated kinase I (PAK I) is a proteolytically activated enzyme that has been identified and isolated from the postribosomal supernatant of rabbit reticulocytes (608). The enzyme phosphorylates histone H4 and other substrates (608); one of the latter is an avian retroviral nucleocapsid protein (609).

It has been determined that three oncogenes, *mil, raf,* and *mos,* encode proteins that appear to have protein serine/threonine kinase activity (reviewed in 2). These enzymes have not been characterized but are of particular interest since the other oncogene-encoded protein kinases are protein tyrosine kinases.

Growing importance is being attached to phosphorylation of the heavy chains of myosin, particularly for protozoal myosins where the phenomenon was discovered. Partially purified *Acanthamoeba* myosin II heavy chain kinase phosphorylates three serines in each heavy chain of this typical myosin at positions 5, 10, and 15 from the carboxyl-terminus (610). The serines occur in repeating peptides of sequence Arg-X-Y-Ser-Z-Arg, where X, Y, and Z are neutral amino acids. Phosphorylation at the tip of the tail of the myosin II rod segments inactivates the actin-activated ATPase activity at sites that are 120,000 daltons distant in the globular head domain, apparently by modifying the conformation of the myosin II filaments (611). Highly purified *Acanthamoeba* myosin I heavy chain kinase phosphorylates a single serine, that probably lies between the catalytic and actin-binding sites in the single heavy chain of the nonfilamentous myosin I, and activates its actin-activated ATPase activity (612). Neither *Acanthamoeba* myosin heavy chain kinase is known to be regulated. Similar effects have been observed upon phosphorylation of the heavy chains of *Dictyostelium* (613) and *Physarum* (614) myosins but the enzymatic details are not as well worked out. Phosphorylation of rabbit alveolar macrophage (615) and bovine brain (616) myosins also occurs at the tip of the tails, analogously to *Acanthamoeba* myosin II, but the consequences of these latter phosphorylations are not known.

Literature Cited

1. Krebs, E. G. 1986. *The Enzymes* 17:3–18
2. Hunter, T., Cooper, J. A. 1985. *Ann. Rev. Biochem.* 54:897–930
3. Beebe, S. J., Corbin, J. D. 1986. *The Enzymes* 17:44–100
4. Krebs, E. G., Blumenthal, D. K., Edelman, A. M., Hales, C. N. 1985. In *Mechanisms of Receptor Regulation*, ed. G. Poste, S. T. Crooke, pp. 159–95. New York: Plenum
5. Taylor, S. S., Saraswat, L. D., Toner, J. A., Bubis, J. 1986. *Ann. NY Acad. Sci.* In press
6. Bramson, H. N., Kaiser, E. T., Mildvan, A. S. 1984. *CRC Crit. Rev. Biochem.* 15:93–124
7. Rickenberg, H. V., Leichtling, B. H. 1987. *The Enzymes* 18:420–56
8. Kato, R., Uno, I., Ishikawa, T., Fujii, T. 1984. *Plant Cell Physiol.* 25:691–96
9. Zoller, M. J., Kerlavage, A. R., Taylor, S. S. 1979. *J. Biol. Chem.* 254:2408–12
10. Uhler, M. D., Chrivia, J. C., McKnight, G. S. 1986. *J. Biol. Chem.* 261:15360–63
11. Erlichman, J., Sarkar, D., Fleischer, N., Rubin, C. S. 1980. *J. Biol. Chem.* 255:8179–84
12. Robinson-Steiner, A. M., Beebe, S. J., Rannels, S. R., Corbin, J. D. 1984. *J. Biol. Chem.* 259:10596–605
13. Jahnsen, T., Lohmann, S. M., Walter, U., Hedin, L., Richards, J. S. 1985. *J. Biol. Chem.* 260:15980–87
14. Erlichman, J., Rubin, C. S., Rosen, O. M. 1973. *J. Biol. Chem.* 248:7607–609
15. Zoller, M. J., Taylor, S. S. 1979. *J. Biol. Chem.* 254:8363–68
16. Titani, K., Sasagawa, T., Ericsson, L. H., Kumar, S., Smith, S. B., et al. 1984. *Biochemistry* 23:4193–99
17. Takio, K., Smith, S. B., Krebs, E. G., Walsh, K. A., Titani, K. 1984. *Biochemistry* 23:4200–206
18. Shoji, S., Ericsson, L. H., Walsh, K. A., Fischer, E. H., Titani, K. 1983. *Biochemistry* 22:3702–709
19. Uhler, M. D., Carmichael, D. F., Lee, D. C., Chrivia, J. C., Krebs, E. G., McKnight, G. S. 1986. *Proc. Natl. Acad. Sci. USA* 83:1300–304
20. Lee, D. C., Carmichael, D. F., Krebs, E. G., McKnight, G. S. 1983. *Proc. Natl. Acad. Sci. USA* 80:3608–12
21. Zoller, M. J., Nelson, N. C., Taylor, S. S. 1981. *J. Biol. Chem.* 256:10837–42
22. Carr, S. A., Biemann, K., Shoji, S., Parmelee, D. C., Titani, K. 1982. *Proc. Natl. Acad. Sci. USA* 79:6128–31
23. Bramson, H. N., Thomas, N., Matsueda, R., Nelson, N. C., Taylor, S. S., Kaiser, E. T. 1982. *J. Biol. Chem.* 257:10575–81
24. Sowadski, J. M., Xuong, N. H., Anderson, D., Taylor, S. S. 1985. *J. Mol. Biol.* 182:617–20
25. Reimann, E. M. 1986. *Biochemistry* 25:119–25

26. Weber, I. T., Steitz, T. A., Bubis, J., Taylor, S. S. 1987. *Biochemistry.* 26: 343–51
27. Lee, J. H., Bechtel, P. J., Phillips, G. N. Jr. 1985. *J. Biol. Chem.* 260:9380–81
28. Glass, D. B., Krebs, E. G. 1980. *Ann. Rev. Pharmacol. Toxicol.* 20:363–88
29. Krebs, E. G., Beavo, J. A. 1979. *Ann. Rev. Biochem.* 48:923–59
30. Nimmo, H. G., Cohen, P. 1977. *Adv. Cyclic Nucleotide Res.* 8:145–266
31. Zetterqvist, O., Ragnarsson, U. 1982. *FEBS Lett.* 139:287–90
32. Kemp, B. E., Graves, D. J., Benjamini, E., Krebs, E. G. 1977. *J. Biol. Chem.* 252:4888–94
33. Kemp, B. E. 1979. *J. Biol. Chem.* 254:2638–42
34. Chessa, G., Borin, G., Marchiori, F., Meggio, F., Brunati, A. M., Pinna, L. A. 1983. *Eur. J. Biochem.* 135:609–14
35. Reed, J., Kinzel, V., Kemp, B. E., Cheng, H.-C., Walsh, D. A. 1985. *Biochemistry* 24:2967–73
36. Rosevear, P. R., Fry, D. C., Mildvan, A. S., Doughty, M., O'Brian, C., Kaiser, E. T. 1984. *Biochemistry* 23:3161–73
37. Walsh, D. A., Krebs, E. G. 1973. *The Enzymes* 8:555–81
38. Flockhart, D. A., Friest, W., Hoppe, J., Lincoln, T. M., Corbin, J. D. 1984. *Eur. J. Biochem.* 140:289–95
39. Mildvan, A. S., Rosevear, P. R., Granot, J., O'Brian, C. A., Bramson, N. H., Kaiser, E. T. 1983. *Methods Enzymol.* 99:93–119
40. Granot, J., Mildvan, A. S., Kaiser, E. T. 1980. *Arch. Biochem. Biophys.* 205: 1–17
41. Kochetkov, S. N., Bulgarina, T. V., Sashenko, L. P., Severin, E. S. 1977. *Eur. J. Biochem.* 81:111–18
42. Gabibov, A. G., Kochtkov, S. N., Sashenko, L. P., Smirnov, I. V., Severin, E. S. 1983. *Eur. J. Biochem.* 135:491–95
43. Bolen, D. W., Stingelin, J., Bramson, N. H., Kaiser, E. T. 1980. *Biochemistry* 19:1176–82
44. Cook, P. F., Neville, M. E., Vrana, K. E., Hartl, F. T., Roskoski, R. Jr. 1982. *Biochemistry* 21:5794–99
45. Whitehouse, S., Feramisco, J. R., Casnellie, J. E., Krebs, E. G., Walsh, D. A. 1983. *J. Biol. Chem.* 258:3693–701
46. Hofmann, F. 1980. *J. Biol. Chem.* 255:1559–64
47. Builder, S. E., Beavo, J. A., Krebs, E. G. 1980. *J. Biol. Chem.* 255:3514–19
48. Granot, J., Mildvan, A. S., Hiyama, K., Kondo, H., Kaiser, E. T. 1980. *J. Biol. Chem.* 225:4569–73
49. Connelly, P. A., Hastings, T. G., Reimann, E. M. 1986. *J. Biol. Chem.* 261:2325–30
50. Rangel-Aldao, R., Rosen, O. M. 1977. *J. Biol. Chem.* 252:7140–45
51. Smith, S. B., White, H. D., Siegel, J. B., Krebs, E. G. 1981. *Cold Spring Harbor Conf. Cell Prolif.* 8:55–65
52. Ogreid, D., Doskeland, S. O., Miller, J. P. 1983. *J. Biol. Chem.* 258:1041–49
53. Robinson-Steiner, A. M., Corbin, J. D. 1983. *J. Biol. Chem.* 258:1032–40
54. Beavo, J. A., Bechtel, P. J., Krebs, E. G. 1975. *Adv. Cyclic Nucleotide Res.* 5:241–51
55. Scott, J. D., Fischer, E. H., Takio, K., Demaille, J. G., Krebs, E. G. 1985. *Proc. Natl. Acad. Sci. USA* 82:5732–36
56. Cheng, H.-C., Van Patten, S. M., Smith, A. J., Walsh, D. A. 1985. *Biochem. J.* 231:655–61
57. Whitehouse, S., Walsh, D. A. 1983. *J. Biol. Chem.* 258:3682–92
58. Cheng, H.-C., Kemp, B. E., Pearson, R. B., Smith, A. J., Misconi, L., et al. 1986. *J. Biol. Chem.* 261:989–92
59. Scott, J. D., Fischer, E. H., Demaille, J. G., Krebs, E. G. 1985. *Proc. Natl. Acad. Sci. USA* 82:4379–83
60. Carmichael, D. F., Geahlen, R. L., Allen, S. M., Krebs, E. G. 1982. *J. Biol. Chem.* 257:10440–45
61. Hemmings, B. A., Aitken, A., Cohen, P., Rymond, M., Hofmann, F. 1982. *Eur. J. Biochem.* 127:473–81
62. Scott, C. W., Mumby, M. C. 1985. *J. Biol. Chem.* 260:2274–80
63. Rangel-Aldao, R., Kupiec, J. W., Rosen, O. M. 1979. *J. Biol. Chem.* 254:2499–508
64. Rangel-Aldao, R., Rosen, O. M. 1976. *J. Biol. Chem.* 251:7526–29
65. Todhunter, J. A., Purich, D. L. 1977. *Biochim. Biophys. Acta* 485:87–94
66. Rangel-Aldao, R., Rosen, O. M. 1976. *J. Biol. Chem.* 251:3375–80
67. Steinberg, R. A., O'Farrell, P. H., Friedrich, U., Coffino, P. 1977. *Cell* 10:381–91
68. Geahlen, R. L., Krebs, E. G. 1980. *J. Biol. Chem.* 255:9375–79
69. Hashimoto, E., Takio, K., Krebs, E. G. 1981. *J. Biol. Chem.* 256:5604–607
70. Geahlen, R. L., Allen, S. M., Krebs, E. G. 1981. *J. Biol. Chem.* 256:4536–40
71. Kuo, J. F., Greengard, P. 1970. *J. Biol. Chem.* 245:2493–98
72. Lincoln, T. M., Corbin, J. D. 1983. *Adv. Cyclic Nucleotide Res.* 15:139–92

73. Gill, G. N., McCune, R. W. 1979. *Curr. Top. Cell. Regul.* 15:1–45
74. Kuo, J. F., Shoji, M., Kuo, W. N. 1978. *Ann. Rev. Pharmacol. Toxicol.* 18:341–55
75. Deleted from proof
76. Lincoln, T. M., Hall, C. L., Park, C. R., Corbin, J. D. 1976. *Proc. Natl. Acad. Sci. USA* 73:2559–63
77. Walter, U. 1981. *Eur. J. Biochem.* 118:339–46
78. Lohmann, S. M., Walter, U., Miller, P. E., Greengard, P., De Camilli, P. 1981. *Proc. Natl. Acad. Sci. USA* 78:653–57
79. Waldmann, R., Bauer, S., Gobel, C., Hofmann, F., Jakobs, K. H., Walter, U. 1986. *Eur. J. Biochem.* 158:203–10
80. Ives, H. E., Casnellie, J. E., Greengard, P., Jamieson, J. D. 1980. *J. Biol. Chem.* 247:2723–28
81. de Jonge, H. R. 1981. *Adv. Cyclic Nucleotide Res.* 14:315–33
82. Gill, G. N., Holdy, K. E., Walton, G. M., Kanstein, C. B. 1976. *Proc. Natl. Acad. Sci. USA* 73:3918–22
83. Lincoln, T. M., Dills, W. L. Jr., Corbin, J. D. 1977. *J. Biol. Chem.* 252:4269–75
84. Flockerzi, V., Speichermann, N., Hofmann, F. 1978. *J. Biol. Chem.* 253:3395–99
85. Gill, G. N., Walton, G. M., Sperry, P. J. 1977. *J. Biol. Chem.* 252:6443–49
86. Takio, K., Wade, R. D., Smith, S. B., Krebs, E. G., Walsh, K. A., Titani, K. 1984. *Biochemistry* 23:4207–18
87. Lincoln, T. M., Corbin, J. D. 1977. *Proc. Natl. Acad. Sci. USA* 74:3239–43
88. Takio, K., Smith, S. B., Walsh, K. A., Krebs, E. G., Titani, K. 1983. *J. Biol. Chem.* 258:5531–36
89. Lincoln, T. M., Flockhart, D. A., Corbin, J. D. 1978. *J. Biol. Chem.* 253:6002–9
90. Gill, G. N. 1977. *J. Cyclic Nucleotide Res.* 3:153–62
91. Monken, C. E., Gill, G. N. 1980. *J. Biol. Chem.* 255:7067–70
92. Hashimoto, E., Takio, K., Krebs, E. G. 1982. *J. Biol. Chem.* 257:727–33
93. Vardanis, A. 1980. *J. Biol. Chem.* 255:7238–43
94. Glass, D. B., Smith, S. B. 1983. *J. Biol. Chem.* 258:14797–803
95. Glass, D. B., El-Maghrabi, M. R., Pilkis, S. J. 1986. *J. Biol. Chem.* 261:2987–93
96. Glass, D. B., Cheng, H.-C., Kemp, B. E., Walsh, D. A. 1986. *J. Biol. Chem.* 261:12166–71
97. Inoue, M., Kishimoto, A., Takai, Y., Nishizuka, Y. 1976. *J. Biol. Chem.* 251:4476–78
98. Kuo, W.-N., Kuo, J. F. 1976. *J. Biol. Chem.* 251:4283–86
99. Glass, D. B. 1983. *Biochem. J.* 213:159–64
100. Schlichter, D. J., Casnellie, J. E., Greengard, P. 1978. *Nature* 273:61–62
101. Aswad, D. W., Greengard, P. 1981. *J. Biol. Chem.* 256:3494–500
102a. Aitken, A., Bilham, T., Cohen, P., Aswad, D., Greengard, P. 1981. *J. Biol. Chem.* 256:3501–506
102b. Hashimoto, E., Takeda, M., Nishizuka, Y., Hamana, K., Iwai, K. 1976. *J. Biol. Chem.* 251:6287–93
103. Glass, D. B., McFann, L. J., Miller, M. D., Zeilig, C. E. 1981. *Cold Spring Harbor Conf. Cell Prolif.* 8:267–91
104. Corbin, J. D., Doskeland, S. O. 1983. *J. Biol. Chem.* 258:11391–97
105. Mackenzie, C. W. III. 1982. *J. Biol. Chem.* 257:5589–93
106. Hofmann, F., Gensheimer, H.-P., Landgraf, W., Hullin, R., Jastorff, B. 1985. *Eur. J. Biochem.* 150:85–88
107. Corbin, J. D., Ogreid, D., Miller, J. P., Suva, R. H., Jastorff, B., Doskeland, S. O., 1986. *J. Biol. Chem.* 261:1208–14
108. Doskeland, S. O., Vintermyr, O. K., Corbin, J. D., Ogreid, D. 1987. *J. Biol. Chem.* In press
109. Foster, J. L., Guttmann, J., Rosen, O. M. 1981. *J. Biol. Chem.* 256:5029–36
110. Hofmann, F., Flockerzi, V. 1983. *Eur. J. Biochem.* 130:599–603
111. Aitken, A., Hemmings, B. A., Hofmann, F. 1984. *Biochim. Biophys. Acta* 790:219–25
112. Landgraf, W., Hullin, R., Gobel, C., Hofmann, F. 1986. *Eur. J. Biochem.* 154:113–17
113. Hofmann, F., Gensheimer, H. P., Gobel, C. 1985. *Eur. J. Biochem.* 147:361–65
114. Pickett-Gies, C. A., Walsh, D. A. 1986. *The Enzymes* 17:396–461
115. Chan, K.-F. J., Graves, D. J. 1984. *Calcium Cell Funct.* 5:1–31
116. Cooper, R. H., Sul, H. S., McCullough, T. E., Walsh, D. A. 1980. *J. Biol. Chem.* 255:11794–801
117. Chrisman, T. D., Jordan, J. E., Exton, J. H. 1982. *J. Biol. Chem.* 257:10798–804
118. Reimann, E. M., Titani, K., Ericsson, L. H., Wade, R. D., Fischer, E. H., Walsh, K. A. 1984. *Biochemistry* 23:4185–92
119. Grand, R. J., Shenolikar, S., Cohen, P. 1981. *Eur. J. Biochem.* 113:359–67
120. Cohen, P., Antoniew, J. F., Davison, M., Taylor, C. 1974. *Metab. Interconvers. Enzymes, Int. Symp. 3rd, 1973*, E. H. Fischer, E. G. Krebs, H. Neurath, E.

R. Stadtman, organizers, pp. 33–42. Berlin/New York: Springer-Verlag
121. Cohen, P. 1978. *Curr. Top. Cell. Regul.* 14:117–96
122. Schramm, H. J., Jennissen, H. P. 1985. *J. Mol. Biol.* 181:503–16
123. Picton, C., Klee, C. B., Cohen, P. 1981. *Cell Calcium* 2:281–94
124. Cohen, P., Burchell, A., Foulkes, J. G., Cohen, P. T. W., Vanaman, T. C., Nairn, A. C. 1978. *FEBS Lett.* 92:287–93
125. Chan, K.-F. J., Graves, D. J. 1982. *J. Biol. Chem.* 257:5956–61
126. Chan, K.-F. J., Graves, D. J. 1982. *J. Biol. Chem.* 257:5948–55
127. Georgoussi, Z., Heilmeyer, L. M. G. 1986. *Biochemistry* 25:3867–74
128. Graves, D. J. 1983. *Methods Enzymol.* 99:268–78
129. Huang, T. S., Bylund, D. B., Stull, J. T., Krebs, E. G. 1974. *FEBS Lett.* 42:249–52
130. Cheng, A., Fitzgerald, T. J., Carlson, G. M. 1985. *J. Biol. Chem.* 260:2535–42
131. Gulyaeva, N. B., Vul'fson, P. L., Severin, E. S. 1977. *Biokhimiya* 43:373–81
132. King, M. M., Carlson, G. M., Haley, B. E. 1982. *J. Biol. Chem.* 257:14058–65
133. Tabatabai, L. B., Graves, D. J. 1978. *J. Biol. Chem.* 253:2196–202
134. Brostrom, C. O., Hunkeler, F. L., Krebs, E. G. 1971. *J. Biol. Chem.* 246:1961–67
135. Burger, D., Stein, E. A., Cox, J. A. 1983. *J. Biol. Chem.* 258:14733–39
136. Cohen, P. 1980. *Eur. J. Biochem.* 111:563–74
137. Picton, C., Klee, C. B., Cohen, P. 1980. *Eur. J. Biochem.* 111:553–61
138. Tam, S. W., Sharma, R. K., Wang, J. H. 1982. *J. Biol. Chem.* 257:14907–13
139. Yoshikawa, K., Usui, H., Imazu, M., Takeda M., Ebashi, S. 1983. *Eur. J. Biochem.* 136:413–19
140. Cohen, P. 1985. *Eur. J. Biochem.* 151:439–48
141. Pickett-Gies, C. A., Walsh, D. A. 1985. *J. Biol. Chem.* 260:2046–56
142. Cohen, P., Watson, D. C., Dixon, G. H. 1975. *Eur. J. Biochem.* 51:79–92
143. Cox, D. E., Edstrom, R. D. 1982. *J. Biol. Chem.* 257:12728–33
144. Hallenbeck, P. C., Walsh, D. A. 1983. *J. Biol. Chem.* 258:13493–501
145a. King, M. M., Fitzgerald, T. J., Carlson, G. M. 1983. *J. Biol. Chem.* 258:9925–30
145b. Cohen, P. 1980. *FEBS Lett.* 119:301–6

146. Silver, P. J., Stull, J. T. 1982. *J. Biol. Chem.* 257:6145–50
147. Slonczewski, J. L., Wilde, M. W., Zigmond, S. H. 1985. *J. Cell Biol.* 101:1191–97
148. Pires, E. M. V., Perry, S. V. 1977. *Biochem. J.* 167:137–46
149. Stull, J. T., Nunnally, M. H., Michinoff, C. H. 1986. *The Enzymes* 17:114–59
150. Sellers, J. R., Adelstein, R. S. 1986. *The Enzymes* 18:382–419
151. Sellers, J. R., Harvey, E. V. 1984. *Biochemistry* 23:5821–26
152. Stull, J. T., Nunnally, M. H., Moore, R. L., Blumenthal, D. K. 1985. *Adv. Enzyme Regul.* 23:123–40
153. Kamm, K. E., Stull, J. T. 1985. *Ann. Rev. Pharmacol. Toxicol.* 25:593–620
154. Hathaway, D. R., Adelstein, R. S., Klee, C. B. 1981. *J. Biol. Chem.* 256:8183–89
155. Serventi, I. M., Coffee, C. J. 1986. *Arch. Biochem. Biophys.* 245:379–86
156a. Walsh, M. P., Vallet, B., Autric, F., Demaille, J. G. 1979. *J. Biol. Chem.* 254:12136–44
156b. Wolf, H., Hoffman, F. 1980. *Proc. Natl. Acad. Sci. USA* 77:5852–55
157. Takio, K., Blumenthal, D. K., Edelman, A. M., Walsh, K. A., Krebs, E. G., Titani, K. 1985. *Biochemistry* 24:6028–37
158. Takio, K., Blumenthal, D. K., Walsh, K. A., Titani, K., Krebs, E. G., 1986. *Biochemistry.* 25:8049–57
159. Mayr, G. W., Heilmeyer, L. M. G. 1983. *Biochemistry* 22:4316–26
160. Crouch, T. H., Holroyde, M. J., Collins, J. H., Solaro, R. J., Potter, J. D. 1981. *Biochemistry* 20:6318–25
161. Edelman, A. M., Takio, K., Blumenthal, D. K., Hansen, R. S., Walsh, K. A., et al. 1985. *J. Biol. Chem.* 260:11275–85
162. Blumenthal, D. K., Takio, K., Edelman, A. M., Charbonneau, H., Titani, K., et al. 1985. *Proc. Natl. Acad. Sci. USA* 82:3187–91
163. Adelstein, R. S., Klee, C. B. 1981. *J. Biol. Chem.* 256:7501–9
164. Walsh, M. P. 1985. *Biochemistry* 24:3724–30
165. Foyt, H. L., Guerriero, V., Means, A. R. 1985. *J. Biol. Chem.* 260:7765–74
166. Lukas, T. J., Burgess, W. H., Prendergast, F. G., Lau, W., Watterson, D. M. 1986. *Biochemistry* 25:1458–64
167. Cole, H. A., Griffiths, H. S., Patchell, V. B., Perry, S. V. 1985. *FEBS Lett.* 180:165–69
168. Ikebe, M., Hartshorne, D. J. 1985. *J. Biol. Chem.* 260:10027–31

169. Perry, S. V., Griffiths, H. S., Levine, B. A., Patchell, V. B. 1985. *Advances in Protein Phosphatases 2*, ed. W. Merlevede, J. Di Salvo, pp. 3–18. Leuven, Belgium: Leuven Univ. Press
170. Barany, K., Csabina, S., Barany, M. 1985. See Ref. 169, pp. 37–58
171. Ikebe, M., Ogihara, S., Tonomura, Y. 1982. *J. Biochem.* 91:1809–12
172. Persechini, A., Hartshorne, D. J. 1983. *Biochemistry* 22:470–76
173. Sellers, J. R., Chock, P. B., Adelstein, R. S. 1983. *J. Biol. Chem.* 258:14181–88
174. Sobieszek, A. 1985. *Biochemistry* 24:1266–74
175. Persechini, A., Kamm, K. E., Stull, J. T. 1986. *J. Biol. Chem.* 261:6293–99
176. Trybus, K. M., Lowey, S. 1985. *J. Biol. Chem.* 260:15988–95
177. Persechini, A., Stull, J. T. 1984. *Biochemistry* 23:4144–150
178. Yazawa, M., Yagi, K. 1978. *J. Biochem.* 84:1259–65
179. Geuss, U., Mayr, G. W., Heilmeyer, L. M. G. 1985. *Eur. J. Biochem.* 153:327–34
180. Kamm, K. E., Stull, J. T. 1985. *Am. J. Physiol.* 249:C238–47
181. Johnson, J. D., Holroyde, M. J., Crouch, T. H., Solaro, R. J., Potter, J. D. 1981. *J. Biol. Chem.* 256:12194–98
182. Malencik, D. A., Anderson, S. R., Bohnert, J. L., Shalitin, Y. 1982. *Biochemistry* 21:4031–39
183. Edelman, A. M., Krebs, E. G. 1982. *FEBS Lett.* 138:293–98
184. Foyt, H. L., Means, A. R. 1985. *J. Cyclic Nucleotide Protein Phosphorylation Res.* 10:143–56
185. deLanerolle, P., Nishikawa, M., Yost, D. A., Adelstein, R. S. 1984. *Science* 223:1415–17
186. Schulman, H. 1984. *J. Cell Biol.* 99:11–19
187. Yamauchi, T., Fujisawa, H. 1983. *Eur. J. Biochem.* 132:15–21
188. Bennett, M. K., Erondu, N. E., Kennedy, M. B. 1983. *J. Biol. Chem.* 258:12735–44
189. McGuiness, T. L., Lai, Y., Greengard, P., Woodgett, J. R., Cohen, P. 1983. *FEBS Lett.* 163:329–34
190. McGuinness, T. L., Lai, Y., Greengard, P. 1985. *J. Biol. Chem.* 260:1696–704
191. Miller, S. G., Kennedy, M. B. 1985. *J. Biol. Chem.* 260:9039–46
192. Fukunaga, K., Yamamoto, H., Matsui, K., Higashi, K., Miyamoto, E. 1982. *J. Neurochem.* 39:1607–17
193. Goldenring, J. R., Gonzalez, B., McGuire, J. S. Jr., DeLorenzo, R. J. 1983. *J. Biol. Chem.* 258:12632–40
194. Kuret, J., Schulman, H. 1984. *Biochemistry* 23:5495–504
195. Ahmad, Z., DePaoli-Roach, A. A., Roach, P. J. 1982. *J. Biol. Chem.* 257:8348–55
196. Payne, M. E., Schworer, C. M., Soderling, T. R. 1983. *J. Biol. Chem.* 258:2376–82
197. Woodgett, J. R., Davison, M. T., Cohen, P. 1983. *Eur. J. Biochem.* 136:481–87
198. Nairn, A. C., Hemmings, H. C. Jr., Greengard, P. 1985. *Ann. Rev. Biochem.* 54:931–76
199. Kennedy, M. B., Greengard, P. 1981. *Proc. Natl. Acad. Sci. USA* 78:1293–97
200. Yamauchi, T., Fujisawa, H. 1981. *FEBS Lett.* 129:117–19
201. Erondu, N. E., Kennedy, M. B. 1985. *J. Neurosci.* 5:3270–77
202. Grab, D. J., Carlin, R. K., Siekevitz, P. 1981. *J. Cell Biol.* 89:440–48
203. Kennedy, M. B., Bennett, M. K., Erondu, N. E. 1983. *Proc. Natl. Acad. Sci. USA* 80:7357–61
204. Ouimet, C. C., McGuinness, T. L., Greengard, P. 1984. *Proc. Natl. Acad. Sci. USA* 81:5604–606
205. Goldenring, J. R., McGuire, J. S. Jr., DeLorenzo, R. J. 1984. *J. Neurochem.* 42:1077–84
206. Kelly, P. T., McGuinness, T. L., Greengard, P. 1984. *Proc. Natl. Acad. Sci. USA* 81:945–49
207. Sahyoun, N., LeVine, H., Cuatrecasas, P. 1984. *Proc. Natl. Acad. Sci. USA* 81:4311–15
208. Edelman, A. M., Hunter, D. D., Hendrickson, A. E., Krebs, E. G. 1985. *J. Neurosci.* 5:2609–17
209. Kelly, P. T., Yip, R. K., Shields, S. M., Hay, M. 1985. *J. Neurochem.* 45:1620–34
210. Kelly, P. T., Vernon, P. 1985. *Dev. Brain Res.* 18:211–24
211. Yamauchi, T., Fujisawa, H. 1982. *Biochem. Biophys. Res. Commun.* 109:975–81
212. Yamamoto, H., Fukunaga, K., Goto, S., Tanaka, E., Miyamoto, E. 1985. *J. Neurochem.* 44:759–68
213. Schworer, C. M., El-Maghrabi, M. R., Pilkis, S. J., Soderling, T. R. 1985. *Biol. Chem.* 260:13018–22
214. Vulliet, P. R., Woodgett, J. R., Cohen, P. 1984. *J. Biol. Chem.* 259:13680–83
215. Deleted in proof
216. Schulman, H. 1984. *Mol. Cell Biol.* 4:1175–78
217. Goldenring, J. R., Vallano, M. L., DeLorenzo, R. J. 1985. *J. Neurochem.* 45:900–5
218. Pearson, R. B., Woodgett, J. R.,

Cohen, P., Kemp, B. E. 1985. *J. Biol. Chem.* 260:14471–76

219a. LeVine, H., Sahyoun, N. E., Cuatrecasas, P. 1986. *Proc. Natl. Acad. Sci. USA* 83:2253–57

219b. Kuret, J., Schulman, H. 1985. *J. Biol. Chem.* 260:6427–33

220. Deleted in proof

221. Shields, S. M., Vernon, P. J., Kelly, P. T. 1984. *J. Neurochem.* 43:1599–609

222. Saitoh, T., Schwartz, J. H. 1985. *J. Cell Biol.* 100:835–42

223. Miller, S. G., Kennedy, M. B. 1986. *Cell* 44:861–70

224. Lai, Y., Nairn, A. C., Greengard, P. 1986. *Proc. Natl. Acad. Sci. USA* 83:4253–57

225. Schworer, C. M., Colbron, R. J., Solderling, T. R. 1986. *J. Biol. Chem.* 261:8581–84

226a. Yamauchi, T., Fujisawa, H. 1985. *Biochem. Biophys. Res. Commun.* 129:213–19

226b. Lou, L. L., Lloyd, S. J., Schulman, H. 1986. *Proc. Natl. Acad. Sci. USA* 83:9497–501

227. LeVine, H., Sahyoun, N. E., Cuatrecasas, P. 1985. *Proc. Natl. Acad. Sci. USA* 82:287–91

228. Lai, Y., Nairn, A. C., Greengard, P. 1986. *Soc. Neurosci. Abstr. XVI Ann. Meet.*, p. 1502

229. Llinas, R., McGuinness, T. L., Leonard, C. S., Sugimori, M., Greengard, P. 1985. *Proc. Natl. Acad. Sci. USA* 82:3035–39

230. Nairn, A. C., Greengard, P. 1987. *J. Biol. Chem.* In press

231a. Nairn, A. C., Bhagat, B., Palfrey, H. C. 1985. *Proc. Natl. Acad. Sci. USA* 82:7939–43

231b. Inoue, M., Kishimoto, A., Takai, Y., Nishizuka, Y. 1977. *J. Biol. Chem.* 252:7610–16

231c. Takai, Y., Kishimoto, A., Iwasa, Y., Kawahara, Y., Mori, T., Nishizuka, Y. 1979. *J. Biol. Chem.* 254:3692–95

232. Takai, Y., Kishimoto, A., Kikkawa, U., Mori, T., Nishizuka, Y. 1979. *Biochem. Biophys. Res. Commun.* 91:1218–24

233. Kishimoto, A., Takai, Y., Mori, T., Kikkawa, U., Nishizuka, Y. 1980. *J. Biol. Chem.* 255:2273–76

234. Kuo, J. F., Andersson, R. G. G., Wise, B. C., Mackerlova, L., Salomonsson, I., et al. 1980. *Proc. Natl. Acad. Sci. USA* 77:7039–43

235. Girard, P. R., Mazzei, G. J., Kuo, J. F. 1986. *J. Biol. Chem.* 261:370–75

236. Kikkawa, U., Takai, Y., Minakuchi, R., Inohara, S., Nishizuka, Y. 1982. *J. Biol. Chem.* 257:13341–48

237. Nishizuka, Y. 1984. *Nature* 308:693–97

238. Schatzman, R. C., Turner, R. S., Kuo, J. F. 1984. *Calcium Cell Funct.* 5:33–66

239. Bell, R. M. 1986. *Cell* 45:631–32

240. Nishizuka, Y. 1986. *Science* 233:305–12

241. Ashendel, C. L. 1985. *Biochim. Biophys. Acta* 822:219–42

242. Ono, Y., Kurokawa, T., Kawahara, K., Nishimura, O., Marumoto, R., et al. 1986. *FEBS Lett.* 203:111–15

243. Parker, P. J., Coussens, L., Totty, N., Rhee, L., Young, S., et al. 1986. *Science* 233:853–59

244. Knopf, J. L., Lee, M.-H., Sultzman, L. A., Kriz, R. W., Loomis, C. R., et al. 1986. *Cell* 46:491–502

245. Coussens, L., Parker, P. J., Rhee, L., Yang-Feng, T. L., Chen, E., et al. 1986. *Science* 233:859–66

246. Huang, K. P., Nakabayashi, H., Huang, F. L. 1986. *Proc. Natl. Acad. Sci. USA,* 83:8535–39

247. Kishimoto, A., Kajikawa, N., Shiota, M., Nishizuka, Y. 1983. *J. Biol. Chem.* 258:1156–64

248. Melloni, E., Pontremoli, S., Michetti, M., Sacco, O., Sparatore, B., et al. 1985. *Proc. Natl. Acad. Sci. USA* 82:6435–39

249. Hoshijima, M., Kikuchi, A., Tanimoto, T., Kaibuchi, K., Takai, Y. 1986. *Cancer Res.* 46:3000–4

250. Huang, K. P., Huang, F. L. 1986. *Biochem. Biophys. Res. Commun.* 139:320–26

251. Kishimoto, A., Nishiyama, K., Nakanishi, H., Uratsuji, Y., Nomura, H., et al. 1985. *J. Biol. Chem.* 260:12492–99

252. Deleted in proof

253. Iwasa, Y., Takai, Y., Kikkawa, U., Nishizuka, Y. 1980. *Biochem. Biophys. Res. Commun.* 96:180–87

254. Hunter, T., Ling, N., Cooper, J. A. 1984. *Nature* 311:480–83

255. Davis, R. J., Czech, M. P. 1985. *Proc. Natl. Acad. Sci. USA* 82:1974–78

256. Gallis, B., Lewis, A., Wignall, J., Alpert, A., Mochizuki, D. Y., et al. 1986. *J. Biol. Chem.* 261:5075–80

257. Ramachandran, C., Yau, P., Bradbury, E. M., Shymala, G., Yasuda, H., Walsh, D. A. 1984. *J. Biol. Chem.* 259:13495–503

258. Turner, R. S., Kemp, B. E., Su, H-D., Kuo, J. F. 1985. *J. Biol. Chem.* 260:11503–507

259. O'Brian, C. A., Lawrence, D. S., Kaiser, E. T., Weinstein, I. B. 1984. *Biochem. Biophys. Res. Commun.* 124:296–302

260. Chan, K-F. J., Stoner, G. L., Hashim,

G. A., Huang, K-P. 1986. *Biochem. Biophys. Res. Commun.* 134:1358–64

261. Ferrari, S., Marchiori, F., Borin, G., Pinna, L. A. 1985. *FEBS Lett.* 184:72–77

262. Wise, B. C., Glass, D. B., Turner, R. S., Kibler, R. F., Kuo, J. F. 1982. *J. Biol. Chem.* 257:8489–95

263. Wise, B. C., Raynor, R. L., Kuo, J. F. 1982. *J. Biol. Chem.* 257:8481–88

264. Kaibuchi, K., Takai, Y., Nishizuka, Y. 1981. *J. Biol. Chem.* 256:7146–49

265. Boni, L. T., Rando, R. R. 1985. *J. Biol. Chem.* 260:10819–25

266. Hannun, Y. A., Loomis, C. R., Bell, R. M. 1986. *J. Biol. Chem.* 261:7184–90

267. Ohkubo, S., Yamada, E., Endo, T., Itoh, H., Hidaka, H. 1984. *Biochem. Biophys. Res. Commun.* 118:460–66

268. McPhail, L. C., Clayton, C. C., Snyderman, R. 1984. *Science* 224:622–25

269. Horn, F., Gschwendt, M., Marks, F. 1985. *Eur. J. Biochem.* 148:533–38

270. Murakami, K., Routtenberg, A. 1985. *FEBS Lett.* 192:189–93

271. Hansson, A., Serhan, C. N., Haeggstrom, J., Ingelman-Sundberg, M., Samuelsson, B. 1986. *Biochem. Biophys. Res. Commun.* 134:1215–122

272. Shoyab, M. 1985. *Arch. Biochem. Biophys.* 236:435–40

273. Wightman, P. D., Raetz, C. R. H. 1984. *J. Biol. Chem.* 259:10048–52

274. Rando, R. R., Young, Y. 1984. *Biochem. Biophys. Res. Commun.* 122:818–23

275. Couturier, A., Bazgar, S., Castagna, M. 1984. *Biochem. Biophys. Res. Commun.* 121:448–55

276. Cabot, M. C., Jaken, S. 1984. *Biochem. Biophys. Res. Commun.* 125:163–69

277. Ganong, B. R., Loomis, C. R., Hannun, Y. A., Bell, R. M. 1986. *Proc. Natl. Acad. Sci. USA* 83:1184–88

278. Mori, T., Takai, Y., Yu, B., Takahashi, J., Nishizuka, Y., Fujikura, T. 1982. *J. Biochem.* 91:427–31

279. Castagna, M., Takai, Y., Kaibuchi, K., Sano, K., Kikkawa, U., Nishizuka, Y. 1982. *J. Biol. Chem.* 257:7847–51

280. Fujiki, H., Tanaka, Y., Miyake, R., Kikkawa, U., Nishizuka, Y., Sugimura, T. 1984. *Biochem. Biophys. Res. Commun.* 120:339–43

281. Brasseur, R., Cabiaux, V., Huart, P., Castagna, M., Baztar, S., Ruysschaert, J. M. 1985. *Biochem. Biophys. Res. Commun.* 127:969–76

282. Wender, P. A., Koehler, K. F., Sharkey, N. A., Dell'Aquila, M. L., Blumberg, P. M. 1986. *Proc. Natl. Acad. Sci. USA* 83:4214–18

283. Parker, P. J., Stabel, S., Waterfield, M. D. 1984. *EMBO J.* 3:953–59

284. Kikkawa, U., Takai, Y., Tanaka, Y., Miyake, R., Nishizuka, Y. 1983. *J. Biol. Chem.* 258:11442–45

285. Niedel, J. E., Kuhn, L. J., Vandenbark, G. R. 1983. *Proc. Natl. Acad. Sci. USA* 80:36–40

286. Konig, B., DiNitto, P. A., Blumberg, P. M. 1985. *J. Cell. Biochem.* 27:255–65

287. Hannun, Y. A., Bell, R. M. 1986. *J. Biol. Chem.* 261:9341–47

288. Sharkey, N. A., Leach, K. L., Blumberg, P. M. 1984. *Proc. Natl. Acad. Sci. USA* 81:607–10

289. Sharkey, N. A., Blumberg, P. M. 1985. *Biochem. Biophys. Res. Commun.* 133:1051–56

290. Uchida, T., Filburn, C. R. 1984. *J. Biol. Chem.* 259:12311–14

291. Jeng, A. Y., Sharkey, N. A., Blumberg, P. M. 1986. *Cancer Res.* 46:1966–71

292. Ashendel, C. L., Staller, J. M., Boutwell, R. K. 1983. *Cancer Res.* 43:4333–37

293. Delclos, K. B., Yeh, E., Blumberg, P. M. 1983. *Proc. Natl. Acad. Sci. USA* 80:3054–58

294. LePeuch, C. J., Ballester, R., Rosen, O. M. 1983. *Proc. Natl. Acad. Sci. USA* 80:6858–62

295. Huang, K. P., Chan, K. F. J., Singh, T. J., Nakabayashi, H., Huang, F. L. 1986. *J. Biol. Chem.* 261:13134–40

296. Hokin, L. E. 1985. *Ann. Rev. Biochem.* 54:205–35

297. Kaibuchi, K., Takai, Y., Sawamura, M., Hoshijima, M., Fujikura, T., Nishizuka, Y. 1983. *J. Biol. Chem.* 258:6701–704

298. Ebeling, J. G., Vandenbark, G. R., Kuhn, L. J., Ganong, B. R., Bell, R. M., Niedel, J. E. 1985. *Proc. Natl. Acad. Sci. USA* 82:815–19

299. Kraft, A. S., Anderson, W. B., Cooper, H. L., Sando, J. J. 1982. *J. Biol. Chem.* 257:13193–96

300. Kraft, A. S., Anderson, W. B. 1983. *Nature* 301:621–23

301. McCaffrey, P. G., Friedman, B. A., Rosner, M. R. 1984. *J. Biol. Chem.* 259:12502–507

302. Shoji, M., Girard, P. R., Mazzei, G. J., Vogler, W. R., Kuo, J. F. 1986. *Biochem. Biophys. Res. Commun.* 135:1144–49

303. Nishihira, J., McPhail, L. C., O'Flaherty, J. T. 1986. *Biochem. Biophys. Res. Commun.* 134:587–94

304. Anderson, W. B., Estival, A., Tapiovaara, H., Gopalakrishna, R.

1985. *Adv. Cyclic Nucleotide Protein Phosphorylation Res.* 19:287–306
305. Wolf, M., Cuatrecasas, P., Sahyoun, N. 1985. *J. Biol. Chem.* 260:15718–22
306. Wolf, M., LeVine, H., May, W. S., Cuatrecasas, P., Sahyoun, N. 1985. *Nature* 317:546–49
307. Drust, D. S., Martin, T. F. J. 1985. *Biochem. Biophys. Res. Commun.* 128:531–37
308. Farrar, W. L., Anderson, W. B. 1985. *Nature* 315:233–35
309. Farrar, W. L., Thomas, T. P., Anderson, W. B. 1985. *Nature* 315:233–37
310. Fearon, C. W., Tashjian, A. H. 1985. *J. Biol. Chem.* 260:8366–71
311. Hirota, K., Hirota, T., Aguilera, G., Catt, K. J. 1985. *J. Biol. Chem.* 260:3243–46
312. Naor, Z., Zer, J., Zakut, H., Hermon, J. 1985. *Proc. Natl. Acad. Sci. USA* 82:8203–207
313. McArdle, C. A., Conn, P. M. 1986. *Mol. Pharmacol.* 29:570–76
314. Liles, W. C., Hunter, D. D., Meier, K. E., Nathanson, N. M. 1986. *J. Biol. Chem.* 261:5307–13
315. Vilgrain, I., Cochet, C., Chambaz, E. M. 1984. *J. Biol. Chem.* 259:3403–406
316. Costa-Casnellie, M. R., Segel, G. B., Lichtman, M. A. 1985. *Biochem. Biophys. Res. Commun.* 133:1139–44
317. Dougherty, R. W., Niedel, J. E. 1986. *J. Biol. Chem.* 261:4097–100
318. Tapley, P. M., Murray, A. W. 1984. *Biochem. Biophys. Res. Commun.* 122:158–64
319. Fabbro, D., Regazzi, R., Costa, S. D., Borner, C., Eppenberger, U. 1986. *Biochem. Biophys. Res. Commun.* 135:65–73
320. Guy, G. R., Gordon, J., Walker, L., Michell, R. H., Brown, G. 1986. *Biochem. Biophys. Res. Commun.* 135:146–53
321. Melloni, E., Pontremoli, S., Michetti, M., Sacco, O., Sparatore, B., Horecker, B. L. 1986. *J. Biol. Chem.* 261:4101–105
322. Rodriguez-Pena, A., Rozengurt, E. 1984. *Biochem. Biophys. Res. Commun.* 120:1053–59
323. Ballester, R., Rosen, O. M. 1985. *J. Biol. Chem.* 260:15194–99
324. Ochoa, S. 1983. *Arch. Biochem. Biophys.* 223:325–49
325. Safer, B. 1983. *Cell* 33:7–8
326. Proud, C. G. 1986. *Trends. Biochem. Sci.* 11:73–77
327. London, I. M., Levin, D. H., Matts, R. L., Thomas, S. B., Petryshyn, R., Chen, J.-J. 1986. *The Enzymes* 18:360–81
328. Siekierka, J., Manne, V., Ochoa, S.

1984. *Proc. Natl. Acad. Sci. USA* 81:352–56
329. Pain, V. M., Clemens, M. J. 1983. *Biochemistry* 22:726–33
330. Trachsel, H., Ranu, R. S., London, I. M. 1979. *Methods Enzymol.* 60:485–95
331. Lundak, T. S., Traugh, J. A. 1980. In *Protein Phosphorylation and Bio-Regulation,* ed. G. Thomas, E. J. Podesta, J. Gordon, pp. 154–61. Basel: Karger
332. Hunt, T. 1979. In *From Gene to Protein: Information Transfer in Normal and Abnormal Cells, Miami Winter Symposium,* ed. T. R. Russell, K. Brew, J. Schultz, H. Harber, pp. 321–46. New York: Academic
333. Farrell, P. J., Balkow, K., Hunt, T., Jackson, R. J. 1977. *Cell* 11:187–200
334. Gross, M., Rynning, J., Knish, W. M. 1981. *J. Biol. Chem.* 256:589–92
335. Traugh, J. A., Del Grande, R. W., Tuazon, P. T. 1981. *Cold Spring Harbor Conf. Cell Prolif.* 8:999–1012
336. Wettenhall, R. E. H., Kudlicki, W., Kramer, G., Hardesty, B. 1986. *J. Biol. Chem.* In press
337. Gonzatti-Haces, M. I., Traugh, J. A. 1982. *J. Biol. Chem.* 257:6642–45
338. Gross, M., Mendelewski, J. 1978. *Biochim. Biophys. Acta* 520:650–63
339. Fagard, R., London, I. M. 1981. *Proc. Natl. Acad. Sci. USA* 78:866–70
340. Gross, M., Kaplansky, D. A. 1983. *Biochim. Biophys.* 740:255–63
341. Hunt, T. 1983. *Philos. Trans. R. Soc. London Ser. B* 302:127–34
342. Jackson, R. J., Herbert, P., Campbell, E. A., Hunt, T. 1983. *Eur. J. Biochem.* 131:313–24
343. Palomo, C., Vicente, O., Sierra, J. M., Ochoa, S. 1985. *Arch. Biochem. Biophys.* 239:497–507
344. Kudlicki, W., Fullilove, S., Kramer, G., Hardesty, B. 1985. *Proc. Natl. Acad. Sci. USA* 82:5332–36
345. Crouch, D., Safer, B. 1984. *J. Biol. Chem.* 259:10363–68
346. Clemens, M. J., McNurlan, M. A. 1985. *Biochem. J.* 226:345–60
347. Samuel, C. E. 1979. *Proc. Natl. Acad. Sci. USA* 76:600–4
348. Ranu, R. S. 1980. *FEBS Lett.* 112:211–15
349. Sen, G. C., Hidenharu, T., Lengyel, P. 1978. *J. Biol. Chem.* 253:5915–21
350. Kimchi, A., Zilberstein, A., Schmidt, A., Shulman, L., Revel, M. 1979. *J. Biol. Chem.* 254:9846–53
351. Berry, M. J., Knutson, G. S., Lasky, S. R., Munemitsu, S. M., Samuel, C. E. 1985. *J. Biol. Chem.* 260:11240–47
352. Samuel, C. E., Duncan, R., Knutson,

G. S., Hershey, J. W. B. 1984. *J. Biol. Chem.* 259:13451–57
353. Morrow, C. D., Gibbons, G. F., Dasgupta, A. 1985. *Cell* 40:913–21
354. Ranu, R. S. 1980. *Biochem. Biophys. Res. Commun.* 97:252–62
355. Grosfeld, H., Ochoa, S. 1980. *Proc. Natl. Acad. Sci. USA* 77:6526–30
356. Petryshyn, R., Levin, D. H., London, I. M. 1983. *Methods Enzymol.* 99:346–62
357. Penn, L. J. Z., Williams, B. R. G. 1985. *Proc. Natl. Acad. Sci. USA* 82:4959–63
358. Berry, M. J., Samuel, C. E. 1985. *Biochem. Biophys. Res. Commun.* 133:168–75
359. Bischoff, J. R., Samuel, C. E. 1985. *J. Biol. Chem.* 260:8237–39
360. Krust, B., Galabru, J., Hovanessian, A. G. 1984. *J. Biol. Chem.* 259:8494–98
361. Das, H. K., Das, A., Ghosh-Dastidar, P., Ralston, R. O., Yaghmai, B., et al. 1981. *J. Biol. Chem.* 256:6491–95
362. Galabru, J., Hovanessian, A. G. 1985. *Cell* 43:685–94
363. Kitajewski, J., Schneider, R. J., Safer, B., Munemitsu, S. M., Samuel, C. E., et al. 1986. *Cell* 45:195–200
364. Whitaker-Dowling, P., Youngner, J. S. 1984. *Virology* 173:171–81
365. Reichel, P. A., Merrick, W. C., Siekerka, J., Mathews, M. B. 1985. *Nature* 313:196–200
366. Schneider, R. J., Safer, B., Munemitsu, S. M., Samuel, C. E., Shenk, T. 1985. *Proc. Natl. Acad. Sci. USA* 82:4321–25
367. Randle, P. J., Fatania, H. R., Lau, K. S. 1984. In *Molecular Aspects of Cell Regulation*, ed. P. Cohen, 3:1–26. Amsterdam: Elsevier
368. Reed, L. J., Yeaman, S. J. 1987. *The Enzymes* 18:77–96
369. Yeaman, S. J., Hutcheson, E. T., Roche, T. E., Pettit, F. H., Brown, J. R., et al. 1978. *Biochemistry* 17:2364–70
370. Mullinax, T. R., Stepp, L. R., Brown, J. R., Reed, L. J. 1985. *Arch. Biochem. Biophys.* 243:655–59
371. Kerbey, A. L., Radcliffe, P. M., Randle, P. J., Sugden, P. H. 1979. *Biochem. J.* 181:427–33
372. Kerbey, A. L., Randle, P. J. 1985. *Biochem. J.* 231:523–29
373. Sugden, P. H., Kerbey, A. L., Randle, P. J., Waller, C. A., Reid, K. B. M. 1979. *Biochem. J.* 181:419–26
374. Kerbey, A. L., Randle, P. J. 1979. *FEBS Lett.* 108:485–88
375. Walsh, D. A., Cooper, R. H., Denton, R. M., Bridges, B. J., Randle, P. J. 1976. *Biochem. J.* 157:41–46

376. Rahmatullah, M., Jilka, J. M., Radke, G. A., Roche, T. E. 1986. *J. Biol. Chem.* 261:6515–23
377. Brandt, D. R., Roche, T. E. 1983. *Biochemistry* 22:2966–71
378. Stepp, L. R., Pettit, F. H., Yeaman, S. J., Reed, L. J. 1983. *J. Biol. Chem.* 258:9454–58
379. Pettit, F. H., Humphreys, J., Reed, L. J. 1982. *Proc. Natl. Acad. Sci. USA* 79:3945–48
380. Linn, T. C., Pettit, F. H., Reed, L. J. 1969. *Proc. Natl. Acad. Sci. USA* 62:234–41
381. Reed, L. J., Pettit, F. H., Yeaman, S. J., Teague, W. M., Bleile, D. M. 1980. In *Enzyme Regulation and Action*, ed. P. Mildner, B. Ries, pp. 47–56. Oxford: Pergamon
382. DeMarucucci, O., Lindsay, J. G. 1985. *Eur. J. Biochem.* 149:641–48
383. Jilka, J. M., Rahmatullah, M., Kazemi, M., Roche, T. E. 1986. *J. Biol. Chem.* 261:1858–67
384. Denyer, G. S., Kerbey, A. L., Randle, P. J. 1986. *Biochem. J.* 239:347–54
385. Parker, P. J., Randle, P. J. 1978. *FEBS Lett.* 95:153–56
386. Odessey, R., Goldberg, A. L. 1979. *Biochem. J.* 178:475–89
387. Hughes, W. A., Halestrap, A. P. 1981. *Biochem. J.* 196:459–69
388. Lau, K. S., Fatania, H. R., Randle, P. J. 1981. *FEBS Lett.* 126:66–70
389. Randle, P. J., Patson, P. A., Espinal, J. 1986. *The Enzymes* 18:97–122
390. Fatania, H. R., Lau, K. S., Randle, P. J. 1981. *FEBS Lett.* 132:285–88
391. Paxton, R., Harris, R. A. 1982. *J. Biol. Chem.* 257:14433–39
392. Odessey, R. 1982. *Biochem. J.* 204:353–56
393. Lawson, R., Cook, K. G., Yeaman, S. J. 1983. *FEBS Lett.* 157:54–58
394. Damuni, Z., Merryfield, M. L., Humphreys, J. S., Reed, L. J. 1984. *Proc. Natl. Acad. Sci. USA* 81:4335–38
395. Cook, K. G., Bradford, A. P., Yeaman, S. J. 1985. *Biochem. J.* 225:731–35
396. Cook, K. G., Bradford, A. P., Yeaman, S. J., Aitken, A., Fearnley, I. M., Walker, J. E. 1984. *Eur. J. Biochem.* 145:587–91
397. Paxton, R., Kuntz, M., Harris, R. A. 1986. *Arch. Biochem. Biophys.* 244:187–201
398. Cook, K. G., Lawson, R., Yeaman, S. J., Aitken, A. 1985. *FEBS Lett.* 164:47–50
399. Cook, K. G., Lawson, R., Yeaman, S. J. 1983. *FEBS Lett.* 157:59–62
400. Lau, K. S., Phillips, C. E., Randle, P. J. 1983. *FEBS Lett.* 160:149–52

401. Fatania, H. R., Patston, P. A., Randle, P. J. 1983. *FEBS Lett.* 158:234–38
402. Fatania, H. R., Lau, K. S., Randle, P. J. 1982. *FEBS Lett.* 147:35–39
403. Yeaman, S. J., Cook, K. G., Boyd, R. W., Lawson, R. 1984. *FEBS Lett.* 172:38–42
404. Espinal, J., Patston, P. A., Fatania, H. R., Lau, K. S., Randle, P. J. 1985. *Biochem. J.* 225:509–16
405. Lau, K. S., Fatania, H. R., Randle, P. J. 1982. *FEBS. Lett.* 144:57–62
406. Paxton, R., Harris, R. A. 1984. *Arch. Biochem. Biophys.* 231:48–57
407. Paxton, R., Harris, R. A. 1984. *Arch. Biochem. Biophys.* 231:58–66
408. Frick, G. P., Blinder, L., Goodman, H. M. 1985. *Fed. Proc.* 44:312–14
409. Harris, R. A., Paxton, R., Goodwin, G. W., Powell, S. M. 1986. *Biochem. J.* 236:209–13
410. Jones, S. M. A., Yeaman, S. J. 1986. *Biochem. J.* 236:209–13
411. Buxton, D. B., Olson, M. S. 1985. *Fed. Proc.* 44:306–7
412. Hathaway, G. M., Traugh, J. A. 1982. *Curr. Top. Cell. Regul.* 21:101–27
413. Pinna, L. A., Meggio, F., Donella-Deana, A., Brunati, A. M. 1985. *Proc. 16th FEBS Congr.* Pt.A, pp. 155–63. VNU Utrecht: Science
414. Hathaway, G. M., Traugh, J. A. 1979. *J. Biol. Chem.* 254:762–68
415. Singh, T. J., Huang, K. P. 1985. *FEBS Lett.* 190:84–88
416. Dahmus, M. E. 1981. *J. Biol. Chem.* 256:3319–25
417. Itarte, E., Huang, K. P. 1979. *J. Biol. Chem.* 254:4052–57
418. Ahmad, Z., Camici, M., DePaoli-Roach, A. A., Roach, P. J. 1984. *J. Biol. Chem.* 259:3420–28
419. Lerch, K., Muir, L. W., Fischer, E. H. 1975. *Biochemistry* 14:2015–23
420. Kudlicki, W., Grankowski, N., Gasior, E. 1978. *Eur. J. Biochem.* 84:493–98
421. Itarte, E., Mor, M. A., Salavert, A., Pena, J. M., Bertomeu, J. F., Guinovart, J. J. 1981. *Biochim. Biophys. Acta* 658:334–47
422. Pierre, M., Loeb, J. E. 1982. *Biochim. Biophys. Acta* 700:221–28
423. Baydoun, H., Hoppe, J., Freist, W., Wagner, K. G. 1982. *J. Biol. Chem.* 257:1032–36
424. Thornburg, W., Gamo, S., O'Malley, A. F., Lindell, T. J. 1979. *Biochim. Biophys. Acta* 571:35–44
425. Plana, M., Guasch, M. D., Itarte, E. 1985. *Biochem. J.* 230:69–74
426. Christmann, J. L., Dahmus, M. E. 1981. *J. Biol. Chem.* 256:3326–31
427. Dahmus, M. E. 1981. *J. Biol. Chem.* 256:3332–39
428. Stetler, D. A., Jacob, S. T. 1985. *Biochemistry* 24:5163–69
429. Tipper, J. P., Bacon, G. W., Witters, L. A. 1983. *Arch. Biochem. Biophys.* 227:386–96
430. Munday, M. R., Hardie, D. G. 1984. *Eur. J. Biochem.* 141:617–27
431. Singh, T. J., Akatsuka, A., Huang, K. P. 1984. *J. Biol. Chem.* 259:12857–64
432. Itarte, E., Plana, M., Guasch, M. D., Martos, C. 1983. *Biochem. Biophys. Res. Commun.* 117:631–36
433. Singh, T. J., Akatsuka, A., Huang, K. P., Murthy, A. S. N., Flavin, M. 1984. *Biochem. Biophys. Res. Commun.* 121:19–26
434. Pendergast, A. M., Traugh, J. A. 1985. *J. Biol. Chem.* 260:11769–74
435. Meggio, F., Donella-Deana, A., Pinna, L. A. 1979. *FEBS Lett.* 106:76–80
436. Tuazon, P. T., Bingham, E. W., Traugh, J. A. 1979. *Eur. J. Biochem.* 94:497–504
437. Kuret, J., Woodgett, J. R., Cohen, P. 1985. *Eur. J. Biochem.* 151:39–48
438. Donella-Deana, A., Grankowski, N., Kudlicki, W., Szyszka, R., Gasior, E., Pinna, L. A. 1985. *Biochim. Biophys. Acta* 829:180–87
439. Cobb, M. H., Rosen, O. M. 1983. *J. Biol. Chem.* 258:12472–81
440. Witters, L. A., Tipper, J. P., Bacon, G. W. 1983. *J. Biol. Chem.* 258:5643–48
441. Akatsuka, A., Singh, T. J., Nakabayashi, H., Lin, M. C., Huang, K. P. 1985. *J. Biol. Chem.* 260:3239–42
442. Thornburg, W., Lindell, T. J. 1977. *J. Biol. Chem.* 252:6660–55
443. Baydoun, H., Hoppe, J., Jacob, G., Wagner, K. G. 1980. *FEBS Lett.* 122:231–33
444. DePaoli-Roach, A. A., Ahmad, Z., Roach, P. J. 1981. *J. Biol. Chem.* 256:8955–62
445. Rigobello, M. P., Jori, E., Carignani, G., Pinna, L. A. 1982. *FEBS Lett.* 144:354–58
446. Hathaway, G. M., Traugh, J. A. 1984. *J. Biol. Chem.* 259:2850–55
447. Cochet, C., Chambaz, E. M. 1983. *Mol. Cell Endocrinol.* 30:247–66
448. Meggio, F., Pinna, L. A. 1984. *Eur. J. Biochem.* 145:593–99
449. Meggio, F., Grankowski, N., Kudlicki, W., Szyszka, R., Gasior, E., Pinna, L. A. 1986. *Eur. J. Biochem.* 159:31–38
450. Huang, K. P., Itarte, E., Singh, T. J., Akatsuka, A. 1982. *J. Biol. Chem.* 257:3236–42
451. Kumon, A., Ozawa, M. 1979. *FEBS Lett.* 108:200–4

452. Cochet, C., Job, D., Pirollet, F., Chambaz, E. M. 1981. *Biochim. Biophys. Acta* 658:191–201
453. Villar-Palasi, C., Kumon, A. 1981. *J. Biol. Chem.* 256:7409–15
454. Boivin, P., Galand, C. 1979. *Biochem. Biophys. Res. Commun.* 89:7–16
455. Dahmus, M. E., Natzle, J. 1977. *Biochemistry* 16:1901–907
456. Rose, K. M., Bell, L. E., Siefken, D. A., Jacob, S. T. 1981. *J. Biol. Chem.* 256:7468–77
457. Glover, C. V. C., Shelton, E. R., Brutlag, D. L. 1983. *J. Biol. Chem.* 258:3258–65
458. Renart, M. F., Sastre, L., Sebastian, J. 1984. *Eur. J. Biochem.* 140:47–54
459. Baydoun, H., Feth, F., Hoppe, J., Erdmann, H., Wagner, K. G. 1986. *Arch. Biochem. Biophys.* 245:504–11
460. Dahmus, G. K., Glover, C. V. C., Brutlag, D. L., Dahmus, M. E. 1984. *J. Biol. Chem.* 259:9001–6
461a. Qi, S. L., Yukioka, M., Morisawa, S., Inoue, A. 1986. *FEBS Lett.* 203:104–8
461b. Padmanabhan, R., Glover, C. V. C. 1987. *J. Biol. Chem.* In press
461c. Glover, C. V. C. 1986. *J. Biol. Chem.* 261:14349–54
462. Hathaway, G. M., Zoller, M. J., Traugh, J. A. 1981. *J. Biol. Chem.* 256:11442–46
463. Feige, J. J., Cochet, C., Pirollet, F., Chambaz, E. M. 1983. *Biochemistry* 22:1452–59
464. Cochet, C., Chambaz, E. M. 1983. *J. Biol. Chem.* 258:1403–406
465. Baydoun, H., Hoppe, J., Freist, W., Wagner, K. G. 1981. *Eur. J. Biochem.* 115:385–89
466. Dabauvalle, M. C., Meggio, F., Creuzet, C., Loeb, J. E. 1979. *FEBS Lett.* 107:193–97
467. DePaoli-Roach, A. A., Roach, P. J., Larner, J. 1979. *J. Biol. Chem.* 254:12062–68
468. Meggio, F., Donella-Deana, A., Pinna, L. A. 1981. *J. Biol. Chem.* 256:11958–61
469. Stetler, D. A., Rose, K. M. 1982. *Biochemistry* 21:3721–28
470. Guasch, M. D., Plana, M., Pena, J. M., Itarte, E. 1986. *Biochem. J.* 234:523–26
471. Ackerman, P., Glover, C. V. C., Osheroff, N. 1985. *Proc. Natl. Acad. Sci. USA* 82:3164–68
472. Pinna, L. A., Donella-Deana, A., Meggio, F. 1979. *Biochem. Biophys. Res. Commun.* 87:114–20
473. Hoppe, J., Baydoun, H. 1981. *Eur. J. Biochem.* 117:585–89
474. Sculley, T. B., Mackinlay, A. G. 1982. *Eur. J. Biochem.* 124:449–55
475. Holmes, C. F. B., Kuret, J., Chisholm, A. A. K., Cohen, P. 1986. *Biochim. Biophys. Acta* 870:408–16
476. Walton, G. M., Spiess, J., Gill, G. N. 1985. *J. Biol. Chem.* 260:4745–50
477. Meggio, F., Donella-Deana, A., Pinna, L. A. 1980. *Biochem. Int.* 1:463–69
478. Meggio, F., Marchiori, F., Borin, G., Chessa, G., Pinna, L. A. 1984. *J. Biol. Chem.* 259:14576–79
479. Kuenzel, E. A., Krebs, E. G. 1985. *Proc. Natl. Acad. Sci. USA* 82:737–41
480. Marin, O., Meggio, F., Marchiori, F., Borin, G., Pinna, L. A. 1986. *Eur. J. Biochem.* 160:239–44
481. Orchinnikov, Y. A. 1982. *FEBS Lett.* 148:179–91
482. Kühn, H. 1984. In *Progress in Retinal Research*, ed. N. Osborne, J. Chader, 3:123–56. Oxford: Pergamon
483. Wilden, U., Kuhn, H. 1982. *Biochemistry* 21:3014–22
484. Aton, B. R., Litman, B. J., Jackson, M. L. 1984. *Biochemistry* 23:1737–41
485. Thompson, P., Findlay, J. B. C. 1984. *Biochem. J.* 220:773–80
486. Rich, D. H., Gross, E., eds. 1981. *Peptides: Synthesis, Structure, Function.* Rockford, Ill: Pierce Chemical Co. 853 pp.
487. Deleted in proof
488. Sitaramayya, A., Liebman, P. A. 1983. *J. Biol. Chem.* 253:1205–209
489. Aton, B., Litman, B. J. 1984. *Exp. Eye Res.* 38:547–59
490. Arshavsky, V. Y., Dizhoor, A. M., Shestakova, I. K., Philippov, P. P. 1985. *FEBS Lett.* 181:264–66
491. Kuhn, H., Hall, S. W., Wilden, U. 1956. *FEBS Lett.* 176:473–78
492. Wilden, U., Hall, S. W., Kuhn, H. 1986. *Proc. Natl. Acad. Sci. USA* 83:1174–78
493. Kuhn, H. 1978. *Biochemistry* 17:4389–95
494. Shichi, H., Somers, R. L. 1978. *J. Biol. Chem.* 253:7040–46
495. Lee, R. H., Brown, B. M., Lolley, R. N. 1981. *Biochemistry* 20:7532–38
496. Lee, R. H., Brown, B. M., Lolley, R. N. 1982. *Biochemistry* 21:3303–307
497. Kuhn, H., Dreyer, W. J. 1972. *FEBS Lett.* 20:1–6
498. Bownds, D., Dawes, J., Miller, J., Stahlman, M. 1972. *Nature New Biol.* 237:125–27
499. Kuhn, H., Mommertz, O., Hargrave, P. A. 1982. *Biochem. Biophys. Acta* 679:95–100
500. Swarup, G., Garbers, D. L. 1983. *Biochemistry* 22:1102–106
501. Miller, J. L., Dratz, E. A. 1984. *Vision Res.* 24:1509–21

502. Somers, R. L., Klein, D. C. 1984. *Science* 226:182–184
503. Benovic, J. L., Strasser, R. H., Caron, M. G., Lefkowitz, R. J. 1986. *Proc. Natl. Acad. Sci. USA* 83:2797–801
504. Strasser, R. H., Benovic, J. L., Caron, M. G., Lefkowitz, R. J. 1986. *Proc. Natl. Acad. Sci. USA* 83:6362–66
505. Dixon, R. A. F., Kobilk, B. K., Strader, D. J., Benovic, J. L., Dohlman, H. G., et al. 1986. *Nature* 321:75–79
506. Kennelly, P. J., Rodwell, V. W. 1985. *J. Lipid Res.* 26:903–14
507. Gibson, D. M., Parker, R. A. 1987. *The Enzymes* 18:180–217
508. Parker, R. A., Miller, S. J., Gibson, D. M. 1984. *Biochem. Biophys. Res. Commun.* 125:629–35
509. Marrero, P. F., Haro, D., Hegardt, F. G. 1986. *FEBS Lett.* 197:183–86
510. Beg, Z. H., Stonik, J. A., Brewer, H. B. Jr. 1985. *J. Biol. Chem.* 260:1682–87
511. Beg, Z. H., Stonik, J. A., Brewer, H. B. Jr. 1987. *Metabolism.* In press
512. Beg, Z. H., Stonik, J. A., Brewer, H. B. Jr. 1979. *Proc. Natl. Acad. Sci.* 76:4375–79
513. Ferrer, A., Hegardt, F. G. 1984. *Arch. Biochem. Biophys.* 230:227–37
514. Keith, M. L., Kennelly, P. J., Rodwell, V. W. 1983. *J. Protein Chem.* 2:209–20
515. Font, E., Sitges, M., Hegardt, F. G. 1982. *Biochem. Biophys. Res. Commun.* 105:705–10
516. Ingebritsen, T. S., Parker, R. A., Gibson, D. M. 1981. *J. Biol. Chem.* 256:1138–44
517. Brown, M. S., Brunsdede, G. Y., Goldstein, J. L. 1975. *J. Biol. Chem.* 250:2502–509
518. Harwood, H. J. Jr., Brandt, K. G., Rodwell, V. W. 1984. *J. Biol. Chem.* 259:2810–15
519. Ferrer, A., Caelles, C., Massot, N., Hegardt, F. G. 1985. *Biochem. Biophys. Res. Commun.* 132:497–504
520. Beg, Z. H., Stonik, J. A., Brewer, H. B. Jr. 1984. *Proc. Natl. Acad. Sci. USA* 81:7293–97
521. Matthews, H. R., Bradbury, E. M. 1982. In *Cell Biology of Physarum and Didymium,* 1:317–69. New York: Academic
522. Lake, R. S., Salzman, N. P. 1972. *Biochemistry* 11:4817–26
523. Matthews, H. R., Huebner, V. D. 1984. *Mol. Cell. Biochem.* 59:81–99
524. Langan, T. A. 1978. *Methods Cell Biol.* 19:127–42
525. Dixon, G. H., Candido, E. P. M., Honda, B. M., Louie, A. J., MacLeod, A. R., Sung, M. T. 1975. *Ciba Found. Symp.* NS 28:229–58
526. Quirin-Stricker, C. 1983. In *Posttranslational Covalent Modifications of Proteins,* pp. 125–42. New York: Academic
527. Lake, R. S. 1973. *J. Cell Biol.* 58:317–31
528. Schlepper, J., Kippers, R. 1975. *Eur. J. Biochem.* 60:209–20
529. Quirin-Stricker, C., Schmitt, M. 1981. *Eur. J. Biochem.* 118:165–72
530. Chambers, T. C., Langan, T. A., Matthews, H. R., Bradbury, E. M. 1983. *Biochemistry* 22:30–37
531. Quirin-Stricker, C. 1984. *Eur. J. Biochem.* 142:317–22
532. Woodford, T. A., Pardee, A. B. 1986. *J. Biol. Chem.* 261:4699–76
533. Mitchelson, K., Chambers, T., Bradbury, E. M., Matthews, H. R. 1978. *FEBS. Lett.* 2:339–42
534. Zeilig, C. E., Langan, T. A. 1980. *Biochem. Biophys. Res. Commun.* 95:1372–79
535. Embi, N., Rylatt, D. B., Cohen, P. 1980. *Eur. J. Biochem.* 107:519–27
536. Goris, J., Defreyn, G., Merlevede, W. 1979. *FEBS Lett.* 99:279–82
537. Vandenheede, J. R., Yang, S.-D., Goris, J., Merlevede, W. 1980. *J. Biol. Chem.* 255:11768–74
538. Hemmings, B. A., Yellowlees, D., Kernohan, J. C., Cohen, P. 1981. *Eur. J. Biochem.* 119:443–51
539. Yang, S.-D., Vandenheede, J. R., Merlevede, W. 1981. *J. Biol. Chem.* 256:10231–34
540. Hemmings, B. A., Resink, T. J., Cohen, P. 1982. *FEBS Lett.* 150:319–24
541. Hemmings, B. A., Aitken, A., Cohen, P., Rymond, M., Hofmann, F. 1982. *Eur. J. Biochem.* 127:473–81
542. Woodgett, J. R., Cohen, P. 1984. *Biochim. Biophys. Acta* 788:339–47
543a. Picton, C., Woodgett, J., Hemmings, B., Cohen, P. 1982. *FEBS Lett.* 150:191–96
543b. DePaoli-Roach, A. A., Ahmad, Z., Camici, M., Lawrence, J. C. Jr., Roach, P. J. 1983. *J. Biol. Chem.* 258:10702–709
544. DePaoli-Roach, A. A. 1984. *J. Biol. Chem.* 259:12144–52
545. Cohen, P., Yellowlees, D., Aitken, A., Donella-Deana, A., Hemmings, B. A., Parker, P. J. 1982. *Eur. J. Biochem.* 124:21–35
546. Cohen, P. 1986. *The Enzymes* 17:461–97
547. Leader, D. P. 1980. In *Molecular Aspects of Cellular Regulation,* ed. P. Cohen, 1:203–33. Amsterdam: Elsevier
548. Deleted in proof

549. Novak-Hofer, I., Thomas, G. 1984. *J. Biol. Chem.* 259:5995–6000
550. Lawen, A., Martini, O. H. W. 1985. *FEBS Lett.* 185:272–76
551. Tabarini, D., Heinrich, J., Rosen, O. 1985. *Proc. Natl. Acad. Sci. USA* 82:4369–73
552. Blenis, J., Erikson, R. L. 1985. *Proc. Natl. Acad. Sci. USA* 82:7621–25
553. Nemenoff, R. A., Gunsalus, J. R., Avruch, J. 1986. *Arch. Biochem. Biophys.* 245:196–203
554. Carracosa, J. M., Wieland, O. H. 1986. *FEBS Lett.* 201:81–86
555. Pelech, S., Olwin, B. B., Krebs, E. G. 1986. *Proc. Natl. Acad. Sci. USA* 83:5968–72
556. Parker, P. J., Katan, M., Waterfield, M. D., Leader, D. P. 1985. *Eur. J. Biochem.* 148:579–86
557. Padel, U., Soling, H.-D. 1985. *Eur. J. Biochem.* 151:1–10
558. Blenis, J., Erikson, R. L. 1986. *Proc. Natl. Acad. Sci. USA* 83:1733–37
559. Gonzatti-Haces, M. I., Traugh, J. A. 1986. *J. Biol. Chem.* 261:15266–72
560. Deleted in proof
561. Blackshear, P. J., Witters, L. A., Girard, P. R., Kuo, J. F., Quamo, S. N. 1985. *J. Biol. Chem.* 260:13304–15
562. Perisic, O., Traugh, J. A. 1985. *FEBS Lett.* 183:215–18
563. Traugh, J. A., Pendergast, A. M. 1986. *Nucleic Acids Res.* 33:195–230
564. Donahue, M. J., Masaracchia, R. A. 1984. *J. Biol. Chem.* 259:434–40
565. Erikson, E., Maller, J. L. 1986. *J. Biol. Chem.* 261:350–55
566. Martin-Perez, J., Rudkin, B. B., Siegmann, M., Thomas, G. 1986. *EMBO J.* 5:725–31
567. Wettenhall, R. E. H., Morgan, F. J. 1984. *J. Biol. Chem.* 259:2084–91
568. Gabrielli, B., Wettenhall, R. E. H., Kemp, B. E., Quinn, M., Bizonova, L. 1984. *FEBS Lett.* 175:219–26
569. Stefanovic, D., Erikson, E., Pike, L. J., Maller, J. L. 1986. *EMBO J.* 5:157–60
570. Nimmo, H. G. 1984. In *Molecular Aspects of Cellular Regulation,* ed. P. Cohen, 3:123–41. Amsterdam/New York/Oxford: Elsevier
571. LaPorte, D. C., Thorsness, P. E., Koshland, D. E. Jr. 1985. *J. Biol. Chem.* 260:10563–68
572. Garnak, M., Reeves, H. C. 1979. *Science* 203:1111–12
573. Wang, J. Y. J., Koshland, D. E. Jr. 1982. *Arch. Biochem. Biophys.* 218:59–67
574. Malloy, P. J., Reeves, H. C., Spiess, J. 1984. *Curr. Microbiol.* 11:37–42
575. LaPorte, D. C., Koshland, D. E. Jr. 1983. *Nature* 305:286–90
576. Nimmo, G. A., Borthwick, A. C., Holms, W. H., Nimmo, H. G. 1984. *Eur. J. Biochem.* 141:401–8
577. LaPorte, D. C., Koshland, D. E. Jr. 1982. *Science* 300:458–60
578. LaPorte, D. C., Chung, T. 1985. *J. Biol. Chem.* 260:15291–97
579. Nimmo, G. A., Nimmo, H. G. 1984. *Eur. J. Biochem.* 141:409–14
580. Ranjeva, R., Boudet, A. M. 1987. *Ann. Rev. Plant Physiol.* 38: In press
581. Yan, T.-F. J., Tao, M. 1982. *J. Biol. Chem.* 257:7037–43
582. Davies, J. R., Polya, G. M. 1983. *Plant Physiol.* 71:489–95
583. Lin, Z.-F., Lucero, H. A., Racker, E. 1982. *J. Biol. Chem.* 257:12153–12156
584. Lucero, H. A., Lin, Z.-F., Racker, E. 1982. 257:12157–60
585. Coughlan, S. J., Hind, G. 1986. *J. Biol. Chem.* 261:11378–85
586. Coughlan, S. J., Hind, G. 1986. *J. Biol. Chem.* 261:14062–68
587. Bennett, J. 1983. *Biochem. J.* 212:1–13
588. Staehelin, L. A., Arntzen, C. J. 1983. *J. Cell Biol.* 97:1327–37
589. Horton, P. 1983. *FEBS Lett.* 152:47–52
590. Allen, J. F. 1983. *CRC Crit. Rev. Plant Sci.* 1:1–12
591. Barber, J. 1986. *Photosyn. Res.* In press
592. Bennett, J. 1980. *Eur. J. Biochem.* 104:85–89
593. Mullet, J. E. 1983. *J. Biol. Chem.* 258:9941–48
594. Black, M. T., Foyer, C. H., Horton, P. 1984. *Biochim. Biophys. Acta* 767:557–62
595. Clark, R. D., Hind, G., Bennett, J. 1985. In *Molecular Biology of the Photosynthetic Apparatus,* ed. K. E. Steinback, S. Bonitz, C. J. Arntzen, L. Bogorad, pp. 259–67. New York: Cold Spring Harbor Lab.
596. Ramakrishna, S., Pucci, D. L., Benjamin, W. B. 1983. *J. Biol. Chem.* 258:4950–56
597. Ramakrishna, S., Benjamin, W. B. 1985. *J. Biol. Chem.* 260:12280–86
598. Sheorain, V. S., Ramakrishna, S., Benjamin, W. B., Soderling, T. R. 1985. *J. Biol. Chem.* 260:12287–92
599. Abdel-Ghany, M., Riegler, C., Racker, E. 1984. *Proc. Natl. Acad. Sci. USA* 81:7388–91
600. Toru-Delbauffe, D., Pierre, M. 1983. *FEBS. Lett.* 162:230–34
601. Usami, M., Takahashi, A., Kadota, K. 1984. *Biochim. Biophys. Acta* 798:306–12
602. Geisow, M. J., Burgoyne, R. D. 1984. *FEBS Lett.* 169:127–32

603. Campbell, C., Squicciarini, J., Shia, M., Pilch, P. F., Fine, R. E. 1984. *Biochemistry* 23:4420–26
604. Pauloin, A., Jolles, P. 1984. *Nature* 311:265–67
605. Schook, W. J., Puszkin, S. 1985. *Proc. Natl. Acad. Sci. USA* 82:8039–43
606. Romhanyi, T., Seprodi, J., Antoni, F., Meszaros, G., Farago, A. 1985. *Biochim. Biophys. Acta* 827:144–49
607. Vilgrain, I., Cochet, C., Chambaz, E. M. 1983. *Cancer Res.* 43:386–91
608. Tahara, S. M., Traugh, J. A. 1981. *J. Biol. Chem.* 256:11558–64
609. Leis, J., Johnson, S., Collins, L. S., Traugh, J. A. 1984. *J. Biol. Chem.* 259:7726–32
610. Cote, G. P., Robinson, E. A., Appella, E., Korn, E. D. 1984. *J. Biol. Chem.* 259:12781–87
611. Kuznicki, J., Albanesi, J. P., Cote, G. P., Korn, E. D. 1983. *J. Biol. Chem.* 258:601–14
612. Albanesi, J. P., Fujisaki, H., Korn, E. D. 1984. *J. Biol. Chem.* 259:14184–89
613. Kuczmarski, E. R., Spudich, J. A. 1980. *Proc. Natl. Acad. Sci. USA* 93:2055–223
614. Ogihara, S., Ikebe, M., Takahashi, K., Tonomura, Y. 1983. *J. Biol. Chem.* 260:14374–78
615. Trotter, J. A., Nixon, C. S., Johnson, M. A. 1985. *J. Biol. Chem.* 260:14374–78
616. Barylko, B., Tooth, P., Kendrick-Jones, J. 1986. *Eur. J. Biochem.* 158:271–82

Ann. Rev. Biochem. 1987. 56:615–49
Copyright © 1987 by Annual Reviews Inc. All rights reserved

G PROTEINS: TRANSDUCERS OF RECEPTOR-GENERATED SIGNALS

Alfred G. Gilman

Department of Pharmacology, Southwestern Graduate School, University of Texas Health Science Center, Dallas, Texas

CONTENTS

PERSPECTIVES AND SUMMARY

To be considered in this discussion is a family of guanine nucleotide–binding proteins (G proteins) that serve as membrane-bound transducers of chemically and physically coded information. Knowledge of this family, particularly that acquired over the past 10 years, permits a rather restrictive definition of the characteristics of its closest members. I present such a definition here in the interest of generalization and describe these characteristics in more detail throughout this review. However, we must remain alert for deviations from the "rules" that may be practiced by as yet undiscovered members of the immediate family or by other related proteins.

0066-4154/87/0701-0615$02.00

The G proteins function as intermediaries in transmembrane signaling pathways that consist of three proteins: receptors, G proteins, and effectors. The receptors that participate in such reactions are legion and include those for a large array of biogenic amine, protein, and polypeptide hormones; autacoids; and neurotransmitters. Best characterized of these receptors are those for β-adrenergic agonists (e.g. epinephrine and the more selective agent isoproterenol) and antagonists. Rhodopsin, too, is a G protein–linked receptor, as, apparently, are those for various odorants. The number of effector molecules known to be controlled by G proteins is more modest: interactions of adenylyl cyclase and a retinal cyclic GMP–specific phosphodiesterase with G proteins are rather well understood. Regulation of the activity of a phosphoinositide phosphodiesterase (phospholipase C) and the function of ion channels by G proteins is strongly suspected, but the details remain unknown.

The G proteins are heterotrimers, with subunits designated α, β, and γ in order of decreasing mass. The α subunits clearly differ among the members of the family and, at least for the moment, define the individual. Common β and γ subunits are probably shared among some α subunits to form the specific oligomers.

The functions of G proteins are regulated cyclically by association of GTP with the α subunit, hydrolysis of GTP to GDP and P_i, and dissociation of GDP. Binding of GTP is closely linked with "activation" of the G protein and consequent regulation of the activity of the appropriate effector. Hydrolysis of GTP initiates deactivation. Dissociation of GDP appears to be rate limiting (or, more precisely, occurs as a result of the rate-limiting process), and this step is accelerated by interaction between G protein and receptor. There is considerable (but not conclusive) evidence that a cycle of dissociation and association of G protein subunits is superimposed on this regulatory GTPase cycle.

G proteins share other unique or unusual characteristics. For example, they are activated by fluoride plus aluminum—the actual ligand probably being AlF_4^-. Also distinctive is that the α subunits of individual G proteins are substrates for ADP-ribosylation catalyzed by bacterial toxins. Best characterized are the reactions carried out by toxins elaborated by *Vibrio cholerae* and *Bordetella pertussis*.

Other recent and related reviews include those by Schramm & Selinger (1), Smigel et al (2), Gilman (3), Levitzki (4), and Stryer (5).

HISTORY

Although space does not permit an extensive historical introduction, I will mention a few of the most important observations, particularly of the 1970s, that serve as background for this review.

The involvement of a G protein in transmembrane signaling was first suggested by the requirement for GTP for hormonal activation of adenylyl cyclase (6). Although knowledge of the biochemistry of the enzyme permitted only speculation on the significance of the phenomenon, this fundamental observation by Rodbell, Birnbaumer, and their colleagues set the stage. Perhaps more confusing than illuminating was the simultaneous finding that GTP interfered with detection of hormone (glucagon) binding to receptors responsible for regulation of adenylyl cyclase activity (7). Maguire et al subsequently found that the effect of guanine nucleotides on receptor binding was specific for agonists and that their affinity for the receptor was reduced (8); although the interpretation was still less than obvious, specificity for agonists lent a strong aura of relevance to function. Crucial, then, were the observations of Cassel & Selinger (9–11), who first assayed catecholamine-stimulated GTPase activity in turkey erythrocyte membranes. These experiments were technically demanding, but their quantitation and interpretation have proven to be essentially correct. Thus, G protein–linked systems are activated on binding of GTP; hydrolysis of GTP initiates or is responsible for deactivation; dissociation of GDP is linked with the rate-limiting step and is controlled by receptor. The latter fact is explained by the negative heterotropic binding interaction between receptor and guanine nucleotide, which must be reciprocal. Schramm's demonstrations that components of the adenylyl cyclase system could be mixed and exchanged by cell fusion (12) presaged their reconstitution in vitro (13–16). The assays that evolved permitted purification of G proteins that are associated with the enzyme (17). The capacity of certain bacterial toxins to ADP-ribosylate specific G proteins was discovered first for cholera toxin (18–20) and subsequently for pertussis toxin (21, 22) and proved to be extraordinarily useful. In the meantime, affinity chromatographic techniques greatly facilitated purification of labile and low abundance molecules such as the β-adrenergic receptor (23) and adenylyl cyclase (24). Identification of analogous systems in the retina (25) and realization of the fact that the basic rules had been worked out before with elongation factor Tu (26) bring us to the present.

INDIVIDUAL G PROTEINS: FUNCTIONS AND MOLECULAR ENTITIES

A chicken and egg problem presents itself in attempts to describe the functions and structure of individual G proteins, since, at the moment, there exist proteins (and "deduced proteins") whose functions are unknown and functions to which a specific molecule has yet to be assigned. I start with a description of those functions that have been implicated as part of the repertoire. Confusion will hopefully be minimized by reference to Table 1.

Table 1 Properties of G protein subunits

Subunit	M_r (kd)	Toxin[a]	Receptor[b]	Effector[c]	Specific peptide[d]
$G_{s\alpha}$	46[e]	C-Arg 201[f]			GEEDPQAARSNSDG
			$\beta>>\alpha$,Rho	AC(+)	KQLQKDKQVYRATHR[g]
$G_{s\alpha}$	44.5	C-Arg 187[f]			TPEPGEDPRVTRAKY[g]
$G_{i\alpha 1}$	40.4	P-Cys 351[f]	M,α,Rho>β	AC(−) Others?	
$G_{i\alpha 2}$	40.5	P-Cys 352[h]	—	—	SKFEDLNKRKDT[i]
$G_{o\alpha}$	39[j]	P-Cys 331[f]	M,α,Rho	—	NLKEDGISAAKDVK
$G_{t\alpha 1}$	40	P-Cys 347 C-Arg 174	Rho>α>>β	PDE	SDLERLVTPGYVPT[k]
$G_{t\alpha 2}$	40.4	P-Cys 351[h] C-Arg 178[h]	—	PDE (?)	LDRITAPDYLPN[k]
$G_\beta{}^l$	37.4				EGNVRVSRELAGHTGY
$G_\gamma{}^l$	8.4				

[a] C = Cholera, P = Pertussis
[b] β = β adrenergic, α = α adrenergic, Rho = rhodopsin, M = muscarinic cholinergic
[c] AC = adenylyl cyclase, PDE = retinal cyclic GMP phosphodiesterase
[d] Peptides utilized successfully for production of specific antisera. Antisera that recognize $G_{s\alpha}$, $G_{i\alpha}$, $G_{o\alpha}$, and $G_{t\alpha}$ have been produced with two peptides, GAGESGKSTIVKQM and HMFDVGGQRDERRK.
[e] Apparent M_r by SDS-PAGE, ~52 kd
[f] Deduced site of ADP-ribosylation. Full $G_{o\alpha}$ sequence not known; residue number refers to Figure 1.
[g] For both forms of $G_{s\alpha}$
[h] Assumed substrate for ADP-ribosylation at this site
[i] Synthesized from sequence of $G_{i\alpha 2}$; $G_{i\alpha 1}$ has amino terminal CQ instead of SK. Antiserum recognizes α_{41}.
[j] Full sequence not available; M_r by SDS-PAGE
[k] Specific for one $G_{t\alpha}$ versus the other
[l] Values for $G_{t\beta\gamma}$

Criteria for Involvement of a G Protein

We now fully appreciate the significance of the early experiments that indicated a role for G proteins in transmembrane signaling reactions, and many of them continue to be repeated frequently in the exploration of other systems. It has become possible to define criteria for involvement of a G protein when a new situation is approached, and it will be useful to list these before consideration of proven or potential functions that are regulated by G proteins.

1. An appropriate ligand for the receptor of interest and GTP are both required to initiate the response in question.
2. The response can be provoked independently of receptor by inclusion of nonhydrolyzable analogues of GTP (GTPγS or Gpp[NH]p) or F⁻ plus Al³⁺. It has been possible to introduce nucleotide analogues into intact cells by injection or perfusion (27, 28) or after permeabilization (29). F⁻ and Al³⁺ have occasionally proven useful with intact cell preparations (30, 31).
3. There is a negative heterotropic interaction between the binding of guanine

nucleotide to a G protein and the binding of agonist to a G protein–linked receptor.

4. Cholera toxin and/or pertussis toxin have characteristic effects on the functions of known G proteins, and they can be utilized with either intact cells or purified components.

5. Certain mutants, particularly of the murine S49 lymphoma, have been extraordinarily useful in the definition of some G protein–regulated functions. It is hoped that novel mutants, deficient in the activities of various G proteins, can be developed.

6. Antibodies with differing reactivities for individual G proteins have recently become available.

7. Purification and reconstitution of individual components of a pathway is the ultimate criterion. This has been achieved with the adenylyl cyclase complex and the retinal phosphodiesterase system.

Functions Regulated by G Proteins

ACTIVATION OF ADENYLYL CYCLASE G_s, named as the stimulatory regulator of adenylyl cyclase, is required for observation of significant levels of adenylyl cyclase activity under physiological conditions (15). The protein was recognized as a discrete entity following its partial resolution from adenylyl cyclase by affinity chromatography on GTP-Sepharose (16) and its functional reconstitution into plasma membranes prepared from an S49 cell mutant (cyc^-) that has subsequently been proven to be devoid of $G_{s\alpha}$ (13–15). The latter assay permitted purification of the protein from several sources (17, 32–35). Mechanisms of regulation of adenylyl cyclase activity and structural properties of G_s are discussed below.

INHIBITION OF ADENYLYL CYCLASE   GTP is also required for receptor-mediated inhibition of adenylyl cyclase. Although a considerable amount of indirect evidence suggested the existence of a distinct G protein to account for this fact, isolation of the species was facilitated greatly by the fruits of an independent line of investigation—studies on the mechanism of action of a toxin from *B. pertussis*. This toxin had been found to abolish hormonal inhibition of adenylyl cyclase and, in some cases, to potentiate stimulation of the enzyme (36, 37). These effects appeared to result from ADP-ribosylation of a 41-kd membrane protein, a reaction catalyzed by the toxin (21, 22). Purification of the substrate for pertussis toxin revealed a guanine nucleotide–binding protein with an obvious resemblance to G_s and G_t (38–41). It was possible to inactivate receptor-mediated inhibition of adenylyl cyclase by treatment of platelet membranes with toxin and, subsequently, to restore hormonal inhibition by reconstitution of membranes with the purified toxin substrate (42, 43). This protein (M_r of α subunit $= 41,000$) has thus been termed G_i (i $=$ inhibitory for adenylyl cyclase).

STIMULATION OF RETINAL CYCLIC GMP PHOSPHODIESTERASE   Light activates a cyclic GMP–specific phosphodiesterase in retinal rod outer segments (44, 45). The observation of a light-activated GTPase activity in the retina (46) and a guanine nucleotide requirement for activation of the phosphodiesterase (47, 48) led to purification of transducin (G_t), another member of the G protein family (25, 49, 50). It thus became clear that the flow of information was from light to rhodopsin, G_t, and the phosphodiesterase in sequence. The concentration of cyclic GMP in retinal rods is a crucial determinant of visual excitation (see 5 for review).

STIMULATION OF PHOSPHOINOSITIDE HYDROLYSIS   Many hormones mobilize intracellular stores of Ca^{2+} by virtue of their ability to stimulate the phosphodiesteratic cleavage of phosphatidylinositol-4,5-diphosphate (PIP_2) to inositol-1,4,5-triphosphate (IP_3) (51). The relevant phosphodiesterase (phospholipase C) is influenced by guanine nucleotides. Thus, GTPγS or Gpp(NH)p (but not GDP, ATP, or ATPγS) stimulate the hydrolysis of polyphosphoinositides by neutrophil membranes (52), and GTP, Gpp(NH)p, or GTPγS is largely required for stimulation of inositol phosphate synthesis by blowfly salivary gland membranes in response to 5-hydroxytryptamine (53). Similar results have been obtained with plasma membrane preparations from rat hepatocytes (54) and human polymorphonuclear leukocytes (55, 56). In these latter two systems the combination of hormone and guanine nucleotide lowered the concentration of Ca^{2+} required to support enzymatic activity to the physiological range. In further support of the notion that a G protein controls phosphoinositide phosphodiesterase activity are the observations that the affinities of agonists for several receptors that stimulate IP_3 synthesis are reduced in the presence of guanine nucleotides (e.g. 57). F^- and Al^{3+} are also able to stimulate the hydrolysis of PIP_2 (31). Beyond this generally consistent set of observations, the situation becomes less clear cut.

    There are clearly discrepant observations on the effects of pertussis toxin on receptor-stimulated synthesis of IP_3. For example, the toxin blocks this effect in polymorphonuclear leukocytes (58), in membranes derived therefrom (55), and in mast cells (59). However, it fails to alter the response in hepatocytes (54), cardiac myocytes (60), astrocytoma cells (60), and fibroblasts (61). These observations suggest participation by different G proteins in different cell types or differential modification of the same G protein. G_i and G_o (see below) are the predominant substrates for pertussis toxin. If G_i or G_o is involved in the response of leukocytes and mast cells, it should be possible to stimulate IP_3 synthesis by addition of the activated (GTPγS-bound) G protein to appropriately prepared membranes; this result has not been reported. The conclusion is that a G protein is likely involved, but its identity is unknown.

Introduction of nonhydrolyzable guanine nucleotide analogues into permeabilized mast cells permits exocytotic secretion in response to addition of extracellular Ca^{2+} (29). Thus, it has been suggested that a G protein may regulate plasma membrane Ca^{2+} channels. However, it is difficult to decide if this is a relatively direct or an indirect response. Treatment of mast cells with pertussis toxin blocks the array of effects that are seen in response to compound 48/80 (a polymeric releaser of histamine), including breakdown of PIP_2, accumulation of inositol polyphosphates, ^{45}Ca influx, generation of arachidonate, and histamine secretion (59). Similar observations have been made with neutrophils (30, 62, 63). The mast cell inositol response to compound 48/80 is not dependent on extracellular calcium, and the Ca^{2+} ionophore A23187 fails to stimulate PIP_2 breakdown. These facts and the observation of guanine nucleotide–mediated IP_3 synthesis in membrane preparations indicate that Ca^{2+} influx does not explain pertussis toxin–sensitive PIP_2 breakdown in neutrophils and mast cells. Arachidonate release and histamine secretion, on the other hand, are largely dependent on extracellular Ca^{2+} and can be evoked by A23187. It is possible that inositol polyphosphates other than IP_3 (e.g. inositol, 1,2,3,4-tetrakisphosphate [IP_4]) may facilitate influx of extracellular Ca^{2+} (64). IP_4 appears to arise by phosphorylation of IP_3. Thus, the scheme might be ordered as follows:

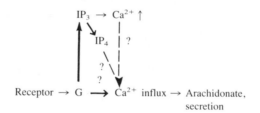

Activation of phosphoinositide breakdown may account for other receptor-mediated effects that likely involve G proteins. Interaction of agonists with a subset of muscarinic receptors of 1321N1 astrocytoma cells (65, 66) or with α_1-adrenergic receptors of rat ventricular myocytes (67) leads to attenuation of cyclic AMP accumulation by stimulation of cyclic nucleotide phosphodiesterase activity. This response is not sensitive to pertussis toxin, is accompanied by hydrolysis of PIP_2, and (at least in the case of astrocytoma cells) is dependent on extracellular Ca^{2+}. It seems most reasonable for the present to assume that the effect on the cyclic nucleotide phosphodiesterase is mediated indirectly by Ca^{2+}.

A pertussis toxin substrate also appears to be involved in reduction of intracellular Ca^{2+} concentrations (68). Somatostatin inhibits K^+-induced prolactin secretion by GH_4C_1 cells and lowers intracellular $[Ca^{2+}]$. These effects are not dependent on cyclic AMP. Pertussis toxin blocks this effect of somatostatin. It is possible that a G protein may be negatively linked to generation of inositol polyphosphates (see 69).

REGULATION OF ION CHANNELS   A few recent reports lend more credence to the exciting possibility that G proteins may exert direct control over the function of ion channels. Pfaffinger et al (27) measured ionic currents in single atrial cells with a whole-cell voltage-clamp technique that permits equilibration of the cytoplasm with a solution of choice. Muscarinic agonists activate an inward rectifying K^+ channel in this preparation in a cyclic nucleotide–independent manner. Observation of the response to acetylcholine required perfusion of the cells with a GTP-containing solution (ATP was also present) and was blocked by prior treatment of cells with pertussis toxin. Breitwieser & Szabo (28) found that this channel could also be activated irreversibly by exposure to acetylcholine after intracellular injection of Gpp(NH)p. This effect was not overcome by addition of isoproterenol, which should stimulate cyclic AMP accumulation. These experiments appear to rule out cyclic nucleotides as mediators of the response. They do not prove direct interaction between G protein and channel. In particular, channel regulation by G protein–mediated alteration of the activity of a protein kinase or a phosphoprotein phosphatase remain as possibilities. Heart is known to contain both G_i and G_o (70); both are ADP-ribosylated by pertussis toxin; either can interact with muscarinic cholinergic receptors in reconstituted systems (71, 72). G_i or G_o has also been suggested as a mediator of neurotransmitter-induced inhibition of voltage-dependent Ca^{2+} channels in chick dorsal root ganglion cells (73).

Mg^{2+} uptake by S49 lymphoma cells is inhibited by β-adrenergic agonists. This response, which is not mediated by cyclic AMP, is absent in cyc^- or UNC S49 cell mutants (which lack $G_{s\alpha}$ or have an altered $G_{s\alpha}$ that cannot interact with receptor, respectively) (74). The implication is that the response requires the β-adrenergic receptor and G_s but not cyclic AMP and, therefore, perhaps not the only effector with which G_s is known to interact, adenylyl cyclase. Beyond these facts, the mechanism of this interesting effect is unknown.

Molecular Entities and Structure

A detailed view of certain aspects of the structures of G proteins is beginning to emerge, thanks in particular to cDNA cloning and sequencing and to the solution of the crystal structure of a related guanine nucleotide–binding protein, the bacterial elongation factor Tu (EF-Tu) (75). Molecular cloning has revealed the primary structures of nearly all of the G proteins that have been purified and has led to an appreciation of at least two additional entities (referred to below as $G_{i\alpha 2}$ and $G_{t\alpha 2}$). To date, however, this approach has not resulted in the discovery of a myriad of novel structures.

$G_{s\alpha}$ G_s, first defined functionally by its ability to activate adenylyl cyclase, was found on purification to be a mixture of two oligomers with differing α subunits (apparent M_r on SDS-polyacrylamide gels, 52,000 and 45,000) and indistinguishable β and γ subunits (17, 32, 76). The relative concentration of the two forms of $G_{s\alpha}$ varies among cells and tissues; functional differences are not yet appreciated.

cDNAs corresponding to $G_{s\alpha}$ have been cloned from bovine brain (77, 78), bovine adrenal (79), and rat brain (80) (Figure 1). The first cDNA was obtained by hybridization with an oligonucleotide probe based on protein sequence obtained from a highly conserved region of $G_{o\alpha}$ and $G_{t\alpha}$ (81). It was identified as $G_{s\alpha}$ by immunoblotting with antibodies generated to peptides synthesized according to sequence deduced from the cDNA (77). This identification was confirmed by failure to find mRNA in the cyc$^-$ ($G_{s\alpha}$-deficient) S49 cell mutant that would hybridize with the cDNA clone (77) and by expression of the cDNA (79, 82). The amino acid sequences that are revealed by the bovine and rat cDNAs differ in only three residues.

The first cDNA for $G_{s\alpha}$ that was isolated encodes a protein of 394 residues and, therefore, an apparent M_r of 46,000. However, upon transient expression in COS-m6 cells, this cDNA was found to direct the synthesis of the 52-kd form of $G_{s\alpha}$ (79). The same result was obtained by expression in *Escherichia coli*, using prokaryotic expression vectors containing either the tac or T7 promoters (M. Graziano, unpublished observation). The implication is that the M_r of the larger form of $G_{s\alpha}$ is actually 46,000 and that its electrophoretic behavior in SDS is anomalous. An alternative cDNA for $G_{s\alpha}$ has also been detected (82). It differs from the first in only 46 contiguous nucleotides, resulting in the alteration of two and the deletion of 14 amino acid residues (residues 73–86 in the larger form). It appears to encode a 44.5-kd protein and directs the synthesis of the 45-kd form of $G_{s\alpha}$ in COS-m6 cells or *E. coli*. Messenger RNA corresponding to each cDNA is detectable by S1 nuclease analysis. In view of their otherwise identical sequence, the two mRNAs are presumed to arise from a single gene for $G_{s\alpha}$ by alternative splicing of internal exons.

$G_{i\alpha}$ First visualized as a 41-kd substrate for ADP-ribosylation by pertussis toxin, oligomeric G_i was purified from rabbit liver (38, 41) and human erythrocytes (35, 39) by techniques nearly identical to those developed for G_s. Although the α subunit of G_i was clearly distinguishable from those of G_s and

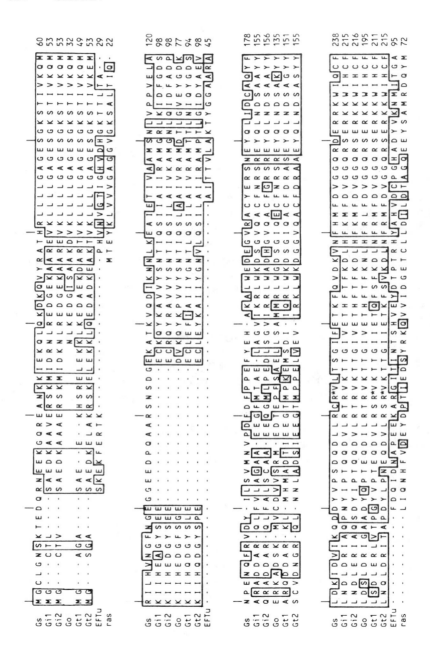

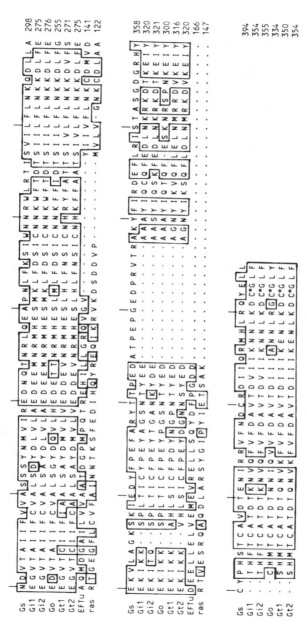

Figure 1 Amino acid sequences of G protein α subunits and the presumptive GTP binding regions of EF-Tu and c-Ha-ras. $G_{s\alpha}$, $G_{i\alpha}1$, $G_{i\alpha}1$, and $G_{i\alpha}2$ are bovine. $G_{i\alpha}2$ and $G_{o\alpha}$ are rat. The amino-terminal sequence of $G_{o\alpha}$ has not yet been determined; the first 11 residues shown are from the bovine protein. Presumed and proven sites of ADP-ribosylation are indicated by*. Boxes surround residues that are identical or conserved among any four of the six G_α subunits and residues in EF-Tu and c-Ha-ras that are identical or conserved with corresponding boxed residues of G_α subunits. Conserved residues are defined as follows: C; S, T, P, A, G; N, D, E, Q; H, R, K; M, I, L, V; F, Y, W.

G_t, the β subunit was apparently identical (40). The small γ subunit, not detected initially because of its poor staining qualities, was detected soon thereafter (41, 76). The functional attributes ascribable to G_i were deduced by reconstitution of the rabbit liver protein into platelet and S49 cell membranes (42, 43, 83). A protein with superficially indistinguishable features has since been purified from bovine (84, 85) and rat (86) brain and is frequently termed G_i. However, there is heterogeneity of substrates for pertussis toxin, which became grossly obvious when brain was studied. Furthermore, inhibition of adenylyl cyclase may not be the exclusive property of "G_i." Caution in nomenclature is thus mandated. (The terminology α_{Mr} serves this purpose.)

Nukada et al (87) purified $G_{i\alpha}$ (α_{41}) from bovine brain and determined the amino acid sequence of several of its tryptic peptides. These sequences are represented faithfully in that deduced from a cDNA clone isolated from a bovine brain cDNA library. The cDNA encodes a protein with 354 amino acid residues and a calculated molecular weight of 40,400. Northern analysis reveals an RNA with approximately 3900 nucleotides.

Itoh et al (80) screened a rat C6 glioma cDNA library with an oligonucleotide probe based on amino acid sequence data obtained with purified rat brain $G_{i\alpha}$ (α_{41}). With the exception of two residues, amino acid sequences deduced from one of the cDNA clones isolated by these investigators matched those determined for seven tryptic peptides derived from the protein. However, the entire sequence of 355 amino acid residues deduced by Itoh et al (80) differs significantly ($\sim 11\%$) from that of Nukada and coworkers (87). Is this difference due to species? The extreme similarity between bovine and rat $G_{s\alpha}$ suggests not, but this argument is hardly definitive. More interesting is the fact that the rat protein sequence obtained by Itoh et al differs from Nukada and associates' bovine cDNA sequence by only three residues in 78. It is suggested that the predominant α_{41} ($G_{i\alpha}$) corresponds to that purified from bovine and rat brain by these two groups and that its sequence is represented by the cDNA of Nukada et al (87). Itoh et al are presumed to have cloned a similar cDNA, but one that encodes a distinct entity. For the moment, we refer to these proteins as $G_{i\alpha 1}$ (Nukada et al) and $G_{i\alpha 2}$ (Itoh et al) or $\alpha_{41,1}$ and $\alpha_{41,2}$.

$G_{o\alpha}$ Sternweis & Robishaw encountered surprising [^{35}S]GTPγS binding activities (10-fold greater than anticipated) during initial attempts to purify G_i from brain. Protein fractionation, ADP-ribosylation with pertussis toxin, and electrophoresis revealed the explanation. Brain contains a plentiful substrate for pertussis toxin in addition to G_i (84). This protein proved to be an obvious member of the G protein family, since, in addition to serving as a substrate for pertussis toxin, it has a guanine nucleotide–binding α subunit ($M_r = 39,000$) and β and γ subunits that are apparently identical to those of G_s and G_i. This new G protein was dubbed G_o (o = other G protein), and evidence was presented that it was not a proteolytic product of the larger G_i (also abundant in brain). Neer and associates detected similar heterogeneity of substrates for pertussis toxin in brain at essentially the same time (85). In addition to α_{41} ($G_{i\alpha}$) and α_{39} ($G_{o\alpha}$), these investigators also noted a 40-kd toxin substrate in their purified preparations. The question of possible proteolytic origin of α_{40} and α_{39} was not settled in this report, and this remains an issue for α_{40}. However, it is possible that α_{40} corresponds to $G_{i\alpha2}$, as defined above.

Although the function of G_o remains to be determined, its discovery has had a major impact. Since bidirectional regulation of adenylyl cyclase activity had presumably been settled with the discovery of G_s and G_i, the existence of another G protein in brain implied a broader role for the family. Furthermore, the abundance of G_o (and G_i) in brain has greatly facilitated experimentation on a number of fronts.

When Itoh and coworkers (80) screened their rat C6 glioma cDNA library for $G_{i\alpha}$ as described above, the first clone detected turned out to correspond to $G_{o\alpha}$. This identification was based on perfect agreement of sequence predicted from the cDNA with that obtained from six tryptic peptides derived from purified rat brain $G_{o\alpha}$ (80). The cDNA for $G_{o\alpha}$ described by Itoh et al lacks nucleotides corresponding to the amino terminus of the protein (probably about 30 amino acid residues); 15 of these are available for the bovine protein (81).

$G_{t\alpha}$ G_t or transducin, purified at about the same time as G_s, has been studied extensively. It is a major component of the disks of the retinal rod outer segment; disks are prepared easily from bovine retina, and mg-quantities of G_t can be purified in one or two days after selective elution of the protein from the disk membrane with GTP (in the absence of detergent) (50). The availability of antibodies to $G_{t\alpha}$ and partial amino acid sequence (81) permitted the essentially simultaneous cloning of cDNAs corresponding to $G_{t\alpha}$ in four laboratories (88–91). Perhaps not surprisingly, at least retrospectively, two sequences, which differ in approximately 20% of the encoded amino acid residues, were elucidated ($G_{t\alpha1}$ and $G_{t\alpha2}$). The clones characterized by three of these groups (89–91), which were selected using expression vectors and antibodies to purified "G_t," encode identical sequences of 350 amino acid residues ($M_r = 40,000$). The odd clone out was selected with an oligonucleotide probe and encodes a protein of 354 residues (88). Antibodies to purified "G_t" detect immunoreactivity in retinal rods but not in cones (92); the same is true of an antipeptide antibody when the peptide sequence was chosen to be specific for the 350-residue protein (93). However, antibodies raised against a peptide synthesized according to sequence specific for the 354-residue protein show reactivity exclusively with cone photoreceptor outer segments (93). It is assumed that there exist two isoforms of G_t—one that activates the photosensitive cyclic GMP–specific phosphodiesterase of rod outer segments (G_{t1}) and one that plays the analogous role in cones (G_{t2}).

THE STRUCTURE OF G_a The availability of essentially complete amino acid sequences for seven G protein α subunits obviously invites comparisons (Figure 1). Overall, the relationship is striking. $G_{i\alpha1}$ and $G_{i\alpha2}$ are most alike (approximately 95% of the residues identical or homologous), in keeping with the tentative designation of both of these molecules as G_is. The two G_t α subunits are also very similar (88% identical or homologous). More surprising is the strength of the relationship between $G_{i\alpha}$s, $G_{t\alpha}$s, and $G_{o\alpha}$ (roughly 80% identical or homologous for all of these comparisons). $G_{s\alpha}$ differs the most (about 50%) from the other α subunits. Its larger mass is due to two discrete "inserts" (residues 72–86 and 324–336) and to additional residues at the amino terminus. A less extensive relationship between G protein α subunits, EF-Tu, and the ras oncogene products is also obvious; the regions of greatest similarity form portions of the guanine nucleotide-binding domain of EF-Tu (75, 94).

Variability among the α subunits is concentrated in three "hot spots": the amino terminus [residues 1–40 of α_{avg}, as defined by Masters et al (95)], residues 120–150, and residues 340–360. Most significant differences between the two $G_{i\alpha}$s are in the second of these regions; nonhomologous differences between the $G_{t\alpha}$s are largely confined to the amino terminal 30 residues. When G_s is compared with any of the other α subunits, the variability that is seen at residues 120–150 extends back to (but not beyond) the region where amino acid residues are inserted in the larger form of $G_{s\alpha}$ (residues 72–86). The variable region near the carboxy terminus of α_{avg} is also immediately adjacent to a G_s-specific insert.

Masters et al (95) have made predictions about the secondary structure of α_{avg}, and the agreement with the crystal structure of the GDP-binding domain of EF-Tu (75) was sufficiently good to inspire a gamble (as defined respectably by one of the authors—see 96). Constraint of the four regions of α_{avg} that are believed to contribute to the guanine nucleotide–binding site (75, 94) with the three-dimensional structure determined for this region of EF-Tu divides α_{avg} into three domains. Two—the amino (1–41) and carboxy (298–396) termini—are obviously mandatory; the third (60–208) results from a long insertion between the first two of the four regions involved in guanine nucleotide binding. Each of the three domains contains one of the "hot spots" mentioned above. Regions of greatest homology are focused around the guanine nucleotide–binding site, and variability increases as one moves away from this core in any direction (with the exception of the extreme carboxy terminus). The authors have speculated on functional roles that might be assigned to these domains. I consider these arguments in the context of protein-protein interactions, below.

SUBUNIT-SPECIFIC AND NONSPECIFIC ANTIBODIES   Elucidation of primary sequence has permitted generation of a number of antipeptide antibodies with predetermined specificity for a given α subunit or for all known α subunits (82, 93, 97). These antibodies have in general been useful for immunoblotting, immunohistochemistry, and immunoprecipitations. There is little information on their reactivities with native subunits or oligomers. Sequences of proven utility are listed in Table 1.

β SUBUNITS   Purification of G_s, G_t, and G_i revealed apparently similar 35-kd polypeptides associated with the more distinctive α subunits. The amino acid composition of β prepared from the three oligomers is indistinguishable; the three proteins yield the same electrophoretic pattern of peptides after proteolysis (40, 98). $\beta\gamma$ subunit complexes are functionally interchangeable: for example, $\beta\gamma$ from G_s or G_i appears to interact identically with $G_{s\alpha}$ (42, 99); $\beta\gamma$ from G_i or G_t can interact with $G_{i\alpha}$ or $G_{t\alpha}$ to reconstitute rhodopsin-stimulated GTPase activity (100). Thus arises the issue (not yet resolved) of the identity or nonidentity of β subunits and, if the latter, their total number. It is a particularly pertinent question in the context of possible dissociation (and mixing) of G protein subunits as part of their mechanism of activation (see below).

It is now clear that there is some level of heterogeneity of β. The "35-kd" subunit of G_s, G_i, and G_o can be resolved into a doublet by SDS-polyacrylamide gel electrophoresis (32, 84). The terminology β_{36}/β_{35} has arisen to define this situation. $G_{t\beta}$, by contrast, displays only one component of this doublet. Its electrophoretic mobility corresponds to that of the upper band, but it is not known if $G_{t\beta}$ is identical to β_{36}. β_{36} and β_{35} are also distinguishable immunologically. Polyclonal antisera to purified $G_{t\beta}$ (β_{36}?) or β_{36}/β_{35} react almost exclusively with β_{36} (97, 101). Antipeptide antibodies prepared against a sequence common to $G_{t\beta}$ and a mixture of β_{36}/β_{35} have great preference for β_{36} (97). This situation is confusing. It is possible that β has very few strong antigenic determinants and that crucial sites may be altered between β_{36} and β_{35}. It is difficult to believe that β_{36} and β_{35} are grossly different.

Evans et al (102) have recently characterized a form of β_{35} from human placental membranes. The protein was resolved (by DEAE) from oligomeric G proteins, wherein the β subunit has the typical β_{36}/β_{35} doublet structure. β_{35} and the β subunit doublet preparation have similar abilities to inhibit adenylyl cyclase activity, presumably by virtue of interaction with $G_{s\alpha}$ (see below). An apparently identical γ subunit is associated with both β_{35} and the β_{35}/β_{36} doublet.

Sugimoto et al (103) have cloned a cDNA for $G_{t\beta}$. The protein has 340 amino acid residues (M_r = 37,400). Two bands were detected when the cDNA was utilized for Northern hybridization with retinal, brain, and liver RNA (~1.8 and 3.3 kb). Two β subunit clones were also isolated from a bovine brain cDNA library. Restriction mapping with seven endonucleases revealed differences only in the 5' noncoding region. It was concluded that the mRNAs for β have the same coding sequence.

An apparently identical cDNA clone for $G_{t\beta}$ was also isolated by Fong and coworkers (104). The authors noted that the entire sequence of $G_{t\beta}$ consists of a reiterated pattern of about 86 amino acid residues; each of these can be divided into two similar 43-residue segments. In addition, there is a resemblance between $G_{t\beta}$ and the carboxy-terminal portion of the yeast CDC4 gene product. (CDC4 is a cell-division-cycle gene of unknown function.) Northern analysis of several tissues revealed 1.8- and 2.9-kb mRNAs.

Robishaw has isolated a cDNA that includes β-subunit-specific sequences from a bovine adrenal library (unpublished observations). The nucleotide sequence of this clone is quite different from those reported previously (103, 104), and the deduced amino acid sequence also differs significantly. This cDNA hybridizes with a 1.8-kb mRNA. Thus there appear to be at least two genes for β. Their relationship to β_{36}/β_{35} is unknown. Nevertheless, there is as yet no reason to believe that there are differences among the β subunits of G_s, G_i, and G_o.

γ SUBUNITS   Ignorance becomes more obvious with regard to γ. This subunit of G_t was recognized early (25, 50), but its detection as a component of G_s and G_i was delayed because of poor avidity for stain (41, 76). β and γ remain tightly associated under nondenaturing conditions. They dissociate as a complex from G_α in the presence of activating ligands (see below).

cDNAs that encode $G_{t\gamma}$ have been cloned and sequenced (105, 106); the protein has been sequenced as well (107). $G_{t\gamma}$ has 74 amino acid residues (M_r = 8400) and is very hydrophilic and acidic. Two-dimensional peptide mapping of γ subunits from human erythrocyte G_s and G_i and from bovine brain failed to reveal differences; $G_{t\gamma}$ (bovine or frog) could be distinguished from these other polypeptides (98). Antibodies to $G_{t\gamma}$ fail to recognize γ subunits from other sources (101, 108). Thus the situation with regard to β and γ may be similar. The specialized retinal rod may have distinct β and γ subunits, while those G proteins that are coincidentally expressed in essentially all (G_s, G_i) or several (G_o) cells may share a common $\beta\gamma$ complex. However, there are hints of greater complexity. There may be multiple γ subunits in evidence in the brain G protein preparations of Sternweis & Robishaw (84). An antibody that recognizes a human placental G protein γ subunit apparently fails to visualize γ in rabbit liver G_i, bovine brain G_i/G_o, bovine G_t, or human platelet G_i (102).

G_p A novel entity, termed G_p by its discoverers (109), has been described recently. Purified from placenta (and visualized in platelets; thus the designation p), G_p may be a member of the immediate G protein family. There are uncertainties, however, which is why its discussion has been postponed to this point.

Purified preparations of G_p contain a GTPγS-binding polypeptide with an apparent molecular weight of 21,000. They also contain approximately equimolar concentrations of an apparently "conventional" $\beta\gamma$ subunit complex. However, evidence for association of putative $G_{p\alpha}$ with $\beta\gamma$ is not yet at hand. G_p is not an obvious substrate for pertussis or cholera toxin, GTPase activity has not yet been demonstrated, and it is not recognized by antibodies to highly conserved domains of the α subunits described above. (It is also not recognized by anti-ras antibodies.) G_p is of obvious interest; given its size and, perhaps, a low affinity for $\beta\gamma$, it may resemble ras more than do the other signal transducing G proteins. The simultaneous choice of the letter p to designate this entity and that hypothetical G protein responsible for regulation of phospholipase C has generated some confusion. The function of G_p, as defined by Evans et al (109), is unknown.

LIGAND–G PROTEIN INTERACTIONS

Studies of the interactions of G protein α subunits with nucleotides have focused particularly on GTPγS (or other nonhydrolyzable triphosphate analogues), GTP, and GDP. The characteristics of the binding reactions are influenced by Mg^{2+}, anions, and proteins that interact with α (particularly receptors and $\beta\gamma$). Given the existence of at least four purified G proteins available for study, the potential for accumulation of data is large.

Binding of GTPγS to oligomeric G proteins or to their resolved α subunits is clearly not a diffusion-controlled process, and it proceeds at a rate that is independent of nucleotide concentration (41, 84, 110, 111). This anomaly is explained by the fact that the proteins, as purified, contain stoichiometric amounts of GDP, obviously bound with high affinity (49, 111). GDP can be removed from G_i or G_o by chromatography in the presence of 1 M $(NH_4)_2SO_4$ and 20% glycerol (111). The kinetics of GTPγS binding to these nucleotide-free α subunits or oligomers is then bimolecular and apparently diffusion-controlled.

There is negative cooperativity of the binding of GTPγS and $\beta\gamma$ to G protein α subunits; thus, GTPγS promotes G protein subunit dissociation (32, 33, 38, 39, 112).

$$G_{\alpha\beta\gamma} + GTP\gamma S \leftrightarrows G_{\alpha}\cdot GTP\gamma S + G_{\beta\gamma} \qquad\qquad 1.$$

Mg^{2+} shifts the equilibrium for this reaction far to the right. The rate of dissociation of GTPγS from G_{α} is slow, but measurable, in the absence of Mg^{2+} (0.4 min^{-1} for $G_{o\alpha}$; 0.2 min^{-1} for $G_{i\alpha}$). $\beta\gamma$ increases the rate of dissociation of GTPγS by about threefold in the absence of the divalent cation (113).

The effect of Mg^{2+} on the binding of GTPγS is striking; the rate of dissociation of the nucleotide from $G_{o\alpha}$ or $G_{i\alpha}$ is reduced to near zero (113). The apparent K_d for interaction of Mg^{2+} with $G_{\alpha}\cdot GTP\gamma S$ is extremely small—about 5 nM. Low concentrations of Mg^{2+} have a similar capability to slow dissociation of GTPγS from oligomeric G_o or G_i. However, the rate of nucleotide dissociation remains measurable until the concentration of Mg^{2+} exceeds 1 mM. At higher Mg^{2+} concentrations subunit dissociation occurs and, as mentioned, $G_{\alpha}\cdot GTP\gamma S\cdot Mg^{2+}$ is extremely stable. The significance of the subunit dissociation reaction will be discussed further below.

The intrinsic fluorescence of G_{α} is enhanced modestly on binding of GTPγS and more dramatically in the presence of nM concentrations of Mg^{2+} (114). F^-, Al^{3+}, and Mg^{2+} cause a similar effect. This change in fluorescence is presumed to reflect the activated state of the G protein α subunit.

The interactions of GDP with G protein α subunits provide a contrast with those of GTPγS (113). Brandt & Ross first reported the differing effect of $\beta\gamma$ on dissociation of GTPγS and GDP from $G_{s\alpha}$ (115); $\beta\gamma$ inhibits the dissociation of GDP (unless the Mg^{2+} concentration is high; see below). This phenomenon has been studied in more detail with G_o and $G_{o\alpha}$. The affinity of GDP for $G_{o\alpha}$ is high (K_d ~40 nM in the absence of Mg^{2+}) and is increased markedly by $\beta\gamma$ (K_d ~0.1 nM). This effect appears to result from both a substantial increase in rate of association of GDP with the protein (surprisingly) and a decrease in the rate of dissociation. In the presence of 10 mM Mg^{2+} the effect of $\beta\gamma$ on the affinity of $G_{o\alpha}$ for GDP is not as great, but it is still substantial (K_d ~100 nM for $G_{o\alpha}$, 10 nM for oligomeric G_o). The effect of Mg^{2+} is to decrease the rate of association of GDP with $G_{o\alpha}$ or G_o and to increase the rate of dissociation from G_o; however, there is no effect of the metal on the dissociation of GDP from $G_{o\alpha}$ (0.3 min^{-1}). It seems probable that the extreme high affinity of Mg^{2+} for the nucleotide-protein complex noted above is a property only of the GTP- (or GTPγS-) bound form of the protein. Thus Mg^{2+} and GTPγS promote dissociation of oligomeric G proteins and the formation of an "activated" state of G_{α}; GDP stabilizes the oligomer and, at modest concentrations of Mg^{2+}, dissociates from it extremely slowly.

The interaction of G proteins with their physiological activator, GTP, is of course more complex, in that nucleotide hydrolysis is involved. The basal

GTPase activity, which is extremely low, has been evaluated for G_s (115), G_i (86, 116, 117), and G_o (85, 86, 113, 118). I will ignore modest quantitative discrepancies between these studies, particularly because most are uninterpretable. A typical molar turnover number is 0.3 min^{-1}; the K_m for GTP is low (0.3 μM), and nM concentrations of Mg^{2+} satisfy the requirement for divalent cation. GTP increases the intrinsic fluorescence of G_α, apparently in the same manner as does GTPγS (118). Fluorescence intensity declines as the bound nucleotide is hydrolyzed. Rate constants and relative steady-state concentrations of G·GTP and G·GDP can thus be measured by quantitation of intrinsic fluorescence (118) or with radioactive nucleotides (115, 119). During steady-state hydrolysis the great majority of the protein exists as the GDP-bound form, since k_{cat} exceeds k_{off} for GDP (0.3 min^{-1}) by an order of magnitude. The rate of dissociation of GDP thus limits the basal GTPase activity. As mentioned, $\beta\gamma$ inhibits the dissociation of GDP from $G_{o\alpha}$ at low concentrations of Mg^{2+} and thereby inhibits GTPase activity (113). As the concentration of Mg^{2+} is increased, the rate of dissociation of GDP from G_o (but not from $G_{o\alpha}$) increases and can exceed the value observed with $G_{o\alpha}$. Under such conditions $\beta\gamma$ activates the GTPase activity of $G_{o\alpha}$ (113). The concentration of Mg^{2+} required for this effect on the dissociation of GDP from oligomeric G proteins is high and is dependent on the protein in question. Although not studied systematically, one can estimate that the effect occurs in the range of 1–10 mM for G_o, 5–50 mM for G_i, and 10–100 mM for G_s (86, 113, 115).

These effects of Mg^{2+} are complicated further by the counter ion, since high concentrations of Cl^- appear to inhibit GTPase activity directly (120). Other effects of relatively modest (mM) concentrations of Cl^- have also been noted, including the ability to inhibit the rate of dissociation of GTPγS and GTP (but not GDP) from $G_{o\alpha}$ (120). Lubrol inhibits the steady-state rate of GTP hydrolysis by interfering with the dissociation of GDP (115). Variations in concentrations of Cl^- and Lubrol account for some of the quantitative discrepancies that are apparent in the literature.

The anomalous ability of F^- to activate adenylyl cyclase was found to be a result of interaction of the anion with G_s (14, 121), and it has since become clear that there is a characteristic effect of F^- on all G proteins (38, 122). Manifestations of these interactions closely resemble those with nonhydrolyzable guanine nucleotide analogues: G proteins become "activated" (i.e. capable of fruitful interaction with their effector molecules), G_α dissociates from $\beta\gamma$, and there is an enhancement of intrinsic fluorescence, at least of G_o (114). Curiously, Al^{3+} (or Be^{2+}) was found to be required for activation of G_s by F^-, and it was suggested that the activating ligand was AlF_4^- (123). Bigay et al (124) have suggested that AlF_4^- interacts only with the GDP-bound form of G_α, and that the anion mimics the role of the γ-phosphate of GTP. It should be possible to verify this very attractive hypothesis by rigorous demonstration of a requirement for bound GDP for AlF_4^--stimulated interaction between G_α and an appropriate effector.

LIGAND-REGULATED PROTEIN-PROTEIN INTERACTIONS

Characteristics of the interactions of ligands with isolated G proteins have been presented above for the sake of simplicity. Their effects on the protein-protein interactions that characterize transmembrane signaling systems are obviously at the heart of mechanism, and these are now being studied in detail with purified reconstituted systems (Figure 2). It should be noted that crucial features of many of these interactions were deduced correctly by study of impure or intact systems, in some cases even before the components had been unambiguously identified. The most important of these deductions were noted above. Of course, this early phase of research was also characterized by many incorrect mechanistic interpretations.

Receptor–G Protein Interactions

The interaction of receptor with a G protein is driven by an appropriate agonist (hormone, photolyzed retinal, etc). This was implied by the comigration of crude β-adrenergic receptors and G_s after solubilization in the presence of agonist (and absence of guanine nucleotide) (125) and by study of the interactions of G_t and rhodopsin. The interaction between R and G is antagonized by guanine nucleotide, either GTP or GDP.

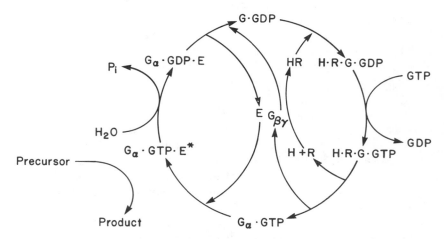

Figure 2 Interactions of receptor, G protein, GTP, and effector. See text for explanation. Modified from Stryer (5).

Pedersen & Ross (126) developed the first successful reconstitution of β-adrenergic receptors and G_s in phospholipid vesicles, and the basic approach has now been utilized extensively for this hormone receptor system and for others. Detailed study of the properties of these interactions has revealed their essential regulatory features (119, 127–132). I will concentrate on the β-adrenergic receptor and G_s in this discussion, since this system has been studied most extensively.

1. HR stimulates dissociation of G·GDP (119). Dissociation of GDP obviously must precede binding of GTP if there is but one site for nucleotide, and only one such site has been detected (however, see 133).
2. HR stimulates nucleotide binding, even when most of the bound GDP has been induced to dissociate by interaction of G·GDP with HR (119). Thus, release of GDP per se is required but is not necessarily sufficient for hormone-stimulated, GTP-mediated activation of G_s. A similar conclusion had been reached by Tolkovsky et al (134), who examined the rate of activation of adenylyl cyclase by Gpp(NH)p and epinephrine in membranes where G_s had hypothetically been cleared of GDP by incubation with hormone. It is perhaps simplest to envision a guanine nucleotide–binding site that is "closed" in the absence of HR and "open" (allowing nucleotide exchange) in its presence.
3. HR stimulates the steady-state GTPase activity of G_s (1–2 min^{-1}) without affecting k_{cat} (4 min^{-1}). This effect is due exclusively to HR-stimulated dissociation of GDP and association of GTP and the resultant accumulation of significant levels of G·GTP.
4. HR functions catalytically (126, 131); one receptor can interact with ~10 molecules of G_s over a period of a few seconds in a single phospholipid vesicle. These observations verify the same conclusion by Tolkovsky & Levitzki (135, 136), who had studied the kinetics of activation of adenylyl cyclase after inactivation of receptor with an irreversible antagonist.
5. HR-stimulated nucleotide exchange requires the $\beta\gamma$ subunit complex of the G protein (131). This observation was made initially with G_t and rhodopsin (137).
6. There are at least two requirements for Mg^{2+} for maximal catecholamine-stimulated GTPase activity: low (nM) concentrations of Mg^{2+} are necessary for nucleotide hydrolysis per se and higher (10 μM) concentrations maximize HR-catalyzed nucleotide exchange (119). This latter requirement is consistent with the initial observation of Iyengar & Birnbaumer (138), who demonstrated that glucagon lowered the concentration of Mg^{2+} necessary for activation of G_s by GTPγS from 25 mM to 10 μM.
7. Reconstitution of R and G results in the establishment of guanine nucleotide–sensitive agonist binding to the receptor (128). Low-affinity agon-

ist states are R and its presumed equivalent, R in the presence of G·GTP or G·GDP (i.e. R and G not associated); the high-affinity state is R·G (nucleotide dissociated). Rojas & Birnbaumer (139) have highlighted the importance of GDP in this negative binding interaction with agonist; participation by GDP seems mandatory. They have also suggested that GTP may not have a similar effect. If true, binding of GTP would apparently not cause dissociation of HR from G, and the active complex of G and effector would then incarcerate HR—drastically reducing its catalytic efficiency.

To summarize, most would agree to the following model. The affinity of HR for G·GDP is sufficient to drive their interaction and to promote dissociation of the nucleotide. HRG is presumably a relatively stable intermediate, but its lifetime is brief in the presence of a normally high concentration of GTP. Binding of GTP causes dissociation of HR (see 133). The lifetime of G·GTP (or G_α·GTP, see below) is many seconds. The catalytic action of HR and the relatively long lifetime of G·GTP provide considerable amplification.

MULTIPLE EFFECTS OF MG^{2+} Effects of Mg^{2+} have been described just above and in the preceding section. It may be useful to summarize these observations and to speculate on their significance. The list of effects and approximate concentrations required is as follows: 1. GTPase, ~5 nM; 2. slow dissociation of GTPγS from G_i or G_o, ~5 nM; 3. fluorescence enhancement of $G_{o\alpha}$, < 100 nM; 4. HR-stimulated G_s activation, ~10 μM; 5. HR-stimulated GTP binding and, by inference, GDP dissociation, ~10 μM; 6. βγ-stimulated GDP dissociation, 1–100 mM; 7. GTPγS-induced subunit dissociation, 1–100 mM.

Once GTP or GTPγS is bound, interaction of Mg^{2+}, presumably with both protein and nucleotide, occurs with extremely high affinity. Effects 1–3 above are all believed to reflect interaction at this site, and this is presumably sufficient to "activate" a resolved G protein α subunit. I speculate that effects 4–7 all reflect interaction of Mg^{2+} at a second site, whose location of G (α, β, or γ) is unknown. The apparent affinity of this site for Mg^{2+} is relatively poor in the absence of HR (1–100 mM). Interaction of Mg^{2+} at this site is necessarry for βγ-facilitated "opening" of the guanine nucleotide–binding site to permit dissociation of GDP, association of GTP or GTPγS, and nucleotide-induced subunit dissociation. HR lowers the concentration requirement for Mg^{2+} at this hypothetical single site; Iyengar & Birnbaumer have stressed the importance of such an interaction (138). Viewed in this context, HR shifts the dependency on Mg^{2+} from a concentration range where βγ stabilizes the binding of GDP to a range where βγ actually facilitates guanine nucleotide exchange.

SPECIFICITY OF R-G INTERACTIONS   The availability of purified G proteins and receptors has permitted tests of specificity of the functional interactions between R and G by reconstitution. Prototypical receptors have been the β-adrenergic (adenylyl cyclase stimulator), α_2-adrenergic (adenylyl cyclase inhibitor), muscarinic cholinergic (cyclase inhibitor or phospholipase C stimulator), and rhodopsin.

G_s appears to be rather specific, in that it interacts selectively with receptors that stimulate adenylyl cyclase activity. The ability of rhodopsin or the α_2-adrenergic receptor to stimulate nucleotide binding to this G protein is minimal (140, 141). Similarly, the interaction between transducin and the β-adrenergic receptor is difficult to detect; however, there is a measurable reaction between transducin and the α_2-adrenergic receptor ($\sim$20% as effective as rhodopsin). G_i and G_o are more promiscuous. It is presumed that their interactions with muscarinic (71, 72, 142) and α_2-adrenergic (141) receptors in vitro reflect their physiological activities. Surprising was the observation of a very significant level of interaction between G_i and β-adrenergic receptors in vitro (143); rhodopsin also stimulates the GTPase activity of G_i and G_o to about the same extent as that of G_t (100, 141).

The unexpected extent of cross-reactivity between receptors and G proteins almost certainly speaks to conservation of structure among the receptor-binding domains of the G proteins and the G protein-binding domains of the receptors. The ability to compare the primary structures of G protein–linked receptors was acquired recently with the cloning of cDNAs for the second such entity, the β-adrenergic receptor (144, 145); the first sequence was, of course, that of rhodopsin (146). The two receptors display an intriguing level of overall similarity, including the fact that both appear to span the bilayer seven times. Interestingly, the most conserved sequences in the two receptors are in the transmembrane spanning regions. It has been suggested that cytoplasmic loop 1-2 of rhodopsin is involved in the interaction with G_t (147).

Masters et al (95) have suggested that the carboxy-terminal domain of G_α is responsible for interaction with receptors. The most compelling argument is that the carboxy-terminal 21 residues of $G_{t\alpha}$ are homologous with an internal region of arrestin (148), a retinal protein that binds to phosphorylated rhodopsin (149). ADP-ribosylation of G_i and G_t on a cysteine residue four removed from the carboxy terminus prevents G protein–receptor interactions (150, 151). Analysis of the UNC mutant of the S49 lymphoma may also reveal a modest amount of information on the receptor-binding domain of G_α. This specific lesion eliminates interaction between G_s and receptors, leaving the other functions of the G protein intact (152).

Lack of the expected specificity for receptor–G protein interaction has stimulated investigation of heretofore unsuspected physiological regulatory mechanisms. Although rhodopsin and G_i presumably never have the opportunity to interact in vivo, the β-adrenergic receptor and G_i presumably do. The questions, therefore, are whether this interaction occurs in vivo; if so, to what purpose; and if not, why not? Ligand-binding studies carried out by Abramson & Molinoff strongly suggest an interaction between the β-adrenergic receptors of cyc⁻ S49 cells and a G protein (153). These cells contain G_i (83, 154, 155); they lack $G_{s\alpha}$ activity, protein, and mRNA (77); they do not appear to contain G_o. Murayama & Ui (156) have suggested that an interaction between β-adrenergic receptors and G_i is responsible for β-adrenergic agonist-mediated inhibition of adenylyl cyclase in adipocyte membranes. Treatment of many cell types with pertussis toxin potentiates the effects of stimulatory hormones on adenylyl cyclase activity (37). Perhaps this is due in part to elimination of an interaction between G_i and stimulatory receptors (143). Also interesting is that pertussis toxin can prevent homologous desensitization of adenylyl cyclase, at least in some systems (157, 158). Cerione et al (159) have attempted to demonstrate a role for G_i in hormonal stimulation of adenylyl cyclase by reconstitution of β-adrenergic receptors with G_s, G_i, and a crude preparation of the cyclase itself. Hormonal stimulation of cyclic AMP synthesis increased as a percentage of basal activity, but absolute activities decreased (basal > hormone stimulated) as G_i was added. This effect is presumably due to $\beta\gamma$ (see below) and probably has little to do with any interaction between the receptor and G_i.

G Protein–Effector Interactions

Two G protein–effector interactions are relatively well defined—G_t-phosphodiesterase and G_s-adenylyl cyclase.

PHOSPHODIESTERASE   The cyclic GMP–specific phosphodiesterase of the rod outer segments is also a heterotrimer (α: 88 kd; β: 84 kd; γ: 11 kd)(160). It is loosely associated with the rod outer segment disks and can be purified in the absence of detergent. The native trimer is essentially inactive, but catalysis is increased markedly after limited tryptic digestion (161). This release from inhibitory constraint is apparently due to proteolysis of the γ subunit (162). γ has a K_d of 0.1 nM for $\alpha\beta$, and the activity of $\alpha\beta$ can be titrated over a broad range (i.e. inhibited) by addition of purified γ. The fully active phosphodiesterase has a ratio of k_{cat}/K_m ($6 \times 10^7 \, \mathrm{M^{-1} \, sec^{-1}}$) equal to those of catalase and carbonic anhydrase—near the diffusion-controlled limit (5). $G_{t\alpha} \cdot \mathrm{Gpp(NH)p}$ (resolved from $\beta\gamma$) activates the phosphodiesterase to the same extent as does trypsin, presumably by alteration of the interactions of the subunits or by displacement of γ (25). A recent study by Sitaramayya et al (163) suggests that the composition of the activated complex may be $\mathrm{PDE}_{\alpha\beta} \cdot G_{t\alpha}$ or $\mathrm{PDE}_{\alpha\beta\gamma} \cdot G_{t\alpha}$. In view of the catalytic prowess of the phosphodiesterase, it is clear that a significant amount of cyclic GMP can be hydrolyzed during the lifetime of $G_{t\alpha} \cdot \mathrm{GTP}$.

Gpp(NH)p causes dissociation of the subunits of G_t (25, 137), and, as noted, $G_{t\alpha} \cdot$Gpp(NH)p activates the phosphodiesterase in the absence of $G_{t\beta\gamma}$. It is not possible to do the same experiment with GTP because of hydrolysis of GTP by $G_{t\alpha}$. However, Fung studied the dependence of the rhodopsin-catalyzed GTPase activity of $G_{t\alpha}$ on the ratio of $G_{t\alpha}$ to $G_{t\beta\gamma}$ (137). The subunits bind to rhodopsin in equimolar quantities, and both α and $\beta\gamma$ are required for rhodopsin-stimulated GTPase activity. However, under conditions where GTPase activity was linearly dependent on $G_{t\alpha}$, the requirement for $\beta\gamma$ was saturated at a $G_{t\alpha} : G_{t\beta\gamma}$ of approximately 20:1. This important experiment indicates that the subunits can be mostly dissociated and function maximally as a receptor-stimulated GTPase. One $G_{t\beta\gamma}$ can catalyze the binding of GTP to many $G_{t\alpha}$ subunits. Thus, subunit dissociation is driven by the binding energy of GTP and Mg^{2+}. The phenomenon is not a unique property of the interaction of G protein α subunits with nonhydrolyzable guanine nucleotide analogues.

ADENYLYL CYCLASE   Adenylyl cyclase exists as multiple molecular species. At least one major form of the enzyme in brain is activated by calmodulin, probably directly (164–166). Most species are also stimulated directly by an unusual diterpene, forskolin, isolated from the roots of the aromatic herb *Coleus forskohlii* (167). The cyclase has been purified from heart (168) and brain (166, 169, 170) using affinity chromatographic techniques with immobilized calmodulin or forskolin, pioneered by Storm and colleagues (171) or Pfeuffer & Metzger (24), respectively. Although some differences among these preparations are apparent, the enzyme appears to be a single polypeptide with a molecular weight of approximately 150,000; it also appears to be a glycoprotein (166, 168). Preparations of the enzyme from Gpp(NH)p-treated membranes (which treatment greatly stabilizes activity) contain $G_{s\alpha}$, but little or no $G_{\beta\gamma}$ (168, 169). Some preparations from untreated membranes appear to be rather free of G protein subunits (166, 170); some apparently are not (169). However, Arad et al (172) have indicated that adenylyl cyclase and G_s copurify during the initial stages of fractionation and appear to be associated even when they have not been exposed to nonhydrolyzable guanine nucleotide analogues. Differences in conditions (particularly in detergent) are presumed to account for these discrepancies (see below).

Levitzki has proposed that adenylyl cyclase is always coupled to G_s in vivo. This argument is based on the work just described (172) and on kinetic analysis, which indicates that the interaction between G_s and C is not rate limiting (134). I find it difficult to conclude that all of the cyclase is always associated with G_s, based on its behavior in detergent. However, this is not to deny the possibility. The suggestion then opens the question of the role of the considerable excess of G_s over adenylyl cyclase and how free G_s interacts with HR compared with the interactions of HR with $G_s \cdot$AC.

Stimulation   Adenylyl cyclase is activated by G_s (Figure 3). In the absence of the regulatory protein catalytic activity is nearly undetectable in the presence of the usual substrate, MgATP (15); some activity is observable with MnATP. A hormone-sensitive adenylyl cyclase activity has been reconstituted in phospholipid vesicles from three purified proteins—the β-adrenergic receptor, G_s, and the cyclase itself. Thus, these three proteins suffice to constitute a primary pathway for hormonal stimulation of cyclic AMP synthesis (173). Since the turnover number of adenylyl cyclase is probably about 1000 min^{-1}, several hundred molecules of cyclic AMP can be made during the lifetime of a single G_s·GTP.

Adenylyl cyclase is activated by $G_{s\alpha}$·GTPγS (112). This interaction is direct, and, as mentioned, a complex of $G_{s\alpha}$·Gpp(NH)p associated with adenylyl cyclase is sufficiently stable to survive purification to homogeneity (168). The capacity of $G_{s\alpha}$ to activate adenylyl cyclase accounts for the activity of oligomeric G_s (112). $G_{\beta\gamma}$ inhibits activation of G_s by GTPγS (110). Activation of G_s by AlF$_4^-$ appears to occur by a similar mechanism: subunits dissociate (32); $G_{\beta\gamma}$ increases the rate of deactivation of F$^-$-activated G_s (99).

Inhibition   Elucidation of the mechanisms of inhibition of adenylyl cyclase by G_i has been less straightforward. Purification and reconstitution of the oligomer indicate that it can indeed mediate hormonal inhibition of the enzyme, and incubation of G_i with GTPγS causes characteristic "activation" of the inhibitory capacity of the protein, accompanied by dissociation of its subunits to $G_{i\alpha}$·GTPγS and $G_{\beta\gamma}$ (42, 43). However, resolution of the subunits, followed by their individual reconstitution with platelet or wild-type

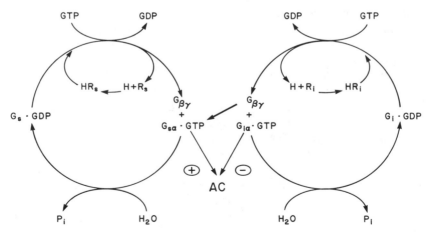

Figure 3   Mechanisms of receptor-mediated stimulation and inhibition of adenylyl cyclase.

S49 cell membranes, revealed only modest inhibitory activity associated with $G_{i\alpha}\cdot$GTPγS; the $\beta\gamma$ subunit complex was the primary source. [The profound inhibitory effect of $\beta\gamma$ (from G_s, G_i, G_o, or G_t) on the adenylyl cyclase activity of normal plasma membranes from a variety of cells has subsequently been observed in several laboratories (174, 175); it is a fact and it has implications (see below).] The inhibitory activity of $\beta\gamma$ is dependent on the presence of G_s (83, 166). Direct inhibition of adenylyl cyclase by $\beta\gamma$ seems unlikely. Although the possibility of this interaction was raised by Katada et al (176), Smigel's data (166) suggest that this effect was due to contamination of crude adenylyl cyclase with G_s. It should be noted that G_i is in considerable excess of G_s in all tissues that have been examined. Thus, G_i can serve as a reservoir of $\beta\gamma$, available to buffer the release of $G_{s\alpha}$.

Based on these observations it was proposed that $\beta\gamma$ can mediate hormonal inhibition of adenylyl cyclase by interaction with $G_{s\alpha}$ (Figure 3):

$$G_{s\alpha} + G_{\beta\gamma} \rightleftarrows G_{s\alpha\beta\gamma} \qquad\qquad 2.$$

Several additional observations support this hypothesis; I will mention one here (43). When platelet membranes are treated briefly with GTPγS and an α_2-adrenergic agonist at low Mg^{2+} concentrations, adenylyl cyclase is "irreversibly" inhibited. This inhibition is of the same magnitude as that produced by maximally effective concentrations of $\beta\gamma$, and it is not additive with the effect of $\beta\gamma$. The inhibition is overcome completely by reconstitution of the membranes with physiological concentrations of $G_{i\alpha}\cdot$GDP. The most reasonable explanation for this fact is interaction between $G_{i\alpha}\cdot$GDP and $G_{\beta\gamma}$ to relieve the inhibition caused by free $\beta\gamma$ in the membrane.

It was never proposed that inhibition of adenylyl cyclase by the indirect action of $\beta\gamma$ was the only possible mechanism, and this notion was untenable from the beginning because of the observation of hormonal inhibition of adenylyl cyclase activity in the cyc$^-$ S49 cell mutant (154). These cells lack all traces of $G_{s\alpha}$, and, logically, $\beta\gamma$ is not inhibitory when reconstituted with cyc$^-$ membranes (83). The relatively modest inhibitory effect of $G_{i\alpha}$ was invoked to explain this situation (83). This inhibitory effect of $G_{i\alpha}\cdot$GTPγS has also been observed by Roof et al (101). It does not appear to be a property of $G_{o\alpha}\cdot$GTPγS (101, 176). The effect of $G_{i\alpha}\cdot$GTPγS on adenylyl cyclase is competitive with that of $G_{s\alpha}$ (176). It should be noted, however, that inhibition of adenylyl cyclase in cyc$^-$ is assayed under unusual conditions; in the absence of $G_{s\alpha}$, forskolin is included to observe a significant level of enzymatic activity.

G PROTEIN SUBUNIT DISSOCIATION   The hypothesis of indirect inhibition of adenylyl cyclase by $G_{\beta\gamma}$ is surrounded by a certain level of controversy, which centers in particular around the issue of dissociation of G protein subunits. When G_s and G_t were the only two G proteins in the picture, the possibility of subunit dissociation was of modest interest to a few. However, there are important implications if an inhibitory, shared subunit is released on activation of any of several G proteins. In particular this would provide a mechanism for coordination of the activity of opposing pathways of trans-membrane signaling (3). Activation of one pathway would inhibit others, depending on the relative strength of signal input and the relative concentrations of the pertinent reactants in a given cell. A corollary is that G proteins that shared a $\beta\gamma$ subunit complex with G_s would all be "G_is" in terms of regulation of adenylyl cyclase. Differences in $\beta\gamma$ among G proteins would be a mechanism to partition their reciprocal interactions. A G protein without a $\beta\gamma$ subunit complex (? ras) would be immune to such regulation. Dissociation of G protein subunits thus provides a literal branch point in pathways for regulation of cell function: G_α initiating certain actions and $G_{\beta\gamma}$ terminating others. In view of these considerations, the issue of the reality of G protein subunit dissociation in the bilayer has assumed some importance.

The subunit dissociation model is based in part on the fact that the phenomenon occurs in solution (and, therefore, usually in the presence of detergent) with all G proteins examined when exposed to GTPγS, Gpp(NH)p, or AlF$_4^-$ and Mg^{2+}. Evidence discussed above indicates that GTP has the same effect on G_t. G protein α subunits are sufficient to activate their effectors. The $\beta\gamma$ subunit complex interferes with these effects. The $\beta\gamma$ subunit complex is distinctly inhibitory to hormone-stimulated adenylyl cyclase activity in normal membranes. This effect does not appear to be exerted directly on adenylyl cyclase and is dependent on the presence of G_s. Thus, it implies that there exists a steady-state concentration of free $G_{s\alpha}$ in the bilayer. Based on these facts, the hypothesis is a reasonable one. Whether it is true is another matter. It has always been recognized that subunit dissociation in the bilayer has not been demonstrated directly. This is an obvious deficiency, although one that is difficult to remedy. Until it is, the question remains open. If such a demonstration is to be taken seriously, I believe that it must occur in a normal membrane, where the concentrations of the reactants are physiological and the environment, although unknown, is relevant.

Trivial criticisms of the subunit dissociation hypothesis include the notion that adenylyl cyclase is always associated with G_s, as discussed above. This possibility does not deny the potential for dissociation of subunits; it is easily accommodated by GTP-induced dissociation of $\beta\gamma$ from the complex of G_s and the enzyme, as noted by Levitzki (4). Similarly, disagreement about the relative abilities of $G_{i\alpha}$ and $G_{\beta\gamma}$ to cause inhibition does not speak to the issue of whether or not subunits actually dissociate.

GTPγS can cause a conformational change of G_s or G_i at 0°C, which was detected as a change in their sedimentation coefficients (177). However, actual subunit dissociation did not occur until the protein was warmed. The authors described the altered form of the G protein as "preactive." Others have mistakenly stated that Codina et al (177) claimed that the ability of G_s to activate adenylyl cyclase preceded subunit dissociation.

It has also been suggested that the lifetime of G·GTP is too short to permit a cycle of subunit dissociation. Recent estimates of k_{cat} suggest a value of approximately 4 min^{-1}. It is difficult to understand how this relatively long lifetime constitutes evidence against the hypothesis without knowledge of the actual rate of dissociation in situ. The status of the proposed $\alpha\beta\gamma \leftrightarrows \alpha + \beta\gamma$ steady state must be measured in the presence and absence of relevant potential perturbants.

G PROTEIN EFFECTOR INTERACTION DOMAIN   Masters et al (95) speculate that the long domain (residues 60–208 of α_{avg}) that is inserted between the first two of the four regions involved in guanine nucleotide binding is involved in G protein–effector interactions. The argument is based particularly on analogy with the corresponding (smaller) regions of EF-Tu and ras. It is also reasonable to suggest that the conformation of this domain is likely to be regulated by GTP. H21a is an S49 cell mutant, wherein G_s is capable of interaction with receptors but incapable of interaction with adenylyl cyclase (178). Elucidation of the molecular basis of this defect may shed light on the domain of G_α necessary for interaction with effectors.

ARF   ADP-ribosylation of $G_{s\alpha}$ by cholera toxin requires the presence of another protein, which has been termed ADP-ribosylation factor or ARF. Distinguishable soluble (179) and membrane-bound (180) forms of this activity have been characterized, and the latter has been purified (181, 182). Although it is certainly intriguing that ARF is a 21-kd GTP-binding protein, it is not ras (182, see also 183). The tight association of purified ARF with endogenous GDP suggests that the protein is a GTPase, although this activity was not detected under the conditions utilized. Evidence indicates that ARF·GTP (or GTPγS) in association with $G_{s\alpha}$·GDP (but not GTPγS) $\pm$ $G_{\beta\gamma}$ is the substrate for cholera toxin. The significance of the apparent association of G_s with ARF is an important (and elusive) question.

G PROTEIN–MEMBRANE INTERACTIONS

This potentially interesting subject has been neglected. Detergent is required for solubilization of G_s, G_i, and G_o and for their behavior as distinct entities in solution. G_t can be eluted from disk membranes with GTP or nonhydrolyzable triphosphate analogues and does not aggregate in the absence of detergent.

Classical transmembrane spanning sequences have not been found for any G protein subunit. It is assumed that the proteins are associated with the inner face of the plasma membrane.

Sternweis (184) has noted that $G_{i\alpha}$·GDP and $G_{o\alpha}$·GDP behave as soluble monomers in the absence of detergent; $G_{\beta\gamma}$ aggregates. $G_{\beta\gamma}$ associates readily with phospholipid vesicles; the α subunits do not. Of interest, the α subunits interact in a saturable fashion with $G_{\beta\gamma}$-containing phospholipid vesicles. The binding of α is essentially stoichiometric with $\beta\gamma$ and is reversed on addition of GTPγS. It is possible that $G_{\beta\gamma}$ serves as the membrane anchor for G_{α} and that G_{α} subunits dissociate from this anchor when activated. If true, their sites of action need not be confined to the plasma membrane. Rodbell has speculated boldly on this subject (185).

COVALENT MODIFICATION OF G PROTEINS

Well-characterized covalent modifications of G proteins are the ADP-ribosylation reactions carried out by toxins elaborated by *V. cholerae* and *B. pertussis* (18–21). Elucidation of the molecular basis of intoxication by these important pathogens is an important landmark, and, of course, the toxins have been of great experimental value. Despite occasional claims to the contrary, there is no convincing evidence for ADP-ribosylation of G proteins as a physiological event.

The complex substrate for ADP-ribosylation of G_s by cholera toxin has been mentioned above. It is not clear if ARF is a requirement for ADP-ribosylation of G_t by the toxin. An arginine residue is the site of modification in G_t (186), and the analogous arginine is presumably ADP-ribosylated in G_s (79). It is not obvious why G_i and G_o are poor substrates for cholera toxin. Differences in sequence surrounding the site of modification or differences in their ability to interact with ARF are the obvious possibilities. The characteristic effect of the ADP-ribosylation is usually described as inhibition of the receptor-stimulated GTPase activity of the G protein (10, 187). This is consistent with the fact that GTP activates receptor-free G_s almost as well as does GTPγS. However, ADP-ribosylation also appears to decrease the affinity of $G_{s\alpha}$ for $G_{\beta\gamma}$, which could account for loss of receptor-dependent GTPase activity (188). The basal GTPase activity of ADP-ribosylated G_s is unimpaired. Further experimentation is necessary.

Pertussis toxin–catalyzed ADP-ribosylation of G_i, G_o, and G_t is somewhat more straightforward, in that ARF is not required; however, resolved α subunits are not substrates. A cysteine residue four removed from the carboxy terminus of the α subunit is the site of modification (81, 189). It is interesting that an analogous cysteine residue in ras is acylated (190). As mentioned above, ADP-ribosylation by pertussis toxin appears to block interactions between G proteins and receptors.

Exposure of cells to phorbol esters causes alterations of their abilities to respond to hormones that interact with G protein–linked receptors. There is, however, no clear picture of mechanism. For example, there are indirect data that are consistent with both enhanced interaction of G_s with adenylyl cyclase (191) and impaired function of the inhibitory pathway (192); these possibilities are obviously not mutually exclusive. Protein kinase C can phosphorylate $G_{i\alpha}$ in vitro, and this reaction is suppressed by $G_{\beta\gamma}$ (193). Unfortunately, there are no data to indicate that this occurs in vivo. This interesting subject is best left for future discussion. Hints of possible covalent modification of G proteins are provided by their isoelectric heterogeneity (194). Variations in the relative quantities of such differing forms have also been noted as a function of development or transformation (195, 196).

Although it is beyond the scope of this review, it is worth mentioning that knowledge of covalent modification of G protein–linked receptors has evolved very impressively in recent years. Rhodopsin is phosphorylated on multiple sites by a rhodopsin kinase, and phosphorylated rhodopsin is then apparently "capped" by interaction with another protein, arrestin (149). The β-adrenergic receptor inspires lavish attention from a hoard of well-known and previously uncharacterized kinases. At least one of these is presumed to be specific for this and related receptors (197). It is predictable that other G protein–linked receptors will be treated similarly. G proteins themselves may be more aloof.

ACKNOWLEDGMENTS

I thank Elliott M. Ross, Paul C. Sternweis, Janet D. Robishaw, Susanne M. Mumby, and Michael P. Graziano for their extremely helpful suggestions. Wendy J. Deaner and Jan Doyle-Argentine provided superb editorial assistance. Work from the author's laboratory was supported by United States Public Health Service Grant GM34497, American Cancer Society Grant CD 225G, and The Raymond and Ellen Willie Chair in Molecular Neuropharmacology.

Literature Cited

1. Schramm, M., Selinger, Z. 1984. *Science* 225:1350–56
2. Smigel, M. D., Ross, E. M., Gilman, A. G. 1984. In *Cell Membranes: Methods and Reviews,* ed. E. L. Elson, W. A. Frazier, L. Glaser, 2:247–94. New York: Plenum
3. Gilman, A. G. 1984. *Cell* 36:577–79
4. Levitzki, A. 1986. *Phys. Revs.* 66:819–54
5. Stryer, L. 1986. *Ann. Rev. Neurosci.* 9:87–119
6. Rodbell, M., Birnbaumer, L., Pohl, S. L., Krans, H. M. J. 1971. *J. Biol. Chem.* 246:1877–82
7. Rodbell, M., Krans, H. M. J., Pohl, S. L., Birnbaumer, L. 1971. *J. Biol. Chem.* 246:1872–76
8. Maguire, M. E., Van Arsdale, P. M., Gilman, A. G. 1976. *Mol. Pharmacol.* 12:335–39
9. Cassel, D., Selinger, Z. 1976. *Biochem. Biophys. Acta* 452:538–51
10. Cassel, D., Selinger, Z. 1977. *Proc. Natl. Acad. Sci. USA* 74:3307–11

11. Cassel, D., Selinger, Z. 1977. *J. Cyclic Nucleotide Res.* 3:11–22
12. Orly, J., Schramm, M. 1976. *Proc. Natl. Acad. Sci. USA* 73:4410–14
13. Ross, E. M., Gilman, A. G. 1977. *Proc. Natl. Acad. Sci. USA* 74:3715–19
14. Ross, E. M., Gilman, A. G. 1977. *J. Biol. Chem.* 252:6966–69
15. Ross, E. M., Howlett, A. C., Ferguson, K. M., Gilman, A. G. 1978. *J. Biol. Chem.* 253:6401–12
16. Pfeuffer, T. 1977. *J. Biol. Chem.* 252:7224–34
17. Northup, J. K., Sternweis, P. C., Smigel, M. D., Schleifer, L. S., Ross, E. M., Gilman, A. G. 1980. *Proc. Natl. Acad. Sci. USA* 77:6516–20
18. Gill, D. M., Meren, R. 1978. *Proc. Natl. Acad. Sci. USA* 75:3050–54
19. Cassel, D., Pfeuffer, T. 1978. *Proc. Natl. Acad. Sci. USA* 75:2669–73
20. Moss, J., Vaughan, M. 1977. *J. Biol. Chem.* 252:2455–57
21. Katada, T., Ui, M. 1982. *J. Biol. Chem.* 257:7210–16
22. Katada, T., Ui, M. 1982. *Proc. Natl. Acad. Sci. USA* 79:3129–33
23. Shorr, R. G. L., Lefkowitz, R. J., Caron, M. G. 1981. *J. Biol. Chem.* 256:5820–26
24. Pfeuffer, T., Metzger, H. 1982. *FEBS Lett.* 146:369–75
25. Fung, B. K.-K., Hurley, J. B., Stryer, L. 1981. *Proc. Natl. Acad. Sci. USA* 78:152–56
26. Kaziro, Y. 1978. *Biochim. Biophys. Acta* 505:95–127
27. Pfaffinger, P. J., Martin, J. M., Hunter, D. D., Nathanson, N. M., Hille, B. 1985. *Nature* 317:536–38
28. Breitwieser, G. E., Szabo, G. 1985. *Nature* 317:538–40
29. Gomperts, B. D. 1983. *Nature* 306:64–66
30. Bokoch, G. M., Gilman, A. G. 1984. *Cell* 39:301–8
31. Blackmore, P. E., Bocckino, S. B., Waynick, L. E., Exton, J. H. 1985. *J. Biol. Chem.* 260:14477–83
32. Sternweis, P. C., Northup, J. K., Smigel, M. D., Gilman, A. G. 1981. *J. Biol. Chem.* 256:11517–26
33. Hanski, E., Sternweis, P. C., Northup, J. K., Dromerick, A. W., Gilman, A. G. 1981. *J. Biol. Chem.* 256:12911–19
34. Hanski, E., Gilman, A. G. 1982. *J. Cyclic Nucleotide Res.* 8:323–36
35. Codina, J., Hildebrandt, J. D., Sekura, R. D., Birnbaumer, M., Bryan, J., et al. 1984. *J. Biol. Chem.* 259:5871–86
36. Hazeki, O., Ui, M. 1981. *J. Biol. Chem.* 256:2856–62
37. Katada, T., Amano, T., Ui, M. 1982. *J. Biol. Chem.* 257:3739–46
38. Bokoch, G. M., Katada, T., Northup, J. K., Hewlett, E. L., Gilman, A. G. 1983. *J. Biol. Chem.* 258:2072–75
39. Codina, J., Hildebrandt, J., Iyengar, R., Birnbaumer, L., Sekura, R. D., Manclark, C. R. 1983. *Proc. Natl. Acad. Sci. USA* 80:4276–80
40. Manning, D. R., Gilman, A. G. 1983. *J. Biol. Chem.* 258:7059–63
41. Bokoch, G. M., Katada, T., Northup, J. K., Ui, M., Gilman, A. G. 1984. *J. Biol. Chem.* 259:3560–67
42. Katada, T., Bokoch, G. M., Northup, J. K., Ui, M., Gilman, A. G. 1984. *J. Biol. Chem.* 259:3568–77
43. Katada, T., Northup, J. K., Bokoch, G. M., Ui, M., Gilman, A. G. 1984. *J. Biol. Chem.* 259:3578–85
44. Chader, G. J., Bensinger, R., Johnson, M., Fletcher, R. T. 1973. *Exp. Eye Res.* 17:483–86
45. Miki, N., Keirns, J. J., Marcus, F. R., Freeman, J., Bitensky, M. W. 1973. *Proc. Natl. Acad. Sci. USA* 70:3820–24
46. Wheeler, G. L., Bitensky, M. W. 1977. *Proc. Natl. Acad. Sci. USA* 74:4238–42
47. Bitensky, M. W., Wheeler, G. L., Aloni, B., Vetury, S., Matuo, Y. 1978. *Adv. Cyclic Nucleotide Res.* 9:553–72
48. Yee, R., Liebman, P. A. 1978. *J. Biol. Chem.* 253:8902–9
49. Godchaux, W. III, Zimmerman, W. F. 1979. *J. Biol. Chem.* 254:7874–84
50. Kuhn, H. 1980. *Nature* 283:587–89
51. Berridge, M. J. 1984. *Biochem. J.* 220:345–60
52. Cockcroft, S., Gomperts, B. D. 1985. *Nature* 314:534–36
53. Litosch, I., Wallis, C., Fain, J. N. 1985. *J. Biol. Chem.* 260:5464–71
54. Uhing, R. J., Prpic, V., Jiang, H., Exton, J. H. 1986. *J. Biol. Chem.* 261:2140–46
55. Smith, C. D., Lane, B. C., Kusaka, I., Verghese, M. W., Snyderman, R. 1985. *J. Biol. Chem.* 260:5875–78
56. Smith, C. D., Cox, C. C., Snyderman, R. 1986. *Science* 232:97–100
57. Lynch, C. J., Charest, R., Blackmore, P. F., Exton, J. H. 1985. *J. Biol. Chem.* 260:1593–600
58. Ohta, H., Okajima, F., Ui, M. 1985. *J. Biol. Chem.* 260:15771–80
59. Nakamura, T., Ui, M. 1985. *J. Biol. Chem.* 260:3584–93
60. Masters, S. B., Martin, M. W., Harden, T. K., Brown, J. H. 1985. *Biochem. J.* 227:933–37
61. Murayama, T., Ui, M. 1985. *J. Biol. Chem.* 260:7226–33

62. Okajima, F., Ui, M. 1984. *J. Biol. Chem.* 259:13863–71
63. Becker, E. L., Kermode, J. C., Naccache, P. H., Yassin, R., Marsh, M. L., et al. 1985. *J. Cell Biol.* 100:1641–46
64. Hansen, C. A., Mah, S., Williamson, J. R. 1986. *J. Biol. Chem.* 261:8100–3
65. Evans, T., Smith, M. M., Tanner, L. I., Harden, T. K. 1984. *Mol. Pharmacol.* 26:395–404
66. Hughes, A. R., Martin, M. W., Harden, T. K. 1984. *Proc. Natl. Acad. Sci. USA* 81:5680–84
67. Buxton, I. L. O., Brunton, L. L. 1985. *J. Biol. Chem.* 260:6733–37
68. Koch, B. D., Dorflinger, L. J., Schonbrunn, A. 1985. *J. Biol. Chem.* 260:13138–45
69. Enjalbert, A., Sladeczek, F., Guillon, G., Bertrand, P., Shu, C., et al. 1986. *J. Biol. Chem.* 261:4071–75
70. Malbon, C. C., Mangano, T. J., Watkins, D. C. 1985. *Biochem. Biophys. Res. Commun.* 128:809–15
71. Florio, V. A., Sternweis, P. C. 1985. *J. Biol. Chem.* 260:3477–83
72. Haga, K., Haga, T., Ichiyama, A., Katada, T., Kurose, H., Ui, M. 1985. *Nature* 316:731–33
73. Holz, G. G. IV, Rane, S. G., Dunlap, K. 1986. *Nature* 319:670–72
74. Maguire, M. E., Erdos, J. J. 1980. *J. Biol. Chem.* 255:1030–35
75. Jurnak, F. 1985. *Science* 230:32–36
76. Hildebrandt, J. D., Codina, J., Risinger, R., Birnbaumer, L. 1984. *J. Biol. Chem.* 259:2039–42
77. Harris, B. A., Robishaw, J. D., Mumby, S. M., Gilman, A. G. 1985. *Science* 229:1274–77
78. Numa, T., Tanabe, T., Takahashi, H., Noda, M., Hirose, T., et al. 1986. *FEBS Lett.* 195:220–24
79. Robishaw, J. D., Russell, D. W., Harris, B. A., Smigel, M. D., Gilman, A. G. 1986. *Proc. Natl. Acad. Sci. USA* 83:1251–55
80. Itoh, H., Kozasa, T., Nagata, S., Nakamura, S., Katada, T., et al. 1986. *Proc. Natl. Acad. Sci. USA* 83:3776–80
81. Hurley, J. B., Simon, M. I., Teplow, D. B., Robishaw, J. D., Gilman, A. G. 1984. *Science* 226:860–62
82. Robishaw, J. D., Smigel, M. D., Gilman, A. G. 1986. *J. Biol. Chem.* 261:9587–90
83. Katada, T., Bokoch, G. M., Smigel, M. D., Ui, M., Gilman, A. G. 1984. *J. Biol. Chem.* 259:3586–95
84. Sternweis, P. C., Robishaw, J. D. 1984. *J. Biol. Chem.* 259:13806–13
85. Neer, E. J., Lok, J. M., Wolf, L. G. 1984. *J. Biol. Chem.* 259:14222–29
86. Katada, T., Oinuma, M., Ui, M. 1986. *J. Biol. Chem.* 261:8182–91
87. Nukada, T., Tanabe, T., Takahashi, H., Noda, M., Haga, K., et al. 1986. *FEBS Lett.* 197:305–10
88. Lochrie, M. A., Hurley, J. B., Simon, M. I. 1985. *Science* 228:96–99
89. Tanabe, T., Nukada, T., Nishikawa, Y., Sugimoto, K., Suzuki, H., et al. 1985. *Nature* 315:242–45
90. Yatsunami, K., Khorana, H. G. 1985. *Proc. Natl. Acad. Sci. USA* 82:4316–20
91. Medynski, D. C., Sullivan, K., Smith, D., Van Dop, C., Chang, F.-H., et al. 1985. *Proc. Natl. Acad. Sci. USA* 82:4311–15
92. Grunwald, G. B., Gierschik, P., Nirenberg, M., Spiegel, A. 1986. *Science* 231:856–59
93. Lerea, C. L., Somers, D. E., Hurley, J. B., 1986. *Science* 234:77–80
94. Halliday, K. R. 1984. *J. Cyclic Nucleotide Res.* 9:435–48
95. Masters, S. B., Stroud, R. M., Bourne, H. R. 1986. *Protein Eng.* 1:47–54
96. Bourne, H. R. 1986. *Nature* 321:814–16
97. Mumby, S. M., Kahn, R. A., Manning, D. R., Gilman, A. G. 1986. *Proc. Natl. Acad. Sci. USA* 83:265–69
98. Hildebrandt, J. D., Codina, J., Rosenthal, W., Birnbaumer, L., Neer, E. J., et al. 1985. *J. Biol. Chem.* 260:14867–72
99. Northup, J. K., Sternweis, P. C., Gilman, A. G. 1983. *J. Biol. Chem.* 258:11361–68
100. Kanaho, Y., Tsai, S.-C., Adamik, R., Hewlett, E. L., Moss, J., Vaughan, M. 1984. *J. Biol. Chem.* 259:7378–81
101. Roof, D. J., Applebury, M. L., Sternweis, P. C. 1985. *J. Biol. Chem.* 260:16242–49
102. Evans, T., Fawzi, A., Fraser, E. D., Brown, M. L., Northup, J. K. 1987. *J. Biol. Chem.* 262:176–81
103. Sugimoto, K., Nukada, T., Tanabe, T., Takahashi, H., Noda, M., et al. 1985. *FEBS Lett.* 191:235–40
104. Fong, H. K. W., Hurley, J. B., Hopkins, R. S., Miake-Lye, R., Johnson, M. S., et al. 1986. *Proc. Natl. Acad. Sci. USA* 83:2162–66
105. Hurley, J. B., Fong, H. K. W., Teplow, D. B., Dreyer, W. J., Simon, M. I. 1984. *Proc. Natl. Acad. Sci. USA* 81:6948–52
106. Yatsunami, K., Pandya, B. V., Oprian, D. D., Khorana, H. G. 1985. *Proc. Natl. Acad. Sci. USA* 82:1936–40
107. Ovchinnikov, Y. A., Lipkin, V. M., Shuvaeva, T. M., Bogachuk, A. P., Shemyakin, V. V. 1985. *FEBS Lett.* 179:107–10

108. Gierschik, P., Codina, J., Simons, C., Birnbaumer, L., Spiegel, A. 1985. *Proc. Natl. Acad. Sci. USA* 82:727–31
109. Evans, T., Brown, M. L., Fraser, E. D., Northup, J. K. 1986. *J. Biol. Chem.* 261:7052–59
110. Northup, J. K., Smigel, M. D., Gilman, A. G. 1982. *J. Biol. Chem.* 257:11416–23
111. Ferguson, K. M., Higashijima, T., Smigel, M. D., Gilman, A. G. 1986. *J. Biol. Chem.* 261:7393–99
112. Northup, J. K., Smigel, M. D., Sternweis, P. C., Gilman, A. G. 1983. *J. Biol. Chem.* 258:11369–76
113. Higashijima, T., Ferguson, K. M., Sternweis, P. C., Smigel, M. D., Gilman, A. G. 1987. *J. Biol. Chem.* 262:762–66
114. Higashijima, T., Ferguson, K. M., Sternweis, P. C., Ross, E. M., Smigel, M. D., Gilman, A. G. 1987. *J. Biol. Chem.* 262:752–56
115. Brandt, D. R., Ross, E. M. 1985. *J. Biol. Chem.* 260:266–72
116. Sunyer, T., Codina, J., Birnbaumer, L. 1984. *J. Biol. Chem.* 259:15447–51
117. Milligan, G., Klee, W. A. 1985. *J. Biol. Chem.* 260:2057–63
118. Higashijima, T., Ferguson, K. M., Smigel, M. D., Gilman, A. G. 1987. *J. Biol. Chem.* 262:757–61
119. Brandt, D. R., Ross, E. M. 1986. *J. Biol. Chem.* 261:1656–64
120. Higashijima, T., Ferguson, K. M., Sternweis, P. C. 1987. *J. Biol. Chem.* In press
121. Howlett, A. C., Sternweis, P. C., Macik, B. A., Van Arsdale, P. M., Gilman, A. G. 1979. *J. Biol. Chem.* 254:2287–95
122. Stein, P. J., Halliday, K. R., Rasenick, M. M. 1985. *J. Biol. Chem.* 260:9081–84
123. Sternweis, P. C., Gilman, A. G. 1982. *Proc. Natl. Acad. Sci. USA* 79:4888–91
124. Bigay, J., Deterre, P., Pfister, C., Chabre, M. 1985. *FEBS Lett.* 191:181–85
125. Limbird, L. E., Gill, D. M., Lefkowitz, R. J. 1980. *Proc. Natl. Acad. Sci. USA* 77:775–79
126. Pedersen, S. E., Ross, E. M. 1982. *Proc. Natl. Acad. Sci. USA* 79:7228–32
127. Brandt, D. R., Asano, T., Pedersen, S. E., Ross, E. M. 1983. *Biochemistry* 22:4357–62
128. Cerione, R. A., Codina, J., Benovic, J. L., Lefkowitz, R. J., Birnbaumer, L., Caron, M. G. 1984. *Biochemistry* 23:4519–25
129. Asano, T., Pedersen, S. E., Scott, C. W., Ross, E. M. 1984. *Biochemistry* 23:5460–67
130. Asano, T., Ross, E. M. 1984. *Biochemistry* 23:5467–71
131. Hekman, M., Feder, D., Keenan, A. K., Gal, A., Klein, H. W., et al. 1984. *EMBO J.* 3:3339–45
132. Pedersen, S. E., Ross, E. M. 1985. *J. Biol. Chem.* 260:14150–57
133. Stryer, L. 1985. *Biopolymers* 24:29–47
134. Tolkovsky, A. M., Braun, S., Levitzki, A. 1982. *Proc. Natl. Acad. Sci. USA* 79:213–17
135. Tolkovsky, A. M., Levitzki, A. 1978. *Biochemistry* 17:3795–810
136. Tolkovsky, A. M., Levitzki, A. 1978. *Biochemistry* 17:3811–17
137. Fung, B. K.-K. 1983. *J. Biol. Chem.* 258:10495–502
138. Iyengar, R., Birnbaumer, L. 1982. *Proc. Natl. Acad. Sci. USA* 79:5179–83
139. Rojas, F. J., Birnbaumer, L. 1985. *J. Biol. Chem.* 260:7829–35
140. Cerione, R. A., Staniszewski, C., Benovic, J. L., Lefkowitz, R. J., Caron, M. G., et al. 1985. *J. Biol. Chem.* 260:1493–500
141. Cerione, R. A., Regan, J. W., Nakata, H., Codina, J., Benovic, J. L., et al. 1986. *J. Biol. Chem.* 261:3901–9
142. Kurose, H., Katada, T., Haga, T., Haga, K., Ichiyama, A., Ui, M. 1986. *J. Biol. Chem.* 261:6423–28
143. Asano, T., Katada, T., Gilman, A. G., Ross, E. M. 1984. *J. Biol. Chem.* 259:9351–54
144. Dixon, R. A. F., Kobilka, B. K., Strader, D. J., Benovic, J. L., Dohlman, H. G., et al. 1986. *Nature* 321:75–79
145. Yarden, Y., Rodriguez, H., Wong, S.K-F., Brandt, D. R., May, D. C., et al. 1986. *Proc. Natl. Acad. Sci. USA.* 83:6795–99
146. Hargrave, P. A., McDowell, H. J., Feldmann, R. J., Atkinson, P. H., Mohans, J. K., Argos, P. 1984. *Vision Res.* 24:1487–99
147. Zuker, C. S., Cowman, A. F., Rubin, G. M. 1985. *Cell* 40:851–58
148. Wistow, G. J., Katial, A., Craft, C., Shinohara, T. 1986. *FEBS Lett.* 196:23–28
149. Wilden, U., Hall, S. W., Kuhn, H. 1986. *Proc. Natl. Acad. Sci. USA* 83:1174–78
150. Van Dop, C., Yamanaka, G., Steinberg, F., Sekura, R., Manclark, C. R., et al. 1984. *J. Biol. Chem.* 259:23–25
151. Okajima, F., Katada, T., Ui, M. 1985. *J. Biol. Chem.* 260:6761–68
152. Haga, T., Ross, E. M., Anderson, H. J., Gilman, A. G. 1977. *Proc. Natl. Acad. Sci. USA* 74:2016–20

153. Abramson, S. N., Molinoff, P. B. 1985. *J. Biol. Chem.* 260:14580–88
154. Jakobs, K. H., Schultz, G. 1983. *Proc. Natl. Acad. Sci. USA* 80:3899–902
155. Hildebrandt, J. D., Sekura, R. D., Codina, J., Iyengar, R., Manclark, C. R., Birnbaumer, L. 1983. *Nature* 302:706–9
156. Murayama, T., Ui, M. 1983. *J. Biol. Chem.* 258:3319–26
157. Heyworth, C. M., Hanski, E., Houslay, M. D. 1984. *Biochem. J.* 222:189–94
158. Wilson, P. D., Dixon, B. S., Dillingham, M. A., Garcia-Sainz, J. A., Anderson, R. J. 1986. *J. Biol. Chem.* 261:1503–6
159. Cerione, R. A., Staniszewski, C., Caron, M. G., Lefkowitz, R. J., Codina, J., Birnbaumer, L. 1985. *Nature* 318:293–95
160. Baehr, W., Devlin, M. J., Applebury, M. L. 1979. *J. Biol. Chem.* 254:11669–77
161. Miki, N., Baraban, J. M., Keirns, J. J., Boyce, J. J., Bitensky, M. W. 1975. *J. Biol. Chem.* 250:6320–27
162. Hurley, J. B., Stryer, L. 1982. *J. Biol. Chem.* 257:11094–99
163. Sitaramayya, A., Harkness, J., Parkes, J. H., Gonzalez-Oliva, C., Liebman, P. A. 1986. *Biochemistry* 25:651–56
164. Salter, R. S., Krinks, M. H., Klee, C. B., Neer, E. J. 1981. *J. Biol. Chem.* 256:9830–33
165. Andreasen, T. J., Heideman, W., Rosenberg, G. B., Storm, D. R. 1983. *Biochemistry* 22:2757–62
166. Smigel, M. D. 1986. *J. Biol. Chem.* 261:1976–82
167. Seamon, K. B., Daly, J. W. 1981. *J. Cyclic Nucleotide Res.* 7:201–24
168. Pfeuffer, E., Drehev, R.-M., Metzger, H., Pfeuffer, T. 1985. *Proc. Natl. Acad. Sci. USA* 82:3086–90
169. Yeager, R. E., Heideman, W., Rosenberg, G. B., Storm, D. R. 1985. *Biochemistry* 24:3776–83
170. Coussen, F., Haiech, J., D'Alayer, J., Monneron, A. 1985. *Proc. Natl. Acad. Sci. USA* 82:6736–40
71. Westcott, K. R., LaPorte, D. C., Storm, D. R. 1979. *Proc. Natl. Acad. Sci. USA* 76:204–8
172. Arad, H., Rosenbusch, J. P., Levitzki, A. 1984. *Proc. Natl. Acad. Sci. USA* 81:6579–83
173. May, D. C., Ross, E. M., Gilman, A. G., Smigel, M. D. 1985. *J. Biol. Chem.* 260:15829–33
174. Bockaert, J., Deterre, P., Pfister, C., Guillon, G., Chabre, M. 1985. *EMBO J.* 4:1413–17
175. Cerione, R. A., Staniszewski, C., Gierschik, P., Codina, J., Somers, R. L., et al. 1986. *J. Biol. Chem.* 261:9514–20
176. Katada, T., Oinuma, M., Ui, M. 1986. *J. Biol. Chem.* 261:5215–21
177. Codina, J., Hildebrandt, J. D., Birnbaumer, L., Sekura, R. D. 1984. *J. Biol. Chem.* 259:11408–18
178. Salomon, M. R., Bourne, H. R. 1981. *Mol. Pharmacol.* 19:109–16
179. Enomoto, K., Gill, M. 1979. *J. Supramol. Struct.* 10:51–60
180. Schleifer, L. S., Kahn, R. A., Hanski, E., Northup, J. K., Sternweis, P. C., Gilman, A. G. 1982. *J. Biol. Chem.* 257:20–23
181. Kahn, R. A., Gilman, A. G. 1984. *J. Biol. Chem.* 259:6228–34
182. Kahn, R. A., Gilman, A. G. 1986. *J. Biol. Chem.* 261:7906–11
183. Gill, D. M., Meren, R. 1983. *J. Biol. Chem.* 258:11908–14
184. Sternweis, P. C. 1986. *J. Biol. Chem.* 261:631–37
185. Rodbell, M. 1985. *Trends Biochem. Sci.* 10:461–64
186. Van Dop, C., Tsubokawa, M., Bourne, H., Ramachandran, J. 1984. *J. Biol. Chem.* 259:696–98
187. Abood, M. E., Hurley, J. B., Pappone, M.-C., Bourne, H. R., Stryer, L. 1982. *J. Biol. Chem.* 257:10540–43
188. Kahn, R. A., Gilman, A. G. 1984. *J. Biol. Chem.* 259:6235–40
189. West, R. E. Jr., Moss, J., Vaughan, M., Liu, T., Liu, T-Y. 1985. *J. Biol. Chem.* 260:14428–30
190. Willumsen, B. M., Norris, K., Papageorge, A. G., Hubbert, N. L., Lowy, D. R. 1984. *EMBO J.* 3:2581–85
191. Bell, J. D., Buxton, I. L. O., Brunton, L. L. 1985. *J. Biol. Chem.* 260:2625–28
192. Jakobs, K. H., Bauer, S., Watanabe, Y. 1985. *Eur. J. Biochem.* 151:425–30
193. Katada, T., Gilman, A. G., Watanabe, Y., Bauer, S., Jakobs, K. H. 1985. *Eur. J. Biochem.* 151:431–37
194. Schleifer, L. S., Garrison, J. C., Sternweis, P. C., Northup, J. K., Gilman, A. G. 1980. *J. Biol. Chem.* 255:2641–44
195. Halvorsen, S. W., Nathanson, N. M. 1984. *Biochemistry* 23:5813–21
196. Woolkalis, W. J., Nakada, M. T., Manning, D. R. 1986. *J. Biol. Chem.* 261:3408–13
197. Benovic, J. L., Strasser, R. H., Caron, M. G., Lefkowitz, R. J. 1986. *Proc. Natl. Acad. Sci. USA* 83:2797–801

Ann. Rev. Biochem. 1987. 56:651–93

THE MOLECULAR BIOLOGY OF THE HEPATITIS B VIRUSES

Don Ganem[1,2] and Harold E. Varmus[1,3]

Departments of Microbiology and Immunology[1], Medicine[2], and Biochemistry and Biophysics[3], University of California Medical Center, San Francisco, California 94143

CONTENTS

0066-4154/87/0701-0651$02.00

PERSPECTIVE

Although the recognition of hepatitis as an infectious disease dates to antiquity, the identification of hepatitis B virus (HBV) as one of its important causes was not achieved until the late 1960s. Over the ensuing 10 years, rapid progress was made in the structural and biological characterization of the virus, but its narrow host range and inability to be propagated in cultured cells stymied early efforts to elucidate the molecular details of viral replication. Within the past decade, however, the development of workable animal models of HBV infection, together with ongoing advances in molecular genetics, has brought these once-refractory areas into full experimental view. The result has been the revelation of a remarkably intricate life cycle, at each step of which unusual strategies are employed. The replication of viral DNA proceeds via reverse transcription of an RNA intermediate, using protein and RNA primers for the generation of the first and second DNA strands. Large fractions of the genome are translated in more than one reading frame; within a frame, proteins from multiple in-phase initiator codons are expressed from overlapping transcripts. The resulting closely related gene products can be posttranslationally processed and assembled into a variety of structures of differing function or subcellular distribution.

The elucidation of these and other remarkable features of the life cycle has offered new insights into the evolutionary origins of these viruses, identified potential targets of antiviral therapy, and provided a new conceptual framework for understanding the mechanisms of viral pathogenesis. In this article we review recent progress in the molecular analysis of the hepatitis B viruses, after a brief overview of their biology. Readers with an interest in a fuller accounting of the biological and clinical aspects of HBV infection are referred to several excellent reviews emphasizing these areas (1–6); other reviews emphasizing virologic aspects of HBV have recently been published (7–10).

THE BIOLOGY OF HBV INFECTION

Parenteral exposure of susceptible hosts to HBV results in primary infection of the liver. Although the classic result of such exposure is acute hepatitis B, a moderately severe illness characterized by hepatocellular injury and inflammation (11), many individuals experience mild or no liver injury despite extensive hepatic infection. Viral replication itself appears not to be cytotoxic (see below): variation in the severity of liver damage between individuals has been attributed instead to differences in host immune responses to virally infected cells (12–15). However, the implicated target antigens and host effector mechanisms remain unidentified.

The cardinal biologic features of HBV infection are species specificity and relative hepatotropism. The host range of human HBV is narrow: to date, productive infections have been established only in human beings and higher primates (16–19). In permissive hosts, viral antigens (20) and DNA are found primarily within liver cells, which harbor abundant quantities of replicative and assembly intermediates as well as mature virions. Recently, viral DNA sequences have been detected in lower copy number in cells other than hepatocytes (28), most commonly in peripheral blood leukocytes (21–27) and bone marrow (22, 23); isolated reports of viral DNA in skin and extrahepatic viscera (29, 30) require confirmation. Taken together, such studies indicate that uptake and persistence of HBV can occur in tissues other than liver and suggest that the tissue tropism of the virus may be broader than once thought. An intriguing feature of extrahepatic HBV infection is that the forms of viral DNA found in these sites often differ in structure from the usual pattern of replicative intermediates seen in productively infected hepatocytes (31; see below). The mechanism(s) by which the genome is replicated and maintained in such sites is unclear, and the biological consequences of extrahepatic viral persistence remain largely unexplored.

Primary infection is usually self-limited, with clearance of viral antigens and infectivity from liver and blood and the development of lasting immunity to reinfection. However, 5–10% of individuals do not resolve primary infection, but develop a persistent, usually lifelong, hepatic infection (4). As in primary infections, such individuals may be asymptomatic or experience varying grades of chronic liver injury (32). Though a minority outcome, persistent infection occupies a central role in the biology of HBV: Chronic HBV carriers, estimated to number over 200 million worldwide (2), represent the reservoir from which infection is spread to other susceptible individuals, either horizontally (chiefly via sexual contact) or vertically (from carrier mothers to newborn babies). Moreover, most of the mortality from HBV infection results from chronic rather than acute disease (33, 34): severe chronic hepatitis B frequently leads to premature death from liver failure (35). But perhaps the most remarkable feature of chronic HBV infection is its association with the development of primary hepatocellular carcinoma (PHC): the risk of PHC development in long-term HBV carriers is over 100-fold that of age-matched noncarriers (36). Understanding the molecular basis of this association represents one of the outstanding challenges of contemporary hepatitis B virology and is considered more fully in the final sections of this review.

THE HEPADNAVIRUS FAMILY

Structural Features

Modern studies of HBV began with the determination of the structural features of the virus (37) and with its recognition as the prototype member of a

family of animal viruses of similar structure and biology (3). Electron microscopy of partially purified preparations of HBV from human serum reveals
three types of particles: 1. 43 nm double-shelled particles, termed "Dane
particles" (after their discoverer) and now known to represent the intact
virion; 2. 20 nm spheres, usually present in 10^3–10^6-fold excess over virions;
and 3. filaments of 20 nm diameter and variable length (37, 38). All three
structures display a common 24 kd glycoprotein antigen [hepatitis B surface
antigen (HBsAg)] on their exterior. Dane particles contain HBsAg only on
their outer shell. The spheres and filaments, on the other hand, are comprised
almost exclusively of this antigen and associated host-derived lipids. Lacking
viral nucleic acid, these subviral lipoprotein particles are noninfectious;
nevertheless, their production in enormous quantities (50–300μg HBsAg/ml
of serum) is a regular feature of infection. The reasons for this enormous
expenditure of synthetic effort are still uncertain, but one attractive possibility
is that these "dummy" particles serve to adsorb neutralizing antisurface
antibodies during the progression of infection.

The identification of the Dane particle as the infectious virion resulted from
the seminal work of Robinson and coworkers (39–44). Removal of the outer
coat of HBsAg with nonionic detergent releases an inner nucleocapsid or core
(45), whose principal structural component is a basic phosphoprotein of 21
kd, hepatitis B core antigen (HBcAg). Cores prepared in this fashion display a
virion polymerase activity that incorporates labeled dNTPs into DNA in the
absence of exogenous primer-template combinations. The radiolabeled product of such reactions, which remains sequestered within the core particle,
proved to be the viral genome; classic studies by Robinson (39–45), Summers
(46), and their coworkers established its structure (cf Figure 1). The genome
is a circular DNA of only 3.2 kilobases in length, the smallest of any animal
DNA virus yet encountered. The molecule displays two remarkable
asymmetries that set it apart from all other viral chromosomes. The first is a
length asymmetry in the two strands: one DNA strand is unit length, while its
complement is less than unit length. The full-length strand is complementary
to the viral mRNAs and by convention is designated to be of minus polarity;
accordingly, the shorter complementary strand is termed the plus strand. The
position of the 5' end of the plus strand is fixed, but its 3' end is variably
situated, differing even between molecules within any given stock. Thus, the
circular genome is only partially duplex (87), possessing a single-stranded
gap region of fixed polarity and variable extent (from 20–80% of unit length);
circularity is maintained by a 5' cohesive terminus (117) of 224 bp. In the
endogenous polymerase reaction, this single-stranded gap is repaired by the
addition of nucleotides to the 3' end of plus strand DNA. The second
asymmetry in the molecule occurs at the 5' termini: minus strand DNA
contains protein in covalent linkage at this position (47), whereas plus strands

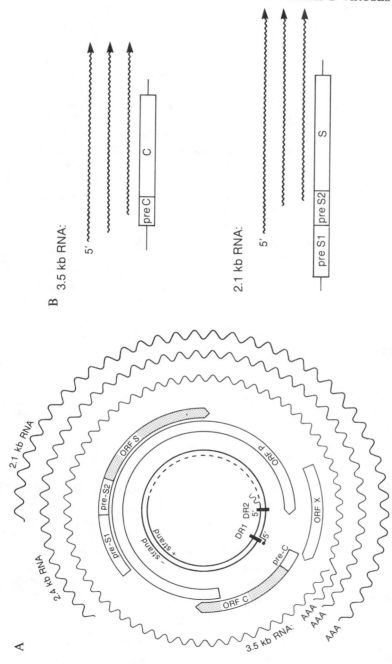

Figure 1 The structural and functional features of hepadnavirus DNA. *A*. The structural and functional features of virion DNA, including the viral direct repeat (DR) sequences (boxed), and the protein (•) and RNA (∿) species found at the 5' ends of the minus and plus DNA strands, respectively. Dashed line indicates the presence of the single-stranded gap. The viral open reading frames (ORFs) are indicated, with arrowheads indicating direction of transcription and translation. Major viral transcripts are indicated as wavy lines in outermost portion of the figure. *B*. Fine structure of the 5' ends of the genomic (*top*) and 2.1 kb subgenomic (*bottom*) RNAs. Wavy lines indicate RNA transcripts; arrows indicate direction of transcription. The positions of coding regions of viral DNA in the vicinity of the 5' ends of each transcript are shown in boxes beneath each RNA set.

contain an attached 5' oligoribonucleotide (48). These asymmetries result from the novel replicative mechanism of the genome, as will be discussed in detail below.

Genomic Organization

Molecular cloning of HBV genomes extracted from Dane particles (71–76) was soon followed by determination of the complete nucleotide sequence of the genome (72, 74, 76, 156). These studies demonstrated a compact coding organization whose key features are summarized in Figure 1. There are four major open reading frames (ORFs), all encoded by minus-strand DNA. The gene for the viral surface antigens revealed unanticipated complexity. The coding region for HBsAg (ORF S) proved to be the 3' portion of a larger coding region: upstream of ORF S is an in-phase reading frame (ORF pre-S) with two conserved in-phase ATG codons that can direct the synthesis of additional HBsAg-related proteins. These codons subdivide the pre-S region into two functional subregions, termed pre-S1 and pre-S2, whose protein products are also found in the virion envelope (see below). Similarly, the coding region for HBcAg (ORF C) is also preceded by a short upstream in-phase ORF (termed ORF pre-C), which could likewise specify a larger HBcAg-related polypeptide. Overlapping these coding regions (and a third, ORF X, whose cognate gene product has not yet been identified) is a large open reading frame, ORF P, which is believed to encode the viral polymerase.

Of course, maps such as those in Figure 1 represent only formal schema derived from sequence inspection; in principle, the final gene products generated in vivo could be substantially different as a result of RNA splicing (92), translational frameshifting (93), or posttranslational processing. The mechanisms by which such regions are expressed, and the structure and function(s) of their resulting gene products, are considered in subsequent sections. Here we make note only of the remarkable parsimony underlying the coding organization: every nucleotide in the genome is in at least one coding region, 50% of the sequence can be read in more than one frame, and even within a frame multiple ATG codons are sometimes used to give rise to multiple related proteins.

Inspection of the sequence has also led to the recognition of conserved cis-acting elements that play important roles in the life cycle. Chief among these is an 11-nucleotide sequence that is represented twice in the genome. These direct repeats (denoted DR1 and DR2 in Figure 1) are located near the 5' ends of the minus and plus DNA strands (respectively) and are critically involved in the initiation of viral DNA synthesis (see below). The other major conserved sequence landmark is the element TATAAA embedded within the 5' end of the core antigen coding sequence. This sequence forms part of the cleavage/polyadenylation signal specifying the common 3' termini of viral mRNAs. The function of these and other *cis*-acting sequences is discussed in detail in subsequent sections.

Animal Hepadnaviruses

The elucidation of the structural features of the HBV virion and its genome was soon followed by the discovery of other naturally occurring animal viruses of similar design (3). The first such agent to be recognized was the woodchuck hepatitis virus (WHV), whose existence was suggested by the astounding incidence of chronic hepatitis and PHC in a colony of Eastern woodchucks *(Marmota monax)* maintained in the Philadelphia Zoo (49). The serum of these animals proved to contain viruslike particles of all three morphologies described above, to display weak serologic crossreactivity with HBsAg, and to contain particle-associated DNA polymerase activity. Viral genomes labeled in the endogenous polymerase reaction possess the overall conformational features of HBV DNA and show significant cross-hybridization to the HBV genome. Viral genomes are found predominantly in the liver, although smaller quantities of viral DNA and RNA are found in peripheral blood leukocytes (50), and viral replicative intermediates are regularly detected in the spleen (B. Korba, J. Gerin, personal communication).

In the ensuing years, similar (but distinct) agents have been recovered from Beechey ground squirrels (ground squirrel hepatitis virus, GSHV) (51) and Pekin ducks (duck hepatitis B virus, DHBV) (52); more recently, less well characterized isolates have been reported in marsupials and tree squirrels (53, 54). On the basis of their common virion morphology, characteristic production of excess surface antigen particles, typical genome size and conformation, and relative hepatotropism, these viruses have been classified as a distinct virus family now termed *hepadnaviruses* (for *hepatotropic DNA viruses*).

Although the members of this family are classified on the basis of their similarities, there are, of course, many biological differences between them. These differences include variation in host range, tissue tropism, pathogenic spectrum, and routes of transmission. In general, the avian virus is the most divergent member of the group, both at the level of nucleotide sequence (see below) and in biological properties. The morphology of DHBV particles is somewhat atypical, and its encapsidated genomes contain a high proportion of molecules that lack a single-stranded gap. Although viral replicative intermediates are found in highest abundance in liver, replication and gene expression also proceed in pancreas, kidney, and spleen (55–58); in animals infected in ovo, the yolk sac is a major site of viral replication (M. Tagawa, W. Robinson, P. Marion, personal communication). Another major distinction concerns routes of viral transmission. In nature, DHBV is transmitted almost exclusively by the vertical route (i.e. from viremic adults to hatchlings via the egg). Indeed, after hatching, ducks rapidly lose susceptibility to exogenous infection: by three weeks of age even large parenteral doses of infectious virus rarely induce detectable viremia (59). While vertical transmission is known to be an important route of virus spread for HBV (and presumed to occur in WHV and GSHV infection), horizontal spread between adults occurs readily for the mammalian hepadnaviruses (34, 60, 61).

The host range of the animal hepadnaviruses is narrow, though perhaps not as restricted as had once been thought. In most cases transmission of virus to hosts closely related to the natural host species is demonstrable: HBV will replicate in several nonhuman primate species (16); GSHV infection has been successfully transmitted to chipmunks (62) and woodchucks (C. Seeger, D. Ganem, H. Varmus, unpublished data), and DHBV to geese (63; R. Sprengel), personal communication). In no case, however, has any hepadnavirus infection yet been transmitted to conventional laboratory animal hosts.

Much interest is now focused on determining the pathologic consequences of infection with the animal hepadnaviruses. The picture is most clear for WHV, since severe chronic hepatitis and hepatoma are frequent in naturally infected populations (49). Recently, controlled inoculation experiments in newborn woodchucks have confirmed the suspected strong correlation between antecedent infection and tumorigenesis (B. Tennant, personal communication). Experience with GSHV is somewhat less extensive, but indicates that while acute infection is rarely severe, chronic hepatitis of variable histologic grades does occur (61). Very recently, hepatocellular carcinomas have been detected among long-term GSHV carriers (66), although similar tumors have been noted in animals that successfully resolved primary infection (66; author's unpublished observations). Primary DHBV infection can also induce liver injury (68), though the long-term consequences of this are less clear. Hepatocellular carcinoma is known to occur in duck flocks in some parts of the world (69, 70), but data on concomitant viral infection is lacking and the frequent presence of dietary aflatoxin, a known hepatic carcinogen, presents additional uncertainties. However, one well-characterized example of PHC arising in a DHBV carrier has been reported (69).

Full-length molecular clones of the genomes of WHV, GSHV, and DHBV have been obtained and their nucleotide sequence determined (77–86). The coding organization of the mammalian viruses is virtually identical to that depicted in Figure 1 for HBV: minor differences include the existence of multiple precore ATG codons in WHV (79) and the presence of 11 noncoding nucleotides in GSHV, situated between the end of ORF X and the pre-C ATG (83). The only substantial divergence in coding organization occurs with DHBV; this species appears to lack an X region, and its core antigen coding region is substantially larger than that of the mammalian viruses (85, 86).

This has led to speculation that the larger core frame may represent the product of a fusion event with an ancestral DHBV X gene, but there is little independent evidence to support or refute such a notion.

The availability of cloned hepadnaviral DNA and susceptible animal hosts has made possible new genetic approaches to the functional dissection of the genome. These approaches devolve from the fact that cloned hepadnaviral genomes are infectious: when injected directly into the liver of the appropriate species, the cloned DNA of all four viruses can initiate a productive infection indistinguishable from authentic viral infection (88–91). Presumably, successful transfection of a few hepatocytes gives rise to progeny virions, which can then spread to other permissive cells. The infectivity of cloned DNA gives assurance (for those clones tested) that no essential viral sequences have been altered during molecular cloning. More importantly, this property allows functional testing of mutations introduced into viral DNA in vitro: inoculation of mutationally altered genomes into susceptible animals allows ready scoring for replication-competence. For replication-proficient mutants, effects of the mutation on host-range, tissue tropism, and pathogenicity can then be assessed. Such approaches have been successfully used to explore the functions of the DR sequences (151) and the precore ORF (see below). Similarly, the construction of WHV-GSHV recombinant genomes has demonstrated the ability of heterologous core and surface proteins to assemble into functional virions (unpublished results of C. Seeger, D. Ganem, H. E. Varmus). As our knowledge of the pathways of viral replication and gene expression grows, it seems likely that such mutational analyses will occupy an increasingly prominent place in hepadnavirology.

REPLICATION OF HEPADNAVIRUS GENOMES

The Central Dogma

The manner in which the peculiar genomes of hepadnaviruses reproduce is perhaps the most fascinating but certainly the most complicated aspect of the molecular biology of these agents. It is therefore helpful to begin with a simple view of the life cycle that can serve as a framework for subsequent discussion of the controversial points.

Four major steps are thought to be fundamental to the replication of hepadnavirus genomes (see Figure 2): 1. conversion of the asymmetric DNA found in virions to covalently closed circular DNA (cccDNA) within the nucleus of infected hepatocytes; 2. transcription of cccDNA by host RNA polymerase to generate an RNA template (the pregenome) for reverse transcription, with encapsidation of the pre-genome into virus cores; 3. synthesis of the first (minus) strand of DNA by copying pregenomic RNA, using a protein primer; and 4. synthesis of the second (plus) strand of DNA by

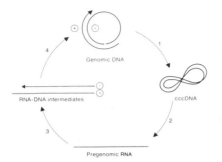

Figure 2 Simplified view of the replication cycle of hepadnaviruses, showing the major steps discussed in the text: 1. conversion of genomic open circular DNA to covalently closed circular (ccc) DNA; 2. transcription of cccDNA to generate the template for reverse transcription (pregenomic RNA); 3. synthesis of the first strand of viral DNA by core-associated reverse transcriptase; and 4. synthesis of the second DNA strand to produce mature genomic DNA.

copying the first DNA strand and using an oligomer of viral RNA as primer. In this scheme, amplification of the viral genome—the step that ensures multiplication, not merely perpetuation, of the genome—occurs during synthesis of pregenomic RNA from cccDNA.

The Nature of the Evidence for Reverse Transcription

Experimental approaches to the replication scheme for hepadnaviruses have benefited from the precedents set by retroviruses, but have lacked their technical advantages: simple cell culture systems for synchronous, high multiplicity infections; temperature-sensitive replication mutants affecting reverse transcriptase; and a readily solubilized, virion-associated polymerase, amenable to enzymological attack. Nevertheless, there is considerable evidence to support the central feature of the virus life cycle, the synthesis of viral DNA from an RNA template.

STRAND ASYMMETRY   The first clue to the unusual mode of genome replication was the asymmetry of the strands of Dane particle DNA. Examination of intrahepatic viral DNA forms (138–143) revealed still more striking asymmetries: infected hepatocytes contain large quantities of minus strands of less than unit length, many of which appear to be unpaired with plus strand DNA. These highly asymmetric forms are located within viral nucleocapsids in the cytoplasm and represent replicative intermediates (144). Such asymmetries are difficult to reconcile with classical semiconservative mechanisms of DNA replication but are compatible with DNA synthesis via reverse transcription. The cytoplasmic location of the asymmetric replicative intermediates likewise pointed to other mechanisms, as most small DNA viral genomes employing semiconservative replication use host machinery localized in the nucleus.

ANALYSIS OF VIRAL DNA SYNTHESIZED BY INTRAHEPATIC PARTICLES
The seminal findings of Summers & Mason (144) have had the greatest

influence on the development of the replication scheme described here. Subviral particles prepared from the livers of DHBV-infected ducks incorporate radioactive precursors into both minus and plus strands of viral DNA in vitro. However, synthesis of the minus strand is insensitive to actinomycin D (implying that the template is not DNA) and generates duplexes that sediment according to density as RNA:DNA hybrids, whereas synthesis of plus strand is sensitive to the drug and generates DNA:DNA duplexes. Thus the minus strand appears to be made from a template of viral RNA and the plus strand from a DNA template, presumably minus strand DNA from which RNA had been removed.

AMINO ACID SEQUENCE RELATIONSHIPS   Computer-based comparisons of deduced amino acid sequences of the longest unassigned reading frame in hepadnavirus genomes with the amino acid sequences of reverse transcriptases encoded by retroviruses (as well as other retrotransposons and the cauliflower mosaic virus) display consistent homologies that strongly support the claim that hepadnaviruses are responsible for the synthesis of their own RNA-directed DNA polymerases (145).

With these general remarks as background, we now turn to a detailed consideration of the mechanism by which reverse transcription proceeds.

The First Step: Production of Covalently Closed Circular DNA

In addition to replicative intermediates, studies of intrahepatic DNA also revealed the presence of covalently closed circular (ccc) forms of the viral genome (138–141); whenever tested, cccDNA has been located in the nucleus (141). A number of observations support the argument that virion DNA undergoes covalent closure to initiate the replicative cycle.

1. When viral DNA in the liver was monitored during the first few days after intravenous infection of duck embryos or newborn ducklings with DHBV, cccDNA was the first novel, virus-specific nucleic acid to appear, preceding the accumulation of viral RNA (115, 116). These findings suggest (though they do not prove) that the virion DNA is directly converted to cccDNA and used as template for synthesis of viral RNA.

2. Since cccDNA is normally encountered in livers of chronically infected animals, it is necessary to account for its persistence. At least three explanations can be proposed: cccDNA has an extremely long half-life; it is able to undergo semiconservative replication to maintain its copy number; or it is continually replenished by conversion of newly synthesized open-circular DNA to cccDNA. By using bromodeoxyuridine to density-label DHBV cccDNA in cultures of infected duck hepatocytes, Tuttleman, Pourcel, &

Summers (146) have shown that the pattern of labeling is compatible with synthesis by reverse transcription, but not by semiconservative replication. Since the open circular DNA found in virus particles is made by reverse transcription, the results imply that the mechanism for maintenance of the cccDNA pool depends on repeated reinfection of hepatocytes. However, it is unclear whether virus particles must exit cells to reinfect them or whether repetition of the life cycle can occur through events that are entirely intracellular, as appears to be the case for certain retroviruses and other retrotransposons (147). In either case, the density labeling experiments provide additional support for the claim that conversion of virion DNA to cccDNA constitutes the initial step in primary infection.

Presumably the conversion to cccDNA is triggered by unidentified events that normally accompany entry of hepadnaviruses into cells, though extracellular virions may contain a small proportion of cccDNA (148). Moreover, covalent circularization requires several steps for which enzymatic activities have not been identified: completion of the plus strand (either by virus-associated or cellular DNA polymerases), removal of primers (protein from the 5' end of the minus strand, ribonucleotides from the 5' end of the plus strand), elimination of a short terminal redundancy of the minus strand (see below), and ligation of DNA ends. It is instructive to recall that most of these events appear to occur correctly even in a bacterial host: fully infectious viral DNA has been repeatedly recovered by molecular cloning of virion DNA pretreated only by partial extension of the plus strands and (probably incomplete) digestion of the terminal protein (88–91).

Step Two: Synthesis, Structure, and Packaging of Pregenomic RNA

It is generally assumed that cccDNA in the host cell nucleus is the template for synthesis of viral RNA presumably by host RNA polymerase II. It is obvious that for the resulting RNA to serve as a template for reverse transcription it must contain complete representation of the viral genome within it. As will be discussed in detail later, hepadnaviruses synthesize several classes of RNA transcripts, but only one class, the 3.5 kb "genomic" RNA, meets this criterion (see Figure 1). This RNA is longer than the length of the viral genome, bearing a terminal redundancy (termed R) that varies from 130 to 270 nt among the hepadnaviruses (118–120). The redundancy arises because transcription is initiated near the start of ORF pre-C, upstream of the only known polyadenylation site in the viral genome; the signals for cleavage and polyadenylation are apparently ignored during the first transit past the site but honored on the second passage.

The 3.5 kb RNAs are heterogeneous, with 5' ends mapping at multiple positions over a 31 nt region that includes the first ATG in the core antigen reading frame (119). (The implications for translation of the 3.5 kb RNAs are considered below.) As judged from close analysis of the events during reverse transcription described below, only the shortest of the 3.5 kb RNAs is likely to serve as template for reverse transcription. This prediction has been confirmed by direct examination of GSHV RNA packaged into the cytoplasmic core particles in which reverse transcription occurs: only the shortest of three 3.5 kb species is found in the particles (94).

Step Three: Synthesis of Minus Strands

The presumptive initiation site for synthesis of minus strand DNA has been determined by mapping its 5' end after removal of the terminal protein with proteases (150, 151). By this criterion, synthesis begins within the short sequence known as DR1. Because DR1 resides within the sequence (R) that is terminally repeated in pregenomic RNA, initiation could occur near either the 5' or the 3' end of the RNA template (see Figure 3). It is simpler to envision initiation near the 3' end (as shown on the right-hand side side of Figure 3), since the minus strand can then be elongated without interruption across the entire genomic sequence, ending with a second copy of the 9 nt that reside

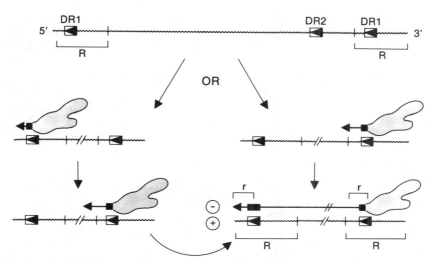

Figure 3 Models for synthesis of the first (minus) strand of hepadnavirus DNA, with priming either within the 5' copy of DR1 in pregenomic RNA (shown on left) or within the 3' copy of DR1 (on the right). Boxed arrowheads represent DR elements, solid rectangles represent copies of the DR elements, thin wavy lines denote RNA, straight lines denote DNA, R indicates the terminal repeat in pregenomic RNA, r indicates the short terminal repeat in full-length minus DNA. The shaded eccentrically shaped symbol represents the putative protein primer for the minus strand. For discussion, see text.

between the initiation site in DR1 and the 5' boundary of R. This predicted terminal redundancy in minus strand DNA has been directly confirmed by structural analysis of virion DNA (48, 151); the redundancy, called r, is likely to have a role in synthesis of the second strand, as described below. It is notable that the length of r is not heterogenous (at least for GSHV), as would be expected if all three of the 3.5 kb RNAs were used as templates for minus strand; the findings are consistent with the observation that only the shortest 3.5 kb RNA is encapsidated.

If the minus strand were initiated near the 5' end of the pregenome, rapid exhaustion of template would require transfer of the nascent minus strand to the 3' end of the same (or another) RNA molecule, presumably in the manner used for a similar step early in retroviral DNA synthesis (147; see left-hand side of Figure 3). However, if an intramolecular transfer between templates occurs, RNAse H cannot be used to expose the 3' end of the growing minus strand, as in retroviral DNA synthesis (147), since that would eliminate part of the template required for synthesis of the redundancy, r.

The obvious unresolved issue in minus strand synthesis is the nature of the presumptive protein primer. Although definitive evidence for protein priming is not yet in hand, the thesis is supported by the precedents of protein primers for other viral DNAs [adenovirus (152) and bacteriophage Φ 29 (153)] and by the linkage of even very short growing minus strands to protein (154). On the other hand, virtually nothing is known about the protein itself: there is no direct support for the presumption that it is virus-coded, and no biochemical properties, even those as fundamental as size, have been determined. Since minus strands appear to be synthesized within virus cores, the protein primer and reverse transcriptase, as well as pregenomic RNA, must be encapsidated. One attractive hypothesis for efficient packaging calls for the polymerase and primer to be generated by proteolysis from a single polypeptide that is encoded by ORF P (and perhaps part of ORF C or ORF X; see below).

Step Four: Synthesis of Plus Strands

Physical mapping of the 5' end of plus strand DNA placed the initiation site at the 3' end of DR2 (151, 155). Covalently linked to the DNA at this position is a short oligomer of viral RNA, suggesting that a fragment of the pregenomic RNA, perhaps generated by RNAse H, primes synthesis of plus strands, as has been described for retroviruses (147). However, when Lien et al (155) sequenced the attached RNA, they found the expected DR sequence flanked by 6 nt from the context of DR1, rather than from the predicted context of DR2. Similar observations were subsequently made with the mammalian hepadnaviruses and supplemented with genetic evidence: when single nucleotide changes were introduced into DR1 or DR2 of GSHV DNA, so that the copies of the repeat were distinguishable, the RNA oligomer found at the 5' end of plus strand DNA bore the sequence characteristic of DR1 rather than

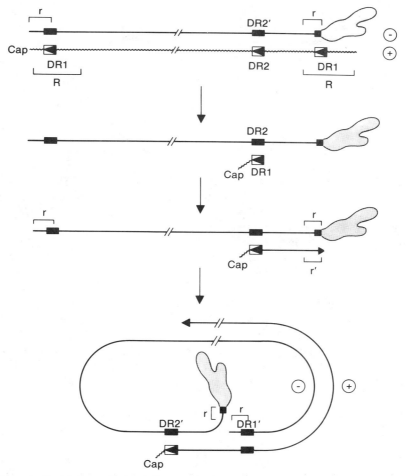

Figure 4 Model for synthesis of the second (plus) strand of hepadnavirus DNA, showing translocation of an oligoribonucleotide containing DR1 to the position of DR2 to initiate DNA synthesis. The symbols are as described in the legend to Figure 3. For clarity of presentation, the orientation of the strands in the open circular form at the bottom of the figure is reversed in relation to the conventional illustration (cf Figures 1 and 2).

DR2 (151). These results indicate that an oligomer containing DR1, from either the 5' or 3' copy of R, is transposed to basepair with the DR2 site in minus strand and to serve as primer at that position (see Figure 4). Provisional evidence for a capped 5' nucleotide on the attached oligomer (155) and the coincidence of the end of the oligomer with the 5' end of the shortest 3.5 kb RNA (119) argue for the hypothesis that the plus strand primer originates from the 5' end of the pregenome.

There is no obvious explanation for what would seem to be a needlessly complex priming mechanism, nor is it known what viral (or cellular) proteins are required for precise cleavage of the pregenome and for transfer of the priming oligomer to the DR2 site. Basepairing between DR1 and DR2 is likely to be important for the translocation, but there has been no further mutagenesis of the DRs to determine the minimum identity required.

Extension of the plus strand, beginning with the sequence downstream of DR2, must soon be stymied when the polymerase reaches the protein-linked 5' end of the minus strand template. Since the nascent plus strand ends with a copy of the short terminal redundancy, r, it can be transferred to the r sequence at the other end of the minus strand template, thereby forming an open circular molecule that approximates the structure of genomic DNA (see Figure 4). The mechanism of this strand transfer is not known but may be facilitated by the high A:T content of r; notably, the A:T composition of r is invariant among the hepadnaviruses (151), though mutations that challenge the significance of this observation have not been made.

Once transferred to the 3' end of the minus strand template, the nascent plus strand could conceivably be fully extended to the 5' end of the template; furthermore, the 3' end of the minus strand is potentially extendable to the 5' end of plus strand DNA. Were these events to occur, hepadnavirus DNA would become a linear molecule with terminal repeats extending from DR2 to DR1 and resembling the LTRs of retroviruses. That this does not happen implies that the hepadnavirus reverse transcriptase is constrained in some unknown fashion; as a result, plus strands are incomplete and heterogeneous, and open circles are the dominant form of virion DNA.

At some point in this process, core particles are selected for export as mature virions, wrapped in surface antigens. Since the predominant form of extracellular DNA is an open circle of duplex DNA, whereas intracellular DNA is dominated by incomplete minus strands, it seems probable that completion of minus strand synthesis and/or conversion to the duplex or circular form is coordinated with export. However, coordination cannot be perfect, since RNA:DNA hybrids have been described in serum-derived particles of HBV (142). In addition, substantial numbers of core particles may devote their DNA to replenishment of the pool of cccDNA (see above); it is not known whether surface antigens affiliate with such particles.

The Hepadnavirus Reverse Transcriptase

Despite strong evidence for reverse transcription of hepadnavirus RNA and considerable curiosity about the protein encoded in the putative *pol* reading frame, a soluble enzymatic activity has not been obtained for any hepadnavirus polymerase. Attempts to prepare the enzyme from virus particles and cores have been unsuccessful, and enzymatically active *pol* protein has yet to be synthesized from cloned DNA in heterologous hosts [though this approach has succeeded for reverse transcriptases encoded by both retroviruses (223, 224) and CaMV (225)]. As a result, nothing is known about the structure, biogenesis, template preferences, and ionic requirements of the hepadnavirus polymerase.

Comparisons With Retroviral and CaMV Reverse Transcription

The central role of reverse transcription in the life cycle of hepadnaviruses has inspired efforts to place these viruses among other viruses and genetic elements that employ RNA-directed DNA synthesis (226). The most useful comparisons have been made with retroviruses and CaMV (227; Figure 5).

The obvious distinction between retroviruses (RNA viruses) and hepadnaviruses (DNA viruses) depends on the timing of virus export: cytoplasmic cores of retroviral particles containing terminally redundant, genomic RNA, primer, and reverse transcriptase bud from the cell membrane prior to reverse transcription, and DNA synthesis normally commences only after the RNA-containing virus particles have entered a new cell. But there are other crucial differences as well (226): retroviruses use a host tRNA as primer for minus strand DNA; initiation of the first strand occurs outside the terminal redundancy (R) in the RNA template, requiring that the nascent minus strand be transferred between templates and affecting the structure of LTRs in the final product; a purine-rich oligomer of genomic RNA, generated by RNAse H at a precise site but not relocated, is used as primer for the plus strand; extension of the plus strand requires another transfer between strands, achieved by copying part of the primer tRNA; the retroviral polymerase fully extends both plus and minus strands, producing LTRs; cccDNA is formed from a linear duplex (rather than open circular) precursor and contains one or two LTRs; integration of viral DNA into the host chromosome is a regular part of the virus life cycle, mediated via an efficient process involving recognition of a specific site in viral cccDNA by the viral integrase function; and synthesis of viral RNA proceeds from an integrated (proviral) template, with little or no transcription of cccDNA.

The life cycle of CaMV, to the extent it is known, appears to incorporate features of both retro- and hepadnaviruses. The extracellular genome, like hepadnavirus DNA, is in the form of open circular DNA, but the primers for the first and second strands are apparently retrovirus-like: a binding site for plant methionine tRNA lies adjacent to the 5' end of the minus strand and polypurine tracts are next to the 5' ends of the multipartite plus strand. CaMV cccDNA is present in chronically infected leaves, and it is the presumptive

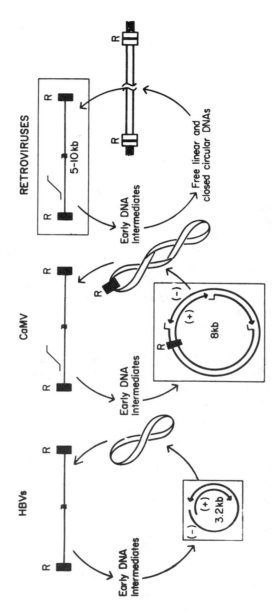

Figure 5 Comparison of the central features of the life cycles of three viruses that use reverse transcriptase: retroviruses, hepadnaviruses, and cauliflower mosaic virus. Boxes indicate the forms of the viral genome packaged into extracellular particles. For discussion, see text.

template for synthesis of viral RNA, since there is no evidence for an integrative mechanism or for LTRs.

VIRAL TRANSCRIPTION

Viral Transcripts: Structure

Genomic replication via reverse transcription implies that viral RNA must serve as both template for genomic DNA synthesis and messenger for the synthesis of viral proteins. These requirements are met by the synthesis of two classes of viral RNAs, genomic and subgenomic (Figure 1). The genomic (3.5 kb) RNAs are the only species that contain the full complement of viral genetic information and hence are the only species that can serve as replication templates. The subgenomic RNAs include two families of transcripts, ca 2.1 and (in HBV and DHBV, at least) 2.4 kb in length. (The latter species have not yet been detected in WHV- and GSHV-infected cells.) All of these RNAs are unspliced, of plus strand polarity, and polyadenylated at a common 3' terminus.

The 5' ends of the genomic RNAs of HBV, GSHV, and WHV are heterogeneous: as shown in Figure 1B, they bracket the pre-C initiator codon. Thus the genomic RNAs could also function as mRNAs for the pre-C and C gene products; consistent with this role, all of these species have been found on polyribosomes in GSHV-infected liver (94). Heterogeneous 5' termini have not yet been clearly discerned for the genomic transcripts of DHBV, although this may relate to the known susceptibility of the longer species to preferential degradation by endogenous nucleases during RNA extraction (94).

The structure of the 2.4 kb RNA (Figure 1A) suggests that its principal role is to serve as mRNA for the expression of the pre-S1 protein. It also includes the genes for the pre-S2, S, and X proteins, but the internal location of these coding regions in this transcript makes it an unlikely template for their translation. The 2.1 kb species are the likely mRNAs for the pre-S2 and S proteins (95, 122, 123). These transcripts, like the genomic RNAs, also display heterogeneous 5' ends, here bracketing the pre-S2 ATG codon (Figure 1). Thus, separate transcripts are produced bearing either the pre-S2 or S ATGs as their first initiator codon; in accord with their presumed roles as pre-S2 and S mRNAs, HBV-transfected cell lines producing only these transcripts synthesize pre-S2 and S polypeptides (195).

Inspection of Figure 1 reveals that the above RNAs could provide candidate messages for many but not all of the viral ORFs. In particular, for the mammalian viruses, no RNAs with ORF X or ORF P at their 5' extremity are apparent. The failure to observe such species could be due to their low abundance or to the existence of other strategies for the expression of these frames (see below). Several groups have attempted to define additional HBV transcripts by analyzing viral RNAs present in cultured cells transfected with cloned viral DNA (95–101) or infected with recombinant viral vectors bearing HBV sequences (102–112). In general, such cells are nonpermissive for HBV replication and do not support correct transcription of the genomic RNA species; most do, however, generate authentic 2.1 kb subgenomic mRNA (95–101), and some produce 2.4 kb mRNA (102, 112). In addition, other HBV-specific transcripts are sometimes observed. COS cells transiently transfected with chimeric SV40-HBV genomes produce a spliced transcript whose 3' splice acceptor site is immediately upstream of ORF X (105). More recently, a large array of novel unspliced transcripts, including a candidate ORF X mRNA, have been identified in cells infected with lytic adenovirus vectors harboring HBV sequences (112). Although such transcripts provide attractive potential solutions to remaining problems in viral gene expression, assessment of their relevance is complicated by their origin in heterologous systems, often in the presence of potent enhancers or transactivators of vector origin.

The Transcriptional Machinery and Its Template

Transcription of hepadnaviral genomes is generally attributed to host RNA polymerases, presumably RNA polymerase II. Although there is no direct evidence for this belief, it is consistent with the observation that correct transcription of subgenomic HBV fragments can occur in nonpermissive cells whose only expressed viral gene product is HBsAg (95–97). In vitro transcription of HBV DNA in nuclear extracts is sensitive to alpha amanitin at concentrations known to inhibit *pol*II (113), but the start sites mapped in vitro do not precisely correspond to those of the 3.5 kb and 2.1 kb RNAs described above. Transcription of HBV sequences can also be observed in *pol*III-dependent in vitro systems and generates a 700 nt RNA of minus polarity that has not been observed in productively infected cells in vivo (114).

The viral DNA form that serves as the transcriptional template is believed to be the closed-circular species. This is the only genomic form present in appreciable quantity in the nucleus, where the host transcriptional machinery resides, and its circularity (and absence of covalent interruptions) could allow transcription of terminally redundant genomic RNAs by readthrough of termination signals. Also consistent with this assignment is the fact that superhelical DNA is the earliest genomic form to arise during infection, preceding the appearance of replicative intermediates derived by reverse transcription of genomic RNA (115, 116).

cis-*Acting Elements*

Little is known about the structure of hepadnaviral promoter elements or of host (or viral) factors that may be involved in their recognition. Inspection of

DNA sequences 5' to the initiation sites of the genomic and 2.1 kb subgenomic RNAs usually reveals no consensus TATAA element within 80–100 nucleotides of these sites (119). Since this element is normally important in eukaryotic promoters for the specification of precise transcription initiation ca 30 nt downstream, its absence may be responsible for the observed heterogeneity of 5' ends; the translational program appears to capitalize on this imprecision to generate additional (functional) gene products. Interestingly, another TATA-less viral promoter, the SV40 late promoter (126), which also generates transcripts with 5' end heterogeneity, shares limited sequence homology with a region upstream of the major HBV subgenomic transcript start sites (122). The absence of TATAA elements may explain the difficulties observed in obtaining correct transcription of viral genes in vitro (113), since transcription in nuclear extracts is strongly dependent on this element (124, 125).

Recently, two independent approaches have revealed a region of HBV DNA to have the properties of a transcriptional enhancer. By screening fragments of the HBV genome for their ability to augment the expression of a linked chloramphenicol acetyl transferase (CAT) gene, Shaul, Rutter, and Laub (129) first identified this element and mapped it to a ~200 nt region some 600 bp upstream of the genomic RNA start site. Tognoni et al (130) identified a similar fragment based on its ability to confer replication-competence on an SV40 replicon driven by an enhancerless T antigen gene. Enhancement is independent of fragment orientation relative to the target promoter and can act on heterologous (non-HBV) promoters. Although the magnitude of enhancement is modest in nonhepatic cells, the effect appears to be substantially greater in several cultured hepatoma lines (129, 131). Recently, a second genomic region, distinct from this enhancer, has been shown to mediate a 3–5-fold augmentation of gene expression in the presence of glucocorticoids; this effect, too, appears to be orientation-independent, but does not display species or tissue-specificity (132).

Other important *cis*-acting elements regulate the termination of transcription. All hepadnaviral RNAs terminate at a common position, some 20 nt downstream of the conserved hexanucleotide TATAAA (AATAAA in DHBV) within the coding region for core antigen. This sequence appears to be the functional homolog of the consensus AATAAA element similarly positioned in other *pol*II transcripts and known to be critical for proper RNA cleavage and polyadenylation (92, 133, 134). As in other transcription units, downstream nucleotides also play a role in correct polyadenylation of HBV transcripts: deletions 3' to the TATAAA hexamer alter the efficiency and precision of polyA addition (105). An interesting feature of hepadnaviral transcription (and one shared by CaMV, retroviruses, and retrotransposons) is the ability to generate overlength transcripts by reading through this element once without cleavage and polyadenylation; successful termination occurs on the "second pass" of the transcription machinery by this site. Such read-through is required both to generate terminally redundant templates for reverse transcription and to allow expression of intact core antigen. The mechanism by which such selective use of this signal is achieved is unknown. Perhaps the close proximity of the hexamer to the 5' ends of the genomic RNAs inhibit its efficient recognition by cellular factors; alternatively, sequences 5' to the genomic RNA start sites might act to enhance recognition of the signal.

Tissue Specificity of Transcription

Production of the 2.1 kb subgenomic HBV mRNAs has been observed in a wide variety of transfected cell lines of diverse tissue and species origin (95–98, 100, 102, 112). This suggests a relative lack of tissue specificity, but does not rule out the possibility that expression of this transcript might be substantially more efficient in hepatocytes. In fact, several [but not all (127)] lines of transgenic mice bearing HBV DNA fragments in the germline do exhibit preferential expression of HBsAg mRNA in the liver (128).

Unlike the 2.1 kb RNA, correct genomic transcripts are not produced in most nonhepatic cultured cell lines transfected with hepadnaviral DNA. Correct genomic transcripts have to date been observed only in 1. productively infected cells in vivo, 2. primary duck hepatocytes infected in culture with DHBV (155a), 3. several well-differentiated human hepatoma cell lines (huH 6, huH 7, and hep G2) transfected with cloned HBV DNA (see below), and 4. an unusual subline of NIH3T3 cells similarly transfected (G. Acs, personal communication). It seems likely that correct transcription from the genomic promoter requires host factors that are present only in certain differentiated target tissues and that are usually not preserved following malignant transformation or upon long-term growth in cell culture. The recent identification of cultured cells capable of such transcription should greatly facilitate tests of these notions and, ultimately, identification of such permissive factors.

VIRAL PROTEINS: SYNTHESIS, PROCESSING, AND ASSEMBLY

Core Antigens

The principal structural protein of the viral nucleocapsid is known as core antigen. Identification of core antigen as a component of the nucleocapsid followed from observations that anticore antibodies could precipitate the viral genome and its associated polymerase activity, but only after removal of the surface envelope with nonionic detergent (39). When viewed in the EM, such immune complexes reveal spherical or icosahedral structures of the same diameter (27 nm) as the inner core of serum virions (37).

As extracted from viral nucleocapsids, core antigen is a basic protein of ca 21 kd (for the mammalian viruses; 35 kd for DHBV). Incubation of viral cores with γ-^{32}P ATP leads to phosphorylation of core antigen on multiple serine and threonine residues (135–137); it is presently unclear whether the responsible kinase activity is intrinsic to core antigen or is due to an associated protein derived from the host cell. HBcAg phosphorylation can also be observed in transfected cultured cells labeled in vivo (137a). No functional role has yet been proposed for core phosphorylation.

The coding region for HBV core antigen was initially identified by its ability to direct the synthesis of core antigenic determinants in *Escherichia coli* recombinants (156, 157). The nucleotide sequence of the core open reading frame (gene C) reveals several interesting features. Its 3' end encodes an extremely basic segment that would be expected to confer nucleic acid binding capacity on the protein. Recent studies (158) confirm that core polypeptides purified from liver do have nonspecific DNA-binding activity; it is highly likely that RNA binding will be demonstrable as well.

At the 5' end of ORF C is a short, in-phase open reading frame that could encode a larger "precore" protein with core antigen determinants at its C-terminus (Figure 1). Candidate mRNAs for such precore proteins have been observed in HBV-, GSHV-, and WHV-infected cells (119, 120). Recent studies by Ou et al (159) have shown that expression of HBV precore sequences in cultured cells profoundly affects the subcellular distribution of core antigen. Normally, core antigen is found in particulate form in the cytoplasm and, depending on the isolate, in the nucleus. (HBV core antigen accumulates predominantly in the nucleus, whereas WHV and DHBV cores are predominantly cytoplasmic. The basis for these differences is not understood.) However, expression of the HBV precore sequences in transfected cell cultures redirects the protein to membranous organelles (presumably ER and Golgi), and some of these precore polypeptides are proteolytically processed and secreted into the medium (159, 160). Antigen in the medium is present predominantly as a 16 kd fragment and reacts preferentially with antisera that recognize epitopes of the conformationally altered form of core antigen known as "e" antigen. Soluble e antigen circulates in the blood during HBV infection, and it seems likely that most of the serum e antigen pool may be derived from expression of the precore region. The inference that the precore region contains a signal sequence that allows the protein to enter the secretory pathway has recently been directly sustained by in vitro translocation experiments using ER vesicles (J. Ou, P. Garcia, P. Walter, W. Rutter, personal communication; V. Bruss, W. Gerlich, personal communication). In this system, cleavage of the N-terminal signal (presumably by signal peptidase in the ER) is observed; the nature of the subsequent processing steps that generate extracellular e antigen is unknown.

The function of this pathway in the viral replicative cycle is an important unresolved problem in hepadnavirology. An interesting analogy exists with the *gag* gene products of certain murine retroviruses, in which upstream "pre-*gag*" coding regions containing signal sequences likewise redirect a subpopulation of *gag* proteins to the cell surface and extracellular space (161). Deletion of these signals does not impair viability of these viruses in cultured cells, though their replicative and pathogenic potential in the whole organism has not been examined (162, 163). Analogous studies of hepadnaviruses have employed a frameshift mutation in the DHBV precore region. When cloned DNA bearing this lesion is used to transfect duck livers in vivo, normal production of viral progeny is observed (C. Chang, G. Enders, H. E. Varmus, D. Ganem, unpublished results). These studies prove that no precursor-product relation exists between precore and core proteins and indicate that, for DHBV at least, the putative precore protein is nonessential. However, as noted previously, the structure of the mammalian and avian core genes differs significantly and caution in extrapolating these findings to the mammalian viruses is warranted.

A final activity of core antigen polypeptides likely to be important in vivo is their ability to assemble into nucleocapsids. HBV core and precore polypeptides have been expressed in both *E. coli* and yeast cells (156, 157, 164). In both systems, core polypeptides can self-assemble into particles morphologically similar to authentic nucleocapsids in the absence of other viral gene products or true pregenomic RNA (164, 165). In yeast cells (though not in bacteria), similar "empty" cores have also been formed by precore polypeptides (164), though precore-derived particles are of lower stability. Precore proteins produced in yeast, however, do not undergo cleavage (164) and their behavior in yeast does not appear to mimic the situation during natural infection. These studies suggest that authentic pregenomic RNA is not required to "nucleate" core assembly, but beyond this permit few inferences about nucleocapsid formation or genome packaging.

The latter is likely to be a process of some complexity. For the mammalian viruses, three different genomic RNAs are produced that differ by at most 31 nucleotides at their 5' ends. Of these species, however, only the shortest is encapsidated into cores (94), where it functions both as template for minus strand DNA synthesis and as the donor of the plus strand primer (see above). Features other than simple primary sequence must be recognized by the packaging apparatus in order to achieve this selectivity, since all the sequences present in the encapsidated molecule are also found in the unencapsidated longer transcripts. What these features are and how they are recognized, however, remain a mystery.

Surface Antigens

STRUCTURE AND FUNCTIONS   Early studies of serum from HBV-infected individuals identified the principal viral surface antigen (HBsAg) as a protein of 24 kd that is present in both unglycosylated (p24^s) and glycosylated (gp27^s) forms (166, 167). As previously noted, these polypeptides are principally found in 20 nm subviral particles containing approximately 100 polypeptide monomers per particle, and host-derived lipid that comprises approximately 25% of the particle mass. The lipid composition of HBsAg particles is relatively invariant and consists of phospholipids (principally phosphatidyl choline), free and esterified cholesterol, and small quantities of triglycerides (168). Little is known about the ultrastructure of these particles despite beginning efforts to apply freeze-fracture electron microscopy and solution X-ray scattering techniques (169). Within the particle, monomeric HBsAg subunits are extensively cross-linked by interchain disulfide bonds; the intact particle is highly immunogenic, and reduction and alkylation of its disulfides markedly reduces both antigenicity and immunogenicity (170–172). However, particle integrity is not entirely dependent upon interchain disulfides: in recombinant yeast strains producing p24^s polypeptides, stable particles assemble prior to the formation of extensive interchain lattices (173).

Surface antigen particles of GSHV and WHV have similar morphologies and buoyant densities to those of HBV, but the DHBV particles are larger (40–50 nm), display a more heterogeneous morphology, and have a buoyant density (1.14 g/cc) suggestive of a higher lipid content (52, 176). The major polypeptides of WHV and GSHV particles are ca 22 kd and are found, as in HBV, in both glycosylated and unglycosylated forms (174, 175). These proteins share extensive serologic crossreactivity. DHBV surface antigen (176) is substantially smaller (17 kd) and, although limited amino acid homology exists with HBsAg, the homologous regions (86) lie outside of its immunodominant epitopes, accounting for the lack of serologic crossreactivity between these molecules.

As previously noted, for the mammalian viruses the ORF for the major viral surface antigen (S) is embedded within a larger coding region that includes the contiguous pre-S ORF; mRNAs exist that can separately encode polypeptides from the pre-S1, pre-S2 and S initiator codons (Figure 1). All three polypeptides (and their glycosylated derivatives) have in fact been detected in the serum of HBV, GSHV, and WHV carriers (177–188). [DHBV encodes only one pre-S protein (183a)]. In all cases, the pre-S proteins represent a minor component of the circulating antigen pool (generally less than 1–10% of the total). Considerable variation exists in their relative proportions, both between virus groups [GSHV carriers, for instance, generally produce very small amounts of pre-S2 proteins (187, 188)] and among individual carriers of a given virus (179).

The structure and function(s) of the pre-S proteins have been studied most extensively in the HBV system. Examination of purified preparations of virions and subviral particles reveals that, while 20 nm particles are composed principally of S polypeptides (with variable quantities of pre-S2 proteins), virions are composed of virtually equimolar proportions of all three species (177). Thus, pre-S1 polypeptides are localized preferentially in virions, a conclusion further supported by the observation that the presence of circulating pre-S1-specific epitopes in HBV infection is strongly correlated with viremia (185). Pre-S2 proteins are likewise found in virions, but they are also present in 20 nm particles in a much higher proportion than their pre-S1 counterparts (177).

Interest in the pre-S proteins has centered around their remarkable immunogenicity and their suspected roles in viral morphogenesis and infectivity. The pre-S2 proteins were discovered first and have been examined in the greatest detail. When mice are immunized with subviral particles bearing both S and pre-S2 determinants, antibody responses to pre-S2 epitopes regularly exceed the anti-S response (189). Studies with congenic mouse strains indicate that this response is regulated by H2-linked genes distinct from those (190) regulating anti-S responses. Interestingly, when S-nonresponder strains are immunized with pre-S2-containing particles, anti-S responses can be evoked; this and other experiments indicate that helper T cells specific for pre-S2 antigens can also provide functional help to B cell clones directed against the S region (191). Antibodies to pre-S2 determinants are readily detected during the course of natural HBV infection (192). Antibody responses to pre-S1 determinants are also elicited by HBV and GSHV infection (185, 187) and have occasionally been observed in hosts lacking demonstrable anti-S antibody (187). The superior immunogenicity of the pre-S2 region has potential implications for the design of future recombinant vaccines. Though HBV vaccines containing only S antigens are generally highly protective (193), some hosts respond poorly to such preparations (194), and it is hoped that inclusion of pre-S2 epitopes in newer vaccines may circumvent this problem.

The functional roles of the pre-S proteins in the viral life cycle have yet to be clearly defined. The pre-S2-specific domain is clearly present on the surface of both HBV virions and subviral particles (182). This region also possesses a binding activity for polymers of serum albumin (181, 195). This binding is species-specific: HBV pre-S2 proteins mediate the binding of human and chimpanzee albumin polymers only. Since this specificity exactly parallels the known host range of HBV, it has been speculated that this interaction may play a role in viral uptake, e.g. by mediating the binding of virus-albumin complexes to putative albumin receptors on the hepatocyte surface. While attractive, this model has not yet been directly tested, and a number of other observations have raised questions about its relevance. First, no binding is observed to monomeric albumin, the principal form of albumin in the circulation. Although smaller quantities of albumin dimers and higher polymers are detectable in the circulation, thus far only polymers produced in vitro by cross-linking with glutaraldehyde are active in the binding assay. Aggregates produced by several other methods, including heat denaturation, prolonged storage, and carbodiimide treatment fail to bind (196). Second, careful attempts to detect analogous binding activities in WHV infection have revealed no such activity in WHsAg particles (197). Finally, there is no direct biochemical evidence for an albumin receptor on the hepatocyte surface, despite active attempts to identify such a molecule.

Still less is known about the function of pre-S1-containing proteins. As noted above, these proteins are preferentially found on virions, and at least some pre-S1 epitopes are displayed on the virion surface, as judged by accessibility to antibodies (188) and trypsin (177). There is mounting evidence that pre-S1 proteins may play a role in viral assembly (see below); recent experiments (198) have suggested that pre-S1 domains may also be involved in hepatocyte binding.

BIOSYNTHESIS AND ASSEMBLY   The assembly of surface antigen polypeptides into particles of diverse structure implies the existence of several pathways of viral morphogenesis and of mechanisms for sorting related components to one pathway or another. Recent studies of the biosynthesis of 20 nm particles have begun to clarify several of the key steps involved in these processes. The S gene product itself contains all the information required for 20 nm particle assembly and secretion: cells transfected with DNA fragments bearing only ORF S sequences (under the control of HBV or heterologous promoters) export morphologically normal subviral particles (102, 103, 195). Thus, pre-S components are not required for either assembly or secretion. Export of particles appears to be by the constitutive secretory pathway (199), though it is slower (200) and less efficient (201) than for conventional secretory or membrane proteins. Particle formation occurs intracellularly: in infected liver or transfected cell lines, 20 nm particles are visible within cisternae of the endoplasmic reticulum (202, 203). Intracellular transport through the Golgi apparatus appears to be the rate-limiting step in particle export, as judged by the predominance of high-mannose N-linked carbohydrates on the intracellular S polypeptides; by contrast, all of the secreted glycoprotein contains processed (endoglycosidase H-resistant) carbohydrate (200).

The first clues to the nature of the intermediates in particle assembly came from studies of S polypeptide synthesis and processing in in vitro translation systems supplemented with microsomal vesicles. In such systems, the product of synthesis is a transmembrane polypeptide as judged by its partial resistance to exogenous proteases and its ability to cosediment with vesicle membranes under alkaline conditions (204). Detailed studies indicate that the transmembrane form of the protein must span the bilayer at least twice (205). To achieve this orientation, the molecule possesses two complex signal sequences, one N-terminal and one internal, whose conjoint action determines the final transmembrane disposition of the polypeptide. Pulse-chase studies confirm that transmembrane S antigen is indeed the initial product of synthesis in vivo (K. Simon, V. Lingappa, D. Ganem, unpublished data). These observations have led to a model for 20 nm particle biogenesis (204). Following the insertion of S polypeptides into the ER bilayer, aggregation of transmembrane monomers occurs, during or following which host proteins are excluded from the aggregate. This aggregate is then transposed into the ER lumen by a process of extrusion or budding. It is noteworthy that the latter steps in this pathway must involve substantial reorganization of the host lipid complement, since the buoyant density of the particles indicates a lipid content much lower than that of most cellular membranes.

This pathway bears some resemblance to morphogenetic events postulated for the envelope glycoproteins of other enveloped animal viruses, particularly in its initial steps. However, in all other virus families, surface glycoproteins are not released as such from the cell. Rather, they remain anchored in the membrane, presumably until interaction with other viral components (matrix or nucleocapsid proteins, for instance) triggers the budding step (207). Such a mechanism ensures that surface glycoproteins exit the cell only after enveloping the viral nucleocapsid. The distinctive feature of S antigen is its ability to carry out the entire process without the requirement for the participation of other viral structures. The result is the liberation of the subviral surface antigen particles that are so characteristic of hepadnaviral infection.

Although the pre-S region is not required for the S protein to enter the pathway, it turns out that the addition of pre-S-encoded domains to the N-terminus of S proteins has a profound effect on their secretory behavior. This is especially true of the pre-S1 domain. When pre-S1 polypeptides are synthesized in the absence of other viral proteins, they are not secreted from cells, despite the presence of a complete S domain in their C-terminal half. Moreover, when coexpressed with S polypeptides at comparable levels, they drastically inhibit the release of S proteins into the medium (201, 208–210). Titration of the two components in vivo indicates that the inhibitory effect of pre-S1 proteins on S secretion is dose-dependent (208, 210). These results suggest that some feature of the pre-S1 protein inhibits its own export, and that this export block can be conferred upon S polypeptides via subunit mixing during particle formation. The site of the export block is known to be distal to the step of membrane insertion, since the pre-S1 proteins are efficiently glycosylated. Recent studies indicate that the pre-S1 polypeptides remain locked in a transmembrane configuration under conditions in which S proteins can progress through the pathway (K. Simon, V. Lingappa, D. Ganem, unpublished data). The structural features of the pre-S1 domain that account for this behavior have not yet been identified with certainty. The pre-S1 proteins are the only surface proteins that are acylated, harboring covalently linked myristic acid at their N-termini (211); such a modification might well restrict the movement of protein domains out of the bilayer. Alternatively, similar consequences might derive from additional putative transmembrane anchors within pre-S1-specific sequences.

The behavior of the pre-S1 proteins in these experiments has led to the suggestion (201) that pre-S1 polypeptides may be the analogues of the conventional envelope glycoproteins of other viruses, moored in the membrane awaiting interaction with nucleocapsids or other viral components (206). In this view, such interactions would then trigger the series of events which for the S proteins occurs constitutively. If so, then the pre-S1 protein would play a key role in directing virion morphogenesis and in differentiating this process from 20 nm particle formation. Although this model of virion assembly has not been directly tested, it is consistent with the experimental fact that serum pre-S1 antigen is found almost exclusively on virions (177).

POLYMERASE   Evidence linking ORF P with the coding region for the viral reverse transcriptase is circumstantial but compelling: as previously reviewed, a segment of the ORF shares homology with the polymerase genes of retroviruses and other genetic elements that employ reverse transcription (145). Beyond this, little is known about the expression of this ORF or the structure of its product(s). Inferences about its synthesis are largely derived from sequence inspection and analogy with the better-understood case of retroviral polymerases. Like retroviral *pol* genes, ORF P overlaps the distal portion of core *(gag)* coding sequences (Figure 1). In retroviruses, both genes are represented in a single (genomic) mRNA, whose translation gives rise to a *gag-pol* fusion protein that is proteolytically processed to the mature *pol* gene products; the protease responsible for the processing is also a component of this polyprotein. For most retroviruses, the fusion protein is generated from the out-of-frame *gag* and *pol* cistrons by ribosomal frameshifting during translation (93).

For the HBVs, the family of genomic RNAs represent analogous mRNAs from which a corresponding core-*pol* polyprotein could be generated. Suspicion that such a mechanism might operate in hepadnaviruses has been heightened by the failure to identify other mRNAs harboring the entire P region but devoid of upstream C coding domains. Immunoblotting of extracts of liver samples from several cases of HBV-related hepatoma and cirrhosis with antisera to bacterial fusion proteins containing core or *pol* domains has demonstrated the presence of 35 kd species with dual immunoreactivity (212). However, the specimens contained only integrated viral DNA forms, raising the possibility that these species might be the product of expression of rearranged genomes. The full-length core-*pol* fusion protein has not been detected and the relationship of such species to the mature *pol* proteins is unknown. It should be noted that hepadnaviral *pol* genes are in the $+1$ frame relative to core, whereas most retroviral *pol* genes are in the -1 frame relative to *gag*.

THE X REGION   Still less is known about the expression of ORF X and the structure and function of its product. That X sequences are indeed expressed as protein during the viral life cycle is shown by the induction of anti-X antibodies during infection. Such antibodies have been detected in human, woodchuck, and squirrel sera by their reactivity with X antigen determinants in synthetic peptides (213), bacterial fusion proteins (214–216), and in products of coupled in vitro transcription-translation of cloned X DNA (187). However, the structure of the X protein in infected cells has not been determined, and nothing is known about its role in viral replication. Preliminary indication that X expression may be essential for viral growth derives from experiments examining the infectivity of mutant GSHV genomes bearing lesions in ORF X: 0/9 animals exposed to genomes bearing frameshifting linker-insertions in X developed infection, while 1/3 recipients of an in-phase insertion became viremic (C. Seeger, D. Ganem, H. E. Varmus, unpublished data). Further experiments with larger numbers of recipients, however, are required to firmly establish this important point.

As previously noted, X sequences are found in both genomic and subgenomic classes of mRNAs, but in both species they are 3' to other abundantly expressed coding regions. In addition, the initial ATG codon of the X region is in a poor sequence context for translation initiation (217). These two features would suggest that if X is expressed from a polycistronic message, its synthesis might be extremely inefficient. It is, of course, possible that minor mRNA species initiating in or splicing to regions just upstream of ORF X may be the true X mRNAs; such RNAs have been observed in heterologous systems (105, 112). Alternatively, X determinants may be expressed as protein from the known viral RNAs via translational frameshifting from overlapping ORF P sequences (Figure 1).

OTHER VIRAL FUNCTIONS

Primer generation   Our current understanding of the viral replicative cycle implies the presence of additional functions whose origins are uncertain. The most obvious of these is the genome-linked protein believed to play a role in the priming of minus-strand DNA synthesis. This protein was originally detected by its ability to affect the behavior of viral DNA during phenol extraction (47); because its molar abundance may be no greater than that of viral DNA, it has thus far not been further characterized structurally. In the absence of such characterization there remains no direct proof that the protein is indeed encoded by the viral genome, though this seems most probable.

It is also likely that viral gene products participate actively in the generation and translocation of the RNA primer for plus-strand DNA synthesis. In influenza virions, which employ an analogous strategy for priming viral transcription, several viral proteins have been implicated in the cap binding and endonucleolytic cleavage steps that generate the primer (218). Since primer generation presumably occurs within viral cores, the required components must be packaged in these structures.

Delta helper functions   In recent years it has become clear that HBV can also serve as a helper virus for the propagation of a replication- defective virus-like agent known as delta (219). Delta infection can only occur in individuals in whom productive HBV infection has been established, either previously or concomitantly. The resulting mixed infection is frequently much more severe than HBV infection alone, with a significant proportion of dually infected individuals developing high-grade liver cell cytotoxicity. The structure of the delta agent has recently been clarified (64, 65, 220–222, 242). The viral genome is a 1650 bp circular single-stranded RNA molecule with extensive self-complementarity; in these regards it resembles the infectious viroid RNAs of higher plants, with which it shares limited but significant sequence homologies. However, unlike viroids, which are naked RNAs, the delta genome is encapsidated by a protein known as delta antigen, composed of polypeptides of 24 and 27 kilodaltons. At least one component of delta antigen is encoded by delta RNA—another important distinction from plant viroids, which lack coding capacity altogether.

The inability of delta agent to be transmitted in the absence of HBV indicates that the latter must supply critical helper functions for delta replication or spread. The principal helper function identified to date is HBsAg: circulating delta RNA-delta antigen complexes are enveloped by an outer coat comprised predominantly of S proteins, with minor amounts of pre-S polypeptides (221). However, it is entirely possible that additional HBV gene products may also be required for complementation of delta agent, either directly or via the activation of host functions.

HEPATOMA AND HEPADNAVIRUS INFECTION

Infection with hepatitis B viruses is one of several chronic inflammatory conditions associated with an increased risk of primary hepatocellular carcinoma (PHC or hepatoma); others include schistosomiasis and alcoholic cirrhosis. The association between PHC and hepadnavirus infection has attracted particular attention, however, because of the documented high risk conferred by chronic infection, the frequency of the neoplastic disease (it is among the most common lethal cancers of man worldwide), and the precedents in tumor virology that offer mechanistic possibilities to explain the association.

At least four kinds of evidence support the connection between hepadnavirus infection and hepatoma. 1. The geographical coincidence of high rates of chronic hepadnavirus infection with high incidence of PHC, particularly in Southeast Asia and parts of equatorial Africa (2). 2. The >100-fold increased risk of hepatoma among carriers of HBsAg in a prospective study of Taiwanese postal workers (36). 3. The appearance of hepatoma in animals chronically infected with hepadnaviruses. [The incidence is particularly high, approaching 100% after two years of infection, in woodchucks (49, 65, 228); tumors in infected ground squirrels (66) and ducks (68, 69) occur more sporadically and have not been unequivocally linked to virus infection.] 4. The presence of integrated hepadnavirus DNA in a large proportion of the tumors and cell lines derived from them (see below).

Efforts to establish whether hepadnaviruses make a specific genetic contribution to tumorigenesis, rather than provoking oncogenic change though nonspecific mechanisms that depend on chronic inflammation and cell regeneration, have been guided by questions that arise from recent progress in the study of other tumor viruses and oncogenes (229). Our discussion is therefore set in the framework of such questions.

Do Hepadnavirus Genomes Contain Viral Oncogenes?

Many RNA and DNA tumor viruses carry genes shown to be directly responsible for neoplastic transformation of cultured cells and for the (generally swift) induction of tumors in appropriate host animals (229, 231). There is little reason, however, to believe that hepadnaviruses harbor viral oncogenes. The latent period between primary infection (generally contracted congenitally in regions of high incidence) and the appearance of hepatoma (at about 40 to years of age) seems inconsistent with the carriage of a hepatic oncogene by the virus. Furthermore, transfection of cultured cells with hepadnavirus DNA has not been reported to induce cellular transformation (though it might be argued that the appropriate target cells have not been used or that any putative oncogene was not properly expressed). Analysis of viral DNA integrated into tumor cell DNA reveals no universally maintained open reading frame, a finding that would be expected for an oncogene required for maintenance of the tumorigenic state. Though one or more reading frames are frequently present in uninterrupted form (232–238), this may simply reflect the preferred use of regions near the DRs as sites of recombination (see below). Finally, hepadnavirus genomes have not been shown to display any homology with normal cellular genes and hence are unlikely to carry genes transduced from host genomes, in the manner of highly oncogenic retroviruses (231).

Although an efficient viral oncogene seems unlikely to exist even in those viruses (HBV, WHV) strongly implicated in hepatocarcinogenesis, the marked difference in incidence of hepatic neoplasms between chronically infected ground squirrels and woodchucks implies that either the viruses differ in oncogenic potency or their hosts respond differently to persistent hepadnavirus infection. The latter seems more likely, since the incidence of PHC also seems to correlate with the more profound inflammatory consequences of infection in the woodchuck. Recent evidence that GSHV can replicate efficiently in woodchucks (63, 65) now provides an opportunity to compare the oncogenic potential of these two hepadnaviruses in a common host.

Does Integration of Hepadnavirus DNA Activate Cellular Proto-Oncogenes?

Some leukemogenic and carcinogenic retroviruses lacking their own oncogenes regularly activate c-*myc* or other proto-oncogenes by insertional mutation, providing promoters or enhancers that augment expression of adjacent genes (239, 240). Though the published evidence is sparse, no simple pattern of this sort can be discerned for hepadnaviruses. The most tantalizing item to date is a single early hepatoma in which the pre-S region of integrated HBV DNA is fused in frame to a cellular sequence that can encode protein closely related to steroid receptor proteins and to the product of the *erb*A proto-oncogene (241). In this case, however, expression of the fusion gene was not documented, and it is not known whether its product is likely to have neoplastic activity in liver cells. Other attempts to find insertionally activated genes, either by testing known proto-oncogenes for virus-induced rearrangements or by seeking chromosomal domains that are interrupted by hepadnaviral DNA insertions in multiple tumors, have been generally unrewarding (Y. K.-T. Fung, H. E. Varmus, unpublished data). The multiple insertions of HBV DNA in one widely studied hepatoma cell line (PLC/PRF/5) do generate some virus-host chimeric transcripts (243), but there is no clear evidence that these are directly related to the neoplastic process, particularly since several integrated units of viral DNA are present in this line.

Does Integration of Hepadnavirus DNA Cause Recessive Mutations?

Some tumors [including retinoblastoma (244), Wilms' tumor (nephroblastoma; 245), osteosarcoma (246), rhabdomyosarcoma, and hepatoblastoma (247)] arise as a consequence of recessive mutations at loci on human chromosomes 13 (q14) and 11 (p13). Though other tumor viruses have not yet been shown to cause tumors in this fashion, it is well established that integration of viral DNA can inactivate resident genes (e.g. 248, 249). Therefore the observation that insertion of HBV DNA in one hepatoma caused a 12 kb deletion mutation in chromosome 11, near or at p13, was initially greeted with considerable interest (235). More recent analysis, however, has shown that the deleted sequences are definitely outside the nearby domain implicated in Wilms' tumor and other neoplasms (250).

Are Other Kinds of Oncogenic Mutations Present in PHC?

Chromosomal translocations, gene amplification and rearrangement, and various lesions, including point mutations, that render proto-oncogenes able to transform cultured cells have been recently described in a variety of human and other tumors (240). Such mutations have also been described in a few virus-associated hepatomas, but there is little or no evidence that hepadnavirus infection is specifically required or even contributory to the lesions.

The c-*myc* proto-oncogene is unequivocally rearranged and aberrantly expressed in three of nine hepatomas in WHV-infected woodchucks, but WHV DNA is not present in the near vicinity of the affected loci (251), implying that a mechanism other than integration of viral DNA is responsible for the arrangement. DNA isolated from occasional hepatomas is capable of transforming NIH3T3 cells to a neoplastic phenotype (253–255), but similar findings have been obtained with hepatic tumors from mice (256, 257) and with many other kinds of human tumors that lack known association with viruses (240). Nevertheless, the identities of the mutant genes and the consequences of the mutations for hepatic carcinogenesis are of great interest, though probably unrelated to virus infection. One of these genes (called *lca*) appears unrelated to all previously isolated oncogenes and has thus far been implicated only in hepatomas (in 2 of 12 HBV-infected patients; 253).

There is one intriguing potential exception to the claim that host mutations are not directly caused by virus infection. Hino et al (258) have described a human hepatoma in which HBV DNA is flanked on one side by sequences from chromosome 17 and on the other by sequences from chromosome 18. These findings suggest that a chromosome 17–18 translocation was mediated by recombination between HBV DNA integrated within each chromosome; however, consequences of the translocation that might be oncogenic (e.g. augmented expression of a nearby proto-oncogene) have not been identified. The finding of a second example of a translocation mediated by HBV DNA (K. Matsubara, personal communication) should focus greater attention on the potential significance of this event.

THE STRUCTURE OF INTEGRATED DNA IN TUMORS AND CHRONICALLY INFECTED LIVERS

Although integrated hepadnavirus DNA has yet to be proved functionally significant in hepatocarcinogenesis, it provides an opportunity to ask how integration has occurred. The strongest impression obtained from the several detailed descriptions of integrated DNA from both primary tumors (235, 236, 241, 259–262) and tumor cell lines (232, 233, 263–270) [and, less commonly, from chronically infected livers (271)] is one of irregularity. The viral DNA may be subgenomic in size or composed of multimers; it may be colinear with the genome or rearranged by one or more deletions or inversions; any or none of the viral open reading frames may remain intact; the cellular sequences that serve as integration sites show no obvious relationship to each other of to sites or recombination within viral DNA; the host sites may be located on any of several chromosomes; and the chromosomal DNA may exhibit large deletions or small duplications at the integration site, implying that cleavage of the chromosome occurs without defined pattern. All of these features are reminiscent of the nonhomologous events that mediate integration of papovavirus DNA or transfected DNA into host chromosomes (230), and they strongly differentiate integration of hepadnavirus DNA from the highly ordered process by which retroviral proviruses are introduced into host chromosomes (147). It is, however, important to recall that deductions about integrative mechanisms may be misleading when they are based on the structure of viral DNA that underwent integration months or years prior to examination of the DNA. The timing of rearrangements is also difficult to estimate. In one case, however, a large inverted repeat was shown to encompass both viral and adjacent host sequences (270), implying that the rearrangement occurred during or after integration, not before.

The only aspect of hepadnavirus integration that appears nonrandom is the frequent use of sequences in or near the DRs as recombination sites (262). Two tumors in which one end of HBV DNA is joined to cellular DNA one or two bp from a 5' boundary of either DR have been used to argue that the integrative mechanism may be endowed with sequence specificity (262), akin to that used to join host DNA to retroviral sequences two bp from the ends of LTRs (147). However, it seems more likely that the frequent use of sequences near the DRs as integration sites is instead a consequence of the preferred use of free ends of replication intermediates in the integration reaction. For example, in the two tumors studied by DeJean et al (262), one junction is near the 3' end of the minus strand and another is near the 5' end of the plus strand in the intermediates. (Curiously, use of the plus strand would require transient formation of a covalent RNA-DNA hybrid to produce the observed junction.) The argument against a site-specific mechanism for integration is further supported by the considerable heterogeneity among the terminal nucleotides in integrated viral DNA with joints near or within DRs. In addition, no simple mechanism can be proposed to generate the observed terminal sequences by orderly cleavage of cccDNA.

NEW DEVELOPMENTS AND FUTURE PROSPECTS

Historically the single greatest impediment to the study of the molecular biology of the hepatitis B viruses has been the inability to observe viral replication in cultured cells. This difficulty has particularly retarded efforts 1. to study early events in infection (uptake and uncoating of virus, delivery of genome to the nucleus), 2. to identify and characterize nonabundant or nonstructural viral proteins, and 3. to develop convenient strategies for examining the phenotypes of replication-defective mutants. In recent months, however, several new culture systems have been developed that support most if not all of the key steps in the viral life cycle. Tuttleman et al have explanted hepatocytes from ducklings into primary culture and shown that they are susceptible to exogenous DHBV infection (155a). Susceptibility is transient: after five to seven days in culture, the cells lose susceptibility to viral infection. Productive infections established during the window of susceptibility are indistinguishable from those occurring in the intact liver: all viral DNA and RNA forms observed in vivo are synthesized in primary culture, and progeny virions are infectious. It is not yet known whether cells prepared in this way can be reproducibly transfected by cloned viral DNA. More recently, several groups (272–274) have demonstrated that three well-differentiated human hepatoma cell lines (Hep G2, huH 6, and huH 7) can produce viral replicative intermediates and Dane particle-like structures following transfection with cloned HBV DNA. Progeny genomes can be seen after transient transfection of mass cultures or following selection of stable transformants. In these hepatoma lines two important differences are apparent from authentic hepatic infection: 1. cccDNA is generally absent, and 2. the viral DNA encapsidated in Dane particles includes forms resembling intrahepatic (replicative) forms as well as mature virion DNA. Presumably the input transfected genomes serve as the transcriptional templates for the genomic and subgenomic RNAs. The failure to produce cccDNA despite seemingly normal genome replication indicates that the normal intra- or extracellular "recycling" pathway by which cccDNA is generated from the pool of progeny genomes derived by reverse transcription is somehow defective. Presumably either the progeny particles are themselves defective or some cellular component involved in their recycling is aberrant or absent. The infectivity of the progeny Dane particles for chimpanzees has not yet been assessed; conversely, the ability of these lines to be infected by authentic HBV virions also remains to be established.

With the development of these culture systems, hepadnaviral research seems destined to enter a new era. Hepadnaviruses should now be susceptible to the same types of genetic and biochemical investigations that have proven so powerful in the analyses of other animal viruses. If experience with these other viruses is a useful guide, we can anticipate that cell culture will complement, but not replace, animal models in the study of viral infection; used in parallel, the two approaches should provide new opportunities to unravel the molecular basis of viral replication, gene expression, and pathogenesis.

ACKNOWLEDGMENTS

We thank our many colleagues whose unpublished data are cited herein for their generous permission to do so. We also thank G. Enders and T. de Lange for helpful comments on the text and J. Marinos and E. Meredith for their invaluable secretarial assistance. Work in the authors' laboratories is made possible by grants from the National Institutes of Health (AI18782 and 22503), to the authors and to the UCSF Liver Center (AM 26743). H. E. V. is an American Cancer Society Professor of Molecular Virology.

Literature Cited

1. Vyas, G. N., Dienstag, J. L., Hoofnagle, J., eds. 1984. *Viral Hepatitis and Liver Disease*. Orlando, Fla: Grune & Stratton
2. Szmuness, W. 1978. *Prog. Med. Virol.* 24:40–69
3. Summers, J. 1981. *Hepatology* 1:179–83
4. Ganem, D. 1982. *Rev. Infect. Dis.* 4:1026–47
5. Marion, P. L., Robinson, W. S. 1983. *Curr. Top. Microbiol. Immunol.* 1983:99–121
6. Tiollais, P., Charnay, P., Vyas, G. H. 1981. *Science* 213:406–11
7. Shafritz, D. A., Lieberman, H. M. 1984. *Ann. Rev. Med.* 35:219–32
8. Standring, D., Rutter, W. 1986. *Prog. Liver Dis.* 8:311–33
9. Tiollais, P., Pourcel, C., DeJean, A. 1985. *Nature* 317:489–94
10. Cattaneo, R., Sprengel, R., Will, H., Schaller, H. 1986. In *Molecular Genetics of Mammalian Development*, Chapter 11. New York: Macmillan
11. Redeker, A. G. 1975. *Am. J. Med. Sci.* 270:9–16
12. Bianchi, L., Gudat, F. 1979. *Prog. Liver Dis.* 6:371–92
13. Chisari, F., Routenberg, J., Anderson, D. 1978. In *Viral Hepatitis*, ed. G. Vyas, S. Cohen, R. Schmid, pp. 245–66. Philadelphia: Franklin Inst.
14. Dienstag, J. L. 1984. See Ref. 1, pp. 135–66
15. Dudley, F., Fox, R., Sherlock, S. 1972. *Lancet* 1:763–66
16. Barker, L. F., Maynard, J. E., Purcell, R. H., Hoofnagle, J. H., Berquist, K. R., et al. 1975. *Am. J. Med. Sci.* 270:189–96
17. Maynard, J. E., Berquist, K. R., Krushak, D. H., Purcell, R. H. 1972. *Nature* 237:514–15
18. Hirschman, R. J., Shulman, N. R., Barker, L. F., Smith, K. O. 1969. *J. Am. Med. Assoc.* 208:1167
19. Barker, L. F., Chisari, F. V., McGrath, P. P., Dalgard, D. W., Kirschstein, R. L., et al. 1973. *J. Infect. Dis.* 127:648–62
20. Berquist, K. R., Peterson, J. M., Murphy, B. L., Ebert, J. W., Maynard, J. E., et al. 1975. *Infect. Immun.* 12:602–5
21. Pontisso, P., Poon, M. C., Tiollais, P., Brechot, C. 1984. *Br. Med. J.* 288:1563–66
22. Romet-Lemonne, J. L., McLane, M. F., Elfassi, E., Haseltine, W. A., Azocar, J., Essex, M. 1983. *Science* 221:667–79
23. Elfassi, E., Romet-Lemonne, J.-L., Essex, M., Frances-McLane, M., Haseltine, W. A. 1984. *Proc. Natl. Acad. Sci. USA* 81:3526–28
24. Lie-Injo, L. E., Balasegaram, M., Lopez, C. G., Herrera, A. R. 1983. *DNA* 2:301–8
25. Yoffe, B., Noonan, C., Melnick, J., Hollinger, F. B. 1986. *J. Infect. Dis.* 153:471–77
26. Noonan, C., Yoffe, B., Mansell, P., Melnick, J., Hollinger, F. B. 1986. *Proc. Natl. Acad. Sci. USA* 83:5698
27. Laure, F., Zagury, D., Saimot, A. G., Gallo, R. C., Hahn, B. H., Brechot, C. 1985. *Science* 229:561–63

28. Blum, H., Stowring, L., Figus, A., Montgomery, C., Haase, A., Vyas, G. N. 1983. *Proc. Natl. Acad. Sci. USA* 80:6685–88
29. DeJean, A., Lugassy, C., Zafrani, S., Tiollais, P., Brechot, C. 1984. *J. Gen. Virol.* 65:651–54
30. Siddiqui, A. 1983. *Proc. Natl. Acad. Sci. USA* 80:4861–64
31. Ganem, D. 1985. *Gastroenterology* 89:1429–30
32. Hoofnagle, J. H., Alter, H. 1984. See Ref. 1, pp. 97–113
33. Francis, D. P., Essex, M., Maynard, J. E. 1981. *Prog. Med. Virol.* 27:127–32
34. Francis, D. P., Maynard, J. E. 1979. *Epidemiol. Rev.* 1:17–31
35. Degroote, J., Fevery, J., Lepoutre, L. 1978. *Gut* 19:510–13
36. Beasley, R. P., Lin, C.-C., Hwang, L.-Y., Chien, C.-S. 1981. *Lancet* 2:1129–33
37. Robinson, W. S., Lutwick, L. I. 1976. *N. Engl. J. Med.* 295:1168–75, 1232–36
38. Dane, D. S., Cameron, C. H., Briggs, M. 1970. *Lancet* 1:695–98
39. Robinson, W. S., Greenman, R. L. 1974. *J. Virol.* 13:1231–36
40. Robinson, W. S. 1977. *Ann. Rev. Microbiol.* 31:357–77
41. Robinson, W. S., Clayton, D. A., Greenman, R. L. 1974. *J. Virol.* 14:384–91
42. Kaplan, P. M., Greenman, R. L., Gerin, J. L., Purcell, R. H., Robinson, W. S. 1973. *J. Virol.* 12:995–1005
43. Hruska, J. E., Clayton, D. A., Rubenstein, J. L. R., Robinson, W. S. 1977. *J. Virol.* 21:666–72
44. Landers, T. A., Greenberg, H. B., Robinson, W. S. 1977. *J. Virol.* 23:368–76
45. Almeida, J. D., Rubenstein, D., Stott, E. J. 1971. *Lancet* 2:1225–27
46. Summers, J., O'Connell, A., Millman, I. 1975. *Proc. Natl. Acad. Sci. USA* 72:4597–601
47. Gerlich, W., Robinson, W. S. 1980. *Cell* 21:801–9
48. Will, H. 1987. *J. Virol.* In press
49. Summers, J., Smolec, J. M., Snyder, R. 1978. *Proc. Natl. Acad. Sci. USA* 75:4533–37
50. Korba, B., Wells, F., Tennant, B., Yoakum, G., Purcell, R., Gerin, J. 1986. *J. Virol.* 58:1–8
51. Marion, P. L., Oshiro, L., Regnery, D. C., Scullard, G. H., Robinson, W. S. 1980. *J. Virol.* 33:795–806
52. Mason, W. S., Seal, G., Summers, J. 1980. *J. Virol.* 36:829–36
53. Cossart, Y., Kiernan, E. 1984. See Ref. 1, p. 647
54. Feitelson, M., Millman, I., Blumberg, R. 1986. *Proc. Natl. Acad. Sci. USA* 83:2994–97
55. Halpern, M. S., England, J. M., Deery, D. T., Petcu, D. J., Mason, W. S., Molnar-Kimber, K. L. 1983. *Proc. Natl. Acad. Sci. USA* 80:4865–69
56. Halpern, M. S., England, J. M., Flores, L., Egan, J., Newbold, J., Mason, W. S. 1984. *Virology* 137:408–13
57. Halpern, M. S., Egan, J., Mason, W. S., England, J. M. 1984. *Virus Res.* 1:213–24
58. Tagawa, M., Omata, M., Yokosuka, O., Uchiumi, K., Imazeki, F., Okuda, K. 1985. *Gastroenterology* 89:1224–29
59. Mason, W. S., Halpern, M. S., England, J. M., Seal, G., Egan, J., et al. 1983. *Virology* 131:375–84
60. Ganem, D., Weiser, B., Barchuk, A., Brown, R. J., Varmus, H. E. 1982. *J. Virol.* 44:366–73
61. Marion, P. L., Knight, S. S., Salazar, F. H., Popper, H., Robinson, W. S. 1983. *Hepatology* 3:519–27
62. Trueba, D., Phelan, M., Nelson, J., Beck, F., Pecha, B. S., et al. 1985. *Hepatology* 5:435–39
63. Marion, P., Cullen, J., Azcarra, R., Van Davelaar, M., Robinson, W. 1987. *Hepatology*. In press
64. Kos, A., Dikema, R., Arnberg, A., Van der Meide, P., Schellekens, H. 1986. *Nature* 323:558–60
65. Chen, P., Kalpana, G., Goldberg, J., Mason, W., Werner, B., et al. 1986. *Proc. Natl. Acad. Sci. USA* 83:8774–78
66. Marion, P., Van Davelaar, M., Knight, S., Salazar, F., Garcia, G., et al. 1986. *Proc. Natl. Acad. Sci. USA* 83:4543–46
67. Deleted in proof
68. Marion, P. L., Knight, S. S., Ho, B. K., Guo, Y. Y., Robinson, W. S., Popper, H. 1984. *Proc. Natl. Acad. Sci. USA* 81:898–902
69. Yokosuka, O., Omata, M., Zhou, Y.-Z., Imazeki, F., Okuda, K. 1985. *Proc. Natl. Acad. Sci. USA* 82:5180–84
70. Omata, M., Uchiumi, K., Ito, Y., Yokosuka, O., Mori, J., et al. 1983. *Gastroenterology* 85:260–67
71. Valenzuela, P., Gray, P., Quiroga, M., Zaldivar, J., Goodman, H. M., Rutter, W. J. 1979. *Nature* 280:815–19
72. Valenzuela, P., Quiroga, M., Zaldivar, J., Gray, P., Rutter, W. J. 1980. In *Animal Virus Genetics: ICN/UCLA Symp. Mol. Cell. Biol.*, ed. B. N. Fields, R. Jaenisch, pp. 57–70. New York: Academic
73. Charnay, P., Pourcel, C., Louise, A.,

Fritsch, A., Tiollais, P. 1979. *Proc. Natl. Acad. Sci. USA* 76:2222–26

74. Galibert, F., Mandart, E., Fitoussi, F., Tiollais, P., Charnay, P. 1979. *Nature* 281:646–50

75. Sninsky, J. J., Siddiqui, A., Robinson, W. S., Cohen, S. N. 1979. *Nature* 279:346–48

76. Ono, Y., Onda, H., Sasada, R., Igarashi, K., Sugino, Y., et al. 1983. *Nucleic Acids Res.* 11:1747–57

77. Cummings, I. W., Browne, J. K., Salser, W. A., Tyler, G. V., Snyder, R. L., et al. 1980. *Proc. Natl. Acad. Sci. USA* 77:1842–46

78. Galibert, F., Chen, T. N., Mandart, E. 1981. *Proc. Natl. Acad. Sci. USA* 78:5315–19

79. Galibert, F., Chen, T. N., Mandart, E. 1982. *J. Virol.* 41:51–65

80. Kodama, K., Ogasawara, N., Yoshikawa, H., Murakami, S. 1985. *J. Virol.* 56:978–86

81. Ganem, D., Greenbaum, L., Varmus, H. E. 1982. *J. Virol.* 44:374–83

82. Siddiqui, A., Marion, P. L., Robinson, W. S. 1981. *J. Virol.* 38:393–97

83. Seeger, S., Ganem, D., Varmus, H. E. 1984. *J. Virol.* 51:367–75

84. Mason, W. S., Aldrich, C., Summers, J., Taylor, J. M. 1982. *Proc. Natl. Acad. Sci. USA* 79:3997–4001

85. Mandart, E., Kay, A., Galibert, F. 1984. *J. Virol.* 49:782–92

86. Sprengel, R., Kuhn, C., Will, H., Schaller, H. 1985. *J. Med. Virol.* 15:323–33

87. Delius, H., Gough, N. M., Cameron, C. H., Murray, K. 1983. *J. Virol.* 47:337–43

88. Will, H., Cattaneo, R., Koch, H.-G., Darai, G., Schaller, H., et al. 1982. *Nature* 299:740–42

89. Will, H., Cattaneo, R., Darai, G., Deinhardt, F., Schellekens, H., Schaller, H. 1985. *Proc. Natl. Acad. Sci. USA* 82:891–95

90. Seeger, C., Ganem, D., Varmus, H. S. 1984. *Proc. Natl. Acad. Sci. USA* 81:5849–52

91. Sprengel, R., Kuhn, C., Manso, C., Will, H. 1984. *J. Virol.* 52:932–37

92. Nevins, J. R. 1983. *Ann. Rev. Biochem.* 52:441–66

93. Jacks, T., Varmus, H. E. 1985. *Science* 230:1237–42

94. Enders, G., Ganem, D., Varmus, H. 1987. *J. Virol.* 61:35–41

95. Standring, D. N., Rutter, W. J., Varmus, H. E., Ganem, D. 1984. *J. Virol.* 50:563–71

96. Dubois, M.-F., Pourcel, C., Rousset, S., Chany, C., Tiollais, P. 1980. *Proc.*

Natl. Acad. Sci. USA 77:4549–53

97. Pourcel, C., Louise, A., Gervais, M., Chenciner, N., Dubois, M.-F., Tiollais, P. 1982. *J. Virol.* 42:100–5

98. Stenlund, A., Lamy, D., Moreno-Lopez, J., Ahola, H., Pettersson, U., Tiollais, P. 1983. *EMBO J.* 2:669–73

99. Stratowa, C., Doehmer, J., Wang, Y., Hofschneider, P. H. 1982. *EMBO J.* 1:1573–78

100. Gough, N. 1983. *J. Mol. Biol.* 165:683–99

101. Siddiqui, A. 1983. *Mol. Cell. Biol.* 3:143–46

102. Laub, O., Rall, L. B., Truett, M., Shaul, Y., Standring, D. N., et al. 1983. *J. Virol.* 48:271–80

103. Liu, C.-C., Yansura, D., Levinson, A. D. 1982. *DNA* 1:213–21

104. Crowley, C. W., Liu, C.-C., Levinson, A. D. 1983. *Mol. Cell. Biol.* 3:44–55

105. Simonsen, C. C., Levinson, A. D. 1983. *Mol. Cell. Biol.* 3:2250–58

106. Will, H., Cattaneo, R., Pfaff, E., Kuhn, C., Roggendorf, M., Schaller, H. 1984. *J. Virol.* 50:335–42

107. Moriarty, A. M., Hoyer, B. H., Shih, J. W. K., Gerin, J. L., Hamer, D. H. 1981. *Proc. Natl. Acad. Sci. USA* 78:2606–10

108. Hsuing, N., Fitts, R., Wilson, S., Milne, A., Hamer, D. 1984. *J. Mol. Appl. Genet.* 2:497–506

109. Shih, M., Arsenakis, M., Tiollais, P., Roizman, B. 1984. *Proc. Natl. Acad. Sci. USA* 81:5867–70

110. Smith, G. L., Mackett, M., Moss, B. 1983. *Nature* 302:490–95

111. Paoletti, E., Lipinskas, B. R., Samsonoff, C., Mercer, S., Panicali, D. 1984. *Proc. Natl. Acad. Sci. USA* 81:193–97

112. Saito, I., Oya, Y., Shimojo, H. 1986. *J. Virol.* 58:554–60

113. Rall, L. B., Standring, D. N., Laub, O., Rutter, W. J. 1983. *Mol. Cell. Biol.* 3:1766–73

114. Standring, D. N., Rall, L. B., Laub, O., Rutter, W. J. 1983. *Mol. Cell. Biol.* 3:1774–82

115. Mason, W. S., Halpern, M. S., England, J. M., Seal, G., Egan, J., et al. 1983. *Virology* 13:575–84

116. Tagawa, M., Omata, M., Okuda, K. 1986. *Virology* 152:477–82

117. Sattler, F., Robinson, W. S. 1979. *J. Virol.* 32:226–33

118. Buscher, M., Reiser, W., Will, H., Schaller, H. 1985. *Cell* 40:717–24

119. Enders, G. H., Ganem, D., Varmus, H. 1985. *Cell* 42:297–308

120. Moroy, T., Etiemble, J., Trepo, C.,

Tiollais, P., Buendia, M. A. 1985. *EMBO J.* 4:1507–14

121. Deleted in proof
122. Cattaneo, R., Will, H., Hernandez, N., Schaller, H. 1983. *Nature* 305:336–38
123. Cattaneo, R., Will, H., Schaller, H. 1984. *EMBO J.* 3:2191–96
124. Breathnach, R., Chambon, P. 1981. *Ann. Rev. Biochem.* 50:349–83
125. Shen, K. T. 1981. *Curr. Top. Microbiol. Immunol.* 93:25–46
126. Reddy, V., Thimmappaya, B., Dhar, R., Subrumanian, K., Zain, S. 1978. *Science* 200:494–500
127. Chisari, F. Y., Pinkert, C. A., Milich, D. R., Filippi, P., McLachlan, A., et al. 1985. *Science* 230:1157–60
128. Babinet, C., Farza, H., Morello, D., Hadchovel, M., Pourcel, C. 1985. *Science* 230:1160–63
129. Shaul, Y., Rutter, W. J., Laub, O. 1985. *EMBO J.* 4:427–30
130. Tognoni, A., Cattaneo, R., Serfling, E., Schaffner, W. 1985. *Nucleic Acids Res.* 13:7457–71
131. Jameel, S., Siddiqui, A. 1986. *Mol. Cell. Biol.* 6:710–15
132. Tur-Kaspa, R., Burk, R., Shaul, Y., Shafritz, D. 1986. *Proc. Natl. Acad. Sci. USA* 83:1627–31
133. Montell, C., Fisher, E. F., Caruthers, M. H., Berk, A. J. 1983. *Nature* 305:600–5
134. Zarkower, D., Stephenson, P., Sheets, M., Wickens, M. 1986. *Mol. Cell. Biol.* 6:2317–23
135. Albin, C., Robinson, W. S. 1980. *J. Virol.* 34:297–302
136. Feitelson, M. A., Marion, P. L., Robinson, W. S. 1982. *J. Virol.* 43:687–96
137. Feitelson, M. A., Marion, P. L., Robinson, W. S. 1982. *J. Virol.* 43:741–48
137a. Roossinck, M., Siddiqui, A. 1987. *J. Virol.* In press
138. Mason, W. S., Aldrich, C., Summers, J., Taylor, J. M. 1982. *Proc. Natl. Acad. Sci. USA* 79:3997–4001
139. Weiser, B., Ganem, D., Seeger, C., Varmus, H. E. 1982. *J. Virol.* 48:1–9
140. Blum, H. E., Haase, A. T., Harris, J. D., Walker, D., Vyas, G. N. 1984. *Virology* 139:87–96
141. Miller, R. H., Robinson, W. S. 1984. *Virology* 137:390–99
142. Miller, R. H., Marion, P. L., Robinson, W. S. 1984. *Virology* 139:64–72
143. Miller, R. H., Tran, C.-T., Robinson, W. S. 1984. *Virology* 139:53–64
144. Summers, J., Mason, W. S. 1982. *Cell* 29:403–15
145. Toh, H., Hayashida, H., Miyata, T. 1983. *Nature* 305:827–29

146. Tuttleman, J., Pourcel, C., Summers, J. 1986. *Cell* 47:451–60
147. Varmus, H. E. 1982. In *Mobile Genetic Elements*, ed. J. Shapiro, pp. 411–503. New York: Academic
148. Ruiz-Opazo, N., Chakrabarty, P. R., Shafritz, D. A. 1982. *Cell* 29:129–38
149. Deleted in proof
150. Molnar-Kimber, K. L., Summers, J. W., Mason, W. S. 1984. *J. Virol.* 51: 181–91
151. Seeger, C., Ganem, D., Varmus, H. E. 1986. *Science* 232:477–83
152. Challberg, M. D., Kelly, T. J. 1982. *Ann. Rev. Biochem.* 51:901–34
153. Escarmis, N., Salas, M. 1981. *Proc. Natl. Acad. Sci. USA* 78:1446–50
154. Molnar-Kimber, K. L., Summers, J. W., Taylor, J. M., Mason, W. S. 1983. *J. Virol.* 45:165–72
155. Lien, J. M., Aldrich, C. E., Mason, W. S. 1986. *J. Virol.* 57:229–37
155a. Tuttleman, J. S., Pugh, J. C., Summers, J. W. 1986. *J. Virol.* 58:17–25
156. Pasek, M., Goto, T., Gilbert, W., Zink, B., Schaller, H., et al. 1978. *Nature* 282:575–79
157. Stahl, S., Mackay, P., Magazin, M., Bruce, S., Murray, K. 1982. *Proc. Natl. Acad. Sci. USA* 79:1606–10
158. Petit, M.-A., Pillot, J. 1985. *J. Virol.* 53:543–51
159. Ou, J., Laub, O., Rutter, W. 1986. *Proc. Natl. Acad. Sci. USA* 83:1578–82
160. Roossinck, M., Jameel, S., Loukin, S., Siddiqui, A. 1986. *Mol. Cell. Biol.* 6:1393–400
161. Edwards, S., Fan, H. 1980. *J. Virol.* 35:41–51
162. Fan, H., Chute, H., Chao, E., Feuerman, M. 1983. *Proc. Natl. Acad. Sci. USA* 80:5965–69
163. Schwartzberg, P., Colicelli, J., Goff, S. 1983. *J. Virol.* 46:538–46
164. Miyanohara, A., Imamura, T., Araki, M., Sugawara, K., Ohtomo, N., Matsubara, K. 1986. *J. Virol.* 59:176–80
165. Cohen, B. J., Richmond, J. E. 1982. *Nature* 296:677–78
166. Peterson, D. L., Roberts, I. M., Vyas, G. N. 1977. *Proc. Natl. Acad. Sci. USA* 74:1530–34
167. Peterson, D. L. 1981. *J. Biol. Chem.* 256:6975–83
168. Gavilanes, F., Gonzalez-Ros, J. M., Peterson, D. L. 1982. *J. Biol. Chem.* 257:7770–77
169. Aggerbeck, L. P., Peterson, D. L. 1985. *Virology* 141:155–61
170. Vyas, G. N., Rao, K. R., Ibrahim, A. B. 1972. *Science* 178:1300–1

171. Sukeno, N., Shirachi, R., Yamaguchi, J., Ishida, N. 1972. *J. Virol.* 9:182–83

172. Imai, M., Gotoh, A., Nishioka, K., Kurashina, S., Miyakawa, Y., Mayumi, M. 1974. *J. Immunol.* 112:416–19

173. Wampler, D., Lehman, E., Boger, J., McAleer, W., Scolnick, E. 1985. *Proc. Natl. Acad. Sci. USA* 82:6830–34

174. Gerlich, W. H., Feitelson, M. A., Marion, P. L., Robinson, W. S. 1980. *J. Virol.* 36:787–95

175. Feitelson, M. A., Marion, P. L., Robinson, W. S. 1981. *J. Virol.* 39:447–54

176. Marion, P. L., Knight, S. S., Feitelson, M. A., Oshiro, L. S., Robinson, W. S. 1983. *J. Virol.* 48:534–41

177. Heermann, K. H., Goldmann, U., Schwartz, W., Seyffarth, T., Baumgarten, H., Gerlich, W. H. 1984. *J. Virol.* 52:396–402

178. Stibbe, W., Gerlich, W. 1983. *J. Virol.* 46:626–28

179. Stibbe, W., Gerlich, W. 1982. *Virology* 123:436–42

180. Neurath, A. R., Kent, S. B. H., Strick, N., Taylor, P., Stevens, C. E. 1985. *Nature* 315:154–56

181. Machida, A., Kishimoto, S., Ohnuma, H., Miyamoto, H., Baba, K., et al. 1983. *Gastroenterology* 85:268–74

182. Machida, A., Kishimoto, S., Ohnuma, H., Baba, K., Ito, Y., et al. 1984. *Gastroenterology* 86:910–18

183. Feitelson, M., Marion, P. L., Robinson, W. S. 1983. *Virology* 130:76–80

183a. Pugh, J., Schaeffer, G., Summers, J., Sninsky, J. 1987. *J. Virol.* In press

184. Pfaff, E., Klinkert, M., Theilmann, L., Schaller, H. 1986. *Virology* 148:15–22

185. Klinkert, M., Theilmann, L., Pfaff, E., Schaller, H. 1986. *J. Virol.* 58:522–25

186. Wong, D., Nath, N., Sninsky, J. 1985. *J. Virol.* 55:223–31

187. Persing, D., Varmus, H., Ganem, D. 1986. *J. Virol.* In press

188. Schaeffer, E., Snyder, R., Sninsky, J. 1986. *J. Virol.* 57:173–82

189. Milich, D. R., Thornton, G. B., Neurath, A. R., Kent, S. B., Michel, M. L., et al. 1985. *Science* 228:1195–98

190. Milich, D. R., Chisari, F. V. 1982. *J. Immunol.* 129:320–24

191. Milich, D. R., McNamara, M. K., McLachlan, A., Thornton, G. B., Chisari, F. V. 1985. *Proc. Natl. Acad. Sci. USA* 82:8168–72

192. Alberti, A., Pontisso, P., Schiavon, E., Realdi, G. 1984. *Hepatology* 4:220–26

193. Szmuness, W., Stevens, C. E., Harley, E. J., Zang, E. A., Oleszko, W. R., et al. 1980. *N. Engl. J. Med.* 303:833–41

194. Hadler, S. 1986. *N. Engl. J. Med.* 315:210–14

195. Persing, D. H., Varmus, H. E., Ganem, D. 1985. *Proc. Natl. Acad. Sci. USA* 82:3440–44

196. Yu, M., Finlayson, J., Shih, J. 1985. *J. Virol.* 55:736–43

197. Pohl, C., Cote, P., Purcell, R., Gerin, J. 1986. *J. Virol.* 60:943–49

198. Neurath, A. R., Kent, S. B. H., Strick, N., Parker, K. 1986. *Cell* 46:429–36

199. Gumbiner, B., Kelly, R. B. 1982. *Cell* 28:51–59

200. Patzer, E. J., Nakamura, G. R., Yaffe, A. 1984. *J. Virol.* 51:346–53

201. Persing, D., Varmus, H., Ganem, D. 1986. *Science* 234:1388–92

202. Gerber, M. A., Hadziyannis, S., Vissoulis, C., Schaffner, F., Paronetto, F., Popper, H. 1974. *Am. J. Pathol.* 75:489–502

203. Patzer, E., Nakamura, G., Simonsen, C., Levinson, A., Brands, R. 1986. *J. Virol.* 58:884–92

204. Eble, B., Lingappa, V., Ganem, D. 1986. *Mol. Cell. Biol.* 6:1454–63

205. Eble, B., MacRae, D., Lingappa, V., Ganem, D. 1987. Submitted for publication

206. Simon, K., Warren, G. 1983. *Adv. Protein Chem.* 36:79

207. Strauss, E., Strauss, J. 1985. In *Virus Structure and Assembly*, ed. S. Casjens, pp. 206–34. Boston: Jones & Bartlett

208. Standring, D., Ou, J., Rutter, W. 1986. *Proc. Natl. Acad. Sci. USA* 83:9338–42

209. Chisari, F., Filippi, P., Milich, D., McLachlan, A., Brinster, R., Palmiter, R. 1986. *J. Virol.* 60:880–87

210. Cheng, K-C, Smith, G. L., Moss, B. 1986. *J. Virol.* 60:337–44

211. Persing, D., Varmus, H., Ganem, D. 1987. *J. Virol.* In press

212. Will, H., Salfeld, J., Pfaff, E., Manso, C., Theilmann, L., Schaller, H. 1986. *Science* 231:594–96

213. Moriarty, A. M., Alexander, H., Lerner, R. A., Thornton, G. B. 1985. *Science* 227:429–33

214. Kay, A., Mandart, E., Trepo, C., Galibert, F. 1985. *EMBO J.* 4:1287–92

215. Meyers, M. L., Trepo, L. V., Nath, N., Sninsky, J. J. 1986. *J. Virol.* 57:101–9

216. Elfassi, E., Haseltine, W. A., Dienstag, J. L. 1986. *Proc. Natl. Acad. Sci. USA* 83:2219–22

217. Kozak, M. 1983. *Microbiol. Rev.* 47:1–45

218. Lamb, R. A., Choppin, P. W. 1983. *Ann. Rev. Biochem.* 52:467–506

219. Rizzetto, M. 1983. *Hepatology* 5:729–37

220. Rizzetto, M., Hoyer, B., Canese, M. G., Shih, J. W. K., Purcell, R. H., Gerin, J. L. 1980. *Proc. Natl. Acad. Sci. USA* 77:6124–28
221. Bonino, F., Heermann, K. H., Rizzetto, M., Gerlich, W. H. 1986. *J. Virol.* 58:945–50
222. Wang, K-S, Choo, Q-L, Weiner, A., Ou, J-H, Najarian, R. C., et al. 1986. *Nature* 323:508–14
223. Tanese, N., Roth, M., Goff, S. P. 1985. *Proc. Natl. Acad. Sci. USA* 82:4944–48
224. Kotewicz, M. L., D'Alessio, J. M., Driftmier, K. M., Blodgett, K. P., Gerard, G. F. 1985. *Gene* 35:249–58
225. Takatsuji, H., Hirochika, H., Fukushi, T., Ikeda, J. 1986. *Nature* 319:240–43
226. Varmus, H. E., Swanstrom, R. 1985. In *Molecular Biology of Tumor Viruses*, Chapter 5S. New York: Cold Spring Harbor Press
227. Varmus, H. E. 1983. *Nature* 304:116–17
228. Snyder, R. L., Summers, L. 1980. *Cold Spring Harbor Conf. Cell Prolif.* 7:447–58
229. Bishop, J. M. 1985. *Cell* 42:23–38
230. Tooze, J. 1980. *DNA Tumor Viruses.* New York: Cold Spring Harbor Press
231a. Bishop, J. M., Varmus, H. E. 1982. In *Molecular Biology of Tumor Viruses*, Chapter 9. New York: CSA
231b. Bishop, J. M., Varmus, H. E. 1985. See Ref. 226, Chapter 9S
232. Shaul, Y., Ziemer, M., Garcia, P. D., Crawford, R., Hsu, H., et al. 1984. *J. Virol.* 51:776–87
233. Ziemer, M., Garcia, P., Shaul, Y., Rutter, W. J. 1984. *J. Virol.* 53:885–92
234. Miyaki, M., Sato, C., Gotanda, T., Matsui, T., Mishiro, S., et al. 1986. *J. Gen. Virol.* 67:1449–54
235. Rogler, C. E., Sherman, M., Su, C. Y., Shafritz, D. A., Summers, J., et al. 1985. *Science* 230:319–22
236. Mizusawa, H., Taira, M., Yaginuma, K., Kobayashi, M., Yoshida, E., Koike, K. 1985. *Proc. Natl. Acad. Sci. USA* 82:208–12
237. Yaginuma, K., Kobayashi, M., Yoshida, E., Koike, K. 1985. *Proc. Natl. Acad. Sci. USA* 82:4458–62
238. DeJean, A., Sonigo, P., Wain-Hobson, S., Tiollais, P. 1984. *Proc. Natl. Acad. Sci. USA* 81:5350–54
239. Varmus, H. E. 1982. *Cancer Surv.* 2:301–19
240. Varmus, H. E. 1984. *Ann. Rev. Genet.* 18:553–612
241. DeJean, A., Bougueleret, L., Grzeschik, K. H., Tiollais, P. 1986. *Nature* 322:70–72

242. Denniston, K., Hoyer, B. H., Smedile, A., Wells, F., Nelson, J., Gerin, J. 1986. *Science* 232:1873–75
243. Ou, J.-H., Rutter, W. J. 1985. *Proc. Natl. Acad. Sci. USA* 82:83–87
244. Caveanee, W. K., Hansen, M. F., Nordenskjold, M., Kock, E., Maumenee, I., et al. 1985. *Science* 228:501–3
245. Orkin, S. H., Goldman, D. S., Sallan, S. E. 1984. *Nature* 309:172–74
246. Hansen, M. F., Koufos, A., Gallie, B. L., Phillips, R. A., Fodstad, O., et al. 1985. *Proc. Natl. Acad. Sci. USA* 82:6216–20
247. Koufos, A., Hansen, M. F., Copeland, N. G., Jenkins, N. A., Lampkin, B. C., Cavenee, W. K. 1985. *Nature* 316:330–34
248. Varmus, H. E., Quintrell, N., Ortiz, S. 1981. *Cell* 25:23–36
249. King, C. R., Kraus, M. H., Aaronson, S. A. 1985. *Science* 229:974–76
250. Glaser, T., Lewis, W. H., Bruns, G. A. P., Watkins, P. C., Rogler, C. E., et al. 1986. *Nature* 321:882–86
251. Moroy, T., Marchio, A., Etiemble, J., Trepo, C., Tiollais, P., Buendia, M-A. 1986. *Nature* 324:276–80
252. Deleted in proof
253. Ochiya, T., Fujiyama, A., Fukushige, S., Hatada, I., Matsubara, K. 1986. *Proc. Natl. Acad. Sci. USA* 83:4993–97
254. Pulciani, S., Santos, E., Lauver, A. V., Long, L. K., Aaronson, S. A., et al. 1982. *Nature* 300:539–42
255. Modali, R., Yang, S. S. 1986. *Prog. Chem. Biol. Res.* 207:147–58
256. Reynolds, S. H., Stowers, S. J., Maronpot, R. R., Anderson, M. W., Aaronson, S. A. 1986. *Proc. Natl. Acad. Sci. USA* 83:33–37
257. Fox, T. R., Watanabe, P. G. 1985. *Science* 228:596–97
258. Hino, O., Shows, T. G., Rogler, C. E. 1986. *Proc. Natl. Acad. Sci. USA* 83:8338–42
259. Ogston, C. W., Jonak, G. J., Rogler, C. E., Astrin, S. M., Summers, J. 1982. *Cell* 29:385–94
260. Shafritz, D. A., Shouval, D., Sherman, H. I., Hadziyannis, S. J., Kew, M. C. 1981. *N. Engl. J. Med.* 305:1067–73
261. Miller, R. H., Lee, S.-C., Liaw, Y.-F., Robinson, W. S. 1985. *J. Infect. Dis.* 151:1081–92
262. DeJean, A., Sonigo, P., Wain-Hobson, S., Tiollais, P. 1984. *Proc. Natl. Acad. Sci. USA* 81:5350–54
263. Monjardino, J. P., Fowler, M. J. F., Thomas, H. C. 1983. *J. Gen. Virol.* 64:2299–304
264. Koshy, R., Koch, S., von Loringhoven,

A. F., Kahmann, R., Murray, K., Hof-schneider, P. H. 1983. *Cell* 34:215–23

265. Koch, S., von Loringhoven, A. F., Hofschneider, P. H., Koshy, R. 1984. *EMBO J.* 3:2185–90

266. DeJean, A., Brechot, C., Tiollais, P., Wain-Hobson, S. 1983. *Proc. Natl. Acad. Sci. USA* 80:2505–9

267. Bowcock, A. M., Pinto, M. R., Bey, E., Juyl, J. M., Dusheiko, G. M., Bern-stein, R. 1985. *Cancer Genet. Cytogenet.* 18:19–26

268. Koike, K., Kobayashi, M., Mizusawa, H., Yoshida, E., Yaginuma, K., Taira, M. 1983. *Nucleic Acids Res.* 11:5391–402

269. Yaginuma, K., Kobayashi, M., Yoshi-da, E., Koike, K. 1985. *Proc. Natl. Acad. Sci. USA* 82:4458–62

270. Mizusawa, H., Taira, M., Yaginuma, K., Kobayashi, M., Yoshida, E., Koike, K. 1985. *Proc. Natl. Acad. Sci. USA* 82:208–12

271. Rogler, C. E., Summers, J. 1984. *J. Virol.* 50:832–37

272. Yaginuma, K., Shirakata, Y., Kobayashi, M., Koike, K. 1987. *Proc. Natl. Acad. Sci. USA* In press

273. Tsurimoto, T., Fujiyama, A., Matsu-bara, K. 1987. *Proc. Natl. Acad. Sci. USA* In press

274. Sureau, C., Romet-Lemonne, J-L., Mullins, J. I., Essex, M. 1986. *Cell* 47:37–47

Ann. Rev. Biochem. 1987. 56:695–726

MOLECULAR GENETICS OF MYOSIN

Charles P. Emerson, Jr.

Department of Biology, University of Virginia, Charlottesville, Virginia 22901

Sanford I. Bernstein

Department of Biology and Molecular Biology Institute, San Diego State University, San Diego, California 92182

CONTENTS

0066-4154/87/0701-0695$02.00

PERSPECTIVES AND SUMMARY

Myosin is a complex multimeric protein that has a central role in contractile processes of eukaryotes. Since its discovery about 50 years ago (1, 2), biochemical studies have provided a detailed understanding of the structure and organization of myosin in muscle and nonmuscle cells and its structural and enzymatic functions in contractile processes. These aspects of myosin protein structure and function have been reviewed extensively (3–7, 7a).

Within the past five years, genes encoding the heavy chain (MHC) subunit and the alkali (alkali MLC) and regulatory (MLC-2) light chain subunits of myosin have been cloned. The structures of these genes have provided a greatly expanded body of primary sequence data on myosin proteins in invertebrates and vertebrates and on myosin isoforms of specialized muscle and nonmuscle cells. Comparative data on myosin proteins from evolutionarily divergent organisms have contributed towards defining important structural and functional domains of MHC and MLC proteins. MHC gene analysis has provided the first complete primary sequences of MHC proteins. The cloning of myosin genes also has led to a molecular genetic understanding of the diversity of MHC and MLC protein isoforms generated by alternative RNA splicing of myosin gene transcripts and by the expression of members of myosin gene families. Myosin genes have been mapped to specific chromosomal loci, and mutations of MHC genes that cause specific muscle dysfunctions have been characterized at the DNA level in the fruit fly, *Drosophila melanogaster,* and in the nematode, *Caenorhabditis elegans.* Cloned rat MLC-2 genes have been introduced transgenically into mouse embryos to identify muscle-specific transcriptional regulatory elements, and a chicken MLC-2 cDNA has been mutated in vitro to identify specific functional domains in the MLC-2 protein.

In this article we review recent progress in the molecular genetics of myosin. The reader is also referred to recent reviews on nematode MHC gene cloning (8) and the alkali MLC genes (9). We have attempted a general overview of the molecular genetics of MHC genes, alkali MLC genes, and regulatory MLC-2 genes, emphasizing the contributions of molecular genetics to our understanding of the biochemistry of myosin and its function in contraction, the evolution of myosin genes and gene families, and the genetic, developmental, and physiological regulation of myosin genes.

OVERVIEW OF MYOSIN STRUCTURE AND FUNCTION

Myosin is a large, hexameric protein composed of two heavy chain subunits (MHC), two alkali light chain subunits (alkali MLC), and two regulatory light chain subunits (regulatory MLC-2) (10–12). Biochemical and physiological studies reveal that myosin in muscles of vertebrates and invertebrates has a complex molecular structure that specifies ATPase activity (13–16), intramolecular conformational changes, intermolecular interactions with actin during contraction (6, 17–21), and assembly into the thick filaments of sarcomeric muscles (21a).

MHC Proteins

The first complete primary structure of a MHC protein was obtained from the structure of a cloned *C. elegans* unc-54 gene that encodes body wall MHC B (22). Interpretation of MHC B structure was facilitated by the partial MHC protein sequence data for rabbit skeletal MHC (23–25). More recently the complete structures of a rat embryonic MHC gene and its MHC protein have been reported (26). These MHC primary sequence data have provided a new general understanding of the structural domains of MHC and its assembly into filaments in muscle (22, 27–29) and a structural basis for characterizing myosin mutations (see GENETICS OF MYOSIN GENES).

The nematode MHC B (Figure 1) is a large protein of 1966 amino acids split into two general structural domains: the S1 globular head in the N-terminal 850 amino acids and an α-helical rod in the C-terminal region (30). The head is attached to the rod at a proteinase-sensitive "hinge" region that is the site of head mobility during crossbridge formation (32, 33). The S1 head has three subdomains separated by two proteinase-sensitive regions. The N-terminal 23K domain includes a highly conserved ATP binding site (Figure 2A) (34, 35) and lysine-88, essential for ATPase activity (36); the 50K and 22K domains include actin-binding sites (32, 37, 38); the 22K domain includes two reactive cysteines (SH_1, SH_2) required for ATPase activity and located within the globular head near the ATP binding site in the 23K domain (39–41), and a MLC binding region (44, 44a–46) that extends into the "hinge" (42, 43) near the proteinase-sensitive junction of the head and the "short S2 region" in the rod.

The rod portion of MHC is a repetitive α-helical sequence, totally devoid of proline residues, that assembles in register with another MHC to form a coiled-coil (47, 27–29). The repetitive rod sequence includes forty 28-amino-acid repeats of seven-residue units with hydrophobic residues at the *a* and *d* positions (Figure 2B). This organization provides a hydrophobic surface for the interactions between the two MHC subunits (27, 28). The 28-amino-acid repeat is interrupted by four invariantly placed (every 197 amino acids) skip residues that interrupt the hydrophobic surface and probably cause local distortions in intramolecular packing of MHC dimers within the thick filament (27, 28, 47a). Pairing between the MHC B monomers likely begins at the S2/LMM boundary at alanine 1185 and continues to alanine 1943, followed by a nonhelical, C-terminal tail piece in *C. elegans* and the *Drosophila*

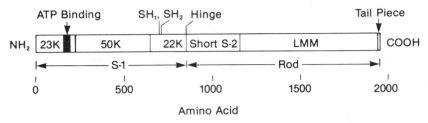

Figure 1   Structural domains of MHC.

muscle MHCs (27) (Figure 2*B*). In sarcomeric muscles, MHC dimers associate in staggered arrays through electrostatic interactions between their coiled-coil rods. Myosin molecules initially assemble in antiparallel orientations to form a polarized thick filament with globular S1 heads protruding at either end of a bare central zone (21a). In the more distal regions of the filament, myosin assembles in parallel (head to tail) orientations, and the protruding heads form crossbridges with a characteristic helical repeat between the thick myosin filaments and the thin actin filaments (48). The pairs of heads from each MHC undergo cyclic cooperative movements at the hinge region (49, 49a) and intramolecular conformational changes (50), leading to ATP binding, ATP hydrolysis, and the generation of tension. The outer surface of the dimer MHC rod has a high charge density that may be significant in the electrostatic interactions that promote filament assembly and result in the periodic 98-residue charge distributions that correlate with observed crossbridge stagger (27).

These structural features of the myosin rod are conserved and likely are general properties important in the assembly and packing of thick filaments. Some isoform-specific features may also be included in the rod sequences as suggested by the fact that nematode body wall muscle has two MHC isoforms, MHC A and MHC B, that assemble only as homodimers (51). MHC A dimers assemble in the central region, and MHC B dimers assemble in the distal region of the thick filament (52). Also the length of thick filaments and number of myosins per filament vary in different muscles and organisms (53). Other proteins also control myosin and thick filament assembly. In *C. elegans, unc-15* paramyosin mutations disrupt assembly (54), and filament core protein(s) provide a molecular scaffold (55). In vertebrates, C protein and proteins H and X are organized in the thick filament and presumably participate in assembly (56).

MLC Proteins

Amino acid sequence data identify two classes of myosin light chains, alkali MLC-1/3 and regulatory MLC-2 (Figures 3*A* and 3*B*) (57–59). Both classes

A

ATP Binding Regions

```
Nem unc-54 MHC    ...MPPHLFAVSDEAYRNMLQDHENQSMLITGESGAGKTENTKK...

Rat emb MHC       ...A***I*SI**N**QF**T*R****I**T********V***R...

Dros 36B MHC      ...V***I**I**G**VD**TN*V****************...

Amoeba myosin II  ...VA**I**I**A***A**NTRQ******T********...
```

B

LMM-COOH Termini

```
                  defgabcdefgabcdefgabcdefgabc
                  :+   + :+   + :+   + :+   + :

Nem unc-54 MHC ...LQDLIDKLQQKLKTQKKQVEEAEELANL

Rat emb MHC    ...****V****V*V*SY*R*A***D*Q**V

Dros 36B MHC   ...M***V*******I**Y*R*I******I*A*

Dict MHC       ...*ETDYKRAKKEAADEQQ*RLTV*NDLRK
```

```
NLQKYKQLTHQLEDAEERADQAENSLSK

H*T*FRKAQ*E**E******I**SQVN*

**A*FRKAQQE**E******L**QAI**

H*SEISL*KDAIDKLQRDH*KTKRE*ET
```

```
MRSKSRASASVAPGLQSSASAAVIRSPSRARASDF

L*A*T*DFT*SRMVVHE*EE

F*A*G**GSVGRGASPA           PRATS*RPQFDGLAFPPRFDLAPENEF
                    I

ETASKIEMQRKMADFFGGFK*
```

Figure 2 Comparisons of ATP binding and LMM-COOH terminal domains of MHC proteins.
   A. ATP binding domains of rat embryonic (26), *Drosophila* 36B (P. T. O'Donnell, S. I. Bernstein, unpublished data), and *Acanthamoeba* myosin II (120) proteins are aligned with the *C. elegans unc-54*, MHC B ATP binding domain, amino acid residues 155–195 (22).
   B. The C-termini of rat embryonic (26), *Drosophila* 36B (124, 125), and *Dictyostelium* (121) MHC proteins are aligned with the *C. elegans unc-54*, MHC B C-terminus (22). The α-helical 28-amino-acid repetitive unit is composed of four hydrophobic amino acid repeats [indicated by colons (:)] with preferred hydrophobic residies (+) at positions d and a, followed by a nonhelical tail piece at the very C-terminus.
   Residues homologous to the nematode sequence are shown as *. Intron positions are indicated by ▼ , intercodon and ↓ , intracodon.

include multiple isoforms. MLCs are members of the super family of calcium-binding proteins that have four EF hand domains composed of an α-helix E, a divalent cation-binding loop, and an α-helix F (60). Some members of this super family (calmodulins and troponin C) have four functional EF hand domains, but MLC-2 has only one functional calcium-binding EF hand, domain I. The other three EF hand domains of MLC-2 and the four domains of alkali MLC have accumulated deletions and nonconservative amino acid substitutions that inactivate their Ca^{2+} binding ability (61).

The globular head of each MHC subunit is associated with one MLC-alkali and one MLC-2 subunit (10–12, 62). Immunoelectronmicroscopy and chemical crosslinking studies show that these MLCs are closely opposed with their N-termini extending from the MHC globular head region back into the neck and hinge region (19, 63, 63a,b). The function of the alkali MLC is not understood, but may be structural. Vertebrate skeletal MHC retains its ATPase activity after removal of the alkali MLC and the regulatory MLC-2 (64). However, MLC-2, but not the alkali MLC, can be reversibly dissociated from scallop myosin (65). Regulatory MLC-2 light chains removed from myosins of various organisms and muscle types also can be reassociated with the scallop MHC/alkali MLC complex to investigate the function of specific isoforms in the regulation of myosin-actin interaction during crossbridge formation (66, 67). The scallop reconstitution system has provided evidence that scallop MLC-2 inhibits myosin-actin interactions and the binding of Ca^{2+} relieves this inhibition (65), whereas smooth muscle and nonmuscle isoforms of MLC-2 regulate myosin-actin interactions through the phosphorylation of a serine in the N-terminal region (Figure 3B) by a Ca^{2+}-regulated, calmodulin-activated kinase (68). Phosphorylation of vertebrate skeletal and cardiac MLC-2 isoforms is not a primary mechanism of control but may modulate their function (69). The regulatory activity of MLC-2 mediated by Ca^{2+} and phosphorylation may induce the profound conformational changes observed in the association of its N-terminus with the alkali MLC and MHC subunits (50).

The function of vertebrate skeletal and cardiac MLC-2 is less well understood, but in vitro mutagenesis studies of a chicken skeletal MLC-2 cDNA establish the functionality of the N-terminal EF hand domain I in Ca^{2+} binding and in the regulation of myosin-actin interactions (70). In these experiments, a full-length chicken cDNA clone was mutated to change amino acid residues in the EF hand domain I. The mutated protein expressed in *Escherichia coli* has lost both its capacity to bind Ca^{2+} and its capacity to inhibit actin-activated myosin ATPase activity in the scallop reconstitution assay. Future mutagenesis studies of other conserved regions of the MLC-2 protein using the scallop myosin reconstitution system should be informative.

A

```
Dros MLC alkali                                                  MADVPKKREVENVEFVFEVMGSPGEGIDAVDLGDALRALN
                                                                ▼                    ──────I──────
Chick MLC alkali 1    MAPKKDVKKPAAAAPAPAPAPAPAPAPAPAKPKEPAIDLKSIKIEFSKEQQDD
Chick MLC alkali 3                                              FK*AFLLFDRTGDAK*TLSQV**IV***G
                                                                             ──────III──────
                                                               *SFS*DEIND
                                                              ▲          ▲

                      LNPTLALIEK-LGGTKKR--NEKKIKLDEFLPIYS-NVKKEKEQGCTEDFIECLKLTDKEENGTMLLAELNHALLALGESL
                              ────────II────────
                      Q***N*E*N*I**NPS*EEM*A***TFE****MLQMAAANN*D**TF***V*G*RVF***G****V-G****R*V*AT***KM
                                              ▲

                                                    FVQRLMSDPVVFE
                      DDEQVETLFADCMDPEDDEGFIPYSQ
                      ────────IV────────       ▼
                                              *LA*MCER*DQLK
                      TE*E**E*MKGQE*SN---*C*NTEA**KHI**V
                                     ▲                 ▲
```

B

```
Dros MLC-2                                                                  Ⓟ
             MADEKKKKVKKKTKEEGGTSETASEAASEAATPAPAATPAPAASATGSKRASGGSRGSRKSKRAGSS-VFSVFSQKQIAE
             ▼
Rat MLC-2    **P--**A*RR-----------------**A*-                        ***N***M*D*T**Q*
(DTNB)      ▲

             FKEAFQLMDADKDGIIGKNDLRAAFDSVGK-IANDKELDAMLGEASGPINFTQLLTLGANRMATSGANDEDEVVIAAFKT
                   ────────I────────                           ────────II────────
             ****TVI*QNR****D*E***DT*AAM*RLNVKNE*******MK*********VF**--MFGEKLK-**DP**-*ITG***V
                  ▲

             FDND--GLIDGDKFREMLMNFGDKFTM*EVDDAYDQMVIDDKNQID-TAALIEMLTGKGEEEEEAA
             ────────III────────                ────────IV────────
             L*PEGK*T*KKQFLE*L*TTQC*R*SQE*IKNMWAAFPP*VGGNV*YKNICY-VI*H-*DAKDQ*
                  ▲                                              ▲
```

Figure 3 Comparisons of alkali MLC and regulatory MLC-2 proteins.

A. *Drosophila* alkali MLC (83, 126) and chicken MLC-1/3 (93) are aligned.

B. *Drosophila* MLC-2 (84) and rat fast skeletal muscle MLC-2 (96) are aligned. The site of phosphorylation of a serine residue is shown Ⓟ

Homologous residues are shown as *. Intron positions are shown as ▼ , intercodon and ↓ , intracodon.

Lines above regions indicate positions of homologous EF hand domains I-IV.

Muscle and Nonmuscle Myosin Isoforms

Electrophoretic and immunological studies show that invertebrates and vertebrates express characteristic MHC and MLC protein isoforms in their functionally specialized muscles (62, 71–73) and at specific embryonic stages and adult physiological states (189, 192, 193, 196, 198a). Myosins from physiologically different muscle types, as well as cardiac myosin isoforms (15, 16), have characteristic ATPase activities (15, 16, 73a), and the abundance of the protein isoform and its mRNA in specific muscles can be modulated in response to thyroid hormone (74, 75, 113, 199a), work hypertrophy (76), or innervation (77). In vertebrates, functionally specialized muscles include fast and slow skeletal muscles, cardiac atrial and ventricular muscles, and vascular and other smooth muscles. In invertebrates, specialized muscles include fibrillar flight muscles, tubular body wall muscles and visceral organ muscles of insects (53), catch muscles in molluscs (65), crustacean claw muscles (77a), and pharyngeal and body wall muscles of nematodes (51, 52). All eukaryotic cells apparently have nonmuscle myosin isoforms (77b, 78). Nonmuscle MHC isoforms are structurally and functionally distinct from isoforms expressed in sarcomeric muscles such as the skeletal and cardiac muscles of vertebrates, but appear biochemically and immunologically related to vertebrate smooth muscle myosins. Nonmuscle myosin is implicated in cellular locomotion, cytoplasmic streaming, and cytokinesis (78).

CLONING OF MYOSIN GENES AND STRUCTURAL COMPARISONS OF MYOSIN PROTEINS

Muscle MHC and MLC genes have been cloned as genomic segments and cDNA clones from the nematode *C. elegans* (22, 79, 80), *Drosophila* (81–84), chicken (47a, 85–93), quail (94, 95), hamster (96a), rat (26, 97, 99, 100–103, 104–107), mouse (108–112), rabbit (113–116), and human (117–119). Nonmuscle MHC genes have been cloned from *Acanthamoeba* (120, 120a) and *Dictyostelium* (121). The complete gene and protein structures of the nematode *unc-54* MHC B gene (22) and the rat embryonic MHC gene (26, 99, 122) have been determined. Complete gene and protein structures have been determined for fast skeletal chicken (93), rat (106), and mouse (123) alkali MLC-1/3 genes and for the *Drosophila* (83) alkali MLC gene (Figure 3A). Complete sequences also have been obtained for a rat MLC-2 fast skeletal gene (96) and the *Drosophila* MLC-2 gene (84) (Figure 3B). Structural data obtained from sequence analyses of these muscle-specific MLC and MHC genes and cDNA clones of their mRNAs have provided an extensive new source of comparative protein and gene structure data for evolutionary analyses. These cloned genes have also been useful as nucleic acid hybridization probes for studies of myosin gene families, the developmental and tissue-specific expression of myosin gene transcripts, and the chromosomal locations of myosin genes and myosin gene mutations.

MHC Genes

Different approaches have been used to clone MHC genes. To isolate the *C. elegans unc-54* (MHC B) gene, cDNA clones were prepared from mRNA enriched for MHC by size selection (79). Putative MHC cDNA clones were identified by hybridization to RNA from *unc-54* mutants that produce reduced amounts or shortened *unc-54* transcripts. DNA sequence analysis established that these cDNAs encode a MHC protein homologous to rabbit skeletal MHC (22). The *unc-54* gene encoding MHC B was isolated by screening a library of nematode genomic DNA with this cDNA clone (123a), and the *unc-54* gene was subsequently used to isolate six additional nonoverlapping MHC genomic clones, designated *myo 1–6* (8). Clones *myo-1, -2,* and *-3* are homologous to both S1 head and HMM rod portions of the *unc-54* gene. DNA fragments encoding segments of the MHC rod of the *myo-1, -2,* or *-3* genes were fused to a β-galactosidase gene, cloned into an expression vector, and then identified with myosin-specific monoclonal antibody probes as encoding MHC isoforms expressed in the body wall, MHC A *(myo-3),* and in the pharynx, MHC C *(myo-2)* and MHC D *(myo-1)* (80). The identities of *myo-4, -5, -6* genes are unknown.

The four nematode sarcomeric MHC genes encode isoforms that share extensive homology in their head and rod regions with rabbit skeletal MHC protein (22–25, 27) and with MHCs encoded by other known sarcomeric MHC genes. One of the most conserved MHC gene and protein sequences is located in the ATP binding domain (35) (Figure 2A). The nematode *unc-54* ATP binding domain is homologous with the very distantly related nonmuscle *Acanthamoeba* myosin I and II (120, 120a). The evolutionary conservation of MHC gene sequences has facilitated the isolation and identification of other MHC genes and gene families.

Other approaches have also been used to isolate MHC genes. A *Dictyostelium* nonmuscle MHC cDNA has been isolated by screening a λgt 11 expression library with a MHC polyclonal antibody (121). Avian and mammalian MHC genes have been isolated by differential screening of cDNA clone libraries of skeletal and cardiac muscle mRNAs using size-selected, muscle-specific cDNA probes or heterologous probes of identified MHC genes (87, 94, 101, 105). Vertebrate sarcomeric MHC genes and gene families have been isolated from genomic DNA libraries screened with MHC cDNA clones.

Some general structural features shared by MHC proteins emerge from DNA sequencing data. The sequences of the globular head are more highly conserved than rod sequences (22, 26). Conserved globular head (S1) sequences include: 1. the ATP binding site in the N-terminal, 23 kd domain (Figure 2A), 2. the region of the two active thiols (cysteines 717 and 727 in the nematode 22 kd domain), and 3. sequences in the 50 kd domain, perhaps involved in actin binding near the 50 kd/22 kd junction. Divergent regions in the head include: 1. the 80-amino-acid N-terminal sequence (122), 2. short sequences around amino acids 210 and 645 (*unc-54* MHC numbering system) that are flexible, charged surface loops that likely are the protease-sensitive sites of the 23 kd, 50 kd, and 22 kd head fragments, and 3. the "swivel" or hinge region, around amino acid 815, that likely provides head mobility during actin-MHC crossbridge formation. The rod region begins at residue 863, the last proline until the C-terminal tail piece. Although the rod sequences are less conserved than the head, sarcomeric MHCs share considerable homology, whereas rod sequences of *Acanthamoeba* and *Dictyostelium* nonmuscle MHCs are divergent (120, 121) (Figure 2B). The rod portions of both muscle and nonmuscle MHCs, however, are α-helical and have a repetitive sequence organization characteristic of α-helical coiled-coil proteins.

The C-terminal sequences of MHCs are divergent (Figure 2B). Nonmuscle MHC, *C. elegans* MHC A and B, and both *Drosophila* 36B MHC isoforms have a nonhelical tail piece sequence at their carboxy termini. Vertebrate MHCs have helical C-termini (27). The *Drosophila* C-terminal sequences are of special interest since two MHC proteins are produced by developmentally regulated, alternative RNA splicing (124, 125) that gives rise to a tubular muscle MHC isoform with a 27-amino-acid terminus and a thorax-specific (indirect flight muscle) MHC isoform with one amino acid and no extended tail piece. Since C-terminal sequences are important for filament assembly in nonmuscle myosins (126a, 127), alternative C-termini of MHC isoforms in different *Drosophila* muscles may be important for determining the different structures of the thick filaments of tubular and thorax muscles (124).

Alkali MLC Genes

Drosophila alkali MLC (83), and chicken (93), rat (106), and mouse (123) alkali MLC-1/3 fast skeletal genes have been cloned, and complete structures determined (Figures 3A and 4A). *Drosophila* alkali MLC and vertebrate MLC-1/3 fast skeletal proteins share regions of local homology in the central EF hand domains II and III (Figure 3A). These regions are thought to be nonfunctional in divalent cation binding, but the presence of conserved sequences suggests that these regions have important functions.

The cloning of vertebrate alkali MLC-1/3 genes has provided a molecular genetic basis, originally suggested by protein (57) and cDNA sequence analyses (128), for the closely related structures of the MLC-1 and MLC-3 proteins. These proteins are products of a single gene, but their RNAs are transcribed from two promoters (Figure 4A) that direct different patterns of alternative exon splicing and result in proteins with variant N-termini and

MYOSIN LIGHT CHAIN – ALKALI

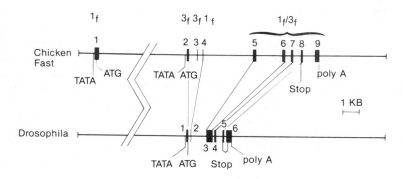

MYOSIN LIGHT CHAIN – 2

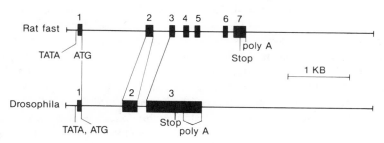

MYOSIN HEAVY CHAIN

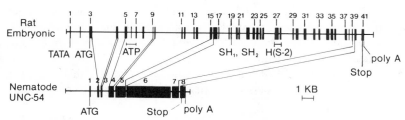

Figure 4 Comparisons of the structures of alkali MLC, regulatory MLC-2, and MHC genes. Exons are shown as boxes and common intron/exon boundaries are shown as lines interconnecting genes.

A. Comparison of the *Drosophila* alkali MLC gene (83) with the chicken alkali MLC-1/3 gene (93). The chicken gene has dual promoters and alternative splicing of exons 3 and 4. The MLC-1 promoter is 10 kb upstream of the MLC-3 promoter. The *Drosophila* gene has regulated inclusion/exclusion of exon 5 (126). (Adapted from 140a.)

B. Comparison of the *Drosophila* MLC-2 gene (84) with the rat MLC-2 fast skeletal muscle gene (96).

C. Comparison of the rat embryonic MHC gene (26) with the nematode *unc-54* gene (22). (Adapted from 26.)

conserved C-termini (Figure 3A). In contrast, the *Drosophila* alkali MLC gene has a single promoter (Figure 4A) and produces a protein similar to vertebrate fast skeletal alkali MLC-3 and to scallop muscle alkali MLC (Figure 3A) (83). However, inclusion or exclusion of exon 5 during *Drosophila* alkali MLC RNA splicing gives rise to two isoforms with different, but structurally similar C-termini (126) (Figure 3A). The functional roles of these two C-termini, if any, are unknown. Like the *Drosophila* alkali MLC isoforms, vertebrate smooth muscle and nonmuscle MLC proteins have a truncated N-terminus and may have alternative C-termini (129). Thus the alkali MLC gene of invertebrate muscles, vertebrate smooth muscle, and nonmuscle cells is similar to the MLC-3 coding portion of vertebrate fast skeletal MLC-1/3 genes, suggesting that MLC-3-type genes are the evolutionary precursors to other alkali MLC genes (9).

The cloning of the mouse fast skeletal alkali MLC-1/3 gene also led to the isolation of a cDNA clone encoding an alkali MLC-1 isoform coexpressed in the cardiac ventricle and slow skeletal muscle (108), and to the isolation of a cDNA clone for an alkali MLC-1 isoform coexpressed in the cardiac atrium and embryonic skeletal muscles (109). In addition, a mouse genomic MLC-1/3 pseudogene has been isolated from *Mus musculus*. This pseudogene appears to be a reverse transcript of an aberrant MLC-1/3 splicing event and is not found in the related species, *Mus spretus,* suggesting that this pseudogene was generated within the past seven million years (123). The alkali MLC-1 ventricle/slow skeletal and the MLC-1 atrial/embryonic skeletal isoforms each are encoded by a single copy gene that produces a protein with an N-terminal extension homologous to alkali MLC-1 fast skeletal protein (130, 131). Neither gene has been shown to produce a MLC-3 type protein by alternative splicing, although minor forms of MLC proteins have been detected in some mammals (131). Amphibian and avian alkali MLC-1 atrial/embryonic skeletal isoforms have not been discovered, suggesting that the mammalian gene is a recent gene duplication product (9). Amphibian and avian ventricular and atrial alkali MLCs are identical (132, 133), and fish cardiac and slow skeletal MLCs are similar if not identical (134), further indicating that alkali MLC gene families are more complex in mammals than in other vertebrates.

Regulatory MLC-2 Genes

MLC-2 regulates interactions between myosin head and actin through the binding of divalent cations or the phosphorylation of a serine residue. The *Drosophila* muscle MLC-2 gene (84) and the rat MLC-2 fast skeletal gene (96) have been cloned and their complete gene structures determined (Figure 4B). The *Drosophila* and rat MLC-2 proteins share homology in the I and II EF-hand domains and in the region of the phosphorylated serine (Figure 3B), suggesting a functional importance of these sequences. Calcium binding in the EF hand domain I apparently is essential for MLC-2 function (70). The *Drosophila* MLC-2 gene encodes an N-terminal extension peptide that is not present in avian, mammalian, or molluscan MLC-2 proteins and that has sequence homology with the vertebrate MLC-1 proteins in the N-terminal proline-alanine-rich region (Figures 3*A* and 3*B*). The N-terminal amino acids of *Drosophila* and rat skeletal MLC-2 proteins (as well as other MLC-2 proteins) are basic. The significance of this conserved feature is unknown, although one possibility is that it is related to the N-terminal conformational changes in MLC-2 during the contraction cycle (50). In vertebrates, several isoforms of MLC-2 exist besides the skeletal muscle protein. Protein sequence data suggest the existence of three other chicken MLC-2 genes that encode a smooth muscle protein, MLC-2SM, and two cardiac proteins, MLC-2A and MLC-2B (58). These chicken isoforms share local regions of homology in EF hand domains I and II, in the region of the phosphorylatable serine, and in the basic N-terminal region, as do the *Drosophila* and rat MLC-2 proteins. These proteins also are divergent in the C-terminal regions thought to be involved in MHC binding (135). A chicken cardiac MLC-2A cDNA and gene (91, 92) and a quail skeletal MLC-2 cDNA (94) also have been cloned and partially characterized. A full-length sequence of a rat cardiac MLC-2 cDNA has been determined (103). The mRNA encoded by this cardiac MLC-2 gene is expressed in both atrial and ventricular tissue (103). Genes encoding avian cardiac MLC-2B and avian and mammalian nonmuscle and smooth muscle MLC-2 have not yet been cloned.

MYOSIN GENE FAMILIES AND ISOFORMS

The evolutionary conservation of MHC, alkali MLC, and regulatory MLC-2 proteins (Figures 2 and 3) is reflected in the structural homology of genes encoding myosin isoforms in widely divergent organisms (Figure 4). Genomic and cDNA clones of myosins have made possible studies of the homology and the complexity of myosin gene families and isoforms in different organisms, the tissue, temporal, and physiological patterns of transcript regulation, and the chromosomal organization of MHC genes.

Drosophila melanogaster has the simplest organization of myosin genes yet discovered. Muscle-specific MHC (81, 82), alkali MLC (83), and MLC-2 (84) genes are single copy in the haploid *Drosophila* genome. In contrast, *Drosophila* actin genes comprise a family of six members that encode highly homologous proteins that have different stage-specific and tissue-specific patterns of expression (136). Despite this simplicity of gene number, the *Drosophila* MHC gene and the alkali MLC gene each encode two protein isoforms with different C-termini produced by a tissue-specific alternative RNA splicing mechanism involving inclusion/exclusion of the penultimate exon. In the pupal thorax, the alkali MLC penultimate exon is excluded (126; S. Falkenthal, personal communication), whereas the MHC penultimate exon is included (124, 125). Despite the opposite tissue-specificity for the inclusion or exclusion, the alternatively spliced exons of the alkali MLC and the MHC gene have similar unusual intron splice junction sequences that may be important in splicing regulation (124, 125). Although other *Drosophila* muscle myosin genes have not been detected, genes that encode divergent nonmuscle MHC and MLC proteins most probably exist. *Drosophila* has a nonmuscle MHC protein that is immunologically distinct from muscle MHC (126b) and may be encoded by a separate gene (136a). Additional muscle MHC isoforms generated by alternative splicing may also be revealed once the complete structure of the 36B gene is determined. Other MHC genes that produce multiple isoforms by alternative splicing have not been discovered in the nematode or vertebrates.

The nematode *(C. elegans)* has a family of MHC genes that encode four muscle-specific MHC isoforms and three unidentified genes with related sequences (8, 80). MLC genes have not yet been identified. MHC A and B isoforms are selectively expressed in body wall muscles in a coordinated 1:2 ratio (137), and MHC C and MHC D isoforms are selectively expressed in the pharyngeal muscles.

Vertebrate MHC gene families are large, and in addition to expression in specific muscle types, vertebrate MHC genes are subject to more complex developmental and physiological regulation. Chickens have as many as 31 MHC genes with homologous ATP binding regions (138). At least five MHC transcripts have been identified in chick and quail by cDNA cloning (88, 89, 94, 95). These transcripts encode embryonic and adult skeletal isoforms and a cardiac isoform also expressed in embryonic skeletal muscle (95). Structural studies of a subset of five chicken genes with closely related sequences reveal that these genes encode MHC proteins with almost identical N-termini (139). This is an unexpected result since N-termini are divergent in other MHC proteins (122). RNA hybridization studies with gene-specific probes from two of these genes indicate they are differentially expressed in different muscles during development (138). Unexpected sequence conservation also has been reported for the rabbit cardiac α- and β-MHCs in their rod regions, suggesting that these genes have undergone recent gene conversion events (116).

The rat has seven sarcomeric MHC genes that have been cloned. These encode an embryonic, neonatal, extraocular, and fast IIA and fast IIB skeletal MHC isoforms, and α-MHC ventricular/atrial and β-MHC ventricular/slow skeletal isoforms (104). These seven MHC genes, which likely represent the entire set of mammalian sarcomeric MHC genes, encode related proteins that have diverse but overlapping patterns of RNA transcript expression in different muscles at different developmental stages, and under different physiological conditions (104, 189).

Alkali MLC and regulatory MLC-2 genes are small gene families in vertebrates (9, 57, 58). The mammalian alkali MLC family includes at least four genes: a MLC-1/3 fast skeletal gene, a MLC-1 ventricular/slow skeletal gene, a MLC-1 atrial/embryonic skeletal gene, and a MLC-3 smooth/nonmuscle gene. The MLC-2 gene family includes at least four genes: a MLC-2 skeletal gene, a MLC-2 cardiac gene (MLC-2A and MLC-2B in chickens), a MLC-2 smooth muscle gene, and a MLC-2 nonmuscle gene. Nonmuscle and smooth muscle MHC and MLC genes have not yet been isolated from higher organisms, so the complexity of these gene families is unknown. *Acanthamoeba* has at least three nonmuscle MHC genes (140), but *Dictyostelium* appears to have only a single gene (121).

CHROMOSOMAL ORGANIZATION OF MYOSIN AND OTHER MUSCLE GENES

Molecular cloning techniques combined with genetic and cytogenetic analyses have permitted chromosomal localization of myosin and other muscle genes in *C. elegans, Drosophila,* and the mouse. These studies reveal that MHC and MLC genes are not closely linked to each other, to other contractile protein genes, or to other muscle-specific gene loci. One exception is the mammalian sarcomeric MHC genes that are clustered at two chromosomal sites, one containing five skeletal MHC genes and the other containing two cardiac genes.

Drosophila melanogaster

Drosophila myosin and other muscle genes have been mapped by in situ hybridization of cloned muscle gene probes to salivary gland chromosomes and by genetic analysis of mutations. A total of 48 muscle gene loci have been mapped, including 36 cloned loci (Figure 5). These muscle genes are scattered over the *Drosophila* chromosomes. The only semblance of a muscle gene cluster is the flight muscle–specific actin gene and the tropomyosin gene family located 150 kb apart at 88F (141). The 36B(2L) muscle MHC gene is unlinked to the MLC-2 and alkali MLC genes, which are at 99E(3R) and 100B(3R), respectively. A nonmuscle MHC gene has been tentatively mapped to 60EF(2R) (D. Kiehart, personal communication).

Caenorhabditis elegans

The majority of the 30 known nematode muscle genes, including the *unc-54* MHC B gene, were identified by analyzing paralyzed or uncoordinated mutants (*unc* mutations) or suppressors of *unc* mutations *(sup)* (Figure 6A). The locations of the dispersed *myo-1, -2,* and *-3* MHC genes and the four actin genes (three of which are clustered on chromosome V) have been mapped at low resolution by in situ chromosome hybridization

DROSOPHILA

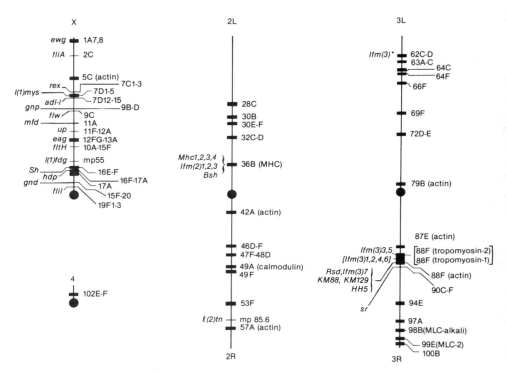

Figure 5   Drosophila muscle gene loci.

Loci identified by genetic studies of muscle-defective mutants are: *ewg, gmp, flw, up, hdp, gnd* (142, 143); *fliA* and *FliI* (144, 145); *rex* and *adl-1* (146); *l(1)mys* (147); *l(1)fdg* (148); *mfd* (149); *fltH* (150); *sh* (151); *sr* (152); *eag* (153); *l(2)tn* (154) and *ifm(3)** (since named *l(3)laker*) (E. Ball, J. Sparrow, personal communication). Recent complementation tests show that *fliI* mutations are allelic to *sdby* (143) and *fltO* (150). Cytological localizations of *fliA, rex, adl-1,* and *hdp* are from unpublished data (T. Homyk, Jr. and C. P. Emerson, Jr.).

DNA clones for the regions including the genes *ewg* (R. Fleming, K. White, personal communication), *Sh* (155), *eag* (156), and *l(1)mys* (157) have been isolated by various molecular genetic strategies.

The chromosomal locations of six actin genes at positions 5C, 42A, 57A, 79B, 87E, and 88F (158, 159), two tropomyosin and a tropomyosin-related gene at 88F (160, 161), myosin light chain genes at 98B and 99E (83), a single myosin heavy chain gene at 36B (81, 82), and a single calmodulin gene at 49A (S. Tobin, personal communication) have been identified by in situ hybridization of cloned genes to salivary gland chromosomes. Dominant flightless mutations mapping to 36B and 88F (162) alter myosin heavy chain gene sequences at 36B (163) and tropomyosin and actin gene sequences at 88F (164, 165, 231).

Further muscle-specific coding sequences at positions 17A, 28C, 30B, 30E–F, 53F, 64C, 64F, 66F, 72D–E, 94E, 97A, 100B, and 102E–F have been identified by in situ chromosome with gene-specific probes (167). *Myo-1, -2,* and *-3* are located in regions of several *unc* and *sup* genes that may represent mutations of these MHC genes.

Mouse

Cloned mouse and other mammalian genes have been mapped to specific chromosomes by analysis of the inheritance of restriction site length polymorphisms in two mouse strains (108, 109, 172), by analysis of chromosome segregation in somatic cell hybrids (172a, 173–176, 178), or by chromosome walking (177) (Figure 6*B*). There are no myosin mutations identified in mammals, although mutations that affect muscle function have been described in the mouse and human, and some of these mutations are linked chromosomally to myosin genes (see GENETICS OF MYOSIN GENES). Mouse alkali MLC-1 and MLC-2 gene family members, although proposed to be products of recent gene duplications, are chromosomally dispersed (108, 109, 172), whearese the mammalian muscle MHC genes are chromosomally linked in two clusters (111). The α-MHC and β-MHC genes are linked in tandem (177), and the skeletal muscle MHC genes are closely linked on mouse chromosome 14 (111) [chromosome 17 in humans (175, 178)]. The proximity and organization of the skeletal genes are not known. The functional and evolutionary significance of the linkage of vertebrate sarcomeric MHC genes remains uncertain in light of the dispersed organization of the four muscle MHC genes of the nematode (167).

REGULATION OF MYOSIN GENE EXPRESSION

During embryonic development of *Drosophila,* birds, and mammals, muscle-specific myosin genes are coordinately activated when myoblast stem cells differentiate and fuse to form multinucleate myotubes (179–181). Heterokaryon experiments involving the fusion of differentiated avian and mammalian skeletal muscle cells with nonmuscle cell types show that myosin and other muscle-specific genes are activated in nonmuscle cell nuclei by trans-acting factors expressed in differentiated muscle cells (181a,b,c).

Invertebrate myosin genes are subject to muscle-specific and developmental stage-specific regulation. In *C. elegans,* the body wall muscle *unc-54* and *myo-3* genes and the pharyngeal *myo-1* and *myo-2* genes are

---

hybridization with late pupal, muscle-specific cDNAs (83). Additional muscle-specific genes at 32C–D, 46D–F, 47F–48D, 62C–D, 63A–C, and 69F (166) and at 49F (P. F. Lasko, A. G. Ayme, M. L. Pardue, personal communication) have been identified by in situ chromosome hybridization.

Genetically identified loci are shown to the left of each chromosome, while cloned chromosomal segments are shown as boxes.

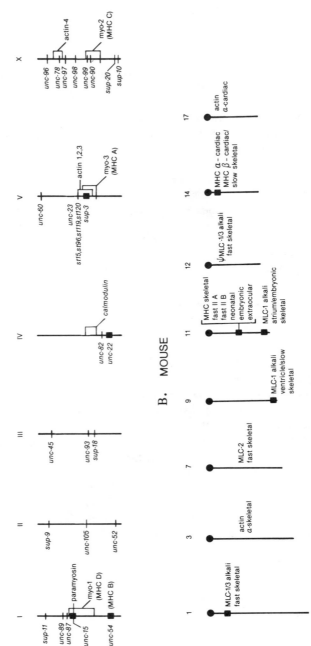

Figure 6 Chromosome maps of *C. elegans* and mouse muscle genes.

A. The nematode chromosome map of muscle genes includes MHC genes *myo-1, 2, 3* and *unc-54*, actin genes 1, 2, 3, and *unc-54*, a calmodulin gene mapped by in situ chromosome hybridization (167, 168). Other genes, identified as *unc* and *sup* mutations, have been mapped genetically (169–171). Brackets and boxes indicate chromosomal regions of specific genes.

B. The mouse chromosome map of muscle genes modified from (9). Mouse skeletal and cardiac MHC genes are in two clusters (111) and are shown linked in tandem as in the rat genome (177; S. Grund, unpublished data cited in Ref. 26). Mouse MHC and MLC genes were mapped by linkage of restriction site polymorphisms between two interbreeding mouse strains (108, 109, 111, 172) and by analysis of species-specific restriction site polymorphisms in somatic cell hybrids (172a, 173–175). Human actin (176) and MHC (175, 178) genes also been mapped using this method. Identified chromosome regions are indicated by boxes. Otherwise, only a linkage group has been assigned.

expressed in different muscle cell types (8), and the *unc-54/myo-3* genes express RNA transcripts in a coordinated 2:1 ratio (137). The *Drosophila* muscle-specific MHC gene, alkali MLC gene, and MLC-2 gene are coactivated at both larval and pupal stages in larval body wall muscles, fibrillar flight muscles, and adult leg muscles (81–84). Duplications and deficiencies of the MHC gene 36B locus cause dosage-dependent changes in MHC expression in all these muscles (81, 163), indicating that this MHC gene is not subject to simple feedback regulation.

Vertebrate muscle myosin genes also show muscle-specific and developmental stage-specific regulation. In addition, vertebrate myosin gene expression is modulated by a variety of factors involved in the embryonic maturation of muscle types and in the maintenance of muscles in adults by hormonal and other physiological signals. When skeletal muscle first differentiates, embryonic MHC genes are activated in myogenic cells in response to altered concentrations of mitogens (183–185) and to the cell cycle–mediated events that accompany myoblast fusion (186, 187). During later development, there are specific transitions in MHC isoform expression. Embryonic mammalian skeletal muscles sequentially express embryonic MHC genes followed by neonatal MHC and then adult MHC gene transcripts (100, 112, 196, 198a). The embryonic to neonatal MHC transition is apparently nerve-independent since it occurs in cultured muscle in the absence of innervation (188). Embryonic and neonatal MHC genes can be expressed in some adult muscles (104, 189). Avian embryonic skeletal muscles coexpress an embryonic MHC, an adult skeletal MHC, and a cardiac MHC gene (88, 95). Immunological observations show that different clonal myoblast cell lineages cultured from early embryonic muscles can express different sets of fast- and slow-specific skeletal muscle MHC isoforms (190), indicating that some patterns of myosin gene expression are determined in myogenic cells. Additional factors such as MHC RNA processing and turnover, RNA translation, protein stability, and assembly into thick filaments also may influence the pattern of MHC isoform accumulation in embryonic skeletal muscle.

In embryonic avian skeletal muscles, MLC genes are coactivated with MHC and encode adult fast isoforms and the adult slow isoforms of alkali MLC-1 and regulatory MLC-2 (191–193). Alkali MLC-3 transcripts, produced by alternative promoter usage of the alkali MLC fast gene (93), are expressed at low levels in early muscle but become predominant prior to hatching (193). By hatching, the adult pattern of mRNAs encoding avian fast and slow skeletal MLC isoforms is established by the selective gene repression and turnover of fast or slow MLC transcripts in inappropriate muscles (194, 195).

Embryonic mammalian muscles (mice and rats) also express adult fast skeletal isoforms of alkali MLC and regulatory MLC-2 (196). Slow skeletal MLC isoforms have not been detected during early development, but an "embryonic" MLC gene is transiently coexpressed in embryonic skeletal muscles (197). This embryonic MLC is encoded by the mammalian-specific alkali MLC-1 embryonic/atrial gene also expressed in the adult cardiac atrium and Purkinje fibers (109).

Gene expression in embryonic avian and mammalian cardiac muscle has not yet been studied in much detail, but early embryonic heart expresses an MHC with epitopes shared by embryonic skeletal muscle in somites (198). Adult cardiac isoforms of alkali MLC and regulatory MLC-2 are expressed in embryonic avian and mammalian hearts (198a), and skeletal fast MLC-1 isoforms also may be coexpressed in the chick heart (199).

The adult pattern of MHC isoform transcript expression can be modulated by thyroid hormone, which can increase or decrease all seven MHC gene-specific transcripts in skeletal (189) and cardiac (75, 113, 199a) muscles. In cardiac muscle, thyroid hormone levels and hypertrophy regulate the transcription of α-MHC and β-MHC genes in opposite directions (113, 200). Thyroid hormone and cardiac hypertrophy induce only quantitative changes in MLC transcripts, indicating that MLCs are not modulated coordinately with MHC isoforms (103). Expression of specific skeletal MLC isoforms and their RNA transcripts, however, is modulated and maintained by innervation and muscle activity (103a,b), signals that also modulate crustacean muscle gene expression (77a).

The apparent complexity of myosin gene regulation in vertebrates is one of the unexpected outcomes of the cloning of myosin and other contractile protein genes (194). Particularly striking is the degree to which cardiac, slow, and fast skeletal genes are coexpressed or have overlapping expression in embryonic and adult muscles (95, 104, 189). Such overlapping transcription suggests either that these specialized muscles share a common muscle gene regulatory mechanism or that coexpressed genes have a complexity of distinct cardiac-specific, fast skeletal–specific, and slow skeletal–specific control elements that modulate their muscle-specific gene expression as well as control elements that coordinate their activation in embryonic muscles. It also has become clear that vertebrate MLC and MHC gene transcripts have different responses to developmental and physiological signals, suggesting that MLC and MHC genes can be modulated independently and that various combinations of vertebrate MLC and MHC protein isoforms can associate and function in different muscles under different physiological conditions.

The dispersal of myosin and other muscle genes in the genomes of invertebrates and vertebrates (Figures 5 and 6) shows that muscle genes are regulated by trans-acting regulatory processes. The possible exceptions to this are the clustered vertebrate skeletal and cardiac MHC genes (111, 177). The opposite responses of the α and β cardiac genes to hypertrophy and thyroid hormone (113, 200), as well as the transition of the skeletal muscle MHC mRNAs during embryogenesis (100, 112, 122), may depend on the clustering of the members of the gene family in mammals. The chromosomal location of a cloned rat MLC-2 fast skeletal gene does not influence its appropriate regulation, as demonstrated by transgenic introduction of this gene into chromosomal sites of mice, and by its activation and expression in skeletal muscle (201). This MLC-2 gene, therefore, must have cis control regions that specify its response to trans-acting regulatory signals. Transfection studies with MHC genes and other MLC genes likely will reveal the nature and complexity of the cis and trans control systems that regulate and coordinate the developmental and physiological expression of myosin genes.

STRUCTURE AND EVOLUTION OF MYOSIN GENES

Alkali MLC Genes

The complete structures of the chicken, mouse, and rat alkali MLC-1/3 fast skeletal genes and the *Drosophila* alkali MLC gene have been determined (Figure 4A). Vertebrate MLC-1/3 genes have conserved exon structures, dual promoters and alternative exon splicing, and regulated expression in fast skeletal muscle (140a, 202). The introns and 5' and 3' untranslated regions of the three vertebrate MLC-1/3 genes are variable in size and have divergent nucleotide sequences, but the MLC-1 and MLC-3 promoters show sequence conservation, perhaps related to transcriptional regulation (202). The MLC-1 promoter is more highly conserved than the MLC-3 promoter (140a). The splice junctions and intron sequences in the 5' regions of exons 2, 3, and 4 do not have unusual features correlated with alternative splicing, but the transcripts produced by these promoters have different primary structures that may direct the alternative splicing choice between exon 3 and exon 4. The homology and positions of these "mini" exons suggest that they arose in the vertebrate lineage by exon duplication (140a, 203). The splice junctions of alternative exons 3 and 4 contain split codons that would produce a protein with frame shifts if both exons were included in a mature transcript, indicating strong selection to maintain alternative splicing.

The *Drosophila* alkali MLC gene is smaller than vertebrate MLC-1/3 fast skeletal genes and has only a single promoter (83, 126). Its exon structure is similar to MLC-3 coding sequences except for exon 3, which is split by an intron in the homologous exons 5 and 6 of the vertebrate gene. There is also an apparent duplication of part of the C-terminal exon that produces two isoforms by regulated alternative splicing (Figures 3A and 4A). Since both alternative exons (5 and 6) have translation stop codons, alternative splicing is not essential to maintain functional protein structure, although the poly A addition and termination sites are encoded exclusively in exon 6. Exon 5 presumably encodes a C-terminus adapted for tubular and visceral muscle function. Similar conjectures can be made for the alternative splicing of *Drosophila* MHC gene transcripts that produce two tissue-specific isoforms with different C-termini by a similar exon inclusion/exclusion process (124, 125).

The *Drosophila* alkali MLC and MLC-2 genes (83, 84) and the vertebrate MLC-3 genes (93, 106, 123), MLC-2 genes (92, 96), and calmodulin genes (204) have similar 5' exon structures with the ATG translation initiation codon located at the end of the first exon. The structure of vertebrate smooth muscle and nonmuscle alkali MLC proteins (58, 129) and invertebrate alkali MLC proteins (65) indicates that these proteins are encoded by genes with MLC-3-like structure and that this gene family is ancient. Genes encoding MLC-1 protein isoforms, therefore, likely originated in the vertebrate lineage. Duplication of an ancestral MLC-3 gene, duplication of the "mini" exon, and addition of an upstream MLC-1 promoter presumably generated the MLC-1/3 fast skeletal muscle gene that in turn was duplicated to form the alkali MLC-1 atrial/embryonic skeletal gene and the alkali MLC-1 ventricle/slow skeletal gene (9). Determination of the exon structure of other MLC-1 genes may reveal the fate of the MLC-3 promoter and of "mini" exons 3 and 4 in these recently evolved gene family members.

Regulatory MLC-2 Genes

The *Drosophila* MLC-2 gene (84) and a rat MLC-2 fast skeletal gene (96) have identical intron locations, separating exons 1 and 2 and 2 and 3, but the rat gene has four additional downstream introns (Figure 4B). A partial sequence of the chicken MLC-2A cardiac gene indicates that it shares identical intron locations between exons 1 and 2, 2 and 3, and 6 and 7 with the rat gene (92). All introns in MLC-2 genes are at different positions from other calcium-binding protein genes except for the first intron that separates the initiation codon from the protein coding region in the second exon (see previous section). None of these intron positions correspond to the EF hand structural domains (Figures 3A and 3B). This suggests that these gene families diverged long ago and that introns were introduced after the duplication of a calcium-binding protein ancestral gene having the four EF hands and the characteristic promoter/ATG first exon (P. Maisonpierre, C. P. Emerson, Jr., in preparation). Alternatively, the ancestral MLC gene may have had a very high density of introns, and different introns were lost in MLC-2 and alkali MLC lineages.

MHC Genes

The structures of the rat embryonic MHC gene (26) and the nematode *unc-54* gene (22) illustrate the degree of evolutionary variation in intron numbers in MHC genes. The rat MHC gene, as well as other vertebrate MHC genes, is much larger than the nematode *unc-54* gene (Figure 4C) and the nematode *myo-1, 2,* and *3* genes (8) because of the interspersion of a larger number of introns in the rat gene. The *C. elegans unc-54* gene and other MHC genes have both gene-specific and conserved intron positions (8, 22).

The positions of introns in the *C. elegans* MHC genes (8, 22) do not correspond to structural and functional domains. Neither the gene region encoding the hinge nor the gene region encoding the 28-amino-acid α-helical repeats in the rod are bounded by introns (Figures 2 and 4*C*), suggesting that these structural domains arose prior to the introduction of introns into an ancestral gene (29). Unlike the *C. elegans* gene, introns of the rat embryonic MHC gene separate three of the five proteinase-sensitive surface loops, regions that have divergent sequences between the rat and *C. elegans* MHC proteins and give rise to the major structural domains of the MHC head (26). This is consistent with the possibility that these domains arose by exon shuffling to form an ancestral gene (205), and that the introns separating the domain coding regions have been lost in the *C. elegans* gene.

Despite the evolutionary variation in MHC gene structure, intron positions are remarkably conserved at the 5' ends of the rat embryonic (122) and the *C. elegans unc-54* MHC genes (8, 22; Figure 2) as well as *Drosophila* (W. A. Kronert, P. T. O'Donnell, S. I. Bernstein, unpublished data), rabbit cardiac (206), and chicken skeletal muscle MHC genes (139). Even the distantly related *Acanthamoeba* myosin II nonmuscle gene (120) has at least one common 5' intron position. The rat embryonic (122), rabbit cardiac (206), chicken skeletal (206a), and *Drosophila* muscle (D. R. Wassenberg, W. A. Kronert, S. I. Bernstein, in preparation) MHC genes have identified 5' transcription start sites preceded by TATA and CAT promoter consensus sequences. One or two upstream exons in each of these genes are entirely composed of nontranslated mRNA sequences. The conservation of intron position in the 5' regions of MHC genes, as well as the conservation of 5' intron positions of MLC-1 and MLC-2 genes (Figures 4*A* and 4*B*), suggests that important regulatory or structural functions are located in these intragenic regions. In addition to conservation of exon structure and protein sequence, vertebrate MHC genes share a 40 bp homologous sequence of unknown function in their 3' nontranslated regions (119). A small translational control ("tc") RNA gene also is located at the 3' end of the chicken skeletal MHC gene (207). Transfection and transgenic studies of myosin genes (201) offer powerful approaches to investigate the functions of the promoters and introns of MHC and MLC genes.

GENETICS OF MYOSIN GENES

Isolation of Muscle Mutants

Mutations of the nematode *unc-54* MHC-B gene (208–210) and the *Drosophila* 36B muscle MHC gene (163) have been identified and characterized at the DNA level. Nematode myosin mutations were recovered from genetic mutagenesis screens for uncoordinated or paralyzed phenotypes (211–213) (Figure 6A), and *Drosophila* myosin mutations were recovered as inducing flightless phenotypes (162, 214) (Figure 5). In *C. elegans,* over 100 mutant alleles of the *unc-54* MHC B locus have been described among 20 different complementation groups that induce the uncoordinated phenotype (215). In *Drosophila,* eight dominant flight muscle mutations have been mapped to the 36B MHC locus among a total of 19 loci that are essential for flight muscle functions (Figure 5).

In *Drosophila* the chromosomal sites of MHC and MLC genes, as well as 26 other muscle genes, have been identified by in situ hybridization with thorax-specific and embryonic muscle-specific DNA clones (Figure 5). The muscle gene products of only some of these genes are known. Mutations that affect actin, MHC, and tropomyosin have been described (Figure 5). In *C. elegans,* 17 muscle-specific *unc* (uncoordinated) and *sup* (suppressor of *unc*) loci have been mapped (Figure 6A), and mutations in MHC B *(unc-54),* paramyosin *(unc-15)* (54), and actin genes (216, 217) have been identified. The other *unc* loci may now be cloned by transposition of the Tc1 repetitive element into the locus of interest, followed by isolation of the interrupted gene using a Tc1 probe (212, 213). The *sup-3* mutation, which enhances *myo-3* gene transcription, has been localized to the structural promoter region of the *myo-3* gene and may represent an upregulated promoter mutation or a *myo-3* gene duplication (218). Two mutations, *unc-90* and *unc-99,* map in the vicinity of the *myo-2* locus (Figure 6A). The lack of mutations in some MHC genes may indicate that these are dominant lethals or are compensated for by the coexpression of the other body wall and pharyngeal MHC genes. Isoform coexpression and isoform compensation also may explain the lack of mouse MHC mutants, since vertebrate MHC genes have overlapping muscle-specific patterns of expression (189).

Mutations causing muscular dysfunction have been described in chicken (219), mouse (221, 222, 222a), and humans (223). The gene products encoded by these mutant loci have not been identified, although in dystrophic chickens, a 3' intron in a fast skeletal MHC gene apparently lacks a small, polymerase III–transcribed gene that encodes a translational control ("tc") RNA (207). Some mouse muscle mutations map to chromosomes having known myosin genes (Figure 6B) (222, 222a). These include *tipsy (tc)* and *shambling (shm)* on chromosome 11, *ducky (du)* on chromosome 9, and *quivering (qu)* on chromosome 7. Other mouse mutations that might be alleles of muscle genes other than myosin include *lethargic (lh)* on chromosome 2, *tottering (tg)* on chromosome 8, and *dystrophic muscularis (dy)* on chromosome 10. A large number of human genetic muscle diseases have been described. These include X-linked, recessive muscle mutations such as Duchenne muscular dystrophy (225), which has been cloned (226), and

Becker, scapuloperoneal, Emery-Dreifuss, and Mabry muscular dystrophies (227). Autosomal recessive and dominant facioscapulohumeral dystrophies and scapuloperoneal syndromes (228) and myotonic dystrophies (229) also have been reported. None of the proteins encoded by these human muscle mutant loci has been identified, although based on our current understanding of muscle mutations in *Drosophila* and *C. elegans,* it seems reasonable to expect that some will be mutations of myosin and other known muscle genes.

MHC mutations in both *Drosophila* and nematode cause major disruptions of myofilament organization (8, 162, 163). Mutations that suppress or enhance mutant MHC phenotypes have been described in the nematode [*sup 9, 10, 11,* and *20* (170, 171)] and in *Drosophila* [*up* and *hdp* (T. Homyk, C. Emerson, in preparation)] (Figures 5 and 6A). The products of these interactive genes may encode contractile proteins that interact structurally or functionally with MHC or they may be genes that regulate the tissue specificity or level of expression of MHC genes. Other *unc* suppressor mutations, such as *sup 5* and *sup 7,* are tRNA suppressors (230).

Drosophila melanogaster *MHC Mutants*

The 36B MHC gene encodes the major MHC of *Drosophila* larval and adult muscles (81, 82). Four dominant mutations, one deletion and three DNA element insertions *(Mhc 1–4),* reduce production of MHC in larval and pupal muscles. These mutations cause selective dysfunction of the highly organized flight muscles in heterozygotes and are embryonic or larval lethals in homozygotes (163). The dominant nature of these mutations, as well as the fact that the 36B(2L) MHC locus is haploinsufficient for flight muscle function (81), establishes that different muscles have different minimum quantitative requirements of MHC protein for function. Other dominant flight muscle mutations *(Ifm(2)1,2,3* and *Bashed)* map to the 36B MHC locus but have not yet been identified at the DNA level. These may include flight muscle–specific MHC gene mutations that affect the thorax-specific exon or the thorax-specific splicing mechanism. An example of this class of mutations is the *Ifm(3)3* tropomyosin mutation that is a DNA element insertion that specifically prevents accumulation of the thorax-specific tropomyosin mRNA, but does not interfere with the use of the alternative exon that is included in the tropomyosin transcripts of nonfibrillar muscles (231).

Caenorhabditis elegans *MHC Mutants*

Over 100 mutant alleles of the *unc-54* MHC B locus have been identified, and 30 have been ordered by fine structure mapping and shown to be distributed throughout the coding region of the *unc-54* locus (215). The molecular cloning of mutant *unc-54* MHC alleles has led to identification at the DNA sequence level of 11 mutations that affect MHC B structure (208) (see OVERVIEW OF MYOSIN STRUCTURE AND FUNCTION).

In general, recessive unc-54 alleles reduce thick filaments in the body wall muscle. These mutants fail to accumulate normal levels of the unc-54 MHC protein because of unstable MHC protein and defective filament assembly (208). Such mutations are deletions of the MHC C-terminal rod region resulting from single base pair changes that introduce termination codons or deletions that cause nonsense mutations. Thus, MHC stability and subsequent thick filament formation require an intact MHC C-terminus. A mutation that deletes only a short region of the C-terminus results in some MHC-B accumulation, suggesting that protein stability is inversely correlated with the number of C-terminal residues deleted. An unexpected observation is that stability of unc-54 mRNA in these C-terminal deletion mutants is apparently reduced because of premature termination of translation (208).

Dominant and semidominant unc-54 alleles that have reduced numbers of thick filaments are apparently DNA deletions or substitutions that result in MHC proteins with conformational distortions in the myosin rod region (8, 208). These mutated proteins fail to assemble into thick filaments. Three such mutations have been sequenced (208). Two are inframe deletions that elimi-nate portions of the rod and disrupt its repeat structure, while the third mutation results in two amino acid substitutions at the extreme N-terminal region of the rod. Both amino acid substitutions may be significant, one preventing the formation of a salt bridge required for stabilization of the hydrophobic core of that region of the molecule and the second resulting in steric hindrance of the rod backbone. The dominance of these mutations apparently results from accumulation of assembly-defective unc-54 MHC protein and interaction of this defective protein with normal MHC proteins produced by the wild-type unc-54 allele and the myo-3 gene. Interestingly, one mutation that maps to the head region of the unc-54 gene also affects thick filament formation, suggesting that the myosin head also is involved in filament assembly.

A third class of unc-54 mutations, originally isolated as a suppressor of unc-22 mutations, results in impaired muscle function but fairly normal muscle structure (232). The unc-22 gene encodes an unknown protein. Homozygotes of suppressor unc-54 mutations, in the absence of an unc-22 mutation, have nearly normal muscle ultrastructure except for slight disorgan-ization of thick filaments in the A band. Although their phenotypes vary, most of these mutants move slowly and stiffly. Genetic fine structure mapping (215) suggested that these mutations are within the head region of the MHC-B protein. Two of these mutant unc-54 genes (s74 and s95) have been sequ-enced (208) and are single base pair changes that cause amino acid sub-stitutions in the 23 kd and 50 kd domains flanking the ATP binding region of the myosin head (Figure 1 and Figure 2A). These mutations may affect either the ATPase activity of the myosin molecule or the binding of actin or MLC.

A large number of *unc-54* mutations remain to be characterized at the DNA sequence level and will likely reveal important structural information about the MHC-B protein and its function and interaction with other muscle proteins. Isolation and characterization of additional nematode and *Drosophila* muscle-specific mutations provide an opportunity to investigate the functions of MHC proteins and to study molecular genetic mechanisms that regulate the developmental and physiological expression of myosin genes.

ACKNOWLEDGMENTS

We thank our colleagues for their cooperation in providing us with results in advance of publication, Dr. Theodore Homyk for his generous help in the assembly of a comprehensive chromosome map of *Drosophila* muscle genes, and Gladys Bryant for her secretarial assistance. Our research was supported by Muscular Dystrophy Association Research Grants to C.P.E., Jr. and S.I.B. and NIH grants GM 26706 (C.P.E., Jr.) and GM 32443 (S.I.B.).

Literature Cited

1. Engelhardt, W. A., Ljubimova, M. N. 1939. *Nature* 144:668–69
2. Szent-Györgyi, A. G. 1945. *Acta Physiol. Scand.* 8(Suppl. 25): 1–115
3. Lowey, S. 1986. In *Myology*, ed. A. G. Engel, B. Q. Banker, pp. 563–88. New York: McGraw–Hill
4. Szent-Györgyi, A. G., Chandler, P. D. 1986. See Ref. 3, pp. 589–612
5. Harrington, W. F., Rodgers, M. E. 1984. *Ann. Rev. Biochem.* 53:35–73
6. Taylor, E. W. 1979. *CRC Crit. Rev. Biochem.* 6:103–64
7. Adelstein, R. S., Eisenberg, E. 1980. *Ann. Rev. Biochem.* 49:921–56
7a. Kendrick-Jones, J., Scholey, J. M. 1981. *J. Muscle Res. Cell Motility* 2:347–72
8. Karn, J., Dibb, N. J., Miller, D. M. 1985. In *Cell and Muscle Motility*, ed. J. W. Shay, pp. 185–235. New York: Plenum
9. Barton, P. J., Buckingham, M. E. 1985. *Biochem. J.* 231:249–61
10. Weeds, A. G. 1969. *Nature* 223:1362–64
11. Weeds, A. G., Lowey, S. 1971. *J. Mol. Biol.* 61:701–25
12. Lowey, S., Risby, D. 1971. *Nature* 234:81–84
13. Barany, M. 1967. *J. Gen. Physiol.* 50:197–216 (Suppl.)
14. Marston, S. B., Taylor, E. W. 1980. *J. Mol. Biol.* 139:573–600
15. Pope, B., Hoy, J. F. Y., Weeds, A. 1980. *FEBS Lett.* 118:205–8
16. Hoh, J. F. Y., McGrath, P. A., Hale, P. T. J. 1978. *J. Mol. Cell Cardiol.* 10: 1053–76
17. Greene, L. E., Eisenberg, E. 1978. *Proc. Natl. Acad. Sci. USA* 75:54–58
18. Trybus, K. M., Huiatt, T. W., Lowey, S. 1982. *Proc. Natl. Acad. Sci. USA* 79:6151–55
19. Vibert, P., Szentkiralyi, E., Hardwicke, P., Szent-Györgyi, A. G., Cohen, C. 1986. *Biophys. J.* 49:131–33
20. Elliott, A., Offer, G. 1978. *J. Mol. Biol.* 123:505–19
21. Squire, J. M. 1981. *The Structural Basis of Muscle Contraction*. New York: Plenum
21a. Huxley, H. E. 1963. *J. Mol. Biol.* 7:281–308
22. Karn, J., Brenner, S., Barnett, L. 1983. *Proc. Natl. Acad. Sci. USA* 80:4253–57
23. Elzinga, M., Collins, J. H. 1979. *Proc. Natl. Acad. Sci. USA* 74:4281–84
24. Elzinga, M., Trus, B. L. 1980. In *Methods in Peptide and Protein Sequence Analysis*, ed. C. Birr, pp. 213–24. New York: Elsevier
25. Capony, J. P., Elzinga, M. 1981. *Biophys. J.* 33:148a
26. Strehler, E. E., Strehler-Page, M.-A., Perriard, J.-C., Periasamy, M., Nadal-Ginard, B. 1986. *J. Mol. Biol.* 190:291–317

27. McLachlan, A. D., Karn, J. 1982. *Nature* 299:226–31
28. McLachlan, A. D., Karn, J. 1983. *J. Mol. Biol.* 164:605–26
29. McLachlan, A. D. 1983. *J. Mol. Biol.* 169:15–30
30. Lowey, S., Slater, H. S., Weeds, A. G., Baker, H. 1969. *J. Mol. Biol.* 42:1–29
31. Deleted in proof
32. Appelgate, D., Reisler, E. 1983. *Proc. Natl. Acad. Sci. USA* 80:7109–112
33. Mornet, D., Bertrand, R., Pantel, P., Audermand, E., Kassab, R. 1981. *Biochemistry* 20:2110–120
34. Szilagyi, L., Bálint, M., Sréter, F. A., Gergely, J. 1979. *Biochem. Biophys. Res. Commun.* 87:936–45
35. Walker, J. E., Saraste, M., Runswick, M. J., Gay, N. J. 1982. *EMBO J.* 1:945–51
36. Hozumi, T., Muhlrad, A. 1981. *Biochemistry* 20:2945–50
37. Labbe, J.-P., Mornet, D., Roseau, G., Kassab, R. 1982. *Biochemistry* 21: 6897–902
38. Sutoh, K. 1982. *Biochemistry* 21:3654–661
39. Wells, J. A., Yount, R. G. 1979. *Proc. Natl. Acad. Sci. USA* 76:4966–70
40. Okamoto, Y., Yount, R. G. 1983. *Biophys. J.* 41:298a
41. Burke, M., Reisler, E. 1977. *Biochemistry* 16:5559–563
42. Lu, R. C. 1980. *Proc. Natl. Acad. Sci. USA* 77:2010–13
43. Lu, R. C., Wong, A. 1982. *Biophys. J.* 37:52a
44. Burke, M., Sivaramakrishnan, M., Kamalakannan, V. 1983. *Biochemistry* 22:3046–53
44a. Bálint, M., Sréter, F. A., Wolf, I., Nagy, B., Gergely, J. 1975. *J. Biol. Chem.* 250:6168–77
45. Szentkiralyi, E. M. 1984. *J. Muscle Res. Cell Motility* 5:147–64
46. Winkelmann, D. A., Almeda, S., Vibert, P., Cohen, C. 1984. *Nature* 307:758–60
47. Parry, D. A. D. 1981. *J. Mol. Biol.* 153:459–64
47a. Kavinsky, C. J., Umeda, P. K., Sinha, A. M., Elzinga, M., Tong, S. W., et al. 1983. *J. Biol. Chem.* 258:5196–205
48. Huxley, H. E. 1985. *J. Exp. Biol.* 115:17–30
49. Huxley, H. E., Brown, W. 1967. *J. Mol. Biol.* 30:383–434
49a. Persechini, A., Hartshorne, D. J. 1981. *Science* 213:1383–85
50. Hardwicke, P. M. D., Wallimann, T., Szent-Györgyi, A. G. 1983. *Nature* 301:478–82
51. Schachat, F. H., Garcea, R. L., Epstein, H. F. 1978. *Cell* 15:405–11
52. Miller, D. M. III, Ortiz, I., Berliner, G. C., Epstein, H. F. 1983. *Cell* 34:477–90
53. Wray, J. S. 1982. In *Basic Biology of Muscles: A Comparative Approach,* ed. B. M. Twarog, R. J. C. Levine, M. M. Dewey, pp. 29–36. New York: Raven
54. MacKenzie, J. M. Jr., Epstein, H. F. 1980. *Cell* 22:747–55
55. Epstein, H. F., Miller, D. M. III, Ortiz, I., Berliner, G. C. 1985. *J. Cell Biol.* 100:904–15
56. Starr, R., Offer, G. 1983. *J. Mol. Biol.* 170:675–98
57. Matsuda, G., Maita, T., Umegane, T. 1981. *FEBS Lett.* 126:111–13
58. Matsuda, G., Maita, T., Kato, Y., Chen, J.-I., Umegane, T. 1981. *FEBS Lett.* 135:232–36
59. Dayhoff, M. 1978. *Atlas of Protein Sequence and Structure,* pp. 276–77. Washington, DC: Natl. Med. Res. Found.
60. Kretsinger, R. H. 1980. *CRC Crit. Rev. Biochem.* 8:119–74
61. Bagshaw, C. R., Kendrick-Jones, J. 1979. *J. Mol. Biol.* 130:317–36
62. Sarkar, S. 1972. *Cold Spring Harbor Symp. Quant. Biol.* 37:14–17
63. Yamamoto, K., Tokunaga, M., Sutoh, K., Wakabayashi, T., Sekine, T. 1985. *J. Mol. Biol.* 183:287–90
63a. Waller, G. S., Lowey, S. 1985. *J. Biol. Chem.* 260:14368–73
63b. Winkelmann, D. A., Lowey, S. 1986. *J. Mol. Biol.* 188:595–612
64. Sivaramakrishnan, M., Burke, M. 1982. *J. Biol. Chem.* 257:1102–105
65. Kendrick-Jones, J., Szentkiralyi, E. M., Szent-Györgyi, A. G. 1976. *J. Mol. Biol.* 104:747–75
66. Simmons, R. M., Szent-Györgyi, A. G. 1978. *Nature* 273:62–64
67. Sellers, J. R., Chantler, P. D., Szent-Györgyi, A. G. 1981. *J. Mol. Biol.* 144:223–45
68. Kendrick-Jones, J., Cande, W. Z., Tooth, P. J., Smith, R. C., Scholey, J. M. 1983. *J. Mol. Biol.* 165:139–62
69. Barany, K., Barany, M., Gillis, J. M., Kushmerick, M. J. 1980. *Fed. Proc.* 39:1547–51
70. Reinach, F. C., Nagai, K., Kendrick-Jones, J. 1986. *Nature* 322:80–83
71. Weeds, A. G., Hall, R., Spurway, N. C. S. 1975. *FEBS Lett.* 49:320–24
72. Gauthier G. F., Lowey, S. 1979. *J. Cell Biol.* 81:10–25
73. Hoh, J. F. Y., Yeoh, G. P. S. 1979. *Nature* 280:321–23

73a. Barany, M. 1967. *J. Gen. Physiol.* 50:197–218
74. Butler-Browne, G. S., Herlicoviez, D., Whalen, R. G. 1984. *FEBS Lett.* 166:71–75
75. Gustafson, T. A., Markham, B. E., Morkin, E. 1985. *Biochem. Biophys. Res. Commun.* 130:1161–67
76. Samuel, J. L., Bertier, B., Bugaisky, L., Marotte, F., Swynghedauw, B., et al. 1984. *Eur. J. Cell. Biol.* 34:300–6
77. Jolesz, F., Sreter, F. A. 1981. *Ann. Rev. Physiol.* 43:531–52
77a. Quigley, M. M., Mellon, D. Jr. 1984. *Dev. Biol.* 106:262–65
77b. Citi, S., Kendrick-Jones, J. 1986. *J. Mol. Biol.* 188:369–82
78. Pollard, T. D. 1981. *J. Cell Biol.* 91(3, part 2):156S–165S
79. MacLeod, A. R., Karn, J., Brenner, S. 1981. *Nature* 291:386–90
80. Miller, D. M., Stockdale, F. E., Karn, J. 1986. *Proc. Natl. Acad. Sci. USA* 83:2305–309
81. Bernstein, S. I., Mogami, K., Donady, J. J., Emerson, C. P. Jr. 1983. *Nature* 302:393–97
82. Rozek, C. E., Davidson, N. 1983. *Cell* 32:23–34
83. Falkenthal, S., Parker, V. P., Mattox, W. W., Davidson, N. 1984. *Mol. Cell. Biol.* 4:956–65
84. Parker, V. P., Falkenthal, S., Davidson, N. 1985. *Mol. Cell. Biol.* 5:3058–68
85. Jakowlew, S. B., Khandekar, P., Datta, K., Arnold, H. H., Narula, S. K., Siddiqui, M. A. Q. 1982. *J. Mol. Biol.* 156:673–82
86. Robbins, J., Horan, T., Gulick, J., Kropp, K. 1986. *J. Biol. Chem.* 261: 6606–12
87. Robbins, J., Freyer, G. A., Chisholm, D., Gilliam, T. C. 1982. *J. Biol. Chem.* 257:549–56
88. Umeda, P. K., Sinha, A. M., Jakovcic, S., Merten, S., Hsu, H. J., et al. 1981. *Proc. Natl. Acad. Sci. USA* 78:2843–47
89. Umeda, P. K., Kavinsky, C. J., Sinha, A. M., Hsu, H. J., Jakovcic, S., Rabinowitz, M. 1983. *J. Biol. Chem.* 258:5206–14
90. Freyer, G. A., Robbins, J. 1983. *J. Biol. Chem.* 258:7149–54
91. Arnold, H. H., Krauskopf, M., Siddiqui, M. A. Q. 1983. *Nucleic Acids Res.* 11:1123–31
92. Winter, B., Klapthor, H., Wiebauer, K., Delius, H., Arnold, H. H. 1985. *J. Biol. Chem.* 260:4478–83
93. Nabeshima, Y., Fujii-Kuriyama, Y., Muramatsu, M., Ogata, K. 1984. *Nature* 308:333–38

94. Hastings, K. E., Emerson, C. P. Jr. 1982. *Proc. Natl. Acad. Sci. USA* 79:1553–57
95. Hallauer, P. 1985. PhD. thesis. Univ. Va.
96. Nudel, U., Calvo, J. M., Shani, M., Levy, Z. 1984. *Nucleic Acids Res.* 12:7175–86
96a. Liew, C.-C., Jandreski, M. A. 1986. *Proc. Natl. Acad. Sci. USA* 83:3175–79
97. Nguyen, H. T., Gubits, R. M., Wydro, R. M., Nadal-Ginard, B. 1982. *Proc. Natl. Acad. Sci. USA* 79:5230–34
98. Deleted in proof
99. Periasamy, M., Wydro, R. M., Strehler-Page, M.-A., Strehler, E. E., Nadal-Ginard, B. 1985. *J. Biol. Chem.* 260: 15856–62
100. Periasamy, M., Wieczorek, D. F., Nadal-Ginard, B. 1984. *J. Biol. Chem.* 259:13573–78
101. Nudel, U., Katcoff, D., Carmon, Y., Zevin-Sonkin, D., Levi, Z., et al. 1980. *Nucleic Acids Res.* 8:2133–46
102. Mahdavi, V., Periasamy, M., Nadal-Ginard, B. 1982. *Nature* 297:659–64
103. Kumar, C. C., Cribbs, L., Delaney, P., Chien, K. R., Siddiqui, M. A. Q. 1986. *J. Biol. Chem.* 261:2866–72
103a. Gauthier, G. F., Burke, R. G., Lowey, S., Hooby, A. W. 1983. *J. Cell Biol.* 97:756–71
103b. Helig, A., Pette, D. 1983. *FEBS Lett.* 151:211–14
104. Wieczorek, D. F., Periasamy, M., Butler-Browne, G. S., Whalen, R. G., Nadal-Ginard, B. 1985. *J. Cell Biol.* 101:618–29
105. Garfinkel, L. I., Periasamy, M., Nadal-Ginard, B. 1982. *J. Biol. Chem.* 257:11078–86
106. Periasamy, M., Strehler, E. E., Garfinkel, L. I., Gubits, R. M., Ruiz-Opazo, N., Nadal-Ginard, B. 1984. *J. Biol. Chem.* 259:13595–604
107. Wydro, R. M., Nguyen, H. T., Gubits, R. M., Nadal-Ginard, B. 1983. *J. Biol. Chem.* 258:670–78
108. Barton, P. J., Cohen, A., Robert, B., Fiszman, M. Y., Bonhomme, F., et al. 1985. *J. Biol. Chem.* 260:8578–84
109. Barton, P. J., Robert, B., Fiszman, M. Y., Leader, D. P., Buckingham, M. E. 1985. *J. Muscle Res. Cell Motility* 6:461–75
110. Robert, B., Weydert, A., Caravatti, M., Minty, A., Cohen, A., et al. 1982. *Proc. Natl. Acad. Sci. USA* 79:2437–41
111. Weydert, A., Daubas, P., Lazaridis, I., Barton, P., Garner, I., et al. 1985. *Proc. Natl. Acad. Sci. USA* 82:7183–87
112. Weydert, A., Daubas, P., Caravatti,

M., Minty, A., Bugaisky, G., et al. 1983. *J. Biol. Chem.* 258:13867–74

113. Sinha, A. M., Umeda, P. K., Kavinsky, C. J., Rajamanickam, C., Hsu, H. J., et al. 1982. *Proc. Natl. Acad. Sci. USA* 79:5847–51

114. Sinha, A. M., Friedman, D. J., Nigro, J. M., Jakovcic, S., Rabinowitz, M., Umeda, P. K. 1984. *J. Biol. Chem.* 259:6674–80

115. Friedman, D. J., Umeda, P. K., Sinha, A. M., Hsu, H. J., Jakovcic, S., Rabinowitz, M. 1984. *Proc. Natl. Acad. Sci. USA* 81:3044–48

116. Kavinsky, C. J., Umeda, P. K., Levin, J. E., Sinha, A. M., Nigro, J. M., et al. 1984. *J. Biol. Chem.* 259:2775–81

117. Appelhans, H., Vosberg, H. P. 1983. *Hum. Genet.* 65:198–203

118. Leinwand, L. A., Saez, L., McNally, E., Nadal-Ginard, B. 1983. *Proc. Natl. Acad. Sci. USA* 80:3716–20

119. Saez, L., Leinwand, L. A. 1986. *Nucleic Acids Res.* 14:2951–69

120. Hammer, J. A. III, Korn, E. D., Paterson, B. M. 1986. *J. Biol. Chem.* 261:1949–56

120a. Hammer, J. A. III, Jung, G., Korn, E. D. 1986. *Proc. Natl. Acad. Sci. USA* 83:4655–59

121. DeLozanne, A., Lewis, M., Spudich, J. A., Leinwand, L. A. 1985. *Proc. Natl. Acad. Sci. USA* 82:6807–10

122. Strehler, E. E., Mahdavi, V., Periasamy, M., Nadal-Ginard, B. 1985. *J. Biol. Chem.* 260:468–71

123. Robert, B., Daubas, P., Akimenko, M.-A., Cohen, A., Garner, I., et al. 1984. *Cell* 39:129–40

123a. Karn, J., Brenner, S., Barnett, L., Cesareni, G. 1980. *Proc. Natl. Acad. Sci. USA* 77:5172–76

124. Bernstein, S. I., Hansen, C. J., Becker, K. D., Wassenberg, D. R. II, Roche, E. S., et al. 1986. *Mol. Cell. Biol.* 6:2511–19

125. Rozek, C. E., Davidson, N. 1986. *Proc. Natl. Acad. Sci. USA* 83:2128–32

126. Falkenthal, S., Parker, V. P., Davidson, N. 1985. *Proc. Natl. Acad. Sci. USA* 82:449–53

126a. Kiehart, D. P., Kaiser, D. A., Pollard, T. D. 1984. *J. Cell. Biol.* 99:1002–14

126b. Kiehart, D. P., Feghali, R. 1986. *J. Cell Biol.* 103:1517–25

127. Kuczmarski, E. R., Spudich, J. A. 1980. *Proc. Natl. Acad. Sci. USA* 77:7292–96

128. Nabeshima, Y., Fujii-Kuriyama, Y., Muramatsu, M., Ogata, K. 1982. *Nucleic Acids Res.* 10:6099–110

129. Nabeshima, Y., Nabeshima, Y. 1984. In *International Cell Biology*, ed. S. Seno, Y. Okada. Japan: Academic

130. Weeds, A. G. 1976. *Eur. J. Biochem.* 66:157–73

131. Whalen, R. G., Butler-Browne, G. S., Gros, F. 1978. *J. Mol. Biol.* 126:415–31

132. Dalla-Libera, L., Sartore, S., Schiaffino, S. 1979. *Biochim. Biophys. Acta* 581:283–94

133. Grandier-Vazeille, X., Tetaert, D., Hemon, B., Biserte, G. 1983. *Comp. Biochem. Physiol. B.* 76:263–70

134. Dinh, T. N.-L., Watabe, S., Ochiai, Y., Hashimoto, K. 1985. *Comp. Biochem. Physiol. B.* 80:203–7

135. Bagshaw, C. R., Kendrick-Jones, J. 1980. *J. Mol. Biol.* 140:411–33

136. Fyrberg, E. A., Mahaffey, J. W., Bond, B. J., Davidson, N. 1983. *Cell* 33:115–23

136a. Kiehart, D. P., Saft, M. S., Laymon, R. A., Goldstein, L. S. B., O'Brien, J. 1986. *J. Cell Biol.* 103:115a

137. Garcea, R. C., Schachat, F., Epstein, H. F. 1978. *Cell* 15:421–28

138. Robbins, J., Horan, T., Gulick, J., Kropp, K. 1986. *J. Biol. Chem.* 261:6606–12

139. Kropp, K., Gulick, J., Robbins, J. 1986. *J. Biol. Chem.* 261:6613–18

140. Hammer, J. A. III, Korn, E. D., Paterson, B. M. 1984. *J. Biol. Chem.* 259:11157–59

140a. Strehler, E. E., Periasamy, M., Strehler-Page, M.-A., Nadal-Ginard, B. 1985. *Mol. Cell. Biol.* 5:3168–82

141. Karlik, C. C., Mahaffey, J. W., Coutu, M. D., Fyrberg, E. A. 1984. *Cell* 37:469–81

142. Deak, I. I. 1977. *J. Embryol. Exp. Morphol.* 40:35–63

143. Deak, I. I., Bellamy, P. R., Bienz, M., Dubuis, Y., Fenner, E., et al. 1982. *J. Embryol. Exp. Morphol.* 69:61–81

144. Homyk, T. Jr. 1977. *Genetics* 87:105–28

145. Homyk, T. Jr., Grigliatti, T. A. 1983. *Dev. Genet.* 4:77–97

146. Homyk, T. Jr., Sinclair, D. A. R., Wong, D. T. L., Grigliatti, T. A. 1986. *Genetics* 113:367–89

147. Newman, S. M. Jr., Wright, T. R. F. 1981. *Dev. Biol.* 86:393–402

148. Newman, S. M. Jr., Wright, T. R. F. 1983. *Dev. Genet.* 4:329–45

149. Hall, J. C. 1982. *Q. Rev. Biophys.* 15:223–479

150. Koana, T., Hotta, Y. 1978. *J. Embryol. Exp. Morphol.* 45:123–43

151. Salkoff, L., Wyman, R. 1981. *Nature* 293:228–30

152. Costello, W. J., Wyman, R. J. 1985. *Wilhelm Roux's Arch. Dev. Biol.* 194:373–76

153. Wu, C.-F., Ganetzky, B., Haugland, F.

N., Liu, A.-X., 1983. *Science* 220: 1076–78

154. Ball, E., Ball, S. P., Sparrow, J. C. 1985. *Dev. Genet.* 6:77–92

155. Han, L. Y., Papazian, D. M., Jan, Y. N., O'Farrell, P. H. 1984. *Soc. Neurosci. Abstr.* 10:1089

156. Ganetzky, B., Wu, C.-F. 1984. *Soc. Neurosci. Abstr.* 10:1090

157. Digan, M. E., Haynes, S. R., Mozer, B. A., Dawid, I. B., Gans, M. 1986. *Dev. Biol.* 114:161–69

158. Fyrberg, E. A., Kindle, K. L., Davidson, N., Sodja, A. 1980. *Cell* 19:365–78

159. Tobin, S. L., Zulauf, E., Sanchez, F., Craig, E. A., McCarthy, B. J. 1980. *Cell* 19:121–31

160. Karlik, C. C., Fyrberg, E. A. 1986. *Mol. Cell. Biol.* 6:1965–73

161. Basi, G. S., Storti, R. V. 1986. *J. Biol. Chem.* 261:817–27

162. Mogami, K., Hotta, Y. 1981. *Mol. Gen. Genet.* 183:409–17

163. Mogami, K., O'Donnell, P. T., Bernstein, S. I., Wright, T. R. F., Emerson, C. P. Jr. 1986. *Proc. Natl. Acad. Sci. USA* 83:1393–97

164. Fyrberg, E. A., Mahaffey, J. W., Karlik, C. C., Coutu, M. D. 1986. In *Molecular Biology of Muscle Development*, ed. C. Emerson, D. Fischman, B. Nadal-Ginard, M. A. Q. Siddiqui, pp. 639–52. New York: Liss

165. Hiromi, Y., Hotta, Y. 1985. *EMBO J.* 4:1681–87

166. Storti, R. V., Szwast, A. E. 1982. *Dev. Biol.* 90:272–83

167. Albertson, D. G. 1985. *EMBO J.* 4:2493–98

168. Salvato, M., Sulston, J., Albertson, D., Brenner, S., 1986. *J. Mol. Biol.* 190:281–90

169. Swanson, M. M., Edgley, M. L., Riddle, D. L. 1984. In *Genetic Maps 1984*, ed. S. J. O'Brian, pp. 286–99. NY: Cold Spring Harbor Lab.

170. Park, E.-C., Horvitz, H. R. 1986. *Genetics* 113:853–67

171. Greenwald, I., Horvitz, H. R. 1986. *Genetics* 113:63–72

172. Robert, B., Barton, P., Minty, A., Daubas, P., Weydert, A., et al. 1985. *Nature* 314:181–83

172a. Czosnek, H., Nudel, U., Shani, M., Barker, P. E., Pravtcheva, D. D., et al. 1982. *EMBO J.* 1:1299–305

173. Czosnek, H., Nudel, U., Mayer, Y., Barker, P. E., Pravtcheva, D. D., et al. 1983. *EMBO J.* 2:1977–79

174. Czosnek, H., Barker, P. E., Ruddle, F. H., Robert, B. 1985. *Somatic Cell Mol. Genet.* 11:533–40

175. Leinwand, L. A., Fournier, R. E., Nad-

al-Ginard, B., Shows, T. B. 1983. *Science* 221:766–69

176. Gunning, P., Ponte, P., Kedes, L., Eddy, R., Shows, T. 1984. *Proc. Natl. Acad. Sci. USA* 81:1813–17

177. Mahdavi, V., Chambers, A. P., Nadal-Ginard, B. 1984. *Proc. Natl. Acad. Sci. USA* 81:2626–30

178. Edwards, Y. H., Parkar, M., Povey, S., West, L. F., Parrington, J. M., Solomon, E. 1985. *Ann. Hum. Genet.* 49:101–9

179. Devlin, R. B., Emerson, C. P. Jr. 1979. *Dev. Biol.* 69:202–16

180. Bernstein, S. I., Donady, J. J. 1980. *Dev. Biol.* 79:388–98

181. Affara, N. A., Daubas, P., Weydert, A., Gros, F. 1980. *J. Mol. Biol.* 140:459–70

181a. Blau, H. M., Pavlath, G. K., Hardeman, E. C., Chiu, C. P., Silberstein, L., et al. 1985. *Science* 230:758–66

181b. Wright, W. E. 1984. *Exp. Cell. Res.* 151:55–69

181c. Wright, W. E. 1984. *J. Cell. Biol.* 98:436–43

182. Deleted in proof

183. Emerson, C. P. Jr., Beckner, S. K. 1975. *J. Mol. Biol.* 93:431–44

184. Linkart, T. A., Clegg, C. H., Haushka, S. D. 1981. *Dev. Biol.* 86:19–30

185. Nguyen, H. T., Medford, R. M., Nadal-Ginard, B. 1983. *Cell* 34:281–93

186. Medford, R. M., Nguyen, H. T., Nadal-Ginard, B. 1983. *J. Biol. Chem.* 258:11063–73

187. Devlin, B. H., Konigsberg, I. R. 1983. *Dev. Biol.* 95:175–92

188. Silberstein, L., Webster, S. G., Travis, M., Blau, H. M. 1986. *Cell* 46:1075–81

189. Izumo, S., Nadal-Ginard, B., Mahdavi, V. 1986. *Science* 231:597–600

190. Miller, J. B., Stockdale, F. E. 1986. *Proc. Natl. Acad. Sci. USA* 83:3860–64

191. Keller, L. R., Emerson, C. P. Jr. 1980. *Proc. Natl. Acad. Sci. USA* 77:1020–24

192. Bandman, E., Matsuda, R., Strohman, R. C. 1982. *Dev. Biol.* 93:508–18

193. Crow, M. T., Olson, P. S., Stockdale, F. E. 1983. *J. Cell. Biol.* 96:736–44

194. Hastings, K. E. M., Emerson, C. P. Jr. 1982. In *Muscle Development, Molecular and Cellular Control*, ed. M. L. Pearson, H. F. Epstein, pp. 215–25. New York: Cold Spring Harbor Lab.

195. Keller, L. R., Emerson, C. P. Jr. 1982. In *Disorders of the Motor Unit*, ed. D. Schotland, pp. 863–75. New York: Wiley

196. Rubinstein, N. A., Kelly, A. M. 1978. *Dev. Biol.* 62:473–85

197. Whalen, R. G., Schwartz, K.,

Bouveret, P., Sell, S. M., Gros, F. 1979. *Proc. Natl. Acad. Sci. USA* 76:5197–201

198. Sweeney, L. J., Clark, W. A., Umeda, P. K., Zak, R., Mamasck, F. J. 1984. *Proc. Natl. Acad. Sci. USA* 81:797–800

198a. Whalen, R. G., Sell, S. M., Eriksson, A., Thornell, L. E. 1982. *Dev. Biol.* 91:478–84

199. Obinata, T., Masaki, T., Takano-Ohmuro, H., Tanaka, T., Shimizu, N. 1983. *J. Biochem.* 94:1025–28

199a. Everett, A. W., Sinha, A. M., Umeda, P. K., Jakovcic, S., Rabinowitz, M., Zak, R. 1984. *Biochemistry* 23:1596–99

200. Lompré, A.-M., Nadal-Ginard, B., Mahdavi, V. 1984. *J. Biol. Chem.* 259:6437–46

201. Shani, M. 1985. *Nature* 314:283–86

202. Daubas, P., Robert, B., Garner, I., Buckingham, M. 1985. *Nucleic Acids Res.* 13:4623–43

203. Frank, G., Weeds, A. G. 1974. *Eur. J. Biochem.* 44:317–34

204. Simmen, R. C. M., Tanaka, T., Ts'ui, K. F., Putkey, J. A., Scott, M. J., et al. 1985. *J. Biol. Chem.* 260:907–12

205. Gilbert, W. 1985. *Science* 228:823–24

206. Umeda, P. K., Levin, J. E., Sinha, A. M., Cribbs, L. L., Darling, D. S., et al. 1986. See Ref. 164, pp. 809–23

206a. Gulick, J., Kropp, K., Robbins, J. 1985. *J. Biol. Chem.* 260:14513–20

207. Zezza, D. J., Heywood, S. M. 1986. *J. Biol. Chem.* 261:7455–60

208. Dibb, N. J., Brown, D. M., Karn, J., Moerman, D. G., Bolten, S. L., Waterston, R. H. 1985. *J. Mol. Biol.* 183:543–51

209. Eide, D. J., Anderson, P. 1985. *Mol. Cell. Biol.* 5:1–6

210. Eide, D., Anderson, P. 1985. *Genetics* 109:67–79

211. Anderson, P., Brenner, S. 1984. *Proc. Natl. Acad. Sci. USA* 81:4470–74

212. Eide, D., Anderson, P. 1985. *Proc. Natl. Acad. Sci. USA* 82:1756–60

213. Moerman, D. G., Benian, G. M.,

Waterston, R. H. 1986. *Proc. Natl. Acad. Sci. USA* 83:2579–83

214. Mogami, K., Fujita, S. C., Hotta, Y. 1982. *J. Biochem.* 91:643–50

215. Waterston, R. H., Smith, K. C., Moerman, D. G. 1982. *J. Mol. Biol.* 158:1–15

216. Waterston, R. H., Hirsh, D., Lane, T. R. 1984. *J. Mol. Biol.* 180:473–96

217. Landel, C. P., Krause, M., Waterston, R. H., Hirsh, D. 1984. *J. Mol. Biol.* 180:497–513

218. Miller, D. M., Maruyama, I. 1986. See Ref. 164, pp. 629–38

219. Asmundson, V. S., Julian, L. M. 1956. *J. Hered.* 47:248–52

220. Deleted in proof

221. Bulfield, G., Siller, W. G., Wright, P. A. L., Monc, K. J. 1984. *Proc. Natl. Acad. Sci. USA* 81:1189–92

222. Roderick, T. H., Davisson, M. T. 1984. See Ref. 169, pp. 343–55

222a. Davisson, M. T., Roderick, T. H. 1981. In *Genetic Variants and Strains of the Laboratory Mouse*, ed. M. C. Green, pp. 283–313. Stuttgart: Gustav Fischer Verlag

223. McKusick, V. A. 1986. *Clin. Genet.* 29:545–88

224. Deleted in proof

225. Engel, A. G. 1986. See Ref. 3, pp. 1185–240

226. Monaco, A. P., Neve, R. L., Colletti-Feener, C., Bertelson, C. J., Kurnit, D. M., Kunkel, L. M. 1986. *Nature* 323:646–50

227. Grimm, T. 1986. See Ref. 3, pp. 1214–50

228. Munsat, T. L. 1986. See Ref. 3, pp. 1251–66

229. Harper, P. S. 1986. See Ref. 3, pp. 1267–96

230. Wills, N., Gesteland, R. F., Karn, J., Barnett, L., Bolten, S., Waterston, R. H. 1983. *Cell.* 33:575–83

231. Karlik, C. C., Fyrberg, E. A. 1985 *Cell* 41:57–66

232. Moerman, D. G., Plurad, S., Waterston, R. H., Baillie, D. L. 1982. *Cell* 29:773–81

Ann. Rev. Biochem. 1987. 56:727–77

INTERFERONS AND THEIR ACTIONS

Sidney Pestka and Jerome A. Langer

Department of Molecular Genetics and Microbiology, University of Medicine and Dentistry of New Jersey, Robert Wood Johnson Medical School, 675 Hoes Lane, Piscataway, New Jersey 08854-5635

Kathryn C. Zoon

Immunology Laboratory, Division of Virology, Office of Biologics Research and Review, Food and Drug Administration, Building 29A, Room 2A17, 8800 Rockville Pike, Bethesda, Maryland 20892

Charles E. Samuel

Department of Biological Sciences, Section of Biochemistry and Molecular Biology, University of California, Santa Barbara, California 93106

CONTENTS

0066-4154/87/0701-0727$02.00

INTRODUCTION

Since the review on interferons that appeared in the *Annual Review of Biochemistry* in 1982 (1), many developments have occurred. These have resulted in a vast expansion of the literature concomitant with the availability of relatively large quantities of purified interferons from natural sources as well as recombinant DNA technology. New insights have been gained in their biochemistry, molecular biology, and clinical application. Because this review is limited in size, it represents a highly selective presentation of the highlights of the past five years. Studies on the numerous biological effects of interferon have only been mentioned in brief. As a whole, this review should provide a scaffold upon which the reader can build. The authors thank our numerous colleagues who were extremely gracious in providing reprints and preprints so that this review could be as up to date as possible.

There appear to be three major classes of interferons (2, 3): leukocyte or alpha[1] interferon (IFN-α); fibroblast or beta interferon (IFN-β); and immune or gamma interferon (IFN-γ). These three classes seem to be present in all mammalian species. There have been a number of suggestions that other kinds of interferon exist (3–6), but it remains to be seen whether these and other molecules with antiviral activity will be classified as interferons.

---

[1]The abbreviations used have followed standard nomenclature as summarized in detail elsewhere (3). In brief, interferon alpha, beta, and gamma are designated IFN-α, IFN-β, and IFN-γ, respectively. The species of origin is designated by a prefix Hu, Mu, Bo, etc for human, murine, or bovine species, respectively, as Hu-IFN-α, Hu-IFN-β, or Hu-IFN-γ, for example. The (2'-5')-oligoadenylate is abbreviated $2,5\text{-}A_n$ where appropriate.

ASSAY AND PRODUCTION OF INTERFERONS

Interferons are measured generally by their antiviral activity in cell culture. Many antiviral assays exist (see Ref. 7 and chapters therein). With the identification of monoclonal antibodies to the interferons (8, 9), radio- and enzyme-immunoassays for the various interferons are now available (10–12). These are relatively rapid (2–4 hours) compared to the antiviral assays that take 16–48 hours. Many of the immunoassays detect only active interferon and are virtually equivalent to the antiviral assays (13). Although it is common to report interferon concentrations in units/ml as defined by international standards, this can be quite deceiving when comparing different interferons (14). Thus, interferon concentrations should be expressed in molarity wherever possible.

Induction of natural interferons has been discussed extensively elsewhere (7, 15, 16). The induction and production procedure is very different from one species and cell to another. Nevertheless, in general double-stranded RNA and viruses prove to be the best inducers of IFN-α and IFN-β. The greatest impact on production of interferons has been made by recombinant DNA technology, by which relatively large amounts of purified interferons are available for basic research and clinical trials (see below).

PURIFICATION AND STRUCTURE OF INTERFERONS

The purification and characterization of interferon for basic science and clinical applications have been recognized as important since the discovery of interferon nearly three decades ago (17). The purification of the natural IFNs was long delayed because of the minute quantities of these molecules produced, their high biological specific activities, and the lack of precision of the antiviral bioassay. Therefore, it was not until the late 1970s that the first natural IFNs were purified and partially characterized (7, 18). Subsequently recombinant DNA technology resulted in the production and purification of large quantities of homogeneous IFNs, which have greatly advanced our knowledge of the chemical and biological properties of these molecules. The methods used to purify IFN have been diverse and in many cases novel; they have included such procedures as affinity chromatography, sodium dodecyl sulfate polyacrylamide gel electrophoresis (SDS PAGE), and high performance liquid chromatography (HPLC) (1, 7, 15). A number of structural properties of both natural and recombinant IFNs have been determined, including their primary amino acid sequence and a limited amount of information on their secondary and tertiary structure.

Natural Interferons

INTERFERON ALPHAS   The purification of natural IFN-α was not only difficult for the reasons mentioned previously but was also complicated by the fact
it was actually a family of structurally related molecules (19–23). The first
purification of Hu-IFN-α's in the late 1970s made use of techniques such as
immunoabsorbant affinity chromatography, SDS PAGE, and HPLC to
achieve pure IFN preparations (19, 20, 24, 25). In general, the specific
activities of the IFN-α's were 2–4 × 10^8 units per mg protein. Subsequently,
multiple species of human interferon alpha have been purified by refinements
of techniques discussed above from virus-induced buffy coats (19, 20, 24–
29), Namalwa cells (55, 30–35), peripheral blood leukocytes from patients
with chronic myelogeneous leukemia (20), and KG-1 cells (36). At present,
the number of species of IFN-α isolated from each cell source ranges from 10
to greater than 16. In general, IFN-α's derived from different sources are
quantitatively and/or qualitatively distinct from each other. The apparent
molecular weights of the IFN-α's range from 16,000 to 27,000 (9, 19, 20,
24–36), and their amino acid compositions are very similar and rich in leucine
and glutamic acid/glutamine (19, 20, 30, 31, 34, 36, 37). In addition,
isoelectric focusing revealed the heterogeneous nature of IFN-α's; multiple
species with pIs ranging from 5.5 to 6.5 have been observed (38). In general,
natural human IFN-α's are polypeptides of 165–166 amino acids (19, 20, 30,
34, 36, 37), although two species have been isolated that lack the 10 COOH-
terminal amino acid residues (39). Their amino acid sequences are quite
homologous with at least 70% sequence identity between individual molecules (20, 22, 23, 26, 30–32, 37, 39–41). Particularly noteworthy is the
conservation of the cysteines at positions 1, 29, 98/99, and 138/139. Based on
studies with recombinant Hu-IFN-αA, these residues are involved in the
formation of disulfide bonds (42–44). Three of the natural IFN-α's appear to
be identical in sequence to two recombinant DNA–derived IFN-α's (21).
Although some controversy existed concerning the presence of carbohydrate
on IFN-α, it is now clear that several species of IFN-α are glycosylated; two
species each from CML and KG-1 cells (45) and three from Namalwa cells
(K. Zoon, unpublished data). The structure and linkage of the carbohydrate
moeity are not known; however, it is postulated that the carbohydrate is linked
to the protein through O-glycoslation, since all but one Hu-IFN-α sequences
lack the sequence Asn-X-Ser/Thr necessary for N-glycosylation and the
carbohydrate composition is consistent with O-glycosylation (45). A population of acid-labile Hu-IFN-α has been described and partially purified (46);
the structure of this IFN is currently unknown.

The purification of Mu-IFN-α from virus-induced L929 cells (47, 48),
C-243 cells (49), and mouse Ehrlich ascites cells (50, 51) was achieved in the

late 1970s by the techniques discussed above, including ion exchange chromatography, gel filtration, controlled pore glass chromatography, immunoadsorbant affinity chromatography, affinity chromatography with ligands such as polynucleotides (Poly U) and octyl groups, isoelectric focusing, and SDS PAGE. The specific activities of the purified IFNs ranged from 4×10^8 to 3×10^9. It must be emphasized, however, that the specific activities are often not comparable from one laboratory to another because of different methods of protein determination. The apparent molecular weight of murine IFN-α is approximately 20,000. It appears to be a glycoprotein (51). As with the Hu-IFN-α's, the Mu-IFN-α's also belong to a family of structurally related proteins (52, 53). The amino acid composition of murine IFN-αC is very similar to that of the human and other murine IFN-α's (21, 53). The first 20 NH$_2$-terminal amino acid residues have been determined for murine IFN-αC (54). This portion of the primary sequence is similar to those derived from the cloned Mu-IFN-α's with 12 out of 20 residues being identical; it is also similar to that of Hu-IFN-α's (21). Monoclonal antibody affinity chromatography, gel filtration, and chromatofocusing have been used successfully to identify a minimum of five species of Mu-IFN-α's from L929 cells that have pIs ranging from 5.6 to 7.5 (55).

INTERFERON BETA   Human IFN-β derived from diploid fibroblasts was the first IFN to be purified (56). Subsequently, improved procedures were developed to purify natural Hu-IFN-β, which included ion exchange chromatography, affinity chromatography (with lectins, phenyl groups, monoclonal antibodies against Hu-IFN-β and Cibacron Blue), and HPLC (57–63). Only one protein species was isolated and identified with these techniques; it exhibited an apparent molecular weight of approximately 20,000 and a specific activity of 2–5×10^8 units/mg protein. The amino acid composition is similar to that observed for Hu- and Mu-IFN-α's and Mu- and Bo-IFN-β (21, 41). The NH$_2$-terminal amino acid sequence for the first 19 residues of natural Hu-IFN-β was determined (59, 62–64) and is in agreement with the sequence derived from the DNA sequence (65). Natural Hu-IFN-β was demonstrated to be a glycoprotein by treatment of the molecule with a mixture of glycosidases (66) and the presence of amino sugars in purified IFN preparations (58, 62). It is believed that the major portion of the carbohydrate moiety is linked by N-glycosylation to the Asn at position 80. Based on the determination of target size for antiviral activity, the functional unit of Hu-IFN-β appears to be a dimer (67). Recently, another human IFN has been reported, which constitutes 5% of the IFN made by poly(I:C)-superinduced diploid fibroblasts, but appears to be structurally distinct from the major form of Hu-IFN-β (6, 68, 69). However, since this molecule shows no structural relationship to any known interferon species, whether it will be considered an interferon remains open (69).

Murine IFN-β was originally purified to homogeneity from L929 cells, C-243 cells, and mouse Ehrlich ascites tumor cells (47–51), described above for Mu-IFN-α. Two forms of Mu-IFN-β have been identified with apparent molecular weights of approximately 35,000 (Class A) and 26,000 (Class B). They have specific activities in the range of 0.5–2 × 10^9 units/mg protein (51). Both appear to be glycoproteins (48, 51). Recently, purification schemes that utilize monoclonal antibody or controlled pore glass, CM-Sephadex, and phosphocellulose (70) have resulted in high yields of Mu-IFN-β. A single species was isolated with an apparent molecular weight of 33,000–40,000. The amino acid compositions of these IFNs are very similar to those of Mu-IFN-α and the Hu-IFN-α's and β. The amino acid sequences of the first 24 residues of Mu-IFN-β (class A) and Mu-IFN-β (class B) were determined to be identical to that deduced from the DNA sequence for murine Mu-IFN-β (21, 54, 65, 70, 71).

The purification of rat and avian interferons has also been described (7). Rat IFN from the cell line RFA-1 induced with Newcastle disease virus was isolated to 88% purity with controlled pore glass and phenyl-Sepharose (72). It has an apparent molecular weight of 18,500 and a specific activity of approximately 4 × 10^8 units/mg protein.

INTERFERON GAMMA   The purification of natural Hu-IFN-γ to homogeneity from human peripheral blood lymphocytes was reported by several groups using a combination of techniques such as controlled pore glass, concanavalin A–Sepharose, Blue-Sepharose, gel filtration, SDS PAGE, and HPLC (73–77). The specific activity of Hu-IFN-γ is 1–25 × 10^7 units/mg protein. Three forms of Hu-IFN-γ have been identified by SDS PAGE with apparent molecular weights of 15,500–17,000, 20,000, and 25,000 (73–78). The differences in molecular weight appear to be due to the extent of glycosylation of a single polypeptide species (75, 78). The amino acid sequences of the 20,000-dalton and the 25,000-dalton species have been determined and were found to be identical to each other and essentially identical to that deduced from the DNA sequence (75). The NH$_2$-terminus was found to be a pyroglutamic acid residue instead of Cys-Tyr-Cys originally predicted from the DNA sequence (75). The 25,000-dalton species was glycosylated at two sites, Asn 25 and Asn 97, while the 20,000-dalton species has carbohydrate linked to only Asn 97. As observed with some natural IFN-α's, natural IFN-γ exhibited heterogeneity at the COOH-terminus in that five truncated forms were identified (75). This heterogeneity is probably related to proteolytic digestion either during or after the secretion of IFN-γ. An NH$_2$-terminal fragment of 45 amino acid residues obtained by CNBr cleavage of Hu-IFN-γ

was found to bind to the IFN-γ receptor but did not elicit antiviral activity or induce HLA antigens; however, it did inhibit the induction of HLA antigens (79). Upon gel filtration natural Hu-IFN-γ yields an apparent molecular weight of 40,000–60,000, thus suggesting under physiological conditions IFN-γ may exist as a dimer (75, 80, 81). In contrast, target analysis of both natural and recombinant Hu-IFN-γ's suggested that the functional unit is a tetramer (67).

Recombinant DNA–Derived Interferons

The application of recombinant DNA technology to the production of IFN (see below) has led to the availability of large quantities of a number of Hu-IFN-α species (18, 82–86), Hu-IFN-β (87, 88), Hu-IFN-γ (89), and Mu-IFN-γ (90) for structural analysis. Techniques such as ion exchange chromatography, affinity chromatography with lectins and monoclonal antibodies, and HPLC have been instrumental in the successful purification of these molecules.

INTERFERON ALPHA   The purification of recombinant Hu-IFN-α's from *Escherichia coli* extracts has been achieved by conventional chromatography (83); by ion exchange, copper chelate chromatography, gel filtration, and crystallization (86); and by ion exchange and monoclonal antibody affinity chromatography (18, 82, 85). Recombinant Hu-IFN-αA and -α2 were purified by these techniques (82–86). They exhibit an apparent molecular weight in the range of 18,500–19,400 and a specific activity of 1.5–2.0×10^8 units/mg protein; each consists of 165 amino acid residues (82–86). The amino acid sequences of recombinant Hu-IFN-αA and -α2 are identical except for residue 23, in which there is a conservative Arg/Lys substitution (21–23, 34, 41, 91). In addition, the amino acid composition, sequence, and molecular weight of Hu-IFN-αA and -α2 correlate with those of several natural Hu-IFN-α species (21, 41). HPLC analysis of tryptic digests of IFN-αA permitted the assignment of the disulfide bonds between residues 1 and 98 and residues 29 and 138 (42). Additional studies with S-sulfonate derivatives of IFN-αA (43) and analogues of IFN-αA prepared by site-specific mutagenesis (44) suggested that the 29-138 disulfide was critical for maximal antiviral activity. However, an NH_2-terminal IFN-α2 fragment consisting of residues 1–110 was shown to exhibit some antiviral activity, thus suggesting that other conformations of the molecule may be active (92). Circular dichroism studies of Hu-IFN-αA estimated the content of alpha-helix to be 45–70% and beta-sheet to be essentially nonexistent (93). In addition, these studies indicated that one tryptophan of IFN-α is in a hydrophobic region of the molecule and is very asymmetric (93). Although crystals of IFN-αA and -α2 have been produced (83, 94), results are not yet available on the tertiary structure of these molecules. Target analysis of IFN-αA suggested that the functional unit is a monomer (67). The recombinant Hu-IFN-αA/D (Bgl II), IFN-αD, IFN-αC, IFN-αI, IFN-αJ, and IFN-αK have also been purified by monoclonal antibody affinity chromatography (18). Multiple species of recombinant DNA–derived IFN-αA and -α2 have been observed by analysis on HPLC and SDS PAGE; they represent variants related to disulfide bond formation (both monomers and oligomers) and presence or absence of methionine at the amino terminus (18, 21, 84, 85).

Bovine IFN-αC was purified from *E. coli* extracts by precipitation with trichloroacetic acid and monoclonal antibody affinity chromatography (95). The IFN has a specific activity of 2×10^8 units/mg protein on bovine cells and was essentially inactive on human and simian cells. The apparent molecular weight of 18,000 was determined by SDS PAGE.

INTERFERON BETA   The purification of recombinant Hu-IFN-β has been achieved primarily by Blue-Sepharose chromatography (87). The recombinant analogue [Ser^{17}]IFN-β has been successfully purified by a combination of steps, including extraction with SDS and 2-butanol, acid precipitation, and gel filtration (88). [Ser^{17}]IFN-β has an apparent molecular weight of approximately 19,000 and a specific activity of 1.1×10^8 units/mg protein, similar to that observed for natural IFN-β. The amino acid composition and sequence is essentially identical (with the exception of residue 17) to that of the natural IFN-β and that derived from the cDNA coding for Hu-IFN-β (88, 96). Three cysteines are found at positions 17, 29, and 141. Since the [Ser^{17}]IFN-β analogue has no apparent loss of biological activity (88, 96) and the [Tyr^{141}]IFN-β has essentially no biological activity (97), it is postulated that the Cys 31 and Cys 141 disulfide bridge is critical for optimal biological activity.

The purification of recombinant Mu-IFN-β was accomplished with controlled pore glass, copper chelate-Sepharose, and affinity chromatography with Matrex gel Blue A and crystallized by vapor diffusion with polyethylene glycol (98). The specific activity was 3.5×10^7 units/mg protein, which was lower than that observed for the natural Mu-IFN-β (0.5–3×10^9 units/mg protein) (47–51). The discrepancy in the specific activities is not understood. The apparent molecular weight of this Mu-IFN-β is 19,900, which is consistent with that predicted from the DNA sequence (71). In addition the NH$_2$-terminal amino acid sequence for the first 24 residues was identical to that predicted from the DNA sequence with the exception of residue 18.

INTERFERON GAMMA   Recombinant Hu-IFN-γ has been successfully purified from *E. coli* (89, 99) and Chinese hamster ovary (CHO) cells (100). Methods for the purification from *E. coli* include extraction with guanidine hydrochloride, silica chromatography, and monoclonal antibody affinity chromatography (89). The specific activity of IFN-γ purified from *E. coli* was 1–2 $\times$ 10^7 units/mg protein, which is in agreement with that observed for purified natural Hu-IFN-γ. The monomer has an apparent molecular weight of 17,000 by SDS PAGE, which is consistent with that derived from the DNA sequence. Upon gel filtration the material migrates as a dimer, which has also been shown for natural IFN-γ (75). However, target size analysis suggests that the functional unit is a tetramer (67). When IFN-γ produced by CHO cells was purified by ion exchange chromatography and concanavalin A-Sepharose (100), two molecular weight forms were isolated with apparent molecular weights of 21,000 and 25,000; these results were in agreement with those derived with the natural IFN-γ's. The specific activity of these species was approximately 1 $\times$ 10^8 units/mg protein, which was 5- to 10-fold higher than observed with natural IFN-γ. Hu-IFN-γ contains no cysteine residues (75).

The purification of Mu-IFN-γ has been accomplished by extraction with polyethyleneimine and chromatography on phenyl-Sepharose, CM-Sepharose, and gel filtration (101). Two variants of IFN-γ, A (136 amino acids) and D (133 amino acids), were purified; they exhibited specific activities of 2 $\times$ 10^6 units/mg protein and 2 $\times$ 10^7 units/mg protein, respectively. They differ in that IFN-γD lacks the Cys-Tyr-Cys at the NH$_2$-terminus. The apparent molecular weight of both species was 15,200. Deletion of 6 residues at the COOH-terminus did not alter the biological activity of the molecule, whereas deletion of 12 residues did (101). Three cysteine residues are present in Mu-IFN-γA and one in -γD; disulfide bonds were not detected in either IFN preparation. The molecular weight of both these IFNs based on gel filtration is approximately 30,000, thus suggesting that the quaternary structure is a dimer.

Crystals of Hu-IFN-γ and Bo-IFN-γ have been obtained that appear suitable for X-ray structure determination (101a; J. R. Rubin, A. A. Kossiakoff, unpublished data). Preliminary results show X-ray diffraction to at least 2.9 Å resolution. Several promising isomorphous derivatives have been obtained with data collection and analysis in progress.

ISOLATION OF GENOMIC AND cDNA INTERFERON CLONES

A similar approach was taken to isolate DNA recombinants containing the human interferon sequences by several laboratories. First, it was necessary to isolate and measure interferon mRNA. This was accomplished several years before cloning was attempted when interferon mRNA was translated in cell-free extracts (102, 103) and in frog oocytes (104–107). A cDNA library was prepared from mRNA isolated from human cells (leukocytes or fibro-blasts) synthesizing the desired interferon. Partially purified mRNA from induced cells was used as a template for cDNA synthesis. The dC-tailed double-stranded DNA obtained was hybridized to dG-tailed pBR322 cleaved at the *Pst*I restriction endonuclease site and the resultant recombinant plasmid introduced into *E. coli* by transformation. Tetracycline-resistant and ampicil-lin-sensitive transformants were screened for the presence of interferon DNA sequences. In our laboratory (108, 109), we first screened all recombinants for their ability to bind to mRNA from cells synthesizing interferon (induced cells), but not to mRNA from uninduced cells (those not producing interferon) by colony hybridization for the presence of induced-specific sequences with ^{32}P-labeled interferon mRNA as a probe in the presence of excess mRNA from uninduced cells (108–110). An analogous approach was used by Tani-guchi et al (111) for initial screening. An mRNA selection procedure was then used to identify pools of positive clones from which individual positive colonies were isolated (108, 111–113). This procedure was used to isolate cDNA clones for human leukocyte (IFN-α) and fibroblast (IFN-β) interferons in several laboratories within a short span of time (108, 111, 112, 114). With protein sequences available for IFN-β, this information was used to obtain positive clones with appropriate probes (115, 116). By analogous procedures, recombinants containing DNA sequences corresponding to human IFN-γ were obtained (117, 118). DNA sequences for murine, rat, bovine, and monkey interferons were obtained with the corresponding recombinants for human interferon as probes (see specific chapters in Refs. 15 and 34 for citations and details).

Interferon Genes

The initial cDNA recombinant plasmids were used as probes to identify the genes and other cDNA clones. In the case of Hu-IFN-α, at least 23 separate loci have been identified corresponding to the major human IFN-α family (15, 34, 119–123). At least 14 of these contain coding sequences for apparently functional proteins (34, 122). These have substantial homology and comprise six linkage groups (122). It is not yet possible to conclude with certainty which DNA sequences are allelic. It appears that IFN-αA and IFN-α2, which differ in one amino acid only, are both represented in each of 14 human DNA samples (123). In addition, two IFN-α loci have identical coding sequences, but different flanking sequences (122). None of these chromosomal genes contains an intron within the coding sequence. Several of the loci represent pseudogenes (119, 120, 122). All mammalian species have large IFN-α gene families (124). None of the human IFN-α species except IFN-αH (IFN-α14) contains an N-glycosylation site.

A related family of IFN-α genes has been described (omega, 125; α_L, 126;

α_{II}, 127). Whereas the major Hu-IFN-α class genes encode proteins of 165 or 166 amino acids, this second class of Hu-IFN-α genes encode 172-amino-acid proteins. Both classes exhibit potent antiviral activity and are coordinately expressed in response to viral induction (127). It appears that both IFN-α classes are present in all mammalian species. The major IFN-α class of human genomic DNA appears to be the minor class in bovine genomic DNA. Both classes of human IFN-α genes are located on chromosome 9 and do not contain any introns (126–128).

The human IFN-β gene is a single gene that also is located on chromosome 9 and codes for a protein of 166 amino acids with about 29% homology to Hu-IFN-αD or IFN-α1 (114–119, 129). As the IFN-α genes, it does not contain any introns. The bovine, equine, and porcine IFN-β gene families contain many genes, whereas humans and most other mammalian species probably contain only one (124, 130); the lion and rabbit genomes contain two IFN-β genes.

Human IFN-γ is specified by a unique gene containing three introns located on chromosome 12 (117, 118, 131, 132). It contains little or no homology to IFN-α or IFN-β (117, 132), although some structural similarity between human IFN-γ and IFN-α or -β has been reported (133). Whereas IFN-α and IFN-β genes exist in most vertebrates examined (124), the existence of IFN-γ genes in nonmammalian vertebrates has not been demonstrated.

As briefly noted above, interferons other than IFN-α, -β, and -γ have been reported (4–6, 68, 69, 134, 135). In particular, it has been suggested that IFN-β represents a family of human genes (135). One of these, initially termed "IFN-β2" has been cloned (6, 68, 69). Although the protein corresponding to "IFN-β2" is neutralized by antiserum to Hu-IFN-β, there is little or no structural homology to Hu-IFN-β or any known interferon or its gene at either the amino acid or nucleotide level; and the protein has little or no antiviral activity compared to Hu-IFN-α, -β, or -γ (68, 69). Thus, it would be useful to designate it by a name other than "IFN-β2." Nevertheless, it remains open whether this is, indeed, a novel interferon.

Expression of Interferons

High-level expression of interferons was achieved in *E. coli* with the use of a strong promoter before the coding region. The general strategy involved removal of the sequence coding for the signal peptide, insertion of an ATG start codon just prior to the coding sequence for the mature secreted protein, and introduction of an operator-promoter and ribosome binding site just upstream from the ATG. In this way, human and animal interferons have been produced at high levels in *E. coli* (15, 34, 110, 114–118, 120, 125, 136). Because of the insertion of a methionine codeword prior to the coding sequence for the mature protein, this methionine must be removed if the protein is to be essentially identical to the natural protein. In the case of Hu-IFN-αA, the methionine is spontaneously removed by *E. coli* so that greater than 99% of the protein lacks this additional terminal methionine (82, 137). However, in the case of some other interferons and many other proteins the terminal methionine remains a part of the protein, so that it is a formidable task to remove it specifically.

It was initially considered that the lack of glycosylation might be a significant problem for proteins produced in *E. coli*. Because most human IFN-α species are not glycosylated (20, 45, 138, 139), however, this was a moot point. In the case of IFN-β and IFN-γ, this is a consideration; nevertheless, most of these interferons that are in clinical trial are the nonglycosylated forms produced in *E. coli*. Glycosylated interferons can be obtained by expressing the proteins in animal cells or in yeast (15, 140–148, and references therein). A structural analysis of the carbohydrate chains of Hu-IFN-γ produced in Chinese hamster ovary cells has been reported (149).

Evolutionary Relationships of Interferon Genes

The homology between human IFN-α and IFN-β genes indicates they originated from a common ancestral gene about 300 million years ago, about the time of the divergence of mammals, reptiles, and birds from fish and amphibia (122, 129, 150). This result suggests that amphibia and fish contain only an IFN-α or IFN-β gene homologous to the mammalian counterparts, but not both, and is in accord with weak hybridization of Hu-IFN-β, but not IFN-α to amphibia and fish DNA (124). The multigene IFN-α family appears to fall into two groups, the major or standard IFN-α and the minor IFN-α (omega, α_L, α_{II}, see above). The major IFN-α species can then be subdivided into two major groups based on the divergence of silent sites (129). Substantial polymorphic variants probably exist among the IFN-α and -β genes (151, 152); in some individuals, a duplicated Hu-IFN-β gene is found. There appears to be little or no clear homology between the IFN-α/β nucleotide sequences on the one hand with that of IFN-γ on the other (131), although some relationships were reported (133, 153).

All three classes of interferon genes (IFN-α, -β, -γ) appear to be evolving rapidly at a rate comparable to that of immunoglobulin genes. In addition, the 3' noncoding sequence of the IFN-α genes shows much greater divergence than the rest of these genes, an observation that remains puzzling (110, 150). Also puzzling is the occurrence of apparent natural hybrids among the human IFN-α genes (34, 122); possible artifacts of cloning, however, have not been ruled out (34). With additional nucleotide sequences of IFN-α, -β, and -γ genes within and between species, many of these questions and enigmas will be resolved. Whereas interferons from a large number of mammalian species have been cloned (34, 52, 53, 70, 71, 95, 130, 148, 154–160), it would be useful to have cloned interferons from some nonmammalian species to enhance our understanding of these evolutionary relationships.

BIOLOGICAL EFFECTS OF INTERFERONS

The interferons exhibit a huge number of biological effects, a few of which are illustrated here. Before homogeneous preparations of interferon were available, it was thought that many of the activities attributed to the crude preparations would disappear on purification. Most of the activities attributed to the crude interferons were reproduced with purified natural and recombinant interferons. The antiviral activity led to the name interferon (17, 161) and serves to define the unit of interferon activity (3). On purifications of the natural human leukocyte interferons (IFN-α), it was found that all fractions that exhibited antiviral activity also exhibited antigrowth activity (162). Surprisingly, however, the ratio of antiviral to antiproliferative activity was not constant from one purified fraction to another (162) or from one interferon to another (163). This observation was confirmed with purified recombinant interferons (14) and extended to other activities as well: stimulation of cytotoxic activities of lymphocytes and macrophages, and of natural killer cell activity (164–167); and increase in expression of some tumor-associated antigens (168–171). In the case of natural killer cell stimulation, IFN-αJ was found to be deficient in this activity while exhibiting other activities (165–167). Recently, Greiner et al (168) observed that Hu-IFN-αD and -αJ lack the ability to stimulate the surface expression of breast and colon tumor–associated antigens, whereas antiviral and antiproliferative activity remain intact. In addition, effects of interferon on lytic viruses such as vesicular stomatitis virus can be dissociated from those on Moloney leukemia virus (172, 173). These and other observations (174–180) indicate that many of the effects of the interferons can be dissociated and are due to different molecular mechanisms. Ultimately, these effects must be explained in terms of the interferon receptors and the biochemical mechanisms involved.

As noted above, the interferons exhibit a wide range of biological and biochemical effects on cells and animals (181 for general references). The antiproliferative activity of interferon was first reported by Paucker et al (182). There are no general rules to predict which cells will be inhibited and which not. In general, empirical results of antiproliferative and antitumor activity have been employed to apply interferon as an antitumor agent (183–185). The interferons also modulate cellular differentiation (186–189).

A major effect of interferons is their modulation of antigens of the major histocompatibility complex (MHC). All interferons (IFN-α, -β, and -γ) induce an increase in surface expression of class I MHC antigens (including β_2-microglobulin), whereas class II antigens are stimulated predominantly by

IFN-γ with little or no effect by IFN-α or -β (190–192). Expression of the Fc receptors is also stimulated by interferon (180, 193). Alterations in surface antigens may be an important mechanism by which interferon can modulate cellular interactions.

Often, IFN-α/β is synergistic with IFN-γ for antiviral or antiproliferative activities (194). In some cases, the interferons may be antagonistic. Mu-IFN-γ stimulates Ia antigen expression. However, treatment of cells with IFN-α or -β blocks stimulation of Ia antigen by IFN-γ (195, 195a). Hu-IFN-α/β blocks the IFN-γ-induced stimulation of H_2O_2 by human mononuclear phagocytes (196); and Mu-IFN-α/β antagonizes down-regulation of mannosyl-fucosyl receptors by Mu-IFN-γ (196a). In addition, one IFN-α species can block the activity of another in some cases: IFN-αJ can block the stimulation of natural killer cell activity by IFN-αA (166).

The interaction of the interferons with their receptors determines the biochemical events and their modulation of cellular functions. This is a complex process that is just beginning to be dissected (see below).

INTERFERON RECEPTORS

That interferon exerts its actions through cell surface–specific receptors (197) was established indirectly through several lines of experimentation (reviewed in 198–201). With the availability of purified interferons and the development of methods for radiolabeling them by iodination and phosphorylation (see articles in Ref. 15), more direct studies of the interferon receptors became possible.

In the current review, the receptor for IFN-α and -β will be considered separately from that of IFN-γ.

IFN-α/β Receptor

Because of the availability of purified IFNs, receptor studies have largely been confined to the human and murine systems. Bovine MDBK cells, on which Hu-IFN-α's are highly active, have often been used in studies of the interaction of Hu-IFN-α with the receptor. In some mouse studies, purified natural IFN, which is a mixture of IFN-α and -β, was used. In the human system the majority of studies have used recombinant IFNs, particularly IFN-αA or -α2, which differ by only a single amino acid. Other recombinant IFN-α's have been used as noted below.

The binding of IFN-α to cells has been studied in both the murine (202–204) and human (e.g. 205–211) systems with a wide variety of cell types (reviewed in 201). Binding to cells is time- and concentration-dependent. In most instances, binding at 4°C reaches its maximum level within 2½ hr.

Binding at 37°C is more complex (202, 206, 212–217). The extent of binding to the cell surface has been evaluated by treating the cells with trypsin or dilute acid at 4°C to eliminate surface-bound IFN (211, 212, 214). Most IFN bound at 4°C is surface bound, whereas the amount of IFN bound at 37°C that is insensitive to trypsin or acid increases with time, suggesting internalization. At 15°C, little internalization was detected. At 22°C the amount of internalization seems to vary with the cell line (J. A. Langer, unpublished data). Binding sites on various human, murine, and bovine cells have also been visualized by electron microscopy (214, 218, 219).

Binding to cells is saturable. In general, cells have been found to have about 2×10^2 to 6×10^3 receptors per cell, characterized by a dissociation constant of 1×10^{-9} to 1×10^{-11} M; i.e. there are a relatively small number of high-affinity receptors (201–205, 207–209, 211, 212, 216, 220–224). This explains the need for ligands of high radiospecific activity. No generalizations have emerged correlating histological or embryological origins with quantity of receptors. It should be noted, however, that naturally occurring cells, e.g. peripheral blood lymphocytes, generally have fewer receptors than many cultured cell lines (207). The characterization of receptors on solid tissue is difficult, presumably because of high nonspecific binding and low specific binding. However, the binding of IFN-α2 to plasma membranes from human placenta and bovine lung has been reported (224a). Many binding curves do not show true saturation, resulting in curvilinear Scatchard plots, prompting some authors to suggest a fairly large number of low-affinity receptors in addition to the high-affinity receptors. Such an interpretation may be possible when measurements are made in the absence of cellular metabolism. However, where secondary events occur so that binding is not in reversible equilibrium, interpretation of curvilinear plots becomes speculative at best. While low-affinity receptors are well documented in a variety of other systems and may exist for IFN, accurate quantitation is difficult because of the low level of radioactivity involved. It is worth noting that the binding of Hu-IFN-α to bovine cells nearly approaches saturation (211), so that the majority of receptors on these cells are, indeed, high affinity. The number of receptors per cell is not necessarily fixed. For example, the number of receptors seems to decrease in T98G and HL-60 cells at higher degrees of confluence or cell density of the culture (224, 225). No general method has been found to amplify or overproduce IFN receptors, although in human HL-60 cells an increase in receptor number accompanies chemically induced myeloid differentiation (224, 226). Other instances of changes in IFN-α receptors linked to differentiation have not yet been reported.

Direct binding studies of human (227–231) and murine (232) IFN-β have been reported. In the case of Hu-IFN-β, the more stable genetically engineered derivative [Ser^{17}]IFN-β, where Cys^{17} was replaced by serine (88, 96), has been especially useful. Binding parameters similar to those measured with IFN-α have been obtained.

The specificity of IFN-α and -β binding has been examined on several levels. Lack of binding to specific receptors seems to be the primary basis for the species specificity of IFN (202, 207, 211, 221). Cells without detectable specific receptors (based on IFN binding) have all been insensitive to the biological effects of IFN (208, 209, 221). However, the lack of receptors is but one cause of insensitivity as demonstrated with mutants selected for their insensitivity to IFN (202, 233–236). Different purified IFN-α's compete for the same binding sites on human and bovine cells (211, 216, 220, 222, 237). There has been no evidence for structural heterogeneity of the receptor. IFN-α and -β share a common receptor, as demonstrated by the ability of IFN-β to compete for the binding of labeled IFN-α (203, 205, 221, 238), and vice versa (229, 231, 232, 239). It should be noted, however, that Hu-IFN-β does not bind to bovine cells, whereas Hu-IFN-α binds well (211, 221). Purified IFN-γ does not compete with IFN-α for its receptor (205, 209, 220, 227, 237, 238, 240, 241), nor the converse (227, 228, 242–247). Most studies have demonstrated that binding of IFN-γ is not competed by IFN-β (227, 228, 243, 245–247) and vice versa (229); however, in two studies IFN-β and IFN-γ compete (231, 242). The reasons for this discrepancy are not apparent.

Two issues have been largely resolved by direct binding studies. Previous reports had suggested that gangliosides might play a major role in defining the IFN receptors (199). While it is now clear that human and murine IFN-β and perhaps Mu-IFN-α can bind to gangliosides and that gangliosides may inhibit the activity of some IFNs (248–253), it is unlikely that gangliosides constitute a major component of the IFN-α/β receptor (240, 254, 255). In addition, despite early reports of "interferon-like" effects on cells by cholera toxin (200), it has been shown by direct binding that cholera toxin does not compete with IFN for binding to the IFN receptor, nor does IFN bind to the cholera toxin receptor (240).

CHARACTERIZATION OF THE IFN-α/β RECEPTOR   The complex formed by covalently binding Hu-IFN-α to its receptor on the cell surface with bifunctional reagents migrates on SDS-polyacrylamide gels with an apparent molecular weight of about 150,000 (209, 256). The molecular weight was not sensitive to reduction by 2-mercaptoethanol, but was decreased by digestion with neuraminidase. Thus crosslinking occurs to a glycoprotein of M_r about 130,000, which is not disulfide-linked to other subunits. A second complex of 300 kd, of which the 150 kd complex is a constituent, has been detected in similar experiments (257). Following binding of [^{125}I]Hu-IFN-α2 to cells at 37°C, a stable noncovalent complex that migrates on Sephacryl S400 with an apparent size of 230 kd can be extracted with digitonin (258). Longer incubation of interferon with cells leads to the appearance of a larger complex (about 650 kd on Superose 6) and a decrease in the amount of the 230 kd complex (217, 258, 259). Within both the 230 kd peak and the 650 kd peak extracted from IFN-treated cells is an intrinsic kinase that phosphorylates a 105 kd protein migrating in these peaks. Possibly related to these complexes is a complex of M_r 500,000 on Sephacryl S500 extracted with Triton X-100 from Daudi cells with covalently crosslinked [^{125}I]Hu-IFN-α (257). When analyzed by SDS-PAGE, this complex contained both the 150 kd and 300 kd covalent complexes described above (257).

Binding activity in extracts of Triton X-100–solubilized cells has been measured by differential precipitation with polyethylene glycol (PEG). With this assay, some physicochemical characterization of the receptor in detergent-solubilized form was determined, suggesting primarily that the molecule is highly asymmetric (256). Some purification of the receptor was achieved by chromatography on wheat-germ agglutinin Sepharose or hydroxyapatite, followed by affinity chromatography on a column of IFN-α linked to CH-Sepharose (260).

An extensive purification of a covalent complex of [^{125}I][Ser^{17}]Hu-IFN-β with its receptor has recently been achieved (239). Following IFN binding and crosslinking with a cleavable reagent, the complex was extracted with Triton, fractionated on wheat-germ agglutinin–Sepharose and then on an immunoaffinity column of polyclonal antihuman IFN-β covalently coupled to protein A–Sepharose. A silver-stained band at 130 kd on an SDS gel coincided with ^{125}I and could be eluted from the gel, radioiodinated, and reanalyzed following cleavage of the crosslinking reagent. This produced a 110 kd ^{125}I-labeled band, presumably corresponding to the receptor binding subunit. It was not possible to demonstrate specific binding of [^{125}I]Hu-IFN-β to the putative receptor.

GENETICS   Linkage of human chromosome 21 to sensitivity of cells to IFN-α and -β is well documented both in somatic cell hybrids and in cells differing in their complement of chromosome 21 (261–267; reviewed in 199, 265–267). Antibodies raised to human-mouse somatic cell hybrids containing human chromosome 21 inhibited the actions of human interferon (265, 268–270), blocked the binding of [^{125}I]Hu-IFN-α to the receptor (271), and were useful in immunoprecipitating the covalent IFN-receptor complex (241). It has therefore been concluded that the gene for the human IFN-α/β receptor is encoded on human chromosome 21. Experiments with hamster-mouse hybrids and selected mouse-human hybrids have established that the murine gene is on chromosome 16 (272, 273). In both species, the IFN-α receptor gene is syntenic with that for soluble superoxide dismutase (SOD-1) (261,

272, 273). In the human case it is possible that the gene maps to the distal end of the long arm of chromosome 21 (21q22) (274–276).

CLONING THE IFN-α RECEPTOR GENE   Two groups have reported progress in cloning the receptor by selecting for murine cells transfected with human DNA that have enhanced sensitivity to human IFN-α and -β. In one case, enhanced protection by Hu-IFN-α against virus killing was used (277, 278), whereas in the other case IFN-induced enhancement of H2 antigens was used as the selection for sorting cells (279). Cells selected in this way have been shown to be responsive to other actions of Hu-IFN-α and to bind [^{125}I]Hu-IFN-α.

ANTIBODIES TO THE IFN-α/β RECEPTOR   Polyclonal antisera that inhibit the antiviral action of Hu-IFN-α have been raised in mice to mouse-human hybrid cells containing human chromosome 21 (268–270). Some antisera partially inhibit the binding of radiolabeled IFN-α to cells (271), and can be used to immunoprecipitate the covalent [^{125}I]Hu-IFN-α-receptor complex (241). Monoclonal antibodies with similar properties have also been described (280).

Two reports of monoclonal anti-idiotypic antibodies produced by immunization with anti-IFN-α antibodies have appeared. One antiserum inhibited the neutralization of IFN-α activity by anti-IFN-α antibodies, exhibited IFN-like activity on human WISH and bovine MDBK cells, and partially competed with [^{125}I]IFN-α for binding to MDBK cells (281). Unfortunately, this hybridoma was lost. The other anti-idiotypic antibody was characterized by its ability to block the activity of IFN-α and -β, by its binding to human cells, which was blocked by IFN-α, and by its ability to immunoprecipitate a protein of M_r about 95,000, similar in size to that expected for the IFN-α receptor (282).

METABOLISM OF CELL-BOUND IFN AND ITS RECEPTOR   When IFN-α is bound to cells at 37°C, the binding increases for about 1 hr, followed by a decrease in bound IFN over the next several hours. Following its binding to cells at 37°C, IFN-α is rapidly internalized via receptor-mediated endocytosis. This was demonstrated both biochemically (206, 212, 214–216, 223, 283, 284) and by following the fate of bound IFN conjugated to electron-dense markers (e.g. colloidal gold or antibody-linked ferritin) with electron microscopy (214, 218, 219, 285). For cells sensitive to the effects of IFN, degradation and secretion of degraded IFN (i.e. TCA-soluble radioactivity) follows. Results with various inhibitors of lysosomal metabolism (e.g. chloroquine, methylamine, monensin) are consistent with this model (212, 214, 216, 223, 283, 284). The reported exception, where internalization and degradation of Mu-IFN-α/β bound to L1210 cells was not apparent, may be due to experimental design or to a true difference in this system (203). A second exception involves Raji cells, which are insensitive to IFN-α, have high-affinity receptors, but do not seem to internalize the bound IFN-α. This may reflect a correlation between internalization and cellular activation by IFN (284). As discussed below, it has not yet been demonstrated that internalization is related to or required for the biological effects of IFN-α.

A second metabolic phenomenon noted for the IFN-α/β receptor, as with other polypeptide receptors, is "down-regulation," the decrease in cell-surface receptors caused by incubation with specific ligand (206, 212, 214–216, 283). This can be induced by low concentrations of either IFN-α or -β, but not by IFN-γ (206, 215). Indirect evidence suggests that the decrease in IFN-α receptors measured on T98G cells at high cell density (225) may result from down-regulation by a very low concentration of endogenous interferon (286). Although prolonged incubation of IFN-γ with T98G neuroblastoma cells led to a decrease in the binding of IFN-α, this effect has no obvious relation to ligand-mediated down-regulation (287). The recovery of normal numbers of cell-surface receptors following removal of IFN-α can be inhibited by cycloheximide, demonstrating a need for protein synthesis (203, 206, 215, 283). On MDBK cells (but not on Daudi cells) a lag of about 4 hr was noted before the number of cell-surface receptors increased following the removal of IFN-α (206, 283). Down-regulation of IFN-α receptors has also been demonstrated on peripheral blood cells from patients with chronic myelogenous leukemia undergoing treatment with Hu-IFN-α (288). Recovery of receptors was noted following removal of IFN.

A receptor half-life of 2–4 hr has been estimated from the loss of about 50% of binding on Daudi, MDBK, or murine L1210 cells during incubation with cycloheximide (203, 206, 283). Following removal of cycloheximide from MDBK cells, a lag of about 2 hr was noted before an increase in receptor number could be observed (283).

FUNCTIONAL ASPECTS OF THE INTERACTION OF IFN-α/β WITH ITS RECEPTOR   Binding of IFN to its receptor is necessary but not sufficient for cellular activation. Thus, cells with undetectable binding of IFN-α or -β (e.g. murine L1210R, human U20S, HT1080, MRC5, or HEC-1) are unresponsive to it (208, 209, 221). Experiments demonstrating that anti-IFN antibodies inhibit the development of an antiviral response in cells producing IFN (289), or that IFN-α or -β microinjected into cells fails to inhibit VSV replication (290, 291), suggest that entry through the receptor is required. No major contradictory experiments have been reported with IFN-α. However, it was found that mouse cells could be activated by Hu-IFN-γ that was delivered via liposomes (292) or by Hu-IFN-γ produced in mouse cells from a vector engineered to preclude secretion of IFN (293). In considering these apparently contradictory results, it is possible that specific internal compartmentation of IFN is crucial for IFN to act internally without interaction with a receptor. Alternatively, it is possible that introduction of IFN to cells by these mechanisms may provide enormous local concentrations that may overcome the species barrier.

Attempts to correlate directly the binding parameters with the degree of activation have proven difficult. Difficulties might be expected a priori from the observation that there is no obvious correlation for any particular IFN-α subtype between its activity in one assay (e.g. antiviral) and its activity in another assay (e.g. antiproliferative) (14, 162). Many biological activities occur at IFN concentrations that correspond to low percentage occupancy of receptors by ligand as measured by receptor binding at 4°C, where metabolic activity is not an issue (203, 207, 217). This has led to the notion that cells have a large excess of "spare receptors" over those necessary for biological response. On murine L1210 cells, however, a close correlation was noted between the concentration of IFN needed for antigrowth activity and the degree of receptor occupancy, although the concentrations needed for antiviral activity were several orders of magnitude lower (203). In a careful but complex study, it was noted that there is a better correlation between antigrowth activity of IFN-α on Daudi cells with an apparent affinity constant measured at 37°C, than with the affinity constant measured at 4°C (213).

Contradictory results exist concerning whether internalization is necessary for IFN action. The strongest evidence comes from experiments that demonstrated that IFN-α/β covalently linked to Sepharose can activate the antiviral response, suggesting that internalization is unnecessary for activity (295, 296). Contradictory, albeit indirect, evidence has been adduced from a correlation between internalization and activation (284). In addition, it was reported by indirect immunoferritin labeling that Mu-IFN-β binds to the nuclear membrane and the nucleoplasm following binding and internalization from the plasma membrane (219). This work is supported by direct binding studies where saturable, specific binding to isolated nuclei of mouse L929 cells was observed. These results suggest the possibility that internalization is followed by transport to the nucleus and might point to a direct role for IFN or an IFN-receptor complex in gene regulation. Confirmatory studies from other laboratories are needed.

A direct signal from the receptor has not yet been demonstrated. Work with mutants of the murine macrophage line J774.2, which are selected either for defects in their cAMP system or for their resistance to the antigrowth effects of IFN-α, has provided evidence that IFN-induced antigrowth activity, but not antiviral activity, may be mediated through a cAMP mechanism (236, 297). In considering other typical second message mechanisms, current evidence does not support a role for rapid alterations in phosphatidylinositide metabolism, cytoplasmic alkalinization, cytoplasmic free calcium concentration (298), or for changes in cGMP levels (299). However, a rapid but transient increase in diacylglycerol concentration following addition of IFN to cells has recently been correlated with both antiviral activity (299a) and inhibition of cell division (299b).

STRUCTURE-FUNCTION STUDIES OF IFN-α AND -β BINDING TO THE RECEPTOR   Features of IFN-α necessary for receptor recognition have not been well defined. Most fragments of IFN-α, whether produced by cyanogen bromide cleavage or by chemical synthesis, have lacked biological activity and binding ability (300–302), with the exception of a fragment of IFN-αA from residue 1 to 110 that retains low biological activity (92). Naturally occurring IFN-α species lacking 10 or 11 residues at the COOH-terminus are active (39), as are genetically engineered IFN-α analogues truncated to residue 151 (44, 301, 303, 304; K. Hotta, S. Pestka, unpublished data). Moreover, a monoclonal antibody that recognizes an epitope within the last 16 residues does not neutralize IFN activity and can bind to IFN-α when it is bound to the receptor (305). Biological activity is also retained when the amino terminus is modified either by extending it with part or all of the signal peptide (112, 306) or by deleting the first four amino acids (307). One of the two disulfides (42), Cys 1 to Cys 98 of IFN-αA, is not required for binding and activity (43). Although internal residues necessary for binding or activity have not been identified, it was shown that the substitution of alanine at conserved positions 30, 32, and 33 of IFN-α2 eliminates antiviral and antiproliferative activity on human cells and antiviral activity on MDBK cells. Moreover, the analogue acted as an antagonist and inhibited the binding of IFN-α2 to MDBK cells, but not to human cells (308, 309). This appears to be another instance where the requirements for high-affinity binding of Hu-IFN-α are less stringent on MDBK cells than on human cells (309a).

Although the IFN-α/β receptor binds the various species of IFN-α and IFN-β, there is functional discrimination whose basis is not known. It was first noted for natural purified IFN-α's (162, 294), and then for recombinant IFN-α's (14), that their relative activities in different assays do not correlate (164–171, 180, 310). That is, an IFN-α with high antiviral activity does not necessarily have high antigrowth and NK activities (165). An extreme example of this is seen when comparing Hu-IFN-αA and Hu-IFN-αJ. Whereas both have high activity in antiviral and antigrowth activity, IFN-αJ is almost devoid of NK activity in a 2-hr incubation (166), despite the fact that its affinity constant for binding to the IFN-α receptor on NK cells is no less than 1/10th that of IFN-αA (311). It is thus clear that receptor occupancy itself is not sufficient for activation.

IFN-γ *Receptor*

Although IFN-γ is generally less stable than IFN-α, Mu- and Hu-IFN-γ have now been radioiodinated and radiophosphorylated to high radiological specific activity for direct binding and receptor studies (227, 245, 247, 312–316). The direct binding of IFN-γ to a variety of human (227, 228, 243–247, 316–318, 318a) and murine cells (313–315, 319–322) has been measured. Most workers have reported a range of 1000 to 10,000 receptors per cell, with dissociation constants ranging from 1×10^{-9} to 1×10^{-11} M, though a larger number of receptors have been reported in a few instances (228, 243, 244, 317). Kinetic constants for binding to several cell lines have been reported (245, 247). Although IFN-γ is macrophage activating factor (MAF), macrophages or macrophage-like cell lines do not have distinctive parameters for binding IFN-γ (244, 245, 247, 316, 319, 322). Although most cells tested have IFN-γ receptors, it was recently shown that several lymphoid cells (from some patients with chronic lymphatic leukemia or hairy cell leukemia, or the T cell line CEM) have undetectable levels of IFN-γ receptors (317). Furthermore, cells isolated from natural sources generally have fewer receptors than established cell lines (317).

The receptors for IFN-γ and IFN-α are distinct based on competitive binding studies, the size of the covalent complex of the ligands and their receptors, segregation of sensitivity in cell mutants and somatic cell hybrids (see 266, 323), and other criteria. This is also consistent with the functional synergy (194, 238, 324) or antagonism (195, 196, 196a) of IFN-α/β and -γ. Most studies have found that IFN-β does not compete with IFN-γ for binding to cells (see above), but contradictory evidence has also been presented (231, 242).

The species specificity of IFN-γ resides in the interaction of IFN-γ with its receptor: Hu-IFN-γ does not bind specifically to mouse, hamster, or bovine cells (245, 247, 316; J. Langer, A. Rashidbaigi, unpublished data) and Mu-IFN-γ does not bind to human or hamster cells (315). Natural and recombinant IFN-γ compete for the same sites (247, 314, 321). Hu-IFN-γ with a tripeptide extension on the amino terminus (Cys-Tyr-Cys) has a slightly decreased affinity for the receptor (245). As with IFN-α, the COOH-terminus of Hu-IFN-γ is not critical for activity (75) or binding to the receptor (J. A. Langer, A. Rashidbaigi, S. Stefanos, T. Mariano, unpublished data). Acid-inactivated Hu-IFN-γ, which retains only about 5% antiviral activity, binds with almost unchanged characteristics to WISH cells compared to untreated IFN-γ, whereas it binds only weakly to peripheral blood monocytes (244). Additionally, acid-inactivated IFN-γ retained its ability to induce HLA-DR antigens on WISH cells, but lost this ability when tested on monocytes. It was therefore suggested that the IFN-γ receptor of WISH cells differs from that of monocytes, but more direct evidence is needed to establish this.

No correlation between binding of Hu-IFN-γ and (2'-5')-oligoadenylate induction was noted on lymphoblastoid and HeLa cells (246). As with IFN-α, however, the IFN-γ activation of cellular function can occur at concentrations at which receptor occupancy is low. For instance, on WEHI-3 cells, the induction of Ia or H2 antigens occurs maximally at 10–100 pM (2–20 units/ml), which corresponds to about 5% receptor occupancy (322).

CHARACTERIZATION OF THE IFN-γ RECEPTOR   Biochemical characterization of the Hu-IFN-γ receptor began with covalent crosslinking. The complex of the receptor and Hu-IFN-γ derivatized with a photoaffinity label migrates on SDS-PAGE at 230 kd (325). Several groups have obtained a 105–130 kd complex after reacting Hu-IFN-γ bound to various cell types with bifunctional reagents (228, 246, 318, 323). The complex was always broad and, in some cases, two bands were resolved (318). Small differences in apparent M_r were noted with different cell types, leading to the suggestion that the receptors on lymphoblastoid cells (which appear unresponsive to IFN-γ) may be slightly different from those on other cells (318). Consistent with the suggestion that the monocyte receptor is distinctive (see above), it was reported that a 160 kd covalent complex is formed with monocytes in contrast to the significantly smaller complexes formed with other cells (326). A third pattern of covalent complexes was seen with four principle bands at about 70, 92, 163, and 325 kd after binding and crosslinking Hu-IFN-γ to plasma membranes prepared from several cell types (317). A model for the receptor as a heterodimer with subunits of 53 kd and 75 kd was proposed. However, on at least several cell lines (including Colo205 and WiDR), four major bands were seen when Hu-IFN-γ was crosslinked to plasma membranes, whereas a single major complex of about 115 kd was formed on intact cells (J. A. Langer, A. Rashidbaigi, unpublished observations).

For various murine cell lines, the crosslinking of Mu-IFN-γ to its receptor results in several bands on SDS polyacrylamide gels, the major one estimated at 90–110 kd (314, 315).

Binding of IFN-γ to detergent-solubilized receptor measured by PEG precipitation has been described (327). Because of the low levels of the receptor and the high nonspecific binding of IFN-γ, this assay has thus far been useful only for partially purified receptor, not for the initial preparation following solubilization of membranes. It has been used to demonstrate a substantial degree of purification of the human IFN-γ receptor following solubilization with Triton X-100 and purification on an IFN-γ affinity column (327).

GENETICS   The chromosomal location of the gene for the human IFN-γ receptor (or its binding subunit) was determined by covalent crosslinking of

Hu-IFN-γ to a panel of human-rodent somatic cell hybrids differing in their complement of human chromosomes (323). The formation of the covalent 115 kd complex described above segregated with chromosome 6, and, more specifically, with the long arm of chromosome 6 (6q). Nevertheless, the presence of human chromosome 6 was not sufficient for protection of cells by Hu-IFN-γ from killing by VSV. Thus, an additional component is necessary for functional expression of the receptor. These results have been independently confirmed by P. Pfizenmaier and coworkers (P. Scheurich, personal communication). Fellous and collaborators identified chromosome 18 as conferring Hu-IFN-γ-dependent enhancement of H2 antigens to human-mouse somatic cell hybrids (328). They therefore proposed that the gene for the Hu-IFN-γ receptor is encoded by chromosome 18. With these same murine-human hybrids (kindly provided by M. Fellous), we have confirmed that human chromosome 6 confers binding of Hu-IFN-γ to these hybrid cells (A. Rashidbaigi, V. Jung, S. Pestka, unpublished data). Experiments are under way to resolve this apparent contradiction. Earlier evidence relating differential IFN-γ sensitivity of human aneuploid cells to the presence of chromosome 21 (reviewed in Ref. 266) does not necessarily imply that the IFN-γ receptor is encoded on that chromosome (329). Recent results demonstrate, however, that human chromosome 21 is required in addition to human chromosome 6 to enable Hu-IFN-γ to induce HLA antigens on murine-human and hamster-human somatic cell hybrids (330). These results demonstrate that two components, encoded by human chromosomes 6q and 21, are necessary and sufficient to generate a functional receptor for Hu-IFN-γ. Both these components are species specific (330). By examining the binding of ^{32}P-labeled Mu-IFN-γ to a panel of hamster-mouse somatic cell hybrids, it was demonstrated that the gene for the murine IFN-γ receptor is on chromosome 10 (330a).

METABOLISM OF IFN-γ AND ITS RECEPTOR   With mouse L1210 cells internalization of IFN-γ was noted within 5–10 min and reached a maximum by 30 min (314). Whereas no release of TCA-soluble ^{125}I was noted within 16 hr, the integrity of the internalized IFN was not analyzed. In these cells IFN-γ dose-dependent down-regulation of up to 90% of the receptors was noted. Recovery of receptors was inhibited by cycloheximide, and the receptor half-life was estimated as 8 hr, a comparatively long time. In human HeLa cells a decrease in binding sites was also apparent when cells were incubated with IFN-γ for 18 hr (246). The addition of cycloheximide to these cells caused a rapid decrease in IFN-γ binding, about 50% in 2 hr. Finally, the kinetics of dissociation of [^{125}I]Hu-IFN-γ seems to vary on different cell lines (316). It has been suggested that this may reflect different fates for IFN-γ and may parallel differences in sensitivity of various cells.

By immunoelectron microscopy and direct binding studies similar to those described above for Mu-IFN-β, MacDonald et al (331) have presented evidence that Mu-IFN-γ is rapidly internalized and translocated to the nucleus of murine cells and that there are specific receptors for Mu-IFN-γ on the nuclear membrane. This result, which raises many issues and possibilities regarding both the metabolism of IFN-γ and its mode of cellular activation, requires independent verification.

Prospects and Problems

In the next few years rapid progress should occur in the purification of IFN receptors, in the cloning of their genes, and in the production of antibodies, which, in turn, will permit studies on structure, function, and gene regulation. The question of structural heterogeneity at the receptor level will be accessible, as will the issue of the existence of high- and low-affinity receptors. Some information relevant to the signal transduction mechanism may be apparent following cloning and purification. Moreover, the generation of three-dimensional models of the IFNs from X-ray crystallographic studies, together with inspired choices for site-directed mutagenesis of the IFNs and their receptors, should contribute to an understanding of the stereochemistry of ligand activation and specificity.

A number of functional and mechanistic questions, similar to those posed for other receptors, need to be answered. These concern the signal transduction mechanisms of the receptor-ligand complexes; the mechanisms for differential activities of various IFN-α's and IFN-β; the biochemical pathways involved; the function of internalization and down-regulation of receptors; the molecular basis of synergy between IFN-α/β and IFN-γ; and the state of IFN receptors in disease (332, 333).

DOUBLE-STRANDED RNA-DEPENDENT ENZYMES AND THEIR POSSIBLE ROLES IN INTERFERON ACTION

Two IFN-induced, double-stranded RNA-dependent enzymes have been identified that may play important roles in the regulation of viral and cellular macromolecular synthesis and degradation (Figure 1). They are the (2'-5')-oligoadenylate [2,5-A_n] synthetase (334) and the P1/eIF-2α protein kinase (335–338). Both the IFN-induced 2,5-A_n synthetase and the IFN-induced P1/eIF-2α protein kinase must be activated following induction by IFN. Both synthetic and natural double-stranded RNAs (dsRNA) can fulfill the activation requirement in vitro (339–341). IFN treatment alone does not lead to significant increases in the intracellular level of 2,5-A_n or the extent of eIF-2α phosphorylation, presumably because neither the induced synthetase nor

ANTIVIRAL ACTIONS OF INTERFERON

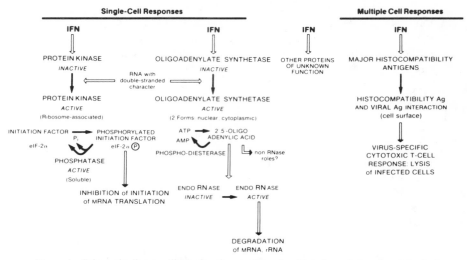

Figure 1 Schematic diagram illustrating the possible role of interferon-induced proteins in the antiviral actions of interferon.

kinase is activated. However, infection of IFN-treated cells in culture with either encephalomyocarditis (EMC) virus (342, 343), reovirus (344, 345), influenza virus (346), vaccinia virus (347), or simian virus 40 (348) causes a significant increase in the intracellular concentration of (2'-5')-oligoadenylates (342, 344, 346–348) and/or the extent of eIF-2α phosphorylation (343, 345, 349).

The (2'-5')-Oligoadenylate Synthetase-Nuclease System

A novel enzyme system that has been extensively characterized in IFN-treated cells is the (2'-5')-oligoadenylate synthetase-nuclease system, some aspects of which have been reviewed in detail (1, 350, 351). Three enzymes that play important roles in the $2,5$-A_n system have so far been identified: a synthetase that catalyzes the formation of oligonucleotides possessing $2',5'$-phosphodiester bonds (334); an endoribonuclease that is activated by certain $2,5$-A_n structures (352, 353); and a phosphodiesterase that catalyzes the hydrolysis of oligonucleotides possessing $2',5'$-phosphodiester bonds (354).

THE $2,5$-A_n SYNTHETASE   The (2'-5')-oligoadenylate synthetase, which is induced by IFN, catalyzes the synthesis of a family of oligonucleotides of the general structure $ppp(A2'p)_nA$ with $n \geq 2$, abbreviated $2,5$-A_n (1, 334). The magnitude of the induction of $2,5$-A_n synthetase by IFN depends on the type of IFN as well as the type and growth state of the cell (1, 350, 351). Induction levels vary from about 10-fold for human HeLa cells that have a high basal enzyme level (355) to about 10,000-fold for chick embryo cells that have a low basal enzyme level (356). IFN-α, -β, and -γ are able to induce 2,5-A_n synthetase expression in mouse and human cells (357–360). However, in addition to IFN treatment and virus infection, the level of 2,5-A_n synthetase has also been observed to be affected by a variety of other parameters, including hormone status (361), glucocorticoid treatment (362), differentiation status (363), and cell-cycle stage (364).

The IFN-induced 2,5-A_n synthetase has been extensively purified from mouse (365) and human (366, 367) cells. A large and a small form of the enzyme have been described (365–368). In Ehrlich ascites cells two mRNAs of 1.5 kb and 3.8 kb are induced by IFN which, when translated in an oocyte system, yield two different enzymically active forms of 2,5-A_n synthetase: one of about 20–30 kd and the other much larger, about 85–100 kd (365). Two 2,5-A_n synthetases of corresponding sizes are induced by IFN in Ehrlich cells, with the bulk of the larger enzyme in the cytoplasmic fraction and the bulk of the smaller enzyme in the nuclear fraction (365). Two similar molecular weight forms of 2,5-A_n synthetase have also been observed in human cells; the two human enzymes appear to have different activation requirements with regard to the dsRNA concentration and pH required for optimal activity (368).

A partial cDNA clone designated E1 of the 2,5-A_n synthetase mRNA was first obtained from human SV80 cells through its ability to select by hybridization an mRNA producing catalytically active enzyme upon translation in oocytes (369). Several laboratories subsequently succeeded in obtaining molecular clones of the human 2,5-A_n synthetase and the complete gene structure of the enzyme has been determined (360, 369–372). There appears to be only a single gene for the human 2,5-A_n synthetase (360, 370, 371) that has been mapped to human chromosome 11 (373). The 13.5 kb gene is comprised of at least eight exons and seven introns (370, 371). Multiple mRNAs encoding active 2,5-A_n synthetase, along with various other RNAs, are produced by cell-specific differential splicing of the primary transcript (360, 370, 371). Differential splicing at the 3' end of the primary transcript yields two mRNAs, a 1.6 kb mRNA coded by seven exons and a 1.8 kb mRNA coded by eight exons; the two mRNAs code for different forms of 2,5-A_n synthetase, a 41.5 kd protein from the 1.6 kb mRNA and a 46 kd protein from the 1.8 kb mRNA, which differ only by their COOH-terminal ends (360, 371). The identity of the mRNA encoding the 100 kd native form of the synthetase that has been described in human cells remains unresolved (366–368); conceivably the larger form of the synthetase is a dimer of the 41.5 kd and/or 46 kd polypeptide.

The amount of the mRNAs encoding 2,5-A_n synthetase activity increases in

IFN-treated cells with kinetics very similar to those observed for the increase in $2,5\text{-}A_n$ synthetase activity itself (369, 374). However, because of differential splicing, various sizes of $2,5\text{-}A_n$ synthetase RNA ranging from 1.5 kb to 3.6 kb are observed with the predominant RNA form dependent on the type of human cell examined (370, 371, 375).

THE $2,5\text{-}A_n$-DEPENDENT ENDORIBONUCLEASE   The only biochemical function of $2,5\text{-}A_n$ so far identified is the activation of an endoribonuclease that is present as a latent protein in both untreated and IFN-treated cells (352, 353, 376–381). This latent endoRNase has been referred to as RNase L (353) or RNase F (352). The activated endoRNase catalyzes the cleavage of both viral and cellular RNA on the 3' side of -UpXp- sequences (predominantly UA, UG, and UU) to yield products with -UpXp 3'-termini (382, 383). Many different types of single-stranded RNA are cleaved, including synthetic polyuridylic acid and various natural RNAs; however, ribosomal RNA is perhaps the best characterized type of RNA observed to be cleaved both in vitro and in vivo (384–386).

The $2,5\text{-}A_n$-dependent RNase has been partially purified from Ehrlich ascites cells (387) and rabbit reticulocytes (388). This RNase activity copurifies with a 75–85 kd protein that binds $2,5\text{-}A_n$; the protein has been identified in cytoplasmic extracts prepared from a variety of mammalian sources (387, 388).

The level of latent $2,5\text{-}A_n$-dependent RNase present in most types of cells in culture differs less than 2-fold between untreated and IFN-treated cells (350, 376, 377, 385). However, culture conditions have been described in which the functional activity of the latent RNase is regulated by parameters other than the simple availability of $2,5\text{-}A_n$. In JLS-V9R cells, IFN treatment causes a 10- to 20-fold increase in $2,5\text{-}A_n$-dependent RNase (389). In NIH 3T3 clone 1 cells, the $2,5\text{-}A_n$-dependent RNase can be independently induced by IFN or by growth arrest (390). The $2,5\text{-}A_n$-dependent RNase is also regulated during cell differentiation; the endoRNase expression is enhanced in differentiated murine embryonal carcinoma (EC) cells and is induced by IFN in differentiated but not undifferentiated EC cells (391). Virus infection in the absence of IFN treatment appears to functionally inactivate the $2,5\text{-}A_n$-dependent RNase, at least in EMC virus–infected HeLa cells (385).

THE (2',5') PHOSPHODIESTERASE   The $2,5\text{-}A_n$ oligonucleotides synthesized in IFN-treated systems are relatively unstable and undergo rapid degradation, which is accompanied by the reversion of the activated $2,5\text{-}A_n$-dependent RNase to an inactive state (392, 393). The enzyme that catalyzes the degradation of $2,5\text{-}A_n$ to yield AMP and ATP, a 2',5'-phosphodiesterase (354), is typically present at comparable levels in untreated cells as well as IFN-treated cells (354, 392–397). However, in some cell lines such as mouse L fibroblasts (354) and human Daudi cells (394), IFN treatment does increase the level of $2',5'$-phosphodiesterase activity about 2- to 4-fold. The $2',5'$-phosphodies-terase has been partially purified from mouse L cells and has an apparent molecular weight of about 40,000 (354). In addition to the hydrolysis of the $2',5'$-phosphodiester bonds of $2,5$-A_n, the phosphodiesterase also apparently catalyzes the degradation of the $3'$ C-C-A terminus of transfer RNA (354).

ANALOGUES OF $2,5$-A_n Because of the rapid degradation of naturally occur-ring $2,5$-A_n oligomers in animal cell systems, several investigators have attempted to prepare analogues with altered stability and biologic activity (398–411). These analogues contained modifications of the sugar moiety, the base, or the phosphodiester linkage. Studies with them revealed important insights into $2,5$-A_n structure-function relationships. A minimum of three adenosine residues are required for optimum binding to and activation of the $2,5$-A_n-dependent endoRNase. The state of the $5'$-terminus phosphorylation markedly influences binding to and activation of the endoRNase, with a minimum of a $5'$-diphosphate required for maximal activity. Modification of the $2'$-terminal residue also has important consequences regarding the stabil-ity and activity of $2,5$-A_n oligomers, as illustrated by the fact that $3'$-O methylation of only the $2'$-terminal adenosine residue generates an analogue more active than the parent $2,5$-A_n oligomer. Finally, the $2',5'$-phosphodiester linkage is crucial because the replacement of a single $2',5'$-bond with a $3',5'$-bond causes a substantial decrease in activity. The stereoconfiguration of the $2',5'$-phosphodiester bonds is also important, as oligomers with the S_p stereoconfiguration are more stable and active than the corresponding R_p isomers.

The P1/eIF-2α Protein Kinase-Phosphoprotein Phosphatase System

Protein phosphorylation-dephosphorylation is an important mechanism by which the functional activity of proteins can be controlled and, hence, biolog-ic processes including IFN action regulated (412, 413). The phosphorylation state of two proteins, P1 and eIF-2α, is acutely sensitive to IFN in many types of animal cells (335–338). Protein P1 is a ribosome-associated protein (337, 414); eIF-2α is the smallest subunit of protein synthesis initiation factor eIF-2 (415). Both protein P1 and eIF-2α are routinely phosphorylated to levels about 5- to 10-fold higher in IFN-treated as compared to untreated cells and derived cell-free extracts (413). Two enzymes, a protein kinase and a phos-phoprotein phosphatase, modulate the phosphorylation state of protein P1 and eIF-2α.

The increase in phosphorylation of P1 and eIF-2α observed in IFN-treated systems is dependent on either virus infection (343, 345, 349) or the addition of dsRNA (337, 416, 417). Within whole cells, about 5–10% of the α subunit of eIF-2 is phosphorylated in untreated cells, whereas 25–30% is phosphorylated within IFN-treated mouse cells upon infection with reovirus (345). The phosphorylation of eIF-2α is also elevated within IFN-treated human cells upon infection with encephalomyocarditis virus (343). The phosphorylation of eIF-2α depends in a similar manner on the multiplicity of infection when catalyzed by extracts prepared from infected cells (345) or, when catalyzed by extracts prepared from uninfected cells, on the concentration of exogenously added dsRNA (418). Several lines of evidence suggest that the dsRNA requirement is involved in the activation of an IFN-induced P1/eIF-2α protein kinase rather than the inhibition of a phosphoprotein phosphatase (419, 420). Activation of the dsRNA-dependent P1/eIF-2α protein kinase may occur in a localized manner (421) mediated by specific mRNA structures (415), thereby leading to a discriminatory inhibitory effect on the translation of selective mRNAs (415, 421).

THE P1/eIF-2α PROTEIN KINASE   Both natural and recombinant IFN-α, -β, and -γ induce the protein P1/eIF-2α protein kinase in murine cells (423, 424). However, in human epithelial-like cells only IFN-α or -β efficiently induce the kinase (425). Recombinant IFN-γ is a very poor inducer of the P1/α kinase in human epithelial-like cells (425); none of the IFNs detectably induce the P1/eIF-2α protein kinase in human fibroblasts (424–427). The synthesis in murine and human cells of an mRNA encoding a polypeptide of the same apparent size of P1 is induced by IFN (428), although it has not yet been directly established whether the induced mRNA actually encodes the kinase; no molecular cDNA clones of the kinase have yet been described. The activity of the protein P1 and eIF-2α kinase, like the induction of an antiviral state, is cAMP-independent as determined with genetic variants of cells deficient in the synthesis or action of cAMP (297, 430).

The IFN-induced dsRNA-dependent P1/eIF-2α protein kinase is ribosome associated (337, 413, 419). The kinase has been purified and characterized, both from IFN-treated and from untreated cells (419, 429, 431–434). Several lines of evidence support the conclusion that IFN treatment induces the synthesis of protein P1, which itself possesses the dsRNA-dependent protein P1/eIF-2α kinase activity (413). First, the dsRNA-dependent phosphorylation of protein P1 is greatly increased in IFN-treated as compared to untreated cells, and the increase in dsRNA-dependent protein kinase activity is blocked by actinomycin D (345, 417–419). Furthermore, the phosphorylation of murine protein P1 in the presence of S10 extract prepared from IFN-treated human cells is observed when the combination of human and mouse extracts includes S10 prepared from IFN-treated mouse cells, but is not observed when the combination includes S10 from untreated mouse cells (419). Second, purification studies have failed to separate the dsRNA-dependent P1/eIF-2α protein kinase activity from protein P1 (418, 419, 432–434). Third, the yield of P1/eIF-2α protein kinase activity obtained following purification from cells treated with IFN is about 5–10 times greater than the yield from an equivalent number of untreated cells (419). Fourth, the amount of P1 is increased by IFN treatment of human cells as measured by Western immunoblot analysis (435, 436). Fifth, the apparent molecular weight of 62 kd observed for native dsRNA-dependent protein kinase by sedimentation analysis (419) is comparable to the molecular weight of 67 kd observed for denatured phosphorylated protein P1 by SDS polyacrylamide gel electrophoresis (414). Sixth, protein P1 contains a nucleotide-binding site as established by the direct photoaffinity labeling of P1 by ATP. The photoaffinity labeling of protein P1 is dsRNA dependent (437).

The purified dsRNA-dependent protein kinase is highly selective for the α subunit of protein synthesis initiation factor eIF-2 and endogenous protein P1 (419, 429). O-Phosphoserine is the major phosphoester linkage both in protein P1 and in eIF-2α (415, 418, 438). Kinase activity is dependent on Mg^{2+}, and the K_m for ATP is 5×10^{-6} M (419). Protein synthesis factors other than the α subunit of eIF-2 (EF-1, EF-2, eIF-3, eIF-4A, eIF-4B, and eIF-5) are not substrates (415, 419). Likewise, calf thymus and *Drosophila* H1 histones are not detectable substrates of the purified kinase, and calf and rabbit thymus histones H2A, H2B, H3, and H4, casein, phosvitin, and tubulin are very poor substrates (419, 429).

The purified P1/eIF-2α protein kinase remains dependent on dsRNA for activation (419). Two forms of P1 have been identified that differ from each other by their phosphorylation patterns; a major phosphopeptide designated X_{ds} is phosphorylated in the presence of dsRNA but not in the absence of dsRNA (418, 438). Adenovirus VAI RNA prevents the activation of the P1/eIF-2α protein kinase by dsRNA (439–441). Although adenovirus VAI RNA possesses significant secondary structure, neither it nor tRNA can substitute for dsRNA as an activator of the kinase (439). The P1/eIF-2α kinase is also inhibited in extracts prepared from vaccinia-infected cells (442–444). A specific kinase inhibitory factor produced by vaccinia has been described that appears to interact with the dsRNA and prevent activation of the P1/eIF-2α protein kinase (422).

A constitutive protein kinase produced by cells from different animal species has been described that catalyzes the phosphorylation of IFN-γ (245, 445, 446). The biologic activity of IFN-γ is not detectably altered by the phosphorylation (245, 445), which occurs on serine residues (315, 446). This kinase is not known to be related to the dsRNA-dependent P1/eIF-2α protein kinase (413).

THE Pl/eIF-2α PHOSPHOPROTEIN PHOSPHATASE   The phosphoprotein phosphatase that catalyzes the dephosphorylation of eIF-2αP is soluble, is dsRNA-independent, and is present at comparable levels in extracts from untreated and IFN-treated cells (420). The phosphatase from IFN-treated cells has not yet been purified; however, a protein phosphatase specific for phosphorylated eIF-2α has been purified and characterized from rabbit reticulocytes (447).

Possible Roles of the dsRNA-Dependent Enzymes in Interferon Action

The extent to which the 2,5-A$_n$ synthetase-endoRNase-phosphodiesterase pathway and the Pl/eIF-2α protein kinase/phosphoprotein phosphatase pathway may be generally involved in the antiviral actions of IFN is yet to be resolved (351, 448). It is clear from studies with cell-free protein synthesizing systems that either 2,5-A$_n$ or the Pl/eIF-2α protein kinase can efficiently inhibit the synthesis of proteins in vitro (1, 350, 413, 449, 450). However, the importance of the 2,5-A$_n$ system or the Pl/eIF-2α protein kinase/phosphatase system in the antiviral action of IFN within intact cells in culture is less clear. Most of the available evidence suggests that a functional 2,5-A$_n$ system is neither sufficient nor required for the antiviral action of IFN against a variety of different viruses that are sensitive to IFN (348, 451–455). By contrast, a functional Pl/eIF-2α protein kinase does appear to correlate with the antiviral action of IFN in at least some virus-cell systems (439, 442, 454). However, it is becoming increasingly clear that the antiviral mechanisms of IFN action are dependent on not only the host cell and challenge virus, but also on the type of IFN (456, 457). Thus, it is possible that the 2,5-A$_n$ system plays an important role in the antiviral action of IFN with certain viruses such as EMC virus in murine cells (407), even though other virus-cell combinations appear more dependent on a functional Pl/eIF-2α protein kinase system for inhibition of virus replication in IFN-treated cells (439, 442, 454). Conceivably a fundamental role of the 2,5-A$_n$ pathway is more closely related to the events associated with cell-cycle progression (364) and cell differentiation (391) rather than antiviral responses.

OTHER MESSENGER RNAs AND PROTEINS INDUCED BY INTERFERON WHOSE FUNCTION IS NOT KNOWN TO DEPEND ON DOUBLE-STRANDED RNA

Overview

IFN treatment of mouse and human cells results in the induction of several proteins in addition to those known to be associated with either the dsRNA-dependent Pl/eIF-2α protein kinase system or the dsRNA-dependent 2,5-A$_n$ synthetase system. The accumulation of about a dozen different polypeptides, ranging in apparent molecular weight from about 15,000 to about 120,000, has been reported to increase in a variety of different types of cells in response to IFN treatment (235, 458–460). These polypeptides, resolved by various gel electrophoresis systems, typically consist of one peptide of about 90–120 kd; several in the range of 65 kd to 80 kd; two in the range of 55 kd to 60 kd; one of about 50 kd; one of about 40 kd; and one of about 15 kd to 20 kd. The functional identity of most of these polypeptides detected by gel electrophoresis techniques, as well as their role in the IFN response, so far is unknown.

A number of cDNAs corresponding to IFN-inducible mRNAs have been isolated (369–371, 461–471). Many of the cDNAs correspond to IFN-inducible mRNAs of unknown identity. However, among the IFN-inducible mRNAs for which molecular clones have been obtained and identified are those that encode the following proteins: $2,5\text{-}A_n$ synthetase (369–371); metallothionein-II (466); thymosin B4 (466); major histocompatibility complex antigens (see below); protein Mx (472); and two proteins of unknown identity, one of 56 kd (461–465) and the other of 15 kd (467). Increased levels of IFN-inducible mRNAs can be detected within 5 min to 2 hr after IFN treatment (463, 466, 473) and, at least in some cases, are due to increased rates of transcription of genes that normally are not transcribed or are transcribed at very low rates in the absence of IFN (462, 463, 471, 474).

Major Histocompatibility Complex Antigens

Among the best characterized of the IFN-induced mRNAs and proteins whose functions are not known to be dependent on dsRNA are the class I, class II, and class III major histocompatibility complex (MHC) antigens and $\beta 2$-microglobulin. IFN-α, -β, and -γ induce the expression of the class I murine H-2 and human HLA-A,B,C major histocompatibility antigens on a variety of different cell types (190, 474–481). In addition, the expression of cell-surface $\beta 2$-microglobulin, a low-molecular-weight subunit associated with the MHC antigens, is also induced by IFN and correlates with the increased synthesis of class I antigens in IFN-treated systems (190, 475, 476, 481).

Both the mRNA for endogenous class I antigens as well as the mRNA transcribed from molecularly cloned heterologous class I genes transfected into cells are increased by IFN treatment (474, 476, 478, 479, 482). In the case of HLA-A mRNA, the increased levels observed in IFN-treated cells is due at least in part to a stimulation in transcription rate (474).

The class II major histocompatibility antigens, the murine I-region–associated Ia antigens, and the human HLA-DR, -DP, and -DQ antigens, are also induced by natural and recombinant IFN-γ (481, 483–488). Induction of class II HLA-D mRNAs and the encoded antigens by IFN-γ occurs both on nonimmune cells (481, 485, 486, 488) as well as on cells of the immune system (483, 484, 487, 488). IFN-γ increases the expression of MHC class II mRNA and gene products, including the second component of complement, C2, and factor B (489). IFN-α, in contrast to IFN-γ, does not significantly induce the expression of either the class II HLA-D antigens (485, 488) or the class III C2 gene product (489). The IFN-induced expression of the MHC antigens may contribute to the antiviral (490) and antiproliferative (491) actions of IFN at the "cell-cell" level, perhaps by enhancement of the antigen-specific lytic effect of cytotoxic T lymphocytes (Figure 1). The cytotoxic T cell response is virus specific and histocompatibility antigen restricted (492, 493). It is assumed that the MHC antigens function in the presentation of foreign antigens, including viral or tumor antigens to T cells. If the basal level of the cell-surface histocompatibility antigens is below the threshold concentration normally required for efficient T cell responses, then the IFN-induced increase in levels of class I and class II antigens could represent an important component of the host defense response to virus infections as well as to tumorigenic cells. It should be noted that increase in tumor-associated antigens (168–171) may also contribute to the antitumor action of IFN.

Protein Mx

Genetic analysis has established that in the mouse the antiviral state against influenza virus induced by IFN is controlled by a host gene designated Mx (457, 494). The gene Mx originates from the inbred mouse strain A2G, which has an inborn resistance to infection by influenza virus (495). An mRNA is induced by IFN in mouse cells bearing the Mx gene, which encodes a polypeptide of about 75 kd (496, 497). The IFN-induced protein Mx, which has been purified to homogeneity, accumulates in the nucleus of mouse cells treated with IFN-α or -β but not IFN-γ (498–500). Antibodies prepared against mouse protein Mx recognize an IFN-induced protein in human cells of about 80 kd, which is induced by IFN-α but not by IFN-γ (501). Unlike the nuclear mouse protein, the human Mx-like protein is localized in the cytoplasm (501).

The mouse Mx gene is located on chromosome 16 (502). Southern analyses indicate that the Mx$^-$ alleles derive from their Mx$^+$ counterpart by deletions. IFN-treated Mx$^+$ mouse cells contain a 3.5 kb Mx mRNA, while untreated Mx$^-$ cells show only traces of a shorter Mx mRNA (472). Mx$^-$ cells transformed with Mx cDNA constitutively express the protein Mx and are resistant to influenza virus (472).

Other Proteins

Among the proteins whose steady-state concentration increases in IFN-treated cells are two polypeptides with guanylate-binding activity of M_r 67,000 and 56,000 that are observed both in IFN-treated human and murine cells (450,

503). The inducibility of the 56 kd guanylate-binding protein (GBP) is inherited as a single autosomal gene (504) located on mouse chromosome 3 (505). The 67 kd GBP is found in the cytoplasm of IFN-treated cells (506); however, the role of GBP in cellular responses to IFN is unknown. The 56 kd human and mouse proteins for which several independent molecular cDNA clones have been obtained (461–465) also appear to be cytoplasmic (465), and in the mouse maps to chromosome 1 (469); however, it has yet to be demonstrated that this protein is equivalent to the 56 kd GBP.

In addition to the enzymes of the P1/eIF-2α protein kinase system and the 2,5-A$_n$ synthetase system, two enzymic activities involved in catabolic reactions also have been reported to be elevated in cells treated with IFN. Xanthine oxidase, which catalyzes the oxidation of the purines hypoxanthine and xanthine as well as a variety of aldehydes, is significantly increased in different organs from mice treated with IFN-α/β or IFN inducers (507–509). Indoleamine 2,3-dioxygenase, which catalyzes the oxidative cleavage of the indole ring that is a constituent of several important regulatory molecules, is increased in mouse lung slices treated with high doses of NDV-induced mouse L cell IFN (510) and in human fibroblasts treated with low doses of recombinant human IFN-γ (511). IFN-γ also induces the increased expression of receptors for tumor necrosis factor (TNF), a protein that mediates cytolytic effects in cells (512, 513).

Different Regulatory Mechanisms for the Expression of IFN-α/β and IFN-γ-Inducible Proteins

Among the initial indications from biologic studies that IFN-α/β and IFN-γ may regulate the expression of different genes were the observations that the activities of IFN-α or -β and IFN-γ were synergistic, both in antiviral assays and in antiproliferative assays (194, 238, 425, 473). Subsequent antiviral studies revealed that the kinetics of development as well as the specific molecular mechanism of the antiviral action of IFN-α against VSV or Sindbis virus in human cells was distinct from that of IFN-γ in the same virus-cell system (452, 453, 513a). Furthermore, the efficiency of induction of specific polypeptides can differ greatly between IFN-α/β and IFN-γ. In certain human cells, for example, polypeptides such as the class I MHC antigens (190, 475–477) and the 2,5-A$_n$ synthetase (359, 360) are induced by all the IFNs; by contrast, the class II MHC antigens (485, 488), the class III MHC C2 product (489), and the 67 kd GBP (506) are efficiently induced by IFN-γ but not by IFN-α/β, whereas the Mx-like protein (501) and the P1/eIF-2α protein kinase (425) are efficiently induced by IFN-α/β but not by IFN-γ.

The extent and length of induction for IFN-inducible mRNAs appear to be regulated differently by IFN-α and IFN-γ (464, 468, 514, 515). Both transcriptional and translational events appear to be important in determining the level of induced mRNA (464, 468, 514). However, very little is known concerning the molecular mechanisms by which IFN initially activates transcription. The consensus IFN response sequence TTCN(G/C)NACCTCNGCAGTTTCTC(C/T)TCT-CT has been derived from homologous sequences identified 5' to the putative transcriptional start sites of IFN-α-inducible genes such as HLA and metallothionein (470). In a transient expression assay, the consensus IFN response sequence is necessary for the induction of the H-2K^b promoter by recombinant IFN-α, -β, and -γ, although the response sequence is only active when associated with the functional enhancer sequence (470a).

For IFN-α there must be either more than one pathway for activation of gene expression from a single IFN-α receptor or functionally distinct IFN-α receptors. The above HLA-metallothionein IFN response consensus sequence has not been found in all IFN-α-inducible genes (371, 465), and some mRNAs inducible by IFN-α in wild-type Daudi cells remain inducible in IFN-resistant Daudi cells whereas others do not (516). By fusion of the 5'-flanking region of a mouse MHC class I gene to the bacterial chloramphenicol acetyltransferase (CAT) gene, it was found that the presence of the interferon consensus sequence is required and sufficient for the MHC promoter to enhance transcription in response to interferons (517). In addition, fusion of the 5' portion of the human IFN-α/β-inducible *ISG-54K* gene to the adenovirus *E1b* gene lacking its own promoter converted the latter into a gene controllable by exogenous IFN-α or -β (517a).

MESSENGER RNAs AND PROTEINS SUPPRESSED BY INTERFERON

Interferon treatment of cells in culture can lead to decreased expression of genes. For example, treatment of Daudi cells with IFN-α (474, 516–518) or IFN-β (474, 519, 520) causes a significant decrease in the level of *c-myc* mRNA. IFN treatment appears not to affect the *c-myc* transcription rate, but rather reduces the half-life of *c-myc* mRNA (474, 520). IFN treatment has also been shown to reduce the amount of *Ha-ras* mRNA and protein in mouse RS485 cells (521) and in human RT4 cells (522), and *c-fos* mRNA in Balb/c 3T3 cells (523).

Some IFNs may act in an autocrine fashion to control cell growth and differentiation (364, 524, 525), and as part of this mechanism may mediate a suppression in the expression of cellular genes. For example, in Balb/c 3T3 cells, platelet-derived growth factor (PDGF) stimulates the rapid expression of a competence gene family including *c-myc* and *c-fos* and the slower expression of IFN-β (526). The IFN-β may function as a feedback inhibitor of growth, as the expression of genes regulated by PDGF, including *c-myc*, *c-fos*, ornithine decarboxylase, and β-actin, are inhibited by IFN treatment

(523). The secretion of a 63 kd polypeptide from Balb/c 3T3 cells is also inhibited by IFN-β (490).

IFN-γ may also suppress gene expression. The synthesis of collagens by human dermal and synovial fibroblasts (527, 528) and human articular and costal chondrocytes (485) is inhibited by IFN-γ. The IFN-γ-mediated inhibition of types I and II collagens is associated with an inhibition in the levels of types I and II procollagen mRNAs (485). Human IFN-α also suppresses collagen synthesis and procollagen mRNA levels to a similar extent as IFN-γ (485).

REGULATION OF INTERFERON EXPRESSION

Virus infection of human or mouse cells induces IFN-α and IFN-β; poly(I)·poly(C) on the other hand induces essentially only IFN-β (104). In uninduced cells, human IFN-β mRNA is not detectable, but appears rapidly after induction by poly(I)·poly(C) (105, 106, 529, 530). Similarly, human interferon genes introduced into cultured cells can be regulated by virus infection or by poly(I)·poly(C) according to the genes, the vectors, and the cell types (142, 143, 145, 531–536). After analysis of IFN and IFN mRNA expression in cells containing deletion mutants of cloned Hu-IFN genes introduced into mouse cells, several distinct DNA sequences were found to be required for their regulation (537–542). In the case of the Hu-IFN-β gene, the -77 to -19 region with respect to the mRNA cap site is required for constitutive and induced expression (537–539); removal of the -210 to -107 region results in an increase of constitutive expressions whereas induced expression is unaffected (537). An enhancer element inducible by poly(I)·poly(C) is located between -77 and -37 from the mRNA cap site and can function upstream or downstream of the IFN-β gene at distances up to one kilobase (539, 540). Sequences related to this element are present in multiple copies in the 5'-flanking regions of both IFN-α and IFN-β genes. Analogous experiments with Hu-IFN-α genes indicated that a region -117 to -74 with respect to the mRNA cap site is necessary for induction by viruses (541, 542).

Further exploration of induction of the human IFN-β gene demonstrated that its induction by poly(I)·poly(C) requires a *trans*-acting factor (543). This factor is inducible by IFN, a phenomenon that explains the priming of IFN production by prior exposure of cells to interferon. Footprinting experiments demonstrate that factors that bind to regions -167 to -94 and -68 to -38 with respect to the mRNA cap site prior to induction with poly(I)·poly(C) are repressor molecules that are released upon induction (544, 545). After induction a transcription factor binds to the region located between -77 and -64. These results indicate that the IFN-β gene is controlled by a negative regulatory mechanism.

CLINICAL STUDIES

In this past year an important landmark was achieved for the therapeutic use of interferon: Hu-IFN-αA and the closely related Hu-IFN-α2 were approved by the Food and Drug Administration in the United States for the treatment of hairy cell leukemia (546). This was a culmination of many years of effort that first began with the treatment of patients with viral diseases and tumors with natural interferons that were relatively impure; then subsequently with purified interferons primarily produced through genetic engineering (7, 15, 547–549). A number of malignancies appear to respond to treatment with either IFN-α, -β, or -γ (547–553). Clinical trials are under way to evaluate the efficacy of interferon in the treatment of cancers, viral diseases, and other disorders (547–553). It is clear that interferons have a place in the therapeutic arsenal. Nevertheless, because animal models have limited use in evaluating agents with substantial species specificity, clinical development will depend on careful trials and imaginative implementation with these potent agents (see 168–171, 186–188, 547–553). The current general availability of virtually unlimited quantities of these agents will catalyze new clinical advancements relatively rapidly. To these ends, it is of utmost importance that basic and preclinical research continue to generate new foundations and insights upon which clinical programs can flourish.

ACKNOWLEDGMENTS

The authors thank Margaret Icangelo for her thoughtful and dedicated assistance in compiling the text from the products of four different word processors; and Graham Cleaves for his efforts in transformation of four incompatible word processing systems into a coherent whole that could be integrated and edited. The work from the laboratory of C. E. S. was supported in part by research grants from the National Institutes of Health and from the American Cancer Society.

Literature Cited

1. Lengyel, P. 1982. *Ann. Rev. Biochem.* 51:251–82
2. Pestka, S., Baron, S. 1981. *Methods Enzymol.* 78:3–14
3. Pestka, S. 1986. *Methods Enzymol.* 119:14–23
4. Duc-Goiran, P., Lebon, P., Chany, C. 1986. *Methods Enzymol.* 119:541–51
5. Wilkinson, M. F., Morris, A. G. 1986. *Methods Enzymol.* 119:96–102
6. Weissenbach, J., Chernajovsky, Y., Zeevi, M., Shulman, L., Soreq, H., et al. 1980. *Proc. Natl. Acad. Sci. USA* 77:7152–56
7. Pestka, S., ed. 1981. *Methods Enzymol.*, Interferons, Part A, Vol. 79. New York: Academic. 677 pp.
8. Secher, D., Burke, D. C. 1980. *Nature* 285:446–50
9. Staehelin, T., Durrer, B., Schmidt, J., Takacs, B., Stocker, J., et al. 1981. *Proc. Natl. Acad. Sci. USA* 78:1848–52
10. Secher, D. S. 1981. *Nature* 290:501–3
11. Staehelin, T., Stahli, C., Hobbs, D. S., Pestka, S. 1981. *Methods Enzymol.* 79:589–95
12. Kelder, B., Rashidbaigi, A., Pestka, S. 1986. *Methods Enzymol.* 119:582–87

13. Pestka, S., Kelder, B., Langer, J. A., Staehelin, T. 1983. *Arch. Biochem. Biophys.* 224:111–16
14. Rehberg, E., Kelder, B., Hoal, E. G., Pestka, S. 1982. *J. Biol. Chem.* 257:11497–502
15. Pestka, S., ed. 1986. *Methods Enzymol.*, Interferons, Part C., Vol. 119. New York: Academic. 845 pp.
16. Torrence, P. F., De Clercq, E. 1977. *Pharmacol. Ther. A* 2:1–88
17. Isaacs, A., Lindenmann, J. 1957. *Proc. R. Soc. London Ser. B* 147:258–67
18. Pestka, S., Tarnowski, S. J. 1985. *Pharmacol. Ther.* 29:299–319
19. Rubinstein, M., Rubinstein, S., Familletti, P. C., Miller, R. S., Waldman, A. A., et al. 1979. *Proc. Natl. Acad. Sci. USA* 76:640–44
20. Rubinstein, M., Levy, W. P., Moschera, J. A., Lai, C.-Y., Hershberg, R. D., et al. 1981. *Arch. Biochem. Biophys.* 210:307–18
21. Langer, J. A., Pestka, S. 1985. *Pharmacol. Ther.* 27:371–401
22. Nagata, S., Mantei, N., Weissmann, C. 1980. *Nature* 287:401–8
23. Goeddel, D. V., Leung, D. W., Dull, T. J., Gross, M., Lawn, R. M., et al. 1981. *Nature* 290:20–26
24. Rubinstein, M., Rubinstein, S., Familletti, P. C., Gross, M. S., Miller, R. S., et al. 1978. *Science* 202:1289–90
25. Zoon, K. C., Smith, M. E., Bridgen, P. J., Zur Nedden, D., Anfinsen, C. B. 1979. *Proc. Natl. Acad. Sci. USA* 76:5601–5
26. Zoon, K. C., Miller, D., Zur Nedden, D., Hunkapiller, M. W. 1982. *J. Interferon Res.* 2:253–60
27. Berg, K., Heron, I. 1980. *Scand. J. Immunol.* 11:489–502
28. Kauppinen, H., Hirvonen, S., Cantell, K. 1986. *Methods Enzymol.* 119:27–35
29. Horowitz, B. 1986. *Methods Enzymol.* 119:39–47
30. Allen, G., Fantes, K. H. 1980. *Nature* 287:408–11
31. Zoon, K. C., Hu, R-Q., Zur Nedden, D., Nguyen, N. Y. 1985. In *The Biology of the Interferon System, 1984,* ed. H. Kirchner, H. Schellekens, pp. 61–67. Amsterdam: Elsevier Science
32. Zoon, K. C., Hu, R.-Q., Zur Nedden, D., Gerrard, T. L., Enterline, J. C., et al. 1986. See Ref. 448, pp. 55–58
33. Allen, G., Fantes, K. H., Burke, D. C., Morser, J. 1982. *J. Gen. Virol.* 63:207–12
34. Pestka, S. 1986. *Methods Enzymol.* 119:3–14
35. Yonehara, S., Yanase, Y., Sano, T., Imai, M., Nakasawa, S., et al. 1981. *J. Biol. Chem.* 256:3770–75
36. Hobbs, D. S., Pestka, S. 1982. *J. Biol. Chem.* 257:4071–76
37. Zoon, K. C., Smith, M. E., Bridgen, P. J., Anfinsen, C. B., Hunkapiller, M. W., et al. 1980. *Science* 207:527–28
38. Bridgen, P. J., Anfinsen, C. B., Corley, L., Bose, S., Zoon, K. C., et al. 1977. *J. Biol. Chem.* 252:6585–87
39. Levy, W. P., Rubinstein, M., Shively, J., Del Valle, U., Lai, C.-Y., et al. 1981. *Proc. Natl. Acad. Sci. USA* 78:6186–90
40. Levy, W. P., Shively, J., Rubinstein, M., Del Valle, U., Pestka, S. 1980. *Proc. Natl. Acad. Sci. USA* 77:5102–4
41. Zoon, K. C., Wetzel, R. 1984. *Handb. Exp. Pharmacol.* 71:79–100
42. Wetzel, R. 1981. *Nature* 289:606–7
43. Morehead, H., Johnston, P. D., Wetzel, R. 1984. *Biochemistry* 23:2500–7
44. DeChiara, T. M., Erlitz, F., Tarnowski, S. J. 1986. *Methods Enzymol.* 119:403–15
45. Labdon, J. E., Gibson, K. D., Sun, S., Pestka, S. 1984. *Arch. Biochem. Biophys.* 232:422–26
46. Chadha, K. C. 1985. In *The Interferon System,* ed. F. Dianzani, G. B. Rossi, pp. 35–41. New York: Raven
47. Iwakura, Y., Yonehara, S., Kawade, Y. 1978. *J. Biol. Chem.* 253:5074–79
48. Knight, E. Jr. 1977. *J. Biol. Chem.* 250:4139–44
49. DeMaeyer-Guignard, J., Tovey, M. G., Gresser, I., DeMaeyer, E. 1978. *Nature* 271:622–25
50. Kawakita, M., Cabrer, B., Taira, H., Rebello, M., Slattery, E., et al. 1978. *J. Biol. Chem.* 253:598–602
51. Cabrer, B., Taira, H., Broeze, R. J., Kempe, T. D., Williams, K., et al. 1979. *J. Biol. Chem.* 254:3681–84
52. Shaw, G. D., Boll, W., Taira, H., Mantei, N., Lengyel, P., et al. 1983. *Nucleic Acids Res.* 11:555–73
53. Daugherty, B. L., Martin-Zanca, D., Kelder, B., Collier, K., Seamans, T. C., Pestka, S. 1984. *J. Interferon Res.* 4:635–43
54. Taira, H., Broeze, R. J., Jayaram, B. M., Lengyel, P., Hunkapiller, M. W., et al. 1980. *Science* 207:528–30
55. Lemson, P. J., Vonk, W. P., Van der Korput, J. A. G. M., Trapman, J. 1984. *J. Gen. Virol.* 65:1365–72
56. Knight, E. Jr. 1976. *Proc. Natl. Acad. Sci. USA* 73:520–23
57. Berthold, W., Tan, C., Tan, Y. H. 1978. *J. Biol. Chem.* 253:5206–12

58. Tan, Y. H., Barakat, F., Berthold, W., Smith-Johannsen, H., Tan, C. 1979. *J. Biol. Chem.* 254:8067–73

59. Stein, S., Kenny, C., Friesen, H.-J., Shively, J., Del Valle, U., Pestka, S. 1980. *Proc. Natl. Acad. Sci. USA* 77:5716–19

60. Knight, E. Jr., Fahey, D. 1981. *J. Biol. Chem.* 256:3609–11

61. Kenny, C., Moschera, J. A., Stein, S. 1981. *Methods Enzymol.* 78:435–47

62. Friesen, H.-J., Stein, S., Evinger, M., Familletti, P. C., Moschera, J., et al. 1981. *Arch. Biochem. Biophys.* 206:432–50

63. Okamura, H., Berthold, W., Hood, L., Hunkapiller, M. W., Inoue, M., et al. 1980. *Biochemistry* 19:3831–35

64. Knight, E. Jr., Hunkapiller, M. W., Korant, B. D., Hardy, R. W. F., Hood, L. E. 1980. *Science* 207:525–26

65. Taniguchi, T., Sakai, M., Fujii-Kuriyama, Y., Muramatsu, M., Kobayashi, S., et al. 1979. *Proc. Jpn. Acad. B* 55:464–69

66. Knight, E. Jr., Fahey, D. 1982. *J. Interferon Res.* 2:421–29

67. Pestka, S., Kelder, B., Familletti, P. C., Moschera, J. A., Crowl, R., et al. 1983. *J. Biol. Chem.* 258:9706–9

68. Revel, M., Ruggieri, R., Zilberstein, A. 1986. See Ref. 448, pp. 207–16

69. Haegeman, G., Content, J., Volckaert, G., Derynck, R., Tavernier, J., Fiers, W. 1986. *Eur. J. Biochem.* 159:625–32

70. Vonk, W. P., Trapman, J. 1983. *J. Interferon Res.* 3:169–75

71. Higashi, Y., Sokawa, Y., Watanabe, Y., Kawade, Y., Ohno, S., et al. 1983. *J. Biol. Chem.* 258:9522–29

72. Kaplan, P., Abreu, S. L. 1985. *J. Interferon Res.* 5:415–22

73. Yip, Y. K., Barrowclough, B. S., Urban, C., Vilcek, J. 1982. *Proc. Natl. Acad. Sci. USA* 79:1820–24

74. Braude, I. A. 1984. *Biochemistry* 23:5603–9

75. Rinderknecht, E., O'Connor, B. H., Rodriguez, H. 1984. *J. Biol. Chem.* 259:6790–97

76. Friedlander, J., Fischer, D. G., Rubinstein, M. 1984. *Anal. Biochem.* 137:115–19

77. Braude, I. A. 1986. *Methods Enzymol.* 119:193–99

78. Braude, I. A., Mehlman, T. 1984. In *Interferon: Research, Clinical Application, and Regulatory Consideration,* ed. K. C. Zoon, P. Noguchi, T.-Y. Liu, pp. 17–25. New York: Elsevier Science

79. Orchansky, P., Fischer, D. G., Novick, D., Rubinstein, M. 1985. In *The 1985 TNO-ISIR Meet. Interferon System, Clearwater Beach, FL.,* 13–18 Oct., 1985, p. 23. Abstr.

80. Falcoff, R. 1972. *J. Gen. Virol.* 16:251–53

81. Yip, Y. K., Pang, R. H. L., Urban, C., Vilcek, J. 1981. *Proc. Natl. Acad. Sci. USA* 78:1601–5

82. Staehelin, T., Hobbs, D. S., Kung, H.-F., Lai, C.-Y., Pestka, S. 1981. *J. Biol. Chem.* 256:9750–54

83. Nagabhushan, T. L., Surprenant, H., Le, H. V., Kosecki, R., Levine, A., et al. 1984. See Ref. 78, pp. 79–88

84. Bodo, G., Fogy, I. 1985. See Ref. 31, pp. 23–27

85. Tarnowski, S. J., Roy, S. K., Liptak, R., Lee, D. K., Ning, R. Y. 1986. *Methods Enzymol.* 119:153–65

86. Thatcher, D. R., Panayotatos, N. 1986. *Methods Enzymol.* 119:166–77

87. Moschera, J. A., Woehle, D., Tsai, K. P., Chen, C.-H., Tarnowski, S. J. 1986. *Methods Enzymol.* 119:177–83

88. Lin, L. S., Yamamoto, R., Drummond, R. J. 1986. *Methods Enzymol.* 119:183–92

89. Kung, H.-F., Pan, Y.-C. E., Moschera, J., Tsai, K., Bekesi, E., et al. 1986. *Methods Enzymol.* 119:204–10

90. Gray, P. W., Goeddel, D. V. 1983. *Proc. Natl. Acad. Sci. USA* 80:5842–46

91. Wetzel, R., Perry, L. J., Estell, D. A., Lin, N., Levine, H. L., et al. 1981. *J. Interferon Res.* 1:381–90

92. Ackerman, S. K., Zur Nedden, D., Heintzelman, M., Hunkapiller, M. W., Zoon, K. 1984. *Proc. Natl. Acad. Sci. USA* 81:1045–47

93. Bewley, T. A., Levine, H. L., Wetzel, R. 1982. *Int. J. Peptide Protein Res.* 20:93–96

94. Miller, D. L., Kung, H.-F., Pestka, S. 1982. *Science* 215:689–90

95. Velan, B., Cohen, S., Grosfeld, H., Leitner, M., Shafferman, A. 1985. *J. Biol. Chem.* 260:5498–504

96. Khosrovi, B. 1984. See Ref. 78, pp. 89–99

97. Shepard, H. M., Leung, D. W., Stebbing, N., Goeddel, D. V. 1981. *Nature* 294:563–65

98. Tanaka, T., Matsuda, S., Kawano, G., Kobayashi, S. 1986. See Ref. 448, pp. 65–72

99. Arakawa, T., Alton, N. K., Hsu, Y.-R. 1985. *J. Biol. Chem.* 280:14435–39

100. Devos, R., Opsomer, C., Scahill, S. J., Van der Heyden, J., Fiers, W. 1984. *J. Interferon Res.* 4:461–78

101. Le, H. V., Mays, C. A., Syto, R.,

Nagabhushan, T. L., Trotta, P. P. 1986. See Ref. 448, pp. 73–80

101a. Vijay-Kumar, S., Senadhi, S., Ealick, S. E., Bugg, C. E., Nagabhushan, T. L., et al. 1987. *J. Biol. Chem.* 262: In press

102. Pestka, S., McInnes, J., Havell, E., Vilcek, J. 1975. *Proc. Natl. Acad. Sci. USA* 72:3898–901

103. Thang, M. N., Thang, D. C., De-Maeyer, E., Montagnier, L. 1975. *Proc. Natl. Acad. Sci. USA* 72:3975–77

104. Cavalieri, R. L., Havell, E. A., Vilcek, J., Pestka, S. 1977. *Proc. Natl. Acad. Sci. USA* 74:3287–91

105. Cavalieri, R. L., Havell, E. A., Vilcek, J., Pestka, S. 1977. *Proc. Natl. Acad. Sci. USA* 74:4415–19

106. Cavalieri, R. L., Pestka, S. 1977. *Tex. Rep. Biol. Med.* 35:117–23

107. Reynolds, F. H. Jr., Premkumar, E., Pitha, P. M. 1975. *Proc. Natl. Acad. Sci. USA* 72:4881–85

108. Maeda, S., McCandliss, R., Gross, M., Sloma, A., Familletti, P. C., et al. 1980. *Proc. Natl. Acad. Sci. USA* 77:7010–13; 78:4648

109. Maeda, S., Gross, M., Pestka, S. 1981. *Methods Enzymol.* 79:613–18

110. Pestka, S. 1983. *Arch. Biochem. Biophys.* 221:1–37

111. Taniguchi, T., Ohno, S., Fujii-Kuriyama, Y., Muramatsu, M. 1980. *Gene* 10:11–15

112. Nagata, S., Taira, H., Hall, A., Johnsrud, L., Streuli, M., et al. 1980. *Nature* 284:316–20

113. McCandliss, R., Sloma, A., Pestka, S. 1981. *Methods Enzymol.* 79:618–22

114. Derynck, R., Content, J., De Clercq, E., Volckaert, G., Tavernier, J., et al. 1980. *Nature* 285:542–47

115. Houghton, M., Stewart, A. G., Doel, S. M., Emtage, J. S., Eaton, M. A. W., et al. 1980. *Nucleic Acids Res.* 8:1913–31

116. Goeddel, D. V., Shepard, H. M., Yelverton, E., Leung, D., Crea, R., et al. 1980. *Nucleic Acids Res.* 8:4057–74

117. Gray, P. W., Leung, D. W., Pennica, D., Yelverton, E., Najarian, R., et al. 1982. *Nature* 295:503–8

118. Devos, R., Cheroutre, H., Taya, Y., Degrave, W., Van Heuverswyn, H., et al. 1982. *Nucleic Acids Res.* 10:2487–501

119. Maeda, S., McCandliss, R., Chiang, T.-R., Costello, L., Levy, W. P., et al. 1981. In *Developmental Biology Using Purified Genes*, ed. D. Brown, C. F. Fox, pp. 85–96. New York: Academic

120. Goeddel, D. V., Yelverton, E., Ullrich, A., Heyneker, H. L., Miozzari, G., et al. 1980. *Nature* 287:411–16

121. Brack, C., Nagata, S., Mantei, N., Weissmann, C. 1981. *Gene* 15:379–94

122. Henco, K., Brosius, J., Fujisawa, A., Fujisawa, J.-I., Haynes, J. R., et al. 1985. *J. Mol. Biol.* 185:227–60

123. Hotta, K., Collier, K. J., Pestka, S. 1986. *Methods Enzymol.* 119:481–85

124. Wilson, V., Jeffreys, A. J., Barrie, P. A., Boseley, P. G., Slocombe, P. M., et al. 1983. *J. Mol. Biol.* 166:457–75

125. Hauptmann, R., Swetly, P. 1985. *Nucleic Acids Res.* 13:4739–43

126. Feinstein, S. I., Mory, Y., Chernajovsky, Y., Maroteaux, L., Nir, U., et al. 1985. *Mol. Cell. Biol.* 5:510–17

127. Capon, D. J., Shepard, H. M., Goeddel, D. V. 1985. *Mol. Cell. Biol.* 5:768–79

128. Owerbach, D., Rutter, W. J., Shows, T. B., Gray, P., Goeddel, D. V., et al. 1981. *Proc. Natl. Acad. Sci. USA* 78:3123–27

129. Taniguchi, T., Mantei, N., Schwarzstein, M., Nagata, S., Muramatsu, M., et al. 1980. *Nature* 285:547–50

130. Leung, D. W., Capon, D. J., Goeddel, D. V. 1984. *Biotechnology* 2:458–64

131. Gray, P., Goeddel, D. 1982. *Nature* 295:503–8

132. Naylor, S. L., Sakaguchi, A. Y., Shows, T. B., Law, M. L., Goeddel, D. V., Gray, P. W. 1983. *J. Exp. Med.* 157:1020–27

133. DeGrado, W. F., Wasserman, Z. R., Chowdhry, V. 1982. *Nature* 300:379–81

134. Dianzani, F., Dolei, A., DiMarco, P. 1986. *J. Interferon Res.* 6: In press

135. Sagar, A. D., Sehgal, P. B., Slate, D. L., Ruddle, F. H. 1982. *J. Exp. Med.* 156:744–55

136. Streuli, M., Nagata, S., Weissmann, C. 1980. *Science* 209:1343–47

137. Staehelin, T., Hobbs, D. S., Kung, H.-F., Pestka, S. 1981. *Methods Enzymol.* 78:505–12

138. Pestka, S., Maeda, S., Hobbs, D. S., Levy, W. P., McCandliss, R., et al. 1981. In *Cellular Responses to Molecular Modulators*, ed. L. W. Mozes, J. Schultz, W. A. Scott, R. Werner, pp. 455–89. New York: Academic. 558 pp.

139. Pestka, S., Maeda, S., Hobbs, D. S., Chiang, T.-R. C., Costello, L. L., et al. 1981. In *Recombinant DNA*, ed. A. G. Walton, pp. 51–74. New York: Elsevier Scientific. 310 pp.

140. Scahill, S. J., Devos, R., Van der Heyden, J., Fiers, W. 1983. *Proc. Natl. Acad. Sci. USA* 80:4654–58

141. Remaut, E., Stanssens, P., Simons, G.,

Fiers, W. 1986. *Methods Enzymol.* 119:366–75

142. Zinn, K., Mellon, P., Ptashne, M., Maniatis, T. 1982. *Proc. Natl. Acad. Sci. USA* 79:4897–901

143. Canaani, D., Berg, P. 1982. *Proc. Natl. Acad. Sci. USA* 79:5166–70

144. Mulcahy, L., Kahn, M., Kelder, B., Rehberg, E., Pestka, S., et al. 1986. *Methods Enzymol.* 119:383–96

145. Innis, M. A., McCormick, F. 1986. *Methods Enzymol.* 119:397–403

146. Schaber, M. D., DeChiara, T. M., Kramer, R. A. 1986. *Methods Enzymol.* 119:416–24

147. Hitzeman, R. A., Chang, C. N., Matteucci, M., Perry, L. J., Kohr, W. J., et al. 1986. *Methods Enzymol.* 119:424–33

148. Dijkema, R., van der Meide, P. H., Dubbeld, M., Caspers, M., Wubben, J., et al. 1986. *Methods Enzymol.* 119:453–64

149. Mutsaers, J. H. G. M., Kamerling, J. P., Devos, R., Guisez, Y., Fiers, W. 1986. *Eur. J. Biochem.* 156:651–54

150. Miyata, T., Hayashida, H., Kikuno, R., Toh, H., Kawade, Y. 1985. In *Interferon 6*, ed. I. Gresser, pp. 1–30. London: Academic. 143 pp.

151. Ohlsson, M., Feder, J., Cavalli-Sforza, L. L., von Gabain, A. 1985. *Proc. Natl. Acad. Sci. USA* 82:4473–76

152. von Gabain, A., Ohlsson, M., Lindstrom, E., Lundstrom, M., Lundgren, E. 1986. *Chim. Scr. R. Acad.,* 26B:357–62

153. Epstein, L. B. 1982. *Nature* 295:453–54

154. Derynck, R. 1983. In *Interferon 5*, ed. I. Gresser, pp. 181–203. London: Academic. 239 pp.

155. Van der Meide, P. H., Dijkema, R., Caspers, M., Vijverberg, K., Schellekens, H. 1986. *Methods Enzymol.* 119:441–53

156. Gold, P., Pestka, S. 1985. *Fed. Proc.* 44:1613

157. Shiroza, T., Nakazawa, K., Tashiro, N., Yamane, K., Yanagi, K. 1985. *Gene* 34:1–8

158. Bowden, D. W., Mao, J.-i, Gill, T., Hsiao, K., Lillquist, J. S., et al. 1984. *Gene* 27:87–99

159. Zwarthoff, E. C., Mooren, A. T. A., Trapman, J. 1985. *Nucleic Acids Res.* 13:791–804

160. Kelley, K. A., Pitha, P. M. 1985. *Nucleic Acids Res.* 13:805–23

161. Lindenmann, J. 1981. *Methods Enzymol.* 78:181–88

162. Evinger, M., Rubinstein, M., Pestka, S.

1981. *Arch. Biochem. Biophys.* 210:319–29

163. Eife, R., Hahn, T., De Tavera, M., Schertel, F., Holtmann, H., et al. 1981. *J. Immunol. Methods* 47:339–47

164. Ortaldo, J. R., Mantovani, A., Hobbs, D., Rubinstein, M., Pestka, S., et al. 1983. *Int. J. Cancer* 31:285–89

165. Ortaldo, J. R., Mason, A., Rehberg, E., Moschera, J., Kelder, B., et al. 1984. *J. Biol. Chem.* 258:15011–15

166. Ortaldo, J. R., Herberman, R. B., Harvey, C., Osheroff, P., Pan, Y.-C., et al. 1984. *Proc. Natl. Acad. Sci. USA* 81:4926–29

167. Pestka, S., Kelder, B., Rehberg, E., Ortaldo, J. R., Herberman, R. B., et al. 1983. In *The Biology of the Interferon System*, ed. E. DeMaeyer, H. Schellekens, pp. 535–49. Amsterdam: Elsevier

168. Greiner, J. W., Fisher, P. B., Pestka, S., Schlom, J. 1986. *Cancer Res.* 46:4984–90

169. Greiner, J. W., Hand, P. H., Noguchi, P., Fisher, P. B., Pestka, S., et al. 1984. *Cancer Res.* 44:3208–14

170. Greiner, J. W., Schlom, J., Pestka, S., Langer, J. A., Giacomini, P., et al. 1987. *Pharmacol. Ther.* In press

171. Giacomini, P., Aguzzi, A., Pestka, S., Fisher, P. B., Ferrone, S. 1984. *J. Immunol.* 133:1649–55

172. Eppstein, D. A., Czarniecki, W., Jacobsen, H., Friedman, R. M., Panet, A. 1981. *Eur. J. Biochem.* 118:9–15

173. Sen, G. C., Herz, R. E. 1983. *J. Virol.* 45:1017–27

174. Lengyel, P. 1981. *Methods Enzymol.* 79:135–48

175. Lengyel, P., Pestka, S. 1981. In *Gene Families of Collagen and Other Proteins*, ed. D. C. Prockop, P. C. Champe, pp. 121–26. Holland/Amsterdam: Elsevier

176. Maheshwari, R. K., Friedman, R. M. 1981. *Methods Enzymol.* 79:451–58

177. Sreevalsan, T., Lee, E., Friedman, R. M. 1981. *Methods Enzymol.* 79:342–49

178. Revel, M., Wallach, D., Merlin, G., Schattner, A., Schmidt, A., et al. 1981. *Methods Enzymol.* 79:149–61

179. Kerr, I. M., Brown, R. E. 1978. *Proc. Natl. Acad. Sci. USA* 75:256–60

180. Yoshie, O., Aso, H., Sakakibara, A., Ishida, N. 1985. *J. Interferon Res.* 5:531–40

181. Stewart, W. E. II. 1979. *The Interferon System*. New York: Springer-Verlag. 421 pp.

182. Paucker, K., Cantell, K., Henle, W. 1962. *Virology* 17:324–34

183. Nishimura, J., Mitsui, K., Ishikawa, T., Tanaka, Y., Yamamoto, R., et al. 1985. *Clin. Exp. Metastasis* 3:295–304
184. Balkwill, F. 1986. *Methods Enzymol.* 119:649–57
185. Brunda, M. J., Wright, R. B. 1986. *Int. J. Cancer.* 37:287–91
186. Fisher, P. B., Grant, S. 1985. *Pharmacol. Ther.* 27:143–66
187. Fisher, P. B., Prignoli, D. R., Hermo, H., Weinstein, I. B., Pestka, S. 1985. *J. Interferon Res.* 5:11–22
188. Rossi, G. B. 1985. See Ref. 150, pp. 31–68
189. Keay, S., Grossberg, S. 1980. *Proc. Natl. Acad. Sci. USA* 77:4099–103
190. Heron, I., Hokland, M., Berg, K. 1978. *Proc. Natl. Acad. Sci. USA* 75:6215–19
191. De Maeyer-Guignard, J., DeMaeyer, E. 1985. See Ref. 150, pp. 69–91
192. Imai, K., Ng, A. K., Glassy, M. C., Ferrone, S. 1981. *J. Immunol.* 127:505–9
193. Fridman, W. H., Gresser, I., Bandu, M. T., Aguet, M., Neauport-Sautes, C. 1980. *J. Immunol.* 124:2436–41
194. Fleischmann, W. R. Jr., Schwarz, L. A. 1981. *Methods Enzymol.* 79:432–40
195. Ling, P. D., Warren, M. K., Vogel, S. N. 1985. *J. Immunol.* 135:1857–63
195a. Inaba, K., Kitaura, M., Kato, T., Watanabe, Y., Kawade, Y., Muramatsu, S. 1986. *J. Exp. Med.* 163:1030–35
196. Garotta, G., Talmadge, K. W., Pink, J. R. L., Dewald, B., Baggiolini, M. 1986. *J. Interferon Res.* In press
196a. Alan, R., Ezekowitz, B., Hill, M., Gordon, S. 1986. *Biochem. Biophys. Res. Commun.* 136:737–44
197. Friedman, R. M. 1967. *Science* 156:1760–61
198. Chany, C. 1976. *Biomedicine* 24:148–57
199. Chany, C. 1984. In *Interferon 3*, ed. R. M. Friedman, pp. 11–32. Amsterdam: Elsevier
200. Aguet, M., Mogensen, K. E. 1984. See Ref. 154, pp. 1–22
201. Zoon, K. C., Arnheiter, H. 1984. *Pharmacol. Ther.* 24:259–78
202. Aguet, M. 1980. *Nature* 284:459–61
203. Aguet, M., Blanchard, B. 1981. *Virology* 115:249–61
204. Aguet, M., Gresser, I., Hovanessian, A. G., Bandu, M.-T., Blanchard, B., et al. 1981. *Virology* 114:585–88
205. Branca, A. A., Baglioni, C. 1981. *Nature* 294:768–70
206. Branca, A. A., Baglioni, C. 1982. *J. Biol. Chem.* 257:13197–200
207. Mogensen, K. E., Bandu, M.-T., Vig-naux, F., Aguet, M., Gresser, I. 1981. *Int. J. Cancer* 28:575–82
208. Baglioni, C., Branca, A. A., D'Alessandro, S. B., Hossenlopp, D., Chadha, K. C. 1982. *Virology* 122:202–6
209. Joshi, A. R., Sarkar, F. H., Gupta, S. L. 1982. *J. Biol. Chem.* 257:13884–87
210. Epstein, C. J., McManus, N. H., Epstein, L. B. 1982. *Biochem. Biophys. Res. Commun.* 107:1060–66
211. Zoon, K., Zur Nedden, D., Arnheiter, H. 1982. *J. Biol. Chem.* 257:4695–97
212. Branca, A. A., Faltynek, C. R., D'Alessandro, S. B., Baglioni, C. 1982. *J. Biol. Chem.* 257:13291–96
213. Mogensen, K. E., Bandu, M.-T. 1983. *Eur. J. Biochem.* 134:355–64
214. Zoon, K. C., Arnheiter, H., Zur Nedden, D., Fitzgerald, D. J. P., Willingham, M. C. 1983. *Virology* 130:195–203
215. Sarkar, F. H., Gupta, S. L. 1984. *Eur. J. Biochem.* 140:461–67
216. Feinstein, S., Traub, A., Lazar, A., Mizrahi, A., Teitz, Y. 1985. *J. Interferon Res.* 5:65–76
217. Uze, G., Mogensen, K. E., Aguet, M. 1985. *Embo J.* 4:65–70
218. Kushnaryov, V. M., Sedmak, J. J., Bendler, J. W., Grossberg, S. E. 1982. *Infect. Immun.* 36:811–21
219. Kushnaryov, V. M., MacDonald, H. S., Sedmak, J. J., Grossberg, S. E. 1985. *Proc. Natl. Acad. Sci. USA* 82:3281–85
220. Yonehara, S., Yonehara-Takahashi, M., Ishii, A., Nagata, S. 1983. *J. Biol. Chem.* 258:9046–49
221. Yonehara, S., Yonehara-Takahashi, M., Ishii, A. 1983. *J. Virol.* 45:1168–71
222. Hannigan, G. E., Gewert, D. R., Fish, E. N., Read, S. E., Williams, B. R. G. 1983. *Biochem. Biophys. Res. Commun.* 110:542–44
223. Evans, T., Secher, D. 1984. *EMBO J.* 3:2975–78
224. Langer, J. A., Pestka, S. 1985. *J. Interferon Res.* 5:637–49
224a. Branca, A. A. 1986. *J. Interferon Res.* 6:305–11
225. Williams, B. R. G., Hannigan, G. E., Saunders, M. E. 1985. In *The 2-5A System*, ed. B. R. G. Williams, R. H. Silverman, pp. 227–36. New York: Liss
226. Yonehara, S., Yonehara, M., Yamaguchi, T., Nagata, S. 1984. *Antiviral Res. Abstr.* 1(No. 3):49
227. Merlin, G., Falcoff, E., Aguet, M. 1985. *J. Gen. Virol.* 66:1149–52
228. Sarkar, F. H., Gupta, S. L. 1984. *Proc. Natl. Acad. Sci. USA* 81:5160–64
229. O'Rourke, E. C., Drummond, R. J.,

Creasey, A. A. 1984. *Mol. Cell. Biol.* 4:2745–49

230. Ruzicka, F. J., Hawkins, M. J., Borden, E. C. 1984. *Fed. Proc.* 43:690. Abstr.

231. Thompson, M. R., Zhang, Z.-Q., Fournier, A., Tan, Y. H. 1985. *J. Biol. Chem.* 260:563–67

232. Jayaram, B. M., Schmidt, H., Yoshie, O., Samanta, H., Floyd-Smith, G., et al. 1983. In *Humoral Factors in Host Defense,* ed. Y. Yamamura, pp. 157–74. New York: Academic

233. Dron, M., Tovey, M. G. 1983. *J. Gen. Virol.* 64:2641–47

234. Tovey, M. G., Dron, M., Mogensen, K. E., Lebleu, B., Mechti, N., et al. 1983. *J. Gen. Virol.* 64:2649–53

235. Dron, M., Tovey, M. G., Eid, P. 1985. *J. Gen. Virol.* 66:787–95

236. Nagata, Y., Rosen, O. M., Makman, M. H., Bloom, B. R. 1984. *J. Cell Biol.* 98:1342–47

237. Aguet, M., Gröbke, M., Dreiding, P. 1984. *Virology* 132:211–16

238. Czarniecki, C. W., Fennie, C., Powers, D., Estell, D. 1984. *J. Virol.* 49:490–96

239. Zhang, Z.-Q., Fournier, A., Tan, Y. H. 1986. *J. Biol. Chem.* 261:8017–21

240. Aguet, M., Belardelli, F., Blanchard, B., Marcucci, F., Gresser, I. 1982. *Virology* 117:541–44

241. Raziuddin, A., Sarkar, F. H., Dutkowski, R., Shulman, L., Ruddle, F. H., et al. 1984. *Proc. Natl. Acad. Sci. USA* 81:5504–8

242. Anderson, P., Yip, Y. K., Vilček, J. 1982. *J. Biol. Chem.* 257:11301–4

243. Orchansky, P., Novick, D., Fischer, D. G., Rubinstein, M. 1984. *J. Interferon Res.* 4:275–82

244. Orchansky, P., Rubinstein, M., Fischer, D. B. 1986. *J. Immunol.* 136:169–73

245. Rashidbaigi, A., Kung, H.-F., Pestka, S. 1985. *J. Biol. Chem.* 260:8514–19

246. Littman, S. J., Faltynek, C. R., Baglioni, C. 1985. *J. Biol. Chem.* 260:1191–95

247. Finbloom, D. S., Hoover, D. L., Wahl, L. M. 1985. *J. Immunol.* 135:300–5

248. Besancon, F., Ankel, H. 1974. *Nature* 252:478–80

249. Vengris, V. E., Reynolds, F. H., Hollenberg, M. D., Pitha, P. M. 1976. *Virology* 72:486–93

250. Kuwata, T., Handa, S., Fuse, A., Morinaga, N. 1978. *Biochem. Biophys. Res. Commun.* 85:77–84

251. Ankel, H., Krishnamurti, C., Besancon, F., Stefanos, S., Falcoff, E. 1980. *Proc. Natl. Acad. Sci. USA* 77:2528–32

252. Belardelli, F., Aliberti, A., Santurbano, B., Antonelli, G. et al. 1982. *Virology* 117:391–400

253. Fuse, A., Handa, S., Kuwata, T. 1982. *Antiviral Res.* 2:161–66

254. MacDonald, H. S., Elconin, H., Ankel, H. 1982. *FEBS Lett.* 141:267–70

255. Gupta, S. L., Raziuddin, A., Sarkar, F. H. 1984. *J. Interferon Res.* 4:305–14

256. Faltynek, C. R., Branca, A. A., McCandless, S., Baglioni, C. 1983. *Proc. Natl. Acad. Sci. USA* 80:3269–73

257. Raziuddin, A., Gupta, S. L. 1985. See Ref. 225, pp. 219–26

258. Eid, P., Mogensen, K. E. 1983. *FEBS Lett.* 156:157–60

259. Eid, P., Mogensen, K. E. 1985. See Ref. 46, pp. 221–25

260. Traub, A., Feinstein, S., Gez, M., Lazar, A., Mizrahi, A. 1984. *J. Biol. Chem.* 259:13872–77

261. Tan, Y. H., Tischfield, J., Ruddle, R. H. 1973. *J. Exp. Med.* 137:317–30

262. Tan, Y. H., Schneider, E. L., Tischfield, J., Epstein, C. J., Ruddle, F. H. 1974. *Science* 186:61–63

263. Tan, Y. H. 1976. *Nature* 260:141–43

264. Slate, D. L., Shulman, L., Lawrence, J. B., Revel, M., Ruddle, F. H. 1978. *J. Virol.* 25:319–25

265. Slate, D. L., Ruddle, F. H. 1981. *Methods Enzymol.* 79:536–42

266. Epstein, C. J., Epstein, L. B. 1983. In *Lymphokines,* ed. E. Pick, 8:277–301. New York: Academic

267. Lubiniecki, A. S., Jones, V., Eatherly, C. 1979. *Arch. Virol.* 60:341–46

268. Revel, M., Bash, D., Ruddle, F. H. 1976. *Nature* 260:139–41

269. Slate, D. L., Ruddle, F. H. 1978. *Cytogenet. Cell Genet.* 22:265–69

270. Slate, D. L., Ruddle, F. H. 1978. *Cytogenet. Cell Genet.* 22:270–74

271. Shulman, L. M., Kamarck, M. E., Slate, D. L., Ruddle, F. H., Branca, A. A., et al. 1984. *Virology* 137:422–27

272. Cox, D. R., Epstein, L. B., Epstein, C. J. 1980. *Proc. Natl. Acad. Sci. USA* 77:2168–72

273. Lin, P.-F., Slate, D. L., Lawyer, F. C., Ruddle, F. H. 1980. *Science* 209:285–87

274. Tan, Y. H., Greene, A. E. 1976. *J. Gen. Virol.* 32:153–55

275. Epstein, L. B., Epstein, C. J. 1976. *J. Infect. Dis.* 133:A56–A62 (Suppl.)

276. Sinet, P. M., Couturier, J., Dutrillaux, B., Poissonier, M., Raoul, U., et al. 1976. *Exp. Cell Res.* 97:47–55

277. Pestka, S., Labdon, J. E., Rashidbaigi, A., Liu, X.-Y., Langer, J. A., et al. 1985. See Ref. 31, pp. 3–11

278. Jung, V., Pestka, S. 1986. *Methods Enzymol.* 119:597–611
279. Chebath, J., Benech, P., Mory, V., Chernajovsky, Y., Horovitz, O., et al. 1985. See Ref. 46, pp. 201–11
280. Shulman, L. M., Ruddle, F. H. 1985. See Ref. 31, pp. 333–37
281. Osheroff, P. L., Chiang, T.-R., Manouses, D. 1985. *J. Immunol.* 135:306–13
282. Yonehara, S., Ishii, A. 1986. See Ref. 448, pp. 167–71
283. Branca, A. A., D'Alessandro, S. B., Baglioni, C. 1983. *J. Interferon Res.* 3:465–71
284. Yonehara, S., Ishii, A., Yonehara-Takahashi, M. 1983. *J. Gen. Virol.* 64:2409–18
285. Kushnaryov, V. M., MacDonald, H. S., Debruin, J., Lemense, G. P., Sedmak, J. J., Grossberg, S. E. 1986. *J. Interferon Res.* 6:241–45
286. Hannigan, G., Williams, B. R. G. 1986. *EMBO J.* 5:1607–13
287. Hannigan, G. E., Fish, E. N., Williams, B. R. G. 1984. *J. Biol. Chem.* 259:8084–86
288. Maxwell, B. L., Talpaz, M., Gutterman, J. U. 1985. *Int. J. Cancer* 36:23–28
289. Vengris, V. E., Stollar, B. D., Pitha, P. 1975. *Virology* 65:410–21
290. Higashi, Y., Sokawa, Y. 1982. *J. Biochem.* 91:2021–28
291. Huez, G., Silhol, M., Lebleu, B. 1983. *Biochem. Biophys. Res. Commun.* 110:155–60
292. Fidler, I. J., Fogler, W. E., Kleinerman, E. S., Saiki, I. 1985. *J. Immunol.* 135:4289–96
293. Sanceau, J., Lewis, J. A., Sondermeyer, P., Beranger, F., Falcoff, R., et al. 1986. *Biochem. Biophys. Res. Commun.* 135:894–901
294. Evinger, M., Pestka, S. 1981. *Methods Enzymol.* 79:362–68
295. Ankel, H., Chany, C., Galliot, B., Chevalier, M. J., Robert, M. 1973. *Proc. Natl. Acad. Sci. USA* 70:2360–63
296. Chany, C., Ankel, H., Galliot, B., Chevalier, M. J., Gregoire, A. 1974. *Proc. Soc. Exp. Biol. Med.* 147:293–99
297. Schneck, J., Rager-Zisman, B., Rosen, O. M., Bloom, B. R. 1982. *Proc. Natl. Acad. Sci. USA* 79:1879–83
298. Mills, G. B., Hannigan, G., Stewart, D., Mellors, A., Williams, B., et al. 1985. See Ref. 225, pp. 357–67
299. Rochette-Egly, C., Tovey, M. G. 1985. *Antiviral Res.* 5:127–35
299a. Yap, W. H., Teo, T. S., Tan, Y. H. 1986. *Science* 234:355–58
299b. Yap, W. H., Teo, T. S., McCoy, E., Tan, Y. H. 1986. *Proc. Natl. Acad. Sci. USA* 83:7765–69
300. Arnheiter, H., Thomas, R. M., Leist, T., Fountoulakis, M., Gutte, B. 1981. *Nature* 294:278–80
301. Wetzel, R., Levine, H. L., Estell, D. A., Shire, S., Finer-Moore, J., et al. 1982. In *Chemistry and Biology of Interferons*, ed. T. Merigan, R. Friedman, pp. 365–76. New York: Academic
302. Ohno, M., Widmer, F., Smith, M. E., Arnheiter, H., Zoon, K. C. 1982. In *Peptides*, ed. K. Blaha, P. Malon. Berlin: de Gruyter
303. Chang, N. T., Kung, H.-F., Pestka, S. 1983. *Arch. Biochem. Biophys.* 221:585–89
304. Franke, A. E., Shepard, H. M., Houck, C. M., Leung, D. W., Goeddel, D. V., et al. 1982. *DNA* 1:223–30
305. Arnheiter, H., Ohno, M., Smith, M., Gutte, B., Zoon, K. C. 1983. *Proc. Natl. Acad. Sci. USA* 80:2539–43
306. Goeddel, D. V., Talmadge, K. W., Pink, J. R. L., Dewald, B., Baggiolini, M. 1980. *Nature* 287:411–16
307. Lydon, N. B., Favre, C., Bove, S., Neyret, O., Benureau, S., et al. 1985. *Biochemistry* 24:4131–41
308. Camble, R., Petter, N. N., Trueman, P., Newton, C. R., Carr, F. J., et al. 1986. *Biochem. Biophys. Res. Commun.* 134:1404–11
309. Marcucci, F., DeMaeyer, E. 1986. *Biochem. Biophys. Res. Commun.* 134:1412–18
309a. Langer, J. A., Pestka, S. 1986. *Methods Enzymol.* 119:305–11
310. Langer, J. A., Pestka, S. 1984. *J. Invest. Dermatol.* 83(Suppl. 1):128s–36s
311. Langer, J. A., Ortaldo, J. R., Pestka, S. 1986. *J. Interferon Res.* 6:97–105
312. Kung, H.-F., Bekesi, E. 1986. *Methods Enzymol.* 119:296–301
313. Wietzerbin, J., Merlin, G., Gaudelet, C., Falcoff, E. 1985. See Ref. 31, pp. 425–28
314. Wietzerbin, J., Gaudelet, C., Aguet, M., Falcoff, E. 1986. *J. Immunol.* 136:2451–55
315. Langer, J. A., Rashidbaigi, A., Pestka, S. 1986. *J. Biol. Chem.* 261:9801–4
316. Ücer, U., Bartsch, H., Scheurich, P., Pfizenmaier, K. 1985. *Int. J. Cancer* 36:103–8
317. Ücer, U., Bartsch, H., Scheurich, P., Berkovic, D., Ertel, C., et al. 1986. *Cancer Res.* 46:5339–43
318. Yonehara, S., Yonehara, M., Fukunaga, R., Nagata, S. 1986. See Ref. 448, pp. 183–87

318a. Nagao, S.-I., Sato, K., Osada, Y. 1986. *Cancer Res.* 46:3279–82

319. Celada, A., Gray, P. W., Rinderknecht, E., Schreiber, R. D. 1984. *J. Exp. Med.* 160:55–74

320. Landolfo, S., Cofano, F., Gandino, L., Gribaudo, G., Cavallo, G. 1985. See Ref. 46, pp. 227–29

321. Cofano, F., Fassio, A., Cavallo, G., Landolfo, S. 1986. *J. Gen. Virol.* 67:1205–9

322. Aiyer, R. A., Serrano, L. E., Jones, P. P. 1986. *J. Immunol.* 136:3329–34

323. Rashidbaigi, A., Langer, J. A., Jung, V., Jones, C., Morse, H. G., et al. 1986. *Proc. Natl. Acad. Sci. USA* 83:384–88

324. Fleischmann, W. R. Jr. 1982. *Cancer Res.* 42:869–75

325. Anderson, P., Nagler, C. 1984. *Biochem. Biophys. Res. Commun.* 120:828–33

326. Rubinstein, M., Fischer, D. G., Orchansky, P. 1986. In *Interferons as Cell Growth Inhibitors and Antitumor Factors,* ed. R. M. Friedman, T. Merigan, T. Sreevalsan, pp. 269–78. New York: Liss

327. Novick, D., Orchansky, P., Israel, S., Rubinstein, M. 1985. See Ref. 79, p. 113. Abstr.

328. Rosa, F. M., Cochet, M. M., Fellous, M. 1986. In *Interferon 7,* ed. I. Gresser, pp. 47–87. London: Academic

329. DeLey, M., Billiau, A. 1982. *Antiviral Res.* 2:97–102

330. Jung, V., Rashidbaigi, A., Jones, C., Tischfield, J., Shows, T., Pestka, S. 1987. *Proc. Natl. Acad. Sci. USA* In press

330a. Mariano, T. M., Kozak, C. A., Langer, J. A., Pestka, S. 1987. *J. Biol. Chem.* In press

331. MacDonald, H. S., Kushnaryov, V. M., Sedmak, J., Grossberg, S. E. 1986. *Biochem. Biophys. Res. Commun.* 138:254–60

332. Mogensen, K. E., Vignaux, F., Gresser, I. 1982. *FEBS Lett.* 140:285–87

333. Boucher, B. J., Temple, R., Toms, G., Mogensen, K. E. 1984. *Clin. Sci.* 67:(Suppl.9) 14

334. Kerr, I. M., Brown, R. E. 1978. *Proc. Natl. Acad. Sci. USA* 75:256–60

335. Lebleu, B., Sen, G. C., Shaila, S., Cabrer, B., Lengyel, P. 1976. *Proc. Natl. Acad. Sci. USA* 73:3107–311

336. Roberts, W. K., Hovanessian, A. G., Brown, R. E., Clemens, M. J., Kerr, I. M. 1976. *Nature* 264:477–80

337. Samuel, C. E., Farris, D. A., Eppstein, D. A. 1977. *Virology* 83:56–71

338. Zilberstein, A., Federman, P., Shulman, L., Revel, M. 1976. *FEBS Lett.* 68:119–24

339. Torrence, P. F., Johnston, M. I., Epstein, D. A., Jacobsen, H., Friedman, R. M. 1981. *FEBS Lett.* 130:291–96

340. Minks, M. A., West, D. K., Benvin, S., Greene, J. J., Ts'o, P. O. P., et al. 1980. *J. Biol. Chem.* 255:6403–7

341. Nilsen, T. W., Maroney, P. A., Robertson, H. D., Baglioni, C. 1982. *Mol. Cell Biol.* 2:154–60

342. Williams, B. R. G., Golgher, R. R., Brown, R. E., Gilbert, C. S., Kerr, I. M. 1979. *Nature* 282:582–86

343. Rice, A. P., Duncan, R., Hershey, J. W. B., Kerr, I. M. 1985. *J. Virol.* 54:894–98

344. Nilsen, T. W., Maroney, P. A., Baglioni, C. 1982. *J. Virol.* 42:1039–45

345. Samuel, C. E., Duncan, R., Knutson, G. S., Hershey, J. W. B. 1984. *J. Biol. Chem.* 259:13451–57

346. Penn, L. J. Z., Williams, B. R. G. 1984. *J. Virol.* 49:748–53

347. Rice, A. P., Kerr, S. M., Roberts, W. K., Brown, R. E., Kerr, I. M. 1985. *J. Virol.* 56:1041–44

348. Hersh, C. L., Brown, R. E., Roberts, W. K., Swyryd, E. A., Kerr, I. M., et al. 1984. *J. Biol. Chem.* 259:1731–37

349. Nilsen, T. W., Maroney, P. A., Baglioni, C. 1982. *J. Biol. Chem.* 257:14593–96

350. Ball, L. A. 1982. *The Enzymes,* 15:281–313

351. Williams, B. R. G., Silverman, R. H. 1985. *Prog. Clin. Biol. Res.* 202:1–478

352. Schmidt, A., Zilberstein, A., Shulman, L., Federman, P., Berissi, H., et al. 1978. *FEBS Lett.* 95:257–64

353. Dougherty, J. P., Samanta, H., Farrell, P. J., Lengyel, P. 1980. *J. Biol. Chem.* 255:3813–16

354. Schmidt, A., Chernajovsky, Y., Shulman, L., Federman, P., Berissi, H., et al. 1979. *Proc. Natl. Acad. Sci. USA* 76:4788–92

355. Baglioni, C., Maroney, P. A., West, D. K. 1979. *Biochemistry* 18:1765–70

356. Ball, L. A. 1979. *Virology* 94:282–96

357. Zerial, A., Hovanessian, A. G., Stefanos, S., Huygen, K., Werner, G. H., et al. 1982. *Antiviral Res.* 2:227–39

358. Baglioni, C., Maroney, P. A. 1980. *J. Biol. Chem.* 255:8390–93

359. Verhaegen-Lewalle, M., Kuwata, T., Zhang, Z-X., De Clercq, E., Cantell, K., et al. 1982. *Virology* 117:425–34

360. Benech, P., Merlin, G., Revel, M., Chebath, J. 1985. *Nucleic Acids Res.* 13:1267–81

361. Stark, G. R., Dower, W. J., Schimke, R. T., Brown, R. E., Kerr, I. M. 1979. *Nature* 278:471–73
362. Krishnan, I., Baglioni, C. 1980. *Proc. Natl. Acad. Sci. USA* 77:6506–10
363. Yarden, A., Shure-Gottlieb, H., Chebath, J., Revel, M., Kimchi, A. 1984. *EMBO J.* 3:969–73
364. Wells, V., Mallucci, L. 1985. *Exp. Cell Res.* 159:27–36
365. St. Laurent, G., Yoshie, O., Floyd-Smith, G., Samanta, H., Sehgal, P. B., et al. 1983. *Cell* 33:95–102
366. Yang, K., Samanta, H., Dougherty, J., Jayaram, B., Broeze, R., et al. 1981. *J. Biol. Chem.* 256:9324–28
367. Wells, J. A., Swyryd, E. A., Stark, G. R. 1984. *J. Biol. Chem.* 259:1363–70
368. Ilson, D. H., Torrence, P. F., Vilcek, J. 1986. *J. Interferon Res.* 6:5–12
369. Merlin, G., Chebath, J., Benech, P., Metz, R., Revel, M. 1983. *Proc. Natl. Acad. Sci. USA* 80:4904–8
370. Saunders, M. E., Gewert, D. R., Tugwell, M. E., McMahon, M., Williams, B. R. G. 1985. *EMBO J.* 4:1761–68
371. Benech, P., Mory, Y., Revel, M., Chebath, J. 1985. *EMBO J.* 4:2249–56
372. Wathelet, M., Moutschen, S., Cravador, A., DeWit, L., Defilippi, P., et al. 1986. *FEBS Lett.* 196:113–20
373. Shulman, L. M., Barker, P. E., Hart, J. T., Messer, P. P. G., Ruddle, F. H. 1984. *Somatic Cell Mol. Genet.* 10:247–57
374. Shulman, L. M., Revel, M. 1980. *Nature* 288:98–100
375. Chebath, J., Benech, P., Mory, Y., Federman, P., Berissi, H., et al. 1985. *Prog. Clin. Biol. Res.* 202:149–61
376. Eppstein, D. A., Samuel, C. E. 1978. *Virology* 89:240–51
377. Ball, L. A., White, C. N. 1979. *Virology* 93:348–56
378. Slattery, E., Ghosh, N., Samanta, H., Lengyel, P. 1979. *Proc. Natl. Acad. Sci. USA* 76:4778–82
379. Baglioni, C., Minks, M. A., Maroney, P. A. 1978. *Nature* 273:684–87
380. Clemens, M. J., Williams, B. R. G. 1978. *Cell* 13:565–72
381. Ratner, L., Wiegand, R. C., Farrell, P. J., Sen, G. C., Cabrer, B., et al. 1978. *Biochem. Biophys. Res. Commun.* 81:947–54
382. Floyd-Smith, G., Slattery, E., Lengyel, P. 1981. *Science* 212:1030–32
383. Wreschner, D. H., McCauley, J. W., Skehel, J. J., Kerr, I. M. 1981. *Nature* 289:414–17
384. Silverman, R. H., Skehel, J. J., James, T. C., Wreschner, D. H., Kerr, I. M. 1983. *J. Virol.* 46:1051–55
385. Silverman, R. H., Cayley, P. J., Knight, M., Gilbert, C. S., Kerr, I. M. 1982. *Eur. J. Biochem.* 124:131–38
386. Wreschner, D. H., James, T. C., Silverman, R. H., Kerr, I. M. 1981. *Nucleic Acids Res.* 9:1571–81
387. Floyd-Smith, G., Yoshie, O., Lengyel, P. 1982. *J. Biol. Chem.* 257:8584–87
388. Wreschner, D. H., Silverman, R. H., James, T. C., Gilbert, C. S., Kerr, I. M. 1982. *Eur. J. Biochem.* 124:261–68
389. Jacobsen, H., Czarniecki, C. W., Krause, D., Friedman, R. M., Silverman, R. H. 1983. *Virology* 125:496–501
390. Krause, D., Panet, A., Arad, G., Dieffenbach, C. W., Silverman, R. H. 1985. *J. Biol. Chem.* 260:9501–7
391. Krause, D., Silverman, R. H., Jacobsen, H., Leisy, S. A., Dieffenbach, C. W., et al. 1985. *Eur. J. Biochem.* 146:611–18
392. Eppstein, D. A., Peterson, T. C., Samuel, C. E. 1979. *Virology* 98:9–19
393. Minks, M. A., Benvin, S., Maroney, P. A., Baglioni, C. 1979. *Nucleic Acids Res.* 6:767–80
394. Goren, T., Kapitkovsky, A., Kimchi, A., Rubinstein, M. 1983. *Virology* 130:273–80
395. Williams, B. R. G., Kerr, I. M., Gilbert, C. S., White, C. N., Ball, L. A. 1978. *Eur. J. Biochem.* 92:455–62
396. Verhaegen-Lewalle, M., Content, J. 1982. *Eur. J. Biochem.* 126:639–43
397. Taira, H., Yamamoto, F., Furusawa, M., Sawai, H., Kawakita, M. 1985. *J. Interferon Res.* 5:583–96
398. Eppstein, D. A., Marsh, Y. V., Schryver, B. B., Larsen, M. A., Barnett, J. W., et al. 1982. *J. Biol. Chem.* 257:13390–97
399. Drocourt, J.-L., Dieffenbach, C. W., Ts'o, P. O. P., Justesen, J., Thang, M. N. 1982. *Nucleic Acids Res.* 10:2163–74
400. Imai, J., Johnston, M. I., Torrence, P. F. 1982. *J. Biol. Chem.* 257:12739–45
401. Doetsch, P., Wu, J. M., Sawada, Y., Suhadolnik, R. J. 1981. *Nature* 291:355–58
402. Haugh, M. C., Cayley, P. J., Serafinowska, H. T., Norman, D. G., Reese, C. B., et al. 1983. *Eur. J. Biochem.* 132:77–84
403. Sawai, H., Imai, J., Lesiak, K., Johnston, M. I., Torrence, P. F. 1983. *J. Biol. Chem.* 258:1671–77
404. Lesiak, K., Imai, J., Floyd-Smith, G., Torrence, P. F. 1983. *J. Biol. Chem.* 258:13082–88

405. Baglioni, C., D'Alessandro, S. B., Nilsen, T. W., den Hartog, J. A. J., Crea, R., et al. 1981. *J. Biol. Chem.* 256:3253–57

406. Imai, J., Lesiak, K., Torrence, P. F. 1985. *J. Biol. Chem.* 260:1390–93

407. Watling, D., Serafinowska, H. T., Reese, C. B., Kerr, I. M. 1985. *EMBO J.* 4:431–36

408. Eppstein, D. A., Van der Pas, M. A., Schryver, B. B., Sawai, H., Lesiak, K., et al. 1985. *J. Biol. Chem.* 260:3666–71

409. Krause, D., Lesiak, K., Imai, J., Sawai, H., Torrence, P. F., et al. 1986. *J. Biol. Chem.* 261:6836–39

410. Eppstein, D. A., Schryver, B. B., Marsh, Y. V. 1986. *J. Biol. Chem.* 261:5999–6003

411. Devash, Y., Gera, A., Willis, D. H., Reichman, M., Pfleiderer, W., et al. 1984. *J. Biol. Chem.* 259:3482–86

412. Krebs, E. G., Beavo, J. A. 1979. *Ann. Rev. Biochem.* 48:923–59

413. Samuel, C. E. 1985. *Prog. Clin. Biol. Res.* 202:247–54

414. Samuel, C. E. 1979. *Virology* 93:281–85

415. Samuel, C. E. 1979. *Proc. Natl. Acad. Sci. USA* 76:600–4

416. Minks, M. A., West, D. K., Benvin, S., Baglioni, C. 1979. *J. Biol. Chem.* 254:10180–83

417. Gupta, S. L. 1979. *J. Virol.* 29:301–11

418. Lasky, S. R., Jacobs, B. L., Samuel, C. E. 1982. *J. Biol. Chem.* 257:11087–93

419. Berry, M. J., Knutson, G. S., Lasky, S. R., Munemitsu, S. M., Samuel, C. E. 1985. *J. Biol. Chem.* 260:11240–47

420. Samuel, C. E., Knutson, G. S. 1982. *J. Interferon Res.* 2:441–45

421. De Benedetti, A., Baglioni, C. 1984. *Nature* 311:79–81

422. Whitaker-Dowling, P., Youngner, J. S. 1984. *Virology* 137:171–81

423. Samuel, C. E., Knutson, G. S. 1982. *J. Biol. Chem.* 257:11789–95

424. Samuel, C. E. 1986. See Ref. 448, pp. 101–10

425. Samuel, C. E., Knutson, G. S. 1983. *Virology* 130:474–84

426. Samuel, C. E., Knutson, G. S., Masters, P. S. 1982. *J. Interferon Res.* 2:563–74

427. Holmes, S. I., Gupta, S. L. 1982. *Arch. Virol.* 72:137–42

428. Colonno, R. J., Pang, R. H. L. 1982. *J. Biol. Chem.* 257:9234–37

429. Samuel, C. E., Knutson, G. S., Berry, M. J., Atwater, J. A., Lasky, S. R. 1986. *Methods Enzymol.* 119:499–516

430. Atwater, J. A., Samuel, C. E. 1982. *Virology* 123:206–11

431. Galabru, J., Hovanessian, A. G. 1985. *Cell* 43:685–94

432. Hovanessian, A. G., Kerr, I. M. 1979. *Eur. J. Biochem.* 93:515–26

433. Kimchi, A., Zilberstein, A., Schmidt, A., Shulman, L., Revel, M. 1979. *J. Biol. Chem.* 254:9846–53

434. Sen, G. C., Taira, H., Lengyel, P. 1978. *J. Biol. Chem.* 253:5915–21

435. Berry, M. J., Samuel, C. E. 1985. *Biochem. Biophys. Res. Commun.* 133:168–75

436. Laurent, A. G., Krust, B., Galabru, J., Svab, J., Hovanessian, A. G. 1985. *Proc. Natl. Acad. Sci. USA* 82:4341–45

437. Bischoff, J. R., Samuel, C. E. 1985. *J. Biol. Chem.* 260:8237–39

438. Krust, B., Galabru, L., Hovanessian, A. G. 1984. *J. Biol. Chem.* 259:8494–98

439. Kitajewski, J., Schneider, R. J., Safer, B., Munemitsu, S. M., Samuel, C. E. et al. 1986. *Cell* 45:195–200

440. O'Malley, R. P., Mariano, T. M., Siekierka, J., Mathews, M. B. 1986. *Cell* 44:391–400

441. Munemitsu, S. M., Berry, M. J., Kitajewski, J., Shenk, T., Samuel, C. E. 1986. See Ref. 326, pp. 87–100

442. Whitaker-Dowling, P., Youngner, J. S. 1983. *Virology* 131:128–36

443. Rice, A. P., Kerr, I. M. 1984. *J. Virol.* 50:229–36

444. Paez, E., Esteban, M. 1984. *Virology* 134:12–28

445. Galliot, B. R., Commoy-Chevalier, M. J., Georges, P., Chany, C. 1985. *J. Gen. Virol.* 66:1439–48

446. Arakawa, T., Parker, C. G., Lai, P-H. 1986. *Biochem. Biophys. Res. Commun.* 136:679–84

447. Crouch, D., Safer, B. 1980. *J. Biol. Chem.* 255:7918–24

448. Stewart, W. E. II, Schellekens, H., eds. 1986. *The Biology of the Interferon System, 1985.* Amsterdam: Elsevier

449. Samuel, C. E. 1982. *Tex. Rep. Biol. Med.* 41:463–70

450. Pfeffer, I. M. 1986. *Mechanisms of Interferon Actions.* Cleveland: Chem. Rubber Co. Press

451. Kingsman, S. M., Samuel, C. E. 1980. *Virology* 101:458–65

452. Masters, P. S., Samuel, C. E. 1982. *J. Biol. Chem.* 258:12026–33

453. Ulker, N., Samuel, C. E. 1985. *J. Biol. Chem.* 260:4319–23

454. Whitaker-Dowling, P., Youngner, J. S. 1986. *Virology* 152:50–57

455. Rice, A. P., Roberts, W. K., Kerr, I. M. 1984. *J. Virol.* 50:220–28

456. Samuel, C. E., Ulker, N., Knutson, G.

S., Zhang, X., Masters, P. S. 1985. See Ref. 31, pp. 131–140

457. Haller, O. 1981. *Curr. Top. Microbiol. Immunol.* 92:25–52

458. Weil, J., Epstein, C. J., Epstein, L. B., van Blerkom, J., Xuong, N. H. 1983. *Antiviral Res.* 3:303–14

459. Guardini, M. A., Schoenberg, M. P., Naso, R. B., Martin, B. A., Gutterman, J. U., et al. 1984. *J. Interferon Res.* 4:67–79

460. Weil, J., Epstein, C. J., Epstein, L. B., Sedmak, J. J., Sabra, J. L., et al. 1983. *Nature* 301:437–39

461. Chebath, J., Merlin, G., Metz, R., Benech, P., Revel, M. 1983. *Nucleic Acids Res.* 11:1213–26

462. Larner, A. C., Jonak, G., Cheng, Y-S. E., Kornat, B., Knight, E., et al. 1984. *Proc. Natl. Acad. Sci. USA* 81:6733–37

463. Engel, D. A., Samanta, H., Brawner, M. E., Lengyel, P. 1985. *Virology* 142:389–97

464. Kusari, J., Sen, G. C. 1986. *Mol. Cell. Biol.* 6:2062–67

465. Wathelet, M., Moutschen, S., Defilippi, P., Cravador, A., Collet, M., et al. 1986. *Eur. J. Biochem.* 155:11–17

466. Friedman, R. I., Manly, S. P., McMahon, M., Kerr, I. M., Stark, G. R. 1984. *Cell* 38:745–55

467. Blomstrom, D. C., Fahey, D., Kutny, R., Korant, B. D., Knight, E. 1986. *J. Biol. Chem.* 261:8811–16

468. Larner, A. C., Chaudhuri, A., Darnell, J. E. 1986. *J. Biol. Chem.* 261:453–59

469. Samanta, H., Pravtcheva, D. D., Ruddle, F. H., Lengyel, P. 1984. *J. Interferon Res.* 4:295–300

470. Friedman, R. I., Stark, G. R. 1985. *Nature* 314:637–39

470a. Israel, A., Kimura, A., Fournier, A., Fellous, M., Kourilsky, P. 1986. *Nature* 322:743–46

471. Luster, A. D., Unkeless, J. C., Ravetch, J. V. 1985. *Nature* 315:672–76

472. Staeheli, P., Haller, O., Boll, W., Lindenmann, J., Weissmann, C. 1986. *Cell* 44:147–58

473. Fleischmann, W. R., Georgiades, J. A., Osborne, L. C., Johnson, H. M. 1979. *Infect. Immun.* 26:248–53

474. Dani, C., Mechti, N., Piechaczyk, M., Lebleu, B., Jeanteur, P., et al. 1985. *Proc. Natl. Acad. Sci. USA* 82:4896–99

475. Basham, T. Y., Bourgeade, M. F., Creasey, A. A., Merigan, T. C. 1982. *Proc. Natl. Acad. Sci. USA* 79:3265–69

476. Fellous, M., Nir, U., Wallach, D., Merlin, G., Rubinstein, M., et al. 1982. *Proc. Natl. Acad. Sci. USA* 79:3082–86

477. Wallach, D., Fellous, M., Revel, M. 1982. *Nature* 299:833–36

478. Yoshie, O., Schmidt, H., Reddy, E. S. P., Weissman, S., Lengyel, P. 1982. *J. Biol. Chem.* 257:13169–72

479. Satz, M. L., Singer, D. S. 1984. *J. Immunol.* 132:496–501

480. Burrone, O. R., Milstein, C. 1982. *EMBO J.* 1:345–49

481. Collins, T., Korman, A. J., Wake, C., Boss, J. M., Kappes, D. J., et al. 1984. *Proc. Natl. Acad. Sci. USA* 81:4917–21

482. Yoshie, O., Schmidt, H., Lengyel, P., Reddy, E. S. P., Morgan, W. R., et al. 1984. *Proc. Natl. Acad. Sci. USA* 81:649–53

483. Wong, G. H. W., Clark-Lewis, I., McKimm-Breschkin, J. I., Schrader, J. W. 1982. *Proc. Natl. Acad. Sci. USA* 79:6989–93

484. King, D. P., Jones, P. P. 1983. *J. Immunol.* 131:315–18

485. Goldring, M. B., Sandell, L. J., Stephenson, M. L., Krane, S. M. 1986. *J. Biol. Chem.* 261:9049–56

486. Wong, G. H. W., Bartlett, P. F., Clark-Lewis, I., Battye, F., Schrader, J. W. 1984. *Nature* 310:688–91

487. Koeffler, H. P., Ranyard, J., Yelton, L., Billing, R., Bohman, R. 1984. *Proc. Natl. Acad. Sci. USA* 81:4080–84

488. Capobianchi, M. R., Ameglio, F., Tosi, R., Dolei, A. 1985. *Human Immunol.* 13:1–11

489. Strunk, R. C., Cole, F. S., Perlmutter, D. H., Colten, H. R. 1985. *J. Biol. Chem.* 260:15280–85

490. Tominaga, S-I., Tominaga, K., Lengyel, P. 1985. *J. Biol. Chem.* 260:16406–10

491. Hayashi, H., Tanaka, K., Jay, F., Khoury, G., Jay, G. 1985. *Cell* 43:263–67

492. Schwartz, R. H. 1985. *Ann. Rev. Immunol.* 3:237–61

493. Hood, L., Steinmetz, M., Malissen, B. 1983. *Ann. Rev. Immunol.* 1:529–68

494. Haller, O., Arnheiter, H., Lindenmann, J., Gresser, I. 1980. *Nature* 283:660–62

495. Lindenmann, J., Iane, C. A., Hobson, D. 1963. *J. Immunol.* 90:942–51

496. Staeheli, P., Colonno, R. J., Cheng, Y-S. E. 1983. *J. Virol.* 47:563–67

497. Horisberger, M. A., Staeheli, P., Haller, O. 1983. *Proc. Natl. Acad. Sci. USA* 80:1910–14

498. Horisberger, M. A., Hochkeppel, H. K. 1985. *J. Biol. Chem.* 260:1730–33

499. Staeheli, P., Dreiding, P., Haller, O., Lindenmann, J. 1985. *J. Biol. Chem.* 260:1821–25
500. Dreiding, P., Staeheli, P., Haller, O. 1985. *Virology* 140:192–96
501. Staeheli, P., Haller, O. 1985. *Mol. Cell. Biol.* 5:2150–53
502. Staeheli, P., Pravtcheva, D., Lundin, L.-G., Acklin, M., Ruddle, F., et al. 1986. *J. Virol.* 58:967–69
503. Cheng, Y-S. E., Colonno, R. J., Yin, F. H. 1983. *J. Biol. Chem.* 258:7746–50
504. Staeheli, P., Prochazka, M., Steigmeier, P. A., Haller, O. 1984. *Virology* 137:135–42
505. Prochazka, M., Staeheli, P., Holmes, R. S., Haller, O. 1985. *Virology* 145: 273–79
506. Cheng, Y-S. E., Becker-Manley, M. F., Chow, T. P., Horan, D. C. 1985. *J. Biol. Chem.* 260:15834–39
507. Ghezzi, P., Bianchi, M., Mantovani, A., Spreafico, F., Salmona, M. 1984. *Biochem. Biophys. Res. Commun.* 119: 144–49
508. Ghezzi, P., Saccardo, B., Bianchi, M. 1986. *J. Interferon Res.* 6:251–56
509. Deloria, L., Abbott, V., Gooderham, N., Mannering, G. J. 1985. *Biochem. Biophys. Res. Commun.* 131:109–14
510. Yoshida, R., Imanishi, J., Oku, T., Kishida, T., Hayaishi, O. 1981. *Proc. Natl. Acad. Sci. USA* 78:129–32
511. Pfefferkorn, E. R., Rebhun, S., Eckel, M. 1986. *J. Interferon Res.* 6:267–79
512. Tsujimoto, M., Yip, Y. K., Vilcek, J. 1986. *J. Immunol.* 136:2441–44
513. Tsujimoto, M., Vilcek, J. 1986. *J. Biol. Chem.* 261:5384–88
513a. Dianzani, F., Baron, S. 1981. *Methods Enzymol.* 78:409–14
514. Faltynek, C. R., McCandless, S., Chebath, J., Baglioni, C. 1985. *Virology* 144:173–80
515. Kelly, J. M., Gilbert, C. S., Stark, G. R., Kerr, I. M. 1985. *Eur. J. Biochem.* 153:367–71
516. McMahon, M., Stark, G. R., Kerr, I. M. 1986. *J. Virol.* 57:362–66
517. Sugita, K., Miyazaki, J., Appella, E., Ozato, K. 1987. Submitted for publication
517a. Levy, D., Larner, A., Chaudhuri, A., Babiss, L. E., Darnell, J. E. Jr. 1986. *Proc. Natl. Acad. Sci. USA* 83:8929–33
518. Einat, M., Resnitzky, D., Kimchi, A. 1985. *Nature* 313:597–600
519. Jonak, G. J., Knight, E. 1984. *Proc. Natl. Acad. Sci. USA* 81:1747–50
520. Knight, E., Anton, E. D., Fahey, D., Friedland, B. K., Jonak, G. J. 1985. *Proc. Natl. Acad. Sci. USA* 82:1151–54
521. Samid, C., Cheng, E. H., Friedman, R. M. 1984. *Biochem. Biophys. Res. Commun.* 119:21–28
522. Soslau, G., Bogucki, A. R., Gillespie, D., Hubbell, H. R. 1984. *Biochem. Biophys. Res. Commun.* 119:941–48
523. Einat, M., Resnitzky, D., Kimchi, A. 1985. *Proc. Natl. Acad. Sci. USA* 82: 7608–12
524. Kohase, M., Henriksen-DeStefano, D., May, L. T., Vilcek, J., Sehgal, P. B. 1986. *Cell* 45:659–66
525. Resnitzky, D., Yarden, A., Zipori, D., Kimchi, A. 1986. *Cell* 46:31–40
526. Zullo, J. N., Cochran, B. H., Huang, A. S., Stiles, C. D. 1985. *Cell* 43:793–800
527. Stephenson, M. I., Krane, S. M., Amento, E. P., McCroskery, P. A., Byrne, M. 1985. *FEBS Lett.* 180:43–50
528. Rosenbloom, J., Feldman, G., Freundlich, B., Jimenez, S. A. 1984. *Biochem. Biophys. Res. Commun.* 123:365–72
529. Sehgal, P. B., Dobberstein, B., Tamm, I. 1977. *Proc. Natl. Acad. Sci. USA* 74:3409–13
530. Raj, N. B. K., Pitha, P. M. 1981. *Proc. Natl. Acad. Sci. USA* 78:7426–30
531. Ohno, S., Taniguchi, T. 1981. *Proc. Natl. Acad. Sci. USA* 78:5305–9
532. Hauser, H., Gross, G., Bruns, W., Hochkeppel, H.-K., Mayr, U., Collins, J. 1982. *Nature* 297:650–54
533. Pitha, P. M., Ciufo, D. M., Kellum, M., Raj, N. B. K., Reyes, G. R., Hayward, G. S. 1982. *Proc. Natl. Acad. Sci. USA* 79:4337–41
534. Tavernier, J., Gheysen, D., Duerinck, F., Van der Heyden, J., Fiers, W. 1983. *Nature* 301:634–36
535. Maroteaux, L., Kahana, C., Mory, Y., Groner, Y., Revel, M. 1983. *EMBO J.* 2:325–32
536. Mitrani-Rosenbaum, S., Maroteaux, L., Mory, Y., Revel, M., Howley, P. M. 1983. *Mol. Cell. Biol.* 3:233–40
537. Zinn, K., DiMaio, D., Maniatis, T. 1983. *Cell* 34:865–79
538. Ohno, S., Taniguchi, T. 1982. *Nucleic Acids Res.* 10:967–77
539. Fujita, T., Ohno, S., Yasumitsu, H., Taniguchi, T. 1985. *Cell* 41:489–96
540. Goodbourn, S., Zinn, K., Maniatis, T. 1985. *Cell* 41:509–20
541. Ragg, H., Weissmann, C. 1983. *Nature* 303:439–42
542. Weidle, U., Weissmann, C. 1983. *Nature* 303:442–46
543. Enoch, T., Zinn, K., Maniatis, T. 1986. *Mol. Cell. Biol.* 6:801–10
544. Zinn, K., Maniatis, T. 1986. *Cell.* 45:611–18

545. Goodbourn, S., Burstein, H., Man-
     niatis, T. 1986. *Cell.* 45:601–10
546. Quesada, J. R., Gutterman, J. U.,
     Hersh, E. M. 1986. *Cancer* 57:1678–80
547. Strander, H. 1986. *Adv. Cancer Res.*
     46:1–265
548. Borden, E. 1987. *Pharmacol. Ther.* In
     press
549. Merigan, T. C. 1981. *Interferons* 3:133–
     54
550. Miescher, P. A., Jaffe, E. R. 1986.
     Semin. Hematol. 23(No. 3, Suppl. 1):1–
     37
551. Krown, S. E., Real, F. X., Vadhan-Raj,
     S., Cunningham-Rundles, S., Krim,
     M., et al. 1986. *Cancer* 57:1662–65
552. Bunn, P. A., Ihde, D. C., Foon, K. A.
     1986. *Cancer* 57:1689–95
553. Neidhart, J. A. 1986. *Cancer* 57:1696–
     99

Ann. Rev. Biochem. 1987. 56:779–827

ras GENES[1]

Mariano Barbacid

Developmental Oncology Section, Frederick Cancer Research Facility, Frederick, Maryland 21701

CONTENTS

PERSPECTIVES AND SUMMARY

Scientists attempting to establish the molecular basis of neoplasia have repeatedly encountered the members of a small gene family known as *ras*. This acronym is derived from the words *rat sarcoma* because these genes were first identified as the transforming principle of the Harvey and Kirsten strains of rat sarcoma viruses (1, 2), two acute transforming retroviruses generated by transduction of the rat H-*ras*-1 and K-*ras*-2 cellular genes, respectively (3, 4). The H-*ras*-1 locus has been transduced into retroviruses on at least two additional occasions during the generation of the Rasheed strain of rat sarcoma virus (5) and the BALB strain of mouse sarcoma virus (6). *ras* genes were inadvertently "rediscovered" when scientists, using gene transfer assays, established the existence of dominant oncogenes in human (7–10) and carcinogen-induced animal tumors (11–14). Most of these transforming genes have been identified as mutated alleles of cellular *ras* genes (11–18). More recently, *ras* oncogenes have been found in tumors induced by retroviruses that lack oncogene sequences. These viruses can integrate in the vicinity of cellular proto-oncogenes, disrupting their normal regulatory elements (19). Activation of *ras* genes by retroviral insertional mutagenesis has recently been described in tumors of avian and mammalian origin (20, 21; J. Ihle, personal communication).

    ras genes have been the focus of intense research since 1982 when their transforming alleles were first identified in human tumors. Unveiling the role of *ras* oncogenes in neoplastic development should have a major impact on our understanding of the pathogenesis of human cancer. However, research on *ras* genes is justified on its own merits. *ras* genes are likely to play a fundamental role in basic cellular functions based on their high degree of conservation throughout eukaryotic evolution. Independently of their phylogenetic origin, they code for proteins that bind guanine nucleotides (22–25), have GTPase activity (25–29) and are associated with the plasma membrane (30–32). These properties, along with their significant sequence homology with G proteins (33–36), suggest that *ras* proteins may participate in the transduction of signals across the cellular membrane.

    Evolution has preserved their basic biological functions. *ras* proteins of yeast can transform mammalian cells (37) and mammalian *ras* proteins can support the growth of mutant yeast cells (37, 38). Yet, *ras* proteins appear to participate in multiple, often distinct, biological processes. In mammals, they have been implicated in cellular proliferation (39) and terminal differentiation (40–42). In yeast, they can be required for survival (43, 44) or mating (45). Therefore, it is likely that *ras* proteins may function at a critical crossroads of signal transduction pathways. The study of *ras* genes and their oncogenic alleles is certainly a most fascinating field that has attracted scientists interested in disciplines as diverse as yeast sporulation, control of cell proliferation, neural differentiation, carcinogen-DNA interactions, and human cancer.

During the last couple of years, reviews dealing with the structure and biochemical properties of *ras* proteins (46) and with the involvement of *ras* genes in human cancer (47–49) and in carcinogen-induced tumors (50) have been published. In this review, I have compiled current information on the different areas of *ras* gene research in an attempt to convey an overall view of the fundamental role that *ras* genes play in normal and neoplastic cellular processes.

PRIMARY STRUCTURE

ras genes are an ubiquitous eukaryotic gene family. They have been identified in mammals, birds, insects, mollusks, plants, fungi, and yeasts. Sequence analysis of these genes and their products has revealed a high degree of conservation, which suggests that they may play a fundamental role in cellular proliferation. In this section, I present an overview of the most significant structural features of this gene family.

Mammalian ras *Genes*

To date, three *ras* genes have been identified in the mammalian genome (3, 4, 15–17, 51, 52). They have been designated H-*ras*-1, K-*ras*-2, and N-*ras*. Two pseudogenes, H-*ras*-2 and K-*ras*-1, have been identified and characterized in rats and humans (3, 4, 53, 54) and are likely to exist in most, if not all, mammals. Several mouse and hamster subspecies possess additional *ras* pseudogenes, probably due to a relatively recent germ line amplification (55). Each of the three functional *ras* genes has been cloned and sequenced in at least two mammalian species. They include human H-*ras*-1 (10, 56–59), K-*ras*-2 (54, 60, 61), and N-*ras* (51, 52, 62, 63); rat H-*ras*-1 (3, 12, 64) and K-*ras*-2 (exons I and II only) (65); and mouse K-*ras*-2 (66) and N-*ras* genes (67, 68). Their location in both human and rodent chromosomes has also been determined (reviewed in 69). N-*ras* has been assigned to the short arm of human chromosome 1 (1p22–p32), whereas H-*ras*-1 and K-*ras*-2 have been assigned to the short arms of chromosomes 11 (11p15.1–p15.5) and 12 (12p12.1–pter), respectively. The chromosomal location of the two human *ras* psuedogenes (H-*ras*-2 maps in the X chromosome and K-*ras*-1 in 6p12–

p23) is also known. In the mouse, H-*ras*-1 has been mapped in chromosome 7, whereas K-*ras*-2 and N-*ras* have been assigned to chromosomes 6 and 3, respectively. Finally, in rats, the H-*ras*-1 gene has been mapped in chromosome 1, K-*ras*-2 in chromosome 4, and the H-*ras*-2 pseudogene in the X chromosome (70).

The three functional *ras* genes code for highly related proteins generically known as p21 (71). The p21 coding sequences of each of these genes are equally distributed in four exons except for the K-*ras*-2 gene, which possesses two alternative fourth coding exons (exons IVA and IVB) that allow the synthesis of two isomorphic p21 proteins of 188 and 189 residues that differ in their carboxy terminal domains (54, 60, 72). Although the spliced junctions of all mammalian *ras* genes correspond precisely, suggesting a common origin from one ancestral gene, their intron structures vary greatly. As a consequence, *ras* genes exhibit distinct genetic complexities ranging from the 4.5 kbp size of H-*ras*-1 to the 50 kbp of K-*ras*-2. Mammalian *ras* genes contain an additional 5' noncoding exon (54, 60, 61, 73) located immediately downstream from their respective promoters (74, 74a). These promoters do not possess the characteristic TATA and CAT boxes commonly found in other eukaryotic genes. Instead, they are rich in G/C boxes, which are presumably involved in the binding of Sp1 proteins, a characteristic of the promoters of housekeeping genes (75).

Comparison of the deduced amino acid sequences of mammalian H-*ras*-1, K-*ras*-2, and N-*ras* p21 proteins has helped to define four domains within these molecules (Figure 1A). The first domain encompasses the amino terminal third of p21 proteins and is a highly conserved region. For instance, the first 85 amino acid residues of mammalian p21 *ras* proteins of known sequence (human H-*ras*-1, K-*ras*-2, and N-*ras,* rat H-*ras*-2, and mouse K-*ras*-2 and N-*ras* genes) are identical (Figure 1A). The next 80 amino acid residues define a second domain where the structures of the different mammalian p21 *ras* proteins diverge slightly from each other (85% homology between any pair of human *ras* genes). A highly variable region encompasses the rest of the molecule except for the last four amino acids, where the sequence Cys[186]-A-A-X-COOH (where A is any aliphatic amino acid) is present in all members of the *ras* gene family (Figure 1A).

ras *Genes of Other Eukaryotes*

ras genes have been highly conserved during evolution (76). They have been identified in chickens (H-*ras* gene) (20), fruit flies (*Drosophila melanogaster* Dras1, Dras2/64B and Dras3 genes) (77–79), mollusks (*Aplysia* Apl-*ras* gene) (80), slime molds (*Dictyostelium discoideum* Ddras gene) (81), plants (*Allium cepa*) (L. Serrano, J. Avila, personal communication), and yeasts (*Saccharomyces cerevisiae* RAS1 and RAS2 genes and *Schizosaccharomyces pombe SPRAS* gene) (82–84). Comparative analysis of the deduced amino acid sequence of the products of these *ras* genes with mammalian p21 *ras* proteins shows a high degree of homology and the same structural domains (Figure 1). Although some of the *ras* gene products of invertebrate species have additional amino terminal residues, they are at least 84% homologous to the highly conserved amino terminal domain of mammalian p21 *ras* proteins. This homology decreases substantially in the second, less conserved domain, and completely disappears in the variable carboxy terminal region with the exception of the conserved carboxy terminal Cys-A-A-X-COOH sequence.

The most striking property of the evolutionary conservation of *ras* genes is their ability to function in heterologous systems. Mammalian *ras* genes under the appropriate control of yeast promoters can complement nonviable $ras1^-ras2^-$ yeast mutants (37, 38). Moreover, mammalian *ras* oncogenes can induce phenotypic alterations in yeast cells (66, 85,). Similarly, chimeric yeast-mammalian *ras* genes and a yeast RAS gene that carries a deletion in its long hypervariable domain are able to efficiently transform mouse NIH3T3 cells in gene transfer assays (37). These results represent the first report of interchangeability between functional genes of yeast and mammals and are the best example to illustrate the high degree of conservation of *ras* genes during evolution.

ras *Oncogenes*

Mammalian *ras* genes acquire transformation-inducing properties by single point mutations within their coding sequences (86–88). Mutations in naturally occurring *ras* oncogenes have been localized in codons 12 (86–88), 13 (89), 59 (90, 91), and 61 (62, 63, 92, 92a). In vitro mutagenesis studies have shown that mutations in codons 63 (93), 116 (94), and 119 (95) can also confer transforming properties to *ras* genes. Missense mutations in the corresponding codons of the *ras* genes of *S. cerevisiae* also induce pronounced phenotypic changes that will be discussed in the section on yeast RAS genes.

The presence of a glycine residue at position 12 appears to be necessary for the normal function of *ras* proteins. Substitution of Gly^{12} by any other amino acid residue (with the exception of proline) results in the oncogenic activation of these molecules (96). A similar affect is observed if Gly^{12} is deleted or if additional amino acids are inserted between Ala^{11} and Gly^{12} (97). Substitution of the neighboring amino acid, Gly^{13}, also has transforming consequences for the harboring cells, although in this case not all substitutions appear to have the same activating effect (89, 93). Whereas Val^{13} and Asp^{13} substitutions clearly yield *ras* oncogenes, replacement of Gly^{13} by Ser^{13} has little effect on the transforming activity of *ras* proteins (93). Miscoding mutations in the domain surrounding codon 61 also play a very important role in the generation of *ras* oncogenes (62, 63, 92). Substitution of Gln^{61} by any

A

Amino acid sequence alignment of ras and ras-related proteins (residues 1–189).

```
                          1                   20                  40                  60                  80                  100
HUMAN/RAT   H-ras-1       MTEYKLVVVGAGGVGKSALTIQLIQNHFVDEYDPTIEDSYRKQVVIDGETCLLDILDTAGQEEYSAMRDQYMRTGEGFLCVFAINNTKSFEDIHQYREQI
CHICKEN     H-ras-1       ----------------------------------------------------------------------------------------------------
HUMAN       K-ras-2A      -----------------------------------------------------------------E------S----------------------------H
MOUSE       K-ras-2A      -----------------------------------------------------------------E------S----------------------------H
HUMAN       K-ras-2B      -----------------------------------------------------------------E------S----------------------------H
MOUSE       K-ras-2B      -----------------------------------------------------------------E------S----------------------------H
HUMAN       N-ras         ------------------------------------RS------------------------------------------------S--A--NL--------
MOUSE       N-ras         ------------------------------------RS------------------------------------------------S--A--NL--------
Drosophila  Dras1         MQ-QT-------G-------F--SY--TD--------------------------------L-------SA----GT-------------------------
Drosophila  Dras2/64B     ------------I---G-----------------------S--D---------------F----E------S-------L--L-DHS--DE-PKFQR------
Dyctiostelium Ddras       -------------I---G------------------------------DKVSI-----------------Q------YS-TSRS-YDE-ASF-----------
S.cerevisiae RAS1         MQGNKSTIR-------G------F-SY-----------------------D-VSI----------E------N----L-YSVTSRN--DELLS-YQ-------
S.cerevisiae RAS2         MPLNKSNIR-------------T-S----------------------------------------E------N----L-YS-TSKS-LDELMT-YQ-------
S.pombe     SPRAS         MRSTYLR-------D-----------S---------------KCE----GA---V-----------------E-------L-YN-TSRS--DE-STFYQ----
```

```
                          120                 140                 160                 180                 189
HUMAN/RAT   H-ras-1       KRVKDSDDVPMVLVGNKCDLAARTVESRQAQDLARSYGIPYIETSAKTRQGVEDAFYTLVREIRQHKLRKLNPPDESGPGCMSCK              CVLS
CHICKEN     H-ras-1       -----------------P-----T---------------------------------------------------N---                      -I-
HUMAN       K-ras-2A      -------E------------PS---DTK--------R----------------------------YR-K-ISKEEKTPGCVKIK-                 -IM
MOUSE       K-ras-2A      -------E------------PS---DTK--------R----------------------------YR-K-ISKEEKTPGCVKIK-                 -IM
HUMAN       K-ras-2B      -------E------------PS---DTK----E------------D---------------------K-EKMSKDGKKKKKKSK  T-               -IM
MOUSE       K-ras-2B      -------E------------PS---DTK----E------------D---------------------K-EKMSKDGKKKKKKSR TR                -TVM
HUMAN       N-ras         -----------------PT---DTK----HE--K----------------------------YRMK---SS-DGTQ---GLP                    -VM
MOUSE       N-ras         -----------------PT---DTK----HE--K----------------------------YR-K---SS-DGTQ---GSP                    -M
Drosophila  Dras1         -H--AEE------A-----SWN-NNE--REV-KQ--------------M--D--------KD-DN-GRRGRKMNKPNCRF-                     KML
Drosophila  Dras2/64B     L---K-REF--LM----KHQQQV-LEE--NTS-NLM----C--L-VN-DQ--HE---IV-KFQIAERPFIEQDYKKGGKR-                    -C-M
Dyctiostelium Ddras       L---K-R--LI----A--DHERQV-VNEG-E--KDSLS  FH-S--S-IN--E---S----KELKGDQSSGKAQKKKKQ                      -LIL
S.cerevisiae RAS1         Q-----YI-V--V---L--ENERQV-YEDGLR--KQLNA-FL----QAIN-DE--S-1-LV-DDGKYNSMNRQLDNTNEIRD(111 aa)            -IIC
S.cerevisiae RAS2         L---T-Y--I-V---S--ENEKQV-Y-DGLNM-KQMNA-FL----QAIN--E----A-LV-DEGGKYNKTLT-NDNSKQTSQ(114 aa)            -II-
S.pombe     SPRAS         L---K-TF-V---A---E-ER-V--REGEQ--K-MHCL-V----L-LN--E---S---T--RYNKSEEKGFQNKQAVQIAQV( 24 aa)            -IC
```

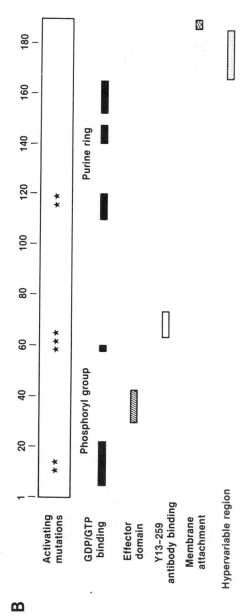

Figure 1 A. Comparative amino acid sequence of *ras* proteins. Residues identical to the human H-*ras*-1 gene product are designated by a dash. *B*. Schematic representation of the structural and functional domains defined within mammalian p21 *ras* proteins.

other amino acid residue, except Pro[61] or Glu[61] (and to a lesser extent Gly[61]), yields *ras* oncogenes (98). Substitutions in residue 59 have only been observed in retroviral (90, 91) or in vitro mutagenized *ras* oncogenes (93). In all cases, substitution of the normal Ala[59] by Thr[59] has been observed. The effect of these mutations on the structure and biochemical properties of *ras* proteins will be discussed in the following section.

Whereas all cellular *ras* oncogenes carry a single activating mutation, each of the four known retroviral *ras* oncogenes exhibits two mutations. The *ras* oncogenes of the Harvey and Kirsten strains have replaced both Gly[12] and Ala[59] residues by Arg[12] and Thr[59] or Ser[12] and Thr[59], respectively (90, 91). The BALB strain exhibits two G → A transitions in codon 12, leading to a Gly[12] (GGA) → Lys[12] (AAA) substitution (99). Finally, the *ras* gene product of the Rasheed strain, p29 *ras*, carries a Gly[12] → Arg[12] substitution as well as 59 additional amino terminal residues derived from the helper virus *gag* gene p15 protein (5). The biological significance of these double mutations is not clear. It could be argued that they reflect selection of those viruses with more malignant properties. However, *ras* oncogenes carrying mutations in codons 12 and 59 are no more transforming than those carrying either mutation alone (93). Moreover, in vitro studies have shown that substitution of Gly[12] by Lys[12] (as seen in BALB-MSV) generates a weakly transforming *ras* oncogene (96). Therefore, it is possible that the observed secondary mutations in retroviral *ras* oncogenes may be an evolutionary safeguard against excessive transforming properties.

ras-*Related Genes*

Emerging evidence suggests that *ras* genes are members of a super gene family. Genes exhibiting limited sequence homology to the *ras* gene family have been identified in a variety of eukaryotic organisms. For instance, a new class of genes, designated *rho*, code for proteins that share 30–40% homology with *ras* p21 proteins (100). *rho* genes have been identified in a variety of species, including such highly diverged organisms as snails *(Aplysia)* and humans. The human and *Aplysia rho* genes share 85% sequence homology, indicating that they are as well conserved in the phylogenetic scale as the *ras* genes (100). At least four additional *ras*-related genes that share 35–55% sequence homology with the *ras* gene family have been identified in humans and rodents (R-*ras*) (100a) primates *(ral)* (100b), fruit flies (D*ras*3) (79), and yeast (YP2) (101). The YP2 yeast gene has been recently shown to be involved in microtubule organization (101a). Whether any of these genes are members of evolutionarily conserved gene families remains to be determined. Finally, some of these *ras*-related genes share significant sequence homology with at least one of the two domains (codons 12–13 and 59–61) frequently involved in the malignant activation of *ras* oncogenes. Whether any of these *ras*-related genes can become oncogenes, either in vivo or by in vitro manipulation, remains to be determined.

BIOCHEMICAL PROPERTIES

ras proteins, independently of their phylogenetic origin, have been shown to bind guanine nucleotides (GTP and GDP) (22–25) and possess intrinsic GTPase activity (25–29). The relevance of these activities to the biological function of *ras* proteins has been demonstrated by three independent lines of evidence: (*a*) microinjection of anti-*ras* antibodies that inhibit guanine nucleotide binding (102) reverses the malignant phenotype of NIH3T3 cells transformed by *ras* oncogenes (103); (*b*) *ras* mutants that have lost their ability to bind guanine nucleotides do not transform NIH3T3 cells (104, 105); and (*c*) the GTPase activity of *ras* genes is severely impaired in their transforming alleles (25–29). In addition to GTP/GDP binding and GTPase activity, *ras* proteins carrying an $Ala^{59} \rightarrow Thr^{59}$ mutation exhibit an autophosphorylating activity of an, as yet, unknown biological significance (23). In all cases, Thr^{59} has been found to be the phosphate receptor site (106). No transphosphorylating activity has been detected with any *ras* protein, including those carrying Thr^{59} mutations.

The biochemical properties of *ras* proteins closely resemble those of the G proteins involved in the modulation of signal transduction through transmembrane signaling systems (107). In fact, certain domains of *ras* proteins exhibit significant sequence homology with the α subunit of G proteins such as G_s, a protein that activates adenylate cyclase in response to β adrenergic stimuli; G_i, which inhibits this enzyme and perhaps activates phospholipase C; G_o, a protein of as yet unknown function; and transducin, a protein that regulates cGMP phosphodiesterase activity in visual signal transduction (33–36). In addition to G proteins, other nucleotide-binding proteins such as the bacterial elongation factor T_u (EF-T_u), the β subunit of ATP-synthase, adenylatekinase, phosphofructokinase, and tubulin also exhibit certain sequence homology to *ras* proteins (108–110).

In this section, the biochemical properties of *ras* proteins will be discussed in conjunction with genetic studies that have made it possible to assign certain functions to specific domains within these molecules.

Guanine Nucleotide Binding

The homology of *ras* genes with G proteins and EF-Tu is basically limited to regions encompassing amino acid residues 5–22 and 109–120 (108, 109). Direct experimental evidence implicating these domains in guanine nucleotide binding has been obtained recently. Antibodies directed against epitopes located within the amino terminal region of *ras* proteins inhibit GTP binding by purified p21 *ras* proteins (102). Conversely, the ability of these antibodies to bind to *ras* proteins is inhibited by preincubation with GTP or GDP (102). X-ray crystallography studies of the GDP binding domain of EF-Tu also support the concept that residues around codon 12 form part of the guanine nucleotide–binding site (109). Based on EF-Tu and *ras* sequence homology, the 12th amino acid residue (Gly) of mammalian *ras* proteins should be located in the phosphoryl binding loop (109, 111). Early studies have predicted that replacement of Gly^{12} by any other amino acid residue (except proline) would disrupt the α-helical structure of the amino terminal domain of *ras* proteins, causing a conformational change that would prevent its proper folding (112–114). Thus, replacement or elimination of Gly^{12} may create a rigid domain that cannot efficiently interact with the phosphoryl region of the GTP molecule, reducing the GTPase activity of *ras* proteins. Two additional residues in this domain, Gly^{15} and Lys^{16}, are present in other guanine nucleotide–binding proteins (109, 111). Substitution of Lys^{16} by Asn^{16} significantly reduces GTP/GDP affinity without affecting base specificity, an observation consistent with the idea that these residues are also part of the phosphoryl group (95).

Crystallographic studies of the $GDP \cdot EF\text{-}T_u$ complex predict that the guanine-binding pocket may interact with two noncontiguous segments of $EF\text{-}T_u$ that are present in *ras* proteins (Asn^{116}-Lys-Cys-Asp^{119} and Ser^{145}-Ala-Lys^{147}) (109, 111). *ras* deletion mutants within residues 109–120 or 130–145 do not bind detectable levels of guanine nucleotides and cannot induce efficient transformation of NIH3T3 cells (104). Moreover, two of three GTP-binding-defective *ras* mutants isolated by random mutagenesis ($Asp^{119} \rightarrow Asn^{119}$ and $Thr^{144} \rightarrow Ile^{144}$) map within these regions (115). Direct biochemical evidence supporting the concept that these regions interact with the guanine ring has been obtained by introducing mutations in residues 116 and 119 (94, 95). Substitution of normal Asn^{116} by a variety of amino acids including isoleucine, lysine, tyrosine, and histidine decreases the binding of guanine nucleotides by several orders of magnitude to below detection levels (94, 115a, 115b). However, only a 10-fold decrease in binding was observed if asparagine was replaced by a related amino acid residue such as glutamine (94). These observations are consistent with the prediction from $EF\text{-}Tu \cdot GTP$ crystallographic studies that Asn^{116} may form a hydrogen bond with the O^6 residue of guanine (116). Substitution of Asp^{119} by alanine leads to a 50-fold increase in the dissociation rate of guanine nucleotides (95). In contrast, the $Asp^{119} \rightarrow Ala^{119}$ mutation has no effect on the dissociation rate of inosine diphosphate, a molecule identical to GDP except for the absence of the amino residue in position 2. These results suggest that Asp^{119} may form a hydrogen bond with the $2\text{-}NH_2$ group of guanine nucleotides (95).

The *ras* protein domain encompassing residues 59–63 does not show significant sequence homology to other guanine nucleotide–binding proteins. However, miscoding mutations in residues 59, 61, or 63 result in the oncogenic activation of *ras* genes (62, 90–93). The $Ala^{59} \rightarrow Thr^{59}$ mutation is of particular interest because Thr^{59} is the substrate for the autophosphorylating activity of those *ras* oncogenes in which it has been identified (106). These observations suggest that residue 59 may be adjacent to the γ-phosphate of the guanine nucleotide, thus favoring the phosphorylation of the hydroxyl group of the mutant Thr^{59} residue (111). However, this hypothesis must await genetic studies in which amino acid substitutions other than Thr are introduced at this position. The interaction of Gln^{61} with guanine nucleotide is not well understood. Substitution of Gln^{61} by 17 different amino acid residues invariably results in decreased GTPase activity (25, 117). However, there is no quantitative correlation between the reduction in this enzymatic activity and the extent of transformation induced by these mutants, suggesting that additional factors might play a role in the malignant activation of *ras* proteins (117).

Another domain implicated in guanine nucleotide binding has been identified by in vitro mutagenesis. Deletion of residues 152–165 completely abolishes guanine nucleotide binding and transforming activity (105). This sequence includes Arg^{164}, a residue conserved in all *ras* genes (Figure 1). Substitution of Arg^{164} by Ala^{164} also results in loss of GTP binding activity (105). The lack of homology between this domain and EF-Tu has made it impossible to predict whether these residues in general, or Arg^{164} in particular, are part of the GDP binding site. Independent studies have shown that substitution of *ras* oncogene residues 164 to 174 by the sequence Pro-Asp-Gln does not result in loss of transforming activity (118). Thus, it is possible that the domain surrounding the conserved Arg^{164} residue plays an indirect role in the binding of guanine nucleotides.

The Effector Domain

Extensive work on deletion mutants of mammalian *ras* genes has led to the definition of five noncontiguous domains (residues 5–63, 77–92, 109–123, 139–165, and the carboxyl terminal sequences Cys^{186}-A-A-X-COOH) that are absolutely required for *ras* function (104). Whereas some of these domains have been clearly implicated in guanine nucleotide binding, others might play a role in effector recognition. Amino acid substitutions at positions 35 (Thr → Ala), 36 (Ile → Ala), 38 (Asp → Ala), and 40 (Tyr → Lys) have been shown to reduce the biological effect of *ras* proteins in assays utilizing both mammalian (NIH3T3 focus formation) and yeast (complementation of growth of *RAS1ras2* mutants on nonfermentable carbon sources and stimulation of adenylate cyclase) cells (119). These mutations do not disrupt the known biochemical activities of these molecules, suggesting that residues

35–40 might be implicated in the effector activity of *ras* proteins (119). Additional support for this hypothesis comes from the observation that all indispensable domains of *ras* proteins lie in internal hydrophic regions except for the domain corresponding to residues 30 to 42, which is hydrophilic and presumably located in the external surface of the molecule (104). These findings suggest that this domain may be involved in the interaction of *ras* proteins with their putative cellular targets (Figure 1B). Recently, a $Gln^{43} \rightarrow Arg^{43}$ substitution has been found to be responsible for the temperature-sensitive phenotype of the 371 strain of Kirsten-MSV (119a). Although the structural consequences of this mutation have not been established, these results add further evidence to the importance of this domain for proper *ras* protein function.

Neutralizing Antibodies

One of the reagents most commonly used to characterize *ras* proteins is a rat monoclonal antibody designated Y13–259 (120). This antibody must bind to a highly conserved domain of *ras* proteins as deduced by its ability to recognize the products of each of the mammalian *ras* genes (62, 120) as well as the *ras* proteins of invertebrate species (24, 121, 122). More importantly, this antibody has been recently shown to specifically block the serum-induced mitogenic response of certain cells in culture and to inhibit morphologic transformation induced by mutated *ras* proteins (39, 123). Binding studies utilizing the products of *ras* deletion mutants have localized the epitope recognized by this antibody within residues 70–89 (124) (Figure 1B). More refined analysis has established that the Y13-259 antibody interacts with the side chains of residues Glu^{63}, Ser^{65}, Ala^{66}, Met^{67}, Gln^{70}, and Arg^{73} (119). Substitution of any of these residues completely abolished the interaction of *ras* proteins with the Y13-259 antibody in immunoblotting assays. In contrast, substitution of residues 61, 62, 64, 68, 71, 72, 75, 78, or any other residue outside the 60–80 region of *ras* proteins had no detectable effect on their affinity for the Y13-259 antibody (119).

Residues critical for Y13-259 binding are presumably located in the exposed, hydrophilic site of an α helix that forms a structure about 20 Å long, a distance consistent with the dimensions of an antibody-binding site (119). Binding of Y13-259 to *ras* proteins does not affect any of their known biochemical properties (124). Moreover, the Y13-259 antibody-binding domain lies in a dispensable hydrophilic region (104, 119). Therefore, it is likely that the biological activities observed by binding of Y13-259 antibodies to *ras* proteins are exerted through indirect conformational changes.

Membrane Attachment

ras proteins have been localized in the inner side of the plasma membrane (30–32). The primary translational product of *ras* oncogenes is synthesized in the cytosol (32, 125). Attachment to the plasma membrane requires a post-translational modification that involves the acylation of Cys^{186} by palmitic acid (32, 126–128). Genetic studies have demonstrated that this posttranslational modification is necessary for the biological function of *ras* proteins. Mutants lacking Cys^{186} code for proteins that remain in the cytosol and cannot induce morphological transformation of NIH3T3 cells (31, 129).

Processed *ras* proteins attached to the plasma membrane exhibit a faster migration rate in SDS-polyacrylamide gel electrophoresis than their un-modified cytosolic counterparts (32, 127, 128). It was presumed that this change in mobility was the result of the acylation of the Cys^{186} residue. However, it is likely that an as yet unidentified posttranslational modification may take place prior to fatty acid acylation. In yeast, RAS proteins are converted to forms with faster electrophoretic mobility before the attachment of palmitic acid (32). Studies with mammalian p21 *ras* proteins indicate that removal of lipids does not restore the mobility of the mature protein to that of the precursor form (127). Moreover, the products of *ras* deletion mutants with a slow rate of acylation exhibit an altered electrophoretic mobility in-dependently of whether they contain palmitate (membrane fraction) or not (cytosolic fraction), indicating that the change in electrophoretic migration is not a consequence of acylation (130). Thus, it is likely that *ras* proteins may undergo two steps of posttranslational modification before they become attached to the plasma membrane.

Dispensable Domains

At least five noncontiguous domains of *ras* proteins have been shown to be dispensable for their biological function. They include the amino terminal end, three internal domains, and the hypervariable region. Evidence that the first few amino acid residues are not important for *ras* function comes from the variability in both the type and number of residues found in this region in the different *ras* gene products (Figure 1A). Moreover, H-*ras* p21 proteins in which the first three amino acids were replaced by unrelated residues retained their biological properties (131).

In vitro mutagenesis studies utilizing viral *ras* oncogene DNA clones have indicated that three internal domains (residues 64–76, 93–108, and 124–138 in mammalian proteins) can be deleted without drastically affecting their transforming properties (104). Each of these domains corresponds to hydrophilic regions that are presumed to be located in the external surface of *ras* proteins (104). It is possible that they represent long hinge regions that can be deleted without serious consequences to the overall structure of the mole-cule. However, in vitro mutagenesis studies must be performed with *ras* oncogenes whose products do not require external signals for activation. Therefore, it is conceivable that a putative receptor domain might be found to be dispensable in these assays. Whether any of these domains are involved in receptor signaling remains to be determined.

Deletion of the entire hypervariable region (residues 166–185 in mammalian *ras* proteins, see Figure 1) does not affect any of the known biochemical properties of *ras* proteins nor their ability to transform NIH3T3 cells (105, 118). It is likely that this long stretch of amino acids, which has been maintained throughout evolution, plays an important role in *ras* function. It is possible that *ras* proteins utilize this domain to interact with other proteins. Its intrinsic variability may serve to confer different functional properties to each of the members of this gene family. Development of biological assays for *ras* proteins other than cellular transformation (e.g. complementation of *S. cerevisiae* ras⁻ mutants) might shed light on whether these apparently dispensable domains play any role in the biological function of *ras* proteins.

A Model for the Function of ras Proteins

The biochemical and biological properties of *ras* proteins, along with their strong resemblance to G proteins, have led to the proposal that *ras* proteins may be involved in signal transduction. A schematic diagram of the currently favored model is shown in Figure 2. This model proposes that *ras* proteins exist in equilibrium between an active and an inactive state. Most of the *ras* molecules in a given cell would exist in their inactive state, which is characterized by a conformation that allows binding of GDP. Normal *ras* proteins will remain in their inactive state until they receive a stimulus from another protein (a receptor?) upstream of a putative pathway of signal transduction. This stimulus would result in the exchange of GDP for GTP followed by a conformational change of the *ras* protein to its active state. Active *ras* proteins would then be able to interact with their putative effector molecules. Once the interaction between the active *ras* proteins and the effector has taken place, they would be immediately deactivated. This can be accomplished by their intrinsic GTPase activity, which would catalyze the hydrolysis of GTP returning the active *ras* protein to the inactive GDP-bound state (Figure 2).

This model takes into account that mutations known to confer transforming properties to *ras* genes must reverse the normal equilibrium between the inactive and active forms. Stabilization of *ras* proteins in their active state would cause a continuous flow of signal transduction, which will result in malignant transformation (Figure 2). Theoretically, this process can be achieved by mutations that inhibit the intrinsic GTPase activity of *ras* proteins (Figure 2a), increase the exchange rate between GDP and GTP (Figure 2b), or induce an active conformational change that does not require binding of guanine nucleotides (Figure 2c). This model also takes into account that normal *ras* genes induce malignant transformation if highly overexpressed (see next section). High levels of normal *ras* proteins may produce enough

ras PROTEINS

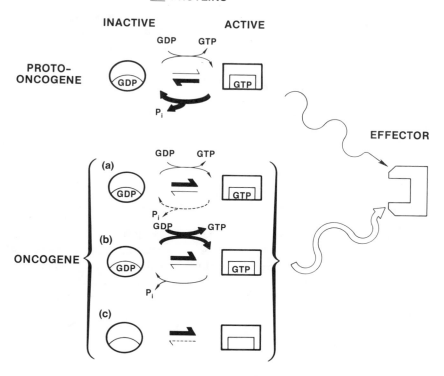

Figure 2   Current model of the mechanism of action of normal and transforming *ras* proteins.

molecules in their GTP-bound active state to induce malignant transformation without affecting the equilibrium between inactive and active forms characteristic of normal *ras* proteins.

Experimental evidence indicates that inhibition of their GTPase activity is the preferred mechanism of activation of oncogenic *ras* proteins (25–29, 117). *ras* oncogenes carrying $Asp^{116} \rightarrow Ile^{116}$ or $Asp^{119} \rightarrow Asp^{119}$ substitutions may exert their transforming properties by mechanisms involving increased dissociation rates of the complex between guanine nucleotides and *ras* proteins (94, 95) (Figure 2*b*). Such an increase in the dissociation rate would result in a significant reduction of the affinity of GDP and GTP for *ras* proteins. This would favor the formation of the active [*ras* protein·GTP] complex due to the higher availability of GTP molecules in the cell. However, in the case of the transforming *ras* gene carrying the $Asp^{116} \rightarrow Ile^{116}$ mutation, its gene product does not bind detectable levels of GTP (94). Whereas this observation might be due to the unusually high dissociation rate (and therefore low affinity) of the [p21·GDP/GTP] complexes, it is conceivable that substitution of aspartic acid by a hydrophobic residue such as isoleucine may induce a conformational change that can activate the *ras* p21 protein without participation of guanine nucleotide binding (Figure 2c).

A discrepancy appears to exist regarding the mechanism by which the Ala[59] → Thr[59] mutation activates *ras* oncogenes. Whereas some authors have reported that these mutants possess low GTPase activity (25, 26), other investigators have found a normal GTPase (132) and increased GDP/GTP dissociation rates (132a). The reasons for these discrepancies remain to be determined.

ras PROTO-ONCOGENES: BIOLOGICAL PROPERTIES

Unveiling the cellular role of *ras* genes requires the use of model systems amenable to genetic analysis. In mammalian cells the extent of genetic manipulation is rather limited. Availability of *ras* DNA clones has allowed a wide range of in vitro mutagenesis studies. However, the only mutants capable of inducing identifiable phenotypes have been the *ras* oncogenes. So far, no experiments involving the genetic manipulation of *Drosophila ras* genes have been reported. Only yeast cells have produced significant genetic information (43–45) that has resulted in the linkage of the *ras* genes to the adenylate cyclase pathway (133). However, these results cannot be extrapolated to other eukaryotic organisms in spite of the conserved structural and functional properties of *ras* genes.

ras *Genes of Yeast*

S. cerevisiae contains two *RAS* loci, *RAS*1 and *RAS*2, that code for proteins of 40,000 (309 amino acids) and 41,000 (322 amino acids) daltons, respectively. *RAS*1 and *RAS*2 are structurally and functionally related to mammalian *ras* genes (82, 83, 134). Genetic analysis of the *RAS*1 and *RAS*2 genes of *S. cerevisiae* has established that neither of them are essential for the viability of these cells (43, 44). Cells containing disruptions in either of these genes can undergo both mitotic and meiotic cell division cycles. However, *ras*1⁻ *ras*2⁻ spores of doubly heterozygous diploids are not capable of vegetative growth (43, 44). These findings clearly illustrate the importance of RAS genes for normal cellular proliferation.

Yeast *RAS* gene products exhibit the same biochemical properties as their mammalian counterparts. They bind GDP and GTP and have an intrinsic GTPase activity (24, 25). RAS proteins carrying Ala[66] → Thr[66] (equivalent to mammalian position 59) and Gln[68] → Leu[68] (equivalent to mammalian position 61) mutations bind normal levels of guanine nucleotides but have an impaired GTPase activity (25). Moreover, the Thr[66] mutant exhibited autophosphorylating activity (25). These findings suggest that *S. cerevisiae* RAS proteins may also function as G proteins according to the model depicted in Figure 2.

Preliminary evidence indicates that *S. cerevisiae* RAS1 and RAS2 genes do not have the same cellular function. Extensive genetic analysis of the RAS2 locus indicates that this gene is necessary for a normal response to nutrient limitation. *ras2⁻* mutants, although viable, have a defect in gluconeogenic growth, accumulate excessive levels of storage carbohydrates, and sporulate prematurely (133, 135, 136). In general, disruption of the RAS2 locus results in an overall premature starvation response. The multiple phenotypes observed in *ras2⁻* mutants may not necessarily be interconnected. For instance, the inability of *ras2⁻* mutants to grow in media containing nonfermentable carbon sources appears to be a consequence of the decrease of RAS1 gene expression under these culture conditions (137). Inhibition of RAS1 gene transcription coupled with the *ras2⁻* mutation may lead to a cell cycle arrest comparable to that observed in *ras1⁻ras2⁻* mutants (43, 44). Support for this hypothesis comes from the observation that suppressors of the *ras2⁻* defect in gluconeogenic growth led to a significant increase in RAS1 mRNA levels (137, 137a). However, direct suppression of the *ras2⁻* phenotype in nonfermentable carbon sources by chimeras capable of constitutively expressing RAS1 proteins remains to be demonstrated. In contrast, the RAS1 product does not seem to play any role in the hypersporulating properties of *ras2⁻* mutants, since the same suppressors that allow these cells to grow on nonfermentable carbon sources have no effect on the hypersporulation phenotype (137). Recent studies have suggested that RAS1 may play an indirect role in the regulation of glucose-induced inositolphospholipid turnover (137b). These observations indicate that the RAS genes of *S. cerevisiae* play different cellular roles in spite of their high sequence homology and genetic complementarity.

Detailed genetic analysis of the RAS2 locus has made it possible to identify the cellular pathways in which this gene is involved. Introduction of a Gly^{19} → Val^{19} substitution (equivalent to mammalian 12 position) in RAS2 results in a dominant gene capable of inducing a series of phenotypic changes in yeast cells (43, 133). RAS2 Val^{19} mutants exhibit a very low sporulating efficiency and can arrest at any phase of the cell cycle upon nutritional stress (43, 133). It it worth noting that this phenotype is the opposite of that induced by elimination of the RAS2 gene (133–136), an observation consistent with the idea that mutations in the critical Gly^{19} codon lead to an enhancement of the normal properties of the RAS2 product. Comparison of the characteristic phenotype of RAS2 Val^{19} mutants with other known yeast mutants has provided fundamental clues regarding the cellular function of the RAS2 gene (133). It was observed that *bcy1*, a suppressor of adenylate cyclase–deficient

(cyr^-) yeasts (138), exhibited an almost identical phenotype to $RAS2$ Val^{19} cells. $bcy1$ is a recessive mutant that affects the regulatory subunit of the cAMP-dependent protein kinase (139). As a consequence, $bcy1$ strains bypass this regulatory step and induce the constitutive activation of the catalytic subunit of this kinase. Genetic crosses between $bcy1$ and several ras mutant strains have conclusively demonstrated that $bcy1$ suppressed the lethality induced by the $ras1^-ras2^-$ genotype (133). Similar results have recently been obtained with an extragenic suppressor mutant designated $sra1$ (137a), which is likely to be an allele of $bcy1$. Other mutations that allow $ras1^-ras2^-$ cells to grow have been identified as alleles of adenylate cyclase that produce elevated levels of cAMP (137a, 139a). These observations indicate that RAS genes must play a role in the adenylate cyclase signal transduction pathway somewhere upstream from the step involved in the regulation of the cAMP-dependent protein kinase activity. As indicated above, adenylate cyclase is regulated by G proteins whose biochemical properties closely resemble those of ras proteins (107). Thus, these studies have raised the interesting possibility that $S.$ $cerevisiae$ RAS genes may be regulatory G proteins that control adenylate cyclase.

Subsequent biochemical studies have supported these genetic observations. $S.$ $cerevisiae$ cells carrying the dominant $RAS2$ Val^{19} mutant exhibit a 4-fold increase in the levels of cAMP (140). In contrast, $RAS1ras2^-$ and $ras1^-ras2^-bcy1$ strains have 4- and 20-fold lower levels of cAMP than wild-type cells, respectively (140). The fact that $ras1^-RAS2$ strains only have slightly depressed cAMP levels adds further support to the concept that $RAS1$ and $RAS2$ genes play different cellular roles. In vitro studies utilizing purified $RAS2$ proteins and crude membrane extracts indicate that the increased levels of cAMP are due to stimulation of adenylate cyclase activity by $RAS2$ proteins (140). This stimulation is dependent on the presence of guanine nucleotides, an expected finding considering the structural and functional homologies between RAS genes and regulatory G proteins. GTP was found to be 50% more efficient than GDP in the $RAS2$-dependent stimulation of cAMP synthesis (140). Moreover, a $RAS2$ Val^{19} protein that has a defective GTPase is about 2-fold more efficient than a wild-type $RAS2$ molecule in stimulating cAMP synthesis. However, if GTP is substituted by a nonhydrolyzable derivative, thus making this reaction independent of the presence of GTPase activity, wild-type $RAS2$ and $RAS2$ Val^{19} mutant proteins exhibit the same high stimulatory effect (140).

Similar experiments were conducted in $Escherichia$ $coli$ cells transformed with $S.$ $cerevisiae$ adenylate cyclase ($CYR1$) and $RAS2$ genes (141). Transformants containing the $CYR1$ gene possess a GTP-independent adenylate cyclase activity, which becomes GTP-dependent upon transformation with the $RAS2$ expression plasmid. Substitution of $RAS2$ by $RAS2$ Val^{19} resulted in both loss of the GTP dependency and increase in the amount of cAMP synthesis. Reconstitution of the GTP-dependent adenylate cyclase was observed by mixing membranes from $CYR1ras1^-ras2^-bcy1$ yeast with *E. coli* cell extracts containing *RAS2* proteins or by mixing membranes of $cyr1RAS1RAS2bcy1$ yeast with *E. coli* extracts containing adenylate cyclase (141). These observations, taken together, indicate that the *RAS2* gene of *S. cerevisiae* participates in the control of adenylate cyclase.

These studies, however, do not resolve whether RAS2 is the regulatory subunit of the catalytic domain of the yeast adenylate cyclase complex. Several lines of evidence argue against this straightforward hypothesis. For instance, yeast cyr^- mutants are viable, whereas $ras1^-ras2^-$ are not. Therefore, *RAS* genes must be involved in additional pathways independent of adenylate cyclase. It is possible that the *RAS2* protein activates adenylate cyclase by interacting with its regulatory subunit, forming a G protein cascade. However, *ras* genes and adenylate cyclase are not linked functionally in other eukaryotic organisms. Mammalian epithelial and fibroblastic cell lines transformed with retroviral *ras* oncogenes exhibit reduced adenylate cyclase activity (142). Moreover, purified p21 *ras* proteins do not activate adenylate cyclase in crude mammalian cell membranes (142). Similarly, human *ras* proteins can induce maturation of frog *(Xenopus)* oocytes without affecting their cAMP levels (143). These observations argue against a direct interaction between *RAS* proteins and the regulatory and/or catalytic subunit of the adenylate cyclase enzymatic complex.

Among those eukaryotic organisms in which *ras* genes do not seem to interact with adenylate cyclase is the fission yeast *S. pombe*. *S. pombe* possesses a single *ras* gene, designated *SPRAS*, which codes for a protein of 219 amino acids whose sequence also conforms to the four structural domains of other *ras* proteins (84) (Figure 1A). The *SPRAS* protein plays a physiological role completely different from those of *S. cerevisiae RAS* products (45, 143a). Disruption of the *SPRAS* locus does not interfere with either growth rates or with the response to nutritional stress conditions. Instead, *S. pombe spras*—mutants completely lose their ability to mate (45, 143a). In addition, these mutations repress rather than stimulate sporulation and they do not affect the intracellular levels of cAMP. So far, there is no genetic or biochemical information regarding the pathways in which the *SPRAS* gene might be involved in *S. pombe*.

The lack of association between *ras* genes and adenylate cyclase in all the eukaryotic organisms studied so far except for the yeast *S. cerevisiae* represents an evolutionary puzzle. It is possible that *ras* proteins function at a pivotal crossroads of signal transduction pathways. If so, selection of one pathway over another may have changed during evolution depending on the particular physiological necessities of each organism. In support of this view is the observation that *S. cerevisiae* depends on its intracellular levels of cAMP to exit from G_1 into the S phase and to switch from G_1 into the sporulation pathway (144). In contrast, there is no evidence for the involvement of cAMP in the control of the life cycle of *S. pombe*. Thus, if *ras* proteins were to be involved in the regulation of cell proliferation it is likely that in *S. Cerevisiae* they do so by interacting with adenylate cyclase. However, other organisms in which cAMP does not play such a role in the control of cell growth may not require that their *ras* products interact with the adenylate cyclase pathway.

ras *Genes of Other Invertebrates*

The function of *ras* genes has been investigated in several invertebrate organisms, including the slime mold *D. discoideum*, the fruit fly *D. melanogaster*, and the mollusk *Aplysia*. The *D. discoideum* Dd*ras* gene was accidentally identified during the course of studies aimed at isolating genes differentially regulated during development (81). A 1.2 kbp Dd*ras* mRNA is expressed at high levels in vegetative cells and disappears at the beginning of differentiation. Two species of Dd*ras* mRNA (0.9 and 1.2 kbp) appear again at the end of the aggregation period (about 12 hours), reaching a maximum after 15 hours, and disappear by 22.5 hours, coinciding with midculmination (81). Protein studies indicate that p23, the Dd*ras* gene product, is also expressed at high levels during vegetative growth and decreases during development (82, 121). Whereas some authors have reported that the decrease in p21 expression is a slow and steady process (82), others have shown a remarkable decrease in the amount of p23 with the onset of differentiation followed by a rise during cellular pseudoplasmodial formation, finally decreasing to the lowest level during the late stages of differentiation (121). These latter results support the concept that Dd*ras* p23 expression correlates with the rate of cell proliferation in *D. discoideum*.

Transformation of *D. discoideum* cells with a series of antisense RNA Dd*ras* constructs linked to a *neo*[R] gene led to a low number of geneticin-resistant survivors, none of which carry the antisense Dd*ras* vectors (145). These results have been interpreted as evidence that *ras* genes are required for cellular proliferation. However, it is not known whether these antisense constructs had any effect on endogenous Dd*ras* expression. More recently, the effect of Gly[19] (equivalent to mammalian Gly[12]) Dd*ras* mutants (Thr[19]) have been shown to induce aberrant developmental phenotypes in this microorganism, suggesting an effect of *ras* genes on cAMP-mediated signal transduction (145a). Unfortunately, none of the biochemical parameters involved in the cAMP-mediated signal transduction pathway appear to be significantly affected by the Dd*ras*-Thr[19] mutant (145a).

In *D. melanogaster*, three D*ras* genes have been identified, mapped, and molecularly cloned (77–79). D*ras*-1 and D*ras*2/64B are the loci most closely related to the human H-*ras*-1 gene, in spite of some discrepancies in the sequence of the latter (77, 78). D*ras*3, an intronless gene, is more distantly related, although chimeric human H-*ras*-1 and D*ras*3 genes were able to transform NIH3T3 cells with low efficiency (79). To date, no genetic studies have been reported suggesting that these loci are located in chromosomal regions not easily accessible to genetic analysis. Each D*ras* locus appears to code for two or three transcripts of different sizes, which are constantly expressed throughout the development of the fly, although the shorter transcripts appear to be more abundant during the early embryonic stages (78, 146). In situ hybridization studies have shown that transcripts from each D*ras* gene exhibit similar distribution at every developmental stage (147). However, a correlation between D*ras* expression and cell proliferation has been observed. Whereas in embryos the D*ras* transcripts are uniformly distributed, in larvae they are restricted to the dividing cells (147).

This picture completely changes in the adult fly. Here, the highest levels of D*ras* RNA are found in fully differentiated, nondividing cells such as the ovaries and the cortex of brain and ganglia (147). Similar results have been recently observed in the marine mollusk, *Aplysia* (80). In this organism, *ras* proteins are most abundantly expressed in nervous tissue, in the ovotestis, and in fertilized eggs. Immunocytochemical analysis of the nervous tissue revealed that *ras* proteins are present in the neuronal cell bodies, as well as in the axons and in the neuropil (80). These studies suggest that *ras* genes may play a role not only in cell proliferation but also in processes involving terminal cell differentiation.

ras *Genes of Mammalian Cells*

The biological function of *ras* genes in mammalian cells is poorly understood. The existence of mutants (oncogenes) capable of inducing transformation-specific phenotypes suggests that these genes may play a role in cell proliferation. However, *ras* oncogenes can induce terminal differentiation of PC12 neural cells (40–42) (see p. 805). Thus, *ras* genes of mammals may also play a dual role in proliferation and in certain differentiation processes.

The difficulty of direct genetic manipulation of *ras* loci within mammalian cells has been partially overcome by microinjection studies using monoclonal antibodies directed against p21 *ras* proteins. Early studies have shown that microinjection of purified mouse *ras* Lys[12], human *ras* Val[12], or retroviral *ras* Arg[12] Thr[59] mutant p21 proteins into NIH3T3 mouse fibroblasts led to transient morphologic transformation and cell proliferation (148–150). Similarly, anti-p21 monoclonal antibodies were able to induce transient reversion of the malignant phenotype of rodent cells transformed by *ras* oncogenes (103, 123). These studies have been taken one step further by examining the effect of Y13-259, a monoclonal antibody capable of recognizing all known *ras* proteins (120), in NIH3T3 cells. Microinjection of this antibody into the cytoplasm of NIH3T3 cells prior to serum stimulation leads to a significant decrease in the number of cells capable of tritiated thymidine uptake (39). These results indicate that cells injected in G_0 were blocked from entering the S phase. Time course experiments suggest that *ras* gene proteins are not required once the cells have entered the S phase. Instead, they appear to be necessary at about eight hours after serum induction, a time that is estimated to coincide with the beginning of the S phase (39). The validity of these observations has been confirmed by showing that NIH3T3 cells transformed by a *ras* mutant that cannot bind the Y13-259 antibody are not prevented from entering the S phase (151). Studies utilizing antibodies specific for each of the mammalian *ras* products may provide additional information on the role of each of these proteins in the proliferation of mammalian cells.

The location of *ras* proteins in the inner surface of the cell membrane (30), along with their similarity to G proteins (107), has raised the possibility that *ras* proteins may participate in the transduction of mitogenic signals. The direct effect of *ras* proteins in the dynamics of cellular membranes has recently been investigated by microinjection experiments (151a). Both normal and transforming *ras* p21 proteins induce the appearance of increased surface ruffles and fluid-phase pinocytosis. Whereas the effect of the normal protein is short-lived, that of its transforming counterpart remains for at least 15 hr after injection (151a). Increased ruffling and pinocytosis are also characteristic membrane responses to certain hormones and mitogenic compounds. These observations support the view that *ras* proteins participate in signal pathways initiated at the cellular surface.

Identification of the receptor and effector systems that interact with *ras* proteins has, so far, been elusive. It has been reported that the GDP binding activity of *ras* proteins in rodent cells transformed by *ras* oncogenes is stimulated by the addition of epidermal growth factor (EGF) (152). These observations, along with the increased production of TGFα, an EGF-like growth factor, by *ras*-transformed cells, have suggested a certain degree of biochemical interaction between *ras* proteins and the EGF receptor pathway.

More recently, it has been suggested that *ras* genes may play a regulatory role in the phosphatidylinositol pathway. Rodent fibroblasts transformed by different *ras* oncogenes exhibit elevated steady-state levels (2- to 3-fold) of phosphatidylinositol-4,5-biphosphate (PIP_2) and their breakdown products, the second messengers 1,2-diacylglycerol (DAG) and inositol-1,4,5-triphosphate (IP_3) (152c). Similar results have been obtained when NIH3T3 cells carrying an inducible N-*ras* proto-oncogene were treated with several growth factors such as bombesin and bradykinin but not with EGF or PDGF (152d). These observations have led to the proposal that N-*ras* p21 proteins may be identical to Gp, the putative G protein that mediates the activation of phospholipase C, the enzyme responsible for the breakdown of PIP_2 into DAG and IP_3 (152d). However, in related experiments in which *ras* p21 proteins were microinjected into REF52 rat fibroblasts, stimulation of phospholipase A_2 was observed instead (151a). Additional experiments are needed to determine whether these observations have physiological relevance or are merely due to an overflow effect caused by the high levels of *ras* proteins present in these cells.

Studies on the effect of Y13-259 antibodies on the proliferative properties of NIH3T3 cells have provided some additional clues. Cells transformed by oncogenes whose normal alleles have been identified as growth factor receptors (*fms*) or that are known to interact with the plasma membrane (*fes* and *src*) cannot enter the S phase if microinjected with Y13-259 antibodies (153). In contrast, this antibody has no effect on cells transformed by two oncogenes whose products are known to be located in the cytoplasm (*mos* and *raf*). These results have been interpreted as evidence that *ras* genes are involved in signal transduction from a variety of receptors and other membrane-associated molecules (153). However, other interpretations are also possible. Further studies with additional oncogenes as well as with antibodies directed against other proto-oncogenes and membrane proteins such as growth factor receptors, protein kinase C, phospholipases, etc, will be necessary to assess the usefulness of this experimental approach.

Studies on the expression of *ras* genes in mammalian cells indicate that they are expressed at low levels in most, if not all, cell lineages. Unlike other proto-oncogenes, *ras* genes are consistently expressed throughout development of the mouse embryo (154, 155). Increased expression (up to eightfold) of *ras* genes has been reported in actively proliferating tissues such as regenerating rat liver (156). However, increased levels of *ras* expression do not always correlate with cellular proliferation. Studies aimed at determining the levels of *ras* proteins in different rat organs have found the highest p21 *ras* expression levels in brain, whereas proliferating tissues only show limited expression (157). A similar study in the mouse revealed the highest levels of expression in heart, another nondividing tissue (158).

More recently, an extensive immunohistochemical survey of p21 *ras* expression in normal fetal and adult human tissue has been conducted (M. Furth, C. Cordon-Cardo, personal communication). In each case, immunoblots of tissue lysates were used to corroborate the immunohistochemical data. The basic conclusions from these studies are (*a*) almost every fetal and adult tissue expresses detectable levels of p21 proteins; (*b*) in most cell lineages, the level of p21 expression was significantly higher in immature than in differentiated cells; and (*c*) certain terminally differentiated cells, including epithelial cells of endocrine glands and the neurons of the central nervous system, expressed high levels of p21 proteins. These findings add further support to the concept that *ras* genes play a dual role in basic cellular proliferation and in certain specific functions of terminally differentiated cells.

ras *ONCOGENES*

Mechanisms of Activation

ras genes can acquire transforming properties by qualitative and quantitative mechanisms. As indicated throughout this review, missense mutations within certain domains yield highly efficient transforming *ras* genes. However, increased expression of normal *ras* proto-oncogenes can also induce certain manifestations of the malignant phenotype. Linkage of normal *ras* genes to retroviral regulatory elements (LTR) results in the malignant transformation of NIH3T3 cells (3, 159, 159a). Similar results have been obtained by integration of multiple copies of a DNA clone of the normal human H-*ras*-1 gene (160). These tumorigenic cells invariably show 30- to 100-fold higher levels of *ras* gene expression than either their normal counterparts or cells transformed by *ras* oncogenes activated by single point mutations (159, 159a, 160). In general terms, the neoplastic properties induced by highly overexpressed *ras* proto-oncogenes are more limited than those induced by their mutated alleles even when driven by their own promoters. Combination of qualitative and quantitative alterations of *ras* genes results in oncogenes (e.g. retroviral *ras* oncogenes) capable of inducing a more complete spectrum of neoplastic pheonotypes.

In vitro Transforming Properties

Cellular *ras* oncogenes transform established rodent cell lines in a dominant fashion (47–49). Neither the resident proto-oncogenes nor cotransfection with their respective normal alleles affects the transformation efficiencies of *ras* oncogenes. However, their levels of expression appear to modulate their transforming potency (161). *ras* oncogenes are necessary not only for initiation, but also for maintenance of the transformed phenotype. This has been demonstrated by the existence of a Kirsten-MSV ts mutant (162) and by in vitro experiments in which *ras* oncogenes were linked to inducible promoters (163) or used to induce reversible transformation of normal rat fibroblasts (161).

Cellular *ras* oncogenes cannot transform rat primary embryo cells (164–166). Tranformation of these cells by *ras* oncogenes requires the cooperation of nuclear oncogenes (167) such as c-*myc* (164), adenovirus E1A (165), polyoma large T (164), N-*myc* (168), or p53 (169, 170). Alternatively, *ras* oncogenes can transform rat embryo cells if the surrounding cells are eliminated, for instance, by cotransfection with the *neo* gene followed by selection with geneticin (171–173). Linkage of *ras* oncogenes to transcriptional enhancers increases the percentage of cells that become transformed under these experimental conditions (171).

The molecular basis for the inhibitory effect of the surrounding cells remains to be elucidated. Reconstitution experiments indicate that normal cells can inhibit the transformed phenotype of *ras*-transfected cells only if plated at high cell density (174). These results explain early observations indicating that *ras*-containing retroviruses can transform rat primary embryo cultures (175) but only at high multiplicity of infection (L. Parada, personal communication). This inhibitory effect can be partially overcome by treating the *ras*-containing primary embryo cells with tumor promoters such as 12-0-tetradecanoyl-phorbol-13-acetate (TPA) (172). However, most TPA-treated or geneticin-resistant rat embryo cells carrying *ras* oncogenes alone will only divide a few times and will not become established as cell lines (172, 173). Quantitative experiments have indicated that primary embryo cells transfected with *ras* oncogenes are one or two orders of magnitude less likely to acquire long-term proliferative properties than those carrying a nuclear (e.g. c-*myc*, p53, adenovirus E1A, etc) oncogene (172, 173). These experiments show that *ras* oncogenes are highly inefficient in rescuing cells from senescence. Moreover, they indicate that induction of malignant transformation by *ras* oncogenes requires the establishment of a proliferative stage in the targeted cells. It is possible, however, that in vitro establishment may not be sufficient criteria to allow the phenotypic expression of *ras* oncogenes. Transformation of rat REF52 cells by these oncogenes requires complementation by adenoviruses E1A (176). Therefore, in vitro transformation of fibroblastic cells by *ras* oncogenes may require additional changes, often associated with, but not necessary for continuous proliferation.

Evaluation of the neoplastic properties of *ras* oncogenes in cells other than cultured fibroblasts has been somewhat hampered by the limited cell types that can be efficiently transfected by these oncogenes. The use of retroviruses carrying *ras* oncogenes has somewhat circumvented this problem. *ras*-containing retroviruses efficiently transform rodent cells of hematopoietic origin, including those of erythroid, lymphoid, and myeloid lineages (177–180). In the case of erythroid cells, the H-*ras* containing Harvey-MSV can induce their malignant transformation without affecting their differentiation program as these cells retain their property to synthesize hemoglobin upon induction (177). Infection of murine mast cells by this virus enhances growth properties of these cells, leading to the establishment of long-term cell lines (178). However, these Harvey-MSV-transformed mast cells require the continuous presence of interleukin-3 (IL-3) for growth, indicating that *ras* oncogenes do not participate in the IL-3 signal transduction pathway (178).

Transformation of myeloid cells results in the generation of both mature macrophage-like and immature myelomonocytic cell lines (179). Both could differentiate either spontaneously in the case of the more mature cells, or by treatment with phorbol esters in the case of the myelomonocytic cells (179). Finally, infection of mouse lymphoid cells with BALB and Harvey-MSVs yields transformed early progenitor cells as determined by the absence of markers specific for the B or T cells (180). Unlike in the case of erythroid and myeloid cells, these early progenitor lymphoid cells cannot be induced to differentiate into more mature lymphoid lineages (180).

Neither cellular nor retroviral *ras* oncogenes have been able to induce stable transformation of normal human fibroblasts (181). However, transfection of normal human bronchial epithelial cells by a DNA clone of the Harvey-MSV *ras* oncogene has yielded colonies of transformed cells (182). These cells are obtained at very low frequency and after careful and long-term culture. They exhibit all the characteristics of malignant cells, including a highly aneuploid karyotype, which suggests that multiple genetic changes have occurred during in vitro selection (182). Other authors have been able to transform human epidermal keratinocytes (183), embryonic kidney cells (184), and normal fibroblasts (185) by *ras*-containing retroviruses. However, transformation of these cells required prior acquisition of indefinite life-span by either infection with DNA tumor viruses such as AD12-SV40 (183) and BK (184) or by treatment with ^{60}Co α-rays (185).

ras oncogenes can also confer additional neoplastic properties to cell lines derived from human tumors. Infection of a human osteosarcoma cell line, HOS, with Kirsten-MSV not only induces drastic morphologic changes but confers on them the property of anchorage-independent growth and of producing tumors in nude mice (186). Transfection of MCF-7 breast carcinoma cells with a DNA clone of Harvey-MSV causes loss of estrogen dependence for neoplastic growth (187). However, bypassing estrogen dependence does not appear to be an intrinsic property of *ras* oncogenes since MCF-7 cells transfected with cellular H-*ras*-1 oncogenes remain hormone dependent (S. Sukumar, personal communication). Thus, it is likely that *ras* oncogenes can only induce the loss of hormone dependence in MCF-7 cells when driven by powerful transcriptional enhancers such as retroviral LTRs.

ras *Oncogenes and Experimental Metastasis*

The ability of *ras* oncogenes to induce metastatic phenotypes has been investigated. NIH3T3 and primary rat embryo cells transfected with *ras* oncogenes can form metastatic nodules in the lung when injected subcutaneously in nude mice (173, 188–190). Other investigators, however, have only observed metastatic foci when *ras*-transformed NIH3T3 cells were retransfected with DNAs isolated from certain metastatic human tumor cell lines (191) or when these cells were injected intravenously (191, 192). In the case of 10T1/2 mouse cells expression of the metastatic phenotype (as determined by intravenous injection) correlates with the levels of expression of transfected H-*ras*-1 oncogenes (192a). In a different system however, transfection of MT1 Cl 5/7 mouse mammary carcinoma cells with the same human H-*ras*-1 oncogene only increased their capacity to metastasize when injected subcutaneously but not intravenously (193). These results suggest that the primary effect of the H-*ras*-1 oncogene may relate to their ability to escape the primary tumor site rather than to colonize new sites (193).

Comparative studies using different mouse cell lines indicate that *ras* oncogenes will confer metastatic properties to NIH3T3 cells, but not C127 cells (190). These results are independent of the levels of expression of p21 *ras* proteins (190). Thus, induction of the metastatic phenotype is not an intrinsic property of *ras* oncogenes. This concept is further supported by the similar incidence of *ras* oncogenes in primary and metastatic tumors (47–49).

ras *Oncogenes and Terminal Differentiation*

In addition to their transforming properties, *ras* oncogenes can also induce terminal differentiation of a rat pheochromocytoma cell line designated PC12. PC12 cells can differentiate into neuron-like cells if infected with *ras*-containing retroviruses or microinjected with oncogenic, but not normal, p21 *ras* proteins (40–42). Treatment of PC12 cells with nerve growth factor (NGF) or cAMP has been shown previously to promote neurite outgrowth (194, 195). Differentiation of PC12 cells induced by *ras* oncogenes clearly proceeds by a mechanism independent of cAMP. However, *ras* oncogenes induce differentiation markers very similar to those induced by NGF. Moreover, both *ras* oncogenes and NGF have a lag of 15–24 hours and require RNA and protein synthesis to induce the neural differentiation of PC12 cells (40–42). Thus, it is possible that *ras* oncogenes may participate in a signal transduction pathway commonly used by NGF. This hypothesis has been strengthened by a recent report showing that microinjection of PC12 cells with antibodies (Y13-259) against p21 *ras* proteins inhibits neurite formation induced by NGF but not by cAMP (196).

At least one other oncogene, v-*src*, has been shown to promote neural differentiation in PC12 cells (197). The effect of v-*src* is highly reminiscent of the effect of *ras* oncogenes. Interestingly, linkage between *src* and *ras* oncogenes has been suggested by two independent lines of evidence. As described above, NIH3T3 cells transformed by v-*src* cannot enter the S phase when microinjected with anti-*ras* Y13-259 antibodies (153). In the next section, I discuss the existence of flat revertants of Kirsten-MSV cells that can suppress the transforming phenotype induced by v-*src* (198). Thus, it is possible that *ras* and *src* genes participate in the same or in very interconnectable pathways. Unfortunately, it is not yet known whether microinjection of anti-p21 antibodies into PC12 cells also inhibits v-*src*-induced neural differentiation. Such experiments may shed light on a possible functional relationship between *ras* and *src* genes in these cells.

Suppressor Genes

One of the advantages of studying *ras* genes in microorganisms is the possibility of identifying genes that can suppress the abnormal phenotypes induced by *ras* genes carrying mutations equivalent to those that activate mammalian *ras* oncogenes. The first of these suppressors (*sup* H) has been recently identified (198a). *sup* H is an allele of *ste* 16, a yeast mutant that does not produce functional a-factor, a mating pheromone. The carboxy terminus of the a-factor is Cys-Val-Ile-Ala-COOH, a sequence similar to the consensus carboxy terminus of *ras* proteins, Cys-A-A-X-COOH (see section on *Membrane Attachment*). Thus, it is likely that *sup* H (now designated *RAM*, for *RAS* protein and a-factor maturation function) represents a mutation in a gene coding for an enzyme responsible for the attachment of fatty acid to the carboxy terminal Cys residue of the *RAS*2 Val^{19} protein (198a).

In mammalian cells there is also genetic evidence for the existence of suppressors of *ras* oncogenes. Flat revertants of NIH3T3 cells transformed with Kirsten-MSV have been described (198, 199). These revertants express high levels of biochemically active p21 protein, have rescuable transforming viruses, and cannot be retransformed by infection with Kirsten-MSV (198, 199). These cells are not tumorigenic and exhibit a very low frequency of retransformation even after long periods of culture. Cell hybridization studies have shown that the revertant phenotype is dominant in hybrids between revertant cells and cells transformed by several *ras* oncogenes (198). These results suggest that the revertant phenotype is not due to the absence of a cellular protein (e.g. a protein located downstream in the putative *ras* signaling pathway). Recently, it has been possible to transmit this suppressor phenotype to other *ras*-transformed NIH3T3 cells by standard gene transfer technology (M. Noda, personal communication).

Hybridization studies between these flat revertants of Kirsten-MSV NIH3T3 cells and cells transformed by other oncogenes indicate that the *ras* suppressor mutation can also suppress transformation induced by the v-*fes* and v-*src* oncogenes but not by v-*mos*, v-*fms*, or v-*sis* (198). These results are in good agreement (with the exception of those for v-*fms*) with those obtained by microinjection of anti-p21 monoclonal antibodies (153). Although the evidence is still mostly circumstantial, it is possible that *ras, fes,* and *src* genes are part of the same or interconnected signaling pathways.

Additional experimental evidence supports the existence of *ras* suppressor genes. Syrian hamster tumors induced by cells transformed by cooperating *ras* and *myc* oncogenes exhibit the nonrandom loss of chromosome 15 (200). Fusion of human tumor cell lines carrying *ras* oncogenes (e.g. HT-1080 and EJ cells) with normal human fibroblasts yields nontumorigenic cells (201, 202). Similarly, fusion of Chinese hamster cells transformed with *ras* oncogenes to nontransformed Chinese hamster cells results in loss of tumorigenicity (203). Most of the hybrids, generated by fusion between normal and tumorigenic cells, are morphologically transformed, indicating that the putative gene(s) involved in the suppression of tumorigenicity must be different from those responsible for the flat phenotype of the Kirsten-MSV revertants. Moreover, some human tumor cell lines containing *ras* oncogenes (e.g. T24 bladder carcinoma cells) are "nontumorigenic" in nude mouse assays (204, 205). Thus, it is possible that those genes involved in the suppression of tumorigenicity of *ras*-transformed cells may not exert their inhibitory action by direct interaction with *ras* oncogenes.

ras ONCOGENES IN ANIMAL TUMORS

ras oncogenes have been implicated in the development of a variety of tumors. They have been found in four strains of acute transforming retroviruses of rodents (3–6) and are occasionally activated in tumors induced by retroviruses that do not carry oncogenes (20; J. Ihle, personal communication). *ras* oncogenes, however, have been more frequently identified in tumors of nonviral etiology. In particular, animal tumors induced by chemical or physical carcinogens have been an abundant source of this class of oncogenes (150). Moreover, *ras* oncogenes are present in about 10% of the most common forms of human neoplasia, thus making them the most frequently identified oncogene family in human cancer (47–49).

As documented in the previous section, the transforming properties of *ras* oncogenes have been extensively documented in in vitro cell culture systems. However, cancer is a multistage disease that probably results from accumulation of independent genetic and epigenetic errors. The progressive generation of aneuploidy during tumor development suggests that most of the aberrations present in tumor cells may occur after malignancy has been irreversibly established. Therefore, the basic question posed by the frequent identification of *ras* oncogenes in spontaneous tumors such as those of humans is whether *ras* oncogenes participate in the induction of neoplastic development or are a consequence of it. Some experimental evidence suggests that *ras* oncogenes can, in fact, become activated after cells have acquired neoplastic properties (206–208). However, recent results obtained in animal model systems indicate that *ras* oncogenes are more likely to participate in the initiation of tumor development. These findings will be reviewed in this section.

Carcinogen-Induced Tumors

ras oncogenes have been found to be reproducibly activated in a variety of carcinogen-induced animal model tumor systems (Table 1). In rats, induction of mammary carcinomas by a single dose of nitroso-methylurea (NMU) during puberty leads to the activation of H-*ras*-1 oncogenes in 86% of the tumors (12, 209). Substitution of this carcinogen by dimethyl-benz(a)anthracene (DMBA) also results in the malignant activation of the H-*ras*-1 locus although in only one fourth of the tumors (209). Activation of the K-*ras*-2 locus has been observed in 40% of kidney mesenchymal tumors induced by a single dose of methyl(methoxymethyl)nitrosamine (DMN) (210) and in 74% of lung tumors (adenocarcinomas and squamous cell carcinomas) by chronic exposure to tetranitromethane (TNM) (J. Stower, M. W. Anderson, personal communication).

In mice (Table 1), H-*ras*-1 oncogenes have been found to be reproducibly activated in 90% of skin papillomas and carcinomas of Sencar mice initiated by DMBA or dibenz(c,h)acridine (DBACR) treatment and followed by promotion with TPA (211–213). Similarly, H-*ras*-1 oncogenes have been found in each of 4 mammary tumors arising from hyperplastic alveolar nodules implanted in mice treated with DMBA (214). Exposure of AKRxRF hybrid mice to X-rays or to repeated NMU treatment leads to the induction of lymphomas, most of which (over 60%) contain activated K-*ras*-2 or N-*ras*

Table 1   Activation of *ras* oncogenes in carcinogen-induced animal tumors

Species	Carcinogen	Tumor	Oncogene	Incidence	Reference
Rat	NMU	Mammary carcinoma	H-*ras*-1	86%	12,209
	DMBA	Mammary carcinoma	H-*ras*-1	23%	209
	DMN	Kidney mesenchymal	K-*ras*-2	40%	210
	TNM	Lung carcinoma	K-*ras*-2	74%	a
Mouse	DMBA	Skin carcinoma	H-*ras*-1	90%	211–213
	DBACR	Skin carcinoma	H-*ras*-1	80%	213
	DMBA	Mammary carcinoma	H-*ras*-1	100%	214
	X-rays	Lymphoma	N-*ras*,K-*ras*-2	57%	14, b
	NMU	Lymphoma	N-*ras*,K-*ras*-2	85%	14, b
	MCA	Thymic lymphoma	K-*ras*-2	83%	214a
	MCA	Fibrosarcoma	K-*ras*-2	50%	215
	HOAFF	Hepatocellular carcinoma	H-*ras*-1	100%	217
	VC	Hepatocellular carcinoma	H-*ras*-1	100%	217
	HODE	Hepatocellular carcinoma	H-*ras*-1	100%	217
	Furfural	Hepatocellular carcinoma	H-*ras*-1	85%	a
	TNM	Lung carcinoma	K-*ras*-2	100%	c

[a] S. Reynolds, M. W. Anderson, personal communication.
[b] A. Pellicer, personal communication.
[c] J. Stowers, M. W. Anderson, personal communication.

oncogenes (14). K-*ras*-2 oncogenes have also been observed in 10 out of 10 lung tumors induced by TNM (J. Stowers, M. W. Anderson, personal communication), and in 10 out of 12 thymic lymphomas (214a) and 2 out of 4 fibrosarcomas induced by MCA (215). Finally, H-*ras*-1 oncogenes have been found in almost 100% of hepatocellular carcinomas of B6C3F$_1$ mice that either arose spontaneously (216) or were induced by carcinogens such as N-hydroxy-2-acetyl-aminofluoride (HOAAF) (217), vinyl carbamate (VC) (217), 1'-hydroxy-2'-3'-dehydroestragole (HODE) (217), or furfural (S. Reynolds, M. W. Anderson, personal communication). The frequent and reproducible activation of *ras* oncogenes in these animal tumors strongly supports the concept that *ras* oncogenes play a causative role in neoplastic development (Table 1).

   ras oncogenes have also been randomly (<10% incidence) identified in certain animal model systems such as rat fibrosarcomas induced by either 1,6- or 1,8-dinitropyrene (217a, M. Nagao, personal communication) or liver carcinomas induced by methyl(acetoxymethyl)nitrosamine (DMN-OAc) (J. Rice, personal communication). Finally, *ras* oncogenes have been identified in certain rodent cell lines transformed by in vitro exposure to chemical carcinogens. Whereas the K-*ras*-2 oncogene was found in mouse cells treated with 3-methyl-cholanthrene (MCA) (218), N-*ras* was found to be activated in four tumorigenic guinea pig cell lines initiated by nitroso compounds and polycyclic hydrocarbons (219; J. Doniger, personal communication).

ras *Genes as Targets of Carcinogens*

It is generally accepted that most carcinogens are mutagens. A significant number of chemical carcinogens are known to form adducts with DNA bases (reviewed in 220). Whereas some of these adducts are highly mutagenic due to their miscoding properties (221, 222), others can lead to mutations due to the generation of apurinic sites or because of the limited fidelity of repair polymerases (reviewed in 223). Whereas most of these mutations will have no consequence to the host, a small number of them can trigger neoplastic development. Identification of these critical mutations has eluded scientists for many years.

   The reproducible activation of *ras* oncogenes in carcinogen-induced tumors has made it possible to correlate their activating mutations with the known mutagenic effects of certain carcinogens (Table 2). Each of the H-*ras*-1 oncogenes present in rat mammary carcinomas induced by NMU but not DMBA became activated by G $\rightarrow$ A transitions are the most common mutations induced by NMU (221, 222). These mutations result from the miscoding properties of O^6-methylguanosine, one of the adducts generated by the methylating activity of NMU. In contrast, DMBA forms large adducts with adenine and guanine residues whose repair very seldom leads to the generation of G $\rightarrow$ A substitutions (220). These findings have led to the proposal that NMU is directly responsible for the malignant activation of H-*ras*-1 oncogenes in these mammary tumors (209).

Although the H-*ras*-1 locus can be activated by G $\rightarrow$ A substitutions in either of the two guanine residues present in the 12th codon (GGA), only those affecting the middle position (G^{35}) have been observed in NMU-induced mammary carcinomas (209). Such a strong positional effect adds further support to the concept that NMU is directly involved in the generation of these mutations. A possible explanation for the reproducible occurrence of $G^{35} \rightarrow A^{35}$ transitions in these tumors has recently been obtained. Analysis of the rate of repair of O^6-methylguanosine residues in the ampicillanase gene of a bacteriophage f1/pBR322 plasmid chimera has revealed a DNA consensus sequence around $_*$unrepaired G_* residues (224). This sequence, GCTGGTCGCCAG$\underline{GA}$GG, where $\overset{*}{G}$ is the unrepaired O^6MeG residue, is 75% homologous (12 out of 16 residues) to the rat H-*ras*-1 sequence around the critical 12th codon (underlined in the consensus sequence). The probability of a random match with this stringency is 0.0017. Moreover, no other sequence in the entire H-*ras*-1 locus exhibits a similar match with this consensus sequence (224). Therefore, it is possible that NMU may induce mammary carcinomas in rats by interacting with a DNA sequence within a critical proto-oncogene domain that cannot be efficiently repaired.

The concept that *ras* oncogenes can be the targets of chemical carcinogens has been further supported by recent experimental evidence obtained with other animal tumor model systems (Table 2). Induction of skin carcinomas in mice by DMBA and phorbol esters involves the specific activation of H-*ras*-1 oncogenes by A $\rightarrow$ T transitions in the second base of codon 61 (212, 213). However, this mutation has not been seen when the initiating carcinogen,

Table 2 Specific mutagenesis of H-*ras*-1 oncogenes in chemically induced tumors

Species	Type of tumor	Carcinogen	Activating mutation	Incidence	Reference
Rat	Mammary carcinoma	NMU	$G^{35} \rightarrow A$	61/61	209
		DMBA	$A^{182}, A^{183} \rightarrow N$	5/5	209
Mouse	Skin papilloma or carcinoma	DMBA	$A^{182} \rightarrow T$	33/37	212,213
Mouse	Hepatocarcinoma	None	$C^{181} \rightarrow A$	6/11	a
			$A^{182} \rightarrow T$	3/11	a
			$A^{182} \rightarrow G$	2/11	a
		HO-AAF	$C^{181} \rightarrow A$	7/7	217
		VC	$A^{182} \rightarrow T$	6/7	217

[a] S. Reynolds, M. W. Anderson, personal communication.

DMBA, was replaced by *N*-methyl-*N*-nitro-*N*-nitroso-guanidine (MNNG), an alkylating carcinogen (212). Similar observations have been reported in hepatomas of male B6C3F$_1$ mice. In this model system, H-*ras*-1 oncogenes present in spontaneous tumors of old age mice (216) are activated by random mutations in the first two nucleotides of codon 61 (S. Reynolds, M. W. Anderson, personal communication). However, when these hepatocellular carcinomas are induced by carcinogenic treatment, the activating mutations conform to a specific pattern (217). Treatment of B6C3F$_1$ mice with a single dose of HO-AAF generates H-*ras*-1 oncogenes activated by C → A transversions in the first base of codon 61 (217). Although the mutagenic specificity of HO-AAF is not known, both *N*-hydroxy-*N*-2-aminofluorene and *N*-acetoxy-acetylaminofluorene (AAAF) are known to induce G-C → T-A transversions (225, 226). Moreover, in vitro treatment of H-*ras*-1 proto-oncogene DNA with AAAF leads to its oncogenic activation by C^{181} → A^{181} transitions (227). Substitution of HO-AAF by VC as the initiating carcinogen results in a completely different mutagenic spectrum. H-*ras*-1 oncogenes present in hepatocellular carcinomas induced by VC consistently exhibit missense mutations (mostly A → T transversions) in the A^{182} residue (217). These findings, taken together, indicate that *ras* oncogenes can be directly activated by the mutagenic properties of the initiating carcinogens.

Multistep Carcinogenesis

Activation of *ras* oncogenes by the initiating carcinogens in tumors induced by a single carcinogenic insult implies that these oncogenes must participate in the initiation of neoplastic development (209, 212, 217). This concept is supported by the identification of transforming H-*ras*-1 genes in the majority of skin papillomas induced by topical application of DMBA or DBACR followed by treatment with tumor promoters (213, 228). Most of these papillomas will regress spontaneously and only a few of them will develop as malignant carcinomas. These results provide independent biological evidence that *ras* genes can be activated during the early stages of tumor development (213, 228).

Activation of *ras* oncogenes is not sufficient to trigger tumorigenesis. Emerging evidence indicates that they must cooperate with secondary genetic events and/or with specific developmental programs in order to elicit neoplastic development. For instance, it has been shown that the active stage of proliferation of the developing mammary gland at the time of the carcinogen insult plays a fundamental role in tumor development (reviewed in 229). Recent results, however, indicate that activation of *ras* oncogenes does not have to be concomitant with mammary gland development (S. Sukumar, personal communication). Treatment of newborn rats with a single dose of NMU leads to the development of mammary carcinomas 2–3 months after the animals have reached sexual development. Most of these tumors carry H-*ras*-1 oncogenes, each activated by the G → A transition diagnostic of NMU-induced mutagenesis. Treatment of these animals with antiestrogen drugs followed by ovariectomy eliminates the occurrence of mammary carcinomas (S. Sukumar, personal communication). These observations suggest that whereas H-*ras*-1 oncogenes might become activated early in life, manifestation of their malignant properties requires the hormone-induced proliferation and/or differentiation that takes place in the mammary gland during sexual maturation.

A similar situation may occur in the skin carcinogenesis model of mice. Initiated cells are known to remain dormant in the skin of DMBA-treated mice for long periods of time until they are treated with tumor promoters (230). Considering that most of the papillomas and carcinomas generated in this model system contain H-*ras*-1 oncogenes (211–213, 228), it is likely that these initiated cells already carry activated *ras* oncogenes. Direct support for this hypothesis has been recently obtained (231). Infection of mouse epidermal skin cells with Harvey-MSV results in the generation of initiated cells that do not express discernible neoplastic properties unless they are treated with TPA. These results indicate that in this model system *ras* oncogenes also require cooperation with cellular proliferation to allow expression of their neoplastic properties.

Induction of skin carcinomas by Harvey-MSV has provided direct experimental evidence for the existence of secondary genetic events in *ras*-induced tumors (231). Whereas Harvey MSV-induced papillomas appear to be polyclonal, the malignant carcinomas are clonal in origin. Papillomas are known to regress to small hyperkeriotic lesions before a few of them begin their invasive growth characteristic of skin carcinomas. Therefore, a second genetic event distinct and independent from *ras* oncogene activation is required for development of skin carcinomas (231). Recently, it has been shown that skin grafts of cultured keratinocytes infected with Harvey-MSV lead to the generation of papillomas but not carcinomas (231a), thus adding further evidence to the concept that *ras* oncogenes are not sufficient for the development of the full malignant phenotype in, at least, this model system.

Similar conclusions have been obtained by independent studies utilizing transgenic mice. Mice carrying in their germ line H-*ras* oncogenes linked to the promoter region of the mouse whey acidic protein (WAP) gene (N. Hynes, B. Groner, personal communication) or to the LTR of the mouse mammary tumor virus (MMTV) (P. Leder, personal communication). These animals develop single mammary tumors after a relatively long latent period, suggesting that a secondary event is necessary for tumor development. Mating transgenic mice carrying MMTV-*ras* chimeras with those containing MMTV-

myc genes (232), results in offspring that develop neoplastic outgrowths in most of their breasts (P. Leder, personal communication). These results suggest that animals carrying two resident oncogenes may not require additional genetic events for neoplastic development.

The requirement of a second genetic event may not be necessary if *ras* oncogenes are activated during the early stages of development. Transgenic mice carrying *ras* oncogenes under the transcriptional control of the elastase promoter die of pancreatic tumors a few days after birth (R. Palmiter, personal communication). In mice, the pancreas evolves from the gut at day 11 of gestation. At day 14 to 15, elastase begins to be expressed in the acimer cells of the pancreas and by day 16 tumors can already be detected. In contrast, the pancreas of transgenic mice carrying the normal *ras* proto-oncogene linked to the elastase promoter developed normally and the mice survived up to one year of age (R. Palmiter, personal communication). These results dramatically illustrate that under certain circumstances expression of a single *ras* oncogene is sufficient to induce tumor development.

Retrovirus-Induced Tumors

To date, four *ras*-containing retroviruses have been identified. They include Harvey-MSV isolated by inoculation of rats with Moloney murine leukemia virus (MuLV) followed by injection of the rat-passaged virus into newborn BALB/c mice (1), Kirsten-MSV derived from rats inoculated with cell-free filtrates from thymic lymphomas of old C3H mice (2), BALB-MSV isolated from an hemangiosarcoma of a BALB/c mouse induced by transmission of a cell-free extract of blood from a spontaneous chloroleukemia of an 18-month-old BALB/c mouse (233, 234), and Rasheed-MSV generated by in vitro cocultivation of a spontaneously transformed rat cell line (235). These viruses primarily cause erythroleukemias and fibrosarcomas when injected into newborn rodents. However, the pathogenic spectrum of these viruses can be altered by manipulations involving changes in dose, route of administration, strains, age of infected animals, etc. These studies have been extensively reviewed in the past (175, 236) and will not be considered here any further.

Introduction of the human H-*ras*-1 oncogene isolated from the T24/EJ bladder carcinoma cell line into appropriate vectors has also yielded transforming retroviruses. These viruses can also induce erythroleukemias and sarcomas when injected into newborn mice (236a). These experiments illustrate the wide malignant spectrum of human *ras* oncogenes. However, they should be interpreted with caution as the levels of expression of these human *ras* oncogenes are much higher in retrovirus-infected cells that they would normally be in human tumors.

Construction of similar retroviruses carrying the normal human H-*ras*-1 proto-oncogene did not yield transforming viruses (236a). These results are somewhat surprising considering the in vitro transforming properties of LTR-*ras* proto-oncogene chimeras (3, 159). In related experiments, the miscoding mutations of codons 12 and 59 of the H-*ras* oncogene present in Harvey-MSV have been eliminated (73). The resulting Gly^{12}/Ala^{59} Harvey-MSV mutant virus exhibits decreased transforming properties, although it retains the capacity to transform NIH3T3 cells and rat embryo cells (73). Other authors have reported that whereas transcriptionally enhanced *ras* proto-oncogenes cannot transform primary rat embryo cultures, they can confer long-term proliferative properties to the transfected cells (171). The reasons for these discrepancies probably lie in the different constructs and assay conditions used in these experiments.

Replication-competent retroviruses that do possess oncogene sequences may also exert their neoplastic properties by transcriptional activation of cellular *ras* genes. Induction of a chicken nephroblastoma by a myeloblastosis-associated retrovirus resulted in elevated levels (25-fold) of a novel H-*ras* transcript that contained retroviral LTR U5 sequences (20). Viral integration did not occur adjacent to the H-*ras* locus but probably next to a distant splicing donor site. Sequence analysis of a DNA clone complementary to the hybrid LTR U5-H-*ras* transcript showed no mutations in any of the critical *ras* codons (20). Thus, it is likely that transcriptional activation of the normal chicken H-*ras* locus contributed to the development of this tumor.

A related event may have accounted for the highly elevated (30-fold) expression of p21 *ras* proteins in a myeloid cell line (416B) derived from a mouse bone-marrow culture infected with the Friend retrovirus complex (237, 238). In this case, a 3.5 kbp segment of the Friend helper virus was inserted within the first intron of the mouse K-*ras*-2 locus (21). As a result, K-*ras*-2 transcription was initiated at a new site (presumably within the retroviral LTR) that excluded the first K-*ras*-2 noncoding exon located upstream from the viral integration site (21). The original 416B cells were nontumorigenic (237, 238). However, these cells have now become malignant, presumably as a consequence of continuous passage in vitro (21). These observations suggest that transcriptional activation of the K-*ras*-2 locus in this mouse hematopoietic cell line may have contributed, but was not sufficient to induce tumorigenic properties in these cells. Similar findings have been recently obtained in DA-2 cells, a cell line derived from a Moloney murine leukemia virus (MLV)–induced tumor. In these cells, Moloney-MLV sequences became integrated 5' of the first coding exon of the H-*ras*-1 locus (J. Ihle, personal communication). This genomic rearrangement is likely to be responsible for the high levels of expression of p21 *ras* proteins observed in DA-2 cells and it is presumed to have contributed to tumor development.

ras ONCOGENES IN HUMAN CANCER

Incidence

ras oncogenes exist in a variety of human cancers. They have been identified in carcinomas of the bladder, breast, colon, kidney, liver, lung, ovary, pancreas, and stomach; in hematopoietic tumors of lymphoid (acute lymphocytic leukemia, B-cell lymphoma, Burkitt's lymphoma) and myeloid (acute and chronic myelogenous leukemias, promyelocytic leukemia) lineage; and in tumors of mesenchymal origin such as fibrosarcomas and rhabdomyosarcomas. Other tumors, including melanomas, teratocarcinomas, neuroblastomas, and gliomas have also been shown to possess *ras* oncogenes. Specific references to these studies can be found in Ref. 49 and will not be included here. Recently, a H-*ras*-1 oncogene has been identified in 1 of 5 keratoacanthomas, a well-characterized human benign tumor of the skin (J. Leon, A. Pellicer, personal communication). Keratoacanthomas exhibit an initial aggressive growth followed by spontaneous regression. These findings resemble those observed in the mouse skin model tumor system (see previous section) in which H-*ras*-1 oncogenes were shown to participate in the early stages of tumor development.

So far, transforming *ras* genes are the oncogenes most frequently identified in human cancer. Their overall incidence is estimated to be around 10–15%. This figure is likely to be lower in certain malignancies such as breast carcinoma, while it may be higher in other tumors such as acute myelogenous leukemia (89). The actual incidence of *ras* oncogenes in human tumors is difficult to assess. Available information has been obtained from transfection assays, which are somewhat insensitive, particularly in detecting large oncogenes such as K-*ras*-2. Moreover, many scientists do not conduct parallel experiments to test the sensitivity of their transfection assays and to control for the intactness of the donor DNA. This is of particular importance in laboratories that lack direct access to reliable tumor material. Finally, some authors only report those tumors or tumor cell lines in which they have obtained positive results. Recent studies utilizing the RNA:RNA hybrid mismatch technology (see next section) have shown the presence of K-*ras*-2 oncogenes carrying codon 12 mutations in 22 of 55 colon carcinomas (M. Perucho, personal communication). Considering that mutations at codon 61 of K-*ras*-2 oncogenes have not yet been analyzed and that N-*ras* oncogenes have also been found in colon carcinomas (49), the overall incidence of *ras* oncogenes in this common type of human cancer may have been underestimated. In any case, available information to date indicates that (*a*) *ras* oncogenes are present in a significant percentage (varying from 5 to 40%) of most human tumors; (*b*) they are not associated with specific types of neoplasia; and (c) their activation does not correlate with the histo-pathological properties of the tumor.

The involvement of *ras* genes in human cancer is not limited to their activation by point mutations. It is likely that expression of abnormally high levels of normal *ras* products may also contribute to malignancy. This could be achieved by perturbation of their regulatory sequences (20, 21) or by gene amplification. In human tumors, there is no evidence for transcriptional activation of *ras* proto-oncogenes by mutations affecting their regulatory elements. In contrast, significant amplification ($\geqslant$10 fold) of *ras* genes has been observed in a variety of human tumors (160, 239–242). However, the overall incidence of *ras* gene amplification in human neoplasia is estimated to be not higher than 1% (160, 240).

Loss of the normal allele in cells carrying *ras* oncogenes has been observed with relative frequency in tumor cell lines, but only rarely in tumor biopsies (243). Interestingly, loss of *ras* alleles also occurs in tumors that do not carry *ras* oncogenes (240, 244, 245). Whether these deletions play any role in the pathogenesis of human cancer remains to be determined.

Expression Studies

Expression of *ras* genes is often increased in tumors relative to normal tissues. Careful quantitative analysis has indicated that the levels of *ras* transcripts in about 50% of human tumors are 2- to 10-fold higher than in control tissue (246–248). Similar results have been obtained by immunoblot analysis of p21 *ras* expression in a variety of primary carcinomas (249–251). Surprisingly, metastatic tissues derived from primary colon carcinomas consistently exhibited low levels of p21 *ras* proteins (249). The expression of *ras* genes in human tumors has also been studied by immunohistochemical techniques. Immunocytochemistry has the advantage of allowing the evaluation of *ras* gene expression in individual cells. These studies have utilized two monoclonal antibodies, Y13-259 (120) and RAP-5 (252), neither of which discriminates between the different p21 *ras* proteins or between their normal and activated forms. Unfortunately, no clear picture has emerged from these studies. Some authors have consistently identified increased levels of immunoreactivity in malignant cells of mammary (252), colon (253), bladder (254), and prostate (255) carcinomas as compared to those of corresponding benign tumors, dysplastic lesions, or normal epithelium. Other authors, however, have not been able to detect any significant differences between normal and malignant tissues (256–258). Moreover, two independent laboratories have reported that the RAP-5 monoclonal antibody may recognize a cytoplasmic cellular component distinct from p21 (258, 258a). In this regard, it is worthwhile to note that a monoclonal antibody that specifically recognizes Val[12] p21 *ras* proteins by immunoblotting or even by immunoperoxidase staining of human tumor cell lines in culture, reacts nonspecifically with a high percentage (over 50%) of formalin-fixed tissue sections (259; see next section). These observations indicate that further work is needed to validate the use of paraffin-embedded formalin-fixed tissue samples in studying the expression of *ras* genes in human tumors.

The biological significance of results obtained by studying the expression of *ras* genes in human tumors is not clear. Enhanced expression of normal *ras* gene products has been implicated in the development of certain animal tumors (20; J. Ihle, personal communication) and in transformation of cells in culture (21, 159, 159a, 160). However, in these cases a 30- to 100-fold increase in the levels of both mRNA and p21 *ras* protein expression has been consistently observed. On the other hand, increases in *ras* transcripts of up to eight-fold can be associated with normal proliferative processes such as regenerating rat liver (156). Therefore, there is not sufficient experimental basis to implicate moderate variation of *ras* proto-oncogene expression in the neoplastic development of human tumors.

Predisposition to Cancer

Extensive genetic evidence suggests the existence of genes responsible for the predisposition of certain individuals to cancer. *ras* oncogenes may not belong to this class of genes. First of all, *ras* oncogenes become activated in somatic rather than in germ line cells (89, 243, 260, 261). Second, normal cells of patients with high cancer risk syndromes do not contain *ras* oncogenes (262). However, a recent population study on the frequency of H-*ras*-1 alleles in normal individuals versus cancer patients has revealed a statistically significant correlation between the latter and certain unusual H-*ras*-1 alleles (244).

The human H-*ras*-1 locus contains a region of around 800 bp of tandemly repeated sequences located 1000 bp downstream from its polyadenylation signal (58). Genetic variability within this stretch of repeated sequences generates a variety of restriction fragment length polymorphisms (RFLP). Up to 20 polymorphic H-*ras*-1 alleles have been identified, of which only four are commonly found in the human population (244). Of a total of 230 samples of white blood cells of normal donors, only 9 (3.9%) exhibited any of the 16 rare alleles. In contrast, of 298 samples of either normal or tumor tissue from cancer patients, 32 (10.7%) contained rare alleles. No significant difference was seen between normal and tumor samples of cancer patients, indicating that the increased frequency of rare alleles in these patients is an inherited property and not a consequence of somatic events associated with tumor development (244). All tumor types, including myelodysplastic lesions, were associated with rare H-*ras*-1 alleles, suggesting that these RFLPs are indicative of genetic predisposition to most, if not all, forms of cancer. Different results have been obtained in an independent study investigating the frequency of rare H-*ras*-1 alleles in lung cancer (245). In this study, 132 patients were examined, 66 with small cell lung carcinomas (SCLC) and 66 with non-SCLC. Rare H-*ras*-1 allelic patterns were found in patients with non-SCLC but not in those with SCLC (245). These observations suggest that large numbers of cancer patients must be examined to fully assess the incidence of rare H-*ras*-1 alleles in specific types of malignancies. The need for such studies has been further indicated by recent reports contesting the possible linkage between rare H-*ras*-1 RFLP and the development of myelodysplasia (263) and melanoma (263a).

It is not clear how these rare H-*ras*-1 alleles may predispose to cancer. It has been speculated that the rare alleles representing specific arrays of tandem-repeats may be in linkage disequilibrium with missense mutations in the H-*ras*-1 coding sequences that may confer weakly oncogenic properties to this locus (244). Recently, it has been shown that these polymorphic repeated sequences have enhancer activity in in vitro CAT assays (263b; S. Ishii, personal communication). Thus, it is possible that a somewhat altered regulation over long periods of time of the H-*ras*-1 gene or of a putative neighboring locus may predispose certain cells to neoplastic growth. However, experimental evidence to support any of these hypotheses has, as yet, to be obtained.

New Diagnostic Methods

So far, identification of *ras* oncogenes in human tumors has relied on tedious and insensitive transfection assays. It is evident that any attempt to utilize information on activation of *ras* oncogenes in clinical studies will require the development of faster and more sensitive diagnostic methods. The first step in this direction has involved the use of oligonucleotides of defined sequence to identify single point mutations in genomic DNA. Hybrids between oligonucleotides that form a perfect match with genomic sequences are more stable than those that exhibit a single mismatch (264). Oligonucleotides corresponding to sequences surrounding the critical 12th and 61st codons of *ras* genes have been successfully used to identify the presence of *ras* oncogenes in certain human neoplasias (89, 265). In spite of the significant advantages of oligonucleotide probing over current gene transfer assays, this technique is still somewhat cumbersome. *ras* genes can theoretically be activated by almost 100 different base substitutions, of which over 30 have already been identified in naturally occurring tumors (49). This heterogeneity can be partially overcome by using oligonucleotide mixtures containing each of the three possible substitutions at a given residue (89, 265). Nevertheless, the significant number of oligonucleotide probes required to identify each possible *ras* oncogene limits the general applicability of this technique.

Several new methods capable of identifying single point mutations over relatively long stretches of DNA sequences may simplify the task of identifying *ras* oncogenes. One of these methods is based on the abnormal electrophoretic migration of DNA heteroduplexes containing single base mismatches in denaturing gradient gels (266). Unfortunately, this method can only identify between 25 and 40% of all possible base substitutions (266). Moreover, considerable technical expertise is required for the preparation of the denaturing gradient gels. The development of [^{32}P]-labeled RNA probes has made possible the identification of single base pair mismatches in RNA:RNA and RNA:DNA hybrids based on their sensitivity to RNAse A digestion (267, 268). Hybridization of tumor DNA or RNA samples with these probes, followed by incubation with RNAse A, cleaves the labeled RNA probes into fragments of defined sizes that can be easily resolved by polyacrylamide gel electrophoresis. These techniques can identify most base substitutions (267, 268). Moreover, the RNA:RNA hybrid method provides useful information regarding the levels of expression of *ras* oncogenes in these tumors. In addition, the relative expression of both normal and transforming alleles can be analyzed (267). Recently, a new method involving the primer-mediated enzymatic amplification of specific genomic sequences has been developed (269). This process, which can easily be automated (K. Mullis, personal communication), should make possible the routine use of oligonucleotide probes to identify *ras* oncogenes in clinical laboratories.

In spite of these technological improvements, the use of nucleic acid hybridization techniques to identify human *ras* oncogenes would always require the use of six different probes (specific for the first and second exons of each of the three *ras* loci) and enough tissue to obtain sufficient amounts of RNA or DNA. In theory, a more practical approach would be to generate monoclonal antibodies capable of discriminating between the putative "active" and "inactive" conformation of p21 *ras* proteins (see Figure 2). Unfortunately, antibodies elicited against peptides containing the missense mutations characteristic of *ras* oncogenes only identify those p21 *ras* proteins carrying the amino acid substitutions present in the immunizing peptide (102, 259, 270). On the other hand, antibodies elicited against bacteria-synthesized p21 *ras* proteins recognize the normal and oncogenic forms with equal efficiency (our unpublished observations). X-ray crystallographic analysis of p21 molecules should help in identifying putative domains that may elicit such diagnostic antibodies.

A potential caveat in using monoclonal antibodies to identify the products of *ras* oncogenes in human tissue sections is the potential lack of specificity (see previous section). Monoclonal antibodies capable of recognizing oncogenic *ras* proteins carrying Val^{12} substitutions by immunoblotting analysis have been shown to be highly specific in immunoperoxidase staining of human carcinoma cell lines in culture (259). However, when this antibody was used to analyze formalin-fixed tumor biopsies, a significant number gave false positive results (259). Therefore, further work will be required to work out the technical challenges posed by the diagnostic identification of *ras* oncogenes in human tumor biopsies.

Future Avenues

In spite of the extensive documentation on the malignant properties of *ras* oncogenes, their presence in human tumors cannot be considered as sufficient evidence of a causative role in the pathogenesis of human cancer. Results obtained with carcinogen-induced tumors and transgenic mice have indicated that *ras* oncogenes can participate in the initiation of carcinogenesis. However, these observations do not preclude the possibility that *ras* oncogenes may also be generated in cells that have already acquired neoplastic properties. To evaluate the involvement of *ras* oncogenes in human cancer it is necessary to determine what percentage of *ras* oncogenes plays a primordial role in tumor development and what percentage becomes activated as a consequence of it. Unfortunately, this question cannot be easily addressed as in human carcinogenesis it is virtually impossible to backtrack to the early events that led to neoplastic development.

Future progress on the involvement of *ras* oncogenes in human cancer depends on results derived from two independent avenues of research. First, we must continue ongoing efforts aimed at unveiling the molecular and biochemical mechanisms by which *ras* oncogenes disrupt the normal proliferative programs. Second, we must develop fast and reliable diagnostic methods to allow the routine identification of *ras* oncogenes in human biopsies. This information, once applied to a large number of tumors, may reveal hidden connections between the activation of these oncogenes and the etiology and/or pathology of human cancer.

It is evident that the molecular pathways that lead to carcinogenesis may not parallel the histo-pathological criteria that have served to develop current cancer treatment programs. We can only hope that understanding the mechanisms by which oncogenes participate in neoplasia may provide a more rational approach to cancer therapy. The presence of *ras* oncogenes in at least 10% of human cancers certainly justifies the intense interest dedicated to this gene family during the last few years.

ACKNOWLEDGMENTS

I would like to thank all those colleagues who sent preprints or shared unpublished information. I am also grateful to those who provided critical comments to this manuscript.

Research sponsored by the National Cancer Institute, DHHS, under contract No. NO1-CO-23909 with Bionetics Research, Inc. The contents of this publication do not necessarily reflect the view or policies of the Department of Health & Human Services, nor does mention of trade names, commercial products, or organizations imply endorsement by the US Government.

Literature Cited

1. Harvey, J. J. 1964. *Nature* 204:1104–5
2. Kirsten, W. H., Mayer, L. A. 1967. *J. Natl. Cancer Inst.* 39:311–35
3. DeFeo, D., Gonda, M. A., Young, H. A., Chang, E. H., Lowy, D. R., et al. 1981. *Proc. Natl. Acad. Sci. USA* 78:3328–32
4. Ellis, R. W., DeFeo, D., Shih, T. Y., Gonda, M. A., Young, H. A., et al. 1981. *Nature* 292:506–11
5. Rasheed, S., Norman, G. L., Heidecker, G. 1983. *Science* 221:155–57
6. Andersen, P. R., Devare, S. G., Tronick, S. R., Ellis, R. W., Aaronson, S. A., Scolnick, E. M. 1981. *Cell* 26:129–34
7. Shih, C., Padhy, L. C., Murray, M., Weinberg, R. A. 1981. *Nature* 290:261–64
8. Krontiris, T. G., Cooper, G. M. 1981. *Proc. Natl. Acad. Sci. USA* 78:1181–84
9. Perucho, M., Goldfarb, M., Shimizu, K., Lama, C., Fogh, J., Wigler, M. 1981. *Cell* 27:467–76
10. Pulciani, S., Santos, E., Lauver, A. V., Long, L. K., Robbins, K. C., Barbacid, M. 1982. *Proc. Natl. Acad. Sci. USA* 79:2845–49
11. Balmain, A., Pragnell, I. B. 1983. *Nature* 303:72–74
12. Sukumar, S., Notario, V., Martin-Zanca, D., Barbacid, M. 1983. *Nature* 306:658–61
13. Eva, A., Aaronson, S. A. 1983. *Science* 220:955–56
14. Guerrero, I., Calzada, P., Mayer, A., Pellicer, A. 1984. *Proc. Natl. Acad. Sci. USA* 81:202–5
15. Parada, L. F., Tabin, C. J., Shih, C., Weinberg, R. A. 1982. *Nature* 297:474–78
16. Santos, E., Tronick, S. R., Aaronson, S. A., Pulciani, S., Barbacid, M. 1982. *Nature* 298:343–47
17. Der, C. J., Krontiris, T. G., Cooper, G. M. 1982. *Proc. Natl. Acad. Sci. USA* 79:3637–40
18. Shimizu, K., Goldfarb, M., Suard, Y., Perucho, M., Li, Y., et al. 1983. *Proc. Natl. Acad. Sci. USA* 80:2112–16
19. Hayward, W. S., Neel, B. G., Astrin, S. M. 1981. *Nature* 290:475–80
20. Westaway, D., Papkoff, J., Moscovici, C., Varmus, H. E. 1986. *EMBO J.* 5:301–9
21. George, D. L., Glick, B., Trusko, S., Freeman, N. 1986. *Proc. Natl. Acad. Sci. USA* 83:1651–55
22. Scolnick, E. M., Papageorge, A. G., Shih, T. Y. 1979. *Proc. Natl. Acad. Sci. USA* 76:5355–59
23. Shih, T. Y., Papageorge, A. G., Stokes, P. E., Weeks, M. O., Scolnick, E. M. 1980. *Nature* 287:686–91
24. Tamanoi, F., Walsh, M., Kataoka, T., Wigler, M. 1984. *Proc. Natl. Acad. Sci. USA* 81:6924–28
25. Temeles, G. L., Gibbs, J. B., D'Alonzo, J. S., Sigal, I. S., Scolnick, E. M. 1985. *Nature* 313:700–3
26. Gibbs, J. B., Sigal, I. S., Poe, M., Scolnick, E. M. 1984. *Proc. Natl. Acad. Sci. USA* 81:5704–8
27. McGrath, J. P., Capon, D. J., Goeddel, D. V., Levinson, A. D. 1984. *Nature* 310:644–49
28. Sweet, R. W., Yokoyama, S., Kamata, T., Feramisco, J. R., Rosenberg, M., Gross, M. 1984. *Nature* 311:273–75
29. Manne, V., Bekesi, E., Kung, H. F. 1985. *Proc. Natl. Acad. Sci. USA* 82:376–80
30. Willingham, M. C., Pastan, I., Shih, T. Y., Scolnick, E. M. 1980. *Cell* 19:1005–14
31. Willumsen, B. M., Christensen, A., Hubbert, N. L., Papageorge, A. G., Lowy, D. R. 1984. *Nature* 310:583–86
32. Fujiyama, A., Tamanoi, F. 1986. *Proc. Natl. Acad. Sci. USA* 83:1266–70
33. Hurley, J. B., Simon, M. I., Teplow, D. B., Robishaw, J. D., Gilman, A. G. 1984. *Science* 226:860–62
34. Tanabe, T., Nukada, T., Nishikawa, Y., Sugimoto, K., Suzuki, H., et al. 1985. *Nature* 315:242–45
35. Lochrie, M. A., Hurley, J. B., Simon, M. I. 1985. *Science* 228:96–99
36. Itoh, H., Kozasa, T., Nagata, S., Nakamura, S., Katada, T., et al. 1986. *Proc. Natl. Acad. Sci. USA* 83:3776–80
37. DeFeo-Jones, D., Tatchell, K., Robinson, L. C., Sigal, I. S., Vass, W. C., et al. 1985. *Science* 228:179–84
38. Kataoka, T., Powers, S., Cameron, S., Fasano, O., Goldfarb, M., et al. 1985. *Cell* 40:19–26
39. Mulcahy, L. S., Smith, M. R., Stacey, D. W. 1985. *Nature* 313:241–43
40. Noda, M., Ko, M., Ogura, A., Liu, D.

G., Amano, T., et al. 1985. *Nature* 318:73–75

41. Bar-Sagi, D., Feramisco, J. R. 1985. *Cell* 42:841–48

42. Guerrero, I., Wong, H., Pellicer, A., Burstein, D. 1986. *J. Cell Physiol.* 129:71–76

43. Kataoka, T., Powers, S., McGill, C., Fasano, O., Strathern, J., et al. 1984. *Cell* 37:437–45

44. Tatchell, K., Chaleff, D. T., DeFeo-Jones, D., Scolnick, E. M. 1984. *Nature* 309:523–27

45. Fukui, Y., Kozasa, T., Kaziro, Y., Takeda, T., Yamamoto, M. 1986. *Cell* 44:329–36

46. Lowy, D. R., Willumsen, B. M. 1986. *Cancer Surv.* In press

47. Marshall, C. 1985. In *RNA Tumor Viruses*, Vol. 2, ed. R. N. Teich, H. Varmus, J. Coffin, pp. 487–558. Cold Spring Harbor, NY: Cold Spring Harbor Lab. 2nd ed.

48. Varmus, H. E. 1984. *Ann. Rev. Genet.* 18:553–612

49. Barbacid, M. 1985. In *Important Advances in Oncology 1986*, ed. V. DeVita, S. Hellman, S. Rosenberg, pp. 3–22. Philadelphia: Lippincott

50. Barbacid, M. 1986. *Trends Genet.* 2:188–92

51. Shimizu, K., Goldfarb, M., Perucho, M., Wigler, M. 1983. *Proc. Natl. Acad. Sci. USA* 80:383–87

52. Hall, A., Marshall, C. J., Spurr, N. K., Weiss, R. A. 1983. *Nature* 303:396–400

53. Miyoshi, J., Kagimoto, M., Soeda, E., Sakayi, Y. 1984. *Nucleic Acids Res.* 12:1821–28

54. McGrath, J. P., Capon, D. J., Smith, D. H., Chen, E. Y., Seeburg, P. H., et al. 1983. *Nature* 304:501–6

55. Chattopadhyay, S. K., Chang, E. H., Lander, M. R., Ellis, R. W., Scolnick, E. M., Lowy, D. R. 1982. *Nature* 296:361–63

56. Goldfarb, M., Shimizu, K., Perucho, M., Wigler, M. 1982. *Nature* 296:404–9

57. Shih, C., Weinberg, R. A. 1982. *Cell* 29:161–69

58. Capon, D. J., Chen, E. Y., Levinson, A. D., Seeburg, P. H., Goeddel, D. V. 1983. *Nature* 302:33–37

59. Reddy, E. P. 1983. *Science* 220:1061–63

60. Shimizu, K., Birnbaum, D., Ruley, M. A., Fasano, O., Suard, Y., et al. 1983. *Nature* 304:497–500

61. Nakano, H., Yamamoto, F., Neville, C., Evans, D., Mizuno, T., Perucho, M. 1984. *Proc. Natl. Acad. Sci. USA* 81:71–75

62. Taparowsky, E., Shimizu, K., Goldfarb, M., Wigler, M. 1983. *Cell* 34:581–86

63. Brown, R., Marshall, C. J., Pennie, S. G., Hall, A. 1984. *EMBO J.* 3:1321–26

64. Ruta, M., Wolford, R., Dhar, R., DeFeo-Jones, D., Ellis, R. W., Scolnick, E. M. 1986. *Mol. Cell. Biol.* 6:1706–10

65. Tahira, T., Hayashi, K., Ochiai, M., Tsuchida, N., Nagao, M., Sugimura, T. 1986. *Mol. Cell. Biol.* 6:1349–51

66. George, D. L., Scott, A. F., Trusko, S., Glick, B., Ford, E., Dorney, D. J. 1985. *EMBO J.* 4:1199–203

67. Guerrero, I., Villasante, A., D'Eustachio, P., Pellicer, A. 1984. *Science* 225:1041–43

68. Guerrero, I., Villasante, A., Corces, V., Pellicer, A. 1985. *Proc. Natl. Acad. Sci. USA* 82:7810–14

69. O'Brien, S. J., ed. 1984. In *Genetics Maps*, pp. 1–584. Cold Spring Harbor, NY: Cold Spring Harbor Lab.

70. Szpirer, J., DeFeo-Jones, D., Ellis, R. W., Levan, G., Szpirer, C. 1985. *Somatic Cell. Mol. Genet.* 11:93–97

71. Shih, T. Y., Weeks, M. O., Young, H. A., Scolnick, E. M. 1979. *Virology* 96:64–79

72. Capon, D. J., Seeburg, P. H., McGrath, J. P., Hayflick, J. S., Edman, U., et al. 1983. *Nature* 304:507–13

73. Cichutek, K., Duesberg, P. H. 1986. *Proc. Natl. Acad. Sci. USA* 83:2340–44

74. Ishii, S., Merlino, G. T., Pastan, I. 1985. *Science* 230:1378–81

74a. Jordano, J., Perucho, M. 1986. *Nucleic Acids Res.* 14:7361–78

75. Ishii, S., Kadonaga, J. T., Tjian, R., Brady, J. N., Merlino, G. T., Pastan, I. 1986. *Science* 232:1410–13

76. Shilo, B. Z., Weinberg, R. A. 1981. *Proc. Natl. Acad. Sci. USA* 78:6789–92

77. Neuman-Silberberg, F. S., Schejter, E., Hoffmann, F. M., Shilo, B. Z. 1984. *Cell* 37:1027–33

78. Mozer, B., Marlor, R., Parkhurst, S., Corces, V. 1985. *Mol. Cell. Biol.* 5:885–89

79. Schejter, E. D., Shilo, B. Z. 1985. *EMBO J.* 4:407–12

80. Swanson, M. E., Elste, A. M., Greenberg, S. M., Schwartz, J. H., Aldrich, T. H., Furth, M. E. 1986. *J. Cell. Biol.* 103:485–92

81. Reymond, C. D., Gomer, R. H., Mehdy, M. C., Firtel, R. A. 1984. *Cell* 39:141–48

82. DeFeo-Jones, D., Scolnick, E. M., Koller, R., Dhar, R. 1983. *Nature* 306:707–9

83. Powers, S., Kataoka, T., Fasano, O.,

Goldfarb, M., Strathern, J., et al. 1984. *Cell* 36:607–12

84. Fukui, Y., Kaziro, Y. 1985. *EMBO J.* 4:687–91

85. Clark, S. G., McGrath, J. P., Levinson, A. D. 1985. *Mol. Cell. Biol.* 5:2746–52

86. Tabin, C. J., Bradley, S. M., Bargmann, C. I., Weinberg, R. A., Papageorge, A. G., et al. 1982. *Nature* 300:143–49

87. Reddy, E. P., Reynolds, R. K., Santos, E., Barbacid, M. 1982. *Nature* 300:149–52

88. Taparowsky, E., Suard, Y., Fasano, O., Shimizu, K., Goldfarb, M., Wigler, M. 1982. *Nature* 300:762–65

89. Bos, J. L., Toksoz, D., Marshall, C. J., Verlaan-de Vries, M., Veeneman, G. H., et al. 1985. *Nature* 315:726–30

90. Dhar, R., Ellis, R. W., Shih, T. Y., Oroszlan, S., Shapiro, B., et al. 1982. *Science* 217:934–36

91. Tsuchida, M., Ohtsubo, E., Ryder, T. 1982. *Science* 217:937–39

92. Yuasa, Y., Srivastava, S. K., Dunn, C. Y., Rhim, J. S., Reddy, E. P., Aaronson, S. A. 1983. *Nature* 303:775–79

92a. Yamamoto, F., Perucho, M. 1984. *Nucleic Acids Res.* 12:8873–85

93. Fasano, O., Aldrich, T., Tamanoi, F., Taparowsky, E., Furth, M., Wigler, M. 1984. *Proc. Natl. Acad. Sci. USA* 81:4008–12

94. Walter, M., Clark, S. G., Levinson, A. D. 1986. *Science* 233:649–52

95. Sigal, I. S., Gibbs, J. B., D'Alonzo, J. S., Temeles, G. L., Wolanski, B. S., et al. 1986. *Proc. Natl. Acad. Sci. USA* 83:952–56

96. Seeburg, P. H., Colby, W. W., Capon, D. J., Goeddel, D. V., Levinson, A. D. 1984. *Nature* 312:71–75

97. Chipperfield, R. G., Jones, S. S., Lo, K.-M., Weinberg, R. A. 1985. *Mol. Cell. Biol.* 5:1809–18

98. Der, C. J., Finkel, T., Cooper, G. M. 1986. *Cell* 44:167–76

99. Reddy, E. P., Lipman, D., Andersen, P. R., Tronick, S. R., Aaronson, S. A. 1985. *J. Virol.* 53:984–87

100. Madaule, P., Axel, R. 1985. *Cell* 41:31–40

100a. Lowe, D. G., Capon, D. J., Delwart, E., Sakaguchi, A. Y., Naylor, S. L., Goeddel, D. V. 1987. *Cell* 48:137–46

100b. Chardin, P., Tavitian, A. 1986. *EMBO J.* 5:2203–8

101. Gallwitz, D., Donath, C., Sander, C. 1983. *Nature* 306:704–7

101a. Schmitt, H. D., Wagner, P., Pfaff, E., Gallwitz, D. 1986. *Cell* 47:401–12

102. Clark, R., Wong, G., Arnheim, N., Nitecki, D., McCormick, F. 1985.

Proc. Natl. Acad. Sci. USA 82:5280–84

103. Feramisco, J. R., Clark, R., Wong, G., Arnheim, N., Milley, R., McCormick, F. 1985. *Nature* 314:639–42

104. Willumsen, B. M., Papageorge, A. G., Kung, H.-F., Bekesi, E., Robins, T., et al. 1986. *Mol. Cell. Biol.* 6:2646–54

105. Lacal, J. C., Anderson, P. S., Aaronson, S. A. 1986. *EMBO J.* 5:679–87

106. Shih, T. Y., Stokes, P. E., Smythers, G. W., Dhar, R., Oroszlan, S. 1982. *J. Biol. Chem.* 257:11767–86

107. Gilman, A. G. 1984. *Cell* 36:577–79

108. Leberman, R., Egner, U. 1984. *EMBO J.* 3:339–41

109. Jurnak, F. 1985. *Science* 230:32–36

110. Gay, N. J., Walker, J. E. 1983. *Nature* 301:262–64

111. McCormick, F., Clark, B. F., la Cour, T. F., Kjeldgaard, M., Norskov-Lauritsen, L., Nyborg, J. 1985. *Science* 230:78–82

112. Santos, E., Reddy, E. P., Pulciani, S., Feldmann, R. J., Barbacid, M. 1983. *Proc. Natl. Acad. Sci. USA* 80:4679–83

113. Pincus, M. R., van Renswoude, J., Harford, J. B., Chang, E. H., Carty, R. P., Klausner, R. D. 1983. *Proc. Natl. Acad. Sci.* 80:5253–57

114. Pincus, M. R., Brandt-Rauf, P. W. 1982. *Proc. Natl. Acad. Sci. USA* 82:3596–600

115. Feig, L. A., Pan, B.-T., Roberts, T. M., Cooper, G. M. 1986. *Proc. Natl. Acad. Sci. USA* 83:4607–11

115a. Clanton, D. J., Hattori, S., Shih, T. Y. 1986. *Proc. Natl. Acad. Sci. USA* 83:5076–80

115b. Der, C. J., Pan, B.-T., Cooper, G. M. 1986. *Mol. Cell. Biol.* 6:3291–94

116. La Cour, T. F., Nyborg, J., Thirup, S., Clark, B. F. 1985. *EMBO J.* 4:2385–88

117. Der, C. J., Finkel, T., Cooper, G. M. 1986. *Cell* 44:167–76

118. Willumsen, B. M., Papageorge, A. G., Hubber, N., Bekesi, E., Kung, H. F., Lowy, D. R. 1985. *EMBO J.* 4:2893–96

119. Sigal, I. S., Gibbs, J. B., D'Alonzo, J. S., Scolnick, E. M. 1986. *Proc. Natl. Acad. Sci. USA* 83:4725–29

119a. Stein, R. B., Tai, J. Y., Scolnick, E. M. 1986. *J. Virology* 60:782–86

120. Furth, M. E., Davis, L. J., Fleurdelys, B., Scolnick, E. M. 1982. *J. Virol.* 43:294–304

121. Pawson, T., Amiel, T., Hinze, E., Auersperg, N., Neave, N., et al. 1985. *Mol. Cell. Biol.* 5:33–39

122. Papageorge, A. G., DeFeo-Jones, D., Robinson, P., Temeles, G., Scolnick, E. M. 1984. *Mol. Cell. Biol.* 4:23–29

123. Kung, H. F., Smith, M. R., Bekesi, E.,

Manne, V., Stacey, D. W. 1986. *Exp. Cell. Res.* 162:363–71

124. Lacal, J. C., Aaronson, S. A. 1986. *Mol. Cell. Biol.* 6:1002–9

125. Shih, T. Y., Weeks, M. O., Gruss, P., Dhar, R., Oroszlan, S., Scolnick, E. M. 1982. *J. Virol.* 42:253–61

126. Sefton, B. M., Trowbridge, I. S., Cooper, J. A., Scolnick, E. M. 1982. *Cell* 31:465–74

127. Chen, Z. Q., Ulsh, L. S., DuBois, G., Shih, T. Y. 1985. *J. Virol.* 56:607–12

128. Buss, J. E., Sefton, B. M. 1986. *Mol. Cell. Biol.* 6:116–22

129. Willumsen, B. M., Norris, K., Papageorge, A. G., Hubbert, N. L., Lowy, D. R. 1984. *EMBO J.* 3:2581–85

130. Lowy, D. R., Papageorge, A. G., Vass, W. C., Willumsen, B. M. 1986. In *UCLA Symposia on Molecular and Cellular Biology: Cellular and Molecular Biology of Tumors and Potential Clinical Applications,* ed. J. Minna, M. Keuhl. In press

131. Lautenberger, J. A., Ulsh, L., Shih, T. Y., Papas, T. S. 1983. *Science* 221:858–60

132. Lacal, J. C., Srivastava, S. K., Anderson, P. S., Aaronson, S. A. 1986. *Cell* 44:609–17

132a. Lacal, J. C., Aaronson, S. A. 1986. *Mol. Cell. Biol.* 6:4214–20

133. Toda, T., Uno, I., Ishikawa, T., Powers, S., Kataoka, T., et al. 1985. *Cell* 40:27–36

134. Dhar, R., Nieto, A., Koller, R., DeFeo-Jones, D., Scolnick, E. M. 1984. *Nucleic Acids Res.* 12:3611–18

135. Tatchell, K., Robinson, L. C., Breitenbach, M. 1985. *Proc. Natl. Acad. Sci. USA* 82:3785–89

136. Fraenkel, D. G. 1985. *Proc. Natl. Acad. Sci. USA* 82:4740–44

137. Breviario, D., Hinnebusch, A., Cannon, J., Tatchell, K., Dhar, R. 1986. *Proc. Natl. Acad. Sci. USA* 83:4152–56

137a. Cannon, J. F., Gibbs, J. B., Tatchell, K. 1986. *Genetics* 113:247–64

137b. Kaibuchi, K., Miyajima, A., Arai, K.-I., Matsumoto, K. 1986. *Proc. Natl. Acad. Sci. USA* 83:8172–76

138. Matsumoto, K., Uno, I., Oshima, Y., Ishikawa, T. 1982. *Proc. Natl. Acad. Sci. USA* 79:2555–59

139. Uno, I., Matsumoto, K., Ishikawa, T. 1982. *J. Biol. Chem.* 257:14110–15

139a. Kataoka, T., Broek, D., Wigler, M. 1985. *Cell* 43:493–505

140. Broek, D., Samiy, N., Fasano, O., Fujiyama, A., Tamanoi, F., et al. 1985. *Cell* 41:763–69

141. Uno, I., Mitsuzawa, H., Matsumoto, K., Tanaka, K., Oshima, T., Ishikawa,

T. 1985. *Proc. Natl. Acad. Sci. USA* 82:7855–59

142. Beckner, S. K., Hattori, S., Shih, T. Y. 1985. *Nature* 317:71–72

143. Birchmeier, C., Broek, D., Wigler, M. 1985. *Cell* 43:615–21

143a. Nadin-Davis, S. A., Nasim, A., Beach, D. 1986. *EMBO J.* 5:2963–71

144. Matsumoto, K., Uno, I., Ishikawa, T. 1983. *Cell* 32:417–23

145. Reymond, C. D., Nellen, W., Firtel, R. A. 1985. *Proc. Natl. Acad. Sci. USA* 82:7005–9

145a. Reymond, C. D., Gomer, R. H., Nellen, W., Theibert, A., Devreotes, P., Firtel, R. A. *Nature* 323:340–43

146. Lev, Z., Kimchie, Z., Hessel, R., Segev, O. 1985. *Mol. Cell. Biol.* 5:1540–42

147. Segal, D., Shilo, B. Z. 1986. *Mol. Cell. Biol.* 6:2241–48

148. Stacey, D. W., Kung, H. F. 1984. *Nature* 310:508–11

149. Feramisco, J. R., Gross, M., Kamata, T., Rosenberg, M., Sweet, R. W. 1984. *Cell* 38:109–17

150. Hyland, J. K., Rogers, C. M., Scolnick, E. M., Stein, R. B., Ellis, R., Baserga, R. 1985. *Virology* 141:333–36

151. Papageorge, A. G., Willumsen, B. M., Johnsen, M., Kung, H.-F., Stacey, D. W., et al. 1986. *Mol. Cell. Biol.* 6:1843–46

151a. Bar-Sagi, D., Feramisco, J. R. 1986. *Science* 233:1061–68

152. Kamata, T., Feramisco, J. R. 1984. *Nature* 310:147–50

152a. Deleted in proof

152b. Deleted in proof

152c. Fleischman, L. F., Chahwala, S. B., Cantley, L. 1986. *Science* 231:407–10

152d. Wakelam, M. J. O., Davies, S. A., Houslay, M. D., McKay, I., Marshall, C. J., Hall, A. 1986. *Nature* 323:173–76

153. Smith, M. R., DeGudicibus, S. J., Stacey, D. W. 1986. *Nature* 320:540–43

154. Muller, R., Slamon, D. J., Adamson, E. D., Tremblay, J. M., Muller, D., et al. 1983. *Mol. Cell. Biol.* 3:1062–69

155. Slamon, D. J., Cline, M. J. 1984. *Proc. Natl. Acad. Sci. USA* 81:7141–45

156. Goyette, M., Petropoulos, C. J., Shank, P. R., Fausto, N. 1983. *Science* 219:510–12

157. Tanaka, T., Ida, N., Shimoda, H., Waki, C., Slamon, D. J., Cline, M. J. 1986. *Mol. Cell. Biochem.* 70:97–104

158. Spandidos, D. A., Dimitrov, T. 1985. *Biosci. Rep.* 5:1035–39

159. Chang, E. H., Furth, M. E., Scolnick, E. M., Lowy, D. R. 1982. *Nature* 297:479–83

159a. McKay, I. A., Marshall, C. J., Cales, C., Hall, A. 1986. *EMBO J.* 5:2617–21

160. Pulciani, S., Santos, E., Long, L. K., Sorrentino, V., Barbacid, M. 1985. *Mol. Cell. Biol.* 5:2836–41

161. Winter, E., Perucho, M. 1986. *Mol. Cell. Biol.* 6:2562–70

162. Shih, T. Y., Weeks, M. O., Young, H. A., Scolnick, E. M. 1979. *J. Virol.* 31:546–60

163. Huang, A. L., Ostrowski, M. C., Berard, D., Hager, G. L. 1981. *Cell* 27:245–55

164. Land, H., Parada, L. F., Weinberg, R. A. 1983. *Nature* 304:596–602

165. Ruley, H. E. 1983. *Nature* 304:602–6

166. Newbold, R. F., Overell, R. W. 1983. *Nature* 304:648–51

167. Weinberg, R. A. 1985. *Science* 230:770–76

168. Schwab, M., Varmus, H. E., Bishop, J. M. 1985. *Nature* 316:160–62

169. Eliyahu, D., Raz, A., Gruss, P., Givol, D., Oren, M. 1984. *Nature* 312:646–49

170. Parada, L. F., Land, H., Weinberg, R. A., Wolf, D., Rotter, V. 1984. *Nature* 312:649–51

171. Spandidos, D. A., Wilkie, N. M. 1984. *Nature* 310:469–75

172. Dotto, G. P., Parada, L. F., Weinberg, R. A. 1985. *Nature* 318:472–75

173. Pozzatti, R., Muschel, R., Williams, J., Padmanabhan, R., Howard, B., et al. 1986. *Science* 232:223–27

174. Spandidos, D. A. 1986. *Anticancer Res.* 6:259–62

175. Harvey, J. J., East, J. 1971. *Int. Rev. Exp. Pathol.* 10:265–360

176. Franza, B. R., Maruyama, K., Garrels, J. I., Ruley, H. E. 1986. *Cell* 44:409–18

177. Hankins, W. D., Scolnick, E. M. 1981. *Cell* 26:91–97

178. Rein, A., Keller, J., Schultz, A. M., Holmes, K. L., Medicus, R., Ihle, J. M. 1985. *Mol. Cell. Biol.* 5:2257–64

179. Pierce, J. H., Aaronson, S. A. 1985. *Mol. Cell. Biol.* 5:667–74

180. Pierce, J. H., Aaronson, S. A. 1982. *J. Exp. Med.* 156:873–87

181. Sager, R., Tanaka, K., Lau, C. C., Ebina, Y., Anisowicz, A. 1983. *Proc. Natl. Acad. Sci. USA* 80:7601–5

182. Yoakum, G. H., Lechner, J. F., Gabrielson, E. W., Korba, B. E., Malan-Shibley, L., et al. 1985. *Science* 227:1174–79

183. Rhim, J. S., Jay, G., Arnstein, P., Price, F. M., Sanford, K. K., Aaronson, S. A. 1985. *Science* 227:1250–52

184. Pater, A., Pater, M. M. 1986. *J. Virol.* 58:680–83

185. Namba, M., Nishitani, K., Fukushima, F., Kimoto, T., Nose, K. 1986. *Int. J. Cancer* 37:419–23

186. Rhim, J. S., Cho, H. Y., Vernon, M. L., Arnstein, P., Huebner, R. J., Gilden, R. V. 1975. *Int. J. Cancer* 16:840–49

187. Kasid, A., Lippman, M. E., Papageorge, A. G., Lowy, D. R., Gelmann, E. P. 1985. *Science* 228:725–28

188. Thorgeirsson, U. P., Turpeenniemi-Hujanen, T., Williams, J. E., Westin, E. H., Heilman, C. A., et al. 1985. *Mol. Cell. Biol.* 5:259–62

189. Greig, R. G., Koestler, T. P., Trainer, D. L., Corwin, S. P., Miles, L., et al. 1985. *Proc. Natl. Acad. Sci. USA* 82:3698–701

190. Muschel, R. J., Williams, J. E., Lowy, D. R., Liotta, L. A. 1985. *Am. J. Pathol.* 121:1–8

191. Bernstein, S. C., Weinberg, R. A. 1985. *Proc. Natl. Acad. Sci. USA* 82:1726–30

192. Bradley, M. O., Kraynak, A. K., Storer, R. D., Gibbs, J. B. 1986. *Proc. Natl. Acad. Sci. USA* 83:5277–81

192a. Egan, S. E., McClarty, G. A., Jarolim, L., Wright, J. A., Spiro, I., et al. 1986. *Mol. Cell. Biol.* In press

193. Vousden, K. H., Eccles, S. A., Purvies, H., Marshall, C. J. 1986. *Int. J. Cancer* 37:425–33

194. Schubert, D., Heinemann, S., Kidokoro, Y. 1977. *Proc. Natl. Acad. Sci. USA* 74:2579–83

195. Greene, L. A., Tischler, A. S. 1976. *Proc. Natl. Acad. Sci. USA* 73:2424–28

196. Hagag, N., Halegoua, S., Viola, M. 1986. *Nature* 319:680–82

197. Alema, S., Casalbore, P., Agostini, E., Tato, F. 1985. *Nature* 316:587–90

198. Noda, M., Selinger, Z., Scolnick, E. M., Bassin, R. H. 1983. *Proc. Natl. Acad. Sci. USA* 80:5602–6

198a. Powers, S., Michaelis, S., Broek, D., Santa Anna-A., S., Field, J., et al. 1986. *Cell* 47:413–22

199. Norton, J. D., Cook, F., Roberts, P. C., Clewley, J. P., Avery, R. J. 1984. *J. Virol.* 50:439–44

200. Oshimura, M., Gilmer, T. M., Barrett, J. C. 1985. *Nature* 316:636–39

201. Stanbridge, E. J., Der, C. J., Doersen, C. J., Nishimi, R. Y., Peehl, D. M., et al. 1982. *Science* 215:252–59

202. Geiser, A. G., Der, C. J., Marshall, C. J., Stanbridge, E. J. 1986. *Proc. Natl. Acad. Sci. USA* 83:5209–13

203. Craig, R. W., Sager, R. 1985. *Proc. Natl. Acad. Sci. USA* 82:2062–66

204. Fogh, W. C., Wright, W. C., Loveless, J. D. 1977. *J. Natl. Cancer Inst.* 58:209–14
205. Fogh, J., Fogh, J. M., Orfeo, T. 1977. *J. Natl. Cancer Inst.* 59:221–26
206. Albino, A. P., Le Strange, R., Oliff, A. I., Furth, M. E., Old, L. J. 1984. *Nature* 308:69–72
207. Tainsky, M. A., Cooper, C. S., Giovanella, B. C., Vande Woude, G. F. 1984. *Science* 225:643–45
208. Vousden, K. H., Marshall, C. J. 1984. *EMBO J.* 3:913–17
209. Zarbl, H., Sukumar, S., Arthur, A. V., Martin-Zanca, D., Barbacid, M. 1985. *Nature* 315:382–85
210. Sukumar, S., Perantoni, A., Reed, C., Rice, J. M., Wenk, M. L. 1986. *Mol. Cell. Biol.* 6:2716–20
211. Balmain, A., Pragnell, I. B. 1983. *Nature* 303:72–74
212. Quintanilla, M., Brown, K., Ramsden, M., Balmain, A. 1986. *Nature* 322:78–80
213. Bizub, D., Wood, A. W., Skalka, A. M. 1986. *Proc. Natl. Acad. Sci. USA* 83:6048–52
214. Dandekar, S., Sukumar, S., Zarbl, H., Young, L. J., Cardiff, R. D. 1986. *Mol. Cell. Biol.* 6:4104–8
214a. Eva, A., Trimmer, R. W. 1986. *Carcinogenesis* 7:1931–33
215. Eva, A., Aaronson, S. A. 1983. *Science* 220:955–56
216. Reynolds, S. H., Stowers, S. J., Maronpot, R. R., Anderson, M. W., Aaronson, S. A. 1986. *Proc. Natl. Acad. Sci. USA* 83:33–37
217. Wiseman, R. W., Stowers, S. J., Miller, E. C., Anderson, M. W., Miller, J. A. 1986. *Proc. Natl. Acad. Sci. USA* 83:5825–29
217a. Ochiai, M., Nagao, M., Tahira, T., Ishikawa, F., Hayashi, K., et al. 1985. *Cancer Lett.* 29:119–25
218. Parada, L. F., Weinberg, R. A. 1983. *Mol. Cell. Biol.* 3:2298–301
219. Sukumar, S., Pulciani, S., Doniger, J., DiPaolo, J. A., Evans, C. H., et al. 1984. *Science* 223:1197–99
220. Singer, B., Kusmierek, J. T. 1982. *Ann. Rev. Biochem.* 51:655–93
221. Eadie, J. S., Conrad, M., Toorchen, D., Topal, M. D. 1984. *Nature* 308:201–3
222. Loechler, E. L., Green, C. L., Essigmann, J. M. 1984. *Proc. Natl. Acad. Sci. USA* 81:6271–75
223. Walker, G. C. 1984. *Microbiol. Rev.* 48:60–93
224. Topal, M. D., Eadie, J. S., Conrad, M. 1986. *J. Biol. Chem.* 261:9879–85
225. Bichara, M., Fuchs, R. P. 1985. *J. Mol. Biol.* 183:341–51
226. Foster, P. L., Eisenstadt, E., Miller, J. H. 1983. *Proc. Natl. Acad. Sci. USA* 80:2695–98
227. Vousden, K. H., Bos, J. L., Marshall, C. J., Phillips, D. H. 1986. *Proc. Natl. Acad. Sci. USA* 83:1222–26
228. Balmain, A., Ramsden, M., Bowden, G. T., Smith, J. 1984. *Nature* 307:658–60
229. Welsch, C. W. 1985. *Cancer Res.* 45:3415–43
230. Van Duuren, B. L., Sivak, A., Katz, C., Seidman, I., Melchionne, S. 1975. *Cancer Res.* 35:502–5
231. Brown, K., Quintanilla, M., Ramsden, M., Kerr, I. B., Young, S., Balmain, A. 1986. *Cell* 46:447–56
231a. Roop, D. R., Lowy, D. R., Tambourin, P. E., Strickland, J., Harper, J. R., et al. 1986. *Nature* 323:822–24
232. Stewart, T. A., Pattengale, P. K., Leder, P. 1984. *Cell* 38:627–37
233. Peters, R. L., Rabstein, L. S., VanVleck, R., Kelloff, G. T., Huebner, R. J. 1974. *J. Natl. Cancer Inst.* 53:1725–29
234. Aaronson, S. A., Barbacid, M. 1978. *J. Virol.* 27:366–73
235. Rasheed, S., Gardner, M. B., Huebner, R. J. 1978. *Proc. Natl. Acad. Sci. USA* 75:2972–76
236. Weiss, R., Teich, N., Varmus, H., Coffin, J., eds. 1984. In *RNA Tumor Viruses,* Vol. 1. Cold Spring Harbor, NY: Cold Spring Harbor Lab. 2nd ed.
236a. Tabin, C. J., Weinberg, R. A. 1985. *J. Virol.* 53:260–65
237. Dexter, T. M., Allen, T. D., Scott, D., Teich, N. M. 1979. *Nature* 277:471–74
238. Ellis, R. W., DeFeo, D., Furth, M. E., Scolnick, E. M. 1982. *Mol. Cell. Biol.* 11:1339–45
239. Fujita, J., Srivastava, S. K., Kraus, M. H., Rhim, J. S., Tronick, S. R., Aaronson, S. A. 1985. *Proc. Natl. Acad. Sci. USA* 82:3849–53
240. Yokota, J., Tsunetsugu-Yokota, Y., Battifora, H., LeFevre, C., Cline, M. J. 1986. *Science* 231:261–65
241. Bos, J. L., Verlaan-de Vries, M., Marshall, C. J., Veeneman, G. H., van Boom, J. H., van der Eb, A. J. 1986. *Nucleic Acids Res.* 14:1209–17
242. Filmus, J. E., Buick, R. N. 1985. *Cancer Res.* 45:4468–72
243. Santos, E., Martin-Zanca, D., Reddy, E. P., Pierotti, M. A., Della Porta, G., Barbacid, M. 1984. *Science* 223:661–64
244. Krontiris, T. G., DiMartino, N. A., Colb, M., Parkinson, D. R. 1985. *Nature* 313:369–74
245. Heighway, J., Thatcher, N., Cerny, T., Hasleton, P. S. 1986. *Br. J. Cancer* 53:453–57

246. Slamon, D. J., deKernion, J. B., Verma, I. M., Cline, M. J. 1984. *Science* 224:256–62
247. Spandidos, D. A., Agnantis, N. J. 1984. *Anticancer Res.* 4:269–72
248. Spandidos, D. A., Kerr, I. B. 1984. *Br. J. Cancer* 49:681–88
249. Gallick, G. E., Kurzrock, R., Kloetzer, W. S., Arlinghaus, R. B., Gutterman, J. U. 1985. *Proc. Natl. Acad. Sci. USA* 82:1795–99
250. Kurzrock, R., Gallick, G. E., Gutterman, J. U. 1986. *Cancer Res.* 46:1530–34
251. Tanaka, T., Slamon, D. J., Battifora, H., Cline, M. J. 1986. *Cancer Res.* 46:1465–70
252. Hand, P. H., Thor, A., Wunderlich, D., Muraro, R., Caruso, A., Schlom, J. 1984. *Proc. Natl. Acad. Sci. USA* 81:5227–31
253. Thor, A., Horan Hand, P., Wunderlich, D., Caruso, A., Muraro, R., Schlom, J. 1984. *Nature* 311:562–65
254. Viola, M. V., Fromowitz, F., Oravez, S., Deb, S., Schlom, J. 1985. *J. Exp. Med.* 161:1213–18
255. Viola, M. V., Fromowitz, F., Oravez, S., Deb, S., Finkel, G., et al. 1986. *N. Engl. J. Med.* 314:133–37
256. Kerr, I. B., Lee, F. D., Quintanilla, M., Balmain, A. 1985. *Br. J. Cancer* 52:695–700
257. Williams, A. R., Piris, J., Spandidos, D. A., Wyllie, A. H. 1985. *Br. J. Cancer* 52:687–93
258. Ghosh, A. K., Moore, M., Harris, M. 1986. *J. Clin. Pathol.* 39:428–34
258a. Robinson, A., Williams, A. R. W., Piris, J., Spandidos, D. A., Wyllie, A. H. 1986. *Br. J. Cancer.* In press
259. Carney, W. P., Petit, D., Hamer, P., Der, C. J., Finkel, T., et al. 1986. *Proc. Natl. Acad. Sci. USA* 83:7485–89
260. Fujita, J., Yoshida, O., Yuasa, Y., Rhim, J. S., Hatanaka, M., Aaronson, S. A. 1984. *Nature* 309:464–66
261. Gambke, C., Hall, A., Moroni, C. 1985. *Proc. Natl. Acad. Sci. USA* 82:879–82
262. Needleman, S. W., Yuasa, Y., Srivastava, S., Aaronson, S. A. 1983. *Science* 222:173–75
263. Thein, S. L., Oscler, D. G., Flint, J., Wainscoat, J. S. 1986. *Nature* 321:84–85
263a. Gerhard, D. S., Dracopoli, N. C., Bale, S. J., Houghton, A. N., Watkins, P., et al. 1987. *Nature* 325:73–75
263b. Colby, W. W., Cohen, J. B., Yu, D., Levinson, A. D. 1986. In *Gene Amplification and Analysis,* Volume 4: *Oncogenes,* ed. T. S. Papas, G. F. Vande Woude, pp. 39–52
264. Conner, B. J., Reyes, A. A., Morin, C., Itakura, K., Teplitz, R. L., Wallace, R. B. 1983. *Proc. Natl. Acad. Sci. USA* 80:278–82
265. Valenzuela, D. M., Groffen, J. 1986. *Nucleic Acids Res.* 14:843–52
266. Myers, R. M., Lumelsky, N., Lerman, L. S., Maniatis, T. 1985. *Nature* 313:495–98
267. Winter, E., Yamamoto, F., Almoguera, C., Perucho, M. 1985. *Proc. Natl. Acad. Sci. USA* 82:7575–79
268. Myers, R. M., Larin, Z., Maniatis, T. 1985. *Science* 230:1242–46
269. Saiki, R. K., Scharf, S., Faloona, F., Mullis, K. B., Horn, G. T., et al. 1985. *Science* 230:1350–54
270. Wong, G., Arnheim, N., Clark, R., McCabe, P., Innis, M., et al. 1986. *Cancer Res.* 46:1–5

Ann. Rev. Biochem. 1987. 56:829–52

BIOSYNTHETIC PROTEIN TRANSPORT AND SORTING BY THE ENDOPLASMIC RETICULUM AND GOLGI

Suzanne R. Pfeffer and James E. Rothman

Department of Biochemistry, Stanford University School of Medicine, Stanford, California 94305

CONTENTS

1. SUMMARY AND PERSPECTIVES

Proteins destined for multiple compartments are synthesized in the cytoplasm by a single class of ribosomes. Therefore, the information or "signals" that

829

0066-4154/87/0701-0829$02.00

specify the localization of each nascent protein must reside in its sequence or structure. The transport machinery that recognizes these signals distributes distinct sets of proteins to the cytosol, the cell surface, lysosomes, secretory storage vesicles (in some cells), mitochondria and chloroplasts, the nucleus, the endoplasmic reticulum (ER), and the Golgi complex. The ER and Golgi function together to process and distribute proteins to the cell surface, secretory storage vesicles, and lysosomes, and their role in this process of biosynthetic protein transport is the major subject of our review.

The broad outline of the pathway of protein localization has been elucidated in the last decade. *Signal peptides,* defined as contiguous blocks of amino acid sequence that contain a sorting signal, direct the first decision that is made: to remain in the cytosol or be translocated across or into a membrane-bound compartment. Proteins containing signal peptides are imported from the cytosol into the nucleus, mitochondria, chloroplasts, or the ER, depending on the type of signal peptide they contain. Proteins lacking such a signal peptide are left behind, permanent residents of the cytosolic compartment by default. The signal peptide–directed import into ER is generally concurrent with translation for kinetic though not mechanistic reasons, giving rise to the ribosome-studded regions of ER known as rough ER.

Selective import from the cytosol is sufficient for nuclear, mitochondrial, or chloroplast localization. In contrast, newly synthesized proteins imported into the ER represent a complex mixture that must be physically transported beyond the ER, as well as sorted according to destination. This mixture includes proteins en route to the cell surface, lysosomes, and secretory storage vesicles, as well as proteins that will remain in the ER or the Golgi.

We refer to the pathway that accomplishes the transport and sorting of newly synthesized proteins imported by the ER as *biosynthetic protein transport,* to distinguish it from *endocytic protein transport.* Both pathways involve the movement of proteins between membrane-bound compartments in the form of transport vesicles. Endocytic transport begins at the cell surface and generally ends in the lysosomes, and represents an ongoing function of the plasma membrane that is not directly related to membrane growth or biosynthesis. Biosynthetic protein transport is the constitutive function of the ER and Golgi complex, and is intimately related to the growth of cellular compartments. It is an extension of the pathway of protein synthesis that supplies several organelles with newly made proteins.

The pathway of biosynthetic protein transport is essentially a generalization of the secretory pathway originally discovered by Palade and coworkers (1) to include destinations other than secretory storage granules. The mixture of proteins imported into the ER is exported to the Golgi, carried forward in transport vesicles. The Golgi is the main distribution center where the majority of the sorting decisions are made. Proteins targeted to the cell surface are carried from the Golgi to the plasma membrane in transport vesicles. Secretory storage vesicles form by budding from the Golgi, and proteins targeted to them are selectively included during the budding. Proteins en route to lysosomes are collected into a specific class of Golgi export vesicles (often referred to as primary lysosomes).

Figure 1 presents the pathway we propose for biosynthetic protein transport, based on the evidence that we review. Transport steps connecting compartments are mediated by the budding and fusion of transport vesicles. Each of these steps is unidirectional and energy-dependent. Open, vertical arrows indicate transport steps that are thought to be signal-mediated; solid, horizontal arrows denote steps that are thought to be signal-independent and to occur by default. The Golgi complex consists of three functionally distinct compartments, termed cis, medial, and trans, which correspond to sequential cisternae in its stack. Signal-dependent sorting to lysosomes and secretory storage vesicles occurs at the trans Golgi. All other transport appears to be nonselective. Retention of proteins by the ER and Golgi appears to be selective.

The trans Golgi is thus the major station for sorting in the Golgi stack. Its appearance reflects this intensive vesicular activity, which deforms the terminal (trans) cisternae into a tubular structure with many buds. These tubulovesicular extensions of the trans Golgi compartment in which sorting occurs have been termed the "trans Golgi network" (see 2 for review).

Given this pathway, consider the fate of a protein that is translocated into the ER because of a signal peptide, but lacks any additional sorting signal. Such a protein will be vectorially transported from the ER to the Golgi, across the Golgi from cis to medial to trans compartments, and on to the cell surface, simply following the default pathway (horizontal arrows, Figure 1). We term this the *bulk flow* because it is nonselective. If the protein is integrated into the membrane of the ER and free to diffuse, it will become a component of the plasma membrane, and move at the same rate as the bulk phase of membrane

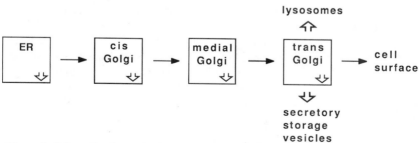

Figure 1   Informational organization and protein flow during biosynthetic protein transport. Horizontal arrows denote steps that we propose are signal-independent and represent the bulk flow. Open, vertical arrows in the boxes indicate retention of resident proteins, which we propose is signal-mediated; open, vertical arrows also represent signal-mediated transport to lysosomes and secretory storage vesicles.

lipid. If the protein is translocated into the lumen of the ER and remains free in solution, it will be externalized from the cell at the same rate as the bulk fluid in the ER lumen, contributing to the constitutive secretion.

For a protein to take any other path requires that it contain a sorting signal. Such signals are likely to be composed of regions on the surface of the protein which we term *signal patches*. Unlike signal peptides, signal patches will in general be formed from noncontiguous regions of the polypeptide chain that are brought together during protein folding. Thus, signal patches will be conformation-dependent. Signal peptides and not patches must be used to direct translocations across membranes, because proteins are unfolded during translocation. By contrast, proteins remain folded as they are sorted in the ER-Golgi system, and surface features (signal patches) can be used to direct their traffic. Signal peptides are easily identified because their function is generally preserved when they are swapped between different proteins, and can be independent of their position in a peptide chain. Signal patches are much more elusive because they cannot easily be transplanted to other proteins, and because attempts to identify them by genetic alteration are hindered by the fact that signal patch function will be eliminated by any change that indirectly affects conformation.

All of the evidence obtained to date suggests that conformation-dependent signal patches are used to direct sorting in the Golgi to lysosomes and secretory granules. Proteins targeted to these locations will move with the bulk flow from the ER through cis, medial, and into the trans Golgi. At the trans Golgi they are removed from the bulk flow in a signal-dependent fashion. These signals must be recognized by a "receptor" system, and in the case of sorting to lysosomes, such a system involving mannose-6-phosphate (man-6-P) receptors is known (see below). The basis for the segregation of secretory proteins to storage vesicles is not yet known.

The bulk movement to the cell surface seems to be mediated by a type of coated vesicle that is nonselective with regard to its content. As discussed below, the coat protein of these bulk carriers has not yet been identified, but it is known that the coat is not composed of clathrin. By contrast, the signal-mediated diversions from the bulk flow at the trans Golgi involve clathrin. Clathrin forms the coat of the Golgi vesicles carrying proteins en route to lysosomes. Also, clathrin coats portions of the secretory storage granules that bud from the trans Golgi compartment. It thus appears that clathrin-coated membranes, both in biosynthetic and endocytic protein transport, execute signal and cognate-receptor-dependent transport steps, while a different type of coat is responsible for the bulk flow to the cell surface. Virtually nothing is known of how either type of vesicle buds, or how its selective fusion is programmed and catalyzed. Nor is it known how clathrin-coated vesicles select their content while rejecting the bulk constituents of the membrane from which they bud, or how residents of the ER and Golgi avoid being carried to the surface by the bulk carriers. Now that the overall organization and flow of proteins and of information in the biosynthetic transport pathway are becoming clear, these mechanistic issues should be the focus of attention in the future.

2. BULK FLOW FROM THE ENDOPLASMIC RETICULUM TO THE CELL SURFACE

Typical plasma membrane proteins and a variety of secretory proteins traverse the secretory pathway in a constitutive manner. Their appearance at the cell surface or in the extracellular medium is independent of any extracellular stimuli as well as any intracellular regulatory barriers. The rate-limiting step in the constitutive transport of most proteins studied to date is their exit from the endoplasmic reticulum (3).

A. *Export from ER to Golgi*

Fitting & Kabat (4) and Lodish et al (3) noted that different membrane and secretory proteins exit the endoplasmic reticulum at distinct rates. For example, in HepG2 cells at 32°C, newly synthesized albumin and α-1-antitrypsin left the ER with a half-time of less than 30 minutes, while 50 minutes were required for C3 complement and α-1-antichymotrypsin export, and transferrin had a half-time of $\sim$150 minutes (3; see also 5, 6). Two viral membrane glycoproteins were also found to exit the ER at different rates in another cell line (4). How can these rate differences be explained?

Lodish and coworkers favor a model in which export from the ER is selective, i.e. signal-dependent (3). Fast-moving proteins would be those bound to a transport receptor; slow-moving ones would move at a slower bulk flow rate. In this type of model, the capture of proteins for departure from the ER might be thought of as an inside-out version of receptor-mediated endocytosis, a process that involves concentration of receptor-ligand complexes in clathrin-coated pits, followed by their removal from the plasma membrane by endocytosis. In the ER, specific receptors might select those polypeptides to be transported on to the Golgi apparatus, leading to their capture by an ER-coated bud. Permanent residents of the ER would be excluded from the buds.

The alternative mechanism would be that there are no signals or receptors for transport out of the ER. Soluble proteins in the lumen of the ER, and membrane proteins free to diffuse in the bilayer to regions where vesicles form, would be transported out at the rate of the bulk phase—the aqueous phase for secretory proteins, the lipid phase for membrane proteins. In this model, the fastest-moving proteins would be those moving at the bulk flow rate. The slower-moving proteins would have to be retarded for one or another reason.

The key fact needed to distinguish these models is the rate of bulk flow out of the ER, which would enable comparison with the rate of protein transport. This determination would require placement of an inert marker inside the ER; the rate of appearance of the marker at the cell surface would then measure the rate of bulk flow. For this purpose, Wieland et al (F. Wieland, M. Gleason, T. Serafini, J. Rothman, in preparation) employed acyl tripeptides of the structure, N-acyl-Asn-[^{125}I]Tyr-Thr-NH$_2$, where acyl is a fatty acid of chain length varying from two to ten carbons. The tripeptides diffused into cells and were trapped in the lumen of the ER upon receipt of a high-mannose oligosaccharide chain at the Asn residue. The rate of secretion of these simple glyco-tripeptides could then be determined. The half-time for transport from the ER to the cell surface was found to be between 5 and 20 min in several cell lines. The fastest membrane and secretory proteins are transported no faster than this bulk flow rate (3). Since the glycopeptides are too small to contain a signal, it appears that no signal is needed for efficient and rapid constitutive transport from ER through the Golgi and to the cell surface.

Given that all proteins in the lumen or the membrane of the ER should, in principle, move to the cell surface at the same bulk flow rate, why do many transiting proteins move more slowly? Differential transport rates might be due to a process similar to adsorption chromatography. Since the bulk of the ER membrane is occupied by immobile, permanent residents, transient adsorption of exported proteins to a large portion of the ER membrane by electrostatic and/or hydrophobic interactions would lead to their retardation, in proportion to the fraction of the time they are adsorbed. The very high ratio of ER membrane surface to aqueous volume and relatively low lumenal protein concentration suggests that adsorption may be a significant factor in ER export. Alternatively, these rate differences may reflect variability in the rates of protein folding and assembly, since proteins do not exit the ER until these processes are completed.

The first indication that unassembled or unfolded proteins are selectively retained in the lumen of the ER came from studies of hetero-oligomeric proteins in instances in which one of the subunits was synthesized without its partner. For example, both the heavy chain of IgM (7), and the gamma chain of the histocompatibility antigen HLA-DR, a transmembrane polypeptide (8), accumulate in the ER unless able to complex with their respective partners, light chains and HLA-DR alpha or beta chains. Yet another striking demonstration that conformation is important for export comes from the observation that retinol-binding protein requires bound ligand for ER exit (9).

This notion has been extended through detailed studies of the assembly of the trimers of influenza virus hemagglutinin in relation to the transport process. Using conformation-specific antibodies, differential protease sensitivity, velocity sedimentation, and chemical crosslinking, two laboratories demonstrated that efficient export of influenza virus hemagglutinin from the endoplasmic reticulum required correct folding and assembly of monomeric subunits into a trimeric form (10, 11). This process had a half-time of about seven minutes, and was followed rapidly by transport to the Golgi apparatus. A similar process has been described for the G glycoprotein of vesicular stomatitis virus (VSV; 12).

Unfolded and/or unassembled chains thus appear to be retained in the ER until folding is completed, and recent evidence suggests that certain of these proteins associate with a 77 kd protein first identified as an immunoglobulin heavy chain *b*inding *p*rotein (BiP; 13, 14). In vitro–generated mutants of influenza hemagglutinin whose transport is arrested in the ER were found to be blocked at different stages in the folding pathway, and unfolded hemagglutinin molecules were found associated with the same ER resident protein shown by others to associate with immunoglobulin heavy chains in the ER of heavy-chain myelomas (11). More importantly, wild-type hemagglutinin appeared to associate with BiP as an intermediate step along its normal pathway of ER export (11). Thus BiP may somehow facilitate oligomerization and/or folding in the lumen of the ER.

Indeed, Munro & Pelham (15) have found that BiP was released from immobilized immunoglobulin heavy chains upon addition of ATP, but not ADP. This would imply for the first time the existence of an ATP pool in the ER lumen, and suggest that BiP may act in ATP-driven cycles to disaggregate incorrectly folded proteins and permit their export (15).

The simplest view is that proteins imported into the ER generally aggregate and/or precipitate before they can properly fold or assemble. We use the term precipitation in this context in a very broad sense. Large, multimolecular aggregates of unfolded proteins (such as those that could be visualized in the electron microscope) as well as individual, unfolded chains adsorbed to the surfaces of resident membrane proteins would be embraced by this term. Such precipitates have not yet been demonstrated, but we infer them from the transient retention of unfolded or partially folded proteins (10, 11). The precipitates would be retained, nonspecifically. A protein would be rapidly transported out of the ER with the bulk flow as soon as it is dissolved from the precipitate and folds. Proteins like BiP and protein disulfide isomerase may facilitate dissolution of such precipitates, which may involve hydrophobic interactions and incorrect (inter- and intra-molecular) S-S bridges.

Why might proteins aggregate or precipitate before they can fold and assemble in the lumen of the ER? Folding in the ER should be more difficult than in the cytoplasm, or inside mitochondria, for two reasons. First, unlike the cytosol, the ER lumen consists almost entirely of freshly imported proteins all trying to fold at once. The ER lumen is also an oxidizing environment designed to favor disulfide bond formation. Any nonspecific contacts among unfolded chains would lead to local precipitation; those proteins bound to BiP would be those being released from this postulated precipitate.

The alternative to precipitation is that some unassembled subunits may be selectively retained until they are fully assembled, after which they are released. BiP could be a part of such a retention machinery.

In any case, binding of unfolded molecules to BiP is not sufficient to explain their transient retention in the ER. First, the quantity of immunoglobulin heavy chain retained greatly exceeds the quantity of BiP bound to it (14). Second, BiP is itself secreted from cells that are synthesizing and secreting antibodies that recognize BiP (14). Thus BiP is intrinsically a soluble protein that is retained in the ER, perhaps as it is operating on precipitates that are themselves nonselectively retained.

It has been reported that BiP is identical with a member of the heat shock family of proteins (15, but see also 11). Yet another perhaps related molecule is egasyn, first identified as a 64,000-dalton ER glycoprotein found associated with microsomal β-glucuronidase (16). The finding that only 10% of egasyn was found complexed with β-glucuronidase suggests that the bulk of egasyn may be complexed with other proteins in the ER (16). Perhaps BiP and egasyn represent members of a family of ER proteins that function to facilitate specific protein oligomerization or to disaggregate incorrectly assembled proteins. Whether egasyn drives the assembly of β-glucuronidase tetramers in an ATP-dependent manner remains to be determined. The unexpected retardation within the ER, of proteins bearing stable terminal glucose residues in deoxynojirimycin (DNM)–treated cells (17), may reflect yet another ER resident that recognizes this abnormal feature of nascent glycoproteins; alternatively, these proteins may not fold correctly.

Thus the selection of proteins for ER export includes not only discrimination between residents and nonresidents, but also an assessment of protein conformation. Unassembled proteins are denied access to departing transport vesicles, possibly by their presence in precipitates. Any mutation that affects protein folding will therefore cause retardation in the ER. This almost certainly explains why a large number of mutations cause transiting proteins to become stuck in the ER. Rather than eliminating a positive signal for transport, these mutations probably slow or eliminate the ability of the protein to fold properly.

B. Transport Through the Golgi Stack

As we have discussed, newly synthesized membrane and secreted proteins are translocated across the endoplasmic reticulum membrane where they fold and also, in many cases, receive high-mannose, asparagine-linked oligosaccharides. They are then transported to the Golgi stack, where they may be further glycosylated, prior to their subsequent delivery to the plasma membrane, secretory vesicles, or lysosomes.

Immunoelectron microscopy experiments monitoring a synchronized wave of transported viral glycoprotein have shown that transported proteins enter the Golgi at the face of the stack adjacent to the endoplasmic reticulum (cis side) and exit at the opposite trans face (18, 19). Histochemical studies have long suggested the distinct chemical and enzymatic nature of individual cisternae (20). Elucidation of the biochemical steps involved in asparagine-linked oligosaccharide processing, and biochemical and immunocytochemical localization of the enzymes that mediate these reactions, have greatly clarified our understanding of the functional organization of the Golgi, and indicate that the Golgi stack is comprised of at least three functionally distinct compartments (see 21, 22 for review).

The cis Golgi is the site where lysosomal hydrolases receive a man-6-P tag, since this compartment is likely to house lysosomal enzyme N-acetylglucosaminyl (GlcNAc) phosphotransferase (23, 24). N-linked oligosaccharides are modified by the action of GlcNAc transferase I in the medial Golgi (25), and subsequent additions of galactose and sialic acid occur in the trans Golgi (26, 27). The requirement of the cis-Golgi GlcNAc phosphotransferase for α-1,2-linked mannose residues present on lysosomal hydrolases (28) would fit well with a model in which Golgi mannosidases I and II, enzymes that destroy such linkages, would be sequestered in a more distal compartment along the transport pathway. Segregation of such competitive reactions may be one function of Golgi stack compartmentalization (21).

Newly synthesized proteins are therefore vectorially transported from ER to cis to medial to trans Golgi, prior to their ultimate distribution from the trans Golgi. The fact that ER $\rightarrow$ Golgi transport could be blocked by energy poisons is inconsistent with a direct continuity of these organelles. The transport of proteins between each of these compartments is therefore thought to be mediated by the abundant, nearby transport vesicles that contain the transported proteins.

The finding that proteins can transfer from one Golgi stack to another in a unidirectional fashion (29, 30) virtually proves the vesicle hypothesis and implies that vesicles have a built-in targeting capacity. Using Chinese hamster ovary (CHO) cells possessing defects in specific glycosyltransferases, it was possible to detect the transfer of the VSV G glycoprotein from the medial Golgi of cells lacking galactosyl transferase to the trans Golgi of wild-type cells, after fusion of these cell types. Analogous experiments using CHO cell lines defective in either GlcNAc transferase I or sialyl transferase showed that transfer was also possible from the cis Golgi of one stack to the medial

Golgi of another, and furthermore, that the likelihood of a forward transfer was at least five times higher than that of a lateral transfer (30).

The cis face of the Golgi stack is usually located next to the transitional region of ER, and the Golgi compartments are placed next to each other in the order of their encounter. This made it seem possible that specific vesicle targeting required this level of cytoplasmic organization. Instead, these "hopping" experiments suggest that vesicle targeting is independent of strict cytoplasmic organization. Therefore each type of vesicle must have its own address marker that specifies its target, and the target membrane receiving this type of vesicle must also have a complementary receptor system that recognizes the address marker and allows vesicle docking.

C. Export from the Golgi to the Cell Surface

Several lines of evidence suggest that newly synthesized proteins depart from the Golgi apparatus by a bulk flow process. First, a variety of secretory proteins traverse the Golgi at the same rate (3), similar to the rate of secretion of the simple glycopeptides (F. Wieland et al, in preparation). Second, when signal-dependent diversion from this flow to lysosomes or secretory granules is saturated or incapacitated, the bulk flow pathway to the surface seems to be the path taken by default. In I-cells that are unable to tag lysosomal hydrolases with man-6-P residues and in cells deficient in one of the man-6-P receptors, a large fraction of lysosomal hydrolases are constitutively secreted (31, 32). Chloroquine-treated AtT-20 cells divert ACTH precursor from its normal secretory vesicle destination to a constitutive secretion pathway (33). Similarly, overexpression of two proteins en route to the yeast lysosome-like vacuole led to the constitutive secretion of a significant fraction of these proteins, presumably by surpassing the capacity of the vacuolar sorting machinery (34, 35). Therefore, there is no apparent barrier for entry into the ER $\rightarrow$ Golgi $\rightarrow$ cell surface pathway: it appears to be nonselective in its choice of cargo and in the rate with which this cargo appears at the cell surface.

In sum, biosynthetic transport in both the ER and the Golgi seems to operate by a bulk flow mechanism. Proteins designated for retention as permanent residents must somehow be selected by "retention signals"; all others move toward the cell surface with the bulk flow. The structure of the ER and the Golgi are well suited for this purpose. The geometry of their flattened cisternae maximizes the surface-to-volume ratio, promoting the transport of lumenal content, since in bulk flow access to a transport vesicle would depend only on the probability of a soluble or membrane protein occupying the internal volume or vesicle membrane surface at the time of vesicle budding.

Bulk flow export need not be entirely random, and indeed must not be in order to permit selective retention to occur (see Section 3C). For example, if resident proteins were somehow confined to regions where vesicle formation could not occur, export would only include transiting proteins and not permanent residents of the ER and the Golgi. The fact that vesicle formation occurs at the rims of Golgi cisternae may be significant in this regard: stacking of Golgi cisternae may itself render central regions incapable of forming vesicles, and one could even speculate that the mechanism of stacking is directly connected with a resident-protein retention mechanism. The vesicles that mediate intra-Golgi transport (cis to medial to trans) need not be discriminate in their choice of cargo, but they must maintain vectoriality and not contain resident proteins. Thus, a binary decision to stay or go forward is sufficient up to the trans Golgi. At this stage, a protein must either stay as a resident, or move to one of multiple forward targets. This necessitates that multiple types of vesicles form at the trans face of the Golgi.

D. The Vesicles Mediating Biosynthetic Protein Transport

The best-studied vesicular transport step has been the process of receptor-mediated endocytosis, in which clathrin-coated vesicles retrieve clusters of receptors and receptor-ligand complexes (36). An endocytic coated pit or vesicle may contain a mixture of different receptor proteins. The fact that different receptors have different probabilities of occupying a coated pit at steady state has been taken to reflect different affinities of receptors for coated pit components (36). Although single-amino-acid substitutions can render a receptor incompetent for coated pit entry (37), it is still possible that endocytic coated pit recognition involves simple receptor-aggregate retrieval.

While the process of clathrin-coated vesicle formation during endocytosis has served as a paradigm for models of other steps in intracellular transport, several sets of findings indicate that the mechanisms by which proteins are collected and captured for biosynthetic transport may be quite different. The JD mutation of the low density lipoprotein (LDL) receptor was originally described as a receptor with normal LDL binding characteristics, yet it was not internalized (38) and did not cluster in clathrin-coated pits (39). Most striking, however, was the finding that the JD LDL receptor was exported to the cell surface at precisely the same rate as the wild-type protein (M. Brown, personal communication; see also 40). Thus, the mechanism by which the JD receptor had access to transport vesicles along the secretory pathway must have been substantially different from that used during endocytosis. This finding would be consistent with a bulk flow model in which constitutively exported proteins do not use clathrin-coated vesicles for their export.

At least three types of vesicles have been found to form from Golgi cisternae. Two of these vesicle types form only at the trans face: one is the secretory storage vesicle, that contains densely packed (and sorted) secretory protein and has a partial clathrin coat (41); the other is a small clathrin-coated vesicle containing acid hydrolases bound to man-6-P receptors (42–46). The third type of vesicle is a small non-clathrin-coated vesicle that may be the bulk carrier (47, 48). These non-clathrin-coated carriers are found at all levels of the Golgi stack. They form from each cisterna and contain the VSV G protein at its prevailing (bulk) concentration in the parental cisterna (47). This property and their localization strongly suggest that these are the bulk carriers responsible for bulk forward movement in biosynthetic protein transport. By contrast, clathrin is found exclusively at the trans face of the Golgi (48, 49).

3. SIGNAL-MEDIATED DIVERSIONS FROM THE BULK TRANSPORT PATHWAY

A. *To Lysosomes*

Proteins en route to lysosomes, and in regulated secretory cells, to secretory vesicles, share a common pathway of intracellular transport with plasma membrane and constitutively secreted proteins. It appears that this common pathway extends as far as the trans Golgi, where these classes of proteins are then in some way sorted to their correct destinations.

Upon arrival in the cis Golgi, lysosomal hydrolases are selectively recognized by a GlcNAc-phosphotransferase that adds a GlcNAc-P to α-1,2-linked mannose residues present on their attached N-linked oligosaccharides. This enzyme has both a protein recognition site that is selective for lysosomal hydrolases and a catalytic site that binds high-mannose chains (50, 51). These sites can be separately mutated, since fibroblasts from certain patients with pseudo-Hurler polydystrophy possess a phosphotransferase with normal activity using α-methylmannoside as an acceptor, but with significantly decreased affinity for lysosomal enzymes (52). The signal on lysosomal enzymes recognized by the phosphotransferase is probably a signal patch since it is exquisitely sensitive to conformation (50, 51).

After modification by the phosphotransferase, a Golgi α-GlcNAc-1-phosphodiester N-acetylglucosaminidase generates man-6-P residues responsible for the sorting of lysosomal hydrolases. The presence of a man-6-P tag enables lysosomal enzymes to bind to one of two man-6-P receptors in the Golgi complex that in some way segregate these proteins for lysosomal delivery (see 53, 54 for review).

Two views exist as to the fate of man-6-P receptor-ligand complexes in the Golgi. While one model involves departure of such complexes from the cis Golgi (43), other immunocytochemical experiments (55, 56; G. Griffiths, personal communication) and biochemical studies suggest that man-6-P receptors leave the Golgi apparatus at the trans face.

Duncan and Kornfeld treated cells with deoxymannojirimycin (DMM), a specific inhibitor of the Golgi enzyme, mannosidase I. Under these conditions, high-mannose oligosaccharide chains present on newly synthesized man-6-P receptors are unable to be processed to complex structures. They then removed the DMM and measured the conversion of receptor oligosaccharides to complex type. If the receptor recycled back through the medial Golgi, it should have been a perfect substrate for the mannosidase residing there, as well as for subsequently acting glycosyltransferases. However, no modification was detected (J. Duncan, S. Kornfeld, personal communication). In other experiments, the man-6-P receptors could be shown to contact sialyl transferase in the trans Golgi, during their intracellular transport (J. Duncan, S. Kornfeld, personal communication). Furthermore, many lysosomal hydrolases contain galactose and sialic acid residues (54), indicative of passage through the trans Golgi.

Taken together, these experiments establish that the bulk of lysosomal hydrolases bound to man-6-P receptors leave the Golgi at the trans face. The transport vesicles carrying these receptor-ligand complexes appear to then fuse with endosomes. Bound ligands are released in the acidic environment of this compartment (eventually converting the endosome to a lysosome) and man-6-P receptors then recycle back to the trans Golgi for another round of transport (53, 54, 57). The existence of a low percentage of cell-surface man-6-P receptors could be readily explained as follows. The volume of endocytic transport between the endosome and plasma membrane is likely to be large relative to the rate of lysosome synthesis. A low level of mis-sorting of the man-6-P receptor, from endosome to plasma membrane rather than back to the trans Golgi, would generate a small pool of cell-surface man-6-P receptor with no detrimental consequences to the cell; on the contrary, this surface receptor provides a scavenger pathway for retrieval of any mis-sorted hydrolases. The fact that the amount of cell-surface man-6-P receptor varies between cell types may simply reflect different volumes of endocytic traffic between these cell lines.

Enzyme cytochemistry (42), as well as immunocytochemistry at the electron microscope level (43), has shown that lysosomal hydrolases and at least one type of man-6-P receptor occupy coated vesicles in the vicinity of the Golgi apparatus. That the coats of these vesicles are comprised of clathrin comes from biochemical fractionation studies that showed that isolated, clathrin-coated vesicles contain lysosomal hydrolases bound to man-6-P receptors (44–46). Double-label immunocytochemistry using antibodies to clathrin and one of the man-6-P receptors is needed to prove the notion that proteins en route to lysosomes leave the Golgi in clathrin-coated vesicles.

B. To Secretory Storage Vesicles

The formation of secretory storage vesicles has long been appreciated to occur at the Golgi trans face (1). These vesicles fuse with the cell surface only upon receipt of an extracellular signal (a hormone or a neurotransmitter, depending on the cell type). This process termed exocytosis (1) is thus regulated, and gives rise to the regulated secretion of the selected contents of the secretory storage vesicle (see 58 for review). By contrast, secretion via bulk flow to the surface is constitutive and continuous. Although it is not yet known how proteins targeted for storage vesicles are collected, it is clear that their content is highly selected, and the evidence suggests that some form of selective precipitation in the trans Golgi is involved.

For example, in AtT-20 cells, the selectivity of this process is demonstrated by the lack of packaging of sulfated glycosaminoglycans (59), laminin (60), viral glycoproteins (61, 62), and immunoglobulin light chains (L. Matsuuchi, R. Kelly, personal communication) into ACTH-containing secretory vesicles. The fact that these cells can package a diverse group of heterologous secretory products, including insulin (63), growth hormone (62), and trypsinogen (60), suggests that the sorting machinery may not in fact recognize specific linear sorting sequences present in each of these proteins, but rather, may employ some more general structural feature for cargo selection, in other words, a signal patch.

Most stored secretory components self-aggregate in the trans Golgi to form morphologically dense cores that are further condensed in secretory vesicles. This type of aggregation is likely to be a pH-dependent process. Most secretory vesicles have acidic interiors (64), and it has recently been shown that the trans Golgi is also a slightly acidic compartment (65), consistent with the first appearance of condensed secretory products.

Recognition of a molecular aggregate could facilitate the sorting process since a single hypothetical receptor molecule could in one step recognize multiple proteins for packaging (58). An aggregate could contain different proteins to generate a complex mixture of secreted proteins. Alternatively, homoaggregates could be targeted to a single secretory vesicle or distinct secretory vesicles. Consistent with the formation of homoaggregates is the recent report of heterogeneity of stored secretory vesicles with respect to their contents (66).

A model in which protein aggregates trigger secretory vesicle formation provides an effective means by which constitutively secreted proteins could be excluded. In this respect, the triggered budding of secretory vesicles might be similar to the triggered budding in phagocytosis, a process that involves aggregate recognition. The diversion of ACTH precursor to the constitutive pathway in chloroquine-treated cells (33) may have merely been due to an inhibition of ACTH aggregate formation at neutral pH.

C. Retention of Selected Proteins by the ER and Golgi as Permanent Residents

How do the ER and Golgi compartments each maintain a unique set of permanent residents in the face of a bulk, apparently nonselective movement toward the cell surface? This is largely a problem for the membranes of these organelles, since at any given time, the vast majority of the proteins in the membranes of the ER and Golgi are permanent residents; their lumenal contents seem to be mostly comprised of transiting proteins. Given that transport forward is by default, then retention in the ER or the Golgi must somehow be programmed by a signal on the retained protein. How this works is not yet known, but poses significant challenges to simple views of intracellular membrane structure.

In order to retain proteins against a bulk, nonselective membrane flow, in a formal sense, there would have to be two continuous but distinct phases in the membranes of the ER and in the Golgi cisternae. The *immobile phase* would consist of regions or domains from which transport vesicles for some reason cannot form. This might include the flat faces of the cisternae of the Golgi stack and the bulk of an ER cisterna. The *mobile phase* would consist of the regions from which vesicles could bud, such as the rims of the Golgi stack of the transitional elements of ER. A protein would need a signal to enter or be retained in the immobile phase; transiting proteins would remain in or diffuse into the mobile phase by default. In this model permanent residents of the ER enter the immobile phase of the ER as soon as they are made. Residents of the Golgi would remain in the mobile phase of the ER and be carried into the Golgi, and continue to move through the stack with the bulk flow in the mobile phase until their target cisterna is reached. There they would somehow segregate into the immobile phase and away from further bulk flow towards the surface.

Each permanent resident of the ER or Golgi must possess a positive retention signal, specific for the appropriate compartment. This organization makes a great deal of genetic sense since it reduces the informational load. Consider the alternative if forward movement, and not retention, were programmed: each protein passing through the ER and each of three Golgi compartments would have a collection of distinct signal patches on its surface, one for each forward transfer that takes place. The number of different proteins passing through the ER-Golgi system en route to all other export destinations is likely to exceed the number of different types of proteins retained. Moreover, the nature of the products exported through the ER-Golgi system differs greatly from cell to cell, whereas the residents of the ER and Golgi are mostly the same. Therefore, many fewer signals need be specified in the genome when proteins are selectively marked for retention, rather than for forward movement. Also, the structure (and function) of transported proteins need not be constrained by a catalogue of multiple signal patches.

How might an immobile phase be generated and maintained? One possibility is that the proteins retained in the immobile phase are locked into place by interaction with a meshwork of a skeletal or matrix protein (like the nuclear lamina), located on either the cytoplasmic or lumenal side. The signal for retention would be a feature on the surface of the protein (a signal peptide or patch) that enables it to bind the matrix. A different kind of matrix would be needed for the ER and each Golgi compartment. Vesicles would not form from the immobile phase because of the rigid and extended matrix. This type of model is attractive because there must be some kind of a matrix in between the cisternae of the Golgi stack to glue them together, and to give the ER and Golgi cisternae their flattened shape. Moreover, a different matrix must already exist between each type of cisterna to enable the Golgi compartments to be attached to each other in their unique order (cis, medial, trans). This model is relatively unattractive because enzymes bound in a meshwork cannot diffuse laterally in the plane of the membrane, and this may be inefficient for their action.

A second type of model does not invoke interactions with a matrix, but envisions that permanent residents form patches by selective, local precipitation in the plane of the bilayer. These patches could be homogeneous or heterogeneous with respect to their polypeptide composition, and need not be large compared to the diameter of transport vesicles (<50 nm). Even patches that are a modest fraction of this size would have to be broken up to fit into the highly curved surface of a transport vesicle. In this model, the energy needed to break up the interactions holding such a patch together would be more than is made available during vesicle budding. The patches would simply not fit into (coated) vesicles of rigidly determined size, and thus be left behind. Proteins would enter these aggregates or "immobile phases" in the target compartment via a signal patch that constitutes a binding site used in the aggregation. Whatever the mechanism of forming immobile phases, it will pose important challenges to our current understanding of membrane structure, and will be key to understanding biosynthetic protein sorting.

D. The Additional Complexities of Polarized Cell Surfaces

The apical and basolateral surfaces of polarized epithelial cells each have a distinct protein composition, and a number of laboratories are investigating the way in which such cells establish and maintain these distinct domains (see 67 for review). The observation of Rodriguez-Boulan & Sabatini (68) that enveloped viruses bud in a polarized manner from cultured Madin-Darby canine kidney (MDCK), as well as the subsequent finding that the constituent viral glycoproteins are themselves delivered to either apical or basolateral plasma membranes, has provided new and excellent tools with which to address this problem.

It is now clear that viral glycoproteins en route to either apical or basolateral domains occupy the same Golgi cisternae and share a common route of transport, up to and including the trans Golgi (69, 70). They are then delivered directly to their appropriate destination (69, 71–73). While plasma membrane proteins are generally exported from the trans Golgi in what appears to be a constitutive, signal-independent process (Figure 1), polarized cells must in some way distinguish proteins destined for either apical or basolateral domains.

Distinct signals might be used for both apical and basolateral transport; alternatively, one of these routes could be signal-mediated while the other is not. Several constitutively secreted endogenous proteins are released exclusively into the apical domain of MDCK cells (74, 75), suggesting that the bulk flow is to this domain. Transport to the apical surface would then be the signal-independent or default pathway, while transport to the basolateral surface would represent signal-mediated diversion from the bulk flow. When a number of exogenous secretory proteins are expressed in MDCK cells, they are found in roughly equal amounts in the apical and basolateral media (74, 75). Considering that the relative surface areas of these domains differ by a factor of about four (76), one might predict that the corresponding volumes of transport to apical and basolateral domains would also differ by this amount. If so, one could argue that the exogenous proteins were secreted preferentially (by a factor of four), into the apical medium.

The polymeric immunoglobulin receptor has the unusual attribute of appearing first at the basolateral surface and then being transported to the apical surface of polarized cells. Mostov and coworkers have recently shown that deletion of the cytoplasmic domain of this receptor leads to its direct targeting to the apical surface, bypassing the basolateral domain (77). This is consistent with the idea that the path to the basolateral surface is signal-mediated, and that the bulk flow (pathway by default) is to the apical surface. A direct experiment using a bulk phase marker is needed to distinguish between models for sorting to apical and basolateral surfaces in the trans Golgi of polarized cells.

4. ELUCIDATING THE TRANSPORT MACHINERY

The pathways of both biosynthetic and endocytic protein transport are now well enough defined for questions of molecular mechanism to be posed. The key to understanding the flow of traffic between membrane compartments is to elucidate the working parts of transport vesicles. How do transport vesicles bud off? How are they targeted? How do they fuse? A combination of biochemical and genetic approaches holds the promise of revealing the responsible machinery in the relatively near future.

A. Biochemical Studies of Transport Reconstituted in Cell-Free Systems

The reconstitution of vesicular transport between compartments has been achieved for transport between the cisternae of the Golgi stack. This system has offered some initial insights into the mechanisms of vesicle budding, targeting, and fusion.

The transport of the VSV-encoded G protein, from the cis compartment of CHO Golgi stacks that lack UDP-GlcNAc transferase I to the medial compartment of wild-type CHO Golgi stacks, could be detected by monitoring the incorporation of GlcNAc as a consequence of transport in a "complementation" assay (78, 79). The unique localization of UDP-GlcNAc transferase I to the medial Golgi (25) permitted biochemical detection of the arrival of G protein into this compartment, and a variety of controls ruled out the possibilities of nonspecific fusion of entire Golgi stacks, or delivery of the transferase to mutant Golgi, rather than G protein to wild-type medial Golgi (79, 80). Transport absolutely required the presence of a crude cytosol fraction, ATP, and proteins on the surface of the Golgi membranes (78, 79, 81).

Electron microscopy revealed that when the isolated Golgi stacks were incubated with cytosol and ATP to initiate transport, coated vesicles of the non-clathrin variety form from their rims (47, 82). These vesicles contain G protein (47) and are thought to be responsible for the transport between the cisternae in the stacks of this cell-free system, resulting in the maturation of glycoprotein as measured in the biochemical assay.

Intercompartmental transport of G protein occurs via a series of sequential intermediates that most likely reflect stages in the budding and fusion of these transport vesicles (82, 83). ATP is required at every stage, and cytosol at all but the last step, which seems to involve the actual fusion of the transport vesicle with the target membrane. The process of targeting a vesicle is probably quite complex, since a series of prefusion complexes is involved.

The budding of transport vesicles requires both ATP and a mixture of proteins from the cytosol. This establishes that vesicle budding is an active process that does not occur in the fashion of a self-assembly. Therefore, even though a coat protein like clathrin can self-assemble to yield empty coats, the budding off of a transport vesicle is apparently a much more complex process. Clathrin represents only the outermost shell of certain vesicle coats (84–86). Therefore, several other proteins are likely to act together with clathrin and clathrin homologues to contribute a substantial portion of the driving force for budding, determine the size of the vesicle, and in the case of signal-mediated transport, possibly provide a scaffold for content selection.

Interestingly, yeast mutants lacking clathrin transport almost normally (87), suggesting that while the outermost clathrin shell of clathrin-coated vesicles is helpful, it is dispensable. Perhaps the 100 kd and 50 kd proteins that form a shell beneath the clathrin lattice (84–86) carry out most of the coat functions outlined above. Clathrin may only provide an extra, dispensable "push" for budding. Once a coated vesicle is formed, the coat must be removed to permit the contact between bilayers needed for fusion with the target membrane. Uncoating is an energy-dependent process (88). A cytosolic uncoating enzyme that couples ATP hydrolysis to the removal of the clathrin type of coat has been identified and studied extensively (see 89 for review).

A cytosolic factor that utilizes fatty acyl–Coenzyme A as a cofactor has recently been identified, raising the possibility that a cycle of acylation and deacylation may promote rounds of transport. The sensitivity of donor and acceptor membranes to N-ethylmaleimide (NEM) treatment (81) led to the identification of this NEM-Sensitive Factor (NSF) that is essential for intra-Golgi transport (90). NSF is present in both cytosol and in Golgi membranes. NSF activity, as measured by the ability of a given fraction to restore activity to NEM-inactivated donor and acceptor Golgi fractions, is abolished by protease, and is larger than 30,000 daltons as judged by ultrafiltration. NSF is present on the surfaces of CHO Golgi membranes and can be extracted from the membranes by high salt concentrations. The ability of fatty acyl coenzyme A to greatly stimulate NSF activity has led to the hypothesis that acyl CoA acts as an acyl chain donor to activate a recipient molecule for transport. Deacylation might recycle this component for another round of transport. The G-protein to be transported is not the acyl acceptor in this cycle, since fatty acyl CoA stimulation of an NSF-dependent transport reaction could be demonstrated using a type of G protein that cannot be acylated.

What might be the function of cytosolic factors that facilitate intracellular transport reactions? In the case of intra-Golgi transport, the reactions that must be reconstituted include collection of transported proteins into a vesicle bud followed by budding, targeting, and fusion of the transport vesicle with the appropriate target membrane. One can envision a series of proteins that act at each of these distinct steps. Remarkably, the cytosol from yeast (91) and from plants (92) will substitute for animal cell cytosol in promoting transport in the Golgi stack and in forming the coated vesicles. This interchangeability means that the transport machinery is extremely similar, even in detail, in all eukaryotes.

The reconstitution of other steps of intracellular protein transport has been recently described. To assay for ER to Golgi transport in vitro, Haselbeck & Schekman (93) made use of a temperature-sensitive yeast strain that accumulates core-glycosylated invertase in the ER at the nonpermissive temperature to devise a complementation assay. A double mutant strain was constructed that also lacked a Golgi α-1,3 mannosyltransferase. The transport of invertase from these donor membranes to acceptor membranes containing α-1,3 mannosyltransferase activity but lacking invertase was detected by monitoring the acquisition of outer chain α-1,3-linked mannose residues on invertase. The reaction required ATP, GDP mannose, and either Mg^{2+} or Mn^{2+} as a divalent cation, but not cytosol. Interestingly, the activity of the donor membranes but not the acceptor membranes was found to be sensitive to NEM as well as trypsin treatments. This is in contrast to the protease sensitivities of both donor and acceptor membranes as well as the cytosol requirement in the cell-free intra-Golgi transport system (79, 81).

Very recently, Balch and coworkers (94) have established a cell-free system that reconstitutes the transport of the VSV G protein from the ER to the Golgi of CHO cells. The trimming of high-mannose oligosaccharides present on G protein by the Golgi enzyme, mannosidase I, was used to monitor this transport. ER $\rightarrow$ Golgi transport required ATP and cytosolic factors, and could only be achieved when homogenates were prepared from mitotic cells.

Woodman & Edwardson have established a cell-free system to study delivery of newly synthesized, influenza virus neuraminidase to the plasma membrane (95). An acceptor fraction is prepared by binding ^{3}H-sialic acid-labeled, Semliki Forest virus to cell surfaces prior to homogenization. Fusion of a neuraminidase-containing, exocytic vesicle with the acceptor preparation leads to release of free ^{3}H-sialic acid. This transport reaction requires ATP, as well as trypsin-sensitive factors present either on the surfaces of the membrane vesicles or free in the cytosol (95).

Crabb & Jackson (96) reconstituted the process of secretory vesicle exocytosis using purified cortical secretory vesicles and plasma membranes from sea-urchin eggs. By attaching sea-urchin eggs to polylysine-coated microscope slides and breaking them open on this surface, these workers were able to obtain stable plasma membrane sheets with the appropriate orientation for secretory granule interaction. Exocytic fusion, as measured by quantitative phase-contrast microscopy, was Ca^{2+} dependent and did not require exogenously added ATP. Whether ATP is somehow involved in this reaction in vivo is not yet known.

Two laboratories have established a cell-free system to investigate the events involved in endocytosis. Davey et al (97) have made use of the susceptibility of sialic acid residues on Semliki Forest virus present in endocytic vesicles in BHK cell extracts (acceptor fraction) to cleavage by the neuraminidase activity of fowl plague virus presented in endocytic vesicles in a donor cell extract, to monitor fusion events along the endocytic pathway in a complementation assay. Like in vitro intra-Golgi transport, organelle fusion as measured in this assay was also ATP dependent, and appeared to be selective as well as efficient (97).

Braell (98) has used the formation of antibody:antigen complexes that would occur only on fusion of donor and acceptor endocytic vesicles to monitor the endocytosis process. The reaction measured in this assay was also ATP- and temperature-dependent, and could further be shown to require cytosolic factors.

Significant progress is being made in the reconstitution of nuclear breakdown and assembly, using either Xenopus eggs (99–103) or synchronized CHO-cell extracts (104, 105). While these processes may at first seem unrelated to our understanding of the secretory pathway, it is important to remember that the ER and Golgi also break down during cell division (see 106 for review). Organelles, like chromosomes, must be distributed to daughter cells, and distinct compartments from the secretory pathway re-established (106). Therefore, factors regulating nuclear formation and breakdown may act upon or be intimately associated with or identical to proteins that maintain compartment boundaries along the secretory pathway (106).

B. Genetic Studies

The secretory pathway in yeast has been defined genetically by a series of temperature-sensitive mutant strains blocked at various stages in protein transport (107, 108). At least 23 complementation groups and thus gene products are required for the transport of secretory proteins from their site of synthesis to the cell surface (107). Perhaps as many as 12 additional genes are necessary for the localization of yeast hydrolases to the lysosome-like vacuole (109, 110). The mis-sorting of carboxypeptidase (CPY; 109) or CPY-invertase fusion protein (110) to the cell surface was used to select for mutant cells that could survive when provided with an amino-blocked peptide or sucrose as carbon source and led to the identification of this second category of complementary groups.

In order to investigate the functions of these gene products, several laboratories are cloning these genes and characterizing the proteins they encode. Secretory pathway mutants can be complemented by unique expression plasmids that contain segments of wild-type yeast DNA. DNA sequence analysis is then used to identify open reading frames, construct gene fusions to β-galactosidase, and generate gene product–specific antibodies.

In this manner, it has been shown that a 29,000-dalton cytosolic protein (sec 53; 111, 112) somehow facilitates assembly of nascent proteins within the ER (113). This may be accomplished by associating with a cytoplasmically oriented ER molecule, or by providing a small molecule needed for the assembly event. The recent establishment of cell-free systems for the study of yeast ER translocation (114–116) may enable a more direct test of the function of this protein.

Several other yeast secretory mutant gene products have been preliminarily characterized. The sec 12 (107) gene product is a lumenally oriented, 55,000-dalton glycoprotein required for ER to Golgi transport (A. Nakano, R. Schekman, personal communication). Interestingly, this protein requires 60 minutes for transport from the ER to Golgi, compared with a half-time of 5 minutes needed for invertase to traverse the entire secretory pathway. The sec 12 protein is thus likely to facilitate transport while residing in the ER. The sec 7 (107) gene product required for Golgi transport is a 220,000-dalton protein that is found tightly associated with the cytoplasmic surfaces of yeast membranes, and requires urea for its extraction from membranes (A. Franzusoff, R. Schekman, personal communication). Perhaps this protein facilitates vesicle formation, vesicle targeting, or maintenance of intracellular compartment boundaries. The sec 7 gene product is also needed for endocytosis (117).

Characterization of these sec mutant gene products has generated several surprises with regard to the localization of the proteins and their known cellular phenotypes. The isolation of suppressor mutations combined with the development of cell-free systems to study these transport steps should enable an elucidation of the role these proteins play in intracellular transport.

ACKNOWLEDGMENTS

We wish to thank Dr. Harvey Lodish for suggesting the terms "mobile" and "immobile" phases. We are also especially grateful to Catherine Shimizu-Haas for expert preparation of the manuscript.

Literature Cited

1. Palade, G. 1975. *Science* 189:347–58
2. Griffiths, G., Simons, K. 1986. *Science* 243:438–43
3. Lodish, H. F., Kong, N., Snider, M., Strous, G. J. A. M. 1983. *Nature* 304:80–83
4. Fitting, T., Kabat, D. 1982. *J. Biol. Chem.* 257:14011–17
5. Ledford, B. E., Davis, D. F. 1983. *J. Biol. Chem.* 258:3304–308
6. Fries, E., Gustafsson, L., Peterson, P. A. 1984. *EMBO J.* 3:147–52
7. Mains, P. E., Sibley, C. H. 1983. *J. Biol. Chem.* 258:5027–33
8. Kvist, S., Wiman, K., Claesson, L., Peterson, P. A., Dobberstein, B. 1982. *Cell* 29:61–69
9. Ronne, H., Ocklind, C., Wiman, K., Rask, L., Obrink, B., Peterson, P. A. 1983. *J. Cell Biol.* 96:907–10
10. Copeland, C. S., Doms, R. W., Bolzau, E. M., Webster, R. G., Helenius, A. 1986. *J. Cell Biol.* 103:1179–91
11. Gething, M.-J., McCammon, K., Sambrook, J. 1986. *Cell* 46:939–50
12. Kreis, T. E., Lodish, H. F. 1986. *Cell* 46:929–37
13. Haas, I. G., Wabl, M. 1983. *Nature* 306:387–89
14. Bole, D. G., Hendershot, L. M., Kearney, J. F. 1986. *J. Cell Biol.* 102:1558–66
15. Munro, S., Pelham, H. R. B. 1986. *Cell* 46:291–300
16. Lusis, A. J., Tomino, S., Paigen, K. 1976. *J. Biol. Chem.* 251:7753–60
17. Lodish, H. F., Kong, N. 1984. *J. Cell Biol.* 98:1720–29
18. Bergmann, J. E., Singer, S. J. 1983. *J. Cell Biol.* 97:1777–87
19. Saraste, J., Hedman, K. 1983. *EMBO J.* 2:2001–6
20. Farquhar, M. G., Palade, G. E. 1981. *J. Cell Biol.* 91:77s–103s
21. Dunphy, W. G., Rothman, J. E. 1985. *Cell* 42:13–21
22. Farquhar, M. G. 1985. *Ann. Rev. Cell Biol.* 1:447–88
23. Goldberg, D. E., Kornfeld, S. 1983. *J. Biol. Chem.* 258:3159–65
24. Pohlmann, R., Waheed, A., Hasilik, A., von Figura, K. 1982. *J. Biol. Chem.* 257:5323–25
25. Dunphy, W. G., Brands, R., Rothman, J. E. 1985. *Cell* 40:463–72
26. Roth, J., Berger, E. G. 1982. *J. Cell Biol.* 93:223–29
27. Roth, J., Taatjes, D. J., Lucocq, J. M., Weinstein, J., Paulson, J. C. 1985. *Cell* 43:287–95
28. Couso, R., Lang, L., Roberts, R. M., Kornfeld, S. 1986. *J. Biol. Chem.* 261:6326–31
29. Rothman, J. E., Urbani, L. J., Brands, R. 1984. *J. Cell Biol.* 99:248–59
30. Rothman, J. E., Miller, R. L., Urbani, L. J. 1984. *J. Cell Biol.* 99:260–71
31. Gonzalez-Noriega, A., Grubb, J. H., Talkad, V., Sly, W. S. 1980. *J. Cell Biol.* 85:839–52
32. Hasilik, A., Neufeld, E. F. 1980. *J. Biol. Chem.* 255:4937–45

33. Moore, H.-P., Gumbiner, B., Kelly, R. B. 1983. *Nature* 302:434–36
34. Stevens, T. H., Rothman, J. H., Payne, G. S., Schekman, R. 1986. *J. Cell Biol.* 102:1551–57
35. Rothman, J. H., Hunter, C. P., Valls, L. A., Stevens, T. H. 1986. *Proc. Natl. Acad. Sci. USA* 83:3248–52
36. Goldstein, J. L., Brown, M. S., Anderson, R. G. W., Russell, D. W., Schneider, W. J. 1985. *Ann. Rev. Cell Biol.* 1:1–39
37. Davis, C. G., Lehrman, M. A., Russell, D. W., Anderson, R. G. W., Brown, M. S., Goldstein, J. L. 1986. *Cell* 45:15–24
38. Brown, M. S., Goldstein, J. L. 1976. *Cell* 9:663–74
39. Anderson, R. G. W., Brown, M. S., Goldstein, J. L. 1977. *Cell* 10:351–64
40. Lehrman, M. A., Goldstein, J. L., Brown, M. S., Russell, D. W., Schneider, W. J. 1985. *Cell* 41:735–43
41. Orci, L., Halban, P., Amherdt, M., Ravazzola, M., Vassalli, J.-D., Perrelet, A. 1984. *Cell* 39:39–47
42. Friend, D. S., Farquhar, M. G. 1967. *J. Cell Biol.* 35:357–76
43. Brown, W. J., Farquhar, M. G. 1984. *Cell* 36:295–307
44. Campbell, C. H., Fine, R. E., Squicciarini, J., Rome, L. H. 1983. *J. Biol. Chem.* 258:2628–33
45. Campbell, C. H., Rome, L. H. 1983. *J. Biol. Chem.* 258:13347–52
46. Schulze-Lohoff, E., Hasilik, A., von Figura, K. 1985. *J. Cell Biol.* 101:824–29
47. Orci, L., Glick, B. S., Rothman, J. E. 1986. *Cell* 46:171–84
48. Griffiths, G., Pfeiffer, S., Simons, K., Matlin, K. 1985. *J. Cell Biol.* 101:949–64
49. Orci, L., Ravazzola, M., Amherdt, M., Louvard, D., Perrelet, A. 1985. *Proc. Natl. Acad. Sci. USA* 82:5385–89
50. Reitman, M. L., Kornfeld, S. 1981. *J. Biol. Chem.* 256:11977–80
51. Lang, L., Reitman, M., Tang, J., Roberts, R. M., Kornfeld, S. 1984. *J. Biol. Chem.* 259:14663–71
52. Lang, L., Takahashi, T., Tang, J., Kornfeld, S. 1985. *J. Clin. Invest.* 76:2191–95
53. Kornfeld, S. 1986. *J. Clin. Invest.* 77:1–6
54. von Figura, K., Hasilik, A. 1986. *Ann. Rev. Biochem.* 55:167–93
55. Willingham, M. C., Pastan, I. H., Sahagian, G. G. 1983. *J. Histochem. Cytochem.* 31:1–11
56. Geuze, H. J., Slot, J. W., Strous, G. J. A. M., Hasilik, A., von Figura, K. 1984. *J. Cell Biol.* 98:2047–54
57. Sahagian, G. G. 1986. *Recent Research on Vertebrate Lectins, Adv. Cell Biol. Monogr.*, ed. B. Parent, K. Olden, pp. 46–64
58. Kelly, R. B. 1985. *Science* 230:25–32
59. Burgess, T. L., Kelly, R. B. 1984. *J. Cell Biol.* 99:2223–30
60. Burgess, T. L., Craik, C. S., Kelly, R. B. 1985. *J. Cell Biol.* 101:639–45
61. Gumbiner, B., Kelly, R. B. 1982. *Cell* 28:51–59
62. Moore, H.-P., Kelly, R. B. 1985. *J. Cell Biol.* 101:1773–81
63. Moore, H.-P., Walker, M. D., Lee, F., Kelly, R. B. 1983. *Cell* 35:531–38
64. Mellman, I., Fuchs, R., Helenius, A. 1986. *Ann. Rev. Biochem.* 55:663–700
65. Anderson, R. G. W., Pathak, R. K. 1985. *Cell* 40:635–43
66. Mroz, E. A., Kechene, C. 1986. *Science* 232:871–73
67. Simons, K., Fuller, S. D. 1985. *Ann. Rev. Cell Biol.* 1:243–88
68. Rodriguez-Boulan, E., Sabatini, D. D. 1978. *Proc. Natl. Acad. Sci. USA* 75:5071–75
69. Rindler, M. J., Ivanov, I. E., Plesken, H., Rodriguez-Boulan, E., Sabatini, D. D. 1984. *J. Cell Biol.* 98:1304–19
70. Fuller, S. D., Bravo, R., Simons, K. 1985. *EMBO J.* 4:297–307
71. Matlin, K., Simons, K. 1984. *J. Cell Biol.* 99:2131–39
72. Misek, D. E., Bard, E., Rodriguez-Boulan, E. 1984. *Cell* 39:537–46
73. Pfeiffer, S., Fuller, S. D., Simons, K. 1985. *J. Cell Biol.* 101:4470–76
74. Kondor-Koch, C., Bravo, R., Fuller, S. D., Cutler, D., Garoff, H. 1985. *Cell* 43:297–306
75. Gottlieb, T., Beaudry, G., Rizzolo, L., Colman, A., Rindler, M., et al. 1986. *Proc. Natl. Acad. Sci. USA* 83:2100–4
76. von Bonsdorff, C.-H., Fuller, S. D., Simons, K. 1985. *EMBO J.* 4:2781–92
77. Mostov, K. E., deBruyn Kops, A., Deitcher, D. L. 1986. *Cell* 47:359–64
78. Fries, E., Rothman, J. E. 1980. *Proc. Natl. Acad. Sci. USA* 77:3870–74
79. Balch, W. E., Dunphy, W. G., Braell, W. A., Rothman, J. E. 1984. *Cell* 39:405–16
80. Braell, W. A., Balch, W. E., Dobbertin, D. C., Rothman, J. E. 1984. *Cell* 39:511–24
81. Balch, W. E., Rothman, J. E. 1985. *Arch. Biochem. Biophys.* 240:413–25
82. Balch, W. E., Glick, B. S., Rothman, J. E. 1984. *Cell* 39:525–36
83. Wattenberg, B. W., Balch, W. E., Rothman, J. E. 1986. *J. Biol. Chem.* 261:2202–207
84. Pearse, B. M. F. 1982. *Proc. Natl. Acad. Sci. USA* 79:451–55

85. Pfeffer, S. R., Drubin, D. G., Kelly, R. B. 1983. *J. Cell Biol.* 97:40–47
86. Zaremba, S., Keen, J. H. 1983. *J. Cell Biol.* 97:1339–47
87. Payne, G., Schekman, R. 1985. *Science* 230:1009–14
88. Patzer, E. J., Schlossman, D. J., Rothman, J. E. 1982. *J. Cell Biol.* 93:230–36
89. Rothman, J. E., Schmid, S. L. 1986. *Cell* 46:5–9
90. Glick, B. S., Rothman, J. E. 1987. *Nature.* 326:309–12
91. Dunphy, W. G., Pfeffer, S. R., Clary, D. O., Wattenberg, B. W., Glick, B. S., Rothman, J. E. 1986. *Proc. Natl. Acad. Sci. USA* 83:1622–26
92. Paquet, M. R., Pfeffer, S. R., Burczak, J. D., Glick, B. S., Rothman, J. E. 1986. *J. Biol. Chem.* 261:4367–70
93. Haselbeck, A., Schekman, R. 1986. *Proc. Natl. Acad. Sci. USA* 83:2017–21
94. Balch, W. E., Wagner, K. R., Keller, D. S. 1987. *J. Cell Biol.* 104:749–60
95. Woodman, P. G., Edwardson, J. M. 1986. *J. Cell Biol.* 103:1829–35
96. Crabb, J. H., Jackson, R. C. 1985. *J. Cell Biol.* 101:2263–73
97. Davey, J., Hurtley, S. M., Warren, G. 1985. *Cell* 43:643–52
98. Braell, W. A. 1987. *Proc. Natl. Acad. Sci. USA.* 84:1137–41
99. Lohka, M., Masui, Y. 1983. *Science* 220:719–21
100. Lohka, M., Maller, J. 1985. *J. Cell Biol.* 101:518–23
101. Miake-Lye, R., Kirschner, M. 1985. *Cell* 41:165–75
102. Newport, J. 1987. *Cell.* 48:205–17
103. Newport, J., Spann, T. 1987. *Cell.* 48:219–30
104. Burke, B., Gerace, L. 1986. *Cell* 44:639–52
105. Suprynowicz, F., Gerace, L. 1986. *J. Cell Biol.* 103:2073–81
106. Warren, G. 1985. *Trends Biochem. Sci.* 10:439–43
107. Novick, P., Field, C., Schekman, R. 1980. *Cell* 21:205–15
108. Schekman, R. 1985. *Ann. Rev. Cell Biol.* 1:115–43
109. Rothman, J. H., Stevens, T. H. 1986. *Cell.* 47:1041–51
110. Bankaitas, V. A., Johnson, L. A., Emr, S. D. 1986. *Proc. Natl. Acad. Sci. USA.* 83:9075–79
111. Ferro-Novick, S., Novick, P., Field, C., Schekman, R. 1984. *J. Cell Biol.* 98:35–43
112. Ferro-Novick, S., Hansen, W., Schauer, I., Schekman, R. 1984. *J. Cell Biol.* 98:44–53
113. Bernstein, M., Hoffmann, W., Ammerer, G., Schekman, R. 1985. *J. Cell Biol.* 101:2374–82
114. Hansen, W., Garcia, P. D., Walter, P. 1986. *Cell* 45:397–406
115. Rothblatt, J. A., Meyer, D. I. 1986. *Cell* 44:619–28
116. Waters, G., Blobel, G. 1986. *J. Cell Biol.* 102:1543–50
117. Riezman, H. 1985. *Cell* 40:1001–9

Ann. Rev. Biochem. 1987. 56:853–79

CYCLIC AMP AND OTHER SIGNALS CONTROLLING CELL DEVELOPMENT AND DIFFERENTIATION IN *DICTYOSTELIUM*

Günther Gerisch

Max-Planck-Institut für Biochemie, D-8033 Martinsried bei München, Federal Republic of Germany

CONTENTS

0066-4154/87/0701-0853$02.00

SUMMARY AND PERSPECTIVES

Because of its unique developmental cycle, *Dictyostelium discoideum* is an excellent organism in which to study cell interactions. Randomly distributed, unconnected cells organize into patterns that give rise to a structure, the fruiting body, that has a defined shape and consists of different types of cells in regular proportions. Since the entire life cycle runs in the haploid state, mutations are easy to detect and to use for the analysis of development.

The primitive nature of the organism does not imply that the signal systems controlling development are simple. *D. discoideum* cells are free-living amoebae that have to be universal in their functions and flexible in their responses, and the signal systems have to accommodate to that. The best-studied signal system in *D. discoideum* development is the cAMP-signal system. This system is organized into networks. Adaptation and desensitization, oscillatory cAMP production, autocatalytic stimulation of early cell development, and probably also proportion regulation of cell types in the multicellular stage are due to these networks. They have, among other functions, a gating effect in the transduction of signals to specific targets. The regulation of genes by periodic pulses, as they are produced by *D. discoideum* cells, is thus separated from the control of genes by continuous cAMP concentrations.

Current research is characterized by the application of molecular techniques to the analysis of *Dictyostelium* development. A key advance was the construction of transformation vectors, which made possible the introduction of antisense constructs (1) and of genes that are modified by site-directed mutagenesis (2, 3). These vectors allow the integration into the genome of cloned DNA as tandem repeats (4, 4a). Some *D. discoideum* strains carry endogenous plasmids (5) that can be used for the construction of extrachromosomal transformation vectors (6). Moreover, these natural plasmids, which reach copy numbers of 100 or more per cell, carry developmentally regulated genes, thus providing an excellent tool for the investigation of factors that program the activity of these genes (7). Further work will profit from these advances as well as from the purification and identification of factors that control development. The goal is to understand how cells switch from one state to another during development, how cells diversify and proportions of cell types are regulated, and how in a multicellular organism patterns of differentiated cells are formed.

In the present review priority will be given to recent publications, even if a phenomenon has first been described by others. This review updates articles in the standard book on *Dictyostelium* development edited by W. F. Loomis (8) and complements recent reviews on receptors (9), chemotaxis (10), phototaxis (11), and control of gene expression during development (12, 13). An authoritative treatise on the systematics and morphology of the *Dictyostelids* has recently been written by K. B. Raper (14).

OVERVIEW OF *DICTYOSTELIUM* DEVELOPMENT

The genus *Dictyostelium* represents a group of eukaryotic microorganisms that form a multicellular body by cell aggregation. During the growth phase cells exist as individual amoebae. These cells respond to factors such as folic acid that may be released from bacteria, their normal food source (15). Upon removal of food sources, intercellular signals produced and recognized by the amoebae play an important role as chemoattractants during cell aggregation. In the multicellular slug stage the cells adhere tightly to each other. In addition they are kept together by a sheath, which like an extracellular matrix surrounds the body, providing it with a smooth, poorly permeable surface (16). Through cell interactions in the slug a pattern is established of cells that are later converted into the cell types constituting the fruiting body. This happens during culmination, a process that involves morphogenetic movements and invagination of cells, similar to gastrulation. In the fruiting body, which marks the end point of asexual development, cells previously located in the anterior region of the slug have formed a stalk, cells from the posterior region have formed a spore head at the tip of the stalk, and cells from the very rear of the slug have been converted into a supporting disc at its base. Germination of the spores is inhibited in the spore head by an adenine derivative called discadenine that is synthesized in late stages of development (17). *Dictyostelium* lives together with other organisms in soil, in an undefined and changing environment. Accordingly, its development is made reversible almost up to the end by mechanisms that erase the accomplishments of developing cells. Erasure occurs in steps, is under the control of several genes, and involves an extracellular low-molecular-weight factor (18, 19).

Up through aggregation the factors acting as intercellular signals are released into the extracellular medium, and the surfaces of responding cells are immediately accessible to them. cAMP, the first chemical discovered to function as a developmental signal in *D. discoideum*, was originally identified as a chemoattractant of aggregating amoebae. Subsequently, cAMP turned out to be a multipurpose signal involved in the control of early development, in slug organization as well as in terminal differentiation of cells into spores and stalk cells. cAMP appears to be also involved in sexual development (20).

This is an alternative to asexual development which, in heterothallic strains, requires interaction of cells of opposite mating types. These cells aggregate together and form zygotes, giant cells that act as centers for the development of macrocysts, the sites of meiosis. Sexual development is controlled by a giant-cell inhibiting factor released by giant cells (21), and by ethylene, which in *D. mucoroides* stimulates macrocyst formation (22).

Asexual development of *Dictyostelium* can be viewed as following the step-by-step activation of genes, whereby the product of one regulatory gene turns on the next gene and so forth (23, 24). But evidence is accumulating for regulatory processes in *Dictyostelium* development that do not fit into a central sequence of consecutively activated genes. These processes include feedback interactions involved in the control of early expressed genes, and regulation of cell-type specification at the multicellular stage. The complexity of these interactions suggests that *Dictyostelium* development resembles more closely the development of higher animals than the sequential assembly of a phage.

REGULATION OF GENE EXPRESSION IN EARLY DEVELOPMENT

Early Expressed Proteins

Development is induced in *D. discoideum* by exhaustion of nutrient bacteria or, in axenically growing laboratory strains, by removal of liquid nutrient medium. After several hours in non-nutrient buffer the amoebae become able to aggregate. Changes between growth and aggregation include increases as well as decreases in the amounts of specific proteins or in the rates of their synthesis (25–27). Some of these changes parallel development only under particular conditions, indicating that they are not strictly coupled to cell development and are not essential for development to proceed (28).

Aggregation-competent cells are distinguished from growth-phase cells by their ability to produce cAMP and to respond chemotactically to it, to assemble into aggregation centers and streams, and to form EDTA-stable contacts. These contacts are characteristic of aggregation-competent cells although they are not essential for aggregation (29). The acquisition of these cell functions is based in part on the developmental regulation of four membrane proteins. During development to aggregation competence the number of cAMP receptors on the cell surface increases about 10-fold to approximately 1×10^5 binding sites per cell, the basal activity of adenylate cyclase increases 20–40-fold, and the activity of cAMP-phosphodiesterase on the cell-surface about 50-fold [cited by Gerisch (30)]. The contact site A glycoprotein, a membrane protein involved in EDTA-stable cell adhesion, is most stringently regulated. It is undetectable in growth phase cells, is maximally expressed in aggregation-competent cells, and is gradually lost during later stages of development (31).

Oscillatory cAMP Production

Aggregation-competent cells of *D. discoideum* produce pulses of cAMP in response to cAMP. These pulses have a half-width of about two minutes and reflect the receptor-mediated activation of adenylate cyclase. After a short phase of accumulation within the cells the newly synthesized cAMP is released. It diffuses to other cells and stimulates them, and is hydrolyzed by extracellular and cell-surface phosphodiesterases. Adenylate cyclase is turned off by adaptation, which means that the cyclase is only activated during a period in which the occupancy of cAMP receptors increases (32, 33). The cycle of receptor-stimulated cAMP synthesis, adaptation, and deadaptation after removal of cAMP by phosphodiesterase underlies the periodic, pulsatile production of cAMP in *D. discoideum* cells. In stirred suspensions cells synchronize their adenylate cyclase activities such that cAMP pulses are periodically produced in phase (34). In a cell layer on agar-plates cAMP is produced in a spatio-temporal pattern of propagated waves (35).

Periodic cAMP production is accompanied by oscillations in cellular cGMP (36) and extracellular H^+ concentrations (36, 37), and in the exchange of extracellular and intracellular Ca^{2+} (36) and K^+ (38). An important question is which of these changes are only coupled to the oscillatory system and which are essential for its function. The receptor/adenylate cyclase complex has been proposed to be the central element of the oscillating system (39). However, experimental work does not provide evidence that periodic cAMP production is essential for oscillations to occur. cAMP pulses are associated in cell suspensions with spikes of decreasing light scattering that seem to be caused by contraction of the cells in response to cAMP. But light scattering oscillations of a more sinusoidal shape continue in suspensions of developing cells after measurable oscillations in cAMP production have ceased (40). Similarly, pulsatile cAMP production is associated with spikes of increased Ca^{2+} uptake into the cells, but sinusoidal oscillations of Ca^{2+} exchange continue without detectable changes in cAMP or cGMP (36). These results can be interpreted to mean that either cAMP production is only coupled to the oscillating system without being intrinsic to its function, or that there are two or more oscillatory systems, one involving cAMP, which may or may not be coupled to the other.

Strain-Specific Differences and Effects of Growth Conditions

Some strains of *D. discoideum* require the cAMP pulses they produce to develop from the growth phase to the aggregation-competent stage. This can be shown by cultivating these strains on agar plates containing μmolar concentrations of a slowly hydrolyzable cAMP analogue, 3', 5'-cyclic adeno-sine-phosphorothioate (cAMPS) (41). This analogue is an agonist of cell-surface cAMP receptors and acts like a continuous cAMP signal that causes adaptation (42, 43). cAMPS inhibits development of wild-type strain NC4 and of the axenically growing strains AX2 and AX3 derived from it. V12M2, which represents another wild-type strain, is resistant to cAMPS, indicating that cAMP pulses are not required for this strain to develop (41).

AX2 and AX3, the most often used axenic strains, differ in their capacity to develop in shaken suspensions. When exponentially growing cells are washed free of nutrients and adjusted to 1×10^7 cells per ml, only AX2 cells will develop to aggregation competence in fast-shaken cultures (44). These cells produce cAMP pulses sufficient for self-stimulation of development. AX3 cells produce cAMP pulses when they are allowed to form clumps (27). Accordingly, they need either cell contact (45) or application of cAMP in pulses of 20 nM size in order to fully acquire aggregation competence under these conditions (46). This is also true for AX2 cells that are harvested from nutrient medium at the late stationary phase rather than during exponential growth. These cells, as well as AX3 cells, can be used for separating the effects of cAMP on early development from those of other factors (44). Differences in strains and culture conditions have to be taken into account when results from different laboratories are compared.

While the contact site A glycoprotein is stringently regulated both in cells grown on bacteria and in axenically grown cells (47), regulation of some other gene products is less precise under axenic conditions. A group of three transcripts, each of them recognized by a specific cDNA-probe, is invariably expressed in axenically grown cells. In cells grown on bacteria these transcripts are strongly expressed during growth, sharply down-regulated after starvation up to the slug stage, and reaccumulating at culmination (48). More often genes that are expressed during axenic growth are suppressed during growth on bacteria and expressed only after starvation. This is the case for the family of lectins called discoidins. Discoidin I accumulates within the cells and is thought to be externalized to act in cell-to-substratum attachment; it contains the sequence gly-arg-gly-asp responsible for the interaction of fibronectin with a cell-surface receptor (49). Cells transfected with discoidin I antisense DNA show no stream formation during aggregation on plastic surfaces (50). Nevertheless, mutants devoid of discoidin I and II develop into fruiting bodies. Nuclear run-off experiments indicate that these mutants do not transcribe the discoidin genes, probably due to the absence of a trans-acting factor. Normal expression of other developmentally regulated proteins in the mutants indicates that the regulation of discoidin is not part of the main stream of developmental changes (51). During normal wild-type development cAMP shuts off the transcription of discoidin I genes (52). Cell-to-cell adhesion has a similar effect and may act through raising the local cAMP concentration in the intercellular space (53).

Classes of Early Expressed Genes

Some of the increases or decreases in gene expression during early development are preferentially enhanced by cAMP pulses. A study using cDNA clones prepared from membrane-bound polysomes has revealed the following classes of transcripts that accumulate during early development (30).

TRANSCRIPTS THAT ACCUMULATE WITHOUT cAMP STIMULATION   These transcripts are present in low amounts during axenic growth and accumulate steadily in suspension cultures during the first six hours of development of AX2 as well as AX3 cells. cAMP pulses have little or no effect on the accumulation of these transcripts. By an unknown mechanism some or all of these transcripts are induced by caffeine in the presence of nutrient medium (54). Caffeine is known to block the receptor-mediated activation of adenylate cyclase (55), but it is unlikely that this is the mechanism by which the transcripts are regulated since adenylate-cyclase activity is low in the untreated growth-phase cells.

TRANSCRIPTS STRONGLY DEPENDENT ON cAMP PULSES   These transcripts accumulate in the AX2 strain sharply between two and six hours of development, and this increase is accelerated by pulses of 20 nM cAMP applied every six minutes. There is little if any increase in cells of the AX3 strain in fast-shaken suspension, but an increase as strong as in the AX2 strain when AX3 cells are exposed to cAMP pulses.

The mRNA encoding the polypeptide moiety of the contact site A glycoprotein belongs to this category of transcripts. Nuclear run-off experiments indicate that induction of transcription is responsible for its accumulation (A. Müller-Taubenberger, personal communication). The glycoprotein is detectable shortly after the onset of transcription. Enhancement of its expression by cAMP pulses is most obvious in suspension cultures of the AX3 strain. Steady-state concentrations of cAMP in the nanomolar range inhibit expression of the glycoprotein, indicating that adaptation occurs in the signal transduction pathway. Caffeine does not induce the contact site A glycoprotein in the presence of nutrient medium (54).

FAMILIES OF DIFFERENTLY REGULATED TRANSCRIPTS   Some cDNA probes recognize several RNA species that show different temporal patterns of regulation. A group of six RNAs is transcribed from ubiquitin genes (56). Some of the ubiquitin transcripts are down-regulated during early development of *D. discoideum*, while the expression of others increases and is enhanced by pulses of cAMP. One of the ubiquitin cDNAs from *D. discoideum* has an open reading frame that extends beyond the 3' end of the ubiquitin encoding region, indicating that the ubiquitin is synthesized as a precursor with a basic polypeptide at its C-terminus. This is similar to a human ubiquitin gene. The 3' terminal region of the human gene encodes a basic polypeptide that is thought to target the ubiquitin precursor into the nucleus where it conjugates to histone 2A (57).

A Network Imposing Autocatalytic Control of Cell Development

In strain AX2 the expression of cell-surface cAMP receptors and adenylate cyclase is enhanced by the stimulation of developing cells with pulses of cAMP. Thus, the expression of components that are responsible for the recognition of cAMP or its production is also under the control of cAMP signals. This means that full expression of these components depends on the functioning of the entire signal system (30).

Pulse-shaped signals are important for expression of these components, which implies that phosphodiesterase is required to limit the length of stimulation and to allow the cells to deadapt between two bursts of cAMP production. *D. discoideum* cells produce not only cell-surface phosphodiesterase but also soluble phosphodiesterase that is released into the medium and is regulated differently from the cell-surface enzyme. The phosphodiesterases are important for the autocatalytic stimulation of cell development by cAMP pulses and also by being part of a servomechanism that limits extracellular cAMP concentrations.

Phosphodiesterase Regulation Involved in the Negative Feedback Control of Extracellular cAMP

Cell-surface phosphodiesterase activity is low in growth phase cells and increases up to the stage of aggregation competence. Its control resembles that of cAMP receptors, adenylate cyclase, and contact sites A in that cAMP pulses are the signals that most efficiently stimulate its expression (58). This is in contrast to the regulation of extracellular phosphodiesterase. This enzyme is already produced during growth phase and its activity is downregulated at the beginning of development by an inhibitor that is a heat-stable, cysteine-rich glycoprotein (59). cAMP induces the extracellular phosphodiesterase and suppresses inhibitor production. Pulses and continuous cAMP signals as well as cAMPS similarly regulate the extracellular phosphodiesterase and its inhibitor, which distinguishes this regulation from the control of early development (42, 60, 61). Apparently extracellular phosphodiesterase and inhibitor regulation is not subject to adaptation.

The membrane-bound and extracellular phosphodiesterases are glycoproteins whose polypeptide moieties appear to be encoded by the same gene (61). Differences in the transduction of pulsatile versus continuous cAMP signals seem to influence the balance between the two forms of the enzyme. The cDNA-derived sequence of phosphodiesterase indicates an unusually long hydrophobic leader of 48 amino acids (62). Possibly this leader is not cleaved off from the cell-surface phosphodiesterase and is responsible for anchoring the protein in the membrane.

Bypass Mutants as Tools to Separate and Uncover Developmental Control Systems

Since the expression of cAMP receptors, adenylate cyclase, and other components of the cAMP-signal system is controlled by the functioning of that system, any defect in one component will suppress the other. This means that mutants defective in a component of the cAMP-signal system will be pleiotropic because they are blocked early in development. Of help in the analysis of developmental controls are mutants in which the requirement for certain control factors is bypassed. Such bypass mutants provide also an optimal genetic background for the selection of mutants defective in specific developmentally regulated proteins, e.g. in the contact site A glycoprotein (29). Mutants in which the expression of cAMP-receptors is no longer under the control of cAMP signals are also important for selecting mutants defective in chemotaxis to cAMP.

The slowly hydrolyzed cAMP analogue, cAMPS, which blocks development in strains NC4, AX2, and AX3, has been employed to select either mutants in which the requirement for cAMP is bypassed, or nonadapting mutants for which signals need no longer to be pulsatile (42, 63). The cAMPS-resistant mutants obtained form small aggregates and fruiting bodies in the presence of cAMPS. Since chemotaxis is largely suppressed by a uniform concentration of cAMPS, but motility of the cells is not, the mutant cells aggregate essentially by random collisions (E. Wallraff, personal communication).

Adenylate cyclase and contact sites A are stringently regulated in the cAMPS-resistant mutants studied, as they are in the wild-type. This result indicates the presence of a second control system that acts either in line with or in parallel to the cAMP-signal system. From a cAMPS-resistant bypass mutant, HG302, a double bypass mutant has been selected that does not require removal of nutrient medium in order to strongly express adenylate cyclase and contact sites A and to aggregate. But the mutant does so only after having finished growth (30). These results indicate a minimum of three controls of early development: one control that acts as a switch between growth and stationary phase in the presence of nutrient medium, and two controls that act after the removal of nutrient medium, one involving cAMP.

POSTAGGREGATIVE DEVELOPMENT

Cell-Type Specification in the Slug Stage

Formation of the tip of the aggregate marks the beginning of postaggregative development. The tip determines the polarity of the slug and emits signals responsible for the spatial arrangement of prestalk and prespore cells, the major cell types constituting the slug. Many new genes are expressed around the time of tip formation (64). This does not mean that these "postaggregative genes" are all expressed in response to the same signal or that the tip itself is required for their expression. Genes specifically expressed in either prestalk or prespore cells are of high interest because they mark the bifurcation of development into two major cell types. Often prestalk-specific genes commence to be expressed late in aggregation; this is earlier than a group of prespore-specific genes that require tight aggregates to be formed (65, 66). Some of the cell-type-specific proteins decline to marginal levels at the end of development (67), while others remain present in the mature stalks or spores (68, 69), or are only induced in one of the mature cell types (70, 71).

Identification of prestalk and prespore cells is primarily based on their position in the slug: prestalk cells are located in the anterior portion which, under normal circumstances, is converted into the stalk of the fruiting body, prespore cells are located in the posterior portion that is destined to form the spore head. The stalk of the fruiting body consists of cells that are lysed with only the cell walls remaining. Accordingly, prestalk cells are rich in lysosomes, which can be stained by neutral red (72). Acid phosphatase (73) and cathepsins (74) are typical prestalk-enriched enzymes. Spores are surrounded in the fruiting body by a mucous coat. In preparation for this, prespore cells express UDP galactose polysaccharide transferase and accumulate spore-coat material in vesicles, the prespore vacuoles (75), which can be identified by electron microscopy and by labeling their contents with antibodies or ^{3}H-fucose (76). Separation of cells in density gradients is widely used for separating prestalk and prespore cells. But prestalk and prespore cells have been successfully separated into two distinct bands on a preformed Percoll gradient only after growth of cells on bacteria and a long period of slug migration (77). Axenically grown cells fractionated in a unimodal distribution rather than into sharply separated bands (78).

There are at least two additional cell types in the slug: anterior-like cells that resemble prestalk cells but are located in the prespore area (72), and cells that are neither labeled with an antiserum against a prestalk antigen nor with an antiserum against a prespore antigen (79). The foot plate of the D. discoideum fruiting body consists of cells resembling stalk cells; their progenitors are located at the rear of the slug, in an area distinguished by its high cAMP-phosphodiesterase activity (80).

Although anterior-like cells resemble prestalk cells, they are not identical with them. Prestalk-specific proteins are present in lower amounts in anterior-like cells and one protein (PSP59) that is synthesized at high rate in prespore cells is also synthesized in anterior-like cells (81). 5'-nucleotidase is thought to be a marker for anterior-like cells (82). In the culmination stage, 5'-nucleotidase activity is particularly high at the surfaces of cells that are located at the border of the prespore and prestalk area (83). These border cells might be anterior-like cells that have moved from the prespore area to enter the prestalk pool (84).

Dissociated slug cells reassemble and sort out according to their origin, and a new tip arises from the restored prestalk area. Prestalk cells show a stronger chemotactic response than prespore cells to cAMP (85) and to oxygen (86), which might be important for sorting out and for the formation of a tip that always points towards the air.

Differences and Variations in the Expression of Prestalk and Prespore Markers

In a comprehensive study, more than 30 proteins were identified in slug cells by two-dimensional electrophoresis to be cell-type specific (67). Two of these proteins are of special interest because early in development they appear to be synthesized in all cells. Synthesis of one of the prespore-specific proteins (PSP59) is preferentially repressed in prestalk cells. Similarly, one of the prestalk-specific proteins (PST71) is apparently produced in all cells before synthesis is shut off in prespore cells. Partial loss of an antigen from the surface of prespore cells has been quantitated by monoclonal antibody-labeling and cell sorter analysis (87). Comparison of nuclear run-off data with levels of transcripts indicates that one of the prestalk-enriched cathepsins, cysteine proteinase 1 (88), is transcriptionally regulated and another one, cysteine proteinase 3, is regulated both transcriptionally and posttranscriptionally (89). The *Dictyostelium ras*-gene is transcribed into two mRNAs. The larger species is expressed during growth and disappears during early development; later both mRNAs accumulate preferentially in prestalk cells (90). A function of the *ras*-protein in morphogenesis is suggested from the aberrant fruiting bodies that are formed by cells transfected with a *ras*-gene in which Gly 12 is mutated into Thr (91). A developmentally regulated modification, probably at the carbohydrate moiety of acid phosphatase, appears to be responsible for the high activity of this lysosomal glycoprotein in prestalk cells (73, 92). Certain carbohydrate epitopes are preferentially found on glycoproteins of prespore cells (93). Earlier in development these epitopes are detected on surface glycoproteins of growth-phase cells and on the contact site A glycoprotein. These epitopes are present on an oligosaccharide moiety called type 2 carbohydrate, which is attached to proteins in the Golgi apparatus (94) and is missing in a class of mutants called mod B, which seem to form smaller fruiting bodies than wild-type (95).

Even if one neglects posttranslational modifications by concentrating on transcripts, differences in the expression among prestalk or prespore-specific gene products and variations among strains or culture conditions are observed. Moreover, with more transcripts being tested the demarcation line becomes diffuse between genes expressed during early development and those expressed later in an often cell-type-specific manner. A number of transcripts tested in axenically grown AX3 cells fall into four classes: 1. Prestalk I messages, which need only removal of nutrient medium to be expressed; 2. prestalk II messages, which require pulsatile stimulation of starving cells with cAMP; 3. prespore I messages that are regulated like prestalk II messages; and 4. prespore II messages, which are expressed only in cells that had formed aggregates (23). In strain NC4, grown on bacteria, the loss of prestalk I and prespore I messages at the culmination stage is more abrupt than in axenically grown AX3 (48).

Single-Cell Systems for Separating Cell Differentiation from Morphogenesis

Difficulties of analyzing interactions between closely connected cells in the slug or culmination stages have been overcome by "in-vitro" systems that permit cell differentiation in monolayers or fast-shaken suspensions of single cells. Under certain conditions single cells differentiate only into prestalk and prespore cells, while under other conditions differentiation proceeds up to mature spores and stalk cells that are both surrounded by a rigid cell wall.

To allow single cells to differentiate, cAMP is normally added to cell cultures at millimolar concentrations. Only nano- to micromolar concentrations are required when a phosphodiesterase inhibitor is supplied (96) or cAMP is replaced by the slowly hydrolyzed analogue cAMPS (97). In strains that require cAMP pulses for early development, constant concentrations of cAMP or cAMPS inhibit development when added to growth-phase cells (41, 46). The strain of choice for investigating stalk-cell and prespore cell differentiation, V12M2, does not require cAMP pulses for early development (41). In the other strains it helps if cells develop up to aggregation competence before cAMP is supplied (48, 66). The stimulation of early development by pulses and of later development by continuous cAMP is clearly seen in a mutant, N7, which does not form stable aggregates because of a defect in cAMP production. Stimulation for six hours with pulses of 10^{-7} M cAMP at intervals of six minutes, followed by incubation with 10^{-4} M repeatedly applied every hour, causes the mutant to develop into small slugs and fruiting bodies (98).

Strategies employed in inducing single cells to differentiate include the replacement of cell-to-cell contact by "conditioning factors" released into the medium, or by contact to purified slug cell membranes (99). A low-molecular-weight factor (CMF), that has been partially purified from medium in which cells have developed, stimulates early development in single AX3 cells attached to a plastic surface (66). After development in the presence of CMF the cells can be induced by cAMP to differentiate into prestalk as well as prespore cells. Another factor, comprised of two low-molecular-weight components, stimulates NC4 cells to differentiate into prespore cells when cAMP and bovine serum albumin are present (100).

This strategy has been complemented by employing mutants in which the requirement is bypassed for certain signals that are normally provided during morphogenesis. A limitation of this approach is that factors that allow differentiation to occur in a particular bypass mutant do not comprise the full repertoire of signals that control differentiation in the wild-type. In the mutant dev1510, cells differentiate precociously into spores or stalk cells (101). Single starved cells of this mutant differentiate predominantly into stalk cells at submicromolar concentrations of cAMP, and into spore-like cells at higher concentrations (102). This result indicates that the effect of cAMP in single-cell systems is not only permissive; it may direct differentiation towards the spore pathway. The finding that both prestalk and prespore markers are expressed in a population of AX3 cells exposed to cAMP and the conditioning factor CMF (66) may be explained by assuming a factor that counterbalances the prespore-directing effect of cAMP under these conditions. However, there are two other caveats against a simplistic view of cell-type determination by cAMP. First, suppression of prestalk-cell formation in aggregates of prespore cells lacks nucleotide specificity. This is shown with prespore cells isolated from slugs. After reaggregation, 40% of these cells convert into prestalk cells. Conversion is prevented by cAMP, but also by 5'-AMP, adenosine, and adenine (103). Second, the direction of differentiation in cAMP-treated cells of V12M2, or of a "sporogenous mutant" derived from it, is influenced by pH conditions. Sporogenous mutants are bypass mutants that form not only mature stalk cells but also spores in the presence of cAMP (104). Weak bases at high extracellular pH favor the spore pathway, while weak acids at low pH favor stalk-cell differentiation (105).

DIF Determines Stalk-Cell Differentiation

At high cell densities, cAMP permits cells of wild-type V12M2 to differentiate into mature stalk cells. Prespore cells are also formed, but no mature spores are obtained. At low cell densities, an additional factor is required for stalk-cell differentiation, the low-molecular weight differentiation inducing factor DIF (106). Cells of a mutant, HM44, which is defective in DIF production, differentiate at all densities only when DIF is supplied together with cAMP. The pattern of proteins synthesized in this mutant shows that DIF not only permits stalk-cell differentiation, but also inhibits prespore-cell differentiation (107). Cells of sporogenous mutants are induced by DIF to differentiate preferentially into stalk cells (104).

DIF consists of several components (106). The major one, DIF-1, is a chlorinated organic compound. DIF activity rises sharply at the tipped aggregate stage that marks the beginning of slug formation, in accord with the proposed role of DIF in cell-type specification (108). But DIF is not enriched in the prestalk area of slugs (108a). Also, the addition of DIF has only little effect on cell-type specific protein synthesis in anterior or posterior portions of a slug (107). These results suggest that cells are primed by other factors to differentiate into one or the other cell type before DIF induces prestalk cells to undergo terminal differentiation. Thus, during normal development differentiation into stalk cells and spores seems to be guaranteed by back-up systems.

Origin of the Prestalk/Prespore Cell Pattern During Normal Development

There are three possible ways that the prestalk/prespore pattern in the slug is created: 1. Cells are different from the beginning of development and sort out in the slug according to their preestablished properties; 2. Aggregated cells interact by short-range signals to differentiate into either prestalk or prespore cells that are distributed like mixed salt and pepper, and sort out later to produce the spatial pattern seen in the slug (109); 3. The pattern is generated after the polarity of the slug is established by intercellular signals that provide each cell in situ with information about its position in the slug. [For a detailed discussion of models see MacWilliams (110).]

In reality, the pattern seems to be generated and maintained by a combination of different mechanisms. Cell-cycle phase at the beginning of development is one factor that determines the fate of a cell. $D.\ discoideum$ has a short S and no G_1 phase (111). Early G_2 cells sort preferentially into the prestalk area and late G_2 cells into the prespore area (112, 113).

Labeling of cells in the tipped aggregate stage with a prespore-specific antibody, mud-1, showed that the antigen recognized by this antibody is at the very beginning expressed in the center of a tipped aggregate near its bottom. Expression of the antigen at the right position has been interpreted to indicate that cell-type specification is based on positional information (114). In contrast, expression of a prestalk antigen in the periphery of a tipped aggregate has been suggested to indicate that cell-type specification occurs first and sorting out next (79). This prestalk antigen is not recognized in the early tip of an aggregate, suggesting that the tip is populated by prestalk cells after its formation. Both reports agree that prespore cells become first detectable in the center of the aggregate. The question of how the pattern is formed needs further work with antibodies and assurance that the antigens labeled are stable markers of individual cells.

Prestalk/prespore patterns in mixtures of wild-type cells with cells of a mutant (Hs2) have been interpreted by assuming two negative feedback loops, one controlling the conversion of anterior-like cells into prestalk cells by an inhibitor, the second controlling conversion of anterior-like cells to prespore cells by another inhibitor (84). Mutant Hs2 shows a reduced prestalk to prespore ratio, but in cell mixtures with wild-type more prestalk cells are formed than in the wild-type alone. This effect is thought to indicate hypersensitivity of the mutant to the inhibitor acting in the first feedback loop. This inhibitor has tentatively been identified with DIF, that of the second loop with cAMP (84). An important role in cell-type conversion is attributed in this hypothesis to anterior-like cells. That role is supported by the high rate of exchange observed between anterior-like cells that migrate within the prespore area and the prestalk pool (115). The model assuming two negative feedback loops has been extended and substantiated by a mathematical study (116).

Regulation of Cell-Type Proportioning

Prespore cells and, to a lesser extent, prestalk cells are still plastic in their capacity to differentiate into either spores or stalk cells. Cell-type specification is sustained and can be modified by signal exchange among cells from tip formation until terminal cell differentiation. Slug cells are capable of binding cAMP to cell-surface receptors and of producing cGMP and cAMP in response to cAMP. Although these activities are reduced to roughly 10% of the activities in aggregating cells, their maintenance suggests that cAMP continues to act as an intercellular signal in the slugs (117, 118).

A mutant that forms only stalks, HL31, shows a normal proportioning of prestalk and prespore cells, indicating that final determination of cell types occurs during fruiting body formation, not earlier than two hours before terminal cell differentiation (67, 119). Reassurance of proportioning in the slug is also indicated by the effect of adenosine deaminase. Adenosine is an inhibitor of cAMP-binding to cell-surface receptors and an endogenous extracellular degradation product of cAMP (120, 121). Addition of adenosine deaminase to slugs causes prestalk cells in the anterior region to form prespore vacuoles indicating cell-type conversion (122).

MECHANISMS OF cAMP ACTION

cAMP Regulates Gene Expression by Interaction with Cell-Surface Receptors

The regulatory subunit of cAMP-dependent protein kinase is thought to be the major intracellular target site of cAMP. Since its specificity differs from that of cell-surface receptors, the site of cAMP action can be probed by the use of cAMP analogues. 2'-deoxy-cAMP has a fairly high affinity for the receptors and a very low affinity for the kinase, while 8-bromo-cAMP is a poor receptor agonist that permeates into the cell and binds to the regulatory subunit. Expression of two postaggregative genes encoding a prespore-specific antigen and glycogen phosphorylase, which is present in prestalk cells as well, is strongly induced by 2'-deoxy-cAMP and only weakly by 8-bromo-cAMP, indicating that the effects of cAMP on postaggregative gene expression are mediated by cell-surface receptors (123). An uncertainty in this approach is the presence of intracellular cAMP-binding proteins that might differ from the specificity of the regulatory subunit.

Heterogeneity of Cell-Surface cAMP-Binding Sites

Aggregation-competent cells of *D. discoideum* possess about 1×10^5 cAMP binding sites on their surface. Computer-assisted curve fitting of cAMP binding data led to the identification of two groups of binding sites, fast dissociating (A) and slowly dissociating sites (B, previously called S) (124). The A sites, representing about 96% of the cAMP-binding sites on the cell surface, exist in two interconvertible forms. Binding of cAMP causes conversion of the high-affinity form A^H (previously called H) to the low-affinity form A^L (previously called L). The conversion of A^H to A^L and the cAMP-induced production of cAMP are inhibited in parallel by Ca^{2+}, suggesting that the A sites mediate the activation of adenylate cyclase (125).

Attempts to fit binding data into a consistent scheme made the following assumptions necessary. The slowly dissociating B sites appear to exist also in a fast dissociating form, designated B^F, to which cAMP initially binds at the surface of living cells (126). Also, two different slowly dissociating sites, B^S and B^{SS}, are assumed to exist. Binding of cAMP is thought to turn B^F into B^S, which is rapidly converted into B^{SS}. Still, there may be some inconsistency in the data that might be solved by further assumptions.

Pretreatment of cells with cAMP desensitizes the cGMP response, i.e. an increase in the intracellular cGMP concentration is no longer observed. In the pretreated cells the conversion of B^S into B^{SS} is blocked, suggesting a role for B-site transition in the activation of guanylate cyclase (126).

In membrane preparations, GTP or GDP appear to convert B^S into B^F and to block the transition of A^H to A^L. These effects suggest that the different forms of A and B sites represent the receptors alone and complexes of them with guanine nucleotide–binding proteins (G-proteins) (127). Substantial activation of adenylate cyclase by GTP has been observed in freshly lysed cells (128). It will be a major effort of future work to identify complexes of receptors with G-proteins and with the catalytic subunits of guanylate and adenylate cyclase.

Desensitization, Receptor Phosphorylation, and Down-Regulation in Response to cAMP

When the second out of two pulsatile stimuli of equal size elicits a weaker response than the first, the term desensitization is used. This is also used as a synonym of adaptation, which is a decrease in the response to a continuing signal, but the equivalence of the two reactions is not strictly proven. In cells stimulated with cAMP, various desensitization phenomena are observed. Guanylate cyclase is desensitized within 10 sec after cAMP application (129). Adenylate cyclase is just starting to be activated at that time and is desensitized about two min later. Even then, cAMP is bound to cell-surface receptors whose total number remains essentially unchanged (130), and signals are still being transmitted. This is shown by the regulation of Ca^{2+} fluxes. If cells are stimulated with cAMP, a net influx of Ca^{2+} is observed and the intracellular Ca^{2+} level remains high as long as cAMP is in the medium. Only after its removal is a net outflux of Ca^{2+} observed that indicates recovery of the resting state (131).

Desensitization may be brought about by modification of components that are specific for certain signal transduction pathways, or by modification of the cAMP receptors that causes them to channel their signals into different pathways of transduction. After desensitization by cAMP of the cGMP response (132), of chemotaxis (133), and of a change in actin polymerization (134), these responses can still be elicited by folate, and vice versa, suggesting a modification of the receptors or of a component closely linked to them. The change in actin polymerization is most likely associated with chemotaxis.

cAMP-receptors have been identified by affinity-labeling of intact cells with ^{32}P-8-azido-cAMP (135, 136) and purified (137). The labeled receptor molecule is a phosphoprotein with M_r 40,000 to 45,000 (138–140). Stimulation of cells with cAMP increases within half a minute the ^{32}P-incorporation and raises the M_r by 2000 to 3000. In cells that produce cAMP periodically, the receptor undergoes oscillatory changes in its M_r (141).

Long exposure of cells to high concentrations of cAMP causes down-regulation of receptors exposed to the cell surface. This response is much slower than desensitization. After exposure of cells for 30 minutes to 1 mM cAMP, fast dissociating type-A receptor sites are reduced by 80–90%. Slowly dissociating B-sites are significantly changed in affinity but not in number. In accord with the proposed role of A-sites in adenylate cyclase activation, the production of cAMP that is induced by saturating cAMP stimuli is drastically decreased. The production of cGMP is not decreased under these conditions, but about 20-fold higher cAMP concentrations are required to induce responses similar to those in control cells (142). This supports the hypothesis that B-sites are involved in guanylate cyclase activation.

Antagonists of cAMP Actions

AMMONIA   Since starving *Dictyostelium* cells consume cellular proteins as an energy source, ammonia is generated as a product of amino acid deamination. Ammonia exerts various effects during the development of *D. discoideum*, e.g. it inhibits EDTA-resistant cell adhesion (143) and repels fruiting bodies rising into the air (144). Ammonia inhibits the cAMP-induced cAMP production. At least some of the developmental effects of ammonia may be caused by this effect (145), but a synergistic effect of cAMP and ammonia on cell-type-specific gene expression also has to be taken into account (146).

Ammonia stimulates spore differentiation in the sporogenous mutant HM18-2 and inhibits stalk-cell differentiation in aggregates of wild-type NC4 (99). The possibility that prestalk and prespore differentiation is determined by effects of ammonia on the intracellular pH (105) has been tested using a pH-sensitive fluorescent dye. No significant difference between cell types has been found (147). It has also been tested whether DIF lowers the intracellular pH and ammonia antagonizes DIF by shifting the pH in the other direction. Determination of cellular pH values by NMR did not reveal a significant change of intracellular pH in response to DIF (148). This is in accord with results showing that pretreatment of cells with ammonia prevents DIF production, but that ammonia is not an antagonist of DIF (149).

ADENOSINE   Extracellular cAMP is in vivo converted into adenosine by two cell-surface enzymes, phosphodiesterase and 5'-nucleotidase. Thus, adenosine production follows each cAMP stimulus. Adenosine delays cell aggregation by interfering with the binding of cAMP to cell-surface receptors (120). Adenosine has been reported to inhibit the binding of cAMP competitively or noncompetitively depending on the developmental stage or strain (121).

Second Messengers

Current work is aimed at unraveling the functional connections between cell-surface receptors and intracellular targets. Signal transduction is complicated by the fact that its pathways are ramified. First, cAMP receptors control not only gene expression but also cell motility. The chemotactic action of cAMP implies that precise information on its spatial distribution in the environment is transmitted to the cell's contractile system. Cells orient in a shallow gradient with not more than a 2% concentration difference between their front and rear ends (150). Second, in addition to cAMP receptors there are folate receptors on the cell surface that project signals to intracellular targets. Although the effects of folate are not fully the same as those of cAMP, the similarity of their actions suggests pathways of signal transduction that join somewhere behind the cAMP and folate receptors (151). Since the contractile system is unlikely to be involved in the control of cell differentiation, the regulation of cell motility is not within the scope of this review. However, since cAMP and folate are both chemoattractants and regulators of cell differentiation, it should be remembered that some of the second messengers involved in chemotaxis might also be involved in the expression of developmentally regulated proteins. This argument holds even though motility responses are observed within seconds after stimulation, while the lag period before the induced synthesis of proteins measures in tens of minutes or in hours.

INOSITOLTRIPHOSPHATE AND Ca^{2+} REGULATION Ca^{2+} redistribution is one of the fastest receptor-mediated responses elicited by cAMP or folate. An increased net influx of Ca^{2+} from the extracellular medium into cells was observed within less than 10 seconds after cAMP application. Before stimulation, cells were equilibrated at low Ca^{2+} concentrations with $^{45}Ca^{2+}$ (152). In other experiments the extracellular Ca^{2+} concentration was suddenly raised to 0.5 mM when a cAMP stimulus was given, and no increased Ca^{2+} influx was observed within one minute in the stimulated cells compared to unstimulated ones (153). Only later did the rate of Ca^{2+} influx become higher in the stimulated cells. This result indicates that an increased Ca^{2+} uptake from the extracellular space is not an essential step in signal transduction.

Treatment of permeabilized cells with putative intracellular messengers is a valuable method for elucidating the sequence of steps in signal transmission. Saponin selectively opens the plasma membrane without emptying intracellular Ca^{2+} storage vesicles. Addition of inositol 1,4,5-triphosphate to permeabilized cells seems to cause within 5 seconds a release of Ca^{2+} from nonmitochondrial stores, suggesting that IP_3 production links receptor activation to Ca^{2+} redistribution (154).

cGMP PRODUCTION   cGMP rises to a sharp peak via receptor-mediated activation of guanylate cyclase, and is primarily degraded by a specific intracellular phosphodiesterase. Treatment of permeabilized cells with Ca^{2+} (155) or IP_3 (156) stimulates cGMP production, suggesting that guanylate cyclase activation is coupled to an increase in the cytoplasmic Ca^{2+} concentration. The *ras*-gene product seems to be involved in the control of guanylate cyclase since cells transformed at high-copy number with a mutated *ras* sequence are defective in desensitization of the cGMP response (157). The activity of cGMP-specific phosphodiesterase increases during development (158). The enzyme is activated by cGMP (159) but is not identical with the major cGMP-binding protein of the cells, the putative mediator of signal transduction. No cGMP-dependent kinase activity has been found to be associated with this cGMP-binding protein (160).

A function of cGMP in the transmission of chemotactic signals is suggested by two findings. First, any compound that acts as a chemoattractant in a cellular slime mold species activates guanylate cyclase. Second, mutants of *D. discoideum* that form particularly long streams ("streamer mutants") are defective in cGMP-specific phosphodiesterase. The delayed cGMP degradation is accompanied by a prolonged chemotactic response of the mutant cells after stimulation with cAMP (161, 162). Involvement of cGMP (163), and of calmodulin (164), in the induction of extracellular cAMP-phosphodiesterase has also been suggested.

INTRACELLULAR CAMP AND PROTEIN KINASE   The presence in *D. discoideum* cells of a cytosolic cAMP-dependent protein kinase and of two other cAMP-regulated enzymes, a phosphoamidase (165) and S-adenosyl-L-homocysteine hydrolase (166), indicates an intracellular function of cAMP in addition to its importance for cell-to-cell signaling. The cAMP-dependent protein kinase of *D. discoideum* differs from mammalian enzymes by having only one cAMP-binding site. The holoenzyme consists of one regulatory and one catalytic subunit (167). The cDNA-derived sequence of the regulatory subunit indicates a molecular weight of 37,000 and shows extended regions of homology with the respective mammalian subunit, particularly at the cAMP-binding domain (168).

Both the catalytic and regulatory subunits show a fourfold increase during growth and the tipped aggregate stage (169, 170). The mRNA for the regulatory subunit accumulates 10–20-fold during early development, and this increase is enhanced by treating AX3 cells with cAMP (171). In accord with its increase during development, cAMP-dependent protein kinase is thought to play a role in the postaggregative stage. A preferential rise of activity in prespore cells suggests a function associated with the differentiation of these cells (172).

The cAMP-dependent protein kinase is unlikely to be important for the control of early development by cAMP pulses. The activation of adenylate cyclase by cAMP pulses is inhibited by caffeine in the wild-type and is constitutively blocked in a mutant, Agip53. Although a transient rise of intracellular cAMP levels is suppressed in these cases, cells can be stimulated by extracellular cAMP pulses to acquire aggregation competence and to express contact sites A and other proteins that are normally expressed during early development (173, 174). These results suggest that second messengers other than intracellular cAMP mediate the effects of extracellular cAMP pulses on early development. The same argument holds for certain prestalk-specific transcripts that are induced by extracellular cAMP under conditions that prevent activation of adenylate cyclate (175).

Reversible translocation of both protein kinase subunits from the cytoplasm to the nucleus has been reported to occur during development. While about 2% of the total soluble activity of these subunits was found in the nuclei of growth-phase cells, 12% of the activity was found in the nuclei of developing cells (176). After dissociation of slug cells, the amount of cAMP-dependent protein kinase in the nucleus decreases, and this decrease is closely correlated with a decay of certain postaggregative mRNA species. Upon cAMP addition to the dissociated cells the kinase activities as well as the amounts of mRNAs recover. Although these correlations may suggest a role for the kinase in the regulation of transcription, the levels of most of the postaggregative mRNAs appear to be regulated by changes in their stability (177).

Targets of Signal Transduction

TRANSCRIPTION REGULATED BY cAMP SIGNALS   cAMP signals have been implicated in the control of transcription as well as stabilization of developmentally regulated mRNA species. Sites responsible for transcriptional control have been identified in the 5' upstream region of a cathepsin gene that is preferentially expressed in prestalk cells. This gene has been designated as pst-cath (2) and the cathepsin as cysteine proteinase 2 (3). Expression of pst-cath can be induced in single cells by cAMP in the presence of the conditioning factor CMF (66). Transformants carrying about 2.5 kb of the 5' flanking region of the pst-cath gene fused to the coding region of an indicator, the *Escherichia coli* β-glucuronidase gene, express the gene fusion in a cAMP-dependent and prestalk-specific manner (178). By linker-scanner mutations, deletions, and point mutations, the importance for transcription of a GC-rich region has been demonstrated (2, 3). This region contains a palindromic repeat and a G-rich box that is homologous to the 3' half of the repeat. Their sequence is:

AACACAGCGGGTGTGTTAAGTTAGGGGTGGGTT.

Lines underneath the sequence indicate the palindrome, lines above a direct repeat including the 3' half of the palindrome and the G box. Mutations in the G box result in expression that is at least 50-fold lower than normal but still regulated by cAMP. The 5' half of the palindrome alone is unable to maintain expression; in conjunction with the G box it confers cAMP-regulated expression to the construct without a need for the 3' half of the repeat (2). The parallel destruction of basal and cAMP-induced expression of the gene might be due to a trans-acting factor, probably a protein, whose amount is increased by stimulating cells with cAMP and whose binding to the GC-rich region is indispensable for transcription.

mRNA STABILITY REGULATED BY cAMP AND CELL CONTACT   Most, but not all, postaggregative mRNA species appear to be regulated by their stability. Before and after cell aggregation, these mRNAs are synthesized at comparable rates, but their lifetimes increase drastically when tight aggregates are formed. Mechanical disruption of the aggregates renders the postaggregative mRNAs unstable except when cAMP is added to the dissociated cells (177).

Stability of the mRNAs in tight aggregates suggests that cell-to-cell contacts are important. Membrane-associated molecules may interact between contiguous cell surfaces to provide signals for RNA stabilization, or high local concentrations of cAMP (179), ammonia (146), or other molecules in the small intercellular space may be responsible for this effect. A convenient technique to study the effect of cell contacts on gene expression is to replace cell-to-cell contact by the contact of cells with an artificial, chemically defined support. Derivatizing a polyacrylamide matrix (180) with glucose provides a surface to which *D. discoideum* cells adhere through glucose-binding proteins on their surface. Despite their adherence the cells are able to move on the glucose-derivatized gels; they form aggregation centers and propagate waves of periodic cAMP signals (181). Development up to the aggregation stage is not affected and contact sites A are synthesized. However, postaggregative mRNA species are not expressed as long as the cells are on the gels. Exposure of tight aggregates, in which these mRNAs are expressed, to glucose-derivatized gels causes rapid spreading of the cells on the gel surface followed by destabilization of the mRNAs. Thus cell-to-substratum adhesion effectively competes with tight cell-to-cell adhesion but cannot replace its RNA-stabilizing effect.

CONCLUDING REMARKS

Acting as a multipurpose signal in *D. discoideum*, cAMP projects to several intracellular targets and mediates functions that vary in the course of development. Not covered in this review is the parallelism between cAMP and folate in eliciting chemotaxis and controlling development (182). *D. discoideum* is a folate auxotroph and therefore unlikely to use folate itself as an intercellular signal (183). But pteridines can be produced by GTP metabolism and are released by *D. discoideum* cells into the medium (184). The induction of folate deaminase by cAMP and the induction of cAMP-phosphodiesterase by folate (42, 151) might be the unavoidable side effects of ramified signal transduction pathways that connect a maximal number of intracellular targets to a few types of cell-surface receptors. However, the attraction of slug cells to folate (185) and the multiple alterations in the development of mutants selected for defects in folate chemotaxis (186) suggest that compounds related to folate play a role in development.

There is much variation in the control of genes between strains and different conditions. Only in some strains are pulsatile cAMP signals required for early development. Variations are possible because control systems are redundant. Bypassing the requirement for cAMP pulses by mutation shows the presence of other systems that alone are able to perform stringent developmental control. Chemotaxis of aggregating cells to cAMP is restricted to a small number of *Dictyostelium* species. Cells of other species produce probably pteridine derivatives in order to aggregate (187). Glorin, the attractant of a related genus, *Polysphondylium,* is a substituted dipeptide (188), but cAMP-dependent protein kinase accumulates during development in *Polysphondylium* as it does in *Dictyostelium* (189). It seems that a general function of cAMP in this group of organisms is the control of morphogenesis at the multicellular stage (190, 191).

Literature Cited

1. Reymond, C. D., Nellen, W., Firtel, R. A. 1985. *Proc. Natl. Acad. Sci. USA* 82:7005–9
2. Datta, S., Firtel, R. A. 1987. *Mol. Cell. Biol.* 7:149–59
3. Pears, C. J., Williams, J. G. 1987. *EMBO J.* 6:195–200
4. Nellen, W., Silan, C., Firtel, R. A. 1984. *Mol. Cell. Biol.* 4:2890–98
4a. Knecht, D., Cohen, S. M., Loomis, W. F., Lodish, H. F. 1986. *Mol. Cell. Biol.*
5. Noegel, A., Welker, D. L., Metz, B. A., Williams, K. L. 1985. *J. Mol. Biol.* 185:447–50
6. Firtel, R. A., Silan, C., Ward, T. E., Howard, P., Metz, B. A., Nellen, W., Jacobson, A. 1985. *Mol. Cell. Biol.* 5:3241–50
7. Noegel, A., Metz, B. A., Williams, K. L. 1985. *EMBO J.* 4:3797–803
8. Loomis, W. F., ed. 1982. *The Development of Dictyostelium discoideum.* New York: Academic
9. Newell, P. C. 1986. *Hormones, Receptors and Cellular Interactions in Plants,* ed. C. M. Chadwick, D. R. Garrod, pp. 155–84. Cambridge: Cambridge Univ. Press
10. McRobbie, S. J. 1986. *CRC Crit. Rev. Microbiol.* 13:335–75
11. Fisher, P. R., Dohrmann, U., Williams, K. L. 1984. *Modern Cell Biol.* 3:197–248
12. Watts, D. J. 1984. *Biochem. J.* 220:1–14
13. Williams, J. G., Pears, C. J., Jermyn, K. A., Driscoll, D. M., Mahbubani, H., Kay, R. R. 1986. *Regulation of Gene Expression,* ed. I. Booth, C. Higgins, pp. 277–98
14. Raper, K. B. 1984. *The Dictyostelids.* Princeton, NJ: Princeton Univ. Press
15. de Wit, R. J. W., Bulgakov, R., Pinas, J. E., Konijn, T. M. 1985. *Biochim. Biophys. Acta* 814:214–26
16. Vardy, P. H., Fisher, L. R., Smith, E., Williams, K. L. 1986. *Nature* 320:526–29
17. Ihara, M., Tanaka, Y., Yanagisawa, K., Taya, Y., Nishimura, S. 1986. *Biochim. Biophys. Acta* 881:135–40
18. Soll, D. R., Mitchell, L. H., Hedberg, C., Varnum, B. 1984. *Dev. Genet.* 4:167–84
19. Hedberg, C., Soll, D. R. 1984. *Differentiation* 27:168–74

20. Abe, K., Orii, H., Okada, Y., Saga, Y., Yanagisawa, K. 1984. *Dev. Biol.* 104:477–83
21. Szabo, S. P., O'Day, D. H. 1984. *Can. J. Biochem. Cell Biol.* 62:722–31
22. Amagai, A. 1984. *J. Gen. Microbiol.* 130:2961–65
23. Chisholm, R. L., Barklis, E., Lodish, H. F. 1984. *Nature* 310:67–69
24. Loomis, W. F. 1985. *Cold Spring Harbor Symp. Quant. Biol.* 50:769–77
25. Finney, R. E., Langtimm, C. J., Soll, D. R. 1985. *Dev. Biol.* 110:171–91
26. Kopachik, W., Bergen, L. G., Barclay, S. L. 1985. *Proc. Natl. Acad. Sci. USA* 82:8540–44
27. Kimmel, A. R., Carlisle, B. 1986. *Proc. Natl. Acad. Sci. USA* 83:2506–10
28. Alexander, S., Cibulsky, A. M., Mitchell, L., Soll, D. R. 1985. *Differentiation* 30:1–6
29. Noegel, A., Harloff, C., Hirth, P., Merkl, R., Modersitzki, M., et al. 1985. *EMBO J.* 4:3805–10
30. Gerisch, G., Hagmann, J., Hirth, P., Rossier, C., Weinhart, U., Westphal, M. 1985. *Cold Spring Harbor Symp. Quant. Biol.* 50:813–22
31. Noegel, A., Gerisch, G., Stadler, J., Westphal, M. 1986. *EMBO J.* 5:1473–76
32. Dinauer, M. C., Steck, T. L., Devreotes, P. N. 1980. *J. Cell Biol.* 86:554–61
33. Klein, P., Fontana, D., Knox, B., Theibert, A., Devreotes, P. 1985. *Cold Spring Harbor Symp. Quant. Biol.* 50:787–99
34. Roos, W., Scheidegger, C., Gerisch, G. 1977. *Nature* 266:259–61
35. Tomchik, K. J., Devreotes, P. N. 1981. *Science* 212:443–46
36. Bumann, J., Malchow, D., Wurster, B. 1986. *Differentiation* 31:85–91
37. Gottmann, K., Weijer, C. J. 1986. *J. Cell Biol.* 102:1623–29
38. Aeckerle, S., Wurster, B., Malchow, D. 1985. *EMBO J.* 4:39–43
39. Goldbeter, A., Martiel, J.-L. 1985. *FEBS Lett.* 191:149–53
40. Gerisch, G., Malchow, D., Roos, W., Wick, U. 1979. *J. Exp. Biol.* 81:33–47
41. Rossier, C., Gerisch, G., Malchow, D., Eckstein, F. 1978. *J. Cell Sci.* 35:321–38
42. Rossier, C., Eitle, E., van Driel, R., Gerisch, G. 1980. *The Eukaryotic Microbial Cell. Cell. Soc. Gen. Microbiol*, ed. G. W. Gooday, D. Lloyd, A. P. J. Trinci, 30:405–24. Cambridge: Cambridge Univ. Press
43. Van Haastert, P. J. M., Kien, E. 1983. *J. Biol. Chem.* 258:9636–42
44. Gerisch, G., Fromm, H., Huesgen, A., Wick, U. 1975. *Nature* 255:547–49
45. Finney, R. E., Langtimm, C. J., Soll, D. R. 1985. *Dev. Biol.* 110:157–70
46. Gerisch, G., Tsiomenko, A., Stadler, J., Claviez, M., Hülser, D., Rossier, C. 1984. *Information and Energy Transduction in Biological Membranes*, ed. C. L. Bolis, E. J. M. Helmreich, H. Passow, pp. 237–47. New York: Liss
47. Murray, B. A., Yee, L. D., Loomis, W. F. 1981. *J. Supramol. Struct. Cell. Biochem.* 17:197–211
48. Oyama, M., Blumberg, D. D. 1986. *Dev. Biol.* 117:550–56
49. Springer, W. R., Cooper, D. N. W., Barondes, S. H. 1984. *Cell* 39:557–64
50. Crowley, T. E., Nellen, W., Gomer, R. H., Firtel, R. A. 1985. *Cell* 43:633–41
51. Alexander, S., Shinnick, T. M. 1985. *Mol. Cell. Biol.* 5:984–90
52. Williams, J. G., Tsang, A. S., Mahbubani, H. 1980. *Proc. Natl. Acad. Sci. USA* 77:7171–75
53. Berger, E. A., Bozzone, D. M., Berman, M. B., Morgenthaler, J. A., Clark, J. M. 1985. *J. Cell. Biochem.* 27:391–400
54. Hagmann, J. 1986. *EMBO J.* 5:3437–440
55. Brenner, M., Thoms, S. D. 1984. *Dev. Biol.* 101:136–46
56. Westphal, M., Müller-Taubenberger, A., Noegel, A., Gerisch, G. 1986. *FEBS Lett.* 209:92–96
57. Lund, P. K., Moats-Staats, B. M., Simmons, J. G., Hoyt, E., D'Ercole, A. J., et al. 1985. *J. Biol. Chem.* 260:7609–13
58. Roos, W., Malchow, D., Gerisch, G. 1977. *Cell Differ.* 6:229–39
59. Franke, J., Kessin, R. H. 1981. *J. Biol. Chem.* 256:7628–37
60. Lappano, S., Coukell, M. B. 1982. *Dev. Biol.* 93:43–53
61. Mullens, I. A., Franke, J., Kappes, D. J., Kessin, R. H. 1984. *Eur. J. Biochem.* 142:409–15
62. Podgorski, G. J., Franke, J., Kessin, R. H. 1986. *J. Gen. Microbiol.* 132:1043–50
63. Wallraff, E., Welker, D. L., Williams, K. L., Gerisch, G. 1984. *J. Gen. Microbiol.* 130:2103–14
64. Cardelli, J. A., Knecht, D. A., Wunderlich, R., Dimond, R. L. 1985. *Dev. Biol.* 110:147–56
65. Saxe, C. L. III, Firtel, R. A. 1986. *Dev. Biol.* 115:407–14
66. Gomer, R. H., Datta, S., Mehdy, M., Crowley, T., Sivertsen, A., et al. 1985. *Cold Spring Harbor Symp. Quant. Biol.* 50:801–12
67. Morrissey, J. H., Devine, K. M.,

Loomis, W. F. 1984. *Dev. Biol.* 103: 414–24

68. Noce, T., Takeuchi, I. 1985. *Dev. Biol.* 109:157–64
69. Dowds, B. C. A., Loomis, W. F. 1984. *Mol. Cell. Biol.* 4:2273–78
70. Wallace, J. S., Morrissey, J. H., Newell, P. C. 1984. *Cell Differ.* 14:205–11
71. Julien, J., Bogdanovsky-Sequeval, D., Felenbok, B., Jacquet, M. 1984. *Cell Differ.* 15:37–42
72. Sternfeld, J., David, C. N. 1981. *Differentiation* 20:10–21
73. Loomis, W. F., Kuspa, A. 1984. *Dev. Biol.* 102:498–503
74. Pears, C. J., Mahbubani, H. M., Williams, J. G. 1985. *Nucleic Acids Res.* 13:8853–66
75. Devine, K. M., Bergmann, J. E., Loomis, W. F. 1983. *Dev. Biol.* 99:437–46
76. Takemoto, K., Yamamoto, A., Takeuchi, I. 1985. *J. Cell Sci.* 77:93–108
77. Ratner, D., Borth, W. 1983. *Exp. Cell Res.* 143:1–13
78. Weijer, C. J., McDonald, S. A., Durston, A. J. 1984. *Differentiation* 28:13–23
79. Gomer, R. H., Datta, S., Firtel, R. A. 1986. *J. Cell Biol.* 103:1999–2015
80. Brown, S. S., Rutherford, C. L. 1980. *Differentiation* 16:173–83
81. Devine, K. M., Loomis, W. F. 1985. *Dev. Biol.* 107:364–72
82. Weijer, C. J. cited by MacWilliams, H., Blaschke, A., Prause, I. 1985. *Cold Spring Harbor Symp. Quant. Biol.* 50:779–85
83. Armant, D. R., Stetler, D. A., Rutherford, C. L. 1980. *J. Cell Sci.* 45:119–29
84. MacWilliams, H., Blaschke, A., Prause, I. 1985. *Cold Spring Harbor Symp. Quant. Biol.* 50:779–85
85. Matsukuma, S., Durston, A. J. 1979. *J. Embryol. Exp. Morphol.* 50:243–51
86. Sternfeld, J., David, C. N. 1981. *J. Cell Sci.* 50:9–17
87. Krefft, M., Voet, L., Gregg, J. H., Williams, K. L. 1985. *J. Embryol. Exp. Morphol.* 88:15–24
88. Williams, J. G., North, M. J., Mahbubani, H. 1985. *EMBO J.* 4:999–1006
89. Presse, F., Bogdanovsky-Sequeval, D., Mathieu, M., Felenbok, B. 1986. *Mol. Gen. Genet.* 203:333–40
90. Reymond, C. D., Gomer, R. H., Mehdy, M. C., Firtel, R. A. 1984. *Cell* 39:141–48
91. Reymond, C. D., Gomer, R. H., Nellen, W., Theibert, A., Devreotes, P., Firtel, R. A. 1986. *Nature* 323:340–43

92. Bennett, V. D., Dimond, R. L. 1986. *J. Biol. Chem.* 261:5355–62
93. Bertholdt, G., Stadler, J., Bozzaro, S., Fichtner, B., Gerisch, G. 1985. *Cell Differ.* 16:187–202
94. Hohmann, H.-P., Gerisch, G., Lee, R. W. H., Huttner, W. B. 1985. *J. Biol. Chem.* 260:13869–78
95. Loomis, W. F., Wheeler, S. A., Springer, W. R., Barondes, S. H. 1985. *Dev. Biol.* 109:111–17
96. Gomer, R. H., Armstrong, D., Leichtling, B. H., Firtel, R. A. 1986. *Proc. Natl. Acad. Sci. USA* 83:8624–28
97. Haribabu, B., Dottin, R. P. 1986. *Mol. Cell. Biol.* 6:2402–8
98. Wang, M., Schaap, P. 1985. *Differentiation* 30:7–14
99. Weeks, G. 1984. *Exp. Cell Res.* 153:81–90
100. Kumagai, A., Okamoto, K. 1986. *Differentiation* 31:79–84
101. Ishida, S. 1980. *Dev. Growth Differ.* 22:143–52
102. Ishida, S. 1980. *Dev. Growth Differ.* 22:781–88
103. Weijer, C. J., Durston, A. J. 1985. *J. Embryol. Exp. Morphol.* 86:19–37
104. Kay, R. R. 1982. *Proc. Natl. Acad. Sci. USA* 79:3228–31
105. Dominov, J. A., Town, C. D. 1986. *J. Embryol. Exp. Morphol.* 96:131–50
106. Kay, R. R., Dhokia, B., Jermyn, K. A. 1983. *Eur. J. Biochem.* 136:51–56
107. Kopachik, W. J., Dhokia, B., Kay, R. R. 1985. *Differentiation* 28:209–16
108. Brookman, J. J., Town, C. D., Jermyn, K. A., Kay, R. R. 1982. *Dev. Biol.* 91:191–96
108a. Brookman, J. J., Jermyn, K. A., Kay, R. R. 1987. *Development.* In Press
109. Meinhardt, H. 1982. *Models of Biological Pattern Formation*, pp. 39–44. London: Academic
110. MacWilliams, H. K. 1984. *Pattern Formation*, ed. G. M. Malacinski, pp. 127–62. New York: Macmillan
111. Weijer, C. J., Duschl, G., David, C. N. 1984. *J. Cell Sci.* 70:111–31
112. McDonald, S. A., Durston, A. J. 1984. *J. Cell Sci.* 66:195–204
113. Weijer, C. J., Duschl, G., David, C. N. 1984. *J. Cell Sci.* 70:133–45
114. Krefft, M., Voet, L., Gregg, J. H., Mairhofer, H., Williams, K. L. 1984. *EMBO J.* 3:201–6
115. Kakutani, T., Takeuchi, I. 1986. *Dev. Biol.* 115:439–45
116. Grinfeld, M., Segel, L. A. 1986. *J. Theor. Biol.* 121:23–44
117. Schaap, P., Spek, W. 1984. *Differentiation* 27:83–87
118. Kesbeke, F., van Haastert, P. J. M.,

Schaap, P. 1986. *FEMS Microbiol. Lett.* 34:85–89

119. Morrissey, J. H., Farnsworth, P. A., Loomis, W. F. 1981. *Dev. Biol.* 83:1–8

120. Newell, P. C. 1982. *FEMS Microbiol. Lett.* 13:417–21

121. Theibert, A., Devreotes, P. N. 1984. *Dev. Biol.* 106:166–73

122. Schaap, P., Wang, M. 1986. *Cell* 45:137–44

123. Schaap, P., van Driel, R. 1985. *Exp. Cell Res.* 159:388–98

124. van Haastert, P. J. M., de Wit, R. J. W. 1984. *J. Biol. Chem.* 259:13321–28

125. van Haastert, P. J. M. 1985. *Biochim. Biophys. Acta* 846:324–33

126. van Haastert, P. J. M., de Wit, R. J. W., Janssens, P. M. W., Kesbeke, F., DeGoede, J. 1986. *J. Biol. Chem.* 261:6904–11

127. Janssens, P. M. W., Arents, J. C., van Haastert, P. J. M., van Driel, R. 1986. *Biochemistry* 25:1314–20

128. Theibert, A., Devreotes, P. N. 1986. *J. Biol. Chem.* 261:15121–25

129. van Haastert, P. J. M., van der Heijden, P. R. 1983. *J. Cell Biol.* 96:347–53

130. Bumann, J., Malchow, D. 1986. *FEMS Microbiol. Lett.* 33:99–103

131. Bumann, J., Wurster, B., Malchow, D. 1984. *J. Cell Biol.* 98:173–78

132. van Haastert, P. J. M. 1983. *Biochem. Biophys. Res. Commun.* 115:130–36

133. van Haastert, P. J. M. 1983. *J. Cell Biol.* 96:1559–65

134. McRobbie, S. J., Newell, P. C. 1984. *Biochem. Biophys. Res. Commun.* 123:1076–83

135. Juliani, M. H., Klein, C. 1981. *J. Biol. Chem.* 256:613–19

136. Theibert, A., Klein, P., Devreotes, P. N. 1984. *J. Biol. Chem.* 259:12318–21

137. Klein, P., Knox, B., Borleis, J., Devreotes, P. 1987. *J. Biol. Chem.* 262:352–57

138. Klein, C., Lubs-Haukeness, J., Simons, S. 1985. *J. Cell Biol.* 100:715–20

139. Devreotes, P. N., Sherring, J. A. 1985. *J. Biol. Chem.* 260:6378–84

140. Klein, P., Vaughan, R., Borleis, J., Devreotes, P. 1987. *J. Biol. Chem.* 262:358–64

141. Klein, P., Theibert, A., Fontana, D., Devreotes, P. N. 1985. *J. Biol. Chem.* 260:1757–64

142. Kesbeke, F., van Haastert, P. J. M. 1985. *Biochim. Biophys. Acta* 847:33–39

143. McConaghy, J. R., Saxe, C. L. III., Williams, G. B., Sussman, M. 1984. *Dev. Biol.* 105:389–95

144. Bonner, J. T., Suthers, H. B., Odell, G. M. 1986. *Nature* 323:630–32

145. Williams, G. B., Elder, E. M., Sussman, M. 1984. *Dev. Biol.* 105:377–88

146. Oyama, M., Blumberg, D. D. 1986. *Dev. Biol.* 117:557–66

147. Ratner, D. I. 1986. *Nature* 321:180–82

148. Kay, R. R., Gadian, D. G., Williams, S. R. 1986. *J. Cell Sci.* 83:165–79

149. Neave, N., Sobolewski, A., Weeks, G. 1983. *Cell Differ.* 13:301–7

150. Mato, J. M., Losada, A., Nanjundiah, V., Konijn, T. M. 1975. *Proc. Natl. Acad. Sci. USA* 72:4991–93

151. Van Ophem, P., van Driel, R. 1985. *J. Bacteriol.* 164:143–46

152. Wick, U., Malchow, D., Gerisch, G. 1978. *Cell Biol. Int. Rep.* 2:71–79

153. Europe-Finner, G. N., Newell, P. C. 1985. *FEBS Lett.* 186:70–74

154. Europe-Finner, G. N., Newell, P. C. 1986. *Biochim. Biophys. Acta* 887:335–40

155. Small, N. V., Europe-Finner, G. N., Newell, P. C. 1986. *FEBS Lett.* 203:11–14

156. Europe-Finner, G. N., Newell, P. C. 1985. *Biochem. Biophys. Res. Commun.* 130:1115–22

157. Van Haastert, P., Kesbeke, F., Reymond, C., Firtel, R. A., Luderus, E., van Driel, R. 1987. *Proc. Natl. Acad. Sci. USA.* In press

158. Coukell, M. B., Cameron, A. M., Pitre, C. M., Mee, J. D. 1984. *Dev. Biol.* 103:246–57

159. Kesbeke, F., Baraniak, J., Bulgakov, R., Jastorff, B., Morr, M., et al. 1985. *Eur. J. Biochem.* 151:179–86

160. Parissenti, A. M., Coukell, M. B. 1986. *Biochem. Cell Biol.* 64:528–34

161. Ross, F. M., Newell, P. C. 1981. *J. Gen. Microbiol.* 127:339–50

162. Coukell, M. B., Cameron, A. M. 1986. *Dev. Genet.* 6:163–77

163. Van Haastert, P. J. M., Pasveer, F. J., van der Meer, R. C., van der Heijden, P. R., van Walsum, H., Konijn, T. M. 1982. *J. Bacteriol.* 152:232–38

164. Hashimoto, F., Hayashi, H. 1985. *Chem. Pharm. Bull.* 33:3023–26

165. Rossomando, E. F., Hadjimichael, J. 1986. *Int. J. Biochem.* 18:481–84

166. Hohman, R. J., Guitton, M. C., Veron, M. 1984. *Arch. Biochem. Biophys.* 233:785–95

167. De Gunzburg, J., Part, D., Guiso, N., Veron, M. 1984. *Biochemistry* 23:3805–12

168. Mutzel, R., Lacombe, M.-L., Simon, M.-N., de Gunzburg, J., Veron, M. 1987. *Proc. Natl. Acad. Sci. USA* 84:6–10

169. Leichtling, B. H., Majerfeld, I. H., Spitz, E., Schaller, K. L., Woffendin,

C., et al. 1984. *J. Biol. Chem.* 259:662–68

170. Part, D., de Gunzburg, J., Veron, M. 1985. *Cell Differ.* 17:221–27
171. De Gunzburg, J., Franke, J., Kessin, R. H., Veron, M. 1986. *EMBO J.* 5:363–67
172. Schaller, K. L., Leichtling, B. H., Majerfeld, I. H., Woffendin, C., Spitz, E., et al. 1984. *Proc. Natl. Acad. Sci. USA* 81:2127–31
173. Wurster, B., Bumann, J. 1981. *Dev. Biol.* 85:262–65
174. Mann, S. K. O., Firtel, R. A. 1987. *Mol. Cell. Biol.* 7:458–69
175. Oyama, M., Blumberg, D. D. 1986. *Proc. Natl. Acad. Sci. USA* 83:4819–23
176. Woffendin, C., Chambers, T. C., Schaller, K. L., Leichtling, B. H., Rickenberg, H. V. 1986. *Dev. Biol.* 115:1–8
177. Mangiarotti, G., Giorda, R., Ceccarelli, A., Perlo, C. 1985. *Proc. Natl. Acad. Sci. USA* 82:5786–90
178. Datta, S., Gomer, R. H., Firtel, R. A. 1986. *Mol. Cell. Biol.* 6:811–20
179. Haribabu, B., Rajkovic, A., Dottin, R. P. 1986. *Dev. Biol.* 113:436–42
180. Schnaar, R., Weigel, P. H., Kuhlenschmidt, M. S., Lee, Y. C., Roseman,

S. 1978. *J. Biol. Chem.* 253:7940–51
181. Bozzaro, S., Perlo, C., Ceccarelli, A., Mangiarotti, G. 1984. *EMBO J.* 3:193–200
182. De Wit, R. J., Bulgakov, R. 1986. *Biochim. Biophys. Acta* 886:76–87
183. Franke, J., Kessin, R. 1977. *Proc. Natl. Acad. Sci. USA* 74:2157–61
184. Tatischeff, I., Klein, R. 1984. *Hoppe-Seyler's Z. Physiol. Chem.* 365:1255–62
185. Mee, J. D., Tortolo, D. M., Coukell, M. B. 1986. *Biochem. Cell Biol.* 64:722–32
186. Segall, J. E., Fisher, P. R., Gerisch, G. 1987. *J. Cell Biol.* 104:151–61
187. Van Haastert, P. J. M., de Wit, R. J. W., Grijpma, Y., Konijn, T. M. 1982. *Proc. Natl. Acad. Sci. USA* 79:6270–74
188. Shimomura, O., Suthers, H. L. B., Bonner, J. T. 1982. *Proc. Natl. Acad. Sci. USA* 79:7376–79
189. Francis, D., Majerfeld, I. H., Kakinuma, S., Leichtling, B. H., Rickenberg, H. V. 1984. *Dev. Biol.* 106:478–84
190. Schaap, P., Wang, M. 1984. *Dev. Biol.* 105:470–78
191. Schaap, P. 1985. *Differentiation* 28:205–8

Ann. Rev. Biochem. 1987. 56:881–914

RECEPTORS FOR EPIDERMAL GROWTH FACTOR AND OTHER POLYPEPTIDE MITOGENS

Graham Carpenter

Departments of Biochemistry and Medicine, Vanderbilt University School of Medicine, Nashville, Tennessee 37232

CONTENTS

PERSPECTIVES AND SUMMARY

The understanding of the biochemistry that underlies the regulation of cell proliferation has been significantly affected by recent advances in the cellular and molecular mechanism of action of peptide growth factors. Although a complete understanding of these processes is still quite distant, biochemical mechanisms and molecular structures are replacing the descriptive body of experimental results available until a few years ago. At this point the number of discrete, well-defined, peptide growth factors is nearly 20 and undoubtedly more remain to be uncovered. Not unlike hormones in general, growth factors may have other biological activities in addition to the stimulation of cell proliferation. Some of the peptide growth factors are known to act, depending

881

0066-4154/87/0701-0881$02.00

on cell type and physiological circumstances, as growth inhibitors or inducers of cell differentiation.

Studies with radiolabeled growth factors have shown clearly that the initial step in the interaction of these extracellular signals with target cells is the rapid and high-affinity binding to receptors present on the plasma membrane. The structural and functional properties of these receptors have been the target of intensive research recently, and at this point there exists a considerable body of data constituting the subject of this review. It is possible to begin individual and comparative analyses of these receptor molecules. Space and time preclude a complete review of the growth factors themselves and their control of the various cellular activities that constitute the mitogenic response. The receptor for epidermal growth factor (EGF) is reviewed in detail and the structures of other receptors are presented for comparative purposes.

EPIDERMAL GROWTH FACTOR RECEPTOR

Epidermal growth factor is a well-characterized mitogen of 53 amino acids that stimulates the proliferation of numerous cell types in vitro and epithelial cells, in particular, in vivo (1–3). The EGF receptor is, with the exception of hemopoietic cells, detectable on a large variety of cell types or tissues. All available evidence indicates that the EGF receptor mediates the biological signals not only of EGF, but also of at least two EGF-like growth factors— transforming growth factor α (TGFβ) (5) and the vaccinia virus growth factor (VGF) (6). The three ligands (7) are approximately 22% identical in amino acid sequence, but bind to the EGF receptor with nearly identical affinities and produce the same responses in target cells. The three ligands are, however, the products of distinct genes that are differentially expressed. In humans, the gene for EGF is located on chromosome 4 (8, 9) and the gene for TGFα on chromosome 2 (10). VGF is encoded by a gene in the vaccinia virus genome. The predominate biological response of target cells to these ligands, both in the intact animal and in cell culture systems, is the enhancement of cell proliferation (1). However, other responses have been described as discussed elsewhere (4).

Receptor Isolation and Structure

The isolation and biochemical characterization of the EGF receptor, including cDNA cloning, have been greatly facilitated by the use of a cell line that overexpresses (by a factor of 20–50 fold) this receptor (11, 12). The cell line is designated A-431 and was derived in 1972 from a human epidermoid carcinoma of the vulva (13). There is no reason to suspect that biochemical information obtained from the EGF receptor of A-431 cells is particularly unusual, but certain details, for example the structure of oligosaccharide side chains, can be expected to differ slightly in other cell types. The long-term physiological responses of A-431 cells to EGF may not be representative of "normal" cellular responses, as EGF provokes a growth inhibition of A-431 cells in most circumstances (29). However, many early responses of A-431 cells to EGF are representative, but perhaps exaggerated, of the normal mitogenic response; for example, the induction of c-*fos* and c-*myc* (30), the turnover of phosphatidylinositol (31), and inositol trisphosphate formation (274), and cytoplasmic alkalinization (32). These cells may prove useful for uncovering some of the intracellular signaling mechanisms employed in the mitogenic response.

The EGF receptor from A-431 cells was first purified to near homogeneity in 1980 by the use of affinity chromatography (14). The purified receptor had an apparent molecular mass of 150,000; however, subsequent purification methods that eliminated calcium from the buffers indicated the actual mass to be 170,000 (15). This corresponded well to mass estimates obtained from earlier studies in which ^{125}I-EGF was chemically crosslinked to the receptor (54, 55, 60). Most cell homogenates contain a calcium-activated protease that cleaves the native molecule of 170,000 daltons to the lower molecular mass (150,000) species (16, 17). The EGF receptor has subsequently been isolated in an active form from mouse liver (61), and active, partially purified preparations have been obtained from human placenta (62, 63).

Prior to the isolation of the EGF receptor, it had been observed that the addition of EGF to receptor-rich membranes from A-431 cells resulted in the rapid activation of protein kinase activity (18, 19), which subsequently was demonstrated to be specific for tyrosine residues (20). Isolation of the EGF receptor resulted in copurification of the growth factor–sensitive tyrosine kinase activity, suggesting that the ligand binding site and kinase might be physically coupled. Consistent with this idea was the demonstration that antibodies to the 170,000-dalton EGF receptor were able to precipitate the growth factor–sensitive kinase activity (15). Treatment of the EGF receptor in membrane vesicles with a radiolabeled ATP affinity reagent demonstrated the presence of an ATP binding site on the 170,000-dalton receptor (21, 22). These studies also demonstrated that if the tyrosine kinase was inactivated by mild heating or *N*-ethylmaleimide exposure, subsequent incubation with the ATP reagent failed to label the receptor. These results all pointed to the somewhat unusual conclusion, at that time, that the EGF receptor molecule was a single polypeptide chain containing both a ligand binding site and a growth factor–sensitive tyrosine kinase activity. Purified preparations of the EGF receptor have been reported to contain two other enzyme activities, DNA topoisomerase and phosphatidylinositol kinase, but these appear to represent contaminating proteins (64–66). Whether or not the association of these proteins with the purified EGF receptor is adventitious remains to be determined. Purified insulin receptor preparations also are reported to contain phosphatidylinositol kinase activity (114, 115).

In 1984, sequence analysis of tryptic peptides derived from the EGF receptor (23) led to cDNA cloning of the receptor and the deduced amino acid sequence of the entire molecule (24). These elegant studies have provided a structural basis on which to understand the ligand binding and tyrosine kinase activities of the EGF receptor as domains of a single polypeptide chain. From this information and the results of subsequent structural studies, a diagrammatic representation of the structural features of the EGF receptor has been constructed (Figure 1). The mature receptor molecule has a polypeptide molecular mass, based on amino acid composition, of 131,600. This is in close agreement with biosynthetic studies carried out in the presence of tunicamycin to inhibit glycosylation (25–27). The mature receptor contains a single polypeptide chain of 1186 amino acid residues (the precursor contains a signal peptide of 24 residues) that is divided into two domains by one hydrophobic membrane anchor sequence. Whether other membrane-spanning regions occur in the folded molecule is not known. One group has, based on more complex modeling programs for membrane protein structure, suggested that additional transmembrane α-helixes might be present in the EGF receptor (28).

EXTRACELLULAR DOMAIN   The sequence data define an extracellular domain in the EGF receptor of 621 amino acid residues (24). This domain, which must fold to accommodate high-affinity ligand binding, is characterized by two structural features: a high content of cysteine (approximately 10%) dispersed in two regions and a relatively large number of canonical sequences for N-linked glycosylation. That many of the extracellular cysteine residues (a total of 51) are in the form of disulfides is likely, but the actual disulfide content of the receptor has not been quantitated. Nevertheless, it is probable that quite thermodynamically stable regions exist in the extracellular region. However, it is not known in what manner these cysteine-rich regions might participate in the actual ligand binding site. Experimental evidence does show that ^{125}I-EGF binding is resistant to preincubation at elevated temperature (10 min at 50°C) (19). While cysteine-rich regions are a feature of several other receptors [insulin, insulin-like growth factor-1 (IGF-1), nerve growth factor (NGF), and low-density lipoprotein (LDL)], these characteristics are not present in other high-affinity receptors [platelet-derived growth factor (PDGF), colony stimulating factor-1 (CSF-1), interleukin-2 (IL-2), and transferrin].

The ligand binding domain of the EGF receptor is also characterized by a relatively high content of carbohydrate. Both indirect (26) and direct (33, 34) studies have failed to detect the presence of O-linked carbohydrate, and it is

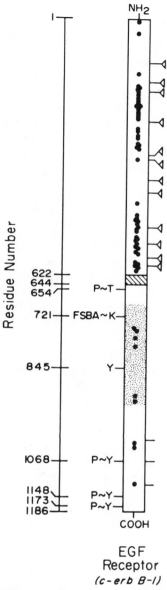

Figure 1 Structural features of the receptor for epidermal growth factor. The following symbols apply: —, canonical sequences for N-linked glycosylations; —◁, probable sites of N-linked oligosaccharide chains; ●, cysteine residues; Y, tyrosine residues; P~Y, phosphotyrosine residues; T, threonine residues; P~T, phosphothreonine residues; K, lysine residues; FSBA~K, lysine residue covalently labeled with *p*-fluorosulfonylbenzoyl adenosine; crosshatched area, membrane-spanning region; stippled area, sequences similar to *src* kinase. Data from references 24, 43–45, 53.

likely that all the carbohydrate on the EGF receptor (approximately 30,000 daltons) is present as N-linked oligosaccharide chains. The sequence data include 12 potential sites for N-linked glycosylation based on the canonical sequence Asn.X.Ser/Thr. Using quite different approaches, two groups (25, 33) estimate that 10–11 of the potential N-linked glycosylation sites are in fact occupied by oligosaccharide chains. Both Con A lectin chromatography of glycopeptides and endoglycosidase H sensitivity of the mature receptor from A-431 cells indicate that the N-linked oligosaccharides include about seven complex-type chains and approximately three high mannose–type chains (25, 33, 34).

High-mannose chains are also detectable on the EGF receptor from other cell types, but in varying amounts (A. M. Soderquist, G. Carpenter, unpublished observations). In other glycoproteins bearing both complex and high mannose–type oligosaccharide chains, the complex chains are usually located proximal to the amino terminus of the polypeptide chain and the high-mannose chains distal to the amino terminus (35). In the EGF receptor, therefore, it would follow, but has not been demonstrated, that the high-mannose chains would be located at the glycosylation sites nearest the transmembrane segment. The detailed structure of the complex-type chains is not available and is likely to vary somewhat in different tissues and cells. What is known is that these oligosaccharides are a mixture of biantennary and multiantennary structures characterized in the outer sugar residues by a low amount of sialic acid, a relatively high amount of fucose (fucosylation may also take place within the core region), and terminal N-acetylgalactosamine residues— an unusual feature of N-linked oligosaccharides (33, 34). The presence of peripheral fucose and/or N-acetylgalactosamine residues is likely to be responsible for the immunologic crossreactivity observed between blood group type A antigens and the EGF receptor from A-431 cells (33–39).

As the extracellular domain has both a high cysteine content (up to 25 disulfides are possible) and a large amount of carbohydrate (approximately 40% of the mass), this segment of the EGF receptor is highly resistant to proteolysis. In intact cells, neither trypsin nor pronase produces a decrease in ^{125}I-EGF binding capacity (G. Carpenter, S. Cohen, unpublished results) or size of the immunoprecipitable receptor (G. Carpenter, A. M. Soderquist, C. Stoscheck, unpublished results). If the cells are disrupted, however, a large tryptic fragment of approximately 120,000 daltons is produced (25, 56–58). This fragment contains the entire extracellular domain, transmembrane segment, and a short portion of the cytoplasmic domain that includes the threonine phosphorylation site at residue 654. If the EGF receptor is pretreated with reducing agents, then the extracellular domain can be more extensively digested by pronase (33) or trypsin (23).

CYTOPLASMIC DOMAIN   The cytoplasmic domain of the EGF receptor has been the subject of intense interest for two reasons: the tyrosine kinase activity encoded in this region of the molecule is considered to be the primary effector system in the transmembrane signaling process, and similar tyrosine kinase activity is present in several other growth factor receptors and in several oncogene products. The reader is referred to a recent review in this series (40) and elsewhere (41, 42) for more extensive discussion of the structural homologies and functional details of this family of proteins.

Immediately adjacent to the membrane-spanning sequence there is, at the cytoplasmic interface, a 13-residue sequence that is highly enriched in basic amino acids. It is likely that this sequence functions during receptor synthesis as a "stop transfer" sequence, following insertion of the amino terminal end of the receptor into the lumen of the endoplasmic reticulum. Whether this type of sequence, which is found in all growth factor receptors, is also important for functions of the mature EGF receptor at the plasma membrane is not known. Interest in a potential functional role of this sequence in the mature receptor has been elicited by the presence of a phosphorylation site for protein kinase C (threonine 654) lying in the midst of this sequence of very basic residues (43, 44). A similar C-kinase phosphorylation site is also a feature of the IL-2 receptor (86, 87).

As indicated by the stippled area within the cytoplasmic domain of the EGF receptor (Figure 1), there is a region of approximately 250 residues that share sequence homology with other members of the tyrosine kinase family, the prototype generally considered to be the *src* kinase of the Rous sarcoma virus. Within this tyrosine kinase region, considered to be the catalytic domain, there is a lysine (residue 721) that has been labeled with an ATP affinity reagent (45). Modeling (47, 59) of the ATP binding site in the EGF receptor predicts that lysine 721 would form a salt bridge with an oxygen atom on the β phosphate of ATP. Approximately 25 residues to the amino side of lysine 721 there is a Gly.X.Gly.X.X.Gly. sequence (residues 695–700), which is highly conserved in nucleotide binding proteins (46). The glycine at residue 695 and the Gly.X.Gly.X.X.Gly. sequence are proposed to be involved, respectively, in recognition of the C2 and N1 atoms on the adenine moiety of ATP and the correct orientation of the ribose and phosphate groups. The modeling studies also suggest that an Asp.Phe.Gly. sequence (residues 831–833), which is highly conserved in protein kinases, also participates in ATP binding. The aspartate residue is proposed to form a hydrogen bond with the amino group at position 6 of the adenine, while hydrophobic interactions occur between the phenylalanine side chain and the adenine ring. The dominant interactions would seem to involve lysine 721 and aspartate 831. ADP and GTP bind to the EGF receptor with an affinity equivalent to that of ATP

(59), and the initial phosphorylation studies (19) reported that the EGF receptor could use either ATP or GTP as phosphate donors.

The essential function of lysine 721 in the tyrosine activity of the EGF receptor is supported by a study of the analogous lysine in the v-*fps* tyrosine kinase of the Fujinami sarcoma virus (48). By site-directed mutagenesis, lysine 950 in the v-*fps* molecule was replaced with either an arginine or glycine residue. In both instances, the altered protein, while metabolically stable, was inactive as a tyrosine kinase and as a transforming gene product. Similar results were obtained when the analogous lysine in the v-*src* tyrosine kinase (lysine 295) was replaced by a methionine residue (49). The failure of arginine to substitute for this lysine residue has been interpreted as evidence supporting a catalytic role of this residue in the ATP hydrolysis and phosphoryl transfer reactions of protein kinases (48). That the tyrosine kinase domain may be able to function catalytically in the absence of other EGF receptor structural features is suggested by proteolysis experiments. Trypsin digestion of EGF receptors has allowed the recovery of a 42,000-dalton fragment that displays growth factor–independent tyrosine kinase activity (50). The protease generation of similar sized, active fragments has been reported for viral tyrosine kinases (51, 52).

Most of the carboxy terminal region of the EGF receptor contains sequences not found in other tyrosine kinases. This C-terminal region is thought to be important for receptor function as it contains the major sites of autophosphorylation. These sites have been identified by sequencing phosphotyrosine-containing tryptic peptides of the EGF receptor derived from A-431 cells labeled with ^{32}P inorganic phosphate in the presence of EGF (53). The phosphorylation sites correspond to tyrosine residues 1173, 1148, and 1068. The tyrosine at residue 1173 seems to be the major site of EGF-induced autophosphorylation of the receptor in intact cells. A fourth tyrosine residue is labeled to a lesser extent in intact cells, but this site has not been identified. When the autophosphorylation reaction is carried out in vitro, three phosphopeptides are identified that correspond to the sites of autophosphorylation in intact cells. Autophosphorylation of the receptor in vitro does not show a similar marked preference for the autophosphorylation of tyrosine 1173 (53). However, studies with peptide substrate corresponding to each of the three autophosphorylation sites of the receptor do show that the peptide corresponding to tyrosine 1173 is preferentially phosphorylated in vitro (95). Whether these autophosphorylation sites are utilized in an ordered and sequential manner has not been demonstrated. To date, there is no evidence that tyrosine 845, within the tyrosine kinase domain, is utilized as an autophosphorylation site in the EGF receptor. This is somewhat curious as this tyrosine is highly conserved in the tyrosine kinase family and is a major autophosphorylation site in the prototype v-*src* molecule.

The autophosphorylation domain at the carboxy terminus of the EGF receptor is often experimentally defined by proteolysis. Treatment of the receptor with proteases, such as trypsin, chymotrypsin, elastase, or calcium-activated neutral protease (calpain), rapidly produces two fragments—a 20,000-dalton fragment containing the carboxy terminus and the three auto-phosphorylation sites and a 150,000-dalton fragment corresponding to the extracellular and tyrosine kinase domains (67, 68). As the carboxy terminal fragment is readily generated by a variety of proteases and two discrete fragments are produced, it is likely that a relatively exposed and flexible region of the receptor, containing these protease sites, connects the more compact tyrosine kinase and autophosphorylation domains. Interestingly, the 150,000-dalton fragment is quite active as a tyrosine kinase. This fragment has diminished activity in autophosphorylation assays (15, 67, 69), which is expected as the primary autophosphorylation sites have been removed, but retains sensitivity to EGF and utilizes new autophosphorylation sites (67, 71) that have not been mapped. The 150,000-dalton receptor species seems to have equivalent activity compared to the intact receptor in the phosphoryla-tion of exogenous substrates (67). As calpain readily generates the carboxy terminal 20,000-dalton fragment in cell homogenates prepared in calcium-containing buffers, this site has been of interest. However, there is no evidence for the generation of the 150,000-dalton fragment of the EGF receptor in intact cells. Based on mapping studies with site-directed antibod-ies (68) and the preferred site of calpain cleavage in model peptides (70), it has been proposed that calpain proteolysis of the EGF receptor occurs at residue 1037 (68). Calpain, interestingly, produces a similar cleavage of the PDGF receptor (88).

Receptor Functions and Regulation

LIGAND BINDING   Available stoichiometry data show that one mole of EGF receptor binds one mole of EGF (72). There are no data in any system to suggest a stoichiometry greater than 1. The ligand binding site in the EGF receptor is somewhat unusual in that at least three EGF-like molecules (EGF, TGF-α, VGF), with little overall sequence conservation (22%), are recog-nized with affinities that are practically indistinguishable. In contrast, the receptors for insulin and IGF-1 (90, for review) bind their homologous ligand with an affinity at least 100-fold greater than that for the heterologous ligand. Insulin and IGF-1 are approximately 50% identical in primary sequence. The ligand binding domain of the PDGF receptor, while not as well characterized, would seem to be more like the EGF binding domain in that three variants of the PDGF molecule (a heterodimer in humans) are recognized—the A:B heterodimer and A:A or B:B homodimers. The A and B chains of the PDGF receptor are approximately 60% identical in amino acid sequence (89).

Parameters of ligand binding to the EGF receptor are most frequently derived following transformation of experimental data to a Scatchard plot (73). This allows the estimation of affinity constants, usually expressed as a K_D value, which for the EGF receptor are reported as 10^{-9} to 10^{-10} M. The Scatchard analysis also allows an estimation of the receptor concentration B_{max}. Most cells display average receptor numbers within the range of 20,000 to 200,000 per cell. The exceptions are the A-431 cell line (11) and several other epithelial carcinoma cell lines (74–76) that have approximately 2×10^6 receptors per cell. Although the binding parameters obtained by Scatchard analysis are generally accepted, it should be noted that these data are subject to considerable error and/or misinterpretation. The K_D values are, in fact, "apparent" K_D values and no attempt has been made to obtain physically valid constants. The apparent K_D may be most seriously in error for those cells that overexpress the EGF receptor. Analyses of ligand binding parameters (77, 78) has shown that unless the receptor concentration is at least 10-fold less than the K_D, the apparent K_D will become a linear function of receptor concentration. Under the conditions generally used for ^{125}I-EGF binding assays, receptor concentrations range from approximately 1 nM for purified, solubilized EGF receptors to 50 μM or more for assays performed with intact cells (see below). In addition, binding parameters obtained with intact cells at temperatures above 10°C are seriously compromised by post-binding events, such as clustering, internalization, and ligand degradation.

B_{max} values are obtained in the Scatchard analysis by extrapolation to the intercept on the y (Bound) axis. If the Scatchard plot is linear this is not complicated, but frequently nonlinear plots are obtained. Extrapolation of the curved plot to the y-axis in these cases can produce a considerable error in the B_{max} value. This is frequently the case with cell lines that overexpress the EGF receptor, and receptor numbers are overestimated.

The nonlinear Scatchard plot presents a more basic difficulty in terms of interpretation. Although alternate interpretations are possible and not disproven, it is generally assumed that these plots reflect the presence of populations of receptors with varying affinities for ^{125}I-EGF. It is equally plausible that the ligand population is heterogeneous and the receptors have different affinities for unlabeled EGF and ^{125}I-EGF (not to mention that ^{125}I-EGF may represent more than one molecular species). In other systems ligand heterogeneity has been shown to produce nonlinear Scatchard plots (79). Nevertheless, these plots are usually analyzed by resolving the curve, for strictly arbitrary reasons, into two straight lines—one representing low-affinity receptors and the other high-affinity receptors.

In studies of ^{125}I-EGF binding, the high-affinity subclass is usually reported to comprise 10% or less of the total receptor population. The difference in apparent K_D values of the high- and low-affinity receptors is generally about 10–20 fold. It is interesting to note the results of computer modeling and Monte Carlo simulations of the resolvability of β-adrenergic receptor subtypes based on relative binding affinities (80). This study indicates that to achieve a 90% successful resolution of a high-affinity receptor subclass representing 10% of the total receptor population, the affinities (K_D values) of the high and low subclasses of receptor need to differ by a factor of approximately 200. A 50% successful resolution requires the affinities to differ by nearly 30-fold. These analyses would indicate that the high-affinity subclass of EGF receptor inferred from Scatchard analyses are not statistically resolvable—if they in fact exist, and there is no biochemical evidence that they do. The evidence for discrete high- and low-affinity receptor subtypes is much stronger for the NGF and IL-2 receptors. In those instances the K_Ds for high- and low-affinity sites differ by 1000-fold and the subtypes can be manipulated biochemically (91–93, for reviews).

Almost all binding studies employ EGF that has been labeled with ^{125}I by the chloramine T procedure. This procedure is known to produce a minor species of ^{125}I-EGF that forms covalent crosslinks, apparently involving oxidized amino acid side chains on the growth factor, with the EGF receptor (82, 83). In this situation the amount of nonphysiological, crosslinked ligand can be 5% of the total cell-bound ^{125}I-EGF (84). This artifact, which may contribute to nonlinear Scatchard plots, does not occur with ^{125}I-EGF prepared with less vigorous oxidizing agents (82).

The analysis of high-affinity binding sites in cells that overexpress the EGF receptor, such as A-431 cells, is additionally compromised in the Scatchard analysis by the fact that at low EGF concentrations (where high-affinity interactions are favored) the amount of bound EGF is a large fraction (over 50%) of the total EGF. In fact, the percent bound can reach 90%. This violates conditions of the Scatchard analysis and, in practice, means that calculation of the free ligand concentration is subject to considerable error. Under these conditions, the author's laboratory has noticed that minor (10%) effects on ^{125}I-EGF binding capacity can, after transformation of the data to a Scatchard analysis, indicate a fourfold difference in the B/F ratio for the apparent high-affinity sites of A-431 cells (M. Wahl, G. Carpenter, unpublished observations).

Interested readers are referred to a recent monograph that describes the difficulties encountered in the design, execution, and interpretation of ligand-receptor binding studies (81).

Although extracellular agents such as lectins and antibodies to the receptor are capable of regulating EGF binding to its receptor, the only known and physiologically relevant mechanism occurs through the activation of protein kinase C. As protein kinase C regulation is intracellular and occurs through modulation of the cytoplasmic domain of the EGF receptor, this mechanism is reviewed in the following section.

TYROSINE KINASE ACTIVITY   The major function of the cytoplasmic domain of the EGF would seem to be intracellular transmission and amplification of the extracellular signal initiated by the binding of EGF. The growth factor–dependent tyrosine kinase activity of the cytoplasmic domain is regarded as the primary, though not necessarily exclusive, mechanism for the generation of intracellular second messages. The following paragraphs are meant to review individual aspects of control of the tyrosine kinase activity of the EGF receptor. Other recent reviews have focused on these topics and the reader is referred to these for additional information and alternate viewpoints (40–42, 85, 94).

General characteristics   Maximal activation of EGF receptor tyrosine kinase activity, in membranes (19), solubilized preparations (14, 63), or purified receptor preparations (69, 95, 96), requires saturating concentrations of EGF, and most concentration curves suggest a linear relationship between the ligand and kinase activation. Growth factor activation is manifested primarily by an increase in the V_{max} of the kinase activity (69, 95). Rather small changes in K_m values, if any, are reported for ATP or exogenous substrates. The reported K_m values for autophosphorylation (approximately 0.25 μM) and for the phosphorylation of exogenous substrates (about 2.5 μM) would not suggest that intracellular ATP levels might regulate the enzyme. The kinase activity also requires either Mn^{2+} or Mg^{2+}, with the former preferred on a molar basis (19, 63). The required cation concentrations are also within normal intracellular levels. A kinetics analysis of EGF-activated exogenous substrate phosphorylation suggests an ordered BiBi mechanism in which ATP binding is preceded by substrate binding and followed by the release of phosphorylated substrate and then ADP (69).

Autophosphorylation mechanism   EGF stimulation of receptor autophosphorylation is generally considered to be intramolecular, based on kinetic analyses of phosphorylation at increasing concentrations of receptor (72, 98). While this mechanism is generally true of tyrosine kinases and protein kinases in general, examples of intermolecular or trans phosphorylation have been reported. In these studies, involving the v-*fps* kinase, inactive kinase molecules (produced by site-directed mutagenesis or the isolation of revertants) were phosphorylated when mixed with active kinase (48, 97). It should be noted that intra- and intermolecular phosphorylation need not be mutually exclusive and that all evidence has been obtained with in vitro assays. Other than a point of enzymological detail, this issue could be significant in a physiological sense. It would be important to know whether occupied and activated receptors are able to phosphorylate unoccupied receptors. The in vitro studies with the EGF receptor do show that under these assay conditions, which include preincubation of the receptor with a saturating concentration of EGF, meaning there are no unoccupied receptors, intramolecular auto-phosphorylation is preferred. In vivo growth factor binding and auto-phosphorylation would begin simultaneously and saturation of surface receptors at 37°C requires 5–10 min, whereas autophosphorylation begins within seconds. Another limitation of the in vitro experiments is that the highest concentration of receptor employed is approximately 50 nM; however, the concentration of EGF receptors in the plasma membrane of a normal cell is approximately 1000 times higher. The calculated concentration in vivo assumes a cell surface area of 1000 μ^2 and 10^5 receptors per cell. The allowed receptor volume is calculated as a shell of 100 Å in width surrounding the cell to approximate the volume of the plasma membrane. For A-431 cells the calculated receptor concentration would be slightly more than 1 mM. (Calculations with assistance of M. Wahl, J. Staros).

Models of kinase activation   In general, two models are considered to explain how ligand binding to the external domain of the EGF receptor can alter the catalytic properties of the kinase domain on the cytoplasmic side of the plasma membrane. The membrane-spanning region is thought to be a rather inflexible α-helical connector whose structure would not be easily changed by ligand binding. The "flush chain" model (42) requires that ligand binding alter the interaction of the external domain with the membrane and that this forces slight changes in the positioning of the membrane-spanning connector, the end result being an indirect alteration of the kinase domain. The model is attractive in that it deals with monomeric receptor species. However, experimental evidence is absent and the energetics would seem to require the plasma membrane, thus not explaining EGF activation in detergent-solubilized receptor preparations.

The second and perhaps most widely cited model is the "cluster" proposal (85, 99). This model proposes that EGF binding induces the aggregation of receptors from monomers and that the process of receptor aggregation, at least dimerization, is intimately involved in the activation of tyrosine kinase activity. The cluster model is attractive physiologically as EGF does produce the rapid formation of receptor clusters on the surface of intact cells (100), and, based on measurements of rotational diffusion at different temperatures, in plasma membrane preparations also (101). The experimental evidence for the cluster model essentially is based on the observations that certain "agonist" monoclonal antibodies to the EGF receptor are able to stimulate DNA synthesis in quiescent cells, to stimulate the tyrosine kinase activity of the receptor (102–104), and to induce an EGF-like stimulation of prolactin secretion by GH_3 cells (105). Less direct (and controversial) experiments with combinations of anti-EGF antibodies and either cyanogen bromide–inactivated EGF or very low EGF concentrations also have been interpreted as evidence in favor of aggregation-induced activation of the EGF receptor (106). As the antibody-induced activation of the EGF receptor required multivalent antibodies (monovalent Fab fragments did not work unless secondary antibodies were added and the active primary antibodies were mostly pentavalent IgM), this mechanism involves the formation of receptor clusters. Experiments in vitro argue that dimerization activates the EGF receptor kinase (85) and that antibody-induced aggregation of the insulin receptor activates its tyrosine kinase activity (127).

There are reservations about the cluster model. A constant objection has been that the model fails to explain the ability of EGF to activate the kinase under conditions where receptor aggregation is difficult to envision; in detergent solutions or when the soluble receptor is "immobilized" on antibody affinity columns. Also, another group has argued, based on in vitro assays, that receptor-receptor interactions decrease kinase activity and that the monomeric EGF receptor is the more active (98). The cluster model, even if correct, still lacks a clear biochemical mechanism to explain the stimulation of kinase activity, particularly if the autophosphorylation reaction is intramolecular.

Also, it seems that other antibodies to either the external or cytoplasmic domains of the EGF receptor can produce enhanced kinase activity with or without receptor clustering and internalization (107–109). These antibodies do not yield a mitogenic response in cells. Other antibodies induce clustering and/or internalization but fail to stimulate the kinase activity (110, 111). It would seem, therefore, that antibody-induced clustering of EGF receptors, activation of the tyrosine kinase, and stimulation of mitogenic responses are dissociable events. In this regard, the seemingly nonspecific capacity of dimethylsulfoxide to enhance the kinase activity of the EGF receptor (112, 113) compounds the difficulty in interrelating these events. None of the proposed "agonist" antibodies have been shown to support the multiplication of an EGF-dependent cell line, which would be the real test of a growth factor agonist. The antibody data may say more about how each of these types of antibodies work than about the mechanism of action of EGF. It should be remembered that ligand-induced receptor aggregation is not a peculiar feature of growth factor receptors, but is a general feature of all ligand-receptor complexes that are subject to endocytosis. Many of these, for example the LDL receptor, have no ligand-dependent enzyme system resident in the receptor. Elements of the "cluster" model, however, remain intriguing and should be testable when the signals for clustering are better understood.

During the early work with the EGF receptor in subcellular preparations, it was noted that the activation of the tyrosine kinase in detergent-solubilized preparation had a temperature requirement (5–10 min at 25°C) that was not necessary in particulate membrane preparations (activation occurring at 0°C) (14). This temperature requirement is also found with the purified EGF receptor and its basis is not understood. Perhaps, the membrane environment provides components that have a facilitating role in the EGF activation of the receptor tyrosine kinase.

Phosphorylation of exogenous substrates   Studies with the insulin receptor have provided evidence that ligand-induced autophosphorylation of the insulin receptor increases the phosphorylation activity of the receptor toward exogenous substrates (116–118). There is also evidence that phosphorylation of the insulin receptor by exogenous protein kinases may enhance the catalytic activity of the insulin receptor (119, 120). In the case of the EGF receptor, however, the evidence is divided. One set of experiments has argued that ligand-induced autophosphorylation of the EGF receptor enhances the molecule's capacity to phosphorylate exogenous substrates (121). Several other reports have not found evidence for this phenomenon and conclude that ligand activation of exogenous substrate phosphorylation is independent of autophosphorylation (95, 69, 68, 123, 140). Considering the differences in experimental design and the fact that synthetic peptide substrates, which have rather high K_m values, were employed, a meaningful conclusion of this point may require analysis with site-directed mutations of the autophosphorylation sites. This technique has been applied to other tyrosine kinases and has shown that alteration of the major tyrosine autophosphorylation of the v-*src* kinase does not alter its catalytic activity toward exogenous substrates (124, 125). However, changing the autophosphorylation site of the v-*fps* kinase (from tyrosine to phenylalanine) does produce a large decrease in that molecule's catalytic activity (126). Similarly, the phosphorylation of the insulin receptor was impaired when autophosphorylation sites were changed from tyrosine to phenylalanine (128).

The significance of autophosphorylation of the EGF receptor remains unclear and, somewhat disappointingly, a similar lack of understanding exists in regard to exogenous substrates for tyrosine phosphorylation. While several proteins are known to be phosphorylated in vitro or in vivo, the identity of some and the functional significance of all are unknown. Table 1 presents a list of those proteins that have been reported to be phosphorylated at tyrosine residues in an EGF-dependent manner—either in vitro and/or in vivo. Of

Table 1 Proteins that are reported tyrosine phosphorylation substrates for the EGF receptor kinase

Substrate	In vivo	In vitro
pp81 (ezrin)	(129, 130)	(18, 131)
pp34–39 (lipocortins)	(130, 132–136)	(137–140)
pp42	(143–148)	—
progesterone receptor	—	(141, 142)
gastrin	—	(149)
growth hormone	—	(150)
middle T antigen	—	(151)
glycolytic enzymes	—	(152)
myosin light chain	—	(153)
erythrocyte band 3	—	(154)

these substrates the lipocortin family, also known as pp34–39 or calpactins
(155–158), and the pp42 molecule are potentially the most interesting. The
lipocortins have been reported to inhibit the activity of phospholipase A_2
(159, 160), and their phosphorylation is reported to inactivate this inhibitory
effect (161). This scheme would suggest that treatment of cells with EGF
should result in the mobilization of free arachidonate and the formation of
eicosanoid products. Several instances have been reported where EGF addition results in increased levels of prostaglandins. However, this mechanism
rests on a rather shaky experimental foundation, particularly in regard to
interaction of lipocortins with phospholipase A_2. A brief review of the
lipocortin-like molecules and their activities has been published recently
(162). Interest in pp42, a protein of unknown structure and function, is
derived mainly from the fact that this molecule is a relatively low-abundance
protein in cells, its stoichiometry of phosphorylation is high, and it is phosphorylated rapidly by several mitogenic agents (145, for summary).

Regulation by protein kinase C   The term "transmodulation" has been applied to those instances in which functions of the EGF receptor are affected by
the action of other proteins. The clearest example of this type of mechanism
involves protein kinase C. The majority of these transmodulation studies have
been carried out in intact cells with a potent activator of protein kinase C—the
tumor promoter 12-O-tetradecanoylphorbol 13-acetate (TPA). The data show
that treatment of cells with TPA results, in most all cell types, in a dramatic
decrease in EGF binding capacity and an attenuation of tyrosine kinase
activity (166, 167). Modulation of the EGF receptor in vitro with purified
protein kinase C has also been reported (95, 163, 164). The natural activator
of kinase C, diacylglycerol, also has been shown to produce similar results

(168–173). Based on Scatchard plot analyses, altered EGF binding in TPA-treated cells was originally reported as a decrease in receptor number. Later studies, also based on Scatchard analyses, indicated the predominant change to be a reduction in receptor affinity for EGF. Recently, it has been concluded from morphological evidence that TPA induces the internalization of unoccupied EGF receptors (165), suggesting an alteration in receptor number. Modulations of EGF receptor functions by protein kinase C have been recently reviewed in more detail elsewhere (41, 94).

In many cases TPA blocks the early responses of cells to EGF. However, the attenuations of EGF receptor function are somewhat difficult to reconcile with reported synergisms between EGF and TPA in the induction of DNA synthesis (174–176), a late event in the mitogenic process. Part of the seeming paradox may be that the inhibitory effects of TPA on receptor function are reportedly transient and not seen in cells that have been pretreated with TPA for several hours. Perhaps, long-term treatment with TPA enhances a step in EGF action or TPA effects DNA synthesis by a mechanism that does not include the EGF receptor and its specific pathway.

It has been known for several years that the EGF receptor contains, in addition to phosphotyrosine, significant levels of phosphothreonine and phosphoserine (130, 177). These nontyrosine phosphorylations are carried out by protein kinases other than the receptor itself. Protein kinase C, activated by TPA in intact cells or added in vitro to EGF receptor preparations, is able to catalyze phosphorylation of the EGF receptor. Phosphopeptide maps indicate several sites in which phosphorylation of the EGF receptor by protein kinase C occurs, but threonine 654 is the dominant phosphorylation site and phosphorylation of this residue is presumed to mediate the attenuations of receptor functions (43, 44, 178). Site-directed mutagenesis of threonine 654 to an alanine residue provides hard evidence for this conclusion (122). The alanine 654 EGF receptors, expressed in recipient cells following transfection, bound ^{125}I-EGF equally well in the presence or absence of TPA.

It is possible that additional transmodulations of the EGF receptors by other growth factors, such as PDGF, may be due to their capacity to activate protein kinase C through diacylglycerol production. Interestingly, treatment of A-431 cells with EGF followed by analysis of tyrosine kinase activity toward exogenous peptide substrates indicated that the kinase activity was reduced by as much as 70% compared to untreated cells (179). Subsequent reports have shown that EGF treatment of intact cells leads to activation of protein kinase C and phosphorylation of a threonine 654 peptide (180, 181). Protein kinase C, therefore, seems to be a critical intermediate in circuits, both within the EGF system itself and through heterologous growth factors, that regulate EGF receptor activity.

Cellular Physiology

RECEPTOR BIOSYNTHESIS   The gene for the EGF receptor is located on the short arm of human chromosome 7 (182–184) in the p14–p12 region (185). Most all cell types, with the exception of hemopoietic cells, express the EGF receptor at levels that vary widely from 20,000 to 200,000 receptors per cell in nontransformed cell populations. In these cells there is no relationship between receptor number and the responsiveness of the cells to EGF. In fact, when cells are grown in the presence of EGF, the receptor number is down-regulated by as much as 90% (230, 231). Therefore, growth responsiveness to EGF can be mediated by as few as 2000 receptors per cell. Several cell lines have been reported in which the EGF receptor is over-expressed—at levels of approximately 1×10^6 receptors per cell (11, 74–76). The most noted of these is the A-431 cell line (11), which was derived from an epidermoid carcinoma. Nearly all of the other receptor-overexpressing cell lines also are derived from squamous carcinomas. Overexpression, or altered expression of the EGF receptor, has been reported for a variety of human tumor (carcinoma) tissues including breast (186–188), liver (189), bladder (190), pancreas (191), glioblastomas (192), sarcomas (193), lung (194, 195), but most frequently for squamous carcinomas (196–201). In some instances of overexpression the EGF receptor gene is amplified, while in other cases the gene is not amplified and increased receptor mRNA may be due to either transcriptional or posttranscriptional mechanisms. A recent study has shown that variants of the A-431 cell line having different levels of EGF receptor vary widely is their tumorigenic potential in nude mice (225). Those variants having a high number of EGF receptors produced larger tumors much more quickly than variants bearing lower numbers of receptors. However, the role of the EGF receptor in tumorigenesis remains unclear.

Transcription   Normal and transformed cells that express the EGF receptor have two mRNA species of 10 kb and 5.6 kb that hybridize with receptor cDNA probes (24, 202, 203). A-431 cells have an additional mRNA species of 2.8 kb that encodes a truncated receptor species that is secreted by the cells (see below). This truncated receptor mRNA also has been detected in some glioblastomas (192). The relationship of the 5.6 and 10 kb mRNAs to each other (products of different splicing reactions?) or the receptor gene (products of different polyadenylation sites?) is not known. Also, it is not clear whether both mRNAs are, in fact, translated in vivo.

There is relatively little information available concerning the structure of 5' or 3' regulatory regions in the EGF receptor gene. Preliminary studies (204) of the 5' flanking region of the EGF receptor gene indicate that the promoter region has neither a TATA box nor a CAAT box. Also, this region is extremely G-C rich (88%), contains five CCGCCC repeats, and in A-431 cells is near a DNAse I hypersensitive site. A discussion of these promoter features relative to similar features of the promoters of other genes concerned with the regulation of cell growth, such as the c-Ha-*ras* 1 promoter, has been presented (205).

The capacity of EGF to increase the synthesis of its receptor has been reported for three cell lines (206–208). This phenomenon is seen at both the mRNA and receptor protein level, but its generality to other cell lines is not known. EGF does not affect the amount of receptor protein synthesized in A-431 cells (209, 210). The mechanism by which EGF can increase EGF receptor mRNA has been reported to occur at the posttranscriptional level (206) and involves accumulation of both the 5.6 and 10 kb mRNAs (206, 207).

Translation and glycosylation   Translation of mRNA for the EGF receptor in vivo is accompanied by cotranslational glycosylation within the lumen of the endoplasmic reticulum at approximately 10 canonical sites for N-linked glycosylation. Cotranslational glycosylation is not unusual, and the evidence, in this instance, is the failure to detect nonglycosylated molecules with brief pulse-labeling protocols (25, 26, 210–214). If the addition of oligosaccharide chains is blocked with an inhibitor such as tunicamycin, then the nonglycosylated molecules (130,000 daltons) are unable to bind EGF when assayed in intact cells or solubilized extracts (26, 215).

While glycosylation is necessary for the receptor to bind its ligand, the addition of carbohydrate per se is not sufficient. Newly synthesized, within 5–10 min of the translation, and glycosylated receptors are unable to recognize EGF, and posttranslational processing events are required before ligand binding capacity is displayed (215, 216). The posttranslational acquisition of ligand binding capacity has been reported for the insulin (217–219) and insulinlike growth factor II (220) receptors, as well as several membrane enzymes. The posttranslational mechanism of activation (i.e. acquisition of EGF binding capacity) of newly synthesized EGF receptors is not known, but seems to occur at late stages of processing in the endoplasmic reticulum, approximately 20–30 min after translation, and may involve the formation of appropriate disulfide bonds within the cysteine-rich ligand binding domain of the receptor (215, 216). As the EGF receptor does not seem to be subject to fatty acid acylation (216, 221, 222), this reaction, which can regulate the biosynthesis of other membrane proteins, would not seem to be involved. While N-linked glycosylation is required in this biosynthetic process of receptor activation, the role of oligosaccharides in ligand binding appears indirect. Enzymatic deglycosylation of the mature EGF receptor does not seem to significantly perturb ligand binding activity (216). There is suggestive evidence (26, 216) that tyrosine kinase activity may also be acquired posttranslationally, but the evidence is not clear due to the lack of EGF binding capacity.

Intracellular biogenesis   Glycosylation does appear necessary for the EGF receptor to be transported from the endoplasmic reticulum, as tunicamycin prevents intracellular transport (26, 216). Certain other receptors, for example the asiologlycoprotein receptor, are transported efficiently without glycosylation (223). Later events in receptor processing within the Golgi do not appear to affect either the acquisition of receptor functions or transport to the cell surface. Inhibition of α-mannosidase II in the Golgi by swainsonine results in the surface appearance of a fully functional EGF receptor having all oligosaccharide chains in the high-mannose form (26). The role of late events in receptor processing, such as the formation of complex oligosaccharide chains in the Golgi, is not evident, at least in tissue culture studies.

The complete biogenesis time for the EGF receptor, that is the average time required from translation to appearance on the cell surface, has been estimated at approximately 3 hr (210, 216). The half-time for conversion of the 160,000-dalton receptor precursor molecule to the mature 170,000-dalton receptor is approximately 1.5 hr. A significant amount of this time (approximately 75 min) is required for translocation of the newly synthesized receptor from the endoplasmic reticulum to the Golgi. The biosynthesis of the EGF receptor has been recently reviewed in more detail elsewhere (172, 224).

INTERNALIZATION AND DEGRADATION   Once EGF receptors have appeared on the cell surface they do not form oligomeric species, or clusters, unless EGF is added (100). An important consideration is the mechanism that leads to the clustering of EGF:receptor complexes under physiological conditions. Since the ligand is not bivalent, this mechanism would seem to involve an EGF-induced allosteric change in the receptor. Analyses of the surface behavior of several "engineered" receptor mutants can be considered. These experiments have not measured clustering directly, but rather have measured EGF-dependent internalization, which rapidly follows the formation of receptor clusters. Phosphorylation of threonine 654 through activation of protein kinase C seems not to be part of this mechanism, as a site-directed mutant containing alanine at residue 654 is efficiently internalized in the presence of EGF (122). The alanine 654 receptor is not internalized by TPA, indicating separate mechanisms of EGF and protein kinase C–induced internalization.

That tyrosine kinase activity of the EGF receptor is not required for ligand-induced internalization has been shown by analysis of insertion and deletion mutants (226, 227). The two insertion mutants, containing four amino acid insertions at different points within the tyrosine kinase domain, are devoid of detectable tyrosine kinase activity and one of the mutants is internalized efficiently in the presence of EGF. The deletion mutants have been constructed by removing progressively larger segments from the carboxy terminus of the EGF receptor. Removal of the last 63 residues, including the two autophosphorylation sites at residues 1148 and 1173, from the carboxy terminus does not significantly perturb the capacity of EGF to induce internalization. Much larger deletions of the cytoplasmic domain, including the tyrosine kinase domain and threonine 654, do prevent EGF-induced internalization. These internalization negative deletion mutants are reported to have a lateral diffusion rate on the cell surface equivalent to that of the wild-type receptor (228) and to bind ^{125}I-EGF efficiently (227).

EGF binding to intact cells occurs rapidly, surface receptors being saturated in 5–10 min at 37°C, and the concurrent clustering and internalization events also proceed very rapidly under physiological conditions. It is clear, however, that persistent occupation of surface receptors for up to 8 hr is required for late-term events such as DNA synthesis to be stimulated by EGF (1). Whether intracellular signaling during this time frame occurs solely from surface EGF:receptor complexes, internalized complexes, or a combination of both is not known. Evidence has been presented to indicate that internalized EGF receptors are active in terms of their capacity to phosphorylate exogenous substrates (140, 228). Similar evidence is reported for the internalized insulin receptor (229). These reports suggest that internalized receptors may be functionally active in terms of intracellular signal generation.

The fate of internalized EGF:receptor complexes is the rapid degradation of both molecules within lysosomes. Initial studies (230) showed that cell-bound ^{125}I-EGF was rapidly degraded by lysosomes. Subsequent experiments that measured directly the half-life of the receptor demonstrated that in the presence of EGF the receptor has a much shorter (up to 10-fold) half-life than in the absence of the ligand (209–211, 230–234). Since these studies of receptor degradation cannot determine whether the receptor is degraded after each internalization event, it is possible that a low level of receptor recycling also occurs and that, on the average, each receptor enters a degradation pathway after a second or third internalization event. Evidence in favor of some recycling has been presented (235, 236). Nevertheless, the intracellular sorting process for the internalized EGF receptor is substantially different than the mechanisms used to sort receptors for LDL or transferrin, which are only recycled in the presence of their ligand (see 237, 238 for reviews).

Receptor-Related Molecules

In addition to the EGF receptor, mammalian cells produce, in different circumstances, several other closely related proteins. The structures of these molecules are depicted in Figure 2 and each is discussed in the following paragraphs.

SECRETED RECEPTOR   A secreted form of the EGF receptor is produced by A-431 cells, but has not been detected in other cells (25, 240). This aberrant receptor species likely arises from the small 2.8 kb EGF receptor mRNA that is thought to be transcribed from a translocated and rearranged EGF receptor gene in A-431 cells (24, 185, 240). The protein sequence deduced from this mRNA would predict the protein product to have a sequence corresponding to most all of the external binding domain of the EGF receptor with approximately nine amino acid residues of unknown origin (including one extra glycosylation site) at the carboxy terminus (24). The transmembrane and cytoplasmic sequences are entirely deleted, hence the molecule is secreted (25, 239). This secreted variant is able to bind EGF, but does not seem to be physiologically significant. Interestingly, the secreted receptor is glycosylated in a manner that is nearly identical to the glycosylation pattern of the intact membrane–anchored receptor, including the presence of high-mannose oligosaccharides, which are rarely found on secreted glycoproteins (A. M. Soderquist, G. Carpenter, manuscript submitted). Membrane anchoring of the EGF receptor within the endoplasmic reticulum and Golgi, therefore, does not seem necessary for correct glycosylation and activation of the external ligand-binding domain of the EGF receptor.

ONCOGENIC *ERB* B MOLECULES   The sequence of the cloned EGF receptor showed that the molecule was, as expected, related to the family of tyrosine kinases (24). However, quite unexpectedly, an extremely high level of sequence homology was observed with the sequence of the *erb* B oncogene of the avian erythroblastosis virus (AEV), which had been reported the previous year (241). Independent chromosomal mapping studies of the EGF receptor (182–185) and v-*erb* B (242) showed that both mapped to the same region on human chromosome 7. It is highly probable, therefore, that the gene for the EGF receptor is the proto-oncogene from which the viral *erb* B gene was derived. It should be noted, however, that the mapping or cloning of the EGF receptor or cellular sequences corresponding to v-*erb* has not been reported in an avian system. Presumably, the v-*erb* B oncogene is derived from sequences in the avian genome. The extent of homology between the human EGF receptor and v-*erb* B gene product is remarkable given the evolutionary distance of the two molecules. A sequence of nearly 400 residues in the cytoplasmic domain of human EGF receptor, including the tyrosine kinase domain, is approximately 95% identical to the v-*erb* B sequence (24). The obvious differences between the two molecules are the large truncation of the EGF receptor extracellular domain in the v-*erb* B molecule and a small

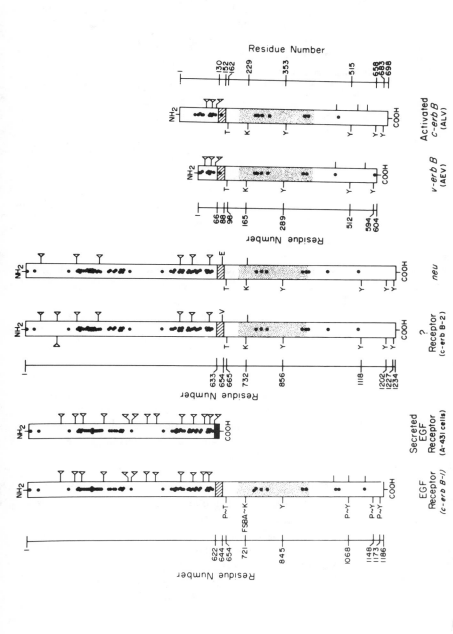

Figure 2 Comparison of the structures of molecules related to the EGF receptor. Symbols are as in Figure 1 plus AEV, avian erythroblastosis virus; ALV, avian leukosis virus; E, glutamic acid residue; V, valine residue. Data from references 24, 250, 252, 253, 257, 241, 246.

truncation at the carboxy terminus, which would include the primary sites of autophosphorylation in the EGF receptor. This has led to predictions that removal of the ligand binding domain and autophosphorylation sites of the EGF receptor would be sufficient to convert the receptor to a transforming protein whose tyrosine kinase is constitutively activated. To date, this idea has not been demonstrated. Three groups recently demonstrated the presence of autophosphorylation activity in the v-*erb* B molecule (243–245), but kinase activity, at least in vitro, is more difficult to assay than that in the EGF receptor.

A second retrovirus, the avian leukosis virus (ALV), also utilizes a transforming protein related to the EGF receptor. This virus, unlike AEV, does not have an oncogene in its genome. ALV activates normal cellular sequences by insertion of provirus into the host genome. Transcription of cellular sequences is controlled then by viral promoters. Induction of lymphomas by ALV involves the transcriptional activation of c-*myc* sequences, while induction of erythroblastosis by the same virus involves activation of c-*erb* B sequences (246). In the latter case, the c-*erb* B sequences that are transcribed correspond to the transmembrane domain and cytoplasmic domain of the EGF receptor. Interestingly and in contrast with the v-*erb* B molecule encoded by AEV, the ALV-induced transcripts encode a protein that does contain the entire carboxy terminal region analogous to the autophosphorylation sites of the EGF receptor.

An analysis of AEV-related viruses of differing oncogenic potential indicates that virally transduced *erb* B sequences encoding the entire carboxy terminus of the EGF receptor have a transforming potential limited to cells of the erythroid lineage (247). In contrast, the viral *erb* B sequences that contain a deletion of the carboxy terminus have a wide potential to transform other cell types, such as fibroblasts. Clearly, the significance of the truncations observed in the v-*erb* B sequences is more complex than was initially anticipated. A recent review is available for a more complete analysis of the biological relationships and functions of *erb* B molecules (248).

THE *NEU* ONCOGENE   More distantly related to the EGF receptor is the *neu* oncogene (249). The product of this oncogene is a glycoprotein of 185,000 daltons whose primary sequence is approximately 50% identical to that of the EGF receptor (250). The *neu* protein crossreacts with certain antibodies to the EGF receptor and v-*erb* B probes hybridize with the *neu* oncogene. There is no evidence, however, that this protein binds EGF or any other known growth factor. The *neu* oncogenehas been cloned, sequenced, and mapped to human chromosome 17 (250, 251, 254). This has allowed the identification and sequencing of the corresponding proto-oncogene (252, 253). The *neu* proto-oncogene is designated c-*erb*-2 as it is clearly distinct from the locus of the

EGF receptor (c-*erb*-1) on chromosome 7. The product of the c-*erb*-2 gene, which has tyrosine kinase activity (255, 256), is likely a receptor for an as yet unknown ligand. Interestingly, the activation of c-*erb*-2 to a transforming protein, i.e. *neu,* can occur by point mutations that bring about a nonconservative replacement (valine to glutamic acid) within the transmembrane domain (257).

STRUCTURES OF OTHER GROWTH FACTOR RECEPTORS

The structures of six other growth factor receptors are now known from cDNA cloning. For comparative purposes these receptors are depicted in Figures 3–5. Of these receptors, the structures of the PDGF (258) and CSF-1 receptors (259), shown in Figure 3, most resemble the structure of EGF receptor in terms of overall organization and size. They differ from the EGF receptor in two rather obvious aspects. These receptors do not have cysteine-rich regions in the extracellular domain and the tyrosine kinase region is interrupted by short sequences that are unrelated to tyrosine kinases. These interrupting sequences are quite different in the PDGF and CSF-1 receptors. However, there is a close resemblance between the overall sequences of these two receptors, and they likely represent members of a receptor family. Each of these receptors is related to a known oncogene. The structural and functional relationships between the CSF-1 receptor and the v-*fms* oncogene are sufficient to conclude that the former is the proto-oncogene for the latter (261). Evidence for low-affinity binding of CSF-1 by the v-*fms* molecule has been reported (263). The sequence similarity between the PDGF receptor (258) and the *kit* oncogene (264) suggests an evolutionary relationship.

The insulin (265, 266) and IGF-1 (267) receptor sequences (Figure 4) are obviously more complex in overall structure, but retain the ligand-activated tyrosine kinase present in the previously described receptors and have a cysteine-rich extracellular domain reminiscent of the EGF receptor. Complexity is introduced in the insulin and IGF-1 receptors by separation of functional activities, ligand binding, and tyrosine kinase activity into different polypeptides, the α and β chains, respectively. Additional complexity is added by the tetrameric structure of the functional receptor molecule.

The structures of the IL-2 and NGF receptors are depicted in Figure 5. The NGF receptor (262; E. Shooter, personal communication) has a cysteine-rich extracellular domain as seen before in the EGF, insulin, and IGF-1 receptors, but has, by comparison, a low carbohydrate content. There is a single transmembrane segment and a smaller, but significant, cytoplasmic domain of 155 residues. There are no sequences related to any protein kinases in this domain. The receptor for IL-2 has an extremely brief cytoplasmic domain of

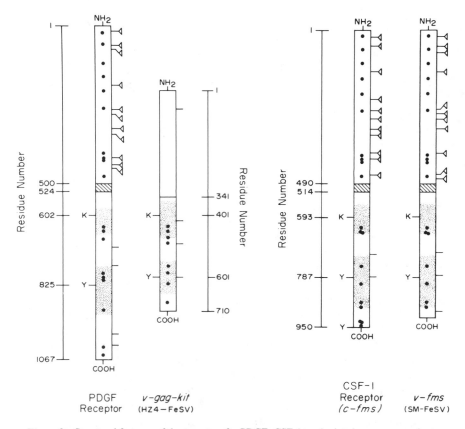

Figure 3 Structural features of the receptors for PDGF, CSF-1, and related oncogene products. Symbols are as for Figure 1 plus H24-FESV, Hardy-Zuckerman 4 feline sarcoma virus; SM-FeSV, McDonough feline sarcoma virus. Data from references 258–260, 264.

13 amino acid residues (268). Most of these residues are basic, and there is in the midst of the cytoplasmic sequence a phosphorylation site for protein kinase C (86, 87). The ligand-binding domain of the IL-2 receptor has an above-average percentage of cysteine residues, but is not as cysteine-rich as the EGF receptor. The external domain is known to contain both N-linked and O-linked carbohydrate. The cytoplasmic domains of these two receptors are of great interest as they must rely on different mechanisms than the other receptors for intracellular signaling.

PROSPECTUS

The structural characteristics of seven growth factor receptors have been described; the properties of the EGF receptor dwelt on in detail. Preliminary

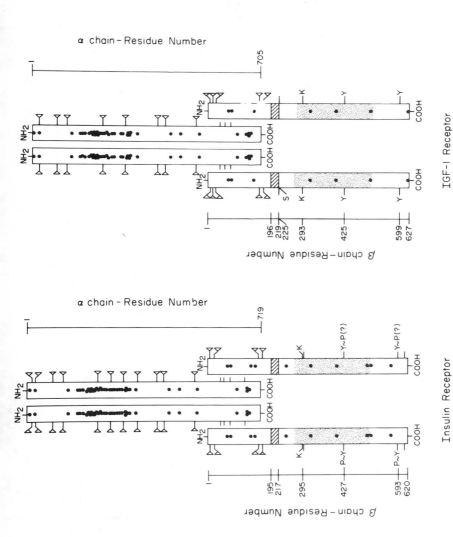

Figure 4 Structural features of the receptors for insulin and IGF-1. Symbols are as in Figure 1. The disulfide bonds between α and α chains and between α and β chains have not been depicted. Data from references 265–267, 272, 273.

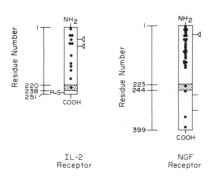

Figure 5 Structural features of the receptors for IL-2 and NGF. Symbols are as in Figure 1 plus P~S, phosphoserine residue. Data from references 86, 87, 262, 268–271, 275).

data on the basic biochemical characteristics of eight additional receptors [IGF-2, acidic and basic forms of fibroblast growth factor (FGF) and endothelial cell growth factor (ECGF), interleukins 1 and 3 (IL-1, IL-3), granulocyte-macrophage colony stimulating factor (GM-CSF), granulocyte colony stimulating factor (G-CSF), and transforming growth factor β (TGF-β)] are available, and their structures may be forthcoming in the near future. There are numerous other growth factors (approximately 7–10) for which receptor information is not presently available. While tyrosine kinase cytoplasmic domains seem to dominate the signaling mechanisms of the receptor structures now known, the examples of the IL-2 and NGF receptors together with preliminary data on other receptors suggest that direct, intramolecular activation of tyrosine kinase activity will not always be a dominant mechanism for the generation of ligand-induced intracellular signals for cell growth and/or differentiation.

ACKNOWLEDGMENTS

This article was written at the time that the 1986 Nobel Prize for Medicine and Physiology was awarded to the two discoverers of growth factors. I am particularly grateful for having had the opportunity to learn from and work with one of them, Stanley Cohen, for the past 13 years. The author would like to thank the many colleagues who supplied manuscripts prior to publication and apologizes that not all valuable information could be cited. The constraints of time, space, and my own ability to absorb all the information in this field place limitations on what can be reasonably reviewed in one article. The author also would like to thank Susan Heaver for typing of the manuscript, several colleagues (Stanley Cohen, James Staros, Matthew Wahl) for reading of the manuscript, and the National Cancer Institute and American Cancer Society for grant support.

Literature Cited

1. Carpenter, G., Cohen, S. 1979. *Ann. Rev. Biochem.* 48:193–216
2. Gospodarowicz, D. 1981. *Ann. Rev. Physiol.* 43:251–63
3. Carpenter, G. 1981. *Handb. Exp. Pharmacol.* 57:89–132
4. Carpenter, G., Goodman, L., Shaver, L. 1986. In *Oncogenes and Growth Control*, ed. T. Graf, P. Kahn, pp. 65–69. Heidelberg: Springer–Verlag
5. Marquardt, H., Hunkapiller, M. W., Hood, L. E., Todaro, G. J. 1984. *Science* 223:1079–82
6. Stroobant, P., Rice, A. P., Gullick, W. J., Cheng, D. J., Kerr, I. M., et al. 1985. *Cell* 42:383–93
7. Carpenter, G., Zendegui, J. G. 1986. *Exp. Cell Res.* 164:1–10
8. Brissenden, J. E., Ullrich, A., Francke, U. 1984. *Nature* 310:781–84
9. Zabel, B. U., Eddy, R. L., Lalley, P. A., Scott, J., Bell, G. I., et al. 1985. *Proc. Natl. Acad. Sci. USA* 82:469–73
10. Brissenden, J. E., Derynck, R., Francke, U. 1985. *Cancer Res.* 45:5593–97
11. Haigler, H., Ash, J. F., Singer, S. J., Cohen, S. 1978. *Proc. Natl. Acad. Sci. USA* 75:3317–21
12. Stoscheck, C. M., Carpenter, G. 1983. *J. Cell. Biochem.* 23:191–202
13. Giard, D. J., Aaronson, S. A., Todaro, G. J., Aronstein, P., Kersey, J. H., et al. 1973. *J. Natl. Cancer Inst.* 51:1417–21
14. Cohen, S., Carpenter, G., King, L. Jr. 1980. *J. Biol. Chem.* 255:4834–42
15. Cohen, S., Ushiro, H., Stoscheck, C., Chinkers, M. 1982. *J. Biol. Chem.* 257:1523–31
16. Cassel, D., Glaser, L. 1982. *J. Biol. Chem.* 257:9845–48
17. Gates, R. E., King, L. E. Jr. 1982. *Mol. Cell. Endocrinol.* 27:263–76
18. Carpenter, G., King, L. Jr., Cohen, S. 1978. *Nature* 276:409–10
19. Carpenter, G., King, L. Jr., Cohen, S. 1979. *J. Biol. Chem.* 254:4884–91
20. Ushiro, H., Cohen, S. 1980. *J. Biol. Chem.* 255:8363–65
21. Buhrow, S. A., Cohen, S., Staros, J. V. 1982. *J. Biol. Chem.* 257:4019–22
22. Buhrow, S. A., Cohen, S., Garbers, D. L., Staros, J. V. 1983. *J. Biol. Chem.* 258:7824–27
23. Downward, J., Yarden, Y., Mayes, E., Scrace, G., Totty, N., et al. 1984. *Nature* 307:521–27
24. Ullrich, A., Coussens, L., Hayflick, J. S., Dull, T. J., Gray, A., et al. 1984. *Nature* 309:418–25
25. Mayes, E. L. V., Waterfield, M. D. 1984. *EMBO J.* 3:531–37
26. Soderquist, A. M., Carpenter, G. 1984. *J. Biol. Chem.* 259:12586–94
27. Decker, S. J. 1984. *Mol. Cell. Biol.* 4:571–75
28. Russo, M. W., Guyer, C. A., Goldman, A., Staros, J. V. 1986. *Biophys. J.* 51:36a
29. Gill, G. N., Lazar, C. S. 1981. *Nature* 293:305–7
30. Bravo, R., Burkhardt, J., Curran, T., Müller, R. 1985. *EMBO J.* 4:1193–95
31. Sawyer, S. T., Cohen, S. 1981. *Biochemistry* 20:6280–86
32. Rothenberg, P., Glaser, L., Schlesinger, P., Cassel, D. 1983. *J. Biol. Chem.* 258:12644–53
33. Cummings, R. D., Soderquist, A. M., Carpenter, G. 1985. *J. Biol. Chem.* 260:11944–52
34. Childs, R. A., Gregoriou, M., Scudder, P., Thorpe, S. J., Rees, A. R., et al. 1984. *EMBO J.* 3:2227–33
35. Pollack, L., Atkinson, P. H. 1983. *J. Cell Biol.* 97:293–300
36. Gooi, H. C., Hounsell, E. F., Picard, J. K., Lowe, A. D., Voak, D., et al. 1985. *J. Biol. Chem.* 260:13218–24
37. Gooi, H. C., Picard, J. K., Hounsell, E. F., Gregoriou, M., Rees, A. R., et al. 1985. *Mol. Immunol.* 22:689–93
38. Gooi, H. C., Hounsell, E. F., Lax, I., Kris, R. M., Libermann, T. A., et al. 1985. *Biosci. Rep.* 5:83–94
39. Fredman, P., Richert, N. D., Magnani, J. L., Willingham, M. C., Pastan, I. 1983. *J. Biol. Chem.* 258:11206–10
40. Hunter, T., Cooper, J. A. 1985. *Ann. Rev. Biochem.* 54:897–930
41. Foulkes, J. G., Rosner, M. R. 1985. In *Molecular Mechanisms of Transmembrane Signalling*, ed. P. Cohen, M. D. Houslay, pp. 217–52. Amsterdam/New York: Elsevier
42. Staros, J. V., Cohen, S., Russo, M. W. 1985. See Ref. 41, pp. 253–78
43. Hunter, T., Ling, N., Cooper, J. A. 1984. *Nature* 311:480–83
44. Davis, R. J., Czech, M. P. 1985. *Proc. Natl. Acad. Sci. USA* 82:1974–78
45. Russo, M. W., Lukas, T. J., Cohen, S., Staros, J. V. 1985. *J. Biol. Chem.* 260:5205–8

46. Wierenga, R., Hol, W. 1983. *Nature* 302:842–44
47. Steinberg, M. J. E., Taylor, W. R. 1984. *FEBS Lett.* 175:387–92
48. Weinmaster, G., Zoller, M. J., Pawson, T. 1986. *EMBO J.* 5:69–76
49. Snyder, M. A., Bishop, M. J., McGrath, J. P., Levinson, A. D. 1985. *Mol. Cell. Biol.* 5:1772–79
50. Basu, M., Biswas, R., Das, M. 1984. *Nature* 311:477–80
51. Levinson, A. D., Courtneidge, S. A., Bishop, J. M. 1981. *Proc. Natl. Acad. Sci. USA* 78:1624–28
52. Brugge, J. S., Darrow, D. 1984. *J. Biol. Chem.* 259:4550–57
53. Downward, J., Parker, P., Waterfield, M. D. 1984. *Nature* 311:483–85
54. Das, M., Miyakawa, T., Fox, C. F., Pruss, R. M., Aharonov, A., Herschman, H. R. 1977. *Proc. Natl. Acad. Sci. USA* 74:2790–94
55. Wrann, M. M., Fox, C. F. 1979. *J. Biol. Chem.* 254:8083–86
56. Fox, C. F., Linsley, P. S., Wrann, M. 1982. *Fed. Proc.* 41:2988–95
57. Chinkers, M., Brugge, J. S. 1984. *J. Biol. Chem.* 259:11534–42
58. Chinkers, M., Garbers, D. 1984. *Biochem. Biophys. Res. Commun.* 123:618–25
59. Vogel, S., Freist, W., Hoppe, J. 1986. *Eur. J. Biochem.* 154:529–32
60. Hock, R. A., Nexø, E., Hollenberg, M. D. 1979. *Nature* 277:403–5
61. Cohen, S., Fava, R. A., Sawyer, S. T. 1982. *Proc. Natl. Acad. Sci. USA* 79:6237–41
62. Hock, R. A., Nexø, E., Hollenberg, M. D. 1980. *J. Biol. Chem.* 255:10737–43
63. Pike, L. J., Kuenzel, E. A., Casnellie, J. E., Krebs, E. G. 1984. *J. Biol. Chem.* 259:9913–21
64. Mroczkowski, B., Mosig, G., Cohen, S. 1984. *Nature* 309:270–73
65. Basu, M., Frick, K., Sen-Majumdar, A., Scher, C. D., Das, M. 1985. *Nature* 316:640–41
66. Thompson, D. M., Cochet, C., Chambaz, E. M., Gill, G. N. 1985. *J. Biol. Chem.* 260:8824–30
67. Gates, R. E., King, L. E. Jr. 1985. *Biochemistry* 24:5209–15
68. Gullick, W. J., Downward, J., Waterfield, M. D. 1985. *EMBO J.* 4:2869–77
69. Erneaux, C., Cohen, S., Garbers, G. L. 1983. *J. Biol. Chem.* 258:4137–42
70. Sasaki, T., Kikuchi, T., Yumoto, N., Yoshimura, N., Murachi, T. 1984. *J. Biol. Chem.* 259:12489–94
71. King, L. E. Jr., Gates, R. E. 1985. *Arch. Biochem. Biophys.* 242:146–56
72. Weber, W., Bertics, P. J., Gill, G. N. 1984. *J. Biol. Chem.* 259:14631–36
73. Scatchard, G. 1949. *Ann. NY Acad. Sci.* 51:660–72
74. Kamata, N., Chida, K., Rikimaru, K., Horikoshi, M., Enomoto, S., et al. 1986. *Cancer Res.* 46:1648–53
75. Cowley, G. P., Smith, J. A., Gusterson, B. A. 1986. *Br. J. Cancer* 53:223–29
76. Filmus, J., Pollak, M. N., Cailleau, R., Buick, R. N. 1985. *Biochem. Biophys. Res. Commun.* 128:898–905
77. Jacobs, S., Chang, K-J., Cuatrecasas, P. 1975. *Biochem. Biophys. Res. Commun.* 66:687–92
78. Chang, K-J., Jacobs, S., Cuatrecasas, P. 1975. *Biochim. Biophys. Acta* 406:294–303
79. Taylor, S. I. 1975. *Biochemistry* 14:2357–61
80. DeLean, A., Hancock, A. A., Lefkowitz, R. J. 1981. *Mol. Pharmacol.* 21:5–16
81. Limbird, L. E. 1986. *Cell Surface Receptors: A Short Course on Theory and Methods.* Boston, Mass: Nijhoff. 196 pp.
82. Baker, J. B., Simmer, R. L., Glenn, K. C., Cunningham, D. D. 1979. *Nature* 278:743–45
83. Linsley, P. S., Blifeld, C., Wrann, M., Fox, C. F. 1979. *Nature* 278:745–48
84. Comens, P. G., Simmer, R. L., Baker, J. B. 1982. *J. Biol. Chem.* 257:42–45
85. Yarden, Y., Schlessinger, J. 1985. *Ciba Found. Symp.* 116:23–45
86. Gallis, B., Lewis, A., Wignall, J., Alpert, A., Mochizuki, D. Y., et al. 1986. *J. Biol. Chem.* 261:5075–80
87. Shackelford, D. A., Trowbridge, I. S. 1986. *J. Biol. Chem.* 261:8334–41
88. Ek, B., Heldin, C.-H. 1986. *Eur. J. Biochem.* 155:409–13
89. Johnsson, A., Heldin, C.-H., Wasteson, Å., Westermark, B., Deuel, T. F., et al. 1984. *EMBO J.* 3:921–28
90. Kahn, C. R. 1985. *Ann. Rev. Med.* 36:429–51
91. Greene, W. C. 1986. *Ann. Rev. Immunol.* 4:69–95
92. Sutter, A., Hosang, M., Vale, R. D., Shooter, E. M. 1984. In *Cellular and Molecular Biology of Neuronal Development,* ed. I. B. Black, pp. 201–14. New York: Plenum
93. Buxser, S. E., Puma, P., Johnson, J. L. 1985. In *Biochemical Actions of Hormones,* ed. G. Litwack, 12:434–60. New York: Academic
94. Thompson, D. M., Gill, G. N. 1985. *Cancer Surv.* 4:767–88
95. Downward, J., Waterfield, M. D., Par-

ker, P. 1985. *J. Biol. Chem.* 260: 14538–46

96. Yarden, Y., Haraiu, I., Schlessinger, J. 1985. *J. Biol. Chem.* 260:315–19

97. Mathey-Prevot, B., Shibuya, M., Samarut, J., Hanafusa, H. 1984. *J. Virol.* 50:325–34

98. Biswas, R., Basu, M., Sen-Majumdar, A., Das, M. 1985. *Biochemistry* 24:3795–802

99. Schlessinger, J., Schreiber, A. B., Levi, A., Lax, I., Libermann, T., et al. 1983. *Crit. Rev. Biochem.* 14:93–111

100. Haigler, H. T., McKanna, J. A., Cohen, S. 1979. *J. Cell Biol.* 81:382–95

101. Zidovetzki, R., Yarden, Y., Schlessinger, J., Jovin, T. M. 1986. *EMBO J.* 5:247–50

102. Schreiber, A. B., Lax, I., Yarden, Y., Eshhar, Z., Schlessinger, J. 1981. *Proc. Natl. Acad. Sci. USA* 78:7535–39

103. Schreiber, A. B., Libermann, T. A., Lax, I., Yarden, Y., Schlessinger, J. 1983. *J. Biol. Chem.* 258:846–53

104. Fernandez-Pol, J. A. 1985. *J. Biol. Chem.* 260:5003–11

105. Hapgood, J., Libermann, T. A., Lax, I., Yarden, Y., Schreiber, A. B., et al. 1983. *Proc. Natl. Acad. Sci. USA* 80:6451–55

106. Schechter, Y., Hernaez, L., Schlessinger, J., Cuatrecasas, P. 1979. *Nature* 278:835–38

107. Defize, L. H. K., Moolenaar, W. H., van der Saag, P. T., de Laat, S. 1986. *EMBO J.* 5:1187–92

108. Beguinot, L., Werth, D., Ito, S., Richert, N., Willingham, M. C., et al. 1986. *J. Biol. Chem.* 261:1801–7

109. Das, M., Knowles, B., Biswas, R., Bishayee, S. 1984. *Eur. J. Biochem.* 141:429–34

110. Gregoriou, M., Rees, A. R. 1984. *EMBO J.* 3:929–37

111. Sunda, H., Magun, B. E., Mendelsohn, J., MacLeod, C. L. 1986. *Proc. Natl. Acad. Sci. USA* 83:3825–29

112. Rubin, R. A., Earp, H. S. 1983. *Science* 219:60–63

113. Rubin, R. A., Earp, H. S. 1983. *J. Biol. Chem.* 258:5177–82

114. Sale, G. J., Fujita-Yamaguchi, Y., Kahn, C. K. 1986. *Eur. J. Biochem.* 155:345–61

115. Machicao, E., Wieland, O. H. 1984. *FEBS Lett.* 175:113–16

116. Rosen, O. M., Herrera, R., Olowe, Y., Petruzzelli, L. M., Cobb, M. H. 1983. *Proc. Natl. Acad. Sci. USA* 80:3237–40

117. Yu, K.-T., Czech, M. P. 1984. *J. Biol. Chem.* 259:5277–86

118. Yu, K.-T., Czech, M. P. 1986. *J. Biol. Chem.* 261:4715–22

119. Yu, K.-T., Werth, D. K., Pastan, I. H., Czech, M. P. 1985. *J. Biol. Chem.* 260:5836–46

120. Graves, C. B., Gale, R. D., Laurino, J. P., McDonald, J. M. 1986. *J. Biol. Chem.* 261:10429–38

121. Bertics, P. J., Gill, G. N. 1985. *J. Biol. Chem.* 260:14642–47

122. Lin, C. R., Chen, W. S., Lazar, C. S., Carpenter, C. D., Gill, G. N., et al. 1986. *Cell* 44:839–48

123. Cassel, D., Pike, L. J., Grant, G. A., Krebs, E. G., Glaser, L. 1983. *J. Biol. Chem.* 258:2945–50

124. Cross, F. R., Hanafusa, H. 1983. *Cell* 34:597–607

125. Snyder, M. A., Bishop, J. M., Colby, W. W., Levinson, A. D. 1983. *Cell* 32:891–901

126. Weinmaster, G., Zoller, M. J., Smith, M., Hinze, E., Pawson, T. 1984. *Cell* 37:559–68

127. Heffetz, D., Zick, Y. 1986. *J. Biol. Chem.* 261:889–94

128. Ellis, L., Clauser, E., Morgan, D. O., Edery, M., Roth, R. A., et al. 1986. *Cell* 45:721–32

129. Gould, K. L., Cooper, J. L., Bretscher, A., Hunter, T. 1986. *J. Cell Biol.* 102:660–69

130. Hunter, T., Cooper, J. A. 1981. *Cell* 24:741–52

131. King, L. E. Jr., Carpenter, G., Cohen, S. 1980. *Biochemistry* 19:1524–28

132. Cooper, J. A., Hunter, T. 1981. *Cell* 91:878–83

133. Erikson, E., Shealy, D. J., Erikson, R. L. 1981. *J. Biol. Chem.* 256:11381–84

134. Sawyer, S. T., Cohen, S. 1985. *J. Biol. Chem.* 260:8233–36

135. Pepinsky, R. B., Sinclair, L. K. 1986. *Nature* 321:81–84

136. Guigni, T. D., James, L. C., Haigler, H. T. 1985. *J. Biol. Chem.* 15081–90

137. Fava, R. A., Cohen, S. 1984. *J. Biol. Chem.* 259:2636–45

138. Ghosh-Dastidar, P., Fox, C. F. 1983. *J. Biol. Chem.* 258:2041–44

139. De, B. K., Misono, K. S., Lukas, T. J., Mroczkowski, B., Cohen, S. 1986. *J. Biol. Chem.* 261:13784–92

140. Cohen, S., Fava, R. A. 1985. *J. Biol. Chem.* 260:12351–58

141. Ghosh-Dastidar, P., Coty, W. A., Griest, R. E., Woo, D. D. L., Fox, C. F. 1984. *Proc. Natl. Acad. Sci. USA* 81:1654–58

142. Woo, D. D. L., Fay, S. P., Griest, R., Coty, W., Goldfine, I., et al. 1986. *J. Biol. Chem.* 261:460–67

143. Cooper, J. A., Bowen-Pope, D., Raines, E., Ross, R., Hunter, T. 1982. *Cell* 31:263–73
144. Nakamura, K. D., Martinez, R., Weber, M. J. 1983. *Mol. Cell. Biol.* 3:380–90
145. Hunter, T., Alexander, C. B., Cooper, J. A. 1985. *Ciba Found. Symp.* 116:188–204
146. Kohno, M. 1985. *J. Biol. Chem.* 260:1771–79
147. Cooper, J. A., Sefton, B. M., Hunter, T. 1984. *Mol. Cell. Biol.* 4:30–37
148. Gilmore, T., Martin, G. S. 1983. *Nature* 306:487–90
149. Baldwin, G. S., Knesel, J., Monckton, J. M. 1983. *Nature* 301:435–37
150. Baldwin, G. S., Grego, B., Hearn, M. T. W., Knesel, J. A., Morgan, F. J., et al. 1983. *Proc. Natl. Acad. Sci. USA* 80:5276–80
151. Segawa, K., Ito, Y. 1983. *Nature* 304:742–44
152. Reiss, N., Kanety, H., Schlessinger, J. 1986. *Biochem. J.* 239:523–30
153. Gallis, B., Edelman, A. M., Casnellie, J. E., Krebs, E. G. 1983. *J. Biol. Chem.* 258:13089–93
154. Shiba, T., Akiyama, T., Kadowaki, T., Fukami, Y., Tsuji, T., et al. 1986. *Biochem. Biophys. Res. Commun.* 135:720–27
155. Glenny, J. R., Tack, B. F. 1985. *Proc. Natl. Acad. Sci. USA* 82:7884–88
156. Glenny, J. 1986. *J. Biol. Chem.* 261:7247–52
157. Saris, C. J. M., Tack, B. F., Kristensen, T., Glenny, J. R., Hunter, T. 1986. *Cell* 46:201–12
158. Kristensen, T., Saris, C. J. M., Hunter, T., Hicks, L. J., Noonan, D. J., et al. 1986. *Biochemistry* 25:4497–503
159. Pepinsky, R. B., Sinclair, L. K., Browning, J. L., Mattaliano, R. J., Smart, J. E., et al. 1986. *J. Biol. Chem.* 261:4239–46
160. Wallner, B. P., Mattaliano, R. J., Hession, C., Cate, R. L., Tizard, R., et al. 1986. *Nature* 320:77–81
161. Hirata, F., Matsuda, K., Notsu, Y., Hattori, T., del Carmino, R. 1984. *Proc. Natl. Acad. Sci. USA* 81:4717–21
162. Brugge, J. S. 1986. *Cell* 46:149–50
163. Cochet, C., Gill, G. N., Meisenhelder, J., Cooper, J. A., Hunter, T. 1984. *J. Biol. Chem.* 259:2553–58
164. Fearn, J. C., King, A. C. 1985. *Cell* 40:991–1000
165. Beguinot, L., Hanover, S. A., Ito, S., Richert, N. D., Willingham, M. C., et al. 1985. *Proc. Natl. Acad. Sci. USA* 82:2774–78
166. Friedman, B. A., Frackelton, A. R., Ross, A. H., Connors, J. M., Fujiki, H., et al. 1984. *Proc. Natl. Acad. Sci. USA* 81:3034–38
167. Davis, R. J., Czech, M. P. 1984. *J. Biol. Chem.* 259:8545–49
168. McCaffrey, P. G., Friedman, B. A., Rosner, M. R. 1984. *J. Biol. Chem.* 259:12502–7
169. Davis, R. J., Ganong, B. R., Bell, R. M., Czech, M. P. 1985. *J. Biol. Chem.* 260:5315–22
170. Davis, R. J., Ganong, B. R., Bell, R. M., Czech, M. P. 1985. *J. Biol. Chem.* 260:1562–66
171. Jetten, A. M., Ganong, B. R., Vandenbark, G. R., Shirley, J. E., Bell, R. M. 1985. *Proc. Natl. Acad. Sci. USA* 82:1941–45
172. Soderquist, A. M., Carpenter, G. 1986. *J. Membr. Biol.* 90:97–105
173. Sinnett-Smith, J. W., Rozengurt, E. 1985. *J. Cell. Physiol.* 124:81–86
174. Matrisian, L. M., Bowden, G. T., Magun, B. E. 1981. *J. Cell. Physiol.* 108:417–25
175. Dicker, P., Rozengurt, E. 1980. *Nature* 287:607–12
176. Frantz, C. N., Stiles, C. D., Scher, C. D. 1979. *J. Cell. Physiol.* 100:413–24
177. Carlin, C. R., Knowles, B. B. 1982. *Proc. Natl. Acad. Sci. USA* 79:5026–30
178. Davis, R. J., Czech, M. P. 1986. *Biochem. J.* 233:435–41
179. Chinkers, M., Garbers, D. L. 1986. *J. Biol. Chem.* 261:8295–97
180. King, C. S., Cooper, J. A. 1986. *J. Biol. Chem.* 261:10073–78
181. Whiteley, B., Glaser, L. 1986. *J. Cell Biol.* 103:1355–62
182. Shimizu, N., Behzadian, A., Shimizu, Y. 1980. *Proc. Natl. Acad. Sci. USA* 77:3600–4
183. Davies, R. L., Grosse, V. A., Kucherlapati, R., Bothwell, M. 1980. *Proc. Natl. Acad. Sci. USA* 77:4188–92
184. Kondo, I., Shimizu, N. 1983. *Cytogenet. Cell Genet.* 35:9–14
185. Merlino, G. T., Ishii, S., Whang-Peng, J., Knutsen, T., Xu, Y-H., et al. 1985. *Mol. Cell. Biol.* 5:1722–34
186. Sainsbury, J. R. C., Malcom, A. J., Appleton, D. R., Farndon, J. R., Harris, A. L. 1985. *J. Clin. Pathol.* 38:1225–28
187. Sainsbury, J. R. C., Sherbert, G. V., Farndon, J. R., Harris, A. L. 1985. *The Lancet* 1:364–66
188. Sainsbury, J. R. C., Farndon, J. R., Harris, A. L., Sherbert, G. V. 1985. *Br. J. Surg.* 72:186–88
189. Kaneko, Y., Shibuya, M., Nakayama, T., Hayashida, N., Toda, G., et al.

1985. *Jpn. J. Cancer Res. (Gann)* 76:1136–40

190. Neal, D. E., Bennett, M. K., Hall, R. R., Marsh, C., Abel, P. D., et al. 1985. *The Lancet* 1:366–68

191. Korc, M., Meltzer, P., Trent, J. 1986. *Proc. Natl. Acad. Sci. USA* 83:5141–44

192. Libermann, T. A., Nusbaum, H. R., Razon, N., Kris, R., Lax, I., et al. 1985. *Nature* 313:144–47

193. Gusterson, B., Cowley, G., McIlhinney, J., Ozanne, B., Fisher, C., et al. 1985. *Cancer* 36:689–93

194. Hendler, F. J., Ozanne, B. 1984. *J. Clin. Invest.* 74:647–51

195. Cerny, T., Barnes, D. M., Hasleton, P., Barber, P. V., Healy, K., et al. 1986. *Br. J. Cancer* 54:265–69

196. Cowley, G., Smith, J. A., Gusterson, B., Hendler, F., Ozanne, B. 1984. *Cancer Cells* 1:5–10

197. Gusterson, B., Cowley, G., Smith, J. A., Ozanne, B. 1984. *Cell Biol. Int. Rep.* 8:649–58

198. Ozanne, B., Shum, A., Richards, C. S., Cassells, D., Grossman, D., et al. 1985. *Cancer Cells* 3:41–49

199. Hunts, J., Ueda, M., Ozawa, S., Abe, O., Pastan, I., et al. 1985. *Jpn. J. Cancer Res. (Gann)* 76:663–66

200. Gullick, W. J., Mardsen, J. J., Whittle, N., Ward, B., Bobrow, L., et al. 1986. *Cancer Res.* 46:285–92

201. Yamamoto, T., Kamata, N., Kawano, H., Shimizu, S., Kuroki, T., et al. 1986. *Cancer Res.* 46:414–16

202. Xu, Y., Ishii, S., Clark, A. J. L., Sullivan, M., Wilson, R. K., et al. *Nature* 309:806–10

203. Xu, Y., Richert, N., Ito, S., Merlino, G. T., Pastan, I. 1984. *Proc. Natl. Acad. Sci. USA* 81:7308–12

204. Ishii, S., Xu, Y., Stratton, R. H., Roe, B. A., Merlino, G. T., et al. 1985. *Proc. Natl. Acad. Sci. USA* 82:4920–24

205. Ishii, S., Merlino, G. T., Pastan, I. 1985. *Science* 230:1378–81

206. Clark, A. J. L., Ishii, S., Richert, N., Merlino, G. T., Pastan, I. 1985. *Proc. Natl. Acad. Sci. USA* 82:8374–78

207. Earp, H. S., Austin, K. S., Blaisdell, J., Rubin, R. A., Nelson, K. G., et al. 1986. *J. Biol. Chem.* 261:4777–80

208. Kudlow, J. E., Cheung, C-Y. M., Bjorge, J. D. 1986. *J. Biol. Chem.* 261:4134–38

209. Krupp, M. N., Connolly, D. T., Lane, M. D. 1982. *J. Biol. Chem.* 257:11489–96

210. Stoscheck, C. M., Soderquist, A. M., Carpenter, G. 1985. *Endocrinology* 116:528–35

211. Decker, S. J. 1984. *Mol. Cell. Biol.* 4:571–75

212. Decker, S. J. 1984. *Mol. Cell. Biol.* 4:1718–24

213. Carlin, C. R., Knowles, B. B. 1984. *J. Biol. Chem.* 259:7902–8

214. Carlin, C. R., Knowles, B. B. 1986. *Mol. Cell. Biol.* 6:257–64

215. Slieker, L. J., Lane, M. D. 1985. *J. Biol. Chem.* 260:687–90

216. Slieker, L. J., Martensen, T. M., Lane, M. D. 1986. *J. Biol. Chem.* 261:15233–41

217. Ronnett, G. V., Lane, M. D. 1981. *J. Biol. Chem.* 256:4704–7

218. Reed, B. C., Lane, M. D. 1981. *Proc. Natl. Acad. Sci. USA* 77:285–89

219. Reed, B. C., Ronnett, G. V., Lane, M. D. 1981. *Proc. Natl. Acad. Sci. USA* 78:2908–12

220. MacDonald, R. G., Czech, M. P. 1985. *J. Biol. Chem.* 260:11357–65

221. Sefton, B. M., Trowbridge, I. S., Cooper, J. A., Scolnick, E. M. 1982. *Cell* 31:465–75

222. Magee, A. I., Courtneidge, S. A. 1985. *EMBO J.* 4:1137–44

223. Breitfeld, P. P., Rup, D., Schwartz, A. L. 1984. *J. Biol. Chem.* 259:10414–21

224. Lane, M. D., Ronnett, G., Slieker, L. J., Kohanski, R. A., Olson, T. L. 1985. *Biochimie* 67:1069–80

225. Santon, J. B., Cronin, M. J., MacLeod, C. L., Mendelsohn, J., Masui, H., et al. 1986. *Cancer Res.* 46:4701–5

226. Prywes, R., Livneh, E., Ullrich, A., Schlessinger, J. 1986. *EMBO J.* 5:2179–90

227. Livneh, E., Prywes, R., Kashles, O., Reiss, N., Sasson, I., et al. 1986. *J. Biol. Chem.* 261:12490–97

228. Kay, D. G., Lai, W. H., Uchihashi, M., Khan, M. N., Posner, B. I., et al. 1986. *J. Biol. Chem.* 261:8473–80

229. Kahn, M. N., Savoie, S., Bergeron, J. J. M., Posner, B. I. 1986. *J. Biol. Chem.* 261:8462–72

230. Carpenter, G., Cohen, S. 1976. *J. Cell Biol.* 71:159–71

231. Stoscheck, C. M., Carpenter, G. 1984. *J. Cell Biol.* 98:1048–53

232. Stoscheck, C. M., Carpenter, G. 1984. *J. Cell. Physiol.* 120:296–302

233. Cooper, J. A., Scolnick, E. M., Ozanne, B., Hunter, T. 1983. *J. Virol.* 48:752–64

234. Beguinot, L., Lyall, R. M., Willingham, M. C., Pastan, I. 1984. *Proc. Natl. Acad. Sci. USA* 81:2384–88

235. Dunn, W. A., Connolly, T. P., Hubbard, A. L. 1986. *J. Cell Biol.* 102:24–36

236. Murthy, V., Basu, M., Sen-Majumdar, A., Das, M. 1986. *J. Cell Biol.* 103:333–42
237. Goldstein, J. L., Brown, M. S., Anderson, R. G. W., Russell, D. W., Schneider, W. J. 1985. *Ann. Rev. Cell Biol.* 1:1–39
238. Wileman, T., Harding, C., Stahl, P. 1985. *Biochem. J.* 232:1–14
239. Weber, W., Gill, G. N., Spiess, J. 1984. *Science* 224:294–97
240. Hunts, J. H., Shimizu, N., Yamamoto, T., Toyoshima, K., Merlino, G. T., et al. 1985. *Somatic Cell Mol. Genet.* 11:477–84
241. Yamamoto, T., Nishida, T., Miyajima, N., Kawai, S., Ooi, T., et al. 1983. *Cell* 35:71–78
242. Spurr, N. K., Solomon, E., Jansson, M., Sheer, D., Goodfellow, P. N., et al. 1984. *EMBO J.* 3:159–63
243. Decker, S. J. 1985. *J. Biol. Chem.* 260:2003–6
244. Gilmore, T., DeClue, J. E., Martin, G. S. 1985. *Cell* 40:609–18
245. Kris, R. M., Lax, I., Gullick, W., Waterfield, M. D., Ullrich, A., et al. 1985. *Cell* 40:619–25
246. Nilsen, T. W., Maroney, P. A., Goodwin, R. G., Rottman, F. M., Crittenden, L. B., et al. 1985. *Cell* 41:719–26
247. Gannett, D. C., Tracy, S. E., Robinson, H. L. 1986. *Proc. Natl. Acad. Sci. USA* 83:6053–58
248. Martin, G. S. 1986. *Cancer Surv.* 5:199–219
249. Schechter, A. L., Stern, D. F., Vaidyanathan, L., Decker, S., Drebin, J. A., et al. 1984. *Nature* 312:513–16
250. Bargmann, C. I., Hung, M-C., Weinberg, R. A. 1986. *Nature* 319:226–30
251. Schechter, A. L., Hung, M-C., Vaidyanathan, L., Weinberg, R. A., Yang-Feng, T. L., et al. 1985. *Science* 976–78
252. Coussens, L., Yang-Feng, T. L., Liao, Y-C., Chen, E., Gray, A., et al. 1985. *Science* 230:1132–39
253. Yamamoto, T., Ikawa, S., Akiyama, T., Semba, K., Nomura, N., et al. 1986. *Nature* 319:230–34
254. Fukushige, S-I., Matsubara, K-I., Yoshida, M., Sasaki, M., Suzuki, T., et al. 1986. *Mol. Cell. Biol.* 6:955–58
255. Stern, D. A., Heffernan, P. A., Weinberg, R. A. 1986. *Mol. Cell. Biol.* 6:1729–40
256. Akiyama, T., Sudo, C., Ogawara, H., Toyoshima, K., Yamamoto, T. 1986. *Science* 232:1644–46
257. Bargmann, C. I., Hung, C-H., Weinberg, R. A. 1986. *Cell* 45:649–57
258. Yarden, Y., Escobedo, J. A., Kuang, W-J., Yang-Feng, T. L., Daniel, T. O., et al. 1986. *Nature* 323:226–32
259. Coussens, L., Van Beveren, C., Smith, D., Chen, E., Mitchell, R. L., et al. 1986. *Nature* 320:277–80
260. Hampe, A., Gobet, M., Sherr, C. J., Gailibert, F. 1984. *Proc. Natl. Acad. Sci. USA* 81:85–89
261. Sherr, C. J., Rettenmier, C. W., Sacca, R., Roussel, M. F., Look, A. T., Stanley, E. R. 1985. *Cell* 41:665–76
262. Johnson, D., Langhan, A., Buck, C. R., Sehgal, A., Morgan, C., et al. 1986. *Cell* 47:545–54
263. Sacca, R., Stanley, E. R., Sherr, C. J., Rettenmier, C. W. 1986. *Proc. Natl. Acad. Sci. USA* 83:3331–35
264. Besmer, P., Murphy, J. E., George, P. C., Qiu, F., Bergold, P. J., et al. 1986. *Nature* 320:415–21
265. Ullrich, A., Bell, J. R., Chen, E. Y., Herrera, R., Petruzzelli, L. M., et al. 1985. *Nature* 313:756–61
266. Ebina, Y., Ellis, L., Jarnagin, K., Edery, M., Graf, L., et al. 1985. *Cell* 40:747–58
267. Ullrich, A., Gray, A., Tam, A. W., Yang-Feng, T., Tsubokawa, M., et al. 1986. *EMBO J.* 5:2503–12
268. Leonard, W. J., Depper, J. M., Crabtree, G. R., Rudikoff, S., Pumphrey, J., et al. 1984. *Nature* 311:626–31
269. Nikaido, T., Shimizu, A., Ishida, N., Sabe, H., Teshigawara, K., et al. 1984. *Nature* 311:631–35
270. Cosman, D., Cerretti, D. P., Larsen, A., Park, L., March, C., et al. 1984. *Nature* 312:768–71
271. Leonard, W. J., Depper, J. M., Kanshisa, M., Krönke, M., Peffer, N. J., et al. 1985. *Science* 230:633–39
272. Stadtmauer, L., Rosen, O. M. 1986. *J. Biol. Chem.* 261:10000–5
273. Herrera, R., Rosen, O. M. 1986. *J. Biol. Chem.* 261:11980–85
274. Wahl, M., Sweatt, J. D., Carpenter, G. 1987. *Biochem. Biophys. Res. Commun.* 142:688–95
275. Radeke, M. J., Misko, T. P., Hsu, C., Herzenberg, L. A., Shooter, E. M. 1987. *Nature.* 324:593–97

Ann. Rev. Biochem. 1987. 56:915–44

PROTEIN GLYCOSYLATION IN YEAST

M. A. Kukuruzinska[1], M. L. E. Bergh[1,2], and B. J. Jackson[1]

[1]Center for Cancer Research, Massachusetts Institute of Technology, Cambridge, Massachusetts 02139

[2]Genzyme Corporation, Boston, Massachusetts 02111

CONTENTS

PERSPECTIVES AND SUMMARY

Chimeras often found themselves in important positions. Centaurs (man/horse) were magical shape-shifters who taught the Hellenic gods. Mermaids (woman/fish) were regarded so highly that, up to the 19th century, English law claimed "all mermaids found in British waters" for the Crown. Surely no less important roles are acknowledged in modern times for glycoproteins

915

0066-4154/87/0701-0915$02.00

(sugar/protein). The diverse processes in which they are involved include: cell-cell recognition, hormone-receptor binding, interactions between microorganisms and their hosts, protein targeting, malignant transformation, and phagocytosis. The role of carbohydrate in these phenomena has been the subject of elaborate investigations, primarily conducted using mammalian and avian systems (1, 2). The inherent complexity of these systems, however, has limited experimental approach. Within the last decade, investigators have begun to recognize an alternative in the yeast systems, which offer straightforward genetics, standard biochemical approaches, and ease of manipulation. It is feasible to isolate mutants, to clone genes, and to disrupt them to determine their essentiality. Increasingly, yeasts have been used as hosts for the expression of heterologous genes, because yeasts and higher eukaryotes share many metabolic pathways. One of these is the asparagine-linked glycosylation pathway, which is similar in yeast and animal cells up to the first stages of glycan processing. Thus, yeast is a good model system in which to study protein glycosylation.

While many excellent reviews on glycoprotein structure, synthesis, and function have appeared, none have focused specifically on glycosylation in yeast. The purpose of this review is to summarize protein glycosylation pathways in yeast with emphasis on genetics, enzymology, and molecular biology. Also, studies on heterologous expression in yeast will be presented.

YEAST GLYCOPROTEINS

Analogous to glycoproteins in animal cells, yeast glycoproteins contain carbohydrate chains linked to proteins at specific amino acid residues: asparagine, in the case of N-linked oligosaccharides, and serine or threonine for O-linked species. In contrast to glycoproteins from higher eukaryotes, the yeast glycans contain outer chains consisting primarily of mannose oligomers, with the exception of some yeasts that add terminal galactose and N-acetylglucosamine residues. In *Saccharomyces cerevisiae* the glycoproteins containing N-linked glycans can be placed in three major categories: cell wall proteins, secreted proteins, and vacuolar proteins. The cell wall mannoproteins serve as structural components and carry antigenic determinants. These glycoproteins contain long carbohydrate chains composed of 100–200 mannose residues, with high sugar to protein ratios. The secreted proteins, as exemplified by invertase and acid phosphatase, function in nutrient acquisition. They contain intermediate-length chains of greater than 50 mannose units. Finally, glycoproteins that are localized in the vacuole, for instance carboxypeptidase Y (CPY) and acid phosphatase, perform digestive roles. Their carbohydrate moieties are quite short, CPY containing around nine mannose units per chain. The O-linked glycans are short, generally not exceeding five mannose residues in length. Often, as in the case of cell wall mannoproteins, both types of glycans, the N-linked and the O-linked, are present on the same protein. Glycoproteins containing only O-linked glycans have not been studied in detail, and their functions, except for the possible involvement in mating, have not been defined.

SYNTHESIS OF N-LINKED GLYCANS

The pathway for the synthesis of N-linked glycans is complex. The current model has been elucidated through studies with animal systems (1, 2). In yeast, the early aspects of the synthesis of N-linked glycans are identical to those found in animal cells. The differences reside in the later 'processing' steps, or the modifications the carbohydrate chains undergo in the Golgi. The biosynthesis originates at the rough endoplasmic reticulum. As proteins are synthesized and translocated into the lumen of the endoplasmic reticulum, oligosaccharides consisting of 14 monosaccharides, $GlcNAc_2Man_9Glc_3$, are assembled on a dolichol-pyrophosphate carrier that is embedded in the membrane of the endoplasmic reticulum (Figure 1). The oligosaccharides are transferred as units to asparagine residues of proteins in the lumen. The initial processing events, the stepwise removal of four monosaccharides, three glucose residues, and one mannose, also begin in the lumen of the endoplasmic reticulum. Once the four monosaccharides are removed, the glycoproteins are transported from the endoplasmic reticulum to the Golgi, in an energy-dependent manner via specific vesicles. In the Golgi the yeast pathway diverges from that in animal cells. In yeast the oligosaccharide chains are elongated through a stepwise addition of mannose residues, although in some species galactose or N-acetylglucosamine residues may be added also. In the Golgi the final sorting events take place. Glycosylated proteins are either secreted, targeted to the cell wall, or targeted to the vacuole. The targeting proceeds via specialized vesicles, and it is energy dependent.

Characterization of intermediates in animal and in yeast cells has been aided by the use of specific enzymes and lectins. The most frequently used enzyme, endoglycosidase H (endo H), is specific for high-mannose structures present on proteins; the oligosaccharides must contain four or more mannose units and the cleavage takes place between two GlcNAc residues of the chitobiosyl moieties linked to asparagine residues. A useful lectin is concanavalin A (ConA), which shows high specificity for oligosaccharides containing three or more mannose residues.

The advantages of the yeast system are evident through the isolation of mutants blocked at specific stages in the biosynthetic pathway. Characterization of mutants defective in the lipid-linked oligosaccharide assembly and monosaccharide removal confirmed the animal model and provided further

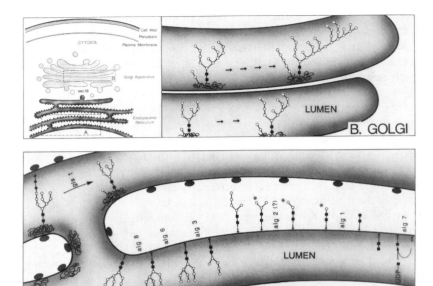

Figure 1   Schematic of glycoprotein biosynthesis in yeast. *A*. Endoplasmic reticulum events: stepwise addition of monosaccharides to lipid-carrier; transfer of the completed oligosaccharide to protein (asparagine) and removal of four monosaccharides from the N-linked carbohydrate chain; addition of the first mannose to protein (serine or threonine) in O-glycosidic linkage. *B*. Golgi events: addition of mannose residues to the N-linked 'core' oligosaccharide on appropriate proteins, resulting in 'outer chain,' which consists of a backbone of $\alpha1,6$-linked mannoses, with $\alpha1,2$-linked side-chains; addition of up to four mannose residues to the O-linked mannose. The mutations in the glycosylation pathway, designated *alg, dpg,* and *gls,* are shown at stages where defects occur. The mutation *sec18* blocks transfer from ER to Golgi. The sugars forming the protein-linked carbohydrate chains are: mannose (○), N-acetylglucosamine (■), and glucose (▲). The arrow represents dolichol-pyrophosphate. The asterisks indicate the structures that may be localized either in the cytoplasm or in the lumen.

insights into specific steps. Moreover, yeast have offered the possibility of mutant constructions that to date are not available in higher eukaryotes. Constructions of double mutants, one in the pathway of interest and another conveying a specific phenotype, facilitate characterization of the defect. For instance, studies of the lipid-linked assembly have been carried out by using a mutation in the secretory pathway, specifically at the point of exit from the endoplasmic reticulum. This defect is referred to as the *sec18* mutation. In this manner the accumulation of intermediates that were transferred to protein but were not extensively processed could be studied.

Isolation and characterization of the individual enzymes involved in the synthesis of glycoproteins from animal cells is difficult and has met with limited success. The enzymes participating in the formation of N-linked glycans represent a small fraction of the total cell mass, are membrane-bound, and are recalcitrant to purification. In yeast this problem is compounded by poor subcellular fractionation techniques. While some of the enzymes have been solubilized, their purification has not been extensive. Many of the studies reported here involved in vitro assays of the enzymatic activities conducted with crude membrane fractions or with partially purified proteins. Because of the limited amount of biochemical data available, this review focuses primarily on the genetic and molecular studies.

Synthesis of Lipid-Linked Oligosaccharides

Lipid-linked oligosaccharides are early intermediates in the biosynthesis of asparagine-linked glycans. Fourteen sugars are transferred stepwise from nucleotide or dolichol-phospho carriers to the polyisoprenoid, Dol-P, to generate $Glc_3Man_9GlcNAc_2$-PP-Dol. The synthesis of this oligosaccharide-lipid donor proceeds in an ordered fashion as evidenced by the isolation of several intermediate lipid-linked oligosaccharides of defined structures (3, 4). Most of the intermediates are identical to those isolated from animal cells (1, 2). The proposed order of events in the lipid-linked oligosaccharide biosynthesis is shown in Figure 2. This process occurs in the endoplasmic reticulum and is completed by the transfer of the oligosaccharide as a unit to an appropriate asparagine residue of the protein.

Much of our knowledge about the individual steps in the lipid-linked assembly has been obtained via isolation and biochemical characterization of mutants defective in this pathway. All of the alg (asparagine-linked glycosylation) mutants described here, except alg7, were isolated through a mannose suicide selection, in which cells defective in asparagine-linked glycosylation survive because they incorporate less radioactive mannose than wild-type cells (5). Seven complementation groups were obtained, and the alg mutants isolated so far are shown in Figure 2. Two of the mutants, alg1 and alg2, exhibited a temperature-sensitive conditional lethal phenotype. The remaining mutants, alg3, alg5, alg6, and alg8, were not affected in the steps necessary for cell viability, although they accumulated incomplete lipid-linked oligosaccharide precursors (5–8). Finally, one complementation group, alg4, was classified as not directly affected in the lipid-linked assembly pathway (6, 9), and its relevance will be addressed later in this section.

UDP-GLCNAC:DOL-P TRANSFERASE   The first step in the assembly of lipid-linked oligosaccharide involves the addition of GlcNAc-P from UDP-GlcNAc to Dol-P. This transfer reaction is inhibited by the antibiotic tunicamycin (10, 11), which is lethal to cells (yeast and higher eukaryotes) after prolonged exposure. The UDP-GlcNAc:Dol-P transferase is a membrane-bound protein

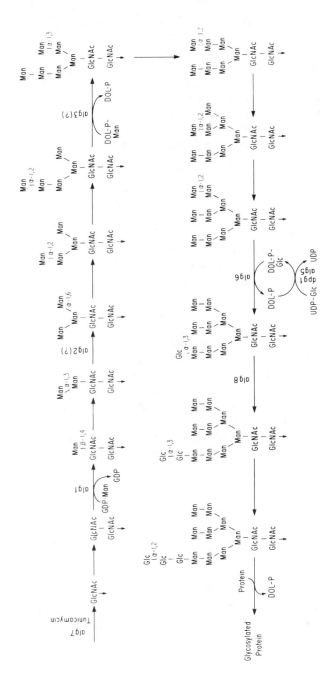

Figure 2 Proposed pathway for the biosynthesis of lipid-linked oligosaccharides in yeast. Abbreviations: GlcNAc, *N*-acetylglucosamine; Man, mannose; Glc, glucose; dol-P, dolichol-phosphate; arrow, dolichol pyrophosphate.

located in the endoplasmic reticulum (12). The transferase gene, *ALG7*, has been cloned on the basis of its ability to rescue cells from lethality associated with tunicamycin (13). Yeast cells were transformed with a multicopy plasmid containing fragments of a yeast genomic library and screened for resistance to the drug. Only cells with multiple copies of the transferase survived. The *ALG7* gene has been sequenced, and it has an open reading frame of 1.34 kb. It encodes a protein with two potential membrane-spanning domains and two potential N-linked glycosylation sites (J. Rine, personal communication). It is essential for cell growth as evidenced by cell lethality observed in strains with null mutations constructed using a standard gene disruption technique (14). The *ALG7* gene is transcribed into two mRNAs of 1.4 and 1.6 kb in size, both of which contain polyadenylation sequences (14). This heterogeneity has been mapped to the 3' untranslated regions on the message. The levels and the ratios of the *ALG7* transcripts change in various *alg* mutants (14, 15). It would be of interest to examine the physiological significance of the 3' message heterogeneity.

The 5' region of the message also displays heterogeneity, albeit the size differences are not as dramatic as those found at the 3' end. There are four transcription initiation sites spanning a 50 bp region; the initiation at the site closest to the ATG codon being the favored one (M. Kukuruzinska, manuscript in preparation). In addition two TATA elements can be identified in the upstream sequence as possible components of the promoter regions.

Dominant mutations that confer resistance to tunicamycin and map in the transferase gene have been isolated. They result in the increased activity of transferase and loss of the ability of the cell to sporulate. The selection of mutants resistant to tunicamycin allowed identification of another gene, *TUN1*, also located on chromosome VII. *Tun1* mutations are recessive and they do not affect glucosamine transport (16).

β1,4 MANNOSYLTRANSFERASE   Cells carrying the *alg1-1* mutation cannot transfer the first mannose to lipid-linked oligosaccharide at the nonpermissive temperature (5). This defect has been identified to be in the β1,4 mannosyltransferase. The gene cloned by complementation of the temperature-sensitive phenotype of the *alg1-1* mutation, *ALG1*, was expressed in *Escherichia coli* and it enabled the bacterial lysate to elongate $GlcNAc_2$-PP-Dol to $Man\beta1,4GlcNAc_2$-PP-Dol (17).

The *ALG1* gene maps to chromosome II (17). The nucleotide sequence of the gene reveals an open reading frame of 1.65 kb and two membrane-spanning domains. There does not appear to be a classical signal sequence at the amino terminus (C. Albright, P. W. Robbins, unpublished results). The *ALG1* gene is transcribed into two messages of 1.7 and 1.85 kb, but the source of heterogeneity has not been determined (M. A. Kukuruzinska, C. Albright, unpublished results).

THE $alg2$ MUTATION   Cells affected by the $alg2$ mutation are temperature-sensitive for growth and accumulate $Man_1GlcNAc_2$-PP-Dol and Man_2-$GlcNAc_2$-PP-Dol at the nonpermissive temperature (6). Two alleles were obtained for this mutation, $alg2-1$ and $alg2-2$, both synthesizing only the two intermediates albeit at different ratios. Although the molecular analysis is not yet complete, three plasmids with a common restriction fragment have been isolated that complement the temperature-sensitivity and the oligosaccharide defect in both alleles (18). Because two intermediates accumulate in these mutant cells, the exact biochemical lesion in the $alg2$ mutants is unclear.

THE $alg3$ MUTATION   The $alg3$ mutants accumulate $Man_5GlcNAc_2$-PP-Dol (6). This mutation may result from a defect in the mannosyltransferase. Alternatively, the accumulation of this intermediate could arise either from a block in the Dol-P-Man synthesis, or from a mutation in the translocating system, as $Man_5GlcNAc_2$-PP-Dol is thought to be synthesized on the cytoplasmic side of the endoplasmic reticulum and elongated on the lumenal side. The data available suggest that the $alg3$ defect is in the transferase gene, because membranes from this mutant synthesize Dol-P-Man (19), and, as will be discussed later, the incomplete lipid-linked precursor is transferred to protein in these cells, an event taking place in the lumen. This mutation is interesting because it suggests that the transferases that catalyze the next steps in the assembly exhibit strict acceptor requirements. As will be described later, some transferases do not have stringent substrate specificities.

GLUCOSYLATION OF LIPID-LINKED OLIGOSACCHARIDES   The final steps in the synthesis of lipid-linked oligosaccharides are glucosylations of the $\alpha1,3$ mannose branch via Dol-P-Glc. Two glucose residues are added with $\alpha1,3$ linkages, while the terminal glucose is added in an $\alpha1,2$ linkage. Three alg mutants with glucosylation defects have been isolated (7, 8). None of the mutants is temperature-sensitive nor do any show striking phenotypes. A slight depression in the rate of Dol-P-Man synthesis in vitro is observed.

Yeast carrying the $alg5$ or $alg6$ mutations accumulate $Man_9GlcNAc_2$-PP-Dol. Cells with an $alg6$ mutation synthesize Dol-P-Glc and are probably defective in the first glucosyltransferase. On the other hand, the $alg5$ mutants are defective in the synthesis of Dol-P-Glc. Membranes prepared from this mutant synthesize glucosylated oligosaccharide only when supplemented with exogenous Dol-P-Glc. In this respect the $alg5$ mutation is similar yet nonallelic to another mutation, $dpg1$, which affects the same reaction. The *dpg1* mutation was isolated as a spontaneous supressor of *gls1* mutation that affects glucosidase I (20).

The *alg8* mutant cannot add the second $\alpha1,3$-linked glucose to lipid-linked oligosaccharide. Instead cells accumulate $Glc_1Man_9GlcNAc_2$-PP-Dol (8). This oligosaccharide is transferred to protein, since endo H treatment of the in vivo [^{3}H]Man-labeled glycoprotein fraction released two oligosaccharides: one had the structure $Glc_1Man_9GlcNAc_1$, the other was its processing product, $Man_8GlcNAc_1$. Pulse-chase studies showed that processing of the transferred oligosaccharide to $Man_8GlcNAc_2$ proceeds more slowly in the *alg8* mutant than in wild-type cells. In order to establish the type of linkage of the glucose residue ($\alpha1,2$ or $1,3$), the *alg8 gls1* double mutant was constructed. Glucosidase I removes the terminal $\alpha1,2$-linked glucose residue from oligosaccharides that have just been transferred to protein. After endo H treatment of *alg8 gls1* membrane proteins the same two oligosaccharides could be isolated, the structure of which were $Glc_1Man_9GlcNAc_1$ and $Man_8GlcNAc_1$, respectively. The presence of $Man_8GlcNAc_1$ indicates that the glucose residue can be removed, therefore it must be in an $\alpha1,3$ linkage (8). However, one other oligosaccharide was released by endo H treatment of protein, which had the size of $Glc_2Man_9GlcNAc_1$. The outer glucose was shown to be in an $\alpha1,2$ linkage. Since processing is slower in the *alg8* mutant, it has been postulated that this oligosaccharide may have been formed after transfer to protein. The enzyme that adds back the outer glucose residue is most likely the $\alpha1,2$ glucosyltransferase. Under normal conditions the enzyme adds the outer glucose to the lipid-linked donor (8). This novel oligosaccharide is probably formed in the *alg8* single mutant, but due to the presence of glucosidase I it is processed too rapidly to be detected.

THE *alg4* MUTATION   The mannose suicide selection provided one complementation group, *alg4*, which contained 16 alleles. All of the isolates are defective in the assembly of the lipid-linked precursor oligosaccharide, and accumulate a spectrum of intermediates, with $Man_{5-8}GlcNAc_2$-PP-Dol being predominant. All but three of the isolates are temperature-sensitive for growth. Because *alg4* mutants are blocked at specific steps in the assembly, and various alleles accumulate different precursors, the *alg4* defect cannot be ascribed to a transferase and is probably of a more global character. A secretory mutation, *sec53*, isolated by a procedure that selected for cells that could not secrete active cell surface enzymes, has been found to be allelic to the *alg4* mutation (6, 21). There are 16 alleles of *sec53* (22), all temperature-sensitive for growth. The *ALG4* gene, isolated by complementation of the temperature-sensitive phenotype of the *alg4* mutation, also complements the alleles in the *sec53* and *alg4* complementation group (23). In most cases, the alleles of both mutations cause defective secretion at the nonpermissive temperature. Some of the *alg4* strains do not show normal secretion at the permissive temperature, while other strains secrete almost as much invertase as the wild-type at the nonpermissive temperature but are arrested in growth. Regardless of the severity of the secretory block, however, the lipid-linked oligosaccharide assembly is defective in these strains (9).

The aberrant lipid-linked glycans are also formed when the membrane fractions are tested for the lipid-linked precursor synthesis at the restrictive temperature, indicating that the glycosylation defect is associated with the endoplasmic reticulum membrane (9).

Diploid strains containing combinations of *alg4* or *alg4/sec53* alleles show, in some cases, complementation of the temperature sensitivity, although their lipid-linked defects are not corrected. This is not due to an unlinked supressor mutation, because all spores from such diploids are temperature-sensitive. This pattern is consistent with intragenic complementation (9).

It seems likely, therefore, that the *ALG4-SEC53* gene product is a multi-functional or a multisubunit protein, which affects the function of the endoplasmic reticulum (9). The in vitro glycosylation data suggest that this protein is associated with endoplasmic reticulum. Recently, the *ALG4-SEC53* gene product has been characterized, however, and it is a 28-kd protein localized in the cytoplasm. It is possible that the protein acts indirectly to facilitate the assembly of the endoplasmic reticulum (24).

Studies on the biosynthesis of the lipid-linked oligosaccharide provided a number of interesting conclusions. First, only early blocks in the lipid-linked assembly pathway result in cell lethality. Second, incomplete lipid-linked precursors, those accumulated in the *alg3, alg5, alg6,* and *dpg1* mutants, contain chains of sufficient length to allow normal growth. Finally, mutant selections identified genes implicated in more generalized functions of the endoplasmic reticulum. The following sections address the question of trans-fer of the lipid-linked precursor oligosaccharide to protein. From the studies with animal systems it was predicted that only the glucosylated lipid-linked precursor could serve as the glycan donor. The availability of yeast mutants accumulating specific intermediates allowed detailed investigation of the structural requirements for the transfer in vivo.

TOPOGRAPHY   An intriguing question in the lipid-linked oligosaccharide assembly concerns the sidedness of the biosynthetic steps. It is not known with certainty on which side of the endoplasmic reticulum membrane, lumen-al or cytosolic, the assembly of the oligosaccharide donor takes place. Most of the studies addressing this question have been carried out using animal cell systems. For instance, treatment of fibroblast microsomal vesicles with ConA showed that some of the lipid-linked intermediates, $Man_{3-5}GlcNAc_2$, were located on the cytoplasmic side of the endoplasmic reticulum (ER) while larger species were lumen-oriented (25). This suggests that $Man_5GlcNAc_2$-PP-Dol is synthesized on the cytoplasmic side of the ER membrane, and then translocated to the lumenal side. The mature precursor, $Glc_3Man_9GlcNAc_2$-lipid, is then completed on the lumenal side where it serves as the donor in polypeptide glycosylation.

Recently, however, Perez & Hirschberg (26) proposed a modification of the model in that the transfer of the two GlcNAc residues might take place on the lumenal rather than the cytoplasmic side of the ER. This implies an extra translocation step as illustrated in Figure 1.

It is not known whether either of the two models applies to yeast, but the observations made with the *alg* mutants may provide further insight. If it is assumed that the oligosaccharyltransferase is oriented toward the lumen, then only donors facing the lumenal side can be transferred to protein. As described in detail below, all donors of the structure $Man_5GlcNAc_2$ and larger are transferred to protein (6–8), an observation consistent with both models. The lipid-linked oligosaccharides $GlcNAc_2$ and $Man_{1-2}GlcNAc_2$, accumulating in *alg1* and *alg2*, respectively, are also transferred in vivo (6). The transfer of the first oligosaccharide is in support of the double-translocation-step model, but transfer of the second one is in disagreement with both models. Perhaps these small species can be found on both sides of the ER membrane. A more detailed description of our current understanding of this problem is given by Hirschberg & Snider in this volume (27).

Transfer of Oligosaccharide Chain from Dolichol-Pyrophosphate to Protein

The transfer of the lipid-linked oligosaccharide to the appropriate asparagine on the polypeptide is catalyzed by oligosaccharyltransferase. Isolation of this enzyme from an animal source was first reported by Das & Heath, who purified the oligosaccharyltransferase 2000-fold from hen oviduct (28). The substrate specificity of this enzyme has been studied extensively. Although species other than $Glc_3Man_9GlcNAc_2$ were transferred to protein acceptors in vitro, evidence from the in vivo studies has led to the hypothesis that only glucosylated lipid-linked glycans could be the substrate for the transfer reaction (reviewed in 1,2). So far, purification of the yeast oligosaccharyltransferase has not progressed beyond the stage of solubilization (29, 30)

PROPERTIES OF DONOR   The characteristics of the transfer of the oligosaccharide from the lipid donor to the peptide acceptor chain in yeast are very similar to those in animal systems. The preferred oligosaccharide species for the transfer has the structure $Glc_3Man_9GlcNAc_2$, and the acceptor sequence on the polypeptide backbone is Asn-X-Ser/Thr.

In vivo labeling experiments suggest that, analogous to the situation in animal cells, a glucosylated donor is involved in protein glycosylation in yeast (31). The identity of this preferred donor was later established as $Glc_3Man_9GlcNAc_2$ (32). Further, in vitro data demonstrate that the glucosylated donor is transferred much more efficiently than the nonglucosylated species (29–31).

Nevertheless, underglucosylated, nonglucosylated, and severely truncated species can be transferred to protein. Several studies have shown that in vitro transfer to protein can take place with a wide variety of lipid-linked intermediates. Even species as small as $GlcNAc_2$-PP-Dol and $Man_1GlcNAc_2$-PP-Dol can serve as donors (33).

More surprising was the observation that the transfer of truncated lipid-linked oligosaccharides also occurred in vivo; in this respect, yeast differs from animal systems. For instance, *alg8* accumulates an underglucosylated donor, which is transferred to protein (8). In *alg5, dpg1,* and *alg6* the nonglucosylated oligosaccharide is used as a substrate, although it has been observed that in *dpg1* cells the amount of carbohydrate transferred to protein is lower than in wild-type (7, 20). The *alg3* mutant accumulates Man_5-$GlcNAc_2$-PP-Dol, which can serve as a donor for transfer (6). In *alg4-sec53* a variety of lipid-linked oligosaccharides accumulate, ranging in size from penta- to decasaccharides. At least some of these species are transferred in vivo (6). Even $GlcNAc_2$ and $Man_{1-2}GlcNAc_2$ can be transferred to protein, since both *alg1* and *alg2* glycosylate invertase in vivo. In the case of *alg1*, a portion of the oligosaccharides on invertase is endo H–resistant, indicating that transfer of $GlcNAc_2$ takes place. A fraction of the oligosaccharides transferred is endo H–sensitive, which indicates that their structure is Man_4-$GlcNAc_2$ or larger. This can be attributed to the leaky nature of the mutation. Invertase made in *alg2* contains mainly endo H–resistant oligosaccharides (6).

PROPERTIES OF ACCEPTOR   The acceptor requirements for yeast N-glycosylation do not seem to differ from those of the animal cell. For all animal oligosaccharyltransferases examined so far, the tripeptide sequence Asn-X-Ser/Thr is a necessary but insufficient prerequisite for N-glycosylation, where X may be any amino acid, with the exception of proline (34–36). Other factors, such as steric inaccessibility, may destroy acceptor properties. Similar results were obtained with yeast in which a variety of synthetic oligopeptides were tested for acceptor activity (37).

TRANSFER   Transfer of the lipid-linked donor to the acceptor protein takes place in the lumen of the rough endoplasmic reticulum. For this reaction to occur the polypeptide chain must be translocated from the cytoplasm into the lumen. The temporal relationships between translation, translocation, and glycosylation reactions are difficult to address. Nevertheless, some data are available from which preliminary conclusions can be drawn.

In animal cells glycosylation has been believed to take place cotranslationally (1, 2), and, therefore, cotranslocationally. Recent studies on the human glucose transporter, however, indicate that translocation of this protein into the microsomal vesicles can take place posttranslationally. Hence, glycosylation may be a posttranslational event. This may apply only to proteins with membrane topology similar to the glucose transporter (38).

Yeast has provided further insight into these events. Reconstitution of the translocation and glycosylation of prepro-α-factor and invertase in a homologous cell-free system showed that for these proteins, translocation could take place posttranslationally (39–41). Moreover, overexpression of the inducible acid phosphatase gene, *PHO5*, in induced, pulse-labeled cells allows detection of the fully translated, yet unglycosylated precursor. The precursor can be found in a processed, glycosylated form after the chase period (42). These data suggest that in some cases translocation and glycosylation may take place posttranslationally. Finally, studies of invertase secretion and glycosylation in the *sec53* mutant suggest that glycosylation and translocation may not have to occur concurrently (R. Feldman, M. Bernstein, and R. Schekman, personal communication). In the *sec53* mutant at the restrictive temperature, 37°C, invertase is bound to the endoplasmic reticulum membrane and is slightly glycosylated, if at all. Proteolytic protection experiments indicate that the protein is translocated. Upon return of the mutant cells to the permissive temperature, invertase is glycosylated and secreted. Therefore, translation, translocation, and glycosylation reactions may not be coupled for all proteins.

Outer-Chain Biosynthesis

In contrast to the preceding steps in the asparagine-linked glycosylation pathway, processing of carbohydrate chains is different in yeast than in higher eukaryotic systems. In animal cells transfer of the $Glc_3Man_9GlcNAc_2$ chain to protein is followed by its conversion to a glycan containing only three mannose residues substituted with sialic acid, galactose, and N-acetylglucosamine. In *S. cerevisiae*, on the other hand, only three glucose and one mannose residues are removed (32, 43). Subsequent elongation takes place through a stepwise addition of mannose residues. The completed carbohydrate chain has the overall structure shown in Figure 3.

REMOVAL OF MONOSACCHARIDES   After the transfer reaction, the stepwise removal of glucose takes place, followed by conversion of $Man_9GlcNAc_2$ to $Man_8GlcNAc_2$. Evidence for this sequence of events is provided by pulse-chase studies of [^{3}H]Man-labeled cells (43). Two glucosidases and one

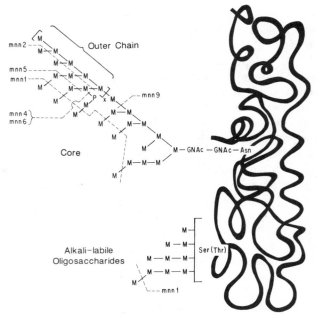

Figure 3   General structure of the protein-linked oligosaccharide chains of *S. cerevisiae* gly-coproteins. The N-linked chain (asparagine) consists of a core unit (8–14 mannose residues) and an outer chain. The length of the outer chain varies, *x* being 0–15. The brackets do not suggest a repeating unit, but indicate the variety of side-chains that can be present on the outer-chain backbone. Linkages are (except for the GlcNAc-bound mannose): $\alpha 1,2$ (—); $\alpha 1,3$ (╱) and $\alpha 1,6$ (╲). Glycoproteins may contain both N-linked and O-linked (serine or threonine) glycans, but can also be of either type. The dashed lines indicate the mannose residues (mono- or dis-accharides, side-chains, or complete outer chain) that are not synthesized in the outer-chain mutations *mnn*.

mannosidase are involved in the removal of the four sugars. The enzymes have been partially purified and characterized (44–46). The subcellular localization of these enzymes has been assigned to the endoplasmic reticulum because endo H treatment of membrane glycoproteins in *sec18* cells, blocked in transport from the endoplasmic reticulum to the Golgi, releases pre-dominantly one oligosaccharide, $Man_8GlcNAc_1$ (7, 47). Glucose removal is not a prerequisite for mannose removal; when the $\alpha 1,2$ glucosidase mutation, *gls1*, is introduced into the *sec18* mutant, $Glc_3Man_8GlcNAc_1$ can be isolated (47).

The inability to remove glucose in *gls1* does not interfere with the forma-tion of the outer chain (48). In contrast, the removal of the terminal mannose in the $Man\alpha 1,2Man\alpha 13Man\alpha 1,6Man\beta 1,4$ sequence (Figure 4), is believed to be essential for initiation of outer-chain synthesis. It has been proposed that this step results in a conformational change of the carbohydrate chain so that it becomes a substrate for the initial elongation steps (43). However, minor quantities of an oligosaccharide containing this particular mannose residue have been isolated from the *mnn1 mnn9* double mutant, which is blocked several steps beyond the $\alpha1,2$ mannosidase step (49). It has been concluded that the removal of the mannose residue is not required for the elongation to occur. However, it may be argued that this mannose has been added back at a later stage (50). Such an aberrant addition would have to be carried out by a Golgi $\alpha1,2$ mannosyltransferase, normally involved in outer-chain synthesis, and not by the endoplasmic reticulum enzyme that uses the lipid-linked oligosaccharide as a substrate. So far, no enzymatic data are available to demonstrate this transfer. In the *alg3* mutant, however, $Man_5GlcNAc_2$ is transferred to invertase in vivo and it is elongated (6). Thus, the specificity of the elongating enzymes may be more relaxed than currently believed. Therefore, it cannot be concluded unequivocally that the removal of the mannose residue is essential for elongation. The availability of a mannosidase mutant would help to resolve this controversy. The mannosidase inhibitor deoxymannojirimycin inhibits the processing mannosidase in vitro (46), but not in vivo (M. L. E. Bergh, P. W. Robbins, unpublished observation).

ELONGATION   Synthesis of the outer chain takes place in the Golgi through a stepwise addition of mannose residues catalyzed by several mannosyltransferases. Evidence for the localization of these reactions was provided by the observation that outer-chain formation does not occur in *sec18* (51). The sugar donor for the Golgi mannosyltransferases is GDP-Man (52). Probably, one of the factors that initially affects the extent of elongation is steric accessibility (53).

The initial steps in the elongation pathway have been elucidated by Trimble & Atkinson (50), who used secreted invertase to isolate and characterize the processing intermediates from Man8 to Man14. The sequence of the events proposed by these authors is shown in Figure 4.

Ballou and coworkers have isolated and characterized many mutants blocked at various stages of outer-chain elongation (reviewed in 54). The carbohydrate chains of the cell wall mannoproteins are the principal immunogens. Antisera obtained to these determinants were used to enrich mutagenized cultures for cells with altered surface antigens. In addition, differential binding to dyes and lectins was used as the basis for mutant screens. Although the *mnn* mutants were originally isolated because of their effects on the cell wall carbohydrate, all yeast glycoproteins are equally affected, since the elongation steps in the N-linked pathway are universal for all cellular components.

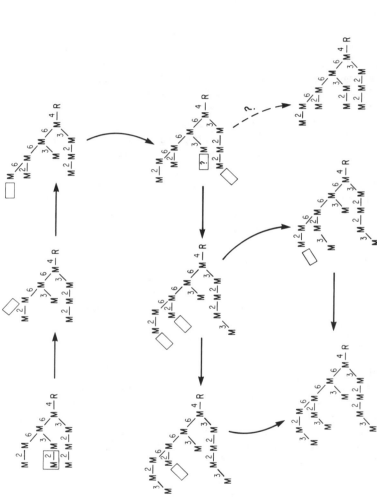

Figure 4 Proposed order of events in the early stage of outer-chain formation. After transfer of Glc$_3$Man$_9$GlcNAc$_2$ to protein, three glucose units are removed (not shown) and one mannose. Then stepwise transfer of mannose residues takes place, along the pathway shown (50). Boxes indicate the localization where the next event (removal or addition) occurs. The question mark indicates a mannose addition that has not yet been demonstrated enzymatically, but that may occur in certain mutants (49, 50).

As can be seen in Figure 3, most of the *mnn* mutations affect the side chains of the outer structures, with the exception of *mnn9*. This mutation blocks synthesis of the outer-chain backbone completely. As a consequence the *mnn9* mutant is much less viable than wild-type, suggesting that the biosynthesis of the outer chain plays an important role in cell growth. Moreover, the *mnn9*, and the *mnn7, mnn8,* and *mnn10* mutants, which carry similar lesions to *mnn9*, are sensitive to lysis, have distorted cell walls, and often form defective spores (55). Thus, the primary phenotype observed in these *mnn* mutants relates to the structure of the cell wall. On the other hand, cells carrying the *mnn2* mutation have no obvious phenotype, even though the side chains are absent. Therefore, while the synthesis of the outer-chain backbone is important, the side-chain biosynthesis can be dispensed with.

In the *alg* mutants incomplete lipid-linked glycans are transferred to protein (see above). To assess which of the lipid-linked precursors were elongated, the secreted invertase isolated from the mutant cultures was treated with endo H (6). In the *alg1* and *alg2* mutants the small oligosaccharide chains that are transferred to protein (GlcNAc$_2$ and Man$_{1-2}$GlcNAc$_2$, respectively) are not substrates for outer-chain formation at the nonpermissive temperature. The *alg3* mutant, on the other hand, transfers a Man$_5$GlcNAc$_2$ oligosaccharide to protein, which apparently is a substrate for elongation (6). The *alg5, alg6,* and *alg8* mutants are also capable of synthesizing outer chain (6, 8).

In summary, the lack of outer-chain elongation of the lipid-linked precursors in the *alg1* and *alg2* mutants suggests that even though such small species can be transferred to protein, some minimal size and/or composition requirements exist for the elongation enzymes in the Golgi. The lack of the core structure and of the outer chain renders these mutants inviable at the restrictive temperature. The specificities of the Golgi enzymes are not stringent, however, because truncated glycans serve as their substrates, as evidenced by the elongation of the incomplete core structure as seen in the *alg3* mutant. This mutant does not exhibit any phenotype and is unaffected in its growth rate. The impaired viability observed with the *mnn9* mutant suggests that the outer-chain addition affects cellular functions. Finally, side chains to the outer-chain backbone, missing in the *mnn2* mutant, are dispensable. The use of *alg mnn* double mutants may provide further insight into the structural requirements that must be met in order to maintain cell viability.

SYNTHESIS OF O-LINKED OLIGOSACCHARIDES

Although the total number of mannose residues in O-linked oligosaccharides approximates that found in N-linked glycans (56), the O-linked chains have been much less thoroughly studied, even though proteins often contain both types of carbohydrate chains (Figure 3).

The biosynthetic pathway for O-glycosylation appears relatively simple. The first mannose residue is transferred, with inversion of configuration, from Dol-P-Man to serine or threonine residues in the endoplasmic reticulum. Up to four additional mannose residues can then be transferred from GDP-Man in the Golgi (37). This contrasts with O-glycosylation in animal cells (see section on HETEROLOGOUS EXPRESSION AND GLYCOSYLATION and Ref. 57). Studies of O-mannosylation of exogenous oligopeptides in the presence of Dol-P-Man and GDP-Man by yeast membranes do not reveal specific sequence requirements on the protein, beyond the need for an acceptor serine or threonine (37).

Secretion, *sec,* mutants have been important in the identification of the subcellular sites at which O-glycosylation occurs (58). The endoplasmic reticulum was determined to be the site at which transfer from Dol-P-Man occurs, because only mannose could be released from the proteins of a *sec18* mutant, which is blocked in the transport of vesicles from the ER to the Golgi at the nonpermissive temperature. At the permissive temperature, mannobiose, not mannose, was the primary β-elimination product. In *sec1* mutants, which appear to be blocked in transport from the Golgi to the plasma membrane, the amounts of each mannose oligomer released are similar in incubations at both permissive and nonpermissive temperatures. Thus the first mannosylation by Dol-P-Man occurs in the endoplasmic reticulum, while subsequent mannosylations, by GDP-Man, occur after the protein has entered the Golgi.

Coordination of O-glycosylation with protein synthesis has not been thoroughly investigated. However, one study has appeared in which yeast protoplasts were labeled with radioactive mannose (59), and the polysomal fraction was isolated. After β-elimination, mannose was the only monosaccharide recovered. It is possible, therefore, that translation and the addition of the first mannose residue may occur simultaneously.

INTERACTION OF THE GLYCOSYLATION PATHWAY WITH OTHER CELLULAR EVENTS

Protein Sorting and Secretion

As discussed in the preceding sections, protein glycosylation comprises transport through the endoplasmic reticulum, where the addition of the N-linked core oligosaccharides takes place, followed by the outer-chain modifications as the protein passes through the Golgi. With the exception of the mitochondrial and perhaps nuclear proteins, synthesis of all proteins destined for noncytoplasmic locations, including those retained in the ER and probably in the Golgi, begins in the ER. The transport events have been studied most extensively in *S. cerevisiae,* where the secretory pathway follows the route:

ER→Golgi body→vesicles→cell surface (60, 61), analogous to that in mammalian cells.

TRANSPORT THROUGH THE ER   Proteins acquire unprocessed carbohydrate chains while they pass through the membrane of the ER. Most are synthesized in precursor forms containing signal peptide sequences. Biogenesis in the ER has been documented through studies involving translation of native yeast mRNAs in the presence of a heterologous microsomal system (62), and most recently in a homologous yeast translation/translocation system where invertase and α-factor mRNAs were used. In the yeast cell-free system, α-factor and invertase both acquired N-linked core oligosaccharides, and the invertase signal sequence was processed. These data have demonstrated that in yeast, glycoprotein synthesis originates in the ER, and that translocation and translation may be uncoupled (39–41). Moreover, the results suggest that glycosylation may be, at times, a posttranslational event.

Much useful information regarding glycosylation and translocation processes has been obtained from studies with temperature-sensitive secretion mutants, sec53 and sec59, which are defective in the import of proteins into the ER (22). In one such mutant, sec53, invertase is only slightly glycosylated at 37°C, while in another, sec59, the protein contains 0–3 out of 10 of the N-linked carbohydrate chains. At the nonpermissive temperature the enzyme is inactive and remains tightly bound to the ER membranes. Upon shift to the permissive temperature, 24°C, the invertase is glycosylated, becomes partially active, and is secreted. The reversibility of this mutation depends on energy, yet it does not require protein synthesis. The effects of tunicamycin treatment of wild-type cells closely resemble the phenotype of the sec53 mutant at 37°C. Thus, inhibition of oligosaccharide synthesis with tunicamycin results in irreversible accumulation of inactive invertase, which is also retained in the ER membrane at 37°C. At 25°C about 50% of unglycosylated invertase is secreted (22). Since the secreted invertase was measured by immunoprecipitation, it is not known whether under these conditions the enzyme is active. A similar result was seen with VSV G protein in infected tissue culture cells treated with tunicamycin (63). It may be speculated that glycosylation is required for proper protein folding, which, in turn, affects secretion. Finally, as noted in the previous section, recent studies on sec53 suggest that glycosylation and translocation may not be coupled (R. Feldman, M. Bernstein, and R. Schekman, personal communication).

As mentioned in the previous section, sec53 is in the same complementation group as alg4, and, in addition to the secretion defect, this mutant carries a lesion in the lipid-linked oligosaccharide assembly pathway. The SEC53-ALG4 gene product is probably a multifunctional protein that may affect the assembly and, therefore, the function of the endoplasmic reticulum (9). The cytoplasmic localization of this protein suggests that it may exert its effect through interaction(s) with some stable component(s) in the endoplasmic reticulum membrane (24). It is possible that the endoplasmic reticulum defect affects otherwise functional transferases, their accessibility to substrates, or the substrate pool sizes. Because many of the endoplasmic reticulum–associated processes are complex, there may exist different levels of regulation. Thus, it would be of interest to investigate gene expression and corresponding enzymatic activities of various transferases in the *sec53* background.

In the *alg* mutants where the assembly of normal lipid-linked oligosaccharides is impaired, the transfer of the truncated carbohydrate units to invertase and its secretion take place (6). The *dpg1* mutation, mentioned previously, has a defect in the synthesis of glucose donor, dolichol phosphoglucose (20). In such a mutant no glucosylated lipid-linked donor is made, resulting in underglycosylation of invertase, which nonetheless is secreted. Furthermore, glucose processing does not affect secretion, since the *gls-1* mutant, defective in glucosidase I, secretes invertase at normal rates (47).

Early steps in the secretory pathway involve sorting events in the endoplasmic reticulum, as some proteins are probably retained in this compartment. Proteins that are destined for the vacuole, the cell surface, or the periplasm are exported from the endoplasmic reticulum to the Golgi. By this point, they have acquired a core oligosaccharide and have undergone the initial processing reactions. Transport from the endoplasmic reticulum to the Golgi is energy-dependent (61, 64), but as described above, does not seem to be affected by glycosylation to a great degree. Conversely, in certain secretion mutants, specifically those blocked at the transport from the endoplasmic reticulum, e.g. *sec18,* (64), normal assembly of lipid-linked oligosaccharide and transfer to protein (invertase) take place.

Therefore, carbohydrate chains are probably not involved as signals in retaining the endoplasmic reticulum components in this compartment, although they may facilitate export. The signals that might prove important seem to reside in the polypeptide portion of the protein. Signal peptide cleavage has been shown to be necessary for expedient transport of invertase (65) and acid phosphatase (42) from the endoplasmic reticulum. A point mutation in the invertase molecule (s2 mutant) resulting from a single base substitution at position 64, (Thr/Ile), and which does not affect the signal peptide cleavage, causes a sevenfold slowing of the transport out of the endoplasmic reticulum (64). This would indicate that, besides signal peptides, there may be other sequences important for sorting in this compartment. Nevertheless, it may be of interest to determine if site-specific mutagenesis of N-linked glycosylation sequences would affect the sorting process.

TRANSPORT THROUGH THE GOLGI   Transport through the Golgi involves completion of the core and the elongation of the man $\alpha 1,6$ branch (see previous section) as well as addition of shorter mannose side-chains in an $\alpha 1,3$ linkage to the backbone. Defects in the processing enzymes, evident in the *mnn* mutants, result in the formation of shorter outer chains on invertase, although secretion is normal (54, 61).

A secretion mutant, *sec7,* blocked in transport from the Golgi, accumulates glycosylated, mature invertase and acid phosphatase at 37°C, indicating that glycosylation is completed in the Golgi (60). It is not clear whether the yeast Golgi is arranged into cis, medial, and trans cisternae, which, in turn, may be enriched with specific processing enzymes, as has been suggested for animal cells (66). Interestingly, the carbohydrate chains present on the vacuolar enzymes are in general shorter than those found on the secreted proteins, suggesting possible organization of the sorting in the Golgi. On the other hand, oligosaccharides present on authentic carboxypeptidase Y and pro- teinase A, vacuolar enzymes, are identical to those found on the misdirected, secreted forms, produced when the proteins' structural genes are present on multicopy plasmids (67, 68). This is consistent with both intra- and ex- tracellular transport from identical cisternae. Furthermore, secretion of in- vertase in the *mnn* mutants indicates that incomplete glycosylation does not affect export from the Golgi (54, 61).

FINAL SORTING EVENTS   Assembly of the plasma membrane is thought to take place via vesicle fusion at the bud portion of the cell. The secretory pathway is also directed to the same side of the membrane (61, 69–72), although it is still unclear if the same vesicles are used for the two processes. Studies of temperature-sensitive actin mutants have implicated this protein in the organization and polarized growth of the yeast cell surface (73). In the actin mutants secretion of invertase is partially inhibited at the restrictive temperature. Moreover, these mutants show delocalized deposition of chitin on the cell surface (73). It has been suggested that polarized transport to the bud portion of the cell requires recognition components at the plasma mem- brane (61). The role of chitin, a $\beta 1,4$ polymer of GlcNAc residues and a major component of the septum between the mother and daughter cell, may prove important here. Mutants in chitin synthase II with less than 10% of the wild-type level of chitin (74) have been isolated recently. It may be of interest to investigate targeting events to the plasma membrane in such mutant backgrounds.

Transport to the vacuole proceeds from the Golgi and has been well defined in the *sec14* mutant with carboxypeptidase Y (75, 76). Many vacuolar enzymes are derived from glycoprotein precursors, and while glycosylation is completed in the Golgi, proteolytic processing takes place in the vacuole. The vacuolar transport appears saturable because cells transformed with a multi-copy plasmid containing a structural gene for carboxypeptidase Y, or with a plasmid where this gene is behind a strong acid phosphatase promoter, secrete as much as 60% of this enzyme (67). The same results were obtained with proteinase A (68). The function of carbohydrate in the targeting and activation of these enzymes remains obscure, as these processes are not affected by glycosylation defects (67, 68). This is in contrast to animal cells where mannose-6-phosphate residues have been shown to participate in proper targeting of the lysosomal enzymes (77).

On the basis of the evidence presented above it appears that the secretory events proceed in the absence of glycosylation, and thus the pathways are not coupled. It cannot be excluded, however, that subtle regulatory mechanisms, on a level of gene activation or transcription, are involved. With the advent of rigorous characterization of mutations affecting various processes and the feasibility of molecular and genetic approaches, such studies are now possible.

Cell Cycle

Proper coordination of cellular processes, including glycosylation, is required for progression of yeast through the cell cycle [for an extensive review of the cell cycle see (78)]. For example, glucans and mannoproteins are synthesized throughout the cell cycle. This finding is consistent with the constant rate of bud growth in log phase cultures.

A more direct role for glycosylation in the cell cycle is supported by mutant studies. Cells carrying the *alg1* mutation arrest unbudded within the first cell cycle at the nonpermissive temperature. Some cell division cycle, *cdc*, mutants exhibit the same phenotype at the nonpermissive temperature. Thus, cells beyond the G1 phase will complete one cycle of growth but will not enter the S phase (78, 79). Treatment with tunicamycin, an inhibitor of the *ALG7* gene product, causes a similar phenotype (80).

Temperature-shift experiments in the presence of *alpha*-factor indicate that the arrests caused by the *alg1-1* mutation and by *alpha*-factor, are temporally indistinguishable (56, 79). In contrast to *alg1*, and despite the closeness of their metabolic blocks in the lipid-linked pathway, the *alg2-1* mutant arrests randomly with a mixture of budded and unbudded cells. The arrest is, however, similar to that of *alg1-1* in being a first cycle arrest, such that cell number does not double (B. Jackson, unpublished results).

These findings may reflect a requirement for synthesis of a glycoprotein upon progression of the cell from G1 to S phase. A possible candidate for such a protein has been examined by Popolo and coworkers (81, 82). Precursor forms of this protein, synthesized specifically during the transition from G1 to S phase, are processed to a form with a molecular weight of 115,000, into which radioactive glucosamine can be incorporated. Synthesis of precursors to the 115,000-dalton protein is especially prominent during recovery from cell cycle arrest of *cdc25*, a mutant that arrests in G1 following exposure to the restrictive temperature (78).

Mating

There is some evidence that O-glycosylated proteins may function in mating. When *a* cells are treated with *alpha*-factor, a protein containing O-linked carbohydrate is synthesized. Its molecular weight is 22,000 and it is extensively O-glycosylated (83). The protein appears to be a mating-type-specific *a* cell agglutinin. The addition of this protein to mixtures of *a* and *alpha* cells, which have been preincubated independently with mating factors, inhibits cellular agglutination.

Additional evidence that O-linked glycosylation may be involved in mating comes from the studies of *fus* (fusion) mutants. A *fus1 fus2* double mutant fails to mate with *fus1, fus2*, or *fus1 fus2* mutants, although diploid formation is normal in crosses to a wild-type strain. These mutants form conjugating pairs; however, subsequent cell wall degradation and cytoplasmic fusion are impaired. The nucleotide sequence of the *FUS1* gene predicts a 390-amino-acid protein with an N-terminal domain consisting of 60% serine and threonine residues (84). This protein is induced 600-fold in cells treated with *alpha*-factor. Given the amino acid composition, the *FUS1* gene product is a candidate for O-glycosylation; however, no structural data are available.

In *S. cerevisiae* the precursor form of the mating pheromone, the prepro-*alpha*-factor, contains three N-linked core oligosaccharides. The carbohydrate chains are located in the polypeptide regions removed during proteolytic processing. The processing occurs late in the secretory pathway, and the mature, secreted form bears no N-linked glycans (85). The function of carbohydrate may lie in facilitating the transit of the pheromone through the secretory pathway, since after treatment of cells with tunicamycin at 25°C, only a fraction of the protein is secreted as compared with untreated cells. Analogous to invertase, the effect of tunicamycin is much more pronounced at 37°C (see section on *Protein Sorting and Secretion*) (85).

GLYCOSYLATION IN OTHER YEASTS

Glycosylation has been investigated most extensively in *S. cerevisiae*. Studies defining the sequence of reactions leading to the formation of mature glycoproteins in other yeasts are less complete. Nevertheless, the major structural components of cell walls have been identified as mannoproteins in *Candida albicans, Candida utilis,* and *Candida parapsilosis* (86–88). The most frequently studied yeast besides *S. cerevisiae* is the fission yeast *Schizosaccharomyces pombe*. Using gold granules labeled with the lectin of *Bandeiraea simplicifolia,* specific for *alpha*-D-galactopyranosyl residues, it has been shown that *S. pombe* cell wall proteoglycan contains nonreducing terminal galactose residues, in addition to mannose (89, 90). In *Kluyveromyces lactis* some of the mannoprotein side-chains are modified by the addition of *N*-acetylglucosamine residues (91). To date, no information is available on the details of the pathways of mannoprotein formation in other yeasts. The cell wall composition, however, suggests that they are similar, if not identical, to those reported for *S. cerevisiae*.

In ascosporogenous yeasts, *Saccharomyces, Hansenula, Saccharomycodes,* and *Pichia,* sexual agglutination occurring between cells of opposite mating type involves glycoprotein components (92). In most cases the agglutination is pronase-sensitive; so far the function of carbohydrate in this process remains unclear. Yet, rigorous characterization of cell surface molecules responsible for sexual agglutination in *Saccharomyces kluyveri* indicates that both the 16-factor and 17-factor, (for 16-cell and 17-cell haploid strains), are glycoproteins. The 16-factor activity is labile to *alpha*-mannosidase and periodate, while the 17-factor activity appears stable to exo-*alpha*-mannosidase digestion (93).

Although difficult to evaluate systematically in other yeasts, many diverse cell functions may be attributed to glycoproteins. Mannose-containing oligosaccharides present in the *C. albicans* cell wall have been shown to be responsible for the infectious nature of this fungus and specifically act in the attachment to human epithelial cells (94). The serotypes A and B of this organism are defined by specific mannose oligosaccharides (88, 95). The membrane-bound precursor of the killer toxin produced by *K. lactis* is a glycoprotein encoded by one of the open reading frames of pGKL1, a linear DNA plasmid (96). In contrast to the small *S. cerevisiae* killer toxin, *K. lactis* toxin is a large, 100-kd protein composed of subunits. As yet, no function has been attributed to the carbohydrate chains.

The presence of galactose residues in the cell surface glycoproteins of *S. pombe* indicates that a galactosyltransferase, not found in *S. cerevisiae,* is involved in posttranslational modifications in this organism. Recently, galactose has also been reported to be present on the *S. pombe* secretory protein, invertase, by reactivity with the *B. simplicifolia* lectin (97). Although structural data are not available, this fission yeast may prove useful in studies of galactosyltransferases found in higher eukaryotes.

HETEROLOGOUS EXPRESSION AND GLYCOSYLATION

As a eukaryotic organism that is easy to manipulate, the yeast *S. cerevisiae* is valuable in the study of expression and function of heterologous genes. Like bacteria, yeast can be cultivated in great quantity. In contrast to bacteria, yeast have a glycosylation system resembling the one found in animal cells, allowing the posttranslational modification often required for biological activity of the introduced proteins. However, this does not imply that yeast is an ideal host for heterologous expression. A variety of problems have already been encountered, such as inefficient translation due to differences in codon usage (98), reduced transcription efficiency (99), low steady-state mRNA levels (100, 101), and proteolysis of the protein synthesized (102). In addition, S. cerevisiae is often not capable of splicing higher eukaryotic RNA (103, 104), although S. pombe is (105). Finally, it is not clear whether yeast is capable of carrying out modifications such as gamma-carboxylation, amidation, and fatty acylation. For example, the VSV and Sindbis virus glycoproteins expressed in yeast are not acylated by the host enzymes (106), whereas some nonviral yeast proteins are acylated (107).

Nevertheless, many foreign genes have been expressed successfully in S. cerevisiae. Some of the foreign glycoproteins produced by yeast include hepatitis B virus surface antigen (108–110), interferon-gamma (99, 111, 112), tissue plasminogen activator (113), monoclonal antibodies (114), a chicken ovalbumin-like protein (115), interleukin-2 (116), and proteins from vesicular stomatitis (106), Sindbis (106), influenza (117, 118), and Semliki Forest viruses (119).

Endogenous forms of many of the above proteins contain N-linked glycans, which can be of the hybrid, mannose, or complex type, depending on the protein and glycosylation site (1, 2). Furthermore, many higher eukaryotic proteins have sialylated O-linked chains of the mucin-type (57). The structures of these glycans are dramatically different from those found in S. cerevisiae. As described in the previous sections, the N-linked chains in yeast are uniformly of the high-mannose type, and the O-linked glycans consist of mannose residues only. It is to be expected, therefore, that when higher eukaryotic glycoproteins are expressed in yeast, their glycosylation pattern will be different.

Structural characterizations of glycans present on foreign proteins in yeast have been conducted in only a limited number of cases. The influenza virus hemagglutinin has been reported to be glycosylated in yeast (107, 108), but no information is available on the exact structure and size of the N-linked glycans. Wood et al reported the N-linked glycosylation of an antibody heavy chain in yeast (114). Surprisingly, a substantial fraction of the protein produced in yeast was not glycosylated, suggesting that the yeast oligosaccharyltransferase might not act as efficiently on some of the foreign glycosylation sites as the homologous enzyme. Similar results were obtained with the prepro-α-factor-somatostatin fusion protein. This protein was underglycosylated when expressed in yeast (120). It has been suggested that this might be due to saturation of the glycosylation machinery: the oligosaccharyltransferase may be present in amounts too low to accommodate the large quantities of substrate produced in an expression system developed to generate high yields of recombinant material. In the case of fusion proteins, however, the protein conformation may be sufficiently different to impair glycosylation sterically. When the hepatitis B surface antigen (HBsA) is expressed in yeast it is not glycosylated at all (109). It is possible that the glycosylation sites of HBsA are not recognized by the yeast transferase, but no studies have been carried out to investigate this possibility.

In contrast, an invertase-prochymosin fusion protein was synthesized in the glycosylated form. The hybrid protein contained three potential glycosylation sites, one from invertase and two from prochymosin. The invertase site was always glycosylated. The two prochymosin sites, which are not a substrate for the calf oligosaccharyltransferase, were also not recognized by the yeast enzyme (121). Although this is a negative result, it is consistent with the interpretation that the specificity of the yeast and calf enzymes is comparable, and/or that the prochymosin folding is similar in the two organisms.

S. cerevisiae recognizes O-linked glycosylation sites on a protein from the fungus *Aspergillus*. Glucoamylase from *Aspergillus awamori* contains both N- and O-linked carbohydrate chains. The structures of the *Aspergillus* O-linked glycans are not of the mucin type; instead, they are similar to those found in *S. cerevisiae*. Both types of glycosylation found on the recombinant protein are similar to those of the native glucoamylase (122).

Little is known about the extent and efficiency of carbohydrate processing on recombinant proteins in yeast. Studies on the processing of the vesicular stomatitis and Sindbis virus proteins expressed in yeast indicate that, in contrast to animal cells, extensive processing does not take place (106). Lack of extensive processing may be due to the same factors that result in underglycosylation.

In view of the information presented above, expression of glycoproteins of pharmacological significance may be complicated. A solution to these problems may lie in the in vitro modification of the carbohydrate chains of the heterologous proteins. In fact, preliminary studies have shown that such an approach may be feasible (123).

RELEVANCE OF GLYCOSYLATION TO YEAST

The most obvious characteristic of glycoproteins in yeast is their localization. They serve as components of cell wall, are membrane-bound, or are localized in separate cellular compartments. Thus they are distinct from the unglycosylated proteins, which remain soluble in the cytoplasm.

Three genes involved in the lipid-linked oligosaccharide assembly, *ALG7*, *ALG1*, and *ALG2*, have been identified as essential for cell growth. This is to be expected in view of the array of functions in which glycoproteins are involved. Interestingly, small lipid-linked glycans that accumulate in the *alg1* and the *alg2* mutants at the restrictive temperature are transferred to protein. These structures do not contain enough information for the Golgi enzymes, however, and elongation is impaired. This results in cell death. The transfer and elongation of the truncated lipid-linked glycan that accumulates in the *alg3* mutant confirms relaxed specificities of the oligosaccharide transferase and of the Golgi enzymes in yeast. The *alg3* mutant does not exhibit a distinct phenotype, and as long as the elongation takes place the viability is not affected. Moreover, neither does incomplete elongation affect cell growth. The *mnn2* mutant lacks the side-chains and the cells show normal doubling times. The importance of the Golgi processes in N-linked glycosylation lies in the addition of the outer backbone, as *mnn9* displays striking growth perturbations.

Glycosylation is necessary for the progression through the cell cycle, as the *alg1* mutant arrests at the G1 phase of the cell cycle. Similar effect is seen after treatment of cells with tunicamycin, a specific inhibitor of the *ALG7* gene product. Recent evidence indicates that the mating type–specific *a* cell agglutinin is an O-glycosylated protein. Glycosylation has been shown to be necessary for the enzymatic activity of some proteins, for instance proteinase A, carboxypeptidase Y, and invertase.

More surprising information stems from the studies of intracellular targeting and secretion, where the absence of glycosyl residues has been shown not to affect proper targeting. In such cases the effects of glycosylation may be more subtle, and may influence rates of processing and turnover or stability.

ACKNOWLEDGMENTS

We thank Drs. P. W. Robbins, C. E. Bulawa, T. C. Huffaker, L. Pillus, K. W. Runge, and R. B. Trimble for critical reading of the manuscript, excellent comments, and helpful discussions. This work was supported by National Institutes of Health Postdoctoral Fellowship Grant GM10367 to B. Jackson, American Cancer Society Postdoctoral Fellowship Grant to M. Bergh, and by the Hoechst Roussel Fellowship of the Life Sciences Research Foundation to M. Kukuruzinska.

Literature Cited

1. Hubbard, S. C., Ivatt, R. J. 1981. *Ann. Rev. Biochem.* 50:555–83
2. Kornfeld, R., Kornfeld, S. 1985. *Ann. Rev. Biochem.* 54:631–44
3. Prakash, C., Vijay, I. K. 1982. *Biochemistry* 21:4810–18
4. Prakash, C., Katial, A., Vijay, I. K. 1983. *J. Bacteriol.* 153:895–902
5. Huffaker, T. C., Robbins, P. W. 1982. *J. Biol. Chem.* 257:3203–10
6. Huffaker, T. C., Robbins, P. W. 1983. *Proc. Natl. Acad. Sci. USA* 80:7466–70
7. Runge, K. W., Huffaker, T. C., Robbins, P. W. 1984. *J. Biol. Chem.* 259:412–17
8. Runge, K. W., Robbins, P. W. 1986. *J. Biol. Chem.* 261:15582–90
9. Runge, K. W., Robbins, P. W. 1986. *Microbiology* 312–16
10. Kuo, S. C., Lampen, J. O. 1974.

Biochem. Biophys. Res. Commun. 58:287–95
11. Lehle, L., Tanner, W. 1976. *FEBS Lett.* 71:167–70
12. Ravoet, A. M., Amar-Costesec, A., Godelaine, D., Beaufay, H. 1981. *J. Cell. Biol.* 91:679–88
13. Rine, J., Hansen, W., Hardeman, E., Davis, R. W. 1983. *Proc. Natl. Acad. Sci. USA* 80:6750–54
14. Kukuruzinska, M. A., Robbins, P. W. 1985. *Proc. Cold Spring Harbor Conf. Mol. Biol. Yeast,* p. 304. New York: Cold Spring Harbor Lab.
15. Kukuruzinska, M. A., Robbins, P. W. 1985. *J. Cell. Biochem.* Suppl. 9C:304
16. Barnes, G., Hansen, W. J., Holcomb, C. L., Rine, J. 1984. *Mol. Cell. Biol.* 4:2381–88
17. Couto, J. R., Huffaker, T. C., Robbins, P. W. 1984. *J. Biol. Chem.* 259:378–82
18. Jackson, B. J., Robbins, P. W. 1986. *Yeast* 2:S166 (Abstr.)
19. Runge, K. W. 1985. *Yeast mutations in the glucosylation steps of the asparagine-linked glycosylation pathway.* PhD thesis. MIT Cambridge
20. Ballou, L., Supal, P., Krummel, B., Markku, T., Ballou, C. E. 1986. *Proc. Natl. Acad. Sci. USA* 83:3081–85
21. Ferro-Novick, S., Hansen, W., Schauer, I., Schekman, R. 1984. *J. Cell. Biol.* 98:44–53
22. Ferro-Novick, S., Novick, P., Field, C., Schekman, R. 1984. *J. Cell. Biol.* 98:35–43
23. Couto, J. 1984. *The molecular biology of a yeast β-mannosyl transferase.* PhD thesis. MIT Cambridge
24. Bernstein, M., Hoffman, W., Ammerer, G., Schekman, R. 1985. *J. Cell. Biol.* 101:2374–82
25. Snider, M. D., Rogers, O. C. 1984. *Cell* 36:753–61
26. Perez, M., Hirschberg, C. B. 1986. *J. Biol. Chem.* 261:6822–30
27. Hirschberg, C. B., Snider, M. D. 1987. *Ann. Rev. Biochem.* 56:63–87
28. Das, R. C., Heath, E. C. 1980. *Proc. Natl. Acad. Sci. USA* 77:3811–15
29. Trimble, R. B., Byrd, J. C., Maley, F. 1980. *J. Biol. Chem.* 255:11892–95
30. Sharma, C. B., Lehle, L., Tanner, W. 1981. *Eur. J. Biochem.* 116:101–8
31. Lehle, L. 1980. *Eur. J. Biochem.* 109:589–601
32. Parodi, A. J. 1981. *Arch. Biochem. Biophys.* 210:372–82
33. Lehle, L., Tanner, W. 1978. *Eur. J. Biochem.* 83:563
34. Bause, E., Hettkamp, H. 1979. *FEBS Lett.* 108:341–44
35. Bause, E. 1983. *Biochem. J.* 209:331–36
36. Ronin, C., Bouchilloux, S., Granier, C., van Rietschoten, J. 1978. *FEBS Lett.* 96:179–82
37. Bause, E., Lehle, L. 1979. *Eur. J. Biochem.* 101:531–40
38. Mueckler, M., Lodish, H. F. 1986. *Cell* 44:629–37
39. Rothblatt, J. A., Meyer, D. I. 1986. *Cell* 44:619–28
40. Waters, M. G., Blobel, G. 1986. *J. Cell. Biol.* 102:1543–50
41. Hansen, W., Garcia, P. D., Walter, P. 1986. *Cell* 45:397–406
42. Haguenauer-Tsapis, R., Hinnen, A. 1984. *J. Mol. Cell Biol.* 4:2669–75
43. Byrd, J. C., Tarentino, A. L., Maley, F., Atkinson, P. H., Trimble, R. B. 1982. *J. Biol. Chem.* 257:14657–66
44. Kilker, R. D., Saunier, B., Tkacz, J. S., Herscovics. A. 1981. *J. Biol. Chem.* 256:5299–303
45. Saunier, B., Kilker, R. D., Tkacz, J. S., Quaroni, A., Herscovics, A. 1982. *J. Biol. Chem.* 257:14155–61
46. Jelinek-Kelly, S., Akiyama, T., Saunier, B., Tkacz, J. S., Herscovics, A. 1985. *J. Biol. Chem.* 260:2253–57
47. Esmon, B., Esmon, P. C., Schekman, R. 1984. *J. Biol. Chem.* 259:10322–27
48. Tsai, P. K., Ballou, L., Esmon, B., Schekman, R., Ballou, C. E. 1984. *Proc. Natl. Acad. Sci. USA* 81:6340–43
49. Tsai, P. K., Frevert, J., Ballou, C. E. 1984. *J. Biol. Chem.* 259:3805–11
50. Trimble, R. B., Atkinson, P. H. 1986. *J. Biol. Chem.* 261:9815–24
51. Esmon, B., Novick, P., Schekman, R. 1981. *Cell* 25:451–60
52. Parodi, A. 1979. *J. Biol. Chem.* 254:8342–52
53. Trimble, R. B., Maley, F., Chu, F. K. 1983. *J. Biol. Chem.* 258:2562–67
54. Ballou, C. E. 1982. *The Molecular Biology of the Yeast Saccharomyces,* pp. 335–60. New York: Cold Spring Harbor Lab.
55. Ballou, L., Cohen, R. E., Ballou, C. E. 1980. *J. Biol. Chem.* 255:5986–91
56. Orlean, P., Schwaiger, H., Appeltauer, U., Haselbeck, A., Tanner, W. 1984. *Eur. J. Biochem.* 140:183–89
57. Sadler, J. E. 1984. *Biology of Carbohydrates,* 2:199–288. New York: Wiley
58. Haselbeck, A., Tanner, W. 1983. *FEBS Lett.* 158:335–38
59. Larriba, G., Elorza, M. V., Villanueva, J. R., Sentandreu, R. 1976. *FEBS Lett.* 71:316–20
60. Novick, P., Ferro, S., Schekman, R. 1981. *Cell* 25:461–69

61. Schekman, R. 1985. *Ann. Rev. Cell Biol.* 1:115–43
62. Bostian, K., Jayachandran, S., Tipper, D. 1983. *Cell* 32:169–80
63. Gibson, R., Schlesinger, S., Kornfeld, S. 1979. *J. Biol. Chem.* 254:3600–7
64. Novick, P., Field, C., Schekman, R. 1980. *Cell* 21:205–15
65. Schauer, I., Emr, S., Gross, C., Schekman, R. 1985. *J. Cell. Biol.* 100:1664–75
66. Dunphy, W. G., Rothman, J. E. 1985. *Cell* 42:13–21
67. Stevens, T. H., Rothman, J. H., Payne, G. S., Schekman, R. 1986. *J. Cell. Biol.* 102:1551–57
68. Rothman, J. H., Hunter, C. P., Valls, L. A., Stevens, T. H. 1986. *Proc. Natl. Acad. Sci. USA* 83:3248–52
69. Tkacz, J., Lampen, J. 1972. *J. Gen. Microbiol.* 72:243–47
70. Tkacz, J., Lampen, J. 1973. *J. Bacteriol.* 113:1073–75
71. Field, C., Schekman, R. 1980. *J. Cell. Biol.* 86:123–28
72. Adams, A. E. M., Pringle, J. R. 1984. *J. Cell. Biol.* 98:934–45
73. Novick, P., Botstein, D. 1985. *Cell* 40:405–16
74. Bulawa, C. E., Robbins, P. 1986. *Yeast* 2:S47 (Abstr.)
75. Hasilik, A., Tanner, W. 1978. *Eur. J. Biochem.* 85:599–608
76. Stevens, T., Esmon, B., Schekman, R. 1982. *Cell* 30:439–48
77. von Figura, K., Hasilik, A. 1986. *Ann. Rev. Biochem.* 55:167–93
78. Pringle, J. R., Hartwell, L. H. 1982. See Ref. 54, pp. 97–142
79. Kibel, F., Huffaker, T., Tanner, W. 1984. *Exp. Cell. Res.* 150:309–13
80. Arnold, E., Tanner, W. 1982. *FEBS Lett.* 148:49–53
81. Popolo, L., Alberghina, L. 1984. *Proc. Natl. Acad. Sci. USA* 81:120–24
82. Popolo, L., Vai, M., Alberghina, L. 1986. *J. Biol. Chem.* 261:3479–82
83. Orlean, P., Ammer, M., Watzele, M., Tanner, W. 1986. *Proc. Natl. Acad. Sci. USA* 83:6263–66
84. Truehart, J., Fink, G. R. 1986. *Yeast* 2:S394 (Abstr.)
85. Julius, D., Schekman, R., Thorner, J. 1984. *Cell* 36:309–18
86. Elorza, M. V., Murgui, A., Sentandreu, R. 1985. *J. Gen. Microbiol.* 131:2209–16
87. Ogawa, T., Yamamoto, H. 1982. *Carbohydr. Res.* 104:271–83
88. Funayama, M., Nishikawa, A., Shinoda, T., Suzuki, M., Fukazawa, Y. 1984. *Microbiol. Immunol.* 28:1359–71
89. Bush, D. A., Horisberger, M., Horman, I., Wursch, P. 1974. *J. Gen. Microbiol.* 81:199–206
90. Horisberger, M., Vonlanthen, M., Rosset, J. 1978. *Arch. Microbiol.* 119:107–11
91. Douglas, R. H., Ballou, C. E. 1980. *J. Biol. Chem.* 255:5979–85
92. Yamaguchi, M., Yoshida, K., Banno, I., Yanagishima, N. 1984. *Mol. Gen. Genet.* 194:24–30
93. Pierce, M., Ballou, C. 1983. *J. Biol. Chem.* 258:3576–82
94. Lee, J. C., King, R. D. 1983. *Infect. Immun.* 41:1024–30
95. Suzuki, M., Fukazawa, Y. 1982. *Microbiol. Immunol.* 26:3878–402
96. Hishinuma, F., Nakamura, K., Hirai, K., Nishizawa, R., Gunge, N., Maeda, T. 1984. *Nucleic Acids Res.* 12:7581–97
97. Moreno, S., Ruiz, T., Sanchez, V., Villanueva, J. R., Rodriguez, L. 1985. *Arch. Microbiol.* 142:370–74
98. Bennetzen, J. L., Hall, B. D. 1982. *J. Biol. Chem.* 257:3026–31
99. Derynck, R., Singh, A., Goeddel, D. V. 1983. *Nucleic Acids Res.* 11:1819–37
100. Mellor, J., Dobson, M. J., Roberts, N. A., Kingsman, A. J., Kingsman, S. M. 1985. *Gene* 33:215–26
101. Chen, C. Y., Oppermann, H., Hitzeman, R. A. 1984. *Nucleic Acids Res.* 12:8951–70
102. Stepien, P. P., Brousseau, R., Wu, R., Narang, S., Thomas, D. Y. 1983. *Gene* 24:287–97
103. Langford, C. J., Gallwitz, D. 1983. *Cell* 33:519–27
104. Pikielny, C. W., Teem, J. L., Rosbash, M. 1983. *Cell* 34:395–403
105. Padgett, R. A., Grabowski, P. J., Konarska, M. M., Seiler, S., Sharp, P. A. 1986. *Ann. Rev. Biochem.* 55:1119–50
106. Wen, D., Schlesinger, M. J. 1986. *Proc. Natl. Acad. Sci. USA* 83:3639–43
107. Wen, D., Schlesinger, M. J. 1984. *Mol. Cell. Biol.* 4:688–94
108. Valenzuela, P., Median, A., Rutter, W. J., Ammerer, G., Hall, B. D. 1982. *Nature* 298:347–350
109. Hitzeman, R. A., Chen, C. Y., Hagie, F. E., Patzer, E. J., Liu, C. C., et al. 1983. *Nucleic Acids Res.* 11:2745–63
110. Miyanohara, A., Toh-E, A., Nozaki, C., Hamada, F., Ohtomo, N., Matsubara, K. 1983. *Proc. Natl. Acad. Sci. USA* 80:1–5
111. Hitzeman, R. A., Leung, D. W., Perry, L. J., Kohr, W. J., Levine, H. L., Goeddel, D. V. 1983. *Science* 2190:620–25

112. Egan, K. M., Koski, R. A., Jones, M., Fieschko, J. C., Bitter, G. A. 1985. See Ref. 14, p. 69
113. Meyhack, B., Hinnen, A. 1984. *Eur. Pat. Appl.* 84810564.9
114. Wood, C. R., Boss, M. A., Kenten, J. H., Calvert, J. E., Roberts, N. A., Emtage, J. S. 1985. *Nature* 314:446–49
115. Mercereau-Puijalon, O., Lacroute, F., Kourilsky, P. 1980. *Gene* 11:163–67
116. Miyajima, A., Otsu, K., Smith, C., Bond, M., Rennick, D., et al. 1985. *J. Cell. Biochem.* Suppl. 9C:156
117. Gething, M. J., Sambrook, J. 1985. *J. Cell. Biochem.* Suppl. 9C:153
118. Jabbar, M. A., Sivasubramanian, N., Nayak, D. P. 1985. *Proc. Natl. Acad. Sci. USA* 82:2019–23
119. Keranen, S., Korpela, K. 1985. *J. Cell. Biochem.* Suppl. 9C:304
120. Green, R., Schaber, M. D., Shields, D., Kramer, R. 1986. *J. Biol. Chem.* 261:7558–65
121. Smith, R. A., Duncan, M. J., Moir, D. T. 1985. *Science* 229:1219–24
122. Innis, M. A., Holland, M. J., McCabe, P. C., Cole, G. E., Wittman, V. P., et al. 1985. *Science* 228:21–26
123. Bergh, M. L. E., Hubbard, S. C., Robbins, P. W. 1986. *Proc. 13th Int. Carbohydr. Symp., Ithaca,* p. 327 (Abstr.)

Ann. Rev. Biochem. 1987. 56:945–93

P450 GENES: STRUCTURE, EVOLUTION, AND REGULATION[1]

Daniel W. Nebert

Laboratory of Developmental Pharmacology, National Institute of Child Health and Human Development, National Institutes of Health, Bethesda, Maryland 20892

Frank J. Gonzalez

Laboratory of Molecular Carcinogenesis, National Cancer Institute, National Institutes of Health, Bethesda, Maryland 20892

CONTENTS

[1]The US Government has the right to retain a nonexclusive royalty-free license in and to any copyright covering this paper.

PERSPECTIVES AND SUMMARY

Cytochromes P450 are enzymes involved in the oxidative metabolism of steroids, fatty acids, prostaglandins, leukotrienes, biogenic amines, pheromones, and plant metabolites. These enzymes also metabolize innumerable drugs, chemical carcinogens, mutagens, and other environmental contaminants. The large degree of overlapping substrate specificities, classes of inducing agents, and drug-drug interactions have caused great difficulty in P450 studies at the level of catalytic activities and protein immunochemistry. P450 enzymes represent the classical "Phase I" metabolism in which the substrate is oxygenated. "Phase II" enzymes often use the oxygen as a site for further metabolism (e.g. glucuronidation, and sulfate, glutathione, or glycine conjugation). Detoxification usually requires both Phase I and Phase II enzymes.

P450 enzymes play a large role in chemical mutagenesis and carcinogenesis. The initiation of cancer by foreign chemicals usually requires at least three steps. 1. Most compounds are inert and require metabolism to the ultimate carcinogenic form. 2. Adducts between DNA and these reactive intermediates escape the usual DNA repair processes and cause nucleotide changes and DNA rearrangements. 3. The result of the first two steps leads to oncogene activation. Without P450 gene expression and, therefore, metabolism of these foreign chemicals by the P450 monooxygenases, the two subsequent steps would often not occur.

The field of P450 molecular biology has literally exploded, and this is the subject of the present review. The first two P450 3' cDNA probes were reported in 1980, and the first P450 full-length cDNA sequence appeared in 1983. Three years later, we have access to 67 complete P450 cDNA or protein sequences in eight eukaryotic species and one prokaryote. Four comparative reviews on the molecular biology of P450 genes have appeared (1–4), but are based on a very limited number of P450 sequences. The reader is referred to other reviews on the subjects of: chemicals that induce P450 activities (5–7), relationship of P450 to cancer (8–11), genetic differences in P450 expression (12), history of P450 research (7), biophysical and biochemical characterization of the P450 monooxygenase reactions (13–15), enzymology of the steroidogenic P450s (16), P450 in yeast (17), marked diminution in P450 activity caused by immunosuppressive agents (18), and recommended nomenclature for the P450 gene superfamily (19).

In the present review we describe the gene structure and evolution of the 10 known P450 gene families, eight of which exist in mammals. Unless otherwise indicated, "percent similarity" always refers to the comparison of full-length amino acid sequences—derived in a few instances from protein sequencing but more than 90% of the time deduced from the cDNA nucleotide sequence. With these data, several conclusions about P450 gene evolution are apparent. 1. The P450 superfamily comprises at least 10 gene families. 2. The P450 superfamily is ancient and has expanded via divergent evolution. 3. The ancestral P450 gene, present probably more than 1.5 billion years ago, had a minimum of 22 exons. 4. Estimation of the unit evolutionary period (UEP; millions of years required for 1% divergence in amino acid sequence) may be difficult due to several instances of gene conversion between homologous P450 genes. 5. Compared with all of the microsomal P450 proteins, two mitochondrial P450 proteins, encoded by nuclear DNA, are more similar to the prokaryotic P450 protein.

Most drugs and combustion products are derived from plants or are similar in chemical structure to plant metabolites (phytoalexins). It is proposed that four of these P450 gene families (I, II, III, and IV) have evolved and diverged in animals due to their exposure to plant metabolites and decayed plant products during the last one billion years. The overlapping substrate specificities of the P450 enzymes are remarkable. There are numerous examples of drugs and other foreign chemicals that are good substrates for the enzymes encoded by two or more P450 gene families—gene families that have diverged so long ago as to be chromosomally nonlinked.

The final portions of this review describe what is known about P450 gene regulation. Striking differences in developmental-, sex- and tissue-specific P450 gene expression have been demonstrated by modern molecular biology techniques. Furthermore, P450 expression vectors have recently been successfully transformed into yeast and transiently transfected into mammalian cell cultures.

The most extensively studied P450 systems to date are the P_1450 gene in mouse hepatoma Hepa-1 cultures and receptor-defective and P_1450 metabolism–deficient mutant cell lines. Upstream P_1450 regulatory sequences include: 1. the TATA box; 2. a tetrachlorodibenzo-p-dioxin (TCDD)–inducible enhancer, which includes 3. an element that augments constitutive gene expression; and 4. a separate control element involved in a negative autoregulatory loop. Metabolism of substrate(s) by the product of the P_1450 gene not only controls its own constitutive expression but regulates the activities of at least two other enzymes having coordinate metabolic functions—UDP glucuronosyltransferase ($UDPGT_1$) and NAD(P)H:menadione oxidoreductase ($NMOR_1$). The P_1450, P_3450, $UDPGT_1$, and $NMOR_1$ genes are under control of the aromatic hydrocarbon (Ah) receptor and are defined as members of the [Ah] gene battery.

HISTORY

Before the end of the 19th century, the concept of "Entgiftung" (detoxification) was already appreciated (20). The correlation of certain clinical effects (e.g. hexobarbital sleeping time) on the intact animal with in vitro drug-metabolizing enzyme activities (20–22) represented a major advance in the field. Subsequently, it was found that certain of these enzyme activities could be enhanced by treatment of rats with polycyclic hydrocarbons such as benzpyrene (5, 23) and with other drugs such as phenobarbital (23–25).

Following the initial reports of a membrane-bound reduced cytochrome·CO complex having an unusual Soret peak at about 450 nm (26, 27), the isolated pigment was characterized and named "cytochrome P-450" (28). The reduced P450·CO complex was found to be photodissociable, permitting its identification as the participant in many chemical reactions such as steroid and drug hydroxylations (29). An unknown paramagnetic hemoprotein, "microsomal Fe_x," was later realized to be one and the same as cytochrome P450, and its unusual spectral properties were suggested to be attributed to a thiolate (mercaptide) cysteinyl sulfur atom bound to the heme iron atom in the fifth coordination site (30). This suggestion has since been confirmed by numerous spectroscopic analyses of purified mammalian and bacterial P450 proteins (4, 13–15).

In liver microsomes from phenobarbital-treated rats, correlations between the progressive rise and fall in P450 content followed spectrophotometrically, and the parallel changes in certain drug-metabolizing activities from these same microsomes (31, 32) provided the important link between inducible biotransformation enzymes and the cytochrome with unusual spectral properties. By 1967 more than 300 substances had been described as having the capacity to induce their own metabolism or the metabolism of chemically related foreign and endogenous compounds (5).

The fact that actinomycin D and cycloheximide blocked P450 induction in cell culture (33, 34) suggested a role for new mRNA and protein synthesis during the induction process. Use of radiolabeled amino acids in phenobarbi-tal- and 3-methylcholanthrene-treated rabbits (35) and α-amanitin (an in-hibitor of RNA polymerase II) in phenobarbital-treated rats (36) led to similar conclusions. In tissue culture and particularly in the intact animal, however, it should be kept in mind that these inhibitors exhibit many broad, nonspecific, and toxic effects.

The purification of multiple P450 proteins (4, 37–41) resulted in polyclonal or monoclonal antibody studies in which the rise and fall of immunoprecipita-ble radioactivity could be correlated with changes in certain drug-metabolizing enzyme activities. Still, measurements of a particular P450 protein were not that much more specific than enzyme assays, with regard to understanding P450 gene expression. With the recent advent of DNA cloning techniques, the regulation of certain P450 genes is now much better un-derstood.

ACTIVITY OF GENE PRODUCTS

Monooxygenases are capable of carrying out a myriad of chemical reactions (Figure 1). The common thread of all oxidative reactions is the insertion of one atom of atmospheric oxygen into the substrate, often producing a highly unstable intermediate, which breaks down to the final product(s). The differ-ent molecular sizes of P450 substrates are quite remarkable: from chromate, carbon disulfide, and ethanol to steroids and 5-ring polycyclic aromatic hydrocarbons. Several of the more unusual P450 reactions include the reduc-tion of chromate (42) and aqueous sulfur dioxide (43), nitrate formation from 2-nitropropane (44), and the (oxygen-sensitive) reduction of tertiary amine N-oxides (45), epoxy derivatives of benzo[a]pyrene (46), and N-hydroxy-2-acetylaminofluorene (47).

The catalytic activity of mammalian P450 drug-metabolizing enzymes is located in the endoplasmic reticulum (4, 12). The multicomponent mem-brane-bound chain (Figure 2) receives electrons from NADPH or NADH; the reducing equivalents are then passed, by way of a flavoprotein oxidoreduc-tase, either via cytochrome b_5 or directly to clusters of P450 proteins. In combination with atmospheric oxygen and any one of innumerable substrates, the P450 forms a trimolecular complex and the substrate is oxygenated (4, 13). Depending on the rate of formation of reactive intermediates, the pres-ence of nearby Phase II enzymes (e.g. epoxide hydrolase, UDP glucuronosyl-transferase, glutathione transferase, sulfotransferase), and the inherent chem-ical stability of the oxygenated intermediate, two possible pathways exist. First, there can be further metabolism (e.g. hydration, or conjugation with glucuronic acid, glutathione, or sulfate) and excretion of innocuous highly polar products from the cell. Second, covalent binding of the oxygenated

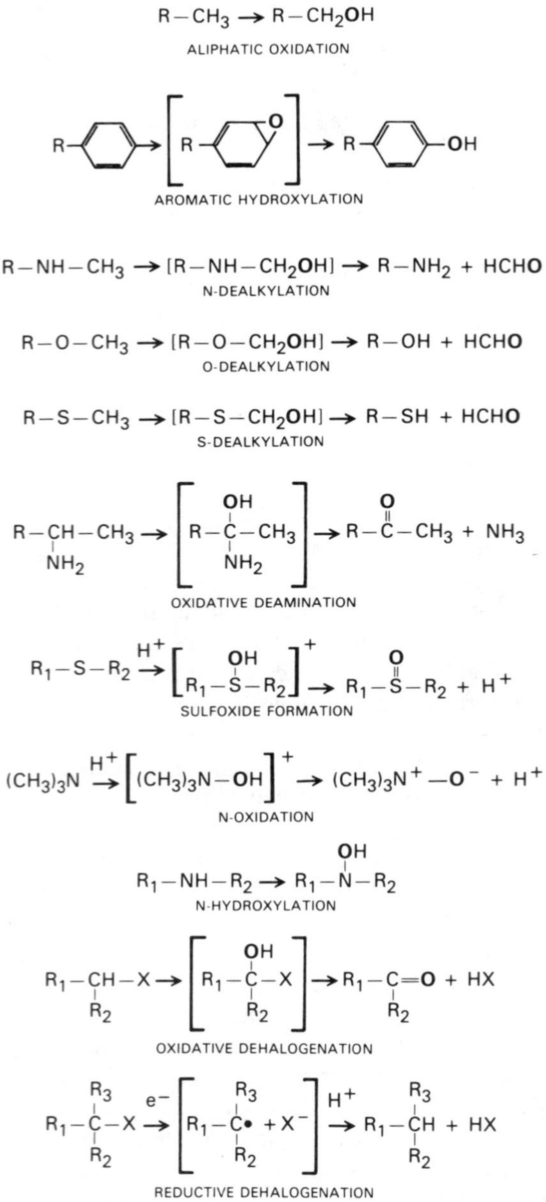

$$R-CH_3 \rightarrow R-CH_2OH$$

ALIPHATIC OXIDATION

AROMATIC HYDROXYLATION

$$R-NH-CH_3 \rightarrow [R-NH-CH_2OH] \rightarrow R-NH_2 + HCHO$$

N-DEALKYLATION

$$R-O-CH_3 \rightarrow [R-O-CH_2OH] \rightarrow R-OH + HCHO$$

O-DEALKYLATION

$$R-S-CH_3 \rightarrow [R-S-CH_2OH] \rightarrow R-SH + HCHO$$

S-DEALKYLATION

$$R-\underset{\underset{NH_2}{|}}{CH}-CH_3 \rightarrow \left[R-\underset{\underset{NH_2}{|}}{\overset{\overset{OH}{|}}{C}}-CH_3\right] \rightarrow R-\overset{\overset{O}{\|}}{C}-CH_3 + NH_3$$

OXIDATIVE DEAMINATION

$$R_1-S-R_2 \overset{H^+}{\rightarrow} \left[R_1-\overset{\overset{OH}{|}}{S}-R_2\right]^+ \rightarrow R_1-\overset{\overset{O}{\|}}{S}-R_2 + H^+$$

SULFOXIDE FORMATION

$$(CH_3)_3N \overset{H^+}{\rightarrow} \left[(CH_3)_3N-OH\right]^+ \rightarrow (CH_3)_3N^+ -O^- + H^+$$

N-OXIDATION

$$R_1-NH-R_2 \rightarrow R_1-\overset{\overset{OH}{|}}{N}-R_2$$

N-HYDROXYLATION

$$R_1-\underset{\underset{R_2}{|}}{CH}-X \rightarrow \left[R_1-\underset{\underset{R_2}{|}}{\overset{\overset{OH}{|}}{C}}-X\right] \rightarrow R_1-\underset{\underset{R_2}{|}}{C}=O + HX$$

OXIDATIVE DEHALOGENATION

$$R_1-\underset{\underset{R_2}{|}}{\overset{\overset{R_3}{|}}{C}}-X \overset{e^-}{\rightarrow} \left[R_1-\underset{\underset{R_2}{|}}{\overset{\overset{R_3}{|}}{C}}\bullet +X^-\right] \overset{H^+}{\rightarrow} R_1-\underset{\underset{R_2}{|}}{\overset{\overset{R_3}{|}}{C}}H + HX$$

REDUCTIVE DEHALOGENATION

Figure 1   Examples of the many diverse monooxygenase activities of the P450 enzymes. The oxygen derived from atmospheric oxygen is denoted by bold print.

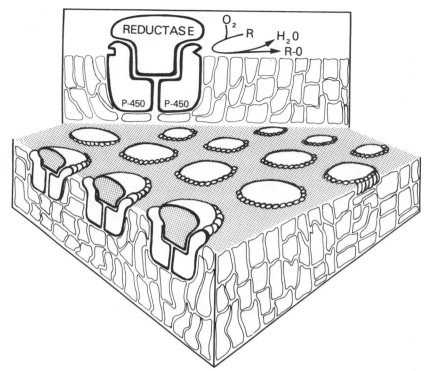

Figure 2   Three-dimensional concept of the monooxygenase system in the endoplasmic reticulum. R, substrate [Reproduced from Ref. 48 with permission from Dr. W. Junk Publishers].

intermediate to proteins [including the P450 protein itself (39, 40)] and nucleic acids can occur; this pathway has been shown in many laboratories to be correlated with chemical carcinogenesis, mutagenesis, drug toxicity, and teratogenesis (8–12).

Various mammalian microsomal preparations have been solubilized with detergents, following which P450 catalytic activity can be reconstituted by means of combining flavoprotein, lipid, and the P450 hemoprotein (49, 50). In such reconstitution experiments, genetic differences in P450 catalytic activity have been shown to be associated only with the P450 protein (51).

More than three fourths of the liver microsomal NADPH-P450 oxidoreductase molecule (Figure 2) is located free of the lipid bilayer (52), whereas P450 protein clusters appear to be more deeply embedded in the membrane—thereby making detergent solubilization of single P450 proteins extremely difficult. The stoichiometry of NADPH-P450 oxidoreductase to P450 molecules in mammalian liver microsomes ranges between 1:10 and 1:100 (50, 53).

For many of the numerous trivial names for the same P450 protein, the reader is referred to several recent reviews and books (1, 2, 4, 10, 19, 39, 40). In the proposal for standardized P450 gene nomenclature (19), there has been an attempt to use Roman numerals and capital letters in a manner that facilitates the matching of previously characterized P450 proteins with the newly defined gene families and subfamilies.

P450 PROTEIN SEQUENCE COMPARISONS AND EVOLUTION

Several conclusions about the P450 gene superfamily can be made (Figure 3 and Table 1). First, the amino acid sequence of a protein from any of the 10 families is ≤36% similar to that from any of the other nine families (19). This definition of gene families, based on percent similarity, is somewhat arbitrary and has been arrived at only after analysis of more than five dozen P450 protein sequences. Those P450 genes considered to be in the same subfamily have ≥68% similarity to other genes in the same subfamily. More than one subfamily in the same family has been found within the P450II and P450XI gene families and represents intermediate divergence: the amino acid sequence of a protein in any one subfamily is about 40% to 65% similar to that in any of the other subfamilies.

Second, because the system is present in all eukaryotes examined and at least some prokaryotes, the superfamily must be an ancient one, probably existing for more than 1.5 billion years. Third, this superfamily represents an example of divergent evolution. Even among proteins with no significant (≤25%) global amino acid alignment, there is sufficient localized similarity (>33%) for those proteins to be considered as gene products of the same superfamily. Figure 4 shows that among all eukaryotic species examined and *Pseudomonas* P450cam, there is a 21-residue cysteinyl fragment associated with the heme iron-binding pocket. It is thus very likely that eukaryotic and prokaryotic genes have evolved from a common ancestor (54). Another possibility is parallel evolution from many origins restricted only by structure-function relationships; before 1.5 billion years ago, for example, there might have existed many primordial oligomers specifying P450-like genes. Based on available data from the P450I, II, and XXI gene families, an ancestral gene most probably had at least 22 exons divided by 21 introns (Figure 5); this conclusion is based on the premise that the introns have not slid more than two amino acids and that intron loss, and not intron addition, occurs during evolution (55).

Fourth, the unit evolutionary period (UEP; millions of years needed for divergence of 1% in amino acid sequence) may not be linear for the P450

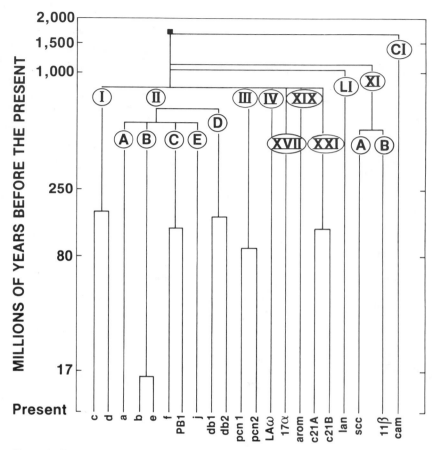

Figure 3 Evolution of the P450 gene family. Data are primarily based on rat for families I through IV, and cow and human for families XI, XVII, XIX, and XXI. The data were compiled from 67 P450 full-length cDNA or protein sequences available as of the end of January, 1987 (19). Whereas all the aforementioned families exist in mammals, the LI family exists in yeast and the CI family is present in *Pseudomonas*.

genes: a UEP value of 4.0 is estimated between the present and the mammalian radiation about 80 million years ago, a UEP value of 9.0 is estimated between 80 million and 400 million years ago, and a UEP value of 14 is estimated between 400 million and 1000 million years ago (19). These estimates are based on single nucleotide exchanges; applying the most commonly used formula, which takes into account the correction for multiple hits during evolution (56), the calculated UEP values are 3.3, 5.7, and 5.0, respectively, i.e. somewhat closer to linearity. At least three examples of

Table 1 P450 gene superfamily[a]

Family, subfamily, and gene designation	Some of the existing names in the literature
P450I (polycyclic aromatic compound–inducible)	
Only one subfamily	
P450IA1	Rat c, rabbit form 6, mouse P_1, human P_1
P450IA2	Rat d, rabbit form 4, mouse P_3, human P_3
P450II (major)	
P450IIA subfamily	
P450IIA1	Rat a
P450IIA2	Human P450(1)
P450IIB subfamily (phenobarbital-inducible)	
P450IIB1	Rat b, rabbit form 2
P450IIB2	Rat e
P450IIC subfamily	
P450IIC1	Rabbit PBc1
P450IIC2	Rabbit PBc2, K
P450IIC3	Rabbit PBc3, form 3b
P450IIC4	Rabbit PBc4, form 1-8
P450IIC5	Rabbit form 1
P450IIC6	Rat PB1
P450IIC7	Rat f
P450IIC8	Human form 1
P450IIC9	Human mp
P450IIC10	Chicken PB15
P450IIC11	Rat M-1, h
P450IID subfamily	
P450IID1	Rat db1, human db1
P450IID2	Rat db2
P450IIE subfamily (ethanol-inducible)	
P450IIE1	Rat j, rabbit form 3a, human j
P450III (steroid-inducible)	
Only one subfamily	
P450IIIA1	Rat pcn1
P450IIIA2	Rat pcn2
P450IV (peroxisome proliferator–inducible)	
Only one subfamily	
P450IVA1	Rat LAω
P450XI (mitochondrial proteins)	
P450XIA subfamily	
P450XIA1	Bovine (and human) scc
P450XIB subfamily	
P450XIB1	Bovine (and human) 11β
P450XVII (steroid 17α-hydroxylase)	
Only one subfamily	
P450XVIIA1	Bovine (and human) 17α

Table 1 *(Continued)*

Family, subfamily, and gene designation	Some of the existing names in the literature
P450XIX	
Only one subfamily	
P450XIXA1	Human arom
P450XXI (steroid 21-hydroxylase)	
Only one subfamily	
P450XXIA1	Bovine (and mouse, human) C21A
P450XXIA2	Bovine (and mouse, human) C21B
P450LI (plant P450)	
Only one subfamily	
P450LIA1	Yeast lan
P450CI (prokaryote P450)	
Only one subfamily	
P450CIA1	*Pseudomonas putida* cam

[a] The sequence data are summarized and cited in Ref. 19.

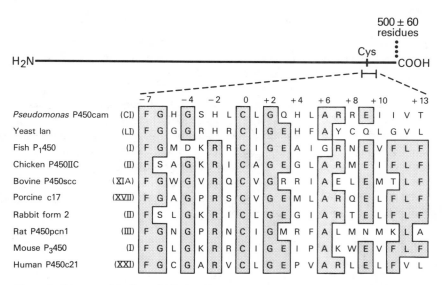

Figure 4 Diagram of the linear P450 protein and approximate location of the highly conserved cysteinyl-containing peptide involved in the heme-binding site among nine eukaryotic species and one prokaryote. Positive and negative numbers refer to amino acid positions downstream and upstream, respectively, from the cysteine that binds the heme iron. These sequences are cited in Ref. 19.

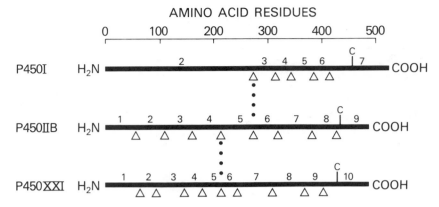

Figure 5   Exon-intron junctions (triangular arrows) of P450 proteins in the I family, the IIB subfamily, and the XXI family. The cysteine (C) in the enzyme active-site is annotated in the COOH-terminal exon 7, exon 9, and exon 10, respectively. It should be noted that the P450I gene, but not the others, has a noncoding first exon. It has been determined that two genes in the rabbit IIC subfamily (111) and rat IID and IIE subfamilies (F. J. Gonzalez, unpublished) have exon-intron junctions identical to those of the rat IIB subfamily. Except for two cases (illustrated by vertical dots) of exon-intron junctions within two residues of one another, all remaining junctions in the three families appear to be unique.

gene conversion between homologous P450 genes in the same subfamily (IA, IIB, and IIIA) have been proposed (57–59). Gene conversion could also contribute to an underestimate of the UEP value.

Lastly, all available linkage data (Table 2) are consistent with the above-mentioned UEP values. The eight mammalian P450 gene families (Figure 3) diverged from one another between 800 and 1100 million years ago. The P450II and P450XI gene families subsequently diverged into at least five and two subfamilies, respectively (Figure 3), between 400 and 600 million years ago, sufficiently long ago so as to result in nonlinked genes representing each subfamily. Except for the two homologous genes in the IA, IID, and XXIA subfamilies (Table 1), it is not yet known but suggested that all genes within any given subfamily will be closely linked.

THE P450I GENE FAMILY

The P450I family in rat, rabbit, mouse, and human has two genes that are inducible by polycyclic hydrocarbons and TCDD (Table 1). Following the initial characterization of polyclonal antibodies to $P_1 450$ and $P_3 450$ from 3-methylcholanthrene-treated C57BL/6N mouse liver (69), 3' specific cDNA clones for P_1 (70, 71) and P_3 (72) were characterized. Full-length cDNA and genomic clones for P_1 and P_3 were subsequently sequenced (73–75). Use of a

Table 2 Chromosomal linkage data for the P450 gene families and subfamilies

P450 gene subfamily	Human			Mouse		
	Locus	Chromosome location	Ref.	Locus	Chromosome location	Ref.
P450IA	*P450C1*	15q22-qter (near *MPI*)	59,60	*P450-1*	Mid-9 (near *Mpi-1*)	61
P450IIA	*P450C2A*	19q13.1–13.2	62			
P450IIB				*P450-2B*	Proximal 7 (*Coh*)	63
P450IIC	*P450C2C*	10	a	*P450-2C*	19	b
P450IID	*P450C2D*	22q11.2-qter	c	*P450-2D*	15	63a
P450IIE	*P450C2E*	10	c	*P450-2E*	7	d
P450IIIA	*P450C3*	7p	c	*P450-3*	6	64
P450XIA	*P450C11A*	15	65,154			
P450XVIIA	*P450C17*	10	66			
P450XXIA	*P450C21*	6 (within *HLA*)	67	*P450-21*	17 (within *H-2*)	68

[a] F. Peter Guengerich, personal communication.
[b] Masa Negishi, personal communication.
[c] Frank J. Gonzalez, O. Wesley McBride, personal communication.
[d] Frank J. Gonzalez, Christine A. Kozak, personal communication.

polyclonal antibody to P_2450 from isosafrole-treated DBA/2N mouse liver (76) led to the isolation and sequencing of the P_2 cDNA; the gene was found to be an allelic variant of the C57BL/6N mouse P_3 gene and differs in only one amino acid residue, plus two nucleotide changes in the 3' nontranslated region (77).

The orthologues of mouse P_1450 and P_3450 are P450c and P450d, respectively, in the rat and forms 6 and 4, respectively, in the rabbit (Table 1). The full-length cDNAs for c (78) and d (79) and the corresponding genes and flanking regions (80, 81) have been sequenced. Although the sequence of the rabbit form 4 protein has appeared (82), only partial cDNA sequences for forms 6 and 4 have been reported to date (83).

The human P_1 full-length cDNA (84) and complete gene and flanking regions (85) have been isolated and sequenced. An increased risk of cigarette smoking–induced bronchogenic carcinoma appears to be associated with the high P_1 inducibility phenotype (11, 86). Although restriction fragment length polymorphisms (RFLPs) have been detected with the human P_1 cDNA, no correlation between enhanced lung cancer risk and any of these RFLP patterns has been found to date (85, 87). The phenotype of P_1 enzyme inducibility can be correlated with the P_1 mRNA content of lymphocytes cultured with mitogen and polycyclic hydrocarbon inducer (85). Interestingly, Epstein-Barr virus–transformed B cell lines from individual patients (88) do not retain their P_1 inducibility phenotype.

The human P_3 full-length cDNA (59, 89) and complete gene and flanking regions (A. K. Jaiswal, K. Ikeya, F. J. Gonzalez, D. W. Nebert, unpublished; 89a) have also been sequenced. Since the P_3 gene is highly correlated with cancer induced by 2-acetylaminofluorene and aminobiphenyls in laboratory animals (90), the human orthologue will also be of great interest in clinical cancer genetics studies.

THE P450II GENE FAMILY

Initial studies with genes in the P450II family were carried out in phenobarbital-treated animals. Whereas this group of genes was originally called the "phenobarbital-inducible family," it now appears that most phenobarbital-inducible genes reside within the IIB and IIC subfamilies.

IIA Subfamily

Rat P450a is highly specific for testosterone 7α-hydroxylase activity (91). The full-length cDNA from nontreated rat liver has been isolated and sequenced and encodes a protein with 52% similarity to rat P450e (91a). Of interest, the rat mRNA is markedly induced by 3-methylcholanthrene but not phenobarbital. When rat e cDNA was used for screening a human liver cDNA library (92), the partial cDNA sequence of the gene called P450(1) was obtained and presumed to be a member of the IIB subfamily. Further analysis shows that this gene belongs to the IIA subfamily (19). The amino acid sequence of P450(1) (331 residues only) is 60% and 51% similar to rat a and e, respectively. All human-rodent orthologues to date exhibit between 71% and 80% similarity in amino acid sequence (19); hence, P450(1) is in the subfamily IIA but is not the orthologue of rat a.

IIB Subfamily

Following the initial characterization of polyclonal antibodies to rat liver P450b and P450e (93), a 3' cDNA clone hybridizing to phenobarbital-induced mRNA was isolated (94, 95), and two overlapping b cDNA clones and a 5' half of the e cDNA, plus some genomic DNA, were sequenced (96). The complex nature of the P450II gene family was first noted by Adesnik and coworkers (97, 98), whose data suggested a minimum of six genes. When the b and e exon sequences were compared (99), only about 40 base substitutions were found over approximately 1.9 kb, resulting in 14 amino acid differences (97% similarity). It is certain that two genes exist because the two gene products (100), as well as both genes (99), have been well characterized. Applying the UEP value of 4.0 (see above), we estimate that the b and e genes diverged no more than 12 million years ago (Figure 3). This conclusion would suggest that the mouse, which has separated from rat about 17 million years ago, would not have orthologues of both b and e. At least one phenobarbital-inducible gene has been partially characterized in the mouse (101). Several other studies of less than full-length cDNA or protein sequences of (presumably) rat e have reported four- to seven-amino-acid differences in the 491-residue protein (96–109, 102–104).

In the rabbit, form 2 is the major phenobarbital-inducible P450 (Table 1). Two independent laboratories have established the amino acid sequence of the protein and found a difference in 11 amino acids out of a total of 491 residues (105–107). This amount of difference probably represents a very extensive polymorphism among rabbits that are not highly inbred. Whether rabbit form 2 is the orthologue of rat b or e has not been established.

Two chicken genes have recently been isolated (108) and one (clone PB15; gene A) has been sequenced and shown to be induced by phenobarbital, allylisopropylacetamide, and 3,5-diethoxycarbonyl-1,4-dihydrocollidine (109). It appears that both genes (A and B) are transcriptionally active, producing mRNAs of 3.2 and 2.5 kb, respectively (B. K. May, personal communication). Of interest, the percent amino acid resemblance between chicken gene A and rat P450II genes is significantly higher for the IIC subfamily (57%) than for the IIB subfamily (49%).

IIC Subfamily

Although most of the genes in this subfamily are constitutively expressed, several appear to be phenobarbital-inducible (110). The nearly complete cDNA sequences of four closely related clones from rabbit liver—PBc1, PBc2, PBc3, and PBc4 (110)—have now been completed (111, 111a; B. Kemper, personal communication). PBc2, also called P450K, has been shown to be responsible for renal ω-1 hydroxylation of lauric acid in turkey kidney (111b). The PBc3 and PBc4 genes were found to be identical with rabbit form 3b (107) and form 1-8 (112), respectively. Rabbit form 1, particularly high for progesterone 21–hydroxylase (113) and benzpyrene hydroxylase (114) activities, has been sequenced (115) and represents a fifth member of the IIC subfamily. Rabbit genes 1 and 1-8 have diverged long after the rabbit-rodent and rabbit-human speciations; therefore, neither the rodent nor the human would be expected to have orthologues to both of these genes (19). Rat full-length cDNAs for PB1 and f have been sequenced (116, 117). Rat PB1 is probably the orthologue of the rabbit genes 1 and 1-8 (19). A P450 gene termed M-1 or P450h, expressed predominantly in the male rat, has recently been sequenced (117a) and found also to reside in the IIC subfamily. Full-length cDNAs have been sequenced for human form 1 (118) and P450mp (F. P. Guengerich, personal communication), responsible for mephenytoin 4-hydroxylation (119). It is presently impossible to determine which laboratory animal P450IIC genes represent the orthologues of human form 1 or mp

(19). Hence, the IIC subfamily is complicated because of its large number of genes.

IID Subfamily

A human debrisoquine 4-hydroxylase polymorphism has been described (120–124) in which the "extensive metabolizer" (EM phenotype) is able to break down the antihypertensive drug 10 to 200 times better than the "poor metabolizer" (PM phenotype). The metabolisms of more than one dozen other drugs appear to be related to this same polymorphism (11, 121). Ethnic variations in the incidence of the PM phenotype vary from nil in Japan and Finland to about 15% in Nigeria (125), whereas the British population, in which the trait was first discovered (120), has about 8% poor metabolizers. The nucleotide sequences of two rat full-length cDNAs, db1 and db2 (Table 1), have recently been determined, and their amino acid sequences are 73% similar (63a). The purified db1 protein has high debrisoquine 4–hydroxylase and bufuralol 1'-hydroxylase activities, whereas the catalytic specificity of db2 is unknown. The antibody to rat db1 can be used to quantitate immunochemically the human db1 enzyme, and a high correlation of immunoprecipitable protein and bufuralol 1'-hydroxylase activity was found among 29 human liver samples (125a). When cDNAs from EM and PM individuals were sequenced, three variants representing aberrant splicing defects were identified: a = retention of intron 5; b = retention of intron 6; b' = loss of first three exons plus the 3' half of exon 6. These splicing errors led to immunologically undetectable, or very low amounts of, human db1 protein (125a). Among a Nigerian population with cancer of the liver and gastrointestinal tract (126) and in a study of 245 cigarette smokers with bronchogenic carcinoma (127), it has been suggested that a disproportionately greater number of the EM phenotype exhibit these malignancies.

IIE Subfamily

At least two P450 proteins appear to be induced by ethanol, imidazole, acetone, trichloroethylene, and pyrazole, and the rank order of potency among this class of inducers varies markedly among laboratory animal species (128–131). The P450 activity exhibiting the greatest induction includes the oxidation of aniline, alcohols (132), and nitrosamines (133). The rat and human P450j full-length cDNAs have been isolated and sequenced (134). Due to the possible correlation of both P450IID and P450IIE genes with malignancy (126, 127, 133), it is anticipated that there will soon be clinical studies attempting to correlate RFLP patterns of these genes with individual risk of cancer (11).

THE P450III GENE FAMILY

The prototype steroid inducer pregnenolone 16α-carbonitrile (PCN) was found to increase total liver microsomal P450 content (135), and a novel P450 protein was purified from PCN-treated rats (136). A cDNA clone was isolated from PCN-treated rats (137); the sequence of rat pcn1 (138) revealed a protein with 88% similarity to that encoded by rat pcn2 (58). The first 17 NH_2-terminal residues of rat pcn1 and pcn2 are identical (58). Among the best drug-metabolizing activities carried out by gene products in this steroid-inducible family are ethylmorphine N-demethylase (136) and testosterone 6β-hydroxylase (58). Interestingly, phenobarbital induces rat pcn2 and pcn1, whereas PCN induces the pcn1 gene but not the pcn2 gene (58).

One human cDNA representing an orthologue of rat pcn1 or pcn2 has been isolated and sequenced in two laboratories. The same gene has been called HLp (139) and P450nf, for nifedipine oxidation (140, 140a). The amino acid sequences of HLp and nf are 97% similar, and they are believed to be allelic variants of the same gene (19). One or more gene conversion events might have made rat pcn1 and pcn2 irregularly homologous (see below). Taking these 100% similarity regions into account, and using a UEP value of 4.0, therefore, we estimate that pcn1 and pcn2 diverged from one another at least 90 million years ago (Figure 3) and that there is a good possibility that both genes will be found in the human.

Macrolide antibiotics, such as triacetyloleandromycin and griseofulvin, appear to induce at least one gene of the P450III family in the rat, P450p (141), and rabbit, form 3c (142). Whether P450p (141) represents rat pcn1 or pcn2 (58) is unknown. Clinical studies of the human orthologues of these P450III genes may be important for predicting macrolide antibiotic response or toxicity caused by such drugs.

THE P450IV GENE FAMILY

The hypolipidemic peroxisome proliferators (e.g. clofibrate) form a novel class of hepatocellular tumor promoters in rodents (143) and induce a protein (P450LAω) having high specificity toward lauric acid ω-hydroxylation (144). The induced P450 protein exhibits a complete lack of immunocrossreactivity with P450s induced by polycyclic hydrocarbons or phenobarbital (145). A P450LAω full-length cDNA has been characterized; the amino acid sequence is $\leq33\%$ similar to that encoded by genes in any of the other seven mammalian P450 gene families (146).

P450 GENE FAMILIES INVOLVED IN STEROIDOGENESIS

There are several forms of congenital adrenal hyperplasia described in which each of the four P450 enzymes involved in adrenal steroidogenesis (Figure 6) is defective (147). The 21-hydroxylase deficiency can be manifest as either a simple virilizing form or a life-threatening salt-wasting form (148). The incidence of 21-hydroxylase deficiency is far greater than the incidence of the other three enzyme deficiencies combined. These rarer forms of congenital adrenal hyperplasia—20,22-desmolase (cholesterol side-chain cleavage; scc) deficiency, 17α-hydroxylase deficiency, and 11β-hydroxylase deficiency— can also be serious and sometimes lethal (147). The reader is referred to Ref. 16 for a detailed review of the history and basic mechanisms of these P450-mediated steroid hydroxylases. It should be noted that P450scc and 11β-hydroxylase are mitochondrial proteins, whereas the 21-hydroxylase and 17α-hydroxylase are microsomal enzymes, and all four are encoded by nuclear genes.

P450XI Family

The sequence of a full-length cDNA for bovine P450scc (149) encodes a protein of 520 amino acids, of which a 39-residue NH_2-terminal peptide is removed posttranslationally to form the mature P450 protein of 481 amino acids. Bovine 11β-hydroxylase and scc proteins have both been shown to undergo this type of posttranslational processing (150). Furthermore, the percent similarity of bovine scc and the prokaryotic P450cam (149) is significantly greater than that of scc and any of the mammalian microsomal P450 proteins (Figure 3). Both of these observations—posttranslational processing and greater resemblance to prokaryotic proteins—are commonly seen with mitochondrial proteins.

Miller and coworkers (151) used a synthetic 72-nucleotide probe identical to a portion of the published scc sequence (149) to isolate a 3' cDNA for the human scc (152). This clone is being used to diagnose patients with congenital adrenal hyperplasia of the 20,22-desmolase-deficiency type (153) and was helpful in sequencing the full-length cDNA for human scc (154). The scc gene is a member of the P450XIA subfamily.

A partial 3' cDNA for bovine adrenocortical 11β-hydroxylase (155) has been used to isolate a partial 3' cDNA encoding human 11β-hydroxylase (156). Southern blot analysis indicated a single copy per haploid genome; the bovine and human probes hybridize to a single mRNA (4.3 kb) from bovine and human adrenal glands (156). The amino acid percent similarity of the 11β-hydroxylase has recently provided the evidence needed for assignment of the corresponding gene to the P450XIB subfamily, while the scc gene is

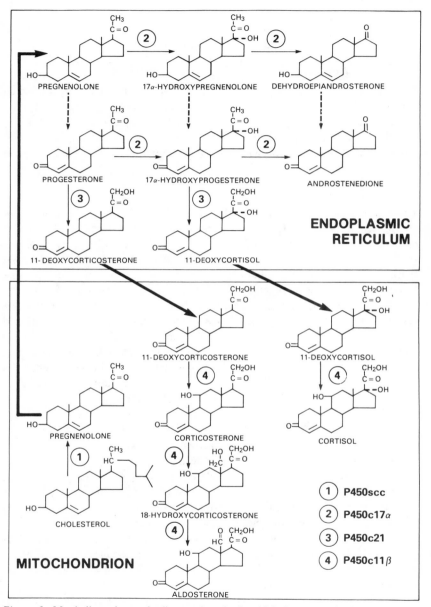

Figure 6 Metabolic pathways leading to the mitochondrial *(bottom)* and microsomal *(top)* biosynthesis of sex steroid and mineralocorticoid hormones and the key steps involving four P450 monooxygenases. The involvement of 11β-hydroxylase in the two steps between corticosterone and aldosterone has recently been established (146a). Estrogens and testosterone are formed by the further metabolism of androstene-3,17-dione. scc, cholesterol side-chain cleavage.

assigned to the P450XIA subfamily (19); both subfamilies appear to comprise single genes (T. Omura, personal communication).

P450XVII Family

The sequence of a full-length cDNA for bovine 17α-hydroxylase confirmed that it is a member of a unique P450 gene family (157). The human 17α-hydroxylase full-length cDNA has been cloned and sequenced, with the use of a porcine cDNA fragment identified via mixed-sequence oligonucleotide probes constructed from knowledge of the porcine nearly full-length protein sequence (158).

P450XIX Family

The "aromatase system" is involved in the synthesis of estrogens from androgens. A polyclonal antibody to the human aromatase P450 (159) was used to clone the 3' portion of the P450arom cDNA (160). This cDNA is sufficiently long to conclude that the arom gene represents its own P450 gene family (E. R. Simpson, personal communication). Because the enzymic activity represents aromatization of the steroid A ring, which includes oxidation and loss of carbon-19, the P450 family has been named XIX (19). The cDNA probe hybridizes to several sizes of mRNA in a number of cells and tissues including ovarian granulosa cells, testicular Sertoli cells, adipose cells, placental syncytiotrophoblasts, hypothalamic cells, and blastocysts (160). The arom cDNA sequence has shown that the cysteinyl fragment T-26 (161) is not the heme-binding peptide and that the cysteinyl fragment T-7 (161) represents the COOH-terminus of the P450arom protein.

P450XXI Family

A partial cDNA for bovine 21-hydroxylase (162) was used to characterize DNA from normal and 21-hydroxylase-deficient patients (67). From detailed segregation analysis, the 21-hydroxylase gene was localized in the HLA region very near the human complement C4A gene (67, 163). With the use of human cosmid clones from the HLA major histocompatibility complex, two 21-hydroxylase genes were found, each located near the 3' end of one of the two C4 genes (164). A tandem arrangement in the human (5'-C4A-C21A-C4B-C21B-3') is similar to that in the mouse, which has Slp (sex-limited protein) in place of the C4A gene (68, 165, 166). Two copies of the C21-hydroxylase genes have also been found in the cow (167). With deletion mutants from patients with various types of 21-hydroxylase deficiency, it was concluded that the human C21B gene is functional, but the C21A gene is not (164, 168). The murine C21A gene, introduced on a cosmid into Y1 adrenocortical tumor cells, is active, while the C21B gene is not (169). One or several small deletions, mostly in exon 3 of the human C21A gene (170, 171), and a deletion of all of exon 2 of the mouse C21B gene (172), account for the nonfunctional gene of the pair of 21-hydroxylase genes in either species. Of interest, the upstream regulatory sequences of the mouse C21B "inactive" gene are functional (172). Thus, the human, mouse, and cow each have a pair of C21 genes, and it is known that one of the two is inactive in the human and mouse. Although a functional bovine C21 gene has been sequenced (173, 174), it remains to be determined whether the second bovine C21 gene is inactive. That there is one active and one inactive C21 gene in close proximity of one another, and that the regulatory region of the inactive gene is functional (172), suggest a possible important role in regulation or other evolutionary survival advantage.

RFLP patterns with the human C21A gene have been characterized (175). Gene rearrangements detected by such a probe may be relevant to a variety of HLA-associated diseases, e.g. juvenile-onset diabetes mellitus (176).

OTHER MAMMALIAN P450 GENES

It is not known how many other unique P450 gene families and subfamilies will be uncovered. Hepatic microsomal cholesterol 7α-hydroxylase (177), renal mitochondrial 25-hydroxy-vitamin D_3 1α-hydroxylase (178), an unusual form of 3-methylcholanthrene-induced "cytochrome P-448" from rat liver (179), and platelet thromboxane synthase (179a) are examples of P450 genes yet to be cloned.

NONMAMMALIAN P450 GENES

The cDNA sequences of fish P_1450 (L. J. Heilmann, D. W. Nebert, unpublished), chicken phenobarbital-induced P450 (108, 109), and yeast P450 (180; J. C. Loper, personal communication) are the only three nonmammalian eukaryotic P450 genes characterized to date. Sequencing of a plant P450 cDNA is under way (M. A. Schuler, personal communication).

P450cam from *Pseudomonas putida* is induced by, and in turn metabolizes, camphor. The deduced protein from the recent cDNA sequence (181) has a two-residue insertion plus three additional amino acid changes, compared with the original protein sequence (182). Reconstitution experiments indicate that prokaryotic P450 metabolism requires three components: the flavoprotein NADH-P450 oxidoreductase, the nonheme iron-sulfur moiety putidaredoxin, and the P450 protein (183, 184). Interestingly, the mitochondrial P450 catalytic activity requires three components: the flavoprotein oxidoreductase, the iron-sulfur moiety adrenodoxin, and the P450 protein (16). Because P450cam is soluble rather than membrane-bound, purification of sufficient quantities for studying the crystal structure has led to a detailed understanding of the substrate-free and substrate-bound forms of this bacterial enzyme (185, 186).

CRITERIA FOR A SINGLE GENE PRODUCT

If one examines the available data regarding the different P450 gene families, it is evident that within a particular family—especially the P450II and P450III families—the enzymes are sometimes highly similar. For example, rat b and e are 97% similar (99), rat db1 and db2 are 73% similar (63a), and rat pcn1 and pcn2 are 88% similar (58). These degrees of resemblance should be taken into consideration when new P450 cDNAs are isolated or when antibodies—even antibodies that react with single bands on Western blots—are used for studying correlations with certain catalytic activities. For example, P450pcn1 and pcn2 are both present in phenobarbital- and PCN-treated rat liver microsomes, both proteins comigrate on sodium dodecyl sulfate-polyacrylamide gels, and antibodies against pcn1 inhibit pcn2 catalytic activity (58). The pcn2 gene was only identified via cDNA cloning and sequencing (58).

Previous studies on microsomal P450-mediated debrisoquine 4-hydroxylase (122–124) never indicated the possible presence of another closely related P450 protein in the P450IID subfamily. Both chromatographic and cDNA sequence data have clearly established that the db1 protein exists in the presence of a second homologous protein in the same subfamily (63a). These proteins can only be resolved on sodium dodecyl sulfate–polyacrylamide gel electrophoresis under unusual running conditions. Moreover, antibodies against the db1 and db2 proteins crossreact with each other, an antibody against db2 inhibits db1 enzymic activity, and the purified proteins comigrate on conventional gels (63a).

In summary, any report in which a P450 protein is quantitated by Western blot analysis, without additional supporting data, must be viewed with caution. A single protein on a gel or immunoblot, or a single NH_2-terminal sequence, should never be used as the sole criterion for a monospecific antibody or a purified protein. The best criteria include isolation, sequencing, and expression of the cDNA. Furthermore, cDNA encoding one P450 can react with at least two mRNAs of identical size on Northern blots, which is known to happen with pcn1 and pcn2 (58) and with db1 and db2 (63a). Probes utilizing specific 3' untranslated regions or specific oligonucleotides should therefore be used to distinguish between highly homologous mRNAs.

P450 EXPRESSION VECTORS

In many instances of P450 cDNA clones so far isolated, the true function of the particular gene product in the intact animal remains uncertain. Expression of the cDNA or genomic clone in yeast, mammalian cultured cells, or transgenic mice (i.e. expression of one enzyme activity and not another) might resolve this uncertainty. Accordingly, the translated region of rat P450c cDNA was placed in the yeast expression vector pAAH5, with (187) and without (188) the rat NADPH-P450 oxidoreductase cDNA on the same vector, and transfected into transformed yeast; benzpyrene hydroxylase activity was found to be expressed. The expression in yeast of a chimeric cDNA encoding the first 518 residues of rat P450c linked to all but the first 56 residues of rat NADPH-P450 oxidoreductase (188a) results in a 130-kd protein with P450c enzymic activity in the absence of exogenously added NADP oxidoreductase. This elegant genetic engineering study (188a) demonstrates the interaction among the FAD and FMN binding sites of the oxidoreductase and the substrate- and heme-binding sites of the P450 protein.

From the yeast expression studies involving rat P450c (188) and a computer analysis of all 67 available P450 protein sequences (A. Puga, D. W. Nebert, unpublished), it appears that the middle third of all P450 enzymes—approximately residues 180 to 320—represents the most critical region for substrate binding and specificity. Consistent with this conclusion is a recent study in which P_1450 cDNA was sequenced from several benzo[a]pyrene-resistant mutant cell lines derived from the mouse hepatoma Hepa-1 parent line: by means of expression of chimeric cDNAs in yeast, it was found that a mutation resulting in a change from arginine to proline at residue 245 produces a two- to three-fold decrease in P_1450 catalytic activity, and that this Pro-245 alteration in concert with a change from leucine to arginine at residue 118 totally abrogrates P_1450 activity (189).

Insertion of the bovine 17α-hydroxylase full-length cDNA into an SV40-derived expression vector, followed by transient transfection of the vector in monkey kidney fibroblast COS cells (190) resulted in both 17α-hydroxylase and 17,20-lyase activities. These activities were determined by both metabolite formation and immunofluorescence of the intact cultured cells with the use of a polyclonal antibody to 17α-hydroxylase (190). Expression of the P450cam protein and enzymic activity in *Escherichia coli* (181, 191) and *Ps. putida* (191) has also been reported.

Enzyme active site–directed inhibitors of enzyme activity (192) or alkylating agents in vitro (193, 194) have been described for the purified mammalian P450 proteins. Interestingly, site-directed mutagenesis studies of P450cam (195) have shown that replacement of Cys-136 with serine—at some (linear) distance in the peptide from the Cys-357 in the enzyme active-site (181)—causes a striking decrease in camphor metabolism. Replacement of Cys-357 with His-357 causes perhaps the most dramatic modifications of P450cam in that the modified apoprotein does not pick up heme from the bacterial cytoplasmic pool; rather, the apoprotein must be reconstituted with heme in vitro (S. G. Sligar, personal communication). It is not yet certain whether the His-357 in this mutant protein binds to the iron.

Analysis of the conserved regions between two proteins encoded by homologous genes in the same subfamily (Figure 7) might also provide insight as to which areas of the protein might be most profitably studied by genetic engineering. For example, in the P450IA subfamily (rat c and d, mouse P_1 and P_3, human P_1 and P_3) one region between residues 50 and 200 and a second region between residues 320 and 500 are conserved. It is likely that the former region interacts with the NADPH-P450 oxidoreductase and the latter region is highly homologous because it includes the heme-binding site. This same biphasic pattern is not seen, however, in any of the other seven gene-pair comparisons (Figure 7).

Gene Conversion

Figure 7 also illustrates the dilemma of homologous genes between 1. what represents recent duplication and divergence with retention of functionally conserved regions and 2. what is the result of gene conversion events. Gene conversion between two homologous P450 genes is likely to have occurred between rat e and another member of the same gene subfamily because of the inclusion of most of one intron (57). The rat b and e genes appear to have a gene conversion event that spans exons 2 to 6 and introns 2 to 5 (57, 99). A gene conversion event within the large exon 2 has been suggested to have occurred between the rodent P_1 and P_3 genes between 80 million and 20 million years ago (59). The one large and two small regions of 100% similarity in amino acid sequence between rat pcn1 and pcn2 (Figure 7) have also been proposed to be the result of gene conversion (58). Disregarding the deleted nucleotides leading to one inactive gene in both C21 gene pairs, we find the degree of conservation of the remaining codons and amino acids between mouse C21A and C21B and between human C21A and C21B (Figure 7) to be highly unusual, especially in view of the likelihood that the C21 gene has undergone duplication more than 80 million years ago (before the human-mouse separation). The fact that the gene duplication event occurred so long ago, and yet regions of the C21A and C21B genes are presently so similar, argues strongly that this represents another case of gene conversion.

CONTROL OF P450 GENE EXPRESSION

The many P450 genes are under complex and distinct control during development, either following exposure of organisms to various foreign compounds or in response to important endogenous signals. Developmental regulation of P450 genes was first suggested by the pioneering studies of Conney et al (196), in which the level of hydroxylated testosterone metabolites generated

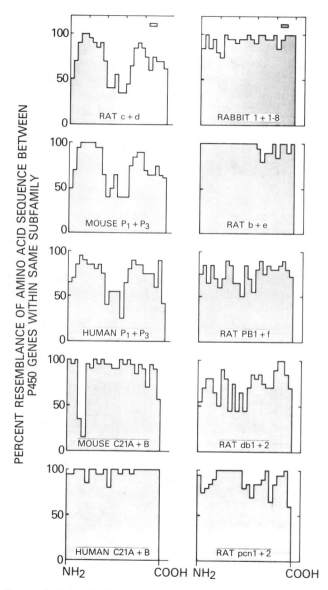

Figure 7   Percent similarity of amino acid sequence between two homologous P450 genes in ten subfamilies. The small boxes between residues 420 and 460 *(top)* denote the mammalian cysteinyl-containing fragments present in the enzyme heme-binding sites. The large decrease in amino acid homology between mouse C21A and C21B in the region of residues 60 to 100 represents the deletion of exon 2 in the inactive C21B gene (172).

from rat liver microsomes was shown to vary as a function of age and sex. That P450 genes are also regulated by a variety of exogenous substances was first recognized more than two decades ago (5). Although numerous studies have documented marked developmental differences in P450 expression and induction by means of enzyme assays or immunochemical techniques, we focus on recent studies in which the mechanisms of developmental- and inducer-specific gene activation have been explored by the use of cDNA probes.

Imprinting and Sex-Specific Expression

The extensive studies by Gustafsson and coworkers on the neonatal "programming" of certain steroid-metabolizing enzyme activities (197, 198) emphasized the importance of androgen receptors in the central nervous system and the role of growth hormone in regulating this process via the hypothalamo-pituitary-liver axis. Several sex-specific P450 proteins associated with neonatal "imprinting" are being characterized by immunochemical techniques in the rat (90, 199–202) and mouse (203–205). For instance, rat P450g and P450h are male-specific while P450i is female-specific (206). Since g, h, and i are highly immunocrossreactive (207) and have similar NH_2-terminal protein sequences (208), it is anticipated that these three sex-specific constitutively expressed P450 genes will reside in a single P450 subfamily. The complete cDNA sequence of P450h from rat liver (117a) has established that the h gene is a member of the IIC family. The availability of mouse cDNA probes to sex-specific P450 genes (117a, 204, 205) should lead to a dissection of the mechanisms underlying the sex-specific imprinting process.

Recent studies in the rat have uncovered an interesting sex-dependent developmental regulation of a gene in the P450III family (58). P450pcn2 is male-specific, and the protein is 88% similar to the PCN-inducible pcn1 (138). During development, the pcn2 protein (and its mRNA) is increased in both male and female rats at 2 weeks of age; by 12 weeks, however, pcn2 increases in males but vanishes in females. During this time the pcn1 gene is inactive. Of further interest is the high correlation of both pcn1 and pcn2 proteins with testosterone 6β-hydroxylase activity, suggesting that these two P450 genes—under distinct regulatory control—may encode isozymes.

Sex-Independent Developmental Regulation

Several P450 proteins and corresponding mRNAs have been found to increase during development in both male and female rodents. In almost every case so far examined, increased mRNA concentrations have been shown to reflect transcriptional activation of the corresponding gene. Expression of the mouse P_1450 gene and its induction by polycyclic aromatic compounds, for example, occurs in the preimplantation embryo of the mouse (209, 210). Early studies that followed enzymic activity (209, 210) have recently been confirmed with mRNA quantitation (211–213). Interestingly, during the 3-week gestation period in mice, the developmental expression of P_3 (the other member of the P450IA subfamily) lags 1 to 2 weeks behind that of P_1 (211–213).

No sex difference has been reported in the P450IIB gene subfamily. However, the rat hepatic P450b and P450e genes are under both constitutive and inducer-dependent developmental control (214). No b or e mRNA is detectable in rats prior to birth, either with or without treatment of the mother with phenobarbital, which is known to cross the placenta; however, constitutive e mRNA appears within one day postpartum, whereas constitutive b mRNA (other than a transient increase at one day of age in the neonate) remains undetectable during the next nine weeks of life (214). In addition, a marked difference was observed in the extent of phenobarbital induction of b and e mRNAs and the ratio between the two mRNAs produced as a function of development (214).

One of the most dramatic developmental gene activiations involves two members of the rat P450IIC subfamily. P450PB1 and P450f are absent at birth and remain at low levels until the rats reach puberty. The developmental increase in PB1 (215) and f (216) proteins have been established via immunochemical quantitation and by mRNA analysis (116). The increases have been shown to represent transcriptional activation of both genes (116). It remains unclear which hormones are responsible for regulating PB1 and f, although the effects of testosterone and gonadectomy have been ruled out in the case of PB1 protein (215).

Finally, the P450IIE protein (and its mRNA) is absent in newborn rats and becomes elevated, via transcriptional activation, to near-maximal levels within two days postpartum (134). In contrast, a fourfold elevation in both enzymic activity and immunodetectable P450j protein in the liver of rats treated with pyrazole or acetone is not accompanied by an increase in the corresponding mRNA (134), suggesting a posttranslational mechanism of control.

The molecular details of how P450 genes are activated during development must await further characterization of these developmentally regulated P450 genes. Likely candidates for important roles in the developmental regulation of these genes include receptors, *trans*-acting transcriptional control factors, and (both positive- and negative-acting) enhancer elements.

Transcriptional Regulation

Phenobarbital treatment was found to cause a marked transcriptional activation of one or both members of the rat P450IIB gene subfamily, b and/or e; the activation is already measurable within 30 min after treatment with the drug (217, 217a). Due to the lack of any known receptor for phenobarbital and the fact that induction of the known IIB genes is absent in all cell culture lines analyzed (M. Adesnik, personal communication; P. S. Guzelian, personal communication), little is known about the mechanism of phenobarbital-induced transcriptional activation of genes.

Of interest is the recent report of phenobarbital induction of P450IA1 (P450c) mRNA in a rat hepatoma cell line (218) but not in intact rat liver (1, 2). This phenomenon, which was found years earlier for mouse P_1450 and rat P450c enzymic activity (219–221), may represent either minor contaminants of P_1450 inducer compounds in phenobarbital preparations (at the millimolar concentrations utilized) or some other tissue culture phenomenon that does not reflect the physiologic condition or the natural inducer response of the P450IA1 gene in the intact animal.

Pregnenolone-16α-carbonitrile (PCN) and dexamethasone also induce a member of the P450III gene family, pcn1 (138). This induction is due to an increase in pcn1 mRNA (137) and has been shown to represent transcriptional activation of the gene in the case of dexamethasone (222). The other member of the P450III gene family, pcn2, the male-specific P450 described earlier, is not induced by PCN or dexamethasone even though its protein exhibits 88% resemblance to the pcn1 protein (58). Furthermore, both pcn1 and pcn2 mRNAs are readily induced by phenobarbital (58). Of interest, transcription of the epoxide hydrolase gene is substantially repressed by dexamethasone treatment (222). These data illustrate a complicated differential regulation of P450III and the epoxide hydrolase genes—at the level of constitutive expression, in response to endogenous steroids, and following exposure to foreign chemicals.

Clofibrate, a peroxisome proliferating agent and inducer of P450LAω, the single known member of the P450IV gene family, elevates LAω mRNA through transcriptional activation (146). Within 1 h after inducer administration, the transcriptional rate of the LAω gene increases dramatically. The kinetics of this increase is similar to that of two other genes coding for peroxisomal enzymes (223). LAω, therefore, may be a member of a gene battery under control of agents responsible for peroxisome proliferation.

The bovine 11β-hydroxylase, scc, 17α-hydroxylase, and 21-hydroxylase P450 genes (224) have been shown to be transcriptionally activated in cultured bovine adrenocortical cells by the peptide hormone corticotropin (ACTH) via cyclic AMP. The mRNA levels of 11β-hydroxylase, scc, and 21-hydroxylase can also be increased by prostaglandins and cholera toxin (225); whether transcriptional activation occurs with these stimuli has not been determined. Gonadotropins (FSH and hCG) regulate scc mRNA in human ovarian granulosa cells via cyclic AMP, while ACTH has no effect (225a); ACTH regulates scc and 17α-hydroxylase mRNAs in human fetal adrenal cells, yet gonadotropins do not (W. L. Miller, personal communication). The level of P450arom mRNA in human adipose stromal cells (226) is regulated by cyclic AMP and its derivatives. Differential expression of 17α-hydroxylase and scc mRNA in the ovary during the bovine estrous cycle (227) and in human ovarian granulosa and theca cells (225a) illustrates the complicated nature of steroidogenic P450 gene expression in endocrine target organs. This expression thus appears to be differentially regulated during fetal development, as well as during hormonal stimulation (227a). It has been postulated (228) that cyclic AMP does not stimulate P450 transcription directly; rather, cAMP may activate the gene encoding a hypothetical steroid hydroxylase-inducing protein (SHIP). The SHIP may then bind to a specific *cis*-acting element of the P450 gene(s). Hence, the steroidogenic P450 gene apparently waits, thinking "some day my SHIP will come." In addition to SHIP, each steroidogenic P450 gene probably has at least one additional *trans*-acting factor responsible for tissue specificity of expression (228).

With cosmid vectors, it has been possible to demonstrate murine C21A gene expression, and its stimulation by ACTH, in mouse Y1 adrenocortical tumor cells but not in mouse L cell fibroblasts (169). A maximum of 230 bp of upstream regulatory sequences is required for efficient expression of the murine C21A gene in Y1 cells (229). As described earlier, the upstream regulatory sequences of the inactive C21B gene are functional and may be important in regulation of the C21A gene in the Y1 cells (172).

Posttranscriptional Regulation

Posttranscriptional regulation plays a major role in the induction of rat "P450p" (139), which is probably the same gene as pcn1 (P. S. Guzelian, personal communication). The enhancement of rat P450p protein, following treatment with macrolide antibiotics, is due, in part, to protein stabilization (230). It must, however, be emphasized that in immunoquantitative studies with an antibody that recognizes both pcn1 and pcn2 proteins, it is not possible to distinguish which gene product (or whether both gene products) is/are being measured. Triacetyloleandromycin (TAO) also induces rabbit form 3c posttranscriptionally (142). Treatment of rabbits with TAO for several days resulted in a fivefold increase in 3c mRNA and no increase in transcription. The mechanism of this mRNA stabilization phenomenon is unknown.

Another example of posttranscriptional regulation has been found in the comparison of rat pcn1, b, and NADPH-P450 oxidoreductase mRNAs following chronic administration of dexamethasone. In addition, marked increases in b and NADPH-P450 oxidoreductase mRNAs were noted in the absence of any increase in transcriptional rates of these genes, suggesting that the steroid inducers may specifically stabilize certain mRNAs (222).

In both rats (133) and rabbits (128, 129) a P450 protein is induced following exposure to ethanol. A cDNA clone of rat P450j, a member of the P450IIE subfamily, was used to demonstrate that the five- to six-fold increase in this protein, following treatment with ethanol, acetone, or 4-methylpyrazole, is not due to an increase in j mRNA (134). The extremely rapid increase in the j protein after exposure to these drugs suggests that this induction process may reflect a posttranslational process (134). In contrast, the P450j protein is increased four- to six-fold in the chemically induced diabetic rat; this is due to a 10-fold elevation in j mRNA in the absence of transcriptional activation, suggesting a specific mechanism of mRNA stabilization (B.-J. Song, F. J. Gonzalez, unpublished). In comparison with these examples of posttranscriptional and posttranslational regulation, the mouse P_1 and P_3 genes have been shown to be differentially controlled by a combination of transcriptional and posttranscriptional events (see below).

Tissue-Specific Regulation

Striking tissue-specific expression of the IIB subfamily genes has been detected with the use of specific oligonucleotide probes (231). Rat b mRNA is absent in liver or kidney of untreated rats but appears to be detectable in lung and testis. In contrast, e mRNA is present in liver but undetectable in the other tissues examined (231). Although these results suggest an interesting tissue-specific regulation of two closely related genes, it cannot presently be determined whether these oligonucleotide probes are sensitive enough to detect low mRNA levels in extrahepatic tissues or whether the probes detect other IIB gene products that are highly homologous to b and e. An earlier study in the rabbit (111a) also uncovered tissue-specific differences in the regulation of genes in the IIC subfamily.

Mouse constitutive P_1 and P_3 mRNAs are detectable in liver, kidney, lung, spleen, and large and small intestine (232). When the transcriptional rate of P_1 and P_3 was compared with mRNA prevalence in liver, lung, and kidney—either in the presence or absence of the inducer TCDD—P_3 mRNA concentrations were found to be 20 to 30 times greater than P_1 mRNA concentrations, and this posttranscriptional regulation exhibited tissue-specific variation. Interestingly, both P_1 and P_3 are found in all tissues examined after TCDD treatment (232).

Expression in Cell Culture

In most rodent hepatoma cell lines and primary hepatocyte cultures examined, the level of P450 is considerably decreased, as compared with normal liver

(233–236). As discussed earlier, the normal P450IIB gene induction response with phenobarbital also appears to be absent in cultured liver–derived cells (234–236). On the other hand, induction of one or more genes in the P450III family by exposure of rat primary hepatocytes to dexamethasone or TAO in the growth medium (230) appears to reflect what is seen in intact liver.

Expression of the mouse P_3 gene, however, is absent in all cell culture lines examined (237) and the same is true for the orthologous rat d gene (234–236). This absence of P_3 expression might be related to the extinction of many genes in cells in culture (233–240). This phenomenon is probably dependent on humoral or diffusible factors, cell-cell contact, or the difference between the differentiated states of cells in the intact animal that are not seen when cells are trypsinized and dispersed in tissue culture. There is no evidence that the mouse P_3 gene is deleted, rearranged, or transcriptionally expressed (237); moreover, 1.8-kb upstream regulatory sequences of the P_3 gene are unable to function in a manner reflecting the intact liver, when expression vectors are transfected transiently or stably into mammalian cell cultures (R. A. Owens, L. A. Neuhold, A. K. Jaiswal, F. J. Gonzalez, D. W. Nebert, unpublished). These negative findings suggest the presence of a strong transcriptional suppressor factor or absence of a *trans*-acting positive control element. In any case, the fact that the P_1 gene is readily induced in cell culture has allowed a detailed molecular analysis of the mechanisms by which the gene expression is affected by TCDD and polycyclic hydrocarbon carcinogens.

$P_1$450 GENE REGULATION

Originating from the discovery of an allelic variant among inbred mouse strains (241), there is presently more known about regulation of the P_1 gene than any other P450 gene (3, 39, 40, 242, 243). The P_1 enzymic activity, aryl hydrocarbon (benzo[a]pyrene) hydroxylase (AHH), is more readily induced in certain mouse strains and not others (12). The lack of AHH inducibility segregates as an autosomal recessive trait between C57BL/6 and DBA/2 mice, and the first inducing chemicals characterized—3-methylcholanthrene and benzo[a]anthracene—were aromatic hydrocarbons; hence, the gene controlling the induction process was named the *Ah* locus. A receptor defect was postulated (244) and subsequently demonstrated (245) to be responsible for the decreased responsiveness of AHH induction in DBA/2 mice.

A series of improved Ah receptor assays with tritiated TCDD as the radioligand (246, 247) and 3'-specific P_1 and P_3 cDNA probes (71, 72) helped establish the chronologic series of events occurring during the induction process (Figure 8). Combustion products (such as benzpyrene and more than a dozen other polycyclic hydrocarbons), TCDD, and plant metabolites similar to β-naphthoflavone bind to the cytosolic Ah receptor with an appar-

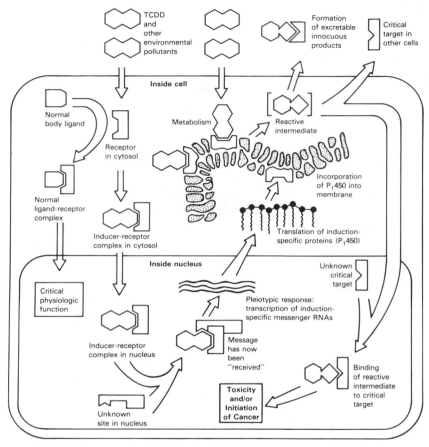

Figure 8   Diagram of P₁450 regulation by the Ah receptor in a cell. That TCDD, as an extremely potent ligand, occupies the Ah receptor such that a critical life function cannot be carried out has been proposed (248) to explain the extreme toxicity (248–250) of this environmental contaminant [Redrawn from Ref. 48 and reproduced with permission from Dr. W. Junk Publishers].

ent K_d < 1.0 nM (246). Although claims have been made that the Ah receptor is normally localized in the nucleus (251), this has been refuted by studies with gel permeation chromatography (247), subcellular biochemical markers (252), and cell enucleation (253). The endogenous ligand (if indeed one exists) for the cytosolic receptor is not known; however, it seems likely that certain plant flavones and/or combustion products have appropriated the Ah receptor for stimulating their own metabolism (242). Numerous other foreign chemicals (e.g. plant opioids and phorbol esters) have been found to bind with high affinity to other cellular receptors. Although a defective Ah receptor is not incompatible with life (12, 242), correlations between the *Ah* locus and decreased binding of epidermal growth factor to its cell-surface receptor (254), immunosuppression (255), longevity and fertility (256), and atherosclerosis (257) have been reported.

Following translocation of the inducer·Ah receptor complex into the nucleus (258)—which is a temperature-dependent step shown to occur in cell culture but not in vitro with isolated cytosol and nuclei (259)—the P_1 and P_3 genes are transcriptionally activated (260), and enhanced mRNA concentrations (258, 261) lead to increased levels of benzpyrene (P_1) and acetanilide (P_3) metabolism in the endoplasmic reticulum (262). Because of this allelic difference in mice and the fact that AHH (P_1) activity converts procarcinogens to ultimate carcinogenic and mutagenic reactive intermediates, genetic differences in cancer risk have been demonstrated in polycyclic hydrocarbon-treated mice (263). The possible correlation between the high AHH inducibility phenotype in the human and cigarette smoking–induced bronchogenic carcinoma (11, 86) was discussed earlier.

Somatic cell genetics has been used to identify genes necessary for the AHH induction response. AHH-inducible cells grown in the presence of benzpyrene generate lethal amounts of toxic metabolites (264, 265). By means of resistance to benzpyrene toxicity, therefore, clones deficient in AHH induction have been selected from the mouse hepatoma line Hepa-1 (266). Such AHH^- clones have been shown to be mutational in origin (267). The recessive mutant clones have been assigned to at least four complementation groups (268) called A, B, C, and D. Group A mutants are heterogeneous (269): some have no detectable P_1 mRNA; others have high constitutive P_1 mRNA levels when grown in the absence of TCDD, yet no AHH activity and no detectable increases in TCDD-inducible P_1 mRNA. These and other data (270) provide strong evidence that the P_1 structural gene is defective in Group A variants (P_1^-). The P_1 cDNAs from two such variant lines have been sequenced and found to be defective (189), further confirming this hypothesis. Groups B, C, and D represent regulatory mutant cell lines. Mutations in Group B result in less than 10% of the Ah receptor levels (271) normally found in the Hepa-1 wild-type *(wt)* and are designated receptorless (r^-). Group B variants probably include mutations in the Ah receptor gene(s). Mutations in Group C affect the normal nuclear translocation of the inducer·receptor complex (271) and are designated nuclear translocation–defective (nt^-). The Group D mutant appears to lack the Ah receptor, which can be restored by 5-azacytidine or sodium butyrate treatment but not mutagen treatment; the D gene may encode a structural component of the Ah receptor or a product required for Ah receptor expression (272). Since intracellular ATP concentration is important for the maintenance of Ah receptor expression in Hepa-1 cells (273), the D gene could even be related to intracellular ATP maintenance or some other event quite far removed from the P_1450 induction process. Miller et al have also isolated and characterized r^- and nt^- benzpyrene-resistant AHH^- mutant lines (274, 275).

A dominant class of AHH^- mutants appears to synthesize a *trans*-acting repressor of P_1 mRNA synthesis (269). Transfection of the dominant class DNA into *wt* Hepa-1 cells has resulted in benzpyrene-resistant AHH^- cells, suggesting that this repressor gene has been transfected (A. J. Watson, O. Hankinson, personal communication).

Expression vectors containing mouse (276, 277) and human (278) P_1 upstream sequences and the chloramphenicol acetyltransferase (CAT) gene in *wt, nt*$^-$, and P_1^- stable transformants have provided evidence for 1. a promoter containing a TATA box, 2. a negative control element between 400 and 800 bases upstream from the mRNA cap site, and 3. a receptor-dependent enhancer more than 900 bases upstream from the cap site (Figure 9). The region of -1647 to -611 from the P_1 mRNA cap site, independent of orientation or distance, has been shown to be active and inducible by TCDD with the heterologous SV40 promoter (277). One TCDD-inducible element can be dissociated from an enhancer of constitutive gene expression (Figure 9), whereas one or more other TCDD-inducible elements cannot (277). Between 1.6 kb upstream and the mRNA cap site, mouse and human P_1 upstream sequences are only 55% similar. However, a computer-aided alignment of 220 bp (-1137 to -918 in the mouse) is 80% similar between the two species (277) and includes one perfectly aligned GGGCGG box [noted for binding the *trans*-acting transcriptional regulatory factor Sp1 (279)].

In cells containing P_1 upstream sequences of 1647 or 800 bp, constitutive CAT gene expression is higher in the transfected P_1^- mutant than in the transfected *wt* line; removal of the -800 to -400 fragment causes a two- to four-fold decrease in constitutive gene expression (Figure 9). These data are consistent with a negative autoregulatory loop (276). Absence of P_1 metabolism leads to the lack of a putative repressor binding to the negative control region, and consequently constitutive P_1 mRNA concentrations are exceedingly high in Group A mutants (269). Interestingly, this metabolism-dependent repression of constitutive activity appears to require the Ah receptor, since no augmentation of control activity is seen in the nt^- mutant containing 400, 800, or 1647 bp of upstream sequences (Figure 9). The interaction of the receptor-dependent, metabolism-dependent, and promoter regions of the P_1 upstream sequences is thus regarded as extremely complicated and depicted by arrows and question marks in Figure 9.

Jones and coworkers (280) have also suggested that P_1 gene transcription is under both positive and negative control. The mouse P_1 gene in that study was isolated from a Hepa-1 "high activity variant" (HAV), exhibiting about 10-fold higher constitutive AHH activity and two-fold higher levels of

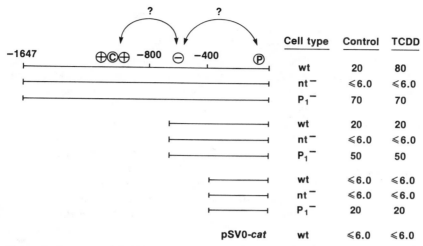

Cell type	Control	TCDD
wt	20	80
nt$^-$	⩽6.0	⩽6.0
P$_1$$^-$	70	70
wt	20	20
nt$^-$	⩽6.0	⩽6.0
P$_1$$^-$	50	50
wt	⩽6.0	⩽6.0
nt$^-$	⩽6.0	⩽6.0
P$_1$$^-$	20	20
pSV0-cat wt	⩽6.0	⩽6.0

Figure 9 Summary of P$_1$450 upstream regulatory sequences driving the chloramphenicol acetyltransferase (CAT) gene in mouse hepatoma Hepa-1 wild-type *(wt)* cells and the nuclear translocation–defective *(nt$^-$)* and P$_1$450 metabolism-defective *(P$_1$$^-$)* mutant cell lines. The −1647, −800, and −400 denote the number of bases upstream from the mRNA cap site. The positive (⊕) and negative (⊖) control regions upstream and downstream from −800, respectively, are illustrated. A DNA element (©) within the TCDD-inducible enhancer that augments constitutive transcription is also shown. The promoter TATA box promoter (Ⓟ) region is also illustrated. Interactions among the TCDD-inducible enhancer, the negative control element, and the promoter region are unknown but depicted as arrows with question marks. The pSV2-*neo* was cotransfected with a 15-fold excess of these various constructs, and stable transformants were selected for their resistance to G418 (276, 277). Numbers at *right* denote average values for CAT activity (pmol/min/mg protein) in pooled transformants treated with control medium or TCDD.

TCDD-inducible P$_1$ mRNA and enzymic activity (281). Furthermore, a large portion of the 5' end of the P$_1$ gene (including the first intron and upstream sequences) was directly inserted into the expression vector, and the transient CAT assay was used (280), in contrast to the pooled stable transformant CAT assay (276). The transcription start site in these constructs was assigned by primer extension of an intron fragment to a position about 630 bp downstream (280) from the cap site of the normal P$_1$ gene that had been established by sequencing, primer extension, and S1 mapping (75). It is conceivable that the P$_1$ gene is altered in HAV cells and that a new promoter exists in the middle of the P$_1$450 first intron (280). Another possibility is that a cryptic promoter in the middle of the first intron has been uncovered by the construction of the expression vector and that the normal promoter does not function in this construct in the transiently transfected cells.

Fujisawa-Sehara et al (282) have reported the effect of rat P450c upstream sequences driving a CAT gene and transiently transfected into mouse Hepa-1 *wt* cells; they demonstrated that a −6.3 to −0.8 kb region can act independently of orientation as an enhancer element with the heterologous SV40 promoter. Also, Jones and coworkers (283) showed that a TCDD-inducible enhancer region (the −1648 to −643 fragment of the normal gene) of the mouse P_1 upstream sequences can act at varying distances and in either orientation, relative to the mouse mammary tumor virus promoter. Moreover, Jones et al (284) proposed at least two discrete TCDD-responsive domains: one between −1299 and −1066, the other between −998 and −890 (in the numbering system of the normal gene).

At least three *cis*-acting dioxin regulatory elements (DREs) with a consensus sequence of $5'$-$^C_C N^{TA}_{GG}$ GCTGGG-$3'$ have been identified in the rat P450c upstream sequences at approximately 1.0, 1.5, and 3.3 kb 5'ward from the mRNA cap site (284a). This consensus sequence exists in the mouse P_1450 upstream sequences at −1020 (amidst the TCDD-responsive enhancer) and at −510 (in the middle of what is believed to be the negative control element), and both DREs are present in the human P_1450 upstream sequences at these same two locations (277).

A potentially very exciting finding (213) suggests a possible interaction between P_1 upstream regulatory sequences and adenovirus-2 E1A-like *trans*-acting factors. P_1 mRNA levels are elevated (in the absence of any foreign chemical inducer) during early mouse embryogenesis and during retinoic acid–induced differentiation of mouse F9 embryonal carcinoma cells in culture (213). The polyoma virus enhancer is inactive in undifferentiated F9 cells and active in differentiated F9 cells (285, 286). Recently it was suggested that undifferentiated F9 cells contain a cellular repressor of the β-globin gene in transfected constructs, and it was shown that E1A products have the same target sequence (within the polyoma virus enhancer element) as the putative cellular repressor in undifferentiated F9 cells (287). The E1A products might require an additional cellular protein for their effect (288). Interestingly, no Ah receptor is detectable in undifferentiated or differentiated F9 cells (H. J. Eisen, unpublished data), so it is possible that the increases in P_1450 mRNA in differentiated F9 cells are not mediated by the Ah receptor. Hence, it is conceivable that the "normal" status of the P_1450 upstream regulatory sequences include bound repressor and that derepression may occur either 1. during differentiation via activation by an unknown endogenous signal or 2. following foreign chemical inducer stimulation.

THE [*Ah*] GENE BATTERY

In addition to the mouse P_1 and P_3 genes, expression of the UDP glucuronosyltransferase (289), NAD(P)H:menadione oxidoreductase (290), and glutathione transferase (291) genes has been rigorously shown by mouse genetics studies to be under Ah receptor control. These five genes are currently

Figure 10 Genes (in boxes) known to be controlled by the Ah receptor and several other enzymes or metabolic pathways known to be stimulated by TCDD and polycyclic hydrocarbons. UDPGT₁, UDP glucuronosyltransferase with 4-methylumbelliferone as substrate. NMOR₁, NAD(P)H:menadione oxidoreductase. GT₁, glutathione transferase with 1-chloro-2,4-dinitrobenzene as substrate.

included in the *[Ah]* battery (Figure 10). Other activities that are known to be increased by TCDD or polycyclic hydrocarbon treatment and may therefore be under Ah receptor control include aldehyde dehydrogenase (292), phosphatidylcholine biosynthesis (293), arachidonic acid metabolism (294), lipid deacylation (295), phospholipase A_2 (296), γ-glutamyltranspeptidase (297), and protein kinase C (298).

Two Group A (P_1^-) mutants exhibit extremely high constitutive $UDPGT_1$ and $NMOR_1$ activities (299). The most plausible explanation for these findings is that the P_1 gene may be subject to autoregulation, which also affects the $UDPGT_1$ and $NMOR_1$ genes (Figure 11). This hypothesis would require that 1. an endogenous substrate be metabolized by the P_1 enzyme, and 2. the hypothetical metabolite, freely diffusible between cells (269), activate a precursor to an active repressor, which interacts with the negative control region and derepresses constitutive P_1 gene transcription. The mouse chromosomal location of the putative repressor gene has been determined (J.-Y. Lee, D. W. Nebert, unpublished data). Another possibility is that the P_1 enzyme directly metabolizes the protein prorepressor to an active repressor; P450 metabolism of proteins is not without precedent (300). Absence of P_1 catalytic activity would lead to no active repressor and, hence, derepression of constitutive P_1 transcription, thereby leading to elevated $P_1$450 mRNA levels in untreated Group A mutant lines (269). Since it has been difficult to separate constitutive P_1 gene expression from TCDD-inducible expression (276, 277), there may be *cis*-acting regulation (shown by an arrow and question mark in Figure 11) between the positive and negative control elements of the P_1 upstream sequences. Alternatively, the unknown endogenous substrate may compete with exogenous inducers as ligands for the Ah receptor, as well as competing for the $P_1$450 enzyme active-site (Figure 11). Furthermore, the repressor activated by the $P_1$450 enzyme normally represses the $UDPGT_1$ and

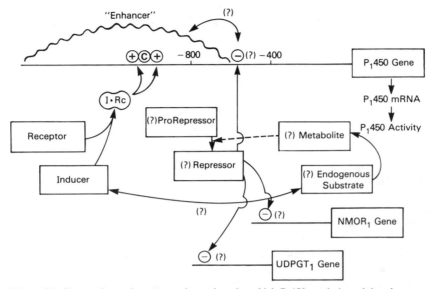

Figure 11 Proposed negative autoregulatory loop by which P_1450 catalytic activity also regulates two other genes in the [*Ah*] gene battery. Arrows with a question mark denote: 1. the apparent interaction between the TCDD-responsive enhancer and regulation of constitutive gene expression via the negative control site ($\ominus$); and 2. potential interaction between the foreign chemical inducer and the unknown endogenous substrate. The $\oplus$, $\copyright$, and $\ominus$ symbols are described in the legend of Figure 9. I, inducer. Rc, Ah receptor. The numbers -800 and -400 represent bp upstream from the P_1 mRNA cap site.

$NMOR_1$ genes; absence of P_1450 metabolism enhances both the enzyme activities and therefore appears to derepress both of these genes.

Rat liver preneoplastic nodules are being studied as a stepwise cancer model system (301, 302). The P_1 enzymic activity is undetectable, while $UDPGT_1$, $NMOR_1$, GT_1, sulfotransferase, and γ-glutamyltranspeptidase activities in preneoplastic nodules are increased (301). It is possible that this phenomenon is somehow associated with the [*Ah*] gene battery.

There appears to be an evolutionary driving force for survival advantage (303) in which drug-metabolizing enzymes convert relatively hydrophobic substrates to very hydrophilic, innocuous, excretable products (Figure 12). The Phase I P450-mediated monooxygenases supply only the first step and, without the conjugating enzymes, can be responsible for toxic, mutagenic, and carcinogenic intermediates. Coordinately linked with the Phase II enzymes, P450-mediated monooxygenation is a necessary and important first step in the conversion of innumerable foreign chemicals to excretable products. The P_1 metabolism–dependent control of the $UDPGT_1$ and $NMOR_1$ genes (Figure 11) is compatible with the scheme illustrated in Figure 12.

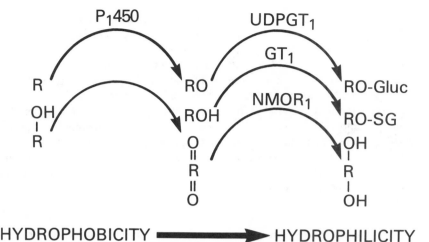

Figure 12   Scheme of coordinate regulation in which genes in the [Ah] battery encode enzymes that are responsible for the metabolism of environmental chemicals.

Absence of all drug-metabolizing enzymes would probably be incompatible with life; the organism would accumulate hydrophobic foreign chemicals and ultimately die from immunosuppression and wasting disease (248–250). A means to turn on the Phase II genes would be important in the case of low or absent P_1 gene function. Hence, absence of P_1 metabolism would result in the derepression, and therefore the highest possible expression, of the Phase II genes, and foreign chemicals could be metabolized and excreted (Figure 12).

SPECULATION ON P450 GENE EVOLUTION

In summary, P450 gene expression represents an exciting field of research from the standpoint of numerous disciplines: chemical carcinogenesis and mutagenesis, pharmacology and toxicology, fundamental molecular biologic studies on gene expression, and speculation about evolution of gene superfamilies. Because P450 proteins participate in the biosynthesis and degradation of steroids and the catabolism of fatty acids, P450 gene expression was probably essential for membrane integrity and turnover in some prokaryotes and the earliest eukaryotes. Another early function of P450 is the use of foreign chemicals as an energy source, which is known to occur in prokaryotes and fungi (303–307). Later in evolution, with the development of excretory organs and digestive tracts, an additional function of P450 became the detoxification of numerous foreign substances, especially plant metabolites.

One of the most intriguing questions facing P450 researchers is whether most P450 enzymes play major roles in physiologic metabolism. Although relatively position-specific monooxygenations of endogenous steroids (e.g. testosterone 6β-, 7α-, and 16α-hydroxylations, estrogen 2-hydroxylation) have been reported for this or that P450 protein, in no case has a clear physiologic function been demonstrated for any of the genes in families P450I through P450IV (Figure 3). One possible exception may be fatty acid ω-hydroxylation by LAω (144–146) in the P450IV gene family. Perhaps evolutionary comparisons would aid in ascribing potential functions for certain P450 genes, such as the potential developmental role for mouse P_1 (213). Clearly if a certain P450 protein is absent from otherwise normal animals, this would suggest that the expression of that P450 gene is not critical. Screening various inbred strains of mice might aid in the identification of important P450 genes (308).

In this connection the human debrisoquine 4-hydroxylase deficiency appears to represent a case in which otherwise normal individuals are deficient in at least one P450 gene. Recent analysis of several "poor metabolizers" (PM phenotype) by use of enzymic activity, immunoblots, and Northern hybridizations has revealed that PM individuals are clearly missing the P450db1 protein (125a). Although correlations between the "extensive metabolizer" (EM) phenotype and increased risk of cancer (126, 127) and between the PM phenotype and Parkinson's disease (309) have been suggested, the fact that a functional db1 enzyme is absent suggests that this P450 has no physiologic role and exists only to metabolize exogenous substances.

It is clear that the enzymes encoded by the P450XI, XVII, XIX, and XXI gene families (Figure 3) carry out functions critical to survival of the organism, e.g. steroid biosynthesis necessary for sexual reproduction. From very early in evolution, certain prokaryotes and simple fungi used P450 enzymes for membrane integrity, as well as to break down food for energy (303–307). We propose that genes in the P450XI, XVII, XIX, or XXI families are most closely related to the earliest P450 gene(s). With the animal-plant divergence at least 1000 million years ago, animals began using plants as a food source and plants began to develop increasingly complicated phytoalexins and other polycyclic oxygen–containing metabolites (310–315) as a means of self-defense against animals. Several hundred million years later, from the time that it became possible for carbon-containing material to burn (e.g. coal and decaying plants), animals were confronted with the need to detoxify combustion products in addition to the innumerable plant metabolites. The enzymes encoded by the P450IA (69) and IIC (114) subfamilies appear to be most specific for the metabolism of combustion products such as benzpyrene. We suggest that most, if not all, of the genes in families P450I through IV have evolved in herbivorous animals first in response to plant metabolites and later in evolution in response to combustion products. The overlapping substrate specificities with which many of these P450 enzymes metabolize flavones, coumarin derivatives, and terpenes is not unlike the overlapping substrate specificities with which these same P450 enzymes metabolize numerous drugs and synthetic and endogenous steroids as well. There are numerous examples of drugs or other foreign chemicals that are good substrates for P450 enzymes encoded by gene families that have diverged so long ago that the genes are not chromosomally linked.

Just as subsets of P450 enzymes are induced by polycyclic hydrocarbons, phenobarbital, pyrazole, steroids, and clofibrate, subsets of UDP glucuronosyltransferases (316) are also induced by these same classes of inducing agents; the induction of other Phase II enzymes has not yet been examined in sufficient detail. The coordinate regulation of Phase I and Phase II genes by the Ah receptor (Figure 12) might therefore be similar to other gene batteries controlled by other receptors. Further studies should provide more insight into this speculation about the evolution of genes encoding drug-metabolizing enzymes.

ACKNOWLEDGMENTS

We thank our colleagues—especially Milt Adesnik, Kathleen Dixon, Cyndi Edwards, John Jones, Byron Kemper, and Walter Miller—for critically reviewing this manuscript. The expert secretarial assistance of Ingrid E. Jordan is greatly appreciated.

Literature Cited

1. Nebert, D. W., Negishi, M. 1982. *Biochem. Pharmacol.* 31:2311–17
2. Adesnik, M., Atchison, M. 1986. *CRC Crit. Rev. Biochem.* 19:247–305
3. Whitlock, J. P. Jr. 1986. *Ann. Rev. Pharmacol. Toxicol.* 26:333–69
4. Black, S. D., Coon, M. J. 1987. *Adv. Enzymol. Relat. Areas Mol. Biol.* In press
5. Conney, A. H. 1967. *Pharmacol. Rev.* 19:317–66
6. Nebert, D. W., Eisen, H. J., Negishi, M., Lang, M. A., Hjelmeland, L. M., et al. 1981. *Ann. Rev. Pharmacol. Toxicol.* 21:431–62
7. Waterman, M. R., Estabrook, R. W. 1983. *Mol. Cell. Biochem.* 53/54:267–78
8. Pelkonen, O., Nebert, D. W. 1982. *Pharmacol. Rev.* 34:189–222
9. Conney, A. H. 1982. *Cancer Res.* 42:4875–917
10. Wolf, C. R. 1986. *Trends Genet.* 2:209–14
11. Gonzalez, F. J., Jaiswal, A. K., Nebert, D. W. 1987. *Cold Spring Harbor Symp.* *Quant. Biol.* 51:879–90
12. Nebert, D. W., Negishi, M., Lang, M. A., Hjelmeland, L. M., Eisen, H. J. 1982. *Adv. Genet.* 21:1–52
13. White, R. E., Coon, M. J. 1980. *Ann. Rev. Biochem.* 49:315–56
14. Jakoby, W. B., Bend, J. R., Caldwell, J., eds. 1982. *Metabolic Basis of Detoxication.* New York: Academic. 357 pp.
15. Anders, M. W., ed. 1985. *Bioactivation of Foreign Compounds.* New York: Academic. 540 pp.
16. Coon, M. J., Koop, D. R. 1983. *The Enzymes* 16:645–77
17. Käppeli, O. 1986. *Microbiol. Rev.* 50:244–58
18. Mannering, G. J., Deloria, L. B. 1986. *Ann. Rev. Pharmacol. Toxicol.* 26:455–515
19. Nebert, D. W., Adesnik, M., Coon, M. J., Estabrook, R. W., Gonzalez, F. J., et al. 1987. *DNA.* 6:1–11
20. Neuberger, A., Smith, R. L. 1983. *Drug Metab. Rev.* 14:559–607
21. Brodie, B. B., Gillette, J. R., La Du, B.

N. 1958. *Ann. Rev. Biochem.* 27:427–54
22. Axelrod, J. 1982. *Trends Pharmacol. Sci.* 3:383–86
23. Brown, R. R., Miller, J. A., Miller, E. C. 1954. *J. Biol. Chem.* 209:211–22
24. Remmer, H. 1959. *Arch. Exp. Pathol. Pharmacol.* 235:279–90
25. Conney, A. H., Davison, C., Gastel, R., Burns, J. J. 1960. *J. Pharmacol. Exp. Ther.* 130:1–8
26. Klingenberg, M. 1958. *Arch. Biochem. Biophys.* 75:376–86
27. Garfinkel, D. 1958. *Arch. Biochem. Biophys.* 77:493–509
28. Omura, T., Sato, R. 1964. *J. Biol. Chem.* 239:2370–78
29. Estabrook, R. W., Cooper, D. Y., Rosenthal, O. 1963. *Biochem. Z.* 338:741–55
30. Mason, H. S., North, J. C., Vanneste, M. 1965. *Fed. Proc.* 24:1172–80
31. Orrenius, S., Ernster, L. 1964. *Biochem. Biophys. Res. Commun.* 16:60–65
32. Remmer, H., Merker, H. J. 1965. *Ann. NY Acad. Sci.* 123:79–97
33. Nebert, D. W., Gelboin, H. V. 1970. *J. Biol. Chem.* 245:160–68
34. Nebert, D. W. 1970. *J. Biol. Chem.* 245:519–27
35. Dehlinger, P. J., Schimke, R. T. 1972. *J. Biol. Chem.* 247:1257–64
36. Jacob, S. T., Scharf, M. B., Vessel, E. S. 1974. *Proc. Natl. Acad. Sci. USA* 71:704–7
37. Lu, A. Y. H., Levin, W. 1974. *Biochim. Biophys. Acta* 344:205–40
38. Lu, A. Y. H., West, S. B. 1980. *Pharmacol. Rev.* 31:277–95
39. Boobis, A., Caldwell, J., DeMatteis, F., Davies, D., eds. 1985. *Microsomes and Drug Oxidations.* London: Taylor & Francis. 428 pp.
40. Ortiz de Montellano, P. R., ed. 1986. *Cytochrome P-450: Structure, Mechanism, and Biochemistry.* New York: Plenum. 539 pp.
41. Thomas, P. E., Bandiera, S., Reik, L. M., Maines, S. L., Ryan, D. E., et al. 1987. *Fed. Proc.* In press
42. Garcia, J. D., Jennette, K. W. 1981. *J. Inorg. Biochem.* 14:281–95
43. Mottley, C., Harman, L. S., Mason, R. P. 1985. *Biochem. Pharmacol.* 34:3005–8
44. Ullrich, V., Hermann, G., Weber, P. 1978. *Biochem. Pharmacol.* 27:2301–304
45. Sugiura, M., Iwasaki, K., Kato, R. 1976. *Mol. Pharmacol.* 12:322–34
46. Sugiura, M., Yamazoe, Y., Kamataki, T., Kato, R. 1980. *Cancer Res.* 40:2910–14
47. Yamazoe, Y., Ishii, K., Yamaguchi, N., Kamataki, T., Kato, R. 1980. *Biochem. Pharmacol.* 29:2183–88
48. Nebert, D. W. 1979. *Mol. Cell. Biochem.* 27:27–46
49. Lu, A. Y. H., Coon, M. J. 1968. *J. Biol. Chem.* 243:1331–32
50. French, J. S., Guengerich, F. P., Coon, M. J. 1980. *J. Biol. Chem.* 255:4112–19
51. Haugen, D. A., Coon, M. J., Nebert, D. W. 1976. *J. Biol. Chem.* 251:1817–27
52. Vermilion, J. L., Coon, M. J. 1978. *J. Biol. Chem.* 253:2694–704
53. Estabrook, R. W., Franklin, M. R., Cohen, B., Shigamatzu, A., Hildebrandt, A. G. 1971. *Metabolism* 20:187–99
54. Nebert, D. W., Kimura, S., Gonzalez, F. J. 1985. See Ref. 39, pp. 145–56
55. Gilbert, W., Marchionni, M., McKnight, G. 1986. *Cell* 46:151–54
56. Jukes, T. H., Cantor, C. R. 1969. *Mammalian Protein Metabolism*, pp. 21–132. New York: Academic
57. Atchison, M., Adesnik, M. 1986. *Proc. Natl. Acad. Sci. USA* 83:2300–304
58. Gonzalez, F. J., Song, B.-J., Hardwick, J. P. 1986. *Mol. Cell. Biol.* 6:2969–76
59. Jaiswal, A. K., Nebert, D. W., McBride, O. W., Gonzalez, F. J. 1987. *J. Exp. Pathol.* 3:1–17
60. Hildebrand, C. E., Gonzalez, F. J., McBride, O. W., Nebert, D. W. 1985. *Nucleic Acids Res.* 13:2009–16
61. Hildebrand, C. E., Gonzalez, F. J., Kozak, C. A., Nebert, D. W. 1985. *Biochem. Biophys. Res. Commun.* 130:396–406
62. Davis, M. B., West, L. F., Shephard, E. A., Phillips, I. R. 1986. *Ann. Hum. Genet.* 50:237–40
63. Simmons, D. L., Kasper, C. B. 1983. *J. Biol. Chem.* 258:9585–88
63a. Gonzalez, F. J., Matsunaga, T., Nagata, K., Meyer, U. A., Nebert, D. W., et al. 1987. *DNA* 6:In press
64. Simmons, D. L., Lalley, P. A., Kasper, C. B. 1985. *J. Biol. Chem.* 260:515–21
65. Miller, W. L., Chung, B., Matteson, K. J., Voutilainen, R., Picado-Leonard, J. 1986. *DNA* 5:61
65a. Chua, S. C., Szabo, P., New, M. I., White, P. C. *Endocrinology* 1987. In press
66. Matteson, K. J., Picado-Leonard, J., Chung, B., Mohandas, T. K., Miller, W. L. 1986. *J. Clin. Endocrinol. Metab.* 63:789–91
67. White, P. C., New, M. I., Dupont, B.

1984. *Proc. Natl. Acad. Sci. USA* 81:7505–509
68. White, P. C., Chaplin, D. D., Weis, J. H., Dupont, B., New, M. I., et al. 1984. *Nature* 312:465–67
69. Negishi, M., Nebert, D. W. 1979. *J. Biol. Chem.* 254:11015–23
70. Lang, M. A., Nebert, D. W., Negishi, M. 1980. *Biochemistry, Biophysics and Regulation of Cytochrome P-450*, pp. 415–22. Amsterdam/New York: Elsevier/North-Holland Biomedical
71. Negishi, M., Swan, D. C., Enquist, L. W., Nebert, D. W. 1981. *Proc. Natl. Acad. Sci. USA* 78:800–4
72. Tukey, R. H., Nebert, D. W. 1984. *Biochemistry* 23:6003–8
73. Gonzalez, F. J., Mackenzie, P. I., Kimura, S., Nebert, D. W. 1984. *Gene* 29:281–92
74. Kimura, S., Gonzalez, F. J., Nebert, D. W. 1984. *J. Biol. Chem.* 259:10705–13
75. Gonzalez, F. J., Kimura, S., Nebert, D. W. 1985. *J. Biol. Chem.* 260:5040–49
76. Ohyama, T., Nebert, D. W., Negishi, M. 1984. *J. Biol. Chem.* 259:2675–82
77. Kimura, S., Nebert, D. W. 1986. *Nucleic Acids Res.* 14:6765–66
78. Yabusaki, Y., Shimizu, M., Murakami, H., Nakamura, K., Oeda, K., et al. 1984. *Nucleic Acids Res.* 12:2929–38
79. Kawajiri, K., Gotoh, O., Sogawa, K., Tagashira, Y., Muramatsu, M. et al. 1984. *Proc. Natl. Acad. Sci. USA* 81:1649–53
80. Sogawa, K., Gotoh, O., Kawajiri, K., Fujii-Kuriyama, Y. 1984. *Proc. Natl. Acad. Sci. USA* 81:5066–70
81. Sogawa, K., Gotoh, O., Kawajiri, K., Harada, T., Fujii-Kuriyama, Y. 1986. *J. Biol. Chem.* 260:5026–32
82. Ozols, J. 1986. *J. Biol. Chem.* 261: 3965–79
83. Okino, S. T., Quattrochi, L. C., Barnes, H. J., Osanto, S., Griffin, K. J., et al. 1985. *Proc. Natl. Acad. Sci. USA* 82: 5310–14
84. Jaiswal, A. K., Gonzalez, F. J., Nebert, D. W. 1985. *Science* 228:80–83
85. Jaiswal, A. K., Gonzalez, F. J., Nebert, D. W. 1985. *Nucleic Acids Res.* 13: 4503–20
86. Kouri, R. E., McKinney, C. E., Slomiany, D. J., Snodgrass, D. R., Wray, N. P., et al. 1982. *Cancer Res.* 42: 5030–37
87. Jaiswal, A. K., Nebert, D. W. 1986. *Nucleic Acids Res.* 14:4376
88. Waithe, W. I., Trottier, Y., Labbe, D., Anderson, A. 1986. *Biochem. Pharmacol.* 35:2069–72
89. Jaiswal, A. K., Nebert, D. W., Gon-

zalez, F. J. 1986. *Nucleic Acids Res.* 14:6773–74
89a. Quattrochi, L. C., Pendurthi, U. R., Okino, S. T., Potenza, C., Tukey, R. H. 1986. *Proc. Natl. Acad. Sci. USA* 83:6731–35
90. Kamataki, T., Maeda, K., Yamazoe, Y., Matsuda, N., Ishii, K., et al. 1983. *Mol. Pharmacol.* 24:146–55
91. Waxman, D. J., Ko, A., Walsh, C. 1983. *J. Biol. Chem.* 258:11937–47
91a. Nagata, K., Matsunaga, T., Gillette, J. R., Gelboin, H. V., Gonzalez, F. J. 1987. *J. Biol. Chem.* 262:2787–93
92. Phillips, I. R., Shephard, E. A., Ashworth, A., Rabin, B. R. 1985. *Proc. Natl. Acad. Sci. USA* 82:983–87
93. Negishi, M., Fujii-Kuriyama, Y., Tashiro, Y., Imai, Y. 1976. *Biochem. Biophys. Res. Commun.* 71:1153–60
94. Fujii-Kuriyama, Y., Taniguchi, T., Mizukami, Y., Sakai, M., Tashiro, Y., et al. 1980. *Proc. Jpn Acad. Ser. B* 56:603–8
95. Fujii-Kuriyama, Y., Taniguchi, T., Mizukami, Y., Sakai, M., Tashiro, Y., et al. 1981. *J. Biochem.* 89:1869–79
96. Mizukami, Y., Sogawa, K., Suwa, Y., Muramatsu, M., Fujii-Kuriyama, Y. 1983. *Proc. Natl. Acad. Sci. USA* 80:3958–62
97. Kumar, A., Raphael, C., Adesnik, M. 1983. *J. Biol. Chem.* 258:11280–84
98. Atchison, M., Adesnik, M. 1983. *J. Biol. Chem.* 258:11285–95
99. Suwa, Y., Mizukami, Y., Sogawa, K., Fujii-Kuriyama, Y. 1985. *J. Biol. Chem.* 260:7980–84
100. Reik, L. M., Levin, W., Ryan, D. E., Maines, S. L., Thomas, P. E. 1985. *Arch. Biochem. Biophys.* 242:365–82
101. Stupans, I., Ikeda, T., Kessler, D. J., Nebert, D. W. 1984. *DNA* 3:129–38
102. Yuan, P., Ryan, D., Levin, W., Shively, J. 1983. *Proc. Natl. Acad. Sci. USA* 80:1169–73
103. Affolter, M., Anderson, A. 1984. *Biochem. Biophys. Res. Commun.* 118: 655–62
104. Affolter, M., Labbe, D., Jean, A., Raymond, M., Noel, D., et al. 1986. *DNA* 5:209–18
105. Heinemann, F. S., Ozols, J. 1983. *J. Biol. Chem.* 258:4195–201
106. Tarr, G. E., Black, S. D., Fujita, V. S., Coon, M. J. 1983. *Proc. Natl. Acad. Sci. USA* 80:6552–56
107. Ozols, J., Heinemann, F. S., Johnson, E. F. 1985. *J. Biol. Chem.* 260:5427–34
108. Mattschoss, L. A., Hobbs, A. A., Steggles, A. W., May, B. K., Elliott, W. H. 1986. *J. Biol. Chem.* 261:9438–43

109. Hobbs, A. A., Mattschoss, L. A., May, B. K., Williams, K. E., Elliott, W. H. 1986. *J. Biol. Chem.* 261:9444–49
110. Leighton, J. K., DeBrunner-Vossbrinck, B. A., Kemper, B. 1984. *Biochemistry* 23:204–10
111. Govind, S., Bell, P. A., Kemper, B. 1986. *DNA.* 5:371–82
111a. Leighton, J. K., Kemper, B. 1984. *J. Biol. Chem.* 259:11165–68
111b. Finlayson, M. J., Kemper, B., Browne, N., Johnson, E. F. 1986. *Biochem. Biophys. Res. Commun.* 141: 728–33
112. Johnson, E. F., Barnes, H., Griffin, K. J., Okino, S., Tukey, R. 1986. *Fed. Proc.* 45:1854
113. Johnson, E. F., Griffin, K. J. 1985. *Arch. Biochem. Biophys.* 237:55–64
114. Raucy, J. L., Johnson, E. F. 1985. *Mol. Pharmacol.* 27:296–301
115. Tukey, R. H., Okino, S. T., Barnes, H. J., Griffin, K. J., Johnson, E. F. 1985. *J. Biol. Chem.* 260:13347–54
116. Gonzalez, F. J., Kimura, S., Song, B.-J., Pastewka, J., Gelboin, H. V., et al. 1986. *J. Biol. Chem.* 261:10667–72
117. Friedberg, T., Waxman, D. J., Atchison, M., Kumar, A., Haaparanta, T., et al. 1987. *Biochemistry.* 25:7975–83
117a. Yoshioka, H., Morohashi, K., Sogawa, K., Miyate, T., Kawajiri, K., et al. 1987. *J. Biol. Chem.* 262:1706–11
118. Tukey, R. H., Quattrochi, L. C., Pendurthi, U. R., Okino, S. T. 1986. *Fed. Proc.* 45:1663
119. Shimada, T., Misono, K. S., Guengerich, F. P. 1986. *J. Biol. Chem.* 261:909–21
120. Idle, J. R., Smith, R. L. 1979. *Drug Metab. Rev.* 9:301–17
121. Eichelbaum, M. 1984. *Fed. Proc.* 43:2298–302
122. Boobis, A. R., Davies, D. S. 1984. *Xenobiotica* 14:151–85
122a. Larrey, D., Distlerath, L. M., Dannan, G. A., Wilkinson, G. R., Guengerich, F. P. 1984. *Biochemistry* 23:2787–95
123. Distlerath, L. M., Reilly, P. E. B., Martin, M. V., Davis, G. G., Wilkinson, G. R., et al. 1985. *J. Biol. Chem.* 260:9057–67
124. Gut, J., Catin, T., Dayer, P., Kronbach, T., Zanger, U., et al. 1986. *J. Biol. Chem.* 261:11734–43
125. Kalow, W. 1984. *Fed. Proc.* 43:2314–18
125a. Gonzalez, F. J., Skoda, R. C., Kimura, S., McBride, O. W., Umeno, M., et al. 1987. *Science.* In press
126. Idle, J. R., Mahgoub, A., Sloan, T. P., Smith, R. L., Mbanefo, C. O., et al. 1981. *Cancer Lett.* 11:331–38
127. Ayesh, R., Idle, J. R., Ritchie, J. C., Crothers, M. J., Hetzel, M. R. 1984. *Nature* 313:169–70
128. Koop, D. R., Coon, M. J. 1984. *Mol. Pharmacol.* 25:494–501
129. Koop, D. R., Crump, B. L., Nordblom, G. D., Coon, M. J. 1985. *Proc. Natl. Acad. Sci. USA* 82:4065–69
130. Ryan, D. E., Ramanathan, L., Iida, S., Thomas, P. E., Haniu, M., et al. 1985. *J. Biol. Chem.* 260:6385–93
131. Ryan, D. E., Koop, D. R., Thomas, P. E., Coon, M. J., Levin, W. 1986. *Arch. Biochem. Biophys.* 246:633–93
132. Morgan, E. T., Koop, D. R., Coon, M. J. 1982. *J. Biol. Chem.* 257:13951–44
133. Yang, C. S., Tu, Y. Y., Koop, D. R., Coon, M. J. 1985. *Cancer Res.* 45: 1140–45
134. Song, B.-J., Gelboin, H. V., Park, S. S., Yang, C. S., Gonzalez, F. J. 1986. *J. Biol. Chem.* 261:16689–97
135. Lu, A. Y. H., Somogyi, A., West, S., Kuntzman, R., Conney, A. H. 1972. *Arch. Biochem. Biophys.* 152:457–62
136. Elshourbagy, N. A., Guzelian, P. S. 1980. *J. Biol. Chem.* 255:1279–85
137. Hardwick, J. P., Gonzalez, F. J., Kasper, C. B. 1983. *J. Biol. Chem.* 258:10182–86
138. Gonzalez, F. J., Nebert, D. W., Hardwick, J. P., Kasper, C. B. 1985. *J. Biol. Chem.* 260:7435–41
139. Molowa, D. T., Schuetz, E. G., Wrighton, S. A., Watkins, P. B., Kremers, P., et al. 1986. *Proc. Natl. Acad. Sci. USA* 83:5311–15
140. Guengerich, F. P., Martin, M. V., Beaune, P. H., Kremers, P., Wolff, T., et al. 1986. *J. Biol. Chem.* 261:5051–60
140a. Beaune, P. H., Umbenhauer, D. R., Bork, R. W., Lloyd, R. S., Guengerich, F. P. 1986. *Proc. Natl. Acad. Sci. USA* 83:8064–68
141. Wrighton, S. A., Maurel, P., Schuetz, E. G., Watkins, P. B., Young, B., et al. 1985. *Biochemistry* 24:2171–78
142. Dalet, C., Blanchard, J. M., Guzelian, P., Barwick, J., Hartle, H., et al. 1986. *Nucleic Acids Res.* 14:5999–6015
143. Reddy, J. K., Azarnoff, D. L., Hignite, C. E. 1980. *Nature* 283:397–98
144. Orton, T. C., Parker, G. L. 1982. *Drug. Metab. Disp.* 10:110–15
145. Tamburini, P. P., Masson, H. A., Bains, S. K., Makowski, R. J., Morris, B., et al. 1984. *Eur. J. Biochem.* 139:235–46

146. Hardwick, J. P., Song, B.-J., Huberman, E., Gonzalez, F. J. 1987. *J. Biol. Chem.* 262:801–10
146a. Yanagibachi, K., Haniu, M., Shively, J. E., Shen, W. H., Hall, P. 1986. *J. Biol. Chem.* 261:3556–62
147. New, M. I., Dupont, B., Grumbach, K., Levine, L. S. 1983. *The Metabolic Basis of Inherited Disease*, pp. 973–1000. New York: McGraw-Hill
148. Speiser, P. W., New, M. I. 1985. *Trends Genet.* 1:275–78
149. Morohashi, K., Fujii-Kuriyama, Y., Okada, Y., Sogawa, K., Hirose, T., et al. 1984. *Proc. Natl. Acad. Sci. USA* 81:4647–51
150. Matocha, M. F., Waterman, M. R. 1985. *J. Biol. Chem.* 260:12259–65
151. Matteson, K. J., Chung, B.-C., Miller, W. L. 1984. *Biochem. Biophys. Res. Commun.* 120:264–70
152. Matteson, K. J., Chung, B.-C., Urdea, M. S., Miller, W. L. 1986. *Endocrinology* 118:1296–305
153. Hauffa, B. P., Miller, W. L., Grumbach, M. M., Conte, F. A., Kaplan, S. L. 1985. *Clin. Endocrinol.* 23:481–93
154. Chung, B.-C., Matteson, K. J., Voutilainen, R., Mohandas, T. K., Miller, W. L. 1986. *Proc. Natl. Acad. Sci. USA* 83:4243–47
155. John, M. E., John, M. C., Simpson, E. R., Waterman, M. R. 1984. *Fed. Proc.* 43:1474
156. Chua, S. C., John, M., White, P. C. 1986. *Pediatr. Res.* 20:262A
157. Zuber, M. X., John, M. E., Okamura, T., Simpson, E. R., Waterman, M. R. 1986. *J. Biol. Chem.* 261:2475–82
158. Chung, B.-C., Picado-Leonard, J., Haniu, M., Bienkowski, M., Hall, P. F., et al. 1987. *Proc. Natl. Acad. Sci. USA* 84:407–11
159. Mendelson, C. R., Wright, E. E., Evans, C. T., Porter, J. C., Simpson, E. R. 1985. *Arch. Biochem. Biophys.* 243:480–91
160. Evans, C. T., Ledesma, D. B., Schulz, T. Z., Simpson, E. R., Mendelson, C. R. 1986. *Proc. Natl. Acad. Sci. USA* 83:6387–91
161. Chen, S., Shively, J. E., Nakajin, S., Shinoda, M., Hall, P. F. 1986. *Biochem. Biophys. Res. Commun.* 135:713–19
162. White, P. C., New, M. I., Dupont, B. 1984. *Proc. Natl. Acad. Sci. USA* 81:1986–90
163. Carroll, M. C., Campbell, R. D., Porter, R. R. 1985. *Proc. Natl. Acad. Sci. USA* 82:521–25
164. White, P. C., Grossberger, D., Onufer, B. J., Chaplin, D. D., New, M. I., et al. 1985. *Proc. Natl. Acad. Sci. USA* 82:1089–93
165. Levi-Strauss, M., Tosi, M., Steinmetz, M., Klein, J., Meo, T. 1985. *Proc. Natl. Acad. Sci. USA* 82:1746–50
166. Amor, M., Tosi, M., Duponchel, C., Steinmetz, M., Meo, T. 1985. *Proc. Natl. Acad. Sci. USA* 82:4453–57
167. Chung, B.-C., Matteson, K. J., Miller, W. L. 1985. *DNA* 4:211–19
168. Carroll, M. C., Palsdottir, A., Belt, K. T., Porter, R. R. 1985. *EMBO J.* 4:2547–52
169. Parker, K. L., Chaplin, D. D., Wong, M., Seidman, J. G., Smith, J. A., et al. 1985. *Proc. Natl. Acad. Sci. USA* 82:7860–64
170. White, P. C., New, M. I., Dupont, B. 1986. *Proc. Natl. Acad. Sci. USA* 83:5111–15
171. Higashi, Y., Yoshioka, H., Yamane, M., Gotoh, O., Fujii-Kuriyama, Y. 1986. *Proc. Natl. Acad. Sci. USA* 83:2841–45
172. Chaplin, D. D., Galbraith, L. G., Seidman, J. G., White, P. C., Parker, K. L. 1986. *Proc. Natl. Acad. Sci. USA* 83:9601–5
173. Yoshioka, H., Morohashi, K., Sogawa, K., Yamane, M., Kominami, S., et al. 1986. *J. Biol. Chem.* 261:4106–9
174. Chung, B.-C., Matteson, K. J., Miller, W. L. 1986. *Proc. Natl. Acad. Sci. USA* 83:4243–47
175. Donohoue, P. A., Jospe, N., Migeon, C. J., McLean, R. H., Bias, W. B., et al. 1986. *Biochem. Biophys. Res. Commun.* 136:722–29
176. Garlepp, M. J., Wilton, A. N., Dawkins, R. L., White, P. C. 1986. *Immunogenetics* 1384:1–6
177. Waxman, D. J. 1986. *Arch. Biochem. Biophys.* 247:335–45
178. Yoon, P. S., Rawlings, J., Orme-Johnson, W. H., DeLuca, H. F. 1980. *Biochemistry* 19:2172–76
179. Seidel, S. L., Shires, T. K. 1986. *Biochem. J.* 235:859–68
179a. Haurand, M., Ullrich, V. 1985. *J. Biol. Chem.* 260:15059–67
180. Kalb, V. F., Loper, J. C., Dey, C. R., Woods, C. W., Sutter, T. R. 1986. *Gene* 45:237–45
181. Unger, B. P., Gunsalus, I. C., Sligar, S. G. 1986. *J. Biol. Chem.* 261:1158–63
182. Haniu, M., Armes, L. G., Yasunobu, K. T., Shastry, B. A., Gunsalus, I. C. 1982. *J. Biol. Chem.* 257:12664–71

183. Peterson, J. A., Basu, D., Coon, M. J. 1966. *J. Biol. Chem.* 241:5162–64
184. Unger, B. P., Sligar, S. G., Gunsalus, I. C. 1986. *The Bacteria*. New York: Academic. 10:557–89
185. Poulos, T. L., Finzel, B. C., Gunsalus, I. C., Wagner, G. C., Kraut, J. 1985. *J. Biol. Chem.* 260:16122–30
186. Poulos, T. L., Finzel, B. C., Howard, A. J. 1986. *Biochemistry.* 25:5314–22
187. Murakami, H., Yabusaki, Y., Ohkawa, H. 1986. *DNA* 5:1–10
188. Oeda, K., Sakaki, T., Ohkawa, H. 1985. *DNA* 4:203–10
188a. Murakami, H., Yabusaki, Y., Sakaki, T., Shibata, M., Ohkawa, H. 1987. *DNA*. 6:In press
189. Kimura, S., Hankinson, O., Nebert, D. W. 1987. *EMBO J*. In press
190. Zuber, M. X., Simpson, E. R., Waterman, M. R. 1986. *Science* 234:1258–61
191. Koga, H., Rauchfuss, B., Gunsalus, I. C. 1985. *Biochem. Biophys. Res. Commun.* 130:412–17
192. Frey, A. B., Kreibich, G., Wadhera, A., Clarke, L., Waxman, D. J. 1986. *Biochemistry.* 25:4797–803
193. Parkinson, A., Ryan, D. E., Thomas, P. E., Jerina, D. M., Sayer, J. M., et al. 1986. *J. Biol. Chem.* 261:11478–86
194. Parkinson, A., Thomas, P. E., Ryan, D. E., Gorsky, L. D., Shively, J. E., et al. 1986. *J. Biol. Chem.* 261:11487–95
195. Unger, B., Jollie, D., Atkins, W., Dabrowski, M., Sligar, S. 1986. *Fed. Proc.* 45:1874
196. Conney, A. H., Levin, W., Jacobson, M., Kuntzman, R., Cooper, D. Y., et al. 1969. *Microsomes and Drug Oxidations*, pp. 279–302. New York: Academic
197. Einarsson, K., Gustafsson, J.-Å., Stenberg, Å. 1973. *J. Biol. Chem.* 248:4987–97
198. Gustafsson, J.-Å., Mode, A., Norstedt, G., Skett, P. 1983. *Ann. Rev. Physiol.* 45:51–60
199. Morgan, E. T., MacGeoch, C., Gustafsson, J.-Å. 1985. *J. Biol. Chem.* 260:11895–98
200. Waxman, D. J. 1984. *J. Biol. Chem.* 259:15481–90
201. Ryan, D. E., Dixon, R., Evans, R. H., Ramanathan, L., Thomas, P. E., et al. 1984. *Arch. Biochem. Biophys.* 233: 636–42
202. Dannan, G. A., Guengerich, F. P., Waxman, D. J. 1986. *J. Biol. Chem.* 261:10728–35
203. Harada, N., Negishi, M. 1985. *Proc. Natl. Acad. Sci. USA* 82:2024–28
204. Burkhart, B. A., Harada, N., Negishi, M. 1985. *J. Biol. Chem.* 260:15357–61
205. Noshiro, M., Serabjit-Singh, C. J., Bend, J. R., Negishi, M. 1986. *Arch. Biochem. Biophys.* 244:857–64
206. Ryan, D. E., Iida, S., Wood, A. W., Thomas, P. E., Lieber, C. S., et al. 1984. *J. Biol. Chem.* 259:1239–50
207. Bandiera, S., Ryan, D. E., Levin, W., Thomas, P. E. 1985. *Arch. Biochem. Biophys.* 240:478–82
208. Haniu, M., Ryan, D. E., Iida, S., Lieber, C. S., Levin, W., et al. 1984. *Arch. Biochem. Biophys.* 235:304–11
209. Galloway, S. M., Perry, P. E., Meneses, J., Nebert, D. W., Pedersen, R. A. 1980. *Proc. Natl. Acad. Sci. USA* 77:3524–28
210. Filler, R., Lew, K. J. 1981. *Proc. Natl. Acad. Sci. USA* 78:6991–95
211. Ikeda, T., Altieri, M., Chen, Y.-T., Nakamura, M., Tukey, R. H., et al. 1983. *Eur. J. Biochem.* 134:13–18
212. Tuteja, N., Gonzalez, F. J., Nebert, D. W. 1985. *Dev. Biol.* 112:177–84
213. Kimura, S., Nebert, D. W. 1987. *J. Exp. Pathol.* 3:61–74
214. Giachelli, C. M., Omiecinski, C. J. 1986. *J. Biol. Chem.* 261:1359–63
215. Waxman, D. J., Dannan, G. A., Guengerich, F. P. 1985. *Biochemistry* 24:4409–17
216. Bandiera, S., Ryan, D. E., Levin, W., Thomas, P. E. 1986. *Arch. Biochem. Biophys.* 248:658–76
217. Hardwick, J. P., Gonzalez, F. J., Kasper, C. B. 1983. *J. Biol. Chem.* 258:8081–85
217a. Adesnik, M., Rivkin, E., Kumar, A., Lippman, A., Raphael, C., Atchison, M. 1982. In *Cytochrome P-450, Biochemistry, Biophysics and Environmental Implications*, ed. E. Hietanen, M., Laitinen, O. Hänninen, pp. 143–48. Amsterdam: Elsevier
218. McManus, M. E., Minchin, R. F., Schwartz, D. M., Wirth, P. J., Huber, B. E. 1986. *Biochem. Biophys. Res. Commun.* 137:120–27
219. Gielen, J. E., Nebert, D. W. 1971. *Science* 172:167–69
220. Gielen, J. E., Nebert, D. W. 1972. *J. Biol. Chem.* 247:7591–602
221. Owens, I. S., Nebert, D. W. 1975. *Mol. Pharmacol.* 11:94–104
222. Simmons, D. L., McQuiddy, P., Kasper, C. B. 1987. *J. Biol. Chem.* 262:326–32

223. Reddy, J. K., Goel, K., Nemali, R., Carring, J., Laffler, G., et al. 1986. *Proc. Natl. Acad. Sci. USA* 83:1747–51
224. John, M. E., John, M. C., Boggaram, V., Simpson, E. R., Waterman, M. R. 1986. *Proc. Natl. Acad. Sci. USA* 83:4715–19
225. Boggaram, V., Simpson, E. R., Waterman, M. R. 1984. *Arch. Biochem. Biophys.* 231:271–79
225a. Voutilainen, R., Tapanainen, J., Chung, B., Matteson, K. J., Miller, W. L. 1986. *J. Clin. Endocrinol. Metab.* 63:202–7
226. Evans, C. T., Corbin, C. J., Saunders, C. T., Simpson, E. R., Mendelson, C. R. 1987. *J. Biol. Chem.* In press
227. Rodgers, R. J., Waterman, M. R., Simpson, E. R. 1986. *Endocrinology* 118:1366–74
227a. Voutilainen, R., Miller, W. L. 1986. *J. Clin. Endocrinol. Metab.* 63:1145–50
228. Waterman, M. R., Simpson, E. R. 1985. See Ref. 39, pp. 136–44.
229. Parker, K. L., Chaplin, D. D., Wong, M., Seidman, J. G., Schimmer, B. P. 1987. *Endocr. Res.* In press
230. Watkins, P. B., Wrighton, S. A., Schuetz, E. G., Maurel, P., Guzelian, P. S. 1986. *J. Biol. Chem.* 261:6264–71
231. Omiecinski, C. J. 1986. *Nucleic Acids Res.* 14:1525–39
232. Kimura, S., Gonzalez, F. J., Nebert, D. W. 1986. *Mol. Cell. Biol.* 6:1471–77
233. Bissell, D. M., Hammaker, L. E., Meyer, U. A. 1973. *J. Cell Biol.* 59: 722–34
234. Owens, I. S., Niwa, A., Nebert, D. W. 1975. *Gene Expression and Carcinogenesis in Cultured Liver*, pp. 378–401. New York: Academic
235. Frey, A. B., Rosenfeld, M. G., Dolan, W. J., Adesnik, M., Kreibich, G. 1984. *J. Cell. Physiol.* 120:169–80
236. Steward, A. R., Wrighton, S. A., Pasco, D. S., Fagan, J. B., Li, D., et al. 1985. *Arch. Biochem. Biophys.* 241:494–508
237. Jaiswal, A. K., Nebert, D. W., Eisen, H. J. 1985. *Biochem. Pharmacol.* 34: 2721–31
238. Lin, R. C., Snodgrass, P. J. 1975. *Biochem. Biophys. Res. Commun.* 64: 725–34
239. Savage, C. R., Bonney, R. J. 1978. *Exp. Cell Res.* 114:307–15
240. Sirica, A. E., Pitot, H. C. 1980. *Pharmacol. Rev.* 31:205–28
241. Nebert, D. W., Gelboin, H. V. 1969. *Arch. Biochem. Biophys.* 134:76–89

242. Eisen, H. J., Hannah, R. R., Legraverend, C., Okey, A. B., Nebert, D. W. 1983. *Biochemical Actions of Hormones*, pp. 227–58. New York: Academic
243. Nebert, D. W., Eisen, H. J., Hankinson, O. 1984. *Biochem. Pharmacol.* 33:917–24
244. Nebert, D. W., Robinson, J. R., Niwa, A., Kumaki, K., Poland, A. P. 1975. *J. Cell. Physiol.* 85:393–414
245. Poland, A. P., Glover, E., Kende, A. S. 1976. *J. Biol. Chem.* 251:4936–46
246. Okey, A. B., Bondy, G. P., Mason, M. E., Kahl, G. F., Eisen, H. J., et al. 1979. *J. Biol. Chem.* 254:11636–48
247. Hannah, R. R., Nebert, D. W., Eisen, H. J. 1981. *J. Biol. Chem.* 256:4584–90
248. Poland, A., Knutson, J. C. 1982. *Ann. Rev. Pharmacol. Toxicol.* 22:517–54
249. Nebert, D. W., Elashoff, J. D., Wilcox, K. R. Jr. 1983. *Am. J. Public Health* 73:286–89
250. McKinney, J. D., Fawkes, J., Jordan, S., Chae, K., Oatley, S., et al. 1985. *Environ. Health Perspect.* 61:41–53
251. Whitlock, J. P. Jr., Galeazzi, D. R. 1984. *J. Biol. Chem.* 259:980–85
252. Denison, M. S., Harper, P. A., Okey, A. B. 1986. *Eur. J. Biochem.* 155:223–29
253. Gudas, J. M., Hankinson, O. 1986. *J. Cell. Physiol.* 128:441–48
254. Kärenlampi, S. O., Eisen, H. J., Hankinson, O., Nebert, D. W. 1983. *J. Biol. Chem.* 258:10378–83
255. Lubet, R. A., Brunda, M. B., Taramelli, D., Dansie, D., Nebert, D. W., et al. 1984. *Arch. Toxicol.* 56:18–24
256. Nebert, D. W., Brown, D. D., Towne, D. W., Eisen, H. J. 1984. *Biol. Reprod.* 30:363–73
257. Paigen, B., Holmes, P. A., Morrow, A., Mitchell, D. 1986. *Cancer Res.* 46:3321–24
258. Tukey, R. H., Hannah, R. R., Negishi, M., Nebert, D. W., Eisen, H. J. 1982. *Cell* 31:275–84
259. Okey, A. B., Bondy, G. P., Mason, M. E., Nebert, D. W., Forster-Gibson, C., et al. 1980. *J. Biol. Chem.* 255:11415–22
260. Gonzalez, F. J., Tukey, R. H., Nebert, D. W. 1984. *Mol. Pharmacol.* 26:117–21
261. Tukey, R. H., Nebert, D. W., Negishi, M. 1981. *J. Biol. Chem.* 256:6969–74
262. Negishi, M., Jensen, N. M., Garcia, G. S., Nebert, D. W. 1981. *Eur. J. Biochem.* 115:585–94

263. Kouri, R. E., Nebert, D. W. 1977. *Origins of Human Cancer*, pp. 811–35. Cold Spring Harbor, NY: Cold Spring Harbor Lab.
264. Gelboin, H. V., Huberman, E., Sachs, L. 1969. *Proc. Natl. Acad. Sci. USA* 64:1188–94
265. Benedict, W. F., Gielen, J. E., Nebert, D. W. 1972. *Int. J. Cancer* 9:435–51
266. Hankinson, O. 1979. *Proc. Natl. Acad. Sci. USA* 76:373–76
267. Hankinson, O. 1981. *Somat. Cell. Genet.* 7:373–88
268. Hankinson, O. 1983. *Somat. Cell. Genet.* 9:497–514
269. Hankinson, O., Andersen, R. D., Birren, B., Sander, F., Negishi, M., et al. 1985. *J. Biol. Chem.* 260:1790–95
270. Montisano, D. F., Hankinson, O. 1985. *Mol. Cell. Biol.* 5:698–704
271. Legraverend, C., Hannah, R. R., Eisen, H. J., Owens, I. S., Nebert, D. W., et al. 1982. *J. Biol. Chem.* 257:6402–407
272. Gudas, J. M., Hankinson, O. 1987. *Somat. Cell Mol. Genet.* In press
273. Gudas, J. M., Hankinson, O. 1986. *J. Cell. Physiol.* 128:449–56
274. Miller, A. G., Whitlock, J. P. Jr. 1981. *J. Biol. Chem.* 256:2433–37
275. Miller, A. G., Israel, D., Whitlock, J. P. Jr. 1983. *J. Biol. Chem.* 258:3523–27
276. Gonzalez, F. J., Nebert, D. W. 1985. *Nucleic Acids Res.* 13:7269–88
277. Neuhold, L. A., Gonzalez, F. J., Jaiswal, A. K., Nebert, D. W. 1986. *DNA* 5:403–11
278. Jaiswal, A. K., Gonzalez, F. J., Nebert, D. W. 1987. *Mol. Endocrinol.* In press
279. Kadonaga, J. T., Jones, K. A., Tjian, R. 1986. *Trends Biochem. Sci.* 11:20–23
280. Jones, P. B. C., Galeazzi, D. R., Fisher, J. M., Whitlock, J. P. Jr. 1985. *Science* 227:1499–502
281. Jones, P. B. C., Miller, A. G., Israel, D. I., Galeazzi, D. R., Whitlock, J. P. Jr. 1984. *J. Biol. Chem.* 259:12357–63
282. Fujisawa-Sehara, A., Sogawa, K., Nishi, C., Fujii-Kuriyama, Y. 1986. *Nucleic Acids Res.* 14:1465–77
283. Jones, P. B. C., Durrin, L. K., Galeazzi, D. R., Whitlock, J. P. Jr. 1986. *Proc. Natl. Acad. Sci. USA* 83:2802–806
284. Jones, P. B. C., Durrin, L. K., Fisher, J. M., Whitlock, J. P. Jr. 1986. *J. Biol. Chem.* 261:6647–50
284a. Sogawa, K., Fujisawa-Sehara, A., Yanrane, M., Fujii-Kuriyama, Y. 1986. *Proc. Natl. Acad. Sci. USA* 83:8044–48
285. Linney, E., Donerly, S. 1983. *Cell* 35:693–99
286. Herbomel, P., Bourachot, B., Yaniv, M. 1984. *Cell* 39:653–62
287. Hen, R., Borrelli, E., Fromental, C., Sassone-Corsi, P., Chambon, P. 1986. *Nature* 321:249–51
288. Kovesdi, I., Reichel, R., Nevins, J. R. 1986. *Science* 231:719–22
289. Owens, I. S. 1977. *J. Biol. Chem.* 252:2827–33
290. Kumaki, K., Jensen, N. M., Shire, J. G. M., Nebert, D. W. 1977. *J. Biol. Chem.* 252:157–65
291. Felton, J. S., Ketley, J. N., Jakoby, W. B., Aitio, A., Bend, J. R., et al. 1980. *Mol. Pharmacol.* 18:559–64
292. Deitrich, R. A., Bludeau, P., Stock, T., Roper, M. 1977. *J. Biol. Chem.* 252:6169–76
293. Ishidate, K., Tsuruoka, M., Nakazawa, Y. 1982. *Biochim. Biophys. Acta* 713:103–11
294. Rifkind, A. B., Muschick, H. 1983. *Nature* 303:524–26
295. Levine, L., Ohuchi, K. 1978. *Cancer Res.* 38:4142–46
296. Bresnick, E., Bailey, G., Bonney, R. J., Wightman, P. 1981. *Carcinogenesis* 2:1119–22
297. Gupta, B. N., McConnell, E. E., Harris, M. W., Moore, J. A. 1981. *Toxicol. Appl. Pharmacol.* 57:99–118
298. Bombick, D. W., Madhukar, B. V., Brewster, D. W., Matsumura, F. 1985. *Biochem. Biophys. Res. Commun.* 127:296–302
299. Robertson, J. A., Hankinson, O., Nebert, D. W. 1987. *Chem. Scripta.* In press
300. Fucci, L., Oliver, C. N., Coon, M. J., Stadtman, E. R. 1983. *Proc. Natl. Acad. Sci. USA* 80:1521–25
301. Farber, E. 1984. *Cancer Res.* 44:5463–74
302. Buchmann, A., Kuhlmann, W., Schwarz, M., Kunz, W., Wolf, C. R., et al. 1985. *Carcinogenesis* 6:513–21
303. Nebert, D. W., Gonzalez, F. J. 1985. *Trends Pharmacol. Sci.* 6:160–64
304. Ferris, J. P., Fasco, M. J., Stylianopoulou, F. L., Jerina, D. M., Daly, J. W., et al. 1973. *Arch. Biochem. Biophys.* 156:97–103
305. Cerniglia, C. E., Gibson, D. T. 1979. *J. Biol. Chem.* 254:12174–80
306. Woods, L. F. J., Wiseman, A. 1979. *Biochem. Soc. Trans.* 7:124–27
307. Wiedmann, B., Wiedmann, M., Kärgel, E., Schunck, W.-H., Müller, H.-G.

1986. *Biochem. Biophys. Res. Commun.* 136:1148–54
308. Nebert, D. W., Felton, J. S. 1976. *Fed. Proc.* 35:1133–41
309. Barbeau, A., Cloutier, T., Roy, M., Plasse, L., Paris, S., et al. 1985. *The Lancet* 2:1213–16
310. Kuc, J., Currier, W. 1976. *Mycotoxins and Other Fungal Related Food Problems,* pp. 356–68. Washington, DC: Am. Chem. Soc.
311. Berenbaum, M., Feeny, P. 1981. *Sci-ence* 212:927–29
312. Maugh, T. H. II. 1982. *Science* 216:722–23
313. Cullis, C. A. 1984. *Nature* 310:366–67
314. Raupp, M. J., Milan, F. R., Barbosa, P., Leonhardt, B. A. 1986. *Science* 232:1408–10
315. McClintock, B. 1984. *Science* 226:792–801
316. Bock, K. W., Burchell, B., Dutton, G. J., Hänninen, O., Mulder, G. J., et al. 1983. *Biochem. Pharmacol.* 32:953–55

AUTHOR INDEX

A

Aaronson, R. P., 382, 538, 540, 547
Aaronson, S. A., 683, 684, 780, 781, 783, 786, 787, 789, 790, 792, 794, 803, 804, 807-9, 811, 813, 816, 817, 882
Abastado, J.-P., 438
Abbott, M. K., 207, 209, 212, 215
Abbott, V., 761
Abdel-Ghany, M., 599
Abdel-Monem, M., 459
Abe, K., 855
Abe, O., 898
Abeijon, C., 78
Abel, P. D., 898
Abeles, R. H., 524
Abelson, J., 150-52, 280, 281
Abood, M. E., 644
Aboul-Ela, F., 447
Abraham, D. J., 516, 517
Abrahams, S. L., 177, 178
Abramson, S. N., 638
Abreu, S. L., 320, 321, 324, 732
Abumrad, N. A., 50
Ackerman, K. E., 171
Ackerman, P., 592
Ackerman, S. K., 733, 747
Acklin, M., 760
Adachi, K., 518
Adachi, T., 337
Adair, W. L., 30
Adair, W. L. Jr., 67, 74, 526
Adamany, A. M., 526
Adamik, R., 629, 637
Adams, A. E. M., 935
Adams, C. E., 518, 519
Adams, J. M., 473, 485, 491
Adamson, E. D., 801
Addison, J. M., 292-94
Adelman, T. G., 295
Adelstein, R. S., 415, 577-80, 696
Adesnik, M., 946, 947, 952, 953, 955, 956, 958-61, 964, 968, 972, 975
Adham, N. F., 340
Adolph, K. W., 551
Adunyah, E. S., 411
Adunyah, S. E., 171, 176, 178
Adya, S., 324
Aeckerle, S., 857
Aerts, R. J., 181
Affara, N. A., 711
Affolter, H., 405

Affolter, M., 959
Agabian, N., 476
Agard, D. A., 548, 554
Aggerbeck, L. P., 675
Agnantis, N. J., 816
Agnew, W., 377
Agol, V. I., 321, 324
Agostini, E., 805
Agranoff, B. W., 171
Agris, P. F., 264, 268, 273, 274
Aguet, M., 740-42, 745, 746, 748-50
Aguilera, G., 584
Agutter, P. S., 538, 545
Aguzzi, A., 739, 747, 760, 764
Aharonov, A., 883
Ahmad, Z., 580-82, 590-92, 596
Ahola, H., 670, 672
Ahrens, D. C., 410
Ailhaud, G., 505
Air, G. M., 366, 376, 385
Aisen, P., 291, 298
Aitio, A., 980
Aitken, A., 571, 573, 574, 589, 592, 595, 596, 599
Aitken, D. M., 100
Aiyer, R. A., 748, 749
Akam, M., 209-12, 222
Akam, M. E., 210, 212
Akatsuka, A., 591, 592
Akerman, K., 422
Akers, R. F., 183
Akimenko, M.-A., 474, 476, 480, 484, 486, 489, 702, 704, 706, 716
Akiyama, T., 896, 903-5, 928, 929
Akusjarvi, G., 329
Alan, R., 740, 748
Albanesi, J. P., 600
Albani, M., 282
Alberghina, L., 936
Albert, K. A., 183
Alberti, A., 676
Alberts, A. W., 16, 46
Alberts, B., 91, 92, 230, 254
Alberts, B. M., 94, 240, 241, 257, 258, 445, 446
Albertson, D. G., 711, 712
Albin, C., 673
Albino, A. P., 807
Aldrich, C., 658, 660, 661
Aldrich, C. E., 664, 665
Aldrich, H., 555
Aldrich, T. H., 782, 783, 786, 789, 799
Alema, S., 805

Alexander, C. B., 896
Alexander, H., 388, 680
Alexander, J., 162
Alexander, R. W., 162, 181, 182
Alexander, S., 388, 856, 858
Alger, B. E., 179, 181
Ali, M. H., 518
Ali, Z., 196
Aliberti, A., 742
Alkonyi, I., 106, 112
Allan, D., 33
Allan, J., 542, 543
Allen, A. K., 28, 29
Allen, E., 443
Allen, G., 730
Allen, J. F., 598
Allen, J. R., 95
Allen, S. M., 571, 572, 592
Allen, T. D., 814
Allfrey, V. G., 542
Allison, V. F., 98
Allred, J. B., 105
Almeda, S., 697
Almeida, J. D., 654
Almers, W., 185
Almoguera, C., 819
Aloni, B., 620
Alonso, M. A., 321, 324
Alpert, A., 583, 887, 906, 908
Alt, F. W., 473, 485, 491
Alter, H., 653
Altieri, M., 971
Altin, J. G., 168
Alton, N. K., 734
Alzner-DeWeerd, B., 265, 279
Amagai, A., 856
Aman, R. A., 111
Amano, T., 619, 638, 781, 799, 805
Amara, S. G., 472, 485
Amar-Costesec, A., 921
Ameglio, F., 759-61
Amento, E. P., 763
Ames, B. N., 153, 265, 273, 280, 282, 283
Amherdt, M., 840
Amiel, T., 790, 798
Amiri, I., 129
Amit, A. G., 387
Amman, D., 403
Ammer, M., 937
Ammerer, G., 849, 924, 934, 939
Amor, M., 964
Amri, E. Z., 505
Amundsen, S. K., 230, 231, 233
Anai, M., 231

995

SUBJECT INDEX

A

Acanthamoeba castellanii
plasma membrane of
sterol:phospholipid ratio in,
47
Acetate
dolichyl-P and, 500
Acetone
cytochrome P450 proteins
and, 960
Acetylcholine receptor
tunicamycin and, 506-7
Acetyl CoA carboxylase
fatty acid synthesis and, 104-
6
N-Acetylgalactosamine
epidermal growth factor receptor and, 886
N-Acetylglucosamine
MDCK cells and, 513
tunicamycin and, 501
β-N-Acetylhexosaminidase
tunicamycin and, 505
N-Acetylneuraminic acid
wheat germ agglutinin and,
29-30
o-Acetylserine sulfhydrase, 103
Aconitase
citrate synthase and, 112
Acrosin, 340
ACTH, 354
cytochrome P450 and, 973
tunicamycin and, 505
Actin
inositol lipid hydrolysis and,
167-68
inositol 1,4,5-trisphosphate
and, 177
β-Actin
interferons and, 762
Actinomycin D
cytochromes P450 and, 948-
49
protein synthesis and, 525
RNA polymerase and, 550
Acute myelogenous leukemia
ras oncogenes in, 815
Acyl carrier protein, 16
Adenine nucleotides
calcium release and, 172
Adenosine
Dictyostelium discoideum and,
870
Adenosine tetraphosphate adenosine
aminoacyl tRNA synthetases
and, 153

Adenoviruses
E2A pre-mRNA of, 471
host-cell translation shut-off
and, 327-29
initiation factor inactivation
and, 319
mRNA translation and, 318
RNA splicing in, 480
Adenylate cyclase
calmodulin and, 400
G proteins and, 163
protein kinase C and, 183
Adenylatekinase
ras proteins and, 787
Adenylyl cyclase
G proteins and, 616-20, 639-
41
pertussis toxin and, 638
ADP
recA protein and, 252
ADP-ribosylation factor
G proteins and, 643
Adrenergic neurotransmitters
calcium channels and, 405
β-Adrenergic receptor
G proteins and, 616
tunicamycin and, 506
β-Adrenergic receptor kinases,
593-94
Agaricus bisporus
sugar-binding lectin of, 35-
36
Alanine
regulation of, 147
Alcohol dehydrogenase, 110
Alcoholic cirrhosis
hepatitis B viruses and, 681
Aldehyde dehydrogenase
cytochrome P450 gene ex-
pression and, 981
Aldolase, 108-10
Aldosterone synthesis, 107
Aleuria aurantia
fucose-binding lectin of, 35
Alkaline phosphatase
tunicamycin and, 504
Alkaloids
aromatic amino acid synthesis
and, 101
N-linked glycoproteins and,
498
pyrrolidine, 521-22
Alkylating reagents
cysteine proteinases and, 337
Aluminum
G proteins and, 616
Amantadine
influenza virus and, 377, 382

Amiloride
sodium/calcium exchanger
and, 410
Amino acid metabolism,
101-4
intermediates in, 91
Amino acids
apoferritin and, 292-94
inositol phosphates and, 163-
64
Aminoacyl tRNA ligases
tRNA and, 263
Aminoacyl tRNA synthetases,
125-54
casein kinase I and, 591
mutations and, 140-45
protein synthesis and, 98
structural organization of,
127-40
tRNA and, 148-52
Aminoacyl tRNA synthetase
synthesis
regulation of, 147-48
p-Aminobenzoic acid synthesis,
101
4-Aminobutyrate, 104
4-Aminobutyrate aminotransferase, 104
Aminodeoxydialdose
tunicamycin and, 501
Aminoimidazole carboxamide
ribonucleotide transformylase, 100
Aminoimidazole ribonucleotide
synthetase, 100
Aminopeptidase N, 355
Aminopeptidases
inhibition of, 337
2-Aminopurine
Escherichia coli dam mutants
and, 441
Ammonia
Dictyostelium discoideum and,
870
glutamine synthesis and, 12
uridine monophosphate synthesis and, 99
Ammonium chloride
influenza virus and, 379
Amoeba proteus
phospholipid transport in, 55
Amphidicolin
thymidylate synthase and, 96
Amphomycin
lipid-linked saccharides and,
509-10
peptidoglycan synthesis and,
509

CUMULATIVE INDEXES

CONTRIBUTING AUTHORS, VOLUMES 52–56

A

Ackers, G., 54:597–629
Albersheim, P., 53:625–63
Ames, G. F.-L., 55:397–426
Amzel, L. M., 52:801–24
Anderson, M. E., 52:711–60
Andreadis, A., 56:467–96

B

Baird, A., 54:403–23
Barbacid, M., 56:779–828
Bass, B. L., 55:599–630
Baughman, G., 53:75–117
Bear, D. G., 53:389–446
Beechem, J. M., 54:43–71
Bennett, V., 54:273–304
Benson, S. A., 54:101–134
Bergh, M. L. E., 56:915–44
Bernstein, S. I., 56:695–726
Berridge, M., 56:159–94
Bessman, S. P., 54:831–62
Bishop, J. M., 52:301–54
Björk, G. R., 56:263–88
Blackburn, E. H., 53:163–94
Bloch, K., 56:1–20
Blumenthal, D. K., 56:567–614
Böhlen, P., 54:403–23
Bond, J. S., 56:333–64
Borst, P., 55:701–32
Bradshaw, R. A., 53:259–92
Brand, L., 54:43–71
Brazeau, P., 54:403–23
Breitbart, R. E., 56:467–96
Breslow, J. L., 54:699–727
Brill, W. J., 53:231–57
Brown, M. S., 52:223–61
Butler, P. E., 56:333–64

C

Cadenas, E., 55:137–66
Campbell, J., 55:733–72
Cantley, L. C., 55:511–38
Carafoli, E., 56:395–434
Caron, M. G., 52:159–86
Carpenter, C. L., 54:831–62
Carpenter, G., 56:881–914
Catterall, W. A., 55:953–86

C (cont.)

Cech, T. R., 55:599–630
Chance, B., 55:137–66
Chase, J. W., 55:103–36
Choppin, P. W., 52:467–506
Chothia, C., 53:537–72
Cimino, G. D., 54:1151–93
Civelli, O., 53:665–715
Clarke, S., 54:479–506
Clayton, D. A., 53:573–94
Cleveland, D. W., 54:331–65
Colman, R. F., 52:67–91
Cooper, A. J. L., 52:187–222
Cooper, J. A., 54:897–930
Cooper, J. A., 55:987–1036
Cox, M. M., 56:229–62
Cushman, S. W., 55:1059–90

D

Darvill, A. G., 53:625–63
Dawidowicz, E. A., 56:43–62
Deininger, P. L., 55:631–62
DeLuca, H. F., 52:411–39
Doerfler, W., 52:93–124
Doolittle, R. F., 53:195–229
Douglass, J., 53:665–715
Duch, D. S., 54:729–64

E

Eckstein, F., 54:367–402
Edelman, A. M., 56:567–614
Edelman, G., 54:135–69
Efstratiadis, A., 55:631–62
Eisenberg, D., 53:595–623
Elbein, A. D., 56:497–534
Emerson, C. P., 56:695–726
Ericson, J. U., 56:263–88
Esch, F., 54:403–23
Evans, R. M., 55:1091–119
Eyre, D. R., 53:717–48

F

Forbes, D. J., 56:535–66
Fraenkel, D., 55:317–38
Fry, S. C., 53:625–63
Fuchs, R., 55:663–700

G

Gallop, P. M., 53:717–48
Gamper, H. B., 54:1151–93
Ganem, D., 56:651–94
Gaull, G. E., 55:427–54
Gerisch, G., 56:853–80
Gilman, A. G., 56:615–50
Glazer, A. N., 52:125–57
Goldstein, J. L., 52:223–61
Gonzalez, F. J., 56:945–94
Gourse, R., 53:75–117
Grabowski, P. J., 55:1119–50
Green, P. J., 55:569–98
Greengard, P., 54:931–76
Griffith, O. W., 55:855–78
Grindley, N. D. F., 54:863–96
Gross, H. J., 54:531–64
Guillemin, R., 54:403–23
Gusella, J. F., 55:831–54
Gustafsson, C. E. D., 56:263–88

H

Habu, S., 54:803–30
Hagervall, T. G., 56:263–88
Hall, M. N., 54:101–134
Hamer, D. H., 55:913–52
Hammarström, S., 52:355–77
Hanahan, D. J., 55:483–510
Harrington, W. F., 53:35–73
Hascall, V. C., 55:539–68
Hasilik, A., 55:167–94
Hassell, J. R., 55:539–68
Hatefi, Y., 54:1015–69
Hayaishi, O., 54:73–100
Hearst, J. E., 54:1151–93
Helenius, A., 55:663–700
Hemmings, H. C. Jr., 54:931–76
Herbert, E., 53:665–715
Hers, H. G., 52:617–53
Hirschberg, C. B., 56:63–88
Höök, M., 53:847–69
Hokin, L. E., 54:205–35
Holmgren, A., 54:237–71
Honjo, T., 54:803–30
Hudson, T. H., 55: 195–224

1079

CHAPTER TITLES, VOLUMES 52–56

Annual Reviews Inc.
A NONPROFIT SCIENTIFIC PUBLISHER

4139 El Camino Way
P.O. Box 10139
Palo Alto, CA 94303-0897 • USA

ORDER FORM

Now you can order
TOLL FREE
1-800-523-8635
(except California)

Annual Reviews Inc. publications may be ordered directly from our office by mail or use our Toll Free Telephone line (for orders paid by credit card or purchase order, and customer service calls only); through booksellers and subscription agents, worldwide; and through participating professional societies. Prices subject to change without notice. ARI Federal I.D. #94-1156476

- **Individuals:** Prepayment required on new accounts by check or money order (in U.S. dollars, check drawn on U.S. bank) or charge to credit card — American Express, VISA, MasterCard.
- **Institutional buyers:** Please include purchase order number.
- **Students:** $10.00 discount from retail price, per volume. Prepayment required. Proof of student status must be provided (photocopy of student I.D. or signature of department secretary is acceptable). Students must send orders direct to Annual Reviews. Orders received through bookstores and institutions requesting student rates will be returned.
- **Professional Society Members:** Members of professional societies that have a contractual arrangement with Annual Reviews may order books through their society at a reduced rate. Check with your society for information.
- **Toll Free Telephone orders:** Call 1-800-523-8635 (except from California) for orders paid by credit card or purchase order and customer service calls only. California customers and all other business calls use 415-493-4400 (not toll free). Hours: 8:00 AM to 4:00 PM, Monday-Friday, Pacific Time.

Regular orders: Please list the volumes you wish to order by volume number.
Standing orders: New volume in the series will be sent to you automatically each year upon publication. Cancellation may be made at any time. Please indicate volume number to begin standing order.
Prepublication orders: Volumes not yet published will be shipped in month and year indicated.
California orders: Add applicable sales tax.
Postage paid (4th class bookrate/surface mail) by **Annual Reviews Inc.** Airmail postage or UPS, extra.

ANNUAL REVIEWS SERIES		Prices Postpaid per volume USA/elsewhere	Regular Order Please send:	Standing Order Begin with:
			Vol. number	Vol. number
Annual Review of ANTHROPOLOGY				
Vols. 1-14	(1972-1985)	$27.00/$30.00		
Vol. 15	(1986)	$31.00/$34.00		
Vol. 16	(avail. Oct. 1987)	$31.00/$34.00	Vol(s). _____	Vol. _____
Annual Review of ASTRONOMY AND ASTROPHYSICS				
Vols. 1-2, 4-20	(1963-1964; 1966-1982)	$27.00/$30.00		
Vols. 21-24	(1983-1986)	$44.00/$47.00		
Vol. 25	(avail. Sept. 1987)	$44.00/$47.00	Vol(s). _____	Vol. _____
Annual Review of BIOCHEMISTRY				
Vols. 30-34, 36-54	(1961-1965; 1967-1985)	$29.00/$32.00		
Vol. 55	(1986)	$33.00/$36.00		
Vol. 56	(avail. July 1987)	$33.00/$36.00	Vol(s). _____	Vol. _____
Annual Review of BIOPHYSICS AND BIOPHYSICAL CHEMISTRY				
Vols. 1-11	(1972-1982)	$27.00/$30.00		
Vols. 12-15	(1983-1986)	$47.00/$50.00		
Vol. 16	(avail. June 1987)	$47.00/$50.00	Vol(s). _____	Vol. _____
Annual Review of CELL BIOLOGY				
Vol. 1	(1985)	$27.00/$30.00		
Vol. 2	(1986)	$31.00/$34.00		
Vol. 3	(avail. Nov. 1987)	$31.00/$34.00	Vol(s). _____	Vol. _____

ANNUAL REVIEWS SERIES	Prices Postpaid per volume USA/elsewhere	Regular Order Please send:	Standing Order Begin with:
		Vol. number	Vol. number

Annual Review of COMPUTER SCIENCE

Vol. 1	(1986) . . . . **$39.00/$42.00**		
Vol. 2	(avail. Nov. 1987) . . . . **$39.00/$42.00**	Vol(s). _____	Vol. _____

Annual Review of EARTH AND PLANETARY SCIENCES

Vols. 1-10	(1973-1982) . . . . **$27.00/$30.00**		
Vols. 11-14	(1983-1986) . . . . **$44.00/$47.00**		
Vol. 15	(avail. May 1987) . . . . **$44.00/$47.00**	Vol(s). _____	Vol. _____

Annual Review of ECOLOGY AND SYSTEMATICS

Vols. 1-16	(1970-1985) . . . . **$27.00/$30.00**		
Vol. 17	(1986) . . . . **$31.00/$34.00**		
Vol. 18	(avail. Nov. 1987) . . . . **$31.00/$34.00**	Vol(s). _____	Vol. _____

Annual Review of ENERGY

Vols. 1-7	(1976-1982) . . . . **$27.00/$30.00**		
Vols. 8-11	(1983-1986) . . . . **$56.00/$59.00**		
Vol. 12	(avail. Oct. 1987) . . . . **$56.00/$59.00**	Vol(s). _____	Vol. _____

Annual Review of ENTOMOLOGY

Vols. 10-16, 18-30	(1965-1971, 1973-1985) . . . . **$27.00/$30.00**		
Vol. 31	(1986) . . . . **$31.00/$34.00**		
Vol. 32	(avail. Jan. 1987) . . . . **$31.00/$34.00**	Vol(s). _____	Vol. _____

Annual Review of FLUID MECHANICS

Vols. 1-4, 7-17	(1969-1972, 1975-1985) . . . . **$28.00/$31.00**		
Vol. 18	(1986) . . . . **$32.00/$35.00**		
Vol. 19	(avail. Jan. 1987) . . . . **$32.00/$35.00**	Vol(s). _____	Vol. _____

Annual Review of GENETICS

Vols. 1-19	(1967-1985) . . . . **$27.00/$30.00**		
Vol. 20	(1986) . . . . **$31.00/$34.00**		
Vol. 21	(avail. Dec. 1987) . . . . **$31.00/$34.00**	Vol(s). _____	Vol. _____

Annual Review of IMMUNOLOGY

Vols. 1-3	(1983-1985) . . . . **$27.00/$30.00**		
Vol. 4	(1986) . . . . **$31.00/$34.00**		
Vol. 5	(avail. April 1987) . . . . **$31.00/$34.00**	Vol(s). _____	Vol. _____

Annual Review of MATERIALS SCIENCE

Vols. 1, 3-12	(1971, 1973-1982) . . . . **$27.00/$30.00**		
Vols. 13-16	(1983-1986) . . . . **$64.00/$67.00**		
Vol. 17	(avail. August 1987) . . . . **$64.00/$67.00**	Vol(s). _____	Vol. _____

Annual Review of MEDICINE

Vols. 1-3, 6, 8-9	(1950-1952, 1955, 1957-1958)		
11-15, 17-36	(1960-1964, 1966-1985) . . . . **$27.00/$30.00**		
Vol. 37	(1986) . . . . **$31.00/$34.00**		
Vol. 38	(avail. April 1987) . . . . **$31.00/$34.00**	Vol(s). _____	Vol. _____

Annual Review of MICROBIOLOGY

Vols. 18-39	(1964-1985) . . . . **$27.00/$30.00**		
Vol. 40	(1986) . . . . **$31.00/$34.00**		
Vol. 41	(avail. Oct. 1987) . . . . **$31.00/$34.00**	Vol(s). _____	Vol. _____